Lecture Notes in Computer Science 3098

Springer
Berlin
Heidelberg
New York
Hong Kong
London
Milan
Paris
Tokyo

Jörg Desel Wolfgang Reisig
Grzegorz Rozenberg (Eds.)

Lectures on Concurrency and Petri Nets

Advances in Petri Nets

Springer

Volume Editors

Jörg Desel
Katholische Universität Eichstätt
Lehrstuhl für Angewandte Informatik
Ostenstr. 14, 85072 Eichstätt, Germany
E-mail: joerg.desel@ku-eichstaett.de

Wolfgang Reisig
Humboldt-Universität zu Berlin
Institut für Informatik
Unter den Linden 6, 10099 Berlin, Germany
E-mail: reisig@informatik.hu-berlin.de

Grzegorz Rozenberg
Leiden University
Leiden Institute of Advanced Computer Science (LIACS)
Niels Bohrweg 1, 2333 CA Leiden, The Netherlands
E-mail: rozenberg@liacs.nl

Library of Congress Control Number: Applied for

CR Subject Classification (1998): F.1, F.2, F.3, C.2.4, C.2, H.3, H.4, C.1

ISSN 0302-9743
ISBN 3-540-22261-8 Springer-Verlag Berlin Heidelberg New York

Springer-Verlag is a part of Springer Science+Business Media

springeronline.com

Printed in Germany

Typesetting: Camera-ready by author, data conversion by Olgun Computergrafik
Printed on acid-free paper SPIN: 11013105 06/3142 5 4 3 2 1 0

Preface

The very first model of concurrent and distributed systems was introduced by C.A. Petri in his seminal Ph.D. thesis in 1964. Petri nets has remained a central model for concurrent systems for 40 years, and they are often used as a yardstick for other models of concurrency. As a matter of fact, many other models have been developed since then, and this research area is flourishing today.

The goal of the 4th Advanced Course on Petri Nets held in Eichstätt, Germany in September 2003 was to present applications and the theory of Petri Nets in the context of a whole range of other models. We believe that in this way the participants of the course received a broad and in-depth picture of research in concurrent and distributed systems.

It is also the goal of this volume to convey this picture. The volume is based on lectures given at the Advanced Course, but in order to provide a balanced presentation of the field, some of the lectures are not included, and some material not presented in Eichstätt is covered here. In particular, a series of introductory lectures was not included in this volume, as the material they covered is well established by now, and well presented elsewhere (e.g., in W. Reisig and G. Rozenberg, eds., "Lectures on Petri Nets," LNCS 1491, 1492, Springer-Verlag, 1997 – these two volumes are based on the 3rd Advanced Course on Petri Nets).

We believe that this volume will be useful as both a reference and a study book for the reader who is interested in obtaining an up-to-date overview of research in concurrent and distributed systems. It will be also useful for the reader who is specifically interested in Petri nets. Although the material presented in this volume is based on the Eichstätt course, the papers included here were written after the course, and therefore they have taken into account numerous comments made by the participants and fellow lecturers during the course. Because of this, and because, to start with, the lecturers were asked to present their material in a tutorial fashion, this volume is very suitable as an auxiliary reading for courses on concurrency and/or Petri nets, and especially useful as the underlying book for a seminar covering this research area.

Acknowledgements

The editors are indebted to the participants and the lecturers of the Advanced Course in Eichstätt for their enthusiastic participation in the course and for their constructive criticism of the presented material both during and after the course. We are also very much indebted to the authors of all the papers for their effort in producing all the material included here. We are grateful for the support of the Katholische Universität Eichstätt, hosting the Advanced Course, and to the Deutsche Forschungsgemeinschaft for its financial support. Many thanks to Springer-Verlag, and in particular to Mr. A. Hofmann, for the pleasant and efficient cooperation in producing this volume.

April 2004

Jörg Desel, Eichstätt
Wolfgang Reisig, Berlin
Grzegorz Rozenberg, Leiden

Table of Contents

Business Process Management Demystified: A Tutorial on Models, Systems and Standards for Workflow Management

Wil M.P. van der Aalst

Department of Technology Management
Eindhoven University of Technology
P.O.Box 513, NL-5600 MB Eindhoven, The Netherlands
w.m.p.v.d.aalst@tm.tue.nl

Abstract. Over the last decade there has been a shift from "data-aware" information systems to "process-aware" information systems. To support business processes an enterprise information system needs to be aware of these processes and their organizational context. Business Process Management (BPM) includes methods, techniques, and tools to support the design, enactment, management, and analysis of such operational business processes. BPM can be considered as an extension of classical Workflow Management (WFM) systems and approaches. This tutorial introduces models, systems, and standards for the design, analysis, and enactment of workflow processes. Petri nets are used for the modeling and analysis of workflows. Using Petri nets as a formal basis, contemporary systems, languages, and standards for BPM and WFM are discussed. Although it is clear that Petri nets can serve as a solid foundation for BPM/WFM technology, in reality systems, languages, and standards are developed in an ad-hoc fashion. To illustrate this XPDL, the "Lingua Franca" proposed by the Workflow Management Coalition (WfMC), is analyzed using a set of 20 basic workflow patterns. This analysis exposes some of the typical semantic problems restricting the application of BPM/WFM technology.

Keywords: Business process management, Workflow management, Workflow management systems, Workflow patterns, XML Process Definition Language (XPDL), Workflow verification.

1 Introduction

This section provides some context for the topics addressed in this tutorial. First, we identify some trends and put them in a historical perspective. Then, we focus on the BPM life-cycle and discuss the basic functionality of a WFM system. Finally, we outline the remainder of this tutorial.

1.1 Historical Perspective

To show the relevance of Business Process Management (BPM) systems, it is interesting to put them in a historical perspective. Consider Figure 1, which shows some of the

J. Desel, W. Reisig, and G. Rozenberg (Eds.): ACPN 2003, LNCS 3098, pp. 1–65, 2004.

ongoing trends in information systems. This figure shows that today's information systems consist of a number of layers. The center is formed by the operating system, i.e., the software that makes the hardware work. The second layer consists of generic applications that can be used in a wide range of enterprises. Moreover, these applications are typically used within multiple departments within the same enterprise. Examples of such generic applications are a database management system, a text editor, and a spreadsheet program. The third layer consists of domain specific applications. These applications are only used within specific types of enterprises and departments. Examples are decision support systems for vehicle routing, call center software, and human resource management software. The fourth layer consists of tailor-made applications. These applications are developed for specific organizations.

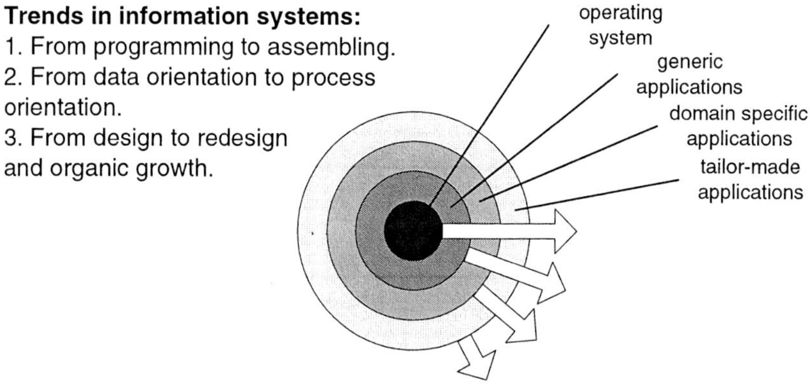

Fig. 1. Trends relevant for business process management.

In the sixties the second and third layer were missing. Information systems were built on top of a small operating system with limited functionality. Since no generic nor domain specific software was available, these systems mainly consisted of tailor-made applications. Since then, the second and third layer have developed and the ongoing trend is that the four circles are increasing in size, i.e., they are moving to the outside while absorbing new functionality. Today's operating systems offer much more functionality. Database management systems that reside in the second layer offer functionality which used to be in tailor-made applications. As a result of this trend, the emphasis shifted from programming to assembling of complex software systems. The challenge no longer is the coding of individual modules but orchestrating and gluing together pieces of software from each of the four layers.

Another trend is the shift from data to processes. The seventies and eighties were dominated by data-driven approaches. The focus of information technology was on storing and retrieving information and as a result data modeling was the starting point for building an information system. The modeling of business processes was often neglected and processes had to adapt to information technology. Management trends such as business process reengineering illustrate the increased emphasis on processes. As a result, system engineers are resorting to a more process driven approach.

The last trend we would like to mention is the shift from carefully planned designs to redesign and organic growth. Due to the omnipresence of the Internet and its standards, information systems change on-the-fly. As a result, fewer systems are built from scratch. In many cases existing applications are partly used in the new system. Although component-based software development still has it problems, the goal is clear and it is easy to see that software development has become more dynamic.

The trends shown in Figure 1 provide a historical context for BPM. BPM systems are either separate applications residing in the second layer or are integrated components in the domain specific applications, i.e., the third layer. Notable examples of BPM systems residing in the second layer are WorkFlow Management (WFM) systems [12, 38, 48, 55, 57, 58, 61] such as Staffware, MQSeries, and COSA, and case handling systems such as FLOWer. Note that leading Enterprise Resource Planning (ERP) systems populating the third layer also offer a WFM module. The workflow engines of SAP, Baan, PeopleSoft, Oracle, and JD Edwards can be considered as integrated BPM systems. The idea to isolate the management of business processes in a separate component is consistent with the three trends identified. BPM systems can be used to avoid hard-coding the work processes into tailor-made applications and thus support the shift from programming to assembling. Moreover, process orientation, redesign, and organic growth are supported. For example, today's WFM systems can be used to integrate existing applications and support process change by merely changing the workflow diagram. Given these observations, the practical relevance of BPM is evident. Although BPM functionality is omnipresent and often hidden in larger enterprise information systems, for clarity we will often restrict the discussion to clear cut "process-aware" information systems such as WFM systems (cf. Section 1.3).

To put the topic of this tutorial in a historical perspective it is worthwhile to consider the early work on office information systems. In the seventies, people like Skip Ellis [32], Anatol Holt [45], and Michael Zisman [78] already worked on so-called office information systems, which were driven by explicit process models. It is interesting to see that the three pioneers in this area independently used Petri-net variants to model office procedures. During the seventies and eighties there was great optimism about the applicability of office information systems. Unfortunately, few applications succeeded. As a result of these experiences, both the application of this technology and research almost stopped for a decade. Consequently, hardly any advances were made in the eighties. In the nineties, there again was a huge interest in these systems. The number of WFM systems developed in the past decade and the many papers on workflow technology illustrate the revival of office information systems. Today WFM systems are readily available [12, 38, 48, 55, 57, 58, 61]. However, their application is still limited to specific industries such as banking and insurance. As was indicated by Skip Ellis it is important to learn from these ups and downs [33]. The failures in the eighties can be explained by both technical and conceptual problems. In the eighties, networks were slow or not present at all, there were no suitable graphical interfaces, and proper development software was missing. However, there were also more fundamental problems: a unified way of modeling processes was missing and the systems were too rigid to be used by people in the workplace. Most of the technical problems have been resolved by now. However, the more conceptual problems remain. Good standards for business process modeling are still missing and even today's WFM systems enforce unnecessary constrains on the process logic (e.g., processes are made more sequential).

1.2 BPM Life-Cycle

As indicated before, *Business Process Management (BPM) includes methods, techniques, and tools to support the design, enactment, management, and analysis of operational business processes.* It can be considered as an extension of classical Workflow Management (WFM) systems and approaches. Before discussing the differences between WFM and BPM, let us consider the *BPM life-cycle*.

The BPM life-cycle has four phases:

- *Process design*
  Any BPM effort requires the modeling of an existing ("as-is") or desired ("to-be") process, i.e., a *process design*. During this phase process models including various perspectives (control-flow, data-flow, organizational, sociotechnical, and operational aspects) are constructed. The only way to create a "process-aware" enterprise information system is to add knowledge about the operational processes at hand.
- *System configuration*
  Based on a process design, the process-aware enterprise information system is realized. In the traditional setting the realization would require a time-consuming and complex software development process. Using software from the second and third layer shown in Figure 1, the traditional software development process is replaced by a configuration or assembly process. Therefore, we use the term *system configuration* for the phase in-between process design and enactment.
- *Process enactment*
  The *process enactment* phase is the phase where the process-aware enterprise information system realized in the system configuration phase is actually used.
- *Diagnosis*
  Process-aware enterprise information system have to change over time to improve performance, exploit new technologies, support new processes, and adapt to an ever changing environment. Therefore, the *diagnosis* phase is linking the process enactment phase to the a new design phase.

Like in software life-cycle models, the four phases are overlapping (cf. Waterfall model) and the whole process is iterative (cf. Spiral model).

As is illustrated in Figure 2, the BPM life-cycle can be used to identify different levels of maturity when it comes to developing process-aware enterprise information systems. In the early nineties and before, most information systems only automated individual activities and where unaware of the underlying process. For the systems that were process-aware, the process logic was hard-coded in the system and not supported in a generic manner. Despite the early work on office automation, the first commercial WFM systems became only practically relevant around 1993 (see Figure 2(a)). The focus of these systems was on "getting the system to work" and support for enactment and design was limited. In the mid-nineties this situation changed and by 1998 many WFM

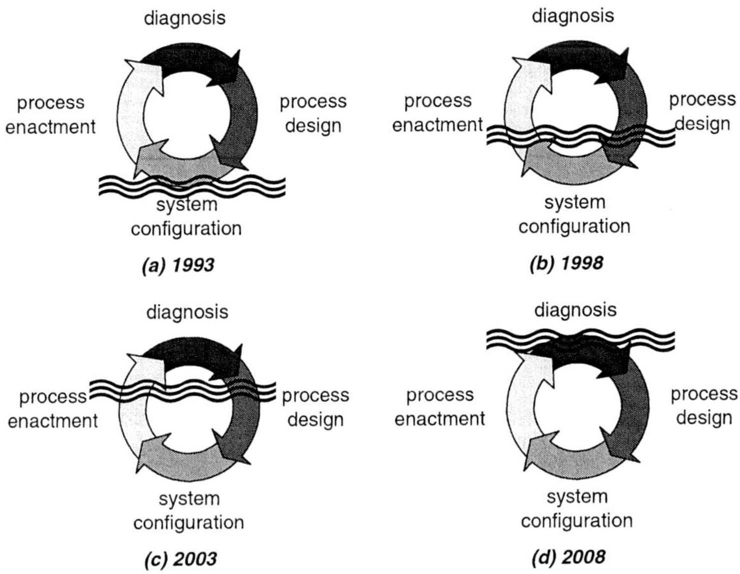

Fig. 2. The BPM life-cycle is used to indicate the maturity of BPM technology over time.

systems had become readily available (see Figure 2(b)). In these systems there was basic support for enactment and design. In the last five years these systems have been further extended allowing for more support during the design and enactment phases (see Figure 2(c)). For example, a case-handling system like FLOWer [22] allows for much more flexibility during the enactment phase than the traditional WFM systems. Today's systems provide hardly any support for the diagnosis phase. Although most BPM software logs all kinds of events (e.g., WFM systems like Staffware log the completion of activities and ERP systems like SAP log transactions), this information is not used to identify problems or opportunities for improvement. In the next five years this situation will probably change when process mining [17, 19] techniques become readily available (see Figure 2(d)).

The BPM life-cycle shown in Figure 2 can also be used to define the difference between WFM and BPM. WFM focusses on the lower half of the BPM life-cycle (i.e., "getting the system to work") while BPM also includes to upper half of the life-cycle. Therefore, BPM also focusses on diagnosis, flexibility, human-centric processes, goal-driven process design, etc. Gartner expects that *Business Process Analysis* (BPA), i.e., software to support the diagnosis phase, will become increasingly important [39]. It is expected that the BPA market will continue to grow. Note that BPA covers aspects neglected by traditional WFM products (e.g., diagnosis, simulation, etc.). *Business Activity Monitoring* (BAM) is one of the emerging areas in BPA. The goal of BAM tools is to use data logged by the information system to diagnose the operational processes. An example is the ARIS Process Performance Manager (PPM) of IDS Scheer [47]. ARIS PPM extracts information from audit trails (i.e., information logged during the execution of cases) and displays this information in a graphical way (e.g., flow times, bottlenecks, utilization, etc.). BAM also includes process mining, i.e., extracting process models from logs [17]. BAM creates a number of scientific and practical challenges (e.g., which processes can be discovered and how much data is needed to provide useful information).

1.3 Workflow Management (Systems)

The focus of this tutorial will be on WFM rather than BPM. The reason is that WFM serves as a basis for BPM and in contrast to BPM it is a mature area with well-defined concepts and widely used software products.

The Workflow Management Coalition (WfMC) defines workflow as: "The automation of a business process, in whole or part, during which documents, information or tasks are passed from one participant to another for action, according to a set of procedural rules." [55]. A Workflow Management System (WFMS) is defined as: "A system that defines, creates and manages the execution of workflows through the use of software, running on one or more workflow engines, which is able to interpret the process definition, interact with workflow participants and, where required, invoke the use of IT tools and applications." [55]. Note that both definitions emphasize the focus on enactment, i.e., the use of software to support the execution of operational processes.

When considering these definitions in more detail it is evident that WFM is highly relevant for any organization. However, at the same time few organizations use a "real" WFM system. To explain this we identify four categories of WFM support:

- *Pure WFM systems*
  At this point in time many WFM systems are available and used in practise. Examples of systems include Staffware Process Suite, FileNET BPM Suite, i-Flow, FLOWer, WebSphere MQ Workflow (formerly known as MQSeries Workflow), TIBCO InConcert, etc.
- *WFM components embedded in other systems*
  Many software packages embed a generic workflow component whose functionality is comparable to the pure WFM systems. For example, most ERP systems provide a workflow component. SAP WebFlow is the workflow component of SAP offering all the functionality typically present in traditional stand-alone WFM products.
- *Custom-made WFM solutions*
  Many organizations, e.g., banks and insurance companies, have chosen not to use a commercially available WFM solution but build an organization-specific solution. These solutions typically only support a subset of the functionality offered by the first two categories. Nevertheless, these systems support the definition and execution of different workflows.
- *Hard-coded WFM solutions*
  The last category refers to the situation were the processes are hard-coded in the applications, i.e., there is no generic workflow support but applications are coupled in such a way that a specific process is supported. The only way to change a process is to change the applications themselves, i.e., unlike the first three categories there is no component that is process-aware. Note that in these hard-coded system an explicit orchestration layer is missing.

At this point in time the majority of business processes are still supported by solutions residing in the third and fourth category. However, the percentage of processes supported by the first two categories is increasing. Moreover, software developers building solutions for the third and fourth category are using the concepts and insights provided by the first two categories. In this context it is interesting to refer to recent developments in the *web services* domain [68]. The functionality of web service composition languages (also referred to as "web service orchestration") like BPEL4WS, BPML, WSCI, WSWSFL, XLANG, etc. is very similar to traditional workflow languages [6, 77].

1.4 Outline and Intended Audience

The goal of this tutorial is to introduce the reader to the theoretical foundations of BPM/WFM using a Petri-net based approach. However, at the same time contemporary systems and languages are presented to provide a balanced view on the application domain.

Section 2 shows the application of Petri nets to workflow modeling. For this purpose, the class of *WorkFlow nets* (WF-nets) is introduced, but also some "syntactical sugaring" is given to facilitate the design of workflows. Section 3 discusses the analysis of workflow models expressed in terms of Petri nets. The focus will be on the verification of WF-nets using classical analysis techniques. Section 4 discusses the typical architecture of a WFM system and discusses contemporary systems. The goal of this section is to show that the step from design to enactment, i.e., the configuration phase (cf. Figure 2), is far from trivial. In Section 5, 20 workflow patterns are used to evaluate the XML Process Definition Language (XPDL), the standard proposed by the Workflow Management Coalition (WfMC). This evaluation illustrates the typical problems workflow designers and implementers are faced with when applying contemporary languages and standards. Section 6 provides an overview of related work. Clearly, only a small subset of the many books and papers on BPM/WFM can be presented, but pointers are given to find relevant material. Finally, Section 7 concludes the tutorial by discussing the role of Petri nets in the BPM/WFM domain.

Note that parts of this tutorial are based on earlier work (cf. [2–6, 12, 15]). For more material the interested reader is referred to [12] and two WWW-sites: one presenting course material (slides, animations, etc.) `http://www.workflowcourse.com` and one on workflow patterns `http://www.workflowpatterns.com`.

This tutorial is intended for people having a basic understanding of Petri nets and interested in the application of Petri nets to problems in the BPM/WFM domain. Sections 2 and 3 are focusing more on the Petri-net side of things while sections 4 and 5 are focusing more on the application domain.

2 Workflow Modeling

In this section, we show how to model workflows in terms of Petri nets. First, we introduce the basic workflow concepts and discuss the various perspectives. Then, we define some basic Petri net notation followed by an introduction to a subclass of Petri nets tailored towards workflow modeling. We conclude this section with an exercise.

2.1 Workflow Concepts and Perspectives

Workflow processes are *case-driven*, i.e., tasks are executed for specific cases. Approving loans, processing insurance claims, billing, processing tax declarations, handling traffic violations and mortgaging, are typical case-driven processes which are often supported by a WFM system. These case-driven processes, also called *workflows*, are marked by three dimensions: (1) the control-flow dimension, (2) the resource dimension, and (3) the case dimension (see Figure 3). The control-flow dimension is concerned with the partial ordering of tasks, i.e., the workflow *process*. The tasks which need to be executed are identified and the routing of cases along these tasks is determined. Conditional, sequential, parallel and iterative routing are typical structures specified in the control-flow dimension. Tasks are executed by resources. Resources are human (e.g., employee) and/or non-human (e.g., device, software, hardware). In the resource dimension these resources are classified by identifying roles (resource classes based on functional characteristics) and organizational units (groups, teams or departments). Both the control-flow dimension and the resource dimension are generic, i.e., they are not tailored towards a specific case. The third dimension of a workflow is concerned with individual cases which are executed according to the process definition (first dimension) by the proper resources (second dimension).

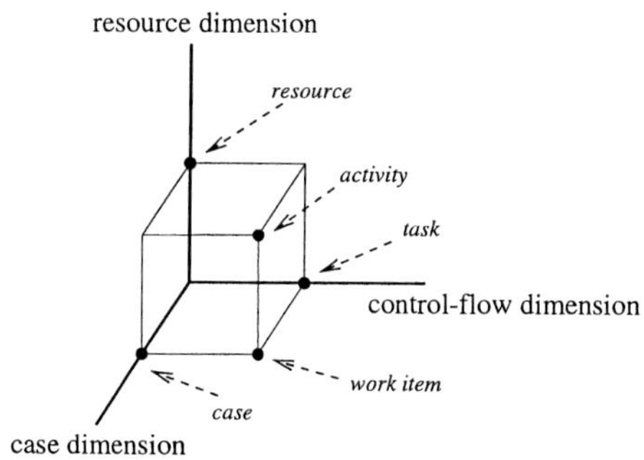

Fig. 3. The three dimensions of workflow.

The primary task of a WFM system is to enact case-driven business processes by joining several perspectives. The following perspectives are relevant for workflow modeling and workflow execution: (1) *control flow* (or process) perspective, (2) *resource* (or organization) perspective, (3) *data* (or information) perspective, (4) *task* (or function) perspective, (5) *operation* (or application) perspective. These perspectives are similar to the perspectives given in [48] and the control flow and resource perspectives correspond to the first two dimensions shown in Figure 3. The third dimension reflects the fact that workflows are case-driven and does not correspond to one of the five perspectives.

In the control-flow perspective, *workflow process definitions* (workflow schemas) are defined to specify which *tasks* need to be executed and in what order (i.e., the routing or control flow). A task is an atomic piece of work. Workflow process definitions are instantiated for specific *cases* (i.e., workflow instances). Since a case is an instantiation of a process definition, it corresponds to the execution of concrete work according to the specified routing. In the *resource* perspective, the organizational structure and its population are specified. The organizational structure describes relations between roles (resource classes based on functional aspects) and groups (resource classes based on organizational aspects). Thus clarifying organizational issues such as responsibility, availability, and authorization. Resources, ranging from humans to devices, form the organizational population and are allocated to roles and groups. The data perspective deals with *control* and *production data*. Control data are data introduced solely for WFM purposes, e.g., variables introduced for routing purposes. Production data are information objects (e.g., documents, forms, and tables) whose existence does not depend on WFM. The task perspective describes the elementary operations performed by resources while executing a task for a specific case. In the operational perspective the elementary actions are described. These actions are often executed using applications ranging from a text editor to custom build applications to perform complex calculations. Typically, these applications create, read, or modify control and production data in the information perspective.

The focus of this tutorial will be on the control-flow perspective. Clearly, this is the most dominant perspective. Moreover, Petri nets can contribute most to this perspective.

2.2 Petri Nets

This section introduces the basic Petri net terminology and notations. Readers familiar with Petri nets can skip this section[1].

The classical Petri net is a directed bipartite graph with two node types called *places* and *transitions*. The nodes are connected via directed *arcs*. Connections between two nodes of the same type are not allowed. Places are represented by circles and transitions by rectangles.

Definition 1 (Petri net). *A Petri net is a triple* (P, T, F)*:*

- P *is a finite set of places,*
- T *is a finite set of transitions* ($P \cap T = \emptyset$),
- $F \subseteq (P \times T) \cup (T \times P)$ *is a set of arcs (flow relation)*

A place p is called an *input place* of a transition t iff there exists a directed arc from p to t. Place p is called an *output place* of transition t iff there exists a directed arc from t to p. We use $\bullet t$ to denote the set of input places for a transition t. The notations $t\bullet$, $\bullet p$ and $p\bullet$ have similar meanings, e.g., $p\bullet$ is the set of transitions sharing p as an input place. Note that we do not consider multiple arcs from one node to another. In the context of workflow procedures it makes no sense to have other weights, because places correspond to conditions.

At any time a place contains zero or more *tokens*, drawn as black dots. The *state*, often referred to as marking, is the distribution of tokens over places, i.e., $M \in P \rightarrow \mathbb{N}$. We will represent a state as follows: $1p_1 + 2p_2 + 1p_3 + 0p_4$ is the state with one token in

[1] Note that states are represented by weighted sums and note the definition of (elementary) (conflict-free) paths.

place p_1, two tokens in p_2, one token in p_3 and no tokens in p_4. We can also represent this state as follows: $p_1 + 2p_2 + p_3$. To compare states we define a partial ordering. For any two states M_1 and M_2, $M_1 \leq M_2$ iff for all $p \in P$: $M_1(p) \leq M_2(p)$

The number of tokens may change during the execution of the net. Transitions are the active components in a Petri net: they change the state of the net according to the following *firing rule*:

(1) A transition t is said to be *enabled* iff each input place p of t contains at least one token.
(2) An enabled transition may *fire*. If transition t fires, then t *consumes* one token from each input place p of t and *produces* one token for each output place p of t.

Given a Petri net (P, T, F) and a state M_1, we have the following notations:

- $M_1 \xrightarrow{t} M_2$: transition t is enabled in state M_1 and firing t in M_1 results in state M_2
- $M_1 \rightarrow M_2$: there is a transition t such that $M_1 \xrightarrow{t} M_2$
- $M_1 \xrightarrow{\sigma} M_n$: the firing sequence $\sigma = t_1 t_2 t_3 \ldots t_{n-1}$ leads from state M_1 to state M_n via a (possibly empty) set of intermediate states $M_2, \ldots M_{n-1}$, i.e., $M_1 \xrightarrow{t_1} M_2 \xrightarrow{t_2} \ldots \xrightarrow{t_{n-1}} M_n$

A state M_n is called *reachable* from M_1 (notation $M_1 \xrightarrow{*} M_n$) iff there is a firing sequence σ such that $M_1 \xrightarrow{\sigma} M_n$. Note that the empty firing sequence is also allowed, i.e., $M_1 \xrightarrow{*} M_1$.

We use (PN, M) to denote a Petri net PN with an initial state M. A state M' is a *reachable state* of (PN, M) iff $M \xrightarrow{*} M'$.

Let us define some standard properties for Petri nets. First, we define properties related to the dynamics of a Petri net, then we give some structural properties.

Definition 2 (Live). *A Petri net (PN, M) is live iff, for every reachable state M' and every transition t there is a state M'' reachable from M' which enables t.*

A Petri net is *structurally live* if there exists an initial state such that the net is live.

Definition 3 (Bounded, safe). *A Petri net (PN, M) is bounded iff for each place p there is a natural number n such that for every reachable state the number of tokens in p is less than n. The net is safe iff for each place the maximum number of tokens does not exceed 1.*

A Petri net is *structurally bounded* if the net is bounded for any initially state.

Definition 4 (Well-formed). *A Petri net PN is well-formed iff there is a state M such that (PN, M) is live and bounded.*

Paths connect nodes by a sequence of arcs.

Definition 5 (Path, Elementary, Conflict-free). *Let PN be a Petri net. A path C from a node n_1 to a node n_k is a sequence $\langle n_1, n_2, \ldots, n_k \rangle$ such that $\langle n_i, n_{i+1} \rangle \in F$ for $1 \leq i \leq k-1$. C is elementary iff, for any two nodes n_i and n_j on C, $i \neq j \Rightarrow n_i \neq n_j$. C is conflict-free iff, for any place n_j on C and any transition n_i on C, $j \neq i-1 \Rightarrow n_j \notin \bullet n_i$.*

For convenience, we introduce the alphabet operator α on paths. If $C = \langle n_1, n_2, \ldots, n_k \rangle$, then $\alpha(C) = \{n_1, n_2, \ldots, n_k\}$.

Definition 6 (Strongly connected). *A Petri net is strongly connected iff, for every pair of nodes (i.e., places and transitions) x and y, there is a path leading from x to y.*

Definition 7 (Free-choice). *A Petri net is a free-choice Petri net iff, for every two transitions t_1 and t_2, $\bullet t_1 \cap \bullet t_2 \neq \emptyset$ implies $\bullet t_1 = \bullet t_2$.*

Definition 8 (State machine). *A Petri net is state machine iff each transition has exactly one input and one output place.*

Definition 9 (S-component). *A subnet $PN_s = (P_s, T_s, F_s)$ is called an S-component of a Petri net $PN = (P, T, F)$ if $P_s \subseteq P$, $T_s \subseteq T$, $F_s \subseteq F$, PN_s is strongly connected, PN_s is a state machine, and for every $q \in P_s$ and $t \in T$: $(q,t) \in F \Rightarrow (q,t) \in F_s$ and $(t,q) \in F \Rightarrow (t,q) \in F_s$.*

Definition 10 (S-coverable). *A Petri net is S-coverable iff for any node there exist an S-component which contains this node.*

See [30, 63] for a more elaborate introduction to these standard notions.

2.3 WF-Nets

In Figure 3 we indicated that a workflow has (at least) three dimensions. The control-flow dimension is the most prominent one, because the core of any workflow system is formed by the processes it supports. In the control-flow dimension building blocks such as the AND-split, AND-join, OR-split, and OR-join are used to model sequential, conditional, parallel and iterative routing [55]. Clearly, a Petri net can be used to specify the routing of cases. *Tasks* are modeled by transitions and causal dependencies are modeled by places and arcs. In fact, a place corresponds to a *condition* which can be used as pre- and/or post-condition for tasks. An AND-split corresponds to a transition with two or more output places, and an AND-join corresponds to a transition with two or more input places. OR-splits/OR-joins correspond to places with multiple outgoing/ingoing arcs. Moreover, in [2] it is shown that the Petri net approach also allows for useful routing constructs absent in many WFM systems.

A Petri net which models the control-flow dimension of a workflow, is called a *WorkFlow net* (WF-net). It should be noted that a WF-net specifies the dynamic behavior of a single case in isolation.

Definition 11 (WF-net). *A Petri net $PN = (P, T, F)$ is a WF-net (Workflow net) if and only if:*

(i) There is one source place $i \in P$ such that $\bullet i = \emptyset$.
(ii) There is one sink place $o \in P$ such that $o \bullet = \emptyset$.
(iii) Every node $x \in P \cup T$ is on a path from i to o.

A WF-net has one input place (i) and one output place (o) because any case handled by the procedure represented by the WF-net is created when it enters the WFM system and is deleted once it is completely handled by the system, i.e., the WF-net specifies the life-cycle of a case. The third requirement in Definition 11 has been added to avoid "dangling tasks and/or conditions", i.e., tasks and conditions which do not contribute to the processing of cases.

Given the definition of a WF-net it is easy derive the following properties.

Proposition 1 (Properties of WF-nets). *Let* $PN = (P, T, F)$ *be Petri net.*

- *If PN is WF-net with source place* i*, then for any place* $p \in P$*:* $\bullet p \neq \emptyset$ *or* $p = i$*, i.e.,* i *is the only source place.*
- *If PN is WF-net with sink place* o*, then for any place* $p \in P$*:* $p\bullet \neq \emptyset$ *or* $p = o$*, i.e.,* o *is the only sink place.*
- *If PN is a WF-net and we add a transition* t^* *to PN which connects sink place* o *with source place* i *(i.e.,* $\bullet t^* = \{o\}$ *and* $t^*\bullet = \{i\}$*), then the resulting Petri net is strongly connected.*
- *If PN has a source place* i *and a sink place* o *and adding a transition* t^* *which connects sink place* o *with source place* i *yields a strongly connected net, then every node* $x \in P \cup T$ *is on a path from* i *to* o *in PN and PN is a WF-net.*

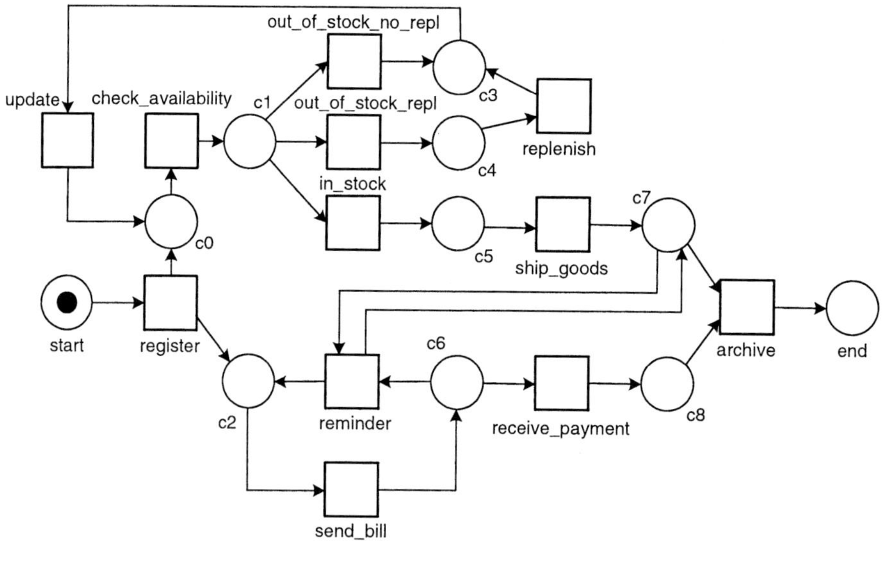

Fig. 4. WF-net.

Figure 4 shows an example of an order handling process modeled in terms of a WF-net. As indicated before cases are represented by tokens and in Figure 4 the token in *start* corresponds to an order. Task *register* is represented by a transition bearing the same name. From a routing point of view it acts as a so-called AND-split (two outgoing arcs) and is enabled in the state shown. If a person executes this task, the token is removed from place *start* and two tokens are produced: one for *c0* and one for *c2*. Then, in parallel, two tasks are enabled: *check_availability* and *send_bill*. Depending on the eagerness of the workers executing these two tasks either *check_availability* or *send_bill* is executed first. Suppose *check_availability* is executed first. Based on the outcome of this task a choice is made. This is reflected by the fact that three arcs are leaving *c1*. If the ordered goods are available, they can be shipped, i.e., firing *in_stock* enables task *ship_goods*. If they are not available, either a replenishment order is issued or not. Firing *out_of_stock_repl* enables task *replenish*. Firing *out_of_stock_no_repl* skips task *replenish*. Note that *check_availability*, place *c1* and the three transitions *in_stock*, *out_of_stock_repl*, and *out_of_stock_no_repl* together form a so-called OR-split: As a result of this construct one token is produced for either *c3*, *c4*, or *c5*. Suppose that not all ordered goods are available, but the appropriate replenishment orders were already issued. A token is produced for *c3* and task *update* becomes enabled. Suppose that at this point in time task *send_bill* is executed, resulting in the state with a token in *c3* and *c6*. The token in *c6* is input for two tasks. However, only one of these tasks can be executed and in this state only *receive_payment* is enabled. Task *receive_payment* can be executed the moment the payment is received. Task *reminder* is an AND-join/AND-split and is blocked until the bill is sent and the goods have been shipped. However, it is only possible to send a reminder if the goods have been actually shipped. Assume that in the state with a token in *c3* and *c6* task *update* is executed. This task does not require human involvement and is triggered by a message of the warehouse indicating that relevant goods have arrived. Again *check_availability* is enabled. Suppose that this task is executed and the result is positive, i.e., the path via *in_stock* is taken. In the resulting state *ship_goods* can be executed. Now there is a token in *c6* and *c7* thus enabling task *reminder*. Executing task *reminder* enables the task *send_bill* for the second time. A new copy of the bill is sent with the appropriate text. It is possible to send several reminders by alternating *reminder* and *send_bill*. However, let us assume that after the first loop the customer pays resulting in a state with a token in *c7* and *c8*. In this state, the AND-join *archive* is enabled and executing this task results in the final state with a token in *end*.

Figure 4 shows some of the limitations of WF-nets. First of all, the construct involving *check_availability*, place *c1* and the three transitions *in_stock*, *out_of_stock_repl*, and *out_of_stock_no_repl* is rather complex for a simple concept as a choice out of three alternatives. Second, the diagram does not show *why* things are happening. The text suggests that some of the tasks are executed by people while others are triggered by external entities or temporal conditions. Unfortunately, this information is missing in Figure 4. Finally, the WF-net does not show the other perspectives. The first two problems can be solved using some "syntactical sugaring" (cf. Figure 5). The third problem will not be addressed in this tutorial. Here we abstract from the other perspectives. The interested reader is referred to [12] for modeling the resource perspective.

Figure 5 shows the same process as the one depicted in Figure 4. However, every task can be an AND/OR-join - AND/OR-split. The semantics of a transition is AND-join - AND-split. Choices can be modeled using places with multiple outgoing arcs.

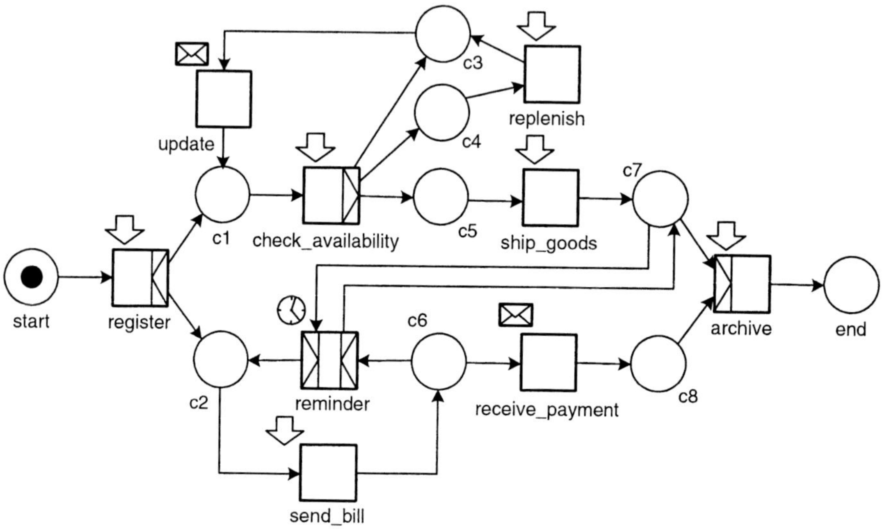

Fig. 5. WF-net extended with some "syntactical sugaring" to denote AND/OR-splits/joins and triggers.

However, the intuition of a task resulting in a choice is better reflected by the notation used in Figure 5: the construct involving *check_availability*, place *c1* and the three transitions *in_stock*, *out_of_stock_repl*, and *out_of_stock_no_repl* is replaced by a single task *check_availability* using the notation for an OR-split. Note that any WF-net with OR-splits can be automatically translated into standard WF-net (i.e., a classical Petri net). Figure 5 also shows three triggers symbols: (1) an arrow denoting a user trigger, (2) an envelope denoting an external trigger, and (3) a clock denoting a time trigger. These three triggers symbols denote that the corresponding tasks need a trigger to be executed, e.g., the tasks bearing an arrow symbol require a user to perform the corresponding activity. Task *receive payment* can only be executed after the payment trigger arrives. Task *reminder* can only be executed after a specified period. Although triggers are extremely important, we will not formalize the concept. For the reader interested in the topic we refer to [28, 34] for a discussion on the reactive nature of WFM systems.

The very simple WF-net shown in Figure 5 shows some of the routing constructs relevant for business process modeling. Sequential, parallel, conditional, and iterative routing are present in this model. There are also more advanced constructs such as the choice between *receive_payment* and *reminder*. This is a so-called *deferred choice* (also referred to as implicit choice) since it is not resolved by the system but by the environment of the system. The moment the bill is sent, it is undetermined whether *receive_payment* or *reminder* will be the next step in the process. Another advanced construct is the fact that task *reminder* is blocked until the goods have been shipped. The latter construct is a so-called *milestone*. The reason that we point out both constructs is that many systems have problems supporting these rather fundamental process patterns. In Section 5.1 we will discuss these patterns in more detail.

2.4 Exercise: Modeling a Complaints Handling Process in Terms of a WF-Net

To conclude this section, we give a small exercise. Model the complaints handling workflow of a travel agency in terms of a WF-net, i.e., construct a diagram similar to Figure 5.

Each year the travel agency has to process many customer complaints. There is a special department for the processing of complaints (department C). There is also an internal department called logistics (department L) which takes care of the registration of incoming complaints and the archiving of processed complaints. The following procedure is used to handle these complaints.

An employee of department L first registers every incoming complaint. After registration a form is sent to the customer with questions about the nature of the complaint. This is done by an employee of department C. There are two possibilities: the customer returns the form within two weeks or not. If the form is returned, it is processed automatically resulting in a report which can be used for the actual processing of the complaint. If the form is not returned on time, a time-out occurs resulting in an empty report. Note that this does not necessarily mean that the complaint is discarded. After registration, i.e., in parallel with the form handling, the preparation for the actual processing is started.

First, the complaint is evaluated by a complaint manager of department C. Evaluation shows that either further processing is needed or not. Note that this decision does not depend on the form handling. If no further processing is required and the form is handled, the complaint is archived. If further processing is required, an employee of the complaints department executes the task "process complaint" (this is the actual processing where certain actions are proposed if needed). For the actual processing of the complaint, the report resulting from the form handling is used. Note that the report can be empty. The result of task "process complaint" is checked by a complaint manager. If the result is not OK, task "process complaint" is executed again. This is repeated until the result is acceptable. If the result is accepted, an employee of the department C executes the proposed actions. After this the processed complaint is archived by an employee of department L.

Give the WF-net, i.e., model the workflow by making a process definition in terms of a Petri net. For the solution to this exercise we refer to [12] or the corresponding WWW site with course material: `http://www.workflowcourse.com`.

3 Workflow Analysis

One of the advantages of using Petri nets for workflow modeling is the availability of many Petri-net-based analysis techniques. In this section, we focus on the verification of WF-nets. The correctness criterion used is the so-called *soundness property*. We will show how this property can be checked and discuss a verification tool specifically designed for workflow analysis.

3.1 Verification, Validation, and Performance Analysis

The correctness, effectiveness, and efficiency of the business processes supported by the WFM system are vital to the organization. A workflow process definition which contains errors may lead to angry customers, back-log, damage claims, and loss of goodwill. Flaws in the design of a workflow definition may also lead to high throughput times, low service levels, and a need for excess capacity. This is why it is important to *analyze* a workflow process definition before it is put into production. Basically, there are three types of analysis:

- *validation*, i.e., testing whether the workflow behaves as expected,
- *verification*, i.e., establishing the correctness of a workflow, and
- *performance analysis*, i.e., evaluating the ability to meet requirements with respect to throughput times, service levels, and resource utilization.

Validation can be done by interactive simulation: a number of fictitious cases are fed to the system to see whether they are handled well. For verification and performance analysis more advanced analysis techniques are needed. Fortunately, many powerful analysis techniques have been developed for Petri nets ([30, 63]). Linear algebraic techniques can be used to verify many properties, e.g., place invariants, transition invariants, and (non-)reachability. Coverability graph analysis, model checking, and reduction techniques can be used to analyze the dynamic behavior of a Petri net. Simulation and Markov-chain analysis can be used for performance evaluation (cf. [59, 63]). The abundance of available analysis techniques shows that Petri nets can be seen as a solver independent medium between the design of the workflow process definition and the analysis of the resulting workflow.

3.2 Verification of the Control-Flow Perspective

In this tutorial we restrict ourselves to *workflow verification*, i.e., we will not discuss techniques for validation and performance analysis. Moreover, we restrict ourselves to the *control flow perspective*. Although each of the perspectives mentioned in Section 2.1 is relevant, the general focus of this tutorial is on control flow perspective, i.e., we use WF-nets as a starting point and demonstrate that Petri-net-based analysis techniques can be used to verify the correctness of a workflow process.

We abstract from the *resource perspective* because, given today's workflow technology, at any time there is only one resource working on a task which is being executed for a specific case. In today's WFM systems it is not possible to specify that several resources are collaborating in executing a task. Note that even if multiple persons are executing one task, e.g., writing a report, only one person is allocated to that task from the perspective of the WFM system: This is the person that selected the work item from the in-basket (i.e., the electronic worktray). Since a person is working on one task at a time and each task is eventually executed by one person (although it may be allocated to a group a people), it is sufficient to check whether all resource classes have at least one resource. In contrast to many other application domains such a flexible manufacturing systems, anomalies such as a deadlock resulting from locking problems are not possible. Therefore, from the viewpoint of verification, i.e., analyzing the logical correctness of a workflow, it is reasonable to abstract from resources. However, if in the future collaborative features are explicitly supported by the workflow management system (i.e., a tight integration of groupware and workflow technology), then the resource perspective should be taken into account.

We partly abstract from the *data perspective*. The reason we abstract from production data is that these are outside the scope of the WFM system. These data can be changed at any time without notifying the WFM system. In fact their existence does not even depend upon the workflow application and they may be shared among different workflows, e.g., the bill-of-material in manufacturing is shared by production, procurement, sales, and quality control processes. The control data used by the WFM system to route cases are managed by the WFM system. However, some of these data are set or updated by humans or applications. For example, a decision is made by a manager based on intuition or a case is classified based on a complex calculation involving production data. Clearly, the behavior of a human or a complex application cannot be modeled completely. Therefore, some abstraction is needed to incorporate the data perspective when verifying a given workflow. The abstraction used in this section is the following. Since control data (i.e., workflow attributes such as the age of a customer, the department responsible, or the registration date) are only used for the routing of a case, we incorporate the routing decisions but not the actual data. For example, the decision to accept or to reject an insurance claim is taken into account, but not the actual data where this decision is based on. Therefore, we consider each choice to be a non-deterministic one. There are other reasons for abstracting from the workflow attributes. If we are able to prove soundness (i.e., the correctness criterion used in this section) for the situation without workflow attributes, it will also hold for the situation with workflow attributes (assuming certain fairness properties). Last but not least, we abstract from triggers and workflow attributes because it allows us to use ordinary Petri nets (i.e., P/T nets) rather than high-level Petri nets. From an analysis point of view, this is preferable because of the availability of efficient algorithms and powerful analysis tools.

For similar reasons we (partly) abstract from the *task* and *operation perspectives*. We consider tasks to be atomic and abstract from the execution of operations inside tasks. The WFM system can only launch applications or trigger people and monitor the results. It cannot control the actual execution of the task. Therefore, from the viewpoint of verification, it is reasonable to focus on the control-flow perspective. In fact, it suffices to consider the life cycle of one case in isolation. The only way cases interact directly is through the competition for resources and the sharing of production data. (Note that control data are strictly separated.) Therefore, if we abstract from resources and data, it suffices to consider one case in isolation. The competition between cases for resources is only relevant for performance analysis.

3.3 Soundness

In this section we summarize some of the basic results for WF-nets presented in [1,3,4].

The three requirements stated in Definition 11 can be verified statically, i.e., they only relate to the structure of the Petri net. However, there is another requirement which should be satisfied:

> *For any case, the procedure will terminate eventually and the moment the procedure terminates there is a token in place o and all the other places are empty.*

Moreover, there should be no dead tasks, i.e., it should be possible to execute an arbitrary task by following the appropriate route though the WF-net. These two additional requirements correspond to the so-called *soundness property*.

Definition 12 (Sound). *A procedure modeled by a WF-net $PN = (P, T, F)$ is sound if and only if:*

(i) For every state M reachable from state i, there exists a firing sequence leading from state M to state o. Formally[2]*:*

$$\forall_M (i \xrightarrow{*} M) \Rightarrow (M \xrightarrow{*} o)$$

(ii) State o is the only state reachable from state i with at least one token in place o. Formally:

$$\forall_M (i \xrightarrow{*} M \ \wedge \ M \geq o) \Rightarrow (M = o)$$

(iii) There are no dead transitions in (PN, i). Formally:

$$\forall_{t \in T} \ \exists_{M,M'} \ i \xrightarrow{*} M \xrightarrow{t} M'$$

Note that the soundness property relates to the dynamics of a WF-net. The first requirement in Definition 12 states that starting from the initial state (state *i*), it is always possible to reach the state with one token in place *o* (state *o*). If we assume a strong notion of fairness, then the first requirement implies that eventually state *o* is reached. Strong fairness means in every infinite firing sequence, each transition fires infinitely often. The fairness assumption is reasonable in the context of WFM: All choices are made (implicitly or explicitly) by applications, humans or external actors. Clearly, they should not introduce an infinite loop. Note that the traditional notions of fairness (i.e., weaker forms of fairness with just local conditions, e.g., if a transition is enabled infinitely often, it will fire eventually) are not sufficient. See [3, 53] for more details. The second requirement states that the moment a token is put in place *o*, all the other places should be empty. The last requirement states that there are no dead transitions (tasks) in the initial state *i*.

The WF-net shown in Figure 5 is sound. This can be verified by checking the three requirements stated in Definition 12. Note that Figure 5 shows triggers and uses syntactic sugaring. For verification we will abstract from this and consider the pure WF-net as shown in Figure 4.

Figure 6 shows a WF-net which is not sound. There are several deficiencies. If *time_out_1* and *processing_2* fire or *time_out_2* and *processing_1* fire, the WF-net will not terminate properly because a token gets stuck in *c4* or *c5*. If *time_out_1* and *time_out_2* fire, then the task *processing_NOK* will be executed twice and because of the presence of two tokens in *o* the moment of termination is not clear.

Given a WF-net $PN = (P, T, F)$, we want to decide whether PN is sound. In [1] we have shown that soundness corresponds to liveness and boundedness. To link soundness to liveness and boundedness, we define an extended net $\overline{PN} = (\overline{P}, \overline{T}, \overline{F})$.

---

[2] Note that there is an overloading of notation: the symbol *i* is used to denote both the *place i* and the *state* with only one token in place *i* (see Section 2.2).

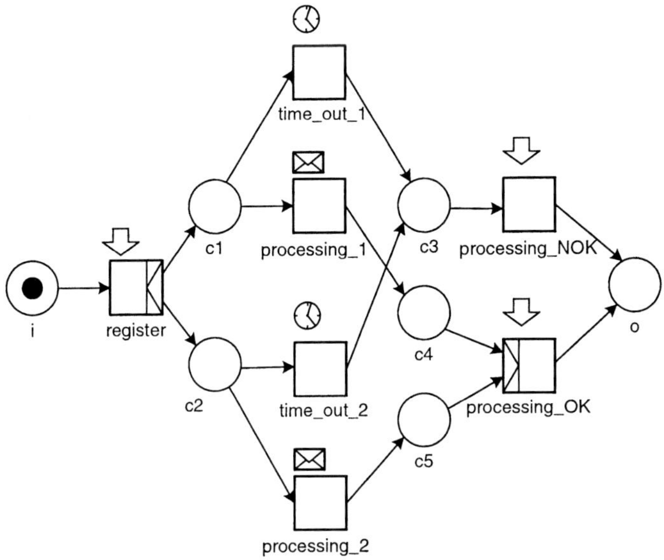

Fig. 6. Another WF-net for the processing of complaints.

$\overline{PN}$ is the Petri net obtained by adding an extra transition t^* which connects o and i. The extended Petri net $\overline{PN} = (\overline{P}, \overline{T}, \overline{F})$ is defined as follows: $\overline{P} = P, \overline{T} = T \cup \{t^*\}$, and $\overline{F} = F \cup \{(o, t^*), (t^*, i)\}$. In the remainder we will call such an extended net the *short-circuited* net of PN. The short-circuited net allows for the formulation of the following theorem.

Theorem 1. *A WF-net PN is sound if and only if* $(\overline{PN}, i)$ *is live and bounded.*

Proof. See [1]. □

This theorem shows that standard Petri-net-based analysis techniques can be used to verify soundness.

In literature there exist many variants of the "classical" notion of soundness used here. Juliane Dehnert uses the notion of relaxed soundness where proper termination is possible but not guaranteed [28, 34]. The main idea is that the scheduler of the workflow system should avoid problems like deadlocks etc. In [54] Ekkart Kindler et al. define variants of soundness tailored towards interorganizational workflows. Kees van Hee et al. [44] define a notion of soundness where multiple tokens in the source place are considered. A WF-net is k-sound if it "behaves well" when there are k tokens in place i, i.e., no deadlocks and in the end there are k tokens in place o. Robert van der Toorn uses the same concept in [71]. In [18, 7] stronger notions of soundness are used and places have to be safe. Another notion of soundness is used in [51, 52] where there is not a single sink place but potentially multiple sink transitions. See [71] for the relation between these variants of the same concept. Other references using (variants of) the soundness property include [41, 60]. For simplicity we restrict ourselves to the classical notion of soundness defined in Definition 12.

3.4 Structural Characterization of Soundness

Theorem 1 gives a useful characterization of the quality of a workflow process definition. However, there are a number of problems:

- For a complex WF-net it may be intractable to decide soundness. (For arbitrary WF-nets liveness and boundedness are decidable but also EXPSPACE-hard, cf. Cheng, Esparza and Palsberg [26].)
- Soundness is a minimal requirement. Readability and maintainability issues are not addressed by Theorem 1.
- Theorem 1 does not show how a non-sound WF-net should be modified, i.e., it does not identify constructs which invalidate the soundness property.

These problems stem from the fact that the definition of soundness relates to the dynamics of a WF-net while the workflow designer is concerned with the static structure of the WF-net. Therefore, it is interesting to investigate structural characterizations of sound WF-nets. For this purpose we introduce three interesting subclasses of WF-nets: *free-choice WF-nets*, *well-structured WF-nets*, and *S-coverable WF-nets*.

Free-Choice WF-Nets. Most of the WFM systems available at the moment, abstract from states between tasks, i.e., states are not represented explicitly. These WFM systems use building blocks such as the AND-split, AND-join, OR-split and OR-join to specify workflow procedures. The AND-split and the AND-join are used for parallel routing. The OR-split and the OR-join are used for conditional routing. Because these systems abstract from states, every choice is made *inside* an OR-split building block. If we model an OR-split in terms of a Petri net, the OR-split corresponds to a number of transitions sharing the same set of input places. This means that for these WFM systems, a workflow procedure corresponds to a free-choice Petri net (cf. Definition 7).

It is easy to see that a process definition composed of AND-splits, AND-joins, OR-splits and OR-joins is free-choice. If two transitions t_1 and t_2 share an input place ($\bullet t_1 \cap \bullet t_2 \neq \emptyset$), then they are part of an OR-split, i.e., a "free choice" between a number of alternatives. Therefore, the sets of input places of t_1 and t_2 should match ($\bullet t_1 = \bullet t_2$). Figure 6 shows a free-choice WF-net. The WF-net shown in Figure 4 is not free-choice; *archive* and *reminder* share an input place but the two corresponding input sets differ.

We have evaluated many WFM systems and only some of these systems (e.g., COSA [66]) allow for a construct which is comparable to a non-free choice WF-net. Therefore, it makes sense to consider free-choice Petri nets in more detail. Clearly, parallelism, sequential routing, conditional routing and iteration can be modeled without violating the free-choice property. Another reason for restricting WF-nets to free-choice Petri nets is the following. If we allow non-free-choice Petri nets, then the choice between conflicting tasks *may* be influenced by the order in which the preceding tasks are executed. The routing of a case should be independent of the order in which tasks are executed. A situation where the free-choice property is violated is often a mixture of parallelism and choice. Figure 7 shows such a situation. Firing transition *t1* introduces parallelism. Although there is no real choice between *t2* and *t5* (*t5* is not enabled), the parallel execution of *t2* and *t3* results in a situation where *t5* is not allowed to occur. However, if the execution of *t2* is delayed until *t3* has been executed, then there is a real choice between *t2* and *t5*. In our opinion parallelism itself should be separated from the choice between two or more alternatives. Therefore, we consider the non-free-choice construct shown in Figure 7 to be improper. In literature, the term *confusion* is often used to refer to the situation shown in Figure 7.

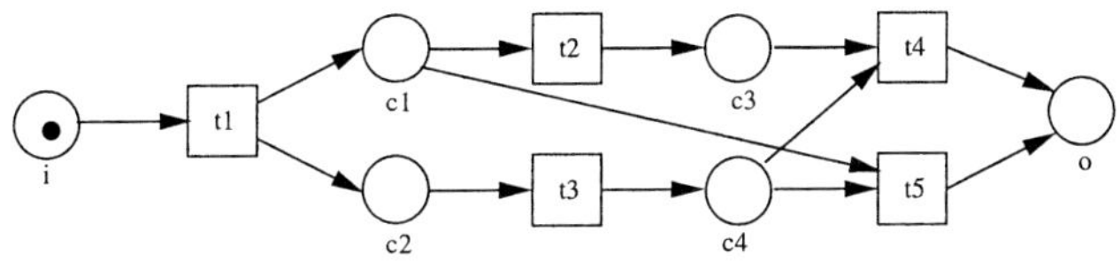

Fig. 7. A non-free-choice WF-net containing a mixture of parallelism and choice.

Free-choice Petri nets have been studied extensively [30, 29, 35, 43], because they seem to be a good compromise between expressive power and analyzability. It is a class of Petri nets for which strong theoretical results and efficient analysis techniques exist. For example, the well-known Rank Theorem [30] enables us to formulate the following corollary.

Corollary 1. *The following problem can be solved in polynomial time. Given a free-choice WF-net, to decide if it is sound.*

Proof. Let PN be a free-choice WF-net. The short-circuited net $\overline{PN}$ is also free-choice. Therefore, the problem of deciding whether $(\overline{PN}, i)$ is live and bounded can be solved in polynomial time (Rank Theorem [30]). By Theorem 1, this corresponds to soundness. □

Corollary 1 shows that, for free-choice nets, there are efficient algorithms to decide soundness. Moreover, a sound free-choice WF-net is guaranteed to be safe (given an initial state with just one token in i).

Lemma 1. *A sound free-choice WF-net is safe.*

Proof. Let PN be a sound free-choice WF-net. $\overline{PN}$ is the Petri net PN extended with a transition connecting o and i. $\overline{PN}$ is free-choice and well-formed. Hence, $\overline{PN}$ is S-coverable [30], i.e., each place is part of an embedded strongly connected state-machine component. Since initially there is just one token $(\overline{PN}, i)$ is safe and so is (PN, i). □

Safeness is a desirable property, because it makes no sense to have multiple tokens in a place representing a condition. A condition is either true (1 token) or false (no tokens).

Although most WFM systems only allow for free-choice workflows, free-choice WF-nets are not a completely satisfactory structural characterization of "good" workflows. On the one hand, there are non-free-choice WF-nets which correspond to sensible workflows (cf. Figure 4). On the other hand there are sound free-choice WF-nets which make no sense. Nevertheless, the free-choice property is a desirable property. If a workflow can be modeled as a free-choice WF-net, one should do so. A workflow specification based on a free-choice WF-net can be enacted by most workflow systems. Moreover, a free-choice WF-net allows for efficient analysis techniques and is easier to understand. Non-free-choice constructs such as the construct shown in Figure 7 are a potential source of anomalous behavior (e.g., deadlock) which is difficult to trace.

Well-Structured WF-Nets. Another approach to obtain a structural characterization of "good" workflows, is to balance AND/OR-splits and AND/OR-joins. Clearly, two parallel flows initiated by an AND-split, should not be joined by an OR-join. Two alternative flows created via an OR-split, should not be synchronized by an AND-join. As shown in Figure 8, an AND-split should be complemented by an AND-join and an OR-split should be complemented by an OR-join.

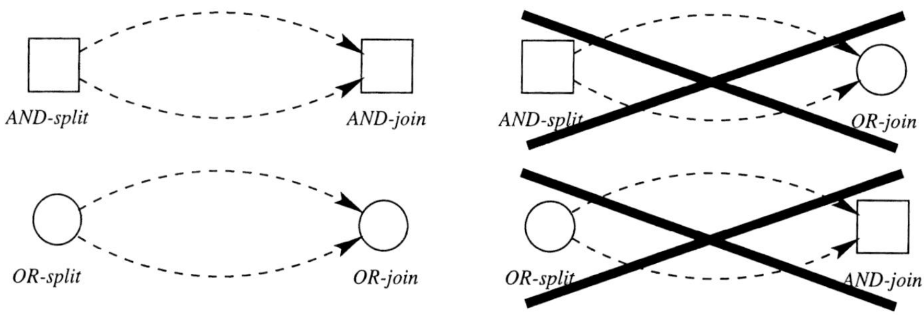

Fig. 8. Good and bad constructions.

One of the deficiencies of the WF-net shown in Figure 6 is the fact that the AND-split *register* is complemented by the OR-join *c3* or the OR-join *o*. To formalize the concept illustrated in Figure 8 we give the following definition.

Definition 13 (Well-handled). *A Petri net PN is well-handled iff, for any pair of nodes x and y such that one of the nodes is a place and the other a transition and for any pair of elementary paths C_1 and C_2 leading from x to y, $\alpha(C_1) \cap \alpha(C_2) = \{x, y\} \Rightarrow C_1 = C_2$.*

Note that the WF-net shown in Figure 6 is not well-handled. Well-handledness can be decided in polynomial time by applying a modified version of the max-flow min-cut technique. A Petri net which is well-handled has a number of nice properties, e.g., strong connectedness and well-formedness coincide.

Lemma 2. *A strongly connected well-handled Petri net is well-formed.*

Proof. Let *PN* be a strongly connected well-handled Petri net. Clearly, there are no circuits that have PT-handles nor TP-handles [36]. Therefore, the net is structurally bounded (See Theorem 3.1 in [36]) and structurally live (See Theorem 3.2 in [36]). Hence, *PN* is well-formed. □

Clearly, well-handledness is a desirable property for any WF-net PN. Moreover, we also require the short-circuited $\overline{PN}$ to be well-handled. We impose this additional requirement for the following reason. Suppose we want to use PN as a part of a larger WF-net PN'. PN' is the original WF-net extended with an "undo-task". See Figure 9. Transition *undo* corresponds to the undo-task, transitions $t1$ and $t2$ have been added to make PN' a WF-net. It is undesirable that transition *undo* violates the well-handledness property of the original net. However, PN' is well-handled iff $\overline{PN}$ is well-handled. Therefore, we require $\overline{PN}$ to be well-handled. We use the term *well-structured* to refer to WF-nets whose extension is well-handled.

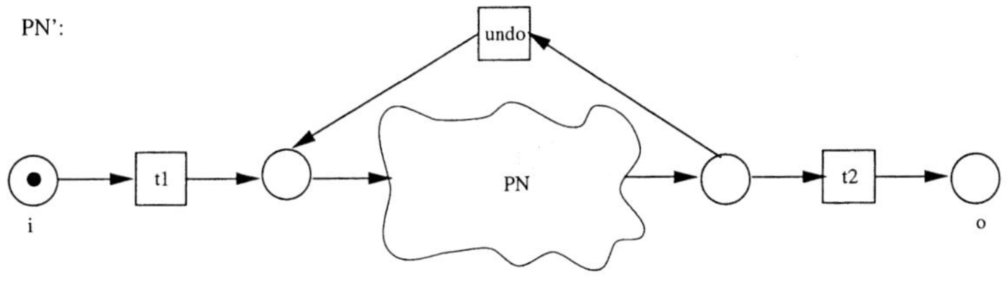

Fig. 9. The WF-net PN' is well-handled iff $\overline{PN}$ is well-handled.

Definition 14 (Well-structured). *A WF-net PN is well-structured iff* $\overline{PN}$ *is well-handled.*

Well-structured WF-nets have a number of desirable properties. Soundness can be verified in polynomial time and a sound well-structured WF-net is safe. To prove these properties we use some of the results obtained for *elementary extended non-self controlling nets.*

Definition 15 (Elementary extended non-self controlling). *A Petri net PN is elementary extended non-self controlling (ENSC) iff, for every pair of transitions* t_1 *and* t_2 *such that* $\bullet t_1 \cap \bullet t_2 \neq \emptyset$*, there does not exist an elementary path* C *leading from* t_1 *to* t_2 *such that* $\bullet t_1 \cap \alpha(C) = \emptyset$.

Theorem 2. *Let PN be a WF-net. If PN is well-structured, then* $\overline{PN}$ *is elementary extended non-self controlling.*

Proof. Assume that $\overline{PN}$ is not elementary extended non-self controlling. This means that there is a pair of transitions t_1 and t_k such that $\bullet t_1 \cap \bullet t_k \neq \emptyset$ and there exist an elementary path $C = \langle t_1, p_2, t_2, \ldots, p_k, t_k \rangle$ leading from t_1 to t_k and $\bullet t_1 \cap \alpha(C) = \emptyset$. Let $p_1 \in \bullet t_1 \cap \bullet t_k$. $C_1 = \langle p_1, t_k \rangle$ and $C_2 = \langle p_1, t_1, p_2, t_2, \ldots, p_k, t_k \rangle$ are paths leading from p_1 to t_k. (Note that C_2 is the concatenation of $\langle p_1 \rangle$ and C.) Clearly, C_1 is elementary. We will also show that C_2 is elementary. C is elementary, and $p_1 \notin \alpha(C)$ because $p_1 \in \bullet t_1$. Hence, C_2 is also elementary. Since C_1 and C_2 are both elementary paths, $C_1 \neq C_2$ and $\alpha(C_1) \cap \alpha(C_2) = \{p_1, t_k\}$, we conclude that $\overline{PN}$ is not well-handled. □

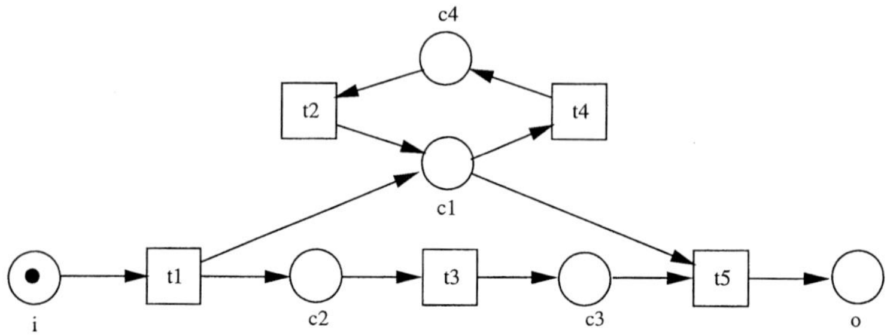

Fig. 10. A well-structured WF-net.

Consider for example the WF-net shown in Figure 10. The WF-net is well-structured and, therefore, also elementary extended non-self controlling. However, the net is not free-choice. Nevertheless, it is possible to verify soundness for such a WF-net very efficiently.

Corollary 2. *The following problem can be solved in polynomial time. Given a well-structured WF-net, to decide if it is sound.*

Proof. Let PN be a well-structured WF-net. The short-circuited net $\overline{PN}$ is elementary extended non-self controlling (Theorem 2) and structurally bounded (see proof of Lemma 2). For bounded elementary extended non-self controlling nets the problem of deciding whether a given marking is live, can be solved in polynomial time (See [23]). Therefore, the problem of deciding whether $(\overline{PN}, i)$ is live and bounded can be solved in polynomial time. By Theorem 1, this corresponds to soundness. □

Lemma 3. *A sound well-structured WF-net is safe.*

Proof. Let $\overline{PN}$ be the net PN extended with a transition connecting o and i. $\overline{PN}$ is extended non-self controlling. $\overline{PN}$ is covered by state-machines (S-components), see Corollary 5.3 in [23]. Hence, $\overline{PN}$ is safe and so is PN (see proof of Lemma 1). □

Well-structured WF-nets and free-choice WF-nets have similar properties. In both cases soundness can be verified very efficiently and soundness implies safeness. In spite of these similarities, there are sound well-structured WF-nets which are not free-choice (Figure 10) and there are sound free-choice WF-nets which are not well-structured. In fact, it is possible to have a sound WF-net which is neither free-choice nor well-structured (Figures 4 and 7).

S-Coverable WF-Nets. What about the sound WF-nets shown in Figure 4 and Figure 7? The WF-net shown in Figure 7 can be transformed into a free-choice well-structured WF-net by separating choice and parallelism. The WF-net shown in Figure 4 cannot be transformed into a free-choice or well-structured WF-net without yielding a much more complex WF-net. Place *c7* acts as some kind of milestone which is tested by the task *reminder*. Traditional WFM systems which do not make the state of the case explicit, are not able to handle the workflow specified by Figure 4. Only WFM systems such as COSA [66] have the capability to enact such a state-based workflow. Nevertheless, it is interesting to consider generalizations of free-choice and well-structured WF-nets: *S-coverable WF-nets* can be seen as such a generalization.

Definition 16 (S-coverable). *A WF-net PN is S-coverable if the short-circuited net* $\overline{PN}$ *is S-coverable.*

The WF-nets shown in Figure 4 and Figure 7 are S-coverable. The WF-net shown in Figure 6 is not S-coverable. The following two corollaries show that S-coverability is a generalization of the free-choice property and well-structuredness.

Corollary 3. *A sound free-choice WF-net is S-coverable.*

Proof. The short-circuited net $\overline{PN}$ is free-choice and well-formed. Hence, $\overline{PN}$ is S-coverable (cf. [30]). □

Corollary 4. *A sound well-structured WF-net is S-coverable.*

Proof. $\overline{PN}$ is extended non-self controlling (Theorem 2). Hence, $\overline{PN}$ is S-coverable (cf. Corollary 5.3 in [23]). □

All the sound WF-nets presented in this tutorial are S-coverable. Every S-coverable WF-net is safe. The only WF-net which is not sound, i.e., the WF-net shown in Figure 6, is not S-coverable. These and other examples indicate that there is a high correlation between S-coverability and soundness. It seems that S-coverability is one of the basic requirements any workflow process definition should satisfy. From a formal point of view, it is possible to construct WF-nets which are sound but not S-coverable. Typically, these nets contain places which do not restrict the firing of a transition, but which are not in any S-component. (See for example Figure 65 in [62].) From a practical point of view, these WF-nets are to be avoided. WF-nets which are not S-coverable are difficult to interpret because the structural and dynamical properties do not match. For example, these nets can be live and bounded but not structurally bounded. There seems to be no practical need for using constructs which violate the S-coverability property. Therefore, we consider S-coverability to be a basic requirement any WF-net should satisfy.

Another way of looking at S-coverability is the following interpretation: S-components corresponds to *document flows*. To handle a workflow several pieces of information are created, used, and updated. One can think of these pieces of information as physical documents, i.e., at any point in time the document is in one place in the WF-net. Naturally, the information in one document can be copied to another document while executing a task (i.e., transition) processing both documents. Initially, all documents are present but a document can be empty (i.e., corresponds to a blank piece paper). It is easy to see that the flow of one such document corresponds a state machine (assuming the existence of a transition t^*). These document flows synchronize via joint tasks. Therefore, the composition of these flows yields an S-coverable WF-net. One can think of the document flows as threads. Consider for example the short-circuited net of the WF-net shown in Figure 4. This net can be composed out of the following two threads: (1) a thread corresponding to logistic subprocess (places $start$, $c0$, $c1$, $c3$,

$c4$, $c5$, $c7$, and end) and (2) a thread corresponding to the actual processing of the complaint (places $start$, $c2$, $c6$, $c8$, and end). Note that the tasks *register* and *archive* are used in both threads.

Although a WF-net can, in principle, have exponentially many S-components, they are quite easy to compute for workflows encountered in practice (see also the above interpretation of S-component as document flows or threads). Note that S-coverability only depends on the structure and the degree of connectedness is generally low (i.e., the incidence matrix of a WF-net typically has few non-zero entries). Unfortunately, in general, it is not possible to verify soundness of an S-coverable WF-net in polynomial time. The problem of deciding soundness for an S-coverable WF-net is PSPACE-complete. For most applications this is not a real problem. Typically, the number of tasks in one workflow process definition is less than 100 and the number of states is less than half a million. Tools using standard techniques such as the construction of the coverability graph have no problems in coping with these workflow process definitions.

Using the Three Structural Characterizations. The three structural characterizations (free-choice, well-structured and S-coverable) turn out to be very useful for the analysis of workflow process definitions. Based on our experience, we have good reasons to believe that S-coverability is a desirable property any workflow definition should satisfy. Constructs violating S-coverability can be detected easily and tools can be build to help the designer to construct an S-coverable WF-net. S-coverability is a generalization of well-structuredness and the free-choice property (Corollary 3 and 4). Both well-structuredness and the free-choice property also correspond to desirable properties of a workflow. A WF-net satisfying at least one one of these two properties can be analyzed very efficiently. However, we have shown that there are workflows that are not free-choice and not well-structured. Consider for example Figure 4. The fact that task *register* tests whether there is a token in *c5*, prevents the WF-net from being free-choice or well-structured. Although this is a very sensible workflow, most WFM systems do not support such an advanced routing construct. Even if one is able to use state-based workflows (e.g., COSA) allowing for constructs which violate well-structuredness and the free-choice property, then the structural characterizations are still useful. If a WF-net is not free-choice or not well-structured, one should locate the source which violates one of these properties and check whether it is really necessary to use a non-free-choice or a non-well-structured construct. If the non-free-choice or non-well-structured construct is really necessary, then the correctness of the construct should be double-checked, because it is a potential source of errors. This way the readability and maintainability of a workflow process definition can be improved.

3.5 Woflan

Few tools aiming at the verification of workflow processes exist. Woflan [73, 72] and Flowmake [64] are two notable exceptions. We have been working on Woflan since 1997. Figure 11 shows a screenshot of Woflan. Woflan combines state-of-the-art scientific results with practical applications [73, 72]. Woflan can interface with leading WFM systems such as Staffware, MQSeries Workflow and COSA but also PNML [24].

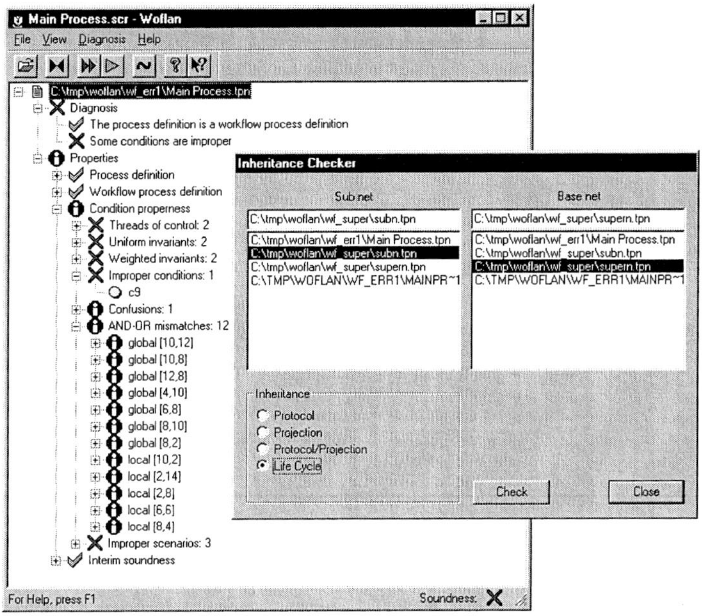

Fig. 11. A screenshot showing the verification and validation capabilities of Woflan.

It can also interface with BPR-tools such as Protos. Workflow processes designed using any of these tools can be verified for correctness. It turns out that the challenge is not to decide whether the design is sound or not. The real challenge is to provide diagnostic information that guides the designer to the error. Woflan also supports the inheritance notions mentioned before. Given two workflow designs, Woflan is able to decide whether one is a subclass of the other. Tools such as Woflan illustrate the benefits of a more fundamental approach.

3.6 Exercise

Consider the solution of the exercise given in Section 2.4. Verify whether the WF-net is sound and make sure that there is an S-cover. A simple verification "web service" is provided via `http://is.tm.tue.nl/research/woflan/`. This web service uses Woflan to verify whether a given process model is sound. Use this web service or download Woflan to check the correctness of your solution.

4 Workflow Management Systems

In this section we provide insight into the functionality of existing WFM systems. First we provide an overview of the workflow market. Then we introduce the typical architecture of a WFM system, followed by an example of a concrete system. Again, we conclude the section with an exercise.

4.1 Overview

In Section 1 we put WFM in a historical perspective and using Figure 2 we discussed the maturity of the BPM market. At this point in time hundreds of WFM/BPM products are available. To illustrate this we use two diagrams of Michael Zur Muehlen [61]. Figure 12 gives a historic overview of office automation and workflow prototypes [61]. Figure 13 provides a historic overview of commercial WFM systems. These two figures show that: (1) workflow management is not something that started in the nineties but already in the seventies with the work of Ellis (*OfficeTalk*, [32]) and Zisman (*Scoop*,[78]) and (2) the number of commercial systems has considerably grown in recent years. Note that given the dynamics of the workflow market, it is difficult to keep diagrams like the one shown in Figure 13 up-to-date. For example, Figure 13 does not show recent systems like FLOWer [22]. Moreover, systems are often named different for commercial reasons. For example, IBM's MQSeries Workflow (formerly known as FlowMark) was recently renamed into WebSphere MQ Workflow.

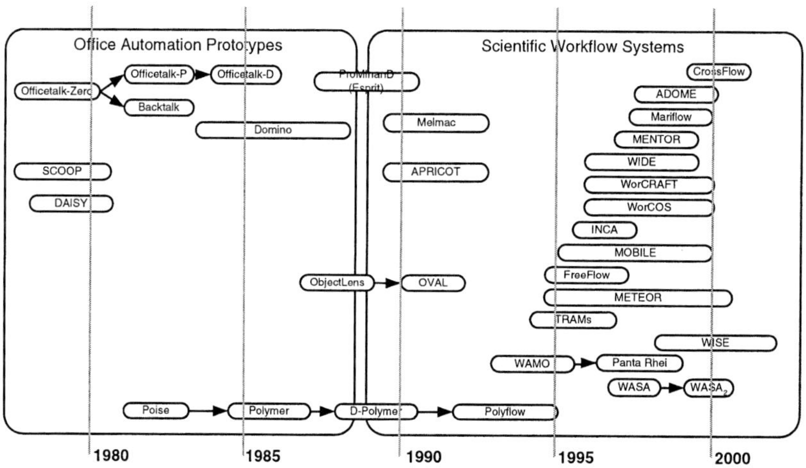

Fig. 12. Historic overview of early systems and research prototypes (Taken from [61]).

Unfortunately, figures 12 and 13 do not show the increased maturity of WFM/BPM products. It also does not show that products target at different types of processes. A well-know classification of WFM systems is given in [40] where the authors distinguish between ad-hoc, administrative, and production workflows and discuss the continuum from human-oriented to system-oriented WFM systems. However, we prefer to use the more recent classification shown in Figure 14 to describe the "workflow spectrum".

Figure 14 shows four types of systems: groupware, production workflow, ad-hoc workflow, and case-handling systems. These systems are characterized in terms of their "focus" (data-driven, process driven, or both) and their "degree of structuredness". Traditional groupware products like Lotos Notes and MS Exchange and production workflow systems like Staffware and MQSeries Workflow form two ends of a spectrum. As

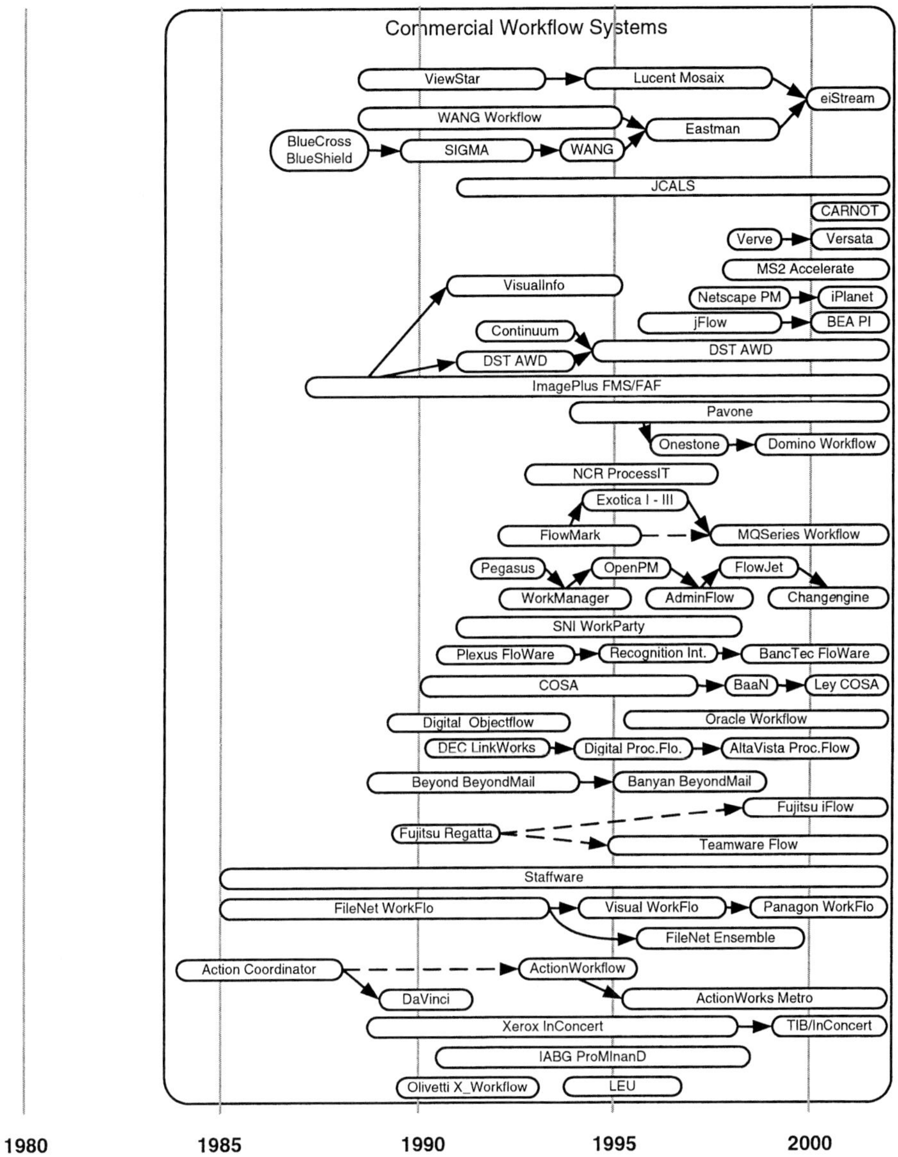

Fig. 13. Historic overview of commercial workflow management systems (Taken from [61]).

Figure 14 shows, traditional groupware products are data-driven (focus on the sharing of information rather than the process) and support only unstructured processes. Note that Lotus Notes and Exchange are not "process-aware" (unless components like Domino Workflow are added). Production workflow systems are process-aware and aim at structured processes. In order to enact a workflow using a production workflow

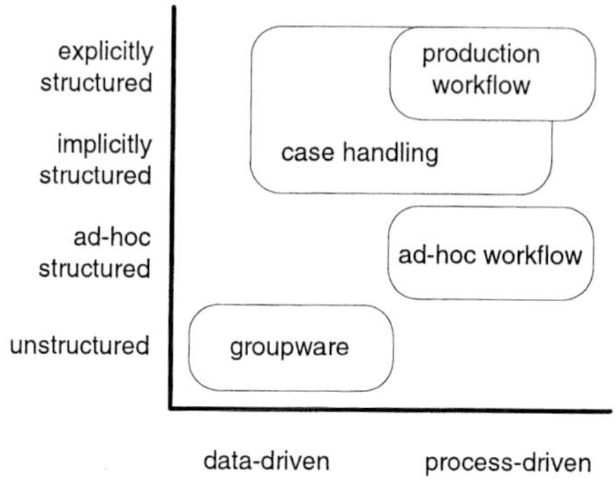

Fig. 14. Classification of systems to support work processes.

system one needs to explicitly specify all possible routes. If something is not explicitly specified at design time, it is not possible. Ad-hoc WFM systems like InConcert (TIBCO), Ensemble (Filenet), and TeamWARE Flow (TeamWARE Group) allow for the creation and modification of workflow processes at execution time. Each case has a private process model and therefore the traditional problems encountered when changing a workflow specification can be avoided. Ad-hoc WFM systems allow for a lot of flexibility. The WFM system InConcert even allows the user to initiate a case having an empty process model. When the case is handled, the workflow model is extended to reflect the work conducted. Another possibility is to start using a template. The moment a case is initiated, the corresponding process model is instantiated using a template. After instantiation, the case has a private copy of the template, which can be modified while the process is running. InConcert also supports "workflow design by discovery": The routing of any completed workflow instance can be used to create a new template. This way actual workflow executions can be used to create workflow process definitions. Figure 14 shows that ad-hoc workflow management systems like InConcert are process-driven and ad-hoc structured. Case-handling systems like FLOWer and Vectus can be positioned in-between groupware, production workflow, and ad-hoc workflow. Unlike in ad-hoc workflow systems the end-users are not expected to change or create process models. Instead the following paradigms are used for case handling [8]:

- avoid context tunneling by providing all information available (i.e., present the case as a whole rather than showing just bits and pieces),
- decide which activities are enabled on the basis of the information available rather than the activities already executed,
- separate work distribution from authorization and allow for additional types of roles, not just the execute role,
- allow workers to view and add/modify data before or after the corresponding activities have been executed (e.g., information can be registered the moment it becomes available).

For more information on case handling we refer to [8,22]. Clearly the classification of systems is not as clear-cut as Figure 14 may suggest. Lotus Notes can be extended with Domino Workflow to join groupware and production workflow functionalities.

Staffware Case Handler and the COSA Activity Manager are extensions of production workflow systems in the direction of case handling (both are based on the generic solution of BPi).

In this tutorial we focus on the classical production workflow systems. However, it is important to understand that they are part of a spectrum and that their application is limited to a specific type of processes (process-driven and explicitly structured).

4.2 Architecture

As indicated by Figure 14, WFM systems target at different processes. Therefore, it is not surprising that there is not one architecture that "fits all systems". Therefore, we present the so-called *reference model* of the Workflow Management Coalition (WfMC) [38, 55]. Figure 15 shows an overview of this reference model. The reference model describes the major components and interfaces within a workflow architecture. The core of any workflow system is the *workflow enactment service*. The workflow enactment service provides the run-time environment which takes care of the control and execution of the workflow. For technical or managerial reasons the workflow enactment service may use multiple *workflow engines*. A workflow engine handles selected parts of the workflow and manages selected parts of the resources. The *process definition tools* are used to specify and analyze workflow process definitions and/or resource classifications. These tools are used at design time. In most cases, the process definition tools can also be used as a BPR-toolset. Most WFM systems provide three process definition tools: (1) a tool with a graphical interface to define workflow processes, (2) a tool to specify resource classes (organizational model), and (3) a simulation tool to analyze a specified workflow[3]. The end-user communicates with the workflow system via the *workflow client applications*. An example of a workflow client application is the well-known *in-basket*. Via such an in-basket work items are offered to the end user. By selecting a work item, the user can execute a task for a specific case. If necessary, the workflow engine invokes applications via Interface 3. The *administration and monitoring tools* are used to monitor and control the workflow. These tools are used to register the progress of cases and to detect bottlenecks. Moreover, these tools are also used to set parameters, allocate people and handle abnormalities. Via Interface 4 the workflow system can be connected to other workflow systems. To standardize the five interfaces shown in Figure 15, the WfMC aims at a common *Workflow Application Programming Interface* (WAPI). The WAPI is envisaged as a common set of API calls and related interchange formats which may be grouped together to support each of the five interfaces (cf. [55]). In Section 5.2 we will describe XPDL, the XML-based language suggested by the WfMC to exchange process definition (i.e., a concrete language for Interface 1).

4.3 Example of a WFM System: Staffware

As indicated in Section 4.1, many WFM systems are available. In this tutorial we only show one system in more detail. Staffware is one of the most widespread WFM systems

[3] In many cases simulation is offered through some export to a standard simulation tool, e.g., COSA supports simulation through an export to ExSpect.

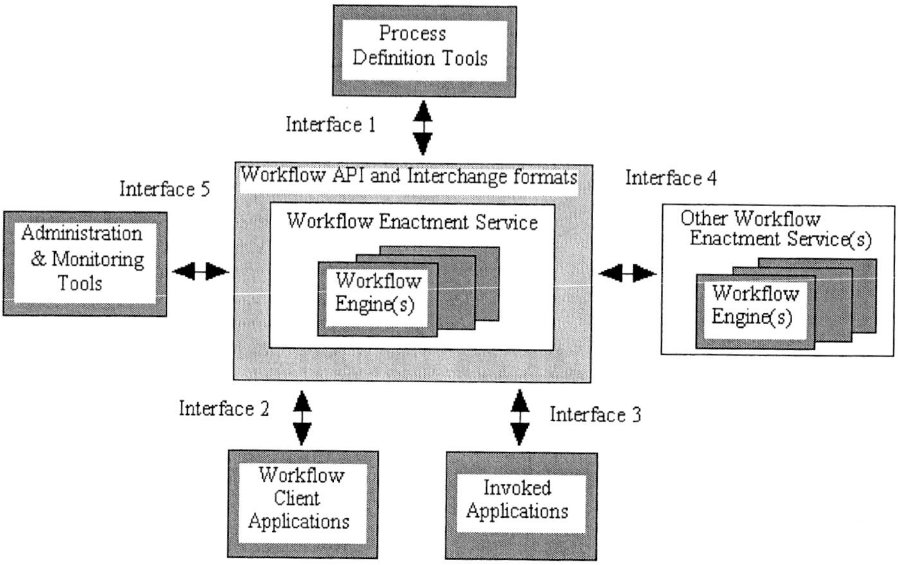

Fig. 15. Reference model of the Workflow Management Coalition (WfMC).

in the world. In 1998, it was estimated by the Gartner Group that Staffware has 25 percent of the global market [25]. Staffware provides the functionality described in the reference model shown in Figure 15. Figure 16 shows some screenshots of the Staffware product. The top window shows the design tool of Staffware while defining a simple workflow process. Work is offered through so-called work queues. One worker can have multiple work queues and one work queue can be shared among multiple workers. The window in the middle shows the set of available work queues (left) and the content of one of these work queues (right). The bottom window shows an audit trail of a case. The three windows show only some of the capabilities offered by Staffware. It is fairly straightforward to map these windows onto the architecture shown in Figure 15.

Let us now consider the modeling language used by Staffware. In Staffware, tasks are called *steps*. There are several kinds of steps: *automatic steps* (offered to an application instead of an end-user), *normal steps* (executed by an end-user), and *event steps* (triggered by some external event). The semantics of a step are OR-join/AND-split, i.e., a step becomes enabled if one of the preceding steps is completed and the completion of step will trigger all subsequent steps. Since the OR-join/AND-split semantics is fixed, two additional building blocks are needed: the *wait step* and the *condition*. The wait step can be used to synchronize flows and has AND-join/AND-split semantics. To model choices, i.e., OR-splits, the condition building block can be used. Staffware only allows for binary choices, i.e., just two possible outcomes (e.g., YES and NO). Staffware processes always start with a start step which is denoted by a symbol representing a traffic light. Termination in Staffware is implicit, i.e., it is possible to start multiple parallel threads which end concurrently. Therefore, there is no need to have one sink node representing the completion of a case. The end of a thread is denoted

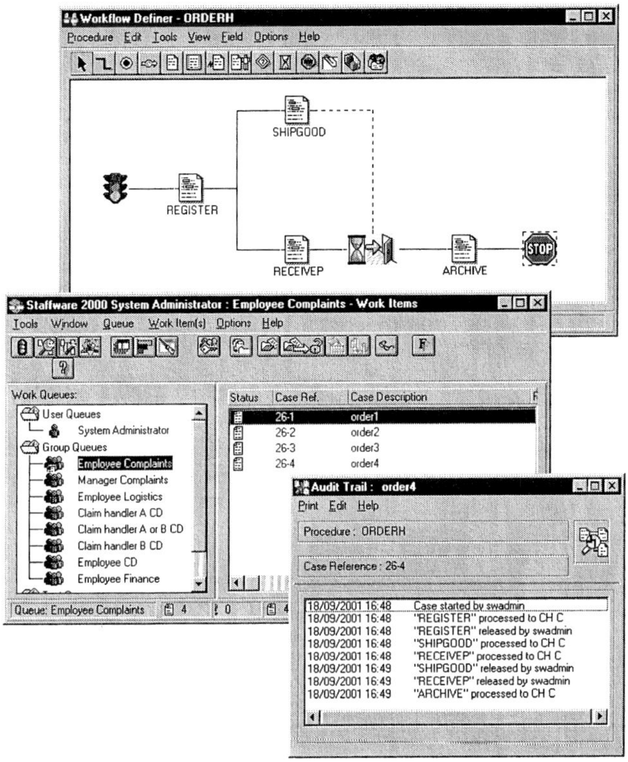

Fig. 16. The Graphical Workflow Definer, Work Queue, and Audit Trail of Staffware.

by a stop symbol. Conditions are modeled by diamond shaped symbols. Wait steps are modeled by symbols in the shape of a sand timer. The basic semantics of a step, a condition, and a wait are shown in Figure 17[4].

Using this translation shown in Figure 17, it is straightforward to map the Staffware model shown in Figure 16 onto a WF-net. The result is shown in Figure 18.

Let us consider now a larger Staffware model also including advanced concepts like time triggers and multiple ending points. For this purpose, we use the model shown in Figure 19. It models a simplified workflow in a travel agency. To organize a trip, a travel agency executes several tasks. First the customer is registered. Then an employee searches for opportunities which are communicated to the customer. Then the customer will be contacted to find out whether she or he is still interested in the trip of this agency and whether more alternatives are desired. There are three possibilities: (1) the customer is not interested at all, (2) the customer would like to see more alternatives, and (3) the customer selects an opportunity. If the customer selects a trip, the trip is booked. In parallel one or two types of insurance are prepared if they are desired. A customer can take insurance for trip cancellation or/and for baggage loss. Note that

[4] Note that the semantics of Staffware steps include a number of particularities not included in the mapping, cf. [72].

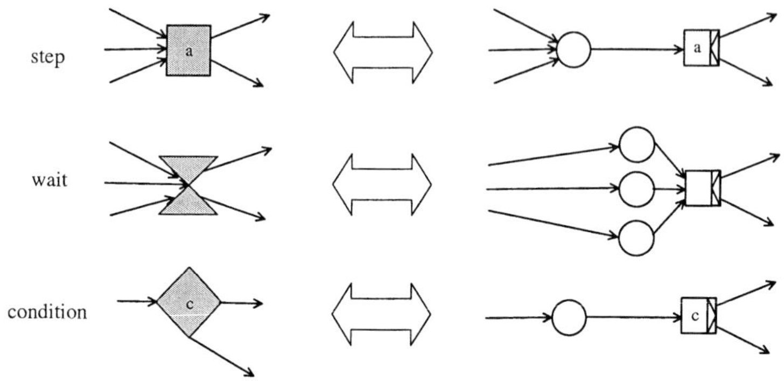

Fig. 17. The semantics of some of the Staffware constructs (left) expressed in Petri nets (right).

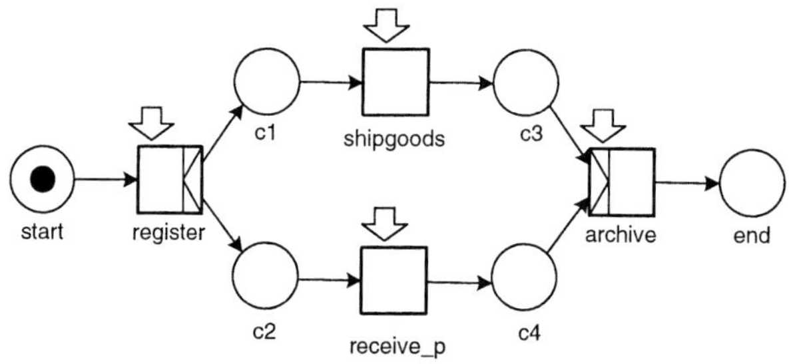

Fig. 18. The Staffware model shown in Figure 16 expressed in terms of a WF-net.

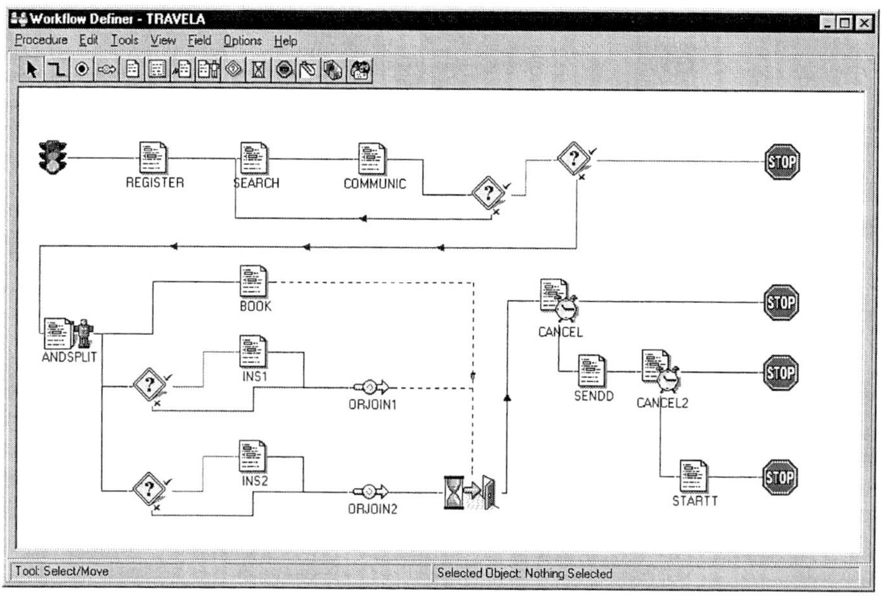

Fig. 19. The workflow of a travel agency modeled in terms of the Staffware language.

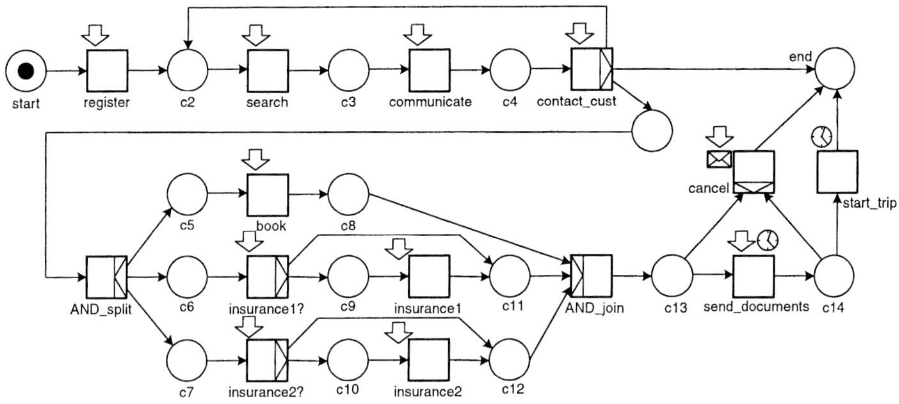

Fig. 20. The travel agency's workflow expressed in terms of a WF-net.

a customer can decide not to take any insurance, just trip cancellation insurance, just baggage loss insurance, or both types of insurance. Two weeks before the start date of the trip the documents are sent to the customer. A trip can be cancelled at any time after completing the booking process (including the insurance) and before the start date. Note that customers who are not insured for trip cancellation can cancel the trip (but will get no refund). Most of the model is self-explanatory. The two OR-join symbols represent "dummy tasks", i.e., Staffware steps not implementing any real task. For the cancellation two steps with a time-out are used: *CANCEL* and *CANCEL2*. The clock symbol is used to indicate steps with a time-out. In such as step, the lower branch is taken if the step is not executed within a given period. For simplicity we did not model all triggers and simplified the choice for both types of insurances.

Let us now translate the model shown in Figure 19 into a WF-net. We do not use the Staffware names but the names used in the original description (Staffware only allows names of up-to 8 characters). The WF-net shown in Figure 20, like any WF-net, has a source place which serves as the start condition (i.e., case creation) and a sink place which serves as the end condition (i.e., case completion). First, the tasks *register*, *search*, *communicate*, and *contact_cust* are executed sequentially. Task *contact_cust* is an OR-split with three possible outcomes: (1) the customer is not interested at all, i.e., a token is put into *end*, (2) the customer would like to see more alternatives, i.e., a token is put into *c2*, and (3) the customer selects an opportunity, i.e., a token is put into *c15* to initiate the booking. Tasks *AND_split* and *AND_join* have just been added for routing purposes. These routing tasks enable the parallel execution of the booking and insurance tasks. The task *book* corresponds to the actual booking of the trip. Tasks *insurance1* and *insurance2* correspond to handling both types of insurance. Since both types of insurance are optional, there is a bypass for each of these tasks. The OR-split *insurance1?* allows for a bypass of task *insurance1* by putting a token in *c11*. After handling the booking and optional insurances the AND-join puts a token in *c13*. The remainder of the process is, from the viewpoint of triggers, very interesting. Note that all tasks executed before this point are either tasks that require a resource trigger or automatic tasks added for routing purposes only. The downward facing arrows denote the resource triggers. If the case is in *c13*, then the normal flow of execution is to first execute task *send_documents* and then execute *start_trip*. Note that task *send_documents* requires both a resource trigger and a time trigger. These two triggers indicate that two weeks before the beginning of the trip a worker sends the documents to the customer. Task *start_trip* has been added for routing purposes and requires a time trigger. Without task *start_trip*, i.e., putting the token in *end* after sending the documents, it would have been impossible to cancel the trip after sending the documents. Task *cancel* is an explicit OR-join and requires both a resource trigger and an external trigger. This task is only executed if it is triggered by the customer. Task cancel can only be executed if the case is in *c13* or *c14*, i.e., after handling the booking and insurance related tasks and before the trip starts.

It is interesting to compare figures 19 and 20. Although the WF-net has more symbols because of the explicit modeling of states (i.e., places), it seems to be a more direct and more elegant way to model the process.

4.4 Exercises

To conclude this section, we provide two small exercises.

- Consider the Staffware process shown in Figure 21. Translate this Staffware model into a WF-net.
- Consider the WF-net shown in Figure 5. Translate this WF-net into a Staffware model, i.e., provide a diagram like the one shown in Figure 21 for the order processing process given in Section 2.3.

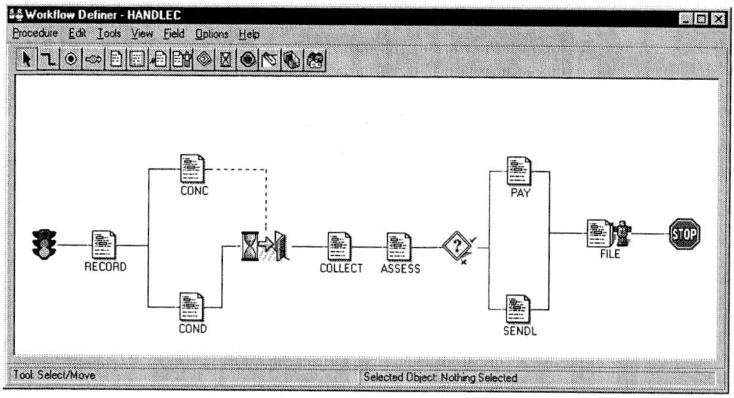

Fig. 21. Exercise: translate this Staffware process into a WF-net.

A solution of the first exercise is given in [12]. A solution for the second exercise is not given and is also far from trivial given the fact that Staffware does not support the Milestone and Deferred choice patterns (cf. Section 5.1 and [15]).

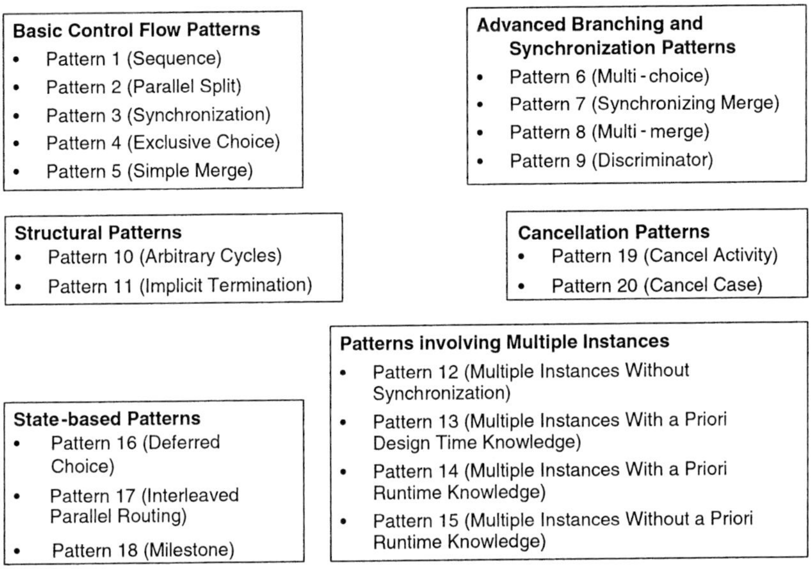

Fig. 22. Overview of the 20 workflow patterns described in [15].

5 Workflow Standards: An Evaluation of XPDL Based on Its Support for Workflow Patterns

There are many languages and standards in the WFM/BPM domain. It is impossible to discuss all of these in any detail. Instead we focus on a single standard. XPDL is not the most important standard but it is typical for many other standards and propriety languages of workflow vendors. For a critical evaluation of XPDL, we use the set of workflow patterns described in [15]. The remainder of this section is structured as follows. First, we introduce the 20 workflow patterns used to evaluate XPDL. Then, in Section 5.2, we provide an overview of the XPDL language. In Section 5.3 the language is analyzed using the 20 workflow patterns. Section 5.4 discusses one of the core semantical problems: The join construct. Finally, Section 5.5 concludes the section by comparing XPDL with WFM systems and other standards such as UML Activity Diagrams, BPEL4WS, BPML, WSFL, XLANG, and WSCI.

5.1 Workflow Patterns

Since 1999 we have been working on collecting a comprehensive set of workflow patterns [15]. The results have been made available through the "Workflow patterns WWW site" `http://www.workflowpatterns.com`. The patterns range from very simple patterns such as sequential routing (Pattern 1) to complex patterns involving complex synchronizations such as the discriminator pattern (Pattern 9). In this tutorial, we restrict ourselves to the 20 most relevant patterns. These patterns can be classified into six categories:

1. *Basic control-flow patterns.* These are the basic constructs present in most workflow languages to model sequential, parallel and conditional routing.
2. *Advanced branching and synchronization patterns.* These patterns transcend the basic patterns to allow for more advanced types of splitting and joining behavior. An example is the Synchronizing merge (Pattern 7) which behaves like an AND-join or XOR-join depending on the context.
3. *Structural patterns.* In programming languages a block structure which clearly identifies entry and exit points is quite natural. In graphical languages allowing for parallelism such a requirement is often considered to be too restrictive. Therefore, we have identified patterns that allow for a less rigid structure.
4. *Patterns involving multiple instances.* Within the context of a single case (i.e., workflow instance) sometimes parts of the process need to be instantiated multiple times, e.g., within the context of an insurance claim, multiple witness statements need to be processed.
5. *State-based patterns.* Typical workflow systems focus only on activities and events and not on states. This limits the expressiveness of the workflow language because it is not possible to have state dependent patterns such as the Milestone pattern (Pattern 18).
6. *Cancellation patterns.* The occurrence of an event (e.g., a customer canceling an order) may lead to the cancellation of activities. In some scenarios such events can even cause the withdrawal of the whole case.

Figure 22 shows an overview of the 20 patterns grouped into the six categories. A detailed discussion of these patterns is outside the scope of this tutorial. The interested reader is referred to [15] and `http://www.workflowpatterns.com`.

We have used these patterns to compare the functionality of numerous WFM systems. The result of this evaluation reveals that (1) the expressive power of contemporary systems leaves much to be desired and (2) the systems support different patterns. Note that we do not use the term "expressiveness" in the traditional or formal sense. If one abstracts from capacity constraints, any workflow language is Turing complete. Therefore, it makes no sense to compare these languages using formal notions of expressiveness. Instead we use a more intuitive notion of expressiveness which takes the modeling effort into account. This more intuitive notion is often referred to as suitability. See [51, 52] for a discussion on the distinction between formal expressiveness and suitability.

The observation that the expressive power of the available WFM systems leaves much to be desired, triggered the question: *How about XPDL as a workflow language?*

5.2 XPDL: XML Process Definition Language

The Workflow Management Coalition (WfMC) was founded in August 1993 as a international non-profit organization. Today there are about 300 members ranging from workflow vendors and users to analysts and university/research groups. The mission of the WfMC is to promote and develop the use of workflow through the establishment of standards for workflow terminology, interoperability and connectivity between workflow products. The WfMC's reference model identifies five interfaces, as shown in Section 4.2. One of the main activities since 1993 has been the development of standards for these interfaces. Interface 1 is the link between the so-called "Process Definition Tools" and the "Enactment Service" (cf. Figure 15). The Process Definition Tools are used to design workflows while the Enactment Service can execute workflows. The primary goal of Interface 1 is the import and export of process definitions. The WfMC defines a process definition as "The representation of a business process in a form which supports automated manipulation, such as modeling, or enactment by a WFM system. The process definition consists of a network of activities and their relationships, criteria to indicate the start and termination of the process, and information about the individual activities, such as participants, associated IT applications and data, etc." [55]. Clearly, there is a need for process definition interchange. First of all, within the context of a single workflow management system there has to be a connection between the design tool and the execution/run-time environment. Second, there may be the desire to use another design tool, e.g., a modeling tool like ARIS or Protos. Third, for analysis purposes it may be desirable to link the design tool to analysis software such as simulation and verification tools. Fourth, the use of repositories with workflow processes requires a standardized language. Fifth, there may be the need to transfer a definition interchange from one engine to another.

To support the interchange of workflow process definitions, there has to be a standardized language [12, 38, 48, 55, 57, 58]. The WfMC started working on such a language soon after it was founded. This resulted in the Workflow Process Definition Language (WPDL) [74] presented in 1999. Although many vendors claimed to be WfMC compliant, few made a serious effort to support this language. At the same time, XML emerged as a standard for data interchange. Since WPDL was not XML-based, the WfMC started working a new language named XML Process Definition Language (XPDL). The starting point for XPDL was WPDL. However, XPDL should not be considered the XML version of WPDL. Several concepts have been added/changed and the WfMC remains fuzzy about the exact relationship between XPDL and WPDL. In October 2002, the WfMC released a "Final Draft" of XPDL [75].

In [75], the authors state "More complex transitions, which cannot be expressed using the simple elementary transition and the split and join functions associated with the from- and to- activities, are formed using dummy activities, which can be specified as intermediate steps between real activities allowing additional combinations of split and/or join operations. *Using the basic transition entity plus dummy activities, routing structures of arbitrary complexity can be specified.* Since several different approaches to transition control exist within the industry, several conformance classes are specified within XPDL. These are described later in the document." The sentence "Using the basic transition entity plus dummy activities, routing structures of *arbitrary complexity* can be specified." triggered us to look into the expressive power of XPDL.

XPDL [75] uses an XML-based syntax, specified by an XML schema. The main elements of the language are: Package, Application, WorkflowProcess, Activity, Transition, Participant, DataField, and DataType. The Package element is the container holding the other elements. The Application element is used to specify the applications/tools invoked by the workflow processes defined in a package. The element WorkflowProcess is used to define workflow processes or parts of workflow processes. A WorkflowProcess is composed of elements of type Activity and Transition. The Activity element is the basic building block of a workflow process definition. Elements of type Activity are connected through elements of type Transition. There are three types of activities: Route, Implementation, and BlockActivity. Activities of type Route are dummy activities just used for routing purposes. Activities of type BlockActivity are used to execute sets of smaller activities. Element ActivitySet refers to a self contained set of activities and transitions. A BlockActivity executes such an ActivitySet. Activities of type Implementation are steps in the process which are implemented by manual procedures (No), implemented by one of more applications (Tool), or implemented by another workflow process (Subflow). The Participant element is used to specify the participants in the workflow, i.e., the entities that can execute work. There are 6 types of participants: ResourceSet, Resource, Role, OrganizationalUnit, Human, and System. Elements of type DataField and DataType are used to specify workflow relevant data. Data is used to make decisions or to refer to data outside of the workflow, and is passed between activities and subflows.

In this section, we focus on the control-flow perspective. Therefore, we will not consider functionality related to the Package, Application, and Participant elements. Moreover, we will only consider workflow relevant data from the perspective of routing. Appendix A shows selected parts of the XPDL Schema [75] relevant for this tutorial. The listing shows the elements Activity, TransitionRestriction, TransitionRestrictions, Join, Split, Transition and Condition. An activity may have one of more "transition restrictions" to specify the split/join behavior. If there is a transition restriction of type Join, the restriction is either set to AND or to XOR. The WfMC defines the semantics of such a restriction as follows: "AND: Join of (all) concurrent threads within the process instance with incoming transitions to the activity: Synchronization is required. The number of threads to be synchronized might be dependent on the result of the conditions of previous AND split(s)." and "XOR: Join for alternative threads: No synchronization is required." [75]. Similarly, there are transition restrictions of type Split that are set to either AND or XOR with the following semantics: "AND: Defines a number of possible concurrent threads represented by the outgoing Transitions of this Activity. If the Transitions have conditions the actual number of executed parallel threads is dependent on the conditions associated with each transition, which are evaluated concurrently." and "XOR: List of Identifiers of outgoing Transitions of this Activity, representing. Alternatively executed transitions. The decision as to which single transition route is selected is dependent on the conditions of each individual transition as they are evaluated in the sequence specified in the list. If an unconditional Transition is evaluated or transition with condition OTHERWISE this ends the list evaluation." [75]. Appendix A also shows the definition of element Transition. A transition connects two activities as indicated by the From and To field and may contain a Condition element.

The WfMC acknowledges the fact that workflow languages use different styles and paradigms. To accommodate this, XPDL allows for vendor specific extensions of the language. In addition, XPDL distinguishes three conformance classes: non-blocked, loop-blocked, and full-blocked. These conformance classes refer to the network structure of a process definition, i.e., the graph of activities (nodes) and transitions (arcs). For conformance class non-blocked there are no restrictions. For conformance class loop-blocked the network structure has to be acyclic and for conformance class full-blocked there has to be a one-to-one correspondence between splits and joins of the same type. These conformance classes correspond to different styles of modeling. Graph based workflow languages like COSA and Staffware correspond to conformance class non-blocked. Languages such as MQSeries, WSFL, and BPEL4WS correspond to conformance class loop-blocked and block-structured languages such as XLANG are full-blocked.

A detailed introduction to XPDL is beyond the scope of this tutorial. For more details we refer to [75].

5.3 The Workflow Patterns in XPDL

In this section, we consider the 20 workflow patterns discussed in Section 5.1, and we show how and to what extent these patterns can be captured in XPDL. In particular, we indicate whether the pattern is directly supported by a XPDL construct. If this is not the case, we sketch a workaround solution. Most of the solutions are presented in a simplified XPDL notation which is intended to capture the key ideas of the solutions while avoiding coding details. In other words, the fragments of XPDL definitions provided here are not "ready to be run".

WP1 Sequence. An activity in a workflow process is enabled after the completion of another activity in the same process. **Example:** After the activity *order registration* the activity *customer notification* is executed.

Solution, WP1. This pattern is directly supported by the XPDL as illustrated in in Listing 1. Within the process Sequence two activities A and B are linked through transition AB.

Listing 1 (Sequence)

```
<WorkflowProcess Id="Sequence">
    <ProcessHeader DurationUnit="Y"/>
    <Activities>
        <Activity Id="A">
            ...
        </Activity>
        <Activity Id="B">
            ...
        </Activity>
    </Activities>
    <Transitions>
        <Transition Id="AB" From="A" To="B"/>
    </Transitions>
</WorkflowProcess>
```

WP2 Parallel Split. A point in the process where a single thread of control splits into multiple threads of control which can be executed in parallel, thus allowing activities to be executed simultaneously or in any order [55, 37]. **Example:** After activity *new cell phone subscription order* the activity *insert new subscription* in Home Location Registry application and *insert new subscription* in Mobile answer application are executed in parallel.

WP3 Synchronization. A point in the process where multiple parallel branches converge into one single thread of control, thus synchronizing multiple threads [55]. It is an assumption of this pattern that after an incoming branch has been completed, it cannot be completed again while the merge is still waiting for other branches to be completed. Also, it is assumed that the threads to be synchronized belong to the same global process instance (i.e., to the same "case" in workflow terminology). **Example:** Activity *archive* is executed after the completion of both activity *send tickets* and activity *receive payment*. Obviously, the synchronization occurs within a single global process instance: the *send tickets* and *receive payment* must relate to the same client request.

Solutions, WP2 & WP3. This pattern directly supported by the XPDL. This is illustrated by the example shown in Listing 2. Within the process Parallel four activities are linked trough four transitions. Transitions AB and AC link the initial activity A to the two parallel activities B and C. Note that the split in activity A is of type AND and no transition conditions are specified. Transitions BD and CD link the two parallel activities B and C to the final activity D. Note that the join in activity D is of type AND and again no transition conditions are specified.

WP4 Exclusive Choice. A point in the process where, based on a decision or workflow control data, one of several branches is chosen. **Example:** The manager is informed if an order exceeds $600, otherwise not.

WP5 Simple Merge. A point in the workflow process where two or more alternative branches come together without synchronization. It is an assumption of this pattern that none of the alternative branches is ever executed in parallel with another one (if it is not the case, then see the patterns Multi Merge and Discriminator). **Example:** After the payment is received or the credit is granted the car is delivered to the customer.

Solutions, WP4 & WP5. XPDL can address the Exclusive choice pattern (WP4) in two ways. In both cases, an activity has a split and multiple outgoing transitions. One way is to use a split of type XOR, i.e., the first transition which as no condition or a condition which evaluates to true is taken. Another way is to use split of type AND and define mutual exclusive transition conditions. Listing 3 shows a solution using the first alternative. Listing 4 shows a solution using the second alternative. In the second solution transitions AB and AC have a condition. In the first solution transitions AB and AC do not have a condition which effectively implies that always the first one (AB) is taken. Besides normal conditions based on workflow relevant data, it is also possible to use conditions of type OTHERWISE (for the default branch to be taken if all other conditions evaluate to false) and of type EXCEPTION (for specifying the branch to be taken after an exception was raised). Listings 3 and 4 also show the direct support for the Simple merge (WP5).

Listing 2 (Parallel Split/Synchronization)

```
<WorkflowProcess Id="Parallel">
    <ProcessHeader DurationUnit="Y"/>
    <Activities>
        <Activity Id="A">
            ...
            <TransitionRestrictions>
                <TransitionRestriction>
                    <Split Type="AND">
                        <TransitionRefs>
                            <TransitionRef Id="B"/>
                            <TransitionRef Id="C"/>
                        </TransitionRefs>
                    </Split>
                </TransitionRestriction>
            </TransitionRestrictions>
        </Activity>
        <Activity Id="B">
            ...
        </Activity>
        <Activity Id="C">
            ....
        </Activity>
        <Activity Id="D">
            ...
            <TransitionRestrictions>
                <TransitionRestriction>
                    <Join Type="AND"/>
                </TransitionRestriction>
            </TransitionRestrictions>
        </Activity>
    </Activities>
    <Transitions>
        <Transition Id="AB" From="A" To="B"/>
        <Transition Id="AC" From="A" To="C"/>
        <Transition Id="BD" From="B" To="D"/>
        <Transition Id="CD" From="C" To="D"/>
    </Transitions>
</WorkflowProcess>
```

WP6 Multi-choice. A point in the process, where, based on a decision or control data, a number of branches are chosen and executed as parallel threads. **Example:** After executing the activity *evaluate damage* the activity *contact fire department* or the activity *contact insurance company* is executed. At least one of these activities is executed. However, it is also possible that both need to be executed.

Solution, WP6. XPDL provides direct support for the Multi-Choice pattern as shown in Listing 5. Depending on the value of amount activity B and/or C is/are executed, e.g.,

Listing 3 (Exclusive Choice/Simple Merge)

```
<WorkflowProcess Id="Choice1">
    <ProcessHeader DurationUnit="Y"/>
    <Activities>
        <Activity Id="A">
            ...
            <TransitionRestrictions>
                <TransitionRestriction>
                    <Split Type="XOR">
                        <TransitionRefs>
                            <TransitionRef Id="AB"/>
                            <TransitionRef Id="AC"/>
                        </TransitionRefs>
                    </Split>
                </TransitionRestriction>
            </TransitionRestrictions>
        </Activity>
        <Activity Id="B">
            ...
        </Activity>
        <Activity Id="C">
              ...
        </Activity>
        <Activity Id="D">
            ...
            <TransitionRestrictions>
                <TransitionRestriction>
                    <Join Type="XOR"/>
                </TransitionRestriction>
            </TransitionRestrictions>
        </Activity>
    </Activities>
    <Transitions>
        <Transition Id="AB" From="A" To="B"/>
        <Transition Id="AC" From="A" To="C"/>
        <Transition Id="BD" From="B" To="D"/>
        <Transition Id="CD" From="C" To="D"/>
    </Transitions>
</WorkflowProcess>
```

if the value of amount is 8 both activities are executed, otherwise just B (amount$>$ 5) or C (amount$<$ 10).

WP7 Synchronizing Merge. A point in the process where multiple paths converge into one single thread. Some of these paths are "active" (i.e. they are being executed) and some are not. If only one path is active, the activity after the merge is triggered as soon as this path completes. If more than one path is active, synchronization of all active paths needs to take place before the next activity is triggered. It is an assumption of this pattern that a branch that has already been activated, cannot be activated again

Listing 4 (Exclusive Choice/Simple Merge)

```
<WorkflowProcess Id="Choice2">
    <ProcessHeader DurationUnit="Y"/>
    <Activities>
        <Activity Id="A">
            ...
            <TransitionRestrictions>
                <TransitionRestriction>
                    <Split Type="AND">
                        <TransitionRefs>
                            <TransitionRef Id="AB"/>
                            <TransitionRef Id="AC"/>
                        </TransitionRefs>
                    </Split>
                </TransitionRestriction>
            </TransitionRestrictions>
        </Activity>
        <Activity Id="B">
            ...
        </Activity>
        <Activity Id="C">
            ...
        </Activity>
        <Activity Id="D">
            ...
            <TransitionRestrictions>
                <TransitionRestriction>
                    <Join Type="XOR"/>
                </TransitionRestriction>
            </TransitionRestrictions>
        </Activity>
    </Activities>
    <Transitions>
        <Transition Id="AB" From="A" To="B">
            <Condition Type="CONDITION">
                choice == "B" </Condition>
        </Transition>
        <Transition Id="AC" From="A" To="C">
            <Condition Type="CONDITION">
                choice == "C" </Condition>
        </Transition>
        <Transition Id="BD" From="B" To="D"/>
        <Transition Id="CD" From="C" To="D"/>
    </Transitions>
</WorkflowProcess>
```

while the merge is still waiting for other branches to complete. **Example:** After either or both of the activities *contact fire department* and *contact insurance company* have been completed (depending on whether they were executed at all), the activity submit report needs to be performed (exactly once).

Listing 5 (Multi Choice/Synchronizing merge)

```
<WorkflowProcess Id="Multi-choice">
    <ProcessHeader DurationUnit="Y"/>
    <Activities>
        <Activity Id="A">
            ...
            <TransitionRestrictions>
                <TransitionRestriction>
                    <Split Type="AND">
                        <TransitionRefs>
                            <TransitionRef Id="AB"/>
                            <TransitionRef Id="AC"/>
                        </TransitionRefs>
                    </Split>
                </TransitionRestriction>
            </TransitionRestrictions>
        </Activity>
        <Activity Id="B">
            ...
        </Activity>
        <Activity Id="C">
            ...
        </Activity>
        <Activity Id="D">
            ...
            <TransitionRestrictions>
                <TransitionRestriction>
                    <Join Type="AND"/>
                </TransitionRestriction>
            </TransitionRestrictions>
        </Activity>
    </Activities>
    <Transitions>
        <Transition Id="AB" From="A" To="B">
            <Condition Type="CONDITION">
                amount &gt; 5 </Condition>
        </Transition>
        <Transition Id="AC" From="A" To="C">
            <Condition Type="CONDITION">
                amount &lt; 10 </Condition>
        </Transition>
        <Transition Id="BD" From="B" To="D"/>
        <Transition Id="CD" From="C" To="D"/>
    </Transitions>
</WorkflowProcess>
```

Solutions, WP7. According to [75] XPDL provides direct support for the Synchronizing merge pattern. Recall the definition of the AND restriction: "AND: Join of (all) concurrent threads within the process instance with incoming transitions to the activity:

Synchronization is required. The number of threads to be synchronized might be dependent on the result of the conditions of previous AND split(s)." [75] which suggests direct support for the Synchronizing merge pattern. If this is indeed the case, then Listing 5 indeed shows an example where activity D either merges or synchronizes the two ingoing transitions depending on the number of threads activated by activity A. Unfortunately, few workflow systems that claim to support XPDL have indeed this behavior. Moreover, XPDL allows for multiple interpretations as discussed in Section 5.4.

WP8 Multi-merge. A point in a process where two or more branches reconverge without synchronization. If more than one branch gets activated, possibly concurrently, the activity following the merge is started for every action of every incoming branch. **Example:** Sometimes two or more branches share the same ending. Two activities *audit application* and *process applications* are running in parallel which should both be followed by an activity *close case*, which should be executed twice if the activities *audit application* and *process applications* are both executed.

Solution, WP8. XPDL only allows for two types of joins: AND and XOR. The semantics of these two joins is not completely clear. A join of type XOR will offer the Simple merge pattern. Recall that the simple merge assumes that precisely one of the incoming transitions will occur. However, XPDL allows for situations where the more incoming transitions will or may occur. Consider Listing 6. Both B and C are executed. Since activity D has a join of type XOR it can already occur when one of these two have been executed. However, it is not clear how many times activity D will occur (and when). In [75] is is stated that "The XOR join initiates the Activity when the transition conditions of any (one) of the incoming transitions evaluates true.". Since it is not specified what should happen if multiple incoming transitions evaluate to true at the same time, we conclude that XPDL does not support the Multi-Merge (WP8). See [15] for typical work-arounds.

WP9 Discriminator. A point in the workflow process that waits for one of the incoming branches to complete before activating the subsequent activity. From that moment on it waits for all remaining branches to complete and "ignores" them. Once all incoming branches have been triggered, it resets itself so that it can be triggered again (which is important otherwise it could not really be used in the context of a loop). **Example:** To improve query response time a complex search is sent to two different databases over the Internet. The first one that comes up with the result should proceed the flow. The second result is ignored.

Solution, WP9. XPDL allows for situations where multiple incoming transitions will or may occur. However, the precise semantics of a join of type XOR is not specified and, similar to WP8, we conclude that the Discriminator (WP9) is not supported.

WP10 Arbitrary Cycles. A point where a portion of the process (including one or more activities and connectors) needs to be "visited" repeatedly *without imposing restrictions* on the number, location, and nesting of these points. Note that block-oriented languages and languages providing constructs such as "while do", "repeat until" typically impose such restrictions, e.g., it is not possible to jump from one loop into another loop.

Listing 6 (Multi-merge?)

```
<WorkflowProcess Id="Parallel">
    <ProcessHeader DurationUnit="Y"/>
    <Activities>
        <Activity Id="A">
            ...
            <TransitionRestrictions>
                <TransitionRestriction>
                    <Split Type="AND">
                        <TransitionRefs>
                            <TransitionRef Id="B"/>
                            <TransitionRef Id="C"/>
                        </TransitionRefs>
                    </Split>
                </TransitionRestriction>
            </TransitionRestrictions>
        </Activity>
        <Activity Id="B">
            ...
        </Activity>
        <Activity Id="C">
            ....
        </Activity>
        <Activity Id="D">
            ...
            <TransitionRestrictions>
                <TransitionRestriction>
                    <Join Type="XOR"/>
                </TransitionRestriction>
            </TransitionRestrictions>
        </Activity>
    </Activities>
    <Transitions>
        <Transition Id="AB" From="A" To="B"/>
        <Transition Id="AC" From="A" To="C"/>
        <Transition Id="BD" From="B" To="D"/>
        <Transition Id="CD" From="C" To="D"/>
    </Transitions>
</WorkflowProcess>
```

Solution, WP10. XPDL distinguishes three conformance classes: non-blocked, loop-blocked, and full-blocked. Conformance class "non-blocked" directly supports this pattern. Note that the transitions basically define a relation and allow for any graph including cyclic ones. For the other conformance classes this is not allowed. For conformance class loop-blocked the network structure has to be acyclic and for conformance class full-blocked there has to be a one-to-one correspondence between splits and joins of the same type.

WP11 Implicit Termination. A given subprocess is terminated when there is nothing left to do, i.e., termination does not require an explicit termination activity. The goal of this pattern is to avoid having to join divergent branches into a single point of termination.

Solution, WP11. XPDL, assuming conformance class "non-blocked", allows for arbitrary graph-like structures. As a result it is possible to have multiple activities without input transitions (i.e., source activities) and multiple activities without output transitions (sink activities). The latter suggests direct support for WP11. Unfortunately, [75] does not clarify the semantics of XPDL in the presence of multiple source and sink activities, e.g., Do all source activities need to be executed or just one? Although XPDL does not specify the expected behavior in such cases, we give it the benefit of the doubt. Note that this illustrates that conformance is still ill-defined in [75] since it refers to syntax rather than semantics.

WP12 MI without Synchronization. Within the context of a single case, multiple instances of an activity may be created, i.e. there is a facility for spawning off new threads of control, all of them independent of each other. The instances might be created consecutively, but they will be able to run in parallel, which distinguishes this pattern from the pattern for Arbitrary Cycles. **Example:** When booking a trip, the activity *book flight* is executed multiple times if the trip involves multiple flights.

Solution, WP12. An activity may be refined into a subflow. The subflow may be executed synchronously or asynchronously. In case of asynchronous execution, the activity is continued after an instance of the subflow is initiated. This way it is possible to "spawn-off" subflows and thus realizing WP12.

WP13-WP15 MI with Synchronization. A point in a workflow where a number of instances of a given activity are initiated, and these instances are later synchronized, before proceeding with the rest of the process. In WP13 the number of instances to be started/synchronized is known at design time. In WP14 the number is known at some stage during run time, but before the initiation of the instances has started. In WP15 the number of instances to be created is not known in advance: new instances are created on demand, until no more instances are required. **Example of WP15:** When booking a trip, the activity *book flight* is executed multiple times if the trip involves multiple flights. Once all bookings are made, an invoice is sent to the client. How many bookings are made is only known at runtime through interaction with the user (or with an external process).

Solutions, WP13-WP15. If the number of instances to be synchronized is known at design time (WP13), a simple solution is to replicate the activity as many times as it needs to be instantiated, and run the replicas in parallel. Therefore, WP13 is supported. However, it is clear that there is no direct support for WP14 and WP15 because any solution will involve explicit bookkeeping of the number of active instances. In fact in [75] is is stated that "Synchronization with the initiated subflow, if required, has to be done by other means such as events, not described in this document." when describing the functionality of asynchronous subflows. Therefore, we conclude that there is no support for WP14 and WP15. Again we refer to [15] for typical workarounds.

WP16 Deferred Choice. A point in a process where one among several alternative branches is chosen based on information which is not necessarily available when this point is reached. This differs from the normal exclusive choice, in that the choice is not made immediately when the point is reached, but instead several alternatives are offered, and the choice between them is delayed until the occurrence of some event. **Example:** When a contract is finalized, it has to be reviewed and signed either by the director or by the operations manager, whoever is available first. Both the director and the operations manager would be notified that the contract is to be reviewed: the first one who is available will proceed with the review.

Solution, WP16. XPDL only allows for choices resulting from conditions on transitions. Hence each choice is directly-based on workflow relevant data and it is not possible offer the choice to the environment. XPDL does not allow for the definition of states (like places in a Petri net) nor constructs like the choice construct in BPML and WSCI and the pick construct in XLANG and BPEL4WS. There is no simple work-around for this omission since it is not possible to shift the moment of decision from the end of an activity to the start of an activity. Moreover, XPDL does not allow for the specification of triggers and/or external events.

WP17 Interleaved Parallel Routing. A set of activities is executed in an arbitrary order. Each activity in the set is executed exactly once. The order between the activities is decided at run-time: it is not until one activity is completed that the decision on what to do next is taken. In any case, no two activities in the set can be active at the same time. **Example:** At the end of each year, a bank executes two activities for each account: *add interest* and *charge credit card costs*. These activities can be executed in any order. However, since they both update the account, they cannot be executed at the same time.

Solution, WP17. Since XPDL does not allow for the definition of states, it is not possible to enforce some kind of mutual exclusion. Hence there is no support for WP17. Even the work-arounds described in [15] are difficult, if not impossible, to apply.

WP18 Milestone. A given activity can only be enabled if a certain milestone has been reached which has not yet expired. A milestone is defined as a point in the process where a given activity has finished and another activity following it has not yet started. **Example:** After having placed a purchase order, a customer can withdraw it at any time before the shipping takes place. To withdraw an order, the customer must complete a withdrawal request form, and this request must be approved by a customer service representative. The execution of the activity *approve order withdrawal* must therefore follow the activity *request withdrawal*, and can only be done if: (i) the activity *place order* is completed, and (ii) the activity *ship order* has not yet started.

Solution, WP18. XPDL does not provide a direct support for capturing this pattern. Therefore, a work-around solution has to be used. Again it is difficult to construct solutions inspired by the ideas in [15]. Since other patterns like WP16 and WP19 are not supported, potential solutions lead to complex process definitions for simply checking the state in a parallel branch.

WP19 Cancel Activity & WP20 Cancel Case. A cancel activity terminates a running instance of an activity, while canceling a case leads to the removal of an entire workflow instance. **Example of WP19:** A customer cancels a request for information. **Example of WP20:** A customer withdraws his/her order.

Solutions, WP19 & WP20. XPDL does not provide explicit constructs for WP19 and WP20. The concept of exceptions seems to be related, but like many other concepts ill-defined. The only construct in XPDL that can raise an exception is the deadline element. Deadlines are used to raise an exception upon the expiration of a specific period of time. A deadline can be raised synchronously or asynchronously: "If the deadline is synchronous, then the activity is terminated before flow continues on the exception path." and "If the deadline is asynchronous, then an implicit AND-SPLIT is performed, and a new thread of processing is started on the appropriate exception transition." [75]. An exception may trigger a transition but cannot be used to cancel activities or cases. Hence, XPDL does not support WP19 and WP20.

5.4 Many Ways to Join

In this section, we evaluated XPDL with respect to the patterns. A more detailed analysis reveals that, not only does XPDL have problems with respect to several patterns, the semantics of many constructs is unclear. To illustrate this we focus on transition restrictions of type Join. The restriction is either set to AND or to XOR and the WfMC defines these settings as follows: "AND: Join of (all) concurrent threads within the process instance with incoming transitions to the activity: Synchronization is required. The number of threads to be synchronized might be dependent on the result of the conditions of previous AND split(s)." and "XOR: Join for alternative threads: No synchronization is required." [75]. To demonstrate that these descriptions do not fully specify the intended behavior, Figure 23 shows seven possible interpretations each expressed in terms of a Petri net (some extended with inhibitor arcs, cf. [63]). Note that Petri nets have formal semantics, and thus, Figure 23 fully specifies the behavior of each construct. Also note that we restrict ourselves to local constructs, i.e., the there are no dependencies other than on the activities directly connected to the join.

The first two constructs correspond to the most straightforward interpretations of the AND-join (Figure 23(a)) and XOR-join (Figure 23(b)). In Figure 23(a), activity C always synchronizes A and B, i.e., if A is never executed, C is never executed[5]. In Figure 23(b), activity C is executed once for each occurrence of A and B. Although Figure 23(a) and Figure 23(b) seem to correspond to straightforward interpretations of the AND-join and XOR-join, few WFM systems actually exhibit this behavior. The other constructs in Figure 23 show other interpretations for both the AND-join and/or XOR-join encountered in contemporary systems. Figure 23(c) shows the situation where activity A is blocked if C was not executed since the last occurrence of A. Similarly, activity B is blocked if C was not executed since the last occurrence of B. Note that this construct uses two inhibitor arcs (i.e., the two connections involving a small circle). Unlike a normal directed arc in Petri net, an inhibitor arc models the requirement that a place has to be empty, i.e., A is only enabled if the input place (not shown) contains a token *and* the output place is empty. Figure 23(d) shows a similar construct but now for the XOR-join, i.e., both activity A and activity B are blocked if C was not executed

[5] Note that this is not the case in XPDL.

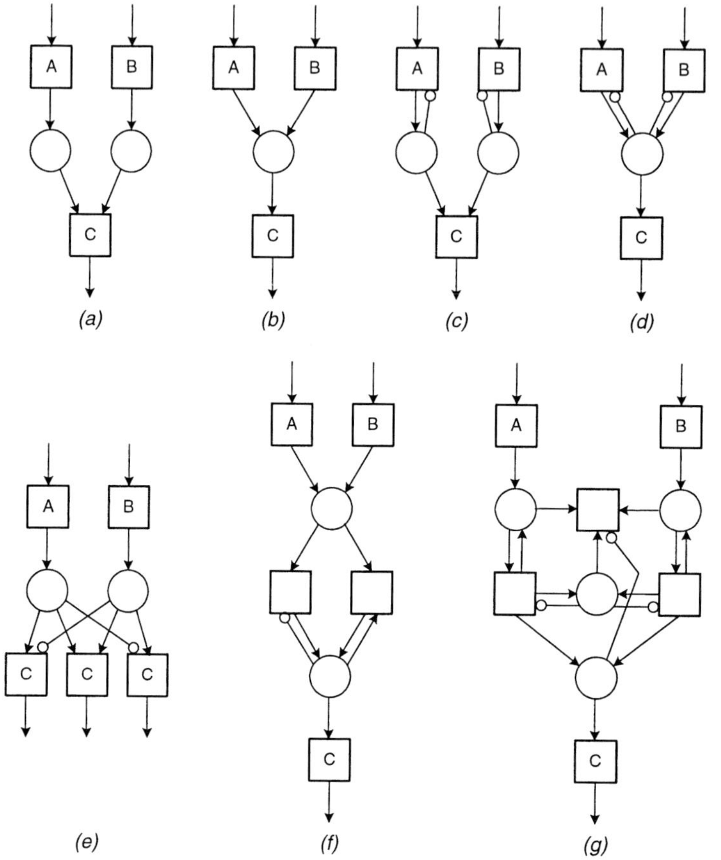

Fig. 23. Seven frequently used ways to join two flows (expressed in terms of Petri nets with inhibitor arcs [63]).

since the last occurrence of A or B. The WFM system COSA [67] uses this interpretation for the AND-join and XOR-join. Figures 23 (c) and (d) use inhibitor arcs to make sure that activity C is only enabled once. This is realized by blocking the preceding activities if needed. An alternative approach is to simply remove additional tokens. Figure 23(e) shows an approach where C synchronizes both flows if both A and B have been executed. If only one of them has been executed, there is no synchronization. Note that there are three instances of C: one for the situation where only A was executed, one for situation where both A and B have been executed, and one where only B was executed. The two inhibitor arcs make sure that the two flows are synchronized if possible. Figure 23(f) shows a similar, but slightly different, approach where simply every attempt to enable C for the second time is ignored. If C is already enabled, then the right transition will occur, otherwise the left one. Consider the scenario where A occurs twice before execution C. In Figure 23(e), C will be executed twice, while in Figure 23(f) C will be executed on ly once. Many systems have a behavior similar to Figure 23(e)/(f), e.g., a normal step in Staffware [69] behaves as indicated by Figure 23(f). (See [72] for a more detailed analysis of Staffware steps.) Although widely supported, the interpretation given in Figure 23(e)/(f) is not very desirable from a modeling point of view since it introduces "race conditions", e.g., the number of times C is executed depends on the interleaving of A, B, and C activities. Figure 23(g) gives yet another interpretation of the AND/XOR-join. C is enabled immediately after the first occurrence of A or B, but after it occurs it is blocked until the other activity has also been executed, i.e., the construct is reset once each of A, B, and C has occurred. Note that this interpretation corresponds to WP9 (Discriminator pattern).

Figure 23 shows that there are many ways to join two flows. In fact, there are many more interpretations. An example is the so-called "wait step" in Staffware [69] which only synchronizes the first time if it is put in a loop (see [72] for more details). Another example is the join in IBM's MQSeries Workflow [46], BPEL4WS [27], and WSFL (Web Services Flow Language, [56]) which decides whether it has to synchronize or not based on the so-called "Dead-Path-Elimination (DPE)" [57]. Given the quote "AND: Join of (all) concurrent threads within the process instance with incoming transitions to the activity: Synchronization is required. The number of threads to be synchronized might be dependent on the result of the conditions of previous AND split(s)." in [75], the latter interpretation seems to be closest to XPDL. Unfortunately, other than IBM-influenced products and standards, no other vendors are using nor supporting this interpretation since it does not allow for Arbitrary cycles (WP10) [9].

The dilemma of joining mixtures of alternative or parallel flows has been discussed in scientific literature. See [9] for pointers to related papers and an elaborate discussion in the context of Event-driven Process Chains (EPC's).

The fact that there are many ways to join and that in [75] the WfMC leaves room for multiple interpretations, brings us to the issue of *conformance*. In [75] it is stated that "A product that claims conformance must generate valid, syntactically correct XPDL, and must be able to read all valid XPDL.". Unfortunately, this quote, but also the rest of [75], does not address the issue of semantics. Note that it is rather easy to generate and read valid XPDL files. The difficult part is to be able to interpret XPDL generated by another tool and execute the workflow as intended.

5.5 Comparing XPDL with Other Languages and Standards

Thus far, we provided a critical evaluation of XPDL based on a set of 20 basic workflow patterns. To conclude this section, we compare XPDL with other standards and 15 workflow products.

Table 1 shows an evaluation of XPDL and six other standards. If a standard directly supports the pattern through one of its constructs, it is rated +. If the pattern is not *directly* supported, it is rated +/-. Any solution which results in spaghetti diagrams or coding, is considered as giving no direct support and is rated -. The rating of XPDL is as explained in this section.

UML activity diagrams [42] are intended to model both computational and organizational processes. Increasingly, UML activity diagrams are also used for workflow modeling. Therefore, it is interesting to analyze their expressiveness using the set of basic workflow patterns as shown in the table. for more information see [31].

Table 1. A comparison of XPDL with other standards such as UML Activity Diagrams, BPEL4WS, BPML, XLANG, WSFL, and WSCI.

pattern	standard						
	XPDL	UML	BPEL4WS	BPML	XLANG	WSFL	WSCI
1 (seq)	+	+	+	+	+	+	+
2 (par-spl)	+	+	+	+	+	+	+
3 (synch)	+	+	+	+	+	+	+
4 (ex-ch)	+	+	+	+	+	+	+
5 (simple-m)	+	+	+	+	+	+	+
6 (m-choice)	+	-	+	-	-	+	-
7 (sync-m)	+[a]	-	+	-	-	+	-
8 (multi-m)	-	-	-	+/-	-	-	+/-
9 (disc)	-	-	-	-	-	-	-
10 (arb-c)	+[b]	-	-	-	-	-	-
11 (impl-t)	+[c]	-	+	+	-	+	+
12 (mi-no-s)	+	-	+	+	+	+	+
13 (mi-dt)	+	+	+	+	+	+	+
14 (mi-rt)	-	+	-	-	-	-	-
15 (mi-no)	-	-	-	-	-	-	-
16 (def-c)	-	+	+	+	+	-	+
17 (int-par)	-	-	+/-	-	-	-	-
18 (milest)	-	-	-	-	-	-	-
19 (can-a)	-	+	+	+	+	+	+
20 (can-c)	-	+	+	+	+	+	+

[a] Although the description of the AND-join suggests support for WP7, XPDL does not specify its precise behavior. In fact, for conformance class "non-blocked", it is unclear how WP7 could be supported

[b] For conformance class "non-blocked", arbitrary graph-like structures are allowed, including arbitrary cycles. For the other conformance classes this is explicitly excluded

[c] For all conformance classes there may be multiple source and/or sink activities. Hence, from a syntactical point of view WP11 is supported. Unfortunately, no semantics are given for this construct

The recently released BPEL4WS (Business Process Execution Language for Web Services, [27]) specification builds on IBM's WSFL (Web Services Flow Language, [56]) and Microsoft's XLANG [70]. XLANG is a block-structured language with basic control flow structures such as sequence, switch (for conditional routing), while (for looping), all (for parallel routing), and pick (for race conditions based on timing or external triggers). In contrast to XLANG, WSFL is not limited to block structures and allows for directed graphs. The graphs can be nested but need to be acyclic. Iteration is only supported through exit conditions, i.e., an activity/subprocess is iterated until its exit condition is met. The control flow part of WSFL is almost identical to the workflow language used by IBM's MQ Series Workflow. See [76, 77] for more information about the evaluation of BPEL4WS, XLANG, and WSFL using the patterns.

BPML (Business Process Modeling language, [21]) is a standard developed and promoted by BPMI.org (the Business Process Management Initiative). BPMI.org is

Table 2. The main results for Staffware, COSA, InConcert, Eastman, FLOWer, Lotus Domino Workflow, Meteor, and Mobile.

pattern	product							
	Staffware	COSA	InConcert	Eastman	FLOWer	Domino	Meteor	Mobile
1 (seq)	+	+	+	+	+	+	+	+
2 (par-spl)	+	+	+	+	+	+	+	+
3 (synch)	+	+	+	+	+	+	+	+
4 (ex-ch)	+	+	+/-	+	+	+	+	+
5 (simple-m)	+	+	+/-	+	+	+	+	+
6 (m-choice)	-	+	+/-	+/-	-	+	+	+
7 (sync-m)	-	+/-	+	+	-	+	-	-
8 (multi-m)	-	-	-	+	+/-	+/-	+	-
9 (disc)	-	-	-	+	+/-	-	+/-	+
10 (arb-c)	+	+	-	+	-	+	+	-
11 (impl-t)	+	-	+	+	-	+	-	-
12 (mi-no-s)	-	+/-	-	+	+	+/-	+	-
13 (mi-dt)	+	+	+	+	+	+	+	+
14 (mi-rt)	-	-	-	-	+	-	-	-
15 (mi-no)	-	-	-	-	+	-	-	-
16 (def-c)	-	+	-	-	+/-	-	-	-
17 (int-par)	-	+	-	-	+/-	-	-	+
18 (milest)	-	+	-	-	+/-	-	-	-
19 (can-a)	+	+	-	-	+/-	-	-	-
20 (can-c)	-	-	-	-	+/-	+	-	-

supported by several organizations, including Intalio, SAP, Sun, and Versata. The Web Service Choreography Interface (WSCI, [20]) submitted in June 2002 to the W3C by BEA Systems, BPMI.org, Commerce One, Fujitsu Limited, Intalio, IONA, Oracle Corporation, SAP AG, SeeBeyond Technology Corporation, and Sun Microsystems. There is a substantial overlap between BPML and WSCI. See [11] for more information about the evaluation of BPML and WSCI using the patterns.

In addition to comparing XPDL to other standards, it is interesting to compare XPDL with contemporary WFM systems. Tables 2 and 3 summarize the results of the comparison of 15 WFM systems in terms of the selected patterns. These tables are taken from [15] and have been added to compare contemporary workflow products with XPDL.

From the comparison it is clear that no tool supports all of the selected patterns. In fact, many of these tools only support a relatively small subset of the more advanced patterns (i.e., patterns 6 to 20). Specifically the limited support for the discriminator, the state-based patterns (only COSA), the synchronization of multiple instances (only FLOWer) and cancellation (esp. of activities), is worth noting.

Please apply the results summarized in tables 1, 2 and 3 with care. First of all, the organization selecting a WFM system/standard should focus on the patterns most relevant for the workflow processes at hand. Since support for the more advanced patterns is limited, one should focus on the patterns most needed. Second, the fact that a pattern is not directly supported by a product does not imply that it is not possible to support the construct at all. As indicated in [15], many patterns can be supported indirectly through mixtures of more basic patterns and coding. Third, the patterns reported in this

Table 3. The main results for MQSeries, Forté Conductor, Verve, Visual WorkFlo, Changengine, I-Flow, and SAP/R3 Workflow.

pattern	product						
	MQSeries	Forté	Verve	Vis. WF	Changeng.	I-Flow	SAP/R3
1 (seq)	+	+	+	+	+	+	+
2 (par-spl)	+	+	+	+	+	+	+
3 (synch)	+	+	+	+	+	+	+
4 (ex-ch)	+	+	+	+	+	+	+
5 (simple-m)	+	+	+	+	+	+	+
6 (m-choice)	+	+	+	+	+	+	+
7 (sync-m)	+	-	-	-	-	-	-
8 (multi-m)	-	+	+	-	-	-	-
9 (disc)	-	+	+	-	+	-	+
10 (arb-c)	-	+	+	+/-	+	+	-
11 (impl-t)	+	-	-	-	-	-	-
12 (mi-no-s)	-	+	+	+	-	+	-
13 (mi-dt)	+	+	+	+	+	+	+
14 (mi-rt)	-	-	-	-	-	-	+/-
15 (mi-no)	-	-	-	-	-	-	-
16 (def-c)	-	-	-	-	-	-	-
17 (int-par)	-	-	-	-	-	-	-
18 (milest)	-	-	-	-	-	-	-
19 (can-a)	-	-	-	-	-	-	+
20 (can-c)	-	+	+	-	+	-	+

tutorial only focus on the process perspective (i.e., control flow or routing). The other perspectives (e.g., organizational modeling) should also be taken into account.

Tables 1, 2 and 3 allow for an objective comparison of the 7 standards and 15 WFM systems. When comparing XPDL to the 6 other standards, it is remarkable to see that XPDL seems to be less expressive than web service composition languages such as BPEL4WS and BPML. An important pattern like the Deferred choice (WP16) is supported by most standards and is vital for practical application of WFM. Nevertheless, it is not even mentioned in [75]. Compared to the 15 WFM systems, XPDL is not as expressive as one would expect. Many systems offer functionality (e.g., the Deferred choice and the Cancel activity patterns), not supported by XPDL. It almost seems that XPDL offers the intersection rather than the union of the functionality offered by contemporary systems. This may have been the initial goal of XPDL. However, if this is the case, two important questions need to be answered.

1. If XPDL offers the intersection rather than the union of the functionality of existing systems, then how to use XPDL in practice? Should workflow designers that want to be able to export only use a subset of the functionality offered by the system? If so, users would not be able to use powerful concepts like the Deferred choice (WP16) and the Cancel activity (WP19) patterns.
2. Why does XPDL support the Synchronizing merge (WP7) while it is only supported by a few systems. Widely-used systems like Staffware do not support this pattern, and therefore, will be unable to interpret the AND-join as indicated in [75].

Note that the issues raised cannot be solved satisfactorily. If XPDL offers the intersection of the functionality of existing systems, it is less expressive than many of the existing tools and standards. If XPDL offers the union of available functionality, it may become impossible to import a process definition into a concrete system and interpret it correctly. (Recall that no system supports all patterns.) Unfortunately, this dilemma is not really addressed by the WfMC [75]. The introduction of extended attributes (i.e., extensions of XPDL for a specific product) and conformance classes (i.e., restrictions to allow the use of specific products) are no solution and only complicate matters.

There have been several comparisons of some of the languages mentioned in this tutorial. These comparisons typically do not use a framework and provide an opinion rather than a structured analysis. A positive example is [65] where XPDL, BPML and BPEL4WS are compared by relating the concepts used in the three languages. Unfortunately, the paper raises more questions than it answers.

Besides the dilemma that XPDL is either not expressive enough or too expressive, there is the problem of semantics. In [75] the WfMC does not give unambiguous specification of all the elements in the language. As a result, many vendors can claim to be compliant while interpreting constructs in a different way. In Section 5.4, we demonstrated that there are many interpretations of seemingly basic constructs like the AND-join and XOR-join. The lack of semantics restricts the application of XPDL and does not allow for a meaningful realization of the topic of conformance. As indicated before, [75] defines conformance as follows: "A product that claims conformance must generate valid, syntactically correct XPDL, and must be able to read all valid XPDL.". Clearly, this inadequate and will not stimulate further standardization in the workflow domain. As a result, web service composition languages like BPML and BPEL4WS may take over the role of XPDL [6].

6 Related Work

There is a lot of literature on WFM and WFM systems. Only some of the books on WFM are referred to in this tutorial [12,37,38,48,55,57,58,61]. There are also many publications reporting on the application of Petri nets to WFM. In this tutorial we mainly referred to papers using WF-nets and soundness (or variants thereof) [1,3,4,28,44,52,54]. For the evaluation of XPDL we relied heavily on the work on workflow patterns. See [15,77] or `http://www.workflowpatterns.com` for more information. The two Springer Lecture Notes in Computing science volumes on BPM can serve as a starting point for finding the state-of-the-art results in this domain [10,16]. Clearly, it is impossible to be complete. Please use the references given in this tutorial for finding more material. Finally, we would like to point out `http://www.workflowcourse.com` as a resource for all kinds of learning material ranging from slides to interactive animations.

7 Conclusion

The goal of this tutorial is to introduce the WFM/BPM domain from a Petri-net point-of-view. The focus of the first part of the tutorial was on the application of Petri nets in this domain. Sections 2 and 3 showed how WF-nets can be used to model and analyze workflow processes. The focus of the second part was more on the application domain itself. Sections 4 and 5 provided information on systems, languages, and standards. To illustrate things we presented a detailed analysis of XPDL using a set of workflow patterns.

To conclude this tutorial we reflect on the role of Petri nets in the WFM/BPM domain. There are at least three good reasons for using Petri nets for workflow modeling and analysis ([2]):

1. *Formal semantics despite the graphical nature*
   On the one hand, Petri nets are a graphical language which allows for the modeling of the workflow primitives identified by the WfMC. On the other hand, the semantics of Petri nets (including most of the extensions) have been defined formally. Many of today's available WFM systems provide ad-hoc constructs to model workflow procedures. Moreover, there are WFM systems that impose restrictions on many of the workflow patterns discussed. Some WFM systems also provide exotic constructs whose semantics are not 100% clear, cf. the join construct in XPDL and many other languages. Because of these problems it is better to use a well-established design language with formal semantics as a solid basis.
2. *State-based instead of event-based*
   In contrast to many other process modeling techniques, the state of case can be modeled explicitly in a Petri net. Process modeling techniques ranging from informal techniques such as dataflow diagrams to formal techniques such as process algebra's are *event-based*, i.e., transitions are modeled explicitly and the states between subsequent transitions are only modeled implicitly. Today's WFM systems are typically event-based, i.e., tasks are modeled explicitly and states between subsequent tasks are suppressed. The distinction between an event-based and a state-based description may appear to be subtle, but patterns like the Deferred choice (WP16) and the Milestone (WP18) show that this is of the utmost importance for workflow modeling.
3. *Abundance of analysis techniques*
   Petri nets are marked by the availability of many analysis techniques. Clearly, this is a great asset in favor of a Petri nets. In this tutorial, we focused on the verification of WF-nets. We have seen that Petri-net-based analysis techniques can be used to determine the correctness of a workflow process definition. The availability of these techniques illustrates that Petri-net theory can be used to add powerful analysis capabilities to the next generation of WFM systems.

However, as indicated in [13] there are also problems when modeling workflows in terms of a Petri nets. For the more advanced routing constructs it is necessary to resort to high-level nets [49, 50]. Moreover, a straightforward application of high-level Petri nets does not always yield the desired result. There seem to be three problems relevant for modeling workflow processes:

1. High-level Petri nets support colored tokens, i.e., a token can have a value. Although it is possible to use this to identify multiple instances of a subprocess, there is no specific support for *patterns involving multiple instances* and the burden of keeping track, splitting, and joining of instances is carried by the designer.

2. Sometimes two flows need to be joined while it is not clear whether synchronization is needed, i.e., if both flows are active an AND-join is needed otherwise an XOR-join. Such *advanced synchronization patterns* are difficult to model in terms of a high-level Petri net because the firing rule only supports two types of joins: the AND-join (transition) or the XOR-join (place).
3. The firing of a transition is always local, i.e., enabling is only based on the tokens in the input places and firing is only affecting the input and output places. However, some events in the workflow may have an effect which is not local, e.g., because of an error tokens need to be removed from various places without knowing where the tokens reside. Everyone who has modeled such a *cancellation pattern* (e.g., a global timeout mechanism) in terms of Petri nets knows that it is cumbersome to model a so-called "vacuum cleaner" removing tokens from selected parts of the net.

Compared to existing WFM languages high-level Petri nets are quite expressive when it comes to supporting the workflow patterns. Recall that we use the term "expressiveness" not in the formal sense. High-level Petri nets are Turing complete, and therefore, can do anything we can define in terms of an algorithm. However, this does not imply that the modeling effort is acceptable. High-level nets, in contrast to many workflow languages, have no problems dealing with state-based patterns. This is a direct consequence of the fact that Petri nets use places to represent states explicitly. Although high-level Petri nets outperform most of the existing languages when it comes to modeling the control flow, the result is not completely satisfactory since the three problems indicated hamper the application in the WFM/BPM domain. This triggered the development of *YAWL (Yet Another Workflow Language)*. YAWL is based on Petri nets but extended with additional features to facilitate the modeling of complex workflows [13, 14]. See `http://www.citi.qut.edu.au/yawl/` for more information or to download the YAWL system.

Acknowledgments

The author would like to thank Lachlan Aldred, Alistair Barros, Twan Basten, Marlon Dumas, Bartek Kiepuszewski, Kees van Hee, Akhil Kumar, Arthur ter Hofstede, Hajo Reijers, Eric Verbeek, Ton Weijters, and Petia Wohed for their collaborative work on the topics discussed in this tutorial.

Disclaimer

The author and the associated institutions assume no legal liability or responsibility for the accuracy and completeness of any information about XPDL or any of the other standards/products mentioned in this tutorial. However, all possible efforts have been made to ensure that the results presented are, to the best of our knowledge, up-to-date and correct.

References

1. W.M.P. van der Aalst. Verification of Workflow Nets. In P. Azéma and G. Balbo, editors, *Application and Theory of Petri Nets 1997*, volume 1248 of *Lecture Notes in Computer Science*, pages 407–426. Springer-Verlag, Berlin, 1997.
2. W.M.P. van der Aalst. Chapter 10: Three Good reasons for Using a Petri-net-based Workflow Management System. In T. Wakayama, S. Kannapan, C.M. Khoong, S. Navathe, and J. Yates, editors, *Information and Process Integration in Enterprises: Rethinking Documents*, volume 428 of *The Kluwer International Series in Engineering and Computer Science*, pages 161–182. Kluwer Academic Publishers, Boston, Massachusetts, 1998.
3. W.M.P. van der Aalst. The Application of Petri Nets to Workflow Management. *The Journal of Circuits, Systems and Computers*, 8(1):21–66, 1998.
4. W.M.P. van der Aalst. Workflow Verification: Finding Control-Flow Errors using Petri-net-based Techniques. In W.M.P. van der Aalst, J. Desel, and A. Oberweis, editors, *Business Process Management: Models, Techniques, and Empirical Studies*, volume 1806 of *Lecture Notes in Computer Science*, pages 161–183. Springer-Verlag, Berlin, 2000.
5. W.M.P. van der Aalst. Making Work Flow: On the Application of Petri nets to Business Process Management. In J. Esparza and C. Lakos, editors, *Application and Theory of Petri Nets 2002*, volume 2360 of *Lecture Notes in Computer Science*, pages 1–22. Springer-Verlag, Berlin, 2002.
6. W.M.P. van der Aalst. Don't go with the flow: Web services composition standards exposed. *IEEE Intelligent Systems*, 18(1):72–76, 2003.
7. W.M.P. van der Aalst and T. Basten. Inheritance of Workflows: An Approach to Tackling Problems Related to Change. *Theoretical Computer Science*, 270(1-2):125–203, 2002.
8. W.M.P. van der Aalst and P.J.S. Berens. Beyond Workflow Management: Product-Driven Case Handling. In S. Ellis, T. Rodden, and I. Zigurs, editors, *International ACM SIGGROUP Conference on Supporting Group Work (GROUP 2001)*, pages 42–51. ACM Press, New York, 2001.
9. W.M.P. van der Aalst, J. Desel, and E. Kindler. On the Semantics of EPCs: A Vicious Circle. In M. Nüttgens and F.J. Rump, editors, *Proceedings of the EPK 2002: Business Process Management using EPCs*, pages 71–80, Trier, Germany, November 2002. Gesellschaft für Informatik, Bonn.
10. W.M.P. van der Aalst, J. Desel, and A. Oberweis, editors. *Business Process Management: Models, Techniques, and Empirical Studies*, volume 1806 of *Lecture Notes in Computer Science*. Springer-Verlag, Berlin, 2000.
11. W.M.P. van der Aalst, M. Dumas, A.H.M. ter Hofstede, and P. Wohed. Pattern-Based Analysis of BPML (and WSCI). QUT Technical report, FIT-TR-2002-05, Queensland University of Technology, Brisbane, 2002.
12. W.M.P. van der Aalst and K.M. van Hee. *Workflow Management: Models, Methods, and Systems*. MIT press, Cambridge, MA, 2002.
13. W.M.P. van der Aalst and A.H.M. ter Hofstede. Workflow Patterns: On the Expressive Power of (Petri-net-based) Workflow Languages. In K. Jensen, editor, *Proceedings of the Fourth Workshop on the Practical Use of Coloured Petri Nets and CPN Tools (CPN 2002)*, volume 560 of *DAIMI*, pages 1–20, Aarhus, Denmark, August 2002. University of Aarhus.
14. W.M.P. van der Aalst and A.H.M. ter Hofstede. YAWL: Yet Another Workflow Language. QUT Technical report, FIT-TR-2002-06, Queensland University of Technology, Brisbane, 2002.
15. W.M.P. van der Aalst, A.H.M. ter Hofstede, B. Kiepuszewski, and A.P. Barros. Workflow Patterns. *Distributed and Parallel Databases*, 14(1):5–51, 2003.

16. W.M.P. van der Aalst, A.H.M. ter Hofstede, and M. Weske, editors. *Busines Process Management*, volume 2678 of *Lecture Notes in Computer Science*. Springer-Verlag, Berlin, 2003.
17. W.M.P. van der Aalst, B.F. van Dongen, J. Herbst, L. Maruster, G. Schimm, and A.J.M.M. Weijters. Workflow Mining: A Survey of Issues and Approaches. *Data and Knowledge Engineering*, 47(2):237–267, 2003.
18. W.M.P. van der Aalst, K.M. van Hee, and R.A. van der Toorn. Component-Based Software Architectures: A Framework Based on Inheritance of Behavior. *Science of Computer Programming*, 42(2-3):129–171, 2002.
19. W.M.P. van der Aalst and A.J.M.M. Weijters, editors. *Process Mining*, Special Issue of Computers in Industry, Volume 53, Number 3. Elsevier Science Publishers, Amsterdam, 2004.
20. A. Arkin, S. Askary, S. Fordin, and W. Jekel et al. Web Service Choreography Interface (WSCI) 1.0. Standards propsal by BEA Systems, Intalio, SAP, and Sun Microsystems, 2002.
21. A. Arkin et al. Business Process Modeling Language (BPML), Version 1.0, 2002.
22. Pallas Athena. *Case Handling with FLOWer: Beyond workflow*. Pallas Athena BV, Apeldoorn, The Netherlands, 2002.
23. K. Barkaoui, J.M. Couvreur, and C. Dutheillet. On liveness in Extended Non Self-Controlling Nets. In G. De Michelis and M. Diaz, editors, *Application and Theory of Petri Nets 1995*, volume 935 of *Lecture Notes in Computer Science*, pages 25–44. Springer-Verlag, Berlin, 1995.
24. J. Billington and et. al. The Petri Net Markup Language: Concepts, Technology, and Tools. In W.M.P. van der Aalst and E. Best, editors, *Application and Theory of Petri Nets 2003*, volume 2679 of *Lecture Notes in Computer Science*, pages 483–506. Springer-Verlag, Berlin, 2003.
25. R. Casonato. Gartner Group Research Note 00057684, Production-Class Workflow: A View of the Market. http://www.gartner.com, 1998.
26. A. Cheng, J. Esparza, and J. Palsberg. Complexity results for 1-safe nets. In R.K. Shyamasundar, editor, *Foundations of software technology and theoretical computer science*, volume 761 of *Lecture Notes in Computer Science*, pages 326–337. Springer-Verlag, Berlin, 1993.
27. F. Curbera, Y. Goland, J. Klein, F. Leymann, D. Roller, S. Thatte, and S. Weerawarana. Business Process Execution Language for Web Services, Version 1.0. Standards propsal by BEA Systems, International Business Machines Corporation, and Microsoft Corporation, 2002.
28. J. Dehnert. *A Methodology for Workflow Modeling: From Business Process Modeling Towards Sound Workflow Specification*. PhD thesis, TU Berlin, Berlin, Germany, 2003.
29. J. Desel. A proof of the Rank theorem for extended free-choice nets. In K. Jensen, editor, *Application and Theory of Petri Nets 1992*, volume 616 of *Lecture Notes in Computer Science*, pages 134–153. Springer-Verlag, Berlin, 1992.
30. J. Desel and J. Esparza. *Free Choice Petri Nets*, volume 40 of *Cambridge Tracts in Theoretical Computer Science*. Cambridge University Press, Cambridge, UK, 1995.
31. M. Dumas and A.H.M. ter Hofstede. UML activity diagrams as a workflow specification language. In M. Gogolla and C. Kobryn, editors, *Proc. of the 4th Int. Conference on the Unified Modeling Language (UML01)*, volume 2185 of *LNCS*, pages 76–90, Toronto, Canada, October 2001. Springer Verlag.
32. C.A. Ellis. Information Control Nets: A Mathematical Model of Office Information Flow. In *Proceedings of the Conference on Simulation, Measurement and Modeling of Computer Systems*, pages 225–240, Boulder, Colorado, 1979. ACM Press.
33. C.A. Ellis and G. Nutt. Workflow: The Process Spectrum. In A. Sheth, editor, *Proceedings of the NSF Workshop on Workflow and Process Automation in Information Systems*, pages 140–145, Athens, Georgia, May 1996.
34. R. Eshuis and J. Dehnert. Reactive Petri nets for Workflow Modeling. In W.M.P. van der Aalst and E. Best, editors, *Application and Theory of Petri Nets 2003*, volume 2679 of *Lecture Notes in Computer Science*, pages 295–314. Springer-Verlag, Berlin, 2003.

35. J. Esparza. Synthesis rules for Petri nets, and how they can lead to new results. In J.C.M. Baeten and J.W. Klop, editors, *Proceedings of CONCUR 1990*, volume 458 of *Lecture Notes in Computer Science*, pages 182–198. Springer-Verlag, Berlin, 1990.
36. J. Esparza and M. Silva. Circuits, Handles, Bridges and Nets. In G. Rozenberg, editor, *Advances in Petri Nets 1990*, volume 483 of *Lecture Notes in Computer Science*, pages 210–242. Springer-Verlag, Berlin, 1990.
37. L. Fischer, editor. *Workflow Handbook 2001, Workflow Management Coalition.* Future Strategies, Lighthouse Point, Florida, 2001.
38. L. Fischer, editor. *Workflow Handbook 2003, Workflow Management Coalition.* Future Strategies, Lighthouse Point, Florida, 2003.
39. Gartner. Gartner's Application Development and Maintenance Research Note M-16-8153, The BPA Market Cathes another Major Updraft. http://www.gartner.com, 2002.
40. D. Georgakopoulos, M. Hornick, and A. Sheth. An Overview of Workflow Management: From Process Modeling to Workflow Automation Infrastructure. *Distributed and Parallel Databases*, 3:119–153, 1995.
41. R.J. van Glabbeek and D.G. Stork. Query Nets: Interacting Workflow Modules that Ensure Global Termination. In W.M.P. van der Aalst, A.H.M. ter Hofstede, and M. Weske, editors, *International Conference on Business Process Management (BPM 2003)*, volume 2678 of *Lecture Notes in Computer Science*, pages 184–199. Springer-Verlag, Berlin, 2003.
42. Object Management Group. *OMG Unified Modeling Language 2.0 Proposal, Revised submission to OMG RFPs ad/00-09-01 and ad/00-09-02, Version 0.671.* OMG, http://www.omg.com/uml/, 2002.
43. M.H.T. Hack. Analysis production schemata by Petri nets. Master's thesis, Massachusetts Institute of Technology, Cambridge, Mass., 1972.
44. K. van Hee, N. Sidorova, and M. Voorhoeve. Soundness and Separability of Workflow Nets in the Stepwise Refinement Approach. In W.M.P. van der Aalst and E. Best, editors, *Application and Theory of Petri Nets 2003*, volume 2679 of *Lecture Notes in Computer Science*, pages 335–354. Springer-Verlag, Berlin, 2003.
45. A. W. Holt. Coordination Technology and Petri Nets. In G. Rozenberg, editor, *Advances in Petri Nets 1985*, volume 222 of *Lecture Notes in Computer Science*, pages 278–296. Springer-Verlag, Berlin, 1985.
46. IBM. *IBM MQSeries Workflow - Getting Started With Buildtime.* IBM Deutschland Entwicklung GmbH, Boeblingen, Germany, 1999.
47. IDS Scheer. ARIS Process Performance Manager (ARIS PPM). http://www.ids-scheer.com, 2002.
48. S. Jablonski and C. Bussler. *Workflow Management: Modeling Concepts, Architecture, and Implementation.* International Thomson Computer Press, London, UK, 1996.
49. K. Jensen. Coloured Petri Nets: A High Level Language for System Design and Analysis. In G. Rozenberg, editor, *Advances in Petri Nets 1990*, volume 483 of *Lecture Notes in Computer Science*, pages 342–416. Springer-Verlag, Berlin, 1990.
50. K. Jensen. *Coloured Petri Nets. Basic Concepts, Analysis Methods and Practical Use. Volume 1.* EATCS monographs on Theoretical Computer Science. Springer-Verlag, Berlin, 1997.
51. B. Kiepuszewski. *Expressiveness and Suitability of Languages for Control Flow Modelling in Workflows.* PhD thesis, Queensland University of Technology, Brisbane, Australia, 2003. Available via http://www.tm.tue.nl/it/research/patterns.
52. B. Kiepuszewski, A.H.M. ter Hofstede, and W.M.P. van der Aalst. Fundamentals of Control Flow in Workflows. *Acta Informatica*, 39(3):143–209, 2003.
53. E. Kindler and W.M.P. van der Aalst. Liveness, Fairness, and Recurrence. *Information Processing Letters*, 70(6):269–274, June 1999.

54. E. Kindler, A. Martens, and W. Reisig. Inter-Operability of Workflow Applications: Local Criteria for Global Soundness. In W.M.P. van der Aalst, J. Desel, and A. Oberweis, editors, *Business Process Management: Models, Techniques, and Empirical Studies*, volume 1806 of *Lecture Notes in Computer Science*, pages 235–253. Springer-Verlag, Berlin, 2000.
55. P. Lawrence, editor. *Workflow Handbook 1997, Workflow Management Coalition*. John Wiley and Sons, New York, 1997.
56. F. Leymann. Web Services Flow Language, Version 1.0, 2001.
57. F. Leymann and D. Roller. *Production Workflow: Concepts and Techniques*. Prentice-Hall PTR, Upper Saddle River, New Jersey, USA, 1999.
58. D.C. Marinescu. *Internet-Based Workflow Management: Towards a Semantic Web*, volume 40 of *Wiley Series on Parallel and Distributed Computing*. Wiley-Interscience, New York, 2002.
59. M. Ajmone Marsan, G. Balbo, and G. Conte et al. *Modelling with Generalized Stochastic Petri Nets*. Wiley series in parallel computing. Wiley, New York, 1995.
60. A. Martens. On Compatibility of Web Services. *Petri Net Newsletter*, 65:12–20, 2003.
61. M. Zur Muehlen. *Workflow-based Process Controlling: Foundation, Design and Application of workflow-driven Process Information Systems*. Logos, Berlin, 2004.
62. W. Reisig. *Petri Nets: An Introduction*, volume 4 of *EATCS Monographs in Theoretical Computer Science*. Springer-Verlag, Berlin, 1985.
63. W. Reisig and G. Rozenberg, editors. *Lectures on Petri Nets I: Basic Models*, volume 1491 of *Lecture Notes in Computer Science*. Springer-Verlag, Berlin, 1998.
64. W. Sadiq and M.E. Orlowska. Analyzing Process Models using Graph Reduction Techniques. *Information Systems*, 25(2):117–134, 2000.
65. R. Shapiro. A Comparison of XPDL, BPML and BPEL4WS (Version 1.4). http://xml.coverpages.org/Shapiro-XPDL.pdf, 2002.
66. Software-Ley. *COSA User Manual*. Software-Ley GmbH, Pullheim, Germany, 1998.
67. Software-Ley. *COSA 3.0 User Manual*. Software-Ley GmbH, Pullheim, Germany, 1999.
68. S. Staab, W. van der Aalst, V.R. Benjamins, A. Sheth, J. Miller, C. Bussler, A. Maedche, D. Fensel, and D. Gannon. Web Services: Been There, Done That? (Trends and Controversies). *IEEE Intelligent Systems*, 18(1):72–85, 2003.
69. Staffware. *Staffware 2000 / GWD User Manual*. Staffware plc, Berkshire, United Kingdom, 2000.
70. S. Thatte. XLANG Web Services for Business Process Design, 2001.
71. R. van der Toorn. *Component-Based Software Design with Petri nets: An Approach Based on Inheritance of Behavior*. PhD thesis, Eindhoven University of Technology, Eindhoven, The Netherlands, 2004.
72. H.M.W. Verbeek. *Verification and Enactment of Workflow Management Systems (submitted)*. PhD thesis, Eindhoven University of Technology, Eindhoven, The Netherlands, 2004.
73. H.M.W. Verbeek, T. Basten, and W.M.P. van der Aalst. Diagnosing Workflow Processes using Woflan. *The Computer Journal*, 44(4):246–279, 2001.
74. WFMC. Workflow Management Coalition Workflow Standard: Interface 1 – Process Definition Interchange Process Model (WFMC-TC-1016). Technical report, Workflow Management Coalition, Lighthouse Point, Florida, USA, 1999.
75. WFMC. Workflow Management Coalition Workflow Standard: Workflow Process Definition Interface – XML Process Definition Language (XPDL) (WFMC-TC-1025). Technical report, Workflow Management Coalition, Lighthouse Point, Florida, USA, 2002.
76. P. Wohed, W.M.P. van der Aalst, M. Dumas, and A.H.M. ter Hofstede. Pattern-Based Analysis of BPEL4WS. QUT Technical report, FIT-TR-2002-04, Queensland University of Technology, Brisbane, 2002.

77. P. Wohed, W.M.P. van der Aalst, M. Dumas, and A.H.M. ter Hofstede. Analysis of Web Services Composition Languages: The Case of BPEL4WS. In I.Y. Song, S.W. Liddle, T.W. Ling, and P. Scheuermann, editors, *22nd International Conference on Conceptual Modeling (ER 2003)*, volume 2813 of *Lecture Notes in Computer Science*, pages 200–215. Springer-Verlag, Berlin, 2003.
78. M.D. Zisman. *Representation, Specification and Automation of Office Procedures*. PhD thesis, University of Pennsylvania, Warton School of Business, 1977.

A XPDL Schema

The listing below shows selected parts of the XPDL Schema given in [75] relevant for this tutorial.

```
<xsd:element name="Activity">
   <xsd:complexType>
      <xsd:sequence>
         <xsd:element ref="xpdl:Description" minOccurs="0"/>
         <xsd:element ref="xpdl:Limit" minOccurs="0"/>
         <xsd:choice>
            <xsd:element ref="xpdl:Route"/>
            <xsd:element ref="xpdl:Implementation"/>
            <xsd:element ref="xpdl:BlockActivity"/>
         </xsd:choice>
         <xsd:element ref="xpdl:Performer" minOccurs="0"/>
         <xsd:element ref="xpdl:StartMode" minOccurs="0"/>
         <xsd:element ref="xpdl:FinishMode" minOccurs="0"/>
         <xsd:element ref="xpdl:Priority" minOccurs="0"/>
         <xsd:element ref="xpdl:Deadline" minOccurs="0"
                      maxOccurs="unbounded"/>
         <xsd:element ref="xpdl:SimulationInformation" minOccurs="0"/>
         <xsd:element ref="xpdl:Icon" minOccurs="0"/>
         <xsd:element ref="xpdl:Documentation" minOccurs="0"/>
         <xsd:element ref="xpdl:TransitionRestrictions" minOccurs="0"/>
         <xsd:element ref="xpdl:ExtendedAttributes" minOccurs="0"/>
      </xsd:sequence>
      <xsd:attribute name="Id" type="xsd:NMTOKEN" use="required"/>
      <xsd:attribute name="Name" type="xsd:string"/>
   </xsd:complexType>
</xsd:element>
...
<xsd:element name="TransitionRestriction">
   <xsd:complexType>
      <xsd:sequence>
         <xsd:element ref="xpdl:Join" minOccurs="0"/>
         <xsd:element ref="xpdl:Split" minOccurs="0"/>
      </xsd:sequence>
   </xsd:complexType>
</xsd:element> <xsd:element name="TransitionRestrictions">
   <xsd:complexType>
      <xsd:sequence>
         <xsd:element ref="xpdl:TransitionRestriction" minOccurs="0"
                      maxOccurs="unbounded"/>
      </xsd:sequence>
   </xsd:complexType>
</xsd:element>
...
<xsd:element name="Join">
```

```
    <xsd:complexType>
        <xsd:attribute name="Type">
            <xsd:simpleType>
                <xsd:restriction base="xsd:NMTOKEN">
                    <xsd:enumeration value="AND"/>
                    <xsd:enumeration value="XOR"/>
                </xsd:restriction>
            </xsd:simpleType>
        </xsd:attribute>
    </xsd:complexType>
</xsd:element>
...
<xsd:element name="Split">
    <xsd:complexType>
        <xsd:sequence>
            <xsd:element ref="xpdl:TransitionRefs" minOccurs="0"/>
        </xsd:sequence>
        <xsd:attribute name="Type">
            <xsd:simpleType>
                <xsd:restriction base="xsd:NMTOKEN">
                    <xsd:enumeration value="AND"/>
                    <xsd:enumeration value="XOR"/>
                </xsd:restriction>
            </xsd:simpleType>
        </xsd:attribute>
    </xsd:complexType>
</xsd:element>
...
<xsd:element name="Transition">
    <xsd:complexType>
        <xsd:sequence>
            <xsd:element ref="xpdl:Condition" minOccurs="0"/>
            <xsd:element ref="xpdl:Description" minOccurs="0"/>
            <xsd:element ref="xpdl:ExtendedAttributes" minOccurs="0"/>
        </xsd:sequence>
        <xsd:attribute name="Id" type="xsd:NMTOKEN" use="required"/>
        <xsd:attribute name="From" type="xsd:NMTOKEN" use="required"/>
        <xsd:attribute name="To" type="xsd:NMTOKEN" use="required"/>
        <xsd:attribute name="Name" type="xsd:string"/>
    </xsd:complexType>
</xsd:element>
...
<xsd:element name="Condition">
    <xsd:complexType mixed="true">
        <xsd:choice minOccurs="0" maxOccurs="unbounded">
            <xsd:element ref="xpdl:Xpression"/>
        </xsd:choice>
        <xsd:attribute name="Type">
            <xsd:simpleType>
                <xsd:restriction base="xsd:NMTOKEN">
                    <xsd:enumeration value="CONDITION"/>
                    <xsd:enumeration value="OTHERWISE"/>
                    <xsd:enumeration value="EXCEPTION"/>
                    <xsd:enumeration value="DEFAULTEXCEPTION"/>
                </xsd:restriction>
            </xsd:simpleType>
        </xsd:attribute>
    </xsd:complexType>
</xsd:element>
```

InterPlay: Horizontal Scale-up and Transition to Design in Scenario-Based Programming

Dan Barak, David Harel, and Rami Marelly

The Weizmann Institute of Science, Rehovot, Israel

Abstract. We describe **InterPlay**, a simulation engine coordinator that supports cooperation and interaction of multiple simulation and execution tools, thus helping to scale-up the design and development cycle of reactive systems. InterPlay involves two main ideas. In the first, we concentrate on the inter-object design approach involving LSCs and the Play-Engine tool, enabling multiple Play-Engines to run in cooperation. This makes possible the distributed design of large-scale systems by different teams, as well as the refinement of parts of a system using different Play-Engines. The second idea concerns combining the inter-object approach with the more conventional intra-object approach, involving, for example, statecharts and Rhapsody. InterPlay makes it possible to run the Play-Engine in cooperation with Rhapsody, and is very useful when some system objects have clear and distinct internal behavior, or in an iterative development process where the design is implementation-oriented and the ultimate goal is to end up with an intra-object implementation.

1 Introduction

The goal of this work is to enrich the scale-up possibilities in the development cycle of reactive systems, when working in an inter-object, scenario-based paradigm, such as that described in [5]. We do this by introducing and implementing a methodology of distributed design, which involves two related ideas. The methodology is intended to supply a new level of flexibility in system development, and to help ensure that the various parts of a system designed by different teams cooperate and integrate into a single working and harmonious system.

The ideas are implemented in what we shall be calling **InterPlay**, a simulation engine coordinator[1] that supports the cooperation and interaction of different simulation and execution tools. These can support different design approaches to the modeling parts of a system or the various levels of abstraction thereof.

There are many proposed approaches to distributed computing, and many feature platform and language independence. This allows connecting applications spanning multiple platforms and operating systems, which have been written by different companies in various languages. Among such solutions are the following: RMI (Remoter Method Invocation) for distributed Java applications [11]; DCOM[2], which is most often associated with Microsoft operating systems but is also supported on Unix, VMS and

[1] In fact, in [5] InterPlay was referred to by the acronym SEC.

[2] Soon to be replaced by .NET [8].

J. Desel, W. Reisig, and G. Rozenberg (Eds.): ACPN 2003, LNCS 3098, pp. 66–86, 2004.

Macintosh [1]; CORBA [9]; and the more recent Web Services using the SOAP communication protocol [10]. While all these approaches apply to the realm of implemented components, there appears to be no solution to the problem of high-level model-driven distributed design that can offer independence of vendors (supporting, e.g., both Rational Rose, and Rhapsody from I-Logix), of overall design philosophy (supporting both an inter-object and an intra-object methodology), and of levels of abstraction. InterPlay can be viewed as an attempt to address these kinds of independence too.

Before discussing the two ideas manifested in InterPlay, we briefly recall the dual approaches to specifying reactive behavior, described, e.g., in [2, 5]. The first approach is an inter-object, scenario-based one, which is based on specifying cross-object scenarios of various modalities, one at a time. This approach is particularly natural for discussing behavior and specifying requirements, and is exemplified by the language of **live sequence charts** (LSCs) [2] and the **play-in/out** method with its supporting **Play-Engine** tool [5]. The second approach is the more conventional intra-object one, which is usually state-based, and is naturally suited for the specification of objects that have clear internal behavior. This approach specifies all possible behaviors for each object in the system, and it leads directly to implementation. It is exemplified by the language of **statecharts** [3] and the **Rhapsody** tool [4, 6], or by conventional object-by-object code. The conceptual duality between these approaches is illustrated visually in Figure 1.

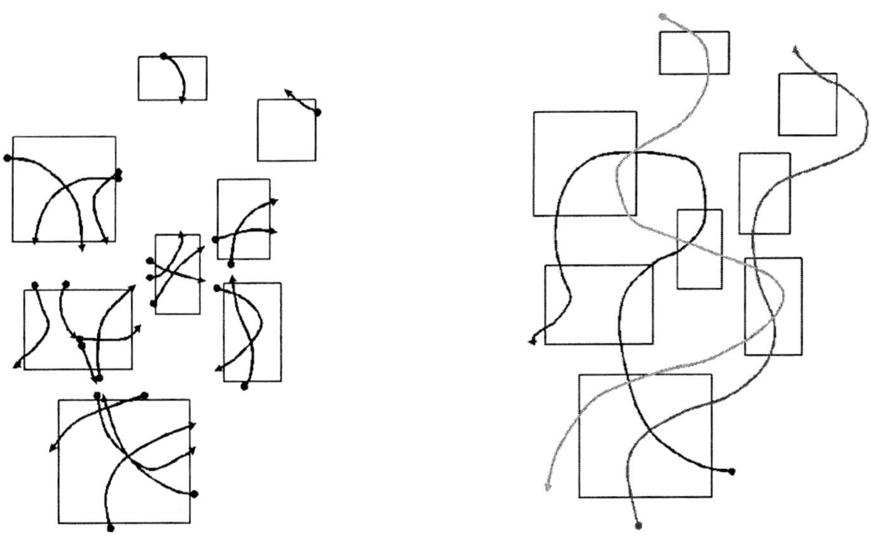

Fig. 1. A visual description of the intra-object and inter-object design approaches respectively.

Let us examine the design and development cycle of a system, observing how the two approaches may be used within it. In the early stages of transforming the client's requirements into a formal specification, the overall functionality of the system is the most important. Here, the main logical components of the system will typically appear, with no specific implementation-related details. This bird's-eye point of view is best described using the inter-object design approach, where we ignore inner mechanisms of system components and focus on the overall behavior of the system, concentrating on interactions among the user, the environment and the system components. Complex systems may have a very large number of objects, practically forcing the distribution of the specification effort – and later also the design and implementation efforts – between multiple teams.

Accordingly, the first ability of InterPlay concentrates on the inter-object approach, and enables multiple Play-Engines to run in cooperation. This makes it possible for different teams to specify the inter-object behavior of different collections of objects, and then run these specifications in a fully cooperative manner. It also makes it possible to refine parts of the system using different Play-Engines. Technically, this is achieved by using **external objects**: each team is assigned some part of the system (actually, a set of objects) to design in detail. A particular team's objects may interact with other objects, to which the team refers as external. These external objects are in fact the *interface* of the other subsystems with respect to the current team's subsystem[3]. All other objects are ignored. The objects with which the team's specification interacts are thus outside the assigned scope and responsibility of the team, yet the team is aware of them, recognizing them as being designed and driven by some other team. The first part of the InterPlay methodology allows these different parts to be executed in tandem, by its ability to have multiple Play-Engines execute together. This distributed design method is illustrated in Figure 2.

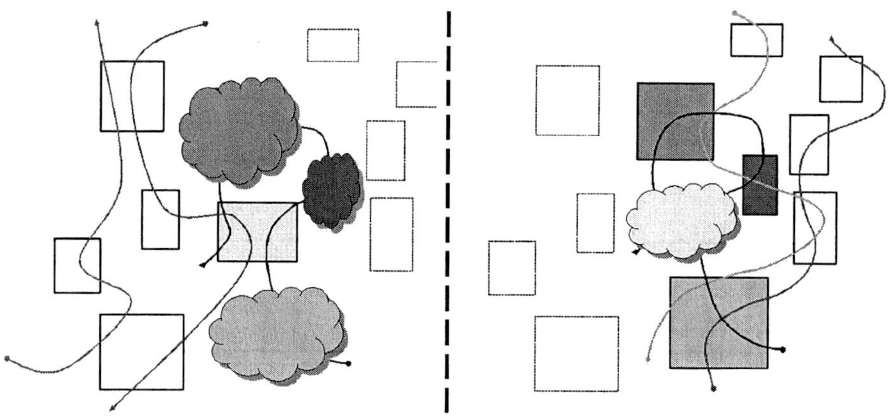

Fig. 2. Distributed design with external objects: External objects are drawn as clouds and each external-internal pair share the same color. Each team specifies a part of the system using the inter-object design approach, and refers to other relevant objects as external.

Let us now turn to the second ability of InterPlay. Following detailed specifications and refinement of requirements, we would like to carry out a transition to design and implementation. While the Play-Engine can indeed execute inter-object specifications,

[3] For more details about external objects, interfaces and distribution to subsystems, see Section 3.

including multiple engines playing together through InterPlay, this is still within the inter-object approach. There will often be objects that have clear and distinct internal behavior which we would like to specify in a more conventional state-based intra-object fashion, using, say, statecharts or code. Moreover, the ultimate goal might be to end up with a complete intra-object implementation, which could be achieved by an iterative development process, during which objects will be gradually provided with intra-object implementation-oriented behavior. The Play-Engine would be useful at the very beginning of this process, and a standard intra-object tool like Rhapsody would be useful at the end, but we want something for the interim, when we have a combination of inter-object and intra-object specifications.

The second feature of InterPlay allows just that: the cooperative execution of a *mixed* system, some parts being specified in a scenario-based fashion, e.g., in LSCs, and others specified in an intra-object state-based fashion, e.g., in statecharts or code. Technically, InterPlay allows the Play-Engine and Rhapsody to execute simultaneously, each taking care of some of the objects. Figure 3 illustrates this, by showing an inter-object specification, with one object designed using the intra-object approach.

The two InterPlay ideas combined enable what we call **horizontal scale-up**, whereby a large system can be split up into parts, each specified in an inter-object or intra-object fashion, at will, and then executed as a whole by Play-Engines cooperating among themselves and/or cooperating with the Rhapsody tool. We view this as a crucial step towards the ability to incorporate the inter-object approach into the development of large and complex systems.

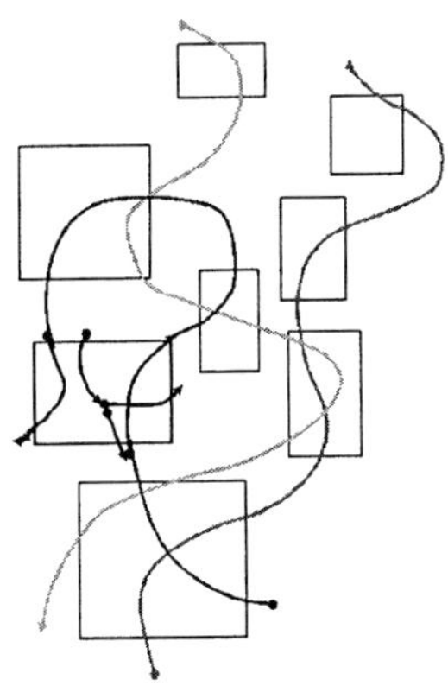

Fig. 3. An inter-object specification with one object designed using the intra-object approach.

The rest of this paper is organized as follows. Section 2 gives a brief overview of the LSC language and the Play-Engine, illustrated using a take-out service system, which serves as a running example throughout the paper. Section 3 discusses the changes introduced in the Play-Engine to support InterPlay and explains their relevance to horizontal scale-up. Section 4 introduces in more detail the InterPlay tool and techniques. Section 5 elaborates on the take-out service example, illustrating the usefulness of InterPlay in integrating the various parts of a system. Section 6 concludes with a discussion of future work, including related research we are carrying out on **vertical scale-up**.

2 The Play-Engine and LSCs

This section provides a short introduction to the language of **live sequence charts** (LSCs) and the *Play-Engine*. The discussion, however, is very brief, and we strongly suggest referring to [5] for more details.

The language of LSCs [2] is a scenario-based visual formalism, which extends classical message sequence charts (MSCs) with logical modalities, thus achieving a far greater expressive power, comparable to that of temporal logic [7]. The Play-Engine supports LSCs, by enabling a system designer to capture behavioral requirements by **playing in** behavior using a graphical interface (GUI) of the target system or an abstract version thereof. As the behavior is played in, the formalized behavior is automatically generated by the Play-Engine, in the form of LSCs.

LSCs have two types of charts, **universal** and **existential**. Universal charts are used to specify restrictions over all possible system runs, and thus constrain the allowed behaviors. A universal chart typically contains a **prechart**, which specifies the scenario which, if successfully executed, forces the system to satisfy the scenario given in the actual chart body. Existential charts, on the other hand, specify sample interactions between the system and its environment, and are required only to be satisfied by at least one system run. They thus do not force the application to behave in a certain way in all cases, and can be used to specify system tests, or simply to illustrate longer (non-restricting) scenarios that provide a broader picture of the behavioral possibilities to which the system gives rise.

We borrow an LSC from our running example, a take-out system described in detail in section 5, to illustrate the main concepts and constructs of the language of LSCs.

In the universal LSC of Figure 4, the prechart (top dashed hexagon) contains the event of the user clicking the `btnOper` button. If this indeed occurs, the chart body then requires the `CustomerConrtol` object to update the occupancy of the restaurant by means of a method call that changes the number of customers in the restaurant. However, we want this update to happen only after a fixed time interval – three clock ticks in our case. The chart body consists of an unbounded loop construct (denoted by '*'), which is repeated infinitely many times, unless interrupted. The loop contains an assignment in which the variable N is assigned the current time. It is important to note that the assignment's variable is local to the containing chart and can be used for the specification of that chart only, as opposed to the system's state variables, which may be used in several charts.

After the assignment comes a **hot** condition, requiring the time to advance 3 ticks before continuing. Hot conditions are mandatory, and must always be true; if not, the requirements are violated and the system aborts. However, when dealing with time, the system simply waits until the specified condition holds. On the other hand, if a **cold** condition is false, the surrounding (sub)chart is exited. This is one example of the way the logical modalities are incorporated into LSCs.

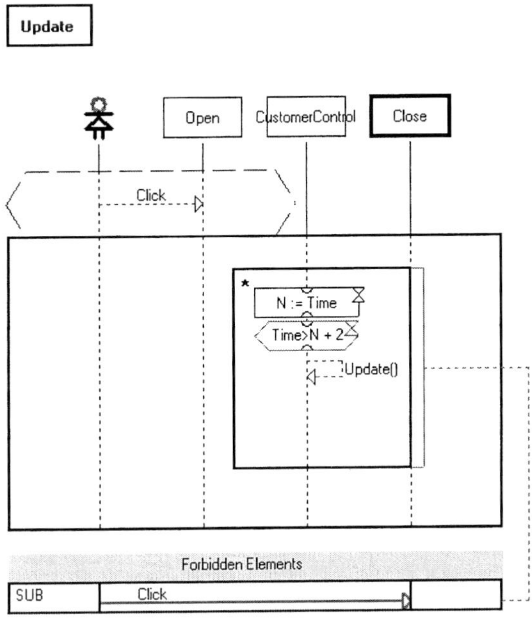

Fig. 4. An LSC example: Updating the occupancy of a restaurant.

An LSC can have **forbidden elements**, listed in a separate area underneath the main chart. Hot and cold elements work similarly there too; e.g., if a hot forbidden condition becomes true, the requirements are violated and the system aborts, whereas a cold one becoming true causes the chart or subchart which is its scope to be exited. In our example in Figure 4, there is a cold forbidden message associated with the loop subchart, the effect being that if the user presses the `btnOper` button again the loop and the chart terminates.

We shall not discuss the play-in process here, but play-out is very relevant. In the play-out phase the user plays the GUI application as he/she would have done when executing a system model (or, for that matter, the final system) but limiting him/herself to 'end-user' and external environment actions only. While doing so, the Play-Engine keeps track of the actions taken, and causes other actions and events to occur as dictated by the LSCs, thus giving the effect of working with a fully operational system or an executable model. It is actually an iterative process, where after each step taken by the user, the play-engine computes a **superstep**, which is a sequence of events carried out by the system as response to the event input by the user. Only those things it is required to do are actually done, while those it is forbidden to do are avoided. This is a minimalistic, but completely safe, way for a system to behave exactly according to the requirements. It is noteworthy that no code needs to be written in order to play out the behavior, nor does one have to prepare a conventional intra-object system model, as is required in most system development methodologies (e.g., using statecharts or some other language for describing the full behavior of each object, as in the UML, for example). We should also emphasize that the behavior played out is up to the user, and need not reflect the behavior as it was played in; the user is not merely tracing scenarios, but is executing the specified behavior freely, as he/she sees fit.

This ability to execute inter-object behavior without building a system model or writing code leads to various improvements in building reactive systems. It enables executable requirements, for example, whereby the Play-Engine becomes a sort of 'universal reactive machine', running the requirements that were played in via a GUI or written directly as LSCs[4]. You provide the global, declarative, inter-object ways you want your system to behave (or to not behave), and the engine simulates these directly. It also allows for executable test-suites, whose executions can then be compared with those of the actual implementation.

As we shall explain later, enabling the cooperation of multiple Play-Engines and these cooperating with conventional tools, allows both distributed design and refinement of such specifications, as well as the gradual introduction of implementation-oriented details in advanced design stages.

3 External Objects in Preparation for InterPlay

Some time ago we introduced external objects into LSCs and implemented them in the Play-Engine along with their respective mechanisms; see Chapter 14 in [5]. However, that introduction was made bearing in mind the idea presented here. In fact, on their own, without InterPlay, external objects are rather hollow, providing little substantial enhancement to the design and development cycle[5]. In this section, we briefly survey the addition of external objects, stressing their role in the scheme we present.

When dealing with reactive systems we distinguish between the system proper and other elements that interact with it, to which we refer as the **environment**. The system's user is separated from the environment and can interact with the system through the GUI, while the other elements of the environment can affect external settings of the system, mainly through changing object properties. Since most reactive systems work in the presence of such external/environmental objects and can affect them and be affected by them, it is necessary to express the interaction with them.

Technically, we have added to the LSCs language and to the Play-Engine a new kind of object, the **external object**, which will be considered as part of the system's environment. External objects are recognized by the system, but are driven externally by another modeling tool, or by code. What will become extremely important, however, is the fact that external objects allow other systems to interact with the one we are working on.

Having external objects within the specification entails more than just breaking up the environment into individual pieces. These pieces are objects in their own right, they have properties, they can be in different states, they can call other objects, etc. However, as we shall see in a moment, in terms of what the Play-Engine knows when 'working on' a particular system with its environment, an external object is abstract; it is not

[4] In principle this could have been done using any other sufficiently powerful scenario-based language, such as timing diagrams or temporal logic.

[5] Without InterPlay no more than two Play-Engines can run cooperatively, and they must always use the exact same system model.

considered to be an ordinary object, and, for example, cannot be triggered (by our Play-Engine specification) to call other objects.

In the LSCs themselves (and also during play-in) external objects are treated much like other objects, and the fact they are external is merely indicated by a little cloud attached to the object-name box. Any object can be made external easily, by flipping the appropriate property in its definition. Thus, objects can be considered internal throughout some portion of the system development process, and then made external later on, whether for refining its design elsewhere, or to implement and test it. We shall see later how this ability can be exploited.

The main difference between internal and external objects occurs during play-out. Usually, property changes of objects, and calls between them, are performed by the Play-Engine as a part of its super-steps. This, however, is not what we want for external objects. The way they are controlled in a simple one-engine use of the Play-Engine is by the system's end-user, but the ultimate goal is for them to be controlled by some other modeling tool, possibly another Play-Engine, or implemented in code. And this is what InterPlay is all about. Consequently, the execution mechanism of the Play-Engine has been modified, so that it does not initiate events that originate from external objects, just as it does not initiate events from the user, or the environment.

Appropriate sets of external objects serve as a commitment between the different teams and their respective parts of the system. They can be compared to an interface in object-oriented programming. The team that sees a specific object as external uses it as a part of its communication mechanism with the outside world. As such, the team relies on this object having certain properties and methods. Hence, the team that 'owns' the object as internal can add properties or methods to it, but not change the original ones. All the added properties and methods added in such a way are for the internal use of that specific team and are not reflected outside on the other external views of the object.

Our methodology is, in a sense, backward compatible, since it can be applied to any existing specification set, even if it was prepared before the introduction of external objects. One of the benefits of this compatibility is that even if two systems have been specified separately, they can later be joined, without any pre-planning. If the two different specifications have referred to some common part, even if slightly differently and by different names, they can still be considered jointly, by choosing the common part to be external in one of the specifications and remaining internal in the other.

In order to support the external objects mechanism, we added to the Play-Engine an **external event manager**, which deals with the technicalities of remote connections to other computers (e.g., IP, ports, etc.) and conveys messages to and from external objects. In fact, once the external manager is activated, the Play-Engine transmits to the outside world the entire sequence of events that occurs among its GUI and internal objects. The Play-Engine also receives via the external manager events and messages from other Play-Engines, or other modeling tools. Since external objects reflect elements specified or implemented outside the scope of the local Play-Engine, events (e.g., property changes or method calls) that originate in those objects also arrive through the external manager. Upon receiving such an event, the Play-Engine acts as if the event originated from the external object itself. In short, the external object is recognized by the local system, but is driven by a remote one.

In order to best serve the InterPlay techniques, the external manager has various operation modes, allowing either cooperation between two Play-Engines or execution by a single Play-Engine and monitoring its run by another. Such a connection was possible between only two Play-Engines having the exact same system model. However, using InterPlay any number of Play-Engines, with different system models, can be connected, as we shall see shortly.

4 InterPlay: Cooperation of Various Design Tools

InterPlay operates in two stages, a preprocessing offline stage, and a main online execution stage. In the first stage a mapping is set up, which associates each internal object with all of its images as external objects in other tools, making them all seem as a single object. During the execution stage InterPlay uses the mapping to translate and transmit messages and events among the connected models and their respective tools, so that whatever happens to an object during play-out is reflected in all its external views.

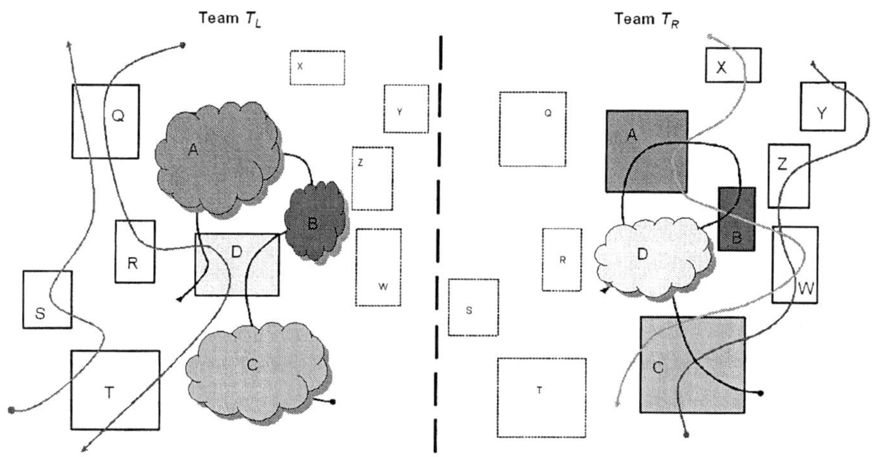

Fig. 5. Inter-object specifications of a system from the points of view of two teams. Objects with thin dotted lines do not actually appear in the relevant specification and are included for better illustration only.

InterPlay's mapping stage is really part of the system's specification, in which one indicates how the different parts of the system fit together. We use the two specifications in Figure 5 throughout this section as a specific example, and concentrate on connecting only multiple Play-Engines.

Consider object D in the figure. It is internal to the left-hand team T_L and external to the right-hand team T_R. Although both teams do deal with this common object, they might refer to it by different names[6] and team T_L might have added to it additional

[6] Had there been another team containing object D as external, it could have referred to it by yet a different name than do teams T_L and T_R.

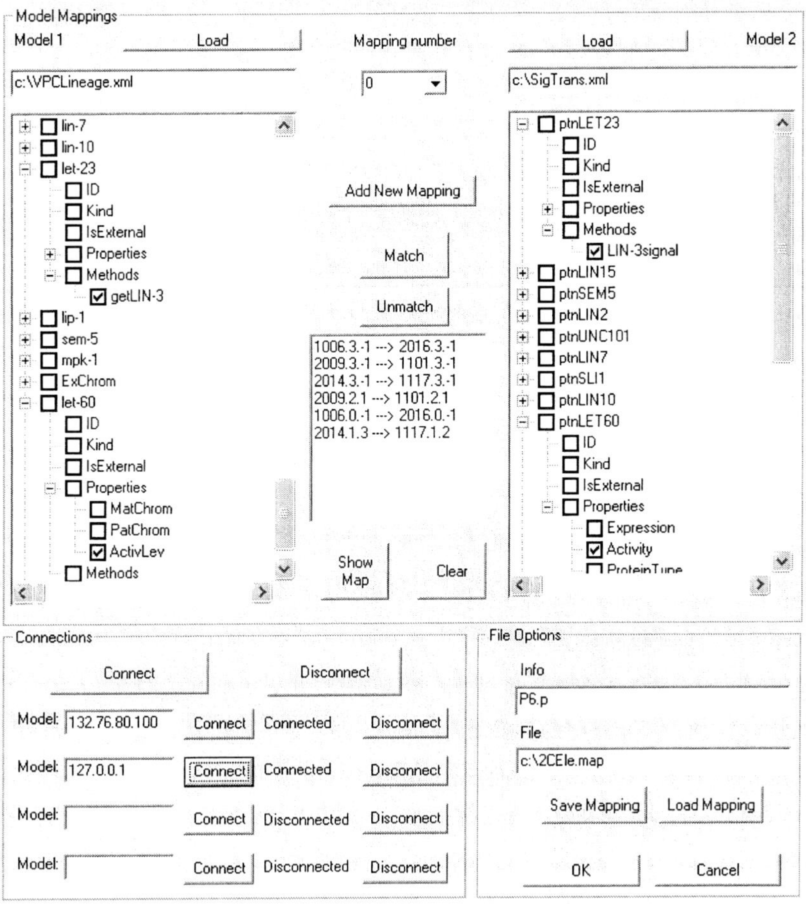

Fig. 6. InterPlay screen shot, mapping two biological models through an interface of a method and a property of two common to both.

properties or methods. Thus InterPlay works on mapping two system parts together, in order to overcome such naming differences while matching an object to its external view. This, of course, does not limit the number of specifications of systems parts and their respective tools that can be fused together. Figure 6 displays a screenshot of InterPlay mapping two system models to each other. These are two parts of a biological system, which communicate using two common proteins Let-23 and Let-60. Although both parts refer to the same proteins, their descriptions are very different in the two models. The common interface is a method in Let-23 and an activity measurement property in Let-60, which are mapped to each other through InterPlay.

When using InterPlay to bridge different levels of abstraction one has to pay particular attention to the specification refinement from coarse to fine. Objects described on the coarse level are **interface objects** for some subsystem that interacts through them.

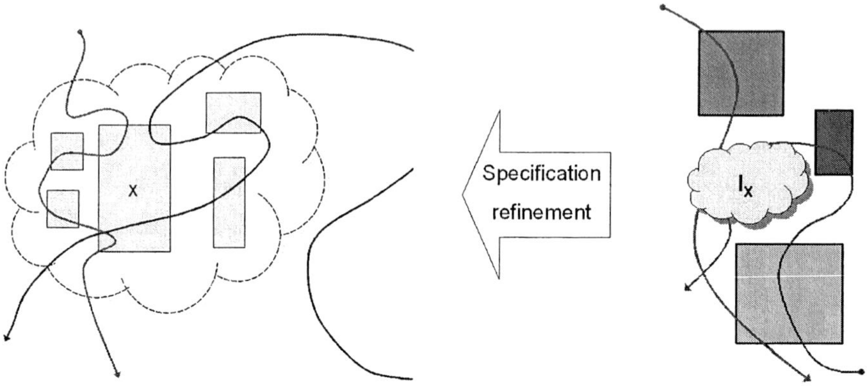

Fig. 7. Multiple levels of abstraction. The left figure represents a specification refinement of the external interface object I_X in the coarse level.

Hence, on a coarse level we describe interactions among interface objects, while when refining the specification we implement[7] the subsystems that interact through them. This rather subtle difference from actually refining an object is further illustrated in Figure 7: The right-hand side of the figure is the coarse level system, in which there is an external interface object I_X. The left-hand portion of the figure is a refinement of the I_X, and within it the internal object X implements the interface object on the coarse level. All other objects on the fine level constitute the subsystem behind the interface X. Thus, all interactions between this subsystem and other subsystems are conducted through object X. When mapping the two specifications through InterPlay, X is matched to I_X, allowing events on the finer abstraction level to be reflected on the coarser level, and vice versa.

Here's how the mapping is set up. InterPlay loads a system model from a Play-Engine specification and displays it to the user. Only the components of the system are loaded (i.e., the GUI and internal objects, with their properties and methods), without any behavior (LSCs) attached. There are several levels of mapping between objects. Assuming object A has not been extended with new methods or properties by team T_R, the mapping can be completed as is, by simply associating (using an appropriate form that pops up) the two versions of A on the object level. This implies that all the object's properties and methods are also mapped.

Assume that object B has been expanded by team T_R. InterPlay allows partial mappings of selected properties and methods, leaving some unmatched. Thus, only the properties and methods common to the two teams will be mapped to each other and we do not allow splitting; e.g., mapping some properties of B in team T_R's specification to object B of team T_L and others to object C therein. This kind of splitting up of an object is closely related to aggregation, and is the central aspect of **vertical scale-up**, which we discuss briefly in Section 6. Nevertheless, InterPlay does allow mapping multiple objects to a single one on the object abstraction level. Going back to Figure 5, it might

[7] By "implementation" in this context we still refer to inter-object design, used to specify in detail a subsystem which has been declared on the coarse level.

be the case that the left-hand team T_L considers objects A, B and C as having the same functionality. For example, D might be a department manager with a direct phone line connection to his/her bosses A, B and C. As only these bosses can call this line, D is impervious to which of them assigns him/her a task. Team T_L can thus use a single external object only, say, A, which will be mapped to the group of objects A, B and C in team T_R's specification. This raises the question of whether any event involving object A in team T_L's specification would have to be reflected in all of its mapped variants on the right. Currently, InterPlay broadcasts such an event to all internal objects mapped to an external one, but other possibilities are mentioned in Section 6.

During play-out, InterPlay carries out the ramifications of the mappings set up in the preliminary phase. Each Play-Engine connects to InterPlay through its external manager. Once connected, played out events (user operations, property changes and method calls) are transmitted to InterPlay, which translates them according to the mappings and sends them to all the relevant Play-Engines. Consider the blue (rightmost) scenario in Figure 5. Play-out starts with team T_R's Play-Engine, involving object C. Since C is internal to T_R's, its Play-Engine performs the necessary events, operating it. InterPlay translates and transmits these events to the T_L's Play-Engine, which traces the scenario as well. The scenario moves on to object D, which is external to T_R's scope, and thus T_R's Play-Engine goes idle. Object D is now 'driven' by the T_L's Play-Engine and through InterPlay the respective events are sent to the T_R's Play-Engine. This initiates an event coming from D, allowing the scenario to proceed. The scenario continues in a similar fashion, with each Play-Engine running and driving its own internal objects, and waiting to receive input from the other one if necessary.

As mentioned above, this description concentrates on several Play-Engines, but a similar process is carried out when the Play-Engine is connected to Rhapsody. More on this later.

5 An Example: The Food Take-Out System

In this section we illustrate InterPlay by a simple example of a food take-out service that enables clients to order food from diverse restaurants through a single ordering center.

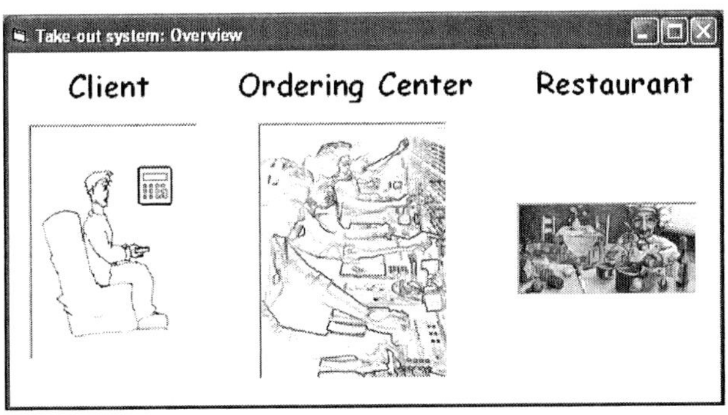

Fig. 8. The three high-level components of the food take-out system.

The development process starts with specifying an inter-object overview of the system's overall functionality. This coarse specification identifies the system's main components – a client, the ordering center and a restaurant component – as illustrated in the GUI of Figure 8. Using the Play-Engine and LSCs, we describe the functionality of the system by interactions among these components, as exemplified in Figure 9. One LSC therein describes the simple process of acquiring a menu from the ordering center, while the other concerns placing a take-out order. Before we explain the latter LSC, note that the `Client` and `Restaurant` were internal at this stage and became external only in later design stages. The prechart contains the event of the `Client` ordering a dish by calling the `Center's Order(Dish)` method. Should this occur, the main chart specifies the `Center` asking for a time estimate on the `Dish` from the `Restaurant`, by calling the `Restaurant's Estimate(Dish)` method. The `Restaurant`'s resulting estimated time is conveyed to the `Center` via the `Time(T)` method. (In accordance to the inter-object design approach we do not specify at this stage how the restaurant calculates this estimated time.) After receiving the estimated time to delivery, the `Client` responds by calling the `Confirm` method with its `ID` and `Decision`. Should the `Client` agree, depicted by the cold `Decision=True` condition, a series of method calls follows, confirming the order to the `Restaurant` and getting an `OrderID` in exchange. If for some reason the `Customer` doesn't wish to order, the chart is simply exited, in effect cancelling the order.

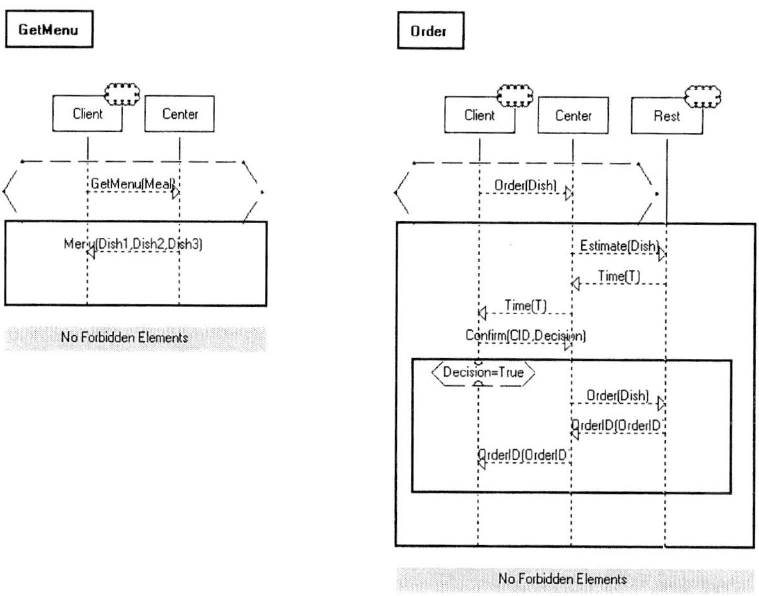

Fig. 9. Two LSCs describing an overview of the behavior of the take-out system.

We now decide to distribute the rest of the specification among three teams, each in charge of one of these components. Each team is required to refine the specification of its assigned subsystem, respecting the interface that was defined on the coarse level. Hence the client has an internal object called `I_Panel`, implementing the interface defined by the `Client` object on the coarse level and serving as its interface with the other system components. It also has an external object called `CommUnit` that implements the ordering center's interface within the client's subsystem. In other words, the entire client subsystem interacts with the rest of the system, represented by `CommUnit`, through its interface, `I_Panel`. Similarly, the restaurant's subsystem has an internal object, `I_Rest`, as its interface with the 'outside world', which in turn is represented by the external `CommUnit`. These objects can be seen in Figures 10 and 12, which show the refined GUIs and additional objects of the client and restaurant subsystems, respectively.

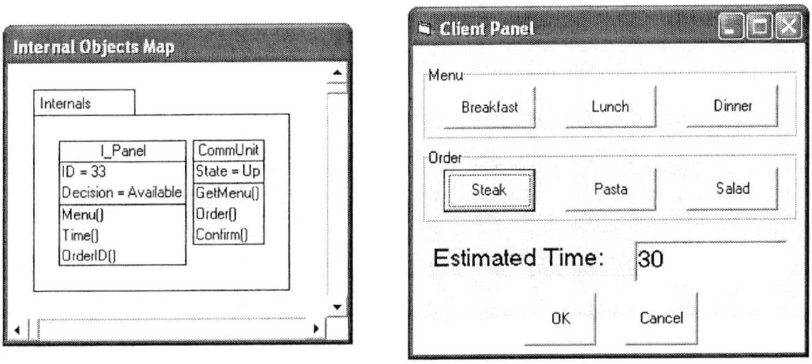

Fig. 10. The client's GUI and internal objects.

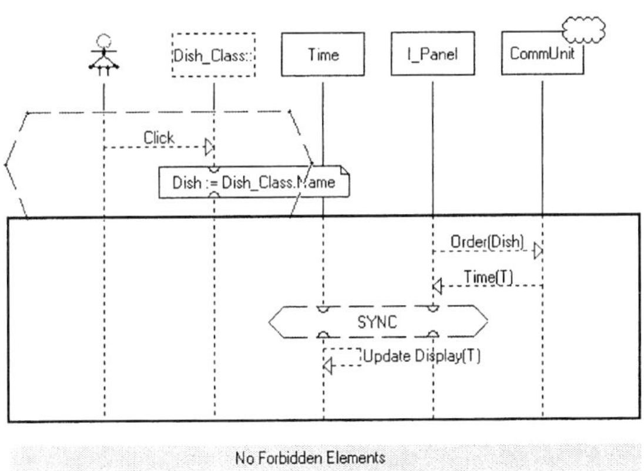

Fig. 11. An LSC that describes the ordering process from the client's point of view. This LSC is invoked by all three buttons referring to dishes (second row in the GUI), which are of class `Dish_Class`.

Having the coarse design level available, we then approach the client subsystem and refine its specification using the aforementioned interface and adding to it further objects and internal behavior. Figures 10 and 11 illustrate this specification refinement, with its GUI and a self-explanatory LSC example that describes the process of the client ordering a dish.

Now that the client's subsystem refinement is complete, we make the `Client` object on the coarse level external. As such, the Play-Engine playing out the coarse level can no longer initiate events from the client. Instead, it waits for them to arrive, having been initiated by another Play-Engine playing out the client subsystem. We played out both specifications, one fine and one coarse, in cooperation, using InterPlay, as we explain shortly. At this stage the restaurant has not been refined yet, so it continued to be 'driven' by the coarse level specification.

Fig. 12. The restaurant's GUI, including cooks and a reflection of its state. Cooks wearing hats are laboring in the kitchen while the others are on break.

The restaurant's team then starts to refine its specification, deciding that the restaurant has to have some cooks to keep the business running, a few customers who sit inside, and two indicator buttons to capture the opening and closing of the restaurant. The team specifies how these parts of the system should behave, independently of, and in ignorance of, how their 'outside world' operates, but still aware of it and interacting with it through the external `CommUnit`. The restaurant's GUI and additional objects are shown in Figure 12, while an LSC example describing part of its internal behavior is shown in Figure 13.

The LSC in the figure specifies how the restaurant calculates the time estimate for a requested dish. It is activated when the `CommUnit` requests an estimate by calling the method `Estimate(Dish)` of the restaurant's interface, `I_Rest`, as defined in the prechart. In the main chart, using a select-case construct, the basic time required for the requested dish is stored. The number of available cooks is also taken into consideration in the if-then-else construct. Finally, the restaurant's interface `I_Rest` returns the preparation time to the ordering center, through the `CommUnit`. The restaurant's specification refinement involved a few other LSCs that deal with its internal behavior, such as one describing the working routine of the cooks in the restaurant, depending on the amount of clientele patronizing it. For lack of space we will not show these here.

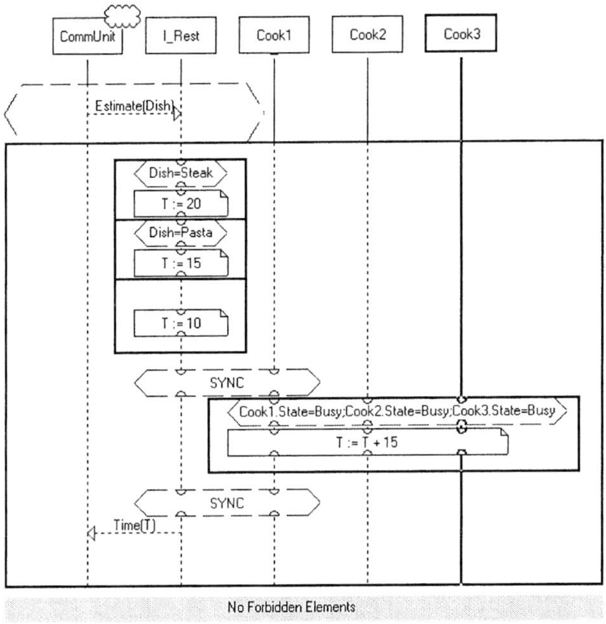

Fig. 13. An LSC describing part of the restaurant's inner behavior.

Recall also Figure 4, which updates the number of clients in the restaurant every 3 clock ticks, by calling the `Update` method.

Having now refined the specifications of the client and the restaurant subsystems, we make both `Client` and `Restaurant` objects external on the coarse level. The three system models, with only the objects and their respective properties and methods, are loaded into InterPlay. We map the refined subsystems to the coarse specification, in turn, by associating their appropriate interface objects: `I_panel` is mapped to `Client` and `I_Rest` is mapped to `Rest`, while both `CommUnits` on the fine level are mapped (separately) to `Center` on the coarse level. Notice that the latter two mappings are made based only on a subset of the methods and properties, while the former two are made on the object level. The mapping of the refined restaurant to the overview of the system is shown in Figure 14.

The entire system can now be run in cooperation by three different Play-Engines, one for each of the two refined subsystems and the third running the coarse specification, providing the functionality of the yet unrefined ordering center and monitoring the entire run. Since the Play-Engine can record a run and later display it as an LSC, we have attached in Figure 15 the three recordings of the respective Play-Engines.

After all of this, and assume we have executed, revised and verified the inter-object specification, we might want to make a transition to design, or in other words, to move towards an intra-object implementation. We could pick the restaurant's interface unit (`I_Rest`), for example, which has clear internal behavior. We would make it external to the inter-object specification of the restaurant, and proceed to define its internal behavior in a state-based fashion using statecharts and Rhapsody. We can now load the

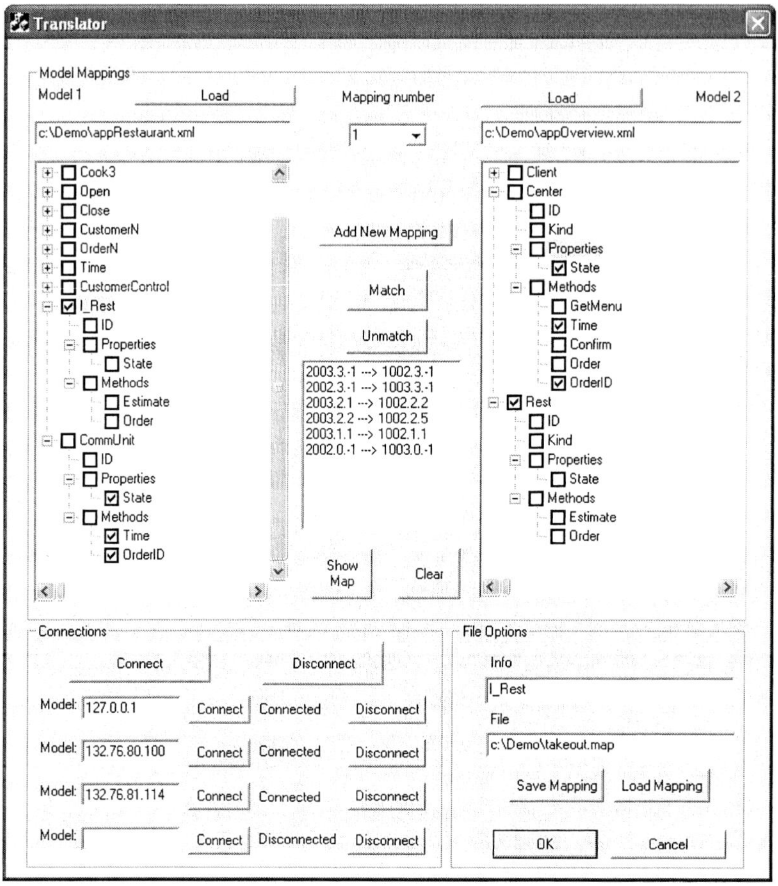

Fig. 14. InterPlay screenshot, showing the mapping of the refined restaurant's subsystem model to the coarse level overview of the take-out system.

unit's system model from Rhapsody into InterPlay and map it to the restaurant's LSCs specification. This would then allow running the intra-object design, or implementation, of the panel both against its specification and in cooperation with the rest of the take-out service system.

Doing this for all the parts of the system that we want to have implemented in an intra-object way would lead to a full implementation. All remaining parts would be played-out in an inter-object manner, with the relevant Play-Engines handing over control whenever an implemented part is to become active.

6 What Next?

In this section we discuss several issues for future research.

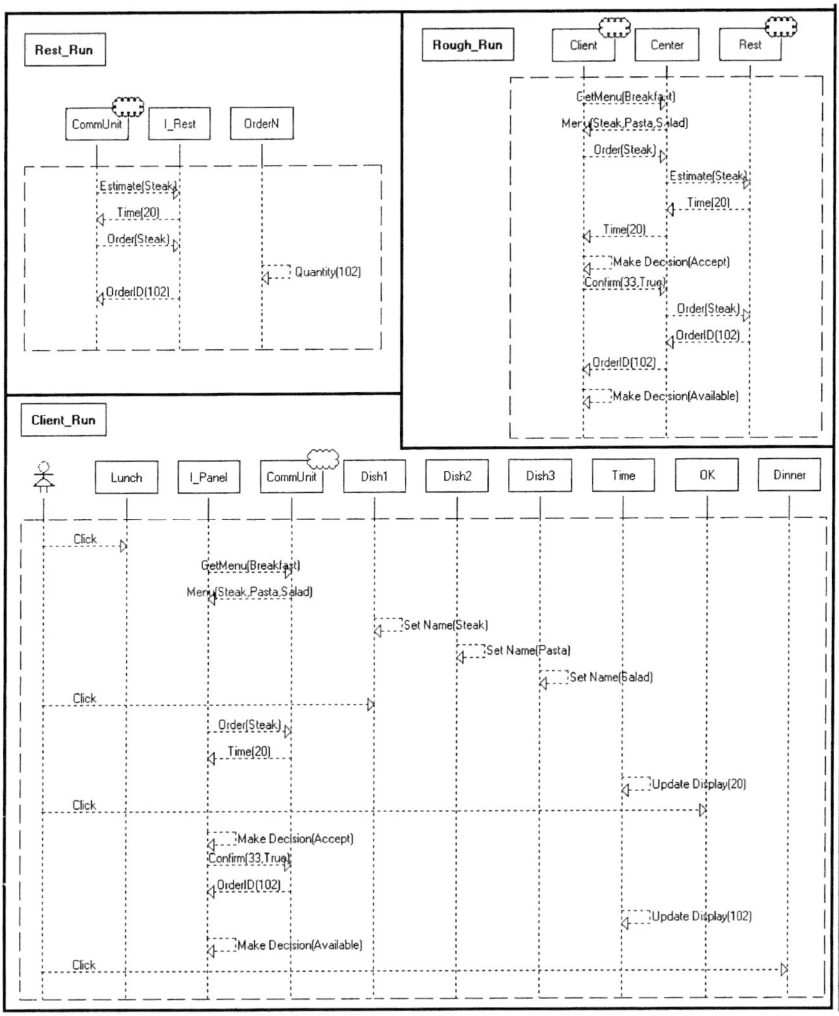

Fig. 15. The results of running the take-out system using three cooperating Play-Engines. Each LSC is the trace of the run from the point of view of one Engine's model.

Connecting to Rhapsody: We have repeatedly stated as our goal not only to connect Play-Engines to each other, but also to allow cooperation between many types of design or modeling tools. We have set up an initial connection between Rhapsody and the Play-Engine. Its present status is that of a feasibility test, and was carried out in a tailored fashion for a particular system model, with very encouraging results. We are now in the final stages of making this connection generic through InterPlay.

We do not make any changes to Rhapsody's framework in order to allow this connection. Instead, we offer an API with which a dll plug-in can be created. The plug-in serves as an observer that receives events of interest from Rhapsody as they take place during the system run, and can also interact with the animation module of Rhapsody, generating events that will impact the animation. We then made it possible for these dll plug-ins to communicate with InterPlay, in both sending and receiving events.

Connecting to Other Implementations: There is clearly much value in allowing the Play-Engine to be connected to other kinds of modeling and implementation tools, including standard programming environments. For example, if a project requires designing a new component that has to fit exactly into an existing complex of implemented components or systems, it could be extremely useful to connect the LSC model we build for it using the Play-Engine via InterPlay directly to the real environment, allowing the composite system to be tested and run as an integrated whole.

Moreover, given such flexible connection abilities, modeling tools like the Play-Engine could be used to conduct integration tests of implemented components even if these were designed using other tools. The implemented system could then be executed with a Play-Engine tracing its runs, making sure they fit the requirements (which would have been predefined as LSCs).

To make such broad connection abilities possible, we intend to construct a simple API for connecting to InterPlay, which most implemented systems will be able to incorporate. Since they would all connect to each other through InterPlay, no changes in any of these tools will be required by this addition.

Synchronous Messages: Synchronous messages, supported by the Play-Engine, raise a whole new level of complexity when one uses InterPlay to carry out truly distributed modeling and implementation. Recall that a synchronous messages is one that flows (for all practical purposes in zero time) from the sending object to the receiving object if and when the former is ready to send it and the latter is free to receive it. When both objects are controlled by a single Play-Engine it is relatively easy to determine whether the message can be sent, and if so to make sure nothing changes in the two objects until the message is delivered. This is far more complicated when the two objects are driven by different Play-Engines, and even worse if they are driven by statecharts or code.

Several possible solutions come to mind, such as using a *two phase commit* protocol, of the kind used in certain kinds of transaction processing. We have not yet dealt with this feature, and doing so would probably require subtle changes both in the Play-Engine and in the InterPlay module.

Centralized Clock Ticks: Another complication that InterPlay gives rise to involves time. Recall that the Play-Engine supports time via a single clock, with a tick event that can be advanced through the host computer's clock or via the model itself (e.g., by the user or by other objects). Clearly, different Play-Engines running different specifications cannot be assumed to advance clock ticks at the same (absolute) rate, and the classical problems of distributed time arise in full force. Even running a single Play-Engine will advance time very differently when run with or without visual animation of the LSCs, not to mention different Play-Engines working in tandem or with other modeling tools.

Without getting into the usual controversies and opinions about how to best deal with time in a distributed environment, it is quite obvious that there are several incentives for supplying a mechanism for centralized clock ticks across InterPlay. (For one, we might be using InterPlay to build an ultimately centralized system in a distributed fashion.) We propose to add the option of receiving clock tick signals from InterPlay through the external event manager. This is relatively easily done. We have also looked closely into Rhapsody, which has a special time mode controlled by the user, and conclude that it too can receive clock ticks from InterPlay through the observer dll without making changes to the main program's framework.

Type Mapping: Currently two objects, or their properties and methods, can be effectively mapped to one another if they are of the same type, or receive parameters of the same type. We plan to consider adding more flexibility to InterPlay through a type-mapping feature, allowing system models to enrich their interaction without having to make further adjustments to the model itself.

Delegating to Multiple Objects: Recall that InterPlay allows mapping multiple objects to a single one on the object abstraction level. However, should an event that involves the single object be necessarily reflected onto all of its multiple images? We do not have enough experience with InterPlay to decide on this quite yet. Other than the obvious approach, currently implemented, of broadcasting each message (and relevant event) to all the objects mapped to the source, we could also implement a scheme that sends it to the latest image to have interacted with the source. We could also have a user-driven mode, letting the user of InterPlay decide at run time how to delegate the message. Recently we have been toying with the idea of allowing asymmetric mappings, which might solve this problem more elegantly, but this is still in preliminary stages only.

Vertical Scale-up: In this paper we have used the term horizontal scale-up to denote the kinds of connections between tools we have discussed. The reason is that what they make possible is the **composition** of collections of objects in a side-by side manner (although in an implicit way a limited kind of refinement can be specified too as we have seen in section 5). Complimentary to this is vertical scale-up, whereby we want to support in LSCs and the Play-Engine the **aggregation**, or **rich refinement** of objects. In other words we want in the large a full notion of hierarchies of objects, complete with multiple-level behavior, even within a single LSC specification. And we want all this related in the play-in and play-out processes. This is a complicated topic, since it is not clear how to best define aggregation in the presence of inter-object behavior. For example, how should scenarios (i.e., LSCs) defined within an object, among its sub-objects, be connected to the scenarios between the parent object and its siblings on the higher level? What kind of mappings should we allow between levels, etc.? We are in the midst of a research project on this, and hope to be able to report on it in a future paper.

References

1. N. Brown and C. Kindel. Distributed Component Object Model Protocol - DCOM/1.0. http://www.microsoft.com/oledev/olecom/draft-brown-dcom-v1-spec-01.txt.
2. W. Damm and D. Harel. LSCs: Breathing Life into Message Sequence Charts. *Formal Methods in System Design*, 19(1), 2001. (Preliminary version in *Proc. 3rd IFIP Int. Conf. on Formal Methods for Open Object-Based Distributed Systems* (*FMOODS'99*), (P. Ciancarini, A. Fantechi and R. Gorrieri, eds.), Kluwer Academic Publishers, 1999, pp. 293–312.).

3. D. Harel. Statecharts: A Visual Formalism for Complex Systems. *Sci. Comput. Prog.,* 8,3, 231-274, 1987. (Preliminary version: Tech. Report CS84-05, The Weizmann Institute of Science, Rehovot, Israel, February 1984.).
4. D. Harel and E. Gery. Executable Object Modeling with Statecharts. *Computer*, July 1997, pp. 31-42.
5. D. Harel and R. Marelly. *Come, Let's Play: Scenario-Based Programming Using LSCs and the Play-Engine*. Springer-Verlag, 2003.
6. I-logix,inc., Website: http://www.ilogix.com/products/rhapsody/index.cfm.
7. Z. Manna and A. Pnueli. *The Temporal Logic of Reactive and Concurrent Systems: Specification*. Springer-Verlag, 1992.
8. Microsoft .NET architecture and resources: http://www.microsoft.com/net/.
9. OMG - Object Management Group. *The Common Object Request Broker: Architecture and Specification*. 2.2 ed, 1992.
10. *Simple Object Access Protocol (SOAP) 1.1*. W3C Note 08 May 2000. http://www.w3.org/TR/2000/NOTE-SOAP-20000508/.
11. A. Wollrath, R. Riggs and J. Waldo. A Distributed Object Model for the Java System. *USENIX Computing Systems*, vol. 9, November/December 1996.

Timed Automata: Semantics, Algorithms and Tools

Johan Bengtsson and Wang Yi*

Uppsala University
{johanb,yi}@it.uu.se

Abstract. This chapter is to provide a tutorial and pointers to results and related work on timed automata with a focus on semantical and algorithmic aspects of verification tools. We present the concrete and abstract semantics of timed automata (based on transition rules, regions and zones), decision problems, and algorithms for verification. A detailed description on DBM (Difference Bound Matrices) is included, which is the central data structure behind several verification tools for timed systems. As an example, we give a brief introduction to the tool UPPAAL.

1 Introduction

Timed automata is a theory for modeling and verification of real time systems. Examples of other formalisms with the same purpose, are timed Petri Nets, timed process algebras, and real time logics [17,42,47,40,8,20]. Following the work of Alur and Dill [5,6], several model checkers have been developed with timed automata being the core of their input languages *e.g.* [50,33]. It is fair to say that they have been the driving force for the application and development of the theory. The goal of this chapter is to provide a tutorial on timed automata with a focus on the semantics and algorithms based on which these tools are developed.

In the original theory of timed automata [5,6], a timed automaton is a finite-state Büchi automaton extended with a set of real-valued variables modeling clocks. Constraints on the clock variables are used to restrict the behavior of an automaton, and Büchi accepting conditions are used to enforce progress properties. A simplified version, namely *Timed Safety Automata* is introduced in [28] to specify progress properties using local invariant conditions. Due to its simplicity, Timed Safety Automata has been adopted in several verification tools for timed automata *e.g.* UPPAAL [33] and Kronos [50]. In this presentation, we shall focus on Timed Safety Automata, and following the literature, refer them as *Timed Automata* or simply automata when it is understood from the context.

The rest of the chapter is organized as follows: In the next section, we describe the syntax and operational semantics of timed automata. The section also addresses decision problems relevant to automatic verification. In the literature, the decidability and undecidability of such problems are often considered to be the fundamental properties of a computation model. Section 3 presents the abstract version of the operational

* Corresponding author: Wang Yi, Division of Computer Systems, Department of Information Technology, Uppsala University, Box 337, 751 05 Uppsala, Sweden. Email: yi@it.uu.se

J. Desel, W. Reisig, and G. Rozenberg (Eds.): ACPN 2003, LNCS 3098, pp. 87–124, 2004.

semantics based on regions and zones. Section 4 describes the data structure DBM (Difference Bound Matrices) for the efficient representation and manipulation of zones, and operations on zones, needed for symbolic verification. Section 5 gives a brief introduction to the verification tool UPPAAL. Finally, as an appendix, we list the pseudo-code for the presented DBM algorithms.

2 Timed Automata

A timed automaton is essentially a finite automaton (that is a graph containing a finite set of nodes or locations and a finite set of labeled edges) extended with real-valued variables. Such an automaton may be considered as an abstract model of a timed system. The variables model the logical clocks in the system, that are initialized with zero when the system is started, and then increase synchronously with the same rate. Clock constraints *i.e.* guards on edges are used to restrict the behavior of the automaton. A transition represented by an edge can be taken when the clocks values satisfy the guard labeled on the edge. Clocks may be reset to zero when a transition is taken.

The first example. Fig. 1(a) is an example timed automaton. The timing behavior of the automaton is controlled by two clocks x and y. The clock x is used to control the self-loop in the location **loop**. The single transition of the loop may occur when $x = 1$. Clock y controls the execution of the entire automaton. The automaton may leave **start** at any time point when y is in the interval between 10 and 20; it can go from **loop** to **end** when y is between 40 and 50, *etc.*

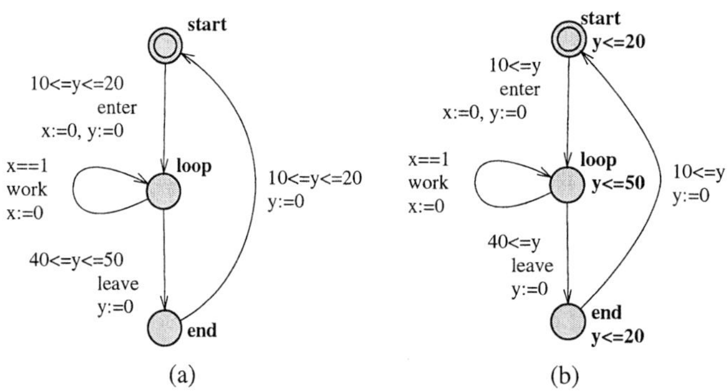

Fig. 1. Timed Automata and Location Invariants.

Timed Büchi Automata. A guard on an edge of an automaton is only an enabling condition of the transition represented by the edge; but it can not force the transition to be taken. For instance, the example automaton may stay forever in any location, just idling. In the initial work by Alur and Dill [5], the problem is solved by introducing Büchi-acceptance conditions; a subset of the locations in the automaton are marked as accepting, and only those executions passing through an accepting location infinitely often are considered valid behaviors of the automaton. As an example, consider again the automaton in Fig. 1(a) and assume that **end** is marked as accepting. This implies that all executions of the automaton must visit **end** infinitely many times. This imposes implicit conditions on **start** and **loop**. The location **start** must be left when the value of y is at most 20, otherwise the automaton will get stuck in **start** and never be able to enter **end**. Likewise, the automaton must leave **loop** when y is at most 50 to be able to enter **end**.

Timed Safety Automata. A more intuitive notion of progress is introduced in *timed safety automata* [28]. Instead of accepting conditions, in timed safety automata, locations may be put local timing constraints called *location invariants*. An automaton may remain in a location as long as the clocks values satisfy the invariant condition of the location. For example, consider the timed safety automaton in Fig. 1(b), which corresponds to the Büchi automaton in Fig. 1(a) with **end** marked as an accepting location. The invariant specifies a local condition that **start** and **end** must be left when y is at most 20 and **loop** must be left when y is at most 50. This gives a local view of the timing behavior of the automaton in each location.

In the rest of this chapter, we shall focus on timed safety automata and refer such automata as *Timed Automata* or simply automata without confusion.

2.1 Formal Syntax

Assume a finite set of real-valued variables $\mathcal{C}$ ranged over by x, y *etc.*standing for clocks and a finite alphabet Σ ranged over by a, b *etc.*standing for actions.

Clock Constraints. A clock constraint is a conjunctive formula of atomic constraints of the form $x \sim n$ or $x - y \sim n$ for $x, y \in \mathcal{C}, \sim \in \{\leq, <, =, >, \geq\}$ and $n \in \mathbb{N}$. Clock constraints will be used as guards for timed automata. We use $\mathcal{B}(\mathcal{C})$ to denote the set of clock constraints, ranged over by g and also by D later.

Definition 1 (Timed Automaton) *A timed automaton $\mathcal{A}$ is a tuple $\langle N, l_0, E, I \rangle$ where*

- *N is a finite set of locations (or nodes),*
- *$l_0 \in N$ is the initial location,*
- *$E \in N \times \mathcal{B}(\mathcal{C}) \times \Sigma \times 2^{\mathcal{C}} \times N$ is the set of edges and*
- *$I : N \longrightarrow \mathcal{B}(\mathcal{C})$ assigns invariants to locations*

We shall write $l \xrightarrow{g,a,r} l'$ when $\langle l, g, a, r, l' \rangle \in E$.

As in verification tools e.g. UPPAAL *[33], we restrict location invariants to constraints that are downwards closed, in the form: $x \leq n$ or $x < n$ where n is a natural number.*

Concurrency and Communication. To model concurrent systems, timed automata can be extended with parallel composition. In process algebras, various parallel composition operators have been proposed to model different aspects of concurrency (see *e.g.*

CCS and CSP [39, 29]). These algebraic operators can be adopted in timed automata. In the UPPAAL modeling language [33], the CCS parallel composition operator [39] is used, which allows interleaving of actions as well as hand-shake synchronization. The precise definition of this operator is given in Section 5.

Essentially the parallel composition of a set of automata is the product of the automata. Building the product automaton is an entirely syntactical but computationally expensive operation. In UPPAAL, the product automaton is computed on-the-fly during verification.

2.2 Operational Semantics

The semantics of a timed automaton is defined as a transition system where a state or configuration consists of the current location and the current values of clocks. There are two types of transitions between states. The automaton may either delay for some time (a delay transition), or follow an enabled edge (an action transition).

To keep track of the changes of clock values, we use functions known as *clock assignments* mapping C to the non-negative reals $\mathbb{R}_+$. Let u, v denote such functions, and use $u \in g$ to mean that the clock values denoted by u satisfy the guard g. For $d \in \mathbb{R}_+$, let $u + d$ denote the clock assignment that maps all $x \in C$ to $u(x) + d$, and for $r \subseteq C$, let $[r \mapsto 0]u$ denote the clock assignment that maps all clocks in r to 0 and agree with u for the other clocks in $C \setminus r$.

Definition 2 (Operational Semantics) *The semantics of a timed automaton is a transition system (also known as a timed transition system) where states are pairs $\langle l, u\rangle$, and transitions are defined by the rules:*

- $\langle l, u\rangle \xrightarrow{d} \langle l, u + d\rangle$ *if* $u \in I(l)$ *and* $(u + d) \in I(l)$ *for a non-negative real* $d \in \mathbb{R}_+$
- $\langle l, u\rangle \xrightarrow{a} \langle l', u'\rangle$ *if* $l \xrightarrow{g,a,r} l', u \in g, u' = [r \mapsto 0]u$ *and* $u' \in I(l')$

2.3 Verification Problems

The operational semantics is the basis for verification of timed automata. In the following, we formalize decision problems in timed automata based on transition systems.

Language Inclusion. A *timed action* is a pair (t, a), where $a \in \Sigma$ is an action taken by an automaton $\mathcal{A}$ after $t \in \mathbb{R}_+$ time units since $\mathcal{A}$ has been started. The absolute time t is called a *time-stamp* of the action a. A *timed trace* is a (possibly infinite) sequence of timed actions $\xi = (t_1, a_1)(t_2, a_2)...(t_i, a_i)...$ where $t_i \leq t_{i+1}$ for all $i \geq 1$.

Definition 3 *A* run *of a timed automaton $\mathcal{A} = \langle N, l_0, E, I\rangle$ with initial state $\langle l_0, u_0\rangle$ over a timed trace $\xi = (t_1, a_1)(t_2, a_2)(t_3, a_3)...$ is a sequence of transitions:*

$$\langle l_0, u_0\rangle \xrightarrow{d_1}\xrightarrow{a_1} \langle l_1, u_1\rangle \xrightarrow{d_2}\xrightarrow{a_2} \langle l_2, u_2\rangle \xrightarrow{d_3}\xrightarrow{a_3} \langle l_3, u_3\rangle \ldots$$

satisfying the condition $t_i = t_{i-1} + d_i$ for all $i \geq 1$.

The timed language $L(\mathcal{A})$ is the set of all timed traces ξ for which there exists a run of $\mathcal{A}$ over ξ.

Undecidability. The negative result on timed automata as a computation model is that the language inclusion checking problem *i.e.* to check $L(\mathcal{A}) \subseteq L(\mathcal{B})$ is undecidable [6, 4]. Unlike finite state automata, timed automata is not determinizable in general. Timed automata can not be complemented either, that is, the complement of the timed language of a timed automaton may not be described as a timed automaton.

The inclusion checking problem will be decidable if $\mathcal{B}$ in checking $L(\mathcal{A}) \subseteq L(\mathcal{B})$ is restricted to the deterministic class of timed automata. Research effort has been made to characterize interesting classes of determinizable timed systems *e.g.* event-clock automata [7] and timed communicating sequential processes [48]. Essentially, the undecidability of language inclusion problem is due to the arbitrary clock reset. If all the edges labeled with the same action symbol in a timed automaton, are also labeled with the same set of clocks to reset, the automaton will be determinizable. This covers the class of event-clock automata [7].

We may abstract away from the time-stamps appearing in timed traces and define the *untimed language* $L_{untimed}(\mathcal{A})$ as the set of all traces in the form $a_1a_2a_3\ldots$ for which there exists a timed trace $\xi = (t_1, a_1)(t_2, a_2)(t_3, a_3)...$ in the timed language of $\mathcal{A}$.

The inclusion checking problem for untimed languages is decidable. This is one of the classic results for timed automata [6].

Bisimulation. Another classic result on timed systems is the decidability of timed bisimulation [19]. Timed bisimulation is introduced for timed process algebras [47]. However, it can be easily extended to timed automata.

Definition 4 *A bisimulation R over the states of timed transition systems and the alphabet $\Sigma \cup \mathbb{R}_+$, is a symmetrical binary relation satisfying the following condition:*

for all $(s_1, s_2) \in R$, if $s_1 \xrightarrow{\sigma} s_1'$ for some $\sigma \in \Sigma \cup \mathbb{R}_+$ and s_1', then $s_2 \xrightarrow{\sigma} s_2'$ and $(s_1', s_2') \in R$ for some s_2'.

Two automata are timed bisimilar iff there is a bisimulation containing the initial states of the automata.

Intuitively, two automata are timed bisimilar iff they perform the same action transition at the same time and reach bisimilar states. In [19], it is shown that timed bisimulation is decidable.

We may abstract away from timing information to establish bisimulation between automata based actions performed only. This is captured by the notion of untimed bisimulation. We define $s \xrightarrow{\epsilon} s'$ if $s \xrightarrow{d} s'$ for some real number d. Untimed bisimulation is defined by by replacing the alphabet with $\Sigma \cup \{\epsilon\}$ in Definition 4. As timed bisimulation, untimed bisimulation is decidable [35].

Reachability Analysis. Perhaps, the most useful question to ask about a timed automaton is the reachability of a given final state or a set of final states. Such final states may be used to characterize safety properties of a system.

Definition 5 *We shall write $\langle l, u\rangle \longrightarrow \langle l', u'\rangle$ if $\langle l, u\rangle \xrightarrow{\sigma} \langle l', u'\rangle$ for some $\sigma \in \Sigma \cup \mathbb{R}_+$. For an automaton with initial state $\langle l_0, u_0\rangle$, $\langle l, u\rangle$, is reachable iff $\langle l_0, u_0\rangle \longrightarrow^* \langle l, u\rangle$.*

More generally, given a constraint $\phi \in \mathcal{B}(C)$ we say that the configuration $\langle l, \phi \rangle$ is reachable if $\langle l, u \rangle$ is reachable for some u satisfying ϕ.

The notion of reachability is more expressive than it appears to be. We may specify invariant properties using the negation of reachability properties, and bounded liveness properties using clock constraints in combination with local properties on locations [38] (see Section 5 for an example).

The reachability problem is decidable. In fact, one of the major advances in verification of timed systems is the symbolic technique [23,46,28,49,34], developed in connection with verification tools. It adopts the idea from symbolic model checking for untimed systems, which uses boolean formulas to represent sets of states and operations on formulas to represent sets of state transitions. It is proven that the infinite state-space of timed automata can be finitely partitioned into symbolic states using clock constraints known as *zones* [12,23]. A detailed description on this is given in Section 3 and 4.

3 Symbolic Semantics and Verification

As clocks are real-valued, the transition system of a timed automaton is infinite, which is not an adequate model for automated verification.

3.1 Regions, Zones and Symbolic Semantics

The foundation for the decidability results in timed automata is based on the notion of *region equivalence* over clock assignments [6,3].

Definition 6 (Region Equivalence) *Let k be a function, called a* clock ceiling, *mapping each clock $x \in C$ to a natural number $k(x)$ (i.e. the ceiling of x). For a real number d, let $\{d\}$ denote the fractional part of d, and $\lfloor d \rfloor$ denote its integer part. Two clock assignments u, v are* region-equivalent, *denoted $u \dot{\sim}_k v$, iff*

1. *for all x, either $\lfloor u(x) \rfloor = \lfloor v(x) \rfloor$ or both $u(x) > k(x)$ and $v(x) > k(x)$,*
2. *for all x, if $u(x) \leq k(x)$ then $\{u(x)\} = 0$ iff $\{v(x)\} = 0$ and*
3. *for all x, y if $u(x) \leq k(x)$ and $u(y) \leq k(y)$ then $\{u(x)\} \leq \{u(y)\}$ iff $\{v(x)\} \leq \{v(y)\}$*

Note that the region equivalence is indexed with a clock ceiling k. When the clock ceiling is given by the maximal clock constants of a timed automaton under consideration, we shall omit the index and write $\dot{\sim}$ instead. An equivalence class $[u]$ induced by $\dot{\sim}$ is called a *region*, where $[u]$ denotes the set of clock assignments region-equivalent to u. The basis for a finite partitioning of the state-space of a timed automaton is the following facts. First, for a fixed number of clocks each of which has a maximal constant, the number of regions is finite. Second, $u \dot{\sim} v$ implies (l, u) and (l, v) are bisimilar w.r.t. the untimed bisimulation for any locaton l of a timed automaton. We use the equivalence classes induced by the untimed bisimulation as symbolic (or abstract) states to construct a finite-state model called the *region graph* or *region automaton* of the original timed automaton. The transition relation between symbolic states is defined as follows:

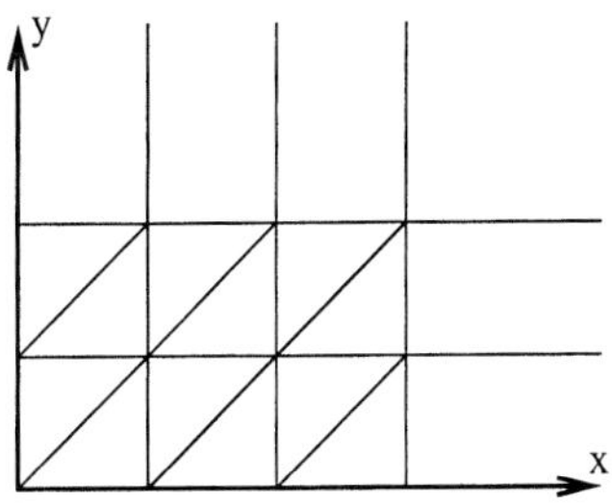

Fig. 2. Regions for a System with Two Clocks.

- $\langle l, [u] \rangle \Rightarrow \langle l, [v] \rangle$ if $\langle l, u \rangle \xrightarrow{d} \langle l, v \rangle$ for a positive real number d and
- $\langle l, [u] \rangle \Rightarrow \langle l', [v] \rangle$ if $\langle l, u \rangle \xrightarrow{a} \langle l', v \rangle$ for an action a.

Note that the transition relation $\Rightarrow$ is finite. Thus the region graph for a timed automaton is finite. Several verification problems such as reachability analysis, untimed language inclusion, language emptiness [6] as well as timed bisimulation [19] can be solved by techniques based on the region construction.

However, the problem with region graphs is the potential explosion in the number of regions. In fact, it is exponential in the number of clocks as well as the maximal constants appearing in the guards of an automaton. As an example, consider Fig. 2. The figure shows the possible regions in each location of an automaton with two clocks x and y. The largest number compared to x is 3, and the largest number compared to y is 2. In the figure, all corner points (intersections), line segments, and open areas are regions. Thus, the number of possible regions in each location of this example is 60.

A more efficient representation of the state-space for timed automata is based on the notion of *zone* and *zone-graphs* [23,27,46,49,28]. In a zone graph, instead of regions, zones are used to denote symbolic states. This in practice gives a coarser and thus more compact representation of the state-space. The basic operations and algorithms for zones to construct zone-graphs are described in Section 4. As an example, a timed automaton and the corresponding zone graph (or reachability graph) is shown in Fig. 3. We note that for this automaton the zone graph has only 8 states. The region-graph for the same example has over 50 states.

A zone is a clock constraint. Strictly speaking, a *zone* is the solution set of a clock constraint, that is the maximal set of clock assignments satisfying the constraint. It is well-known that such sets can be efficiently represented and stored in memory as DBMs (*Difference Bound Matrices*) [12]. For a clock constraint D, let $[D]$ denote the maximal set of clock assignments satisfying D. In the following, to save notation, we shall use D to stand for $[D]$ without confusion. Then $\mathcal{B}(\mathcal{C})$ denotes the set of zones.

A symbolic state of a timed automaton is a pair $\langle l, D \rangle$ representing a set of states of the automaton, where l is a location and D is a zone. A symbolic transition describes all the possible concrete transitions from the set of states.

Definition 7 *Let D be a zone and r a set of clocks. We define $D^{\uparrow} = \{u + d \mid u \in D, d \in \mathbb{R}_+\}$ and $r(D) = \{[r \mapsto 0]u \mid u \in D\}$. Let $\leadsto$ denote the symbolic transition relation over symbolic states defined by the following rules:*

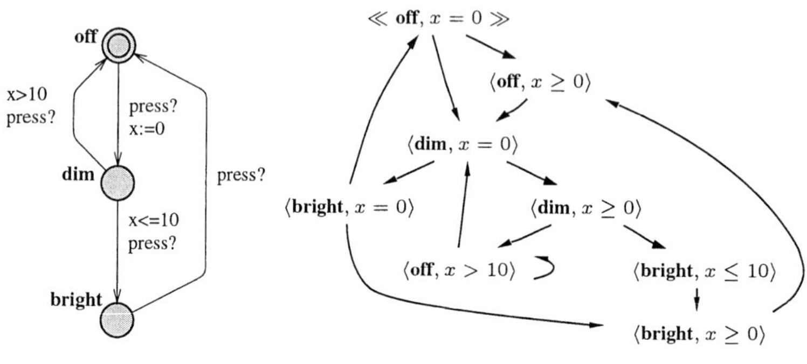

Fig. 3. A Timed Automaton and its Zone Graph.

- $\langle l, D\rangle \rightsquigarrow \langle l, D^{\uparrow} \wedge I(l)\rangle$
- $\langle l, D\rangle \rightsquigarrow \langle l', r(D \wedge g) \wedge I(l')\rangle$ *if* $l \xrightarrow{g,a,r} l'$

We shall study these operations in details in Section 4 where $D^{\uparrow}$ is written as $\mathtt{up}(D)$ and $r(D)$ as $\mathtt{reset}(D, r{:=}0)$. It will be shown that the set of zones $\mathcal{B}(\mathcal{C})$ is closed under these operations, in the sense that the result of the operations is also a zone. Another important property of zones is that a zone has a canonical form. A zone D is *closed under entailment* or just closed for short, if no constraint in D can be strengthened without reducing the solution set. The canonicity of zones means that for each zone $D \in \mathcal{B}(\mathcal{C})$, there is a unique zone $D' \in \mathcal{B}(\mathcal{C})$ such that D and D' have exactly the same solution set and D' is closed under entailment. Section 4 describes how to compute and represent the canonical form of a zone. It is the key structure for the efficient implementation of state-space exploration using the symbolic semantics.

The symbolic semantics corresponds closely to the operational semantics in the sense that $\langle l, D\rangle \rightsquigarrow \langle l', D'\rangle$ implies for all $u' \in D'$, $\langle l, u\rangle \rightarrow \langle l', u'\rangle$ for some $u \in D$. More generally, the symbolic semantics is a correct and full characterization of the operational semantics given in Definition 2.

Theorem 1 *Assume a timed automaton with initial state* $\langle l_0, u_0\rangle$.

1. *(soundness)* $\langle l_0, \{u_0\}\rangle \rightsquigarrow^* \langle l_f, D_f\rangle$ *implies* $\langle l_0, u_0\rangle \rightarrow^* \langle l_f, u_f\rangle$ *for all* $u_f \in D_f$.
2. *(Completeness)* $\langle l_0, u_0\rangle \rightarrow^* \langle l_f, u_f\rangle$ *implies* $\langle l_0, \{u_0\}\rangle \rightsquigarrow^* \langle l_f, D_f\rangle$ *for some* D_f *such that* $u_f \in D_f$

The soundness means that if the initial symbolic state $\langle l_0, \{u_0\}\rangle$ may lead to a set of final states $\langle l_f, D_f\rangle$ according to $\rightsquigarrow$, all the final states should be reachable according to the concrete operational semantics. The completeness means that if a state is reachable according to the concrete operational semantics, it should be possible to conclude this using the symbolic transition relation.

Unfortunately, the relation $\rightsquigarrow$ is infinite, and thus the zone-graph of a timed automaton may be infinite, which can be a problem to guarantee termination in a verification procedure. As an example, consider the automaton in Fig. 4. The value of clock y drifts away unboundedly, inducing an infinite zone-graph.

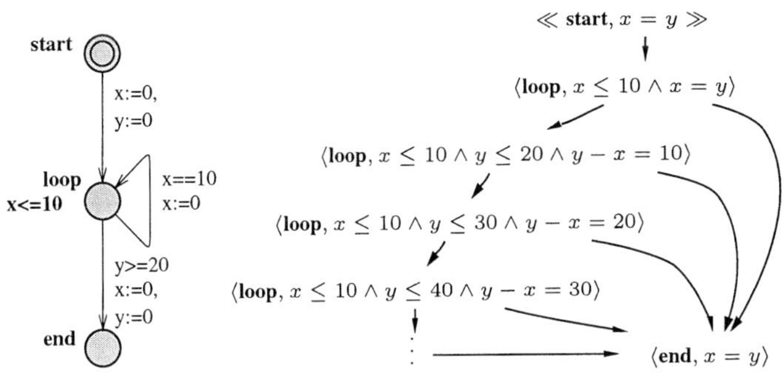

Fig. 4. A Timed Automaton with an Infinite Zone-Graph.

The solution is to transform (*i.e.* normalize) zones that may contain arbitrarily large constants to their representatives in a class of zones whose constants are bounded by fixed constants *e.g.* the maximal clock constants appearing in the automaton, using an abstraction technique similar to the widening operation [26]. The intuition is that once the value of a clock is larger than the maximal constant in the automaton, it is no longer important to know the precise value of the clock, but only the fact that it is above the constant.

3.2 Zone-Normalization for Automata without Difference Constraints

In the original theory of timed automata [6], difference constraints are not allowed to appear in the guards. Such automata (whose guards contain only atomic constraints in the form $x \sim n$) are known as diagonal-free automata in the literature in [18]. For diagonal-free automata, a well-studied zone-normalization procedure is the so-called *k-normalization* operation on zones [43, 41]. It is implemented in several verification tools for timed automata *e.g.* UPPAAL to guarantee termination.

Definition 8 (k-Normalization) *Let D be a zone and k a clock ceiling. The semantics of the k-normalization operation on zones is defined as follows:*

$$\mathsf{norm}_k(D) = \{u | u \dot{\sim}_k v, v \in D\}$$

Note that the normalization operation is indexed by a clock ceiling k. According to [43, 41], $\mathsf{norm}_k(D)$ can be computed from the canonical representation of D by

1. removing all constraints of the form $x < m$, $x \leq m$, $x - y < m$ and $x - y \leq m$ where $m > k(x)$,
2. replacing all constraints of the form $x > m$, $x \geq m$, $x - y > m$ and $x - y \geq m$ where $m > k(x)$ with $x > k(x)$ and $x - y > k(x)$ respectively.

Let $[D]_k$ denote the resulted zone by the above transformation. This is exactly the normalized zone as defined in Definition 8, that is, $[D]_k = \{u | u \dot{\sim}_k v, v \in D\}$

As an example, the normalized zone-graph of the automaton in Fig. 4 is shown in Fig. 5 where the clock ceiling is given by the maximal clock constants appearing in the automaton.

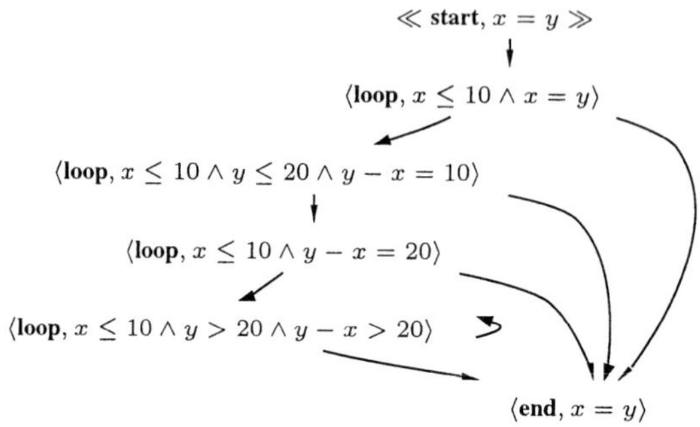

Fig. 5. Normalized Zone Graph for the Automaton in Fig. 4.

Note that for a fixed number of clocks with a clock ceiling k, there can be only finitely many normalized zones. The intuition is that if the constants allowed to use are bounded, one can write down only finitely many clock constraints. This gives rise to a finite characterization for $\rightarrow$.

Definition 9 $\langle l, D\rangle \rightsquigarrow_k \langle l', \mathit{norm}_k(D')\rangle$ *if* $\langle l, D\rangle \rightsquigarrow \langle l', D'\rangle$.

For the class of diagonal-free timed automata $\rightsquigarrow_k$ is sound, complete and finite in the following sense.

Theorem 2 *Assume a timed automaton with initial state $\langle l_0, u_0\rangle$, whose maximal clock constants are bounded by a clock ceiling k. Assume that the automaton has no guards containing difference constraints in the form of $x - y \sim n$.*

1. *(soundness) $\langle l_0, \{u_0\}\rangle \rightsquigarrow_k^* \langle l_f, D_f\rangle$ implies $\langle l_0, u_0\rangle \longrightarrow^* \langle l_f, u_f\rangle$ for all $u_f \in D_f$ such that $u_f(x) \le k(x)$ for all x.*
2. *(Completeness) $\langle l_0, u_0\rangle \longrightarrow^* \langle l_f, u_f\rangle$ with $u_f(x) \le k(x)$ for all x, implies $\langle l_0, \{u_0\}\rangle \rightsquigarrow_k^* \langle l_f, D_f\rangle$ for some D_f such that $u_f \in D_f$*
3. *(Finiteness) The transition relation $\rightsquigarrow_k$ is finite.*

Unfortunately the soundness will not hold for timed automata whose guards contain difference constraints. We demonstrate this by an example. Consider the automaton shown in Fig. 6. The final location of the automaton is not reachable according to the operational semantics. This is because in location S_2, the clock zone is ($x - y > 2$ and $x > 2$) and the guard on the outgoing edge is $x < z+1 \wedge z < y+1$ which is equivalent to $x - z < 1 \wedge z - y < 1 \wedge x - y < 2$. Obviously the zone at S_2 can never enable the guard, and thus the last transition will never be possible. However, because the maximal constants for clock x is 1 (and 2 for y), the zone in location S_2: $x - y > 2 \wedge x > 2$ will be normalized to $x - y > 1 \wedge x > 1$ by the maximal constant 1 for x, which enables the guard $x - z < 1 \wedge z - y < 1$ and thus the symbolic reachability analysis based on the above normalization algorithm would incorrectly conclude that the last location is reachable.

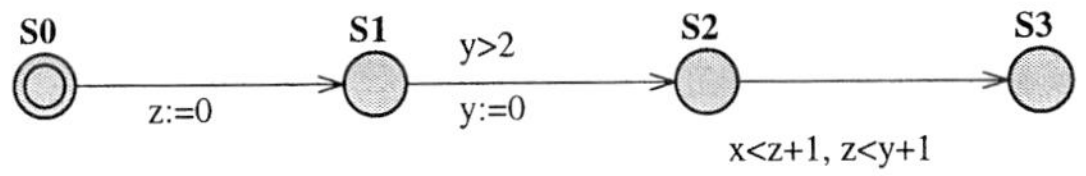

Fig. 6. A counter example.

(a) Zones without normalization

$$
\begin{aligned}
S_0 &: \begin{cases} x - y = 0 \\ y - z = 0 \\ z - x = 0 \end{cases} \\
S_1 &: \begin{cases} x - y = 0 \\ z - x \leq 0 \\ z - y \leq 0 \end{cases} \\
S_2 &: \begin{cases} y - x < -2 \\ y - z \leq 0 \\ z - x \leq 0 \\ \mathbf{0} - x < -2 \end{cases}
\end{aligned}
$$

(b) Zones normalized with k-normalization

$$
\begin{aligned}
S_0 &: \begin{cases} x - y = 0 \\ y - z = 0 \\ z - x = 0 \end{cases} \\
S_1 &: \begin{cases} x - y = 0 \\ z - x \leq 0 \\ z - y \leq 0 \end{cases} \\
S_2 &: \begin{cases} y - x < -1 \\ y - z \leq 0 \\ z - x \leq 0 \\ \mathbf{0} - x < -1 \end{cases} \\
S_3 &: \begin{cases} y - x < -1 \\ y - z < 0 \\ z - x < 0 \\ \mathbf{0} - x < -1 \\ \mathbf{0} - z < 0 \\ x - z < 1 \end{cases}
\end{aligned}
$$

Fig. 7. Zones for the counter example in Fig. 6.

The zones in canonical forms, generated in exploring the state-space of the counter example are given in Fig. 7. The implicit constraints that all clocks are non-negative are not shown.

Note that at S_0 and S_1, the normalized and un-normalized zones are identical. The problem is at S_2 where the intersection of the guard (on the only outgoing edge) with the un-normalized zone is empty and non-empty with the normalized zone.

3.3 Zone-Normalization for Automata with Difference Constraints

Our definition of timed automata (Definition 1) allows any clock constraint to appear in a guard, which may be a difference constraint in the form of $x - y \sim n$. Such automata are indeed needed in many applications *e.g.* to model scheduling problems [24]. It is shown that an automaton containing difference constraints can be transformed to an equivalent diagonal-free automaton [18]. However, the transformation is not applicable since it is based on the region construction. Besides, it is impractical to implement such an approach in a tool. Since the transformation modifies the model before analysis, it is difficult to trace debugging information provided by the tool back to the original model.

In [14, 16], a refined normalization algorithm is presented for automata that may have guards containing difference constraints. The algorithm transforms DBMs according to not only the maximal constants of clocks but also difference constraints appearing in the automaton under consideration. Note that the difference constraints correspond to the diagonal lines which split the entire space of clock assignments. A finer partitioning is needed.

We present the semantical characterization for the refined normalization operation based on a refined version of the region equivalence from Definition 6.

Definition 10 (Normalization Using Difference Constraints) *Let $\mathcal{G}$ stand for a finite set of difference constraints of the form $x-y \sim n$ for $x, y \in \mathcal{C}$, $\sim \in \{\leq, <, =, >, \geq\}$ and $n \in \mathbb{N}$, and k for a clock ceiling. Two clock assignments u, v are equivalent, denoted $u \dot{\sim}_{k,\mathcal{G}} v$ if the following holds:*

- *$u \dot{\sim}_k v$ and*
- *for all $g \in \mathcal{G}$, $u \in g$ iff $v \in g$.*

The semantics of the refined k-normalization operation on zones is defined as follows:

$$\mathsf{norm}_{k,\mathcal{G}}(D) = \{u | u \ \dot{\sim}_{k,\mathcal{G}} \ v, v \in D\}$$

Note that the refined region equivalence is indexed by both a clock ceiling k and a finite set of difference constraints $\mathcal{G}$, and so is the normalization operation.

Since the number of regions induced by $\dot{\sim}_k$ is finite and there are only finitely many constraints in $\mathcal{G}$, $\dot{\sim}_{k,\mathcal{G}}$ induces finitely many equivalence classes. Thus for any given zone D, $\mathsf{norm}_{k,\mathcal{G}}(D)$ is well-defined in the sense that it contains only a finite set of equivalence classes though the set may not be a convex zone, and it can be computed effectively according to the refined regions. In general, $\mathsf{norm}_{k,\mathcal{G}}(D)$ is a non-convex zone, which can be implemented as the union of a finite list of convex zones. The next section will show how to compute this efficiently.

The refined zone-normalization gives rise to a finite characterization for $\rightarrow$.

Definition 11 $\langle l, D\rangle \rightsquigarrow_{k,\mathcal{G}} \langle l', \mathsf{norm}_{k,\mathcal{G}}(D')\rangle$ *if* $\langle l, D\rangle \rightsquigarrow \langle l', D'\rangle$.

The following states the correctness and finiteness of $\rightsquigarrow_{k,\mathcal{G}}$.

Theorem 3 *Assume a timed automaton with initial state $\langle l_0, u_0\rangle$, whose maximal clock constants are bounded by a clock ceiling k, and whose guards contain only a finite set of difference constraints denoted $\mathcal{G}$.*

1. *(soundness) $\langle l_0, \{u_0\}\rangle$ $(\rightsquigarrow_{k,\mathcal{G}})^*$ $\langle l_f, D_f\rangle$ implies $\langle l_0, u_0\rangle \rightarrow^* \langle l_f, u_f\rangle$ for all $u_f \in D_f$ such that $u_f(x) \leq k(x)$ for all x.*
2. *(Completeness) $\langle l_0, u_0\rangle \rightarrow^* \langle l_f, u_f\rangle$ with $u_f(x) \leq k(x)$ for all x implies $\langle l_0, \{u_0\}\rangle$ $(\rightsquigarrow_{k,\mathcal{G}})^*$ $\langle l_f, D_f\rangle$ for some D_f such that $u_f \in D_f$*
3. *(Finiteness) The transition relation $\rightsquigarrow_{k,\mathcal{G}}$ is finite.*

3.4 Symbolic Reachability Analysis

Model-checking concerns two types of properties *liveness* and *safety*. The essential algorithm of checking liveness properties is loop detection, which is computationally expensive. The main effort on verification of timed systems has been put on safety

Algorithm 1 Reachability analysis.

```
PASSED = ∅, WAIT = {⟨l_0, D_0⟩}
while WAIT ≠ ∅ do
  take ⟨l, D⟩ from WAIT
  if l = l_f ∧ D ∩ φ_f ≠ ∅ then return "YES"
  if D ⊈ D' for all ⟨l, D'⟩ ∈ PASSED then
    add ⟨l, D⟩ to PASSED
    for all ⟨l', D'⟩ such that ⟨l, D⟩ ⇝ k,G⟨l', D'⟩ do
      add ⟨l', D'⟩ to WAIT
    end for
  end if
end while
return "NO"
```

properties that can be checked using reachability analysis by traversing the state-space of timed automata.

Reachability analysis can be used to check properties on states. It consists of two basic steps, computing the state-space of an automaton under consideration, and searching for states that satisfy or contradict given properties. The first step can either be performed prior to the search, or done *on-the-fly* during the search process. Computing the state-space on-the-fly has an obvious advantage over pre-computing, in that only the part of the state-space needed to prove the property is generated. It should be noted though, that even on-the-fly methods will generate the entire state-space to prove certain properties, *e.g.* invariant properties.

Several model-checkers for timed systems are designed and optimized for reachability analysis based on the symbolic semantics and the zone-representation (see Section 4). As an example, we present the core of the verification engine of UPPAAL (see Algorithm 1).

Assume a timed automaton $\mathcal{A}$ with a set of initial states and a set of final states (*e.g.* the bad states) characterized as $\langle l_0, D_0\rangle$ and $\langle l_f, \phi_f\rangle$ respectively. Assume that k is the clock ceiling defined by the maximal constants appearing in $\mathcal{A}$ and ϕ_f, and $\mathcal{G}$ denotes the set of difference constraints appearing in $\mathcal{A}$ and ϕ_f. Algorithm 1 can be used to check if the initial states may evolve to any state whose location is l_f and whose clock assignment satisfies ϕ_f. It computes the normalized zone-graph of the automaton on-the-fly, in search for symbolic states containing location l_f and zones intersecting with ϕ_f.

The algorithm computes the transitive closure of $\leadsto_{k,\mathcal{G}}$ step by step, and at each step, checks if the reached zones intersect with ϕ_f. From Theorem 2, it follows that the algorithm will return with a correct answer. It is also guaranteed to terminate because $\leadsto_{k,\mathcal{G}}$ is finite. As mentioned earlier, for a given timed automaton with a fixed set of clocks whose maximal constants are bounded by a clock ceiling k, and a fixed set of diagonal constraints contained in the guards, the number of all possible normalized zones is bounded because a normalized zone can not contain arbitrarily large or arbitrarily small constants. In fact the smallest possible zones are the refined regions. Thus the whole state-space of a timed automaton can only be partitioned into finitely many symbolic states and the worst case is the size of the region graph of the automaton, induced by the refined region equivalence. Therefore, the algorithm is working on a finite structure and it will terminate.

Algorithm 1 also highlights some of the issues in developing a model-checker for timed automata. Firstly, the representation and manipulation of states, primarily zones, is crucial to the performance of a model-checker. Note that in addition to the operations to compute the successors of a zone according to $\rightsquigarrow_{k,\mathcal{G}}$, the algorithm uses two more operations to check the emptiness of a zone as well as the inclusion between two zones. Thus, designing efficient algorithms and data-structures for zones is a major issue in developing a verification tool for timed automata, which is addressed in Section 4. Secondly, PASSED holds all encountered states and its size puts a limit on the size of systems we can verify. This raises the research challenges *e.g.* state compression [14], state-space reduction [15] and approximate techniques [9].

4 DBM: Algorithms and Data Structures

The preceding section presents the key elements needed in symbolic reachability analysis. Recall that the operations on zones are all defined in terms of sets of clock assignments. It is not clear how to compute the result of such an operation. In this section, we describe how to represent zones, compute the operations and check properties on zones. Pseudo code for the operations is given in the appendix.

4.1 DBM Basics

Recall that a clock constraint over C is a conjunction of atomic constraints of the form $x \sim m$ and $x - y \sim n$ where $x, y \in C$, $\sim \in \{\leq, <, =, >, \geq\}$ and $m, n \in \mathbb{N}$. A zone denoted by a clock constraint D is the maximal set of clock assignments satisfying D. The most important property of zones is that they can can be represented as matrices *i.e.* DBMs (Difference Bound Matrices) [12, 23], which have a canonical representation. In the following, we describe the basic structure and properties of DBMs.

To have a unified form for clock constraints we introduce a reference clock $\mathbf{0}$ with the constant value 0. Let $C_0 = C \cup \{\mathbf{0}\}$. Then any zone $D \in \mathcal{B}(C)$ can be rewritten as a conjunction of constraints of the form $x - y \preceq n$ for $x, y \in C_0$, $\preceq \in \{<, \leq\}$ and $n \in \mathbb{Z}$.

Naturally, if the rewritten zone has two constraints on the same pair of variables only the intersection of the two is significant. Thus, a zone can be represented using at most $|C_0|^2$ atomic constraints of the form $x - y \preceq n$ such that each pair of clocks $x - y$ is mentioned only once. We can then store zones using $|C_0| \times |C_0|$ matrices where each element in the matrix corresponds to an atomic constraint. Since each element in such a matrix represents a bound on the difference between two clocks, they are named *Difference Bound Matrices* (DBMs). In the following presentation, we use D_{ij} to denote element (i, j) in the DBM representing the zone D.

To construct the DBM representation for a zone D, we start by numbering all clocks in C_0 as $0, \ldots, n$ and the index for $\mathbf{0}$ is 0. Let each clock be denoted by one row in the matrix. The row is used for storing lower bounds on the difference between the clock and all other clocks, and thus the corresponding column is used for upper bounds. The elements in the matrix are then computed in three steps.

- For each constraint $x_i - x_j \preceq n$ of D, let $D_{ij} = (n, \preceq)$.
- For each clock difference $x_i - x_j$ that is unbounded in D, let $D_{ij} = \infty$. Where ∞ is a special value denoting that no bound is present.
- Finally add the implicit constraints that all clocks are positive, *i.e.* $\mathbf{0} - x_i \leq 0$, and that the difference between a clock and itself is always 0, *i.e.* $x_i - x_i \leq 0$.

As an example, consider the zone $D = x - \mathbf{0} < 20 \wedge y - \mathbf{0} \leq 20 \wedge y - x \leq 10 \wedge x - y \leq -10 \wedge \mathbf{0} - z < 5$. To construct the matrix representation of D, we number the clocks in the order $\mathbf{0}, x, y, z$. The resulting matrix representation is shown below:

$$M(D) = \begin{pmatrix} (0,\leq) & (0,\leq) & (0,\leq) & (5,<) \\ (20,<) & (0,\leq) & (-10,\leq) & \infty \\ (20,\leq) & (10,\leq) & (0,\leq) & \infty \\ \infty & \infty & \infty & (0,\leq) \end{pmatrix}$$

To manipulate DBMs efficiently we need two operations on bounds: comparison and addition. We define that $(n, \preceq) < \infty$, $(n_1, \preceq_1) < (n_2, \preceq_2)$ if $n_1 < n_2$ and $(n, <) < (n, \leq)$. Further we define addition as $b_1 + \infty = \infty$, $(m, \leq) + (n, \leq) = (m + n, \leq)$ and $(m, <) + (n, \preceq) = (m + n, <)$.

Canonical DBMs. Usually there are an infinite number of zones sharing the same solution set. However, for each family of zones with the same solution set there is a unique constraint where no atomic constraint can be strengthened without losing solutions.

To compute the canonical form of a given zone we need to derive the tightest constraint on each clock difference. To solve this problem, we use a graph-interpretation of zones. A zone may be transformed to a weighted graph where the clocks in $\mathcal{C}_0$ are nodes and the atomic constraints are edges labeled with bounds. A constraint in the form of $x - y \preceq n$ will be converted to an edge from node y to node x labeled with $(n, \preceq)$, namely the distance from y to x is bounded by n.

As an example, consider the zone $x - \mathbf{0} < 20 \wedge y - \mathbf{0} \leq 20 \wedge y - x \leq 10 \wedge x - y \leq -10$. By combining the atomic constraints $y - \mathbf{0} \leq 20$ and $x - y \leq -10$ we derive that $x - \mathbf{0} \leq 10$, *i.e.* the bound on $x - \mathbf{0}$ can be strengthened. Consider the graph interpretation of this zone, presented in Fig. 8(a). The tighter bound on $x - \mathbf{0}$ can be derived from the graph, using the path $\mathbf{0} \rightarrow y \rightarrow x$, giving the graph in Fig. 8(b). Thus, deriving the tightest constraint on a pair of clocks in a zone is equivalent to finding the shortest path between their nodes in the graph interpretation of the zone. The conclusion is that a canonical, *i.e.* closed, version of a zone can be computed using a shortest path algorithm. Floyd-Warshall algorithm [25] (Algorithm 2) is often used to transform zones to canonical form. However, since this algorithm is quite expensive (cubic in the number of clocks), it is desirable to make all frequently used operations preserve the canonical form, *i.e.* the result of performing an operation on a canonical zone should also be canonical.

Minimal Constraint Systems. A zone may contain redundant constraints. For example, a zone contains constraints $x - y < 2, y - z < 5$ and $x - z < 7$. The constraint $x - z < 7$ is obviously redundant because it may be derived from the first two. It is

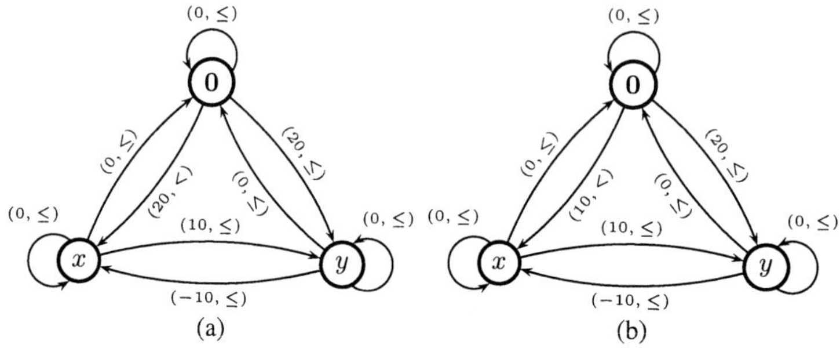

Fig. 8. Graph interpretation of the example zone and its closed form.

desirable to remove such constraints to store only the minimal number of constraints. Consider, for instance, the zone $x - y \leq 0 \land y - z \leq 0 \land z - x \leq 0 \land 2 \leq x - \mathbf{0} \leq 3$. This is a zone in a minimal form containing only five constraints. The closed form of this zone contains more than 12 constraints. It is known, *e.g.* from [31], that for each zone there is a minimal constraint system with the same solution set. By computing this minimal form for all zones and storing them in memory using a sparse representation we might reduce the memory consumption for state-space exploration. This problem has been thoroughly investigated in [31,41,36].

The following is a summary of the published work on the minimal representation of zones. We present an algorithm that computes the minimal form of a closed DBM. Closing a DBM corresponds to computing the shortest path between all clocks. Our goal is to compute the minimal set of bounds for a given shortest path closure. For clarity, the algorithm is presented in terms of directed weighted graphs. However, the results are directly applicable to the graph interpretation of DBMs.

First we introduce some notation: we say that a cycle in a graph is a *zero cycle* if the sum of weights along the cycle is 0, and an edge $x_i \xrightarrow{n_{ij}} x_j$ is *redundant* if there is another path between x_i and x_j where the sum of weights is no larger than n_{ij}.

In graphs without zero cycles we can remove all redundant edges without affecting the shortest path closure [31]. Further, if the input graph is in shortest path form (as for closed DBMs) all redundant edges can be located by considering alternative paths of length two.

As an example, consider Fig. 9. The figure shows the shortest path closure for a zero-cycle free graph (a) and its minimal form (b). In the graph we find that $x_1 \xrightarrow{9} x_2$ is made redundant by the path $x_1 \xrightarrow{7} x_4 \xrightarrow{2} x_2$ and can thus be removed. Further, the edge $x_3 \xrightarrow{15} x_2$ is redundant due to $x_3 \xrightarrow{5} x_1 \xrightarrow{9} x_2$. Note that we consider edges marked as redundant when searching for new redundant edges. The reason is that we let the redundant edges represent the path making them redundant, thus allowing all redundant edges to be located using only alternative paths of length two. This procedure is repeated until no more redundant edges can be found.

This gives the $O(n^3)$ procedure for removing redundant edges presented in Algorithm 3. The algorithm can be directly applied to zero-cycle free DBMs to compute the minimal number of constraints needed to represent a given zone.

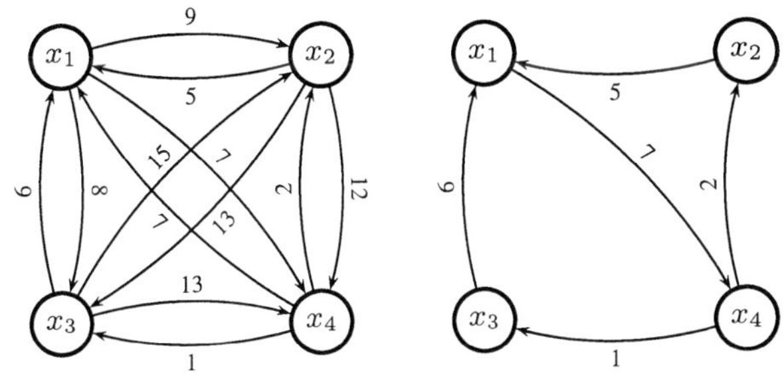

(a) (b)

Fig. 9. A zero cycle free graph and its minimal form.

However, this algorithm will not work if there are zero-cycles in the graph. The reason is that the set of redundant edges in a graph with zero-cycles is not unique. As an example, consider the graph in Fig. 10(a). Applying the above reasoning on this graph would remove $x_1 \xrightarrow{3} x_3$ based on the path $x_1 \xrightarrow{-2} x_2 \xrightarrow{5} x_3$. It would also remove the edge $x_2 \xrightarrow{5} x_3$ based on the path $x_2 \xrightarrow{2} x_1 \xrightarrow{3} x_3$. But if both these edges are removed it is no longer possible to construct paths leading into x_3. In this example there is a dependence between the edges $x_1 \xrightarrow{3} x_3$ and $x_2 \xrightarrow{5} x_3$ such that only one of them can be considered redundant.

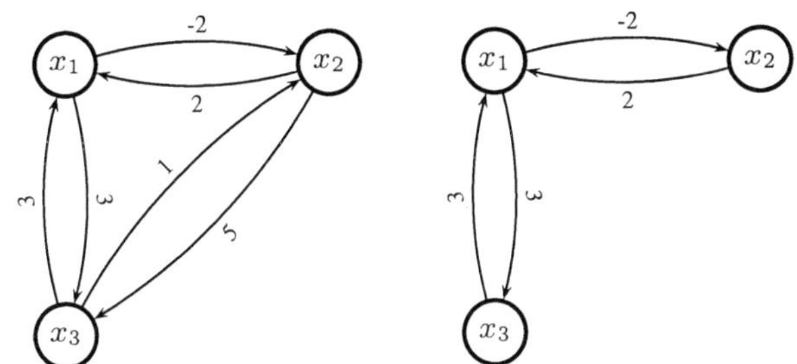

(a) (b)

Fig. 10. A graph with a zero-cycle and its minimal form.

The solution to this problem is to partition the nodes according to zero-cycles and build a super-graph where each node is a partition. The graph from Fig. 10(a) has two partitions, one containing x_1 and x_2 and the other containing x_3. To compute the edges in the super-graph we pick one representative for each partition and let the edges between the partitions inherit the weights from edges between the representatives. In our example, we choose x_1 and x_3 as representatives for their equivalence classes. The edges in the graph are then $\{x_1, x_2\} \xrightarrow{3} \{x_3\}$ and $\{x_3\} \xrightarrow{3} \{x_1, x_2\}$. The super-graph is clearly zero-cycle free and can be reduced using Algorithm 3. This small graph can not be reduced further. The relation between the nodes within a partition is uniquely defined by the zero-cycle and all other edges may be removed. In our example all edges within each equivalence class are part of the zero-cycle and thus none of them can be removed. Finally the reduced super-graph is connected to the reduced partitions. In our example we end up with the graph in Fig. 10(b). Pseudo-code for the reduction-procedure is shown in Algorithm 4.

Now we have an algorithm for computing the minimal number of edges to represent a given shortest path closure that can be used to compute the minimal number of constraints needed to represent a given zone.

4.2 Basic Operations on DBMs

This subsection presents all the basic operations on DBMs except the ones for zone-normalization, needed in symbolic state space exploration of timed automata, both for forwards and backwards analysis. The two operations for zone-normalization are presented in the next subsection.

First note that even if a verification tool only explores the state space in one direction all operations are still useful for other purposes, *e.g.* for generating diagnostic traces. The operations are illustrated graphically in Fig. 11.

To simplify the presentation we assume that the input zones are consistent and in canonical form. The basic operations on DBMs can be divided into two classes:

1. **Property-Checking:** This class includes operations to check the consistency of a DBM, the inclusion between zones, and whether a zone satisfies a given atomic constraint.
2. **Transformation:** This class includes operations to compute the strongest postcondition and weakest precondition of a zone according to conjunction with guards, time delay and clock reset.

Property-Checking

`consistent`(D) The most basic operation on a DBM is to check if it is consistent, *i.e.* if the solution set is non-empty. In state-space exploration this operation is used to remove inconsistent states from the exploration.

For a zone to be inconsistent there must be at least one pair of clocks where the upper bound on their difference is smaller than the lower bound. For DBMs this can be checked by searching for negative cycles in the graph interpretation. However, the most efficient way to implement a consistency check is to detect when an upper bound is set to lower value than the corresponding lower bound and mark the zone as inconsistent by setting D_{00} to a negative value. For a zone in canonical form this test can be performed locally. To check if a zone is inconsistent it will then be enough to check whether D_{00} is negative.

`relation`(D, D') Another key operation in state space exploration is inclusion checking for the solution sets of two zones. For DBMs in canonical form, the condition that $D_{ij} \leq D'_{ij}$ for all clocks $i, j \in C_0$ is necessary and sufficient to conclude that $D \subseteq D'$. Naturally the opposite condition applies to checking if $D' \subseteq D$. This allows for the combined inclusion check described in Algorithm 5.

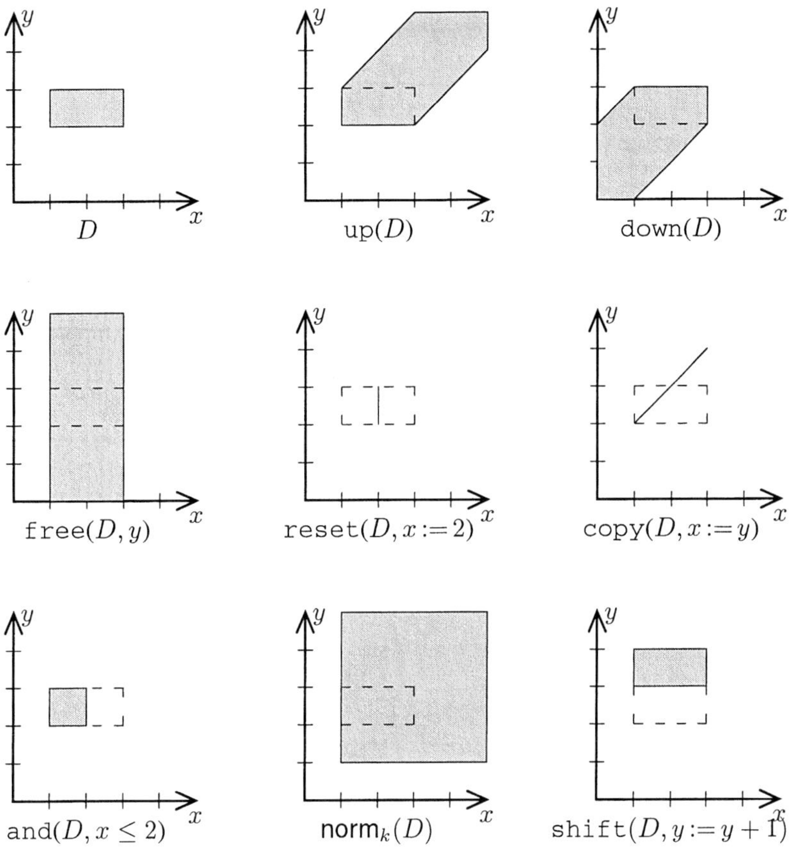

Fig. 11. DBM operations applied to the same zone where for $\mathsf{norm}_k(D)$, k is defined by $k(x) = 2$ and $k(y) = 1$.

satisfied($D, x_i - x_j \preceq m$) Sometimes it is desirable to non-destructively check if a zone satisfies a constraint, *i.e.* to check if the zone $D \wedge x_i - x_j \preceq m$ is consistent without altering D. From the definition of the consistent-operation we know that a zone is consistent if it has no negative cycles. Thus, checking if $D \wedge x_i - x_j \preceq m$ is non-empty can be done by checking if adding the guard to the zone would introduce a negative cycle. For a DBM on canonical form this test can be performed locally by checking if $(m, \preceq) + D_{ji}$ is negative.

Transformations

up(D) The up operation computes the strongest postcondition of a zone with respect to delay, *i.e.* up(D) contains the clock assignments that can be reached from D by delay. Formally, this operation is defined as $\mathrm{up}(D) = \{u + d \mid u \in D, d \in \mathbb{R}_+\}$.

Algorithmically, up is computed by removing the upper bounds on all individual clocks (In a DBM all elements D_{i0} are set to ∞). This is the same as saying that any clock assignment in a given zone may delay an arbitrary amount of time. The property that all clocks proceed at the same speed is ensured by the fact that constraints on the differences between clocks are not altered by the operation.

This operation preserves the canonical form, *i.e.* applying up to a canonical DBM will result in a new canonical DBM. The up operation is also presented in Algorithm 6.

down(D) This operation computes the weakest precondition of D with respect to delay. Formally $\mathtt{down}(D) = \{u \mid u+d \in D, d \in \mathbb{R}_+\}$, *i.e.* the set of clock assignments that can reach D by some delay d. Algorithmically, down is computed by setting the lower bound on all individual clocks to $(0, \leq)$. However due to constraints on clock differences this algorithm may produce non-canonical DBMs. As an example, consider the zone in Fig. 12(a). When down is applied to this zone (Fig. 12(b)), the lower bound on x is 1 and not 0, due to constraints on clock differences. Thus, to obtain an algorithm that produce canonical DBMs the difference constraints have to be taken into account when computing the new lower bounds.

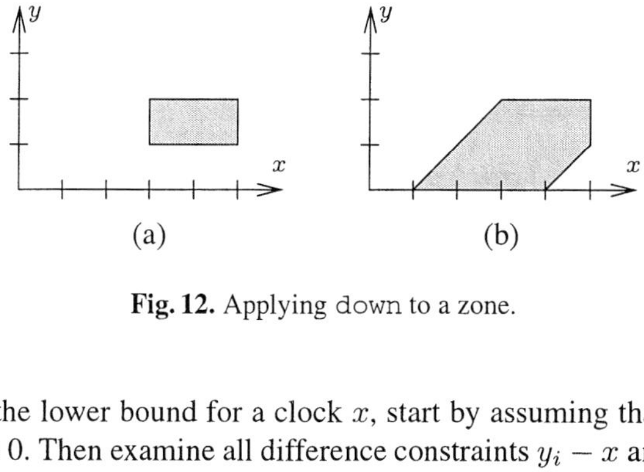

(a) (b)

Fig. 12. Applying down to a zone.

To compute the lower bound for a clock x, start by assuming that all other clocks y_i have the value 0. Then examine all difference constraints $y_i - x$ and compute a new lower bound for x under this assumption. The new bound on $\mathbf{0} - x$ will be the minimum bound on $y_i - x$ found in the DBM. Pseudo-code for down is presented in Algorithm 7.

and($D, x_i - y_j \preceq b$) The most used operation in state-space exploration is conjunction, *i.e.* adding a constraint to a zone. The basic step of the and operation is to check if $(b, \preceq) < D_{ij}$ and in this case set the bound D_{ij} to $(b, \preceq)$. If the bound has been altered, *i.e.* if adding the guard affected the solution set, the DBM has to be put back on canonical form. One way to do this would be to use the generic shortest path algorithm, however for this particular case it is possible to derive a specialization of the algorithm allowing re-canonicalization in $O(n^2)$ instead of $O(n^3)$.

The specialized algorithm takes the advantage that D_{ij} is the only bound that has been changed. Since the Floyd-Warshall algorithm is insensitive to how the nodes in the graph are ordered, we may decide to treat x_i and x_j last. The outer loop of Algorithm 2 will then only affect the DBM twice, for $k = i$ and $k = j$. This allows the canonicalisation algorithm to be reduced to checking, for all pairs of clocks in the DBM, if the path via either x_i or x_j is shorter than the direct connection. The pseudo code for this is presented in Algorithm 8.

free(D, x) The free operation removes all constraints on a given clock, *i.e.* the clock may take any positive value. Formally this is expressed as $\texttt{free}(D, x) = \{u[x = d] \mid u \in D, d \in \mathbb{R}_+\}$. In state-space exploration this operation is used in combination with conjunction, to implement reset operations on clocks. It can be used in both forwards and backwards exploration, but since forwards exploration allows other more efficient implementations of reset, free is only used when exploring the state-space backwards.

A simple algorithm for this operation is to remove all bounds on x in the DBM and set $D_{0x} = (0, \leq)$. However, the result may not be on canonical form. To obtain an algorithm preserving the canonical form, we change how new difference constraints regarding x are derived. We note that the constraint $\mathbf{0} - x \leq 0$ can be combined with constraints of the form $y - \mathbf{0} \preceq m$ to compute new bounds for $y - x$. For instance, if $y - \mathbf{0} \leq 5$ we can derive that $y - x \leq 5$. To obtain a DBM on canonical form we derive bounds for D_{yx} based on D_{y0} instead of setting $D_{yx} = \infty$.In Algorithm 9 this is presented as pseudo code.

reset($D, x := m$) In forwards exploration this operation is used to set clocks to specific values, *i.e.* $\texttt{reset}(D, x := m) = \{u[x = m] \mid u \in D\}$. Without the requirement that output should be on canonical form, reset can be implemented by setting $D_{x0} = (m, \leq)$, $D_{0x} = (-m, \leq)$ and remove all other bounds on x. However, if we instead of removing the difference constraints compute new values using constraints on the other clocks, like in the implementation of free, we obtain an implementation that preserve the canonical form. Such an implementation is presented in Algorithm 10.

copy($D, x := y$) This is another operation used in forwards state-space exploration. It copies the value of one clock to another. Formally, we define $\texttt{copy}(D, x := y)$ as $\{u[x = u(y)] \mid u \in D\}$. As reset, copy can be implemented by assigning $D_{xy} = (0, \leq)$, $D_{yx} = (0, \leq)$, removing all other bounds on x and re-canonicalize the zone. However, a more efficient implementation is obtained by assigning $D_{xy} = (0, \leq)$, $D_{yx} = (0, \leq)$ and then copy the rest of the bounds on x from y. For pseudo code, see Algorithm 11.

shift($D, x := x + m$) The last reset operation is shifting a clock, *i.e.* adding or subtracting a clock with an integer value, *i.e.* $\texttt{shift}(D, x := x + m) = \{u[x = u(x) + m] \mid u \in D\}$. The definition gives a hint on how to implement the operation. We can view the shift operation as a substitution of $x - m$ for x in the zone. With this reasoning m is added to the upper and lower bounds of x. However, since lower bounds on x are represented by constraints on $y - x$, m is subtracted from all those bounds. This operation is presented in pseudo-code in Algorithm 12.

4.3 Zone-Normalization

The operations for zone-normalization are to transform zones which may contain arbitrarily large constants to zones containing only bounded constants in order to obtain a finite zone-graph.

norm$_k(D)$ For a timed automaton and a safety property to be checked, that contain no difference constraints, the k-normalization $\mathsf{norm}_k(D)$ is needed, and it can be computed based on the canonical form of D (see Section 3). It is to remove all upper bounds higher than the maximal constants and lower all lower bounds higher than the maximal constants down to the maximal constants. The result of $\mathsf{norm}_k(D)$ is illustrated graphically in Fig. 11.

In the canonical DBM representation of a zone, the operation consists of two steps: first, remove all bounds $x - y \preceq m$ such that $(m, \preceq) > (k(x), \leq)$ and second, set all bounds $x - y \preceq m$ such that $(m, \preceq) < (-k(y), <)$ to $(-k(y), <)$. Pseudo-code for k-normalization is given in Algorithm 13 where k_i denotes $k(x_i)$.

The k-normalization will not preserve the canonical form of a DBM, and the best way to put the result back on canonical form is to use Algorithm 2.

norm$_{k,\mathcal{G}}(D)$ For automata containing difference constraints in the guards, it is more complicated and expensive to compute the normalized zones. Assume an automataon $\mathcal{A}$ containing the set of difference constraints $\mathcal{G}$ and the maximal clock constants bounded by a clock ceiling k. Assume a zone D of $\mathcal{A}$ to be normalized. According to the semantical characterization for $\mathsf{norm}_{k,\mathcal{G}}(D)$ in Definition 10 we know that if a difference constraint is not satisfied by any assignment in the zone D, it should not be satisfied by any assignment in the normalized one, $\mathsf{norm}_{k,\mathcal{G}}(D)$, and if all assignments in D satisfy a difference constraint then so should all assignments in $\mathsf{norm}_{k,\mathcal{G}}(D)$. This leads to a core normalization algorithm consisting of three steps.

1. Collect all difference constraints g used as guards in $\mathcal{A}$ such that
   (a) $g \wedge D$ is empty. This is the case when g is outside of D.
   (b) $\neg g \wedge D$ is empty. That is the case when g contains D completely.
   Let $\mathcal{G}_{\mathrm{unsat}} = \{g | g \wedge D = \emptyset\} \cup \{\neg g | \neg g \wedge D = \emptyset\}$
2. Compute $\mathsf{norm}_k(D)$, that is, to run the k-normalization without considering the difference constraints.
3. Subtract (or cut) the k-normalized zone of D by all difference constraints in $\mathcal{G}_{\mathrm{unsat}}$, that is to compute $\mathsf{norm}_k(D) \wedge \neg\mathcal{G}_{\mathrm{unsat}}$.

The last step is to make sure that none of the collected difference constraints are satisfied after the k-normalization. In Algorithm 14, the core normalization is given as pseudo code. The set $\mathcal{G}_{\mathrm{d}}$ used in the algorithm is the set of difference constraints appearing in the automaton under consideration with the maximal clock constants bounded by a given clock ceiling k as input.

It appears to be the case that $\mathsf{norm}_k(D) \wedge \neg\mathcal{G}_{\mathrm{unsat}}$ is the normalized zone we are looking for. Unfortunately this is not. The core normalization does not handle the case when a difference constraint splits the zone D to be normalized. That is, there is a guard g such that $g \wedge D \neq \emptyset$ and $\neg g \wedge D \neq \emptyset$. In this case, we need to split D by g using Algorithm 15, and then apply the core normalization algorithm to the parts of D separately, which are the sub-zones of D resulted from the splitting. The union of the normalized sub-zones is what we are looking for, that is $\mathsf{norm}_{k,\mathcal{G}}(D)$.

The complete normalization procedure is presented in Algorithm 16. The splitting, denoted by split in the description, is used as a preprocessing step and then the core normalization algorithm (*i.e.* Algorithm 14) is applied to all the resulted sub-zones resulted from the splitting.

Finally, the symbolic transition relation $\leadsto_{k,\mathcal{G}}$ can be computed as follows: If $\langle l, D\rangle \leadsto \langle l', D'\rangle$, $\langle l, D\rangle \leadsto_{k,\mathcal{G}} \langle l', D''\rangle$ for all $D'' \in Q$ used in Algorithm 16, *i.e.* the algorithm for $\mathsf{norm}_{k,\mathcal{G}}(D')$.

To demonstrate the normalization procedure we apply it to the zone for location S_2 in our counter example. The difference constraints in the example are $g_1 \equiv x - z < 1$ and $g_2 \equiv z - y < 1$. The zone contains both clock assignments satisfying g_1 and assignments satisfying its negation, and thus we have to split the zone with respect to this constraint prior to normalization, giving the zones below.

$$\begin{cases} y - x < -2 \\ y - z < -1 \\ z - x \leq 0 \\ \mathbf{0} - x < -2 \\ \mathbf{0} - z < -1 \\ x - z < 1 \end{cases} \quad \begin{cases} y - x < -2 \\ y - z \leq 0 \\ \mathbf{0} - x < -2 \\ z - x \leq -1 \end{cases}$$

(a) satisfying g_1 (b) satisfying $\neg g_1$

Zone (a) above does not contain any clock assignments satisfying g_2 and thus it will not be split further. Zone (b) however needs to be split into two parts satisfying g_2 and $\neg g_2$. This gives us the following zones to normalize by the core normalization procedure.

$$\begin{cases} y - x < -2 \\ y - z < -1 \\ z - x \leq 0 \\ \mathbf{0} - x < -2 \\ \mathbf{0} - z < -1 \\ x - z < 1 \end{cases} \quad \begin{cases} y - x < -2 \\ y - z \leq 0 \\ \mathbf{0} - x < -2 \\ z - x \leq -1 \end{cases} \quad \begin{cases} y - x < -2 \\ y - z \leq -1 \\ z - x \leq -1 \\ \mathbf{0} - x < -2 \\ \mathbf{0} - z \leq -1 \end{cases}$$

(a) g_1 and $\neg g_2$ (b) $\neg g_1$ and g_2 (c) $\neg g_1$ and $\neg g_2$

The sets of difference constraints not satisfied by the zones (a), (b) and (c) shown above are: $\mathcal{G}_{\text{unsat}}^{(a)} = \{\neg g_1, g_2\}$, $\mathcal{G}_{\text{unsat}}^{(b)} = \{g_1, \neg g_2\}$, $\mathcal{G}_{\text{unsat}}^{(c)} = \{g_1, g_2\}$ respectively. We apply k-normalization to each of them, giving:

$$\begin{cases} y - x < -1 \\ y - z < -1 \\ z - x \leq 0 \\ \mathbf{0} - x < -1 \\ \mathbf{0} - z < -1 \\ x - z < 1 \end{cases} \quad \begin{cases} y - x < -1 \\ y - z \leq 0 \\ \mathbf{0} - x < -1 \\ x - z \geq 1 \end{cases} \quad \begin{cases} y - x < -1 \\ y - z \leq -1 \\ z - x \leq -1 \\ \mathbf{0} - x < -1 \\ \mathbf{0} - z \leq -1 \end{cases}$$

(A) g_1 and $\neg g_2$ (B) $\neg g_1$ and g_2 (C) $\neg g_1$ and $\neg g_2$

Since the k-normalized zones (A), (B) and (C) shown above do not enable any constraint in $\mathcal{G}_{\text{unsat}}$, we need not to subtract the corresponding difference constraints from the zones. Finally, we note that, as the un-normalized zones (a), (b) and (c), none of the normalized zones (A), (B) and (C) intersects with $g_1 \wedge g_2$; the transition from S_2 to S_3 is not enabled by the normalization procedure.

4.4 Zones in Memory

This section describes a number of techniques for storing zones in computer memory. The section starts by describing how to map DBM elements on machine words. It continues by discussing how to place two-dimensional DBMs in linear memory and ends by describing how to store zones using a sparse representation.

Storing DBM Elements. To store a DBM element in memory we need to keep track of the integer limit and whether it is strict or not. The range of the integer limit is typically much lower than the maximum value of a machine word and the strictness can be stored using just one bit. Thus, it is possible to store both the limit and the strictness in different parts of the same machine word. Since comparing and adding DBM elements are frequently used operations it is crucial for the performance of a DBM package that they can be efficiently implemented for the chosen encoding. Fortunately, it is possible to construct an encoding of bounds in machine words, where checking if b_1 is less than b_2 can be performed by checking if the encoded b_1 is smaller than the encoded b_2.

The encoding we propose is to use the least significant bit (LSB) of the machine word to store whether the bound is strict or not. Since strict bounds are smaller than non-strict we let a set (1) bit denote that the bound is non-strict while an unset (0) bit denote that the bound is strict. The rest of the bits in the machine word are used to store the integer bound. To encode ∞ we use the largest positive number that fit in a machine word (denoted `MAX_INT`).

For good performance we also need an efficient implementation of addition of bounds. For the proposed encoding Algorithm 17 adds two encoded bounds b_1 and b_2. The symbols & and | in the algorithm are used to denote bitwise-and and bitwise-or, respectively.

Placing DBMs in Memory. Another issue is how to store two-dimensional DBMs in linear memory. In this section we present two different techniques and give a brief comparison between them. The natural way to put matrices in linear memory is to store the elements by row (or by column), *i.e.* each row of the matrix is stored consequently in memory. This layout has one big advantage, its good performance. This advantage is mainly due to the simple function for computing the location of a given element in the matrix: $loc(x, y) = x*(n+1)+y$. This function can (on most computers) be computed in only two instructions. This is important since all accesses to DBM elements use this function. How the different DBM elements are place in memory with this layout if presented in Fig. 13(a).

The second way to store a DBM in linear memory is based on a layered model where each layer consists of the bounds between a clock and the clocks with lower index in the DBM. In this representation it is cheap to implement local clocks, since all information about the local clocks are localised at the end of the DBM. The drawback with this layout is the more complicated function from DBM indices to memory locations. For this layout we have:

$$loc(x, y) = \begin{cases} y * (y + 1) + x & \text{if } x \leq y \\ x * x + y & \text{otherwise} \end{cases}$$

This adds at least two instructions (one comparison and one conditional jump) to the transformation. This may not seem such a huge overhead, but it is clearly noticeable. The cache performance is also worse when using this layout than when storing the DBMs row-wise. This layout is illustrated in Fig. 13(b).

The conclusion is that unless the tool under construction supports adding and removing clocks dynamically the row-wise mapping should be used. On the other hand, if the tool supports local clocks the layered mapping may be preferable since no reordering of the DBM is needed when entering or leaving a clock scope.

$$\begin{pmatrix} 0 & 1 & 2 & 3 \\ 4 & 5 & 6 & 7 \\ 8 & 9 & 10 & 11 \\ 12 & 13 & 14 & 15 \end{pmatrix} \quad \begin{pmatrix} 0 & 2 & 6 & 12 \\ 1 & 3 & 7 & 13 \\ 4 & 5 & 8 & 14 \\ 9 & 10 & 11 & 15 \end{pmatrix}$$

(a) Row wise (b) Layered

Fig. 13. Different layouts of DBMs in memory.

Storing Sparse Zones. In most verification tools, the majority of the zones are kept in the set of states already visited during verification. They are used as a reference to ensure termination by preventing states from being explored more than once. For the states in this set we may benefit from storing only the minimal number of constraints using a sparse representation.

A straight forward implementation is to store a sparse zone as a vector of constraints of the form $\langle x, y, b \rangle$. We may save additional memory by omitting implicit constraints, such as $\mathbf{0} - x \leq 0$. A downside with using sparse zones is that each constraint require twice the amount of memory needed for a constraint in a full DBM, since the sparse representation must store clock indices explicitly. Thus, unless half of the constraints in a DBM are redundant we do not gain from using sparse zones.

A nice feature of the sparse representation is that checking whether a zone D_f represented as a full DBM is included in a sparse zone D_s may be implemented without computing the full DBM for D_s. It suffices to check for all constraints in D_s that the corresponding bound in D_f is tighter. However, to check if $D_s \subseteq D_f$ we have to compute the full DBM for D_s.

5 UPPAAL

In the last decade, there have been a number of tools developed based on timed automata to model and verify real time systems, notably Kronos [50] and UPPAAL [33]. As an example, we give a brief introduction to the UPPAAL tool (www.uppaal.com).

UPPAAL is a tool box for modeling, simulation and verification of timed automata, based on the algorithms and data-structures presented in previous sections. The tool was released for the first time in 1995, and since then it has been developed and maintained in collaboration between Uppsala University and Aalborg University.

5.1 Modeling with UPPAAL

The core of the UPPAAL modeling language is networks of timed automata. In addition, the language has been further extended with features to ease the modeling task and to guide the verifier in state space exploration. The most important of these are shared integer variables, urgent channels and committed locations.

Networks of Timed Automata. A *network of timed automata* is the parallel composition $A_1|\cdots|A_n$ of a set of timed automata $A_1, \ldots, A_n$, called processes, combined into a single system by the CCS parallel composition operator with all external actions hidden. Synchronous communication between the processes is by hand-shake synchronization using input and output actions; asynchronous communication is by shared variables as described later. To model hand-shake synchronization, the action alphabet Σ is assumed to consist of symbols for input actions denoted $a?$, output actions denoted $a!$, and internal actions represented by the distinct symbol τ.

An example system composed of two timed automata is shown in Fig. 14. The network models a time-dependent light-switch (to the left) and its user (to the right). The user and the switch communicate using the label *press*. The user can press the switch (press!) and the switch waits to be pressed (press?). The product automaton, *i.e.* the automaton describing the combined system is shown in Fig. 15.

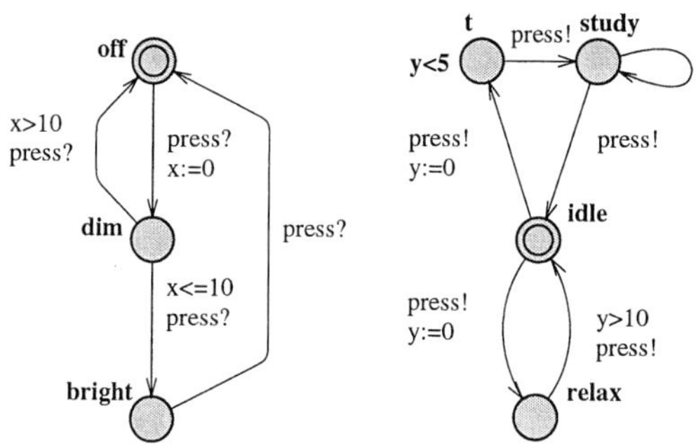

Fig. 14. Network of Timed Automata.

The semantics of networks is given as for single timed automata in terms of transition systems. A state of a network is a pair $\langle l, u \rangle$ where l denotes a vector of current locations of the network, one for each process, and u is as usual a clock assignment remembering the current values of the clocks in the system. A network may perform two types of transitions, delay transitions and discrete transitions. The rule for delay transitions is similar to the case of single timed automata where the invariant of a location vector is the conjunction of the location invariants of the processes. There are two rules for discrete transitions defining local actions where one of the processes makes a move on its own, and synchronizing actions where two processes synchronize on a channel and move simultaneously.

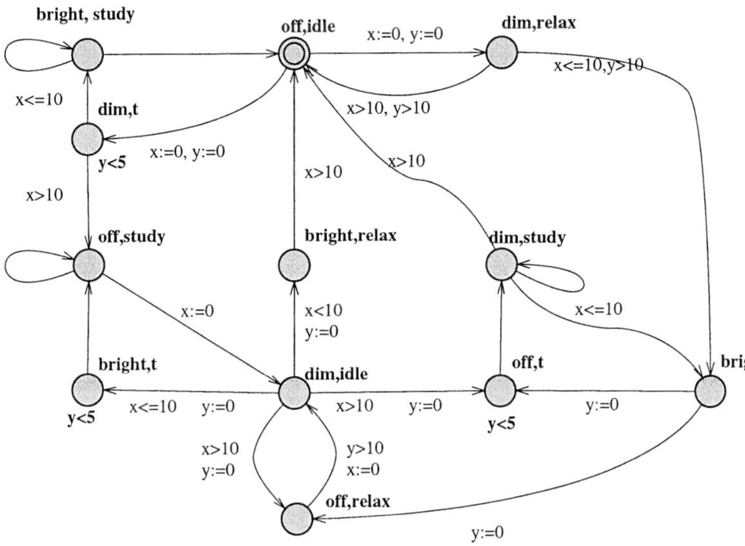

Fig. 15. Product Automaton for the Network in Fig. 14.

Let l_i stand for the ith element of a location vector l and $l[l_i'/l_i]$ for the vector l with l_i being substituted with l_i'. The transition rules are as follows:

- $\langle l, u\rangle \xrightarrow{d} \langle l, u+t\rangle$ if $u \in I(l)$ and $(u+d) \in I(l)$, where $I(l) = \bigwedge I(l_i)$
- $\langle l, u\rangle \xrightarrow{\tau} \langle l[l_i'/l_i], u'\rangle$ if $l_i \xrightarrow{g,\tau,r} l_i'$, $u \in g$, $u' = [r \mapsto 0]u$, $u' \in I(l[l_i'/l_i])$
- $\langle l, u\rangle \xrightarrow{\tau} \langle l[l_i'/l_i][l_j'/l_j], u'\rangle$ if $l_i \xrightarrow{g_i,a?,r_i} l_i'$, $l_j \xrightarrow{g_j,a!,r_j} l_j'$, $i \neq j$, $u \in g_i \wedge g_j$, $u' = [r_i \cup r_j \mapsto 0]u$ and $u' \in I(l[l_i'/l_i][l_j'/l_j])$.

Note that a network is a closed system which may not perform any external action. In fact, the CCS hiding operator is embedded in the above rules.

Shared Integer Variables. Clocks may be considered as typed variables with type *clock*. In UPPAAL, one may also use integer variables and arrays of integers, each with a bounded domain and an initial value. Predicates over the integer variables can be used as guards on the edges of an automaton process and the integer variables may be updated using resets on the edges. In the current version of UPPAAL, the syntax related to integer variables resembles the standard C syntax. Both integer guards and integer resets are standard C expressions with the restriction that guards can not have side-effects.

The semantics of networks can be defined accordingly. The clock assignment u in the state configuration $\langle l, u\rangle$ can be extended to store the values of integer variables in addition to clocks. Since delay does not affect the integer variables, the delay transitions are the same as for networks without integer variables. The action transitions are extended in the natural way, *i.e.* for an action transition to be enabled the extended clock assignment must also satisfy all integer guards on the corresponding edges and when a transition is taken the assignment is updated according to the integer and clock resets.

There is a problem with variable updating in a synchronizing transition where one of the processes participating in the transition updates a variable used by the other. In UPPAAL, for a synchronization transition, the resets on the edge with an output-label is performed before the resets on the edge with an input-label. This destroys the symmetry of input and output actions. But it gives a natural and clear semantics for variable updating. The transition rule for synchronization is modified accordingly:

- $\langle l, u\rangle \xrightarrow{\tau} \langle l[l_i'/l_i][l_j'/l_j], u'\rangle$ if $l_i \xrightarrow{g_i,a?,r_i} l_i', l_j \xrightarrow{g_j,a!,r_j} l_j', i \neq j, u \in g_i \wedge g_j, u' = [r_i \mapsto 0]([r_j \mapsto 0]u)$ and $u' \in I(l[l_i'/l_i][l_j'/l_j])$

Urgent Channels. To model urgent synchronizing transitions, which should be taken as soon as they are enabled, UPPAAL supports a notion of urgent channels. An urgent channel works much like an ordinary channel, but with the exception that if a synchronization on an urgent channel is possible the system may not delay. Interleaving with other enabled action transitions, however, is still allowed. In order to keep clock constraints representable using convex zones, clock guards are not allowed on edges synchronizing on urgent channels.

To illustrate why this restriction is necessary, consider the network shown in Fig. 16. Both processes may independently go from their first state to their second state. In the second state, the first process must delay for at least 10 time units before it is allowed to synchronize on the urgent channel u. In the second state, the other process must delay for at least 5 time units before it is allowed to synchronize on the urgent channel u. As soon as both processes have spent the minimal time periods required in their second state, they should synchronize and move to their third state. The problem is in [**S2,T2**] where the zone may be for example $(x \geq 10 \wedge y = 5) \vee (y \geq 5 \wedge x = 10)$ which is a non-convex zone.

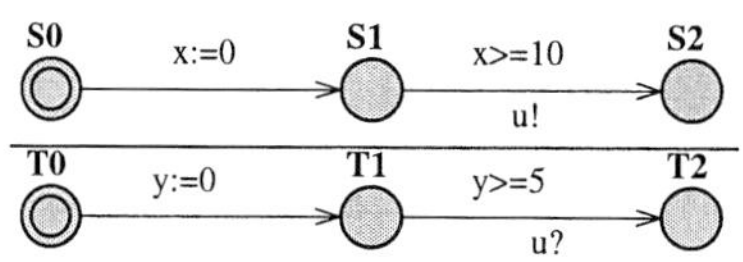

Fig. 16. An example of a network with non convex timing regions.

For this example, the problem can be solved by splitting the non-convex zone into two convex ones. But in general, the splitting is a computationally expensive operation. In UPPAAL, we decided to avoid such operations for the sake of efficiency. So only integer guards are allowed on edges involving synchronizations on urgent channels.

Committed Locations. To model atomic sequences of actions, *e.g.* atomic broadcast or multicast, UPPAAL supports a notion of *committed locations*. A committed location is a location where no delay is allowed. In a network, if any process is in a committed location then only action transitions starting from such a committed location are allowed. Thus, processes in committed locations may be interleaved only with processes in a committed location.

Syntactically, each process A_i in a network may have a subset $N_i^C \subseteq N_i$ of locations marked as committed locations. Let $C(l)$ denote the set of locations in l, that are committed. For the same reason as in the case of urgent channels, as a syntactical restriction, no clock constraints but predicates over integer variables are allowed to appear in a guard on an outgoing edge from a committed location.

The transition rules are given in the following, where $\rightarrow_c$ denotes the transition relation for a network with committed locations and $\rightarrow$ denotes the transition relation for the same network without considering the committed locations.

- $\langle l, u\rangle \xrightarrow{d}_c \langle l, u+d\rangle$ if $\langle l, u\rangle \xrightarrow{d} \langle l, u+d\rangle$ and $C(l) = \emptyset$
- $\langle l, u\rangle \xrightarrow{\tau}_c \langle l[l_i'/l_i], u'\rangle$ if $\langle l, u\rangle \xrightarrow{\tau} \langle l[l_i'/l_i], u'\rangle$ and either $l_i \in C(l)$ or $C(l) = \emptyset$
- $\langle l, u\rangle \xrightarrow{\tau}_c \langle l[l_i'/l_i][l_j'/l_j], u'\rangle$ if $\langle l, u\rangle \xrightarrow{\tau} \langle l[l_i'/l_i][l_j'/l_j], u'\rangle$ and either $l_i \in C(l)$, $l_j \in C(l)$ or $C(l) = \emptyset$

5.2 Verifying with UPPAAL

The model checking engine of UPPAAL is designed to check a subset of TCTL formula [2] for networks of timed automata. The formulas should be one of the following forms:

- `A[]` ϕ — Invariantly ϕ.
- `E<>` ϕ — Possibly ϕ.
- `A<>` ϕ — Always Eventually ϕ.
- `E[]` ϕ — Potentially Always ϕ.
- ϕ `-->` ψ — ϕ always leads to ψ. This is a shorthand for $\forall\Box(\phi \Rightarrow \forall\Diamond\psi)$.

where ϕ, ψ are local properties that can be checked locally on a state, *i.e.* boolean expressions over predicates on locations and integer variables, and clock constraints in $\mathcal{B}(C)$.

The transition system of a network may be unfolded into an infinite tree containing states and transitions. The semantics of the formulas are defined over such a tree. The letters A and E are used to quantify over paths. A is used to denote that the given property should hold for all paths of the tree while E denotes that there should be at least one path of the tree where the property holds. The symbols `[]` and <> are used to quantify over states within a path. `[]` denotes that all states on the path should satisfy the property, while <> denotes that at least one state in the execution satisfies the property. In Fig. 17 the four basic property types are illustrated using execution trees, where the dashed arrows are used to denote repetitions in the trees. The states satisfying ϕ are denoted by filled nodes and edges corresponding to the paths are highlighted using bold arrows.

The two types of properties most commonly used in verification of timed systems are `E<>`ϕ and `A[]`ψ. They are dual in the sense that `E<>`ϕ is satisfied if and only if `A[]`$\neg\phi$ is not satisfied. This type of properties are often classified as safety properties,

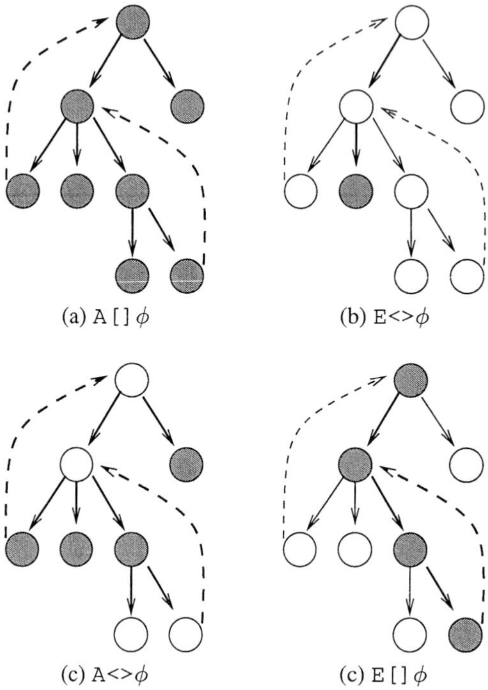

Fig. 17. (T)CTL-formulae.

i.e. meaning that the system is safe in the sense that a specified hazard can not occur. It is also possible to transform so called bounded liveness properties, *i.e.* properties stating that some desired state will be reached within a given time, into safety properties using observer automata [1] or by annotating the model [37]. For example, to check if an automaton will surely reach a location l within 10 time units, we use one clock x (set to 0 initially) and introduce a boolean variable l_b (set to false initially). For each incoming edge to l in the automaton, set l_b to true. Then if the automaton satisfies the invariant property "$x \leq 10 \vee l_b$", it will surely reach l within 10 time units provided that the automaton contains no zeno loops which stop time to progress.

The other three types of properties are commonly classified as unbounded liveness properties, *i.e.* they are used to express and check for global progress. These properties are not commonly used in UPPAAL case-studies. It seems to be the case that bounded liveness properties are more important for timed systems.

5.3 The UPPAAL Architecture

To provide a system that is both efficient, easy to use and portable, UPPAAL is split into two components, a graphical user interface written in Java and a verification engine written in C++. The engine and the GUI communicate using a protocol, allowing the verification to be performed either on the local workstation or on a powerful server in a network.

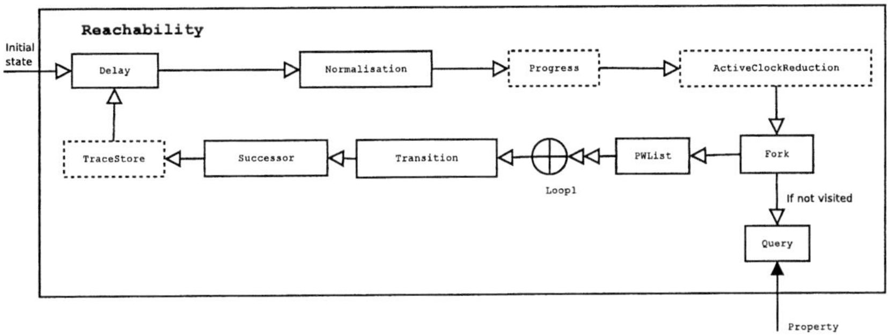

Fig. 18. Schematic view of the reachability pipeline in UPPAAL.

To implement the reachability analysis algorithm 1, the UPPAAL verification engine is organized as a pipeline that incarnates the natural data flow in the algorithm. A sketch of this pipeline is shown in Fig. 18. This architecture simplifies both activating and deactivating optimizations at runtime by inserting and removing stages dynamically, and introducing new optimizations and features in the tool by implementing new or changing existing stages.

In addition to the zone-manipulation algorithms described in Section 4 and the pipeline architecture, in UPPAAL a number of optimizations have been implemented:

- Minimal constraint systems [31] and CDDs [32, 11], to reduce memory consumption by removing redundant information in zones before storing them.
- Selective storing of states in PASSED [31], where static analysis is used to detect states that can be omitted safely from PASSED without losing termination.
- Compression [13] and sharing [10, 21] of state data, to reduce the memory consumption of PASSED and WAIT.
- Active clock reduction [22], that use live-range analysis to determine when the value of a clock is irrelevant. This does not only reduce the size of individual states but also the perceived state-space.
- Supertrace [30] and Hash Compaction [45, 44] where already visited states are stored only as hash signatures, and Convex-hull approximation [9] where convex hulls are used to approximate unions of zones, for reducing memory consumption at a risk of inconclusive results.

Acknowledgements

We would like to thank Pavel Krcal for pointing out an error in an early version of this document.

Appendix: Pseudo-Code for Operations on DBMs

Algorithm 2 `close`(D): Floyds algorithm for computing shortest path

```
for k := 0 to n do
  for i := 0 to n do
    for j := 0 to n do
      D_ij := min(D_ij, D_ik + D_kj)
    end for
  end for
end for
```

Algorithm 3 Reduction of Zero-Cycle Free Graph G with n nodes

```
for i := 1 to n do
  for j := 1 to n do
    for k := 1 to n do
      if G_ik + G_kj ≤ G_ij then
        Mark edge i → j as redundant
      end if
    end for
  end for
end for
Remove all edges marked as redundant.
```

Algorithm 4 Reduction of negative-cycle free graph G with n nodes

```
for i := 1 to n do
  if Node_i is not in a partition then
    Eq_i = ∅
    for j := i to n do
      if G_ij + G_ji = 0 then
        Eq_i := Eq_i ∪ {Node_i}
      end if
    end for
  end if
end for
Let G' be a graph without nodes.
for each Eq_i do
  Pick one representative Node_i ∈ Eq_i
  Add Node_i to G'
  Connect Node_i to all nodes in G' using weights from G.
end for
Reduce G'
for each Eq_i do
  Add one zero cycle containing all nodes in Eq_i to G'
end for
```

Algorithm 5 `relation`(D, D')

```
f_{D⊆D'} := t
f_{D⊇D'} := t
for i := 0 to n do
  for j := 0 to n do
    f_{D⊆D'} := f_{D⊆D'} ∧ (D_ij ≤ D'_ij)
    f_{D⊇D'} := f_{D⊇D'} ∧ (D_ij ≥ D'_ij)
  end for
end for
return ⟨f_{D⊆D'}, f_{D⊇D'}⟩
```

Algorithm 6 `up`(D)

```
for i := 1 to n do
  D_i0 := ∞
end for
```

Algorithm 7 `down`(D)

```
for i := 1 to n do
  D_0i = (0, ≤)
  for j := 1 to n do
    if D_ji < D_0i then
      D_0i = Dji
    end if
  end for
end for
```

Algorithm 8 `and`(D, g)

```
if D_yx + (m, ≼) < 0 then
  D_00 = (−1, ≤)
else if (m, ≼) < D_xy then
  D_xy = (m, ≼)
  for i := 0 to n do
    for j := 0 to n do
      if D_ix + D_xj < D_ij then
        D_ij = D_ix + D_xj
      end if
      if D_iy + D_yj < D_ij then
        D_ij = D_iy + D_yj
      end if
    end for
  end for
end if
```

Algorithm 9 `free`(D, x)

```
for i := 0 to n do
  if i ≠ x then
    D_xi = ∞
    D_ix = D_i0
  end if
end for
```

Algorithm 10 `reset`$(D, x := m$

```
for i := 0 to n do
  D_xi := (m, ≤) + D_0i
  D_ix := D_i0 + (−m, ≤)
end for
```

Algorithm 11 `copy`$(D, x := y)$

```
for i := 0 to n do
  if i ≠ x then
    D_xi := D_yi
    D_ix := D_iy
  end if
end for
D_xy := (0, ≤)
D_yx := (0, ≤)
```

Algorithm 12 `shift`$(D, x := x + m)$

```
for i := 0 to n do
  if i ≠ x then
    D_xi := D_xi + (m, ≤)
    D_ix := D_ix + (−m, ≤)
  end if
end for
D_x0 := max(D_x0, (0, ≤))
D_0x := min(D_0x, (0, ≤))
```

Algorithm 13 norm$_k(D)$)

```
for i := 0 to n do
  for j := 0 to n do
    if D_ij ≠ ∞ and D_ij > (k_i), ≤) then
      D_ij = ∞
    else if D_ij ≠ ∞ and D_ij < (−k_j, <) then
      D_ij = (−k_j, <)
    end if
  end for
end for
close(D)
```

Algorithm 14 Core normalization: $\mathsf{Core\text{-}Norm}_k(D)$.

```
𝒢_unsat := ∅
for all g ∈ 𝒢_d do
  if D ∧ g = ∅ then
    𝒢_unsat := 𝒢_unsat ∪ {g}
  end if
  if D ∧ ¬g = ∅ then
    𝒢_unsat := 𝒢_unsat ∪ {¬g}
  end if
end for
D := norm_k(D)
for all g ∈ 𝒢_unsat do
  D := D ∧ ¬g
end for
return D
```

Algorithm 15 Zone splitting: $\mathsf{split}(D)$.

```
Q := {D}, Q' := ∅
for all g ∈ 𝒢_d do
  for all D' ∈ Q do
    if D' ∧ g and D' ∧ ¬g then
      Q' := Q' ∪ {D' ∧ g, D' ∧ ¬g}
    else
      Q' := Q' ∪ {D'}
    end if
  end for
  Q := Q', Q' := ∅
end for
return Q
```

Algorithm 16 Normalization: $\mathsf{norm}_{k,\mathcal{G}}(D)$.

```
Q := ∅
for all D' ∈ split(D) do
  Q := Q ∪ {Core-Norm_k(D')}
end for
return Q
```

Algorithm 17 Algorithm for adding encoded bounds.

```
if b1 = MAX_INT or b2 = MAX_INT then
  return MAX_INT
else
  return b1 + b2 − ((b1&1)|(b2&1))
end if
```

References

1. Luca Aceto, Augusto Bergueno, and Kim G. Larsen. Model checking via reachability testing for timed automata. In *Proceedings, Fourth Workshop on Tools and Algorithms for the Construction and Analysis of Systems*, volume 1384 of *Lecture Notes in Computer Science*. Springer-Verlag, 1998.
2. Rajeev Alur, Costas Courcoubetis, and David L. Dill. Model-checking for real-time systems. In *Proceedings, Seventh Annual IEEE Symposium on Logic in Computer Science*, pages 414–425. IEEE Computer Society Press, 1990.
3. Rajeev Alur, Costas Courcoubetis, and David L. Dill. Model-checking in dense real-time. *Journal of Information and Computation*, 104(1):2–34, 1993.
4. Rajeev Alur, Costas Courcoubetis, and Thomas A. Henzinger. The observational power of clocks. In *International Conference on Concurrency Theory*, pages 162–177, 1994.
5. Rajeev Alur and David L. Dill. Automata for modeling real-time systems. In *Proceedings, Seventeenth International Colloquium on Automata, Languages and Programming*, volume 443 of *Lecture Notes in Computer Science*, pages 322–335. Springer-Verlag, 1990.
6. Rajeev Alur and David L. Dill. A theory of timed automata. *Journal of Theoretical Computer Science*, 126(2):183–235, 1994.
7. Rajeev Alur, Limor Fix, and Thomas A. Henzinger. Event-clock automata: a determinizable class of timed automata. *Theoretical Computer Science*, 211(1–2):253–273, 1999.
8. Rajeev Alur and Thomas A. Henzinger. A really temporal logic. *Journal of the ACM*, 41(1):181–204, 1994.
9. Felice Balarin. Approximate reachability analysis of timed automata. In *Proceedings, 17*th *IEEE Real-Time Systems Symposium*. IEEE Computer Society Press, 1996.
10. Gerd Behrmann, Alexandre David, Kim G. Larsen, and Wang Yi. Unification & sharing in timed automata verification. In *Proceedings, Tenth International SPIN Workshop*, volume 2648 of *Lecture Notes in Computer Science*. Springer-Verlag, 2003.
11. Gerd Behrmann, Kim G. Larsen, Justin Pearson, Carsten Weise, and Wang Yi. Efficient timed reachability analysis using clock difference diagrams. In *Proceedings, Eleventh International Conference on Computer Aided Verification*, volume 1633 of *Lecture Notes in Computer Science*, pages 341–353. Springer-Verlag, 1999.
12. Richard Bellman. *Dynamic Programming*. Princeton University Press, 1957.
13. Johan Bengtsson. Reducing memory usage in symbolic state-space exploration for timed systems. Technical Report 2001-009, Department of Information Technology, Uppsala University, 2001.
14. Johan Bengtsson. Clocks, dbms and states in timed systems. Ph.D. Thesis, ACTA Universitatis Upsaliensis 39, Uppsala University, 2002.
15. Johan Bengtsson, Bengt Jonsson, Johan Lilius, and Wang Yi. Partial order reductions for timed systems. In *Proceedings, Ninth International Conference on Concurrency Theory*, volume 1466 of *Lecture Notes in Computer Science*, pages 485–500. Springer-Verlag, 1998.
16. Johan Bentsson and Wang Yi. On clock difference constraints and termination in reachability analysis of timed automata. In *Formal Methods and Software Engineering, 5th International Conference on Formal Engineering Methods, ICFEM 2003, Singapore, November 5-7, 2003, Proceedings*, volume 2885 of *Lecture Notes in Computer Science*. Springer, 2003.
17. Bernard Berthomieu and Michel Diaz. Modeling and verification of timed dependent systems using timed petri nets. *IEEE Transactions on Software Engineering*, 17(3):259–273, 1991.
18. Beatrice Bérard, Volker Diekert, Paul Gastin, and Antoine Petit. Characterization of the expressive power of silent transitions in timed automata. *Fundamenta Informaticae*, 36:145–182, 1998.

19. Karlis Cerans. Decidability of bisimulation equivalences for parallel timer processes. In *Computer Aided Verification*, volume 663 of *LNCS*. Springer, 1992.
20. Zhou Chaochen. Duration calculus, a logical approach to real-time systems. *Lecture Notes in Computer Science*, 1548:1–7, 1999.
21. Alexandre David, Gerd Behrmann, Kim G. Larsen, and Wang Yi. A tool architecture for the next genreation of UPPAAL. Technical Report 2003-011, Department of Information Technology, Uppsala University, 2003.
22. Conrado Daws and Sergio Yovine. Reducing the number of clock variables of timed automata. In *Proceedings, 17*th *IEEE Real-Time Systems Symposium*. IEEE Computer Society Press, 1996.
23. David L. Dill. Timing assumptions and verification of finite-state concurrent systems. In *Proceedings, Automatic Verification Methods for Finite State Systems*, volume 407 of *Lecture Notes in Computer Science*, pages 197–212. Springer-Verlag, 1989.
24. Elena Fersman, Paul Pettersson, and Wang Yi. Timed automata with asynchronous processes: Schedulability and decidability. In J.-P. Katoen and P. Stevens, editors, *Proc. of the 8*th *International Conference on Tools and Algorithms for the Construction and Analysis of Systems*, number 2280 in Lecture Notes in Computer Science, pages 67–82. Springer–Verlag, 2002.
25. Robert W. Floyd. Acm algorithm 97: Shortest path. *Communications of the ACM*, 5(6):345, June 1962.
26. N. Halbwachs. Delay analysis in synchronous programs. In *Fifth Conference on Computer-Aided Verification*, Elounda (Greece), July 1993. LNCS 697, Springer Verlag.
27. Thomas A. Henzinger, Xavier Nicollin, Joseph Sifakis, and Sergio Yovine. Symbolic model checking for real-time systems. In *Proceedings, Seventh Annual IEEE Symposium on Logic in Computer Science*, pages 394–406, 1992.
28. Thomas A. Henzinger, Xavier Nicollin, Joseph Sifakis, and Sergio Yovine. Symbolic model checking for real-time systems. *Journal of Information and Computation*, 111(2):193–244, 1994.
29. C.A.R. Hoare. Communicating sequential processes. *Communications of the ACM*, 21(8):666–676, August 1978.
30. Gerard J. Holzmann. *Design and Validation of Computer Protocols*. Prentice-Hall, 1991.
31. Kim G. Larsen, Fredrik Larsson, Paul Pettersson, and Wang Yi. Efficient verification of real-time systems: Compact data structure and state space reduction. In *Proceedings, 18*th *IEEE Real-Time Systems Symposium*, pages 14–24. IEEE Computer Society Press, 1997.
32. Kim G. Larsen, Justin Pearson, Carsten Weise, and Wang Yi. Clock difference diagrams. *Nordic Journal of Computing*, 1999.
33. Kim G. Larsen, Paul Petterson, and Wang Yi. UPPAAL in a nutshell. *Journal on Software Tools for Technology Transfer*, 1997.
34. Kim G. Larsen, Paul Pettersson, and Wang Yi. Compositional and Symbolic Model-Checking of Real-Time Systems. In *Proc. of the 16*th *IEEE Real-Time Systems Symposium*, pages 76–87. IEEE Computer Society Press, December 1995.
35. Kim Guldstrand Larsen and Yi Wang. Time-abstracted bisimulation: Implicit specifications and decidability. *Information and Computation*, 134(2):75–101, 1997.
36. Fredrik Larsson. Efficient implementation of model-checkers for networks of timed automata. Licentiate Thesis 2000-003, Department of Information Technology, Uppsala University, 2000.
37. Magnus Lindahl, Paul Pettersson, and Wang Yi. Formal Design and Analysis of a Gear-Box Controller. In *Proceedings, Fourth Workshop on Tools and Algorithms for the Construction and Analysis of Systems*, number 1384 in Lecture Notes in Computer Science, pages 281–297. Springer-Verlag, 1998.

38. Magnus Lindahl, Paul Pettersson, and Wang Yi. Formal Design and Analysis of a Gearbox Controller. *Springer International Journal of Software Tools for Technology Transfer (STTT)*, 3(3):353–368, 2001.
39. Robin Milner. *Communication and Concurrency*. Prentice Hall, 1989.
40. Xavier Nicollin and Joseph Sifakis. The algebra of timed processes, ATP: Theory and application. *Journal of Information and Computation*, 114(1):131–178, 1994.
41. Paul Pettersson. *Modelling and Verification of Real-Time Systems Using Timed Automata: Theory and Practice*. PhD thesis, Uppsala University, 1999.
42. G. M. Reed and A. W. Roscoe. A timed model for communicating sequential processes. *Theoretical Computer Science*, 58(1-3):249–261, 1988.
43. Tomas Gerhard Rokicki. *Representing and Modeling Digital Circuits*. PhD thesis, Stanford University, 1993.
44. Ulrich Stern and David L. Dill. Improved probabilistic verification by hash compaction. In *Correct Hardware Design and Verification Methods: IFIP WG10.5 Advanced Research Working Conference Proceedings*, 1995.
45. Pierre Wolper and Dennis Leroy. Reliable hashing without collision detection. In *Proceedings, Fifth International Conference on Computer Aided Verification*, volume 697 of *Lecture Notes in Computer Science*, pages 59–70. Springer-Verlag, 1993.
46. Mihalis Yannakakis and David Lee. An efficient algorithm for minimizing real-time transition systems. In *Proceedings, Fifth International Conference on Computer Aided Verification*, volume 697 of *Lecture Notes in Computer Science*, pages 210–224. Springer-Verlag, 1993.
47. Wang Yi. CCS + time = an interleaving model for real time systems. In *Proceedings, Eighteenth International Colloquium on Automata, Languages and Programming*, volume 510 of *Lecture Notes in Computer Science*. Springer-Verlag, 1991.
48. Wang Yi and B. Jonsson. Decidability of timed language-inclusion for networks of real-time communicating sequen tial processes. In *Proc.* 14th *Conf. on Foundations of Software Technology and Theoretical Computer Science*, volume 880 of *Lecture Notes in Computer Science*. Springer-Verlag, 1994.
49. Wang Yi, Paul Petterson, and Mats Daniels. Automatic verification of real-time communicating systems by constraint-solving. In *Proceedings, Seventh International Conference on Formal Description Techniques*, pages 223–238, 1994.
50. Sergio Yovine. Kronos: a verification tool for real-time systems. *Journal on Software Tools for Technology Transfer*, 1, October 1997.

Petri Nets and Dependability

Simona Bernardi[1], Andrea Bobbio[2], and Susanna Donatelli[1]

[1] Dipartimento di Informatica, Università di Torino, Torino, Italy
{bernardi,susi}@di.unito.it
[2] Dipartimento di Informatica, Università del Piemonte Orientale, Alessandria, Italy
{bobbio}@di.unipmn.it

Abstract. Dependability evaluation main objective is to assess the ability of a system to correctly function over time. There are many possible approaches to the evaluation of dependability: in these notes we are mainly concerned with dependability evaluation based on probabilistic models. Starting from simple probabilistic models with very efficient solution methods we shall then come to the main topic of the paper: how Petri nets can be used to evaluate the dependability of complex systems.

1 Introduction

The term *dependability* is normally used to refer to the ability of an element (hardware or software component, plant or whatever complex system) to correctly perform its intended function, or *mission*, over time.

In this paper we are interested in the *quantitative evaluation* of dependability, a research field that has many practical implications, as: 1) the analysis of risks and safety; 2) the specification and contract document of a system - it is usually the case that the definition of a new system also includes requirements about dependability, that is to say on how much we can rely on the system being built, whether this is a software product, an automation system, or a bridge; 3) incidence of maintenance in the life cycle of a system - being able to estimate the dependability of an object allows to predict how often it will break down, with the consequence of additional costs on maintenance, and to take decision on the balance between investing more time and money on the construction of the system and having bigger maintenance costs; 4) dimensioning of technical assistance sector: being able to predict how often a component of a car will break down allows to estimate the number of spare components needed over a certain time period, the costs of the repairs during the warranty period, and the planning of the preventive maintenance.

To study the evolution over time of the dependability of a system it is necessary to be able to foresee when and how its component will be subject to malfunctioning, and how the malfunctioning of a component may affect the system behavior. There are two major approaches: measuring of physical, existing systems, and evaluation of abstract models of the (existing or planned) systems.

For what concerns measures, we can distinguish the following classes:

◇ *Field measures.*
It assumes that the system is already operational, and that is possible to measure the quantities of interest without altering the system behavior. Field measures should be collected for a "statistically significant" period.

J. Desel, W. Reisig, and G. Rozenberg (Eds.): ACPN 2003, LNCS 3098, pp. 125–179, 2004.

◇ *Single component measures.*
Only a subset of the components are placed under test to collect measures. This is usually accompanied by acceleration technique, that generate, in a short period of time, the same conditions that the component will encounter along a much longer period.
◇ *Prototype measures.*
A prototype of the system is built, and measures are taken on the prototype under normal operational condition. This technique is very expensive, and it can be used only for goods of large consume or that involve an expected high safety risk.

The approach based on models requires first the construction of abstract mathematical models that describe the behavior of the system: the quantity of interest is the computed through the analysis and solution of the model. Modeling for dependability suffers of the same problems as any other modeling approach: the choice of the right level of abstraction for the quantity/property we want to evaluate and the complexity of the model solution.

Models are usually distinguished according to the following characteristics:

◇ *Modeling language.*
A model can be described in terms of basic quantities and mathematical expressions that relate the overall behavior to the basic quantities, as well as through an high level language with a well defined semantics, as for example Petri nets or queueing networks, or application-specific languages, in which the language elements have a direct counterpart on the system basic components. In general, the higher level is the language, the easier is to define the model, and usually the harder is to solve it. By model solution we mean the evaluation of the quantity of interest, in our case it is usually a direct definition of the dependability of the system.
◇ *Solution methods.*
There are two large classes of solution methods: analytical techniques and simulation. Analytical techniques assume that it is possible to derive from the model a set of equations that describe the quantities under evaluation, and that there are mathematical techniques, exact or approximate, to solve the equations. Simulation consists instead in executing the model on a computer a number of times that is sufficient to provide a statistically acceptable estimation of the quantities of interest. There are high level modeling languages for which simulation is the only viable analysis technique, as it is usually the case for application-oriented languages, in which the semantics of the basic elements of the language is not defined in mathematical terms, but through a piece of program code.

Topic of these notes is indeed to describe various modeling approaches to the evaluation of dependability, with a large emphasis on Petri nets.

Measures and models should by no means be considered as competitors. The evaluation of dependability requires the synergic use of both: measures can indeed be used to set the model parameters, while models can be used to drive the, usually expensive, measuring activity.

With the inherent complexity of modern systems it is extremely difficult to precisely and deterministically describe the physical, technological, and environmental factors and interactions that provoke a system malfunctioning. It is instead very much accepted that the time to failure of a system (how long does it take for the system to go into a faulty state) is not a deterministic quantity, but a random variable of a, generally unknown, distribution. A similar rationale supports also the idea that, if a system can be repaired, then the time to repair (how long it takes for the system to be repaired) is also a random variable.

The above observations led to the development of a probabilistic approach to the quantitative evaluation of system dependability, that is the main topic of this paper.

Let us now introduce the terminology used in this paper.

Faults, errors and failures. We adhere to the terminology discussed in [43], and we say that when the delivered service of a system deviates from fulfilling the system intended function, then the system has a *failure*. A failure is due to a deviation from the correct state of the system, known as *error*. Such a deviation is due to a given cause, for instance related to the physical state of the system, or to a bad system design. This cause is called a *fault*. We shall generically refer to fault, errors, and failures as the "FEF elements."

Systems and components. We view a system as built out of elementary components. We shall first discuss dependability of a component in isolation (that can be seen as a single-component system), to then introduce the dependability of a system as a function of the dependability of its components. When dealing with a single component we do not distinguish the FEF elements, so we equivalently refer to a component as being faulty/non-faulty, failed/not-failed, not-working/working, down/up, which are current terms in the literature. For a system with a simple structure like whose presented in Sub-Section 5.1 we distinguish between fault in a component and failure of the system (provoked by one or more faulty components). The reader should nevertheless be aware that it is also common of the literature to use the generic term failure, so that the failure of one or more component provokes the failure of the system.

Net classes. Since the main topic of the paper is on quantitative analysis based on probabilistic approach, we shall consider Petri nets with timed transitions. The time associated to transition is either zero (immediate transition) or it is a delay described by an exponentially distributed random variable. We shall use two specific net classes: Generalized Stochastic Petri Nets (GSPN) [1], and their colored counterpart Stochastic Well-formed Nets (SWN) [18].

The paper is organized as follows: in Section 2 the basic concepts of dependability are introduced. Section 3 presents two combinatorial techniques for the dependability analysis of systems consisting of a number of independent components: the reliability block technique and fault tree analysis. Section 4 describes state enumeration techniques that can be applied also for the dependability analysis of systems in which the independence assumption among components does not hold. The material presented in Sections 2,3 and 4 derives from [13]. Section 5 introduces dependability modeling using Petri Nets. Two examples are presented: a simple one, representing a system with two independent components, and a more complex example with several interacting components that is a simplified version of the case study analyzed in [10]. Section 6 describes a systematic, compositional approach to the construction of Stochastic Petri Net models for dependability and Section 7 describes the application of such approach to the automation system domain. The material presented in Sections 6 and 7 has been taken from [8, 7, 4]. Finally, conclusions are written in Section 8.

2 Basic Concepts of Dependability

A first step towards quantitative evaluation of dependability is the definition of the dependability quantity. The definition of dependability takes different flavors depending on whether we consider a system that, once broken, stays broken forever, or a system that, once broken, it is repaired and goes through cycles of correct functioning and repairs. In the first case the measure to be considered is the *reliability*, in the second case is the *availability*, as we shall see in the following.

2.1 Dependability of Non-repairable Components: Reliability

The first case considered is that of non-repairable components, that is to say the system under study is seen as a monolithic component that, once it is broken/malfunctioning, it will stay in that state forever. In this case the dependability of the system is characterized by the reliability quantity. A commonly accepted definition of reliability [60] is:

> *The reliability of a component at time t is the probability that the component correctly fulfills the assigned mission during the interval* $[0,t]$*, given its environmental conditions.*

Observe that the definition relates the reliability of a component to its environment: it is therefore possible that the same component will have a very different reliability depending on the environment in which it is placed.

Let τ be the random variable that represents the *time to failure* of the component under study, τ being a time quantity, it is defined only for non-negative values. The *probability distribution function* of the variable τ is:

$$F(t) = Prob\{\tau \leq t\} \tag{1}$$

that defines the probability that the system is malfunctioning at time t. The following properties hold true for $F(t)$:

$$\begin{cases} F(0) = 0 \\ \lim_{t \to \infty} F(t) = 1 \\ F(t) \text{ not decreasing in } t \end{cases} \tag{2}$$

The reliability function is defined as the complement of $F(t)$:

$$R(t) = Prob\{\tau > t\} = 1 - F(t) \tag{3}$$

that defines the probability that the system is still working properly (still up) at time t (and since the system is not repairable, being up at time t it means that no fault has taken place). The following properties hold true for $R(t)$:

$$\begin{cases} R(0) = 1 \\ \lim_{t \to \infty} R(t) = 0 \\ R(t) \text{ not increasing in } t \end{cases} \tag{4}$$

If $F(t)$ is derivable, the probability density function of the variable τ is:

$$f(t) = \frac{dF(t)}{dt} = -\frac{dR(t)}{dt} \tag{5}$$

where $f(t)\,dt$ is the probability that the value of τ falls in between t and $t+dt$, that is to say, the fault takes place in between t and $t+dt$. Moreover:

$$\int_a^b f(t)\,dt = Prob\{a < \tau \leq b\} = F(b) - F(a)$$

represents the probability that the fault takes place in the interval $[a,b]$.

The expected value of the variable τ, $E[\tau]$, is called *Mean Time To Failure*, and is indicated by the acronym *MTTF*.

The hazard rate. The hazard (or failure) rate represents the probability that a component gets faulty between t and $t+dt$, given that it was correctly functioning up to time t (that is to say, the hazard rate is equal to the probability density function of the τ variable, *conditioned* on the fact that the component was still working correctly at time t [51]).

$$h(t) = Prob\{t < \tau \leq t+dt \,|\, \tau > t\} = \frac{Prob\{t < \tau \leq t+dt\,, \tau > t\}}{Prob\{\tau > t\}} \tag{6}$$

From (6), using (5), we can derive:

$$h(t) = \frac{f(t)}{R(t)} = -\frac{1}{R(t)}\frac{dR(t)}{dt} \tag{7}$$

and solving for $R(t)$ we get:

$$R(t) = exp\left[-\int_0^t h(x)\,dx\right] \tag{8}$$

that is the fundamental equation that relates reliability and hazard rate.

The classic shape of the failure rate when plotted over the time axis is that of a bath-tube: the failure rate is high at the beginning of the life of the object, it then remains stable for a significant period, and it finally increases. In terms of behavior of a system, it means that most systems have a very high probability of breaking when they are new, this probability decreases while experiencing a correct functioning of the system, up to a point in time in which the failure rate is constant, that is to say the failure rate does not depend on the particular time instant that we are considering, and up to a certain time barrier after which the aging of the system is predominant and the probability of breaking down increases as the time passes by. The bath-tube shape is particularly evident for manufacturing products, while for electronic components the aging effect is less evident.

In the technical literature it is often the case that the failure rate is a single constant value, which implies that we are assuming that the system is in its period of life corresponding to the bottom of the bath-tube: a constant failure rate is therefore equivalent to saying that the system has no memory of its past.

Which distribution for τ? Given that the time of correct functioning of a system is a random variable, what is an adequate distribution for it? We shall present two candidate distributions: exponential and Weibull.

The main characteristic of the exponential distribution is that the failure rate is constant, and, vice-versa, any distribution with a constant failure rate is exponential [63]. Given a constant failure rate, $h(t) = cost = \lambda$, from (7) and (5) we can derive:

$$\begin{aligned} F(t) &= 1 - e^{-\lambda t} \\ R(t) &= e^{-\lambda t} \\ f(t) &= \lambda\, e^{-\lambda t} \\ h(t) &= \lambda \end{aligned} \tag{9}$$

The mean value is of τ is $MTTF = 1/\lambda$, that is to say, the failure rate has a clear physical meaning: it is the inverse of the failure rate.
The exponential distribution is known as *memoryless* since the reliability conditioned on the fact that the component has been working correctly already for a duration $t = a$, is equal to the reliability at time $t = 0$.

The Weibull distribution is:

$$F(t) = 1 - exp\left[-(t/\eta)^{\beta}\right]$$

where $\eta > 0$ is the scaling parameter (displacement on the x-axis), and $\beta > 0$ is the shaping parameter. Changing β we get a different characterization of the failure rate:

$$\beta < 1 \Longrightarrow h(t) \text{ decreasing}$$

$$\beta = 1 \Longrightarrow h(t) \text{ constant}$$

$$\beta > 1 \Longrightarrow h(t) \text{ increasing}$$

Observe that the exponential distribution can be seen as a Weibull with $\beta = 1$. The ability to represent various behaviors of the failure rate is the major appeal of the Weibull distribution for dependability modeling.

2.2 Dependability of Repairable Components: Availability

We now consider the case of a component that, once broken, can be repaired. The behavior of a repairable component over time is therefore determined not only by the way in which it fails, but also by the way in which it is repaired, and we can consider the life of a system as an alternation between two states: Up (system is working) and Down (system is not working and it is under repair).

We assume that the intervals of correct functioning (time to failure), and the intervals of incorrect functioning (time to repair) are described by random variables. Let $\tau_1, \tau_2, \tau_3, \ldots$ be the random variables of the successive duration of the up times, and $\theta_1, \theta_2, \theta_3, \ldots$ the associated repair times, under the hypothesis that the repair is "regenerative", that is to say that after the repair the component is "good as new", then all the τ_i have the same distribution $F(t)$, and all the θ_i have the same distribution $G(t)$, and we can describe the behavior of the system with only two random variables τ, duration of the Up times, and θ, duration of the Down times. $G(t)$ represents the probability that the component is repaired in $[0,t]$, and it is called Maintainability. Similarly to what we have done for $F(t)$, we get for $G(t)$ the following expressions:

$$g(t) = \frac{dG(t)}{dt}$$
$$h_g(t) = \frac{g(t)}{1 - G(t)}$$
$$MTTR = \int_0^{\infty} t\,g(t)\,dt$$

where MTTR is the Mean Time To Repair, and $h_g(t)$, the repair rate, is the probability that the repair is terminated in the interval $[t, t+dt]$, given that the component was still unrepaired at time t. If we assume that the repair rate $h_g(t)$ is time-independent, $h_g(t) = cost = \mu$, then the maintainability is an exponential function:

$$G(t) = 1 - e^{-\mu t} \qquad \text{and} \qquad MTTR = \frac{1}{\mu} \tag{10}$$

The assumption of time-independence is not very realistic since, in general, the time it takes to finish a repair does depend on how long the repair has already taken, but this assumption is nevertheless often taken in the literature and in the practice, for the advantages that it offers in the solution process.

It is clear that if a system is subject to failures and repairs, the reliability function $R(t)$ is not particularly informative since for any t greater than the time of the first failure, the value of $R(t)$ is always going to be zero.

A new quantity is therefore defined, and it is called *availability*, indicated as $A(t)$. $A(t)$ is the probability that the system is Up at time t.

$$A(t) = Prob\{at\ time\ t,\ state = \mathbf{Up}\} \tag{11}$$

The *unavailability* $U(t)$ is instead the probability that the system is Down at time t.

$$U(t) = Prob\{at\ time\ t,\ state = \mathbf{Down}\} \tag{12}$$

and, since we assume that the system is either Up or Down , we have:

$$A(t) + U(t) = 1$$

The computation of $A(t)$ and $U(t)$ of a system starts from the observation that $A(t)$ ($U(T)$) is equivalent to the probability of being in the Up (Down) state at time t. But the probability of being in the Up state can be computed writing equilibrium equation, since the variation over time of the probability of being in state Up is equal to the difference between the probability of entering the Up state and the probability of leaving the state, that, assuming a fixed failure (repair) rate equal to λ (μ), amounts to the following equations:

$$\begin{cases} \dfrac{dA(t)}{dt} = -\lambda A(t) + \mu U(t) \\ \\ \dfrac{dU(t)}{dt} = \lambda A(t) - \mu U(t) \end{cases} \tag{13}$$

Assuming that at time 0 the system is working properly, we can set $A(0) = 1$, we can solve the equations (13), and obtain:

$$\begin{aligned} A(t) &= \frac{\mu}{\lambda+\mu} + \frac{\lambda}{\lambda+\mu} e^{-(\lambda+\mu)t} \\ \\ U(t) &= \frac{\lambda}{\lambda+\mu} - \frac{\lambda}{\lambda+\mu} e^{-(\lambda+\mu)t} \end{aligned} \tag{14}$$

And we get:

$$A(0) = 1 \quad ; \quad \lim_{t\to\infty} A(t) = A_\infty = \frac{\mu}{(\lambda+\mu)} \tag{15}$$

The typical shape of the availability function is made up of a transient term that exhibits an exponential decay, and a time independent term that constitute the horizontal asymptote.

Since in a repairable system $MTTF \gg MTTR$, and hence $\lambda \ll \mu$, the contribution of the transient term decays very quickly, and therefore the availability is often identified by its asyntotic behavior in (15).

If A_∞ is the asymptotic availability, we can write

$$A_\infty = \frac{\mu}{\lambda+\mu} = \frac{1/\lambda}{1/\lambda + 1/\mu} = \frac{MTTF}{MTTF + MTTR} \tag{16}$$

Although the above expression has been derived under the hypothesis of constant failure and repair rate, it has been proven [24] that it holds for any distribution $F(t)$ and $G(t)$, given that MTTF is the mean value of $F(t)$ and MTTR is the mean value of $G(t)$.

3 Combinatorial Methods for System Dependability

Assuming that we are able to characterize the failure and repair distribution of a component, we have seen how to predict its availability. But if a system is a complex aggregate of components, it may be difficult to associate directly to the system a failure and repair distribution. As usual in computer science, when a problem is too complex, a divide and conquer technique may lead to viable solution, and this is indeed the approach that we shall discuss next: given a number of independent components, and a well defined way of combining them into a *system configuration*, we shall see how to compute the reliability (availability) of a (repairable) system. We present two techniques: *reliability blocks* [60] and *fault trees* [59].

3.1 Reliability of Non-repairable Systems

In reliability blocks we assume that the system under study is built out of two basic configuration schemes: series configuration and parallel configuration.

A system is the series configuration of two components if the failure of one of them provokes the failure of the whole system. If we assume that the components are statistically independent, and letting $R_1(t), R_2(t), \ldots, R_n(t)$ be the reliability of the n components at time t, then the reliability of the system at time t, $R_s(t)$ is [60]:

$$R_s(t) = R_1(t) \cdot R_2(t) \cdots R_n(t) \tag{17}$$

The reliability of the system is the product of the reliability of the components, and since the $R_i(t)$ are positive values less than 1, it implies that more components we have in a series the less reliability we have.

If the failure rate of the components is constant (and therefore the $R_i(t)$ are exponential distributions of parameter λ_i), we derive from (17):

$$R_s(t) = e^{-\lambda_s t} \qquad \text{with :} \quad \lambda_s = \sum_{i=1}^{n} \lambda_i \qquad \text{and} \qquad MTTF = \frac{1}{\lambda_s} \tag{18}$$

meaning that the whole system fails according to an exponential distribution.

If we want to build a more robust system, we can use the concept of parallel redundancy: the system is built out of n components, and the whole system fails only if all the components fail. Again, assuming that the components are independent, and for the simple case of $n = 2$, we can state that the unreliability $F_s(t)$ can be computed as:

$$F_s(t) = F_1(t) \cdot F_2(t) \tag{19}$$

from which we derive:

$$R_s(t) = R_1(t) + R_2(t) - R_1(t) \cdot R_2(t) \tag{20}$$

Equation (20) implies that the reliability of the system is greater that the maximum reliability of its components. In the case of constant failure rate we get:

$$R_s(t) = e^{-\lambda_1 t} + e^{-\lambda_2 t} - e^{-(\lambda_1+\lambda_2)t} \quad \text{e} \quad MTTF = \frac{1}{\lambda_1} + \frac{1}{\lambda_2} - \frac{1}{\lambda_1 + \lambda_2} \tag{21}$$

The expression for $F_s(t)$ of (19) can be generalized to the n case, leading to:

$$[1 - R_s(t)] = \prod_{i=1}^{n} [1 - R_i(t)] \tag{22}$$

from which $R_s(t)$ can be computed.

The equations above allow the computation of the reliability function of systems that have a complex parallel-series composition scheme, indeed each $R_i(t)$ could be the reliability of a complex subsystem, that can be expressed with a formula that depends on whether the subsystem is the serial or parallel composition of sub-subsystems, and so on.

A special type of parallel redundancy is the so-called k out of n (indicated as $k{:}n$): the system has n redundant components placed in parallel, and the system works fine when at least k out of the n components works properly. An example of such a system can be a redundant water pipeline in which the required water flow is ensured as far as k pipelines are correctly transporting their flows.

The probability that exactly i components out of n work fine, in the hypothesis that all components have the same failure rate R, is given by the binomial distribution:

$$Pr\{i:n\} = \binom{n}{i} R^i (1-R)^{(n-i)} \tag{23}$$

since the system is working in all cases in which i components, with $i > k$, are correctly working, we get:

$$R_{k:n} = \sum_{i=k}^{n} \binom{n}{i} R^i (1-R)^{(n-i)} \tag{24}$$

3.2 Availability of Repairable Systems

Let us now consider the case of a system built out of *repairable* components. Again we shall consider the failure process and the repair process of the components totally independent: this hypothesis is less realistic than in the case of a single component, since it is often the case that there is a limited number of resources allocated for repairs, shared among all components, so that the computed availability may be an upper bond of the real value.

If there are n components configured in series, then the availability of the system A_s is given by:

$$A_s = A_1 \cdot A_2 \cdots A_n$$

where the $A_1, A_2, \ldots, A_n$ and A_s are evaluated for the same time instant t, or at infinity. If the constant rate hypothesis applies, then the first equation in (14) can be used.

Again, the availability of the system is smaller than the availability of the worst component, and therefore to increase the availability of a system it is advisable to act on the worst component.

For what concerns the redundant parallel systems, considering the simplifying hypothesis of two components, we have that the unavailability U_s is given by:

$$U_s = U_1, \cdot U_2 \tag{25}$$

from which we can derive

$$A_s = (1 - U_s) = A_1 + A_2 - A_1 \cdot A_2 \tag{26}$$

showing that the availability of the system is greater than the availability of the most available component.

3.3 Fault Trees

Another approach to the combinatorial study of systems built out of independent components is that of *fault trees* [59]: logical trees for the representation and analysis of the critical conditions whose combined occurrence causes a specific event, called the *Top Event (TE)*, to occur.

When the TE is one particular undesired event, then the analysis of the combination of elementary events that lead to the occurrence of the TE assumes the name of *Fault-tree analysis (FTA)*. Elementary events are of the type: component X is working properly, or component X is not working properly.

The TE is the root of the tree and the construction of the tree is usually "*top / down*" from general to specific. The FTA is particularly suited to the analysis of complex systems comprising several subsystems or components which are connected in various configurations, with a high level of redundancy. FTA is commonly used by reliability engineers dealing with aircraft, space, chemical and nuclear systems, and it is also considered in the IEC-1025 standard [36]. The interested reader can find a full treatment of the topic in [3, 31, 32, 22, 59].

The methodological approach to dependability based on FTA consists of the following steps: definition of the Top Event, construction of the Fault Tree, qualitative analysis, and qualitative analysis.

Definition of the Top Event. TE candidates are events whose occurrence may lead to unsafe operating conditions, catastrophic failure or malfunction, unaccomplishment of the assigned mission, and so on.

If more TE's need to be investigated a different tree for each one the TE's must be generated and analyzed.

Construction of the fault-tree. Once the TE has been defined, the construction of the FT proceeds by identifying the *immediate causes* for the occurrence of the TE, and their logical relationship (for example, whether the immediate causes must occur separately or simultaneously for the TE to occur).

The immediate, necessary and sufficient causes for the TE constitute the first level of the tree. Each immediate cause is now treated as a sub-top event, and the analysis proceeds to determine their immediate causes. In this way, the construction evolves iteratively from events to their causes, continuously approaching finer resolution, until a desired level of detail is reached.

Interactions between causes at each level of the iterative construction are represented by means of logic gates (usually OR and AND gates, but more complex gates can be defined, as, for example, k out of n gates), while the output of the logical gates represents the occurrence of the higher level of the tree. The events at which the construction of the tree is ended are called *terminal events*.

Qualitative Analysis of a FT. The qualitative analysis is aimed at identifying all the combinations of events that cause the top event to occur, as a function of the terminal events. Combinations are ranked according to the number of events, since the smaller the number of events that cause the TE the less resilient to failure it is likely to be our system. The qualitative analysis of an FT consists in deriving a logical expression of the TE, in such a way that all the combinations of events whose simultaneous occurrence provokes the TE are evidenced. A combination of events whose simultaneous occurrence provokes the TE is called a *cut set* (CS) for the system. A CS that does not contain any subset which is again a CS , is minimal and is called a *minimal cut set* (MCS) or *mincut.*

Definition. *A* CS *is a set of terminal events whose simultaneous occurrence forces the occurrence of the* TE. *A* CS *is an* MCS *if it does not contain any subset of terminal events that is still a* CS.

Suppose the FT has m MCS denoted by $K_1, K_2, \ldots, K_m$. According to the above definition, the occurrence of any K_i $(i = 1, 2, \ldots, m)$ implies the occurrence of the TE, hence:

$$TE = K_1\, or\ K_2\, or\ \ldots or\ K_m = OR_{i=1}^{m}\, K_i \tag{27}$$

The list of all the MCS provides a very valuable information to the analyst since it provides all the minimal sets of failure events that can provoke the TE to occur, that is the system failure event, and allows the analysts to identify the potential weak points of the system and to initiate corrective actions.

The determination of the CS proceeds iteratively in a top down fashion, starting from the TE and applying the rules of the logic algebra, guided by the gate typology, until all the terminal nodes are reached. If the FT does not contain repeated events, the above search directly provides the MCS, otherwise if the FT contains repeated events the list of the MCS must be further extracted from the obtained CS.

Quantitative Analysis. The quantitative analysis has the objective of evaluating the probability of occurrence of the TE, of the MCS and of any other intermediate event of the FT in terms of the probability of occurrence of the basic events. FTA assumes that the failures of the basic components are statistically independent. According to this assumption, the properties of the FT are completely specified if an individual probability is assigned to each single basic event.

Let A_i be a terminal event and denote $Q_i = P(\overline{A}_i)$, where $\overline{A}_i$ stays for "component A_i not working". To compute the probability values in either cases a mission time must be fixed. Denote the mission time T_M, hence

$$Q_i = P(\overline{A}_i, T_M) \tag{28}$$

If a component is non-repairable, then Q_i is the component unreliability computed at time T_M, if it is repairable Q_i is the component unavailability computed at time T_M.

Many FT tools accept as input parameter for each basic even only a constant failure and a constant repair rate, thus implicitly assuming that the failure and repair times of each component are exponentially distributed and restricting the analysis to this case, only. By denoting with λ_i and μ_i, respectively, the constant failure and a repair rate of component A_i, formula (28) becomes [3, 64]:

$$Q_i = \frac{\lambda_i}{\lambda_i + \mu_i}\left(1 - e^{-(\lambda_i+\mu_i)T_M}\right) \tag{29}$$

from which the value for a non-repairable component (the usual unreliability expression for a component with constant failure rate) is obtained by setting $\mu_i = 0$.

Probability of the TE. Since the TE can always be expressed in disjunctive normal form in terms of the MCS, then:

$$P(TE) = P(K_1 + K_2 +, \ldots, K_m)$$

where each CS K is the AND of a number (called the CSorder) of terminal events, then

$$K = \mathcal{A}_1 \, \mathcal{A}_2 \cdots \mathcal{A}_\ell$$

and

$$Pr(K) = Pr(\mathcal{A}_1) Pr(\mathcal{A}_2) \cdots Pr(\mathcal{A}_\ell) \tag{30}$$

If there are m MCS's $(K_1, K_2, \cdots, K_m)$, since the occurrence of a single MCS implies the TE, then the probability of TE is given by

$$Pr(TE) = Pr(K_1 + K_2 + \cdots + K_m) \tag{31}$$

Recall that, if A and B are two events, then

$$Pr(\mathcal{A} \, or \, \mathcal{B}) = Pr(\mathcal{A}) + Pr(\mathcal{B}) - Pr(\mathcal{A} \, \mathcal{B}) \tag{32}$$

and that the OR of m events requires an expansion into $(2^m - 1)$ terms involving the computation of the probability of the AND of groups of j events, $(j = 1, 2, \ldots, m)$.

The computation of the probability of the TE can therefore be quite complex, but we should consider that the probability of the single events are in general quite low (they are failure probabilities), and that all terms that compute the product of a significant number of basic events can be quite low. It is therefore a widely accepted practice to truncate the computation, especially considering that upper and lower bounds on $Pr(TE)$ can be computed using the probability of the single CS's and the probability of the AND of pair of CS's, as follows.

$$Pr(TE) \leq \sum_i Pr(K_i) \tag{33}$$

$$Pr(TE) \geq \sum_i Pr(K_i) - \sum_{i>j} Pr(K_i \, K_j) \tag{34}$$

The computation of the probability of occurrence of the TE and of the MCS is, usually, the main concern of a FTA. However, several other useful measures can be defined and evaluated, like the expected number of failed components, the main failure equivalence, and the Mean Time To Failure.

FTA is a widespread practice for the availability analysis of systems, and there are indeed a number of tools that support it, among them we cite [55, 62].

4 State Enumeration Techniques

The work of the previous section is here extended to consider systems of independent components but with arbitrary configurations, and systems in which the independence assumption does not hold. Both techniques are based on the idea of enumerating all the possible states of the system, and to classify them as "good" or "bad". The probability of being in a state (at time t or in steady-state) is then computed, and the reliability of the system is then obtained summing up the probability of all the good states.

If the components are independent we can provide a closed form expression for the probability of each state, thus adopting a combinatorial approach, as for reliability blocks, while if dependencies among components are to be taken into account, more complex quantitative analysis techniques need to be considered, based on the solution of the associated stochastic process.

In this section we shall first consider the problem of state enumeration, to then separately discuss the quantitative analysis for the independent case and for the dependent one, limited to the simpler case in which the associated stochastic process is a Markov chain.

Consider a system with n components and any configuration among them. The usual hypothesis is to assume that each single component can be represented by two mutually exclusive conditions or states referred to as *working* (or *Up*) and *failed* (or *Down*), identified by the state indicator variable x_i associated to the i-th component, with the following encoding:

$$x_i = \begin{cases} 1 & \text{component } i \text{ Up} \\ 0 & \text{component } i \text{ Down} \end{cases}$$

The state of the system is identified by the vector $\mathbf{x} = (x_1, x_2, \ldots, x_n)$ [3]. The state space of the system Ω is the set of all the possible values of $\mathbf{x}$, i.e. the set of all the possible combinations of the n components being working or failed, leading to $N = |\Omega| = 2^n$.

The state space, that we shall call RG for similarity with Petri nets terminology, is a labeled directed graph whose nodes are the states of the system and each edge represents the transition between states due to failure or repair. If we assume that no multiple failures or repairs can take place at the same time, which is indeed the case for independent components working in continuous time, then there is a direct arc labelled i between state $\mathbf{x}$ and $\mathbf{x}'$ only if the two states differ only in the value of variable x_i.

We assume that the system as a whole can be classified according to a binary behavior: *working* or *failed*. Hence, we introduce a binary indicator variable y for the system [3,41]:

$$y = \begin{cases} 1 & \text{system is in a working state} \\ 0 & \text{system is in a failed state} \end{cases} \tag{35}$$

The value of y is a function of the state, and we can define $y = \varphi(\mathbf{x})$, or $y = \varphi(x_1, x_2, \ldots, x_n)$

The state space Ω can be partitioned in two exhaustive and mutually exclusive subsets Ω_u and Ω_d.

$$\Omega_u = \{\Omega : \varphi(\mathbf{x}) = 1\} \qquad ; \qquad \Omega_d = \{\Omega : \varphi(\mathbf{x}) = 0\}$$

Let $N_u = |\Omega_u|$ be the cardinality of Ω_u, and $N_d = |\Omega_d|$ the cardinality of Ω_d, then

$$\Omega = \Omega_u \cup \Omega_d \qquad ; \qquad \Omega_u \cap \Omega_d = 0 \qquad ; \qquad N = N_u + N_d \tag{36}$$

Observe that the system configuration is totally identified by the function $y = \varphi(\mathbf{x})$. For example, in the 2-parallel connection configuration the only system failed state is $\mathbf{x} = (0,0)$, that is to say, both components have to be down for the whole system to fail.

The failure process defined on the state space. The evolution of the system in time can be represented by means of the succession of states passed through by the system due to failure or repair events of its components.

Denoting by $Z(t)$ the function of the time that represents the state occupied by the system at time t. For any value of t, $Z(t)$ is a random variable that assumes non negative values in the states of Ω. The probability that the system is in state $\mathbf{x}$ at time t is denoted by $p_{\mathbf{x}}(t)$ and is defined as:

$$p_{\mathbf{x}}(t) = Pr\{Z(t) = \mathbf{x}\} \tag{37}$$

under the normalization condition:

$$\sum_{\mathbf{x}\in\Omega} p_{\mathbf{x}}(t) = 1 \qquad \text{for any} \quad t \geq 0\,.$$

On the failure process the following measures con be defined:
Reliability: since there are many states (all whose in Ω_u) in which the system is considered as working properly, then the Reliability of the system is obtained summing up the probability of each of these states, leading to:

$$R_S(t) = \sum_{\mathbf{x}\in\Omega_u} p_{\mathbf{x}}(t) \tag{38}$$

Availability:

$$A(t) = \sum_{\mathbf{x}\in\Omega_u} p_{\mathbf{x}}(t) \tag{39}$$

Mean sojourn time spent in a state up to time t:

$$\ell_{\mathbf{x}}(t) = \int_0^t p_{\mathbf{x}}(z)\, dz \tag{40}$$

System MTTF

$$MTTF = \int_0^\infty R_S(z)\, dz = \sum_{\mathbf{x}\in\Omega_u} \int_0^\infty p_{\mathbf{x}}(z)\, dz = \lim_{t\to\infty} \sum_{\mathbf{x}\in\Omega_u} \ell_i(t)$$

Average interval availability:

$$A_I(t) = \frac{1}{t} \sum_{\mathbf{x}\in\Omega_b} \ell_i(t)$$

$Z(t)$ is a stochastic process defined over the discrete state space Ω and with continuous time parameter t. The quantitative evaluation of the state probabilities expressed by equation (37), completely determines the stochastic process $Z(t)$ and, hence, the behavior of the system. If the components are statistically independent, evaluation of expression (37) can be performed by resorting to combinatorial formulas, presented in the following subsection, while the case of statistically dependent components will be treated next.

4.1 Independent Components

If the components of a system are statistically independent, the probability $p_{\mathbf{x}}(t)$ of being in a generic state $\mathbf{x}$ with characteristic vector $\mathbf{x} = (x_1, x_2, \ldots, x_n)$ at time t can be expressed as the product of the probabilities of the individual variables:

$$p_{\mathbf{x}}(t) = Pr\{(x_1(t)\} \cdot Pr\{(x_2(t)\} \cdots Pr\{(x_n(t)\} \tag{41}$$

where, thanks to the independent component assumption, each term $Pr\{(x_i(t)\}$ is given by:

$$\begin{cases} Pr\{x_i(t) = 1\} = R_i(t) \\ Pr\{x_i(t) = 0\} = 1 - R_i(t) \end{cases} \tag{42}$$

where $R_i(t)$ is the probability that component i is in working condition at time t, and coincides with the reliability of component i in case of non-repairable components or with the availability of component i in case of repairable components.

In the usual case in which the time to failure distribution of each individual component is considered exponentially distributed with failure rate λ_i, equation (42) takes the form:

$$Pr\{x_i(t)\} = \begin{cases} R_i(t) = e^{-\lambda_i t} & \text{if} \quad x_i(t) = 1 \\ 1 - R_i(t) = 1 - e^{-\lambda_i t} & \text{if} \quad x_i(t) = 0 \end{cases} \tag{43}$$

4.2 Markovian Methods for Dependent Components

The analysis above relies on the hypothesis that the components are independent, but this is not always the case. In the previous section we have shown how state enumeration techniques can be used to cheaply compute the reliability of a system built out of independent components. The hypothesis of independence is not always reasonable, for example the failure of a component may induce a larger load on the remaining components, thus increasing their failure rate, or two or more components have a common cause of failure (for example a computer and a video can be seen as independent, but if they use the same source of power their failure are not independent). If the component are statistically dependent, i.e. the failure or repair process of any one of them is dependent on the state of the other(s), more sophisticated techniques are necessary, able to incorporate the conditional dependencies of each component with respect to the state of the other ones.

Consider a system with two components whose state space is the Cartesian product of the state spaces of the components and is depicted in Figure 1; if $\lambda_1 \neq \lambda_1'$ then the failure rate of the first component depends on the state of the second component (for example the failure rate of the first component increases if the second component is not working properly). This dependency does not allow to compute the reliability of the system in the simple product form of Equation (41), and we need to solve the associated stochastic process. Continuous Time Markov Chains (CTMC) are stochastic process with a good tradeoff between expressiveness and solution cost.

A very large literature exists on the topic, the interested reader may refer, for example, to [51, 25, 42, 64]. Reliability analysis through CTMC is also dealt with in the international standard IEC1165 [37].

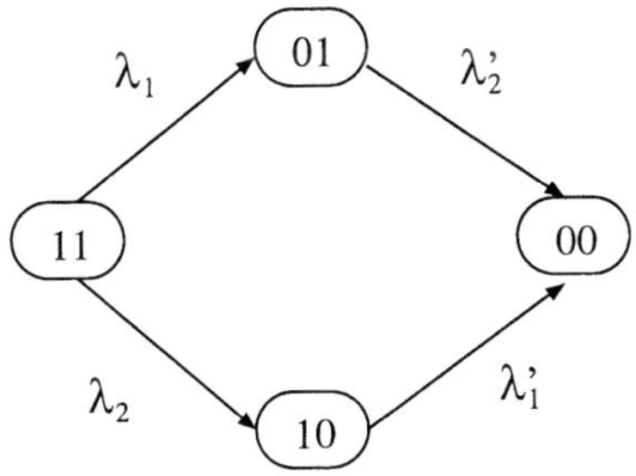

Fig. 1. The state space of a two components system

Let $Z(t)$ be a stochastic process defined over the discrete state space Ω. $Z(t)$ is a Continuous Time Markov Chain (CTMC) [51,64] if, given any ordered sequence of time instants $(0 < t_1 < t_2 < \ldots < t_m)$, the probability of being in state $\mathbf{x}^{(m)}$ at time t_m depends only on the state occupied by the system at the previous instant of time t_{m-1} and not on the complete sequence of state occupancies. This property, that is usually referred to as the *Markov property*, can be rephrased by saying that the future evolution of the process only depends on the present state and not on the past. More formally the Markov property may be written as:

$$P\{Z(t_m) = \mathbf{x}^{(m)} | Z(t_{m-1}) = \mathbf{x}^{(m-1)}, , \ldots, Z(t_1) = \mathbf{x}^{(1)}\}$$
$$= P\{Z(t_m) = \mathbf{x}^{(m)} | Z(t_{m-1}) = \mathbf{x}^{(m-1)}\} \tag{44}$$

If we number the states from 1 to N, for example taking the lessicographical order, then we can define the *transition probability matrix* of the process:

$$\mathbf{P}(u,v) = [p_{ij}(u,v)]$$

of dimension $(N \times N)$ whose entries are the transition probabilities $p_{ij}(u,v)$ defined as:

$$p_{ij}(u,v) = P\{Z(v) = j | Z(u) = i\} \qquad (u \leq v) \tag{45}$$

$p_{ij}(u,v)$ represents the probability that the Markov chain $Z(t)$ is in state j at time v, given it was in state i at time u, and it is called the *transition probability* between state i and j. For the transition probabilities $p_{ij}(u,v)$, the following initial conditions hold:

$$p_{ii}(v,v) = 1 \qquad ; \qquad p_{ij}(v,v) = 0 \tag{46}$$

Further, let $p_i(t)$ be the (unconditional) probability that the system is in state i at time t. $p_i(t)$ is the state occupancy probability, or simply the *state probability*, and is defined as:

$$p_i(t) = Pr\{Z(t) = i\} \tag{47}$$

and $\mathbf{p}(t) = [p_i(t)]$ denotes the row vector of dimension $(1 \times N)$ whose entries are the state probabilities $p_i(t)$ defined in (47). $\mathbf{p}(t)$ is called the *state probability vector* of the process.

If the Markov process is homogeneous[1] we can derive the following equation:

$$\frac{d\mathbf{p}(t)}{dt} = \mathbf{p}(t)\cdot\mathbf{Q} \qquad \text{with initial condition} \qquad \mathbf{p}(0) = \mathbf{p_0} \tag{48}$$

where $\mathbf{p_0}$ is a probability vector describing the initial conditions, and $\mathbf{Q}$ is a matrix, called infinitesimal generator, whose elements q_{ij} are the conditional probabilities of jumping in state j in a small interval Δt, given that the CTMC was in state i at the beginning of the interval. The quantities q_{ij} are called the transition rates of the process, and they are a simple function of the p_{ij} values.

Equation (48) is the fundamental equation for CTMC: it consists of a set of N first order differential equations with constant coefficients, that provide the state occupancy probabilities at time t, from which the required performance models can be computed. Various analytical and numerical solution techniques are available for the fundamental CTMC Equation ([61]).

If the Markov chain is irreducible, that is to say each state is reachable from any other state, then the limit

$$\lim_{t\to\infty} p_i(t) = \pi_i$$

always exists and it is independent of the initial state; equation 48, when t goes to infinity, simplifies to:

$$\pi\cdot\mathbf{Q} = \mathbf{0} \qquad \text{with} \qquad \sum_{i=1}^{N} \pi_i = 1 \tag{49}$$

where the normalizing condition on the right is necessary to impose that the solution is a probability vector. Equation (49) is a linear set of homogeneous equations, and can be solved with numerical solution techniques [61].

Observe that, if the components are non-repairable, then the associated MC will never be irreducible (from a state in which there is a failed component is never possible to come back to the initial state in which all components are working properly), and therefore the only reasonable measures to be computed are the probabilities at time t, and derived quantities.

From what concerns complexity, we can observe that most techniques used for the solution "at time t" or in steady state are based on iterative methods: at each iteration the most expensive operation is a vector-matrix multiplication (vector of size equal to the number of states and matrix with a number of non-null elements equal to the number of arcs in the RG). The iteration procedure is stopped when a certain (estimated) convergence towards the actual solution is reached.

The number of iteration is usually the major factor affecting the solution cost.

5 Dependability Modeling Using Petri Nets

Specifying systems at the state space level can be an error-prone, low level activity. Petri nets have been widely recognized in the literature as an effective way to specify systems

[1] A Markov process is said to be homogeneous when the transition probabilities in matrix $\mathbf{P}(u,v)$ depend only on the length time interval $(v-u)$ and not on the values of the time instants v and u.

using a reasonably high-level formalism, while at the same time having a precise operational semantics that allows the derivation of the associated state-space. In particular the class of Stochastic Petri Nets [47, 2] (SPN) has a semantics defined through Markov chains, so they are considered a natural language to use when the stochastic process underlying the system is a Markov chain. With the term SPN we actually indicate the more general class of Petri nets with stochastic delays associated to transitions, that include various extensions like ESPN [28], GSPN [1], Stochastic Reward nets [48], and colored/parametric extensions like Stochastic Activity Networks [57], and Stochastic Well-formed Nets [18].

Indeed SPN have been widely used not only for the study of generic performance indices, but also specifically for dependability studies, as testified by the available SPN tools that allow the computation of some pre-defined dependability quantities, like SURF-2 [30], UltraSAN [23], and SPNP [20]. Among the initial works on the use of SPNs for dependability modeling and analysis we can mention [33, 35, 48, 58, 39, 46, 14]. In [29] Extended SPNs are used for carrying out sensitivity analysis of the system reliability and availability when the error coverage probability varies. The work [49] presents an overview of different classes of non Markovian Petri Nets used for dependability analysis. Net-compositionality has been adopted in [40, 56, 16, 11] to cope with the complexity of dependability modeling. Concerning works on the translation of Fault Trees into SPNs we mention the works [34, 45, 15]. In [34] and in [45] the translation into Petri Nets allows to model the dependences among system components, while in [15] SWNs are used as a target formalism of the translation to exploit the symmetries of the parametric fault trees.
But what do we gain from the use of Petri nets?

- An high level language to describe the system.
- The possibility of reusing the tools and the solution methods available for the SPN tools, including the possibility of validating qualitative properties (like liveness or deadlock freeness) that can be an important issue when the system being modeled as a complex behavior.

What do we loose by using SPN? The solution associated to an SPN is usually produced solving the associated CTMC, but the CTMC is, in principle, a "flat" structure, in which all the information on independency between components is basically lost, thus forcing the solution of the whole Markov chain with numerical methods, even in whose cases, like that of independent components, in which the solution is just the product of the solution of the components.

The section is organized as follows: we shall first show a very simple model, equivalent to a 2-parallel configuration, and then a slightly more complex case, in which a model of a system is modified to include the presence of faults. We conclude the section with the presentation of a (simplified version) of a case study of a "dependability mechanisms".

5.1 Simple PN Models of Dependability

The very simple SPN model of Figure 2 represents a system with two independent components. Each SPN is a simple sequence place - transition - place, where the tran-

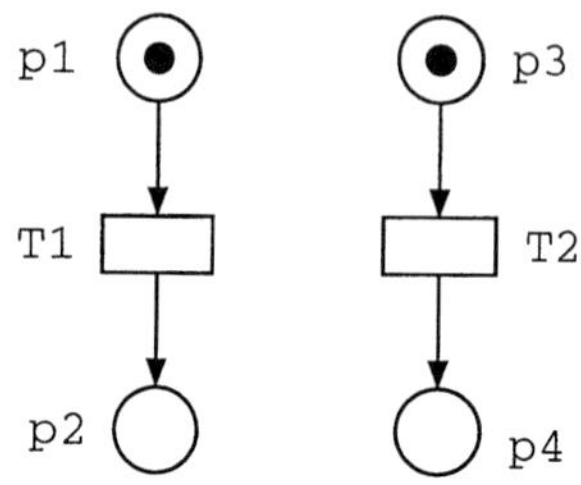

Fig. 2. A simple SPN model of a two component system

sition represents the failure. The Markov chain of this SPN is exactly that of Figure 1 if $\lambda_i = \lambda'_i$ and it is equal to the rate of transition T_i. If $\lambda_i \neq \lambda'_i$ then it is necessary to resort to marking dependent rates.

How do we compute the reliability/unreliability of the system? Again, as for CTMC based approaches, we can sum up the probability of the Up and Down states, where the Up and Down states depend on the system configuration. But how can we specify a system configuration? There are two possible approaches: an implicit one, in which it is the definition of the measure that encodes the configuration, and an explicit one, in which the configuration is reported in the net.

If we assume the implicit approach, and we want to express a series configuration, then to compute the unreliability of the system we have to sum up the probability of all states in which either place p_2 is marked, or place p_4 is marked, or both. For the parallel configuration we sum over a single state: that with a token each in places p_2 and p_4.

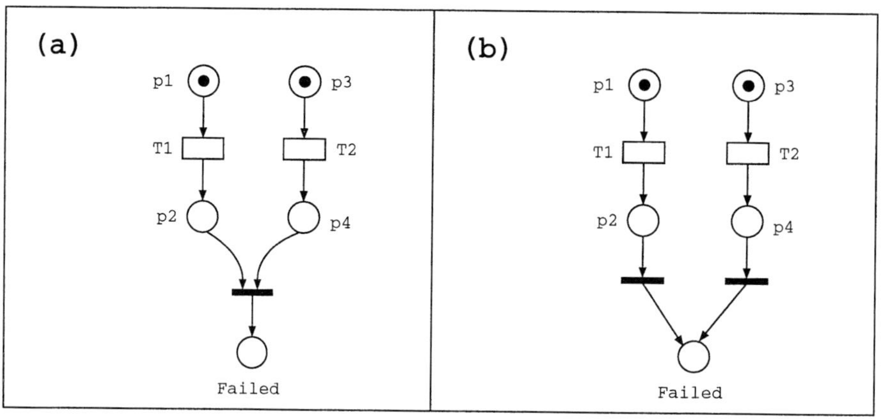

Fig. 3. Explicit modeling of the failure state

If we want to make the configuration explicit we can add a place *Failed*, as it is done in Figure 3. The place is connected to places p_2 and p_4 in different manners, to reflect the parallel configuration in Figure 3(a), and the series one in Figure 3(b).

The simple SPN model of Figure 2 can be modified so as to take into account repairs, leading to the SPN model of Figure 4 where one repair transition per component (named T_3 and T_4) has been added. Again, we can define the configuration in an implicit manner through the definition of the availability/unavailability metrics, or we can explicitly define the configuration in the model. Indeed the modification of the net is not as simple as in the unrepairable components case, since we need to put a token in place *Failed* without removing the tokens from the places P_2 and P_4, moreover we need to model also the fact that, in consequence of a component repair, the whole system can be working again, and this may not be trivial.

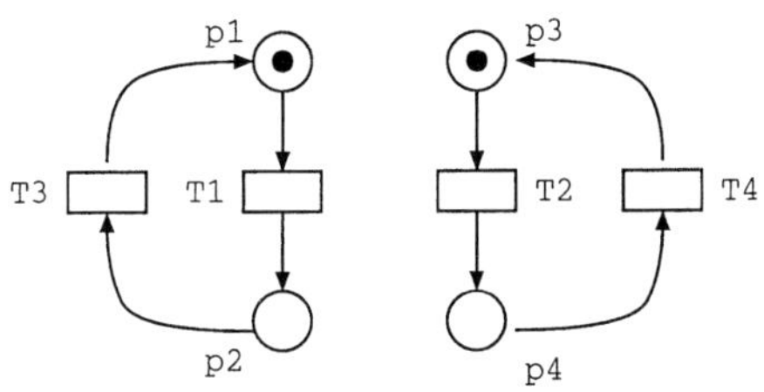

Fig. 4. Modeling of repair actions

The approach based on reliability blocks can be translated into SPN with a limited effort, nevertheless since the SPN solution requires the solution of the Markov chain, then this translation makes sense only if we need to insert dependencies between the components (for example a simple failure rate dependency).

A similar argument holds true also for fault trees. There has been a number of translation defined from fault trees of various flavors to various flavors of SPN, always with the objective of being able to include dependencies [33, 45], or to exploit symmetries of the fault tree also in the solution process [15].

In the previous SPN models we have assumed that a component has only two states, Up and Down, represented as distinct places, but a model can be a much more complicated net. In the next section we shall introduce a simplified version of the model of a piece of software that provides a sort of parallel redundancy. Although being a simplified version of the real code, it has indeed a more articulated structure than the two-states approach discussed in these SPN introductory examples, moreover it is a good example of the problem related to the modeling of the restart of normal operation after a failure.

5.2 A More Complex PN Model of Dependability: The Local Voter

The Local Voter mechanism (LV) presented in this sub-section is a simplified version of a fault-tolerance mechanism designed and implemented within the TIRAN EEC project [17] and studied in [10]. A fault-tolerance mechanism is basically a piece of code aimed at improving the reliability of a complex software system. LV aims at masking occurrences of faults during the execution of the code of an application process.

Fault masking is achieved by the adoption of a spatial redundancy (as in the $k : n$ case that we have seen for reliability blocks) of the execution of the piece of code and by the voting on the results coming from the replicas. Depending on the voting technique adopted in the LV and on the spatial redundancy, a limited number of faults may be masked; for instance, by using a majority voting algorithm and by running concurrently K copies, up to $[\frac{K-1}{2}]$ faults can be made transparent for an application process.

In [10] the purpose of the modeling activities was to evaluate the "goodness"of the mechanism both from a qualitative (i.e., correctness with respect to the design specification) and from a quantitative point of view (i.e., performance and dependability). In particular, concerning the quantitative analysis of the LV two important issues were addressed: first, the amount of overhead induced by the use of the mechanism with respect to not to use it in the application execution; second, the probability of voting success.

Observe that the first measure is not a dependability measure in a strict sense, but it is aimed at evaluating the cost of "dealing with faults", in particular the cost of masking them whenever possible. The second measure can be considered instead a reliability measure, since we can consider as the "assigned mission" of our system the ability to vote on an agreed value.

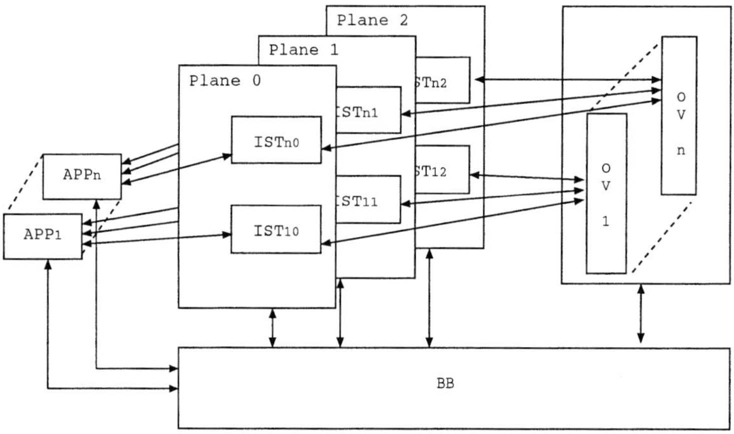

Fig. 5. Representation of the local voter mechanism

Figure 5 shows a graphical representation of the simplified LV; the LV can be used concurrently by several application processes and three replicas are considered per application.

The replicas are executed on separate "planes", that naturally correspond to separate processing nodes. The application process APP_i that uses the LV mechanism is split in two parts, a part that does not require a replicated execution, and a part that requires it. If there are n applications that use LV, then each application has its distinct piece of code to be executed.

The three replicas of an application i, called IST_{i0}, IST_{i1} and IST_{i2} in the figure, receive the same input data from the application.

When a replica IST_{ij} ends its computation, it sends its output data to the appropriate voting OV_i; there is one voting task per application.

The components of the local voter interact with the backbone BB, which is a sort of run-time support for the TIRAN library of mechanisms which handles all exceptions as well as the recovery actions. All interactions among tasks are based on communication through mailbox.

Table 1 lists the acronyms used for the different tasks, and for each task lists how many copies of that task there are in a LV that serves n applications.

Table 1. Acronyms

Acr.	description	no. of copies
APP	application	n
IST	replicated software to vote upon	$3*n$
OV	output voter	n
BB	backbone	1

Each OV_i is characterized by a timer which is set and starts to count-down for expiration as soon as OV_i receives the first output from one of the replicas of the corresponding application APP_i. If either all the three replicas of APP_i or two of them are received before the time-out expiration then the timer is disabled and a voting on the available replicas is carried out. In any case, OV_i will send a message to the BB to notify the voting outcome on the available replicas and, if it is the case, the time-out expiration. The BB is in charge of notifying the termination of the elaboration to the application and of restarting the system in case of a time-out occurrence.

The SWN Model of the Local Voter. The following assumptions were made to model LV: tasks communicate in an asynchronous manner via mailboxes, and there is one mailbox for each ordered pair of tasks; time required to prepare a message is in general negligible, while the time to actually transmit it from the task output buffer to the recipient mailbox is not. With respect to the graphical representation, we have used cross-lined places to emphasize mailboxes and shadowed boxes to delimit portions of the nets that correspond to "recovery actions", and that will be explained in a second step. Moreover, we have adopted the SWN syntax of the GreatSPN [5] tool: net objects, i.e., places and transitions, are denoted as *name*|*label* where *name* is the name of the object and *label* is the label, τ labels are omitted. Labels are used for net composition.

Colors have been used to identify applications, planes and to distinguish the output value as "termination with normal operation" or "termination under abnormal conditions". Three color classes have been defined:

AP is the color class of applications that can request a replicated execution of their code, and it is defined as $AP = \{ap1, .., apn\}$;

P is the color of the planes, and there are always three planes, therefore $P = \{pl1, pl2, pl3\}$;

Exc is the color used to distinguish the positive or negative outcome of a LV activity, and it is built out of two static subclasses $Exc = Ecx1 \cup Ecx2$, where $Exc1 = \{e1\}$ means that there has been a time-out expiration, while $Exc2 = \{e2\}$ means that there was no time-out expiration.

Since the system is specified compositionally, it is a very natural choice to model each component of Figure 5 as an isolated SWN, to then compose them. This approach simplifies the model construction and allows model reuse, but it might make more complex the modeling of whose activities that require the knowledge of the global state, as for the restart activity after a failure.

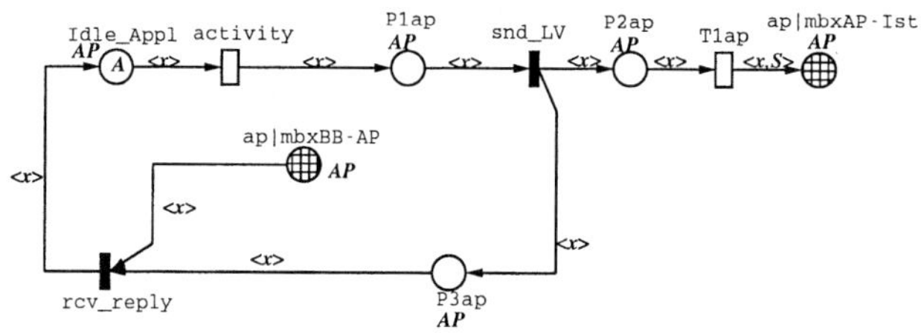

Fig. 6. The application model

Figure 6 shows the SWN model of the APP_i, that cyclically execute their own activity, broadcast the input to their replicas (tasks IST_{ij}), and wait for a message of termination of elaboration coming from the backbone BB.

Figure 7 shows the SWN model of a copy of the code to be executed on the different planes: it is assumed that all replicas are activated at the beginning and then suspend themselves waiting for a message from the APP tasks. There are $|AP| \times |P|$ replicas, i.e., one for each application and for each plane. Each replica (x,y) waits for the input message (x,y) from the application x. When such message is received the replica of application x on plane y starts its activity, modeled by timed transition comp, and then sends the result of the computation to OV.

Figure 8 shows the SWN model of the output voter OV: there is an OV for each application that can use LV.

Each OV executes the voting algorithm (majority voting 2 out of 3) on replicas of the same application, independently from the others. OV waits for the replicas outcome from the three different planes. As soon as the first outcome is received, a timeout for reception of the other two replicas outcome is set (transition setTOforx). Then three situations may occur:

C1 all the three outcomes are received before the time-out expiration (transition recv3noTO fires) and voting on the three outcomes takes place;

C2 the time-out has expired and two of the three outcomes have been received (transition recv2&TO fires), and a vote on the two replicas takes place;

C3 the time-out has expired and only one of the three outcomes has been received by OV (transition recv1&TO fires).

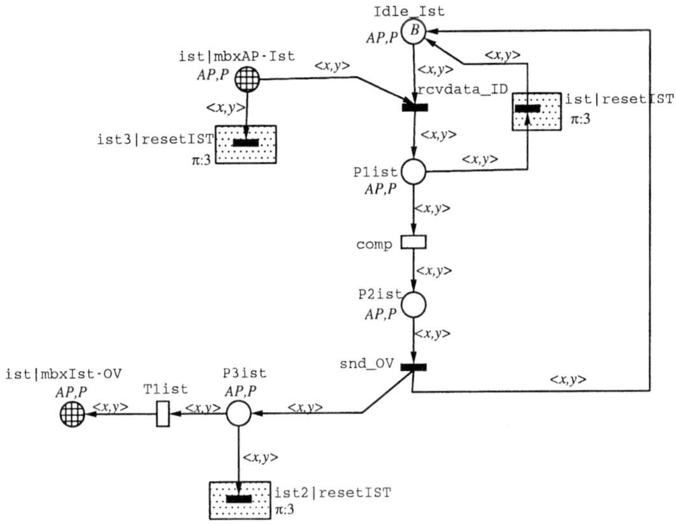

Fig. 7. The model of the replicated code

Under condition C1 an exception message of type $e2$ (no time-out has occurred) is sent to the the backbone BB; in cases C2 and C3 a message of exception of type $e1$ (a time-out has occurred) is sent to BB. Observe that we are not passing on to the backbone the information on whether the vote was successful or not, although this will be a trivial extension, since the success or failure of the 2-out-of-3 algorithm is modeled in detail in the SWN of Figure 8.

When the message is sent to BB, OV waits for an acknowledge from BB to return back into its idle state. Observe that we are assuming that no direct answer goes back directly from OV to APP, not even in the case of a "normal" 3-out-of-3 voting, since we impose that all restarted are caused by BB.

Figure 9 shows the SWN model of the BB task, or, more precisely, of that part of BB devoted to interactions with LV. BB is in an idle state until it receives an exception message coming from OV. If the exception is of type $e2$, i.e., no time-out has occurred, then BB sends an acknowledge to OV and to the application. If instead the exception is of type $e1$, then a time-out has occurred, and therefore a reset operation is needed, before sending back the messages to OV and to APP.

Local Voter without Recovery Actions: An Open Model. A first analysis was performed for the case of a "single run" for each application. In order to obtain the complete model the single nets have to be composed using the program *algebra* [10], a program associated to the GreatSPN tool that allows superposition of nets over places and transitions. The nets used are the one without shadowed portion and, since in the non-shadowed portion no message is passed from OV to BB, each application is executed only once. The resulting SWN net has been solved, for the single application case, using the reachability graph construction of GreatSPN, that produces 68 tangible states.

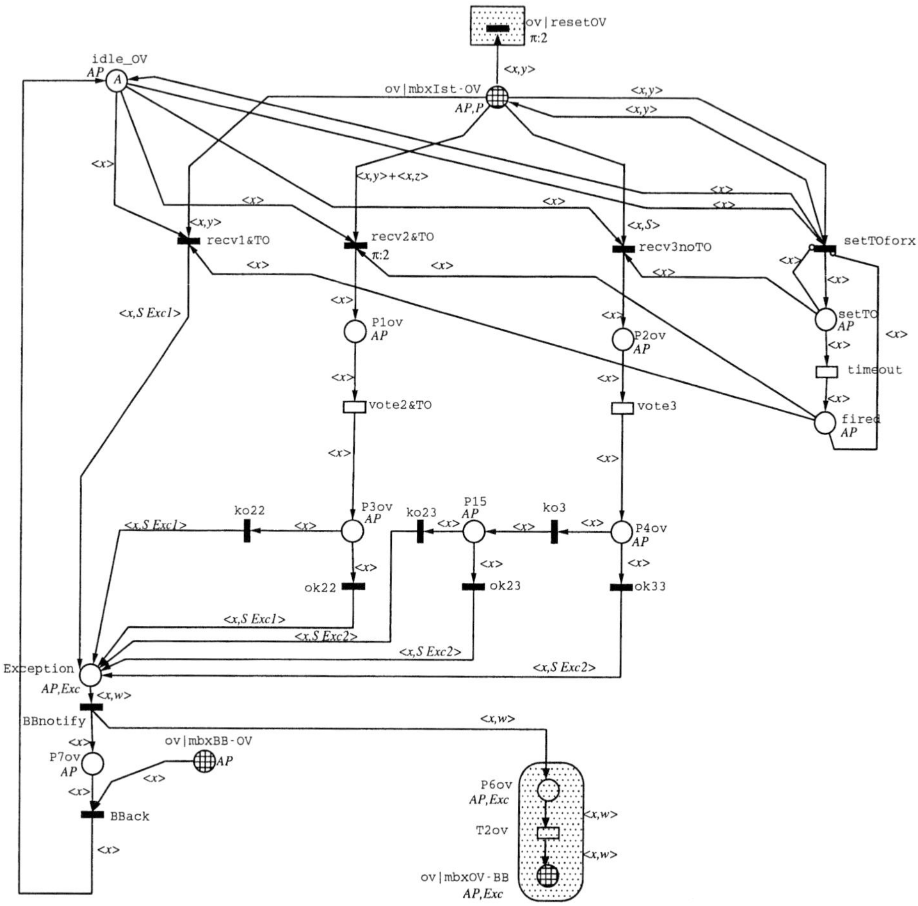

Fig. 8. The model of the output voter

There are 7 dead markings. Three of them correspond to the case of time-out expiration after OV receives the results coming from one replica and is waiting for the results to be sent by the other two replicas. The three markings differ from the identity of the replica that has sent the results to the OV before the time-out expiration. All components, except OV and APP, are in their initial states (idle state), APP and OV are both waiting for a message from BB, that will, of course, never arrives. Three deadlocks correspond to the case of time-out expiration after the results coming from any two replicas have been received. The last deadlock represents the case of reception of all the three replicas before the time-out occurrence. The qualitative behavior was judged correct by the system designer, and the modeling activity could move on to the subsequent step.

Local Voter and Recovery Actions: An Ergodic Model In a second step the recovery actions due to the time-out expiration have been added to the model. The recovery action taken by BB is:

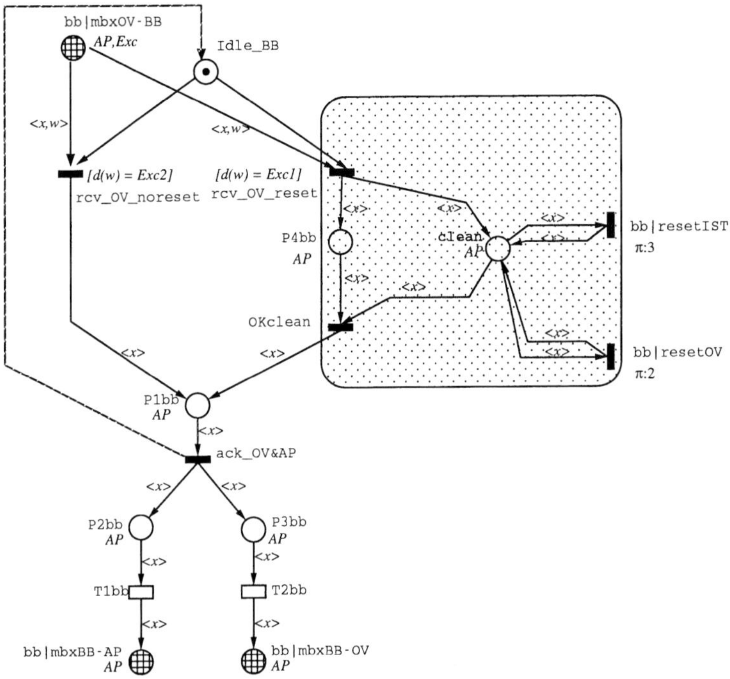

Fig. 9. The model of the backbone

- to remove messages from mailboxes that refer to the application that has caused the exception;
- to take the corresponding tasks back to the their initial states.

To accomplish this BB enables a number of immediate transitions, one per model component, and they are labelled in such a way as to superpose with the resetting transitions in the model components. Observe that these transitions are assigned a different priority, mainly to avoid the generation of useless inter leavings of immediate transitions, that could significantly slow down the state space generation.

The model is obtained by composing all nets, including also the shadowed portions. The resulting SWN is ergodic (since there is a single strongly connected component in the reachability graph and only exponential and immediate transitions are present).

The reachability graph for the single application case has 106 tangible states and the initial marking is a home state. The generation takes a few seconds on a 256Mbyte Pentium 4 machine.

Local Voter without Recovery Actions and Explicit Faults In the models considered up to now no fault is explicitly included in the model, so that a time-out can expire only due to a delay in the completion of one of the replicas. In order to consider the effect of explicit faults the model of IST has been modified to include a timing transition that models the fault and that takes IST into an error state place: the modified model is depicted in Figure 10.

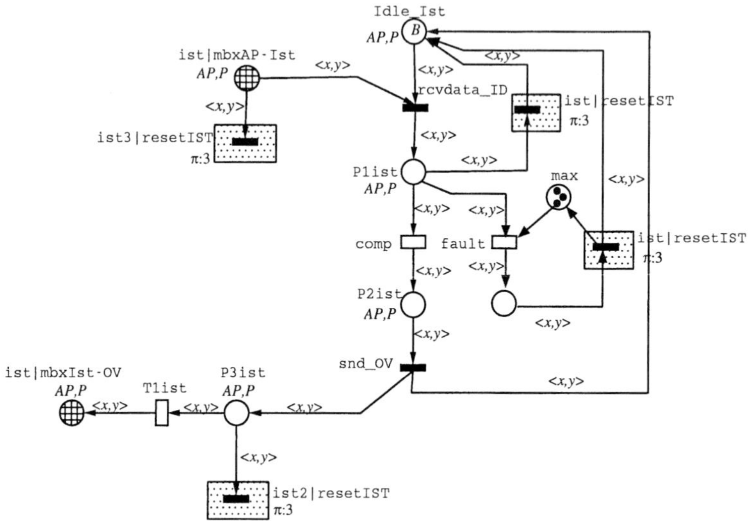

Fig. 10. The modified model of the replicated code

The model assumes that:

- only the replicas IST_{ij} can be affected by a fault and only during their computation phase;
- faults are independent.

The resulting model, for the single application case, has 119 tangible markings and there are 20 dead markings. Among them a very interesting one is the marking that represents the state of the model where all the replicas are in an error state, and this corresponds to a case in which no replica will ever reach OV, so no time-out will be set. This case was, up to the modeling phase, overlooked by the specification document and it is an example of use of Petri Nets for the correctness analysis of the mechanism.

Quantitative Results. The mechanism overhead may be analyzed using the ergodic model with a single application and the time-out deactivated, and computing the mean time to execute a computation through the local voter (inverse of the throughput of transition activity of Figure 6) divided by the mean time spent by a single replica to perform the operation (inverse of the throughput of transition comp of Figure 7).

The probability of different voting outcomes for one application cycle is given by the relative throughput of transitions recv1&TO,recv2&TO, recv3&noTO of Figure 8. The value of the metric depends on the length of the time-out as well as on the probability of matching of the results produced by the replicas. Observe that, not having explicit faults in the ergodic model, the time-out can expire only due to excessive delays of the computations on the planes and/or of the communications between the model components. This implies that, assuming that the results produced by the replicas always match, the probability of 3-out-of-3 voting converges to one as the length of the time-out goes to infinity. This behavior cannot be observed, instead, if we consider the effect of explicit faults and we compute the same metric on the third model. A detailed description of the quantitative analysis of the LV can be found in [10].

6 Dependability of Complex Systems Using Petri Nets

A system can be a complex aggregation of components, and the ways in which the up and down states of a component influence the up and down states of the whole system may not be so straightforward as what we have seen in the previous sections. As we have seen in the introduction (Section 1), when the delivered service of a system deviates from fulfilling the system intended function, we say that the system has a *failure*. A failure is due to a deviation from the correct state of the system, known as *error*. Such a deviation is due to a given cause, for instance related to the physical state of the system, or to a bad system design. This cause is called a *fault*.

But if we consider a system as a set of interacting components, which is pretty much in the line with the way systems are designed nowadays, then we should consider, as pointed out in [44], that the failure of a sub-component (deviation from its intended functionality) may be perceived by the other sub-components as an external fault, thus giving rise to the so-called Fault-Error-Failure (FEF) chain.

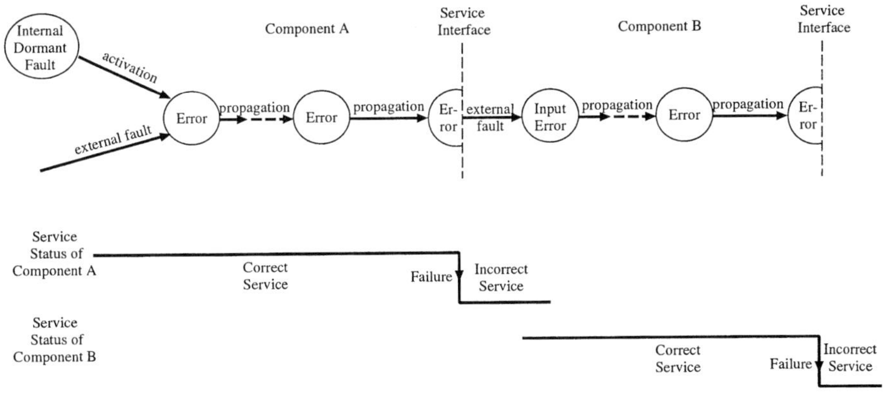

Fig. 11. Relationships between faults, errors and failures

Error propagation within and among system components is explicitly shown in Figure 11 extracted from [44]: internal propagation is caused by the computation process of the faulty component A while the external propagation from component A to component B, that receives service from A, occurs when, through internal error propagation, an error reaches the service interface of the faulty component A. Then, the service delivered by component B to component A becomes incorrect provoking the failure of component A that appears as an external fault to component B and propagates the error into B.

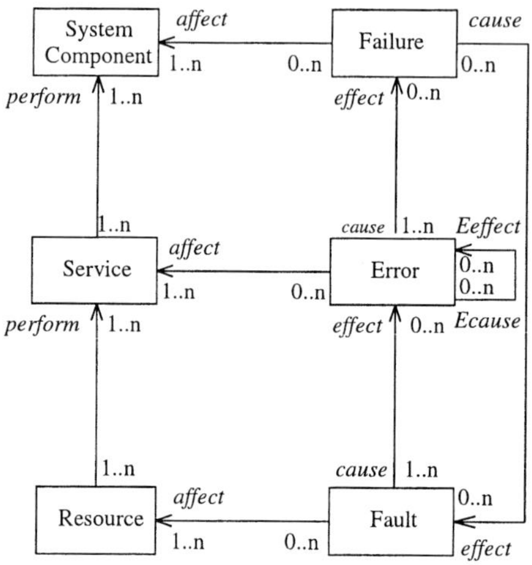

Fig. 12. The CD system description

To be able to model the FEF chain we need a more articulated vision of the system, that allows to clearly identify the components that can be affected by a fault, the system elements on which faults can induce erroneous behaviors, and how these behaviors may lead to a (sub)system failure. Although with a different aim in mind, the work in [27] introduces a view of the system as a layered structure in which a system component realizes a certain function by using a set of resources. Since the pattern of usage of resources can be quite complicated, an intermediate level is added, called services. This point of view on system behavior leads to a three layer structure, represented as an UML Class Diagram (CD) in the left portion of Figure 12.

The right portion of the CD shows instead the FEF chain: relationships between faults, errors and failures as well as their propagation among the system entities is captured by the *cause-effect* associations. Once customized on a specific application the CD shows which faults provoke which errors and which (set of) errors provoke a failure, that is to say a deviation from the function delivered by the system. The diagram also connects each type of fault, error and failure with the corresponding system entity affected by it, so a fault may affect a resource, an error may affect a service performed on one or more faulty resources, and a failure may affect the whole system if errors are not recovered in due time.

Moreover, if a service is affected by an error, the error can be propagated to another service either performed by the same resource (i.e., internal propagation) or by another resource communicating with the former (i.e., external propagation). This error propagation is represented by the *Ecause-Eeffect* association.

Since a failure of a system component can be perceived by another component as an external fault (as described in [44]), an association exists between the failure and the fault classes. This is an aspect that adds additional complexity to the modeling and is out of the scope of these notes.

We would like to use the high-level information provided by the CD of Figure 12 to drive the construction of PN models. To do this we need first to introduce the PSR modeling approach that has been presented in [27], and then to modify it to allow for the treatment of the FEF aspects.

6.1 A Layered Approach to Modeling: The PSR

The PSR is a model construction approach in which the PN model is organized into three levels: resources, services and processes. Resources are at the bottom level, and they provide operations for the services, where a service is basically a complex pattern of use of the resources. Services are then requested by the application model placed at the highest level, called process level (and that we have called "System Component" in the CD, since it is more intuitive, although it is not the original term used in [27]).

PSR provides a schema of how the resource, service, and process levels nets should look like, and a compositional operator to compose them.

Figure 13, bottom part, depicts a model of a resource: a resource can be idle, and it can offer one or more operations *op_i* through the sequence of actions start operation, operation, end operation. Each transition of the sequence has an associated label (shown in italics in the figure), that is used for composition with the service level, and that is derived from the association *perform(Resource,Service)* of the CD diagram of Figure 12, prefixing it with an *S_* or with an *E_* to indicate Start and End of operation, and postfixed with the operation index. Depending on the type of operation it may be necessary to acquire a lock (transition *lockRes* and *unlockRes*). Figure 13, in the middle, models a service. A service can be requested by a process through the pair of labels of start and end service (*S_perform(Service, System Component), E_perform(Service, System Component)*). Once activated the service can request resource operation via the *S_perform(Resource, Service)_i, E_perform(Resource, Service)_i* labels.

The upper part of Figure 13 depicts a skeleton of the process model that uses services: the request of a service is performed through the label *perform(Service,System Component)* and through a matching function that maps the label into the pair of labels *S_perform(Service,System Component), E_perform(Service,System Component)*.

Each level is defined through net composition operators based on transition superposition ("horizontal composition"), then resource level is composed with the service level, and the resulting net is composed with the process level also through transition superposition ("vertical composition"). Transition superposition of nets is based on transition labels: two transitions of equal label in two separate nets are fused into a single one: the formal definition of the superposition operator for GSPN can be found in [27], and the SWN extension is instead presented in [10]. Therefore if $\{R_i\}, \{S_k\}, \{P_m\}$, are the sets of GSPN (or SWN) models representing resources, services, and processes, respectively, and if $||$ is the transition superposition operator, then the full model of the system is given by:

$$
\begin{aligned}
R &= R_1 || R_2 || \cdots || R_{n_r} \\
S &= S_1 || S_2 || \cdots || S_{n_s} \\
P &= P_1 || P_2 || \cdots || P_{n_p} \\
PSR &= R || S || P
\end{aligned}
$$

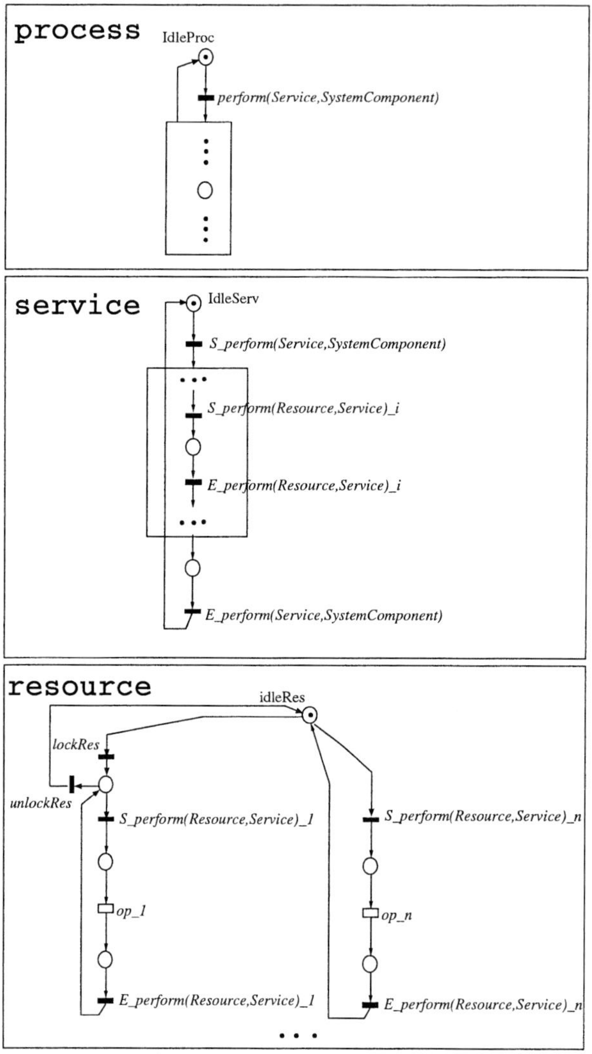

Fig. 13. Resource, service and process models

From now on the term PSR will refer to a structure of the models according to Figure 13.

6.2 PSR and FEF Elements

The PSR originally defined in [27] is not adequate for dependability modeling, since it does not take into consideration the interactions with the FEF elements. In [7, 4] the PSR has been modified by changing the basic models of the resources, services, and process so as to allow interactions with the FEF elements, and by extending the formula to compose the $\{R_i\}$, $\{S_k\}$, and $\{P_m\}$ with models of the FEF elements.

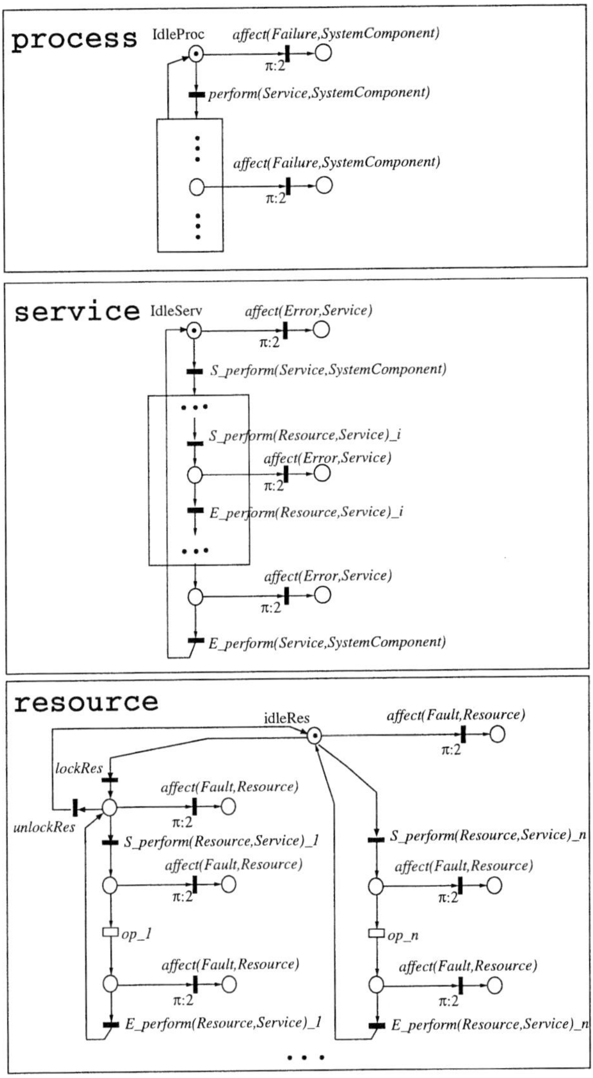

Fig. 14. The modified models of resources, services, and processes

Figure 14 presents the modification to the basic PSR models. Following the CD scheme of Figure 12, we assume that faults affect only the behavior of the resources, errors are perceived at the service level, while failure are a concern of System Components, and therefore of the process level.

Figure 14, bottom part, models a resource. From each state in which a fault can be perceived by the resource a transition labeled *affect(Fault,Resource)* has been added (to be used for synchronization with the fault model) which takes the resource into a faulty state. Again, *affect(Fault,Resource)* is the association that, in the CD diagram of a specific application, relates a specific type of fault to a specific type of resource affected by that fault.

Also for service and process models (Figure 14 in the middle part and at the bottom part, respectively), transitions to be used for synchronization with an error and a failure model, respectively, have been added which take the services/processes into an anomalous state.

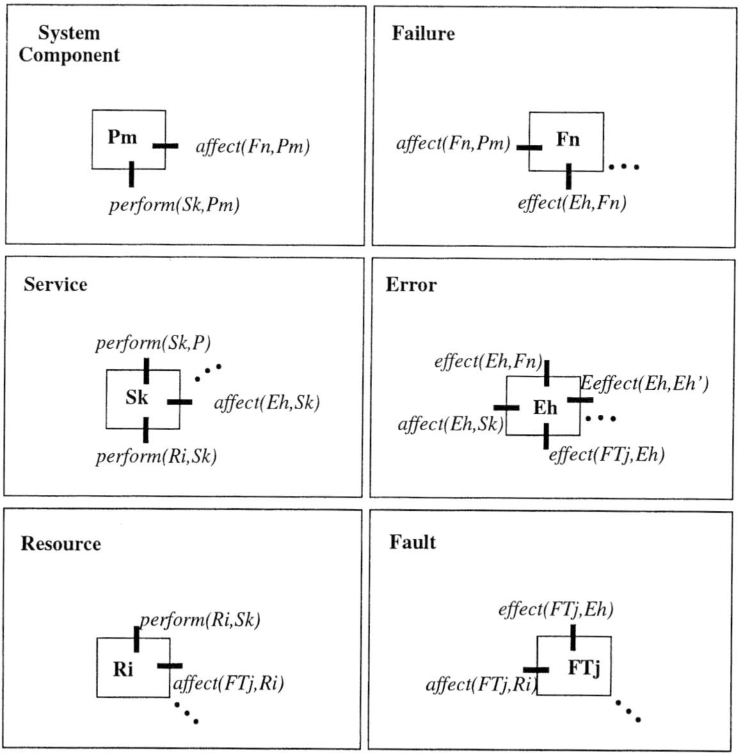

Fig. 15. Organization of the Petri net models in the layered approach

Now that we have the modified resource, service, and process models we need to modify the PSR construction: indeed the $\{R_i\}$, $\{S_k\}$, and $\{P_m\}$ models have to be integrated with the models of the FEF elements. Figure 15 assumes that there are a number of models for faults, errors, and failures, called $\{FT_j\}$, $\{E_h\}$, and $\{F_n\}$, respectively, and it provides a schematic view of how they are organized in three levels: fault models are placed at the resource level, error models at the service level and failure models at process level. Each model is depicted as a box with explicit interface transitions that are labelled according to the associations defined in the CD of Figure 12.

Each level is obtained through horizontal composition, and the global model is obtained through vertical composition of the levels, as for the original PSR. Therefore if $\{R_i\},\{FT_j\},\{S_k\},\{E_h\},\{P_m\},\{F_n\}$ are the sets of GSPN (or SWN) models representing resources, faults, services, errors, processes, and failure, and if $||$ is the transition superposition operator, then the full model of the system is given by

$$\begin{aligned} R &= R_1 \,||\, R_2 \,||\cdots||\, R_{n_r} \,||\, FT_1 \,||\, FT_2 \,||\cdots||\, FT_{n_r} \\ S &= S_1 \,||\, S_2 \,||\cdots||\, S_{n_s} \,||\, E_1 \,||\, E_2 \,||\cdots||\, E_{n_e} \\ P &= P_1 \,||\, P_2 \,||\cdots||\, P_{n_p} \,||\, F_1 \,||\, F_2 \,||\cdots||\, F_{n_f} \\ PSR &= R \,||\, S \,||\, P \end{aligned}$$

The proposed approach to the construction of Petri net models for dependability shares similarities with other approaches. In [54] there is an example of organization of dependability GSPN model into layers to separate the architecture model, the service model and the failure modes model, although without explicitly modeling the FEF chain. The modeling of the FEF is made more explicit instead in [12], although the approach taken is that of a top down hierarchical approach more than flat compositional as in the PSR.

6.3 PN Models of the FEF Elements

The compositional approach depicted in Figure 15 requires the definition of models for the FEF elements. In the following we present a library of FEF element models that respect the transition interface of the boxes of Figure 15. The original definition of the library can be found in [8]. The library is built starting from a classification of faults, errors, and failures into a hierarchy of classes that has been devised in the DepAuDE [26] EEC project, and whose complete description can be found in [9]. The classification of the FEF elements in DepAuDE was heavily inspired by the work in [43], customized on the automation system field that was the application target of DepAuDE.

The hierarchy of the classes has a counterpart in a hierarchy of PN components. To set the field to the description of the hierarchy of the FEF models we need first to define the notion of PN component and that of hierarchy for PN components.

PN component is a GSPN system [1] (or an SWN one [18]) labeled over transitions, parametric with respect to transition rates (weights) and/or initial marking and with an associated list $RESULTS$ of performance results to be computed and/or verified and a list $CONSTR$ of constraints to be verified.

Hierarchy of PN models. Hierarchy in a Class Diagram involves a notion of *inheritance*, that involves the structure of a class (attributes, operation names, and associations) as well as behavior (operations). In GSPN the inheritance of the structure of a class is reflected into 1) inheriting parameters (rate/weight, initial marking), results to be computed, constraints to be verified, and possibly adding new ones; 2) inheriting, and in case modifying, the labels associated to either places or transitions. Inheritance of the dynamic behavior of the super-class is reflected in either maintaining the same net structure in the sub-class or modifying it by applying transformation rules that preserve the *behavioral inheritance* [66]. Two main notions of behavioral inheritance are introduced in [66]: *protocol inheritance* and *projection inheritance*. Although they have been defined for labeled transition systems, it is rather straightforward to use them for the reachability graphs (RGs) of SPN models. Intuitively, let p and q be two SPN models representing the behavior of a class P and of its super-class Q, respectively; protocol inheritance can be verified by not allowing to fire transitions that are present in p and not in q (i.e., *blocking new actions*) and by checking whether the RGs of p and q are equivalent. Projection inheritance can be verified, instead, by considering not observable the transitions that are present in p and not in q (i.e., *hiding the effect of new actions*) and by checking whether the RGs of p and q are equivalent. In both cases, branching bisimulation [53] is used as equivalence relation. Branching bisimulation belongs to the class of observational equivalence, in which two systems are equivalent if an external observer cannot discriminate between them. Of course two systems may or may not be equivalent depending on what an observer is allowed to see. In our context an observer is allowed to see all transition labels, unless otherwise stated, that is to say the labels that are used to compose models.

The two basic notions of inheritance are combined [66] in order to obtain a stronger and a weaker notion. *Stronger* inheritance is preserved if both protocol and projection inheritance are satisfied. *Life cycle* inheritance is the weaker notion: the set of transitions present in p and not in q is partitioned into "not-observable" and "not-allowed to fire" such that the observable behavior of P equals the behavior of Q.

Observe that the proposed rules for inheritance only consider the net functional behavior, the stochastic behavior may not be preserved, and usually it is not.

Fault Models. Figure 16(A) is the classification of faults: the root class of the inheritance tree describes a generic fault; the first level of the inheritance tree distinguishes *Physical Fault, Design Fault, Interaction Fault* and *Malicious Logic*. Physical faults are characterized by two input attributes that allow to specify the maximum time during which the fault is active and can be perceived by the system (*duration*) and the frequency of its occurrence (*fault_rate*). Attribute *fault_dormancy*, the length of time between the occurrence of a fault and the appearance of the corresponding error, is considered instead as a metric to be evaluated. The different type of usage of class attributes is denoted by prefixing the attribute name with a specific symbol (i.e., "$" = input attribute, "/" = metric to be evaluated, "/$" = metric to be evaluated and validated).

Physical Faults may be either considered permanent or temporary: their discrimination depends on the values assigned to the input attributes *min-duration* and *max-duration* as emphasized by the constraint written in the note symbol. Permanent physical faults and temporary physical faults are further specialized by several sub-classes. For example, temporary physical faults are discriminated in *DevTemp Physical Faults*, that is internal faults due to the development phase; *Transient Physical Faults*, that is faults induced by environmental phenomena; *Intermittent Physical Faults*, i.e., internal physical defects that become active depending on a particular point-wise condition.

Transient and intermittent physical faults classes are enriched with some input parameters, such as: *latency_rateN* and *latency_rateB*, representing the rate of transient fault activation in case of normal conditions and burst conditions, respectively, *persistence_rate*, representing the rate of fault disactivation and *latency_rate*, representing the rate of intermittent fault activation.

Dotted boxes in Figure 16(A) represent classes that are not described here: the interested reader can find a complete description in [9].

GSPN component models for faults have been built according to the hierarchy view of Figure 16(A): each GSPN model is an elaboration of previous generic Petri net models of fault generator proposed in [50] where only physical faults are considered and they are classified with respect to their persistence in permanent and temporary, the latter being further specialized in transient and intermittent.

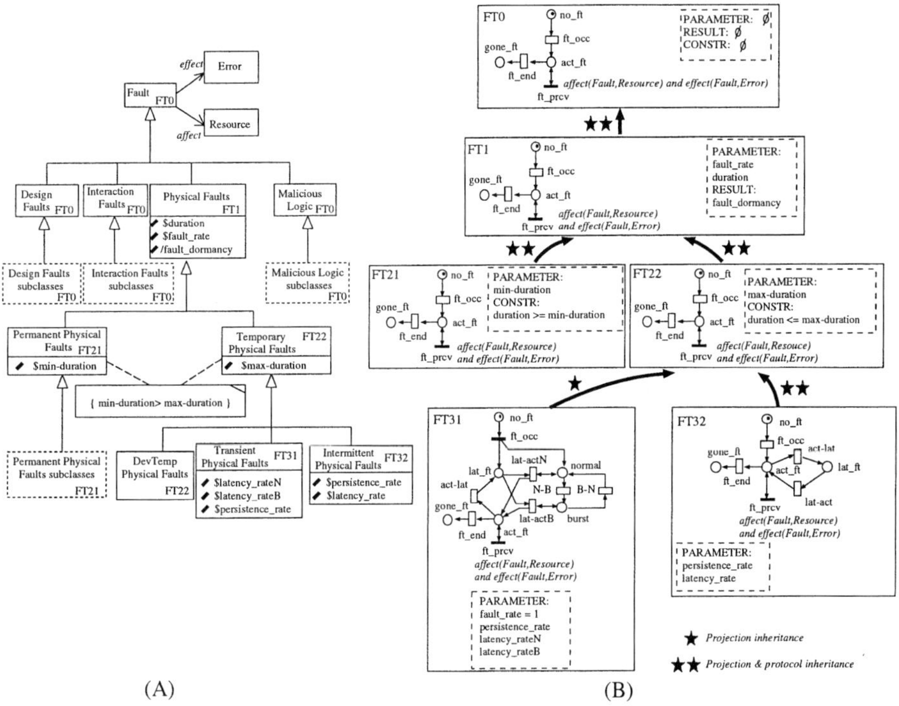

Fig. 16. GSPN component models of fault classes

In Figure 16(B) each box is a labelled GSPN component: a GSPN net with a set of parameters, results to be computed, and constrains to be verified. For sake of graphical clarity the rates of transitions are not shown in the figure, they are listed in the text when needed. Labels are associated to transitions, they determine the observable part of the net behavior and they identify the transitions that can be used for the composition with another model.

The behavior of *Fault* super-class of Figure 16(A) corresponds to the GSPN component model FT_0 of Figure 16(B) characterized by three states: the fault is not present (place *no_ft*), the fault is active and it may be perceived by a system entity (place *act_ft*) and the fault is terminated (place *gone_ft*). The fault occurrence is represented by the firing of transition *ft_occ*, and when the fault is active it can be perceived (transition ft_prcv) by a system entity, causing an error situation. Transition ft_prcv is labeled so as to allow synchronization with the affected system entity (i.e., the resource model of Figure 14) and with an error model. The fault termination is represented by the firing of transition ft_end. Since neither attributes nor constraints are specified for the *Fault* super-class, the lists of parameters, results and constraints of the corresponding GSPN model FT_0 are empty.

Classes *Design Faults, Interaction Faults, Malicious Logic*, and their corresponding sub-classes, present the same behavior of the more general class *Fault*, so that the GSPN model FT_0 is reused to represent these classes also.

The class *Physical Faults* is associated with the GSPN model FT_1 that inherits from FT_0 and adds to the parameter list *fault_rate* and *duration* and to the result list *fault_dormancy*. The parameters and the result correspond to the homonyms attributes defined in the *Physical Fault* class.

The net structure of FT_0 has been maintained, but rates of transitions ft_occ and ft_end have been defined as functions of the added parameters, i.e., $w(ft\_occ) = fault\_rate$ and $w(ft\_end) = 1/duration$.

The behavior of *Permanent Physical Faults* and of *Temporary Physical Faults* classes is represented by the GSPN models FT_{21} and FT_{22}, respectively. Both the models inherit from model FT_1, add a parameter (the parameter *min-duration* for model FT_{21} and the parameter *max-duration* for model FT_{22}) and maintain the same net as FT_1. A fault is classified permanent if it lasts more than min-duration, and it is classified temporary if it lasts less than max-duration, with the constraints, derived from the note symbol of the CD of Figure 16(A), that min-duration is greater than max-duration. The interaction of the models with the corresponding resource and error models (that amounts to the labels associated to transitions) is also inherited from FT_1.

With respect to the fault models proposed in [50], where permanent faults remain always active while temporary faults once occurred after a certain amount of time eventually disappear, both the fault models FT_{21} and FT_{22} are characterized by a termination state (i.e., place $gone\_ft$) and the represented fault classes are discriminated by the fault duration.

Temporary faults can still be distinguished into intermittent and transient faults. Intermittent faults, once occurred, are characterized by alternating periods in which they are active, and they can be perceived by the system entity, and periods in which they are latent and hence they do not cause any error. Transient faults, instead, disappear a certain amount of time after their activation; however, unlike generic temporary faults, they are characterized by a complex mechanism of activation that depends on the condition of the external environment.

The behavior of *Transient Physical Faults* class is represented by the GSPN model FT_{31} in which a fault moves from the latent state to the active state with a different rate depending on the environment conditions. Under normal condition, represented by the place *normal* marked, transition *lat-actN* with rate parameter equal to *latency_rateN* will fire, while under "burst" condition, represented by the place *burst* marked, transition *lat-actB* with rate parameter equal to *latency_rateB* will fire.

The behavior of *Intermittent Physical Faults* class is represented by the GSPN model FT_{32}, in which firing of transition *act-lat* (with rate parameter equal to *per-*

sistence_rate) brings the state of the fault from active to latent and, vice-versa, firing of transition *lat-act* (with rate parameters equal to *latency_rate*) changes the fault state from latent to active.

GSPN models FT_{31} and FT_{32} inherit from FT_{22}: for the structure, new parameters have been added with respect to the parameter list of FT_{22} and for FT_{31} the parameter *fault_rate* is now not relevant (and it has been set to the default value of 1), since the fault activation depends upon the two transitions *lat-actN* and *lat-actB*. From the behavioral point of view the GSPN model FT_{32} strongly inherits from FT_{22}, i.e., it preserves both the projection and the protocol inheritance, while the GSPN model FT_{31} preserves only the projection inheritance, that is to say, if any of the transitions $act - lat$, *lat-actN*, and *lat-actB* is used in a synchronization with another model, then it may be the case that FT_{31} is not able to act as FT_{22}.

Finally, the sub-classes *DevTemp* of temporary physical faults and sub-classes of permanent physical faults inherit the behavior of their super-classes and they have been represented by the same GSPN models associated to the latter.

All the fault GSPN models described above can have more than one label for each transition; in particular, transition *ft_prcv* is characterized by two labels: one is used to interact with the resource model affected by the fault and the other is used to interact with the corresponding error model.

Error Models. Errors are deviations from the correct state of the system that may cause a subsequent failure [44]; they are caused by faults affecting the resources of the system and they are related to the services performed by the faulty resources. A classification of errors is given by the CD of Figure 17(A), taken from [9], that considers only errors caused by physical faults, and discriminate them depending on which type of resource has been affected. The type of resources considered in DepAuDE are processing, memory, and communication.

The super-class *Error* of the hierarchy/logical view is modeled by the GSPN ER_0 - shown in Figure 17(B). The class is characterized by two attributes that are mapped in two results to be computed on ER_0: *error_latency*, the length of time between the occurrence of an error and the appearance of the corresponding failure, and *PE*, the probability of error. Note that for some results, as *PE*, it is already possible to give their definitions, since their computation is based only on local information; the definition of other results, as *error_latency*, requires instead information on the whole system.

The GSPN model ER_0 is characterized by four states: the error is not present (place *no_err*), the error is generated (place *pot_err*), the error is occurred (place *error*) and the error has been detected (*detected*). Places *error* and *detected* are used to define the result *PE* as the probability that one of the places is marked. The error can be caused by either a fault occurred in a resource or by the error propagation effect: the error generation is represented by the firing of transition *cause* that is labeled so as to ensure synchronization with caused fault or error model. The labels are derived from the associations *effect(Fault, Error)* and association *Eeffect(Error,Error)* of the CD of Figure 17(A). In general, ER_0 contain as many transition "cause" (i.e., with input place *no_err* and with output place *pot_err*) as the number of GSPN models representing potential causes of the error. The occurrence of the error in the corresponding service is represented by the firing of transition *err_occ*, properly labeled to ensure synchronization with the service

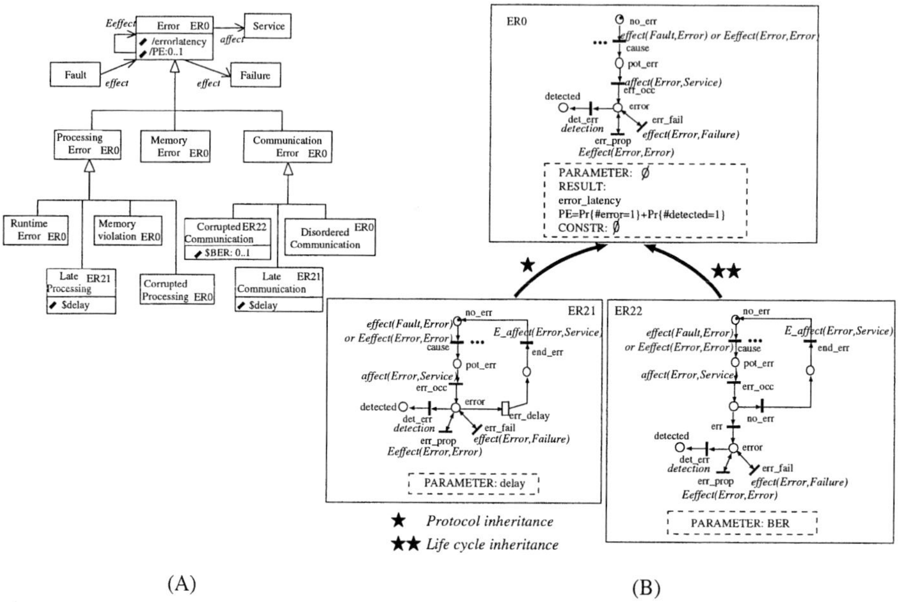

Fig. 17. GSPN component models of error classes

model. Transition *det_err* represents error detection carried out by some other model, synchronized through the label *detection*.

Test transitions *err_prop* and *err_fail* are instead interface transitions for an error model and for a failure mode model, respectively.

Classes *Processing Error, Memory Error, Communication Error, Runtime Errors, Memory Violation, Corrupted Processing* and *Disordered Communication* present the same behavior of the super-class *Error*, so that GSPN model ER_0 is reused to represent the behavior of these classes also.

Late Processing and *Late Communication* classes are characterized by an input attribute, *delay*, whose values indicate the delay caused in the execution of the corresponding erroneous function. Their behavior is modeled by the GSPN ER_{21} where *delay* has been added to their parameter list. Moreover, the model contains a pair of causal connected transitions: *err_delay*, that represents the delay caused by the error, and *end_err* that brings the error model to its initial state (*no_err*). Timed transition *err_delay* is characterized by a rate equal to $w(err\_delay) = 1/delay$, immediate transition *end_err* is, instead, an interface transition and has to be synchronized with the service model in order to bring it from an erroneous state to a normal state. Protocol inheritance is preserved for model ER_{21}; indeed if transition *err_delay* is not allowed to fire the RG of ER_{21} is equal to the RG of ER_0.

Finally, GSPN model ER_{22} has been associated to the *Corrupted Communication* class characterized by the input attribute *BER* (Bit Error Rate). *BER* has been added to the parameter list of ER_{22} and it has been assigned to the weight of the immediate transition *err*. For model ER_{22} life cycle inheritance is preserved, when considering transition *err* non observable and transitions *no_err* and *end_err* not allowed to fire.

Failure Models. Failures are deviations of the service delivered by the system with respect to the system intended function. The CD shown in Figure 18(A) associates to a generic failure mode two metrics to be computed and verified: *PF*, i.e., the probability of failure, and *RF*, rate of failure. The CD represents a classification of failures with respect to the their impact on the system, that is whether their occurrences are considered acceptable or not depending on the criticality level associated to the system process they affect. The different failure mode assumptions are represented by the sub-classes: *Halting Failure, Degrading Failure* and *Repairing Failure*. Halting failures cause the system activity not to be any longer perceptible by the user. Depending whether the absence of system activity takes form of a frozen output or of silence, they are further classified in passive failures and in silent failures, respectively. Degrading failures still allow the system to provide a subset of its specified behavior. Repairing Failure requires instead that faulty resources originating the failure be replaced or repaired before the system activity continues. Repairing actions are undertaken during the failure treatment phase and are performed by proper mechanisms (association *address*).

The failure hierarchy of Figure 18(A), defined in [9] can be exploited to construct GSPN model components representing different failure modes. The main purpose of GSPN models representing failure modes is to synthesize in a unique place the set of (erroneous) states that have equivalent consequences on the system. These models correspond to the failure mode layer described in [54] that allows to arrange an SPN model in a manner suitable for the analysis of different levels of service degradation. In Figure 18(B) two skeletons of GSPN models representing a generic failure mode and a repairing failure mode, respectively, are depicted. The model F_0 is characterized by three main states: *no_fail*, *pot_fail* and *fail*, respectively meaning the absence of failure, the occurrence of the error conditions causing it and the failure occurrence. Several error conditions may cause the occurrence of a failure: the firing of transition *cond_i* represents the occurrence of one of such conditions; since, in general, the failure occurrence is caused by a combination of errors, *cond_i* is a multi-labeled transition with labels derived from association *effect(Error, Failure)* for synchronization with the error models. Transition *fail_occ* has to be synchronized with a System Component model, so that its label is derived from association *affect(Failure, SystemComponent)*.

Concerning the result list, the GSPN model is characterized by two metrics derived from the homonyms attributes of *Failure* class: *PF*, defined as the probability the place *fail* is marked, and *RF* defined as the throughput of transition *fail_occ*.

Model F_1, representing a repairing failure mode, contains one transition more with respect to F_0: *fail_repair* that is an interface transition to be synchronized with reconfiguration mechanism models. Model F_1 respects protocol inheritance.

7 A Methodological Approach to the Construction of Petri Nets Models for Dependability in the Automation System Domain

The Petri Net component models and their organization into a three-layered structure described in Section 5 constitute a first support in the construction of a Petri Net model suitable to be analyzed through either numerical or simulation techniques. But a number of points are still open: how does the modeler identifies, in the system being modeled,

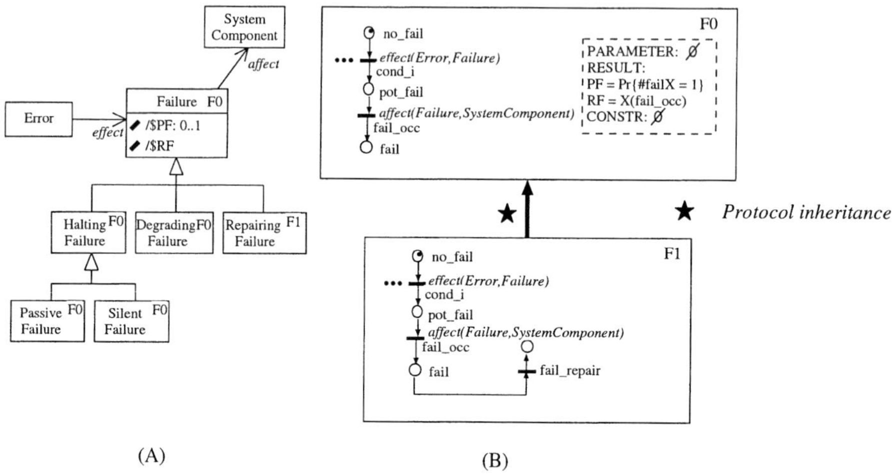

Fig. 18. GSPN component models of failure mode classes

the resources operations, the services and the system components required by the PSR approach? And while modeling a system that includes some FT mechanisms, where should the mechanisms models be placed in the context of Figure 15? In this section we try to give an answer to these questions by restricting the scope to a particular domain, that of automation systems. A larger treatment of the topic can be found in [7, 4].

This work was developed in the EEC-IST project DepAuDE [26] as part of a methodological effort to support the analyst from the early phases of the project (collection of dependability requirements) down to the definition, validation and dependability evaluation of fault tolerance strategies adopted for automation systems [9].

The DepAuDE methodology follows the approach of integrating different notations during the dependability process as suggested by emerging standard like IEC 60300 [21]. In particular, UML Class Diagrams (CDs) and SPN are used in the methodology with different roles, but the information contained in the CDs is exploited to drive the SPN modeling process.

CDs are meant as a support for the requirements collection and/or for structuring and/or reviewing for completeness already available requirements. A set of predefined CDs for the automation system domain (called *generic* CD scheme) and guidelines on how to produce from it the *customized* one that refers to the target application are provided.

Stochastic Petri nets - in particular GSPN and SWN – are used to support dependability design validation and evaluation through modeling. The methodology supports the construction of *PN evaluation scenarios*. A PN scenario consists of a set of PN model components and of their interactions, plus the set of model parameters and the set of performance and validation properties of interest.

The methodology helps the analyst in the construction of a PN scenario by providing a set of predefined reusable PN models for some of the UML classes, a suggested structure of interaction of the model components, guidelines on how to extract information from Class Diagrams, and in particular on the Class Diagram instantiated on the specific application, automatic translation from UML State-Charts and Sequence Diagrams into PN, and a suggested approach to the dependability analysis.

The issue of re-use of high level information available from the UML design has been a major concern also for the European Esprit project HIDE [16], that has devised an integrated environment supporting dependability analysis of UML-based system design from the early stages, based on the automatic generation of PN from a number of UML diagrams that encode specific dependability aspects in a rather abstract form. The goal being the evaluation from the early stages of the design the resulting model is obviously rather abstract (since a limited amount of information is available), which is an advantage from a computational point of view, but may be not sufficiently detailed to allow also qualitative properties to be checked.

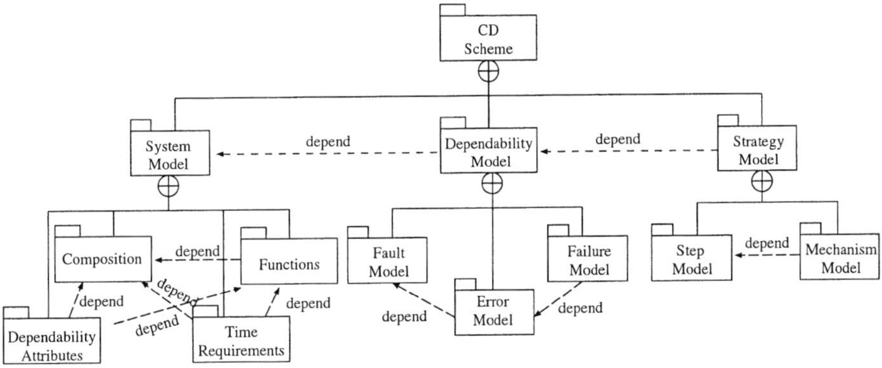

Fig. 19. Scheme hierarchy

UML Class Diagrams. The CDs of the generic CD scheme of the DepAuDE methodology are grouped into the hierarchical structure of *packages* represented in Figure 19, where each non-leaf package of the structure encapsulates a set of inner packages together with their definition dependency relationships. The scheme is therefore constituted by a set of CDs that describe the system in terms of automation components, automation functions, dependability attributes, and timing requirements (left branch in Figure 19), a set of CDs that describe the dependability model in terms of the FEF chain (central branch), and a set of CDs devoted to the strategy model (right branch) that is seen as a set of dependability actions/steps, that can be achieved through a number of software “mechanisms”. The *Fault Model*, *Error Model* and *Failure Model* packages contain, respectively, the hierarchy views of fault, errors and failures described in sub-Section 6.3.

Class attributes are used to represent either parameters, whose values have to be provided as input to the specifications, or measures to be computed or upper/lower bounds whose values have to be provided as input to the specification and to be validated at later stages of the development. We have chosen to discriminate these different types of usage of class attributes by prefixing the name of the attributes with a specific symbol ("$" , "/" or "/$", respectively).

Elements of a generic CD can be customized on a specific application with the help of a set of guidelines [9]. In the customized CD the value of the class attributes and of the association multiplicities have been set and new classes and associations are added.

The customized CD still refers to classes, and not to objects, but certain classes and associations have been made more specific using information from the application. We now illustrate a few Class Diagrams of the generic CD scheme, trimmed so as to simplify explanation, while still, we hope, containing enough information for what will be discussed later.

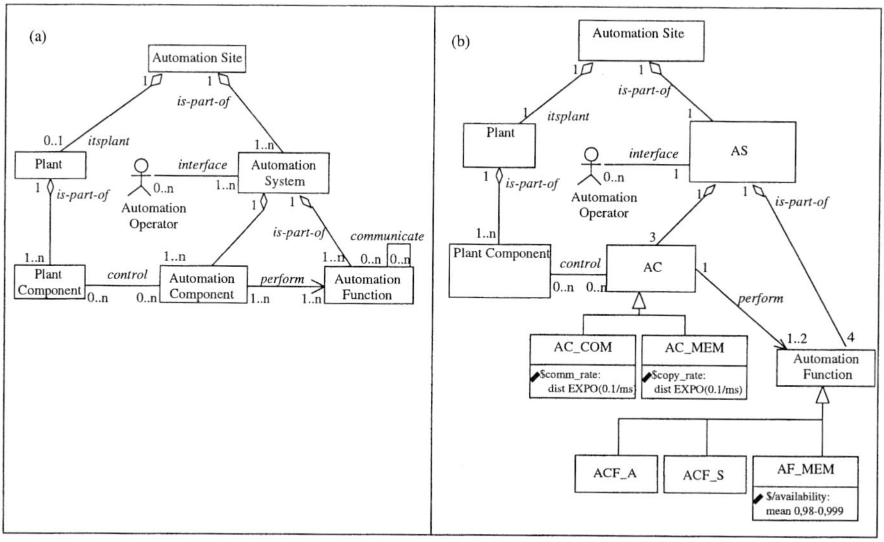

Fig. 20. The CD *Structure* of *Composition* (a) and its instantiation to the running example (b)

Figure 20(a) describes a portion of the system: an automation site is a plant (optional) together with one or more automation systems. An automation system is composed by a set of automation functions, that can *communicate* among them, and by a set of automation components that can be used to *perform* one of more automation functions. An automation component *controls* zero or more plant components. An *instantiated* version of the CD of Figure 20(a) will contain application specific information.

Let us consider, for example, a cyclic application that activates two concurrent processes: each process reads a sample input from a plant, elaborates the future state, saves the new state in memory and produces the new output for the plant. The memory units can be affected by physical faults that may cause errors in the automation functions. Communication units are instead assumed not affected by faults. To increase the dependability of the automation system a fault-tolerance strategy has been devised consisting of error detection, error diagnosis and error recovery. The error detection step uses a standard watchdog mechanism while error diagnosis and recovery steps are implemented by a recovery mechanism. If the watchdog expires, it sends a notification message to a software recovery mechanism, that provides to terminate the watchdog and check the status of the automation system: if no error is present then it is a false alarm, and the watchdog is simply reinitialized. If instead an error is present then a recovery action is carried out. Measures to be computed are the availability of the automation functions and the probability of failure of the automation system.

Figure 20(b) shows an instantiated version of the CD depicted in Figure 20(a) to the example: there are two types of automation components dealing with communication (AC_{COM}) and memory (AC_{MEM}), three types of automation functions dealing with asynchronous communication (ACF_A), synchronous communication (ACF_S), and memory (AF_{MEM}), and a single type of automation system AS.

Classes AC_{COM} and AC_{MEM} are characterized, respectively, by the input attributes *comm_rate*, representing the communication rate, and *copy_rate*, representing the rate of the copy operation. The values assigned to these attributes are exponential probability distribution functions with parameter $\lambda = 0.1$. Class AF_{MEM} is characterized, instead, by a metric to be evaluated and verified against the requirements that is the mean availability of the automation functions performed by the memory components whose value has to be included in the interval $[0.98, 0.999]$.

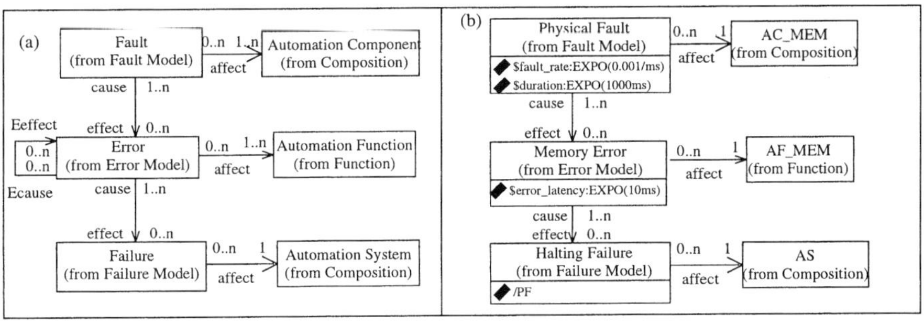

Fig. 21. The generic *FEF chain* (a) and its instantiation to the running example (b)

Figure 21(a) is a trimming of the CD of the FEF chain. Once customized on a specific application it shows which faults cause which errors, how errors propagates, and which (set of) errors cause a failure, that is to say a deviation from the service delivered by the system. The diagram also connects each type of fault, error and failure with the corresponding system components affected by it. The *instantiated* version is shown in Figure 21(b): it contains only one type of faults (*Physical Faults*), a single type of error (*Memory Error*) and a failure (*Halting Failure*). The instantiation of the *affect* relationship relates the fault to the AC_{MEM} component only, the error to function AF_{MEM}, and failure to AS. Values are set to the input attributes of *Physical Fault* and *Memory Error* classes (i.e., the attributes prefixed with the "$" symbol), while the attribute *PF* (prefixed with the "/" symbol) emphasizes that the probability of failure of the automation system is a measure of interest to be computed.

The CD description of a system contains a lot of useful information for the construction of PN evaluation scenarios, in particular we have observed the following relationships: (1) the *package structure* provides indication on the organization of PN component models; (2) the *aggregations* provides information that allows to identify PN components and the composition formulae; (3) *binary general associations* (associations from now on) among classes indicate interactions and can therefore be used to identify labels for PN model composition; (4) *classes* are rich of *attributes* that are useful to set rate parameters (input and/or upper-lower bound attributes) and to define the performance/dependability indices (output measures and upper-lower bound attribute to be checked); (5) information on the *FEF chain* is fundamental to set the relationship among the PN models of faults, errors, failures, and system components; (6) *hierarchies* can indicate reuse of PN components through inheritance [65]; (7) the *Strategy Model* package allows to identify which mechanisms are used for which fault, error, or failure (dependability strategy).

Composition Scheme of PN Models. To use the PSR in the automation system domain we need to identify the main PN models involved and their interactions. This identification is again driven by the CDs. The organization of PN models into layers is described by Figure 22. The package structure with the three branches of Figure 19 is reflected in the organization into three columns of Figure 22, while to decide in which level to place the various PN models we have considered the FEF chain first (Figure 21). Faults are at the lowest level of the chain, and they have therefore been placed at the resource level. Consequently also automation components (AC), that are affected by faults, have been placed at the same level. With a similar reasoning errors and automation functions (AF) have been placed at the service level, and failure models and automation system (AS) have been placed at the process level.

This is depicted in Figure 22 by the set of boxes AC_i for automation components, FT_j for the fault models, AF_k for the automation functions, ER_h for the error models, AS for the automation system, and $FAIL_n$ for the failure models.

The composition of automation components with faults requires a proper assignment of labels to interface transitions for horizontal composition, that can be derived from the associations *affect(Fault, Automation Component)* of the CD scheme in Figure 21.

The labels for the composition of automation functions with errors and for error propagation are derived from the association *affect(Error, Automation Function)* and *Eeffect(Error, Error)*, from the CD of Figure 21.

The inter-level interaction between service and resource is given by the association *perform(Automation Component, Automation Function)* of the CD of Figure 20 and by the cause-effect association between faults and errors of Figure 21.

The labels for the composition of automation system with failure are derived from the association *affect(Failure, Automation System)*, while the propagation from error to failure is based on the association *effect(Error, Failure)*.

The software mechanisms are instead placed either at the service or at the process level, depending on whether they address errors or failures.

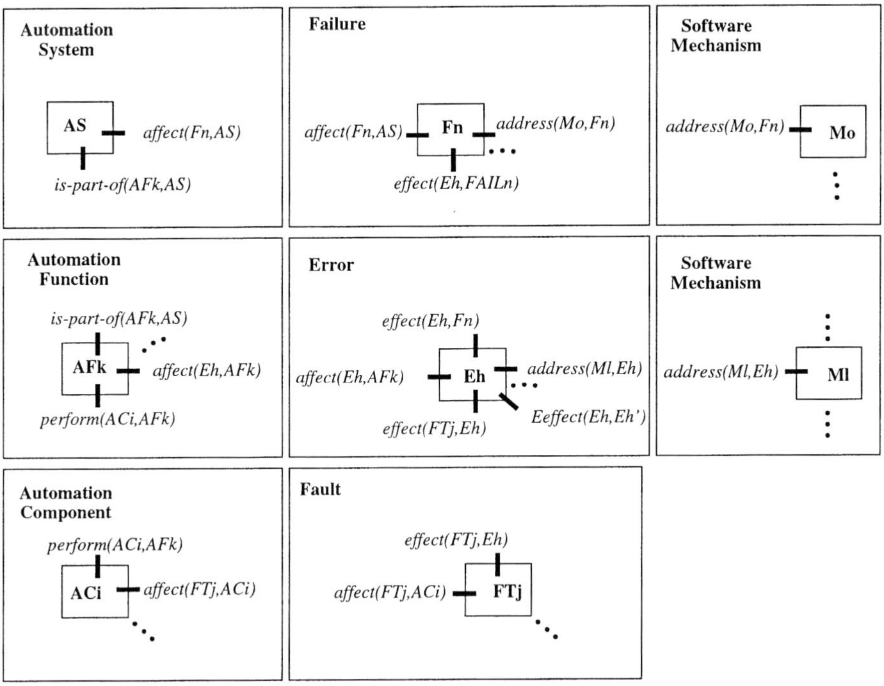

Fig. 22. Organization of the PN models in the DepAuDE methodology

Observe that the relationship between software mechanisms and automation functions is through the error models, although there are cases, like the Local Voter presented in Sub-Section 5.2, in which it seems more natural to have a direct relationship between the mechanisms and the functions, but unfortunately, no information of the subject is contained in the CDs, so that no general guidelines can be derived.

Getting an Executable PN Model. Once the basic structure of the PN models has been identified it is necessary to complete them with a number of information related to the specific application. To reach this goal we have identified a number of steps that will be illustrated through an example.

Step 1. Select the concrete classes of the customized CD scheme amenable to a PN model

The set of PN component models can be identified by examining the customized CD scheme to select the classes that are relevant from a quantitative point of view. Good candidates are classes of the customized CD scheme that contain attributes specifying input parameters, metrics to be computed and/or to be verified.

If we assume a class level specification, then all GSPNs are initialized with a single token, while in case of object level specification the identities of the objects come into play. In the context of modeling of distributed object software, an in-depth treatment of the specification level is reported in [67], where a formalism derived from Colored Petri Nets (CPN) [38] has been defined. Having used GSPN for the class models, SWN [18] is a natural choice at the object level to keep track of the object identities.

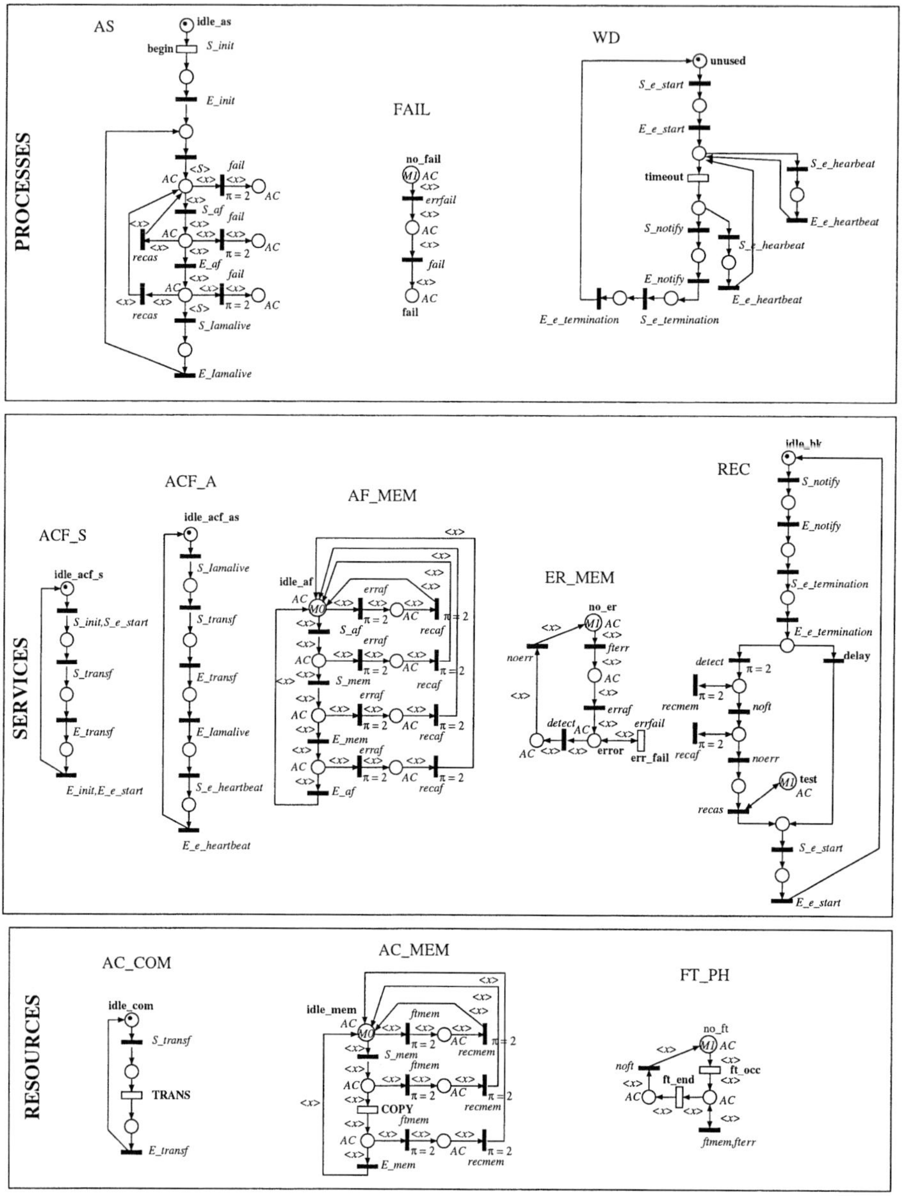

Fig. 23. Example of analyzable PN model: the set of PN components

Example. The customized CD scheme allows the identification of the PN components shown in Figure 23.

At the resource level, left to right, we have: the communication unit, the memory units, and the physical fault models. For AC_{COM} and AC_{MEM} we have reused the resource models of Figure 13 and of Figure 14, respectively, with transmission and copy as basic operations, while adding reset transitions in AC_{MEM}. For FT_{PH} the predefined physical fault model FT_0 of Figure 16 has been used (with reset).

At service level, left to right, there are: the synchronous communication function used by the automation system to initialize the watchdog, the asynchronous communication function used by the automation system to send signals to the watchdog, the memory automation functions, the errors affecting the memory automation functions, and the recovery mechanism. Models ACF_A and ACF_S are simplified models of communication presented in [6]. AF_{MEM} follows the basic skeleton of service model of Figure 14 with "reset" transitions. ER_{MEM} is an error memory model obtained by refining the PN model component ER_0 of Figure 17. Model *REC* has been produced from the high level design specification of the recovery mechanism.

At process level there are: the automation system, the halting failure mode of the automation system, and the watchdog mechanism. Model *AS* has been produced from the high level design specification of the automation system. Model *FAIL* is a customization of the predefined failure mode model F_0 of Figure 18. *WD* is a simplification of the watchdog model resulting from the automatic translation of the State-Chart specification of the watchdog [11].

Colors have been used to keep track of the multiple copies of AC_{MEM}, FT_{PH}, and ER_{MEM}, as well as to model two parallel subprocesses of *AS*.

Step 2. Customize the composition rules using the associations of the customized CDs. The names/rolenames of the binary general associations defined in the CDs have been used to characterize the labels of the PN components in Figure 23.

If object level specification is assumed, and therefore SWN models are used, further "control" SWN component models may be necessary, to do the right association between colors, as, for example, in the case of AC and AF model, to associate to each *Automation Component* the corresponding *Automation Function*.

Example. The associations named *affect* allow to define three synchronization labels: *ftmem*, to synchronize the memory model and the faults, *erraf*, to synchronize the automation functions and the memory, and finally, *fail*, to synchronize the automation system and the halting failure model. New labels are introduced for the interaction between the recovery mechanism and the memory error model (*detect, noerr*) and between the recovery mechanism and the automation functions model (*recaf*). We can then define the sets of labels for the horizontal compositions of the resource level ($L_{res} = \{ftmem\}$), of the service level ($L_{srv} = \{erraf\}$, and $L'_{srv} = \{detect, noerr, recaf\}$), and the process level ($L_{pr} = \{fail\}$).

The resource layer model *R*, the service layer model *S* and process layer model *P* are then obtained by applying the composition operator $||$ over transition labels:

$$R = \left(AC_{COM} \underset{\emptyset}{||} AC_{MEM}\right) \underset{L_{res}}{||} FT_{PH},$$

$$S = \left\{\left[\left(ACF_A \underset{\emptyset}{||} ACF_S\right) \underset{\emptyset}{||} AF_{MEM}\right] \underset{L_{srv}}{||} ER_{MEM}\right\} \underset{L'_{srv}}{||} REC,$$

$$P = \left(AS \underset{L_{pr}}{||} FAIL\right) \underset{\emptyset}{||} WD$$

A similar procedure allows to identify the labels for the vertical composition of layers based on the associations *perfom* between automation components and automation functions, and on the associations *effect* between physical faults and memory errors, and between memory errors and halting failure. New labels are also added to represent: the interactions among the recovery mechanism model and the component models laying at resource level and at process level, the interactions among the automation system model and the automation (communication) functions models and, finally, the interactions among the watchdog model and the automation communication functions models.

The final PN model *PSR* is then obtained using the parallel composition of the various levels upon the identified labels, according to the PSR methodology.

Step 3. Define the initial marking of the composed PN.
A complete definition of the initial marking is possible only when system design specification is available, by composing the initial marking of the components, that may require information on the object identities.

Example. The initial marking is based on the assumptions that there is one communication unit, and that physical faults affects only one the first memory. The marking parameter M_0 has one token per color of the class $C = C_1 \cup C_2 = \{c_1\} \cup \{c_2\}$ and it is used for AC_{MEM} and AF_{MEM}. The marking parameter M_1, defined as the single color c_1, is used for FT_{PH},ER_{MEM} and *FAIL*.

Step 4. Initialize the rate/weight parameters of the PN composed model and define the results.
The rate/ weight parameters and performance/ dependability indices should be defined according to the values set to the input and output attributes of the customized classes. The remaining ones are added and initialized by the modeler.

Example. From the customized CDs of Figure 20(b) and of Figure 21(b) we can identify the following input parameters for the PN model: *comm_rate*, *copy_rate*, *fault_rate* and *duration* representing the communication rate, the rate of copy operation, the rate of the fault occurrence and its duration, respectively.

The metrics to be evaluated and/or validated are specified by three attributes: *availability*, defined in class AF_{MEM} of Figure 20(b), and *PF*, defined in the *Halting Failure* class of Figure 21(b), specifying the probability of failure. These information extracted from the CDs are only indications, no formal definition is associated to them, and they have to be defined by the modeler.

Step 5. Perform the analysis.
To perform the analysis it may not be a straightforward task, since it may require a modification of the PN model: as exemplified by the Local Voter mechanism in Sub-Section 5.2 we point out that for certain types of indices it may be necessary to perform a transient analysis in which the states representing a failure are made absorbing, while if instead a recovery strategy is being evaluated it is likely that the model should be made ergodic.

Example. We have used GreatSPN tool [52] to construct the PN component models depicted in Figure 23 and the program `algebra` [10] to carry out their composition. The reachability graph of the final SWN model contains 115 tangible markings, 778 vanishing markings and 4 dead markings (failure of the automation system). The model can be used for the computation of the probability of failure defined as the probability that place *fail* becomes marked within time t, i.e., $Pr\{M[fail](x) = 1, x \leq t\}$. The modified ergodic model (in which a restart from failure has been modeled) is characterized by 103 tangible markings and 1051 vanishing markings. The ergodic model can be used for the computation of the mean availability of the memory automation function that can be affected by error that is defined as the probability that place *error* is empty.

8 Conclusions

In this paper we have introduced the quantitative evaluation of dependability based on a probabilistic approach, following an order of presentation that somehow reflects also be the historical development of the dependability field.

Starting from the dependability of simple systems, that can be expressed as relatively simple formulas of the dependability of the elementary components of the system, following a divide and conquer approach, we have then discussed the role of state enumeration techniques for dependability, and in particular of state enumeration techniques based on Continuous Time Markov Chain.

Since state enumeration is a low-level, error-prone activity, the researchers have looked with interest into higher level formalisms as Petri nets, and in particular to whose classes of Petri nets that have an underlying stochastic process semantics as either the simple Markov Chain, like SPN, or as the more complex form of Markov Regenerative Process, like Markov Regenerative SPN [19].

When building a model of complex systems for dependability, the interplay between the system components and the FEF elements plays a central role, we have therefore also presented a systematic, compositional approach to the construction of SPN models for dependability. The approach has been exemplified with an example taken from the automation system domain.

Acknowledgements

We would like to thank Andras Horváth that has been the co-author of the work [10], part of which has been reused in this paper, as well as the person who implemented the program *algebra* for net-composition in the *GreatSPN* tool.

References

1. M. Ajmone Marsan, G. Balbo, G. Conte, S. Donatelli, and G. Franceschinis. *Modelling with Generalized Stochastic Petri Nets*. J. Wiley, 1995.
2. M. Ajmone Marsan, A. Bobbio, and S. Donatelli. Petri Nets in performance analysis, an introduction. In W. Reisig and G. Rozenberg, editors, *Lectures in Petri Nets: basic models*, pages 211–256, Berlin, Germany, 1998. Springer Verlag. LNCS, Vol 1491.
3. R.E. Barlow and F. Proschan. *Statistical Theory of Reliability and Life Testing*. Holt, Rinehart and Winston, New York, 1975.

4. S. Bernardi. *Building Stochastic Petri Net models for the verification of complex software systems.* PhD thesis, Dipartimento di Informatica, Università di Torino, April 2003.
5. S. Bernardi, C. Bertoncello, S. Donatelli, G. Franceschinis, G. Gaeta, M. Gribaudo, and A. Horváth. GreatSPN in the new millenium. Technical report, In Tools of Aachen 2001, International MultiConference on Measurement, Modelling and Evaluation of Computer-Communication System, 2001.
6. S. Bernardi and S. Donatelli. Performance Validation of Fault-Tolerance Software: A Compositional Approach. In *Proc. of the International Conference on Dependable Systems and Networks, DSN'01*, pages 379–388, Göteborg, Sweden, July 2001. IEEE Computer Society ed.
7. S. Bernardi and S. Donatelli. Building Petri net scenarios for dependable automation systems. In *Proc. of the* 10th *International Workshop on Petri Nets and Performance Models (PNPM2003)*, pages 72–81, Urbana-Champain, Illinois (USA), September 2003. IEEE Computer Society ed.
8. S. Bernardi and S. Donatelli. Stochastic Petri nets and inheritance for dependability modelling. In *Proc. of the* 10th *Pacific Rim International Symposium on Dependable Computing (PRDC04)*, Papeete, tahiti (French Polynesia), March 2004. IEEE C.S. to be published.
9. S. Bernardi, S. Donatelli, and G. Dondossola. Methodology for the generation of the modeling scenarios starting from the requisite specifications and its application to the collected requirements. Deliverable D1.3b - DepAuDE IST Project 25434, June 2002.
10. S. Bernardi, S. Donatelli, and A. Horváth. Special section on the pratical use of high-level Petri Nets: Implementing Compositionality for Stochastic Petri Nets. *Journal of Software Tools for Technology Transfer (STTT)*, 3(4):417–430, August 2001.
11. S. Bernardi, S. Donatelli, and J. Merseguer. From UML Sequence Diagrams and Statecharts to analysable Petri Net models. In *Proceedings of the* 3rd *International Workshop on Software and Performance*, pages 35–45, Rome (Italy), July 2002.
12. C. Betous-Almeida and K. Kanoun. Stepwise Construction and Refinement of Dependability Models. In *Proc. of the International Conference on Dependable Systems and Networks, DSN'02*, pages 515–524, Washington, D.C., USA, June 2002. IEEE Computer Society ed.
13. A. Bobbio. Teoria e Metodi di affidabilità. Dispense COREP - Dipartimento di Informatica, Università del Piemonte Orientale, Alessandria, Italia (in italian).
14. A. Bobbio. Petri Nets Generating Markov Reward Models for Performance/Reliability Analysis of Degradable Systems. In Puigjaner, R. et al., editors, *Modeling Techniques and Tools for Computer Performance Evaluation. Proceedings of the Fourth International Conference, 1988, Palma, Spain*, pages 353–365, New York, NY, USA, 1989. Plenum.
15. A. Bobbio, G. Franceschinis, R. Gaeta, and L. Portinale. Parametric Fault Tree for the Dependability Analysis of Redundant Systems and Its High-Level Petri Net Semantics. *IEEE Trans. Software Eng.*, 29(3):270–287, 2003.
16. A. Bondavalli, M. Dal Cin, D. Latella, I. Majzik, A. Pataricza, and G. Savoia. Dependability analysis in the early phases of UML-based system design. *International Journal of Computer Systems Science & Engineering*, 16(5):265–275, September 2001.
17. O. Botti, V. De Florio, G. Deconinck, F. Cassinari, S. Donatelli, A. Bobbio, A. Klein, H. Kufner, R. Lauwereins, E. Thurner, and E. Verhulst. TIRAN: Flexible and Portable Fault Tolerance Solutions for Cost Effective Dependable Applications. In P. Amestoy, P Berger, M. Daydé, I. Duff, V. Frayssé, L. Giraud, and D. Ruiz, editors, *Proceedings 5th Int. Euro-Par Conference on Parallel Processing (Europar'99) - Lecture Notes in Computer Science Vol. 1685*, pages 1166–1170. Springer Verlag, Berlin, Germany, Toulouse, France, Aug. 31-Sep. 3 1999.
18. G. Chiola, C. Dutheillet, G. Franceschinis, and S. Haddad. Stochastic Well-Formed coloured nets for symmetric modelling applications. *IEEE Transaction on Computers*, 42(11):1343–1360, November 1993.

19. H. Choi, V.G. Kulkarni, and K. Trivedi. Markov Regenerative Stochastic Petri Nets. *Performance Evaluation*, 20:337–357, 1994.
20. Gianfranco Ciardo and Kishor S. Trivedi. SPNP: The Stochastic Petri Net Package (Version 3.1). In *Proc. 1st Int. Workshop on Modeling, Analysis and Simulation of Computer and Telecommunication Systems (MASCOTS'93)*, pages 390–391. IEEE Comp. Soc. Press, 1993.
21. International Electrotechnical Commission. IEC-60300-3-1: Dependability Management. IEC, 3 rue de Varembé CH 1211 Geneva, Switzerland, 2001.
22. S. Contini and A Poucet. Advances on fault tree and event tree techniques. In In A.G. Colombo and A. Saiz de Bustamante, editor, *System Reliability Assessment*, pages 77–102. Kluwer Academic P.G., 1990.
23. J.A. Couvillion, R. Freire, R. Johnson, W. Douglas Obal II, M.A. Qureshi, M. Rai, W.H. Sanders, and J.E. Tvedt. Performability Modeling with UltraSAN. *IEEE Software*, 8(5):69–80, 1991.
24. D.R. Cox. *Renewal theory*. Chapman & Hall, London, 1962.
25. D.R. Cox and H.D. Miller. *The theory of stochastic processes*. Chapman and Hall, London, 1965.
26. DepAuDE EEC-IST project 2000-25434. http://www.depaude.org.
27. S. Donatelli and G. Franceschinis. The PSR methodology: integrating hardware and software models. In *Proc. of the* 17th *International Conference in Application and Theory of Petri Nets, ICATPN '96*, Osaka, Japan, June 1996. Springer Verlag. LNCS, Vol 1091.
28. J. B. Dugan, K. S. Trivedi, R. M. Geist, and V. F. Nicola. Extended Stochastic Petri Nets: Applications and Analysis. In Gelenbe, E, editor, *PERFORMANCE'84: Models of Comput. System Performance, Proc. of the 10th Int. Symp., Paris*, pages 507–519, Amsterdam, 1984. Elsevier.
29. J.B. Dugan and K.S. Trivedi. Coverage modelling for dependability analysis of fault tolerant systems. *IEEE Transaction on Computers*, 38(6):775–787, 1989.
30. C. Béounes et al. SURF-2: A Program for Dependability Evaluation of Complex Hardware and Software Systems. In 23rd *Int. Symp. on Fault-Tolerant Computing*, pages 668–673, Toulouse (France), 1993.
31. N.B. Fuqua. *Reliability Engineering for Electronic Design*. Marcel Dekker Inc., New York, 1987.
32. E.J. Henley and H Kumamoto. *Reliability Engineering and Risk Assessment*. Prentice Hall, Englewood Cliffs, 1981.
33. G. S. Hura. A Petri Net Approach to Enumerate all System Success Paths for Reliability Evaluation of a Complex System. *Microelectron. Reliab. (GB)*, 22(3):427–428, 1982.
34. G. S. Hura and J. W. Atwood. The Use of Petri Nets to Analyze Coherent Fault Trees. *IEEE Trans. Reliab. (USA)*, 37(5):469–474, 1988.
35. O. Ibe, A. Sathaye, R. Howe, and K. S. Trivedi. Stochastic Petri Net Modeling of VAXcluster Availability. In *Proc. Third Int. Workshop on Petri Nets and Performance Models (PNPM89)*, pages 112–121, Kyoto (Japan), 1989.
36. IEC-10125. *Fault Tree Analysis*. IEC-Standard-No. 10125, 1990.
37. IEC-61165. *Application of Markov techniques*. IEC-Standard-No. 61165, 1995.
38. Jensen K. *Coloured Petri Nets. Basic Concepts, Analysis Methods and Practical Use*, volume 1. Springer-Verlag, 1997. Monographs in Theoretical Computer Science, ISBN:3-540-60943-1.
39. K. Kanoun, M. Borrel, T. Moreteveille, and A. Peytavin. Modeling the Dependability of CAUTRA, a Subset of the French Air Traffic Control System. In *Proceedings of the* $26^t h$ *Int. Symp. Fault-Tolerant Computing (FTCS-26)*, pages 95–515, Sendai (Japan). LAAS-REPORT.

40. K. Kanoun and Borrel M. Dependability of fault-tolerant systems. Explicit modeling of the interactions between hardware and software. In *Proc. of the 2nd Annual IEEE International Computer Performance and Dependability Symposium (IPDS'96)*, pages 252–261, Urbana Champain, USA, Sept 1996. IEEE-CS Press.
41. A. Kaufmann, D. Grouchko, and R. Cruon. *Mathematical Models for the Study of the Reliability of Systems*. Academic Press, 1977.
42. V. G. Kulkarni. *Modeling and Analysis of Stochastic Systems*. Chapman & Hall, 1995.
43. J. C. Laprie. Dependability – Its attributes, impairments and means. In B. Randell, J.C. Laprie, H. Kopetz, and B. Littlewood, editors, *Predictably Dependable Computing Systems*, pages 3–24. Springer Verlag, 1995.
44. Randell B. Laprie J.C. and Avizienis A. Fundamental Concepts of Dependability. Technical report, LAAS - NewCastle University - UCLA, 2001. LAAS Report no. 01-145, NewCastle University Report no. CS-TR-739, UCLA CSD Report no. 010028.
45. M. Malhotra and K.S. Trivedi. Dependability Modelling using Petri net based models. *IEEE Transactions on Reliability*, 44(3):428–440, 1995.
46. M. Ajmone Marsan, A. Bobbio, G. Conte, and A. Cumani. Performance analysis of degradable multiprocessor systems using Generalized Stochastic Petri Nets. *Distributed Processing Technical Committee Newsletter*, 6(SI-1):47–54, 1984.
47. M.K. Molloy. Performance analysis using Stochastic Petri Nets. *IEEE Transaction on Computers*, 31(9):913–917, September 1982.
48. J. Muppala, G. Ciardo, and K. Trivedi. Stochastic reward nets for reliability prediction. *Communications in Reliability, Maintainability and Serviceability*, 1(2):9–20, July 1994.
49. J. Muppala, R. Fricks, and K. S. Trivedi. Techniques for System Dependability Evaluation. In W. Grassman, editor, *Computational Probability*, pages 445–480, The Netherlands, 2000. Kluwer Academic.
50. I. Mura, S. Chiaradonna, and A. Bondavalli. Modelli teorici e pratici per la rappresentazione del processo di guasto. Progetto di ricerca PDCC-ENEA: Aspetti specifici e tecniche di tolleranza ai guasti. (in italian).
51. A. Papoulis. *Probability, Random Variables and Stochastic Processes*. Mc Graw Hill, New York, 1965.
52. Performance Evaluation group of Torino. The GreatSPN tool. http://www.di.unito.it/ greatspn.
53. L. Pomello, G. Rozenberg, and C. Simone. A Survey of Equivalence Notions for Net Based Systems. *Lecture Notes in Computer Science; Advances in Petri Nets 1992*, 609:410–472, 1992.
54. M Rabah and K. Kanoun. Performability evaluation of multipurpose multiprocessor systems: the "separation of concerns" approach. *IEEE Transactions on Computers. Special Issue on Reliable Distributed Systems*, 52(2):223–236, February 2003.
55. R. A. Sahner and K.S. Trivedi. Reliability Modeling using SHARPE. *IEEE Transactions on Reliability*, R-36(2):186–193, June 1987.
56. W.H. Sanders and L.M. Malhis. Dependability Evaluation Using Composed SAN-Based Reward Models. *Journal of Parallel and Distributed Computing*, 15(3):238–254, 1992.
57. W.H. Sanders and J.F. Meyer. Stochastic Activity Networks: Formal Definitions and Concepts. *Lecture Notes in Computer Science*, 2090:315–??, 2001.
58. W. Schneeweiss. *Petri Nets for Reliability Modelling*. LiLoLe-Verlag GmbH, Hagen (Germany), 1999.
59. W.G. Schneeweiss. *The Fault Tree Method*. LiLoLe Verlag, 1999.
60. M.L. Shooman. *Probabilistic reliability: an engineering approach*. Mc Graw Hill, 1968.
61. W.J. Stewart. *Introduction to the Numerical Solution of Markov Chains*. Princeton University Press, 1994.

62. K.J. Sullivan, J.B. Dugan, and D. Coppit. The Galileo Fault Tree Analysis Tool. In *Proceedings of the 29th Annual International Symposium on Fault-Tolerant Computing*, Madison, Wisconsin, 1999. IEEE.
63. K. Trivedi. *Probability & Staticstics with Reliability, Queueing & Computer Science applications*. Prentice Hall, 1982.
64. K. Trivedi. *Probability & Statistics with Reliability, Queueing & Computer Science applications*. Wiley, II Edition, 2001.
65. W.M.P. Van der Aalst. Inheritance of Dynamic Behavior in UML. In Daniel Moldt ed., editor, *Proc. of the* 2th *Workshop on Modelling of Objects, Components and Agents, MOCA'02*, Aarhus, Denmark, August 2002. Technical Report ISSN 0105-8517, Dept. of Computer Science, University of Aarhus.
66. W.M.P. Van der Aalst and T. Basten. Life-cycle inheritance: A Petri-net based approach. In P. Azéma and G. Balbo, editors, *Application and Theory of Petri Nets 1997*, volume 1248, pages 62–81. Springer Verlag, Berlin, 1997.
67. X. Xie and S.M. Shatz. Development of Class-level and Instance-level Design Model for Distributed Systems. *International Journal of Informatica, special issue on Component Based Software Development*, 25:465–474, 2001.

Process Algebra
A Petri-Net-Oriented Tutorial

Eike Best[1] and Maciej Koutny[2]

[1] Parallel Systems, Faculty of Computing Science
Carl von Ossietzky Universität Oldenburg
D-26111 Oldenburg, Germany
Eike.Best@informatik.uni-oldenburg.de

[2] School of Computing Science, University of Newcastle
Newcastle upon Tyne NE1 7RU, United Kingdom
Maciej.Koutny@newcastle.ac.uk

Abstract. Process algebras aim at defining algebraic calculi for concurrency and communication between concurrent processes. This paper describes some of the issues that would seem to be worth discussing when process algebraic ideas are related to Petri net theoretical concepts.

Keywords: Petri nets, process algebras.

1 Introductory Remarks

Many research monographs and textbooks have been written on the subject of *process algebras* (see, for instance, [3, 6, 19, 20, 24–26]), and even a comprehensive handbook of 1342 pages was published three years ago [4]. Thus, even when the topic is restricted to the relationship between process algebras and Petri nets, there is a vast amount of work that cannot be described in short a tutorial. We therefore chose to write an account of *selected* thoughts that might occur to someone attempting to form a link between process algebras and Petri nets; moreover, while aiming at an easy-to-be-followed presentation, we included several pointers to the relevant papers and results. We will discuss a well-known concrete process algebra (the reader should be aware that there are several of them, just as there are several classes of Petri nets), and highlight some of the key issues both in the design of such an algebra, and in its translation into Petri nets.

Process algebras have been studied for about 25 years, but some ideas behind them go back even further; indeed, even regular expressions can be viewed as a very simple process algebra equivalent to finite state automata. First seeds for what has by now become a standard approach to process algebra have been sown by Robin Milner [24], based on his work on flow graphs, and Tony Hoare [18, 19], based on his ideas about designing concurrent programs. Also, at around the same time, the relationship between the path notation, which originates from operating systems design (Roy Campbell and Nico Habermann [8]), and Petri nets has been pioneered by Peter Lauer and others [20]. Soon after that, research groups in Amsterdam started to investigate an axiomatic approach to process algebra [3].

J. Desel, W. Reisig, and G. Rozenberg (Eds.): ACPN 2003, LNCS 3098, pp. 180–209, 2004.

We here concentrate on the relationship between process algebras and Petri nets which, in particular, could allow one to apply Petri-net-based analysis methods (such as the S-invariant method) in the process algebraic framework, and to transfer concepts of concurrency and causality that are well-developed in Petri net theory into the domain of process algebras. Apart from the already mentioned work on path expressions, some early influential research in this area has been carried out by Ursula Goltz [17], Gerard Boudol and Ilaria Castellani [7], Rob van Glabbeek and Frits Vaandrager [16], Ugo Montanari and his research group in Pisa [11], and Ernst-Rüdiger Olderog [26]. Much of this work aimed at giving a faithful translations for the existing process algebras, the majority of which had been designed without Petri nets in mind. By contrast, but also building upon this earlier work, the 'box algebra' approach of [6] on which this paper is based, aimed at designing a 'Petri nets-friendly' process algebra.

2 Basic CCS

As a basis for our discussion, we have chosen the original *Calculus of Communicating Systems* by Robin Milner [24, 25], referred to as the *(basic) CCS*. The word 'basic' here reflects the fact that there exist numerous extensions (e.g., [2, 9]) as well as interesting restrictions (e.g., [10]) of CCS and related process algebras. CCS is *action-based*, which means that it is built upon a set of (*atomic*) 'observable' actions $A = \{a, \widehat{a}, b, \widehat{b}, c, \widehat{c}, \ldots\}$ together with a distinct 'unobservable' (or 'silent') action τ. Every observable action a is coupled with another observable action, denoted by $\widehat{a}$ and called its *conjugate* (and $\widehat{a}$ has a as its conjugate, i.e., $a = \widehat{\widehat{a}}$). (Though Milner's CCS uses over-barring, such as $\overline{a}$, to denote conjugates, we changed this notation to over-hatting in order to avoid potential confusion with another over-barring, to be introduced later on.)

CCS provides a method for structuring a collection of actions in such a way that some overall coordinated behaviour is described in order, e.g., to specify the desired behaviour of some still-to-be-designed system, or to describe crucial aspects of the behaviour of some existing system. Technically, this is achieved by imposing a syntax on top of the set of actions, as follows:

$$E \ ::= \ \mathsf{nil} \ \big| \ E + E \ \big| \ z.E \ \big| \ E \,\mathsf{|}\, E \ \big| \ E \backslash a \ .$$

In the above, a ranges over A, and z over $A \cup \{\tau\}$. The syntax defines CCS *process expressions* (or just *processes*) E, and it should be read in the usual Backus-Naur style. Thus, for example, nil is a process, and if E and F are processes then so is $E + F$. Within the syntax, symbols '$+$' and '$\mathsf{|}$', as well as constructs '$z.$' and '$\backslash a$' denote *process algebraic operators*.

There is a single basic process nil without any operator at all which can be viewed, in the usual way, as a nullary operator, or a *constant*. There is, furthermore, an infinite family of unary operators, parameterised by actions (the *prefixing operators*, $z.$). Another family of unary operators, written in postfix style, are the *restrictions* $\backslash a$ (one for each $a \in A$). The only binary operators, written in infix notation, are the *choice* '$+$' and *composition* '$\mathsf{|}$'. To avoid an excessive use of the parentheses, it is assumed that prefixing binds more tightly than choice, and that choice binds more tightly than composition; thus, for instance, $a.b.\mathsf{nil} + c.\mathsf{nil} \mid d.\mathsf{nil}$ stands for $((a.(b.\mathsf{nil})) + (c.\mathsf{nil})) \mid (d.\mathsf{nil})$.

Process variables, such as X, are also considered to be basic processes, each such variable being associated with a unique defining expression of the form $X \stackrel{\mathrm{df}}{=} E$. Variables provide support for different levels of abstraction within a CCS specification, as well as for recursion: e.g., X with $X \stackrel{\mathrm{df}}{=} a.X$ is a recursive expression.

Every process has an associated set of behaviours. It is one of the main objectives of *formal semantics* of a process algebra to make this notion precise. The behaviour of process expressions is often given in terms of 'can make a move' statements, which are similar to 'an action can be executed' statements in general concurrent systems terminology, or 'a transition can occur' statements in Petri net terminology (for example, the process $a.\mathsf{nil}$ can make move a, and thereafter no further move). Referring to CCS syntactic constructs, we have that the basic process nil can make no move at all; $E + F$ behaves either like E or like F; $z.E$ can make the move z and then behaves like E; $E \mid F$ behaves like both E and F, together with some further (synchronised) activities which will be described later; $E \backslash a$ behaves like E, except that the actions a and $\widehat{a}$ are blocked; and, finally, X behaves like the process E in the equation $X \stackrel{\mathrm{df}}{=} E$.

A straightforward attempt to capture the behaviour associated with a process expression could be to emulate the approach taken when the language of a regular expression is defined. However, this might lead to identifying processes which may behave rather differently in certain contexts (see the discussion in Section 4). To address this problem, CCS uses more refined scheme for describing the possible moves of a process, namely that offered by structural operational semantics (SOS), pioneered by Gordon Plotkin [28]. In SOS, the concept of *inference rules* of formal logic is applied to processes, using the fact that if a process E makes a move, then the 'remaining' behaviour of E after making this move can be described by another process expression E'. For example, $E = a.\mathsf{nil} + b.c.\mathsf{nil}$ can make move b, and the remaining behaviour can be described by process $E' = c.\mathsf{nil}$ (note that the left-hand side branch of the + together with the + itself, has disappeared because it was not selected, and that the b has disappeared because it was executed). In general, we say that 'E makes a move z and becomes E'', and write this formally as $E \xrightarrow{z} E'$. For processes, then, the SOS rules have the following format: *if* some expression E can make a move and become E' (the rule's *premise*), *then* some expression D (syntactically related to E) can also make a move and become D' (the rule's *conclusion*). A rule is divided by a horizontal bar, above which its premise, and below which its conclusion are given. For example, the two rules for choice are:

$$\frac{E \xrightarrow{z} E'}{E + F \xrightarrow{z} E'} \quad \text{and} \quad \frac{F \xrightarrow{z} F'}{E + F \xrightarrow{z} F'}$$

In this vein, one may go through the syntax providing suitable rule(s) for each syntactic clause, i.e., for each operator (or family of operators). For example, in case of restriction we have a single rule:

$$\frac{E \xrightarrow{z} E' \ , \ z \neq a \ , \ z \neq \widehat{a}}{E \backslash a \xrightarrow{z} E' \backslash a}$$

Thus $E\backslash a$ can make a z-move if E can and z is different from a and $\widehat{a}$. It is also worth noting that there are no rules at all for the process nil.

3 From Process Expressions to Petri Nets

We now will go through the basic CCS process algebra and discuss several of the concepts contained in it, as well as some of the issues pertaining to its connection to Petri nets. The discussion of infinite behaviour is deferred to the last section of this paper.

Sequential Composition. The basic CCS means of generating sequential behaviour is through the prefixing operator. For instance, $a.b.\mathsf{nil}$ can 'make an a-move', thereafter 'make a b-move', and then terminate:

$$a.b.\mathsf{nil} \xrightarrow{a} b.\mathsf{nil} \xrightarrow{b} \mathsf{nil}.$$

It may be observed here that $a.b.\mathsf{nil}$, $b.\mathsf{nil}$ and nil are expressions with different *structure*, indicating that a CCS expression may be subjected to a structural change through behaviour (in general, the structure and the behaviour of CCS expressions are closely intertwined). Moreover, it is fairly clear that a Petri net equivalent of $a.b.\mathsf{nil}$ should be something like the transition-labelled net shown on the left-hand side of Figure 1. Note that we do need the labelling as one might write a process such as $a.a.\mathsf{nil}$, giving rise to two different transitions with the same label a.

Fig. 1. Net for $a.b.\mathsf{nil}$ before and after the occurrence of a.

There is a conceptual discrepancy in thinking about structure and behaviour in CCS and in Petri nets. After a has occurred in the net, the transition labelled a is still visible in its structure (see the right-hand side of Figure 1), though it cannot be activated anymore, whilst the expression $a.b.\mathsf{nil}$ changes to $b.\mathsf{nil}$ after the occurrence of a, and thus the a is 'lost' (this way of viewing behaviour is actually quite particular to the standard SOS semantics rather than to process algebra as such).

What would happen if we wished to put more complicated partial expressions into sequence, rather than just two actions a and b? The CCS syntax tells us that having a in sequence with something arbitrarily complicated is possible. However, it explicitly disallows anything more complicated than just an action to be written *in front* of the prefix operator symbol. For instance, writing $(a + b).\mathsf{nil}$ is forbidden. Theoretically speaking, this is not too restrictive, because the general sequential composition can be 'emulated', using in an essential way the still-to-be-explained composition operator and restriction. Nevertheless, it may be useful to have a more general sequential operation at one's disposal, and there are some process algebras with such an operator as a basic feature, e.g., [3].

Choice Composition. This seems to be a rather innocent operation. However, as we will see, some care needs to be taken when it is applied in combination with other constructs. Let us first see how a simple choice between a and b, such as $a.\mathsf{nil} + b.\mathsf{nil}$, can be expressed in terms of Petri nets. We haven not yet considered the nil by itself, but after looking at its role in the previous example, it should indicate 'the end of a net (or of net's execution)'. We emphasised this by naming the place to which nil corresponds x (for 'exit', as opposed to the place where the initial token resides, which is called e, for 'entry').

Given $a.\mathsf{nil}+b.\mathsf{nil}$, should the corresponding net have only one end, or two, or even three ends? All three possibilities are shown in Figure 2, and all of them can be found in the literature. Moreover, when the choice is embedded in a context, we might wish to model it as being an *internal* one, by which is meant that the process performing the choice should be free to 'make up its mind' without being hindered by any external influence. A good way of modelling internal choice is shown on the right-hand side of Figure 2. The choice between a and b is 'pre-poned' there, in the sense that it depends on a prior choice between two silent τ actions (there are process algebras in which one may distinguish syntactically between external and internal choice, such as TCSP [19]). However, such a translation is inappropriate for CCS: Indeed, CCS has another expression corresponding to this net, $\tau.a.\mathsf{nil} + \tau.b.\mathsf{nil}$, which needs to be distinguished carefully (although we have not yet explained exactly why) from the original expression $a.\mathsf{nil} + b.\mathsf{nil}$.

Now, which of the three possibilities (a/b/c) in Figure 2 is to be preferred as a translation of $a.\mathsf{nil} + b.\mathsf{nil}$? Cases can be made for all of them: Figure 2(b) [11] seems to be the closest Petri net correspondent of that expression, since there are two nil's in it and two x places in the net; Figure 2(c) has strong category theoretical (if not categorical) arguments in its favour [33]; and Figure 2(a) [6] treats choice as a forward/backward symmetric sequential construct (almost as a nondeterministic **if–fi** statement in programming [12], or like $a+b$ in a regular expression). Moreover, the latter allows the net to be used as a building block in the following sense. An expression such as $(a + b); c$ where the semicolon denotes sequential composition, would give, using Figure 2(a), the left-hand side

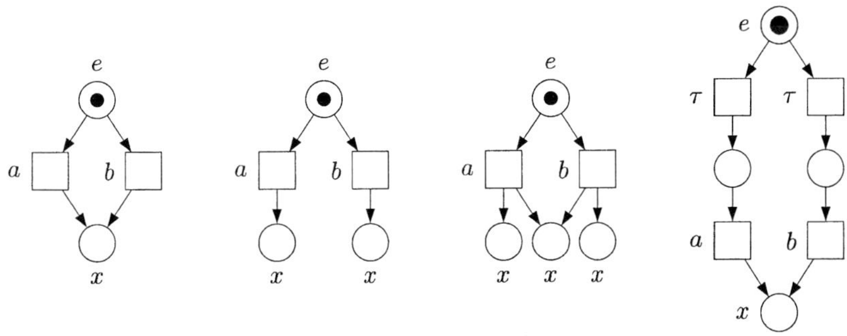

Fig. 2. Nets for $(a.\mathsf{nil}) + (b.\mathsf{nil})$: (a) one-ended, (b) two-ended, (c) three-ended, and (d) with internal choice.

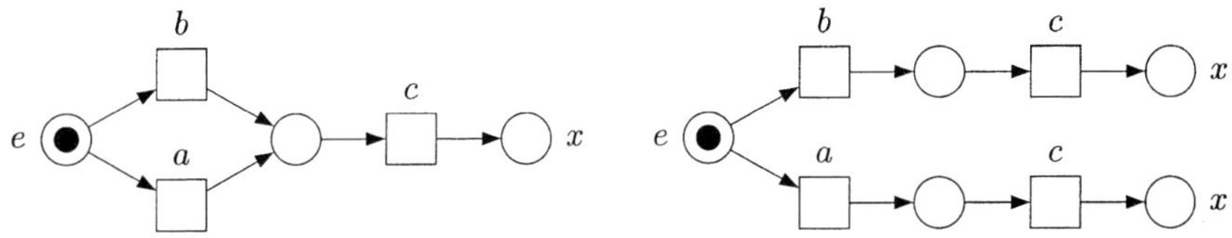

Fig. 3. Nets for $(a + b); c$ constructed from Figure 2(a) and Figure 2(b).

of Figure 3. Figure 2(b) would not quite as easily fit such an extension; the net shown on the right-hand side of Figure 3 is close to $(a + b); c$ and has two x places, but also contains a duplication of the c action which is not present in the process expression.

Parallel Composition. Putting processes in parallel for performance reasons, or for architectural reasons, has long been a challenge for theoreticians, as well as for practitioners, because the behaviour of such compound processes seems to be much more difficult to describe and to master than sequential behaviour. For instance, if two parallel processes access common variables, one needs to be able to control these accesses and their orderly interaction. Numerous special programming constructs have been invented in order to manage this, and other kinds of, interaction between parallel processes. One that plays a particularly important role in process algebra is the concept of a *handshake communication*. This is a form of a symmetric synchronisation, by which an action that is shared between, say, n processes can be executed only if *all n of them* are ready to do so. This is much like in Petri net theory, where a transition with n input places can occur only if all of these places carry a token.

Basic CCS has a particular incarnation of the handshake, in that it is always two-partnered (i.e., n=2), and it always results in an internal action. Moreover, handshakes can occur only between a actions and their conjugates, $\widehat{a}$ actions. The idea is, quite literally, that two conjugates fit together like two hands about to engage in a handshake, and that the resulting τ synchronisation describes the handshake itself. Let us have a look at the three SOS rules for CCS composition:

$$\frac{E \xrightarrow{a} E'}{E \mid F \xrightarrow{a} E' \mid F} \qquad \frac{F \xrightarrow{a} F'}{E \mid F \xrightarrow{a} E \mid F'} \qquad \frac{E \xrightarrow{a} E' \; , \; F \xrightarrow{\widehat{a}} F'}{E \mid F \xrightarrow{\tau} E' \mid F'}$$

First, note carefully the difference between the first two rules and the rules for choice composition: the processes 'remaining' after move a still contain parts of the original processes which had not been involved in the move, while in the rule for choice these have disappeared. This models parallel (rather than nondeterministically selected) behaviour. Second, note the appealing simplicity of the third, handshake rule: if E can make an a-move and F can make an $\widehat{a}$-move, then their composition $E \mid F$ can make a τ-move, creating as a new process the composition of whatever remains from E and F after making their respective moves.

For instance, what are the sequences of moves, or in technical terms, the *interleavings*, that the process $a.a.\mathsf{nil} \mid \widehat{a}.\widehat{a}.\mathsf{nil}$ could make? If we applied only the first two rules of $\mid$, we would have move sequences $aa\widehat{a}\widehat{a}$, $a\widehat{a}a\widehat{a}$, $a\widehat{a}\widehat{a}a$, $\widehat{a}\widehat{a}aa$, $\widehat{a}a\widehat{a}a$, $\widehat{a}aa\widehat{a}$ (and all their prefixes as well). But if we now also take the third rule into account, there are still more behaviours of that expression, such as $\tau a\widehat{a}$. What happened here is that we let the first a on the left-hand side of $a.a.\mathsf{nil} \mid \widehat{a}.\widehat{a}.\mathsf{nil}$ be synchronised with the first $\widehat{a}$ on the right-hand side, and the rest got executed in an unsynchronised way. Of course, $\tau\widehat{a}a$ is also a possible behaviour, and so are $a\widehat{a}\tau$, $\widehat{a}a\tau$ and $\tau\tau$. Still more behaviour is possible if we let the left-hand side process execute its first a and then synchronise its second a with the first $\widehat{a}$ on the right-hand side, thus: $a\tau\widehat{a}$. And, by symmetry, we may have $\widehat{a}\tau a$. What we just witnessed suggests that a parallel composition can give rise to a bewildering amount of behaviour. In fact, using only prefixing and composition, it is possible to construct processes that give rise to exponentially many interleavings (in terms of the size of an expression). What is so bewildering about this (or perhaps, what so bewildered early researchers on the subject) is not the sheer size of the set of potential interleavings, but the wide variety of possible interferences of one process by another one at unpredictable points in time, some of which may be quite unwanted in real applications. In practice, one is interested in such a global point of view, though, in order to prove assertions about the possible set of *all* behaviours. Of course, the $\mid$ rule does not really help here, being (as all SOS rules are, and also as the transition rule in Petri nets is) local in nature. However, such a rule is a solid formal foundation for further applications of more advanced and global methods, just as the transition rule in Petri nets is necessary for, e.g., the state equation, and the latter, for instance, for the structural analysis of nets.

Other process algebras have different parallel operators and different action synchronisation schemes. COSY and, more particularly, the path notation [8, 20], have top-level parallelism, which means that one may express a collection of parallel processes and paths, inside which the parallel operator may not occur;

i.e., no nesting of parallel composition is allowed. Synchronisation occurs through *equality of action names*, that is, if an action a occurs in two or more different processes (or, to be precise, paths), it may be executed only if all of them are ready to do so. TCSP [19] also has synchronisation on common action names – with some subtle differences to COSY – but, on the other hand, allows nested parallelism. Hoare's original CSP [18] can be characterised as a mixture of top-level parallelism and conjugation-based two-way synchronisation. Conjugation is syntactically expressed in CSP by !-actions (send actions) and ?-actions (receive actions); see [5] for an account of its translation into Petri nets. The PBC [6] generalises conjugation to n-way synchronisation, where $n \geq 2$.

Let us now turn to the translation of processes using the composition operator into Petri nets. Actually, it is not very hard to define such a translation. Given $E \mid F$, one first draws separately the nets for E and F and lays them side by side (or, in technical terms, forms their *disjoint union*). In a second step, one collects all pairs of transitions that are labelled by a in one of the components and by $\widehat{a}$ in the other component and inserts a common transition, labelled by τ, that acts in the net of each component like the transition it originates from does. We need to do this not just for $a, \widehat{a}$-pairs, but also for all other pairs of conjugates. For the $a.a.\mathsf{nil} \mid \widehat{a}.\widehat{a}.\mathsf{nil}$ example, we actually have *four* matching $a, \widehat{a}$-pairs, because there are two a's on the left-hand side and two $\widehat{a}$'s on the right-hand side, and any two of them can be matched. The construction thus yields the labelled Petri net shown in Figure 4. Virtually all translations of CCS into Petri nets that exist in the literature will yield this net; there is not really any alternative way of the sort we discussed in the case of the choice operator.

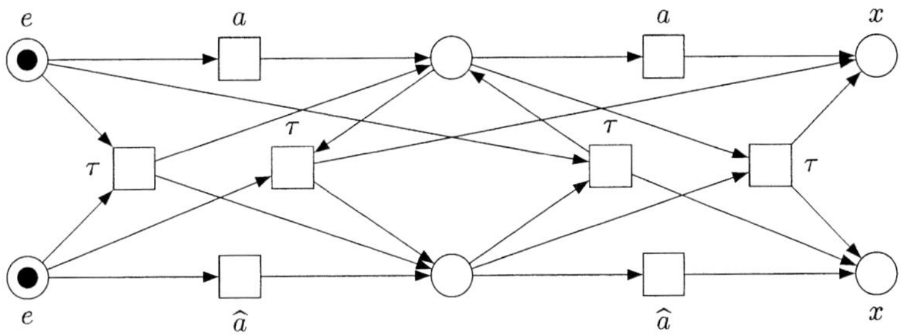

Fig. 4. Net for $a.a.\mathsf{nil} \mid \widehat{a}.\widehat{a}.\mathsf{nil}$.

Restriction. To analyse its effect, consider the process $(b.a.\mathsf{nil})\backslash a$. It can make a b-move, and then no further move, because the a has been restricted away. Note carefully that the same is true for the process $b.\mathsf{nil}$, but that there is still an important difference between these two processes: the latter 'terminates properly', while the former 'gets stuck' (or, more technically, it *deadlocks* without reaching a proper final state). This difference is not visible in terms of the behavioural sequences of the two terms, unless we add some information about termination.

The Petri net of a restricted process $E\backslash a$ is simply obtained by omitting from the net of E all a-labelled and all $\widehat{a}$-labelled transitions together with the adjacent arcs. Thus, the net corresponding to $(b.a.\mathsf{nil})\backslash a$ looks like that on the left-hand side of Figure 5. On the right-hand side of Figure 5, the net corresponding to $b.\mathsf{nil}$ is shown. The isolated x place on the left-hand side may appear superfluous, but it (and the fact that place s is *not* an exit place) captures the fact that this net cannot terminate properly, and thus describes the main distinction between processes $(b.(a.\mathsf{nil}))\backslash a$ and $b.\mathsf{nil}$. It is therefore good to keep that place around, or at least keep the information somewhere that place s is not an exit place.

Fig. 5. Nets for $(b.a.\mathsf{nil})\backslash a$ and $b.\mathsf{nil}$.

Restriction is typically used in combination with composition $|$, the main idea being that a pair of conjugated actions is used in order to create a desired synchronisation, after which the pair is no longer needed and can be restricted away. For instance, we may consider the process $((a.(a.\mathsf{nil}))\,|\,(\widehat{a}.(\widehat{a}.\mathsf{nil})))\backslash a$, coming from the expression discussed in the last section. Restricting away the a's (and $\widehat{a}$'s) implies that only the τ-behaviour of that expression still remains possible; that is, it can make two τ moves, and that is all. Let us have a look at the corresponding net in Figure 6 to check whether this is correctly reflected. Only the first τ-transition can occur, and then the fourth, putting both tokens on their respective x places (and thus terminating execution properly). Rightfully, the second and third τ-transitions have become 'dead' (in the sense that they cannot be executed anymore), since they were both associated with a prior execution of a or, respectively, of $\widehat{a}$, and both executions are no longer possible. One might be tempted to delete them from the net. However, such an optimisation is in general very hard to detect, and it is thus better not to incorporate it in any basic constructions, but to add it, if desired, as an appendix to the systematic construction.

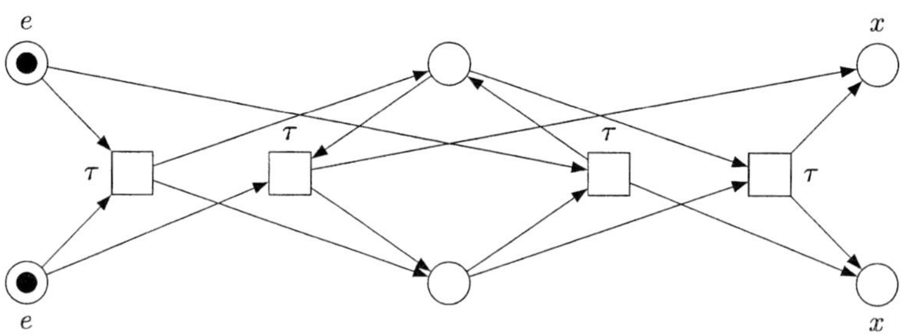

Fig. 6. Net for $(a.a.\mathsf{nil}\,|\,\widehat{a}.\widehat{a}.\mathsf{nil})\backslash a$.

A similar argument as for the difference between $(b.a.\mathsf{nil})\backslash a$ and $b.\mathsf{nil}$ can be used to justify why processes $\tau.a.\mathsf{nil} + \tau.b.\mathsf{nil}$ and $a.\mathsf{nil} + b.\mathsf{nil}$ should not be treated as equivalent. For suppose that either is put into the context $(\bullet\backslash a)$, that is, suppose that the 'hole' in that expression (the bullet, '$\bullet$') is first replaced by $\tau.a.\mathsf{nil} + \tau.b.\mathsf{nil}$ and then by $a.\mathsf{nil} + b.\mathsf{nil}$. We obtain, respectively:

$$(\tau.a.\mathsf{nil} + \tau.b.\mathsf{nil})\backslash a \quad \text{and} \quad (a.\mathsf{nil} + b.\mathsf{nil})\backslash a \ .$$

The internal choice in the first of these expressions can be resolved in such a way that the τ just in front of the a is selected, and after that, the entire expression can make no further move and is 'deadlocked', similarly to the expression $(b.(a.\mathsf{nil}))\backslash a$ after executing b. By contrast, in the second expression, no τ can be selected, and the a cannot be selected either, because it is restricted away; thus, the b remains executable, and after executing it, a final state has been reached, so that no such deadlock is possible. Thus, the two expressions behave differently in the same context, and one should, as a rule, be careful to distinguish them. Any formal semantics that does *not* distinguish two expressions which behave differently in a given (desired) context is technically called *not fully abstract.* More precisely, if a formal semantics distinguishes two expressions whenever they behave differently in at least one context out of a predefined set of contexts (and equates them if such a context cannot be found), then it is called *fully abstract* (with respect to that set of contexts).

4 Algebraic Laws, and Bisimilarity

Looking at the two SOS rules for the choice operator, it is clear that processes like $E+F$ and $F+E$ should be semantically the same. Now, if we looked at their nets, such a conclusion would be easy to defend as the nets are just isomorphic. However, no such notion of isomorphism suggests itself easily for expressions. Here is another case: $E\backslash a$ and $(E\backslash a)\backslash a$. Clearly, restricting a two times in a row will give the same result as restricting a just once; technically speaking, restriction is *idempotent.* Yet the two expressions are syntactically quite different. But if we consider their nets, then it is easy to see that they are again isomorphic, since the outer restriction in $(E\backslash a)\backslash a$ does not find any transitions to be removed anymore. As another example, consider $a.(b.\mathsf{nil}+b.\mathsf{nil})$ and $a.b.\mathsf{nil}+a.b.\mathsf{nil}$. It seems very reasonable that these two expressions could, or even should, be treated as equivalent. Syntactically speaking, this could be achieved by something like a distributivity law for prefixing (over choice). However, one needs to be careful, because we *do not* want a general left-distributivity law to hold in the process algebra context. The argument is supplied by the following two expressions:

$$a.(b.\mathsf{nil} + c.\mathsf{nil}) \quad \text{and} \quad a.b.\mathsf{nil} + a.c.\mathsf{nil}.$$

There is a context in which they behave differently, viz. $(((\bullet \mid \widehat{a}.\widehat{c}.\mathsf{nil})\backslash a)\backslash b)\backslash c$. Put first $a.(b.\mathsf{nil}+c.\mathsf{nil})$ in place of the bullet and observe that there is no deadlock: two τ's can be executed in succession, and the execution terminates normally.

Then put $a.b.\mathsf{nil}+a.c.\mathsf{nil}$ in place of the bullet and observe that now there *can* be a deadlock. It arises when the $\widehat{a}$ is synchronised with the a in front of the b, rather than with the a in front of the c, and executed. If this is not seen immediately from the SOS rules, we would encourage the reader to draw the two nets and check that they differ in an additional τ-labelled transition in the second net, which is the one creating the deadlock. We conclude that distributivity of prefix over choice (from the left) is not, in general, a good idea. Still, it seems that we *ought* to have some means of equating expressions, because there are just too many of them around: It is easy to find more examples that one might intuitively call equivalent, e.g., $a.\mathsf{nil}$ and $a.\mathsf{nil}+a.\mathsf{nil}$, or even (we are talking 'interleaving') $a.\mathsf{nil} \mid b.\mathsf{nil}$ and $a.b.\mathsf{nil}+b.a.\mathsf{nil}$.

The discussion suggests that a *behavioural*, rather than a structural, or a syntactic, equivalence relation is sought for. One of the fundamental contributions of early CCS [24, 27] is the development of a very important notion of this kind, called *bisimulation equivalence*, or just *bisimulation*. The basic idea is to relate sets of states of the two processes that are subject to comparison, in such a way that from two related states, if one of them can make a move, the other one can make a similar move such that the new states are again related. By a 'state' of a process, we mean any process that can be reached after a number of moves (including the initial process which can be reached after zero moves). The set of all reachable states can be viewed as the nodes of a directed graph, whose arrows correspond to the moves from one state to another. This edge-labelled graph is usually called the *transition system* (of the process). (The terminology is due to [21].) If an initial state is present (as is usually the case), we may use a special symbol for it – below, we use a circle which is not entirely black. These notions are actually similar to the *reachability graphs*, which comprises the set of reachable markings of a Petri net in graphical form.

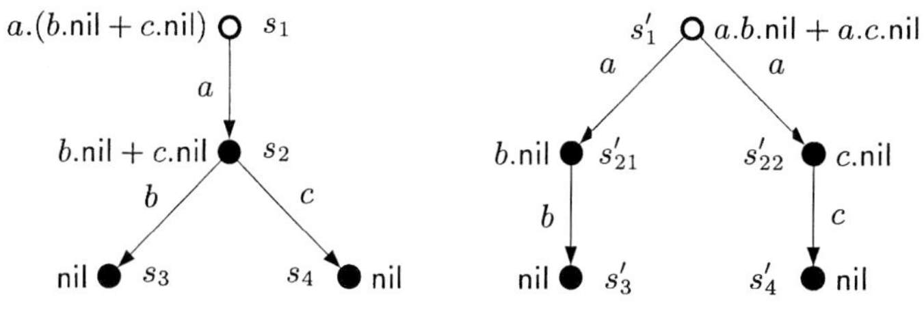

Fig. 7. Transition systems for $a.(b.\mathsf{nil}+c.\mathsf{nil})$ and $a.b.\mathsf{nil}+a.c.\mathsf{nil}$.

Figure 7 explains why no bisimulation can be found between $a.(b.\mathsf{nil}+c.\mathsf{nil})$ and $a.b.\mathsf{nil}+a.c.\mathsf{nil}$: On the right-hand side, neither state s'_{21} nor state s'_{22} allow *both* b and c actions to occur, and thus neither of them corresponds to state s_2. Intuitively, the moment of choice between b and c is pre-poned in the process $a.(b.\mathsf{nil})+a.(c.\mathsf{nil})$, choosing here between the first or the second a already determines the choice between b and c. Note that, by contrast, a bisimulation *can* be found between the two expressions $a.((b.\mathsf{nil})+(b.\mathsf{nil}))$ and $a.(b.\mathsf{nil})+a.(b.\mathsf{nil})$. Their two transition systems are exactly the same as the ones shown in Figure 7, except that the c is on both sides replaced by a b. In that case, $\{(s_1, s'_1), (s_2, s'_{21}), (s_2, s'_{22}), (s_3, s'_3), (s_4, s'_4)\}$ is a bisimulation.

Because bisimulation is defined on transition systems, rather than on expressions or nets, it can be defined in a similar way for any formal model from which a transition system can be derived. A transition system over a given set of symbols Σ is a pair $(S, \rightarrow)$ such that S is some set of states (usually with a designated initial state) and $\rightarrow$ is a set of triples $(s, x, \tilde{s})$, such that s and $\tilde{s}$ are states and x is a symbol from Σ. A triple $(s, x, \tilde{s})$ indicates that a move from state s to state $\tilde{s}$ (not necessarily different from s) is possible by executing symbol x. One can also write $s \xrightarrow{x} \tilde{s}$ instead of $(s, x, \tilde{s}) \in \rightarrow$. For example, the left-hand side of Figure 7 shows a transition system over $\Sigma = \{a, b, c\}$ with four states $S = \{s_1, s_2, s_3, s_4\}$ and three triples: (s_1, a, s_2), (s_2, b, s_3) and (s_2, c, s_4). Note that a transition-labelled Petri net gives rise to two transition systems: (i) its reachability graph where the set of symbols is the set of transitions, and (ii) the graph obtained from (i) by replacing each transition with its label (the set of symbols is now the set of transition labels).

Let us have a look at the formal definition of bisimulation: Suppose $(S_1, \rightarrow_1)$ and $(S_2, \rightarrow_2)$ are two transition systems with initial states s_{01} and s_{02}, respectively. Then a relation $\beta \subseteq S_1 \times S_2$ is called a bisimulation *if and only if* the initial states are related by β (i.e., $(s_{01}, s_{02}) \in \beta$), and whenever two states are related, then the moves in $(S_1, \rightarrow_1)$ can be mirrored by moves in $(S_2, \rightarrow_2)$ and vice versa, i.e., for each pair $(s_1, s_2) \in \beta$ we have:

- If $s_1 \xrightarrow{z} s'_1$, then there is $s'_2 \in S_2$ such that $s_2 \xrightarrow{z} s'_2$ and $(s'_1, s'_2) \in \beta$.
- If $s_2 \xrightarrow{z} s'_2$, then there is $s'_1 \in S_1$ such that $s_1 \xrightarrow{z} s'_1$ and $(s'_1, s'_2) \in \beta$.

We have been careful to give the Σ in the definition of a transition system a different name than A, even though one usually takes $\Sigma = A$ or $\Sigma = A \cup \{\tau\}$, in the case of basic CCS. The resulting notion of bisimulation is called *strong bisimulation*. Often, one wants to 'neglect τ-moves', whenever possible. For instance, the processes $a.\tau.\mathsf{nil}$ and $a.\mathsf{nil}$ are not strongly bisimilar, because the τ is treated like a normal action. It is possible (and often desirable) to consider a weaker notion of bisimulation that ignores internal τ-moves whenever this not lead to a loss of vital semantical information. For instance, it would be too hasty to equate $a.\mathsf{nil} + b.\mathsf{nil}$ and $a.\mathsf{nil} + \tau.b.\mathsf{nil}$, because executing the τ implies a 'silent' but firm commitment to executing b (rather than a) as the next action. A version of bisimulation called *weak bisimulation* [25] arises out of the previous definition thus: β is called a weak bisimulation between the two transition systems *if and only if* the initial states are related by β, and whenever two states are related, then the moves in $(S_1, \rightarrow_1)$ can be mirrored by (possibly empty) *sequences* in $(S_2, \rightarrow_2)$ and vice versa, i.e., for each pair $(s_1, s_2) \in \beta$ we have:

- If $s_1 \xrightarrow{z} s'_1$, then there is $s'_2 \in S_2$ such that $s_2 \xrightarrow{w} s'_2$ and $(s'_1, s'_2) \in \beta$, where: w is a sequence of τ's if $z = \tau$, and z surrounded by sequences of τ's if $z \neq \tau$; and $s_2 \xrightarrow{w} s'_2$ means that s'_2 is reached from s_2 by successive executions of the actions in w.

- If $s_2 \xrightarrow{z} s_2'$, then there is $s_1' \in S_1$ such that $s_1 \xrightarrow{w} s_1'$ and $(s_1', s_2') \in \beta$, where w is similar as above.

Let us write $E \cong E'$ and $E \cong_w E'$ if E and E' are *strongly bisimilar* and, respectively, *weakly bisimilar*, i.e., if there exists a strong (weak, respectively) bisimulation between their transition systems. It is fairly clear that these relations are indeed equivalence relations, i.e., they are reflexive, symmetric, and associative. Obviously, a strong bisimulation is also a weak bisimulation, so that $E \cong E'$ implies $E \cong_w E'$ (i.e., $\cong \subseteq \cong_w$). For instance, $a.b.\mathsf{nil}$ and $a.\tau.b.\mathsf{nil}$ are weakly bisimilar (when checking this be aware that the empty sequence is also a sequence), since the two intermediate states in the latter can be related to the single intermediate state of the former, but $a.\mathsf{nil} + b.\mathsf{nil}$ and $a.\mathsf{nil} + \tau.b.\mathsf{nil}$ are not.

Strong and weak bisimilarity of process expressions are actually quite strong equivalence notions, short of actual syntactic equality of process terms. There are also much weaker notions of equality between processes. A particularly important one is *trace equality*, defined as follows: E and E' are trace-equivalent if the set of all sequences of possible moves of E and E' (i.e., their interleavings) coincide. It is easy to prove that whenever E and E' are (strongly or weakly) bisimilar, then E and E' are also trace equivalent. However, $a.(b.\mathsf{nil} + c.\mathsf{nil})$ and $a.b.\mathsf{nil}+a.c.\mathsf{nil}$ are trace equivalent without being bisimilar. Between bisimulation and trace equality, there is a spectrum of interesting equivalence notions, as described in [14]. Although bisimulation is a strong equivalence notion, there are still stronger ones. A next to useless one, already mentioned, is syntactic equality. Another one is *transition system isomorphism*, defined as follows: E and E' are *ts-isomorphic* if there exists an isomorphism between their transition systems. For instance, $E+F$ and $F+E$ are ts-isomorphic, but $a.a.\mathsf{nil}$ and $a.\mathsf{nil} \mid a.\mathsf{nil}$ are not (but they are bisimilar). There also are desirable identifications that require only a slight strengthening of the notion of bisimulation. For instance, one might wish to distinguish $(b.a.\mathsf{nil})\backslash a$ and $b.\mathsf{nil}$, because (as discussed earlier on) the latter, but not the former, terminates properly; and yet, they are (strongly and weakly) bisimilar.

Now that we have defined strong and weak bisimilarity and sketched their characteristics, some real *calculations* can be started, in particular, we may now set out to *prove* a host of equivalences. For example, the following algebraic properties of the choice operator become actually provable:

$$
\begin{array}{lll}
E + \mathsf{nil} & \cong E & \text{(nil is neutral w.r.t. choice, cf. the next line)} \\
E + F & \cong F + E & \text{(choice is symmetric)} \\
(E + F) + G & \cong E + (F + G) & \text{(choice is associative),}
\end{array}
$$

for all E, F, and G. Similarly, the composition operator $\mid$ is symmetric, associative, and has nil as a neutral element, with respect to strong bisimulation. The associativity of $\mid$ requires a trifle more thought than the other properties. It holds because of the careful design of basic CCS. In many practical cases, a composition of two processes is immediately followed by restriction, such as in $(E \mid F)\backslash a$. Suppose that we had defined $\mid$ in such a way that it *included* restriction. Then we would have lost associativity. To see this, compare the following two expressions:

$$(((a.\mathsf{nil} \mid a.b.\mathsf{nil})\backslash a) \mid \widehat{a}.\mathsf{nil})\backslash a \quad \text{and} \quad (a.\mathsf{nil} \mid ((a.b.\mathsf{nil} \mid \widehat{a}.\mathsf{nil})\backslash a))\backslash a\ .$$

The expression on the left-hand side is bisimilar to nil, while the expression on the right-hand side can make a τ-move followed by a b-move, and therefore, associativity (of the hypothetical composition operator) does not hold. The reason why such an example cannot occur for the basic CCS composition operator, is that all conjugate pairs are still around after synchronisation, unless they are *explicitly* restricted away by a *different* operator.

Restriction, too, has several useful algebraic properties, such as:

$$\begin{array}{lll}(E\backslash a)\backslash b & \cong (E\backslash b)\backslash a & \text{(restriction is symmetric)}\\ (E+F)\backslash a & \cong (E\backslash a)+(F\backslash a) & \text{(restriction distributive over choice)}\end{array}$$

and more (for instance, the reader might wish to find out under which condition(s) restriction distributes over composition).

We also have some *congruence* properties, that is, we may prove that $\cong$ is preserved under some operations. Just as one particular example, the following holds true:

$$E \cong E' \quad \Longrightarrow \quad E \mid F \ \cong\ E' \mid F\ ,$$

for all processes E, E' and F. There are many more algebraic properties of this kind, but this is not the right place to list them all – well, if 'listing them all' is at all possible. Suppose that we had a finite set of such properties, such that for *any* two expressions E and E', if E and E' are bisimilar, then this can be proved using only those properties. If such a set could be found, then we would call bisimilarity *finitely axiomatisable*. However, it has been shown that bisimilarity is *not* finitely axiomatisable for process algebras including basic CCS [30]. Thus, the list of 'interesting algebraic properties' analogous to the above ones, has to be virtually endless. Let us just list two of them. The first one gives a general property for weak bisimulation:

$$a.(E+\tau.F)+a.F \ \cong_w\ a.(E+\tau.F)\,,$$

which is taken from [25]. It is a good exercise to prove this for simple E and F, for instance $b.\mathsf{nil}$ and $c.\mathsf{nil}$, and also to check why $\cong_w$ cannot be replaced here by the stronger $\cong$. Another property is the *expansion property.* It allows composite expressions to be 'developed' into large sums (comprising the possible 'first actions' that can be executed in the composite expression followed by smaller composite expressions). We give here only an application of this property:

$$\begin{array}{l}a.\mathsf{nil} \mid \widehat{a}.\mathsf{nil}\\ \quad \cong\ a.(\mathsf{nil} \mid \widehat{a}.\mathsf{nil}) + \widehat{a}.(a.\mathsf{nil} \mid \mathsf{nil}) + \tau.(\mathsf{nil} \mid \mathsf{nil})\\ \quad \cong\ a.\widehat{a}.(\mathsf{nil} \mid \mathsf{nil}) + \widehat{a}.a.(\mathsf{nil} \mid \mathsf{nil}) + \tau.(\mathsf{nil} \mid \mathsf{nil})\\ \quad \cong\ a.\widehat{a}.\mathsf{nil} + \widehat{a}.a.\mathsf{nil} + \tau.\mathsf{nil}\ .\end{array}$$

In this chain of bisimilarities, the first two come from the expansion property, and the last one from nil | nil $\cong$ nil. We have managed to reduce an expression with | operator into one without. The expansion property can *always* be used in this way, showing that (at least with respect to bisimulation as defined above) the | operator is redundant for recursion-free expressions[1]. This is strongly related to the fact that, so far, we *only* consider interleaving semantics. In fact, the validity of the expansion property can be viewed as a direct *formalisation of interleaving* (concurrency can be replaced by nondeterministic interleaving).

We have called the above equalities 'properties' rather than 'laws', because they are consequences of the definitions, rather than being postulated. It is possible to turn the approach upside down, as exemplified by [3]. Let us start with just the set A of actions (and possibly τ) and simply *postulate* a set of reasonable algebraic equations for a set of operators that are interesting in that they make sense for a process algebra, such as ' | ' for choice (as before) and ';' for sequential composition. We quote the first five postulates (also called *axioms*) from [3]:

$$p+q=q+p \qquad (p+q)+r=p+(q+r) \qquad p+p=p$$
$$(p+q);r=p;r+q;r \qquad (p;q);r=p;(q;r)\ .$$

The axioms in the first row are already known from basic CCS's choice operator, except that here, p, q and r range over an unknown 'set of processes', and the equality is not a concrete one, but a postulated one (which may, or may not[2], later be substituted by a concrete one). The next one is the 'wrong' distributivity, or rather the 'right' one, both by nature and since it can safely be assumed to hold (as opposed to left distributivity, which was rejected earlier). Note that it cannot even be formulated in basic CCS, since a compound process such as $p+q$ cannot be placed in front of sequential composition there. For the same reason, associativity of sequential composition (the last one in the above list of axioms) cannot be formulated in basic CCS. Thus, it is already clear that with the above axioms, one aims at a process algebra that it somewhat different from basic CCS. Eventually, through the development and after adding more and more axioms, processes satisfying these axioms can be constructed via terms and graphs as transition systems, i.e.: transition systems are a *model* of those equations, after the appropriate operations have been defined on them and bisimilarity is taken as equality.

What is the advantage of postulating axioms and looking for models that satisfy them, over defining a particular model such as basic CCS and proving properties about it? Well, sometimes one wishes to be somewhat at liberty with the adopted axioms and, in particular, wants to investigate their interplay. Searching for interesting models and proving that they satisfy the axioms could,

[1] However, the practical use of this insight is very limited indeed; for example, [3] gives the expansion of ($a.a.a$.nil | $b.b.b$.nil) | $c.c.c$.nil. It takes two full pages in small print!

[2] For instance, we might be as careless as to postulate *contradictory* laws.

however, be a none too trivial matter. There are also other angles by which process algebras can be viewed and classified. For instance, one might be interested in categorising them in terms of different SOS semantics formats [1]. Another approach is *parametric*, in the sense that one considers a whole class of concrete process algebras and proves properties that hold for each one. We will turn to one of these approaches in the next section.

5 B.TOYA: A Basic Toy Algebra

We are now prepared to consider process algebras with operations that slightly deviate from, or generalise, those of basic CCS. We'll discuss another concrete algebra and its semantics, which are designed specifically with their relationship to Petri nets in mind. Inevitably, through such a relationship, we need to consider Petri nets as *composable* entities, and the intended translation will be *compositional*, in the sense that the net of a composite process is the composition of the nets associated with the composed processes.

Let us start by decomposing the basic CCS composition operator. As described above, it consists of a first step (corresponding to forming the disjoint union of Petri nets) and a second step (calculating the synchronisations of whatever conjugate pairs can be found in the constituent processes, or respectively, their nets). Now suppose that we separated disjoint union from synchronisation. In CCS, composition forces synchronisation for *all* pairs $a, \widehat{a}$, $b, \widehat{b}$, etc., in a single step. By analogy to the unary restriction operator, let us contemplate a *unary* operator (parametrised by a and written in postfix style) that effects synchronisation just on pairs $a, \widehat{a}$. Let us denote that operator $\mathsf{sy}\ a$, and let us use $\mathsf{rs}\ a$, instead of the previous $\backslash a$, to denote restriction. Suppose that we can prove the following properties for the sy operator:

$$
\begin{array}{lcll}
E\ \mathsf{sy}\ a & \cong & E\ \mathsf{sy}\ \widehat{a} & \text{(insensitivity to conjugation)} \\
(E\ \mathsf{sy}\ a)\ \mathsf{sy}\ a & \cong & E\ \mathsf{sy}\ a & \text{(idempotence)} \\
(E\ \mathsf{sy}\ a)\ \mathsf{sy}\ b & \cong & (E\ \mathsf{sy}\ b)\ \mathsf{sy}\ a & \text{(symmetry),}
\end{array}
$$

with some (strong) equivalence $\cong$. This would, in fact, make synchronisation algebraically quite similar to restriction (though almost 'opposite' in meaning), since restriction satisfies the same properties as well. The three properties allow us, by the way, to extend both operators uniquely to finite sets B of action names, $\mathsf{sy}\ B$ and $\mathsf{rs}\ B$, and even to $B = A$, i.e., the entire set of action names. Using $\mathsf{sy}\ A$ then comes quite close to the original CCS intention behind the (second step in the) composition operator.

The point of the last paragraph was not to advertise yet another synchronisation operator, but to argue that the set of (elementary) manipulations in basic CCS (and, as it turns out, in similar concrete process algebras as well) can be divided – in terms of the nets associated with expressions – into two classes:

- Those manipulating *places*. Examples are choice, prefixing (or, more generally, sequential composition), and disjoint parallel composition (i.e., the first step in CCS's $\mathsf{|}$ operator).

– Those manipulating *transitions*. Examples are restriction and synchronisation (i.e., the second step in CCS's $\mid$ operator).

Let us first have a look at the latter class. Clearly, transitions will be manipulated depending on their labels. In case of restriction $\mathsf{rs}\ a$, for example, all transitions labelled a or $\widehat{a}$ are simply removed. In case of synchronisation $\mathsf{sy}\ a$, all transitions are left as they are, but some new ones are added, which can be viewed as the sums of two old ones (one labelled by a, the other by $\widehat{a}$).

The above suggests to search for a uniform and general way of describing these transition-(label-)based operations all at once. In [6], *general relabellings* were introduced for this purpose. The idea is that a set of transitions (by virtue of their labels) can be specified to give rise to a new transition, which is the net-theoretic sum of the old ones. Because different transitions may have the same label, we need to consider *multisets of labels* as the arguments of general relabelling. Therefore, a general relabelling ρ is defined as a relation $\rho \subseteq \mathsf{mult}(\Sigma) \times \Sigma$, where in case of basic CCS, Σ could be $A \cup \{\tau\}$, and $\mathsf{mult}(\Sigma)$ denotes the set of multisets whose elements are in Σ. As an example, if we wished to describe restriction $\mathsf{rs}\ a$ to have the same effect as restriction $\backslash a$ in basic CCS, we could consider:

$$\rho_{\,\mathsf{rs}\ a} = \{\, (\{z\}, z) \mid z \in A \backslash \{a, \widehat{a}\} \,\}.$$

That is, all transitions with labels not in $\{a, \widehat{a}\}$ are left intact, while those with label a or $\widehat{a}$ have no entry in $\rho_{\,\mathsf{rs}\ a}$ and are therefore omitted. As another example, synchronisation $\mathsf{sy}\ a$ can be described by:

$$\rho_{\,\mathsf{sy}\ a} = \{\, (\{z\}, z) \mid z \in A \cup \{\tau\} \,\} \cup \{(\{a, \widehat{a}\}, \tau)\}.$$

That is, *all* transitions are left intact, but some new τ-labelled ones are added, as they arise from pairs of transitions, of which one is labelled by a and the other by $\widehat{a}$.

For the sake of using the concept of general relabelling in a slightly non-standard way, let us design a 'toy' process algebra with a unary synchronisation operator that is parametrised by three action names, $y, y', z \in \Sigma$, and that takes any pair y, y', creates a synchronised action z out of them (like previously the τ out of a and $\widehat{a}$), and deletes all the y's and y''s afterwards. Denote this operator by $\bullet[y, y' \to z]$. Furthermore, in our toy algebra, we wish to have forward/backward symmetric choice, sequential, and parallel operators, the latter meaning disjoint parallelism. Basic TOYA is thus defined as follows (we assume $y, y', z \in \Sigma$):

$$\mathsf{B.TOYA}: \ E ::= z \ \big| \ E[y, y' \to z] \ \big| \ E + E \ \big| \ E ; E \ \big| \ E \,\|\, E \ ,$$

where we have chosen the '$\|$' symbol for parallel composition, which is different from the composition in CCS, and the semicolon to indicate sequential composition, which is more general than prefixing in CCS.

By calling the operators forward/backward symmetric, we mean that their (intended) Petri nets should look symmetric, whether viewed from the entry

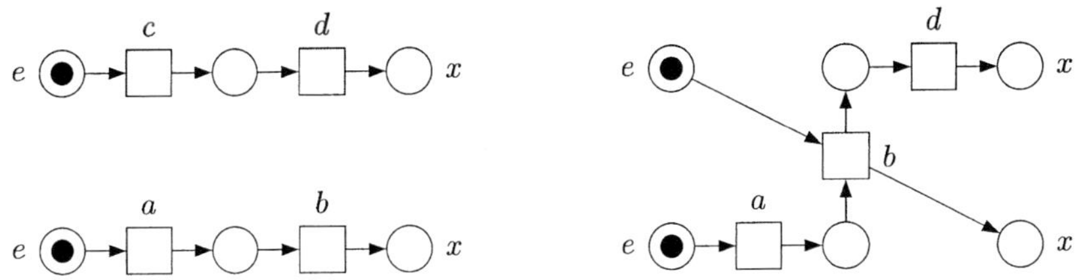

Fig. 8. Nets for $((a;b) \parallel (c;d))$ and $((a;b) \parallel (c;d))[b,c \to b]$.

places or seen from the exit places (in Figure 2, only the nets shown on the left-hand side and on the right-hand side are forward/backward symmetric, the others are not). For instance, consider the expression:

$$((a;b) \parallel (c;d))[b,c \to b].$$

Its net *before* applying the $\bullet[b,c \to b]$ operator is shown on the left-hand side of Figure 8, and the net *after* applying it is shown on the right-hand side of the figure. Consider, by contrast, the expression:

$$((a;b) + (c;d))[b,c \to b]$$

which is like the previous one, except that the $\parallel$ has been replaced by a choice operator. Its net is shown on the right-hand side of Figure 9. Observe that it is intuitively correct that the b action in the middle can never occur, because the b in $(a;b)$ requires a to occur first, while executing a disables c, which would be needed for the (resulting) b to occur.

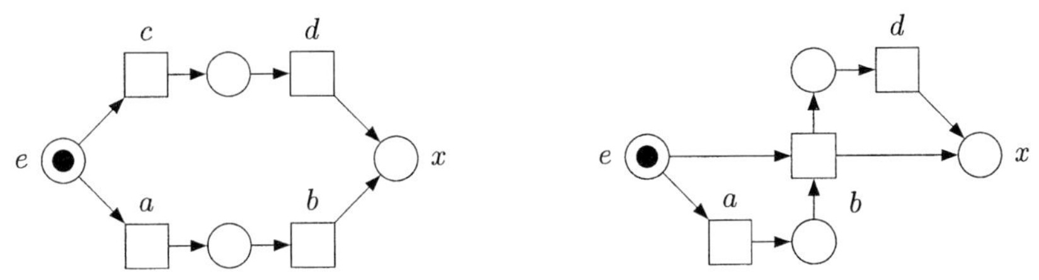

Fig. 9. Nets for $(a;b) + (c;d)$ and $((a;b) + (c;d))[b,c \to b]$.

Let us have a look at the net translation of B.TOYA expressions in which the parallel operator is nested, such as $(a \parallel b) + (c \parallel d)$ and $(a \parallel b);(c \parallel d)$. In the former, as soon as a is chosen, neither c nor d can be chosen anymore. In the latter, c or d have to wait until both a and b have been completed. It is well-known how this can be achieved by means of Petri nets, and the two translations are shown in Figure 10. Both of these translations involve manipulations on the e and x places of the constituent nets, known as *place multiplication*. On the left-hand side of Figure 10, the e places of the first constituent, $(a \parallel b)$, are multiplied with the e places of the second constituent, $(c \,\|\, d)$, and the same is done (for reasons of forward/backward symmetry) with their x places. On the right-hand side of Figure 10, the x places of the first constituent are multiplied with the e places of the second – as befits a sequential composition.

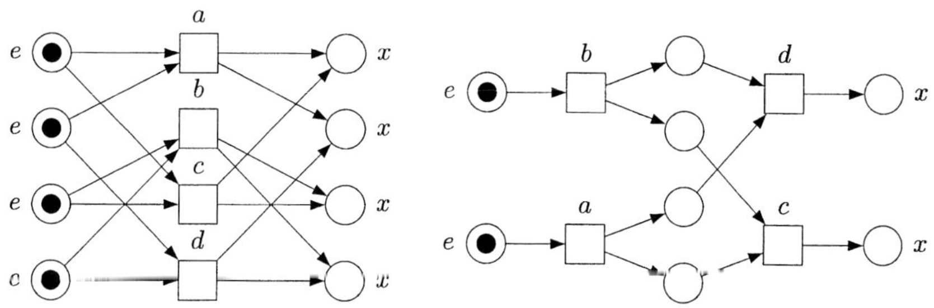

Fig. 10. Nets for $(a \,\|\, b) + (c \,\|\, d)$ and $(a \,\|\, b); (c \,\|\, d)$.

A systematic way of unifying place multiplication and its use in the Petri net semantics of process algebras is to use *transition refinement* [15]. In transition refinement, a transition in a Petri net is replaced by a whole (sub)net (i.e., a net which, after refinement, becomes a subnet). If the net by which transition t is replaced (called the *refining* net), carries e and x tags on some of its places, then one can combine the former with the input places of t, and the latter with the output places of t, pairwise, to create appropriate new places by place multiplication. If care is exercised, then properties of the resulting *refined* net can be inferred from properties of the refining net and the net in which t is embedded. In particular, the S-invariants of the refined net can be calculated from those of the original net and those of the refining net [6][3]. Moreover, transition refinement is symmetric, i.e., refining the first transition t_1 and then t_2 is the same as refining the two transitions in the other order. Hence one may employ, without ambiguity, a *simultaneous refinement* of a set of transitions. And finally, for a vast class of cases, the safeness (1-boundedness) of the resulting net can be inferred from the safeness of the original net and the safeness of the refining nets.

Let us see how simultaneous transition refinement can help in unifying the treatment of choice and sequential composition, and also of disjoint concurrent composition. Figure 11 shows three nets which correspond to the operations just mentioned. Let us look first at the net for choice, on the left-hand side of Figure 11: Refining the transition labelled ρ_1 by the net corresponding to $(a \,\|\, b)$, and simultaneously refining the transition labelled ρ_2 by the net corresponding to $(c \,\|\, d)$, yields the net shown on the left-hand side of Figure 10. Let us then consider the net for sequence in the middle of Figure 11: Refining the transition labelled ρ_1 by the net corresponding to $(a \,\|\, b)$, and simultaneously the transition

[3] This result is due to R. Devillers.

labelled ρ_2 by the net corresponding to $(c \parallel d)$, yields the net shown on the right-hand side of Figure 10.

Thus, the two operations can quite similarly be described by simultaneous transition refinement. They are distinguished only by the original Petri net to which transition refinement is applied. Disjoint concurrent composition can also be described by simultaneous transition refinement, with the only distinction that the to-be-refined-net is the one shown on the right-hand side of Figure 11. Thus, these three nets can be viewed as binary *operator nets*: They each take two nets as arguments and create a new net out of them through simultaneous transition refinement.

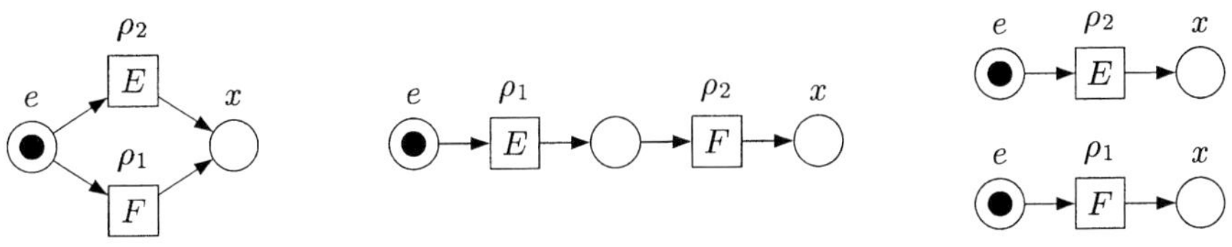

Fig. 11. Operator nets for choice, sequence and parallel composition.

Actually, general relabellings can be integrated seamlessly into this idea of simultaneous transition refinement. For, suppose that every transition in an operator net is labelled by a general relabelling ρ. Then one can define simultaneous refinement in such a way that during its application, the respective relabellings are *also* carried out. In Figure 11, we have already provided every transition with a general relabelling ρ_i $(i = 1, 2)$, but since each of these operations does *not* create any new transitions, nor destroy old ones, every ρ_i should be defined as the 'identity',

$$\rho_i \;=\; \{(\{z\}, z) \mid z \in \Sigma\},$$

which leaves every transition intact, and unmodified. An operator net for the unary operation $[y, y' \to z]$ has one transition and it carries, by contrast, the following general relabelling:

$$\rho_{[y,y'\to z]} \;=\; \{(\{v\}, v) \mid v \in \Sigma \setminus \{y, y'\}\} \cup \{(\{y, y'\}, z)\}\,,$$

as shown on the left-hand side Figure 12. Finally, we need a standard Petri net for each of the constants of the algebra, i.e., for an action $z \in \Sigma$. It is shown on the right-hand side of Figure 12.

Fig. 12. Operator net for $E[y, y' \to z]$, and Petri net for constant z.

As an example, we consider the systematic derivation of the net corresponding to the expression $((a;b) \parallel (c;d))[b, c \rightarrow b]$, cf. Figure 8. To get this net, we will proceed uniformly by structural induction: first, we create four nets for a, b, c, d (using four times the right-hand side of Figure 12). Then, we use simultaneous refinement twice on the operator net for sequential composition, to get nets for $(a;b)$ and for $(c;d)$, respectively. We use these two to simultaneously refine the operator net for parallel composition and derive a net for $((a;b) \parallel (c;d))$. Finally, we refine the net shown on the left-hand side of Figure 12 by the latter (which involves an application of $\rho_{[y,y' \rightarrow z]}$), to get the end result shown on the right-hand side Figure 8.

This approach can be called parametric, because we could, if we so wished, consider many other operators of the algebra, not just the ones shown in Figures 11 and 12. However, it is surely not a very good idea to use just *any* Petri net in the role of an operator net. In particular, we wish to still be able to define decent SOS rules for the process algebra(s) so obtained.

6 SOS Rules for B.TOYA

The standard SOS approach does not match exactly the idea of marking changes in a Petri net, because a net's structure is *not* changed by transition occurrences. However, it is possible to modify SOS in such a way that it does not affect the underlying structure of a process algebraic expression. To this end, we may introduce markings into process algebraic expressions. A (perhaps deceptively) simple (but, as we will see, sufficient) way of doing this is by overbarring and/or underbarring subexpressions. For example, we may write $\overline{a}$ to describe the fact that a token is present before a, using which, a may be executed. Similarly, $\underline{a}$ can be used to describe the fact that a token is present after a and it *may* have arrived there by executing a (although it might have arrived there in some other way).

Let us look at all possible over- and underbarrings of all subexpressions of a simple expression such as $a;b$, i.e.: $\overline{a;b}$, $\overline{a};b$, $\underline{a};b$ $a;\overline{b}$, $a;\underline{b}$ and $\underline{a;b}$ (there are no other syntactic possibilities). But these are more than enough: The first describes the initial marking of the entire expression $a;b$. The second describes the initial marking of its first part, a. The third describes the intermediate marking reached after executing a. The fourth describes the initial marking of its second part, b. The fifth and sixth describe the marking reached after executing b, and the marking reached after executing the entire expression, respectively. These expressions (that arise from B.TOYA expressions by over- or underbarring subexpressions) will be called *dynamic expressions* (as opposed to non-barred ones, which will be called *static*).

Clearly, due to the nature of sequential composition, some of the above should be identified, because they correspond to the same marking in the associated net: $\overline{a;b} \;\equiv\; \overline{a};b$, $\underline{a};b \;\equiv\; a;\overline{b}$ and $a;\underline{b} \;\equiv\; \underline{a;b}$. We use here the symbol $\equiv$ to mean (loosely speaking): 'giving rise, in the intended semantics, to the same marked Petri net'. Let us have a closer look at $\underline{a};b \;\equiv\; \overline{b};a$: Both dynamic expressions correspond to the intermediate marking reached after executing the transition labelled ρ_1 (which is the one corresponding to a in the expression $a; b$) in the operator net for sequential composition, see Figure 11 (middle). In general, simply by looking at the operator net (and not worrying about how arbitrarily complex a concrete sequentially composed expression may become after refinement), we can find all the identifications to be made by $\equiv$. Before we give the list of identifications for B.TOYA, we extend the syntax to include dynamic expressions. We denote them by G and H, rather than D, E or F, which continue to stand for static expressions:

$$\mathsf{DB.TOYA}: \quad G \ ::= \ \overline{E} \ \big| \ \underline{E} \ \big| \ G[y, y' \to z] \ \big| \ G + E \ \big| \ E + G \ \big| \\ G; E \ \big| \ E; G \ \big| \ G \,\|\, H.$$

Note how the fact that choice is distinguished from concurrent composition is reflected in this syntax: in a dynamic expression, only one of the two sides of a choice may be dynamic, while both sides of a concurrent composition are dynamic. Here are the general identifications of DB.TOYA expressions, as derived from the corresponding operator nets:

$$\begin{array}{ll}
(1) & \text{(a)}\ \overline{E[y,y' \to z]} \equiv \overline{E}[y,y' \to z] \ , \ \text{(b)}\ \underline{E}[y,y' \to z] \equiv \underline{E[y,y' \to z]} \\
(2) & \text{(a)}\ \overline{E+F} \equiv \overline{E}+F \equiv E+\overline{F} \ , \ \text{(b)}\ \underline{E}+F \equiv E+\underline{F} \equiv \underline{E+F} \\
(3) & \text{(a)}\ \overline{E;F} \equiv \overline{E};F \ , \ \text{(b)}\ \underline{E};F \equiv E;\overline{F} \ , \ \text{(c)}\ E;\underline{F} \equiv \underline{E;F} \\
(4) & \text{(a)}\ \overline{E \,\|\, F} \equiv \overline{E} \,\|\, \overline{F} \ , \ \text{(b)}\ \underline{E} \,\|\, \underline{F} \equiv \underline{E \,\|\, F} \ .
\end{array}$$

Line (3) is the general version of what has already been explained on an example. Line (2a) states that the initial marking of a choice initialises both of its constituents (but only one of them at the same time). Line (4a) states, by contrast, that the initial marking of a concurrent composition initialises both of its constituents at the same time. The other identifications have similar justifications.

We need a general context rule: suppose that $\mathcal{C}(\bullet)$ is some valid dynamic context, i.e., that in place of the $\bullet$ some dynamic expression G could be written, such that the resulting string $\mathcal{C}(G)$ is again a valid dynamic expression. Then,

$$\text{(CR)} \quad \text{If } G \equiv H \text{ , then } \mathcal{C}(G) \equiv \mathcal{C}(H) \ .$$

To mimic the marking changes in the (intended) Petri net semantics, we need an equivalent of the Petri net transition rule in terms of dynamic expressions. Let us generalise this by defining *concurrent occurrences*, i.e., *steps*, of actions. Thus, for instance, in Figure 10, on the left-hand side, a step $\{a, b\}$ can occur, since a and b are concurrently enabled, and, for similar reasons, a step $\{c, d\}$, but also a step $\{a\}$ or a step $\{b\}$ (or also the empty step, making no marking change), but *not*, of course, a step $\{a, c\}$. The rules are provided by the next list:

$$(\alpha) \quad \overline{a} \xrightarrow{\{a\}} \underline{a} \qquad (A) \; \frac{G \xrightarrow{\gamma+k\cdot\{y,y'\}} H\,,\; k\in\mathbb{N}}{G[y,y'\to z] \xrightarrow{\gamma+k\cdot\{z\}} H[y,y'\to z]}$$

$$(B)\;(1)\; \frac{G \xrightarrow{\gamma} H}{G+E \xrightarrow{\gamma} H+E} \qquad (2)\; \frac{G \xrightarrow{\gamma} H}{E+G \xrightarrow{\gamma} E+H}$$

$$(C)\;(1)\; \frac{G \xrightarrow{\gamma} H}{G;E \xrightarrow{\gamma} H;E} \qquad (2)\; \frac{G \xrightarrow{\gamma} H}{E;G \xrightarrow{\gamma} E;H}$$

$$(D)\quad \frac{G \xrightarrow{\gamma} G'\,,\; H \xrightarrow{\delta} H'}{G \,\|\, H \xrightarrow{\gamma+\delta} G' \,\|\, H'}\,,$$

where we write $\xrightarrow{\gamma}$ for a step $\gamma \in \mathsf{mult}(\Sigma)$, and γ and δ are, in general, multisets of actions[4]. Also, $k\cdot\{y,y'\}$ denotes the multiset where y and y' have multiplicity k (and all other actions multiplicity 0). Note that (α) is the only rule which allows something (namely, a) to actually 'happen'. All other rules are only 'context rules', in the sense that they allow to infer some global occurrence from something that happens locally. This is reinforced by the observation that, in fact, the rules for choice and for sequential composition, (B) and (C), are *exactly* the same, differing only in the operator used. In fact, sequence and choice are distinguished *only* by their different $\equiv$ relations in lines (2) and (3). Concurrent composition and synchronisation $[y,y' \to z]$ again have the same context rule, and are distinguished only by their different shape in terms of the syntax of DB.TOYA.

The rule for choice may be contrasted with the corresponding rule for basic CCS, given in section 2. The main difference is (and that is, of course, what was desired) that the underlying structure of a dynamic expression (that is, the structure obtained by removing all over- and underbars) does not change during the execution of a choice.

Let us see how one would derive the step sequence $\xrightarrow{\{a\}\{b\}}$, which is (or should be, by inspecting the intended Petri net semantics shown on the right-hand side Figure 8) derivable as a valid execution from the initial state of the expression:

$$((a;b) \,\|\, (c;d))[b,c\to b]\,.$$

In the derivation that follows, we list on the right-hand side the rules used in each derivation step.

$$\begin{aligned}
\overline{((a;b) \,\|\, (c;d))[b,c\to b]} &\equiv \overline{((a;b) \,\|\, (c;d))}[b,c\to b] && (\text{ 1a })\\
&\equiv (\overline{(a;b)} \,\|\, \overline{(c;d)})[b,c\to b] && (\text{ 4a, CR })\\
&\equiv ((\overline{a};b) \,\|\, (\overline{c};d))[b,c\to b] && (\text{ 2·3a, CR })
\end{aligned}$$

[4] They have to be multisets, rather than a sets, since two transitions with the same label could occur concurrently.

$$\begin{array}{lll} \xrightarrow{\{a\}} & ((\underline{a};b)\parallel(\overline{c};d))[b,c\to b] & (\ \alpha,\ \mathrm{C1},\ \mathrm{D},\ \mathrm{A}\ ) \\ \equiv & ((a;\overline{b})\parallel(\overline{c};d))[b,c\to b] & (\ \mathrm{3b},\ \mathrm{CR}\ ) \\ \xrightarrow{\{b\}} & ((a;\underline{b})\parallel(\underline{c};d))[b,c\to b] & (\ 2{\cdot}\alpha,\ \mathrm{C2},\ \mathrm{C1},\ \mathrm{D},\ \mathrm{A}\ ) \\ & \ldots & \end{array}$$

In the line with $\xrightarrow{\{a\}}$, rule (D) has been applied with $\gamma = \{a\}$ and $\delta = \emptyset$, and rule (A) with $\gamma = \{a\}$ and $k = 0$. In the line with $\xrightarrow{\{b\}}$, rule (D) has been applied with $\gamma = \{b\}$ and $\delta = \{b\}$, and rule (A) with $\gamma = \emptyset$ and $k = 1$.

For historic (and also self-explanatory) reasons, semantics such as these SOS rules are called *operational*, and semantics such as given by the Petri net translation (through operator nets and simultaneous transition refinement, cf. section 5) are called *denotational*, and it is desirable to prove an equivalence between them (see, e.g., [26]). In the case of B.TOYA, we have a strong equivalence: *consistency*, i.e., every step sequence derivable by the SOS rules can also be derived in the net by means of the transition rule, and *completeness*, i.e., for every step sequence derivable in the net, there is *some* possibility of deriving it from the SOS rules. In proving these equivalence properties, it is essential that all operator nets satisfy certain appealing and desirable properties (for instance, 1-safeness, but also, a special property called *factorisability* [6]), and it is also essential that the relationship between the process algebra on the one hand, and the Petri nets on the other hand, is tight and one-to-one. In particular, it may be dangerous to use intermediate τ transitions too liberally. For instance, the translation of choice shown on the right-hand side of Figure 2 has to be rejected.

7 Recursion

From the theory of computability, it is known that Turing machines can be simulated by 2-stack pushdown automata. Using basic CCS with recursion, an arbitrary number of stacks can be modelled. Thus – and this is how an argument in [25] goes – basic CCS with recursion is Turing-powerful. Therefore, it is impossible to find finite 1-safe (basic, i.e., place/transition) Petri nets corresponding to basic CCS expressions in general, since such nets have the computing power of finite automata. Even when the restriction to 1-safeness is lifted, we will not be able to find a translation to finite nets, because finite, unbounded place/transition nets are still not Turing-powerful, even though their computing power extends beyond that of finite automata [29]. If a finite Petri net translation is desired, one needs to use high-level Petri nets. A translation of basic CCS and TCSP into finite high-level nets has been described in [32].

In this section, we will show what may happen if the *finiteness* restriction, rather than the 1-boundedness restriction, is lifted, i.e., we will search for a translation of basic CCS with recursion into (possibly infinite) nets. And we will strive to keep the other restriction, 1-safeness, intact. Actually, rather than basic CCS, we will consider B.TOYA, augmented by a facility to express recursion:

$$\mathsf{R.TOYA}: \quad E \ ::= \ a \ \big| \ E[y,y' \to z] \ \big| \ E+F \ \big| \ E;F \ \big| \ E \,\|\, F \ \big| \ X\,,$$

where every variable X is, by definition (and as in basic CCS), associated with a defining equation $X \stackrel{\mathrm{df}}{=} E$. In such a defining equation, E is called the *body*. Since basic CCS can essentially be expressed in R.TOYA, save for the differences in synchronisation and parallel composition discussed above, the explanation of R.TOYA below could (albeit with some effort) be carried over to basic CCS. With the $X \stackrel{\mathrm{df}}{=} E$ device, we can write mutually recursive systems of equations, such as $X_1 \stackrel{\mathrm{df}}{=} a; X_2$ & $X_2 \stackrel{\mathrm{df}}{=} X_2 \,\|\, X_1$. However, there are standard (well, almost standard) ways of reducing the number of equations in such systems, and we will concentrate, in the following, on just a single variable, X, and its unique defining equation, $X \stackrel{\mathrm{df}}{=} E$.

For instance, let us consider the equation $X \stackrel{\mathrm{df}}{=} a; X$, with body $a; X$. Clearly, we should be able to derive sequences of a-moves of arbitrary length from the initial state of X, but how can this be done formally? Well, let us start as follows:

$$\overline{X} \ \stackrel{\emptyset}{\longrightarrow} \ \overline{a;X} \ \ \equiv \ \ \overline{a};X \ \stackrel{a}{\longrightarrow} \ \underline{a};X \ \ \equiv \ \ a;\overline{X}.$$

In this derivation, all steps except the first one are applications of ideas already mentioned. The second and the last one, for instance, are applications of rules (3a) and (3b), which belong to the operator net for sequential composition. The first step, $\overline{X} \stackrel{\emptyset}{\longrightarrow} \overline{a;X}$, is an application of the *recursion principle*, which states the following:

> In any dynamic context, a recursion variable X may be (syntactically) rewritten by the associated body E of its defining equation $X \stackrel{\mathrm{df}}{=} E$.

A dynamic context, here, is any $\mathcal{C}(\bullet)$ such that, when R.TOYA expressions X and E are inserted into the place of the bullet, $\mathcal{C}(X)$ and $\mathcal{C}(E)$ yield dynamic expressions[5]. The relation of 'rewriting' is here formally denoted by $\stackrel{\emptyset}{\longrightarrow}$, because in the (intended) Petri net semantics, this should be considered as an empty step, i.e., a non-move or a non-change[6]. In the first step of the derivation shown above, we have the context $\overline{(\bullet)}$ with the bullet replaced by X, and the recursion principle allows us to rewrite X to the body E, that is, to $a;X$, and thus, in the given context, to $\overline{a;X}$.

Well, the same recursion principle may be applied at the end of the previous derivation, the context being $a;\overline{(\bullet)}$, and we will get as a result that another a can occur, thus:

$$a;\overline{X} \ \stackrel{\emptyset}{\longrightarrow} \ a;\overline{a;X} \ \ \equiv \ \ a;(\overline{a};X) \ \stackrel{a}{\longrightarrow} \ a;(\underline{a};X) \ \ \equiv \ \ a;(a;\overline{X})\,,$$

and the recursion principle can again be applied at this point. Clearly, the process may be repeated indefinitely, and so we may derive an infinite sequence of a-moves from $\overline{X}$.

[5] Dynamic R.TOYA expressions are defined similarly to dynamic B.TOYA expressions.

[6] This is to be distinguished from an *actual* move, however 'silent' it may be. For instance, a τ-move would, as a step containing only one move, be written as $\stackrel{\{\tau\}}{\longrightarrow}$.

Let us have a look at another, not so obvious, example: $X \stackrel{\mathrm{df}}{=} a + (X; a)$. We claim that from $\overline{X}$, two a-moves can be made in succession. Here is a derivation:

$$\overline{X} \xrightarrow{\emptyset} \overline{a + (X; a)} \equiv^2 a + (\overline{X}; a) \xrightarrow{\emptyset} a + (\overline{a + (X; a)}; a) \equiv a + ((\overline{a} + (X; a)); a)$$
$$\xrightarrow{a} a + ((\underline{a} + (X; a)); a) \equiv^2 a + ((a + (X; a)); \overline{a}) \xrightarrow{a} a + ((a + (X; a)); \underline{a})$$

where $\equiv^2$ abbreviates 'two $\equiv$-steps'. Looking carefully at this derivation reveals that, (i), a sequence of a-moves of any arbitrary finite length can be derived from $\overline{X}$, but also: (ii), no *infinite* sequence of a-moves can be derived from $\overline{X}$, in contrast to the previous example. Property (i) can easily be seen by observing that the above derivation can be made longer by repeating some intermediate applications of the recursion principle, and (ii) is true because in order for any move to be made at all from $\overline{X}$, the *first* a in $a + (X; a)$ has to be chosen first. Or, put differently, the above derivation cannot be extended at its tail, as the previous one could (we may continue to replace X in the dynamic expression $a + ((a + (X; a)); \underline{a})$ by its body $a + (X; a)$, but we may never derive another move from it, because the underbar can never be 'promoted' to an overbar).

Now let us discuss a perhaps weird example: $X \stackrel{\mathrm{df}}{=} X$. Surely, the recursion principle lets us rewrite $\overline{X}$ indefinitely often into $\overline{X}$, but the expression does not change and no actual *move* will ever result from this, not even a tiny little τ-move. This is particularly strange when we consider what could be the Petri net associated with such an X. Apparently, *any* Petri net satisfies an equation such as $X = X$, under any arbitrary (reasonable) notion of equality. However, consider any Petri net that has some transition that *can* occur. A move of such a transition cannot be derived from the above recursion principle alone, and thus, a reasonable *principle of economy*[7] allows us to exclude such nets from consideration as the semantics of $X \stackrel{\mathrm{df}}{=} X$. This creates a huge simplification: we will admit as solutions of $X \stackrel{\mathrm{df}}{=} X$ only such Petri nets that have no activated transitions, and, by extension of the principle of economy, as solutions of a general equation $X \stackrel{\mathrm{df}}{=} E$ only those nets with 'minimalistic' behaviour[8]. Of course, a 'most minimal' Petri net with no activated transitions, is one which has no transitions at all. Nevertheless, our nets should have at least an e-labelled place and an x-labelled place, because otherwise they cannot easily be used as building blocks for larger nets. Thus, in the end, the Petri net corresponding to $X \stackrel{\mathrm{df}}{=} X$ has, by definition (and by the principle of economy, applied to places as well), exactly two places (one e-labelled and the other x-labelled), and no transition.

How, then, should the Petri nets associated with the other examples, $X \stackrel{\mathrm{df}}{=} a; X$ and $X \stackrel{\mathrm{df}}{=} a + (X; a)$, look like? First of all, it is a good idea to construct nets for their bodies, with X treated as a simple transition. Figure 13 shows, on its left-hand side, the net corresponding to the body, i.e. $a; X$, of $X \stackrel{\mathrm{df}}{=} a; X$.

---

[7] Which states, more precisely, that the recursion principle is the *only* means, in addition to the rules for the other operators, of deriving moves for a recursive equation.

[8] This notion, which we slightly ironised here, can actually be made quite precise.

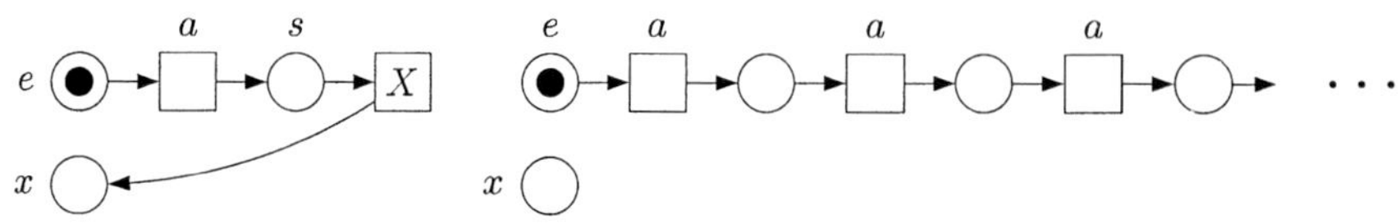

Fig. 13. Nets for $X \overset{\text{df}}{=} a; X$: the body and solution.

As one can see, it has been constructed just in the same way as other previous examples. However, it is drawn in a slightly twisted way. Why? Well, look at what happens if we refine the X transition by the net shown on the right-hand side, identifying the place called s with the e-labelled place of the right-hand side, and also identifying the two x-labelled places: we will get back 'the same' net as already shown on the right-hand side. The refinement into X just serves as a prolongation of the net to the left, or, seen in another way, 'pushes' the chain of a-labelled transitions one position further to the right, without changing anything about the structure. Actually, it is possible (and done in [6]) to define the refinement operator in such a way that we may omit the apostrophes around 'the same' above: *exactly* the same net arises when the right-hand side net is refined into the X on the left-hand side: visually and formally, the net shown on the right-hand side is a *solution* of $X \overset{\text{df}}{=} a; X$, with $\overset{\text{df}}{=}$ replaced by actual equality, and it is thus a *fixpoint* of the (Petri net) mapping $f(X) = a; X$.

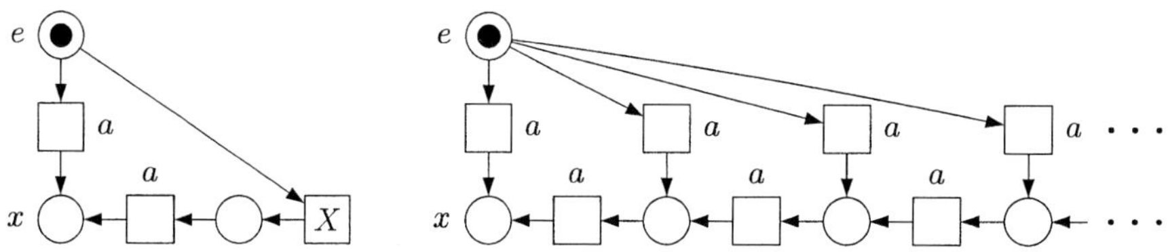

Fig. 14. Nets for $X \overset{\text{df}}{=} a + (X; a)$: the body and solution.

Exactly the same happens, in fact, in the other example, shown in Figure 14. Note that this net exhibits precisely the same behaviour as the one we derived from the expression $X \overset{\text{df}}{=} a + (X; a)$ by SOS arguments: an arbitrarily long finite chain of a-moves is possible, but no infinite one. These two examples can be generalised: refinement works in all cases of recursion expressible in R.TOYA, and yields solutions with exactly the same step sequences as derivable from the SOS rules[9]. Most meaningful recursions will, in some way, be guarded, that is, be such that recursion variables do not occur immediately at the beginning of an execution of the right-hand side of a defining equation (thus, $a; X$ is guarded while $a + (X; a)$ is not). Without guards, however, more strange cases need to be

[9] Actually, one can prove strong (in the sense that not only moves, but steps, are considered) transition system isomorphism.

considered, for instance $X \stackrel{\text{df}}{=} (a \parallel b) + X$. A little reflection shows that it should have an uncountably large solution – but does it also have countable solutions? *With* guards, however, one will usually have the unique solution property (see, e.g., [3]), and this is also true in 'minimal' Petri net models.

The recursion principle is present, in one form or another, in most process algebras. For instance, basic CCS recursion is explained in [25] thus[10]:

> The recursion rule for $X \stackrel{\text{df}}{=} E$ says, informally, that any action which may be inferred for the body E, 'unwound' once (by substituting itself for its bound variable) may be inferred for X itself.

In [3], a model for a (guardedly) recursive equation is explicitly constructed by repeatedly substituting the body, E, of $X \stackrel{\text{df}}{=} E$ into X's that (re)appear on the right-hand side, to get better and better approximations of the solution which is taken as a (projective) limit of these approximations.

8 Concluding Remarks

We have discussed some of the issues arising in giving Petri net semantics to process algebras, as well as some of the issues arising in process algebra in general, not specifically in the context of Petri nets. Why should one want to define a translation from process algebra to Petri nets? Generally speaking, this may be useful as soon as one wishes to attempt exploiting Petri-net based techniques and/or concepts for the analysis and/or the design of process-algebra based systems. In [24], a translation from a concurrent language (close to the ones in actual use) into basic CCS was given. The PEP system (Programming Environment based on Petri nets [31]) uses a similar translation, as well as a translation from concurrent programs (and inputs described in other concurrent systems notations) into M-nets ([23], a process algebra with a high-level Petri net semantics), and from there into low-level (basic) nets like the ones considered in this paper, and further into their unfoldings [22]. Travelling back and forth on this chain of translations, unfolding-based (and more generally, Petri-net based) verification methods (e.g., [13]), can be applied to concurrent programs.

Finally, we would like to discuss very briefly an approach which is directly opposite to that presented in this paper. Roughly speaking, one might be interested how can a process algebraic expression be derived from a given Petri net? A first point to note is that such an expression (if any) is not likely to be unique, and there may even be several systematic constructions which yield different results. For instance, in B.TOYA one may write two very dissimilar expressions

$$(a \parallel b); \tau; (a \parallel b) \quad \text{and} \quad ((a; (c; a)) \parallel (b; (\widehat{c}; b))) \ \mathsf{sy}\ c\ \mathsf{rs}\ c$$

which have the same Petri net semantics. Another point to note is that a Petri net is not, in general, structured. Thus, outside information has to be used when

[10] Page 57. This is not a literal quote, but hopefully conveys the intended meaning.

searching for a good translation into a process algebra. Having said that, in special cases, systematic constructions can be given. For instance, if the Petri net is 1-bounded and covered by S-components, then, using COSY, a translation of such a net into a path expression (and further into an expression of CCS or of TCSP) may be devised. And, if a Petri net is such that every place has only one input arc, other systematic translations can also be found.

Acknowledgment

The first author gratefully acknowledges financial support by the research project UK/EPSRC BESST – Behavioural Synthesis of Systems with Heterogenous Timing (grant GR/R16754).

References

1. L. Aceto, W.J. Fokkink, C. Verhoef: Structured operational semantics. In [4], 197-292.
2. J.C.M. Baeten, C.A. Middelburg: Process algebra with timing: Real time and discrete time. In [4], 627-684.
3. J.C.M. Baeten, W.P. Weijland: **Process Algebra**. Cambridge Tracts in Theoretical Computer Science 18, Cambridge University Press, 1990.
4. J.A. Bergstra, A. Ponse, S.A. Smolka (eds): **Handbook of Process Algebra**. Elsevier Science B.V., Amsterdam, 2001.
5. E. Best: **Semantics of Sequential and Parallel Programs**. Prentice Hall, 1996.
6. E. Best, R. Devillers, M. Koutny: **Petri Net Algebra**. EATCS Monographs on Theoretical Computer Science, Springer-Verlag, 2001.
7. G. Boudol, I. Castellani: Flow models of distributed computations: Three equivalent semantics for CCS. Information and Computation 114(2), 247-314, 1994.
8. R.H. Campbell, A.N. Habermann: The specification of process synchronization by path expressions. Symposium on Operating Systems, 89-102, 1974.
9. I. Castellani: Process algebras with localities. In [4], 945-1045.
10. S. Christensen, Y. Hirshfeld, F. Moller: Bisimulation is decidable for basic parallel processes. Proc. of CONCUR'93, Lecture Notes in Computer Science 715, Springer-Verlag, 143-157, 1993.
11. P. Degano, R. De Nicola, U. Montanari: A partial ordering semantics for CCS. Theoretical Computer Science 75(3), 223-262, 1990.
12. E.W. Dijkstra: **A Discipline of Programming**. Prentice Hall, 1976.
13. J. Esparza: Model checking using net unfoldings. Science of Computer Programming 23, 151-195, 1994.
14. R. van Glabbeek: The linear time – branching time spectrum I. In [4], 3-99.
15. R. van Glabbeek, U. Goltz: Refinement of actions in causality based models. REX Workshop on Stepwise Refinement of Distributed Systems: Models, Formalism, Correctness. Lecture Notes in Computer Science 430, Springer-Verlag, 267-300, 1990.
16. R. van Glabbeek, F.W. Vaandrager: Petri net models for algebraic theories of concurrency. Proc. of PARLE, Parallel Architectures and Languages, Vol. II, Lecture Notes in Computer Science 259, Springer-Verlag, 224-242, 1987.

17. U. Goltz: **Über die Darstellung von CCS-Programmen durch Petrinetze.** Dissertation, Gesellschaft für Mathematik und Datenverarbeitung, 1988.
18. C.A.R. Hoare: Communicating sequential processes. Comm. of the ACM 21, 666-677, 1978.
19. C.A.R. Hoare: **Communicating Sequential Processes**. Prentice Hall, 1985.
20. R. Janicki and P.E. Lauer: **Specification and Analysis of Concurrent Systems – the COSY Approach**. EATCS Monographs on Theoretical Computer Science, Springer-Verlag, 1992.
21. R. Keller: Formal verification of parallel programs. Comm. of the ACM 19, 371-384, 1976.
22. V. Khomenko, M. Koutny, W. Vogler: Canonical prefixes of Petri net unfoldings. Acta Informatica 40, 95-118, 2003.
23. H. Klaudel, F. Pommereau. M-nets: a survey. Manuscript, 2003 (submitted).
24. R. Milner: **A Calculus of Communicating Systems**. Lecture Notes in Computer Science 92, Springer-Verlag, 1980.
25. R. Milner: **Communication and Concurrency**. Prentice Hall, 1989.
26. E.R. Olderog: **Nets, Terms and Formulas**. Cambridge Tracts in Theoretical Computer Science 23, Cambridge University Press, 1991.
27. D. Park: Concurrency and automata on infinite sequences. Proc. 5th GI Conference, Karlsruhe, Lecture Notes in Computer Science 104, Springer-Verlag, 167-183, 1981.
28. G.D. Plotkin: A structural approach to operational semantics. Report FN-19, Computer Science Department, University of Aarhus, 1981.
29. L. Priese, H. Wimmel: **Petri-Netze**. Springer-Verlag, Theoretische Informatik, 2003.
30. P. Sewell: Nonaxiomatisability of equivalences over finite state processes. Annals of Pure and Applied Logic, 90(1-3), 163-191, 1997.
31. Ch. Stehno: Real-time systems design with PEP. Proc. TACAS'02, Lecture Notes in Computer Science 2280, Springer-Verlag, 476-480, 2002.
32. D. Taubner: **Finite Representation of CCS and TCSP Programs by Automata and Petri Nets**. Lecture Notes in Computer Science 369, Springer-Verlag, 1989.
33. G. Winskel: Petri nets, algebras, morphisms and compositionality. Information and Control 72, 197-238, 1987.

A Coloured Petri Net Approach to Protocol Verification*

Jonathan Billington, Guy Edward Gallasch, and Bing Han

Computer Systems Engineering Centre
University of South Australia
Mawson Lakes Campus, SA 5095, Australia
jonathan.billington@unisa.edu.au,
{guy.gallasch,bing.han}@postgrads.unisa.edu.au

Abstract. The correct operation of communication and co-operation protocols, including signalling systems in various networks, is essential for the reliability of the many distributed systems that facilitate our global economy. This paper presents a methodology for the formal specification, analysis and verification of protocols based on the use of Coloured Petri nets and automata theory. The methodology is illustrated using two case studies. The first belongs to the category of data transfer protocols, called Stop-and-Wait Protocols, while the second investigates the connection management part of the Internet's Transmission Control Protocol (TCP). Stop-and-Wait protocols (SWP) incorporate retransmission strategies to recover from data transmission errors that occur on noisy transmission media. Although relatively simple, their basic mechanisms are important for practical protocols such as the data transfer procedures of TCP. The SWP case study is quite detailed. It considers a class of protocols characterized by two parameters: the maximum sequence number (MaxSeqNo) and the maximum number of retransmissions (MaxRetrans). We investigate the operation of the protocol over (lossy) in-sequence (FIFO) channels, and then over (lossy) re-ordering media, such as that provided by the Internet Protocol. Four properties are considered: the bound on the number of messages that can be in the communication channels; whether or not the protocol provides the expected service of alternating sends and receives; (unknowing) loss of messages (i.e. data sent but not received, and not detected as lost by the protocol); and the acceptance of duplicates as new messages. The model is analysed using a combination of hand proofs and automatic techniques. A new result for the bound of the channels (2MaxRetrans+1) is proved for FIFO channels. It is further shown that for re-ordering channels, the channels are unbounded, loss and duplication can occur, and that the SWP does not provide the expected service. We discuss the relevance of these results to the Transmission Control Protocol and indicate the limitations of our approach and the need for further work. The second case study (TCP) illustrates the use of hierarchies to provide a compact and readable CPN model for a complex protocol. We advocate an incremental approach to

* Partially supported by an Australian Research Council (ARC) Discovery Grant (DP0210524).

J. Desel, W. Reisig, and G. Rozenberg (Eds.): ACPN 2003, LNCS 3098, pp. 210–290, 2004.

both modelling and analysis. The importance of stating the assumptions involved is emphasised and we illustrate how they affect the abstractions that can be made to simplify the model. The incremental approach to analysis allows us to validate the model against the TCP definition and to show how errors in the connection establishment procedures can be found. Finally we provide some observations and tips on the how the methodology can be used based on many years of experience. The emphasis of the paper is on providing a tutorial style introduction to the methodology, examining case studies in depth, rather than breadth, and giving some insight into the process while noting its limitations.

1 Introduction

The global economy is becoming more and more dependent on distributed systems. An important example is the Internet which connects millions of computers all over the world via the interconnection of thousands of networks. It provides the infrastructure for the world wide web and email and the development of new information services such as electronic commerce, GRID computing, web services and mobile data services. At the heart of distributed systems are the communication and co-operation protocols that ensure that the required services are provided to their users. It is thus vitally important that these protocols operate correctly.

A protocol needs to satisfy a set of properties defined for the communication service it is meant to provide (e.g. data is neither lost nor duplicated and arrives in sequence, and there are no deadlocks). Proving that a protocol satisfies its required properties is known as protocol verification. Protocol verification [40,70] is a difficult problem due to inherent complexity [77].

This paper summarises a protocol verification methodology set in the context of the broader field of protocol engineering [9]. The paper does not attempt to compare the merits of this approach with other approaches. For a comparison of the main techniques for protocol specification and analysis, including the Petri net approach, the reader is referred to a recent survey by Babich and Deotto [4].

Coloured Petri Nets [48–50,53] have been used successfully for the modelling and analysis of a wide range of concurrent and distributed systems [50] including communication systems and protocols [11,14,23,24,31,32,51,83]. Thus the methodology uses Coloured Petri Nets for the specification of protocol behaviour.

In order to prove that the *protocol specification* satisfies the requirements of its users, a higher level specification, known as the *service specification* is also modelled with Coloured Petri Nets (CPNs). We then wish to prove that the protocol specification is a refinement of the service specification, in that it complies with the sequencing constraints on user observable events (known as service primitives) that are embodied in the service specification. These constraints can be expressed as the *service language*: the set of sequences of service primitives at each of the user interfaces. In principle, the service language can be derived from the CPN service specification by generating its occurrence graph (reachability graph, state space), converting it to an automaton by nominating halt states and labelling events that are not observable (by users) as empty, and then reducing this automaton to its minimised deterministic form using standard automata reduction techniques.

We can follow the same approach with the protocol specification, masking out internal events such as retransmissions, to obtain the *protocol language*: the set of sequences of service primitives generated by the protocol. We then compare the protocol and service languages. If they are the same, then we can say that the protocol specification satisfies the service sequence constraints. If the protocol language is a subset of the service language, then the protocol satisfies the constraints, but may not provide all the desired features of the service. Finally if the protocol language is not a subset of the service language, then it contains sequences that are erroneous (if the service language has been defined correctly).

We are also interested in other behaviour of the protocol, such as whether or not it will deadlock or livelock in various circumstances. In general, we need to define (a priori) a set of properties that the protocol needs to fulfill, such as correct termination or transparent delivery of data. These properties can be expressed in some language (often a temporal logic) and model checking techniques used over the occurrence graph (OG) to prove their existence or otherwise. Thus our methodology comprises two parts: the first checks sequencing constraints are satisfied at the user interface; the second proves general properties (such as boundedness or absence of deadlock) and specific protocol properties by hand proofs on the CPN model or by investigating the occurrence graph.

After presenting the methodology in some detail in Section 2, this paper formalises the methodology in Section 3 and illustrates it with two case studies. The first of these is concerned with data transfer procedures and comprises the class of Stop-and-Wait protocols [73] (see Sections 4 to 8). This is motivated by their basic mechanisms being important for practical protocols such as the Internet's Transmission Control Protocol (TCP). The second case study (Sections 9 to 12) investigates the connection management part of protocols, using the rather complex 3-way handshake of TCP. The paper also attempts to provide some guidelines for modelling and analysing protocols using high-level Petri net techniques based on twenty-five years experience of the first author and the work of his colleagues and students. The paper does not aim to be complete, it does however aim to give some insight and detailed illustrations of a Coloured Petri Net approach to protocol verification.

2 Protocol Verification Methodology

Our first attempts to develop a protocol verification methodology were published in [19]. Since then we have used it with some success to verify that industrial scale protocols do or do not meet their service specifications [34–37, 39, 55, 58, 59, 74, 78, 80].

The main steps of the methodology are:

1. Service specification: specify the service to be provided to the users of the protocol under investigation;
2. Protocol specification: specify a protocol entity for each party involved in communication;
3. Underlying service specification: specify the characteristics of the communication medium by which the different protocol entities communicate, by defining the communication service provided by the underlying service in the protocol architecture;
4. Composite specification: combine the protocol specification with the specification of the medium to obtain a composite specification of the protocol entities communicating over the underlying service;
5. Analysis: analyse the composite specification using reachability analysis and/or theorem proving to investigate desired properties of the system; and
6. Comparison: compare the service specification with the composite specification to see if the composite specification is a faithful refinement of the service.

2.1 Service Specification

This first step of the methodology has led to the development of formal service specifications using high-level nets for a number of protocols. The first of these was for the ISO Open Systems Interconnection (OSI) Transport Service [7]. The OSI standardisation effort had strongly supported the notion of service specification and promulgated guidelines for their development, known as the OSI service conventions [46]. This has greatly assisted the development of formal service specifications. In the case of OSI and other protocol development forums, such as ITU-T [41] and the Wireless Application Protocol Forum [81], this has led to the inclusion of service definitions as an integral part of developing protocol specifications. In contrast, this has not been the case in the development of Internet drafts and Request for Comments (RFCs) used in the Internet community.

Integral to the development of service specifications is the notion of a *service primitive*. A service primitive represents an interaction between the user of the service (often another protocol entity in a higher layer) and the provider of the service (i.e. the protocol operating over its underlying service). It corresponds to some feature of the service, such as a request by a user to establish or release a connection or to transfer data, or an indication by the provider that a connection has been requested by a remote user. Primitives are meant to be implementation independent, allowing them to be implemented in various ways such as message passing or procedure calls. They are also considered to be atomic events in service specifications, and are readily modelled by labelling transitions in a Coloured Petri net with the name of the primitive.

In an attempt to verify industrial protocols we have recently developed a number of service specifications. These have included the Wireless Application

Protocol (WAP) transaction layer [35] and the capability exchange signalling (CES) protocol of ITU-T recommendation H.245 [56], where the service definitions exist in the standards documents. We have also attempted to create service specifications for Internet Request for Comments such as the Resource Reservation Protocol (RSVP) [78,79],the Internet Open Trading Protocol (IOTP) [58,60] and the Internet's Transmission Control Protocol (TCP) [15,16]. We have found that the specification of services has ranged from relatively straightforward (WAP) to requiring significant ingenuity.

Although the CES protocol and service are very simple, ensuring that the service specification properly reflected the sequences of service primitives required complex synchronisation mechanisms [56]. The work for RSVP and IOTP was much more complicated, firstly due to the complex nature of the protocols and secondly that no service definitions had been written as part of the standardisation process. With IOTP there was the added complexity of catering for 4 interfaces, due to there being 4 user roles (Consumer, Merchant, Payment Handler and Delivery Handler). This complexity has led to the development of local automata which express the sequencing constraints at each one of the interfaces separately, before trying to define the global sequences by converting the automata into Coloured Petri nets (CPNs) and synchronising them via queues. We considered that attempting to directly build the correct CPN (covering all interfaces at once) would be too error prone, and that a divide and conquer approach of specifying local interface sequences first, as is done in the service conventions, would be an easier task. This has led to defining a validation step when specifying services this way. The validation step comprises proving that the CPN service specification does conform to the local sequences as defined by the local automata. This is done using reachability analysis and automata reduction techniques [5]. Interested readers can consult [60] for the details. For complex service specifications (such as IOTP), this validation step has proved to be of significant value, as now an iterative approach is used to remove errors from the CPN specification of the service.

Our experience with both the CES protocol and TCP has shown that the reachability graph for many service specifications is not finite. This is due to the service provider (e.g. the Internet) having an unknown storage capacity (number of buffers) and that the service allows an arbitrary number of service data units to be accepted by the service provider, before they are delivered to a peer user. This has important ramifications for our comparison step, which we shall return to below in section 2.3.

2.2 Protocol, Underlying Service and Composite Specifications

Currently we tend to consider steps 2 to 4 together, whereas step 1 is quite independent, and could be performed by a separate member of the protocol verification team, concurrently with steps 2 to 4. The reasons for considering these 3 steps together are that:

- it is important that the level of abstraction used for modelling the underlying service and the protocol entities is the same;

- for verification, the precise architectural location where the underlying service is considered, may not correspond to a strict layer boundary; and
- separate specifications for the protocol and underlying service tend to generate a larger state space when combined, due to there being ways of optimising the specification at the boundaries when the composite specification is considered.

Regarding the first point, normally we consider protocol data units (packets) as being transferred through the underlying medium, rather than service data units.

We illustrate the second point by considering how to model the error detection mechanism in protocols that need to recover from transmission errors. To detect transmission errors, packets contain redundant bits known as a checksum. The operation of checksums is very well known and either does not need to be verified or can be verified separately. We thus can assume that the checksum works correctly and then use a non-deterministic approach to model the *effect* of the checksum: a packet is accepted as correct (passes the checksum) or is discarded (fails the checksum). The effect is that one possible action is to discard (or lose) the packet. The processing of a checksum is an operation that occurs on the contents of the whole packet, and thus it is done before the details of the main protocol mechanisms are considered. Thus checksum processing, although part of the protocol, can be considered as a preliminary layer that can be combined with the characteristics of the medium (underlying service). This loss of a packet (due to the checksum) can then be combined with a lossy medium - such as that provided by the Internet Protocol (IP) [61] - where packets can be dropped at routers due to congestion. Hence we only need to model the loss of a packet once. It may be due to a bit error, or due to congestion, but from the point of view of the major protocol mechanisms, it does not matter. Thus when modelling a transport protocol such as TCP, we only model *above* the checksum procedures of TCP, and hence the boundary for verification is not strictly at the TCP and Internet Protocol boundary.

When using hierarchical CPNs it is natural to model the medium (underlying service) between protocol entities as a hierarchical substitution transition as in [53]. This makes a lot of sense from a specification point of view, as the details of the operation of the medium can be hidden and specified separately. Further, the medium can be changed separately as we change our view of its characteristics. However, this can have a penalty when considering verification and state space explosion. The use of a substitution transition requires that there is both an input place (perhaps representing a sending entity buffer) and an output place (representing a receiving entity buffer). This can lead to a combinatorial explosion of states of the buffers in both directions of information flow in the medium. To avoid this component of combinatorial state space explosion, we can often provide a coarser model, where the storage of the sending buffer, the medium and the receiving buffer are combined and modelled by one place. Thus we do not model any of the details of transferring packets from one buffer to another, which is not relevant to most protocol mechanisms, and hence avoid these different buffer states and the consequential state space explosion that results. However, this does mean that we can not hide the medium in the specification using hierarchical CPNs, as no hierarchical place is provided. Thus it is important to keep in mind sources of state space explosion when creating a model that is to be verified using model checking approaches.

2.3 Analysis and Comparison

The methodology in [19] proposed to concentrate on comparing sequences of service primitive events at each of the interfaces, between the service specification and the composite specification. It was assumed that other properties, such as absence of *deadlock* or *livelock*, or boundedness properties could be decided by querying the reachability graph.

Deadlocks can be determined by examining the dead markings of the reachability graph of the composite specification, to see if they are desired or not. (Dead markings correspond to the leaf nodes of the reachability graph.) Desired dead markings correspond to required terminal states, such as in a connection establishment protocol, where both protocol entities perform an initialisation sequence (for example to synchronise sequence numbers or to set flow control window sizes) before data is transferred. The connection establishment protocol needs to place each protocol entity in the *data transfer* state, and have no packets left in the underlying medium. This would be a dead marking corresponding to the desired terminal state of correct establishment. Dead markings that are not desired, we then call deadlocks. These may correspond to *unspecified receptions*, where a packet is left in the medium because the protocol does not define a procedure for processing the packet in some circumstances, or to when the different protocol entities are not synchronised (e.g. one is established, while the other is closed).

Livelocks occur when the protocol entities are involved in the exchange of control information (such as a reset) but no progress is made with respect to the aim of the protocol, which is to transfer data (for example), and that there is no way out of this cyclic behaviour. Livelocks can be detected by calculating the strongly connected components (SCC) of the reachability graph. Each terminal SCC can then be examined. It may be a dead marking or it may be a component that involves cycles. Each of these cycles needs to be checked to see if it is desired or not. In a data transfer protocol, the main loop sending and receiving data is an expected component. However, other components may not be desired (such as each end constantly sending each other resets) and they would be considered to be livelocks. Non-terminal SCCs that contain cycles are not livelocks, but may correspond to *tempo blocking* behaviour, such as repeated loss and retransmission. The difference is that there is always the possibility of escaping from this cyclic behaviour.

Also of interest are bounds on the maximum number of messages in the communication channels, as this may affect buffer and network dimensioning or the need for congestion control procedures. These properties are generic for protocols. There may also be many other properties that are specific to a particular protocol that we wish to prove, and this can be done using a model checking approach on the reachability graph.

Given that we can determine *divergent* behaviour (deadlocks and livelocks) from the reachability graph of the composite specification, it is not important if this information is lost when comparing the service and composite specifications. Once we have determined the presence or absence of deadlocks or livelocks in our protocol we are interested in whether or not the sequences of primitives that occur in the composite specification are compatible with the service specification. Thus we can use the notion of *language equivalence*[1] or *language inclusion* to check this compatibility. We refer to the set of sequences of service primitives that occur in the service specification as the *service language* and similarly, those that occur in the composite specification as the *protocol language*.

If the protocol language is the same as the service language, then the protocol is a faithful refinement of the service specification. If the protocol language is included in the service language, then we may be able to define conditions under which the protocol is also a faithful refinement, for example, if there are some options which the protocol does not include, or some concurrent behaviour that is acceptable but not essential, see [58,59]. It may also be the case that the protocol does not implement an essential part of the service, in which case the protocol needs to be revised to include it (for example, it is unlikely that the empty set would be an acceptable subset of service primitive sequences!). However, if the protocol language is not a subset of the service language, then there is at least one sequence in the protocol that does not exist in the service specification. This means that there is an error, either in the service specification or in the protocol specification. If the error is in the service specification, then it is normally readily fixed by inspecting the sequences and understanding how they occur. If it is in the protocol specification (which is more usually the case) then the sequence of service primitives needs to be traced back to protocol behaviour (in the reachability graph) to see how the sequence was generated, normally a more difficult task, as there are usually many more epsilon transitions in the protocol specification.

Obtaining the Service Language. The service language is obtained from the service specification by generating its occurrence graph and using automata reduction techniques [5]. Normally, transitions in the service specification are labelled by service primitive names, except for some transitions that may only relate to synchronisation transitions or garbage collection. Once the OG is generated, it contains all sequences of transitions that can occur in the CPN for the initial marking. To obtain the service language, any non-primitive transition is mapped to an internal transition (epsilon transition), and acceptance (halt)

[1] There are many other equivalence notions defined in the literature (155 reported in 1993 by van Glabbeek) [77], including observational, failures, testing and Valmari's Chaos-Free Failures Divergences (CFFD). The problem with language equivalence is that progress properties, such as the absence of deadlock and livelock are not preserved. However, since these properties can be obtained directly from the protocol's state space [49] language equivalence is sufficient.

states are designated. Designation of halt states may not be trivial and requires some experience of protocols on the part of the verifier. In connection management and transaction protocol services halt states will often correspond to dead markings, while in data transfer services there may only be one halt state corresponding to the initial marking. The OG can then be considered as a Finite State Automaton (FSA), that encodes a language. This FSA is then transformed into an equivalent deterministic and minimum FSA that preserves the sequences of primitives while removing the epsilon transitions. This uses 5 algorithms and is implemented in tools such as FSM from AT&T [33]. This minimum deterministic FSA is the service language.

Obtaining the Protocol Language. We use the same procedure to obtain the protocol language from the composite specification. An important difference is that the service specification has been created with service primitives in mind. In the composite specification, there may be few guidelines as to which transitions correspond to service primitives, because the protocol may have been developed without a service specification, as is the case for Internet protocols. In this case, significant judgement is required to label transitions correctly. It is worse than that because if the protocol is modelled first (a natural way to proceed to get a good understanding of the protocol before trying to retrofit its service), it may be that decisions have been made in the protocol model which mean that for a particular transition, a service primitive will only correspond to some of its modes and not others. Thus the creation of the FSA corresponding to the OG needs to map transition modes (rather than transitions) to primitives or epsilons.

Comparing Languages. Languages can be compared if they are represented by deterministic automata. We use the FSM tool to obtain the difference between the service language and the protocol language and vice versa. For details of how this is done see [34, 58, 78].

Infinite State Systems. We have assumed in this section that the systems we are dealing with are finite state and for physical systems, this seems to be a reasonable assumption. Unfortunately, there are times when we do not know the range of a parameter, even though we may be sure it is finite. An example of this is the storage capacity of the Internet. In this case, we would like our results to be general, that is, to apply to any arbitrary value of the storage capacity. Our approach to this problem [17,56] has been to introduce a parameter representing the storage of the Internet (as the length of a queue) into the model. We can then obtain results for small values of this capacity using standard reachability analysis. In the case of the CES protocol service, we find that the OG grows linearly with the length of the queue [56]. We can thus derive recursive formulae for the OG for any value of the length of the queue. Further, we have been able to show that the corresponding deterministic automaton (DFSA) also grows linearly in the size of the queue and have derived a recursive formula for it [57]. The hope is that we shall be able to derive a similar recursive formula for the

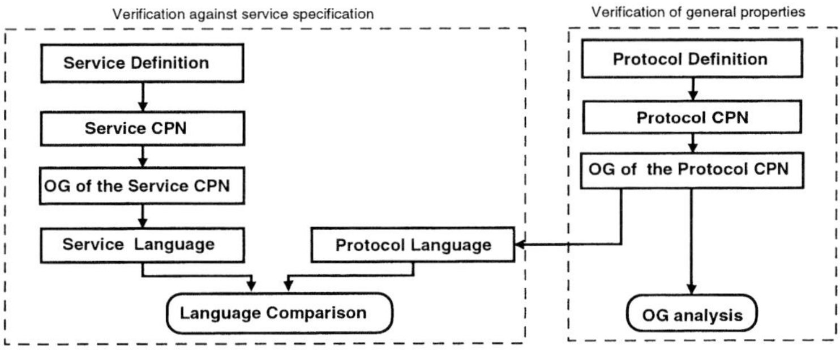

Fig. 1. Protocol verification methodology.

protocol language DFSA and then be able to extend language comparison to recursive FSAs. In the case of TCP [17], we have been able to show that the OG of the service grows exponentially with the size of the queue and have been able to derive expressions for the nodes and arcs of the OG directly, without the need for recursive formulae.

2.4 Summary

The verification methodology is summarised in Fig. 1. The dashed box on the right shows the procedures for verifying properties of a protocol. We start with the protocol definition (often provided by an international standard) and model it with CPNs. From the CPN model, we use a software package for CPNs called Design/CPN [29] to generate its OG. By analysing the OG we can obtain information about the dynamic behaviour and properties of the protocol. This may be proving correct termination (e.g. absence of deadlock and livelock), investigating boundedness properties and message sequences, or more specific properties that could be written in a temporal logic or other technique suitable for querying OGs.

The dashed box on the left of Fig. 1 illustrates the steps required to verify a protocol against its service language [19]. We do this by comparing the sequences of *service primitives* that occur as a result of the protocol's operation (the *protocol language*), with the sequences specified in the service specification (the *service language*). We firstly create a CPN model of the service specification, in which service primitives are associated with CPN transitions. The OG of the CPN model is calculated. The OG includes all the possible occurrence sequences of CPN transitions. The CPN model may include transitions that do not model service primitives, but rather internal events of the service provider required to ensure correct operation. We need to eliminate these internal transitions while preserving service primitive sequences. To do this, we treat the OG as a FSA and use standard FSA reduction techniques [5]. This minimised and deterministic FSA embodies the *service language*. In a similar way, we generate the *protocol language*. These are compared to see if all the sequences of service primitives in the protocol language are sequences specified in the service language, to discover if there are any inconsistencies between the protocol definition and the service definition. The automata reductions and comparison algorithms are automated in tools such as FSM [33].

3 Definitions

3.1 High-Level Petri Nets

We use Coloured Petri nets and Design/CPN for the specification of protocols and services, and refer readers to [48–50, 53] for an introduction to CPNs and their definitions. However, for some of the proofs in this paper, it is useful to use the definition of the High-level Petri Net (HLPNs) semantic model presented in clause 5 of international standard ISO/IEC 15909-1 [44]. It is drawn from similar work published in [8, 10] and earlier work in [47].

In order to be self-contained, we begin by presenting definitions concerning multisets and vectors, followed by the semantic model from [44]. We then define occurrence sequences for high-level nets and provide some propositions that we require in the analysis of the Stop-and-Wait protocol.

Multisets

Definition 1. *A multiset, $B : A \rightarrow \mathbb{N}$, is a function that associates a natural number, known as the multiplicity, with each of the elements of a non-empty basis set, A.*

The multiplicity of $a \in A$ in B, is given by $B(a)$. The set of all multisets over A is denoted by μA.

Multiset Operations

Definition 2. Equality:
Two multisets, $B1, B2 \in \mu A$, are equal, $B1 = B2$, iff $\forall a \in A, B1(a) = B2(a)$.

Definition 3. Comparison:
$B1$ is less than or equal to $B2$, $B1 \leq B2$, iff $\forall a \in A, B1(a) \leq B2(a)$ and $B1$ is greater than or equal to $B2$, $B1 \geq B2$, iff $\forall a \in A, B1(a) \geq B2(a)$.

We define addition and subtraction on multisets $B1, B2 \in \mu A$ as follows.

Definition 4. Addition and Subtraction:
$B = B1 + B2$ *iff* $B(a) = B1(a) + B2(a)$
$B = B1 - B2$ *iff* $B(a) = B1(a) - B2(a)$ *if* $B1(a) \geq B2(a)$

There are times when we wish to subtract one multiset from another when the above restriction on multiset subtraction does not apply. We then need to consider multisets as vectors.

Vectors

Definition 5. *A vector V over a basis set A is a function $V : A \rightarrow \mathbb{Z}$ where $\mathbb{Z}$ is the set of integers.*

The set of all vectors over A is denoted by νA. Subtraction is a closed operation for vectors, defined componentwise as follows.

Definition 6. Vector Subtraction:
For $V1, V2 \in \nu A$, $V = V1 - V2$ iff $\forall a \in A, V(a) = V1(a) - V2(a)$.

High-Level Petri Net

We now define a High-level Petri net (*HLPN*) [8,44].

Definition 7. *$HLPN = (P, T, D; Type, Pre, Post, M_0)$ where*

- *P is a finite set of Places.*
- *T is a finite set of Transitions, disjoint from P ($P \cap T = \emptyset$).*
- *D is a non-empty finite set of non-empty domains where each element of D is called a type.*
- *$Type : P \cup T \longrightarrow D$ is a function used to assign types to places and to determine transition modes.*
- *$Pre, Post : TM \longrightarrow \mu PLACE$ are the pre and post mappings with*
  - *$TM = \{(t, m) | t \in T, m \in Type(t)\}$, the set of transition modes; and*
  - *$PLACE = \{(p, g) | p \in P, g \in Type(p)\}$, the set of* elementary *places.*
- *$M_0 \in \mu PLACE$ is a multiset called the initial marking of the net.*

Marking of a HLPN

Definition 8. *A* Marking *of the HLPN is a multiset, $M \in \mu PLACE$.*

Enabling of Transition Modes

Definition 9. *A single transition mode, $tm \in TM$, is* enabled *at a marking M iff $Pre(tm) \leq M$.*

We can also define the concurrent enabling of a finite multiset of transition modes (see [10,44]) but this is not required for this paper.

Transition Rule

Definition 10. *The* transition rule *for a single transition mode $tm \in TM$ enabled at a marking M is given by*

$$M' = M - Pre(tm) + Post(tm)$$

where an occurrence of tm results in the new marking M'.

The occurrence of a single transition mode $tm \in TM$ in marking M is denoted by $M[tm\rangle M'$ or $M \xrightarrow{tm} M'$.

Occurrence Sequences and Reachability

In addition to that defined in [44], we define the following in relation to occurrence sequences.

Definition 11. *Let M be a marking of the HLPN. A finite sequence of transition modes, $tm_1, tm_2, \ldots, tm_k \in TM, k \in \mathbb{N}^+$, is called a* finite occurrence sequence *able to occur from M, if there are markings $M_1, M_2, \ldots, M_k$ such that*

$$M \xrightarrow{tm_1} M_1 \xrightarrow{tm_2} M_2 \ldots \xrightarrow{tm_k} M_k$$

We denote a finite occurrence sequence by $\sigma_k = tm_1 tm_2 \ldots tm_k$ and write $M \xrightarrow{\sigma_k} M_k$.

Definition 12. *A marking M' is* reachable *from a marking M if there is a finite occurrence sequence σ_k leading from M to M', i.e. $M \xrightarrow{\sigma_k} M'$.*

Definition 13. *An infinite sequence of transition modes, $\sigma = tm_1 tm_2 tm_3 \ldots$ is called an* infinite occurrence sequence, *able to occur from a marking M, if there are markings $M_1, M_2, \ldots$ such that*

$$M \xrightarrow{tm_1} M_1 \xrightarrow{tm_2} M_2 \xrightarrow{tm_3} \ldots$$

Following directly from these definitions are Propositions 1 and 2. They are HLPN extensions of the propositions given in [28] and are used in the proof of Theorem 4 in Section 6.

Proposition 1. *An infinite occurrence sequence σ of transition modes can occur from a marking M if and only if every finite prefix of σ can also occur from M.*

Proposition 2. *If M and L are markings of a HLPN and for a finite occurrence sequence σ_k, $M \xrightarrow{\sigma_k} M'$ and $L \xrightarrow{\sigma_k} L'$ then (using vector subtraction) $M' - M = L' - L$.*

The following proposition is useful for proving that a finite occurrence sequence of transition modes can be repeated indefinitely.

Proposition 3. *If M and L are markings satisfying $M \geq L$ then every occurrence sequence that can occur from L can also occur from M.*

Proof. Consider the infinite occurrence sequence $\sigma = tm_1 tm_2 tm_3 \ldots$ that can occur from L. Now all finite prefixes of σ (i.e. $\sigma_k = tm_1 tm_2 \ldots tm_k, k \in \mathbb{N}^+$) can occur from L according to Proposition 1, i.e.

$$L \xrightarrow{tm_1} L_1 \xrightarrow{tm_2} \ldots \xrightarrow{tm_k} L_k$$

Using the enabling condition (and transition rule) as defined above, this means that $L \geq Pre(tm_1)$. Occurrence of tm_1 at marking L leads to a marking L_1 in which the next transition mode in the sequence, tm_2, is enabled, and from which the rest of the sequence $tm_2 tm_3 \ldots tm_k$ can occur.

Now consider the marking M. We know that $M \geq L$. If $L \geq Pre(tm_1)$ then this means that $M \geq Pre(tm_1)$ also. Thus transition mode tm_1 is enabled in M.

Occurrence of tm_1 from M will lead to a new marking M_1, i.e. $M \xrightarrow{tm_1} M_1$, and by Proposition 2 we know that $M_1 - M = L_1 - L$ (again using vector subtraction). Knowing $M \geq L$ and substituting L for M we obtain

$$M_1 - L \geq L_1 - L$$

$$\Rightarrow M_1 \geq L_1$$

We know that tm_2 is enabled in L_1, and by the above arguments, tm_2 is also enabled in M_1. By repeated application of the above arguments, we obtain that for every intermediate marking $L_n (1 \leq n \leq k)$ during execution of the occurrence sequence σ_k there is a corresponding marking M_n such that $M_n \geq L_n$. So all finite prefixes σ_k of the infinite occurrence sequence σ can occur from M, and by Proposition 1, σ can occur from M. Thus any occurrence sequence that can occur from L can also occur from M. □

3.2 Definitions of Occurrence Graphs and Associated Automata

Part of the methodology requires the generation of a CPN's occurrence graph (OG) and its transformation to an appropriate finite state automaton (FSA). This section provides the definitions that are useful for this purpose.

Occurrence Graphs

We consider that an OG can be defined as a labelled directed graph, where the nodes of the graph represent markings of the CPN, and the directed arcs represent the transition modes that can occur in all executions of the net from the initial marking. The arcs are thus labelled by the transition mode. We thus start by defining a labelled directed graph.

Definition 14. *A labelled directed graph is a triple $G = (V, L, E)$ where*

- *V is the set of vertices or nodes;*
- *L is a set of labels; and*
- *$E \subseteq V \times L \times V$ is a set of labelled directed edges.*

Definition 15. *An occurrence graph of a HLPN with an initial marking M_0, is a labelled directed graph $OG = (V, TM, A)$ where*

- *$V = [M_0\rangle$ is the set of markings reachable from M_0 (the reachability set);*
- *TM is the set of transition modes of the HLPN; and*
- *$A = \{(M, tm, M') \in V \times TM \times V | M[tm\rangle M'\}$ is the set of arcs (directed edges) labelled by transition modes.*

Abstract OGs

When we only consider sequences of primitives, there are two abstractions of the OG which are useful. The first removes the modes (variable bindings) from the transition modes to give transitions only. The second removes the details of the markings, so that they are just represented by integers. Both abstractions may be used together. We formalise these abstractions in the following definitions.

For an occurrence graph where we are only interested in the transition names, rather than transition modes, e.g. in the case of service primitives where we are not concerned with service primitive parameter values, but just the primitive name, then an abstract OG with respect to transitions, AOG^T, is defined as follows.

Definition 16. *An abstract OG, with respect to transitions, of a HLPN with an initial marking M_0 and a set of transition modes, TM, is a labelled directed graph $AOG^T = (V, T, A)$ where*

- *$V = [M_0\rangle$ is the set of markings reachable from M_0;*
- *T is the set of transitions of the HLPN; and*
- *$A = \{(M, t, M') \in V \times T \times V | (t, m) \in TM and M[(t, m)\rangle M'\}$ is a set of arcs labelled with transition names.*

In the case where we are only interested in the identification of the markings for the nodes, and not the details of the markings, we introduce an injection, I, mapping the set of reachable markings into the set of positive integers:

$$I : [M_0\rangle \longrightarrow \mathbb{N}^+$$

where $I(M_0) = 1$ represents the initial marking.

Definition 17. *An abstract OG, with respect to markings, of a HLPN with an initial marking M_0 and a set of transition modes TM, is a labelled directed graph $AOG^M = (V, TM, A)$ where*

- *$V = \{I(M) | M \in [M_0\rangle\}$ is the set of nodes;*
- *TM is the set of transition modes of the HLPN; and*
- *$A = \{(I(M), tm, I(M')) \in V \times TM \times V | M[tm\rangle M'\}$ is the set of arcs labelled with transition modes.*

This definition is useful for the analysis of the Stop-and-Wait protocol (see section 6).

Finally we combine the two abstractions to obtain an abstract OG with respect to markings and transitions, AOG^{MT}.

Definition 18. *An abstract OG, with respect to markings and transitions, of a HLPN with an initial marking, M_0, and a set of transition modes, TM, and given an injection, I, mapping markings to positive integers, is a labelled directed graph $AOG^{MT} = (V, T, A)$ where*

– $V = \{I(M) | M \in [M_0\rangle\}$ *is the set of nodes;*
– T *is the set of transitions of the HLPN; and*
– $A = \{(I(M), t, I(M')) \in V \times T \times V | (t, m) \in TM and M[(t, m)\rangle M'\}$ *is the set of arcs labelled with transition names.*

This last definition is useful when we are only interested in sequences of transitions, e.g., sequences of service primitive names.

FSA Associated with the OG
The next step is then defining the mapping from an abstract OG to its FSA. In our methodology, the FSAs are only used to determine language equivalence (or inclusion) and thus we are not concerned with the details of the markings, and hence we can use an abstract OG that does not include the marking details. Hence we just represent the nodes by positive integers.

When we construct the CPN model of the protocol, we normally do so with primitive events in mind, and thus label transitions in the CPN with service primitive names. However, in some cases (such as for the stop-and-wait protocol) it is convenient to have a more general mapping from transition modes to service primitives. (In the following, we restrict our attention to a mapping to service primitive names, as that is our current focus, but in general the mapping could be to service primitives in general, i.e. where service primitive parameters are included as well as the name.) We thus need a function that maps each transition mode in the abstract OG to either a service primitive name, or to an epsilon. Lets call this function *Prim* as it returns a primitive name (or epsilon). Thus we have

$$Prim : TM' \longrightarrow SP \cup \{\epsilon\}$$

where

– $TM' \subseteq TM$ is the set of transition modes used to label arcs in the abstract OG; and
– SP is the set of service primitive names in the system we are describing.

Given an abstract OG with respect to markings $AOG^M = (V, TM, A)$ (see Definition 17) we can formulate its corresponding FSA as

Definition 19. $FSA_{AOG^M} = (V, SP, A_{SP}, v_0, F)$ *where*

– V *is the set of nodes of the abstract OG (the states of the FSA);*
– SP *is the set of service primitive names of the system of interest (the alphabet of the FSA);*
– $A_{SP} = \{(v, Prim(tm), v') | (v, tm, v') \in A\}$ *is the set of transitions labelled by service primitives or epsilons for internal events (the transition relation of the FSA);*
– $v_0 = 1$ *corresponds to the abstract initial marking (initial state of the FSA); and*
– $F \subseteq V$ *is the set of final states.*

4 Stop-and-Wait Protocols

This part of the paper illustrates our approach to the verification of data transfer protocols by investigating the class of Stop-and-Wait protocols (SWP) [66,73]. The work presented here is a major eleboration and extension of that recently published by the first two authors [12] and is based on [13].

We choose the SWP class because the protocol mechanisms are readily understood and because they are the simplest representative class of data transfer protocols since they include sequence numbers and retransmission counters. The class of SWP protocols is characterised by two parameters: the maximum sequence number and the maximum number of retransmissions.

Stop-and-Wait is an elementary form of flow control [66,73] between a sender and a receiver. The sender stops after transmitting a message and waits until it receives an acknowledgement indicating that the receiver is ready to receive the next message. Stop-and-Wait Protocols often operate over noisy channels and combine flow control with error recovery using a timeout and retransmission scheme, known as Automatic Repeat ReQuest (ARQ) [73]. In this case, a checksum [73] is included to detect transmission errors. Messages that pass the checksum are acknowledged as received correctly. A message that fails the checksum is discarded by the receiver. In this case, the sender of the message will not receive an acknowledgement within its specified timeout period, and thus retransmits the message. This works well if the cause of not receiving the acknowledgement is due to the message being discarded (due to errors). However, the acknowledgement is also error protected by a checksum and it could have been discarded. In this case the retransmitted message is an unnecessary duplicate of the original message that has already been received correctly. To prevent duplicate messages being accepted as new messages a sequence number is appended to each message.

The class of SWPs are important because many practical data transfer protocols use *sliding window* mechanisms that have their foundations based on Stop-and-Wait principles. Sliding Window protocols [66, 73] improve the efficiency of SWPs by allowing many messages (rather than one) to be sent before requiring an acknowledgement. The number of messages that can be sent before the sender must stop and wait to receive an acknowledgement is known as the *window*. Cumulative acknowledgements and more sophisticated error retransmission schemes (such as Selective Reject) [66] can also improve efficiency. These schemes are used in many practical protocols such as TCP [62]. The underlying principles of ARQ used in sliding window protocols are the same as those used in SWPs, so that a window size of 1 corresponds to a Stop-and-Wait protocol. Thus it is essential that the stop-and-wait mechanisms work correctly if the more advanced protocols are also to be correct.

It is well known [73] that for sliding window protocols to work properly in detecting and discarding duplicates, the sequence number space needs to be one greater than the number of unacknowledged messages (the window). In the case of Stop-and-Wait protocols which have just one outstanding unacknowledged message, the sequence number space can be just two numbers, usually {0,1}.

When a SWP uses the sequence numbers {0,1} it is called an Alternating Bit protocol (ABP) [6], because the sequence number can be implemented using just one bit in the header of the message, and the sequence number value alternates between 0 and 1. Acknowledgement messages in this case serve a dual purpose: that of flow control (indicating that the receiver is ready to receive another message) and transmission error recovery (informing the sender not to retransmit as the data has been received correctly). This is the simplest class of SWPs where the maximum sequence number is instantiated to 1. We consider Stop-and-Wait protocols with an arbitrary maximum sequence number as this takes us a step closer to sliding window protocols, where the window size is arbitrary, and hence the sequence number space (which must be at least one greater than the window size) is also arbitrary. It may also be the case that SWPs with larger sequence number spaces can work correctly over media with a limited amount of re-ordering (see [52]), but we do not consider this situation in this paper.

A number of papers, articles and books [1,3,6,18,30,63,68,71–73,76] have been written about the ABP. Many demonstrate that the ABP will work perfectly over an underlying medium that behaves in a FIFO (First-In First-Out) manner and that may also include loss. The ABP is often used as a case study when developing a new modelling language or a derivation from an existing modelling language, to demonstrate the use or effectiveness of the new language. This is the case in [68] where the ABP is used as an example to illustrate a new Timed Rewriting Logic (TRL) for capturing the static and dynamic aspects of SDL (Specification and Description Language) [45]. Another example of this is in [71] and [72] where the ABP is formally modelled and analysed using Temporal Petri nets (derivations of Petri nets with restrictions on the firing of transitions based on formulae containing temporal operators.) The ABP is used to illustrate modelling and analysis of protocols using Petri Nets in [30]. The Abracadabra Service and Protocol Example [76] describes a protocol using Alternating Bit sequence numbers, Retransmissions on timeout, Acknowledgements, Connection And Disconnection (ABRACAD), and is one of a graded set of examples used to provide guidelines for the application of three standardised formal description techniques, namely Estelle [21], LOTOS (Language Of Temporal Ordering Specifications) [20] and SDL [45]. Billington et al [18] use a variant of the ABP [22] to demonstrate a software tool. Reisig [63] develops the ABP in a series of steps as part of a case study on acknowledged messages, developed incrementally using simple Petri net models to illustrate the principles and operation of the ABP over FIFO (First-In First-Out) communication channels.

Some of the above papers demonstrate that the ABP will work as expected over FIFO channels that may also include loss. It appears, however, that the situation in which messages may be re-ordered by the medium has not been considered. The ABP was originally designed to provide a reliable data link service over an unreliable point-to-point physical link. In this situation overtaking of messages does not occur. However, the same ARQ mechanisms are used in transport level protocols, such as TCP [62], that operate over a medium that does not guarantee in-sequence delivery and may also lose messages [73]. It is therefore useful to investigate this situation for SWPs.

In Section 5 we present and explain our Coloured Petri Net (CPN) [48, 53] model of the SWP and discuss some of the modelling decisions made during its construction. The model is then analysed in Section 6 using a combination of hand proofs and language analysis. A discussion of the impact of the analysis results on the Transmission Control Protocol is given in Section 7, as well as the identification and discussion of a limitation of our approach. Finally some concluding remarks are presented in Section 8.

5 The Stop-and-Wait Protocol Model

We model a Stop-and-Wait protocol that includes error recovery using retransmissions operating over a lossy re-ordering medium. The CPN model of our SWP is given in Figs. 2 and 3. Figure 2 presents the graphical representation of the system, while Fig. 3 defines all the constants, sets and functions required and declares the types of the variables used in the annotations associated with the graphical representation. The software tool, Design/CPN [29], was used for the construction of the model. Design/CPN has four main facilities: an editor, a simulator, a state space tool and a performance tool. The simulation engine and state space tool are built using CPN ML [27], a variant of the functional programming language Standard ML of New Jersey (SML/NJ) [67]. CPN ML is used for the net annotations in Fig. 2 and the declarations in Fig. 3.

In Fig. 3 colour sets (types) are defined using the keyword color and enumerated types (Sender, Seq, Retranscounter) are created with the set constructor with. Variables are declared using the keyword var and are typed by a colour set, e.g. the variables sn and rn of type Seq (sequence number). Functions are defined using the keyword fun and values (e.g. constants) are defined using the keyword val.

We now describe the CPN model of the SWP in detail. The model comprises three main parts: the *Sender* (on the left), the *Receiver* (on the right) and an underlying bidirectional communication medium, *Network*, in the middle.

5.1 Sender

The sender consists of four places, four transitions and their interconnecting arcs. The places, sender_ready and wait_ack, represent the two states of the sender (either ready to send a new data message or awaiting an acknowledgement) and are typed by the colour set Sender, representing a single sender. The place, sender_ready, has an initial marking of one s token, indicating that the Sender is initially in the ready state. The seq_no place stores the sender sequence number, which is either the number of the message just sent (an unacknowledged message) or if acknowledged, the number of the next message to be sent. It is typed by the colour set Seq (sequence number) and has an initial marking of a single 0 token, indicating that the first message to be sent will have sequence number 0. The current number of retransmissions is recorded in place retrans_counter, typed by the colour set RetransCounter and is initially 0.

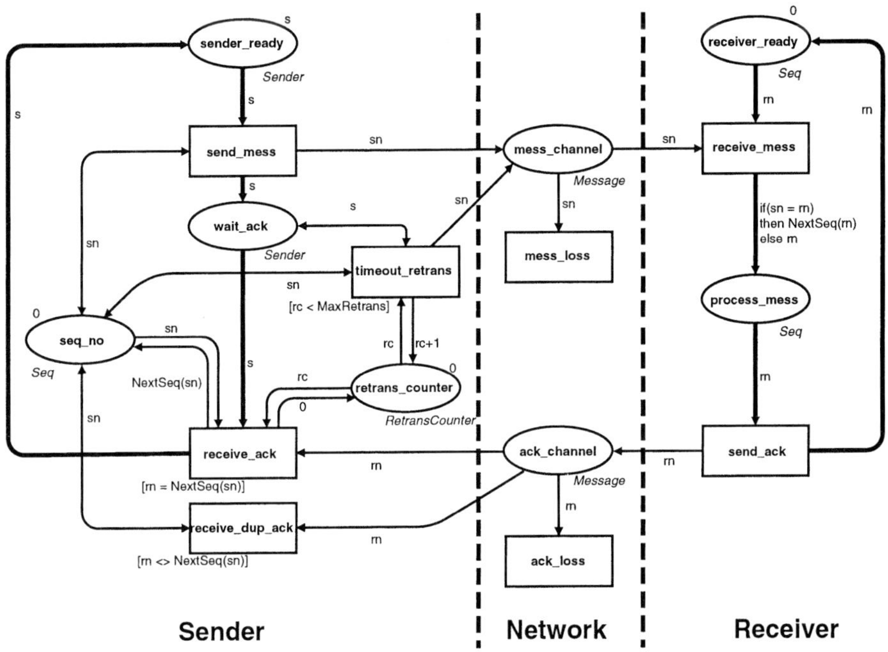

Fig. 2. The CPN of the Stop-and-Wait Protocol operating over a lossy re-ordering channel.

```
val MaxRetrans = 1;
val MaxSeqNo = 1;

color Sender = with s;
color Seq = int with 0..MaxSeqNo;
color RetransCounter = int with 0..MaxRetrans;
color Message = Seq;

var sn,rn   : Seq;
var rc : RetransCounter;

fun NextSeq(n) = if(n = MaxSeqNo) then 0 else n+1;
```

Fig. 3. Global Declarations for the Stop-and-Wait Protocol CPN.

Transition send_mess models the sending of a message to the receiver. Message content is not represented as no protocol operations involve it (the protocol behaves the same way irrespective of content). The same is true for the addresses of sender and receiver, as we only have one of each in this model. Consequently, a message (or an acknowledgement) can be modelled by just its sequence number.

When the sender is ready send_mess may occur. It writes its sequence number (as the message) to the message channel and changes state to waiting for an acknowledgement.

The timeout_retrans transition models the expiry of the retransmission timer and the retransmission of the currently unacknowledged message. This transition can only occur if the sender is waiting for an acknowledgement and there have been less than MaxRetrans retransmissions of this message (see the guard). When timeout_retrans occurs, the retransmission counter is incremented by 1 and the retransmitted message is placed into the message channel.

Transition receive_ack models the receipt of expected acknowledgements from the receiver, i.e. those that acknowledge the currently outstanding message. Duplicate acknowledgements are received and discarded by transition receive_dup_ack. These may result from acknowledged retransmissions, where delay rather than loss was the cause of the retransmission. The complementary guards on these transitions identify the acknowledgement as being expected or a duplicate. An expected acknowledgement will have a sequence number one greater than the sender sequence number. The function NextSeq is used to increment the sequence number modulo (MaxSeqNo + 1), as shown in Fig. 3. An occurrence of receive_ack will remove the acknowledgement from the channel, return the sender to the ready state, reset the retransmission counter to 0 and increment the sequence number stored in seq_no using modulo arithmetic. The transition receive_dup_ack discards duplicate acknowledgements irrespective of the state of the sender.

5.2 Receiver

The receiver consists of two places and two transitions. The places receiver_ready and process_mess model the states of the receiver and are typed by the colour set Seq. A sequence number token present on one of these places indicates that the receiver is in that state (either ready to receive a message, or processing a message.) The receiver_ready place has an initial marking of one 0 token, indicating that initially the receiver is in the ready state and expecting a message with sequence number 0.

Transition receive_mess models the receipt of a message from the sender. The annotation on the arc from receive_mess to process_mess compares the sequence number of the message (sn) with the sequence number expected by the receiver (rn). If they match, then the message is the one expected (and is passed onto the user, a process that is not modelled) and the sequence number is incremented modulo (MaxSeqNo + 1) by the NextSeq function and placed in process_mess. If they don't match, a duplicate is detected (and discarded) and the receive sequence number is placed in process_mess unchanged. Transition send_ack occurs when the receiver has finished processing the message, indicating that enough buffer space is available to receive another message. This transition sends an acknowledgement containing the next sequence number expected by the receiver and returns the receiver to the ready state. Sending an acknowledgement when a duplicate message is received is necessary because if an acknowledgement of a (new) message is lost and subsequent retransmissions of the same message are not acknowledged, the system will fail to progress as no acknowledgement will ever be delivered to the sender.

5.3 Underlying Medium

The underlying communication medium is modelled as a bidirectional channel consisting of one place and one transition for each direction of communication. The mess_channel place models the message channel while the ack_channel place models the acknowledgement channel. Both channel places are typed by the colour set Message (a sequence number, see Fig. 3) and are both initially empty. This models the overtaking behaviour of the communication medium. The two transitions mess_loss and ack_loss model the loss of messages and acknowledgements respectively. This corresponds to either loss in the network (due to congestion and buffer overflow in a router), or to discarding messages (and acknowledgements) due to checksum failures.

5.4 Discussion of Modelling Decisions

A straightforward way to begin modelling a system such as this is to use one place for each state of an entity, one place for each data item, and one transition for each action. This is evident in the Sender, where we have one place for each state (i.e. ready, waiting for an acknowledgement), one place for each item of data (i.e. sequence number, retransmission counter) and one transition for each action. A representation such as this gives a clearer visual indication of the control flow within the Sender than if the sender states were folded and represented by a single place typed by the set of states. However, this is normally only possible for protocols with very few states. As the number of states increases, so do the number of arcs, which leads to a visually complex diagram with many arc crossings, distracting from the major flows. This is alleviated to some extent by the use of thicker lines for arcs to emphasise major flows, as is illustrated for control flow in Fig. 2.

The receiver has been modelled using a different style, where there has been a folding of the receiver sequence number and receiver state. The transition receive_mess represents both the receipt of an expected message and the discarding of duplicate messages, requiring a complex arc annotation. This provides a more compact representation of the receiver clearly highlighting the control flow loop. This demonstrates the versatility of CPNs in being able to illustrate visually control flow and data flow. To do this well requires significant experience, especially for complex protocols that require a hierarchical approach. We used different styles for the sender and receiver to illustrate the different approaches. When modelling a complex protocol this would rarely be done, and a consistent style throughout the whole model is advocated.

We may want the model structure to reflect the structure of the real life system, given a certain amount of abstraction. For example, we have illustrated this by modelling the sequence number at the sender in the net structure as a separate place. This is to reflect the fact that in an implementation, the sequence number as an entity of data may exist separately from, and regardless of, the state of the sender. It also simplifies modelling of the sender, because duplicate acknowledgement messages can be received and discarded regardless of the state of the sender. Conversely, the meaning of the value of the sequence number (either the next to be sent when in state sender_ready, or the message to be acknowledged when in state wait_ack) is dependent on the state, and hence this would favour folding the sequence number into the state places, as in the receiver. Modelling the loss of messages and acknowledgements by separate transitions (instead folding them into the receive transitions) allows for more flexibility in analysis, as to analyse a system without a lossy channel requires nothing more than adding a [false] guard to each loss transition. Thus various trade-offs present themselves to the modeller, even in models as simple as this.

6 SWP CPN Model Analysis

6.1 Properties of Interest

As described previously, with the SWP it is usual to place an upper bound (MaxRetrans) on the number of retransmissions that are allowed per message. When this limit is reached, the communication medium is considered to be down. In practice, an indication is given to a management entity that invokes a procedure to deal with the fault. This interaction (and procedure) is not modelled as it is not part of the SWP. In our model, the protocol will just terminate in a state where the retransmission counter has reached its maximum value (MaxRetrans). This is an expected terminal state, indicating that the network is down, and that the last message sent may have been lost.

Thus we are not particularly concerned with terminal states. Instead we focus on properties that are quintessential for correct operation of the SWP. The first concerns bounds on the channels, the second that duplicates are not accepted as new messages, the third that messages are not lost unknowingly and the fourth that the protocol conforms to the Stop-and-Wait service of alternating sends and (correct) receives, ensuring that messages are received in the same order as they were sent.

6.2 FIFO Channels

Our first step is to consider the SWP operating over communication channels that preserve sequence. This corresponds to the SWP operating over a physical link (as is the case for data link protocols) and is thus important in its own right. It is also important from the point of view of incremental analysis of the SWP operating over re-ordering channels. This is because most networks will provide a FIFO channel most of the time. Thus it is important to ensure that the SWP will operate correctly over FIFO channels, before we investigate the re-ordering case.

The CPN in Fig. 4 and associated declarations in Fig. 5 show our Stop-and-Wait protocol operating over a lossy FIFO channel. Places mess_channel and ack_channel are modified to operate as FIFO queues by altering their colour set from Message to MessList (a list of messages), giving them an initial marking of the empty list and modifying appropriate arc expressions on incoming and outgoing arcs to manipulate the list as a FIFO queue. All arcs placing a message into one of the channels do so by appending it to the *end* of the message list using the infix append operator (^^). All arcs removing a message from one of the channels do so by removing a message from the *beginning* of the list using the infix 'cons' operator (::). Loss in the medium is modelled by transitions mess_loss or ack_loss. This loss behaviour includes the discarding of corrupted messages (due to failing the checksum) and loss due to routers dropping packets (if applicable).

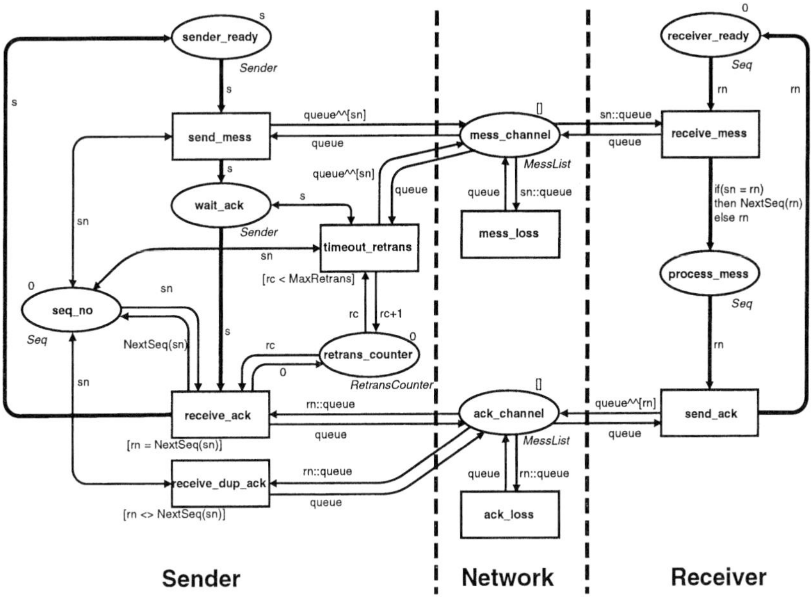

Fig. 4. A CPN of the Stop-and-Wait Protocol operating over an in-order medium.

We now consider the first property of interest, that of the bounds on the channels, and then investigate the other three properties (loss, duplication and SWP service).

Channel Bounds. In this section we state and prove two theorems regarding the maximum length of each of the message queues in places mess_channel and ack_channel.

```
val MaxRetrans = 1;
val MaxSeqNo = 1;

color Sender = with s;
color Seq = int with 0..MaxSeqNo;
color RetransCounter = int with 0..MaxRetrans;
color Message = Seq;
color MessList = list Message;

var sn,rn   : Seq;
var rc : RetransCounter;
var queue   : MessList;

fun NextSeq(n) = if(n = MaxSeqNo) then 0 else n+1;
```

Fig. 5. Declarations of the CPN shown in Fig. 4.

Theorem 1. *For the Stop-and-Wait CPN of Figs. 4 and 5 with* MaxRetrans $\geq$ *0 and* MaxSeqNo $\geq$ *1, the message queue length in place* mess_channel *is bounded by* (2MaxRetrans + 1), *i.e.* $\forall M \in [M_0\rangle, |queue| \leq (2\mathsf{MaxRetrans} + 1)$ *where* $M(\mathsf{mess\_channel}) = 1`queue$, $queue \in \{0, ..., \mathsf{MaxSeqNo}\}^*$ *and* $|queue|$ *is the length of the list 'queue'.*

Proof. We use the notation $\oplus$ for addition modulo (MaxSeqNo + 1) and 'n-message' as shorthand for 'a message with sequence number n'.

We firstly examine the case where MaxRetrans = 0. From the initial marking of the CPN as shown in Fig. 4, the only transition that can occur is send_mess, which inserts a 0-message into the message queue, so $|queue| = 1$. Transition timeout_retrans will never occur because of its guard. The only enabled transitions are mess_loss and receive_mess, both of which remove the 0-message, resulting in $|queue| = 0$. An occurrence of mess_loss leads to a dead marking and hence the theorem holds in this case. An occurrence of receive_mess with the variable bindings sn = rn = 0 indicates this is the message expected by the receiver. The receiver sequence number is incremented by the NextSeq function on the arc from receive_mess to process_mess. Now the only enabled transition is send_ack, inserting an acknowledgement (rn = 1) into the acknowledgement queue on place ack_channel. The only possible action now is to remove this acknowledgement from the queue, either through the occurrence of ack_loss (leading to a dead marking and hence the theorem holds) or through receive_ack as rn = NextSeq(sn) = 1. An occurrence of receive_ack will return the model to a state identical to the initial state except that all sequence numbers are now 1 instead of 0. The behaviour described above is the *only* behaviour of the system. This behaviour repeats, each time leading to the 'same' state except for the sequence numbers that have been incremented modulo MaxSeqNo. When the sequence number wraps back to zero, the CPN returns to the initial state. This demonstrates that for any MaxSeqNo$\geq$ 1, $|queue| \leq 1$ for all markings when MaxRetrans = 0. This proves the theorem for MaxRetrans = 0.

For the more general case of MaxRetrans ≥ 1, the situation is complicated by the possibility of duplicate messages and duplicate acknowledgements. Consider that we start with empty queues. The maximum number of messages with a given sequence number n that can be inserted into the message queue is the original (send_mess occurs) plus MaxRetrans duplicates (by MaxRetrans occurrences of timeout_retrans) giving $|queue| =$ (MaxRetrans+1) (given the queue was empty). At this point the sender must stop and wait until it receives an acknowledgement (receive_ack) for the n-message. The minimum number of n-messages that need to be received and acknowledged (i.e. when no loss occurs) is one, leaving MaxRetrans n-messages in the message queue. When this acknowledgement is received, the retransmission counter is reset to zero and (MaxRetrans+1) $(n \oplus 1)$-messages can be sent, giving a queue with MaxRetrans n-messages followed by (MaxRetrans+1) $(n \oplus 1)$-messages and $|queue| =$ (2MaxRetrans+1). Because of the FIFO property of the communication channels, the remaining MaxRetrans n-messages must be removed (by loss or receipt) before the first $(n \oplus 1)$-message can be received and acknowledged allowing messages with sequence number $n \oplus 2$ to be placed in the message channel. Thus before any new message can be sent, the length of the queue can be no more than MaxRetrans. As already discussed, only (MaxRetrans+1) new messages can be added to the queue, giving a maximum queue length of 2MaxRetrans+1. □

A similar theorem to that stated in Theorem 1 holds for the acknowledgement channel.

Theorem 2. *For the Stop-and-Wait CPN of Figs. 4 and 5 with MaxRetrans ≥ 0 and MaxSeqNo ≥ 1, the acknowledgement queue in place ack_channel is bounded by $(2MaxRetrans + 1)$, i.e. $\forall M \in [M_0\rangle, |queue| \leq (2MaxRetrans + 1)$ where $M(ack\_channel) = 1`queue$, $queue \in \{0, ..., MaxSeqNo\}^*$ and $|queue|$ is the length of the list 'queue'.*

Proof. From Theorem 1 at most (2MaxRetrans+1) messages can be in the message queue. Also, from the proof of Theorem 1 when there are (2MaxRetrans+1) messages in the message queue the acknowledgement queue is empty. We know that exactly one acknowledgement is generated for each message accepted by the receiver, by the occurrence of receive_mess followed by send_ack, which implies that a message is removed from mess_channel for every acknowledgement generated by the receiver. Further, from the proof of Theorem 1, the sum of messages and acknowledgements in the channel places can be no more than MaxRetrans before a new message can be sent. Thus although the removal of one acknowledgement can result in the addition of (MaxRetrans + 1) messages, this can only happen when the sum of messages and acknowledgements in the channels is MaxRetrans. Thus (taking loss into account) the sum of the messages and acknowledgements in any marking must be $\leq$ (2MaxRetrans + 1). Therefore the maximum number of acknowledgements that can be in ack_channel is (2MaxRetrans + 1) (when all the messages in mess_channel have been received and acknowledgements deposited in ack_channel). □

Loss, Duplication and Stop-and-Wait Property

We firstly prove that the SWP satisfies the SW property for a range of parameter values and then consider whether or not loss and duplication can occur. We would like to prove the following theorem.

Theorem 3. *The Stop-and-Wait CPN of Figs. 4 and 5 where* MaxRetrans $\geq$ *1 and* MaxSeqNo $\geq$ *1, satisfies the Stop-and-Wait property.*

We use language analysis to prove this theorem for a significant range of values of MaxRetrans and MaxSeqNo.

For the SWP service, we define two primitives: a *send* at the sender entity interface; and a *receive* at the receiver entity interface. We can then define the *service language* as 0 or more repetitions of the sequence (send, receive). This can be represented by the regular expression $(send\ receive)^*$ or by the Finite State Automaton (FSA) shown in Fig. 6.

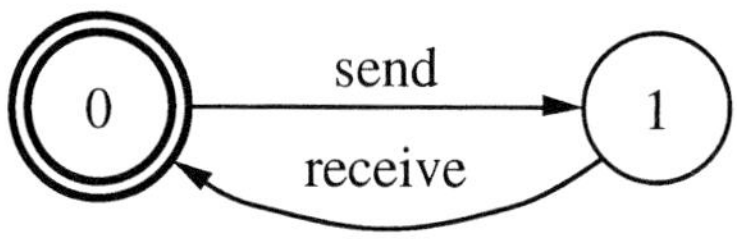

Fig. 6. FSA for the SWP service.

The next step is to generate the *protocol language* from the *protocol specification*. The protocol language just contains service primitive events. Our protocol specification is the CPN model of Figs. 4 and 5. In this CPN we can consider that the send primitive occurs when the send_mess transition occurs and that the receive primitive occurs when receive_mess occurs when the bindings of sn and rn are the same (sn = rn). (Otherwise the occurrence of receive_mess represents the discarding of duplicates, which does not correspond to a receive primitive.)

Following the verification methodology, the protocol language is obtained from the CPN's reachability graph by treating it as a FSA. All non-service primitive transitions (i.e. those associated with sending and receiving acknowledgements, with loss, retransmission or discarding duplicates) are replaced by empty (ϵ) transitions and the resulting FSA minimised [5] to produce the minimum deterministic FSA. This FSA represents all possible sequences of service primitives, generated from the protocol, and is thus the protocol language. We use the suite of tools available in the FSM package [33] for FSA minimisation and comparison.

Reachability graphs of the CPN defined in Fig. 4 and Fig. 5 were generated using Design/CPN [29] for both lossy and lossless media. (Loss is disabled by adding a guard of [false] to each of the two loss transitions.) We generated the OGs for a range of values of the parameters MaxRetrans and MaxSeqNo. Table 1 gives the statistics for some selected values of the parameters for lossy FIFO channels. Each OG is generated on a 2.4 Ghz PC with 1 GByte of memory.

Table 1. OG Results for SWP operating over lossy FIFO channels.

SeqNo. Bits	MaxSeqNo	MaxRetrans	Nodes	Arcs	Dead	Channel Bound	Time (hh:mm:ss)
1	1	0	12	12	4	1	00:00:00
1	1	1	80	194	4	3	00:00:00
1	1	2	264	834	4	5	00:00:00
1	1	3	640	2278	4	7	00:00:00
1	1	4	1300	4956	4	9	00:00:00
2	3	0	24	24	8	1	00:00:00
2	3	1	160	388	8	3	00:00:00
2	3	2	528	1668	8	5	00:00:00
2	3	3	1280	4556	8	7	00:00:00
2	3	4	2600	9912	8	9	00:00:00
9	511	0	3072	3072	1024	1	00:00:01
9	511	1	20480	49664	1024	3	00:00:29
9	511	2	67584	213504	1024	5	00:03:22
9	511	3	163840	583168	1024	7	00:16:34
9	511	4	332800	1268736	1024	9	00:55:32
10	1023	0	6144	6144	2048	1	00:00:04
10	1023	1	40960	99328	2048	3	00:01:32
10	1023	2	135168	427008	2048	5	00:11:49
10	1023	3	327680	1166336	2048	7	00:57:07
10	1023	4	665600	2537472	2048	9	04:02:14

The first two columns give the value of the number of bits required to encode the sequence number and the corresponding maximum sequence number. The next column records MaxRetrans. The next three columns list the numbers of nodes (markings), arcs, and dead markings in each OG, respectively. The second last column indicates the maximum number of messages in the message queue and the maximum number of acknowledgements in the acknowledgement queue, confirming Theorems 1 and 2. The last column records the time it took to generate the OG in hours, minutes and seconds.

The results indicate that the state space is linear in the size of the sequence number space (MaxSeqNo + 1) and the number of dead markings is given by 2((MaxSeqNo + 1). We expect the number of dead markings to be independent of the number of retransmissions, and for there to be a dead marking for each sequence number for two cases: firstly, when all messages are lost; and secondly, when all acknowledgements are lost.

Generating the reachability graph and answering our analysis questions is readily achieved with Design/CPN for small values of the MaxSeqNo and MaxRetrans parameters. We can obtain results for some practical values of sequence numbers. For example, the X.25 protocol allows the use of 3 bit, 7 bit and 15 bit sequence numbers. For MaxSeqNo = 127 (7 bit sequence numbers) and MaxRetrans = 3, we find the reachability graph contains 40960 states and takes 94 seconds to generate. Increasing MaxSeqNo to 1023 (10 bit sequence numbers) and for MaxRetrans = 4 we find that there are 665600 reachable states, taking over 4 hours to generate. At around 1.5 million states the generation time becomes too slow to be feasible. (1.5 million states takes Design/CPN days to generate and will exhaust the memory on a PC with 1Gb of RAM.) Thus obtaining a result for 15 bit sequence numbers is problematic. Further, TCP, which uses 32 bit sequence numbers (MaxSeqNo = 4294967295), would require the generation of a reachability graph containing over 10^{12} states (for MaxRetrans = 3). Clearly this is not feasible with Design/CPN.

We generated a similar set of statistics for the case without loss. In this case there are no dead markings and the size of the state space is smaller but still is linear in the size of the sequence number space. For example, without loss, the reachability graph for MaxRetrans=1 and MaxSeqNo=1 comprises 48 nodes and 86 arcs, as opposed to 80 nodes and 194 arcs in the lossy case. This reachability graph is shown in Fig. 7.

We have abbreviated the names of transitions (S for send, R for receive and T for timeout), included the sequence number and indicated when a duplicate is received. We can see that the behaviour is quite complex even for this simple case. The usual behaviour when there are no retransmissions is given by the cycle of nodes 1,2,3,5,7,9,12,17,1. The rest of the graph depicts the behaviour when retransmissions occur, leading to the need to receive duplicate acknowledgements. It is worth noting that this graph is strongly connected (all markings are mutually reachable from each other).

The suite of tools available in the FSM package [33] was used for FSA generation and manipulation. A mapping was provided for the reachability graph to distinguish between the transition occurrences of interest and internal events (ϵ transitions). Final (halt) states were chosen to be those states in which the sender and receiver are both in their ready states with the same sequence numbers, as the protocol can terminate after sending an arbitrary number of messages, not necessarily a multiple of the modulo value. All the reachability graphs that were generated in both the lossy and lossless cases were then converted into a format understandable by the FSM tools. For the lossless FIFO medium, all the minimum deterministic FSA produced were identical to that shown in Fig. 6. Thus for the range of parameters tested (MaxSeqNo up to 1023 and MaxRetrans up to 4), the SWP operating over a lossless in-order medium is language equivalent to its service. For the lossy case, the FSA in Fig. 8 was produced for each combination of parameters tested. It shows that there can be sequences of alternating sends and receives that may end after either a send or a receive. This is expected, as the sequence may end after a send if a message and all its retransmissions are lost, ending in a dead marking, where the medium is declared down. (In this case the sender cannot tell if the last message was lost or successfully received, as it could have been that the acknowledgements were lost instead.) Given that this is inevitable for a finite number of retransmissions, we consider that the SWP satisfies the stop-and-wait property in this case.

We have thus shown that the SWP satisfies the stop-and-wait property for the values of the parameters tested. We also conjecture that this result implies

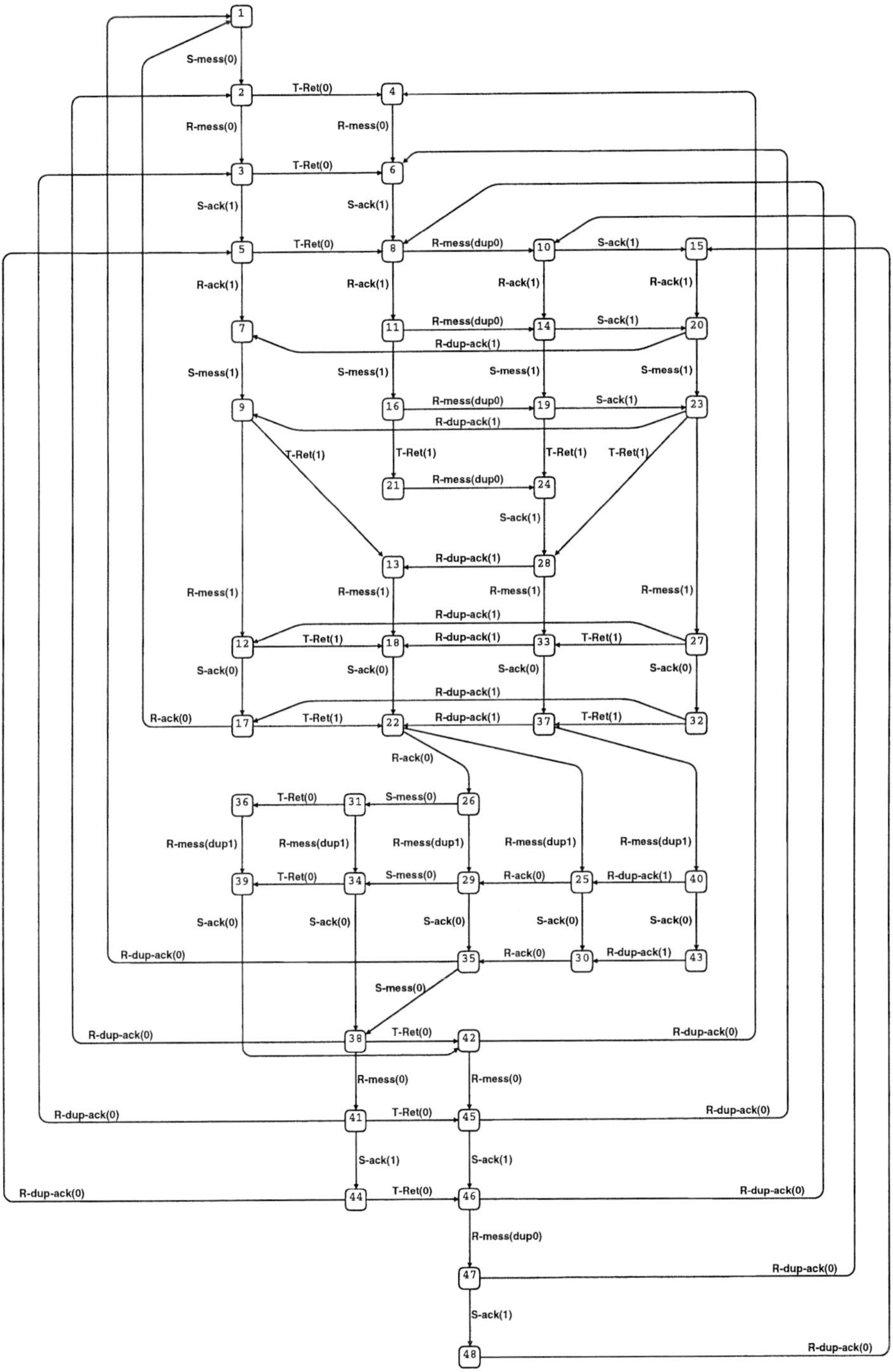

Fig. 7. OG of the Stop-and-Wait protocol, with MaxRetrans=1, MaxSeqNo=1, operating over a FIFO channel.

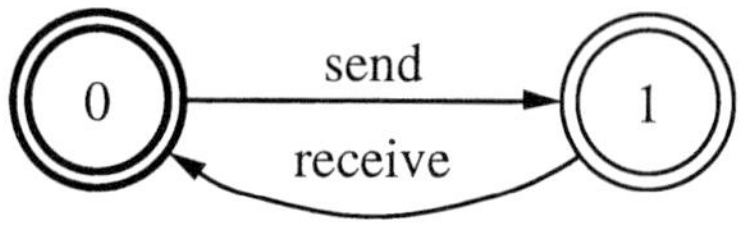

Fig. 8. A FSA showing sequences of send and receive primitives when the Stop-and-Wait protocol operates over a lossy FIFO channel.

that no duplicates are accepted as new messages by the receiver, and no messages are lost, except for possibly the last message in the case of a lossy medium as discussed above. We have further confidence that these conjectures are true because the Alternating Bit Protocol is widely accepted as being correct over lossy FIFO channels [2].

We thus conclude that the SWP operates as expected over lossy FIFO channels. If we allow messages in the channel to be re-ordered, will these properties still be satisfied? The next section shows they are not.

6.3 Re-ordering Channels

Channel Bounds. We wish to prove that the number of messages in the communication channel has the potential to grow without bound. This is formally captured in the following theorem.

Theorem 4. *For the Stop-and-Wait CPN of Figs. 2 and 3 where* MaxRetrans $\geq$ *1 and* MaxSeqNo $\geq$ *1, the message channel (place* mess_channel*) is unbounded.*

Proof. To prove this theorem, we show that a cycle of transition occurrences exists where the total effect of each cycle is to increase the number of messages in the message channel by one, and that this cycle can be repeated indefinitely. The following lemma is used in our proof.

Lemma 1. *Let σ_k be a finite occurrence sequence that can occur from a marking L, i.e. $L \xrightarrow{\sigma_k} L'$. If $L' \geq L$, then the occurrence sequence σ_k can be repeated indefinitely from marking L.*

Proof. Given $L \xrightarrow{\sigma_k} L'$ and $L' \geq L$, from Proposition 3 we know that σ_k can also occur from L', i.e. $L' \xrightarrow{\sigma_k} L''$. From Proposition 2, we know that $L'' - L' = L' - L$ and because $L' \geq L$ we know that $L'' \geq L'$. Thus from Proposition 3 we know that σ_k can occur from L''. Thus by repeated application of Propositions 2 and 3, σ_k can repeat indefinitely from marking L. □

All that is left to do to complete the proof of Theorem 4 is to identify a finite occurrence sequence σ_k of transitions in our CPN model, such that σ_k can be repeated indefinitely (i.e. $L \xrightarrow{\sigma_k} L'$ with $L' \geq L$), and that the total effect of an occurrence of the sequence σ_k is to increase the number of messages in the communication channel.

Consider our CPN model from Fig. 2 with declarations as shown in Fig. 3 but with MaxRetrans ≥ 1 and MaxSeqNo ≥ 1. From the initial marking, M_0, only the send_mess transition is enabled (with the variable sn bound to 0) indicating that the sender is ready to begin a transmission. The occurrence of this transition leads to a new marking M_1 in which there is a message (a '0' token) in the mess_channel place.

Transitions timeout_retrans, mess_loss and receive_mess are all enabled at marking M_1. We note that the medium is lossy, but that this does not mean that messages *must* be lost. An occurrence of receive_mess, with variables sn and rn bound to 0, models the receipt of this message and leads to a marking M_2. The arc inscription of the output arc from transition receive_mess to place process_mess determines that this message is not a duplicate (sn = rn = 0) and indicates this by placing a '1' token (through evaluation of NextSend(0)) into this place.

Transition send_ack is enabled in marking M_2 and when fired returns the receiver to the ready state and places an acknowledgement message into the acknowledgement channel (place ack_channel). This results in marking M_3.

The timeout_retrans transition now occurs, with rc and sn bound to 0. The guard on timeout_retrans evaluates to true, because MaxRetrans $\geq$ 1. This leads to a marking M_4 in which the retransmission counter has been incremented (a '1' token on retrans_counter) and a duplicate message '0' is in the message channel (place mess_channel).

Transition receive_ack rather than receive_dup_ack is enabled in M_4, due to the complimentary guards, with a binding of rn = 1, sn = 0 and rc = 1. Occurrence of receive_ack removes the acknowledgement message from the acknowledgement channel, returns the sender to the ready state, increments the sequence number (NextSeq(0) = 1), and resets the retransmission counter to 0. The resulting marking is M_5, with

$$
\begin{array}{ll}
M_5(\text{sender\_ready})=1\text{`s} & M_5(\text{retrans\_counter})=1\text{`0} \\
M_5(\text{seq\_no})=1\text{`1} & M_5(\text{receiver\_ready})=1\text{`1} \\
M_5(\text{ack\_channel})=\emptyset & M_5(\text{wait\_ack})=\emptyset \\
M_5(\text{process\_mess})=\emptyset & M_5(\text{mess\_channel})=1\text{`0}
\end{array}
$$

M_0 and M_5 are similar in many respects, but $M_5 \not\geq M_0$ as the sequence numbers stored at the sender and at the receiver have been incremented by one. Note that there is an additional message ('0') left in the message channel. Let us refer to the above sequence of transition occurrences as σ_0 where σ_0 = send_mess <sn=0>, receive_mess <rn=0, sn=0>, send_ack <rn=1>, timeout_retrans <rc=0, sn=0>, receive_ack <rc=1, rn=1, sn=0> and $M_0 \xrightarrow{\sigma_0} M_5$. The binding of variables for each transition occurrence is written inside angular brackets.

For illustration purposes, we firstly consider alternating bit sequence numbers (MaxSeqNo = 1). Consider the sequence of transition modes σ_1 where σ_1 = send_mess <sn=1>, receive_mess <rn=1, sn=1>, send_ack <rn=0>, timeout_retrans <rc=0, sn=1>, receive_ack <rc=1, rn=0, sn=1>

σ_1 is very similar to σ_0, with the exception of the bindings of sn and rn. In all instances, the values to which sn and rn are bound have been incremented, modulo (MaxSeqNo+1) (modulo 2 in this case). σ_1 can occur from M_5, resulting in a marking M_{10}, with

M_{10}(sender_ready)=1‘s M_{10}(retrans_counter)=1‘0
M_{10}(seq_no)=1‘0 M_{10}(receiver_ready)=1‘0
M_{10}(ack_channel)=∅ M_{10}(wait_ack)=∅
M_{10}(process_mess)=∅ M_{10}(mess_channel)=1‘0 + 1‘1

We note that $M_{10} = M_0 + \{((\text{mess\_channel}, 0), 1), ((\text{mess\_channel}, 1), 1)\}$. Thus M_{10} is a covering marking of M_0, i.e. $M_{10} \geq M_0$. The sequence numbers at sender and receiver have wrapped back to their original value of 0 and M_{10} is identical to M_0 with the addition of the two extra messages in the message channel. According to the transition rule for HLPNs, additional tokens on a place will not disable any transitions that were previously enabled. From Lemma 1, the occurrence sequence $\sigma_0\sigma_1$ can repeat indefinitely, increasing the number of tokens in mess_channel by two each cycle. Thus mess_channel is unbounded and we have proved Theorem 4 for MaxSeqNo = 1.

Generalising for MaxSeqNo ≥ 1, we have MaxSeqNo+1 sequence numbers and thus require $\sigma_0, \sigma_1, \ldots, \sigma_{\text{MaxSeqNo}}$, defined in the same way as σ_0 and σ_1 above. For $0 \leq j \leq$ MaxSeqNo, σ_j = send_mess <sn=j>, receive_mess <rn= sn=j>, send_ack <rn=(j⊕1)>, timeout_retrans <rc=0, sn=j>, receive_ack <rc=1, rn=(j⊕1), sn=j>.

The occurrence of $\sigma_0\sigma_1 \ldots \sigma_{\text{MaxSeqNo}}$ in marking M_0 leads to a marking M_m, where m = 5MaxSeqNo and

M_m(sender_ready)=1‘s M_m(retrans_counter)=1‘0
M_m(seq_no)=1‘0 M_m(receiver_ready)=1‘0
M_m(ack_channel)=∅ M_m(wait_ack)=∅
M_m(process_mess)=∅
M_m(mess_channel)=1‘0 + 1‘1 + ... + 1‘MaxSeqNo

M_m covers marking M_0 so that $\sigma_0\sigma_1 \ldots \sigma_{\text{MaxSeqNo}}$ can repeat indefinitely from marking M_0, resulting in MaxSeqNo additional messages in the message channel for each repetition. Thus the message channel is unbounded. □

A similar theorem holds for the acknowledgement channel.

Theorem 5. *For the Stop-and-Wait CPN of Figs. 2 and 3 where MaxRetrans ≥ 1 and MaxSeqNo ≥ 1, the acknowledgement channel (place ack_channel) is unbounded.*

Proof. The proof is similar to that of Theorem 4, hence we just provide a sketch. Consider the transition sequence send_mess, receive_mess<rn=sn>, send_ack, timeout_retrans, receive_ack, receive_mess<sn≠rn> and send_ack. (Binding elements have been omitted where they are not important.) Transition occurrence sequences $\sigma_0, \sigma_1, \ldots \sigma_{\text{MaxSeqNo}}$ are defined in a similar way. The occurrence sequence $\sigma_0, \sigma_1, \ldots \sigma_{\text{MaxSeqNo}}$ can be repeated indefinitely from M_0, resulting in MaxSeqNo additional acknowledgements in the acknowledgement channel for each repetition. □

Loss, Duplication and Stop-and-Wait Property. As previously discussed, when the Stop-and-Wait protocol operates as required, one message will be received correctly at the receiver for every original message sent by the sender. It turns out that this is not always the case for the Stop-and-Wait protocol operating over a medium that reorders messages. This demonstrates that the protocol does not satisfy the Stop-and-Wait service. Further we can show that sequences of sends and receives exist where there are more receives than sends, indicating that duplicates are accepted. Finally we can also demonstrate that there are sequences in which there are more sends than receives, indicating that messages can be lost.

We summarise these results in the following theorems.

Theorem 6. *The Stop-and-Wait CPN of Figs. 2 and 3 where* MaxRetrans ≥ 1 *and* MaxSeqNo ≥ 1*, does not satisfy the Stop-and-Wait service.*

Theorem 7. *For the Stop-and-Wait CPN of Figs. 2 and 3 where* MaxRetrans ≥ 1 *and* MaxSeqNo ≥ 1*, the receiver may incorrectly accept duplicate messages as new messages.*

Theorem 8. *For the Stop-and-Wait CPN of Figs. 2 and 3 where* MaxRetrans $\geq$ *1 and* MaxSeqNo ≥ 1*, messages can be lost without the sender or receiver being aware of it.*

Proof. We use language analysis to prove the above theorems. Due to the unbounded communication channels in our original model shown in Fig. 2, the resulting reachability graph is infinite. However, to prove our theorems, we only need to demonstrate that it is possible for the system to malfunction. We therefore limit the capacity of the communication channels to two. The rationale behind this is that if the protocol operates incorrectly with a channel capacity of two messages, the same incorrect behaviour will also be present in a channel with capacity greater than two. Capacities of 0 and 1 are not appropriate, as a capacity of 0 results in no communication and a capacity of 1 prohibits overtaking. Thus a capacity of two is the minimum needed to show interesting behaviour.

To obtain the smallest reachability graph of interest, we also set MaxRetrans $= 1$ and MaxSeqNo $= 1$. We argue that if incorrect behaviour is evident when MaxRetrans $= 1$ then the same behaviour can occur for MaxRetrans ≥ 1 (as it includes MaxRetrans $= 1$) and similarly for MaxSeqNo (as sequence numbers always wrap, but the sequences illustrating the incorrect behaviour will be longer).

Channel capacity has been implemented as shown in Fig. 9 with declarations shown in Fig. 10. The initial marking of the mess_channel and ack_channel places has been modified so that each place contains a certain number of empty tokens, in this case two each, representing empty buffers. Each time a message is placed in the channel, an empty token must be removed, and whenever a message is removed, an empty token must be put back. This is shown on the arc expressions connecting the mess_channel and ack_channel places to the surrounding transitions.

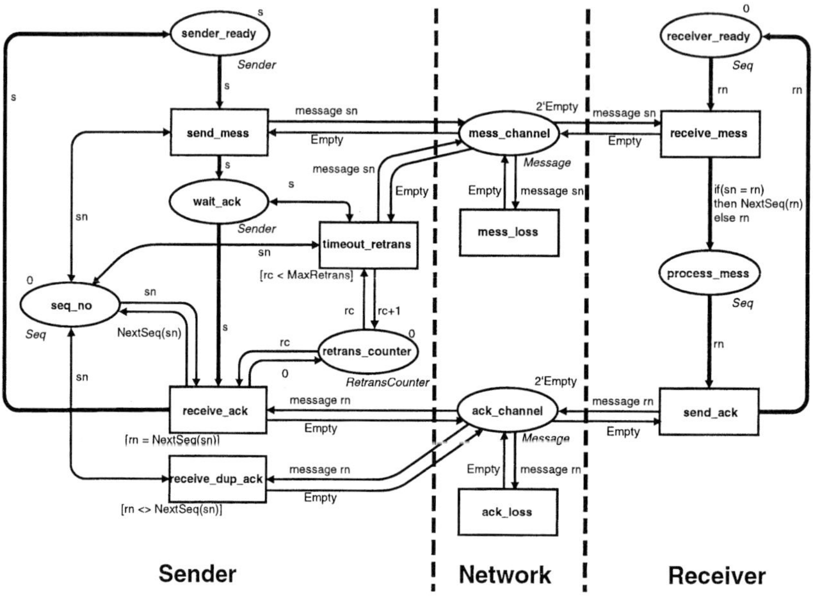

Fig. 9. A CPN of the Stop-and-Wait Protocol operating over a reordering medium with finite capacity.

```
val MaxRetrans = 1;
val MaxSeqNo = 1;

color Sender = with s;
color Seq = int with 0..MaxSeqNo;
color RetransCounter = int with 0..MaxRetrans;
color Message = union message : Seq + Empty;

var sn,rn   : Seq;
var rc : RetransCounter;

fun NextSeq(n) = if(n = MaxSeqNo) then 0 else n+1;
```

Fig. 10. Declarations of the CPN shown in Fig. 9.

Design/CPN [29] was used to generate the reachability graph of this CPN (Fig. 9) for the configuration shown in Fig. 10, without loss in the channel. The reachability graph contains 410 nodes and 848 arcs. After interpreting this as a FSA, the FSM package was used to obtain the equivalent minimum deterministic FSA as shown in Fig. 11. We have replaced send with s and receive with r in the figure due to size constraints.

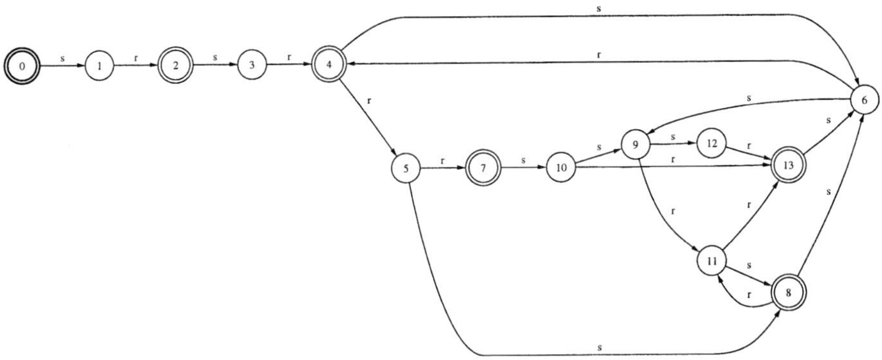

Fig. 11. FSA showing erroneous sequences of send and receive primitives for the Stop-and-Wait protocol operating over a lossless reordering medium.

This FSA shows that the SWP operating over a medium that reorders messages does not satisfy its service. For example, we see that the sequence $4 \xrightarrow{r} 5 \xrightarrow{s} 8 \xrightarrow{s} 6 \xrightarrow{r} 4$ violates the Stop-and-Wait service of alternating send and receive events. There are other more interesting sequences also. There are incorrect sequences of send and receive primitives, indicating that the receiver can mistakenly accept duplicate messages as new messages. For example, the cycle $7 \xrightarrow{s} 10 \xrightarrow{r} 13 \xrightarrow{s} 6 \xrightarrow{r} 4 \xrightarrow{r} 5 \xrightarrow{r} 7$ shows that it is possible for the system to enter a loop where the receiver accepts four messages as legitimate messages for every two sent by the sender. Another such loop is $5 \xrightarrow{s} 8 \xrightarrow{r} 11 \xrightarrow{r} 13 \xrightarrow{s} 6 \xrightarrow{r} 4 \xrightarrow{r} 5$. To illustrate how duplication can happen in the protocol, we have included a protocol trace corrresponding to the initial sequence $0 \xrightarrow{s} 1 \xrightarrow{r} 2 \xrightarrow{s} 3 \xrightarrow{r} 4 \xrightarrow{r} 5$ as shown in Fig. 12.

In Fig. 12, event 1 corresponds to sending a message (send primitive) with sequence number 0 (mess(0)), which is received (event 2: receive) and acknowledged (event 3) by sending ack(1). The timer then expires at the sender, and mess(0) is retransmitted (event 4) giving rise to a duplicate (mess(0)[dup]) which is delayed in the medium. The sender then receives ack(1) (event 5) and sends out its next message, mess(1), at event 6 (send). At this stage there are two messages in the channel. Because the retransmitted mess(0) is delayed, it is overtaken by mess(1) which is expected and received normally by the receiver (event 7:receive) who acknowledges it with ack(0) at event 8. At this point, the primitive events have been as expected: send, receive, send, receive. Next the sender retransmits mess(1) (mess(1)[dup]) at event 9 (not relevant to this discussion). Then at event 10 (receive), the receiver is expecting a mess(0) and receives it. However, it is a duplicate of the first message, and not a new message. The receiver wrongly interprets it as a new message and a receive primitive occurs, giving the sequence send, receive, send, receive, receive.

We now consider a third cycle in Fig. 11, given by $13 \xrightarrow{s} 6 \xrightarrow{s} 9 \xrightarrow{s} 12 \xrightarrow{r} 13$. This cycle shows that for every 3 messages sent, only one is received, demon-

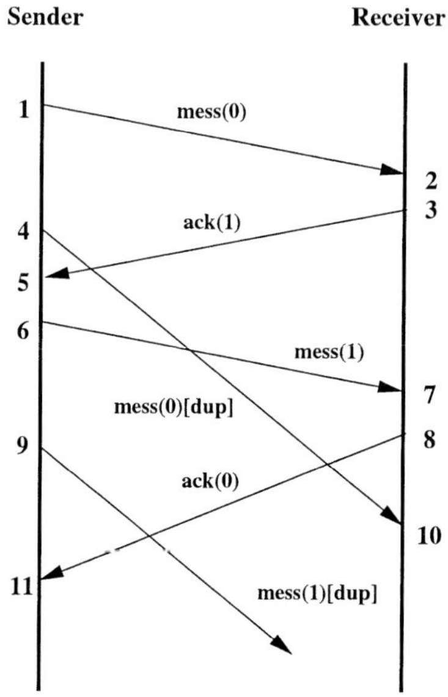

Fig. 12. Time Sequence diagram showing the acceptance of a duplicate for the Stop-and-Wait protocol operating over a lossless reordering medium.

strating message loss even though there is no loss in the medium! We illustrate this behaviour with the protocol trace shown in Fig. 13.

The sequence starts as expected with the first two messages (mess(0) and mess(1)) sent (events 1 and 8) and received (2 and 9) correctly. Retransmissions occur for both mess(0) (event 4) and mess(1) (event 11), but these are correctly discarded as duplicates (events 6 and 13). However, they give rise to duplicate acknowledgements (events 7 and 14) which the sender incorrectly interprets as acknowledgements (events 16 and 18) for the new messages sent (events 15 and 17). For example, this is due to ack(0) overtaking ack(1) and mess(0) being sent before ack(1) arrives. Now mess(1) (the fourth message sent) overtakes mess(0) (the third message sent) and is misinterpreted by the receiver as a duplicate and discarded (event 19) because the receiver is expecting mess(0). An acknowledgement (ack(0)) is thus sent (event 20) indicating that the receiver is expecting a message with a sequence number 0. Meanwhile, the sender has received duplicate ack(0) (event 18) which it interprets as a good acknowledgement for the fourth message (which is discarded by the receiver, as already discussed) and transmits (event 21) the fifth message (mess(0)). This gives us the primitive sequence: send, receive, send, receive, send, send, send. The receiver now receives the third message (mess(0)) correctly (event 22) and acknowledges it (event 23). The receiver is thus expecting to receive a message with sequence number 1, but receives the new fifth message (event 25) and discards it as a duplicate. However, the sender now receives ack(1) (event 24) and believes that the fifth message has been received correctly. This sequence demonstrates how two mes-

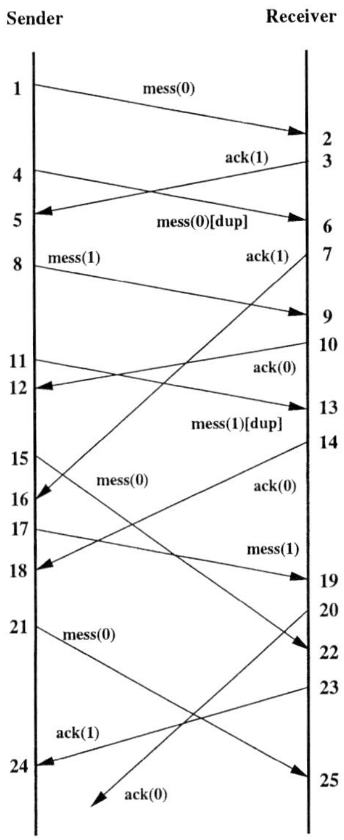

Fig. 13. Time Sequence diagram showing how loss can occur for the Stop-and-Wait protocol operating over a lossless reordering medium.

sages (the fourth and fifth) can be lost due to incorrect protocol mechanisms, while both the sender and receiver believe that there is no problem. Note that this sequence only requires two retransmissions to occur.

It is interesting to note that problems with acceptance of incorrect messages do not occur until the sequence numbers wrap, i.e. at node 4 in Fig. 11.

We also considered the case when the channel was lossy (in addition to reordering). The same parameter settings were used. The reachability graph contained 624 nodes and 2484 arcs. The reduced FSA showing the protocol language for this configuration contains 29 nodes and 47 arcs, and is shown in Fig. 14. There are many incorrect sequences in this language also. □

7 Discussion

7.1 Practical Relevance

In order to understand the relevance of these results to practical protocols, let us consider the error recovery and flow control strategies implemented in TCP [62]. TCP uses retransmission on timeout to recover from packet loss and a sliding

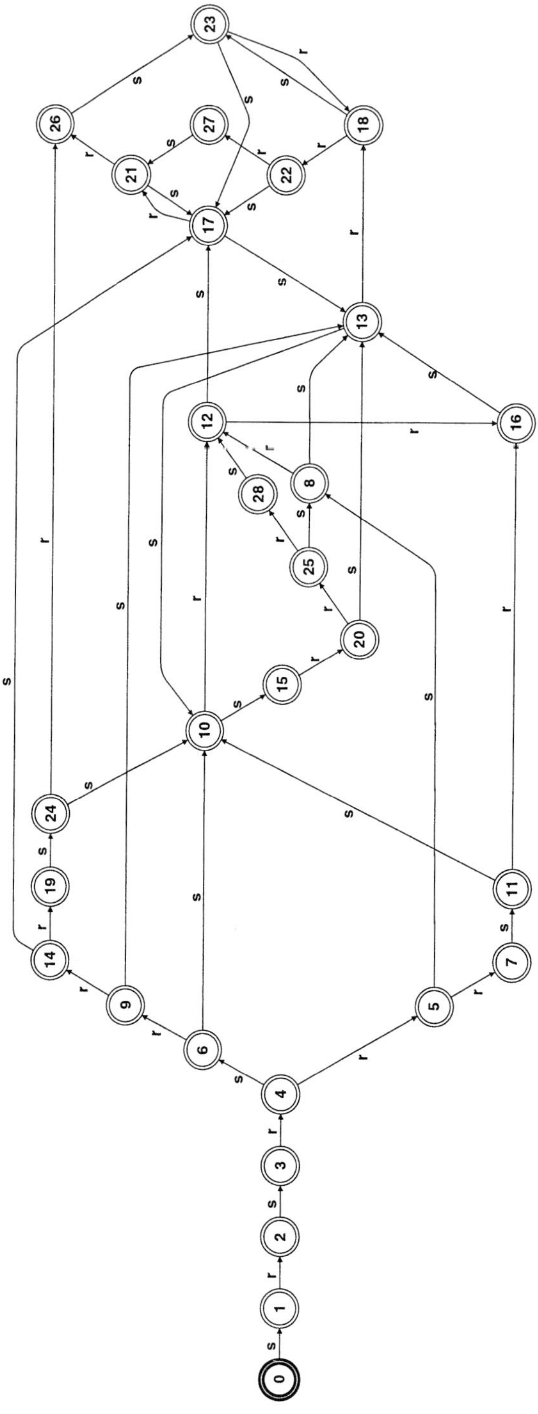

Fig. 14. FSA showing erroneous sequences of send and receive primitives for the Stop-and-Wait protocol operating over a lossy reordering medium.

window mechanism for flow control, which includes dynamic window changes. TCP operates over IP (the Internet Protocol), which allows packets (known as segments) to be dropped or reordered. The correctness of TCP's data transfer procedures can thus be related to the correctness of the Stop-and-Wait protocol operating over a medium that allows reordering.

It is necessary to distinguish between 'old' duplicates, those left from a previous connection, and duplicates caused by retransmissions within a connection. TCP uses a 32 bit sequence number, giving $2^{32} = 4294967296$ sequence numbers. Each sequence number is associated with 1 byte of data. Apart from unboundedness, the problems associated with the Stop-and-Wait protocol only arise after sequence numbers wrap, so that delayed duplicates can disrupt the acknowledgement mechanism, resulting in loss of messages or mistaken acceptance of duplicates as new messages. This will only happen in TCP after 4Gbytes of data have been transmitted and duplicates still remain in the network.

The threat posed by old duplicates was recognised by the designers of the Internet. They introduced the concept of a life-time for a packet in the IP layer, known as *time-to-live*. (This is implemented as a 'hop count' in practice.) The idea is that duplicate packets left floating around a network will be discarded once their time-to-live expires. TCP also implements a 3-way handshake for connection establishment, such that the starting sequence number for a connection can be chosen carefully for each new connection. In this way, (time-to-live combined with the 3-way handshake) TCP tried to avoid the problem caused by old duplicate packets being accepted due to wrapping sequence numbers.

RFC (Request for Comment) 793 [62], the protocol specification for TCP maintained by the Internet Engineering Task Force (IETF) [42], states that the maximum lifetime for a segment of data (MSL) is two minutes. Thus there will not be a problem with duplicate packets if they are destroyed before sequence numbers can wrap. Every byte is given a sequence number, thus for a transmission rate of 1 megabit per second (125 kBytes/sec) and ideal conditions for data transfer, the sequence number space will be exhausted in approximately $2^{32}/(1.25 * 10^5) \approx 9.5$ hours. Clearly this is not a problem. For a transmission rate of 100 megabits/sec (12.5 megabytes/sec) we see that the sequence numbers will wrap after $2^{32}/(1.25 * 10^7) \approx 5$ minutes and 45 seconds. This is getting close to the maximum packet lifetime, but should not pose a problem unless the hop count mechanism takes longer than 2 minutes to quash packets. With the introduction of Gigabit networks [73] the sequence numbers of TCP could wrap after only 34 seconds of data transfer at 1 gigabit/second. Although the maximum throughput of a network rarely approaches the theoretical maximum, it would not be unreasonable to assume that with a very large window size and very large data transfers, wrapping of sequence numbers would occur after about one minute, allowing for the possibility of duplicates being in the channel at the same time as new packets with the same sequence numbers.

This is the condition necessary for packet loss and the acceptance of duplicates (as a new packet) to occur. However, to get duplicates, there must be retransmissions caused by additional delay due to network congestion or lack of responsiveness in the receiver (e.g. an overloaded web server) which will reduce throughput. This delay, however, does not need to be very great to cause retransmissions, and hence the effect on throughput may not be significant. Another factor limiting throughput is TCP's window size and the round trip delay (RTD). In standard TCP implementations, the maximum window size is 2^{16} bytes, limiting the throughput to (2^{16}/RTD) bytes/sec. The speed of light propagation delay contribution to RTD will then provide a limit irrespective of the transmission speed. However, to allow users to take advantage of high-speed networks, RFC 1323 [43] proposes to increase the maximum window size to 2^{30} bytes or 1Gbyte, in which case the speed of light delays are no longer a limiting factor.

It is unlikely that duplicates are a problem for TCP with the current speed of networks, however these problems may become more probable if network speed were to increase by another order of magnitude, i.e. 10 gigabit/second. There are additional ramifications to be considered if incorrect acceptance of duplicates or loss of data becomes a problem. For safety critical applications operating over the Internet the consequences could be catastrophic.

There are a number of suggested ways in which this problem could be solved, or at least alleviated. RFC 1323 [43] specifies a number of TCP extensions for high performance. The extension for a larger window size has already been mentioned. Another extension is Protect Against Wrapped Sequence Numbers (PAWS) which proposes a solution to wrapping sequence numbers within a connection, by including a 32 bit time-stamp in every segment. Another solution involves extending the sequence number space, to 2^{64}, i.e. 64 bit sequence numbers. Even at 10 gigabit/second, a 64 bit sequence number field would take 470 years to wrap. The procedure of sequence numbering may also be reviewed, as currently every byte is given a sequence number. Providing a sequence number for every packet would extend the usefulness of the existing 32-bit sequence numbers.

How likely is it that unbounded growth of messages in the communication channels will actually occur? The unbounded growth is caused by retransmissions due to delayed acknowledgements. Given the variability of the round trip delay (due to the unpredictability of network congestion or overloaded servers) it is not uncommon for these delays to occur. This is countered to some extent by TCP measuring round trip delay and setting its retransmission timeout period accordingly. However, due to transients, unnecessary retransmissions will always occur. The unbounded growth, however, only occurs because the duplicates are not received by the receiver. This is highly unlikely. Also those that are delayed in the network will be expunged after their time-to-live limit has expired. Thus TCP already has mechanisms in place to prevent unbounded growth. TCP has also developed sophisticated techniques to cope with network congestion [73], so we don't see that our unboundedness result for re-ordering media will cause major difficulties with protocols such as TCP. Nonetheless, as network speeds increase the problem will get worse, particularly if the time to live value is maintained at 2 minutes. In general we can say that the contribution to congestion over FIFO channels is well contained (determined by the maximum number of retransmissions allowed) whereas over channels that re-order messages, it requires other mechanisms to contain congestion.

One avenue of further study is to generalise these results for other flow control mechanisms, such as the Sliding Window mechanism used in TCP and other protocols.

7.2 Shortcomings of Our Approach

The language analysis performed in Section 6.3 uncovered a number of errors, but was not able to detect all errors. We discovered many scenarios in which the Stop-and-Wait protocol, operating over a reordering channel (bounded or unbounded), can generate sequences of alternating send and receive events in which duplicate data is accepted and messages can be lost.

The following sequence of events is just one of many in which the above two problems are evident. This particular sequence was chosen because it is one of the shortest. Starting from the initial state of the CPN as depicted in Figs. 9 and 10, it illustrates the possibility of acceptance of duplicate data and message loss. This scenario represents a loop in the reachability graph of the model and exists independently of the boundedness and lossy properties of the channel. In the following event sequence, we use 'message n' as shorthand for 'the message with sequence number n'. The corresponding transition and binding of variables is written after each action.

1. Send message 0 (send_mess <sn=0>)
2. Receive message 0 (receive_mess <rn=0, sn=0>)
3. Send acknowledgement for message 0 (send_ack <rn=1>)
4. Timeout and retransmit message 0 (timeout_retrans <rc=0, sn=0>)
   - A duplicate message 0 remains in the channel.
5. Receive the ack of message 0 (receive_ack <rc=1, rn=1, sn=0>)
6. Send message 1 (send_mess <sn=1>)
7. Receive message 1 (receive_mess <rn=1, sn=1>)
   - Message 1 has overtaken message 0.
8. Send acknowledgement for message 1 (send_ack <rn=0>)
9. Receive the ack of message 1 (receive_ack <rc=0, rn=0, sn=1>)
10. Send message 0 (send_mess <sn=0>)
    - Now there are two message 0's in the channel, a new message 0 and the duplicate of the previous message 0 from line 4.
11. Receive message 0 (receive_mess <rn=0, sn=0>)
    - Duplicate data accepted (unless the new message 0 overtakes the old message 0.) The receiver believes this to be the correct (new) message 0.
12. Send acknowledgement for message 0 (send_ack <rn=1>)
13. Receive and discard a duplicate message 0 (receive_mess <rn=1, sn=0>)
    - Loss of the data in the new message 0 (unless overtaking occurred in step 11, in which case we are discarding the duplicate message 0 from step 4.)
14. Send a duplicate acknowledgement of message 0 (send_ack <rn=1>)

15. Receive the ack of message 0 (receive_ack <rc=0, rn=1, sn=0>)
16. Receive the duplicate ack of message 0 (receive_dup_ack <rn=1, sn=1>)
17. Send message 1 (send_mess <sn=1>)
18. Receive message 1 (receive_mess <rn=1, sn=1>)
19. Send acknowledgement of message 1 (send_ack <rn=0>)
20. Receive ack of message 1 (receive_ack <rc=0, rn=0, sn=1>)
21. Repeat from the beginning.

The problem arises in steps 11 and 13 of the above sequence. Because we do not include message data in our model, we can not determine which message is being received at step 11 (the original message from step 10 or the duplicate from step 4) and which is being discarded at step 13. Note that the global sequence of alternating send and receive events, as defined in our service, still holds.

Language equivalence is sufficient to prove or disprove that the correct *sequences* of events were occurring in the protocol, as defined in our service specification. From it we were able to detect numerous errors relating to the sequences of events, and to infer from those incorrect sequences that loss or duplication was occurring. However, given our existing service and protocol specifications, we were unable to detect the incorrect data acceptance and loss problems identified above, when the event sequences were as expected. The fault lies not with language analysis itself but with what we are applying language analysis to. The service specification is incomplete. Indeed, the service only specifies that there must be alternating send and receive events, i.e. that every send event (for a new message) is followed by a receive event (for what the receiver believes is the same new message). It abstracts from the data that is sent and received. What appears to be a reasonable abstraction (that is, the data to be sent is not required as it does not affect the operation of the protocol) is adequate for proving or disproving properties such as deadlocks and livelocks, and for determining the sequences of events, but says nothing about the data delivered on the occurrence of such events.

The assumption we made in defining the service is related to the notion of *data independence* [54, 64, 65, 77, 82]. Conceptually, a system is said to be data independent if the operation of the system is independent of the specific data it is operating on. Sabnani [65] uses data independence principles to define and verify properties about the Alternating Bit Protocol operating over a link-layer channel with a capacity of one. For example, to prove that data is passed up to the receiving user in the same order in which it is supplied by the sending user, both Wolper [82] and Sabnani [65] tell us that we need a minimum of three distinct data values. Incorporating these ideas into our service and protocol specifications may provide a solution.

Another avenue for investigation is the work presented in [52]. Knuth essentially derives design rules for appropriate bounds on sequence numbers for data transfer protocols operating over FIFO channels. He then provides a generalisation of these rules for channels that are basically FIFO in nature but may exhibit limited (bounded) reordering. Investigating the derivation of these rules may provide insight into our use of sequence numbers in the analysis of the protocol specification.

8 Concluding Remarks on the Stop-and-Wait Protocol

The class of Stop-and-Wait protocols (SWPs) include message acknowledgements and retransmission on time-out procedures to recover from transmission errors or from dropped packets in a communication medium such as the Internet. They form the basis of the data transfer procedures for both data link layer protocols and transport level protocols. The retransmission procedure can result in duplicate messages due to acknowledgements being lost or delayed. To detect duplicates a SWP inserts sequence numbers into messages and keeps track of the sequence numbers at both ends. However, sequence numbers need to be from a finite sequence number space and the number of times that a message can be retransmitted is also limited. We thus characterise SWPs using two parameters: the maximum sequence number (MaxSeqNo) and the maximum number of retransmissions (MaxRetrans). We demonstrate how a simple CPN model can be built that is parameterised by MaxSeqNo and MaxRetrans and discuss some of the modelling decisions. The first model operates over a lossy re-ordering medium. We also show how the CPN model can be modified to operate over (lossy) FIFO channels (applicable to data link protocols) and that this is an important starting point for analysing the re-ordering case. We consider four properties that we believe are important for stop-and-wait protocols: the bound on the channels; (unknowing) loss of messages; acceptance of duplicates as new messages; and the stop-and-wait property of alternating sends and receives. In the case of lossy FIFO channels, we manually prove that the communication channels are bounded by one more than twice the maximum number of retransmissions (2MaxRetrans + 1). We believe this is a new result. This illustrates an approach to proving properties of protocols for arbitrary parameter values using manual proofs. Using the protocol verification methodology, i.e., by generating the occurrence graph of the protocol and using automata reduction, we also show that the stop-and-wait property holds for small values of the parameters and conjecture that this implies that no loss or duplication occurs.

Protocols (such as TCP) operating over the Internet Protocol have to contend not only with loss due to transmission errors and packets dropped at routers, but also with the possibility that the order of packets is not maintained. Since TCP can behave as a Stop-and-Wait protocol under certain conditions it is interesting to investigate the behaviour of SWPs over a reordering medium. We analysed our CPN model with re-ordering channels and proved that: the communication channels are unbounded; messages can be lost, although the sender believes they have been confirmed by the receiver; duplicates can be accepted as new messages by the receiver; and that the SWP does not satisfy its service of alternating sends and receives. We provided a manual proof that the channels were unbounded so long as both parameters were positive giving a general result. The last 3 properties were obtained using our automatic verification method for the case when the channel capacity was 2 and MaxRetrans and MaxSeqNo were both 1. We then argued that this would also imply that these error conditions would occur for any channel capacity greater than one, and any positive values of the parameters. We also noted that the first result is independent of sequence number wrap, while the last three results depend on sequence numbers wrapping before the problems occur.

We discuss the practical relevance of these results to TCP. We conclude that sequence number wrap is possible in Gigabit networks, particularly if the extended window size option is used. RFC 1323 discusses this problem and suggests a mechanism (PAWS) using 32 bit time stamps to reject old duplicates, which hopefully will eliminate the problems associated with sequence number wrap. The problem with unbounded channels is not serious, but could add to congestion problems as the speed of networks increases. Our discussion is at a high-level and does not investigate in detail TCP's procedures for data transfer, nor the suggested PAWS scheme.

Our language analysis results allow us to detect incorrect sequences of events in our protocol specification and to deduce that loss and duplication are occurring. However, we illustrate that our data abstraction assumption (that data is not required as it does not affect the operation of the protocol) prevents us from detecting data loss and duplication in situations where the sequences of events are correct (i.e. correspond to the SWP service). Thus, with the data abstraction used, language analysis does not provide a method for verifying correct operation of the protocol in terms of absence of loss and duplication. To solve this problem we plan to apply data independence principles and techniques in order to define service and protocol specifications that capture the required information.

9 Transmission Control Protocol

The purpose of this part of the paper is to provide an example of an application of the methodology to an important complex protocol of the Internet, the Transmission Control Protocol (TCP). Our concern here will be to illustrate the first steps in modelling and analysing a complex protocol in the hope that this experience will help others to tackle other complex protocols. We concentrate on the connection management aspects of the protocol (especially establishment), rather than data transfer, which has already been discussed in detail for the Stop-and-Wait protocol.

TCP is specified in Internet Request For Comments (RFC) number 793 [62]. Its goal is to establish, maintain and close point-to-point connections between host computers attached to the Internet. Its main purpose is the reliable transfer of data between host computers. It also provides facilities for many connections to be running simultaneously to support multiple Internet application sessions such as those related to the World Wide Web and Email.

9.1 TCP Messages

In order to establish and release connections and to transfer data, RFC 793 defines a set of messages that are exchanged between the two computers. A TCP message, known as a segment, comprises a header field and a data field that carries application data. The TCP header field provides control information for handling multiple connections, their management (the opening and closing of connections) and reliable data transfer including end-to-end flow control. The format of a TCP segment is given in Fig. 15.

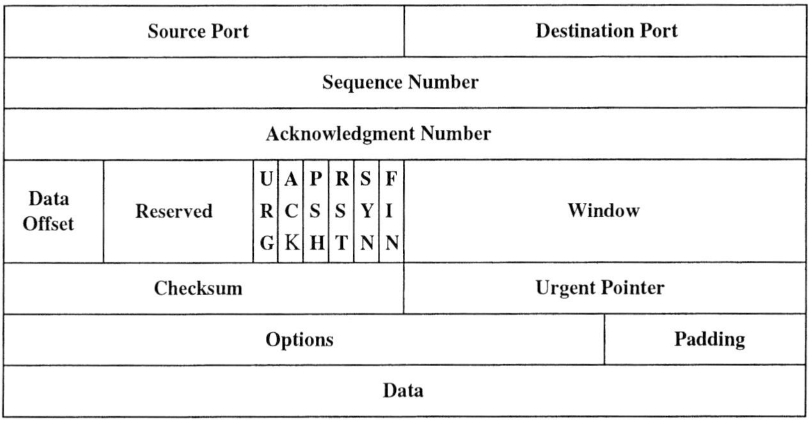

Fig. 15. TCP Segment Format.

The 16 bit source and destination port fields are used to identify the application that is going to use the connection and allow multiple connections to be running simultaneously. On a particular connection, a sequence number is associated with every octet of data that is to be sent from one host computer to another. When transmitting data, the 32 bit sequence number field specifies the sequence number of the first data octet in the segment. The 32 bit acknowledgement number field indicates the successful receipt of data octets and contains the next sequence number of the data octet that the sender of the acknowledgement segment is expecting to receive. The four bit data offset field contains the header length (in 32 bit words). The next 6 bits are reserved, then there are a set of 6 control bits that are vitally important for TCP connection management. We describe them in detail in the next paragraph. A 16 bit window field is used for flow control. It signals to the receiver of the segment the number of data octets that the sender of the segment is prepared to receive, beginning with the acknowledgement number in the segment.

As already mentioned the header contains six 1-bit control flags: URG (urgent), ACK (acknowledgement), PSH (push), RST (reset), SYN (synchronisation) and FIN (finish). The URG flag if set, is used in conjunction with an urgent pointer field, to indicate to the receiver the position of data in the octet stream that needs priority when being delivered to the user. A segment with the ACK flag set indicates that the acknowledgement number field is valid. A set PSH flag indicates to the receiving TCP process that all queued data, including that just received, must be immediately delivered to the user. When set, the RST flag informs the receiver of the segment to reset the connection. This normally results in the connection being aborted. A TCP connection is initiated by setting the SYN bit in a segment. The SYN carries an initial sequence number which indicates that the first octet of data to be sent will carry the next sequence number (initial sequence number plus one). The initial sequence number is chosen according to the value of a clock that the host runs, instead of always being set to zero. This is to reduce the probability of delayed old duplicate SYNs interfering with the connection. Finally, a segment with the FIN bit set indicates that the sender of the segment has no more data to send. It is used to gracefully close the connection (i.e. without data loss). The sequence number of the FIN is that of the last data octet sent plus one. A TCP segment is usually named after the control bits that are set. For example, a SYNACK segment refers to the segment which has both the SYN and ACK bits set.

The remaining header fields comprise a 16 bit checksum used to detect transmission errors, a 16 bit urgent pointer required for urgent data (already discussed above) and an options field (allowing, for example, a maximum segment size to be indicated in a SYN segment).

9.2 TCP Connection Management Procedures

A TCP state diagram [26,62,69] is included in the RFC to illustrate TCP connection management procedures. It defines TCP's 11 states and a core set of state changes related to processing user calls and connection management segments. It is incomplete in that it does not include TCP's state variables nor does it incorporate reset processing and error handling. A much more comprehensive pseudo-code like description of TCP's procedures is given in Section 3.9 of RFC 793. In this section we just illustrate the procedures using message sequence diagrams.

Figure 16 (a) is a message sequence diagram for normal connection setup and tear down. On the left is the client (the initiator of the connection) and on the right is the server. Time progresses down the page. The client's states (e.g. CLOSED, SYN_SENT, ESTABLISHED) are written to the left of the vertical line representing the client. A similar convention is adopted for the server side. User commands (i.e. *active open*, *passive open*, and *close*) are written in parentheses on top of some states, indicating when they occur. TCP uses a "three-way handshake" [75] to establish a connection, i.e., three segments are used by the two communicating hosts to open the connection. In Fig. 16, the sequence number and the acknowledgement number (when relevant) are included with the segment name. The procedure is initiated by the TCP entity on one host (client), and responded to by the TCP entity on the other (server). In Fig. 16 (a), after receiving an active open command from its user, the TCP client sends out a SYN segment with a sequence number ISS1, its initial sequence number for the connection. The client also changes state from CLOSED to SYN_SENT. The TCP server enters LISTEN after receiving a passive open command from its user. To respond to the SYN from the client, it sends out a SYNACK segment with an acknowledgement number ISS1+1 as well as a sequence number ISS2, the server's initial sequence number for the connection. After sending the SYNACK segment, the TCP server changes state from LISTEN to SYN_RCVD. To respond to the SYNACK, the TCP client sends an ACK with sequence number ISS1+1 and acknowledgement number ISS2+1 and changes state from SYN_SENT to ESTABLISHED. After receiving the ACK, the server goes into ESTABLISHED from SYN_RCVD. The connection is now set up between the client and the server.

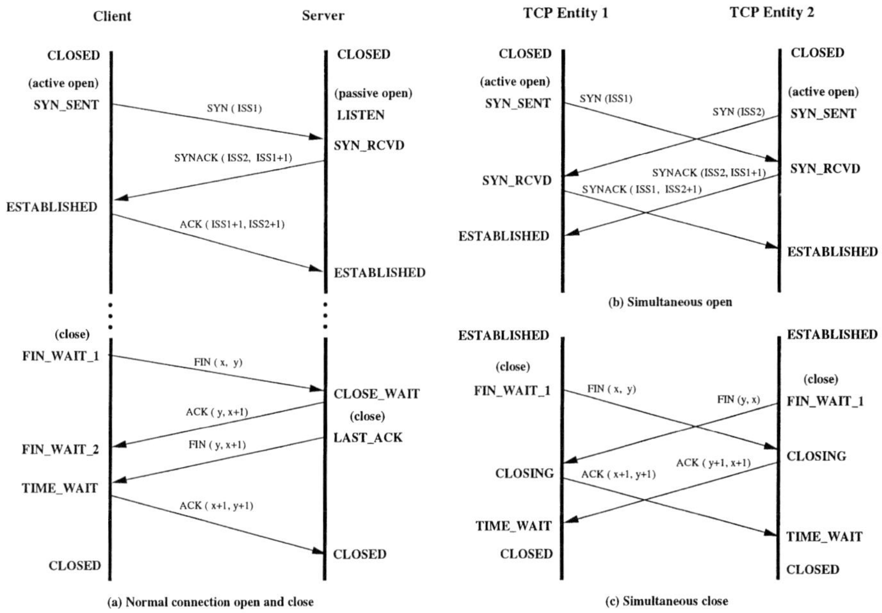

Fig. 16. Message sequences for TCP connection management.

Graceful connection termination is illustrated at the bottom of Figure 16 (a). Firstly, the TCP client user indicates that it has no further data to send by issuing a close command. The TCP client sends a FIN segment to the server and enters the FIN_WAIT_1 state. On receipt of the FIN, the server acknowledges it and informs its user that the client is closing the connection. When the server user has no more data to send, it issues a close command to the server TCP entity, which sends a FIN to the client, closing the connection from the server to the client. Finally the FIN is acknowledged by the client. The sequence number (x) and acknowledgement number (y) used in this message sequence diagram assume that the server had no more data to send.

The TCP connection management protocol also incorporates procedures for simultaneously opening and simultaneously closing connections by both TCP entities. Figs. 16 (b) and (c) show the sequences respectively. For more information about them, see [62, 69].

As well as the major states defined in the state diagram, TCP state variables are used to maintain the state of a TCP connection and are stored in a record called the *Transmission Control Block* (TCB). Important TCP variables for connection management are: SND_UNA, SND_NXT, ISS and RCV_NXT. SND_UNA records the oldest sequence number of a segment that has been sent but has yet to be acknowledged. SND_NXT stores the sequence number of the next segment to be sent. ISS represents the initial send sequence number of the initiating TCP entity, while RCV_NXT stores the sequence number of the next expected incoming segment.

10 CPN Model of TCP Connection Management

In this section we explore building a CPN model of TCP's connection management procedures. We start by discussing the assumptions and abstractions used.

10.1 Modelling Assumptions and Abstractions

We make five assumptions when modelling TCP connection management.

1. The communication channel does not lose, corrupt or duplicate packets, but may delay and re-order packets.

   The reason for starting from a non-lossy channel is that a lossy channel may hide possible deadlocks in the protocol, such as unspecified receptions which the channel can conveniently lose, but mostly will not! Thus this anomaly would be missed when inspecting the leaf nodes of the reachability graph if arbitrary channel loss is included, but nonetheless it would be a problem. Excluding loss initially allows these imperfections to be detected by inspecting dead markings.

2. There is no retransmission.

   We would like to ensure that the procedures work without the complication of retransmissions, which we know will cause state space explosion. Thus, when no loss is involved, retransmissions are not necessary for the procedures to operate correctly. Later, we shall need to investigate the effect of loss and the use of retransmissions, once the basic behaviour is confirmed to be correct. This assumption allows us to ignore the retransmission procedures for SYN and FIN segments.

3. We consider a single instance of a TCP connection.

   Because all instances of a connection operate the same way, we can just consider one instance. It is only when we wish to consider contention for resources (such as buffer space in a host) between a number of running connections, that we need to consider multiple connections. This is ruled out of the scope of the initial analysis of TCP. This assumption allows us to ignore the source and destination port numbers.

4. The receive buffer is big enough to store all the incoming segments.

   Since we are only considering connection management, and not data transfer, it is reasonable to assume that the receiver will be able to store the connection management segments. This allows us ignore the flow control window and the calculations and comparisons based on it. This simplifies the model and thus reduces the size of the state space.

5. The user issues four commands to the TCP entity: active open, passive open, send and close.

   The TCP user interface also allows for three other calls: abort, receive and status. We do not model the abort call at this stage, as we wish to investigate TCP's basic behaviour of establishing and gracefully releasing connections, before including arbitrary aborts. Once we know the core behaviour is satisfactory, then we can start to investigate the rarer and more complex behaviour that includes user aborts. The receive and status calls do not affect the operation of the protocol and can be considered as local interface matters. They are also not modelled.
   The TCP RFC says little about feedback to the user (such as receipt of data, or indications that the connection has been requested or is established) and thus at this stage of the investigation, we do not consider it. It is however of vital importance with respect to whether or not TCP satisfies its service, where we must be explicit about such interactions. Since RFC 793 does not define a service, we firstly investigate the operation of the protocol without it. Once more experience is gained, then we are in a much better position to define the service [15–17].

We model TCP segments at the level of detail needed for analysing the connection management procedures, given the above assumptions. A TCP segment is thus modelled by including: the sequence number, the acknowledgement number and four control flags: SYN, ACK, RST and FIN. Other fields in the TCP header can be omitted because they do not affect the operation of the connection management procedure. For example, we do not need to model the checksum as discussed in Section 2.2, and the flags PSH and URG, the urgent pointer field and the window are only concerned with the data transfer procedures. We also do not consider options.

10.2 Architecture

Our CPN model comprises 6 places and 87 transitions. It is organised into three hierarchical levels, as shown in Fig. 17.

The first level has one page called TCP_Overview. The second level also has one page, named TCP_Entity. Since TCP is symmetrical (both ends implement the same procedures) we only need to define the procedures once, and then instantiate them for each end, using *page instances* [53]. The third level has eleven pages, one for each TCP state. This is standard practice in many protocol definitions and has been used in the modelling of other communication protocols,

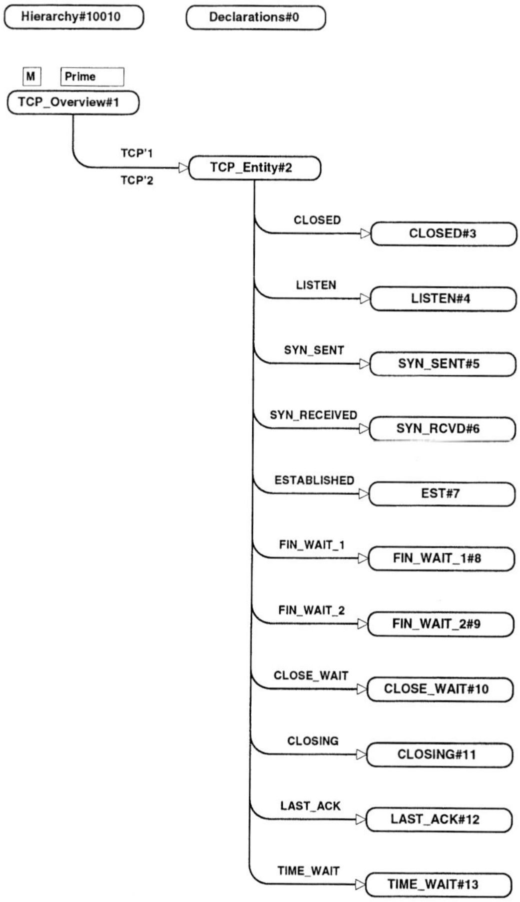

Fig. 17. Hierarchy page for the CPN model.

for example, [34]. This approach is also consistent with the way TCP is specified in Section 3.9 of RFC 793. (It turns out that other structures are possible which take advantage of common procedures in different states ([37]). However, the use of a state-based approach provides a clean and readily followed structure when first starting to model protocols, especially if they include state diagrams or state tables in their definitions. Once this experience has been obtained, further important optimisations of the structure can be undertaken.)

10.3 Declarations for the TCP Model

The declarations shown in Fig. 18 define the colour sets and any associated variables for User Commands, TCP segments and the TCB and include values for the initial sequence numbers.

```
1 (* User Commands *)
2 color COMMAND = with A_Open | P_Open | Send | Close;
3
4 (* TCP Segments *)
5 color Int = int;
6 color CTLbit = with SYN | RST | ACK | FIN;
7 color ACKflag = with on | off;
8 color SEG_CTL = product CTLbit*ACKflag;
9 color SEG = record
10              SEQ: Int *
11              ACK: Int *
12              CTL: SEG_CTL;
13 var seg: SEG;
14
15 (* Transmission Control Block *)
16 color STATE = with CLOSED | LISTEN | SYN_SENT | SYN_RCVD | EST |
17              CLOSE_WAIT | LAST_ACK | FIN_W1 | FIN_W2 |CLOSING | TIME_WAIT;
18 color SV = record
19            RCV_NXT: Int *
20            SND_NXT: Int *
21            SND_UNA: Int *
22            ISS: Int;
23 var v: SV;
24 color LISTENstat = with lis | cls;
25 var i: LISTENstat;
26 color TCB = product STATE*SV*LISTENstat;
27
28 (* ISS *)
29 val ISS_tcp1 = 10;
30 val ISS_tcp2 = 20;
```

Fig. 18. Declarations for TCP user commands, segments and TCB.

The colour set COMMAND (line 2) defines four commands: A_Open (active open); P_Open (passive open); Send; and Close, that are issued by users. The colour set SEG (lines 9–12) defines a TCP segment as a record with three entries: SEQ, ACK and CTL. SEQ is used to model the sequence number and ACK, the acknowledgement number. Both numbers are defined as integers. Note that the actual sequence number and acknowledgement number space is finite, ranging from 0 to $2^{32} - 1$. Because we are only dealing with connection management and not data transfer, only a very small portion of the sequence number space is used for establishing and releasing a connection. Thus we can also assume that the sequence numbers do not wrap (i.e. cycle back to zero), so that modulo arithmetic is not required. (Note that including modulo arithmetic is essential when modelling the data transfer procedures.)

CTL (line 12), typed by the colour set SEG_CTL, is used to model TCP's control flags. SEG_CTL (line 8) is a product of CTLbit (line 6) and ACKflag (line 7) that model the four segment types and the ACK flag respectively. The ACKflag indicates the status (on or off) of the Acknowledgement Number field. If the ACK flag is on it indicates that the Acknowledgement field is valid. Variable seg (line 13), the variable for TCP segments, is declared as type SEG. An example of a SYNACK using ML syntax is $\{SEQ = 20, ACK = 11, CTL = (SYN, on)\}$.

The Transmission Control Block (lines 15–30) defines at line 26 a colour set, TCB, as a product of STATE (line 16), SV (lines 18–22) and LISTENstat (line 24). STATE comprises the 11 TCP states. SV defines the four TCP variables explained in Section 9.2: RCV_NXT, SND_NXT, SND_UNA and ISS. We use integers for state variable values (instead of 32 bit integers) for the same reason as given above for sequence numbers. LISTENstat is a Boolean used to keep track of whether or not the TCP entity has previously been in the LISTEN state. If it has, variable i, typed by LISTENstat (line 25), will take the value lis, otherwise it will take the value cls, indicating that the entity has not previously been in LISTEN but CLOSED. This is used to determine the next state that TCP enters from states SYN_SENT and SYN_RCVD on receipt of a RESET segment (see RFC 793, Section 3.4). Variable v (line 23), the variable for the record of TCP state variables, can take any value belonging to SV, such as $\{RCV\_NXT = 21, SND\_NXT = 11, SND\_UNA = 11, ISS = 10\}$. ISS_tcp1 and ISS_tcp2 (lines 29–30) define the initial send sequence numbers for each TCP entity.

We also define six functions that create TCP segments, depending on the value of state variables or incoming segments, as shown in Fig. 19. All the functions are similar, so we just illustrate in detail the creation of a segment by using the function SYNseg (lines 32–35) that creates a SYN segment, as an example. The initial sequence number, stored in TCB as the fourth component of the state variable record and specified in ML as $\#ISS(v)$, is assigned to the sequence number field of the SYN segment. By convention, the acknowledgement number field is assigned the value 0 (for null), because the SYN segment is used to initiate a connection and therefore cannot carry a valid acknowledgement number. In the SYN segment, the control bit SYN is on and ACK is off (indicating that the acknowledgement number is not valid), so entry CTL of the segment record is assigned (SYN,off). Other TCP segments are modelled in a similar way. Note that functions RSTackon and RSTackoff take incoming segments as their argument rather than TCB's state variables. They are used to model the RST segment with ACK on and off respectively.

10.4 The Top and Second Level Pages

The top level page is shown in Fig. 20, which provides an abstract view of the protocol.

There are 6 places in Fig. 20. Places User_1 and User_2, typed by COMMAND, model TCP user commands. Changing the initial marking of a user place will change the command issued to TCP, resulting in modelling different cases. Places TCB_1 and TCB_2, typed by TCB, model TCP state information. A token in either place represents a local TCP (compound) state. Places H1_H2

```
(* Functions for TCP Segments *)
fun SYNseg(v: SV): SEG =
    {SEQ = #ISS(v),
     ACK = 0,
     CTL = (SYN,off)};

fun SYNACKseg(v: SV): SEG =
    {SEQ = #ISS(v),
     ACK = #RCV_NXT(v),
     CTL = (SYN,on)};

fun ACKseg(v: SV): SEG =
    {SEQ = #SND_NXT(v),
     ACK = #RCV_NXT(v),
     CTL = (ACK,on)};

fun FINseg(v: SV): SEG =
    {SEQ = #SND_NXT(v),
     ACK = #RCV_NXT(v),
     CTL = (FIN,on)};

fun RSTackon(seg: SEG): SEG =
    {SEQ = 0,
     ACK = #SEQ(seg)+1,
     CTL = (RST,on)};

fun RSTackoff(seg: SEG): SEG =
    {SEQ = #ACK(seg),
     ACK = 0,
     CTL = (RST,off)};
```

Fig. 19. Functions for creating TCP Segments.

and H2_H1, typed by SEG, each model a unidirectional communication channel. H1_H2 indicates the transmission direction is from host 1 to host 2, whereas H2_H1 indicates the opposite direction. A token in the communication channel place represents that a segment is in transit from one TCP entity to its peer TCP entity, and may be anywhere in the network or in a buffer of either local entity.

Also in Fig. 20, are two *substitution transitions* named TCP'1 and TCP'2. Each represents a TCP connection management process that implements both the establishment and termination procedures. A substitution transition may be viewed as a macro that is linked with another CPN page (known as a *subpage*) that is called when the CPN executes. TCP'1 and TCP'2 are both linked with the second level page (Fig. 21), which serves as a page instance.

The places associated with a substitution transition need to be assigned to places on the subpage. For example, places User_1 and User_2 in Fig. 20 are both

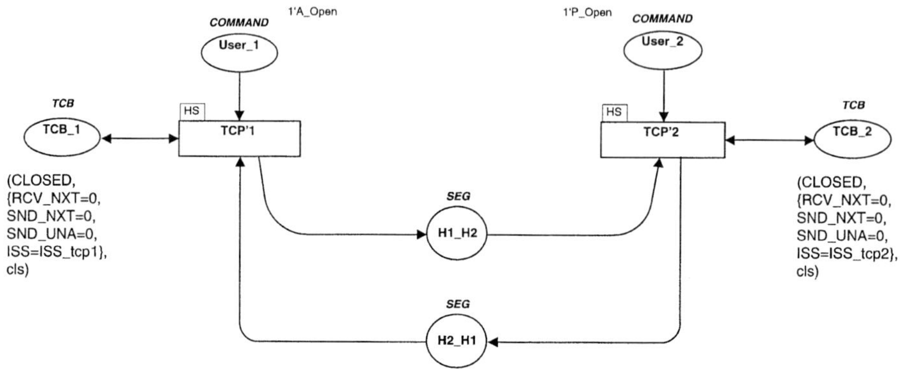

Fig. 20. Top level CPN page: TCP_Overview.

assigned to place User in Fig. 21. In contrast, Place H1_H2 in Fig. 20 is assigned to place Out for TCP'1 and In for TCP'2 (see Fig. 21) and vice versa for H2_H1.

The second level page structures the TCP connection management process into eleven substitution transitions. Each transition is named by a TCP state and is linked with a page at the third level, which models TCP's behaviour for that state.

10.5 Third Level Pages

We illustrate the TCP model at the executable level by considering four pages at the third level: CLOSED, LISTEN, SYN_SENT and SYN_RCVD.

The CLOSED Page. The CLOSED page (Fig. 22) models TCP's behaviour for the CLOSED state. It has 4 transitions and 4 places that have already been described.

When a server wishes to receive connection requests, it issues a passive open to its TCP entity. This is modelled by transition Passive_Open. TCP enters the LISTEN state (with state variables unchanged) after receiving a passive open (P_Open) command. When transition Passive_Open occurs, the value of the LISTENstat flag of the token in TCB is changed from *cls* to *lis*, indicating that TCP has been in LISTEN. Transition Active_Open models the expected behaviour of TCP sending out a SYN after receiving an active open (A_Open) command from its user. TCP enters SYN_SENT and updates its state variables as shown on the output arc to place TCB.

When TCP is CLOSED, it is not expecting any incoming segments. If it receives one, then it needs to inform the sender that it is closed, by sending a reset. However, if the incoming segment is a reset, then it is discarded, as the receiver does not need to inform its peer to close. As indicated by its guard, transition Rcv_noRST models TCP's response to any incoming segment that is not a reset (i.e. the RST bit is not set). TCP sends a RST segment to its peer

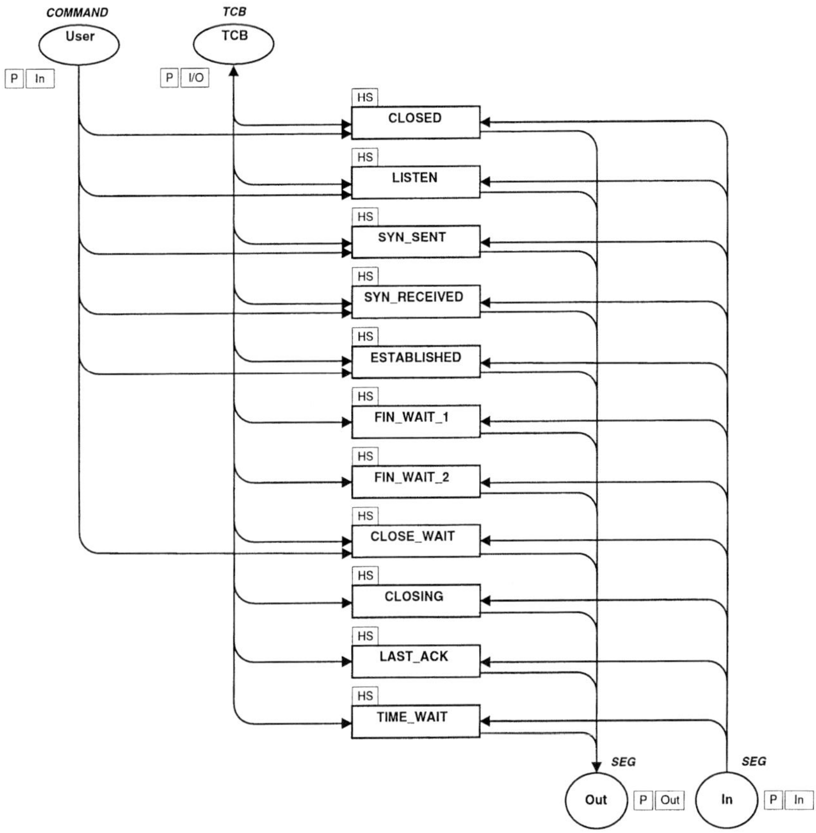

Fig. 21. Second level CPN page: TCP_Entity.

and remains CLOSED with its state variables unchanged. As indicated by the output arc expression, the RST sent by TCP may carry an acknowledgement number or not, depending on the status of the ACK bit in the incoming segment. Transition Rcv_RST models TCP's behaviour on receipt of a segment with RST on, that is, discarding the incoming segment and remaining in CLOSED.

The LISTEN Page. The actions taken by TCP when in the LISTEN state are shown in Fig. 23.

The LISTEN page has 5 transitions: Close, Send, Rcv_ACK, Rcv_RST and Rcv_SYN. Transitions Close and Send model TCP's behaviour on the receipt of user commands *Close* and *Send* respectively. A passive open command followed by *Send* is equivalent to a two stage active open, resulting in the sending of a SYN segment and TCP entering the SYN_SENT state. Receiving a *Close* command while in LISTEN returns TCP to CLOSED, with no need to send any segment, as no peer has requested a connection.

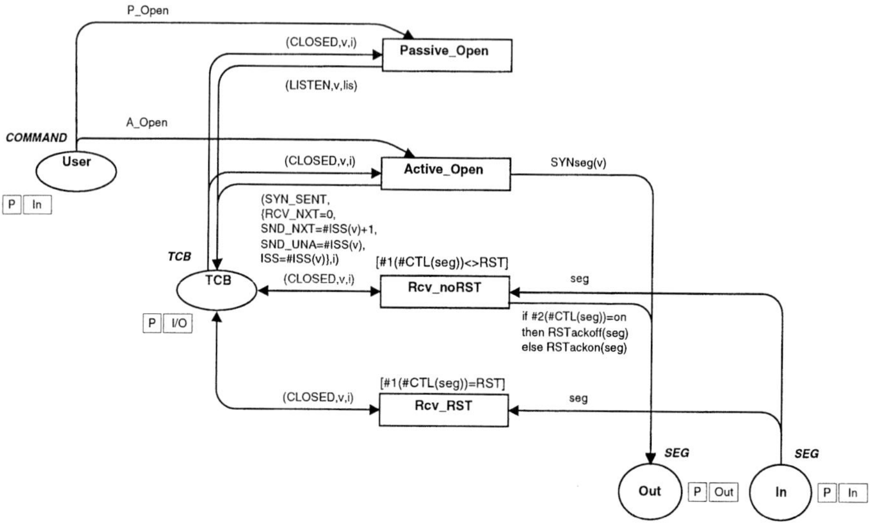

Fig. 22. The CLOSED page.

Transition Rcv_ACK and Rcv_RST model unexpected behaviour. Transition Rcv_ACK specifies TCP's response to any incoming segment with its ACK on that is not a reset. It replies with a RST segment and remains in LISTEN. Transition Rcv_RST models TCP remaining in LISTEN after receiving and discarding a RST segment.

When in LISTEN, TCP expects to receive a connection request. Transition Rcv_SYN models this situation by receiving a SYN segment. TCP returns a SYNACK and enters SYN_RCVD with its state variables updated.

The SYN_SENT Page. The SYN_SENT page (Fig. 24) has 9 transitions that are created in accordance with the TCP specification for the SYN_SENT state.

Transition Close models TCP entering CLOSED from SYN_SENT in response to a Close command. Transitions Rcv_uACK and Rcv_uACK_RST model TCP's response to an unacceptable ACK segment with the RST bit off and on respectively. Transitions RST_LISTEN and RST_CLOSED are used to respond to a RST with an acceptable ACK number based on TCP's previous state information. That is, if it has previously been in LISTEN, indicated by the LISTENstat flag having the value *lis*, TCP changes state to LISTEN from SYN_SENT (modelled by transition RST_LISTEN), Otherwise, it goes into CLOSED (modelled by transition RST_CLOSED). Transition Rcv_RSTnoACK is used to respond to a RST without an ACK. Transitions Rcv_SYNACK and Rcv_SYN model TCP's response to an incoming segment with the SYN bit on under different conditions. If the incoming segment acknowledges the SYN, TCP goes into ESTABLISHED from SYN_SENT and sends out an ACK (modelled by Rcv_SYNACK). Otherwise, it enters SYN_RCVD and sends out a SYNACK (modelled by Rcv_SYN). If the incoming segment has neither SYN nor RST set (normally an acceptable

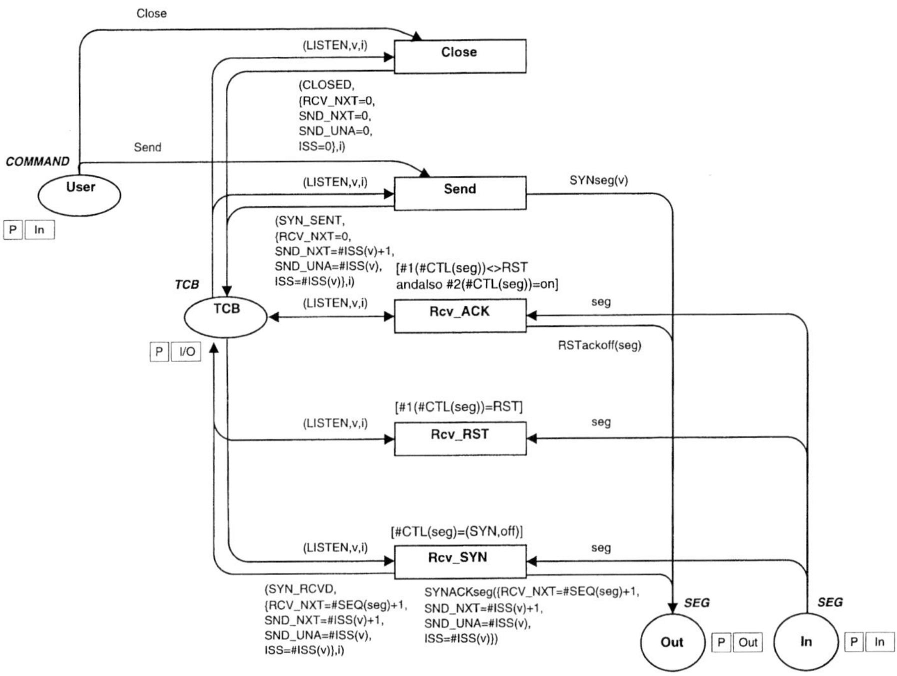

Fig. 23. The LISTEN page.

ACK), TCP drops the segment. This is modelled by transition noSYN_noRST. Note that if the ACK in an incoming segment is not acceptable, TCP sends out a RST, as specified by transition Rcv_uACK.

The SYN_RCVD Page depicted in Fig. 25 has 11 transitions. Transition Close models TCP sending a FIN segment and entering FIN_WAIT_1 after receiving a *Close* command from its user. Rcv_uSeq and Rcv_uSeq_RST are used to respond to an incoming segment with an unacceptable sequence number (first item of the guard) with RST off or on respectively.

Other transitions in Fig. 25 require that the sequence number of the incoming segment is acceptable (first item of the guard). Transitions RST_LISTEN and RST_CLOSED model TCP's response to a RST with an acceptable sequence number based on TCP's previous state. That is, if it has previously been in LISTEN, TCP enters LISTEN from SYN_RCVD, otherwise, it changes state to CLOSED. Also depending on TCP's previous state, transitions SYNinw_LISTEN and SYNinw_CLOSED model TCP sending out a RST after receiving an acceptable SYN (i.e. its sequence number is in the receive window). If the SYN's sequence number is not in the window, an ACK is sent as a result of the sequence number check (see transition Rcv_uSeq). Transition Rcv_ACKoff models TCP dropping the incoming segment and remaining in SYN_RCVD in response to a segment with ACK off. Transitions Rcv_ACKonA and Rcv_ACKonU model TCP's behaviour on receipt of an ACK under different conditions. If the

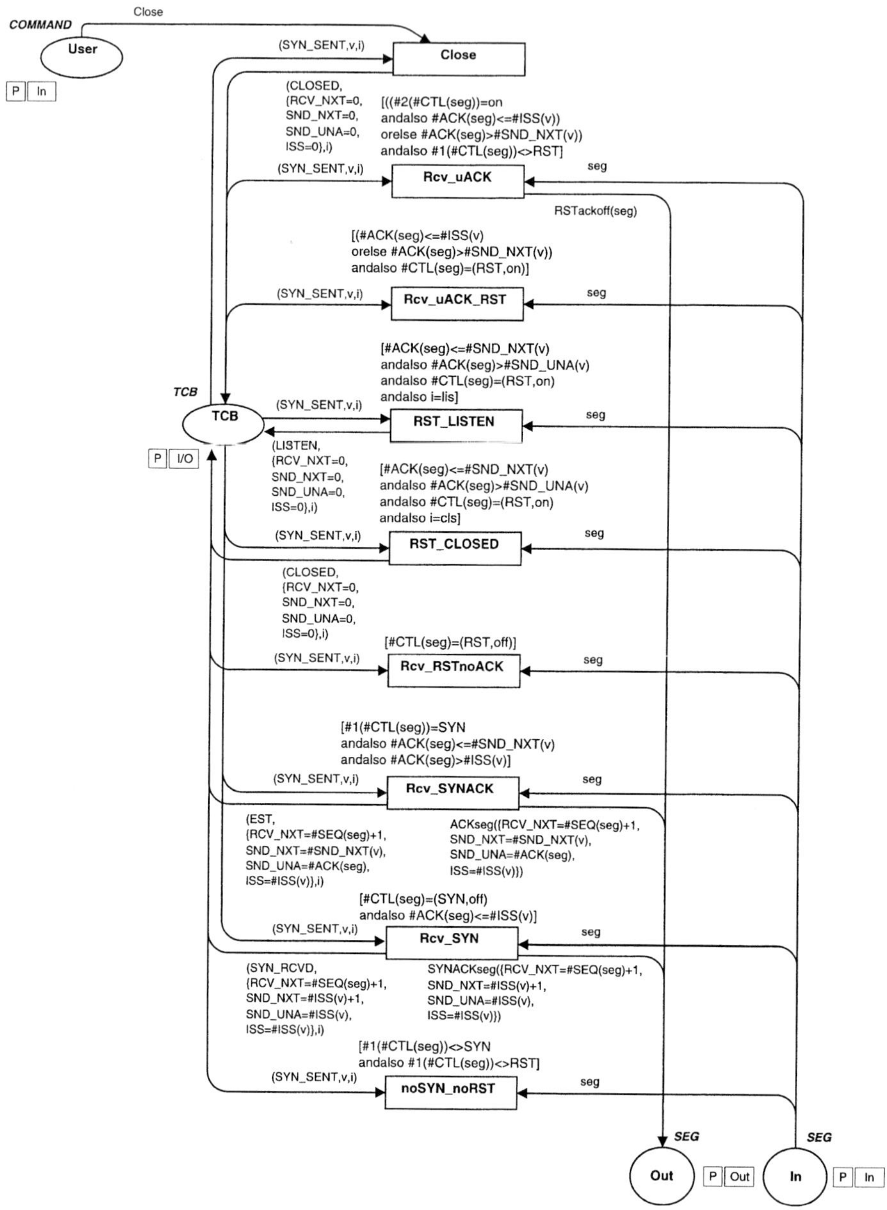

Fig. 24. The SYN_SENT page.

ACK is acceptable, TCP enters ESTABLISHED (modelled by Rcv_ACKonA), otherwise, TCP sends out a RST (modelled by Rcv_ACKonU). Finally, transition Rcv_FIN models TCP entering CLOSE_WAIT and sending an ACK on receipt of a FIN segment.

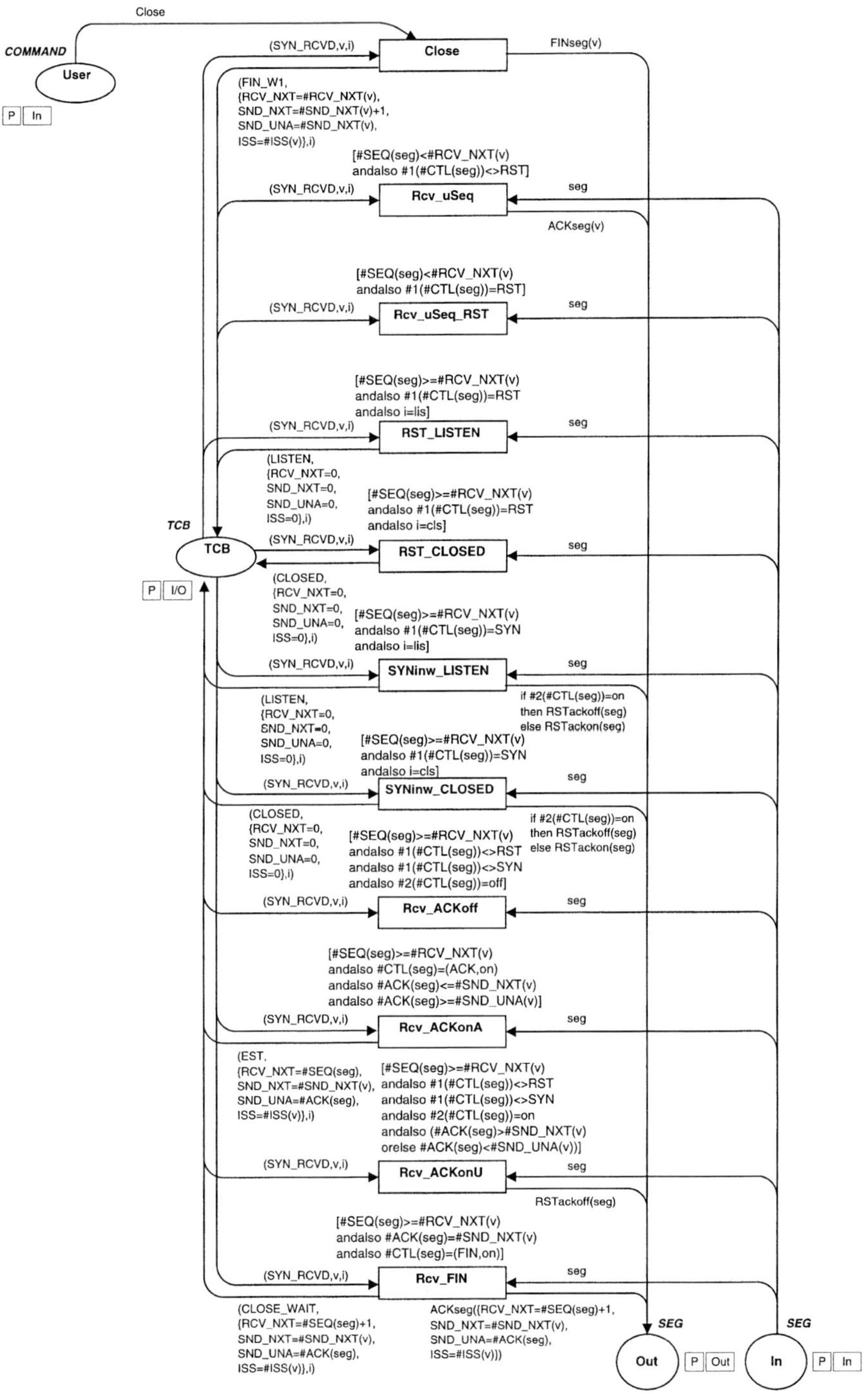

Fig. 25. The SYN_RCVD page.

11 Functional Analysis of TCP Connection Management

The TCP connection management protocol is analysed by generating occurrence graphs (OGs) using Design/CPN 4.0.5 running under Linux on a 800MHZ Intel Celeron laptop computer with 128 MB RAM.

11.1 Initial Configurations

We can analyse many different connection management scenarios by choosing a number of different initial markings of places User_1 and User_2 in Fig. 20. This allows us to start simply by just considering the connection establishment protocol, before analysing the effect of releasing connections. This incremental approach has two advantages. Firstly it allows us to gain confidence in the model by providing the simplest analysis results which can be checked against the specification (i.e. RFC 793). This is an important step in model validation. Secondly, we may discover errors in the protocol, which we can more easily debug in a simpler reachability graph. Once we gain confidence that the model is correct, we can then look at more complex scenarios which exercise all parts of the specification, by choosing the initial marking of places User_1 and User_2 accordingly. For example, we could include an active open and a close command in one user place, and a passive open, send and close in the other.

In this paper we shall just illustrate the approach by considering two cases that just relate to connection establishment. In Case 1, place User_1 has an A_Open (active open) command as the initial marking, while place User_2 has a P_Open (passive open) command as its initial marking. This represents the usual client-server opening scenario that would occur, for example, when requesting a web page. In Case 2, both user places have an A_Open as the initial marking, which allows for the simultaneous opening of a connection, that may occur in peer to peer applications. The initial markings of the remaining places are the same for the different cases. The channel places are empty, while the TCB places are as shown in Fig. 20. Since the initial send sequence (ISS) number for each TCB can be chosen arbitrarily within the 32 bit integer range, we arbitrarily select ISS=10 for TCB_1 and ISS=20 for TCB_2.

11.2 Analysis Results

When we analysed our CPN model for the above two cases, we found that there was a problem with the simultaneous open procedures (Case 2). We modified the original model (Model A) twice to remove the problems. We refer to the first modified model as Model B and to its modification as Model C.

Analysis of Model A

OGs for Model A are obtained for both cases. The results are summarised in Table 2 and the OGs for Cases 1 and 2 are given in Figs. 26 and 27.

The table shows the number of nodes, arcs and dead markings in the OG. The final column indicates whether or not the dead markings are considered to

Table 2. Results for Model A.

Model A	Nodes	Arcs	Dead Markings	Deadlocks
Case 1	11	12	2(0)	0
Case 2	42	60	2(1)	0

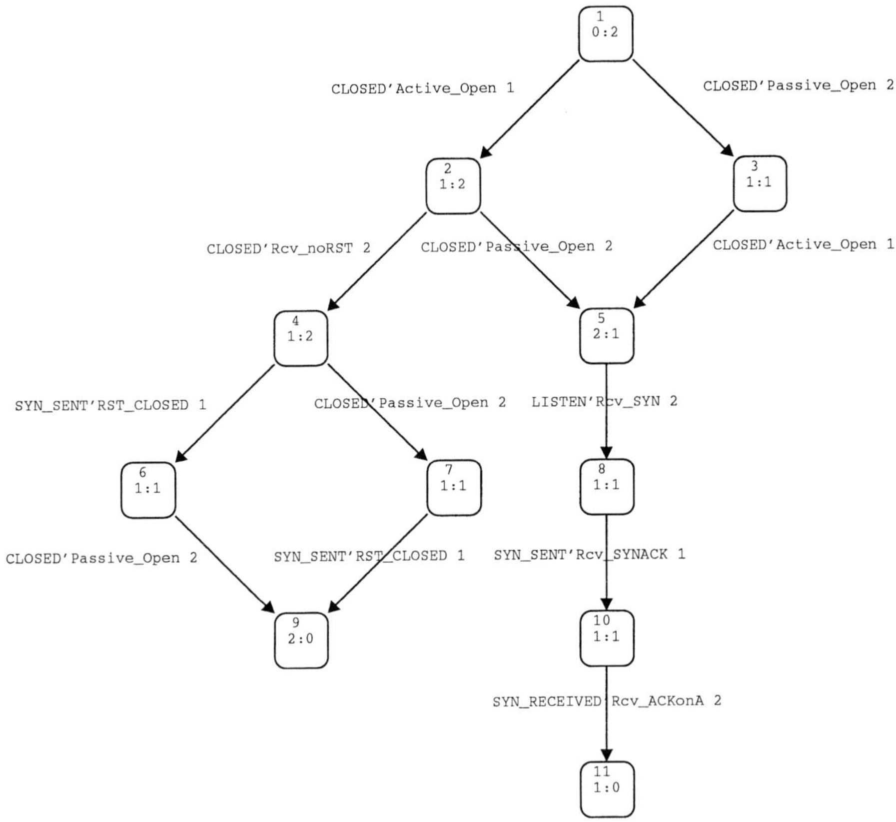

Fig. 26. OG of Model A for Case 1.

be deadlocks (i.e. undesired terminal markings). The integer in parenthesis in column Dead Markings shows the number of dead markings reached through undesired transition sequences. Both OGs are small and are generated in less than 1 second.

Case 1 has two dead markings, nodes 9 and 11 in Fig. 27. The details of the dead markings are given in Fig. 28. Node 9 has one TCP entity in CLOSED, the other in LISTEN, and nothing in the channel. This is an expected terminal state because when the TCP server is in CLOSED (the passive open command has yet to be issued) and receives a SYN, it sends out a RST that changes the state of the TCP client from SYN_SENT to CLOSED. The server TCP entity then

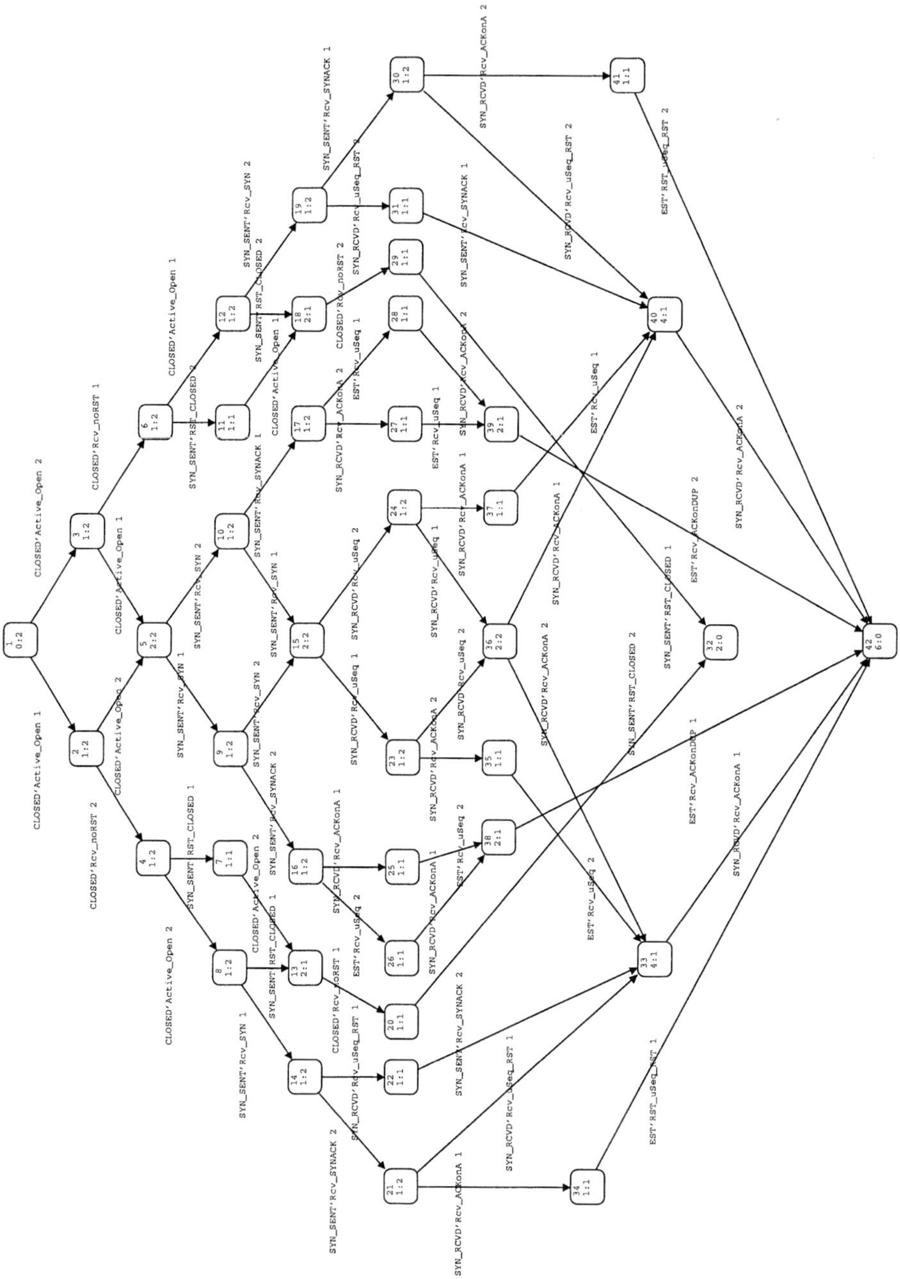

Fig. 27. OG of Model A for Case 2.

goes into LISTEN after receiving the passive open command. The other dead marking (Node 11) has both TCP entities in the ESTABLISHED state and nothing in the channel. Also, the sequence numbers are synchronised at both ends, that is, the send next (SND_NXT) and the send oldest unacknowledged (SND_UNA) numbers are equal on one side and also equal the receive next number (RCV_NXT) at the other end of the connection. This is thus the desired terminal state in which the connection is properly established at both ends. Examining the sequences leading to the two dead markings (see Fig. 26) indicates that they are reached through desired transition sequences.

```
 9      9
 2:0    User 1: empty
        User 2: empty
        H1_H2 1: empty
        H2_H1 1: empty
        TCB 1: 1'(CLOSED,{RCV_NXT = 0,SND_NXT = 0,SND_UNA = 0,ISS = 0},cls)
        TCB 2: 1'(LISTEN,{RCV_NXT = 0,SND_NXT = 0,SND_UNA = 0,ISS = 20},lis)

 11     11
 1:0    User 1: empty
        User 2: empty
        H1_H2 1: empty
        H2_H1 1: empty
        TCB 1: 1'(EST,{RCV_NXT = 21,SND_NXT = 11,SND_UNA = 11,ISS = 10},cls)
        TCB 2: 1'(EST,{RCV_NXT = 11,SND_NXT = 21,SND_UNA = 21,ISS = 20},lis)
```

Fig. 28. The dead markings for Case 1 (Model A).

In Case 2 there are two dead markings, nodes 32 and 42 as shown in Fig. 27. Both are desired terminal states as can be seen from the marking details given in Fig. 29. Node 42 has both TCP entities in the ESTABLISHED state with synchronised state variables and nothing in the channel. The other dead marking (node 32) has both TCPs in the CLOSED state and nothing in the channel.

```
 32     32
 4:0    User 1: empty
        User 2: empty
        H1_H2 1: empty
        H2_H1 1: empty
        TCB 1: 1'(CLOSED,{RCV_NXT = 0,SND_NXT = 0,SND_UNA = 0,ISS = 0},cls)
        TCB 2: 1'(CLOSED,{RCV_NXT = 0,SND_NXT = 0,SND_UNA = 0,ISS = 0},cls)

 42     42
 6:0    User 1: empty
        User 2: empty
        H1_H2 1: empty
        H2_H1 1: empty
        TCB 1: 1'(EST,{RCV_NXT = 21,SND_NXT = 11,SND_UNA = 11,ISS = 10},cls)
        TCB 2: 1'(EST,{RCV_NXT = 11,SND_NXT = 21,SND_UNA = 21,ISS = 20},cls)
```

Fig. 29. The dead markings for Case 2 (Model A).

There are 38 sequences of transitions from node 1 to node 42, of length 7 or 8. Thus we do not find the sequence corresponding to the scenario shown in Fig. 16(b), which is of length 6. Instead, we find sequences similar to the trace obtained from the OG and shown in Fig. 30. This scenario is drawn as a message

Fig. 30. A sequence of the OG for Case 2 (Model A).

sequence diagram in Fig. 31(a), which shows that an ACK is sent in response to each SYNACK before each TCP entity enters the ESTABLISHED state. This behaviour is not desired because it adds delay when there is a simultaneous open and also is not according to TCP's intent as described by the message sequence diagrams in [62].

By examining the trace in Fig. 30, we see that it is transition Rcv_uSeq in Fig. 25 that sends out the redundant ACK, on receipt of segment SYNACK (see the transition from node 24 to node 36 in Fig. 30). Transition Rcv_uSeq

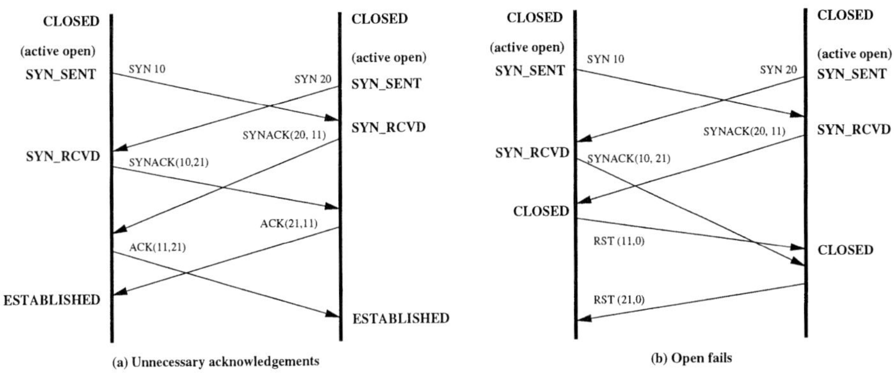

Fig. 31. Problematic simultaneous open scenarios.

sends out an ACK in response to any incoming segment that satisfies the two inequalities on its transition guard, according to the specification on page 69 of RFC 793.

> "If an incoming segment is not acceptable, an acknowledgment should be sent in reply (unless the RST bit is set, if so drop the segment and return) ..."

This means that the sequence number of the incoming SYNACK is less than the value of RCV_NXT. In addition, the second inequality is satisfied because the SYNACK is not a RST segment. As the inequalities are applicable not only to the expected SYNACK, but also to other segments (including unexpected SYNACKs), we do not revise them to remove the problematic ACK for this case. Instead, we question the value of RCV_NXT on whether or not it reflects the sequence number of the next segment that TCP is expecting after receiving the SYN segment. The next segment that TCP is expecting is a SYNACK which has the same sequence number as that of the SYN (see Fig. 8, page 32 of RFC 793 and Fig. 16(b)). Examining the marking immediately prior to the firing of transition Rcv_uSeq, we find that the value of RCV_NXT is updated to SEQ(seg)+1 by transition Rcv_SYN on the SYN_SENT page (Fig. 24), according to processing a SYN in SYN_SENT on page 68 of RFC 793. To remove this inconsistency, we propose not to update RCV_NXT in the case of simultaneous open, keeping its value as SEQ(seg). Therefore, $\#SEQ(seg) < \#RCV\_NXT(v)$ will be false, and hence transition Rcv_uSeq will not be enabled.

To test our assertion, we make the corresponding modification, assigning $\#SEQ(seg)$ to RCV_NXT, on the output arc of transition Rcv_SYN to place TCB (see Fig. 24). This gives us Model B.

Analysis of Model B

The analysis results of Model B are again generated in less than 1 second and shown in Table 3. The results for Case 1 are the same as those of Case 1 for Model A.

Table 3. Results for Model B.

Model B	Nodes	Arcs	Dead Markings	Deadlocks
Case 1	11	12	2(0)	0
Case 2	44	62	2(1)	0

Examining the sequences leading to the dead markings of Case 2 shows that this undesired sequence is removed. However, it reveals another problematic sequence, which involves the generation of RSTs by each TCP entity in response to a SYNACK, as shown in Fig. 32. The corresponding scenario is drawn in Fig. 31(b), where the connection is unnecessarily terminated.

Examining the trace in Fig. 32, we see that it is transition SYNinw_CLOSED in Fig. 25 that sends out the RST. SYNinw_CLOSED rejects the SYN in the window by sending out a RST, as specified on page 71 of RFC 793. However, the SYNACK is mistakenly treated as an old duplicate, because RFC 793 specifies that for state SYN_RCVD, all segments with the SYN bit on (hence for a SYNACK) are rejected by sending out a RST. It is worth mentioning that our proposed solution for the first problem helps to reveal this. To remove the second problem, we must check the ACK bit while checking the SYN bit for state SYN_RCVD in the case of simultaneous open.

We make the corresponding modifications to the guards of transitions SYNinw_CLOSED and Rcv_ACKonA in Fig. 25. Firstly, we replace $\#1(\#CTL(seg)) = SYN$ with $\#CTL(seg) = (SYN, off)$ for transition SYNinw_CLOSED to exclude the SYNACK. Then we replace $\#CTL(seg) = (ACK, on)$ with $\#1(\#CTL(seg)) <> RST$ and $\#1(\#CTL(seg)) <> FIN$ for transition Rcv_ACKonA to accept the SYNACK. We also replace the annotation on the arc from Rcv_ACKonA to TCB by function. It is the same as the previous annotation except that it updates the value of RCV_NXT with #SEQ(seg)+1 if a SYNACK is received and with #SEQ(seg) if an ACK is received. We thus get Model C.

Analysis of Model C

The analysis results of Model C are shown in Table 4.

Table 4. Results for Model C.

Model C	Nodes	Arcs	Dead Markings	Deadlocks
Case 1	11	12	2(0)	0
Case 2	39	54	2(0)	0

The results of Case 1 are the same as those of Case 1 for Models A and B. Analysis of Case 2 shows that there are two dead markings that are desired and reached through desired sequences. One dead marking (Node 35 in Fig. 33) has TCP entities in ESTABLISHED, state variables synchronised and nothing in the channel. The other (Node 32) has both TCPs in CLOSED and nothing in the channel.

Fig. 32. A sequence of the OG for Case 2 (Model B).

Based on the sequences from the reachability analysis, three typical simultaneous open scenarios that are expected are obtained. They are presented in Fig. 16(b), and Fig. 34 (a) and (b). The scenarios shown in Fig. 34 are not discussed in RFC 793, nor in the text books (e.g., [26, 69]).

12 Concluding Remarks on TCP

In this part of the paper we have illustrated two parts of the verification methodology for a complex practical protocol. Firstly we have provided some guidelines

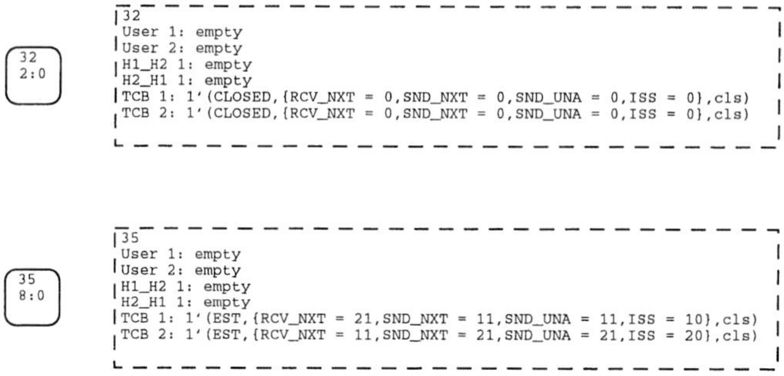

Fig. 33. The dead markings for Case 2 (Model C).

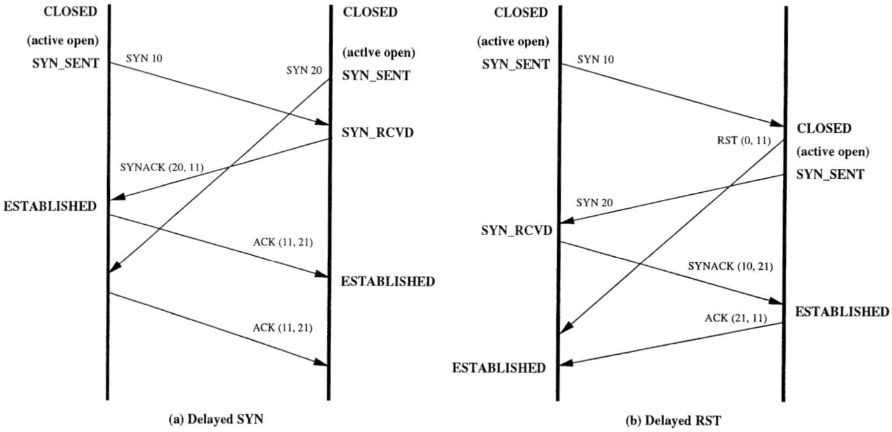

Fig. 34. Two expected simultaneous open scenarios.

for taking the first steps when modelling a complex protocol. The protocol is divided into its connection management part and its data transfer part. We firstly model the connection management part and importantly list the assumptions made. We then illustrate the use of hierarchical CPNs to structure the specification, taking advantage of symmetry, to just specify the connection management procedures once, but call them for each TCP entity by using page instances. This reduces the complexity of the model and eases maintenance. The detailed part of the model is structured into the processing that occurs per state which is a standard way of specifying protocols and a useful starting point for organising specifications. Secondly, we have illustrated the process of analysing a connection management protocol using reachability analyis with the help of Design/CPN.

The analysis has revealed some problems with the procedures for simultaneously opening a connection as specified in Section 3.9 of RFC 793. One problem causes delay and additional traffic due to an unnecessary exchange of acknowledgements when establishing the connection. A solution to this was incorporated and analysed revealing a more serious problem where the procedure can fail to establish the connection. This is due to another error in the functional specification of RFC 793. A further correction to the model is made which removes the problem. Note that the second error also occurs in the finite state machine (FSM) diagram of RFC 793. So the FSM and narrative descriptions of Section 3.9 of RFC 793 are consistently in error. No deadlocks or other subtle errors were found. These results were reported in [38, 39].

This part has only touched the surface regarding the analysis of TCP connection management, where we have presented results just for the case of opening a connection. Further work addresses closing of the connection, relaxes some of the modelling assumptions, for example, allowing segments to be lost in the channel and recovered by retransmissions, develops a service specification [15–17] and compares the connection management protocol language with its service language. This work is consolidated in [37].

13 Some Observations Concerning Specification and Verification

13.1 Modelling Assumptions

After reading the definition of the protocol or service in a standards document (or other primary reference) it is vitally important to write down the assumptions that are made when modelling the protocol (or service). This is to ensure that the analysis results obtained are with respect to the set of assumptions made and that this is firmly in the mind of both the verifier and the reader of the results. The assumptions can be with respect to scope or restrictions within the scope, or abstractions that are made. Illustrations of these for complex protocols are given in [34, 37, 58, 78].

13.2 Specification Structure

There are several ways to structure a specification using hierarchical nets such as Coloured Petri Nets. One of the most popular ways is to structure specifications according to the (major) states of the protocol entities. For each major state, each action in that state is modelled by a transition, for example, the processing of an incoming message or a command from a user. Many international standards are structured this way by using state tables, or Specification and Description Language (SDL) [45] processes where there is one SDL diagram for each major state. If the CPN structure matches the structure in the international standard, then this aids in validating the CPN model against the standard, to ensure that the CPN accurately reflects the standard.

However, there are some specifications where this approach has significant drawbacks and can lead to specifications that have a lot of redundancy. This is the case when the processing of various input messages are treated the same way in a *set of states*. For example in TCP, the state machine comprises 11 states, and processing of some actions such as reset is essentially the same for

8 of these states. Thus the specifier needs to be aware of commonalities that exist for various states and to structure the specification according to processing actions (such as opening connections, resetting, closing connections and dealing with timeouts and retransmissions) rather than slavishly following a state-based approach. This leads to a more elegant specification that has fewer transitions, where actions that are common are clearly seen to be the same, and where maintenance is facilitated. This is following the usual rules of good programming and good writing, which is at the core of the object-based approach. The reduction in the specification is achieved by the folding of transitions that have the same actions for different states, and using a variable that runs over states, with the appropriate restrictions placed in a guard (see [37]).

13.3 Specification Validation

The validation of a CPN specification against the definition of a protocol provided in a standards document or in a proprietary definition is a very important step in making sure that the analysis results do apply to the system of interest. Validation may involve several steps. Firstly developing specifications incrementally and checking that each transition does accurately reflect the intent of the system is essential. This may include stepping through the model using interactive simulation on a tool such as Design/CPN. The next step is to incrementally analyse the model using the OG. Errors discovered may be in the model or in the protocol definition. This is detected by carefully checking if the error is in fact in the system or introduced into the model due to some misinterpretation or inaccurate assumption. Errors in the model need to be removed and then the model re-analysed, iteratively removing inaccuracies. In circumstances where the protocol definition can be discussed with its inventors, this is a vital step in resolving assumptions made in the model.

13.4 Specification versus Verification

Various concerns are more important when developing specifications for implementation rather than for verification. A specification for implementation needs to be readable to ease understanding, and complete so that it contains all essential details to ensure that implementations will interwork correctly. Due to complexity, the verifier will want to modify the specification to just concentrate on essential aspects that need verification. Otherwise the verification task becomes impossible. The main approach concerns making the right abstractions and dividing the specification into manageable and separable components that can be verified independently.

When considering the development aspects of protocol engineering in full, the protocol engineering team will need to develop the protocol architecture and complete service and protocol specifications. The art of the protocol verifier is then to have sufficient insight into how the protocol specification can be divided into manageable parts for verification. This may then lead to proof obligations that the independence assumptions are valid for the properties that require verification.

Let us briefly look at some of the techniques that are used.

Independence of Protocol Mechanisms. Transaction protocols, such as the Internet Open Trading Protocol, define a set of transactions that may be carried out by the different parties. In this case, it is possible to treat each transaction separately. This greatly reduces the complexity of verification. It is only when transactions interact in some way (such as when a refund is dependent on a previous purchase), that we need to consider transactions together. Hence a lot of insight can be obtained and errors can be detected by considering each transaction independently (see [58]).

For connection-oriented protocols, we can divide the operation of the protocol into phases: connection establishment; data transfer; and connection release, and verify each phase separately, before considering their interactions. The next step will normally be to consider the release or abortion of connections at any time during the life of the connection, considering connection establishment and termination together.

Data Abstraction. Protocol messages may include fields that are not used or affected by the protocol mechanisms under investigation. For example, address and multiplexing fields (such as port numbers in TCP) allow multiple connections to be in progress at the same time between a large number of end systems. However, the operation of each connection is the same, and as long as we are not concerned about resources that are shared between connections (such as buffer space), we can just consider one connection as a representative and analyse its behaviour. This means that we can greatly reduce the address space and multiplexing fields. For example, in the case of TCP, we do not need to include these fields at all, and can just consider two protocol entities interacting over a medium representing the operation of IP, where every message that is sent is destined for the peer entity. This is easily achieved by having two places for the medium, one for each direction of information flow. Further, if our objective is to verify TCP connection management procedures, then we can safely ignore fields that are concerned with data transfer, such as windows for flow control, and urgent pointers and flags that are concerned with urgent data transfer. Moreover, we can omit the checksum (used to detect transmission errors) and model its effect using non-determinism as discussed in Section 2.2. We also do not need to model the header length or options (such as the maximum packet size), nor the data that may be transferred. This then reduces the message to a small tuple of values. In the case of TCP a message can be reduced to a triple where we model the message type (such as a connect request (SYN) or reset (RST)), its sequence number and an acknowledgement number. This also reduces the amount of state information that needs to be modelled in the TCP entities, such as the window size. An illustration of the approach is given in [39] and explained earlier in this paper.

With data transfer protocols it is normally sufficient to consider the sender and receiver parts of a protocol entity separately. Thus we just consider one way flow of data across the medium between a sender and receiver, with a return path for control information such as acknowledgements. However, ignoring the data field in data transfer protocols can lead to situations where although service primitive sequences are satisfied, duplicate data and data loss may be occurring, as discussed in Section 7.2.

13.5 Modelling the Underlying Medium

The behaviour of the protocol depends on the medium over which it operates. It is important to start with the simplest medium that makes sense. This is often a FIFO queue. For example, although the Internet Protocol allows for reordering, loss, delay and duplication of messages, most of the time it behaves like a FIFO queue. Thus as a first step, it is important to verify that the protocol will operate correctly over this perfect medium. The reason for this is that media that can lose messages or reset or disconnect connections, can hide undesirable behaviour such as the occurrence of deadlocks. However, that the medium may occassionally lose a message or reset a connection is cold comfort to the users of the protocol who may have to wait a long time for such an event to occur to remove the deadlock. Once the protocol is shown to operate correctly over this friendly medium, then medium imperfections can be introduced and the protocol re-analysed.

13.6 Incremental Approach

To gain confidence in the CPN model and with its verification it is important to use an incremental approach in modelling and analysing the system under investigation. This is illustrated in [78] where RSVP mechanisms are examined one at a time. A similar approach is taken for IOTP [58] where error free and successful transactions are examined first and then arbitrary cancellation and error handling procedures are added.

13.7 Parameters

Complex protocols include several parameters of significance such as the size of the sequence number space, flow control and congestion control window sizes, the maximum number of retransmissions, and the data that is to be sent. Another parameter may also be the capacity of the medium over which the protocol operates. For example, in the Stop-and-Wait protocol we have two parameters of interest, the maximum sequence number and the maximum number of retransmissions.

Reachability analysis requires that parameters of the system be instantiated with particular values before the reachability graph can be generated. Thus we have to generate the OG and its equivalent deterministic automaton for every parameter value. The first step is to start with the smallest values that make sense (such as a medium capacity of one or two, no retransmissions, and maximum sequence number of one) to obtain results to give a feel for how the system operates.

As the values of these parameters increase, reachability analysis suffers from the state explosion problem [77] and becomes intractable. It also means that we can only obtain results for a (small) subset of values, rather than a general result for any value. It may be that results can be obtained for the values of interest of the parameters involved, such as the maximum values of the retransmission counters in the transaction service of WAP [34]. However, in general we would like a result that is valid for any value of the parameters involved. In this case we resort to theorem proving, quite often based on the results obtained from reachability analysis for small parameter values.

In some cases we can invoke the notion of *data independence* [82] to reduce what could be an infinite set (e.g. data items) to a finite and possibly very small number (such as 3), when protocol operations do not affect the data. This can be the case for the data that is sent, when only read/write or assignment operations are involved [65].

Recently, we have managed to obtain symbolic expressions (possibly recursive) for the reachability graphs and their deterministic automata in terms of the medium capacity for the Capability Exchange Signalling service and TCP's data transfer service as discussed under infinite state systems in Section 2.3.

14 Conclusions

This paper summarises the steps of a protocol verification methodology and discusses them in some detail based on several years of attempting to use the methodology for the verification of industrial scale protocols. The methodology uses Coloured Petri Nets (CPNs) to specify both the service provided by the protocol to is users, and the composite specification of the protocol entities interacting over a communication medium or channel. The composite specification is analysed using tool supported reachability analysis to discover behavioural properties, such as undesirable terminal states, livelocks, channel bounds or sequences of events which are inefficient, without recourse to the service specification. Some of this behaviour (inefficient sequences or bounds) are not visible to the users and are thus not captured in service specifications. However, because they have impact on the use of network resources they are worthy of investigation.

Most protocols are characterised by a set of parameters (such as window sizes, sequence number range and maximum number of retransmissions) which need to be instantiated for automatic reachability analysis. For parameters with a small range, such as the maximum number of retransmissions for the Wireless Application Protocol transaction layer (where no more than 8 is required) repeated automated runs may be sufficient to obtain the desired results [34]. The range of results obtained may be extended by using advanced state space techniques, such as the sweep-line method [25, 36], partial orders or equivalence classes or some combination [14]. However, for many industrial protocols, the number of parameters and their significant ranges, preclude obtaining results for all but very small values of the parameters using automated analysis. When this is the case, the automated approach may be able to be supplemented by hand proofs to obtain general results for all values of the parameters. Quite often the intuition behind these hand proofs arises from the results obtained from using automated reachability analysis for small values of parameters.

It is also valuable to compare the behaviour of the composite specification with that of the service specification. For some protocols the service has also been defined as part of the process of defining the protocol. This greatly facilitates the creation of a CPN model of the service. However, the service definitions invariably do not completely specify the global sequences of user observable events (known as service primitives) concentrating instead on defining the local interface sequences. The CPN modeller must then use their intuition to obtain a consistent set of global sequences. Sometimes this involves complex structures in the CPN to ensure that the correct sequences occur, while in other cases it is quite straightforward. However, as far as we are aware, none of the Internet protocols have included service definitions, and thus a service specification must be created, based on the protocol definition, interface definitions if they exist and the experience of the modeller. In this circumstance, it is recommended that the protocol be modelled and analysed first, and then the service developed, based on the experience gained. Once a service model is obtained, the sequences of service primitives embedded in the service model (the service language) can be compared with the service primitive sequences that occur in the composite specification (the protocol language). This can be done automatically for finite state systems using well-known reachability analysis and automata reduction and comparison algorithms. This approach also suffers from the state explosion problem. For infinite state systems, or where parameter values are unknown, we briefly discuss an approach using recursive techniques to provide symbolic representations of the occurrence graphs and associated automata. The intuition behind these symbolic representations has been obtained from using reachability analysis for small values of parameters. In some cases it is possible to derive the symbolic representations directly without the need for recursive formulations.

The methodology is illustrated by two case studies. One considers the class of Stop-and-Wait protocols (SWP) as a representative of the class of data transfer protocols. This involves the inclusion of two parameters: the maximum sequence number and the maximum number of retransmissions. We define 4 properties of interest (queue bounds, data loss, duplication and the Stop-and-Wait property) and prove that the SWP operates correctly over FIFO channels. The channel bound depends the maximum number of retransmisions ($\mathsf{MaxRetrans}$) and is given by $2\mathsf{MaxRetrans} + 1$ for both channels. We believe this to be a new result. This is obtained using hand proofs, which we develop in detail to illustrate the process. To prove that there is no loss or duplication and that the Stop-and-Wait property holds (i.e. that the protocol satisfies its service of alternating sends and receives) we use automatic reachability and language analysis. We do this for a significant range of parameter values e.g. up to 10 bit sequence numbers for up to 4 retransmissions, which gives confidence that the results are general. However, we have no general proof. We also prove that the properties do not hold when the SWP operates over lossy (or lossless) re-ordering channels. Again hand proofs are used to prove that the channels are unbounded, giving a general result when both the maximum number of retransmissions and the maximum sequence number are any positive integer. We use language analysis for the other proofs, and argue that they are generally applicable for medium capacities of two or greater, and for the other two parameters (MaxRetrans and MaxSeqNo being positive integers. The results for loss and duplication are illustrated with time sequence diagrams. We discuss the practical relevance of these results by considering their impact on the Transmission Control Protocol. We conclude that the problems can occur once transmission rates are at about 10 Gbit/s. We also discuss a limitation of our approach. We have shown that loss and duplication can occur in the SWP by considering sequences where there are more sends than receives (loss) or more receives than sends (duplication). We also illustrate that both loss and duplication can occur even when the Stop-and-Wait property of alternating sends and receives holds. This is due to our data abstraction assumption, which is too strong. We note that using the notion of data independence may overcome this limitation.

The second case study examines the connection management procedures of the Transmission Control Protocol, as a representative example of connection management procedures as opposed to the data transfer procedures of the SWP. It also illustrates the application of the methodology to a practical protocol of major significance. We exemplify the process of writing down assumptions, regarding the creation of the model, that simplify the analysis task. We also stress the importance of this step. We build a model of significant complexity and analyse the connection establishment protocol. This allows us to discover two problems with the simultaneous open procedures, and provide a solution which we then verify to be correct. This is easily handled by brute force reachability analysis. However, we have only illustrated the use of the methodology for the analysis of the very simplest of procedures. Further work is in hand to provide a comprehensive verification of the connection management procedures, including release, abort and the use of retransmissions [37].

Finally we end the paper with some observations and recommendations regarding the use of the methodology. Better tool support is required for the methodology to allow the seamless integration of reachability analysis and language analysis and to allow language equivalence or inclusion to be done on-the-fly using advanced techniques such as the sweep-line method. Promising areas of future work include the incorporation of the notion of data independence into the methodology and the use of recursive techniques to obtain general results for parameteric verification.

Acknowledgements

We gratefully acknowledge the contributions of many people to this work. First and foremost Jonathan greatly appreciates the guidance and encouragement of Professor Fred Symons who introduced the work to Telecom Australia Research Laboratories (TRL) in the late 1970s. His former colleagues in TRL, especially Geoffrey Wheeler, Michael Wilbur-Ham, Tim Batten, John Gilmore and Greg Findlow have provided important contributions and insights. Past and present CSEC postgraduate students have contributed significantly to the area, in particular: Lin Liu for her work on methodology, including recursive formulae for occurrence graphs and automata; Chun Ouyang for work on methodology, complex protocol specification and defining acceptable subsets of service specifications using IOTP; Dr. Maria Villapol for her work on incremental specification and verification with RSVP; Dr. Steven Gordon for applying the methodology, including the sweep-line technique, to WAP; and Mat Farrington who pointed out the relevance of RFC 1323 to our investigations. We would also like to thank Dr. Lars Kristensen, a post doctoral fellow with CSEC for two years from 2000 to 2002, for his many valuable suggestions regarding the methodology, the use of Design/CPN and his work on the development of the sweep-line technique.

References

1. P. Aziz Abdulla and B. Jonsson. Verifying Programs with Unreliable Channels. *Information and Computation*, 127(2):91–101, June 1996.
2. Y. Afek, H. Attiya, A. Fekete, M. Fischer, N. Lynch, Y. Mansour, D. Wang, and L. Zuck. Reliable Communication Over Unreliable Channels. *Journal of the ACM*, 41(6):1267–1297, Nov. 1994.
3. Y. Afek and G.M. Brown. Self-Stabilization of the Alternating Bit Protocol. In *Proceedings of the 8th Symposium on Reliable Distributed Systems*, pages 80–83. IEEE Comput. Soc. Press, 1989.
4. F. Babich and L. Deotto. Formal Methods for the Specification and Analysis of Communication Protocols. *IEEE Communications Surveys*, 4(1):2–20, Third Quarter 2002.
5. W.A. Barrett and J.D. Couch. *Compiler Construction: Theory and Practice*. Science Research Associates, 1979.
6. K.A. Bartlett, R.A. Scantlebury, and P.T. Wilkinson. A Note on Reliable Full-Duplex Transmission over Half-Duplex Links. *Communications of the ACM*, 12(5):260–261, May 1969.
7. J. Billington. *Abstract Specification of the ISO Transport Service Definition using Labelled Numerical Petri Nets*, volume III of *Protocol Specification, Testing, and Verification*, pages 173–185. Elsevier Science Publishers, 1983.
8. J. Billington. Extensions to Coloured Petri Nets. In *Petri Nets and Performance Models, The Proceedings of the Third International Workshop, PNPM '89, Kyoto, Japan, December 11-13*, pages 61–70. IEEE Computer Society, 1989.
9. J. Billington. Formal specification of protocols: Protocol Engineering. In *Encyclopedia of Microcomputers*, volume 7, pages 299–314. Marcel Dekker, New York, 1991.

10. J. Billington. *Protocol Specification using P-Graphs, a Technique based on Coloured Petri Nets*, volume 1492 of *Lecture Notes in Computer Science, Lectures on Petri Nets II: Applications*, pages 293–330. Springer-Verlag, 1998.
11. J. Billington, M. Diaz, and G. Rozenberg, editors. *Application of Petri Nets to Communication Networks*, volume 1605 of *Lecture Notes in Computer Science*. Springer-Verlag, 1999.
12. J. Billington and G. E. Gallasch. How Stop and Wait Protocols Can Fail Over The Internet. In *Proceedings of FORTE'03*, volume 2767 of *Lecture Notes in Computer Science*, pages 209–223. Springer-Verlag, 2003.
13. J. Billington and G. E. Gallasch. An Investigation of the Properties of Stop-and-Wait Protocols over Channels which can Re-order messages. Technical Report 15, Computer Systems Engineering Centre, School of Electrical and Information Engineering, University of South Australia, Australia., 2004.
14. J. Billington, G.E. Gallasch, L.M. Kristensen, and T. Mailund. Exploiting equivalence reduction and the sweep-line method for detecting terminal states. *IEEE Transactions on Systems, Man and Cybernetics, Part A: Systems and Humans*, 34(1):23–37, January 2004.
15. J. Billington and B. Han. Formalising the TCP Symmetrical Connection Management Service. In *Proceedings of the Design, Analysis and Simulation of Distributed Systems Conference, Orlando, Florida, USA*, pages 178–184, March 2003.
16. J. Billington and B. Han. On Defining the Service Provided by TCP. In *Proceedings of the 26th Australasian Computer Science Conference, Adelaide, Australia*, volume 16 of *Conferences in Research and Practice in Information Technology*, pages 129–138, February 2003.
17. J. Billington and B. Han. Closed Form Expressions for the State Space of TCP's Data Transfer Service Operating over Unbounded Channels. In *Proceedings of the 27th Australasian Computer Science Conference, Dunedin, New Zealand*, volume 26 of *Conferences in Research and Practice in Information Technology*, pages 31–39, January 2004.
18. J. Billington, G.R. Wheeler, and M.C. Wilbur-Ham. PROTEAN: A High-level Petri Net Tool for the Specification and Verification of Communication Protocols. *IEEE Transactions on Software Engineering*, 14(3):301–316, March 1988.
19. J. Billington, M.C. Wilbur-Ham, and M.Y. Bearman. Automated Protocol Verification. In *Protocol Specification, Testing and Verification, V*, pages 59–70. North Holland, Amsterdam, 1986.
20. T. Bolognesi and E. Brinksma. Introduction to the ISO Specification Language LOTOS. *Comput. Networks and ISDN Sys.*, 14(1):25–59, 1987.
21. S. Budkowski and P. Dembinski. An Introduction to Estelle: A Specification Language for Distributed Systems. *Comput. Networks and ISDN Sys.*, 14(1):3–23, 1987.
22. CCITT. ISDN user-network interface data link layer specification. Technical report, Draft Recommendation Q.921, Working Party XI/6, Issue 7, Jan. 1984.
23. S. Christensen and L. O. Jepsen. Modelling and Simulation of a Network Management System Using Hierarhical Coloured Petri Nets. In *Proceedings of the 1991 European Simulation Multiconference*, pages 47–52. Society for Computer Simulation, 1991.
24. S. Christensen and J.B. Jørgensen. Analysis of Bang and Olufsen's BeoLink Audio/Video System Using Coloured Petri Nets. In *Proceedings of ICATPN'97*, volume 1248 of *Lecture Notes in Computer Science*, pages 387–406. Springer-Verlag, 1997.

25. S. Christensen, L.M. Kristensen, and T. Mailund. A Sweep-Line Method for State Space Exploration. In *Proceedings of TACAS 2001*, volume 2031 of *Lecture Notes in Computer Science*, pages 450–464. Springer-Verlag, 2001.
26. D.E. Comer. *Internetworking with TCP/IP: Principles, Protocols, and Architecture*, volume 1. Prentice Hall, Upper Saddle River, NJ, 2000.
27. CPN ML: An Extension of Standard ML. `http://www.daimi.au.dk/designCPN/sml/cpnml.html`.
28. J. Desel and W. Reisig. *Place/Transition Petri Nets*, volume 1491 of *Lecture Notes in Computer Science, Lectures on Petri Nets I: Basic Models*, pages 122–173. Springer-Verlag, 1998.
29. Design/CPN Online. `http://www.daimi.au.dk/designCPN/`.
30. M. Diaz. Modelling and Analysis of Communication and Co-operation Protocols Using Petri Net Based Models. In *Protocol Specification, Testing and Verification*, pages 465–510. North-Holland, 1982.
31. G. A. Findlow, G. S. Gerrand, J. Billington, and R. J. Fonc. Modclling ISDN Supplementary Services Using Coloured Petri Nets. In *Proceedings of Communications '92*, pages 37–41, Sydney, Australia, 1992.
32. D. J. Floreani, J. Billington, and A. Dadej. Designing and Verifying a Communications Gateway Using Colored Petri Nets and Design/CPN. In *Proceedings of ICATPN'96*, volume 1091 of *Lecture Notes in Computer Science*, pages 153–171. Springer-Verlag, 1996.
33. FSM Library, AT&T Research Labs. `http://www.research.att.com/sw/tools/fsm/`.
34. S. Gordon. *Verification of the WAP Transaction Layer using Coloured Petri Nets.* PhD thesis, School of Electrical and Information Engineering, University of South Australia, 2001.
35. S. Gordon and J. Billington. Analysing the WAP class 2 Wireless Transaction Protocol using Coloured Petri nets. In *Proceedings of 21st International Conference on Application and Theory of Petri Nets*, volume 1825 of *Lecture Notes in Computer Science*, pages 207–226. Springer-Verlag, 2000.
36. S. Gordon, L.M. Kristensen, and J. Billington. Verification of a Revised WAP Wireless Transaction Protocol. In *Proceedings of 23rd International Conference on Application and Theory of Petri Nets*, volume 2360 of *Lecture Notes in Computer Science*, pages 182–202. Springer-Verlag, 2002.
37. B. Han. Formal Specification of the TCP Service and Verification of TCP Connection Management. Draft PhD Thesis, University of South Australia, April 2004.
38. B. Han and J. Billington. An Analysis of TCP Connection Management Using Coloured Petri Nets. In *Proceedings of the 5th World Multi-Conference on Systemics, Cybernetics and Informatics (SCI'2001)*, pages 590–595, Orlando, Florida, July 2001.
39. B. Han and J. Billington. Validating TCP Connection Management. In *Proceedings of the Workshop on Software Engineering and Formal Methods, Adelaide, Australia*, volume 12 of *Conferences in Research and Practice in Information Technology*, pages 47–55, June 2002.
40. G.J. Holzmann. *Design and Validation of Computer Protocols.* Prentice-Hall, 1991.
41. International Telecommunication Union. `http://www.itu.int/home/`.
42. Internet Engineering Task Force. `http://www.ietf.org`.
43. The Internet Engineering Task Force. TCP Extensions for High Performance. RFC 1323, 1992.

44. ISO/IEC. *Software and Systems Engineering – High-level Petri Nets – Part 1: Concepts, Definitions and Graphical Notation.* ISO/IEC FDIS 15909-1, Final Draft International Standard, International Organisation for Standardization, February 2004.
45. ITU-T. *Recommendation Z.100: Functional Specification and Description Language (SDL).* International Telecommunications Union, 2002.
46. ITU-T. *Recommendation X.210, Information Technology - Open Systems Interconnection - Basic Reference Model: Conventions for the Definition of OSI Services.* International Telecommunications Union, Nov. 1993.
47. K. Jensen. Coloured Petri Nets and the Invariant Method. *Theoretical Computer Science*, 14:317–336, 1981.
48. K. Jensen. *Coloured Petri Nets. Basic Concepts, Analysis Methods and Practical Use. Vol. 1, Basic Concepts.* Springer-Verlag, 2nd edition, 1997.
49. K. Jensen. *Coloured Petri Nets. Basic Concepts, Analysis Methods and Practical Use. Vol. 2, Analysis Methods.* Springer-Verlag, 2nd edition, 1997.
50. K. Jensen. *Coloured Petri Nets. Basic Concepts, Analysis Methods and Practical Use. Vol. 3, Practical Use.* Springer-Verlag, 1997.
51. J.B. Jørgensen and K. H. Mortensen. Modelling and Analysis of Distributed Program Execution in BETA Using Coloured Petri Nets. In *Proceedings of ICATPN'96*, volume 1091 of *Lecture Notes in Computer Science*, pages 249–268. Springer-Verlag, 1996.
52. D. E. Knuth. Verification of Link-Level Protocols. *BIT*, 21:31–36, 1981.
53. L.M. Kristensen, S. Christensen, and K. Jensen. The Practitioner's Guide to Coloured Petri Nets. *International Journal on Software Tools for Technology Transfer*, 2(2):98–132, 1998.
54. R. Lazic and D. Nowak. *A Unifying Approach to Data-independence*, volume 1877 of *Lecture Notes in Computer Science*, pages 581–595. Springer-Verlag, 2000.
55. L. Liu and J. Billington. Modelling and Analysis of the CES Protocol of H.245. In *Proc. of the Third Workshop and Tutorial on Practical Use of Coloured Petri Nets and the CPN Tools*, pages 95–114, Aarhus, Denmark, 29-31 August 2001.
56. L. Liu and J. Billington. *Tackling the Infinite State Space of a Multimedia Control Protocol Service Specification*, volume 2360 of *Lecture Notes in Computer Science*, pages 273–293. Springer-Verlag, 2002.
57. L. Liu and J. Billington. Obtaining the Service Language for H.245's Multimedia Capability Exchange Signalling Protocol: the Final Step. In *Proc. of the 10th International Multi-Media Modelling Conference*, Brisbane, Australia, 5-7 January 2004.
58. C. Ouyang. *Formal Specification and Verification of the Internet Open Trading Protocol using Coloured Petri Nets.* PhD thesis, School of Electrical and Information Engineering, University of South Australia, Australia, March 2004. submitted.
59. C. Ouyang and J. Billington. On verifying the Internet Open Trading Protocol. In *Proceedings of 4th International Conference on Electronic Commerce and Web Technologies*, volume 2738 of *Lecture Notes in Computer Science*, pages 292–302, Prague, Czech Republic, 1-5 September 2003. Springer-Verlag.
60. C. Ouyang, L. M. Kristensen, and J. Billington. A formal service specification of the Internet Open Trading Protocol. In *Proceedings of 23rd International Conference on Application and Theory of Petri Nets*, volume 2360 of *Lecture Notes in Computer Science*, pages 352–373, Adelaide, Australia, 24-30 June 2002. Springer-Verlag.
61. J. Postel. Internet Protocol - DARPA Internet Program Protocol Specification. RFC 791, IETF, September 1981.

62. J. Postel. Transmission Control Protocol. RFC 793, 1981.
63. W. Reisig. *Elements of Distributed Algorithms: Modelling and Analysis with Petri Nets.* Springer-Verlag, 1998.
64. A.W. Roscoe. *The Theory and Practice of Concurrency.* Prentice Hall International Series in Computer Science, 1998.
65. K. Sabnani. An Algorithmic Technique for Protocol Verification. *IEEE Transactions on Communications*, 36(8):924–931, 1988.
66. W. Stallings. *Data and Computer Communications.* Prentice Hall, 6th edition, 2000.
67. Standard ML of New Jersey. `http://cm.bell-labs.com/cm/cs/what/smlnj/`.
68. L.J. Steggles and P. Kosiuczenko. A Timed Rewriting Logic Semantics for SDL: a case study of the Alternating Bit Protocol. *Electronic Notes in Theoretical Computer Science*, 15, 1998.
69. W.R. Stevens. *TCP/IP Illustrated*, volume 1. Addison-Wesley, Reading, MA, November 1994.
70. C.A. Sunshine. Formal Techniques for Protocol Specification and Verification. *IEEE Computer*, pages 346–350, September 1979.
71. I. Suzuki. Formal Analysis of the Alternating Bit Protocol by Temporal Petri Nets. *IEEE Transactions on Software Engineering*, 16(11):1273–1281, 1990.
72. I. Suzuki. Specification and Verification of the Alternating Bit Protocol by Temporal Petri Nets. In *Proceedings of the 32nd Midwest Symposium on Circuits and Systems*, pages 157–160. IEEE Press, 1990.
73. A. Tanenbaum. *Computer Networks.* Prentice Hall, 4th edition, 2003.
74. A. Tokmakoff and J. Billington. An Approach to the Analysis of Interworking Traders. In *Proc. of ICATPN'99*, volume 1639 of *Lecture Notes in Computer Science*, pages 127–146. Springer-Verlag, 1999.
75. R.S. Tomlinson. Selecting sequence numbers. In *Proc. of SIGCOMM/SIGOPS Interprocess Commun. Workshop*, ACM, pages 11–23, 1975.
76. K. J. Turner (Ed.). *Using Formal Description Techniques: An Introduction to Estelle, Lotos and SDL.* Wiley Series in Communication and Distributed Systems. John Wiley & Sons, 1993.
77. A. Valmari. The State Explosion Problem. In *Lectures on Petri Nets I: Basic Models*, volume 1491 of *Lecture Notes in Computer Science*, pages 429–528. Springer-Verlag, 1998.
78. M. E. Villapol. *Modelling and Analysis of the Resource Reservation Protocol.* PhD thesis, Electrical and Information Engineering, University of South Australia, Australia, November 2003.
79. M. E. Villapol and J. Billington. Generation of a Service Language for the Resource Reservation Protocol using Formal Methods. In *Proc. INCOSE2001, the 11th Annual International Symposium of the International Council on Systems Engineering*, CD-ROM, Melbourne, Australia, 1-5 July 2001.
80. M. E. Villapol and J. Billington. *Analysing Properties of the Resource Reservation Protocol*, volume 2679 of *Lecture Notes in Computer Science*, pages 377–396. Springer-Verlag, 2003.
81. WAP Forum. `http://www.wapforum.org/`.
82. P. Wolper. Expressing Interesting Properties of Programs in Propositional Temporal Logic. In *Proceedings of the 13th Annual ACM Symposium on Principles of Programming Languages (POPL)*, pages 184–193. ACM, 1986.
83. J. Xu and J. Kuusela. Analyzing the Execution Architecture of Mobile Phone Software with Coloured Petri Nets. *Int. Journal on Software Tools for Technology Transfer*, 2(2):133–143, 1998.

Extending the Zero-Safe Approach to Coloured, Reconfigurable and Dynamic Nets*

Roberto Bruni, Hernán Melgratti, and Ugo Montanari

Dipartimento di Informatica, Università di Pisa, Italia
{bruni,melgratt,ugo}@di.unipi.it

Abstract. As web applications become more and more complex, primitives for handling interaction patterns among independent components become essential. In fact, distributed applications require new forms of transactions for orchestrating the progress of their negotiations and agreements. Still we lack foundational models that accurately explain the crucial aspects of the problem. In this work we explore how to model transactions in *coloured*, *reconfigurable* and *dynamic nets*, (i.e., high-level/high-order Petri nets that can express mobility and can extend themselves dynamically during their execution). Starting from *zero-safe* nets – a well-studied extension of Place/Transition Petri nets with a transactional mechanism based on a distinction between consistent (observable) and transient (hidden) states – we show how the zero-safe approach can be smoothly applied to a hierarchy of nets of increasing expressiveness.

1 Introduction

To some extent, *place/transition Petri nets* (P/T nets) [27, 28] are for Concurrency Theory what finite automata are for the Theory of Computation: their rigorous theories have been consolidated in pioneering work; they are foundational models for many other languages and calculi; they have been enriched in a number of ways (e.g. time, stochastic, data type, high-order and reflection) for taking into account particular features demanded by real case studies and scenarios; they have been applied with success interdisciplinarily and even in industrial applications; they have been a constant reference for comparing emerging paradigms with; they admit intuitive graphical presentations; they are widely used in software engineering and in system specification and verification.

Nowadays, one of the main challenges for researchers with interest in Concurrency is the definition of adequate models for global computing applications, where aspects like distribution, name and code mobility, security, quality of services, and coordination are stretched to the very limit. Several of these aspects have been investigated separately and sometimes combined especially with

* Research supported by the MSR Cambridge Project NAPI, by the FET-GC Project IST-2001-32747 AGILE, by the MIUR Project COFIN 2001013518 COMETA, and by the MURST-CNR 1999 Project, *Software Architectures on Cooperative WAN*.

J. Desel, W. Reisig, and G. Rozenberg (Eds.): ACPN 2003, LNCS 3098, pp. 291–327, 2004.

the help of suitable *process calculi* (e.g. π-calculus [26], join calculus [21], spi-calculus [1], ambient calculus [19]), where new primitives can be easily introduced and experimented with. Although such calculi are much more expressive than P/T nets, it is possible to recover their spirit by progressively enriching the basic P/T net model with high-level and high-order features, like exemplified in [15]. The *Petri Box* calculus [4] is a different approach for reconciling both worlds.

In this paper, the aspect we want to focus on is *orchestration*. In fact, as more and more complex global computing applications are developed, then more primitives for handling common interaction patterns between independent components become essential. Academy and Industry are showing renewed interest in the orchestration of distributed applications via programming languages and calculi with primitive transactional mechanisms for managing electronic negotiations and contracts carried on among independent components. Although some solutions have been proposed in the literature (see the section on *related work*), still there is no complete agreement on the foundational models that better expose the crucial points of the problem.

The solution we propose in the paper relies on the so-called *zero-safe approach*, that is shown to span along a hierarchy of concurrent models (of increasing expressiveness), from P/T nets to *dynamic nets*, (i.e., high-level petri nets that can express dynamic network reconfigurability and reflection), stepping through *coloured nets* and *reconfigurable nets*. The hierarchy is indeed the one proposed in [15], where it is also shown that each net flavors correspond to a typeable fragment of join calculus. The straight consequence is that the zero-safe approach can be transferred also to those (sub)calculi at no additional cost.

Zero-safe approach. Zero-safe nets (ZS *nets*) have been introduced to model transactions in concurrent systems [11]. The basic model extends P/T *nets* with a mechanism for expressing serializable concurrent (multiway) transactions. In ZS nets there are two kinds of places (and, consequently, two kinds of tokens), called stable and zero-safe. Roughly, a transaction on a ZS net is a concurrent computation that departs from and arrives to a multiset of stable tokens. Recently, they have been used in [8] to encode short-running transactions of Microsoft Biztalk®, a commercial workflow management system [30]. ZS nets additionally provides a "dynamic" specification of transactions boundaries (as opposed to the "static" one of BizTalk) supporting multiway transactions, which retain several entry and exit points, and admit a number of participants which is statically unknown. Nevertheless, ZS nets are not suitable to express some interesting aspects of negotiations in global computing, such as value passing, dynamic reconfiguration of communication, name mobility, programmable compensations and nesting. Also their expressive power is limited as e.g. reachability is decidable [16].

ZS nets offer a two-level view of the modeled system: (1) the concrete operational view where transient places and the coordination mechanism between activities participating to a transaction are fully exposed; (2) the abstract view, where transactions are seen as atomic activities, and the user is aware of stable places only, while transient places are transparent. In fact, the abstract view is given by an ordinary P/T net, whose places are the stable places of the ZS net and whose transitions are the transactions of the ZS net. Moreover, the correspondence between the two views admits a rigorous mathematical characterization as coreflection between suitable model categories. It is worth remarking that ZS nets with a finite number of transitions can yield abstract P/T nets with infinitely many transitions. From the system designer viewpoint, this means that the combinatorial features of ZS nets can be exploited to keep small the size of the architecture.

The zero-safe approach has been extended to more expressive frameworks such as nets with read and inhibitor arcs [13], which have been shown expressive enough to give a concurrent operational semantics to the language TraLinda (an extension of Linda with transactional primitives).

From the implementation point of view, a distributed interpreter for ZS nets has been proposed in [10] that is based on the ordinary unfolding construction for Petri nets, while both centralized and distributed interpreters have been proposed in [7, 8], which are written in (distributed) join calculus. In particular, while the centralized implementation closely corresponds to the spirit of BizTalk's Transaction Manager and can be written in the join fragment corresponding to coloured nets, the distributed implementation exploits a novel commit protocol, called *Distributed 2-Phase Commit* (D2PC) and exploits reflection for dynamic creation of local transaction managers. Given the correspondence in [15], the distributed interpreter can be directly translated in dynamic nets, but neither in reconfigurable nets, nor in coloured nets.

A hierarchy of transactional frameworks. In this paper, we progressively enrich ZS nets by adding: (1) the value passing mechanism of coloured nets; (2) the dynamic interconnection mechanism of reconfigurable nets; (3) the high-order features of dynamic nets.

In most cases, it is shown that that two-level view of the zero-safe approach is fully preserved, in the sense that, e.g. the abstract net of a coloured ZS net is a coloured P/T net, and so on. Moreover, most constructions are consistent with the obvious embedding derived from the hierarchy, in the sense that, e.g. if we regard a coloured ZS net as a reconfigurable ZS net and take the corresponding abstract reconfigurable P/T net, then we get a coloured P/T net. In other words, the diagram in Figure 1 commutes (vertical arrows are the obvious embedding, while horizontal arrows stand for the construction of abstract nets). We used the word "most", because although we conjecture that the construction of abstract nets can be extended to the dynamic case, at the moment the problem is still open and therefore the *tower* in Figure 1 misses the roof.

For each *layer* of the tower we give several examples for illustrating the main features of the corresponding model. The two main case studies which are presented are the *mobile lessees problem* and the *mailing list*. Regarding the mobile lessees problem, first it is shown that an instance of the problem can always be represented as a ZS net, then it is shown that colours allow for modeling all the instances with a unique coloured ZS net. Regarding the mailing list example, first it is shown that reconfigurable arcs are needed for modeling

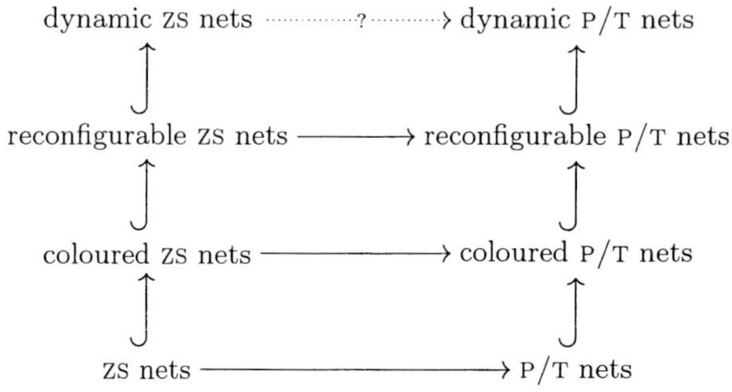

Fig. 1. The hierarchy of transactional frameworks.

dynamic message delivery, and then it is shown that the example can be extended with dynamic creation of new mailing lists by exploiting reflection in dynamic ZS nets.

Structure of the paper. In Sections 2 and 3 are background sections, where we recall the basics of P/T nets and ZS nets. In particular, we define the operational semantics of such models and the notion of a *causal net*, the notion of a *process* and the notion of an *abstract net*, which are later extended to account for colours, reconfiguration and high-order. The modeling of (instances of) the mobile lessees problem is instead original.

Section 4 and 5 contains the original proposals for extending the zero-safe approach to coloured and reconfigurable nets. In both cases, the operational and abstract semantics are defined and related by strong correspondence theorems.

Section 6 attempts to extend the zero-safe approach to dynamic nets. The operational semantics of dynamic ZS nets is presented and discussed on the basis of the mailing list example, whereas the abstract semantics is just informally discussed to put in evidence the difficulties in completing the tower in Figure 1.

Some concluding remarks and future work are in Section 7.

Related work. This part collects pointers to recent approaches to formal methods applied to negotiations for distributed systems. It can be skipped without compromising the reading of the rest of the paper.

Recent works have addressed the extension of the coordination language Linda [23] to express transactions. In particular, the serializability of transactions in *JavaSpaces* [31] have been studied in [17] by adding new primitives for handling traditional flat transactions to Linda. An alternative extension with multiway negotiations is proposed in [14], called TraLinda. The semantics of TraLinda relies on a zero-safe extension of contextual nets. Contextual nets have been previously used in [29] to study the serializability on database transactions.

While aforementioned works are closely related to the classical notion of transactions – "all or none" effect of a transaction is observable – web services languages, such as BPEL [6], WSFL[25], XLANG [32], and its graphical representation Biztalk[30], are particular aware of primitives for handling long-lived negotiations. Such languages do not guarantee atomicity of transactions, but provide programmable *compensations* for undoing actions performed by failed negotiations. The compensation mechanism has been introduced in [22] for designing long-running transactions in database applications.

A compensation language, called StAc, has been proposed in [18]. Processes and compensations in StAc are written in terms of atomic activities. Nevertheless, the interaction among activities is reduced to data sharing and is not described at the top-level (they are hidden on the detailed description of the atomic activities).

In the spirit of process description languages, an extension of the π-calculus with nested compensation contexts has been introduced in [5]. Nevertheless, the extension accounts only for compensation and there is no mechanism to restrict the interactions of transactional processes: the communication capabilities of a process do not change when it runs inside a transactional context. A different approach is taken in cJoin [9] – an extension of the Join calculus with nested, compensatable negotiations – where processes in different transactions can interact by joining their original negotiations into a larger one. Finally, [20] introduces the pike calculus based on *conclaves* (i.e., set of dependent processes) as main abstractions for programming fault-tolerant applications. Different notions of transactions can be modelled in pike by combining such abstractions.

2 Petri Nets

In Petri nets, *places* are repositories of *tokens* (i.e. resources, messages) and *transitions* fetch and produce tokens. We consider an infinite set of resource names $\mathcal{P}$. Given $S \subseteq \mathcal{P}$, we denote with $\wp_{\mathrm{f}}(S)$ the set of all finite subsets of S.

Definition 1 (Net). *A* net N *is a 4-tuple* $N = (S_N, T_N, \delta_{0N}, \delta_{1N})$ *where* $S_N \subseteq \mathcal{P}$ *is the (nonempty) set of places,* $a, a', \ldots$, T_N *is the set of transitions,* $t, t', \ldots$ *(with* $S_N \cap T_N = \emptyset$*), and the functions* $\delta_{0N}, \delta_{1N} : T_N \to \wp_{\mathrm{f}}(S_N)$ *assign respectively, source and target to each transition.*

We will denote $S_N \cup T_N$ by N, and omit subscript N whenever no confusion arises. We abbreviate a transition $t \in T$ such that $\delta_0(t) = s_1$ and $\delta_1(t) = s_2$ as $s_1[\rangle s_2$, where s_1 is usually referred to as the *preset* of t (written ${}^\bullet t$) and s_2 as the *postset* of t (written $t^\bullet$). Similarly for any place a in S, the preset of a (written ${}^\bullet a$) denotes the set of all transitions with target in a (i.e., ${}^\bullet a = \{t | a \in t^\bullet\}$, and the postset of a (written $a^\bullet$) denotes the set of all transitions with source in a (i.e., $a^\bullet = \{t | a \in {}^\bullet t\}$. Moreover, let ${}^\circ N = \{x \in N | {}^\bullet x = \emptyset\}$ and $N^\circ = \{x \in N | x^\bullet = \emptyset\}$ denote the sets of *initial* and *final elements* of N respectively. A place a is said to be *isolated* if ${}^\bullet x \cup x^\bullet = \emptyset$.

Remark. We consider only nets whose transitions have a non-empty preset, i.e. such that ${}^\circ N \subseteq S$.

$$\frac{m\,[\rangle\, m' \in T \quad m'' \in \mathcal{M}_S}{m \oplus m'' \rightarrow_T m' \oplus m''}\ (\text{FIRING}) \qquad \frac{m_1 \rightarrow_T m_1' \quad m_2 \rightarrow_T m_2'}{m_1 \oplus m_2 \rightarrow_T m_1' \oplus m_2'}\ (\text{STEP})$$

Fig. 2. Operational semantics of P/T nets.

Note that in a net, the target and the source of a transition is a set of states, and thus transitions can consume and produce at most one token in each state. More generally in P/T nets, a transition can fetch and produce several tokens in a particular place, i.e., the pre and postset of a transition are multisets.

Definition 2 (Multiset). *Given a set S, a* multiset *over S is a function $m : S \rightarrow \mathbb{N}$. Let $dom(m) = \{a \in S \mid m(a) > 0\}$. The set of all finite multisets (i.e., with finite domain) over S is written $\mathcal{M}_S$. The empty multisct (i.c., with $dom(m) = \emptyset$) is written $\emptyset$. The multiset union $\oplus$ is defined as $(m_1 \oplus m_2)(a) = m_1(a) + m_2(a)$.*

Note that $\oplus$ is associative and commutative, and $\emptyset$ is the identity for $\oplus$. Hence, $\mathcal{M}_s$ is the free commutative monoid $S^{\oplus}$ over S. We write a for a singleton multiset m such that $dom(a) = \{a\}$ and $m(a) = 1$.

Definition 3 (p/t net). *A* marked place / transition Petri net *(P/T net) is a tuple $N = (S_N, T_N, \delta_{0N}, \delta_{1N}, m_{0N})$ where $S_N \subseteq \mathcal{P}$ is a set of places, T_N is a set of transitions, the functions $\delta_{0N}, \delta_{1N} : T_N \rightarrow \mathcal{M}_{S_N}$ assign respectively, source and target to each transition, and $m_{0N} \in \mathcal{M}_{S_N}$ is the initial marking.*

The notions of pre and postset, initial and final elements, and isolated places are straightforwardly extended to consider multisets instead of sets.

The operational semantics of P/T nets is given by the inference rules in Figure 2. Given a net N, the proof $m \rightarrow_T m'$ means that a marking m evolves to m' under a *step*, i.e., the concurrent firing of several transitions. Rule FIRING describes the evolution of the state of a net (represented by the marking $m \oplus m''$) by applying a transition $m[\rangle m'$, which consumes the tokens m corresponding to its preset and produces the tokens m' corresponding to its postset. The multiset m'' represents idle resources, i.e. the tokens that persist during the evolution. Rule STEP stands for the parallel composition of computations, meaning that several transitions can be applied in parallel as far as there are enough tokens to fire all of them. We omit the subscript T whenever it is clear from the context. The sequential composition of computations is indicated $\rightarrow^*$, i.e. $m \rightarrow^* m'$ denotes the evolution of m to m' under a (possibly empty) sequence of steps.

Example 1 (A simple P/T net). Let N be a P/T net s.t. $S = \{\mathtt{a}, \mathtt{b}, \mathtt{c}, \mathtt{d}\}$, $T = \{\mathtt{t}_1, \mathtt{t}_2\}$, $\delta_0(\mathtt{t}_1) = \delta_0(\mathtt{t}_2) = \{\mathtt{a}, \mathtt{b}\}$, $\delta_1(\mathtt{t}_1) = \{\mathtt{c}\}$, $\delta_1(\mathtt{t}_2) = \{\mathtt{d}\}$, $m_0 = \{\mathtt{a}, \mathtt{b}\}$. Figure 3(a) shows the graphical representation of N. As usual, places are represented with circles, transitions with boxes, tokens with dots, and the pre and postset functions are represented with arcs. Figure 3(b) shows a possible computation in N for the initial marking $\{\mathtt{a}, \mathtt{a}, \mathtt{b}, \mathtt{b}\}$, which corresponds to the concurrent firing of $\mathtt{t}_1$ and $\mathtt{t}_2$.

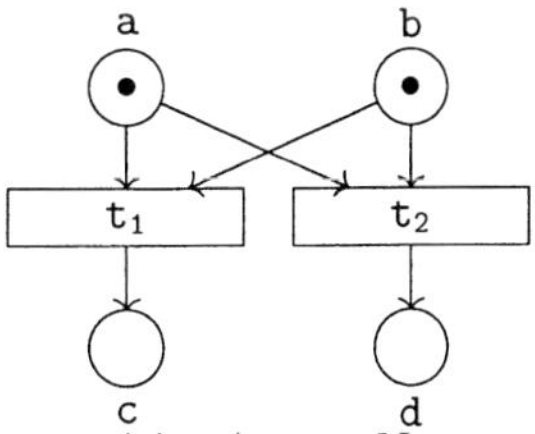

(a) P/T net N.

$$\dfrac{\dfrac{\mathtt{t}_1 = \mathtt{a} \oplus \mathtt{b}[\rangle \mathtt{c} \in T}{\mathtt{a} \oplus \mathtt{b} \to_T \mathtt{c}}\,\text{FIRING} \qquad \dfrac{\mathtt{t}_2 = \mathtt{a} \oplus \mathtt{b}[\rangle \mathtt{d} \in T}{\mathtt{a} \oplus \mathtt{b} \to_T \mathtt{d}}\,\text{FIRING}}{\mathtt{a} \oplus \mathtt{a} \oplus \mathtt{b} \oplus \mathtt{b} \to_T \mathtt{c} \oplus \mathtt{d}}\,\text{STEP}$$

(b) A computation in N for $\mathtt{a} \oplus \mathtt{a} \oplus \mathtt{b} \oplus \mathtt{b}$

Fig. 3. A simple P/T net.

3 Zero-Safe Nets

In this section we recall the basics of the zero-safe approach by following the presentation given in [11]. Zero-safe nets are an extension of Petri nets suitable to express transactions. Differently from P/T nets, the places of zero-safe nets are partitioned into ordinary and transactional ones (called *stable* and *zero*, respectively). Accordingly to the ordinary terminology, in a '0-safe' net all places cannot contain any token in all reachable markings. *Zero-safe net* – note the word 'zero' instead of the digit '0' – is used to denote that the net contains zero places that cannot contain any token in any observable marking. The role of zero places is to coordinate the atomic execution of complex collections of transitions.

Definition 4 (ZS net). *A* Zero-Safe net *(ZS net) is a 6-tuple $B = (S_B, T_B, \delta_{0B}, \delta_{1B}, m_{0B}, Z_B)$ where $N_B = (S_B, T_B, \delta_{0B}, \delta_{1B}, m_{0B})$ is the underlying P/T net and the set $Z_B \subseteq S_B$ is the set of* zero places. *The places in $S_B \backslash Z_B$ (denoted by L_B) are called* stable places. *A stable marking m is a multiset of stable places (i.e., $m \in \mathcal{M}_{L_B}$), and the initial marking m_{0B} must be stable.*

Note that markings $m \in \mathcal{M}_{S_B}$ can be seen as pairs (s, z) with $m = s \oplus z$, where $s \in \mathcal{M}_{L_B}$ is a stable marking and $z \in \mathcal{M}_{Z_B}$ is the multisets of zero resources, because $\mathcal{M}_{S_B} \simeq \mathcal{M}_{L_B} \times \mathcal{M}_{Z_B}$. Transitions are written $m[\rangle m'$, with m and m' multisets of stable and zero places. A transaction goes from a stable marking to another stable marking. The key point is that stable tokens produced during a transaction are made available only at commit time, when no zero tokens are left. As usual, we omit subscripts when referring to components of a ZS net if they are clear from the context.

The operational semantics of ZS nets is defined by the two relations $\Rightarrow_T$ and $\to_T$ in Figure 4. Rules FIRING and STEP are the ordinary ones for Petri nets, for the execution of one/many transition(s). However, sequences of steps differ from the ordinary transitive closure of $\to_T$: The rule CONCATENATION composes zero tokens in series but stable tokens in parallel, hence stable tokens produced by the first step cannot be consumed by the second step. Transactions are step sequences from stable markings to stable markings, when CLOSE can be applied. The moves $(s, \emptyset) \Rightarrow_T (s', \emptyset)$ define all the atomic activities of the net, and hence they can be performed in parallel and sequentially as the transitions

(FIRING)
$$\frac{s \oplus z[\rangle s' \oplus z' \in T \quad s'' \in \mathcal{M}_{L_B} \quad z'' \in \mathcal{M}_{Z_B}}{(s \oplus s'', z \oplus z'') \rightarrow_T (s' \oplus s'', z' \oplus z'')}$$

(STEP)
$$\frac{(s_1, z_1) \rightarrow_T (s'_1, z'_1) \quad (s_2, z_2) \rightarrow_T (s'_2, z'_2)}{(s_1 \oplus s_2, z_1 \oplus z_2) \rightarrow_T (s'_1 \oplus s'_2, z'_1 \oplus z'_2)}$$

(CONCATENATION)
$$\frac{(s_1, z) \rightarrow_T (s'_1, z'') \quad (s_2, z'') \rightarrow_T (s'_2, z')}{(s_1 \oplus s_2, z) \rightarrow_T (s'_1 \oplus s'_2, z')}$$

(CLOSE)
$$\frac{(s, \emptyset) \rightarrow_T (s', \emptyset)}{(s, \emptyset) \Rightarrow_T (s', \emptyset)}$$

Fig. 4. Operational semantics of ZS nets.

of an ordinary net. It is worth noting that a step $(s, \emptyset) \Rightarrow_T (s', \emptyset)$ can be itself the parallel composition of several concurrent transactions (by rule STEP).

One of the main advantages of the zero safe approach is that it prevents combinatorial explosion at the specification level. In fact, atomic activities can be defined in terms of several subactivities, which keeps the description of the system small, tractable and modular.

Example 2 (The free choice problem). Suppose the net introduced in Example 1 (see Figure 3(a)) to code the assignment of two resources a and b either to the activity c or d. By firing t_1 the resources are assigned to c, and by t_2 to d. The nondeterministic choice encoded by the net corresponds to a centralized coordination mechanism that guarantees that both resources are assigned atomically to the same activity. Nevertheless, if one wants to model the system using a *free choice* net, where all decisions are made locally (i.e., by looking just one place) the situation is different. Consider the free choice net shown in Figure 5. It models the system with two independent decisions: one for the assignment of a, the other for the assignment of b.

Nevertheless, the free choice net admits computations not allowed in the abstract system in Figure 3(a). In fact, the free choice net has deadlocks: consider the firing of $\mathtt{assign}_{\mathsf{a,c}}$ and $\mathtt{assign}_{\mathsf{b,d}}$. In this case, the net cannot evolve to either b or c, which is a computation not possible in the original net.

ZS nets can be used to overcome this problem, by defining intermediate places as zero places. The assignment problem can be modelled as the ZS net in Figure 6, where smaller circles stand for zero places. The ZS net avoids deadlocks because computations ending in markings containing zero tokens are recoverable and not observable.

Example 3 (Mobile lessees). The general problem that we want to model consist of a set of apartments that can be rented immediately, a group of people looking for an apartment, and people that want to change their apartments, i.e., they are willing to move to another apartment if somebody else can rent their actual apartments. Consider an instance of the problem with three apartments A, B, C and four people P, Q, R, S. The initial state can be represented as in Figure 7(a), where the apartment A is available for rent, P and S are searching

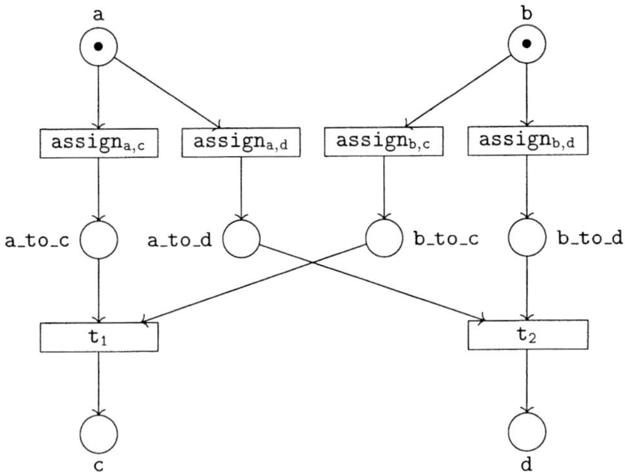

Fig. 5. Free choice net for the assignment problem.

for an apartment, Q wants to leave B, and R wants to leave C. Figure 7(b) shows the preferences of people on apartments.

The formulation of the problem as a ZS net is shown in Figure 8. Note that there is a place for any apartment available for immediate rent in the initial state (`A_free`), a place for any person looking for an apartment (`S_wants` and `P_wants`), and a place for any person willing to change apartment (`Q_changes_B` and `R_changes_C`). There is also a place for any possible rent (i.e., accordingly to the preference matrix in 7(b)). For instance, the transition `S_takes_A` states that person S can rent the apartment A whenever A is free and S is searching for an apartment, and a token in `S_moves_A` means that the person S has rented the apartment A. The more interesting transitions are `Q_leaves_B` and `R_leaves_C`, each of them starts a transaction. In fact, they describe the activity of changing an apartment as the orchestration of two different activities, one in which a person finds a new apartment, and other in which the apartment is rented. For instance, when `Q_leaves_B` is fired a token is produced in `Q_search`, and another in `B_avail`. Note this transaction can finish only when both tokens are consumed, meaning that both Q has rented a new apartment and B has been rented. The initial marking denotes the initial state of the problem.

In Figure 9 we show a proof for a transaction in which Q leaves B and takes A, R leaves C and takes B; and P takes B, while S remains without apartment. For space reason, we abbreviate the name of places (i.e. $\mathtt{A_f}$ for `A_free`, $\mathtt{QB_c}$ for `Q_changes_B`, and similarly for the rest). Moreover, we write stable places with capital letters, while zero places are written with lower case. The computation corresponds to the parallel begin of two transactions (where `Q_changes_B` and `R_changes_C` are decomposed into two subactivities) followed by the parallel execution of `Q_takes_A`, `P_takes_C` and `R_takes_B`

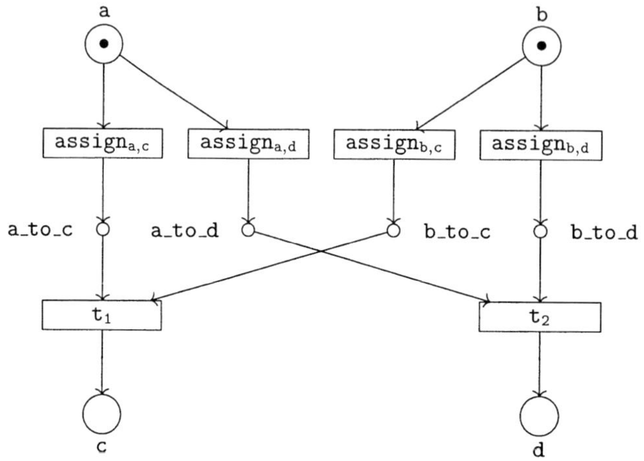

Fig. 6. Free choice net for the assignment problem.

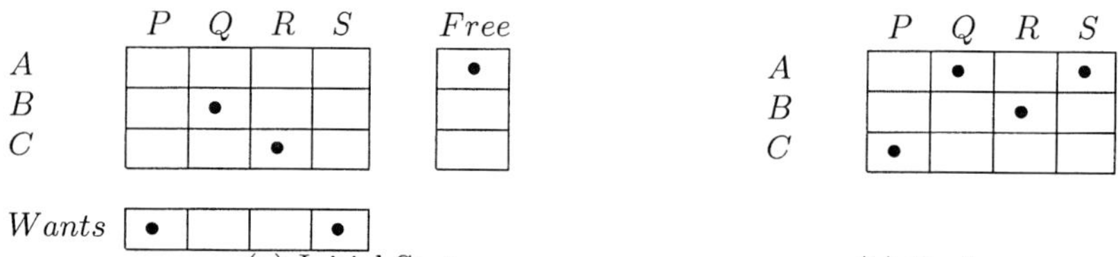

(a) Initial State. (b) Preferences.

Fig. 7. An instance of the mobile lessees.

3.1 Abstract Semantics

As stated by the operational semantics of ZS nets (Figure 4), the observable states of a system are those represented by stable markings, while the meaningful computations (i.e., the atomic activities) are the *stable steps* of the net, i.e. the steps consuming and producing stable markings (relation $\Rightarrow$). Since stable steps can be composed in sequence and parallel, a stable step can be thought of as the execution of several *basic* transactions, i.e., stable steps that cannot be decomposed into other stable steps. Consequently, all the correct behaviours of the system can be derived from the set of basic transactions of the net. The abstract semantics of ZS net is intended to capture the behaviour of a ZS net in terms of its basic transactions.

In this context, a transaction denotes an activity of the system that might be composed by many, possibly concurrent, coordinated subactivities. Since the concurrent semantics of an operational model is usually defined by considering as equivalent all the computations where the same concurrent events are executed in different orders, it follows that we should quotient out those transactions which are equivalent from a concurrent point of view, in such a way that the

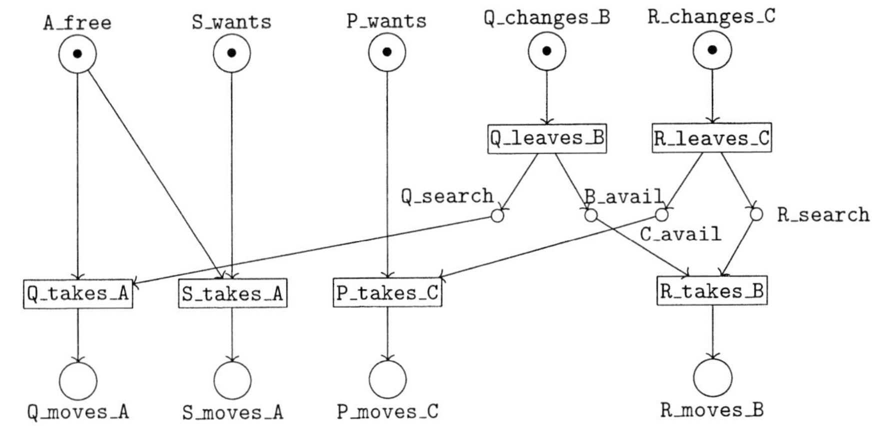

Fig. 8. ZS net of the mobile lessees example.

A:

$$\dfrac{\dfrac{\dfrac{A_f \oplus q_s \,[\rangle\, QA_m \in T}{(A_f, q_s) \rightarrow_T (QA_m, \emptyset)}\,(F) \quad \dfrac{b_a \oplus r_s \,[\rangle\, RB_m \in T}{(\emptyset, b_a \oplus r_s) \rightarrow_T (RB_m, \emptyset)}\,(F)}{(A_f, q_s \oplus b_a \oplus r_s) \rightarrow_T (QA_m \oplus RB_m, \emptyset)}\,(S) \quad \dfrac{P_w \oplus c_a \,[\rangle\, PC_m \in T}{(S_w \oplus P_w, c_a) \rightarrow_T (S_w \oplus PC_m, \emptyset)}\,(F)}{(S_w \oplus P_w \oplus A_f, q_s \oplus r_s \oplus b_a \oplus c_a) \rightarrow_T (S_w \oplus QA_m \oplus RB_m \oplus PC_m, \emptyset)}\,(S) \quad (A)$$

$$\dfrac{\dfrac{\dfrac{\dfrac{QB_c \,[\rangle\, q_s \oplus b_a \in T}{(QB_c, \emptyset) \rightarrow_T (\emptyset, q_s \oplus b_a)}\,(F) \quad \dfrac{RC_c \,[\rangle\, r_s \oplus c_a \in T}{(RC_c, \emptyset) \rightarrow_T (\emptyset, r_s \oplus c_a)}\,(F)}{(QB_c \oplus RC_c, \emptyset) \rightarrow_T (\emptyset, q_s \oplus r_s \oplus b_a \oplus c_a)}\,(S) \quad A}{(S_w \oplus P_w \oplus A_f \oplus QB_c \oplus RC_c, \emptyset) \rightarrow_T (S_w \oplus QA_m \oplus RB_m \oplus PC_m, \emptyset)}\,(\text{CONCAT})}{(S_w \oplus P_w \oplus A_f \oplus QB_c \oplus RC_c, \emptyset) \Rightarrow_T (S_w \oplus QA_m \oplus RB_m \oplus PC_m, \emptyset)}\,(\text{CLOSE})$$

Fig. 9. A proof for the execution of a transaction in the mobile lessees ZS net.

actual order of execution of concurrent transitions in the ZS net is invisible in the abstract net.

In order to identify the equivalent executions from a concurrent point of view there are two main approaches: the *collective token philosophy* (CTph) and the *individual token philosophy* (ITph). The net semantics under the CTph does not distinguish among different instances of the idealized resources (i.e., tokens). This is a valid interpretation of the behaviour of a system only when any such instance is *operationally* equivalent to all the others. Nevertheless, tokens may have different origins and histories, thus the *causality* information carried on by different tokens is disregarded when identifying equivalent computations w.r.t. CTph, which turns to be the main drawback of this approach. Alternatively, the ITph takes into account the causal dependencies arising in concurrent executions.

The abstract semantics of zero-safe nets has been largely studied under both philosophies. In particular, in both cases the abstract net is characterized by an adjunction on suitable categories [11]. We summarise here the basics of the abstract semantics under the ITph, which is the most significant one. In particular, the distinction between tokens with different origins and history relies on the notion of a deterministic process. A deterministic process denotes a particular computation in a net, and therefore it explicitly carries the causal information between firings. To define processes we need two other concepts: net morphisms and causal nets.

Definition 5 (Net morphism). *Let N, N' be* P/T *nets. A pair $f = (f_S : S_N \to S_{N'}, f_T : T_N \to T_{N'})$ is a* net morphism *from N to N' (written $f : N \to N'$) if $f_S(\delta_{iN}(t)) = \delta_{iN'}(f_T(t))$.*

We usually omit subscripts when they are clear from the context. With abuse of notation we apply functions (i.e., f_S) over (multi)sets, meaning the multiset obtained by applying the function element-wise: $f_S(\{m_0, \ldots, m_n\}) = f_S(m_0) \oplus \ldots \oplus f_S(m_n)$.

Definition 6 (Causal Net and Process). *A net $K = (S_K, T_K, \delta_{0K}, \delta_{1K})$ is a* causal net *(also called* deterministic occurrence net*) if it is* acyclic *and $\forall t_0 \neq t_1 \in T_K, \delta_{iN}(t_0) \cap \delta_{iN}(t_1) = \emptyset$, for $i = 0, 1$.*

A (Goltz-Reisig) process *for a* P/T *net N is a net morphism P from a causal net K to N.*

Two processes P and P' of N are *isomorphic* and thus equivalent if there exists a net isomorphism $\psi : K_P \to K_{P'}$ such that $\psi; P' = P$.

Given a process $P : K \to N$, the set of *origins* and *destinations* of P are defined as $O(P) = {}^\circ K$ and $D(P) = K^\circ \cap S_K$, respectively. We write $pre(P)$ and $post(P)$ for the multisets denoting the initial and final markings of the process, i.e. $pre(C) = P(O(P))$ and $post(C) = P(D(P))$. Moreover, as isomorphisms respect initial and final markings, we say that $O(\xi) = pre(P)$, $D(\xi) = post(P)$, for $\xi = [\![P]\!]_\approx$. Finally, the set of *evolution places* of a process P is the set $E_P = \{P(a) | a \in K, |{}^\bullet a| = |a^\bullet| = 1\}$.

Definition 7 (Connected transaction). *Given a* ZS *net B, let P be a process of the underlying* P/T *net N_B. The equivalence class $\xi = [\![P]\!]_\approx$ is a* connected transaction *of B if:*

- *$pre(P)$ and $post(P)$ are stable markings, i.e., the process starts by consuming stable tokens and produces only stable tokens;*
- *$E_P \subseteq Z_B$, i.e. stable tokens produced during the transaction cannot be consumed during in the same transaction;*
- *P is connected, i.e. the set of transitions T_K is non-empty, and for all $t_0, t_1 \in T_K$ there exists an undirected path connecting t_0 and t_1; and*
- *P is full, i.e., it does not contain idle (i.e., isolated) places (i.e., $\forall a \in S_K, |{}^\bullet a| + |a^\bullet| \geq 1$).*

We denote by Ξ_B (ranged by ς) the set of connected transaction of B.

A connected transaction can be executed when the state of the net contains enough stable tokens to enable all the transitions independently. At the end of

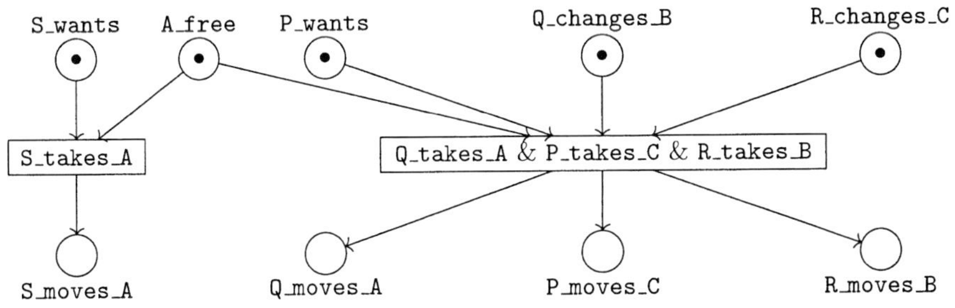

Fig. 10. Abstract net of the mobile lessees example.

its execution no token may be left on zero places (nor may be found on them at the beginning of the step). This means that all the zero tokens produced by a transaction are also consumed by the same transaction. Moreover, in a connected transaction no intermediate marking is stable.

Definition 8 (Causal abstract net). *Let* $B = (S_B, T_B, \delta_{0B}, \delta_{1B}, m_{0B}, Z_B)$. *The net* $I_B = (S_B \backslash Z_B, \Xi_B, \delta_{0I}, \delta_{1I}, m_{0B})$, *with* $\delta_{0I}(\varsigma) = pre(\varsigma)$ *and* $\delta_{0I}(\varsigma) = post(\varsigma)$, *is the* causal abstract net *of* B *(we recall that* $pre(\varsigma)$ *and* $post(\varsigma)$ *denote the multisets* $P_\varsigma(O(\varsigma))$ *and* $P_\varsigma(D(\varsigma))$, *respectively, and that* Ξ_B *is the set of all the connected transactions of* B*).*

Example 4 (Abstract Net for the Mobile lessees problem). Figure 10 shows the abstract net corresponding to the ZS net in Figure 8. In the abstract net there are only two transitions, each of them representing an abstract transaction of the ZS net: `S_takes_A`, corresponding to the homonymous transition in the ZS net; `Q_takes_A` & `P_takes_C` & `R_takes_B`, for the atomic negotiation in which Q leaves B and takes A, R leaves C and takes B; and P takes B. These two transitions are enough to model the abstract behaviour of the system. In fact, any other combination is not possible because it would imply that some exchanged apartment (i.e., B or C) remains available for rent or a person willing to change apartment (Q or R) remains without apartment, which is an inconsistent state (with pending negotiations).

Note that one of the main advantages of the approach is that it allows to fully specify the behaviour of a system without analyzing all possible global combinations. Consider an instance of the lessees problem with a larger number of apartments and people, and a more complicated set of preferences. It could be tedious to figure out which are all the possible combinations that correspond to consistent transformations in the system. Moreover, it is possible to describe an infinite abstract net with a finite ZS net, as the multicasting system presented in [11] or the generalized version of the mobile lessees problem analyzed in Section 7.

The correspondence between the concrete and the abstract view is stated by the following theorem [12].

Theorem 1. *Let* B *be a* ZS *net and* I_B *its abstract net. Then* $m \rightarrow_{T_{I_B}} m'$ *iff* $m \Rightarrow_{T_B} m'$.

4 Adding Colours to zs Nets

4.1 Coloured p/t Nets

In coloured P/T nets [24] (known also as *high-level* nets), tokens carry on information, which is given by their *colours*. Actually, colours are values/data associated with a particular instance of a resource. Hence, the state and transitions of a net exploit also the information present in tokens, i.e. their colours.

We consider an infinite set of constant colour names $\mathcal{B}$, ranged over by $x, x_1, \ldots$ and an infinite set of colour variables $\mathcal{V}$, ranged over by $v, w, \ldots$. We denote the set of constants and variables with $\mathcal{C} = \mathcal{B} \cup \mathcal{V}$. Moreover, we require constant and variable colours to be disjoint ($\mathcal{B} \cap \mathcal{V} = \emptyset$) and different from place names, i.e. $\mathcal{C} \cap \mathcal{P} = \emptyset$. Let S be a set, S^* stands for the set of all finite (possible empty) sequences on S, i.e. $S^* = \{(s_1, \ldots, s_n) | n \geq 0 \wedge s_i \in S\}$. The empty sequence is denoted by $\bullet$, and the underlying set of a sequence $(s_1, \ldots, s_n)$ by:

$$\overline{(s_1, \ldots, s_n)} = \bigcup_i \{s_1\}$$

Definition 9 (Coloured net). *A coloured net N is a 5-tuple $N = (S_N, C_N, T_N, \delta_{0N}, \delta_{1N})$ where $S_N \subseteq \mathcal{P}$ is the (nonempty) set of places, $C_N \subseteq \mathcal{C}$ is the set of colours, T_N is the set of transitions (with $S_N \cap T_N = \emptyset$), and the functions $\delta_{0N}, \delta_{1N} : T_N \to \wp_f(S_N \times C_N^*)$ assign respectively, source and target to each transition. To assure that a transition fetches and produces at most one token in a place we require that $\forall t \in T_N, if (s, c_1), (s, c_2) \in \delta_{iN}(t)$ then $c_1 = c_2$, for $i = 0, 1$.*

The pre and postset of a transition are defined similarly to Section 2, but taking into account that they are coloured sets instead of sets. Analogously, for any place a in S_N, the preset of a (written ${}^\bullet a$) denotes the set of all transitions with target in a (i.e., ${}^\bullet a = \{t | (a, c) \in t^\bullet\}$, and the postset of a (written $a^\bullet$) denotes the set of all transitions with source in a (i.e., $a^\bullet = \{t | (a, c) \in {}^\bullet t\}$. The definitions for the sets of *initial* and *final elements*, and *isolated* place are identical to those given in Section 2.

Note that in coloured nets a transition $m_1 [\rangle m_2$ denotes a pattern that should be matched/instantiated with appropriated colours in order to be applied. In particular, constant colours appearing in m_1 act as values that should be matched in order to fire the transition, while variables should be instantiated with appropriate colours. Variables are binders of colours occurring in m_2. For instance, the transition $t = (a_1, v), (a_2, v), (a_3, x_1) [\rangle (a_1, v), (a_4, x_2)$ denotes a pattern stating that whenever a_1 and a_2 contain tokens with the same colour (but they can be of any constant colour because of the variable v) and a_3 contains a token with constant colour x_1, the transition can be fired. When t is fired, the tokens matching the preset are consumed, and a new token is put in a_1, whose colour corresponds to the consumed tokens in a_1 and a_2, and a token with colour x_2 is produced in a_4. Consequently, the firing of t over $m = (a_1, x_3), (a_2, x_3), (a_3, x_1)$

will produce $m' = (a_1, x_3), (a_4, x_2)$. From a functional point of view, colour variables occurring in the preset of a transition act as its parameters, which are called *received colours.*

Definition 10 (Received colours of a transition). *The* colour of a set $s \subseteq \mathcal{S} \times \mathcal{C}^*$ *is defined as* $col(s) = \cup_{(a,c) \in s} \overline{c}$, *the* set of constants *is* $col_{\mathcal{B}}(s) = col(s) \cap \mathcal{B}$, *and the* set of variables $col_{\mathcal{V}}(s) = col(s) \cap \mathcal{V}$. *Given a transition* $t = m[\rangle m'$, *the set of* received colours *(also* received names*) of* t *is given by* $rn(t) = col_{\mathcal{V}}(m)$.

Remark. As variables are used to describe parameters in a transition, we will consider only coloured nets in which each transition $t = m[\rangle m'$ satisfies $col_{\mathcal{V}}(m') \subseteq rn(t)$. This restriction states that all variables occurring in the postset of a transition are bound to some variable in the preset.

Clearly, previous definitions can be straightforwardly extended to consider coloured multisets instead of sets.

Definition 11 (Coloured Multiset). *Given two sets* S *and* C, *a* coloured multiset *over* S *and* C *is a function* $m : S \to C \to \mathbb{N}$. *Let* $dom(m) = \{(s, c) \in S \times C \mid m(s)(c) > 0\}$. *The set of all finite multisets over* S *and* C^* *is written* $\mathcal{M}_{S,C}$. *The multiset union is defined as* $(m_1 \oplus m_2)(s)(c) = m_1(s)(c) + m_2(s)(c)$. *We write* $s(c)$ *for a multiset* m *such that* $dom(m) = \{(s, c)\}$ *and* $m(s)(c) = 1$. *Additionally,* $(s, c) \in m$ *is a shorthand for* $(s, c) \in dom(m)$, *while* $s \in m$ *means* $(s, c) \in m$ *for some* c.

Definition 12 (c-p/t net). *A* coloured marked place / transition net *(*C-P/T *net) is a 6-tuple* $N = (S_N, C_N, T_N, \delta_{0N}, \delta_{1N}, m_{0N})$ *where* $S_N \subseteq \mathcal{P}$ *is the set of places,* $C_N \in \mathcal{C}$ *is the set of colours,* T_N *is a set of transitions, the functions* $\delta_{0N}, \delta_{1N} : T \to \mathcal{M}_{S_N, C_N}$ *assign respectively, source and target to each transition, and* $m_{0N} \in \mathcal{M}_{S_N, C_N}$ *is the initial marking. Moreover,* $\forall t \in T_N, col_{\mathcal{V}}(t^\bullet) \subseteq rn(t)$ *(i.e., variables in the postset are bound to received names), and* $col(m_{0N}) \subseteq \mathcal{B}$ *(i.e., tokens in the initial marking are coloured with constants).*

As aforementioned, the firing of a transition t in a coloured net requires to instantiate t with appropriate colours, i.e., those corresponding to tokens present in places. Consequently, the instantiation of a transition corresponds to a substitution on colour variables.

Definition 13 (Substitution on colours). *Let* $\sigma : \mathcal{V} \to \mathcal{B} \cup \mathcal{V}$ *be a partial function. The* substitution $v\sigma$ on a colour variable v *is* c *if* $\sigma(v) = c$, *otherwise it is* v, *i.e., it is the identity when* σ *it is not defined. Instead, the* substitution $x\sigma$ on a constant colour x *produces* x, *i.e., it has no effect. The* substitution on a colour sequence *is the simultaneous substitution on the names appearing in the sequence, i.e.,* $(c_1, \ldots, c_n)\sigma = (c_1\sigma, \ldots, c_n\sigma)$. *The* colour substitution on a multiset $m \in \mathcal{M}_{S,C}$ *is given by* $(m \star \sigma)(s)(c) = \sum_{d\sigma = c} m(s)(d)$.

The operational semantics of coloured nets is given by replacing the rule FIRING in Figure 2 by the following version:

(COLOURED-FIRING)

$$\frac{t = m\,[\rangle\, m' \in T \quad m'' \in \mathcal{M}_{S,C}}{m \star \sigma \oplus m'' \rightarrow_T m' \star \sigma \oplus m''} \quad dom(\sigma) = rn(t) \text{ and } \sigma(v) \in \mathcal{B} \text{ for } v \in dom(\sigma)$$

Remark: α-equivalence on defined names. Note that the variables chosen to denote colours in the preset of a transition are meaningless. Actually, they act as binders whose scope is just *that* transition, and consequently the can be changed without modifying the meaning of a transition. Therefore we define the following relation over transitions, called α-conversion on received colours.

Two transitions $t_1 = m_1[\rangle m_1'$ and $t_2 = m_2[\rangle m_2'$ are α-convertible if there exists an injective substitution $\sigma : \mathcal{V} \rightarrow \mathcal{V}$, where $dom(\sigma) \subset rn(t_1)$, such that $m_1 \star \sigma = m_2$ and $m_1' \star \sigma = m_2'$. The α-conversion is an equivalence relation, which is denoted by $\equiv_\alpha$. We usually talk about transitions up-to α-equivalence.

4.2 Coloured zs Nets

The zs version of a coloured net is obtained also by distinguishing stable places from zero ones.

Definition 14 (c-zs net). *A* coloured zs net *(c-zs net for short) is a 7-tuple* $B = (S_B, C_B, T_B, \delta_{0B}, \delta_{1B}, m_{0B}, Z_{0B})$ *where* $N_B = (S_B, C_B, T_B, \delta_{0B}, \delta_{1B}, m_{0B})$ *is the underlying c-p/t net and the set* $Z_B \subseteq S_B$ *is the set of* zero places. *The places in* $S_B \backslash Z_B$ *(denoted by* L_B*) are called* stable places. *A stable marking m is a coloured multiset of stable places (i.e.,* $m \in \mathcal{M}_{L_B,C_B}$*), and the initial marking* m_{0B} *must be stable and satisfy* $col(m_{0B}) \subseteq \mathcal{B}$.

The operational semantics of c-zs nets is a straightforward extension of rules given in Figure 4, where the firing rule is replaced by the following version:

(COLOURED-FIRING)

$$\frac{t = s \oplus z\,[\rangle\, s' \oplus z' \in T \quad s'' \in \mathcal{M}_{L,C} \quad z'' \in \mathcal{M}_{Z,C}}{(s \star \sigma \oplus s'', z \star \sigma \oplus z'') \rightarrow_T (s' \star \sigma \oplus s'', z' \star \sigma \oplus z'')} \quad \begin{array}{l} dom(\sigma) = rn(t), \text{ and} \\ \sigma(v) \in \mathcal{B} \text{ for } v \in dom(\sigma) \end{array}$$

We still write a marking $m = s \oplus z$ as a pair (s, z) to denote that $s \in \mathcal{M}_{L,C}$ and $z \in \mathcal{M}_{Z,C}$.

Example 5 (c-zs net for the mobile lessees problem). A more general representation for the mobile lessees problem introduced in the Example 3 can be given in terms of c-zs nets. Consider the net in Figure 11, where label on arcs corresponds to the colours of the pre and postset of a transition. Tokens present in the place `free` represent apartments that are available for being rented immediately. The actual identity of the apartment is given by the colour of the token. Similarly people looking for an apartment are represented as coloured tokens in `wants`, and those willing to change apartment as tokens (w', v') in `changes`,

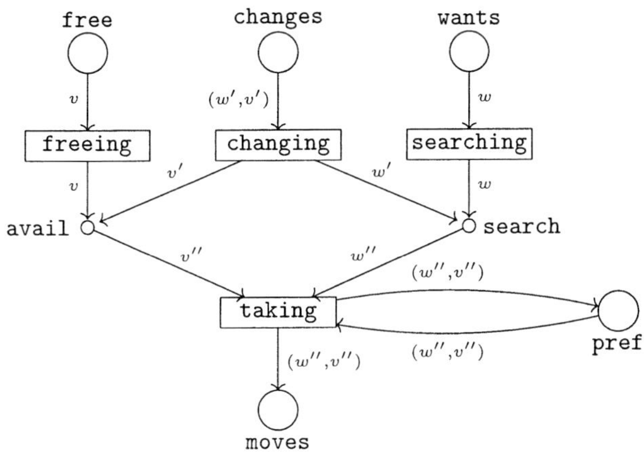

Fig. 11. Coloured ZS net of the mobile lessees example.

meaning that the person w' changes the apartment v'. The transition `freeing` initiates a transaction by making available for rent an offered apartment. Analogously, `searching` initiate a transaction in which a person is looking for an apartment. Transition `changing` starts a transaction by rendering available the offered apartment and putting a token in the place of persons looking for an apartment. Finally, the transition `taking` states that a person w'' searching for an apartment can take the available apartment v'' if she likes it (i.e., a token with colour (w'', v'') is in the set of preferences). A token with colour (w'', v'') produced in the place `moves` means that the person w'' has moved to the apartment v''. It is worth noting that tokens are actually produced on place `moves` when no token is left in the zero places (i.e., `avail` or `search`).

While in the ZS net different instances of the problem (i.e., different set of apartments, people or preferences) correspond to different structures of the net (i.e., states, transitions and flow function), in the coloured version the structure is the same for every instance of the problem, the only thing that changes is the initial marking. In fact, all the information about a particular instance of the problem is represented by colours.

Contextual nets. The self-loops introduced to model the preference sets in the Example 5 can be better modelled as *read arcs.* Nets with read arcs allow for modelling "read without consume", where many readers can access concurrently the same resource. Consider transition `taking` in Figure 11. Actually, there is no need to consume the token in `pref`. To fire `taking` it is enough to check the presence of a token with suitable colours on `pref`. The extension of the ZS model to contextual nets have been studied in [13].

ZS *nets can be encoded as* C-P/T *nets.* ZS nets have been encoded in [7] into a fragment of the join calculus corresponding to the coloured nets. In such encoding (which is called flat) the transactional mechanism of ZS nets is implemented through a centralized coordinator, which is aware of the zero tokens present in the net. Roughly, stable tokens produced during a transaction are kept frozen by the coordinator, which will release them when no zero token is left in the net.

4.3 Abstract Net under the ITph Approach

In order to define the abstract net associated to a C-ZS net we revise the notion of causal net and processes in order to take into account colours.

Definition 15 (Coloured Causal Net). *A coloured net* $K = (S_K, C_K, T_K, \delta_{0K}, \delta_{1K})$ *is a* causal net *if it is* acyclic *and transitions do not share places in their pre and postsets, i.e. if* $a \in \delta_{iN}(t_0)$ *and* $a \in \delta_{iN}(t_1)$ *then* $t_0 = t_1$.

When viewing coloured causal nets as descriptions of runs, it should be clear that differently from causal nets, in the coloured version a causal net can be blocked because the bindings between the different colours are not consistent. Consider a simple causal net with the following transitions $t_0 = (a_0, \bullet)[\rangle(b_0, x_0)$, $t_1 = (a_1, \bullet)[\rangle(b_1, x_1)$, and $t_2 = (b_0, v), (b_1, v)[\rangle(a, \bullet)$. Starting with the marking $a_0(\bullet) \oplus a_1(\bullet)$, it cannot execute completely because t_0 produces a token with the constant colour x_0 on b_0 and t_1 a token with colour x_1 on b_1, but t_2 requires tokens in b_0 and b_1 with the same colour.

In general, causal nets can execute completely when the colours used to label transitions are sequences of variables of the same length.

Definition 16 (Plain nets). *A causal net* K *is a* plain net *if* $\exists k \in \mathbb{N}$ *s.t.* $\forall t \in T, (s, c) \in {}^\bullet t \cup t^\bullet : \overline{c} \subseteq \mathcal{V} \wedge |c| = k$, *with* $|c|$ *denoting the length of* c.

As for P/T net, we define a notion of morphism between C-P/T nets.

Definition 17 (Coloured net morphism). *Let* N, N' *be* C-P/T *nets. A tuple* $f = (f_S : S_N \rightarrow S_{N'}, f_T : T_N \rightarrow T_{N'}, \sigma = \{\sigma_t\}_{t \in T_N})$ *is said a* coloured net morphism *from* N *to* N' *(written* $f : N \rightarrow_\sigma N'$*) if* $f_S({}^\bullet t)[\rangle f_S(t^\bullet) = {}^\bullet f_T(t) \star \sigma_t[\rangle f_T(t)^\bullet \star \sigma_t$.

Note that a morphism explains also the correspondence between colours used by transitions. Each σ_t relates each colour appearing in t with a colour in $f_T(t)$. Moreover, transitions in N are required to be a particular case of those in N'.

Definition 18 (Process of a coloured net). A (Goltz-Reisig) process *for a* C-P/T *net* N *is coloured net morphism* $P : K \rightarrow_\sigma N$, *from a coloured causal net* K *to* N, *s.t.* $\forall \sigma_t \in \sigma$, σ_t *is injective and* $\sigma_t : \mathcal{V} \rightarrow \mathcal{V}$.

Two coloured processes P and P' are *isomorphic* and thus equivalent if there exists a net isomorphism $\psi : K_P \rightarrow_\sigma K_{P'}$ such that $\psi; P' = P$.

A process P associates a coloured causal net K to a C-P/T net N. As K is itself a coloured net, its transitions can be fired for any suitable substitution of colours. Therefore, a process describes several runs that start from initial markings with different colours. (Our approach is similar to that presented in [3]).

Nevertheless, this does not mean that a process can be instantiated for any possible combination of colours. In fact, a process stands for executions where just one token is produced and/or consumed from a particular place, and consequently the colours appearing in the preset and postset of a place must coincide. Consequently, a process implicitly defines a relation among colours in admissible markings. A compatible execution of a process is an instantiation of colours that satisfies such constraints.

For simplicity, we defined the notion of compatible execution for the equivalence class of P, i.e. $[\![P]\!]_{\approx}$.

Definition 19 (Compatible execution of $[\![P]\!]_{\approx}$). *Let $\varsigma \in [\![P]\!]_{\approx}$ such that $\forall t_1, t_2 \in T_{\varsigma}, rn(t_1) \cap rn(t_2) = \emptyset$, i.e., transitions do not share variables. A substitution σ is said a* compatible execution *of ς if $\forall a \in S_{\varsigma}, {}^{\bullet}a \star \sigma = a^{\bullet} \star \sigma$. If such σ exists, we say that $[\![P]\!]_{\approx}$ is compatible. A process P is* compatible *if there exists a compatible execution for $[\![P]\!]_{\approx}$.*

A compatible execution captures the notion of unification that takes place when computing in a coloured net.

Example 6. Two simple processes for the C-ZS net in Figure 11 are presented in Figure 12. The first one represents a person looking for an apartment that takes a free apartment, while the second shows the process in which two people exchange their apartments.

The first process can be used as representative of its equivalence class because its transitions do not share variables. The substitution $\sigma = \{v/v', w/w'\}$ is a compatible execution for the first process. Note that w/w' in σ captures the idea that the token consumed from the state `wants` refers to the same person of the token used from the preference set (similarly, v/v' relates the different tokens referring to the same apartment). Observe also that the substitution $\sigma' = \{v/x, v'/x, w/x, w'/x\}$ is a compatible execution for the same process, which requires all names to be equals to the constant x. Clearly, σ' imposes more restrictive constraints than σ. Moreover, σ' is a particular case of σ.

On the other hand, the second process cannot be taken as representative of its equivalence class because some transitions, such as `changing'` and `taking`, share variables. Nevertheless, a representative can be obtained by applying α-conversion on transitions.

Proposition 1. *Any process P of a plain net is compatible.*

As we are interested on capturing the most general definition for equivalent executions, we will associate particular cases to instantiations of more general ones. Consequently, we are interested on the less restrictive constraints on colours implied by a process, which is called the *most general compatible execution.*

Definition 20 (mgce). *A compatible execution σ is said the* most general compatible execution *(shorten as* mgce*) of ς, written σ_{ς}, if for every other compatible execution σ' there exist a substitution γ s.t. $\forall t \in T_{\varsigma}, (t \star \sigma_{\varsigma}) \star \gamma = t \star \sigma'$.*

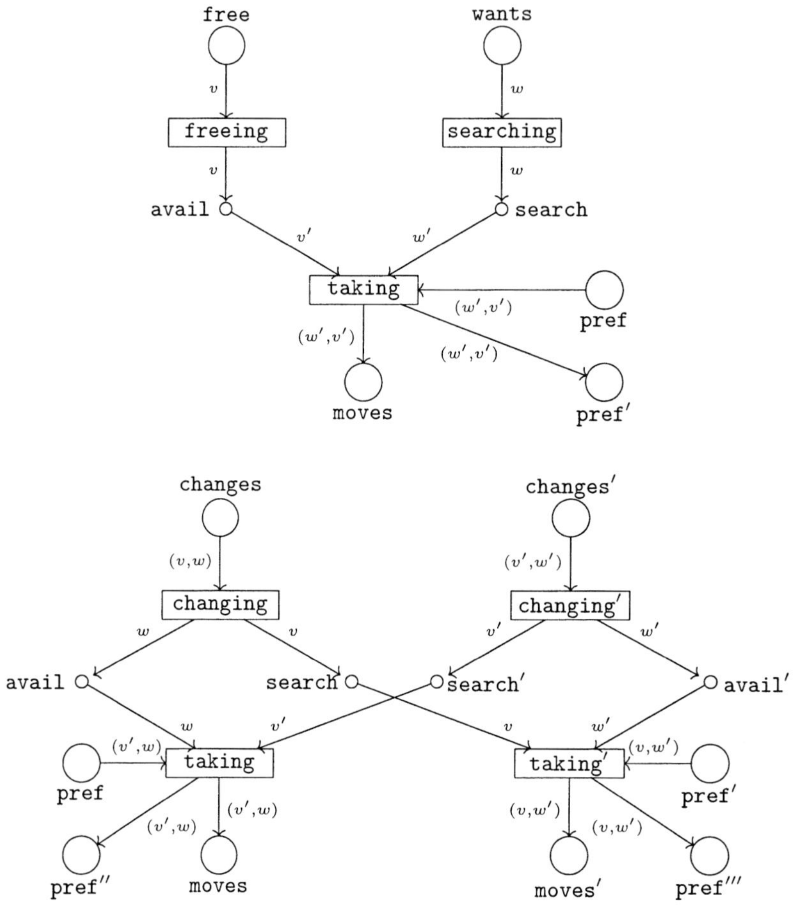

Fig. 12. Two coloured processes for the mobile lessees example.

We write by Ξ_B (ranged by ς) the set of connected transaction of B.

The definition for connected transactions is identical to Definition 7, but requiring processes to be compatible. Consequently, the definition of the abstract net is immediate, the only difference is that abstract transitions are defined in terms of the processes and their mgce.

Definition 21 (Causal abstract coloured net). *Let* $B = (S_B, C_B, T_B, \delta_{0B}, \delta_{1B}, m_{0B}, Z_B)$. *The net* $I_B = (S_B \backslash Z_B, C_B, \Xi_B, \delta_{0I}, \delta_{1I}, m_{0B})$, *with* $\delta_{0I}(\varsigma) = pre(\varsigma) \star \sigma_\varsigma$ *and* $\delta_{0I}(\varsigma) = post(\varsigma) \star \sigma_\varsigma$, *is the* causal abstract net *of* B *(we recall that* σ_ς *is the mgce for* ς*, that* $pre(\varsigma)$ *and* $post(\varsigma)$ *denote the multisets* $P_\varsigma(O(\varsigma))$ *and* $P_\varsigma(D(\varsigma))$*, respectively, and that* Ξ_B *is the set of all the compatible connected transactions of* B*).*

Example 7 (Abstract coloured net for the generalized mobile lessees problem). Figure 13 shows a partial view of the abstract net corresponding to the mobile

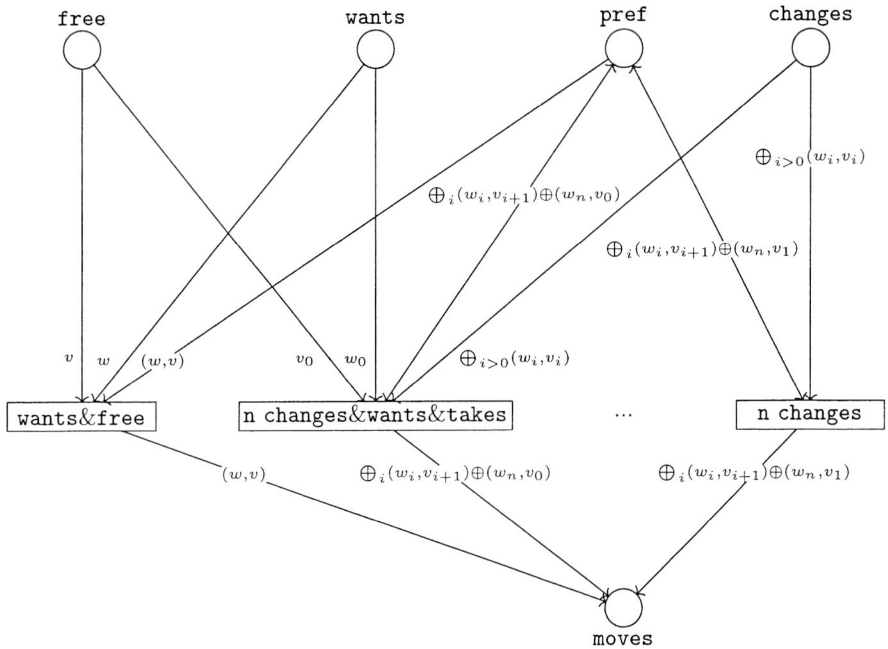

Fig. 13. Partial abstract net of the C-ZS net of the mobile lessees example.

lessees example. Transition `wants&free` corresponds to the atomic step in which a person who is searching for an apartment rents an available apartment. Transition `n changes` corresponds to the case in which n people interchange their apartments. Note there are infinite transitions of this kind, one for any $n \geq 2$. Similarly, the transition `n changes&wants&free` stands for the atomic step in which n people change their apartments, but one of them takes a free apartment and one person looking for an apartment participates in the exchange. Observe that this infinite abstract net is modelled with a finite concrete ZS net.

Finally, the correspondence between the two different views provided by the concrete ZS net and the abstract net is guaranteed by the following result.

Theorem 2. *Let B be a* C-ZS *net and I_B its abstract net. Then $m \rightarrow_{T_{I_B}} m'$ iff $m \Rightarrow_{T_B} m'$.*

Proof (sketch). $\Rightarrow$) By induction on the structure of the proof $m \rightarrow_{T_{I_B}} m'$. (i) When the reduction is obtained by applying rule FIRING, then there is a transition $m_1[\rangle m_1'$ in I_B s.t. $m_1 \star \sigma \oplus m'' = m$ and $m_1' \star \sigma \oplus m'' = m'$, i.e., m_1 is consumed, m_1' is produced, and m'' denotes idle resources. Consequently, by the construction of the abstract net there is a connected transaction (a compatible process) ς with a mgce σ_ς s.t. $pre(\varsigma)\star\sigma_\varsigma = m_1$ and $post(\varsigma)\star\sigma_\varsigma = m_1'$. We can build a proof for $m_1\star\sigma_\varsigma\star\sigma \Rightarrow_B m_1'\star\sigma_\varsigma\star\sigma$ using ς in the following way: at each step use rules COLOURED-FIRING and STEP for firing all enable transitions, then combine steps with rule CONCATENATION. The concatenation rule can always be applied because the evolution places of a concatenable transaction are zero-places. The family of substitutions σ_t used by the morphism explains how variables appearing in the transitions (of the net B) are used in the process. Consequently, any substitution used in the proof to fire a transition t is defined as $\sigma_t \star \sigma_\varsigma \star \sigma$ restricted to $rn(t)$. Note that proof obtained by adding idle resources in the application of FIRING is also a valid rule. Consequently, the idle resources m'' can always be added to the computation described by the process. (ii) When the reduction is obtained by applying rule STEP the proof is immediate by using inductive hypothesis on premises and by noting that steps in the abstract net corresponds also to steps in the ZS net.

$\Leftarrow$) Note that it is possible to define a process P describing the computation $m \Rightarrow_{T_B} m'$, s.t. $pre(P) = m$, $post(P) = m'$. Moreover, the causal net used by P contains a place for each produced token in the proof, and a transition for any application of FIRING. Note that by rule CONCATENATION all evolution places corresponds to zero places. If two independent computations are combined with rule STEP, then the process has independent subnets, each of them is a process from a stable marking to a stable marking. So, they can be considered independently. Obviously, each independent process is compatible, because it is a possible computation of the net (the compatible execution can be built from the substitutions used during the proof). Therefore the abstract net contains a transition representing this process. This is guaranteed because transitions in the abstract net are defined in terms of the mgce (i.e., any compatible instantiation of the processes can be obtained as an instantiation the mgce). Consequently, there is a firing corresponding to any independent subnet in the process. The entire computation in the abstract net can be obtained by using rule STEP to combine concurrent firings. Isolated places in the causal net are idle resources and can be added to any firing in the proof.

ZS *nets as* C-ZS *nets.* P/T (and ZS) nets can be seen as a particular case of C-P/T (resp., C-ZS) nets where tokens are coloured with the empty sequence $\bullet$.

Definition 22 (Coloured version of a p/t net). *Let* $N = (S_N, T_N, \delta_{0N}, \delta_{1N}, m_{0N})$ *be a* P/T *net. The coloured net* $C_N = (S_N, \emptyset, T_N, \delta_{0C_N}, \delta_{1C_N}, m_{0C_N})$, *with* $\delta_{iC_N}(t)(a, \bullet) = \delta_{iN}(t)(a)$ *for* $i = 1, 2$ *and* $m_{0C_N}(a, \bullet) = m_{0N}(a)$ *is the* coloured version *of* N. *Given a* ZS *net* B, *the* C-ZS *net* C_B *is coloured version* B *if its underlying* C-P/T *net* N_{C_B} *is the coloured version of the underlying* P/T *net* N_B *of* B, *and* $Z_B = Z_{C_B}$.

It should be noted that the construction of the abstract nets under these two different views is consistent. That is, the coloured abstract net for a ZS net coincides with the abstract (non-coloured) net.

Theorem 3. *Let* B *a* ZS *net,* I_B *its abstract* P/T *net,* C_B *and* C_{I_B} *their coloured versions, and* I_{C_B} *the abstract net of* C_B *(i.e., the colored version of* B*). Then* $C_{I_B} \approx I_{C_B}$.

Proof (sketch). The proof follows from the fact that the coloured version C_P of a process P of a net N is a process of the coloured net C_N. Moreover, if $P \approx P'$ then $C_P \approx C'_P$. On the other hand, the coloured version of P is always a plain net, because all colours are sequences of length 0. Therefore any coloured process C_P is compatible, and consequently $C_{I_B} \approx I_{C_B}$.

5 Reconfigurable zs Nets

5.1 Reconfigurable Nets

The idea behind reconfigurable nets (R-P/T nets) is that basic colours are names of places in the net, and consequently the postset of a transition is not static, but depends on the colours of the consumed tokens. For instance, a transition $t = a(v)[\rangle v(a)$ denotes a pattern that consumes a token from a and generates a token in the place corresponding to the colour v of the consumed message. In fact, if t is applied to $m_1 = a(b)$ it produces $m_2 = b(a)$. Instead, when applied to $m'_1 = a(b')$, it generates $m'_2 = b'(a)$.

Consequently, the definitions of nets and of place/transitions nets can be extended in order to allow received names to appear as places in the postsets of transitions. We consider an infinite set of variable names $\mathcal{V}$, ranged over by $v, w, \ldots$. We require also variable names be different from place names, i.e., $\mathcal{V} \cap \mathcal{P} = \emptyset$. Moreover, the constant colours are names of places, hence $\mathcal{C} = \mathcal{P}$.

Definition 23 (r-p/t net). *A* Reconfigurable marked place / transition net *is a 5-tuple $N = (S_N, T_N, \delta_{0N}, \delta_{1N}, m_{0N})$ where $S_N \subseteq \mathcal{P}$ is a set of places, T_N is a set of transitions, the functions $\delta_{0N} : T_N \to \mathcal{M}_{S_N, S_N \cup \mathcal{V}}$ and $\delta_{1N} : T_N \to \mathcal{M}_{S_N \cup \mathcal{V}, S_N \cup \mathcal{V}}$ assign respectively, source and target to each transition, and $m_{0N} \in \mathcal{M}_{S_N, S_N}$ is the initial marking. Moreover, for every t in T_N we require $\delta_{1N}(t) \subseteq \mathcal{M}_{S_N \cup rn(t), S_N \cup rn(t)}$, i.e., variables occurring in the postset of a transition are received names.*

Note that we allow variables to occur in the preset of a transition just in colour positions, while they can also occur in place positions in the postsets. Variables are used analogously to variables in the coloured model, i.e., they are the parameters of a transition that should be instantiated in order to fire the transition. As usual, we consider transitions up-to α-conversion.

The main difference between coloured and reconfigurable nets is that when a transition t is fired in a R-P/T net, the variables in the postset of t should be substituted also when they appear in place position. The following definition introduces the substitution of names occurring both in colour and place position.

Definition 24 (Substitution). *Let $\sigma : \mathcal{V} \to \mathcal{V} \cup \mathcal{P}$ be a partial function. The* substitution *σ on a multiset $m \in \mathcal{M}_{\mathcal{V}\cup\mathcal{P}, \mathcal{V}\cup\mathcal{P}}$ is given by $(m\sigma)(s)(c) = \sum_{r\sigma = s \wedge d\sigma = c} m(s)(d)$.*

The operational semantics for reconfigurable nets can be defined by replacing the rule (FIRING) in Figure 4 by the following (RECONF-FIRING) rule:

(RECONF-FIRING)

$$\frac{t = m \,[\rangle\, m' \in T \quad m'' \in \mathcal{M}_{S,S}}{m \star \sigma \oplus m'' \rightarrow_T m'\sigma \oplus m''} \quad \begin{array}{l} dom(\sigma) = rn(t), \text{ and} \\ \sigma(v) \in S \text{ for } v \in dom(\sigma) \end{array}$$

Comparing rule (COLOURED-FIRING) of C-P/T nets and (RECONF-FIRING) of R-P/T nets, it should be clear that in both cases a transition t can be fired on m only when m contains an instance of the preset obtained by renaming only colours (i.e., $m \star \sigma$). That is, a transition in both C-P/T nets and R-P/T nets consumes messages from a fixed set of places. Differently, in C-P/T nets, tokens generated by firing t corresponds to an instance of the postset of t obtained by substituting only colours accordingly to σ, whereas in R-P/T nets the renaming also affects names appearing in place position. For this reason, in C-P/T nets a transition produces messages in a fixed set of places (although their colour can be different for each firing). Instead, in R-P/T nets two different firings of the same transition can produce messages in different places, i.e. the postset of a transition changes dynamically depending on the colours of the consumed messages.

5.2 Reconfigurable ZS Nets

The first consideration is that in R-P/T net there is no difference between places and colours. Taking into account that places in ZS nets can be either stable or zero, also colours are zero and stable. Consequently, the distinction between stable and zero markings must also consider colours present inside places. Actually, a stable marking should contain only stable names, therefore we write s to indicate $s \in \mathcal{M}_{L,L}$ and we write z for denoting a zero marking. At a first glance, it could appear that any other marking denotes a zero marking. This is not the case for a non-empty marking $m \in \mathcal{M}_{L,Z}$. Markings of this kind contain stable places with tokens coloured with zero names, which is somehow contrary to the ZS approach. Note that in ZS nets, tokens in stable places produced during a transaction are released only at commit, when all zero tokens have been consumed. Consequently, we will restrict zero markings to $z \in \mathcal{M}_{Z,S}$. We denote the set of well-defined markings as $\mathcal{W}_{L,Z} = \mathcal{M}_{L,L} \cup \mathcal{M}_{Z,L \cup Z}$. Additionally, we consider the set of variable names $\mathcal{V}$ as partitioned into sets: $\mathcal{V}_L$, the set of stable variables $V, W, \ldots$, and $\mathcal{V}_Z$ the set of zero variables $v, w, \ldots$.

Definition 25 (r-zs net). *A* Reconfigurable ZS net *is a 6-tuple* $B = (S_B, T_B, \delta_{0B}, \delta_{1B}, m_0B, Z_B)$ *where* $N_B = (S_B, T_B, \delta_{0B}, \delta_{1B}, m_0B)$ *is the underlying* R-P/T *net and the set* $Z_B \subseteq S_B$ *is the set of* zero places. *The places in* $S_B \backslash Z_B$ *(denoted by* L_B*) are called stable places. A stable marking* m *is a coloured multiset of* stable places *(i.e.,* $m \in \mathcal{M}_{L_B,L_B}$*), and the initial marking* m_0B *must be stable. Moreover, we impose the pre and postset functions to be defined over well-defined markings, i.e.,* $\forall t \in T_B, \delta_{iN}(t) \in \mathcal{W}_{L_B \cup \mathcal{V}_L, Z_N \cup \mathcal{V}_Z}$*, for* $i = 1, 2$.

We require transitions to be fired with appropriate names, that is zero variables are substituted by zero places and stable variables by stable places.

Definition 26 (Type preserving substitution). *A substitution* σ *is* type preserving *if* $\forall V \in \mathcal{V}_L, \sigma(V) \in (L_B \cup \mathcal{V}_L)$ *and* $\forall v \in \mathcal{V}_Z, \sigma(v) \in (Z_B \cup \mathcal{V}_Z)$.

In what follows we assume all substitutions to be type preserving. Then, firing rule for R-ZS nets can be written as follows.

(RECONF-FIRING)

$$\frac{t = s \oplus z \; [\rangle \; s' \oplus z' \in T \quad s'' \in \mathcal{M}_{L,L} \quad z'' \in \mathcal{M}_{Z,S}}{(s \star \sigma \oplus s'', z \star \sigma \oplus z'') \rightarrow_T (s'\sigma \oplus s'', z'\sigma \oplus z'')} \quad \begin{array}{l} dom(\sigma) = rn(t), \text{ and} \\ \sigma(v) \in S \text{ for } v \in dom(\sigma) \end{array}$$

Example 8 (Mailing list). Consider a data structure that allows to send atomically a message to a list of subscribers (in the sense that it is either sent to all or to none). Figure 14 shows a ZS net corresponding to such structure. **Nil** is a stable constant colour, all other colours used for labelling arcs are stable variables.

The stable place `newSubs` contains the tokens corresponding to the agents that want to be subscribed to the list. Their colours are the places in which they expect to receive a new message. Place `top` contains the element on top of the list (the latest subscriber). We assume that an empty list is denoted with a token coloured with the constant colour **Nil**. A list is encoded with several tokens in place `subscList`, where each token carries on the information corresponding to one subscriber and the next subscriber in the list, hence their colours are pairs.

By firing the transition `add` a new subscriber N is added on top of the list. The token corresponding to the previous subscriber on top of the list (whose colour is T) is replaced with a new token of colour N, i.e., the new subscriber becomes the top of the list. Also, a new token is produced in `subscList` whose colour is (N, T), meaning that the subscriber that follows N in the list is the previous element on top of the list T.

Transition `tell` allows to send a message M to every subscriber in the list. When `tell` is fired a new transaction is initiated, because a new token is generated in place `sending`, which is a zero place. Note that the top of the list is consumed, and a new token with the same colour is produced in `top`, but it will be released only when the transaction finishes. Therefore, transitions `add` and `tell` will not be enabled until the current transaction finishes.

The zero token present in `sending` contains the information of the subscriber to notify (i.e., the first colour of the pair), and the message to send (i.e., the second colour). Transition `notify` is a reconfigurable transition. In fact, it consumes from `sending` the token (T, M), and sends M to the subscriber T (nevertheless this token will be available actually when the transaction finishes). Additionally, `notify` takes from `subscList` the subscriber F that follows T in the list, and update the state of the transaction by putting in the zero place `sending` a token to notify the next subscriber with M.

The transaction finishes when the end of the list is reached. That is, when the token in `sending` is addressed to the receiver **Nil**. At this point, the transition `end` can be fired to consume the zero token present in the net, which will release all the stable tokens produced during the transaction. At this moment

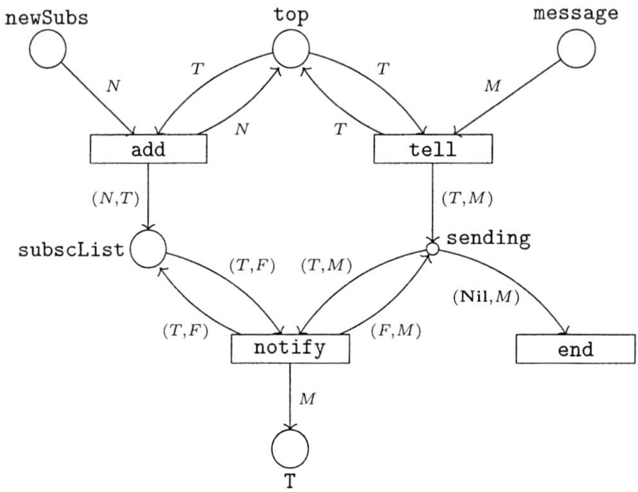

Fig. 14. A R-ZS net for the mailing list example.

all subscribers atomically receive message M, and the top of the list is available for executing new activities.

5.3 Abstract Net under the ITph Approach

The definition of the abstract semantics of R-ZS net also relies on the identification of the basic atomic computations of the net, and for the ITph approach on the notion of Goltz-Reisig processes. The interesting point here is that during a computation on a reconfigurable net some transitions are instantiated in a particular way. Consider the reconfigurable net shown in 15(a), consisting of two transitions $t_1 = a(u,v)[\rangle u(v)$ and $t_2 = c(w)[\rangle w(\bullet)$, where a, b and c are places, and u, v and w variables. Figure 15(b), shows a possible execution in the net where the transition $\mathtt{t}_1'$ is an particular case of t_1, where the received name u has been used as colour c. Consequently, our notion of processes of a reconfigurable net is based on this idea of instantiation.

Definition 27 (Instance of a transition). *Let $t = m[\rangle m'$ be a transition. A transition i is an* instance *of t for a substitution σ if $dom(\sigma) \subseteq rn(t)$ and $i = m \star \sigma[\rangle m'\sigma$.*

Definition 28 (Reconfigurable net morphism). *Let N, N' be* R-P/T *nets. A tuple $f = (f_S : S_N \to S_{N'}, f_T : T_N \to T_{N'}, \rho = \{\rho_t\}_{t \in T_N}, \sigma = \{\sigma_t\}_{t \in T_N})$ is a* reconfigurable net morphism *from N to N' (written $f : N \to_{\sigma,\rho} N'$) if $\forall t \in T_N$:*

- *$\rho_t : \mathcal{V} \to \mathcal{P}$ (i.e., substitutes variables by constants);*
- *$f_S({}^\bullet t \star \rho_t)[\rangle f_S(t^\bullet \star \rho_t)$ is an instance of ${}^\bullet f_T(t)[\rangle f_T(t)^\bullet$ for σ_t.*

Substitution ρ_t are referred to as proper substitutions *or* instantiations.

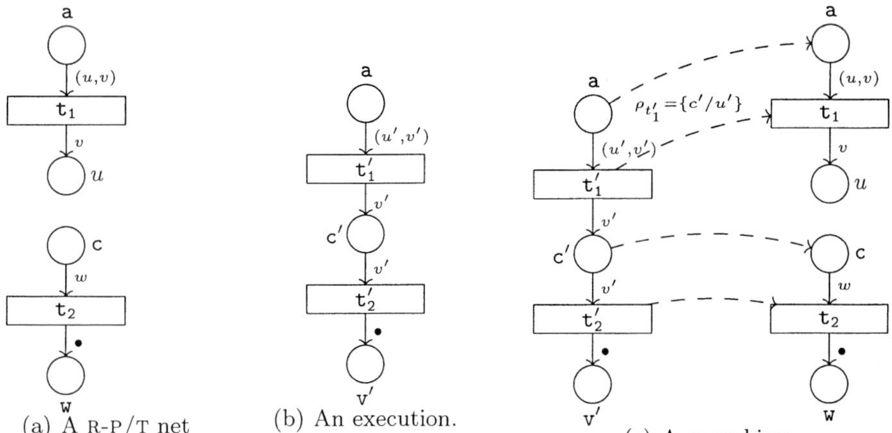

Fig. 15. Reconfigurable net morphism.

For the particular case of coloured transitions (i.e., transitions without reconfigurable capabilities), the condition required on the mapping is analogous to that on coloured net morphisms (Definition 17). In fact, no proper instantiations are needed. Moreover, if used they correspond to instantiations of colours.

Figure 15(c) shows a morphism between reconfigurable nets. Note that the received name u' of $\mathtt{t}_1'$ has been instantiated as c' (i.e., the proper substitution is $\rho_{t_1} = \{c'/u'\}$), because t_2 consumes messages from c. Nevertheless, the whole net is still a reconfigurable net. In fact, the place v' in which the final transition will produce the token depends on the colour of the token consumed from a'.

Definition 29 (Process of a reconfigurable net). *A* (Goltz-Reisig) process *for a* R-P/T *net N is a reconfigurable net morphism $P : K \rightarrow_{\sigma,\rho} N$, from a reconfigurable causal net K to N, s.t. every ρ_{t_k} is minimal (i.e., for every other ρ'_{t_k} that satisfies the morphism conditions $\rho'_{t_k} \subseteq \rho_{t_k}$ holds) and every σ_{t_k} is injective and $\sigma_{t_k} : \mathcal{V} \to \mathcal{V}$.*

As done for coloured nets, we also define a notion of compatible execution of a process to capture the relation between the different colours appearing in the causal net.

Definition 30 (Compatible execution of $[\![P]\!]_\approx$). *Let $\varsigma \in [\![P]\!]_\approx$ s.t. $\forall t_1, t_2 \in T_\varsigma, rn(t_1) \cap rn(t_2) = \emptyset$, i.e., transitions do not share variables. A substitution σ is said a* compatible execution *of ς if:*

- *if $\forall a \in S_\varsigma, {}^\bullet a \star \sigma = a^\bullet \star \sigma$.*
- *σ is consistent with any proper instantiation ρ_t in ς, i.e., $\forall t \in T_\varsigma$, $\rho_t \subseteq \sigma$.*

If such σ exists, we say that $[\![P]\!]_\approx$ is compatible. A process P is compatible *if there exists a compatible execution for $[\![P]\!]_\approx$.*

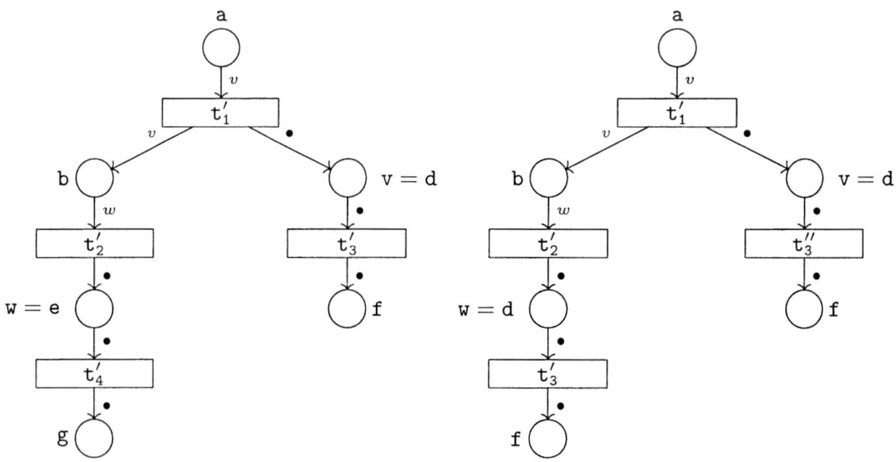

Fig. 16. Two reconfigurable processes.

The first condition is similar to that for compatible executions of coloured processes. The second one assures that in a compatible execution a name is not instantiated in different ways.

Example 9. Consider a reconfigurable net consisting of the following transitions: $t_1 = a(v)[\rangle b(v) \oplus v(\bullet)$, $t_2 = b(w)[\rangle w(\bullet)$, $t_3 = d(\bullet)[\rangle f(\bullet)$ and $t_4 = e(\bullet)[\rangle g(\bullet)$. Figure 16 shows two process of the net (both can be taken as representative of their equivalence class because their transitions do not share variable names). For simplicity, we call transitions in the causal net t' if they are mapped to t, while corresponding places have the same name. Places with name such as $\mathtt{w} = \mathtt{e}$ denote the proper instantiations used by the morphism. Consider the first process, the place $\mathtt{v} = \mathtt{d}$ can be mapped only into d, because $\mathtt{t}_3'$ in the net corresponds to t_3 in the original net. Consequently, $\mathtt{t}_1'$ is an instance of t_1 for the proper instantiation $\{d/v\}$.

Note that the first process does not admit a compatible execution σ. By the first condition of a compatible execution, σ should include substitutions $\{v/w\}$, $\{w/v\}$ or $\{u/v, u/w\}$. By the second, as $\mathtt{t}_1'$ is a proper instantiation for $\{d/v\}$ then $\{d/v\}$ should be in σ. Similarly, by considering $\mathtt{t}_2'$, $\{e/w\}$ should be in σ. Hence, all conditions together are inconsistent because a substitution is a function and cannot assign two different substitutions for the same variable.

The second process admits $\sigma = \{d/v, d/w\}$ as a compatible execution.

The definitions for the most general compatible execution (mgce), compatible process, connected transactions, and causal abstract net are analogous to those presented in Section 4.3.

Example 10 (Abstract net for the mailing list example). In Figure 17 we (partially) show the abstract net corresponding to the R-ZS net in Figure 14. Transition `add` is identical to transition `add` in Figure 14. The transition `tell n` sends

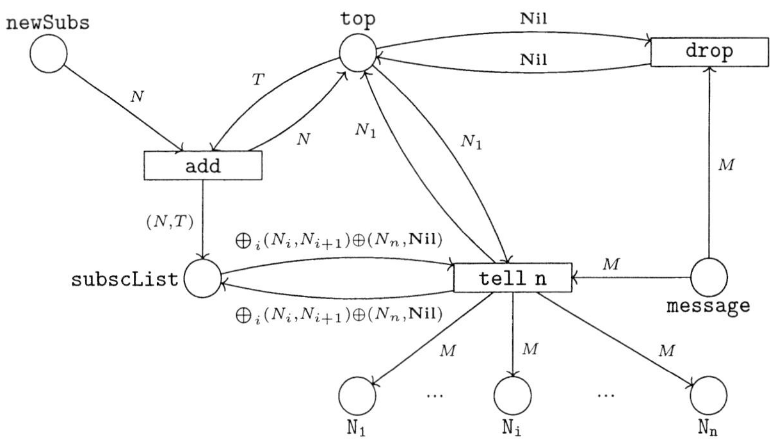

Fig. 17. A partial view of the abstract net for the mailing list example.

atomically a message M to n subscribers in the list whose top is N_1 and finishes in **Nil**. There is one such transition for any $n \geq 2$. The transition drop handles the case in which the list is empty. In such situations the message sent is simply lost (consumed).

Also for the reconfigurable case, the following theorem assures the correspondence between the abstract and the concrete view.

Theorem 4. *Let B be a* R-ZS *net and I_B its abstract net. Then $m \rightarrow_{T_{I_B}} m'$ iff $m \Rightarrow_{T_B} m'$.*

Proof. The proof follows as in Theorem 2 (considering also proper instantiations).

C-ZS *nets as* R-ZS *nets.* C-ZS nets are a particular case of R-ZS net, where no transition uses received names as places in their postset. Thus, given a C-ZS net B, it is possible to construct its abstract C-P/T net C_B and its abstract R-P/T net R_B. The following results assure that both constructions are isomorphic.

Proposition 2. *Let N be a coloured net. If P is a compatible coloured process, then P is a compatible reconfigurable process.*

Theorem 5. *Let B be a* C-ZS *net, C_B its abstract* C-P/T *net, and R_B its abstract* R-P/T *net R_B. Then $R_B \approx C_B$.*

Proof (sketch). The proof follows from Proposition 2. First noting that equivalence classes are the same under both views. Finally, as proper instantiations are not used for coloured transition, a compatible execution under the coloured view is also compatible under the reconfigurable view.

6 Towards Dynamic zs Nets

6.1 Dynamic Nets

While in reconfigurable nets the sets of states and transitions remains unchanged during computations, *dynamic nets* can create new components while executing: new places and transitions may be added to the net when a transition is fired. The main idea is that the firing of a transition may allocate a new subnet, which is parametric on the actual values of the received names. Nevertheless, it is not possible to modify existing transitions: they always consume tokens from a fixed multiset of places and the postset is always the same expression (multiset of places or nets) parametric on the received values. Moreover, it is not possible to attach new transitions with preset in a place after the net has been instantiated (i.e., there is not input capability). The definition given here of dynamic nets follows the presentation given in [2].

Definition 31 (DN). *The set* DN *is the least set satisfying the following equation:*

$$\begin{aligned}\mathcal{N} = \{(S_N, T_N, \delta_{0N}, \delta_{1N}, m_{0N}) \mid \\ S_N \subseteq \mathcal{P} \ \wedge \ \delta_{0N} : T_N \rightarrow \mathcal{M}_{S_N,\mathcal{C}} \ \wedge \ \delta_{1N} : T_N \rightarrow \mathcal{N} \ \wedge \ m_{0N} \in \mathcal{M}_{\mathcal{P},\mathcal{C}}\}\end{aligned}$$

If $(S_N, T_N, \delta_{0N}, \delta_{1N}, m_{0N}) \in$ DN: S_N is the set of places, T_N is the set of the transitions, δ_{0N} and δ_{1N} are the functions assigning the pre and postset to every transition, and m_{0N} is the initial marking. Note that while in previous nets the initial marking is required to be a multiset over the places of the net, here we allow a net to fixed a marking over states that are not defined by it. In fact, the initial marking m_{0N} is a multiset over $\mathcal{P}$ (i.e., $m_{0N} \in \mathcal{M}_{\mathcal{P},\mathcal{C}}$) and not over the places of the net S_N ($\mathcal{M}_{S_N,\mathcal{C}}$). A trivial example is the way in which a coloured transition $a(v)[\rangle b(\bullet)$ is written: $a(v)[\rangle(\emptyset, \emptyset, \emptyset, \emptyset, b(\bullet))$, where b clearly does not belong to the new subnet. In what follows, we write coloured and reconfigurable transitions as in the previous sections, and use verbose notation just for transitions that allocate new components. Also the postset of transitions defined in a new subnet can produce tokens in places not defined by it. Nevertheless, well-defined subnets cannot use places that do not belong to the net they are in.

Names defined in S_N act as binders on N. Therefore, nets are considered up-to α-conversion on S_N. Specially, the creation in N of a new subnet N_1 means the creation of a α-equivalent net N_1' s.t. all names in $S_{N_1'}$ are guaranteed to be different from any other place in N (i.e., they are fresh).

Example 11 (A simple dynamic net). Consider the net N represented in Figure 18(a). The double-lined arrow indicates the dynamic transition $t = a(\bullet)[\rangle N_1$, which creates an instance of the subnet N_1 when fired. We allow the initial marking of N_1 and the postset of transitions in T_{N_1} to generate tokens in a. Therefore, the following is a valid definition for N_1: $S_{N_1} = \{d\}, T_{N_1} = \{t_1\}, m_1 = a(\bullet) \oplus d(\bullet)$ and $t = d(\bullet)[\rangle a(\bullet)$. A firing of t will produced the net shown in 18(b). A new place d and a transition t (whose pre and postset are $d(\bullet)$ and $a(\bullet)$, resp.) have

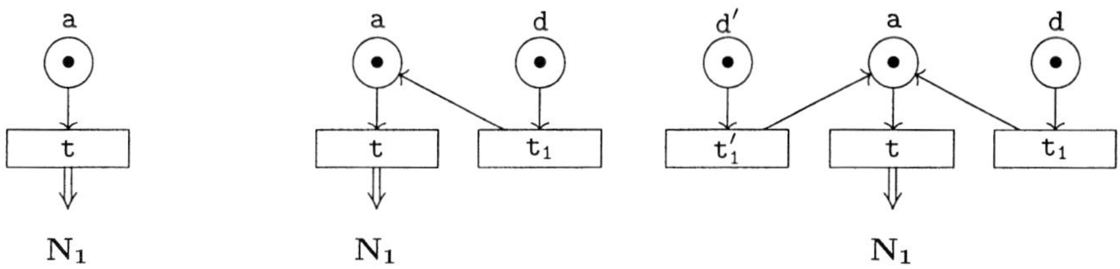

(a) Initial state. (b) After firing t. (c) After two firings of t.

Fig. 18. A simple dynamic net.

been added to the net. Also two tokens have been produced: one in a and the other in d, accordingly to the initial marking of N_1. In this marking t is enabled and can be fired again. The intended meaning of the new activation of t is to create a new subnet: a new place and a new transition whose names are different from others in the net (Figure 18(c)).

Definition 32 (Defined and Free names). *The set of* defined names in a marking *m is $dn(m) = \{a | a \in m\}$, i.e. names appearing in place position. Given $N = (S_N, T_N, \delta_{0N}, \delta_{1N}, m_{0N}) \in$ DN, the* set of defined *(dn) and* free *(fn)* names *of transitions, sets of transitions, and nets are defined as follow:*

$dn(m_1[\rangle N_1) = dn(m_1)$
$dn(N) = dn(T_N) = \bigcup_{t \in T_N} dn(t)$
$fn(m_1[\rangle N_1) = dn(m_1) \cup col_{\mathcal{B}}(m_1) \cup (fn(N_1) \setminus rn(m_1))$
$fn(T_N) = \bigcup_{t \in T_N} fn(t) \setminus dn(T_N)$
$fn(N) = fn(T_N) \setminus S_N$

Definition 33 (Dynamic Net). *$N \in DN$ is a* dynamic net *if $fn(N) = \emptyset$.*

As mentioned above, a well-defined net N does not generate tokens in places that do not belong to it. Note that the condition on the free names imposed for dynamic nets (i.e., $fn(N) = \emptyset$) assures that tokens are always generated in the same net. In this case, any name is bound to a particular place defined in the net, which is guaranteed to be different to any other place.

As for coloured and reconfigurable nets, the firing of a transition t requires the postset to be instantiated with the received colours of t, i.e., the parameters of the t ($rn(t)$). Hence, we need a suitable notion of substitution on nets.

Definition 34 (Instantiation of a net). *Let $\sigma : \mathcal{V} \to \mathcal{P} \cup \mathcal{V}$ be a substitution. The* instantiation of a transition *$t = m_1[\rangle N_1$ with σ s.t. $rn(t) \cap dom(\sigma) = \emptyset$ is defined as $t\sigma = m_1[\rangle N_1\sigma$. Given a dynamic net $N = (S_N, T_N, \delta_{0N}, \delta_{1N}, m_{0N})$, the* instantiation of *N with σ s.t. $dom(\sigma) \cap S_N = \emptyset$ is defined as $N\sigma = (S_N, T_N, \delta_{0N}, \delta_{1N}\sigma, m_{0N}\sigma)$, where $\delta_{1N}\sigma(t) = (\delta(t)_{1N})\sigma$.*

The condition imposed on the substitution used to instantiate a net (or a transition) avoids the capture of free names appearing in the substitution. If this side condition is not satisfied, an α-conversion on the places of the net (or on the received names of the transition) can be applied before.

(DYN-FIRING)

$$\frac{t = m\,[\rangle\, N_1 \in T \qquad m'' \in \mathcal{M}_{S_N,C}}{(S,T,m \star \sigma \oplus m'') \to (S,T,m'') \circledcirc N_1\sigma} \quad \begin{array}{l} dom(\sigma) = rn(t), \text{ and} \\ \sigma(v) \in S \text{ for } v \in dom(\sigma) \end{array}$$

(DYN-STEP)

$$\frac{(S,T,m_1) \to (S,T,m_1') \circledcirc N_1 \quad (S,T,m_2) \to (S,T,m_2') \circledcirc N_2}{(S,T,m_1 \oplus m_2) \to (S,T,m_1' \oplus m_2') \circledcirc (N_1 \oplus N_2)}$$

Fig. 19. Operational semantics of dynamic nets.

Definition 35 (Composition of nets). *Let N_1 and N_2 be dynamic nets. The* addition *of N_2 to N_1 (written $N_1 \circledcirc N_2$) defined as $N_1 \circledcirc N_2 = (S_{N_1} \cup S_{N_2}, T_{N_1} \cup T_{N_2}, \delta_{0N_1} \cup \delta_{0N_2}, \delta_{1N_1} \cup \delta_{1N_2}, m_{0N_1} \oplus m_{0N_2})$ provided with the fact that $N_1 \cap N_2 = \emptyset$ and $fn(N_1) \cap S_{N_2} = \emptyset$. The addition $N_1 \circledcirc N_2$ is said the* parallel composition *of N_1 and N_2 (written $N_1 \oplus N_2$) if also $fn(N_2) \cap S_{N_1} = \emptyset$.*

Observe that side conditions for the parallel composition avoid free names in one net to be captured by the transitions defined by of the other. Nevertheless, when a subnet N_2 is added to a net N_1 ($N_1 \circledcirc N_2$) we allow the free names of N_2 to be capture by the definitions in N_1. We remind that we are considering nets up-to α-conversion in the name of the places, thus it is always possible to rename places in order to satisfy the side conditions mentioned above.

In order to provide the operational semantics for dynamic nets, we remark that the state of a net is not given just in terms of the markings, but also in the structure of the net. The operational semantics is presented in Figure 19. For simplicity we write (S,T,m) as a shorthand for $(S,T,\delta_0,\delta_1,m)$. Rule DYN-FIRING stands for the firing of t when the marking contains an instance of the preset of t (for a suitable substitution on colours σ). The resulting net consists of the original net, where the consumed tokens have been removed, and a new instance of N_1 (i.e., the postset of t). Note that the composition $\circledcirc$ of nets assures that the names of the added components are fresh. Rule DYN-STEP stands for the parallel composition of computations when the initial marking contains enough tokens to execute them independently. By requiring $(N_1 \oplus N_2)$, the components added by concurrent activities are guaranteed to be disjoint.

It is worth noting that reconfigurable nets are a particular case of dynamic nets. In fact when t is a reconfigurable rule, i.e. $N_1 = (\emptyset, \emptyset, m_1)$, the expression $(S,T,m) \circledcirc N_1 = (S,T,m \oplus m_1)$ corresponds to the RECONF-FIRING rule.

6.2 Applying the zs Approach to Dynamic Nets

The evolving structure of dynamic nets opens several possibilities when applying the ZS approach. The more obvious option is to provide transactions by allowing any net to define stable and zero places, as done for the other kind of nets. Nevertheless, other options can take advantage of the possibility of creating

(DYN-FIRING)
$$\frac{t = s \oplus z[\rangle N_1 \in T \qquad s'' \in \mathcal{M}_{L,L} \qquad z'' \in \mathcal{M}_{Z,S}}{(S,T,(s{\star}\sigma \oplus s'', z{\star}\sigma \oplus z''),Z) \to (S,T,(s'',z''),Z) \otimes N_1\sigma} \quad \begin{array}{l} dom(\sigma) = rn(t), \text{ and} \\ \sigma(v) \in S \text{ for } v \in dom(\sigma) \end{array}$$

(DYN-STEP)
$$\frac{(S,T,m_1,Z) \to (S,T,m_1',Z) \otimes N_1 \quad (S,T,m_2,Z) \to (S,T,m_2',Z) \otimes N_2}{(S,T,m_1 \oplus m_2,Z) \to (S,T,m_1' \oplus m_2',Z) \otimes (N_1 \oplus N_2)}$$

(DYN-CONCATENATION)
$$\frac{(S,T,s_1 \oplus z_1,Z) \to (S'',T'',z'',Z'') \otimes s_1' \quad (S'',T'',s_2 \oplus z'',Z'') \to (S',T',s_2' \oplus z',Z')}{(S,T,(s_1 \oplus s_2,z),Z) \to (S',T',(s_1' \oplus s_2',z'),Z)}$$

(DYN-CLOSE)
$$\frac{(S,T,s_1,Z) \to (S',T',s_1',Z')}{(S,T,s_1,Z) \Rightarrow (S',T',s_1',Z')}$$

Fig. 20. Operational semantics of flat dynamic SZ nets.

subnets to specify subactivities that should be executed atomically, providing in this way a hierarchy of atomic activities, i.e., nested transactions. On the rest of this section we describe the operational semantics for the first case, called *flat dynamic* ZS nets. We left as interesting problems to be investigated in the future the characterization of the abstract net, and the different possibilities for applying the ZS approach to dynamic nets.

Flat Dynamic zs Nets. As mentioned above, flat dynamic ZS nets correspond to a direct application of the ZS approach where the places of a net B are either stable, i.e. in $L_B = S_B \backslash Z_B$, or zero, i.e., in Z_B. As for reconfigurable nets, we rely on two disjoint set of variables: $\mathcal{V}_{L_B}$, ranged over by $V, W, \ldots$ for stable variables, and $\mathcal{V}_{Z_B}$ for zero variables $v, w, \ldots$. Similarly, we use $s \in \mathcal{M}_{L,L}$ for denoting a stable marking, and $z \in \mathcal{M}_{Z,S}$ for zero markings. Moreover $\mathcal{W}_{L,Z} = \mathcal{M}_{L,L} \cup \mathcal{M}_{Z,L\cup Z}$ stands for the set of well-defined markings.

Definition 36 (Flat d-zs net). *A* flat dynamic ZS net *is a 6-tuple* $B = (S_B, T_B, \delta_{0B}, \delta_{1B}, m_{0B}, Z_B)$ *where* $N_B = (S_B, T_B, \delta_{0B}, \delta_{1B}, m_{0B})$ *is the underlying dynamic net and the set* $Z_B \subseteq S_B$ *is the set of* zero places. *The places in* $S_B \backslash Z_B$ *(denoted by* L_B*) are called* stable places. *A stable marking* m *is a coloured multiset of stable places (i.e.,* $m \in \mathcal{M}_{L_B,L_B}$*), and the initial marking* m_{0B} *must be stable. Moreover, we impose the pre and postset functions to be defined over well-defined markings.*

Rules in Figure 20 shows the operational semantics of flat D-ZS nets. The rules are the straightforward extension of rules corresponding to R-ZS nets for the case of dynamic transitions.

Example 12 (Private Mailing Lists). Consider the mailing list problem presented in Example 8. Suppose there are n users u_i, each of them needs to send atomically messages present in m_i to listeners whose names are in s_i. (Every user has its own list of subscribers and messages). The system can be modelled as a flat dynamic ZS net by reusing the mailing list structure in Figure 14. Consider the dynamic net in Figure 21 for the case of two users. The net N_1 (appearing in the postset of $t = \mathtt{new}(V, W)[\rangle N_1)$ corresponds exactly to the net in Figure 14 plus the initial marking $m_{0N_1} = V(\mathtt{newSubs}) \oplus W(\mathtt{message}) \oplus \mathtt{top}(\mathbf{Nil})$.

There are n transitions $\mathtt{subsc_i}$ and $\mathtt{dist_i}$, i.e., a pair for each user u_i. The listeners for the user u_i are in place s_i, while the messages are in m_i. For instance, the listeners for u_1 are j_1 and j_2, and it has only the message l_1 to send. To create a list for u_1 it is necessary to put a token on $\mathtt{new}$ with colour $(\mathtt{a_{s_1}}, \mathtt{d_{m_1}})$ (we omitted it in Figure 14 for space limitations) . The obtained net after firing $\mathtt{new}$ with colours $(\mathtt{a_{s_1}}, \mathtt{d_{m_1}})$ is shown in Figure 22. Note that a new instance of N_1 has been created. For convenience in the graphical representation we renamed $\mathtt{newSubs}$ with 1_s and $\mathtt{message}$ with 1_m. Observe that tokens corresponding to the initial marking of N_1 have been produced: token **Nil** in $\mathtt{top}$, 1_s in $\mathtt{a_{s_1}}$, and 1_m in $\mathtt{d_{m_1}}$. Now, listeners in $\mathtt{s_1}$ can be subscribed to the list by firing the reconfigurable transition $\mathtt{subsc_1}$. Note that any coloured token W in s_1 is forwarded to the place 1_s, which will enable the transition $\mathtt{add}$ of the mailing list structure. Similar is the case for tokens in $\mathtt{m_1}$, which are forwarded by $\mathtt{dist_1}$ to the place 1_m.

Suppose that t is fired again for a token $(\mathtt{a_{s_2}}, \mathtt{d_{m_2}})$ in $\mathtt{new}$. In this case, a new mailing list structure is created, which is guaranteed to be independent of the first structure.

About the abstract view of flat dynamic ZS *nets.* The main difficulty when defining the abstract dynamic net describing the atomic movements of the concrete ZS nets is to figure out a suitable notion for a process. A process, viewed as morphism from a causal net into a P/T net, identifies elements of the causal net as particular instances of elements in P/T net. For P/T, C-P/T, and R-P/T nets, where the elements of the net are fixed, the correspondence between instances and general elements is quite clear. In particular, states are mapped into states and transitions (instance of some pattern) to transitions (representing a general pattern). Instead, when describing the execution of a dynamic net D it could be necessary to talk about states and transitions that are not present in D (although D describes how to create them). This question is still open and remains as an interesting problem that bears further investigation.

7 Conclusions

In this paper we have extended the zero-safe approach along the hierarchy of increasingly expressive models characterized in [15]. The results are summarized in Figure 1. Although the more general case of dynamic nets presents some technical difficulties in reconciling the operational and abstract view, the zero-safe approach has been shown somehow orthogonal to the whole hierarchy. Notably, when defining the operational semantics of C-ZS nets and R-ZS nets only the rule describing the firing of a transition is modified. Instead, for dynamic nets all rules are rewritten to consider also the structure of the net as part of the state.

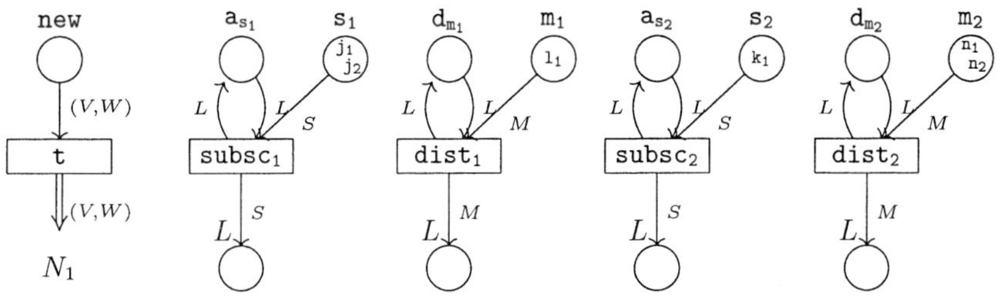

Fig. 21. Private Mailing Lists.

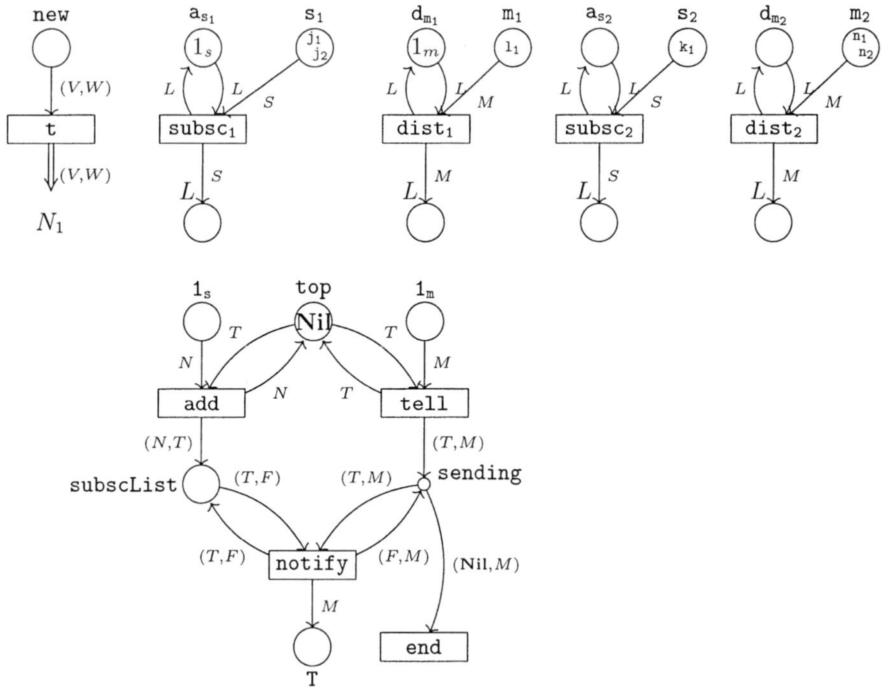

Fig. 22. Private Mailing Lists after firing t with colours $(\mathtt{a}_{\mathtt{s}_1}, \mathtt{d}_{\mathtt{m}_1})$.

Regarding the abstract semantics, it is clear that the description of the abstract view associated to a dynamic ZS net – in particular the characterization of a process in such an evolving structure – remains as an open problem.

On the other hand, the extensions proposed here account only for flat transactions. We plan to investigate alternative extensions of dynamic nets for modelling nested transactions. In particular, by exploiting the capability of creating new subnets to describe sub-transactions. Moreover, the description of compensations in this framework is an ambitious goal that we leave to future work.

Finally, we think that the distributed two phase commit protocol proposed in [8] (used to encode ZS nets in Join) can be reused or extended to implement dynamic ZS nets.

References

1. M. Abadi and A.D. Gordon. A calculus for cryptographic protocols: The spi calculus. *Inform. and Comput.*, 148(1):1–70, 1999.
2. A. Asperti and N. Busi. Mobile petri nets. Technical Report UBLCS96-10, University of Bologna, May 1996. 1996.
3. P. Baldan, H. Ehring, R. Heckel, K. Hoffmann and H. Ehrig. High-level net processes. In W. Brauer, H. Ehrig, J. Karhumäki, and A. Salomaa, editors, *Formal and Natural Computing: Essays Dedicated to Grzegorz Rozenberg*, Volume 2300 of *Lect. Notes in Comput. Sci.*, pages 191–219, Springer Verlag, 2002.
4. E. Best, R. Devillers, and J. Hall. The Petri Box Calculus: A new causal algebra with multi-label communication. 609:21–69, 1992.
5. L. Bocchi, C. Laneve, and G. Zavattaro. A calculus for long-running transactions. In *Proc. of Sixth IFIP International Conference on Formal Methods for Open-Object Based Distributed Systems (FMOODS'03)*, Lect. Notes in Comput. Sci. Springer Verlag, 2003. To appear.
6. BPEL Specification. version 1.1. `http://www-106.ibm.com/developerworks/webservices/library/ws-bpel/`, May 2003.
7. R. Bruni, C. Laneve, and U. Montanari. Centralized and distributed orchestration of transactions in the jo in calculus. Technical Report TR-02-12, Computer Science Department, University of Pisa, 2002.
8. R. Bruni, C. Laneve, and U. Montanari. Orchestrating transactions in join calculus. In L. Brim, P. Jancar, M. Kretinsky, and A. Kucera, editors, *Proceedings of CONCUR 2002, 13th International Conference on Concurrency Theory*, volume 2421 of *Lect. Notes in Comput. Sci.*, pages 321–336. Springer Verlag, 2002.
9. R. Bruni, H. Melgratti, and U. Montanari. Nested commits for mobile calculi: extending Join, 2003. Submitted.
10. R. Bruni and U. Montanari. Executing transactions in zero-safe nets. In M. Nielsen and D. Simpson, editors, *Proceedings of ICATPN 2000, 21st Int. Conf. on Application and Theory of Petri Nets*, volume 1825 of *Lect. Notes in Comput. Sci.*, pages 83–102. Springer Verlag, 2000.
11. R. Bruni and U. Montanari. Zero-safe nets: Comparing the collective and individual token approaches. *Inform. and Comput.*, 156(1-2):46–89, 2000.
12. R. Bruni and U. Montanari. Transactions and zero-safe nets. In H. Ehrig, G. Juhás, J. Padberg, and G. Rozenberg, editors, *Advances in Petri Nets: Unifying Petri Nets*, volume 2128 of *Lect. Notes in Comput. Sci.*, pages 380–426. Springer Verlag, 2001.
13. R. Bruni and U. Montanari. Zero-safe net models for transactions in Linda. In U. Montanari and V. Sassone, editors, *Proceedings of ConCoord 2001, International Workshop on Concurrency and Coordination*, volume 54 of *Elect. Notes in Th. Comput. Sci.*, 2001.
14. R. Bruni and U. Montanari. Concurrent models for linda with transactions. Math. Struct. in Comput. Sci., 2003. To appear.

15. M. Buscemi and V. Sassone. High-level Petri nets as type theories in the Join calculus. In F. Honsell and M. Miculan, editors, *Proceedings of FoSSaCS 2001, Foundations of Software Science and Computation Structures*, volume 2030 of *Lect. Notes in Comput. Sci.*, pages 104–120. Springer Verlag, 2001.
16. N. Busi. On zero safe nets. Private communication, April 1999.
17. N. Busi and G. Zavattaro. On the serializability of transactions in javaspaces. In U. Montanari and V. Sassone, editors, *Elect. Notes in Th. Comput. Sci.*, volume 54. Elsevier Science, 2001.
18. M. Butler, M. Chessell, C. Ferreira, C. Griffin, P. Henderson, and D. Vines. Extending the concept of transaction compensation. *IBM Systems Journal*, 41(4):743–758, 2002.
19. L. Cardelli and A.D. Gordon. Mobile ambients. In M. Nivat, editor, *Proceedings of FoSSaCS'98, Foundations of Software Science and Computational Structures*, volume 1378 of *Lect. Notes in Comput. Sci.*, pages 140–155. Springer Verlag, 1998.
20. D Duggan. An architecture for secure fault-tolerant global applications. *TCS*. To appear.
21. C. Fournet and G. Gonthier. The reflexive chemical abstract machine and the Join calculus. In *Proceedings of POPL'96, 23rd Annual ACM SIGPLAN - SIGACT Symposium on Principles of Programming Languages*, pages 372–385. ACM Press, 1996.
22. H. Garcia-Molina and K. Salem. Sagas. In U. Dayal and I.L. Traiger, editors, *Proceedings of the ACM Special Interest Group on Management of Data Annual Conference*, pages 249–259. ACM Press, 1987.
23. D. Gelernter. Generative communication in Linda. *ACM Transactions on Programming Languages and Systems*, 7(1):80–112, 1985.
24. K. Jensen. *Coloured Petri Nets. Basic Concepts.* EATCS Monographs on Theoretical Computer Science. SV, 1992.
25. F. Leymann. WSFL Specification. version 1.0. `http://www-306.ibm.com/software/solutions/webservices/pdf/WSFL.pdf`, May 2001.
26. R. Milner, J. Parrow, and J. Walker. A calculus of mobile processes, I and II. *Inform. and Comput.*, 100(1):1–40,41–77, 1992.
27. C.A. Petri. *Kommunikation mit Automaten.* PhD thesis, Institut für Instrumentelle Mathematik, Bonn, 1962.
28. W. Reisig. *Petri Nets: An Introduction.* EATCS Monographs on Theoretical Computer Science. Springer Verlag, 1985.
29. G. Ristori. *Modelling Systems with Shared Resources via Petri Nets.* PhD thesis, Computer Science Department, University of Pisa, 1994.
30. U. Roxburgh. Biztalk orchestration: Transactions, exceptions, and debugging, 2001. Microsoft Corporation. Available at `http://msdn.microsoft.com/library/en-us/dnbiz/html/bizorchestr.asp`.
31. Sun Microsystem, Inc. JavaSpaces^{TM} service specifications, v.1.1, 2000.
32. S. Thatte. XLANG: Web Services for Business Process Design. `http://www.gotdotnet.com/team/xml_wsspecs/xlang-c/default.htm`, 2001.

A Survey on Non-interference with Petri Nets

Nadia Busi and Roberto Gorrieri

Dipartimento di Scienze dell'Informazione, Università di Bologna
Mura A. Zamboni, 7, 40127 Bologna, Italy

Abstract. Several notions of non-interference have been proposed in the literature to study the problem of confidentiality in nondeterministic and concurrent systems. Here we rephrase some of them – notably *SNNI* and *BNDC* – over the model of safe Place/Transition Petri Nets. The common feature of these non-interference properties is that they are all defined as extensional properties based on some notion of behavioural equivalence on systems. Here we also address the problem of defining non-interference by looking at the structure of the net systems under investigation. We define *structural* non-interference properties based on the absence of particular places in the net. We characterize structural properties that are slight refinement of well-known properties such as *SNNI* and *SBNDC*. We then argue that, in order to capture all the intuitive interferences at the structural level, it is necessary to consider the net originated by the region construction, yielding the property *RBNI* we advocate.

1 Introduction

Non-interference has been defined in the literature as an extensional property based on some observational semantics: the high part of a system is non-interfering with the low part if whatever is done at the high level produces no visible effect on the low part of the system. The original notion of non-interference in [8] was defined, using trace semantics, for system programs that are deterministic. Generalized notions of non-interference were then designed to include (nondeterministic) labeled transition systems and finer notions of observational semantics such as bisimulation (see, e.g., [12, 6, 11, 13, 7]). Relevant properties in this class are the trace-based properties *SNNI* and *NDC*, as well as the bisimulation-based properties *BSNNI*, *BNDC* and *SBNDC* proposed by Focardi and Gorrieri some years ago [6, 7] on a CCS-like process algebra. In particular, *SNNI* states that a system R is secure if the two systems $R \setminus H$ (all the high level actions are prevented) and R/H (all the high level actions are permitted but are unobservable) are trace equivalent. *BNDC* intuitively states that a system R is secure if it is bisimilar to R in parallel with any high level process Π w.r.t. the low actions the two systems can perform. And *SBNDC* tells that a system R is secure if, whenever a high action h is performed, the two instances of the system before and after performing h are bisimilar from a low level point of view.

The first part of the paper is devoted to show that these non-interference properties, originally proposed on the Security Process Algebra, can be naturally

J. Desel, W. Reisig, and G. Rozenberg (Eds.): ACPN 2003, LNCS 3098, pp. 328–344, 2004.

defined also on Petri Nets; in particular – to keep the presentation as simple as possible – we use 1-safe Place/Transition Petri Nets [10]. The advantage of this proposal is the import in the Petri Net theory of security notions that makes possible the study of security problems. Technically, what we do is to introduce two operations on nets, namely parallel composition (with synchronization in TCSP-like style) and restriction, and suitable notions of observational equivalences on the low part of the system (low trace equivalence and low bisimulation); then, five security properties are defined and compared in a rather direct way. In particular, the two properties based on low trace semantics, namely *SNNI* and *NDC*, are equivalent. On the contrary, in the bisimulation case, *BSNNI* is weaker than *BNDC*, which turns out to be equivalent to *SBNDC*.

In this approach, the security property is based on the dynamics of systems; they are all defined by means of one (or more) equivalence check(s); hence, non-interference checking is as difficult as equivalence checking, a well-studied hard problem in concurrency theory.

In the second part of the paper, instead, we address the problem of defining statically non-interference by looking at the structure of the net systems under investigation:

– in order to better understand the causality and conflict among different system activities, hence grounding more firmly the intuition about what is an interference, and
– in order to find more efficiently checkable non-interference properties that are sufficient conditions for those that have already received some support in the literature.

We define structural non-interference properties based on the absence of particular places in the net. We identify two special classes of places: *causal places*, i.e., places for which there are an incoming high transition and an outgoing low transition; and, *conflict places*, i.e. places for which there are both low and high outgoing transitions. Intuitively, causal places represent potential source of interference (*hilo* flow for *high input – low output*), because the occurrence of the high transition is a prerequisite for the execution of the low transition. Similarly, conflict places represent potential source of interference (*holo* flow for *high output – low output*), because the occurrence of a low event tells us that a certain high transition will not occur.

The first result of the paper is that when causal places are absent, we get a non-interference property which is slightly finer than *SNNI*. More precisely, if N has no causal places, then N satisfies *SNNI*. We present an example that shows that this structural notion is actually finer than *SNNI*.

The second result is that when also conflict places are absent, we get a property, called *Place-Based Non-Interference* (*PBNI* for short), which is slightly finer than *SBNDC*. More precisely, if the net N has no causal and no conflict places, then N satisfies *SBNDC*. A relevant counterexample shows that the inclusion is strict. This counterexample also hints that *PBNI* may still miss some potentially dangerous interferences.

In order to capture all the intuitive interferences at the structural level, we argue that it is necessary to consider nets that are *saturated* w.r.t. the region construction [4, 1]. Intuitively, given the marking graph $MG(N)$ of a net N, another net N' is obtained by adding to N all the possible (useful) places such that $MG(N')$ is isomorphic to $MG(N)$. The final property we propose is called *Region-Based Non-Interference* (*RBNI* for short) that we advocate as the most intuitive non-interference notion in this setting.

The paper is organised as follows. In Section 2 we recall the basic definitions about transition systems and Petri Nets. In Section 3 we recast the behavioural approach to non-interference properties, originally defined in a process algebraic setting, on Petri Nets. The original structural property *PBNI* is introduced in Section 4, while *RBNI* is presented in Section 5. Finally, some conclusive remarks are drawn.

2 Basic Definitions

Here we recall the basic definition about transition systems and safe Place/Transition Petri Nets we will use in the following.

2.1 Transition Systems

Definition 1. *A transition system is a triple* $TS = (St, E, \rightarrow)$ *where*

- St *is the set of states*
- E *is the set of events*
- $\rightarrow \subseteq St \times E \times St$ *is the transition relation.*

In the following we use $s \xrightarrow{e} s'$ *to denote* $(s, e, s') \in \rightarrow$.

A rooted *transition system is a pair* (TS, s_0) *where* $TS = (St, E, \rightarrow)$ *is a transition system and* $s_0 \in St$ *is the* initial state.

Definition 2. *Let* $TS_1 = (St_1, E_1, \rightarrow_1, s_1)$ *and* $TS_2 = (St_2, E_2, \rightarrow_2, s_2)$ *be two rooted transition systems. An* isomorphism *is a bijection* $f : St_1 \rightarrow St_2$ *such that*

- $s \xrightarrow{e} s'$ *iff* $f(s) \xrightarrow{e} f(s')$
- $s_2 = f(s_1)$.

If there exists an isomorphism between TS_1 *and* TS_2 *then we say that* TS_1 *and* TS_2 *are* isomorphic.

2.2 Petri Nets

Definition 3. *Given a finite set* S, *a* multiset over S *is a function* $m : S \rightarrow \omega$. *The set of all multisets over* S *is denoted by* $\mathcal{M}(S)$ *The* multiplicity *of an element* s *in* m *is the natural number* $m(s)$. *We write* $m \subseteq m'$ *if* $m(s) \leq m'(s)$ *for all* $s \in S$. *The operator* $\oplus$ *denotes* multiset union: $(m \oplus m')(s) = m(s) + m'(s)$ *for all* $s \in S$. *The operator* $\setminus$ *denotes* multiset difference: $(m \setminus m')(s) = max\{m(s) - m'(s), 0\}$. *We say that* $s \in m$ *if* $m(s) > 0$. *If* $X \subseteq S$, *with abuse of notation we use* X *to denote the multiset* $X(s) = 1$ *if* $s \in X$ *and* $X(s) = 0$ *otherwise.*

Definition 4. *A net is a tuple* $N = (S, T, F)$*, where*

- S *and* T *are the (finite) sets of* places *and* transitions, *such that* $S \cap T = \emptyset$
- $F \subseteq (S \times T) \cup (T \times S)$ *is the flow relation*

A multiset over the set S of places is called *marking*. Given a marking m and a place s, we say that the place s contains $m(s)$ tokens.

Let $x \in S \cup T$. The *preset* of x is the set ${}^\bullet x = \{y \mid F(y, x)\}$. The *postset* of x is the set $x^\bullet = \{y \mid F(x, y)\}$. The preset and postset functions are generalized in the obvious way to set of elements: if $X \subseteq S \cup T$ then ${}^\bullet X = \bigoplus_{x \in X} {}^\bullet x$ and $X^\bullet = \bigoplus_{x \in X} x^\bullet$. A transition t is enabled at marking m if ${}^\bullet t \subseteq m$. The firing (execution) of a transition t enabled at m produces the marking $m' = (m \backslash {}^\bullet t) \oplus t^\bullet$. This is usually written as $m[t\rangle m'$.

A *net system* is a pair (N, m_0), where N is a net and m_0 is a marking of N, called *initial marking*. With abuse of notation, we use (S, T, F, m_0) to denote the net system $((S, T, F), m_0)$.

The set of *markings reachable from* m, denoted by $[m\rangle$, is defined as the least set of markings such that

- $m \in [m\rangle$
- if $m' \in [m\rangle$ and there exists a transition t such that $m'[t\rangle m''$ then $m'' \in [m\rangle$.

The set of *firing sequences* is defined inductively as follows:

- m_0 is a firing sequence;
- if $m_0[t_1\rangle m_1 \dots [t_n\rangle m_n$ is a firing sequence and $m_n[t_{n+1}\rangle m_{n+1}$ then $m_0[t_1\rangle m_1 \dots [t_n\rangle m_n[t_{n+1}\rangle m_{n+1}$ is a firing sequence.

Given a firing sequence $m_0[t_1\rangle m_1 \dots [t_n\rangle m_n$, we call $t_1 \dots t_n$ a *transition sequence*. The set of transition sequences of a net N is denoted by $TS(N)$. We use σ to range over $TS(N)$. Let $\sigma = t_1 \dots t_n$; we use $m[\sigma\rangle m_n$ as an abbreviation for $m[t_1\rangle m_1 \dots [t_n\rangle m_n$.

The *marking graph* of a net N is

$$MG(N) = ([m_0\rangle, T, \{(m, t, m') \mid m \in [m_0\rangle \wedge t \in T \wedge m[t\rangle m'\})$$

A net is *pure* if ${}^\bullet t \cap t^\bullet = \emptyset$ for all transitions $t \in T$. A net is *simple* if the following condition holds for all $x, y \in S \cup T$: if ${}^\bullet x = {}^\bullet y$ and $x^\bullet = y^\bullet$ then $x = y$.

A net system is *safe* if each place contains at most one token in any marking reachable from the initial marking, i.e., $m(s) \leq 1$ for all $s \in S$ and for all $m \in [m_0\rangle$. A net system is *reduced* if each transition can occur at least one time: for all $t \in T$ there exists $m \in [m_0\rangle$ such that $m[t\rangle$.

In the following we consider safe net systems. To lighten the definitions, in Sections 4 and 5 we we consider safe net systems that are pure, simple and reduced.

3 A Behavioural Approach to Non-interference

In this section we want to recast some basic properties, proposed by Focardi and Gorrieri some years ago [6, 7], in our setting. Our aim is to analyse systems that can perform two kinds of actions: high level actions, representing the interaction of the system with high level users, and low level actions, representing the interaction with low level users. We want to verify if the interplay between the high user and the high part of the system can affect the view of the system as observed by a low user. We assume that the low user knows the structure of the system, and we check if, in spite of this, he is not able to infer the behavior of the high user by observing the low view of the execution of the system.

Hence, we consider nets whose set of transitions is partitioned into two subsets: the set H of high level transitions and the set L of low level transitions. To emphasize this partition we use the following notation. Let L and H be two disjoint sets: with (S, L, H, F, m_0) we denote the net system $(S, L \cup H, F, m_0)$.

The non-interference properties we are going to introduce are based on some notion of *low* observability of a system, i.e., what can be observed of a system from the point of view of low users. The low view of a transition sequence is nothing but the subsequence where high level transitions are discarded.

Definition 5. *Let $N = (S, L, H, F, m_0)$ be a net system. The* low view *of a transition sequence of N is defined as follows:*

$$\Lambda_N(\varepsilon) = \varepsilon$$

$$\Lambda_N(\sigma t) = \begin{cases} \Lambda_N(\sigma)t & \text{if } t \in L \\ \Lambda_N(\sigma) & \text{otherwise} \end{cases}$$

The definition of Λ_N is extended in the obvious way to sets of transitions sequences: $\Lambda_N(\Sigma) = \{\Lambda_N(\sigma) \mid \sigma \in \Sigma\}$ for $\Sigma \subseteq (L \cup H)^*$.

Definition 6. *Let N_1 and N_2 be two net systems. We say that N_1 is* low-view trace equivalent *to N_2, denoted by $N_1 \overset{\Lambda}{\approx}_{tr} N_2$, iff $\Lambda_{N_1}(TS(N_1)) = \Lambda_{N_2}(TS(N_2))$.*

We define the operations of parallel composition (in TCSP-like style) and restriction on nets, that will be useful for defining some non-interference properties.

Definition 7. *Let $N_1 = (S_1, L_1, H_1, F_1, m_{0,1})$ and $N_2 = (S_2, L_2, H_2, F_2, m_{0,2})$ be two net systems such that $S_1 \cap S_2 = \emptyset$ and $(L_1 \cup L_2) \cap (H_1 \cup H_2) = \emptyset$. The parallel composition of N_1 and N_2 is the net system*

$$N_1 \mid N_2 = (S_1 \cup S_2, L_1 \cup L_2, H_1 \cup H_2, F_1 \cup F_2, m_{0,1} \oplus m_{0,2})$$

Definition 8. *Let $N = (S, L, H, F, m_0)$ be a safe net system and let U be a set of transitions. The restriction on U is defined as $N \backslash U = (S, L', H', F', m_0)$, where*

$$L' = L \setminus U$$

$$H' = H \setminus U$$

$$F' = F \setminus (S \times U \cup U \times S)$$

Strong Nondeterministic Non-Interference (*SNNI* for short) is a trace-based property, that intuitively says that a system is secure if what the low-level part can see does not depend on what the high-level part can do.

Definition 9. *Let $N = (S, L, H, F, m_0)$ be a net system. We say that N is SNNI iff $N \overset{\Lambda}{\approx}_{tr} N \backslash H$.*

The intuition is that, from the low point of view, the system where the high level transitions are prevented should offer the same traces as the system where the high level transitions can be freely performed. In essence, a low-level user cannot infer, by observing the low view of the system, that some high-level activity has occurred.

As a matter of fact, this non-interference property captures the information flows from high to low, while admits flows from low to high. For instance, the net N' of Figure 1 is *SNNI* while the net N'' is not *SNNI*.

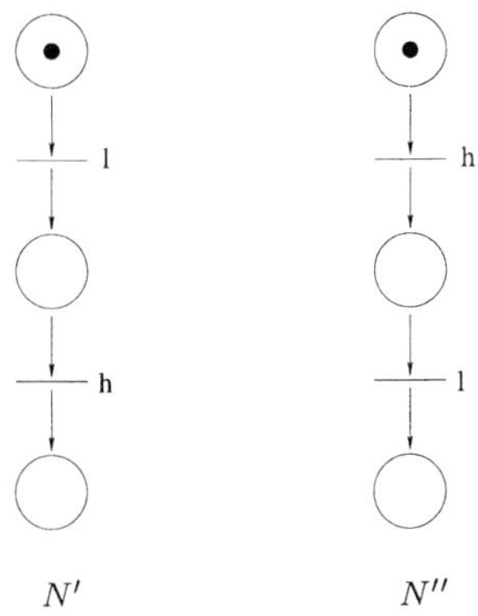

Fig. 1. The net system N' is *SNNI* while N'' is not *SNNI*.

An alternative notion of non-interference, called *Nondeducibility on Composition* (*NDC* for short), says that the low view of a system N in isolation is not to be altered when considering each potential interaction of N with the high users of the external environment.

Definition 10. *Let $N = (S, L, H, F, m_0)$ be a net system. We say that N is a high-level net if $L = \emptyset$.*

Definition 11. *Let $N = (S, L, H, F, m_0)$ be a net system. N is NDC iff for all high-level nets $K = (S_K, \emptyset, H_K, F_K, m_{0,K})$: $N \backslash H \overset{\Lambda}{\approx}_{tr} (N \mid K) \backslash (H \setminus H_K)$.*

The left-hand term represents the low view of the system N in isolation, while the right-hand term expresses the low view of N interacting with the high environment K (note that the activities resulting from such interactions are invisible by the definition of low bisimulation). *NDC* is a very intuitive property: whatever high level system K is interacting with N, the low effect is unobservable. However, it is difficult to check this property because of the universal quantification over high systems. Luckily enough, we will then prove that *SNNI* and *NDC* are actually the same non-interference property.

Theorem 1. *Let $N = (S, L, H, F, m_0)$ be a net system. N is SNNI if and only if N is NDC.*

The two properties above are based on (low) trace semantics. It is well-known [7] that bisimulation semantics is more appropriate than trace semantics because it captures also some indirect information flows due to, e.g., deadlocks. For this reason, we now consider non-interference properties based on bisimulation. To this aim, we first need to introduce a notion of low–view bisimulation.

Definition 12. *Let $N_1 = (S_1, L_1, H_1, F_1, m_{0,1})$ and $N_2 = (S_2, L_2, H_2, F_2, m_{0,2})$ be two net systems. A* low–view bisimulation *from N_1 to N_2 is a relation on $\mathcal{M}(S_1) \times \mathcal{M}(S_2)$ such that if $(m_1, m_2) \in R$ then for all $t \in \bigcup_{i=1,2} L_i \cup H_i$:*

- *if $m_1[t\rangle m_1'$ then there exist σ, m_2' such that $m_2[\sigma\rangle m_2'$, $\Lambda_{N_1}(t) = \Lambda_{N_2}(\sigma)$ and $(m_1', m_2') \in R$*
- *if $m_2[t\rangle m_2'$ then there exist σ, m_1' such that $m_1[\sigma\rangle m_1'$, $\Lambda_{N_2}(t) = \Lambda_{N_1}(\sigma)$ and $(m_1', m_2') \in R$*

If $N_1 = N_2$ we say that R is a low–view bisimulation on N_1.

We say that N_1 is low–view bisimilar *to N_2, denoted by $N_1 \stackrel{\Lambda}{\approx}_{bis} N_2$, if there exists a low–view bisimulation R from N_1 to N_2 such that $(m_{0,1}, m_{0,2}) \in R$.*

The first obvious variation on the theme is to define the bisimulation based version of *SNNI*, yielding *BSNNI*.

Definition 13. *Let $N = (S, L, H, F, m_0)$ be a net system. We say that N is BSNNI iff $N \stackrel{\Lambda}{\approx}_{bis} N \backslash H$.*

Obviously, *BSNNI* $\subseteq$ *SNNI*. The converse is not true: the net N in Figure 2 is *SNNI* but not *BSNNI*. Note that *SNNI* misses to capture the indirect information flow present in this net: if the low transition l is performed (and hence low observed), the low user can infer that the high transition h has not been performed, hence deducing one piece of high knowledge.

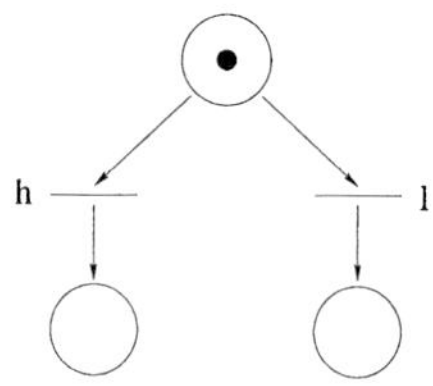

Fig. 2. A net system that is *SNNI* but not *BSNNI*.

Similarly, *BNDC* can be defined from *NDC*, yielding a rather appealing security property, which is finer than *BSNNI*.

Definition 14. *Let $N = (S, L, H, F, m_0)$ be a net system. N is BNDC iff for all high-level nets $K = (S_K, \emptyset, H_K, F_K, m_{0,K})$: $N \backslash H \stackrel{\Lambda}{\approx}_{bis} (N \mid K) \backslash (H \setminus H_K)$.*

Theorem 2. *Let $N = (S, L, H, F, m_0)$ be a net system. If N is BNDC then N is BSNNI.*

Unfortunately, the converse is not true: Figure 3 reports a net that is *BSNNI* but not *BNDC*; the reason why can be easily grasped by looking at their respective marking graphs in Figure 4.

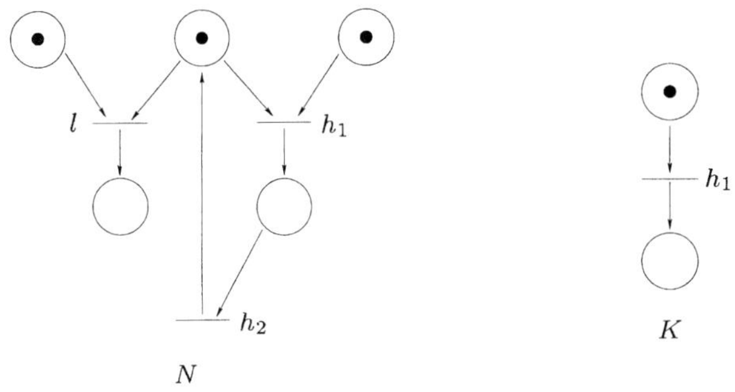

Fig. 3. A net system that is *BSNNI* but not *BNDC*.

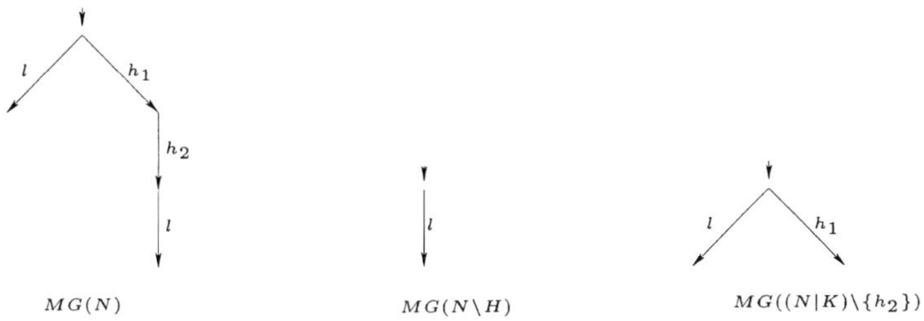

Fig. 4. The marking graphs of the net systems N, $N\backslash H$ and $(N \mid K)\backslash\{h_2\}$.

BNDC is quite appealing but, because of the universal quantification on all poossible high level systems, it is difficult to check. The next property, called *Strong Bisimulation Non Deducibility on Composition* (*SBNDC* for short), is actually an alternative characterization of *BNDC* which is easily checkable.

Definition 15. *Let $N = (S, L, H, F, m_0)$ be a net system. N is SBNDC iff for all markings $m \in [m_0\rangle$ and for all $h \in H$ the following holds:*

if $m[h\rangle m'$ then there exists a low–view bisimulation R on $N\backslash H$ such that $(m, m') \in R$.

Theorem 3. *Let $N = (S, L, H, F, m_0)$ be a net system. N is BNDC if and only if N is SBNDC.*

The theorem above holds because we are in an unlabeled setting: transitions are not labeled. In [6, 7] it is proved that – for the security Process Algebra – *SBNDC* is strictly finer than *BNDC*.

4 Place-Based Non-interference in Petri Nets

In this section we define a non-interference property based on the absence of some kinds of places in a net system. Consider a net system $N = (S, L, H, F, m_0)$.

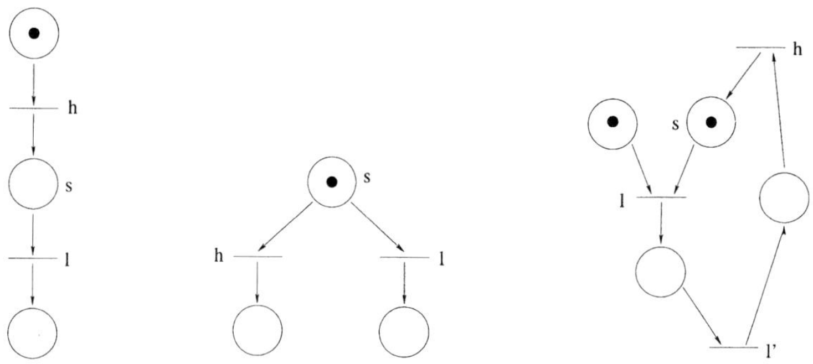

N_1 N_2 N_3

Fig. 5. Examples of net systems containing conflict and (potentially) causal places.

Consider a low level transition l of the net: if l can fire, then we know that the places in the preset of l are marked before the firing of l; as the nets under investigation are pure nets, we also know that such places become unmarked after the firing of l. If there exists a high level action h that produces a token in a place s in the preset of l (see the system N_1 in Figure 5), then the low level user can infer that h has occurred if he can observe the occurrence of the low level action l. We note that there exists a causal dependency between the transitions h and l, because the firing of h produces a token is consumed by l. Consider now the situation illustrated in the system N_2 of Figure 5: in this case, place s is in the preset of both l and h, i.e., l and h are competing for the use of the resource represented by the token in s. Aware of the existence of such a place, a low user knows that no high-level action h has been performed, if he observes the low-level action l. Place s represents a conflict between transitions l and h, because the firing of l prevents h from firing.

Our idea is to consider a net system secure if it does not contain places of the kinds illustrated above.

In order to avoid the definition of a security notion that is too strong, and that prevents systems with no flow of information to be considered secure, we need to refine the concept of causal place. Let s be a place such that $s \in h^\bullet \cap {}^\bullet l$. If s is empty in the initial state of the system, then the low user can infer that h has occurred from the occurrence of l. On the other hand, if s is marked in the initial state, then the first occurrence of l can happen even if h has not fired; thus, the low level user can infer that h has occurred by observing two occurrences of l. Hence, in this last case, such a place s is a source of a flow of information only if transition l can be fired at least two times. For example, consider the net system N_3 reported in Figure 5. Place s is a potentially causal place, but the system has to be considered secure, as the only (maximal) transition sequence is $ll'h$.

Definition 16. *Let $N = (S, L, H, F, m_0)$ be a net system. Let s be a place of N such that $s^\bullet \cap L \neq \emptyset$.*

The place $s \in S$ is a potentially causal place *if $^\bullet s \cap H \neq \emptyset$. A potentially causal place s is a* causal place *if the following condition holds: if $m_0(s) > 0$ then there exists a transition sequence $t_1 \ldots t_n$ and $i < n$ s.t. $t_i, t_n \in s^\bullet \cap L$.*

The place $s \in S$ is a conflict place *if $s^\bullet \cap H \neq \emptyset$.*

Definition 17. *Let $N = (S, L, H, F, m_0)$ be a net system. We say that N is PBNI (*Place Based Non-Interference*) if, for all $s \in S$, s is neither a causal place nor a conflict place.*

Now we show that the absence of causal places implies *SNNI*. We need the following preliminary lemma.

Lemma 1. *Let $N = (S, L, H, F, m_0)$ be a net system without causal places. if $m_0[\sigma\rangle m_1$ then there exists m_2 s.t. $m_0[\Lambda_N \sigma\rangle m_2$ and $m_2(s) \geq m_1(s)$ for all $s \in {}^\bullet L$.*

Theorem 4. *Let $N = (S, L, H, F, m_0)$ be a net system. If N has no causal places then N is SNNI.*

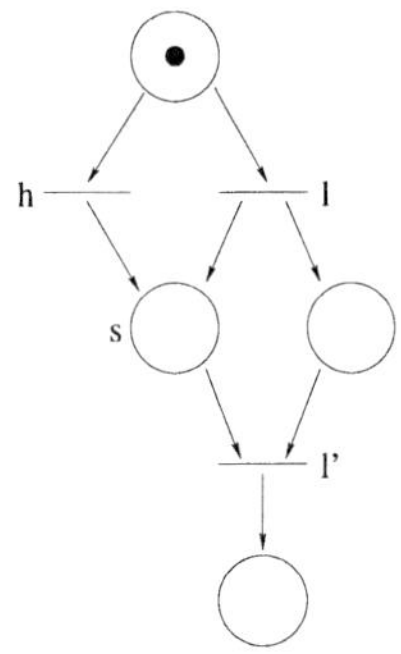

N_4

Fig. 6. A net system containing a causal place, whose marking graph is *SNNI*.

The converse is not true. For example, consider the net system N_4 in Figure 6: place s is a causal place, but N_4 is *SNNI* (but not *SBNDC*). However, as we will see in Section 5, in absence of any form of conflicts in the system, *SNNI* implies the absence of causal places.

As *SBNDC* can reveal the presence of conflicts between high-level transitions and low-level transitions, the absence of causal places in a system is not sufficient to guarantee *SBNDC*. Consider for example the system N_2 in Figure 5, and its marking graph $MG(N_2)$ reported in Figure 8. The system N_2 has no causal places, but N_2 is not *SBNDC*. In fact, $m_1 \xrightarrow{h} m_2$ and the markings m_1 and m_2 have different low-level behaviours, because m_1 can perform l whereas m_2 cannot perform any action.

If we take into account also conflict places, we obtain that the absence of both causal and conflict places is a sufficient condition for *SBNDC*.

Theorem 5. *Let $N = (S, L, H, F, m_0)$ be a net system. If N is PBNI then N is SBNDC.*

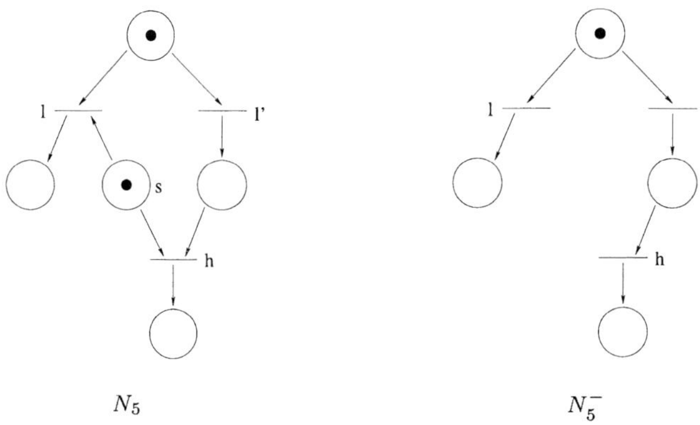

Fig. 7. Two examples of net systems that illustrate the inadequacy of *SBNDC* and *PBNI*.

On the other hand, the absence of causal and conflict places is not a necessary condition for *SBNDC*. Consider the system N_5 reported in Figure 7: the system contains a conflict place, s, hence N_5 is not *PBNI*. However, N_5, whose marking graph is reported in Figure 8, is *SBNDC*: in fact, the only high-level transition is $m_3 \xrightarrow{h} m_4$, and m_3 and m_4 are behaviourally equivalent because both markings have no low outgoing moves.

In our opinion, the system N_5 is not secure, because the occurrence of the low-level transition l permits to a low-level user to deduce that no high-level action has been (and will be) performed. We note that the same kind of information flow is exhibited by the system N_2 of Figure 5, which, on the contrary, is not *SBNDC*.

Hence, *SBNDC* fails to capture some kinds of interference, concerned with the presence of a conflict between a low-level transition and a high-level one. Indeed,

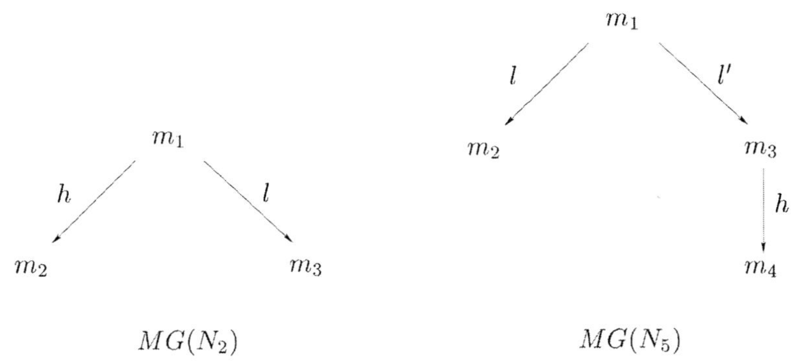

Fig. 8. The marking graphs of the systems N_2 (Figure 5) and N_5 (Figure 7).

also the absence of conflict places, hence *PBNI*, is not sufficient to ensure the absence of the kind of interference discussed above. Consider for the example the system N_5^- of Figure 7, obtained by removing the conflict place s from N_5. The two systems N_5 and N_5^- have the same behaviour, as their marking graphs are isomorphic, but N_5^- is *PBNI*. The example above suggests us to look for conflict places not only in the system under investigation, but in all the systems exhibiting the same marking graph.

5 Region-Based Non-interference in Petri Nets

In this section we enhance *PBNI* to capture the kind of interference we envisaged in system N_5^-. We learned from the previous section that in order to capture some kinds of information flows – arising from conflicts among high and low transitions – it is necessary to look for the presence of conflict places in all the systems whose marking graph is isomorphic to the marking graph of the analyzed system. To construct all such places, we exploit the notion of region, introduced in [4] and investigated, e.g., in [1, 2] for the synthesis of Petri nets[1]. A region is a set of states in the marking graph of a net, corresponding to a real or a potential place of the net. After recalling some basic notions and results on regions (see, e.g., [2]), the non-interference notion based on regions is introduced.

5.1 Theory of Regions

Given the marking graph G of a safe net system N, a region of G is basically a set of markings corresponding to the states where a real or potential place of N is marked. In other words, a region r groups together all the states of the graph in which a place r contains a token. Let r be a region of $MG(N)$. Consider a place s that is necessary for a transition t to happen, i.e., $s \in {}^\bullet t$. Let $m \in r$

[1] The restriction to safe Place/Transition nets is essential to keep the presentation of the region construction as simple as possible.

and assume that $m[t\rangle$; then, s is marked in m; as we consider pure nets, s is no longer marked after the firing of t. Thus, we have a transition $m \xrightarrow{t} m'$ in the marking graph, and $m'(s) = 0$; hence, $m' \notin r$. So, for each state in r, if a t-labelled transition exits from it then that transition enters a state that is not in r. Moreover, if a state m is outside r, then t cannot happen in m, because the place s in the preset of t is empty; so we do not have t- labelled transitions exiting from s. To summarize, if $s \in {}^\bullet t$, then each t-labelled transition of the graph starts inside r and ends outside r. Analogously, if a transition t produces a token in s, i.e., $s \in t^\bullet$, then each t-labelled transition in the graph has source outside r and target inside r.

Suppose now that place s is unrelated to transition t, i.e., $s \notin {}^\bullet t \cup t^\bullet$. If t fires in a state where s is marked, then place s is marked also after the firing of t; that is, if a t-labelled transition starts inside r, then it also ends inside r. Analogously, if t happens in a state where s is empty, then s remains empty also after the firing of t, i.e., t-labelled transitions that start outside r also end outside r.

From the above discussion we deduce that t-labelled transitions have a uniform behaviour w.r.t. r: either all of them cross r exiting, or all of them cross r entering, or none of them cross r.

We recall here the notion of region and some relevant results that will be used later.

Definition 18. *Let $TS = (St, E, \rightarrow)$ be a transition system, a set $r \subseteq St$ is said to be a* region *if and only if $\forall s_1 \xrightarrow{e} s_1', s_2 \xrightarrow{e} s_2'$ the following conditions hold:*

- *if $s_1 \in r$ and $s_1' \notin r$ then $s_2 \in r$ and $s_2' \notin r$;*
- *if $s_1 \notin r$ and $s_1' \in r$ then $s_2 \notin r$ and $s_2' \in r$.*

It is easy to see that both St and $\emptyset$ are regions, and they are called the *trivial* regions. The set of *non-trivial* regions of a transition system TS will be denoted with $Reg(TS)$.

The complementary set of a region is itself a region:

Proposition 1. *Let $TS = (St, E, \rightarrow)$ be a transition system. If r is a region of TS, then also $St \setminus r$ is a region of TS.*

As t-labelled arcs have a uniform behaviour w.r.t. a region, we can define the analogous of preset and postset for events and regions

Definition 19. *Let $TS = (St, E, \rightarrow)$ be a transition system and $e \in E$. The preregionset and the postregionset of e are the sets of regions defined as follows:*

$$\begin{aligned} {}^\circ e &= \{r \in Reg(TS) \mid \forall (s, e, s') \in \rightarrow : s \in r \wedge s' \notin r\} \\ e^\circ &= \{r \in Reg(TS) \mid \forall (s, e, s') \in \rightarrow : s \notin r \wedge s' \in r\} \end{aligned}$$

Given a region r of TS, ${}^\circ r = \{e \in E \mid r \in e^\circ\}$ and $r^\circ = \{e \in E \mid r \in {}^\circ e\}$.

The following proposition explains the relation between the places of a net system and the regions of its marking graph.

Definition 20. *Let $N = (S, T, F, m_0)$ be a net system and let $s \in S$. With r_s we denote the set of states of $MG(N)$ where s is marked: $r_s = \{m \in [m_0\rangle \mid m(s) = 1\}$.*

Proposition 2. *Let $N = (S, T, F, m_0)$ be a net system and let $s \in S$. The set r_s is a region of $MG(N)$.*

Proposition 3. *Let $N = (S, T, F, m_0)$ be a net system and let $s \in S$. We have that ${}^\bullet s = {}^\circ r_s$ and $s^\bullet = r_s^\circ$.*

On the other hand, a region not always corresponds to a place of the net, but may represent a potential place. The addition of such a potential place to the net system has no influence on its behaviour.

Definition 21. *Let $N = (S, T, F, m_0)$ be a net system and let r be a region of $MG(N)$ s.t. the following holds: $\forall s \in S : {}^\circ r \neq {}^\bullet s$ or $r^\circ \neq s^\bullet$. Let s_r be a place s.t. $s_r \notin S$. We net system $N^{+r} = (S', T, F', m_0')$ is defined as follows:*

$$
\begin{aligned}
S' &= S \cup \{s_r\} \\
F' &= F \cup \{(s_r, t) \mid r \in {}^\circ t\} \cup \{(t, s_r) \mid r \in t^\circ\} \\
m_0' &= \begin{cases} m_0 \oplus \{s_r\} & \text{if } m_0 \in r \\ m_0 & \text{otherwise} \end{cases}
\end{aligned}
$$

Proposition 4. *Let N be a net system and r be a region of $MG(N)$. Then $MG(N)$ is isomorphic to $MG(N^{+r})$.*

Given a net sytem N, we can construct the *saturated* version of (the marking graph of) N, obtained by using all the nontrivial regions of $MG(N)$ as places. Note that the set $Reg(MG(N))$ is finite, as the set of nontrivial regions of a transition system is a subset of the powerset of the set of states of the transition system, and the set of states of the marking graph of a safe Petri net is finite.

Definition 22. *Let $TS = (St, E, \rightarrow, s_0)$ be the marking graph of a net system. The net system $Sat(G) = (S, T, F, m_0)$ is defined as follows:*

$$
\begin{aligned}
S &= Reg(G) \\
T &= E \\
F &= \{(r, e) \mid r \in {}^\circ e\} \cup \{(e, r) \mid r \in e^\circ\} \\
m_0(r) &= \begin{cases} 1 & \text{if } s_0 \in r \\ 0 & \text{otherwise} \end{cases}
\end{aligned}
$$

Proposition 5. *Let N be a net system. Then $MG(N)$ is isomorphic to $MG(Sat(MG(N)))$.*

5.2 Region-Based Non-interference

We introduce a non-interference property based on the absence of some kinds of regions in the marking graph of a net system.

Definition 23. *Let $N = (S, L, H, F, m_0)$ be a net system. Let r be a region in $Reg(MG(N))$ such that $r^\circ \cap L \neq \emptyset$.*

The region $r \in Reg(MG(N))$ is a potentially causal region *if $^\circ r \cap H \neq \emptyset$. A potentially causal region r is a* causal region *if the following condition holds: if $m_0 \in r$ then there exists a transition sequence $t_1 \ldots t_n$ and $i < n$ s.t. $t_i, t_n \in r^\circ \cap L$.*

The region r is a conflict region *if $r^\circ \cap H \neq \emptyset$.*

Definition 24. *Let $N = (S, L, H, F, m_0)$ be a net system. We say that N is RBNI (*Region-Based Non-Interference*) if, for all regions $r \in Reg(MG(N))$, r is neither a causal region nor a conflict region.*

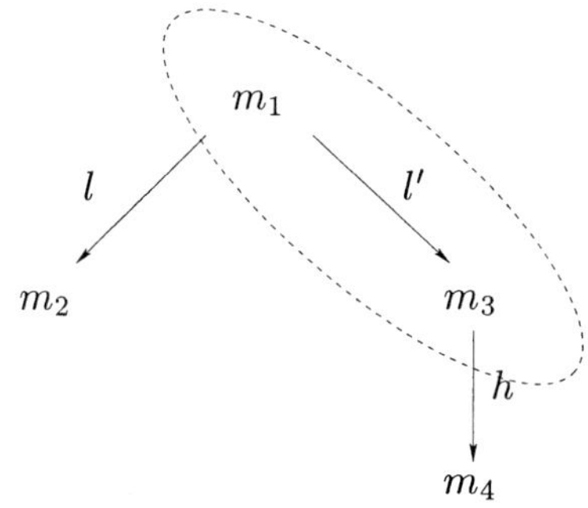

Fig. 9. A conflict region of net N_5^- (Figure 7).

Consider the net system N_5^- in Figure 7. We have that the region $r = \{m_1, m_3\}$ illustrated in Figure 9 is a conflict region, as $l, h \in r^\circ$. Hence, N_5^- is *PBNI* but it is not *RBNI*.

Proposition 6. *Let $N = (S, L, H, F, m_0)$ be a net system. If N is RBNI then N is also PBNI.*

Instead of looking for causal (resp. conflict) regions in the marking graph of a net system N, we can equivalently check for presence of causal (resp. conflict) places in the saturated version of N.

Proposition 7. *Let $N = (S, L, H, F, m_0)$ be a net system. The system N is RBNI if and only if the system $Sat(MG(N))$ is PBNI.*

In Section 4 we argued that the absence of causal regions is not a necessary condition for *SNNI*, because of the existence of places that contain both causal and conflict relations. Now we show that if no conflict is present, i.e., there exist no conflict region in the marking graph of the system, then *SNNI* is equivalent to the absence of causal places.

Theorem 6. *Let $N = (S, H \cup L, F, m_0)$ be a net system such that $MG(N)$ has no conflict regions. Then N has no causal places if and only if N is SNNI.*

A consequence of the above result is that, in absence of conflicts, *PBNI* is equivalent to *RBNI*.

Corollary 1. *Let* $N = (S, L, H, F, m_0)$ *be a net system such that* $MG(N)$ *has no conflict regions. Then N is PBNI if and only if N is RBNI.*

6 Conclusions

A survey is presented on five behavioural non-interference properties, as well as on two new structural ones, *PBNI* and *RBNI*, that we propose to firm more strongly the intuition about the nature of interferences and to obtain more efficiently checkable property. With the help of many examples, we have shown that *RBNI* seems to capture all the intuitive interferences that are possible due to causality and conflict. Moreover, *PBNI* is a sufficient condition for *SNNI* and *SBNDC*, hence offering a very efficient way to check these observational non-interference properties.

The two properties *PBNI* and *RBNI* are structural because no notion of observational equivalence is considered in their definition; however, to be precise, the definition of *RBNI* requires an exploration of the state space (marking graph), hence it is in some sense a *behavioural* property.

The current investigation was conducted for safe Place/transition Petri nets. The choice of such a restrictive class is due to the fact the we wanted to introduce our security properties, in particular *RBNI* with the minimal technical overhead. The results presented here scales smoothly to elementary net systems [5] as well as safe nets with self-loops.

The current investigation was conducted in an unlabeled setting: transitions in the Petri nets are unlabeled. A natural extension of this approach is to consider labeled systems, also equipped with the unobservable action ε. Labels can be used to represent an abstraction of the system where different transitions are considered as equivalent (from the observational point of view). Therefore, we can model situations where the low user is not able to recognize precisely the low transition in execution but only its equivalence class w.r.t. observation. Similarly, label ε is used to model transitions that the low user cannot observe and which is not interested to. Such an extension would also permit to export our approach to process algebras, because it is well-known (see e.g., [3]) how to map (some) process algebras to safe Petri nets.

References

1. E. Badouel and Ph. Darondeau. Theory of regions. *Lectures on Petri Nets I: Basic Models*, Springer LNCS 1491:529:586, 1998.
2. J. Desel and W. Reisig. The synthesis problem of Petri nets. *Acta Informatica*, 33:296–315, 1996.
3. P. Degano, R. De Nicola, U. Montanari, "A Distributed Operational Semantics for CCS based on C/E Systems", *Acta Informatica* 26, 59-91, 1988.

4. A. Ehrenfeucht and G. Rozenberg. Partial (set) 2-structures; I and II. *Acta Informatica*, 27:315–368, 1990.
5. J.Engelfriet and G. Rozenberg. Elementary Net Systems *Lectures on Petri Nets I: Basic Models*, Springer LNCS 1491, 1998.
6. R. Focardi, R. Gorrieri, *"A Classification of Security Properties"*, Journal of Computer Security 3(1):5-33, 1995
7. R. Focardi, R. Gorrieri, *"Classification of Security Properties (Part I: Information Flow)"*, Foundations of Security Analysis and Design - Tutorial Lectures (R. Focardi and R. Gorrieri, Eds.), Springer LNCS 2171:331-396, 2001
8. J.A. Goguen, J. Meseguer, *"Security Policy and Security Models"*, Proc. of Symposium on Security and Privacy, IEEE CS Press, pp. 11-20, 1982
9. C. A. Petri, *Kommunikation mit Automaten*, PhD Thesis, Institut für Instrumentelle Mathematik, Bonn, Germany, 1962.
10. W. Reisig, "Petri Nets: An Introduction", EATCS Monographs in Computer Science, Springer, 1985.
11. A.W. Roscoe, *"CSP and Determinism in Security Modelling"*, Proc. of IEEE Symposium on Security and Privacy, IEEE CS Press, pp. 114-127, 1995
12. P.Y.A. Ryan, *"Mathematical Models of Computer Security"*, Foundations of Security Analysis and Design - Tutorial Lectures (R. Focardi and R. Gorrieri, Eds.), Springer LNCS 2171:1-62, 2001
13. P.Y.A. Ryan, S. Schneider, *"Process Algebra and Noninterference"*, Proc. of 12th Computer Security Foundations Workshop, IEEE CS Press, pp. 214-227, 1999

Synthesis of Asynchronous Hardware from Petri Nets

Josep Carmona[1], Jordi Cortadella[1], Victor Khomenko[2], and Alex Yakovlev[2]

[1] Universitat Politècnica de Catalunya, Barcelona, Spain
jcarmona@ac.upc.es, jordicf@lsi.upc.es
[2] University of Newcastle, Newcastle upon Tyne NE1 7RU, UK
{Victor.Khomenko,Alex.Yakovlev}@ncl.ac.uk

Abstract. As semiconductor technology strides towards billions of transistors on a single die, problems concerned with deep sub-micron process features and design productivity call for new approaches in the area of behavioural models. This paper focuses on some of recent developments and new opportunities for Petri nets in designing asynchronous circuits such as synthesis of asynchronous control circuits from large Petri nets generated from front-end specifications in hardware description languages. These new methods avoid using full reachability state space for logic synthesis. They include direct mapping of Petri nets to circuits, structural methods with linear programming, and synthesis from unfolding prefixes using SAT solvers.

1 Introduction

1.1 Semiconductor Technology Progress

The International Technology Roadmap for Semiconductors (ITRS) [1] predicts the end of this decade will be marked by the appearance of a System-on-a-Chip (SoC) containing four billion 50-nm transistors that will run at 10GHz. With a steady growth of about 60% in the number of transistors per chip per year, following the famous Moore's law, the functionality of a chip doubles every 1.5 to 2 years. Such a SoC will inevitably consist of many separately timed communicating domains, regardless of whether they are internally clocked or not [1]. Built at the deep sub-micron level, where the effective impact of interconnects on performance, power and reliability will continue to increase, such systems present a formidable challenge for design and test methods and tools.

The key point raised in the ITRS is that *design cost* is the greatest threat to the continued phenomenal progress in microelectronics. The only way to overcome this threat is through improving the productivity and efficiency of the design process, particularly by means of design automation and component reuse. The cost of design and verification of processing engines has reached the point where thousands of man-years are spent to a single design, yet processors reach the market with hundreds of bugs [1].

J. Desel, W. Reisig, and G. Rozenberg (Eds.): ACPN 2003, LNCS 3098, pp. 345–401, 2004.

1.2 Self-timed Systems and Design Tools

Getting rid of global clocking in SoCs offers potential added values, traditionally quoted in the literature [60]: greater operational robustness, power savings, electro-magnetic compatibility and self-checking. While the asynchronous design community continues its battle for the demonstration of these features to the semiconductor industry investors, the issue of design productivity may suddenly turn the die to the right side for asynchronous design. Why?

One of the important sub-problems of the productivity and reuse problem for globally clocked systems is that of timing closure. This issue arises when the overall SoC is assembled from existing parts, called Intellectual Property (IP) cores, where each part has been designed separately (perhaps even by a different manufacturer) for a certain clock period, assuming that the clock signal is delivered accurately, at the same time, to all parts of the system. Finding the common clocking mode for SoCs that are built from multiple IP cores is a very difficult problem to resolve.

Self-timed systems, or less radical, globally asynchronous locally synchronous (GALS) systems [11,70], are increasingly seen by industry as a natural way of composing systems from predesigned components without the necessity to solve the timing closure problem in its full complexity. As a consequence, self-timed systems highlight a promising route to solving the productivity problem as companies begin to realise. But they also begin to realise that without investing into design and verification tools for asynchronous design the above promise will not materialise. For example, Philips, whose products are critical to the time-to-market demands, is now the world leader in the exploitation of asynchronous design principles [27]. Other microelectronics giants such as Intel, Sun, IBM and Infineon, follow the trend and gradually allow some of their new products involve asynchronous parts. A smaller 'market niche' company Theseus Logic has been successful in down-streaming the results of their recent investment in asynchronous design methods (Null-Convention Logic) [26].

1.3 Design Flow Problem

The major obstacle now is the absence of a flexible and efficient design flow, which must be compatible with commercial CAD tools, such as for example the Cadence toolkit. A large part of such a design flow would be typically concerned with mapping the logic circuit (or sometimes macro-cell) netlist onto silicon area using place and route tools. Although hugely important this part is outside our present scope of interest, as it is essentially the same as in the traditional design flow. What we are concerned with is the stage in which the behavioural specification of a circuit is converted into the logic netlist implementation.

The pragmatic approach to this stage suggests that the specification should appear in the form of a high-level Hardware Description Language (HDL). Examples of such languages are the widely known VHDL and VERILOG, as well as TANGRAM [2] or BALSA [22] that are more specific for asynchronous design. The latter are based on the concepts of processes, channels and variables, similar to Hoare's CSP.

We can in principle be motivated by the success of behavioural synthesis achieved by synchronous design in the 90s. However, for synchronous design the task of translating an HDL specification to logic (see, e.g., [47]) is fairly different from what we may expect in the asynchronous case.

Its first part was concerned with the so-called architectural synthesis, whose goal was the construction of a register-transfer level (RTL) description. This required extracting a control and data flow graph (CDFG) from the HDL, and performing scheduling and allocation of data operations to functional data path units in order to produce an FSM for a controller or sequencer. The FSM was then constructed using standard synchronous FSM synthesis, which generated combinational logic and rows of latches.

Although some parts of architectural synthesis, such as CDFG extraction, scheduling and allocation, might stay unchanged for self-timed circuits, the development of the intermediate level, an RTL model of a sequencer, and its subsequent circuit implementation, would be quite different.

1.4 How Can Petri Net Help?

Two critical questions arise at this point. Firstly, what is the most adequate formal language for the intermediate (still behavioural) level description? Secondly, what should be the procedure for deriving logic implementation from such a description?

The present level of development of asynchronous design flow suggests the following options to answer those questions:

(1) Avoid (!) answering them altogether. Instead, follow a syntax-driven translation of the HDL directly into a netlist of hardware components, called handshake circuits. This sort of silicon-compilation approach was pursued at Philips with the TANGRAM flow [2]. Many computationally hard problems involving global optimisation of logic were also avoided. Some local 'peephole' optimisation was introduced at the level of handshake circuit description. Petri nets were used for that in the form of Signal Transition Graphs (STGs) and their composition, with subsequent synthesis using the PETRIFY tool [52, 18]. Similar sort of approach is currently followed by the designers of the BALSA flow, where the role of peephole optimisation tools is played by the FSM-based synthesis tool MINIMALIST [12]. The problem with this approach is that, while being very attractive from the productivity point of view, it suffers from the lack of global optimisation, especially for high-speed requirements, because direct mapping of the parsing tree into a circuit structure may produce very slow control circuits.

(2) Translate the HDL specification into a STG for controller part and then synthesise this it using PETRIFY. This approach was employed in [4], where the HDL was VERILOG. This option was attractive because the translation of the VERILOG constructs preserved the natural semantical execution order between operations (not the syntax structure!) and PETRIFY could apply logic optimisation at a fairly global level. If the logic synthesis stage was not constrained by the state space explosion inherent in PETRIFY, this would have been an ideal situation.

However, the state space explosion becomes a real spanner in the works, because the capability of PETRIFY to solve the logic synthesis problem is limited by the number of logic signals in the specification. STGs involving 40–50 binary variables can take hours of CPU time. The size of the model is critical not only for logic minimisation but, more importantly, for solving state assignment and logic decomposition problems. The state assignment problem often arises when the STG specification is extracted automatically from an HDL. This forces PETRIFY into solving Complete State Coding (CSC) using computationally intensive procedures involving calculation of regions in the reachability graph.

While the logic synthesis powers of PETRIFY should not be underestimated, one should be realistic *where* they can be applied efficiently. Thus the solution lies where the design productivity similar to that of (1) can be achieved together with the circuit optimality offered by (2). We believe that the way to such a solution is through finding more efficient ways of logic synthesis in the framework of the design flow shown in Fig. 1.

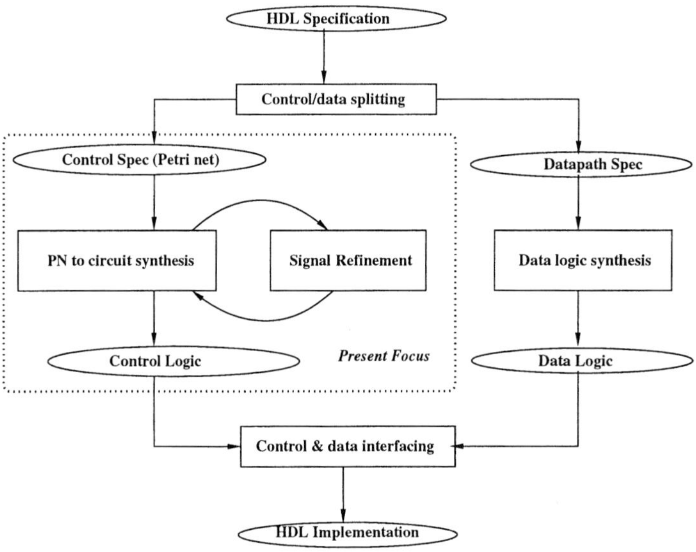

Fig. 1. Design Flow with Logic Synthesis from Petri nets.

The original HDL specification is syntactically and semantically analysed, giving rise to control and data path specifications. Data path can be synthesised using standard RTL-based (synchronous) design flow, applied to the main fragments of the data path, namely combinational logic and registers. There exist methods of converting such logic to self-timed implementations, e.g., [43]. This aspect of design is outside our scope here. The control specification is assumed to be extracted from the HDL in the form of a Petri net, which will thus act as the intermediate behavioural representation. Such an extraction is in general non-trivial and relies on rigorous semantic relationship between control-flow constructs used in typical behavioural HDLs and their equivalents in Petri nets. For example, if one uses BALSA, such constructs basically include sequencing, parallelisation, two-way and multi-way selection, arbitration and (forever, while and for) loops, as well as macro and procedure calls. Those can be translated into Petri nets quite efficiently as done for example in PEP [3] for the translation of basic high-level programming language notation, $B(PN)^2$, into Petri nets.

1.5 Methods for Logic Synthesis from Petri Nets

The question of what kind of Petri nets is appropriate for subsequent logic synthesis of control depends on the method used for synthesis. Roughly, synthesis methods are split into two main categories. The first category comprises techniques of direct mapping of Petri net constructs to logic. In various forms it appeared in [51, 20, 32, 68, 74, 6, 58]. In the framework of 1-safe Petri nets and speed-independent circuits this problem was solved in [68], however only for autonomous (no inputs) specification where all operations were initiated by the control logic specified by a labelled Petri net. Another limitation was that the technique did not cover nets with arbitrary dynamic conflicts. Hollaar's one-hot encoding method [32] allowed explicit interfacing with the environment but required fundamental mode timing conditions, use of internal state variables as outputs and could not deal with conflicts and arbitration in the specifications. Patil's method [51] works for the whole class of 1-safe nets. However, it produces control circuits whose operation uses 2-phase (non-return-to-zero) signalling. This results in lower performance than what can be achieved for 4-phase circuits used in [68].

The second category considers the Signal Transition Graph refinement of the Petri net control specification. These methods usually perform an explicit logic synthesis, by deriving Boolean equations for the output signals of the controller using the notion of next state functions obtained from the STG [14, 18]. It should be noted that sometimes the STG specification for control can be obtained directly from the original specifications, e.g., if those are provided in the form of Timing Diagrams.

In this paper we will not concentrate on the problem of synthesis of Petri nets for logic synthesis of controllers and refer the reader to most recent literature, such as [4].

Our focus will be on the most recent advances in logic synthesis from Petri nets and Signal Transition Graphs. These methods try to avoid using the state space generated by the Petri net model directly. They follow two possible approaches. The first one, called a *structural* approach, performs graph-based transformations on the STG and deals with the approximated state space by means of linear algebraic representations. The second one, called an *unfolding-based* method, represents the state space in the form of true concurrency (or partial order) semantics provided by Petri net unfoldings.

The remaining structure of the paper is as follows. Section 2 introduces the problem of synthesis of control circuits from Petri net based specifications. It will

```
y := 0;
loop
   x := READ (IN);
   WRITE (OUT, (x + y)/2);
   y := x;
end loop
```

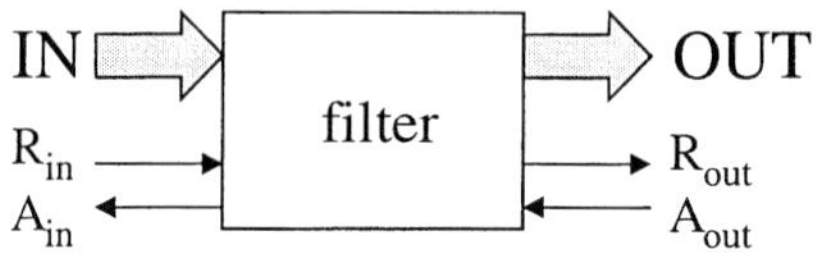

Fig. 2. High-level specification of a filter.

do it in an informal way by considering two characteristic examples of control logic to be designed by this sort of methodology. Section 3 provides an overview of the traditional state-based synthesis, which is currently implemented in the PETRIFY tool. Section 4 describes structural methods and use of integer linear programming in logic synthesis. Section 5 presents how Petri nets unfoldings and Boolean satisfiability problem (SAT) solvers can be used in the synthesis of asynchronous control logic. Section 6 briefly overviews some other related methodologies and outlines the important current and future research directions.

2 Synthesis Problem: Simple Examples and Signal Transition Graph Definition

We shall introduce the problem of synthesis of control circuits from Petri nets specifications using two simple but realistic design examples. This will also help us to present the two main types of control hardware that can be designed with the methods described in this paper. The first example, a simple data processing controller, will illustrate the design flow starting from an algorithmic, HDL-based, specification. The second one, an interface controller, will show the design starting from a waveform, Timing Diagram based, specification. Algorithmic and waveform specifications are most popular forms of behavioural notation amongst hardware designers. While describing the second example we will introduce our main specification model, Signal Transition Graph (STG).

2.1 A Simple Filter Controller

We illustrate a typical design flow by means of the example shown in Fig. 2. The algorithm describes a simple filter that reads data items from an input channel (IN) and writes the filtered data into an output channel (OUT) by averaging the last two samples, x and y. (Note that the first output value in this case may be invalid and should be ignored by the environment.) The interaction with the environment is asynchronous, using a four-phase protocol implemented by a pair of ⟨*Request,Acknowledge*⟩ signals, as shown in Fig. 3.

One of the possible implementations of the filter is depicted in the block diagram of Fig. 4. It contains two level-sensitive latches, x and y, and one adder (the averaging of x and y is achieved simply by a one-bit right shift of the bits

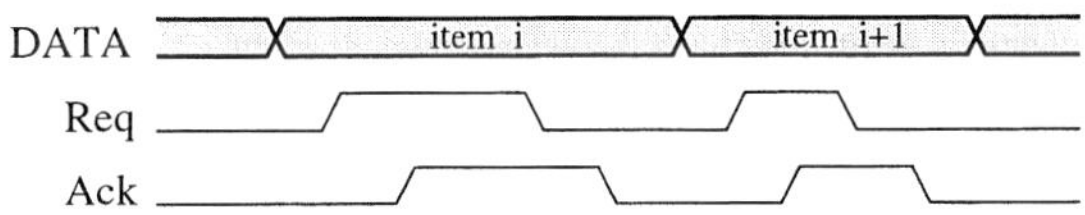

Fig. 3. Four-phase handshake protocol.

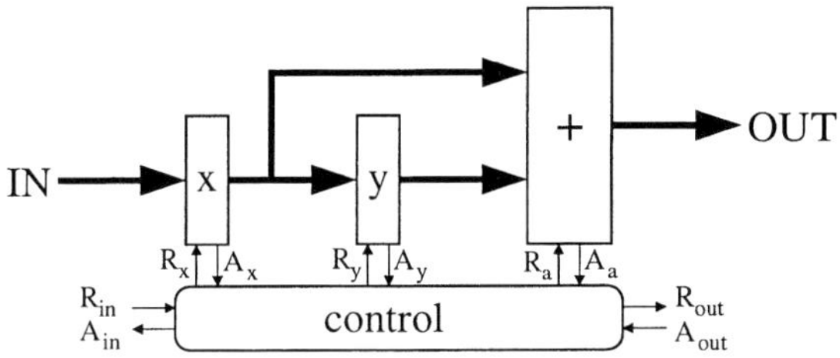

Fig. 4. Block diagram for the filter.

of the sum $x + y$). Each of the components operates according to a four-phase protocol as follows:

- The latches are transparent when R is high and opaque when low. A being high indicates that the data transfer through the latch has been completed.
- The adder starts its operation when R goes high. After a certain delay, signal A will be asserted, indicating that the addition has been finished and the output is valid. After that, R and A go low to complete the four-phase protocol.

The acknowledge signals of the latches and the adder can be implemented in many different ways, depending on how the blocks are designed. One way of doing that is by simply inserting a delay between R and A that mimics the worst-case delay of the corresponding block, as typically done for bundled-data components in micropipelines [64].

The signals $\langle R_{in}, A_{in} \rangle$ and $\langle R_{out}, A_{out} \rangle$ perform the synchronisation of the IN and OUT channels, respectively. R_{in} indicates the validity of IN. After A_{in} goes high, the environment is allowed to modify IN. On the other side, R_{out} and A_{out} should be able to control a level-sensitive latch in a similar way as described above for the latches x and y.

Synthesis of control. The synchronisation of the functional units depicted in Fig. 4 is performed by the *control* block, which is responsible for circulating the data items in the data-path in such a way that the required computations are performed as specified by the algorithm.

In this paper, we use a specially interpreted Petri nets, called Signal Transition Graphs (STGs), to specify the behaviour of asynchronous controllers. The transitions represent signal events (i.e., rising or falling edges of signals), whereas the arcs and places represent the causality relations among the events.

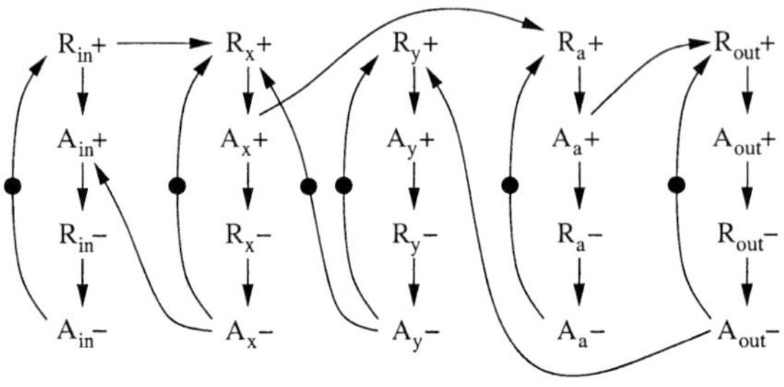

Fig. 5. Behavioural specification of the control.

Fig. 5 describes one possible behaviour of the control that results in a correct operation of the circuit. In this cases, the behaviour can be described by a *marked graph*, a subclass of Petri nets without choice. Marked graphs are often represented by omitting the places between transitions.

Each pair of req/ack signals commit a four-phase protocol, determined by the arcs $R^+ \to A^+ \to R^- \to A^- \to R^+$. The rest of the arcs are the ones that define how data items move along the data-path. For the sake of brevity, only a couple of them are discussed.

The arc $R_{in}^+ \to R_x^+$ indicates that the latch x can become transparent when there is some valid data at the IN channel. Moreover, the data can only be read once the latch y has captured the previous data from x. This is guaranteed by the arc $A_y^- \to R_x^+$.

On the other hand, the adder will start a new operation every time the latch x has acquired new data. This is indicated by the arc $A_x^+ \to R_a^+$. The result will be sent to the OUT channel when the addition has completed (arc $A_a^+ \to R_{out}^+$).

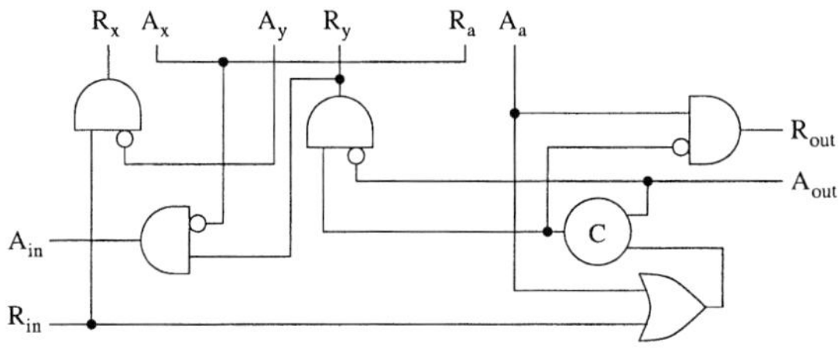

Fig. 6. Asynchronous controller for the filter.

From the specification of the control, a logic circuit can be synthesised. The circuit shown in Fig. 6 has been obtained by the PETRIFY tool.

2.2 VME Bus Controller

Our second example is a fragment of a VME bus slave interface [75]. It will help us to illustrate how the STG specification of an asynchronous controller can be derived from its original Timing Diagram specification. Fig. 7(a) depicts the interface of a circuit that controls data transfers between a VME bus and a device. The main task of the bus controller is to open and close the data transceiver through signal *d* according to a given protocol to read/write data from/to the device.

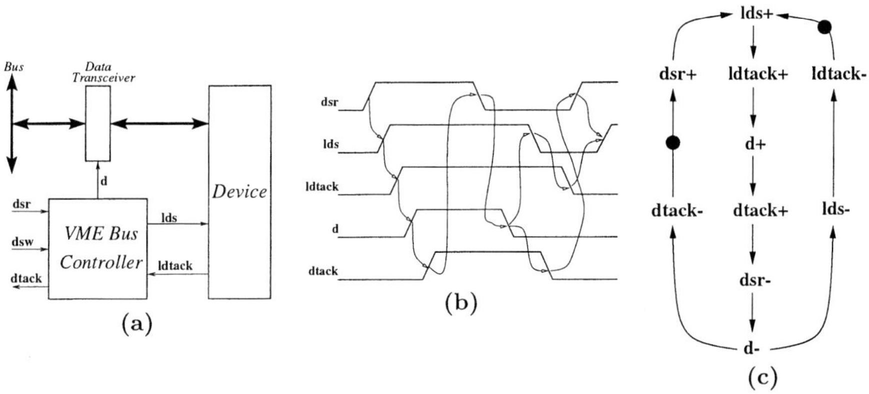

Fig. 7. VME bus controller: interface **(a)**, the timing diagram for the read cycle **(b)** and the STG for the read cycle **(c)**.

The input and output signals of the bus controller are as follows:

- *dsr* and *dsw* are input signals that request to do a read or write operation, respectively.
- *dtack* is an output signal that indicates that the requested operation is ready to be performed.
- *lds* is an output signal to request the device to perform a data transfer.
- *ldtack* is an input signal coming from the device indicating that the device is ready to perform the requested data transfer.
- *d* is an output signal that enables the data transceiver. When high, the data transceiver connects the device with the bus. The direction of the transfer (read or write) is defined by the high or low level of a special (RW) signal, which is part of the address/data bundle.

Fig. 7(b) shows a timing diagram of the read cycle. In this case, signal *dsw* is always low and not depicted in the diagram. The behaviour of the controller is as follows: a request to read from the device is received by signal *dsr*. The controller transfers this request to the device by asserting signal *lds*. When the device has the data ready (*ldtack* high), the controller opens the transceiver to transfer data to the bus (*d* high). Once data has been transferred, *dsr* will become low indicating that the transaction must be finished. Immediately after, the controller will lower signal *d* to isolate the device from the bus. After that, the transaction will be completed by a return-to-zero of all interface signals, seeking for a maximum parallelism between the bus and the device operations.

Our controller also supports a write cycle with a slightly different behaviour. For the sake of simplicity, we have described in detail only the read cycle.

The model that will be used to specify asynchronous controllers is based on Petri nets [53, 49]. It is called *Signal Transition Graph* (STG) [55, 13]. Roughly speaking, an STG is a formal model for *timing diagrams*. Now we explain how to derive an STG from a timing diagram.

From timing diagrams to signal transition graphs. A timing diagram specifies the events (signal transitions) of a behaviour and their causality relations. An STG is a formal model for this type of specifications. In its simplest form, an STG can be considered as a causality graph in which each node represents an event and each arc a causality relation. An STG representing the behaviour of the read cycle for the VME bus is shown in Fig. 7(c). Rising and falling transitions of a signal are represented by the superscripts + and −, respectively.

Additionally, an STG can also model all possible dynamic behaviours of the system. This is the rôle of the tokens held by some of the causality arcs. An event is *enabled* when it has at least one token on each input arc. An enabled event can *fire*, which means that the event occurs. When an event fires, a token is removed from each input arc and a token is put on each output arc. Thus, the firing of an event produces the enabling of another event. The tokens in the specification represent the initial state of the system.

The initial state in the specification of Fig. 7(c) is defined by the tokens on the arcs $dtack^- \rightarrow dsr^+$ and $ldtack^- \rightarrow lds^+$. In this state, there is only one event enabled, viz. dsr^+. It is an event on an input signal that must be produced by the environment. The occurrence of dsr^+ removes a token from its input arc and puts a token on its output arc. In that state, the event lds^+ is enabled. In this case, it is an event on an output signal, that must be produced by the circuit modelled by this specification.

After firing the sequence of events $ldtack^+$, d^+, $dtack^+$, dsr^- and d^-, two tokens are placed on the arcs $d^- \rightarrow dtack^-$ and $d^- \rightarrow lds^-$. In this situation, two events are enabled and can fire in any order independently from each other, i.e., these events are *concurrent*, which is naturally modelled by the STG.

Choice in signal transition graphs. In some cases, alternative behaviours, or modes, can occur depending on how the environment interacts with the system. In our example, the system will react differently depending on whether the environment issues a request to read or a request to write.

Typically, different behavioural modes are represented by different timing diagrams. For example, Fig. 8(a) and 8(b) depict the STGs corresponding to the read and write cycles, respectively. In these pictures, some arcs have been split and circles inserted in between. These circles represent *places* that can hold tokens. In fact, each arc going from one transition to another has an implicit place that holds the tokens located in that arc.

By looking at the initial markings, one can observe that the transition dsr^+ is enabled in the read cycle, whereas dsw^+ is enabled in the write cycle. The combination of both STGs models the fact that the environment can non-deterministically choose whether to start a read or a write cycle.

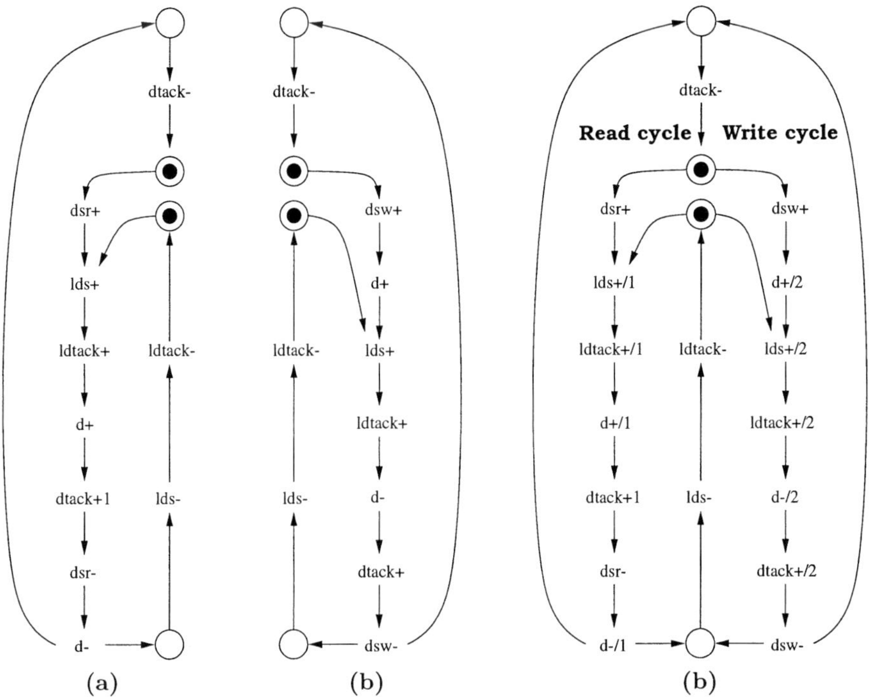

Fig. 8. VME bus controller: read cycle **(a)**, write cycle **(b)**, read and write cycles **(c)**.

This combination can be expressed by a single STG with a choice place, as shown in Fig. 8(c). In the initial state, both transitions, dsr^+ and dsw^+, are enabled. However, when one of them fires, the other is disabled since both transitions are competing for the token in the choice place. This type of choice is called *free choice* because the transitions, dsr^+ and dsw^+, connected to the choice place have no other input places that could affect the process of choice making.

Here is where one can observe an important difference between the expressiveness of STGs and timing diagrams: the former are capable of expressing non-deterministic choices while the latter are not.

2.3 More Formal Definition of Signal Transition Graphs

To be able to introduce the methods of synthesis of asynchronous circuits in subsequent sections, we will need a more formal definition of an STG. STGs are a particular type of labelled Petri nets, where transitions are associated with the changes in the values of binary variables. These variables can for example be associated with wires, when modelling interfaces between blocks, or with input, output and internal signals in a control circuit.

A *net* is a triple $N \stackrel{\mathrm{df}}{=} (P, T, F)$ such that P and T are disjoint sets of respectively *places* and *transitions*, and $F \subseteq (P \times T) \cup (T \times P)$ is a *flow relation.* A *marking* of N is a multiset M of places, i.e., $M : P \to \{0, 1, 2, \ldots\}$. We adopt the standard rules about representing nets as directed graphs, viz. places are represented as circles, transitions as rectangles, the flow relation by arcs, and markings are shown by placing tokens within circles. As usual, ${}^\bullet z \overset{\text{df}}{=} \{y \mid (y, z) \in F\}$ and $z^\bullet \overset{\text{df}}{=} \{y \mid (z, y) \in F\}$ denote the *pre-* and *postset* of $z \in P \cup T$. We will assume that ${}^\bullet t \neq \emptyset$, for every $t \in T$. A *net system* is a pair $\Sigma \overset{\text{df}}{=} (N, M_0)$ comprising a finite net N and an *initial* marking M_0. We assume the reader is familiar with the standard notions of the theory of Petri nets, such as the *enabledness* and *firing* of a transition and marking *reachability*, as well as other standard notions and classification associated with Petri nets [49].

A *Signal Transition Graph (STG)* is a quadruple $\Gamma \overset{\text{df}}{=} (N, M_0, Z, \lambda)$, where

- $\Sigma = (N, M_0)$ is a Petri net (PN) based on a net $N = (P, T, F)$,
- Z is a finite set of binary signals, which generates a finite alphabet $Z^\pm = Z \times \{+, -\}$ of signal transitions
- $\lambda : T \to Z^\pm$ is a labelling function.

Labelling λ does not need to be 1-to-1 (some signal transitions may occur several times in the PN), and it may be extended to a partial function, in order to allow some transitions to be "dummy" ones (denoted by ϵ), that is to denote "silent events" that do not change the state of the circuit.

When talking about individual signal transitions, the following meaning will be associated with their labels. A label x^+ is used to denote the transition of signal x from 0 to 1 (rising edge), while x^- is used for a 1 to 0 transition (falling edge). In the following it will often be convenient to associate STG transitions directly with their labels, "bypassing" their Petri net identity. In such cases if the labelling is not 1-to-1 (so called multiple labelling), we will also use a subscript or an index separated by slash denoting the instance number of the $x^\pm$.

Sometimes, when reasoning on a pure event-based level, it will also be convenient to hide the direction of a particular edge and use $x^\pm$ to denote either a x^+ transition or an x^- transition.

An STG inherits the basic operational semantics from the behaviour of its underlying Petri net. In particular, this includes: (i) the rules for transition enabling and firing, (ii) the notions of reachable markings, traces, and (iii) the temporal relations between transitions (precedence, concurrency, choice and conflict). Likewise, STGs also inherit the various structural (marked graph, free-choice, etc.) and behavioural properties (boundedness, liveness, persistency, etc.), and the corresponding classification of PNs. Namely:

- **Choice place.** A place is called a choice (or conflict) place if it has more than one output transition.
- **Marked graph and State machine.** A PN is called a *marked graph* (MG) if each place has exactly one input and one output transition. Dually, a PN is called a *state machine* (SM) if each transition has exactly one input and one output place. MGs have no choice. Safe SMs have no concurrency.

- **Free-choice.** A choice place is called *free-choice* if every its output transition has only one input place. A PN is *free-choice* if all its choice places are free-choice.
- **Persistency.** A transition $t \in T$ is called *non-persistent* if some reachable marking enables t together with another transition t', and t becomes disabled after firing t'. Non-persistency of t with respect to t' is also called a *direct conflict* between t and t'. A PN is *persistent* if it does not contain any non-persistent transition.
- **Boundedness and safeness.** A PN *k-bounded* if for every reachable marking the number of tokens in any place is not greater than k (a place is called k-bounded if for every reachable marking the number of tokens in it is not greater than k). A PN is *bounded*, if there is a finite k for which it is k-bounded. A PN is *safe* if it is 1-bounded (a 1-bounded place is called a safe place).
- **Liveness.** A PN is *live* if for every transition t and every reachable marking M there is a firing sequence that leads to a marking M' enabling t.

The signal transition labelling of an STG may sometimes differentiate between input and non-input signals, thus forming two disjoint subsets, Z_I (for inputs) and Z_O (for non-inputs, or simply outputs), such that $Z = Z_I \cup Z_O$. An STG is called *autonomous* if it has no input signals (i.e., $Z_I = \emptyset$).

Graphically, an STG can either be represented in the standard form of a labelled PN, drawing transitions as bars or boxes and places as circles, or in the so-called STG shorthand form. The latter, as was first shown in the above examples, designates transitions directly by their labels and omits places that have only one input and one output transition.

Examples of STGs, in their shorthand notation, were shown in Fig. 8, describing a simple VME bus controller example. It was assumed in them that $Z_I = \{dsr, dsw, ldtack\}$ and $Z_O = \{lds, dtack, d\}$. The first two STGs, in Fig. 8(a) and 8(b), are marked graphs (they do not have choice on places). The third one, in Fig. 8(c), modelling both read and write operation cycles, is not a marked graph because it contains places with multiple input and output transitions. It is not a free-choice net either, because one of its choice places, the input to transitions $lds^+/1$ and $lds^+/2$, is not a free-choice place. The latter is however a *unique choice place* because whenever one of the above two transitions is enabled the other is not, which is guaranteed by the other choice place, which is a free-choice one. Thus, behaviourally, this net does not lead to dynamic conflicts (arbitration) or confusion, as it is free from any interference between choice and concurrency.

3 State-Based Synthesis from Signal Transition Graphs

The main purpose of this section is to present a state-based method to design asynchronous control circuits, i.e., those circuits that synchronise the operations performed by the functional units of the data-path through handshake protocols. The method uses the STG model of a circuit as its initial specification. The key

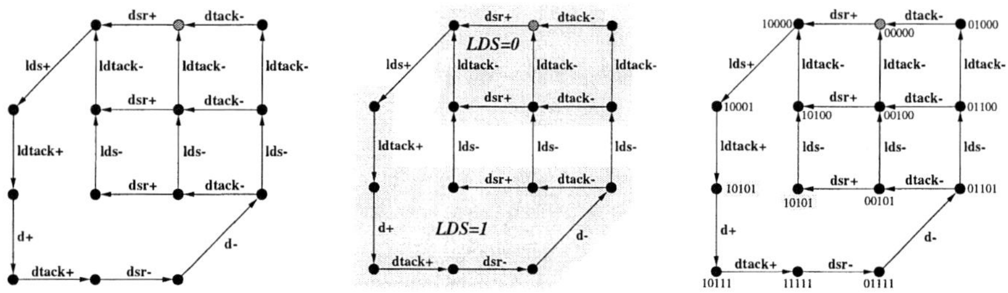

(a) (b) (c)

Fig. 9. Reachability graph of read cycle **(a)**, its binary partitioning for signal *lds* **(b)**, and the encodings of the reachable states **(c)**. The order of signals in the binary encodings is: *dsr*, *dtack*, *ldtack*, *d*, *lds*.

steps in this method are the generation of a state graph, which is a binary encoded reachability graph of the underlying Petri net, and deriving Boolean equations for the output signals via their next state functions obtained from the state-graph. This method is surveyed here very briefly and informally, using our VME bus controller example. For more details the reader is referred to the book and the PETRIFY tool [18].

3.1 State Graphs

State space. An STG is a succinct representation of the behaviour of an asynchronous control circuit that describes the causality relations among the events. However, the state space of the system must be derived by exploring all possible firing orders of the events. Such exploration may result in a state space much larger than the specification.

Unfortunately, the synthesis of asynchronous circuits from STGs requires an exhaustive exploration of the state space. Finding efficient representations of the state space is a crucial aspect in building synthesis tools. Other techniques based on direct translation of Petri Nets into circuits or on approximations of the state space exist [42, 50], but usually produce circuits with area and performance penalty.

Going back to our example of the VME bus controller, Fig. 9(a) shows the reachability graph corresponding to the behaviour of the read cycle. The initial state is depicted in gray.

For simplicity, the write cycle will be ignored in the rest of this section. Thus, we will consider the synthesis of a bus controller that only performs read cycles.

Binary interpretation. The events of an asynchronous circuit are interpreted as rising and falling transitions of digital signals. A rising (falling) transition represents a switch from 0 (1) to 1 (0) of the signal value. Therefore, when considering each signal of the system, a binary value can be assigned to each state for that signal. All those states visited after a rising (falling) transition and before a falling (rising) transition represent situations in which the signal value is 1 (0).

In general, the events representing rising and falling transitions of a signal induce a partition of the state space. As an example, let us take signal *lds* of the bus controller. Fig. 9(b) depicts the partition of states. Each transition from `LDS=0` to `LDS=1` is labelled by lds^+ and each transition from `LDS=1` to `LDS=0` is labelled by lds^-.

It is important to notice that rising and falling transitions of a signal must alternate. The fact that a rising transition of a signal is enabled when the signal is at 1 is considered a specification error. More formally, a specification with such problem is said to have an *inconsistent state coding.*

After deriving the value of each signal, each state can be assigned a binary vector that represents the value of all signals in that state. A transition system with a binary interpretation of its signals is called a *state graph* (SG). The SG of the bus controller read cycle is shown in Fig. 9(c).

3.2 Deriving Logic Equations

In this section we explain how an asynchronous circuit can be automatically obtained from a behavioural description. We have already distinguished two types of signals in a specification: inputs and outputs. Further, some of the outputs may be observable and some internal. Typically, observable outputs correspond to those included in the specification, whereas internal outputs correspond to those inserted during synthesis and not observable by the environment. Synthesising a circuit means providing an implementation for the output signals of the system.

This section gives an overview of the methods used for the synthesis of asynchronous circuits from an SG.

System behaviour. The specification of a system models a protocol between its inputs and outputs. At a given state, one or several of these two situations may happen:

- The system is waiting for an input event to occur. For example, in the state `00000` of Fig. 9(c), the system is waiting for the environment to produce a rising transition on signal *dsr*.
- The system is expected to produce a non-input (output or internal) event. For example, the environment is expecting the system to produce a rising transition on signal *lds* in state `10000`.

In concurrent systems, several of these things may occur simultaneously. For example, in state `00101`, the system is expecting the environment to produce dsr^+, whereas the environment is expecting the system to produce lds^-. In some other cases, such as in state `01101`, the environment may be expecting the system to produce several events concurrently, e.g., $dtack^-$ and lds^-.

The particular order in which concurrent events will occur will depend on the delays of the components of the system. Most of the synthesis methods discussed

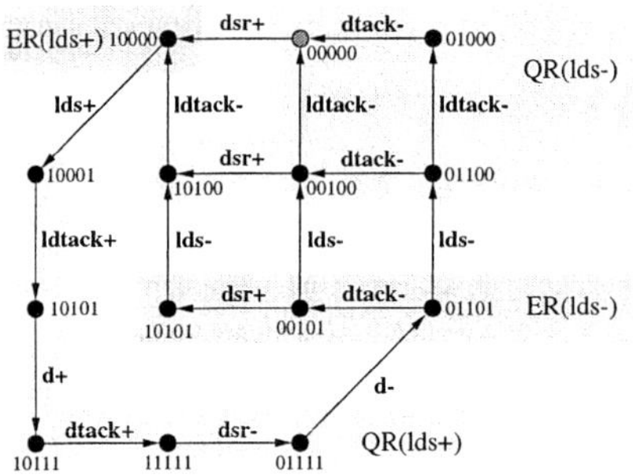

(a)

State region	current value of *lds*	next value of *lds*
ER(*lds*$^+$)	0	1
QR(*lds*$^+$)	1	1
ER(*lds*$^-$)	1	0
QR(*lds*$^-$)	0	0

(b)

Fig. 10. Excitation and quiescent regions for signal *lds* **(a)** and the corresponding next-state function **(b)**.

here aim at synthesising circuits whose correctness does not depend on the actual delays of the components. These circuits are called *speed-independent.*

A correct implementation of the output signals must be in such a way that signal transitions on those signals must be generated *if and only if* the environment is expecting them. Unexpected signal transitions, or not generating signal transitions when expected, may produce circuit malfunctions.

Excitation and Quiescent Regions. Let us take one of the output signals of the system, say *lds*. According to the specification, the states can be classified into four regions:

- The *positive excitation region*, ER(*lds*$^+$), includes all those states in which a rising transition of *lds* is enabled.
- The *negative excitation region*, ER(*lds*$^-$), includes all those states in which a falling transition of *lds* is enabled.
- The *positive quiescent region*, QR(*lds*$^+$), includes all those states in which signal *lds* is at 1 and *lds*$^-$ is not enabled.
- The *negative quiescent region*, QR(*lds*$^-$), includes all those states in which signal *lds* is at 0 and *lds*$^+$ is not enabled.

Fig. 10(a) depicts these regions for signal *lds*. It can be easily deduced that ER(*lds*$^+$) $\cup$ QR(*lds*$^-$) and ER(*lds*$^-$) $\cup$ QR(*lds*$^+$) are the sets of states in which signal *lds* is at 0 and 1, respectively.

Next-State Functions. Excitation and quiescent regions represent sets of states that are behaviourally equivalent from the point of view of the signal for which they are defined. The semantics of these regions are the following:

- ER(*lds*$^+$) is the set of states in which *lds* is at 0 and the system must change it to 1.
- ER(*lds*$^-$) is the set of states in which *lds* is at 1 and the system must change it to 0.

- QR(lds^+) is the set of states in which lds is at 1 and the system must not change it.
- QR(lds^-) is the set of states in which lds is at 0 and the system must not change it.

According to this definition, the behaviour of each signal can be determined by calculating the *next value* expected at each state of the SG. This behaviour can be modelled by Boolean equations that implement the so-called *next-state* functions (see Fig. 10(b)).

Let us consider again the bus controller and try to derive a Boolean equation for the output signal lds. A 5-variable Karnaugh map for Boolean minimisation is depicted in Fig. 11. Several things can be observed in that table. There are many cells of the map with a *don't care* (–) value. These cells represent binary encodings not associated to any of the states of the SG. Since the system will never reach a state with those encodings, the next-state value of the signal is irrelevant.

lds=0

ldtack,d \ dsr,dtack	00	01	11	10
00	0	0	-	1
01	-	-	-	-
11	-	-	-	-
10	0	0	-	0

lds=1

ldtack,d \ dsr,dtack	00	01	11	10
00	-	-	-	1
01	-	-	-	-
11	-	1	1	1
10	0	0	-	1 / 0

CSC conflict

Fig. 11. Karnaugh map for the minimisation of signal lds.

The shadowed cells correspond to states in the excitation regions of the signal. The rest of cells correspond to states in some of the quiescent regions. If we call f_{lds} the next-state function for signal lds, here are some examples on the value of f_{lds}:

$$
\begin{array}{ll}
f_{lds}(10000) = 1 & \text{state in } \mathsf{ER}(lds^+) \\
f_{lds}(10111) = 1 & \text{state in } \mathsf{QR}(lds^+) \\
f_{lds}(00101) = 0 & \text{state in } \mathsf{ER}(lds^-) \\
f_{lds}(01000) = 0 & \text{state in } \mathsf{QR}(lds^-)
\end{array}
$$

3.3 State Encoding

At this point, the reader must have noticed a peculiar situation for the value of the next-state function for signal lds in two states with the same binary encoding: 10101. This binary encoding is assigned to the shadowed states in Fig. 9(c).

Unfortunately, the two states belong to two different regions for signal lds, namely to ER(lds^-) and QR(lds^+). This means that the binary encoding of

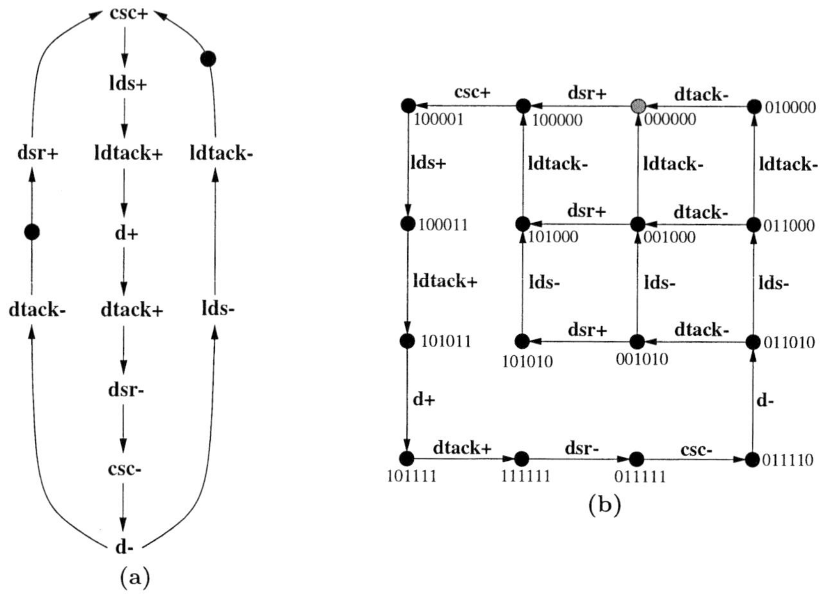

Fig. 12. An STG (a) and its SG (b) satisfying the CSC property.

the SG signals alone cannot determine the future behaviour of *lds*. Hence, an ambiguity arises when trying to define the next-state function. This ambiguity is illustrated in the Karnaugh map of Fig. 11.

Roughly speaking, this phenomenon appears when the system does not have enough memory to "remember" in which state it is. When this occurs, the system is said to violate the *Complete State Coding* (CSC) property. Enforcing CSC is one of the most difficult problems in the synthesis of asynchronous circuits.

Fig. 12 presents a possible solution for the SG of the VME bus controller. It consists of inserting a new signal, *csc*, that adds more memory to the system. After the insertion, the two conflicting states are disambiguated by the value of *csc*, which is the last value in the binary vectors of Fig. 12.

Now Boolean minimisation can be performed and logic equations can be obtained (see Fig. 13). In the context of Boolean equations representing gates we shall liberally use the "=" sign to denote "assignment", rather than mathematical equality. Hence *csc* on the left-hand side of the last equation stands for the *next* value of signal *csc*, while *csc* on the right-hand side corresponds to its current value. The resulting circuit contains cycles: the combinational feedbacks play the rôle of local memory in the system.

The circuit shown in Fig. 13 is said to be speed-independent, i.e., it works correctly regardless of the delays of its components. For this to be true, it is required that each Boolean equation is implemented as one *complex gate*. This roughly means that the internal delays within each gate are negligible and do not produce any externally observable spurious behaviour. However, the external delay of the gates can be arbitrarily long.

$$
\begin{aligned}
lds &= d + csc \\
dtack &= d \\
d &= ldtack \cdot csc \\
csc &= dsr \cdot (csc^{+}\overline{ldtack})
\end{aligned}
$$

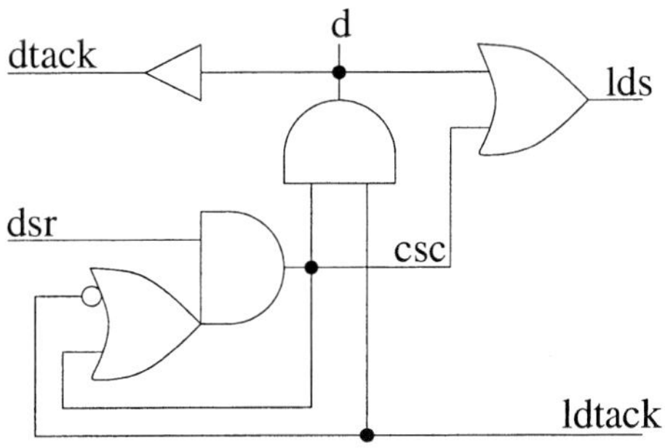

Fig. 13. Logic equations and implementation of the VME bus controller.

Note that signal *dtack* is merely implemented as a buffer, and a wire is enough to preserve that behaviour. But note that the specification indicates that the transitions of *dtack* must occur always *after* the transitions of d. For this reason, the resulting equation is $dtack = d$ and not vice versa. Thus, the buffer introduces the required delay to enforce the specified causality.

3.4 Properties for Implementability

In conclusion to this section let us summarise the main properties required for the STG specification to be implementable as a speed-independent circuit [18]:

- *Boundedness* of the STG that guarantees the SG to be finite.
- *Consistency* of the STG, that ensures that the rising and falling transitions of each signal alternate in all possible runs of the specification.
- *Completeness of state encoding* (CSC) that ensures that there are no two different states with the same signal encoding but different behaviour of the output or internal signals.
- *Persistency* of signal transitions in such a way that no signal transition can be disabled by another signal transition, unless both signals are inputs. This property ensures that no short glitches, known as *hazards*, will appear at the disabled signals. (Arbitration is implemented by 'factoring out' the arbiter into the environment and using a special circuit able to resolve metastability.)

4 Synthesis Using Structural Methods, Linear Programming and STG Decomposition

4.1 Rationale

Structural methods provide a way to avoid the state space explosion problem, given that they rely on succinct representations of the state space. The main benefit of using structural methods is the ability to deal with large and highly concurrent specifications, that cannot be tackled by state-based methods. On the other hand, structural methods are usually conservative and approximate, and can only be exact when the behaviour of the specifications is restricted in some sense. For instance, in [66] structural methods for the synthesis of asynchronous circuits are presented for the class of *marked graphs*, a very restricted class of Petri nets where choices are not allowed. In this section we present structural methods to solve some of the main problems in the synthesis of asynchronous control circuits from well-formed specifications.

As it was explained in previous sections, the synthesis of asynchronous circuits from an STG can be separated into two steps [18]: (i) checking and (possibly) enforcing implementability conditions and (ii) deriving the next-state function for each signal generated by the system. Most of the existing CAD tools for synthesis perform steps (i) and (ii) at the underlying state graph level, thus suffering from the state space explosion problem.

In order to avoid the state explosion problem, structural methods for steps (i) and (ii) have been proposed in the literature. Approaches like the ones presented in [66, 50, 9, 8] can be considered purely structural. Among the methods applied by these approaches, graph theoretic-based and linear algebraic are the essential techniques. The work presented in this section uses both linear algebraic methods and graph theoretic-based methods.

Regarding step (i), in this section an encoding technique to ensure implementability is presented. It is inspired by the work of René David [20]. The main idea is to insert a new set of signals in the initial specification in a way that unique encoding is guaranteed in the transformed specification.

To the best of our knowledge, the results reported in [38, 29] are the first ones that use linear algebraic techniques to approach the encoding problem. In the former approach, a complete characterisation of the encoding problem is presented, provided that, like in Section 5, unfoldings are used to represent the underlying state space of the net. Linear algebraic methods to verify the encoding are presented in this section, where the computation of the unfolding is not performed, at the expense of checking only sufficient conditions for synthesis. However, the experimental results indicate that this approach is highly accurate and often provides a significant speed-up compared with [38, 39]. One can imagine a design flow where the methods presented in this section are used to pre-process the specifications, and complete methods like the ones presented in Section 5 are only used when purely structural methods fail.

Another alternative to alleviate the state space explosion problem is by using decomposition techniques. We apply them when performing step (ii). More specifically, in this section an algorithm for computing the set of signals needed to synthesise a given signal is presented, which also uses linear algebraic techniques. This allows to project the behaviour into that set of signals and perform the synthesis on the projection.

In summary, this section covers the two important steps (i) and (ii) in the synthesis of asynchronous circuits: it proposes powerful methods for checking CSC/USC and a method for decomposing the specification into smaller ones *while preserving the implementability conditions.*

4.2 Structural Technique to Ensure a Correct Encoding

The first example of use of structural methods is presented in this section. The technique is inspired by previous work on using a special type of cells, called *David cells.* This type of cells, first introduced in [20], were used in [69] to mimic the token flow of a Petri net. Fig. 14 depicts a very simple example on how these cells can be abutted to build a distributor that controls the propagation of activities along a ring.

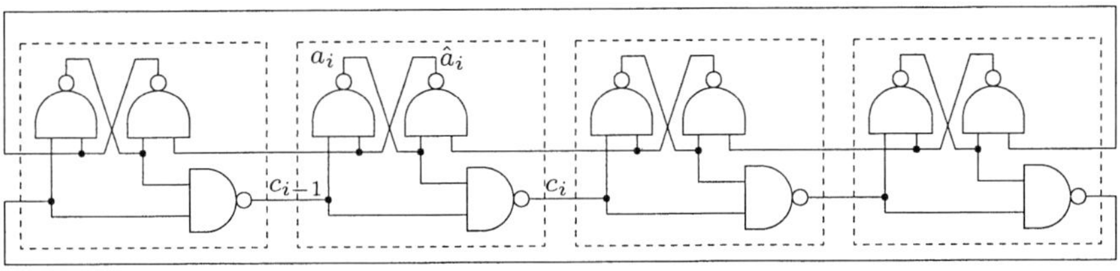

Fig. 14. Distributor built from David cells [42].

The behaviour of one of the cells in the distributor can be summarised by the following sequence of events:

$$\cdots \to \underbrace{c_{i-1}^-}_{\substack{i\text{-th cell}\\ \text{excitation}}} \to \underbrace{a_i^+ \to \hat{a}_i^-}_{i\text{-th cell setting}} \to$$

$$\to \underbrace{\hat{a}_{i-1}^+ \to a_{i-1}^- \to c_{i-1}^+}_{(i-1)\text{-th cell resetting}} \to \underbrace{c_i^-}_{\substack{(i+1)\text{-th cell}\\ \text{excitation}}} \to \cdots$$

Let us explain how one can use David Cells to ensure a correct encoding of the system specified by a given STG. The main idea is to add a new signal for each place of the original net. The semantics of the new signal inserted is to mimic the token flow of the corresponding place in the original net. The technique is shown in Fig. 15.

For instance place p_4 induces the creation of signal sp_4. Moreover, provided that when transition $dtack^+$ is enabled, it adds a token to p_4, in the transformed net it will induce that near (preceding) $dtack^+$ there must be an sp_4^+. A similar reason makes to have sp_4^- near (following) dsr^-. New transitions are inserted in a special way: internal signal transitions must not be inserted in front of an input signal transition. The reason for that is to try to preserve the I/O interface (see more on this in [7]).

The derived STG is guaranteed to have a correct encoding (in the example, the right STG is guaranteed to satisfy the USC property). The theory underlying the technique can be found in [9]. It can be applied for any STG with the underlying free-choice live and safe Petri net (FCLSPN).

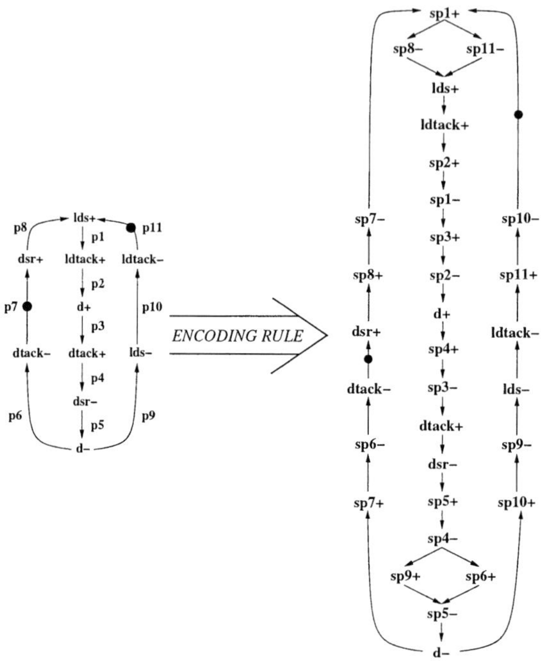

Fig. 15. Encoding rule applied to the VME Bus Controller example.

4.3 ILP Models for Fast Encoding Verification

The main drawback of this technique is that a correct encoding is ensured at the expense of inserting a lot of new signals into the net, and thus the final implementation can be very inefficient in terms of area and/or performance. Therefore it would be nice to have an *oracle* that could tell us when the application of the encoding technique is needed, provided that the computation of the state space and subsequent checking cannot be done for the specification at hand due to efficiency reasons.

This section presents Integer Linear Programming (ILP) models as oracles that we can use to verify the encoding in an STG. The good news is that when we query such oracles, usually it takes short time for them to answer, even for very large STGs. The bad news is that they are not perfect oracles: we can only trust them when they say "Yes, your STG is correctly encoded".

In this section we assume some basic knowledge of *Linear Programming* (see, e.g., [56]). The rest of this section has three main parts: first it is shown how to use linear algebraic techniques for deciding whether a given marking M is reachable in a Petri net. Second, using this technique, models for finding encoding conflicts are presented, and third, experimental results are shown.

Approximation of the reachability set of a PN. Computing the reachability graph from a given PN is a very hard problem, because the size of the reachability graph may grow exponentially with respect to the size of the PN, or it even can be infinite. The main reason is that the concurrency in the PN leads to a blow up in the reachability graph. The reader can find in [65] an in-depth discussion on the rôle of concurrency in relation to the size of the reachability graph.

Therefore, it is interesting to approach the problem of reachability using other models or techniques. In this section we describe how to use ILP techniques to compute approximations of reachable markings of a PN.

Given a firing sequence $M_0 \xrightarrow{\sigma} M$ of a PN N, the number of tokens for each place p in M is equal to the number of tokens of p in M_0 plus the number of tokens added by the input transitions of p appearing in σ minus the tokens removed by the output transitions of p appearing in σ, which can be expressed as the following *token conservation equation:*

$$M(p) = M_0(p) + \sum_{t \in {}^\bullet p} \#(\sigma, t) F(t, p) - \sum_{t \in p^\bullet} \#(\sigma, t) F(p, t) \ .$$

Definition 1 (Incidence matrix of a PN). *The matrix* $\mathbf{N} \in \{-1, 0, 1\}^{|P| \times |T|}$ *defined by* $\mathbf{N}(p, t) \stackrel{\text{df}}{=} F(p, t) - F(t, p)$ *is called the* incidence matrix *of* N.

Definition 2 (Parikh vector). *Let* σ *be a feasible sequence of* N. *The vector* $\boldsymbol{\sigma} \stackrel{\text{df}}{=} (\#(\sigma, t_1), ..., \#(\sigma, t_n))$ *is called the* Parikh vector *of* σ.

Using the previous definitions, the token conservation equations for all the places in the net can be written in the following matrix form:

$$M = M_0 + \mathbf{N} \cdot \boldsymbol{\sigma} \ .$$

This equation allows to approximate the reachability set of a Petri net by means of an ILP:

Definition 3 (Marking Equation). *If a marking* M *is reachable from* M_0, *then there exists a sequence* σ *such that* $M_0 \xrightarrow{\sigma} M$, *and the* marking equation

$$M = M_0 + \mathbf{N} \cdot X$$

has at least one solution $X \in \mathbb{N}^{|T|}$.

Note that the marking equation provides only a necessary condition for reachability. If the marking equation is infeasible, then M is not reachable from M_0, but the inverse does not hold in general: there are markings satisfying the marking equation which are not reachable. Those markings are said to be *spurious* [59]. Fig. 16(a,b,c) presents an example of spurious marking: the Parikh vector $\boldsymbol{\sigma} = (320011)$ and the marking $M = (00020)$ are a solution of the marking equation shown in Fig. 16(b) for the Petri net in Fig. 16(a)[1]. However, M is not reachable: only sequences visiting *negative* markings can lead to M. Fig. 16(c) depicts the graph containing the reachable markings and the spurious markings (shadowed). This graph is called the *potential reachability graph.* The initial marking is represented by the state (10000).

[1] Both in the figure and the explanation, we abuse the notation and skip the commas in the definition of Parikh vectors and markings.

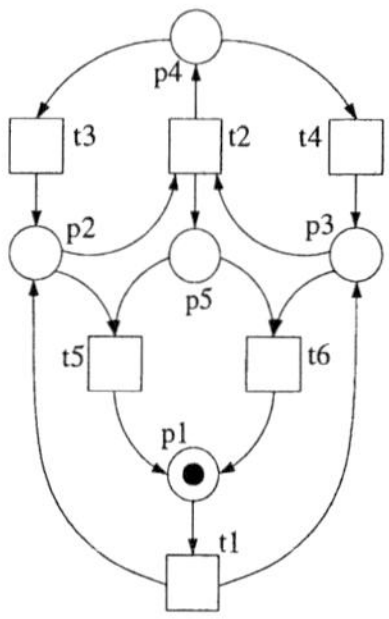

(a)

$$\mathrm{m} = \mathrm{m}_0 + \mathrm{N}\,\vec{\sigma}$$

$$\begin{bmatrix}0\\0\\0\\2\\0\end{bmatrix} = \begin{bmatrix}1\\0\\0\\0\\0\end{bmatrix} + \begin{bmatrix}-1 & 0 & 0 & 0 & +1 & +1\\ +1 & -1 & +1 & 0 & -1 & 0\\ +1 & -1 & 0 & +1 & 0 & -1\\ 0 & +1 & -1 & -1 & 0 & 0\\ 0 & +1 & 0 & 0 & -1 & -1\end{bmatrix} \begin{bmatrix}3\\2\\0\\0\\1\\1\end{bmatrix}$$

(b)

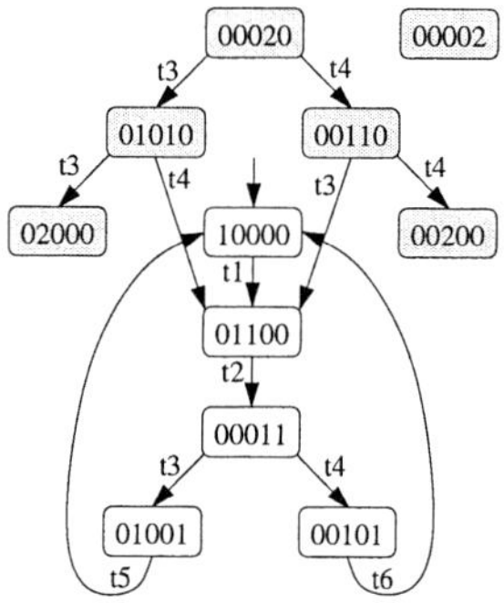

(c)

Fig. 16. A Petri net **(a)**, a spurious solution $M = (00020)^T$ **(b)**, and the potential reachability graph **(c)**.

ILP models to find encoding conflicts. Let us explain with the example of the VME Bus Controller how to derive an ILP formulation that detects USC/CSC conflicts in a given STG.

The incidence matrix of the STG corresponding to the VME example is as as follows:

	lds^+	dsr^+	$ldtack^+$	$ldtack^-$	d^+	$dtack^-$	$dtack^+$	lds^-	drs^-	d^-
p_1	+1	0	-1	0	0	0	0	0	0	0
p_2	0	0	+1	0	-1	0	0	0	0	0
p_3	0	0	0	0	+1	0	-1	0	0	0
p_4	0	0	0	0	0	0	+1	0	-1	0
p_5	0	0	0	0	0	0	0	0	+1	-1
p_6	0	0	0	0	0	-1	0	0	0	+1
p_7	0	-1	0	0	0	+1	0	0	0	0
p_8	-1	+1	0	0	0	0	0	0	0	0
p_9	0	0	0	0	0	0	0	-1	0	+1
p_{10}	0	0	0	-1	0	0	0	+1	0	0
p_{11}	-1	0	0	+1	0	0	0	0	0	0

The initial marking of the underlying Petri net is $M_0 \stackrel{\text{df}}{=} (00000010001)$, and the vector $x = (1110000000)$ is a solution of the marking equation ($M_1 = M_0 + \mathbf{N}x$). It means that the sequence of transitions corresponding to the Parikh vector x is fireable at M_0, and it leads to M_1, where $M_1 = (01000000000)$. From M_1, the vector $z = (0100111011)$ is a solution of the marking equation, ($M_2 = M_1 + \mathbf{N}z$), where $M_2 = (00000001100) \neq M_1$. The non-zero positions of vector z correspond to transitions d^+, $dtack^+$, dsr^-, d^-, $dtack^-$ and dsr^+. Looking at vector z, one can realise that for each signal appearing in it, the same number of rising and falling transitions of the signal appear (for instance, d^+ and d^- occur once). This type of sequences are called *complementary sequences.* The importance of finding complementary sequences is due to the fact that they connect two markings (M_1 and M_2 in the example) that have the same encoding, since that each signal appearing in the sequence *ends up with the same value that it had at the beginning.* The reader can assign any meaningful value to each signal in marking M_1 and check that M_2 will have the same encoding.

So, according to the marking equation, there are two different markings, M_1 and M_2, such that M_2 is reachable from M_1 by firing a complementary sequence, i.e., both markings have the same encoding. We found an USC conflict. The corresponding ILP model is:

ILP model for USC checking:

$$
\boxed{\begin{array}{l}\text{Reachability conditions:}\\ \quad M_1 = M_0 + \mathbf{N}x \\ \quad M_2 = M_1 + \mathbf{N}z \\ \quad M_1, M_2, x, z \geq 0, x, z \in \mathbb{Z}^{|T|}\end{array}}
$$

$$
\begin{array}{l}\texttt{z is complementary seq.}\\ M_1 \neq M_2\end{array} \tag{1}
$$

Note, that as it was said in the previous section, the marking equation provides only sufficient conditions for a marking to be reachable. Therefore the markings M_1 and M_2, that are solution for model (1), can indeed be spurious, and the corresponding model will incorrectly use them as example of encoding conflicts. This is why in the introduction we said that our ILP models are non-perfect oracles: only when the model finds no solution (a conflict between two markings) one can be sure that the STG is free of conflicts. On the contrary, when they find a conflict, only for very restricted classes of nets (marked graphs or live, safe and cyclic free-choice nets [21]) one can be sure that the conflict is a real one.

Now let us show how to find CSC conflicts using ILP techniques. Informally, for a given signal a of the STG, a CSC conflict exists for a if the following conditions hold: let $a_i^{\pm}$ be a transition of signal a. Then, a CSC conflict exists if: (i) M_2 is reachable from M_1, (ii) M_1 and M_2 have the same code, (iii) $a_i^{\pm}$ is enabled in M_1 and (iv) for every transition $a_j^{\pm}$ of signal a, $a_j^{\pm}$ is not enabled in M_2. For safe systems, the enabledness of a transition x at a marking M can be characterised by the sum of tokens of the places in ${}^{\bullet}x$ at M: x is enabled at M

if and only if the sum of tokens of the places in $^\bullet x$ is equal to the number of places in $^\bullet x$:

ILP model for CSC checking:

$$\begin{array}{ll} (i) & \boxed{\text{Reachability conditions (same as in (1))}} \\ (ii) & \texttt{z is complementary seq.} \\ (iii) & \sum_{p \in {}^\bullet a_i^\pm} M_1(p) = |{}^\bullet a_i^\pm| \\ (iv) & \forall a_j^\pm : \sum_{p \in {}^\bullet a_j^\pm} M_2(p) < |{}^\bullet a_j^\pm| \end{array} \tag{2}$$

Note that the constraint $M_1 \neq M_2$ is not needed in (2). If we continue with the example of the VME Bus Controller, it can be shown that the USC conflict described in the previous section is also a CSC conflict for signal d. Given that it has been shown before that the assignments $x = (1110000000)$ and $z = (0100111011)$ satisfy the first two constraints, now we show that constraints (iii) and (iv) are also satisfied by x and z. The former constraint is satisfied because

$$\sum_{p \in {}^\bullet d^+} M_1(p) = M_1(p_2) = 1 = |\{p_2\}| = |{}^\bullet d^+| ,$$

and constraint (iv) is also satisfied since

$$\sum_{p \in {}^\bullet d^+} M_2(p) = M_2(p_2) = 0 < 1 = |\{p_2\}| = |{}^\bullet d^+| .$$

Note that constraint (iv) is not verified for transition d^-, because the consistency of the STG is assumed. Thus, a CSC conflict has been detected in the VME Bus Controller example.

Experimental Results on Using ILP to Verify the Encoding. The ILP methods presented have been implemented in MOEBIUS, a tool for the synthesis of speed-independent circuits. The experiments have been performed on a *Pentium*^{TM} 4/2.53 GHz and 512M RAM.

The experiments for CSC/USC detection are presented in Tables 1 and 2. Each table reports the CPU time of each approach in seconds. We use 'time' and 'mem' to indicate that the algorithm had not completed within 10 hours or produced a memory overflow, respectively. The following tools were compared:

- CLP: the approach presented in [38] for the verification of USC/CSC. It uses non-linear integer programming methods and works on STG unfolding.
- SAT: the approach presented in [39] for the verification of CSC[2]. It uses a satisfiability solver and works on STG unfolding.
- ILP: the approach presented in this section.

From the results one can conclude, as it was expected, that checking USC is easier than checking CSC, given the different nature of the two problems: for

[2] Checking for USC is not implemented in our version of SAT

Table 1. CSC detection for well-structured STGs.

Benchmark	$\|P\|$	$\|T\|$	$\|Z\|$	CLP	SAT	ILP
PpWk(2,9)	71	38	19	< 1	< 1	< 1
PpWk(2,12)	95	50	25	< 1	< 1	< 1
PpWkCsc(2,9)	72	38	19	3	< 1	< 1
PpWkCsc(2,12)	96	50	25	246	1	< 1
PpWk(3,6)	70	38	19	< 1	< 1	< 1
PpWk(3,9)	106	56	28	11	< 1	< 1
PpWk(3,12)	142	74	37	933	< 1	< 1
PpWkCsc(3,6)	72	38	19	3	< 1	< 1
PpWkCsc(3,9)	108	56	28	2075	< 1	< 1
PpWkCsc(3,12)	144	74	37	time	1	< 1
PpArb(2,9)	86	48	23	< 1	< 1	< 1
PpArb(2,12)	110	60	29	< 1	< 1	< 1
PpArbCsc(2,9)	88	48	23	41	< 1	< 1
PpArbCsc(2,12)	112	60	29	1022	16	< 1
PpArb(3,6)	92	54	25	< 1	< 1	< 1
PpArb(3,9)	128	72	34	< 1	< 1	< 1
PpArb(3,12)	164	90	43	< 1	< 1	< 1
PpArbCsc(3,6)	95	54	25	61	< 1	< 1
PpArbCsc(3,9)	131	72	34	time	2	< 1
PpArbCsc(3,12)	167	90	43	time	16	1
TangramCsc(3,2)	142	92	38	< 1	< 1	1
TangramCsc(4,3)	321	202	83	< 1	< 1	9
Art(10,9)	216	198	99	< 1	< 1	< 1
Art(20,9)	436	398	199	5	10	< 1
Art(30,9)	656	598	299	38	82	< 1
Art(40,9)	876	798	399	138	265	< 1
Art(50,9)	1096	998	499	377	630	1
ArtCsc(10,9)	752	630	315	time	861	182
ArtCsc(20,9)	1532	1270	635	time	mem	1623
ArtCsc(30,9)	2312	1910	955	time	mem	5413
ArtCsc(40,9)	3092	2550	1275	time	mem	12602
ArtCsc(50,9)	3872	3190	1595	time	mem	25210

verifying USC only one ILP model is needed to be solved, whereas for verifying CSC n models are needed, where n is the number of non-input signals in the STG. Moreover, when some encoding conflict exists, the ILP solver can find it in short time. This is explained by the fact that proving the absence of encoding conflicts requires an exhaustive exploration of the *branch-and-bound* tree visited by ILP solvers.

The speed-up shown by ILP with respect to the unfolding approach of SAT or CLP are because in ILP approximations of the state space are used, whereas SAT or CLP (as will be explained in the next section) are exact. However, our conservative approach has proven to be highly accurate in the experimental results.

4.4 Computing the Necessary Support for a Given Signal

In this section we are going to adapt model (2) to derive a decomposition method for the synthesis of a given signal. The main idea is to try to compute those signals in the STG that are needed to ensure that a given signal will be free of encoding conflicts, if we abstract away of the rest of the STG. We will call such a set of signals a *support*.

Table 2. USC detection for well-structured STGs.

Benchmark	$\|P\|$	$\|T\|$	$\|Z\|$	CLP	ILP
PPWK(3,9)	106	56	28	10	< 1
PPWK(3,12)	142	74	37	876	< 1
PPWKCSC(3,9)	108	56	28	2002	< 1
PPWKCSC(3,12)	144	74	37	time	1
PPARB(3,9)	128	72	34	< 1	< 1
PPARB(3,12)	164	90	43	< 1	< 1
PPARBCSC(3,9)	131	72	34	time	1
PPARBCSC(3,12)	167	90	43	time	1
TANGRAM(3,2)	142	92	38	< 1	1
TANGRAM(4,3)	321	202	83	< 1	6
ART(40,9)	876	798	399	146	1
ART(50,9)	1096	998	499	328	2
ARTCSC(40,9)	3092	2550	1275	time	851
ARTCSC(50,9)	3872	3190	1575	time	1387

Let us use as example the STG shown in Fig. 12(a), where a new signal (csc) has been inserted in the original STG of the VME Bus Controller to solve the encoding conflict. A possible support for signal d is $\{ldtack, csc\}$. Fig. 17(a) shows the projection induced by this support, and the final implementation of d is shown in Fig. 17(b). The rest of this section is devoted to explaining how to compute efficiently a support for a given output signal a.

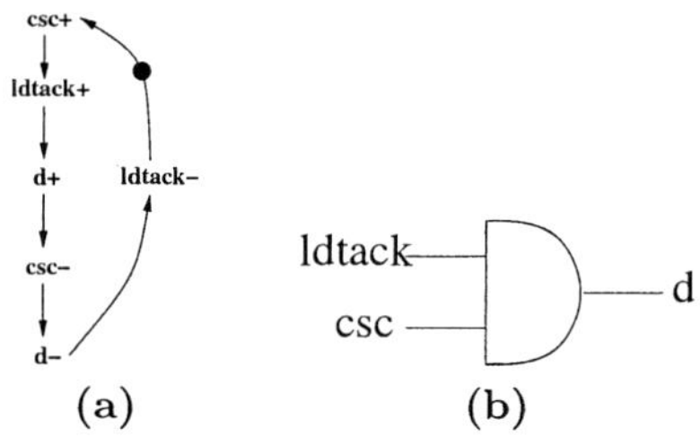

Fig. 17. Projection of the STG in Fig. 12 for signal d **(a)** and a circuit implementing d **(b)**.

The computation of a support can be performed iteratively: starting from an initial assignment, ILP techniques can be used to guide the search. Suppose we have an initial candidate set of signals $Z' \subseteq Z$, candidate to be a support of a given signal a. A way of determining whether Z' is a support for signal a is by solving the following ILP problem:

ILP model for checking support:

$$\boxed{(i), (iii) \text{ and } (iv) \text{ from } (2)} \quad\quad (3)$$
$$\texttt{z is complementary seq. for signals in } Z'$$

If (3) is infeasible, then Z' is enough for implementing a. Otherwise the set Z' must be augmented (from signals in $Z \setminus Z'$) with more signals until (3) is infeasible. Moreover if (3) is feasible, adding a complemented signal b from $Z \setminus Z'$ will not turn the problem infeasible because z is still a complementary sequence for signals in $Z' \cup \{b\}$. On the contrary, adding a uncomplemented signal will assign a different code to markings M_1 and M_2 of (3). Therefore, the uncomplemented signals in z will be the candidates to be added to Z'. The algorithm for finding a support set for a non-input signal a is the following:

Algorithm for the calculation of support:

Support (STG S, Signal a) **returns** support of a

> $Z' :=$ Trig$(a) \cup \{a\}$
> **while** (3) is infeasible **do**
> > Let b be an uncomplemented signal in z
> > $Z' := Z' \cup \{b\}$
>
> **endwhile**
> **return** Z'

where $Trig(a)$ is the set of signals that directly cause the switching of signal a. In the next section we are going to present an example of using this algorithm for the synthesis of the VME Bus Controller STG specification.

4.5 Synthesis of the VME Bus Controller Using Structural Methods

Let us show how to use the structural methods to synthesise the VME example. In addition to the methods presented in this section, we use Petri net transformations for stepwise transformation and projection. For a formal presentation of the kit of transformations used in the example, see [9].

First, as shown in Section 4.3, we can use the ILP model (1) to realise that the original STG of the VME Bus controller has encoding conflicts. Consequently we apply the encoding technique presented in Section 4.2 to enforce CSC.

Afterwards, in order to derive a efficient implementation, we can try to eliminate as many signals inserted by the encoding technique as possible, while keeping a correct encoding. The idea is to eliminate a signal and only accept the removal if the transformed STG still has a correct encoding. Fig. 18 shows how the removal of the first five signals is done, using the USC ILP model (1) as an oracle.

The process can be iterated until no more signals can be removed. The final STG is shown in the centre of Fig. 19. From that STG, the algorithm for support computation is run for every output signal, and the corresponding projection is found. This is shown also in Fig. 19.

And finally, from each projection the corresponding circuit is obtained. Given that the projections are usually small (the support for a given signal is often very small in practice, and the corresponding projections are usually quite small), state-based algorithms for synthesis introduced in Section 3 can be applied. The final synthesis of each projection is shown in Fig. 20.

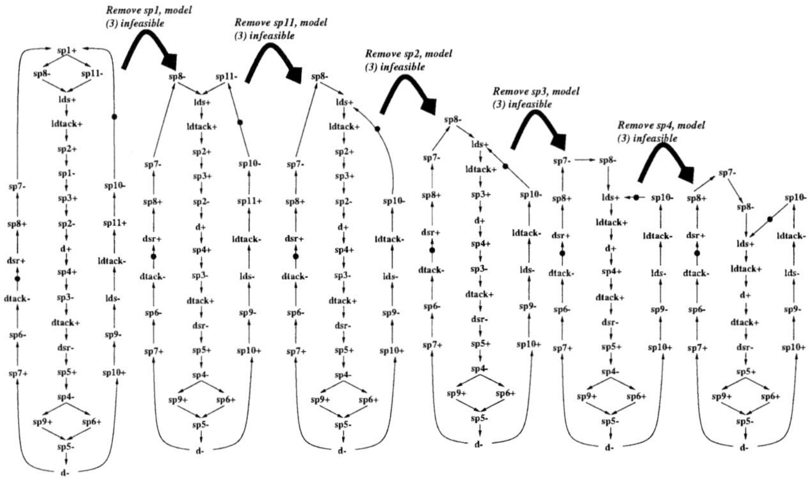

Fig. 18. Greedy removal of signals sp_1, sp_{11}, sp_2, sp_3 and sp_4.

Table 3 shows the results of experiments on synthesis to check the quality of the generated circuits. The column 'Lit' reports the number of literals, in factored form, of the netlist. The results are compared with the circuits obtained by PETRIFY [18], a state-based synthesis tool, on the same controllers. From the reported CPU time, the time needed for computing a support and the corresponding projection was negligible compared with the time needed for deriving logic equations. Table 3 shows that the quality of the circuits obtained by the ILP-based technique is comparable to that of the circuits obtained by PETRIFY. Moreover it is clear that the structural approach can deal with larger specifications.

Table 3. Support computation, projection and synthesis compared to state-based approach.

benchmark	states	$\|P\|$	$\|T\|$	$\|Z\|$	Lit. Pfy	Lit. ILP	CPU Pfy	CPU ILP
PpWkCsc(2,6)	8192	47	26	19	57	57	5	1
PpWkCsc(2,9)	524.288	71	38	19	87	87	49	2
PpWkCsc(3,9)	2.7×10^7	106	56	28	mem	130	mem	3
PpWkCsc(3,12)	2.2×10^{11}	142	74	37	time	117	time	3
PpArbCsc(2,6)	61440	62	36	17	77	77	21	83
PpArbCsc(2,9)	3.9×10^6	110	60	29	107	107	185	59
PpArbCsc(3,9)	3.3×10^9	131	72	34	163	165	10336	289
PpArbCsc(3,12)	1.7×10^{12}	167	90	43	time	210	time	608
TangramCsc(3,2)	426	142	92	38	97	103	56	146
TangramCsc(4,3)	9258	321	202	83	mem	247	mem	7206

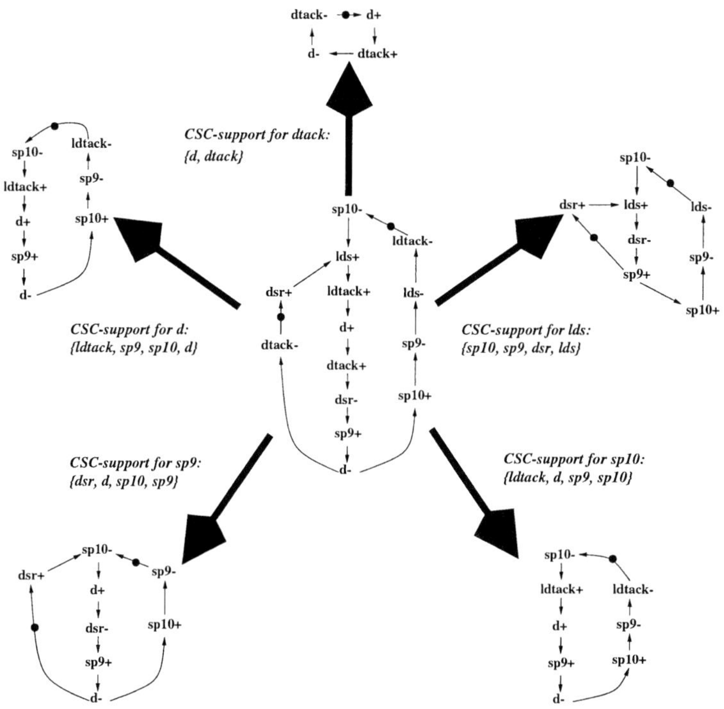

Fig. 19. Support computation and projection for the VME Bus Controller example.

4.6 Conclusions

Several examples of using structural methods are presented in this section. We have given intuition on how these methods are used for the problem of synthesis of control circuits from STG specifications. Although in some cases they can provide only sufficient conditions, in general those methods are highly accurate and provide a significant speed-up with respect to other approaches, as has been demonstrated by the experimental results shown for the problem of verifying the encoding.

In conclusion, structural methods are necessary for being able to handle large and concurrent specifications. We advocate for their use, either isolated or in combination with approaches like the one presented in the next section.

5 Synthesis Using Petri Net Unfoldings

While the state-based approach is relatively simple and well-studied, the issue of computational complexity for highly concurrent STGs is quite serious due to the state space explosion problem. This puts practical bounds on the size of control circuits that can be synthesised using such techniques, which are often restrictive, especially if the STG models are not constructed manually by a designer but rather generated automatically from high-level hardware descriptions.

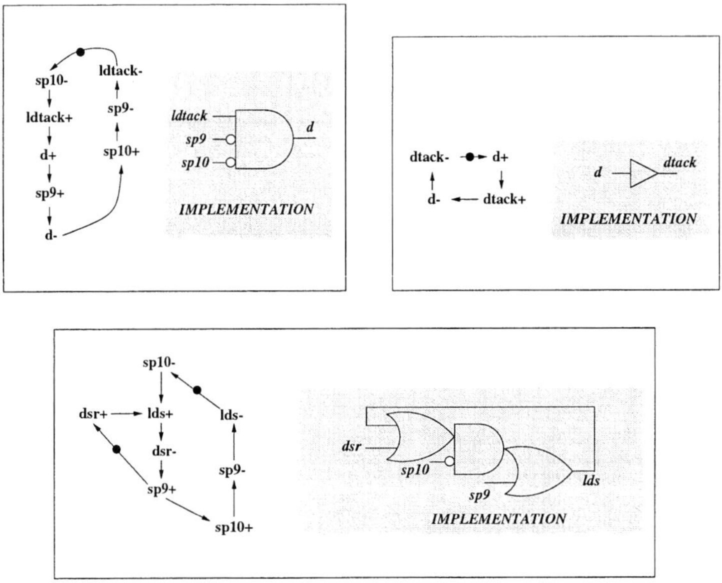

Fig. 20. Speed-independent synthesis of the VME Bus Controller.

In order to alleviate this problem, Petri net analysis techniques based on causal partial order semantics, in the form of Petri net unfoldings, are applied to circuit synthesis. In particular, the following tasks are addressed: (i) detection of encoding conflicts; (ii) resolution of encoding conflicts; and (iii) derivation of Boolean equations for output signals. We show that the notion of an encoding conflict can be characterised in terms of satisfiability of a Boolean formula (SAT), and the resulting algorithms solving tasks (i) and (iii) achieve significant speedups compared with methods based on state graphs. Moreover, we propose a framework for resolution of encoding conflicts (task (ii)) based on *conflict cores.*

5.1 STG Unfoldings

A *finite and complete unfolding prefix* π of an STG Γ is a finite acyclic net which implicitly represents all the reachable states of Γ together with transitions enabled at those states. Intuitively, it can be obtained through *unfolding* Γ, by successive firings of transition, under the following assumptions: (a) for each new firing a fresh transition (called an *event*) is generated; (b) for each newly produced token a fresh place (called a *condition*) is generated. The unfolding is infinite whenever Γ has an infinite run; however, if Γ has finitely many reachable states then the unfolding eventually starts to repeat itself and can be truncated (by identifying a set of *cut-off* events) without loss of information, yielding a finite and complete prefix. We denote by B, E and $E_{cut} \subseteq E$ the sets of conditions, events and cut-off events of the prefix, respectively. Fig. 21(b) shows a finite and complete unfolding prefix (with the only cut-off event is depicted as a double box) of the STG shown in Fig. 21(a).

Efficient algorithms exist for building such prefixes [25, 31, 36, 37], which ensure that the number of non-cut-off events in a complete prefix can never exceed the number of reachable states of Γ. However, complete prefixes are often exponentially smaller than the corresponding state graphs, especially for highly concurrent Petri nets, because they represent concurrency directly rather than by multidimensional 'diamonds' as it is done in state graphs. For example, if the original Petri net consists of 100 transitions which can fire once in parallel, the state graph will be a 100-dimensional hypercube with 2^{100} nodes, whereas the complete prefix will coincide with the net itself.

Due to its structural properties (such as acyclicity), the reachable markings of Γ can be represented using *configurations* of π. A configuration C is a downward-closed set of events (being downward-closed means that if $e \in C$ and f is a causal predecessor of e, then $f \in C$) without structural conflicts (i.e., for all distinct events $e, f \in C$, $^\bullet e \cap {}^\bullet f = \emptyset$). Intuitively, a configuration is a partial-order execution, i.e., an execution where the order of firing of concurrent events is not important.

After starting π from the implicit initial marking (whereby one puts a single token in each condition which does not have an incoming arc) and executing all the events in C, one reaches the marking denoted by $Cut(C)$. We denote by $Mark(C)$ the corresponding marking of Γ, reached by firing a transition sequence corresponding to the events in C. It is remarkable that each reachable marking of Γ is $Mark(C)$ for some configuration C, and, conversely, each configuration C generates a reachable marking $Mark(C)$. This property is a primary reason why various behavioural properties of Γ can be re-stated as the corresponding properties of π, and then checked, often much more efficiently (in particular, one can easily check the consistency and deadlock-freeness of Γ [57, 35]). The experimental results in Table 4 demonstrate that high levels of compression are indeed achieved in practice.

For the unfolding of a consistent STG we define by $Code_z(C)$ the value (0 or 1) corresponding to signal z in the encoding of the state $Mark(C)$; we also define $Out_z(C)$ to be 1 if $z \in Out(M)$ and 0 otherwise, and $Nxt_z(C) \stackrel{\mathrm{df}}{=} Code_z(C) \oplus Out_z(C)$, where '$\oplus$' is the 'exclusive or' operation.

5.2 Visualisation and Resolution of State Encoding Conflicts

A number of methods for resolution of CSC conflicts have been proposed so far (see, e.g., [16] for a brief review). The techniques in [67, 75] introduce constraints within an STG, called *lock relation* and *coupledness relation*, which provide some guidance. These techniques recognise that if all pairs of signals in the STG are 'locked' using a chain of handshaking pairs then the STG satisfies the CSC property. The synthesis tool PETRIFY uses the theory of regions [16] for this purpose.

The above techniques work reasonably well. However, they may produce sub-optimal circuits or fail to solve the problem in certain cases, e.g., when

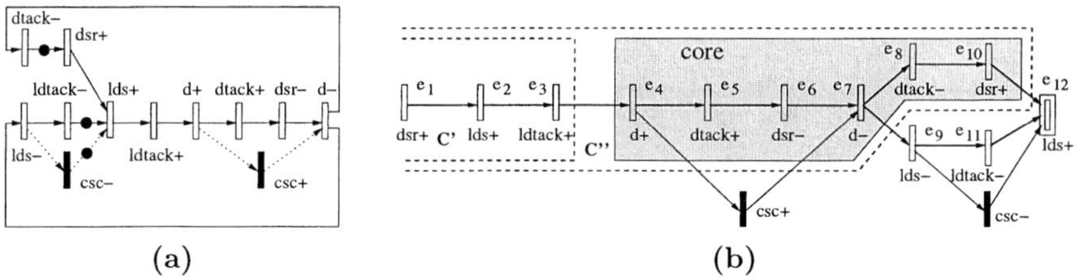

inputs: *dsr*, *ldtack*; **outputs:** *lds*, *d*, *dtack*; **internal:** *csc*

Fig. 21. VME bus controller: STG for the read cycle, with the dotted lines appearing in the final STG satisfying the CSC property **(a)** and unfolding prefix with a conflict pair of configurations and a new signal *csc* resolving the CSC conflict **(b)**. The order of signals in the binary encodings is: *dsr*, *dtack*, *lds*, *ldtack*, *d*.

a controller specification is defined in a compact way using a small number of signals. Such specifications often have CSC conflicts that are classified as irreducible by PETRIFY. Therefore, manual design may be required for finding good synthesis solutions, particularly in constructing interface controllers, where the quality of the solution is critical for the system's performance.

According to a practising designer [54], the synthesis tool should offer a way for the user to understand the characteristic patterns of a circuit's behaviour and the cause of each encoding conflict and allow one to interactively manipulate the model by choosing where in the specification to insert new signals. The visualisation method presented here is aimed at facilitating a manual refinement of an STG with CSC conflicts, and works on the level of unfolding prefixes. In order to avoid the explicit enumeration of encoding conflicts, they are visualised as *cores*, i.e., sets of transitions 'causing' one or more of them. All such cores must eventually be eliminated by adding new signals that resolve the encoding conflicts to yield an STG satisfying the CSC property. Optionally, our method can also work in a completely automatic or semi-automatic manner, making it possible for the designer to see what is going on and intervene at any stage during the process of CSC conflict resolution.

5.3 Encoding Conflicts in a Prefix

A CSC conflict can be represented as an unordered *conflict pair* of configurations $\langle C', C'' \rangle$ whose final states are in CSC conflict, as shown if Fig. 21(b). In Section 5.6 a SAT-based technique for detecting CSC conflicts is described. Essentially, it allows for efficiently finding such conflict pairs in STG unfolding prefixes.

Note that the set of all conflict pairs may be quite large, e.g., due to the following 'propagation' effect: if C' and C'' can be expanded by the same event e then $\langle C' \cup \{e\}, C'' \cup \{e\} \rangle$ is also a conflict pair (unless these two configurations enable the same set of output signals). Therefore, it is desirable to reduce the number of pairs needed to be considered, e.g., as follows. A conflict pair $\langle C', C'' \rangle$ is called *concurrent* if $C' \not\subseteq C''$, $C'' \not\subseteq C'$ and $C' \cup C'$ is a configuration.

Proposition 1 ([45]). *Let $\langle C', C'' \rangle$ be a concurrent CSC conflict pair. Then $C \stackrel{\text{df}}{=} C' \cap C''$ is such that either $\langle C, C' \rangle$ or $\langle C, C'' \rangle$ is a CSC conflict pair.*

Thus concurrent conflict pairs are 'redundant' and should not be considered. The remaining conflict pairs can be classified as follows:

Conflicts of type I are such that $C_1 \subset C_2$ (Fig. 21(b) illustrates this type of CSC conflicts).
Conflicts of type II are such that $C_1 \setminus C_2 \neq \emptyset \neq C_2 \setminus C_1$ and there exist $e' \in C_1 \setminus C_2$ and $e'' \in C_2 \setminus C_1$ such that $e' \# e''$.

The following notion is crucial for the proposed approach:

Definition 4. *Let $\langle C', C'' \rangle$ be a conflict pair. The corresponding* complementary set *is defined as $\mathcal{CS} \stackrel{\text{df}}{=} C' \triangle C''$, where $\triangle$ denotes the symmetric set difference. $\mathcal{CS}$ is a* core *if it cannot be represented as the union of several disjoint complementary sets. A complementary set is of type I/II if the corresponding conflict pair is of type I/II, respectively.* ◊

For example, the core corresponding to the conflict pair shown in Fig. 21(b) is $\{e_4, \ldots, e_8, e_{10}\}$ (note that for a conflict pair $\langle C', C'' \rangle$ of type I, such that $C' \subset C''$, the corresponding core is simply $C'' \setminus C'$).

One can show that every complementary set $\mathcal{CS}$ can be partitioned into $C_1 \setminus C_2$ and $C_2 \setminus C_1$, where $\langle C', C'' \rangle$ is a conflict pair corresponding to $\mathcal{CS}$. Moreover, if $\mathcal{CS}$ is of type I then one of these parts is empty, while the other is $\mathcal{CS}$ itself. An important property of complementary sets is that for each signal $z \in Z$, the difference between the numbers of z^+– and z^-–labelled events in $\mathcal{CS}$ is the same in these two parts (and is 0 if $\mathcal{CS}$ is of type I). This suggests that a complementary set can be eliminated by introduction of a new internal signal and insertion of its transition into this set, as this would violate the stated property.

It is often the case that the same complementary set corresponds to different conflict pairs, so the designer can save time by analysing the cores rather than the full list of CSC conflicts, which can be much longer.

5.4 Framework for Visualisation and Resolution of Encoding Conflicts

The visualisation is based on showing the designer the cores in the STG's unfolding prefix. Since every element of a core is an instance of the STG's transition, the cores can easily be mapped from the prefix to the STG. For example, the core $\{e_4, \ldots, e_8, e_{10}\}$ in Fig. 21(b) can be mapped to the set of transitions $\{d^+, dtack^+, dsr^-, d^-, dtack^-, dsr^+\}$ of the original STG shown in Fig. 21(a).

Cores are important for resolution of encoding conflicts. By introducing an additional internal signal and insertion of its transition, say csc^+, one can destroy a core eliminating thus the corresponding encoding conflicts. To preserve the consistency of the STG, the signal transition's counterpart, csc^-, must also be added to the specification *outside the core,* in such a way that it is neither concurrent to nor in structural conflict with csc^+. It is sometimes possible to insert csc^- into another core thus eliminating it also, as shown in Fig. 22(b). Another restriction is that an inserted signal transitions cannot trigger an input signal transition (the reason is that this would impose constraints on the environment which were not present in the original STG, making it 'wait' for the newly inserted signal). More about the formal requirements for the correctness of inserting a new transition can be found in [18].

The core in Fig. 21(b) can be eliminated by inserting a new signal, csc^+, somewhere in the core, e.g., concurrently to e_5 and e_6 between e_4 and e_7, and by inserting its complement outside the core, e.g., concurrently to e_{11} between e_9 and e_{12}. (Note that concurrent insertion of these two transitions avoids an increase in the latency of the circuit, where each transition is assumed to contribute a unit delay.) The final STG satisfying the CSC property is shown in Fig. 21(a) with dotted lines taken into account.

It is often the case that cores overlap. In order to minimise the number of inserted signals, and thus the area and latency of the circuit, it is advantageous to insert a signal in such a way that as many cores as possible are eliminated by it. That is, a signal should be inserted into *the intersection of several cores* whenever possible.

To assist the designer in exploiting core overlaps, another key feature of our method, viz. the *height map* showing the quantitative distribution of the cores, is employed in the visualisation process. The events located in conflict cores are highlighted by shades of colours. The shade depends on the *altitude* of an event, i.e., on the number of cores it belongs to. (The analogy with a topographical map showing the altitudes may be helpful here.) The greater the altitude, the darker the shade. 'Peaks' with the highest altitude are good candidates for insertion of a new signal, since they correspond to the intersection of maximum number of cores.

Using this representation, the designer can select an area for insertion of a new signal and obtain a local, more detailed description of the cores overlapping with the selection. When an appropriate core cluster is chosen, the designer can decide how to insert a new signal transition optimally, taking into account the design constraints and his/her knowledge of the system being developed.

The overview of the process of resolution of CSC conflicts is shown in Fig. 22(a). Given an STG, a finite and complete prefix of its unfolding is constructed, and the cores are computed. If there are none, the process stops. Otherwise, the height map is shown to the designer, who chooses a set of overlapping cores. In phases one and two, an additional signal transition splitting the core is inserted together with its counterpart. The inserted transitions are then transferred to the STG, and the process is repeated. Depending on the number of conflict cores, the resolution process may involve several cycles.

After completion of phase one, the height map is updated. The altitudes of the events in the core cluster where the new signal transition has been inserted are made negative, to prompt the designer that if the counterpart transition is

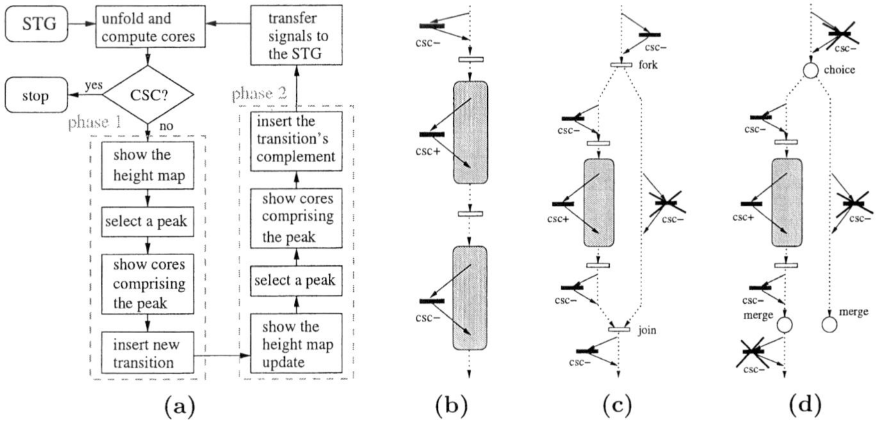

Fig. 22. The process of resolution of encoding conflicts **(a)** and strategies for eliminating conflict cores **(b–d)**. Several possibilities are shown for insertion of csc^-, but only one of them should be used. (The positions where csc^- is *not* allowed are shown as transitions that are crossed out.)

inserted there, some of the cores in the cluster will reappear. Moreover, in order to ensure that the insertion of the counterpart transition preserves consistency, the areas where it cannot be inserted (in particular, the events concurrent to or in structural conflict with this transition) are faded out.

Typical cases in STG specifications are schematically illustrated in Fig. 22(b–d). Cores 'in sequence', can be eliminated in a 'one-hot' manner as depicted in Fig. 22(b). Each core is eliminated by one signal transition, and its complement is inserted outside the core, preferably, into another non-adjacent one[3].

An STG that has a core in one of the concurrent branches can also be tackled in a 'one-hot' way, as shown in Fig. 22(c). Note that in order to preserve the consistency the transition's counterpart cannot be inserted into the concurrent branch, but can be inserted before the fork transition or after the join one. In a branch which is in a structural conflict with another branch, the transition's counterparts must be inserted in the same branch somewhere between the choice and the merge points, as shown in Fig. 22(d).

Obviously, the described cases do not cover all possible situations and all possible insertions (e.g., one can sometimes insert a new signal transition before the choice point and its counterparts into *each* branch, etc.), but we hope they do give an idea how the cores can be eliminated. [45] presents this method of resolution of CSC conflicts using STG unfoldings in more detail.

5.5 Boolean Satisfiability

Boolean satisfiability problem (SAT) has great theoretical interest as the canonical $\mathcal{NP}$-complete problem. Though it is very unlikely that it can be solved in

[3] The union of two adjacent cores is usually a complementary set which will not be destroyed if both the transition and its counterpart are inserted into it.

polynomial time, there are algorithms which can solve many interesting SAT instances quite efficiently. SAT solvers have been successfully applied to many practical problems such as AI planning, ATPG, model checking, etc. The research in SAT has led to algorithms which routinely solve SAT instances generated from industrial applications with tens of thousands or even millions variables [78].

Thus it is often advantageous to re-state the problem at hand in terms of SAT, and then apply an existing SAT solver. In this paper, the SAT approach will be used for detection of CSC conflicts in Section 5.6 and derivation of equations for logic gates of the circuit in Section 5.7.

The *Boolean satisfiability problem (SAT)* consists in finding a *satisfying assignment*, i.e., a mapping $A : Var_\varphi \to \{0,1\}$ defined on the set of variables Var_φ occurring in a given Boolean expression φ such that φ evaluates to 1. This expression is often assumed to be given in the *conjunctive normal form (CNF)* $\bigwedge_{i=1}^{n} \bigvee_{l \in L_i} l$, i.e., it is represented as a conjunction of *clauses*, which are disjunctions of *literals*, each literal l being either a variable or the negation of a variable. It is assumed that no two literals in the same clause correspond to the same variable.

Some of the leading SAT solvers, e.g., zCHAFF [48], can be used in the *incremental mode,* i.e., after solving a particular SAT instance the user can modify it (e.g., by adding and/or removing a small number of clauses) and execute the solver again. This is often much more efficient than solving these related instances as independent problems, because on the subsequent runs the solver can use some of the useful information (e.g., learnt clauses, see [78]) collected so far. In particular, such an approach can be used to compute *projections* of assignments satisfying a given formula, as described in sequel.

Let $V \subseteq Var_\varphi$ be a non-empty set of variables occurring in a formula φ, and $Proj_V^\varphi$ be the set of all restricted assignments (or projections) $A|_V$ such that A is a satisfying assignment of φ. Using the incremental SAT approach it is possible to compute $Proj_V^\varphi$, as follows.

Step 0: $\mathcal{A} := \emptyset$.
Step 1: Run the SAT solver for φ.
Step 2: If φ is unsatisfiable then return $\mathcal{A}$ and terminate.
Step 3: Add $A|_V$ to $\mathcal{A}$, where A is the computed satisfying assignment.
Step 4: Append to φ a new clause $\bigvee_{v \in V \wedge A(v)=1} \neg v \vee \bigvee_{v \in V \wedge A(v)=0} v$.
Step 5: Go back to Step 1.

Suppose now that we are interested in finding only the minimal elements of $Proj_V^\varphi$, assuming that $A|_V \leq A'|_V$ if $(A|_V)(v) \leq (A'|_V)(v)$, for all $v \in V$. The above procedure can then be modified by changing Step 4 to:

Step 4': Append to φ a new clause $\bigvee_{v \in V \wedge A(v)=1} \neg v$.

Similarly, if we were interested in finding all the maximal elements of $Proj_V^\varphi$, then one could change Step 4 to:

Step 4": Append to φ a new clause $\bigvee_{v \in V \wedge A(v)=0} v$.

Moreover, in the latter two cases, before terminating an additional pass over the elements stored in $\mathcal{A}$ should be made in order to eliminate any non-minimal (or non-maximal) projections.

5.6 Detection of State Encoding Conflicts Using SAT

Let C' and C'' be two configurations of the unfolding of a consistent STG and z be an output signal. C' and C'' are in *Complete State Coding conflict for z* (CSC^z *conflict*) if $Code_x(C') = Code_x(C'')$ for all $x \in Z$ and $Nxt_z(C') \neq Nxt_z(C'')$. This notion is very similar to the notion of a CSC conflict; in particular, each CSC^z conflict is a CSC conflict, and each CSC conflict is a CSC^z conflict for some output signal z, i.e., the problem of detection of CSC conflicts is easily reducible to the problem of detection of CSC^z conflicts, and we will mostly concentrate on the latter problem. A CSC^z conflict can be represented as an unordered *conflict pair* of configurations $\langle C', C'' \rangle$ whose final states are in CSC^z conflict; for example, the conflict pair of configurations shown in Fig. 21(b) is in CSC^{lds} and CSC^d conflict.

Constructing a SAT Instance. We adopt the following naming conventions. The variable names are in the lower case and names of formulae are in the upper case. Names with a single prime (e.g., conf'_e and $\mathcal{CONF}'$) are related to C', and ones with double prime (e.g., conf''_e) are related to C''. If there is no prime then the name is related to both C' and C''. If a formula name has a single prime then the formula does not contain occurrences of variables with double primes, and the counterpart double prime formula can be obtained from it by adding another prime to every variable with a single prime. The subscript of a variable points to which element of the STG or the prefix the variable is related, e.g., conf'_e and conf''_e are both related to the event e of the prefix. By a name without a subscript we denote the list of all variables for all possible values of the subscript, e.g., conf' denotes the list of variables conf'_e, where e runs through the set $E \setminus E_{cut}$.

The following Boolean variables will be used in the proposed translations:

- For each event $e \in E \setminus E_{cut}$, we create two Boolean variables, conf'_e and conf''_e, tracing whether $e \in C'$ and $e \in C''$ respectively.
- For each signal $x \in Z$, we create a variable code_x to trace the value of x. Since the values of all the signals must match at the final states of C' and C'', we use the same set of variables for both configurations.
- For each condition $b \in B \setminus E^{\bullet}_{cut}$ which is an instance of a place from P^1_Z (defined later), we create two Boolean variables, cut'_b and cut''_b, tracing whether $b \in Cut(C')$ and $b \in Cut(C'')$ respectively.
- For each event $e \in E$ which is an instance of the output signal z for which the CSC^z condition is being checked, we create two Boolean variables, en'_e and en''_e, tracing whether e is 'enabled' by C' and C'' respectively. Note that unlike conf' and conf'', such variables are also created for the cut-off events.

Our aim is to build a Boolean formula $\mathcal{CSC}^z$ such that: (i) $\mathcal{CSC}^z$ is satisfiable iff there is a CSC^z conflict; and (ii) for every satisfying assignment, the two sets of non-cut-off events of the prefix, $C' \stackrel{\text{df}}{=} \{e \in E \setminus E_{cut} \mid \mathsf{conf}'_e = 1\}$ and $C'' \stackrel{\text{df}}{=} \{e \in E \setminus E_{cut} \mid \mathsf{conf}''_e = 1\}$, constitute a conflict pair $\langle C', C'' \rangle$ of configurations. $\mathcal{CSC}^z$ will be the conjunction of constraints described below.

For example, these variables will assume the following values for the CSC^d conflict depicted in Fig. 21(b) (the order of signals in the binary codes is: *dsr*, *dtack*, *lds*, *ldtack*, *d*): $\mathsf{conf}' = 111000000000$, $\mathsf{conf}'' = 111111110100$, $\mathsf{code} = 10110$, $\mathsf{en}'_{e_4} = 1$, $\mathsf{en}'_{e_7} = 0$, and $\mathsf{en}''_{e_4} = \mathsf{en}''_{e_7} = 0$ (the values of cut' and cut'' are not shown).

Configuration constraints. The role of first two constraints, $\mathcal{CONF}'$ and $\mathcal{CONF}''$, is to ensure that C' and C'' are both legal configurations of the prefix (not just arbitrary sets of events). $\mathcal{CONF}'$ is defined as the conjunction of the formulae

$$\bigwedge_{e \in E \setminus E_{cut}} \bigwedge_{f \in {}^\bullet({}^\bullet e)} (\mathsf{conf}'_e \Rightarrow \mathsf{conf}'_f) \quad \text{and} \quad \bigwedge_{e \in E \setminus E_{cut}} \bigwedge_{f \in E_e} \neg(\mathsf{conf}'_e \wedge \mathsf{conf}'_f) ,$$

where $E_e \stackrel{\text{df}}{=} (({}^\bullet e)^\bullet \setminus \{e\}) \setminus E_{cut}$. The former formula ensures that if $e \in C'$ then all the direct causal predecessors of e are also in C', which in turn ensures that C' is a downward closed set of events. The latter one ensures that C' contains no structural conflicts. (One should be careful to avoid duplication of clauses when generating this formula.) $\mathcal{CONF}''$ is defined similarly.

$\mathcal{CONF}'$ and $\mathcal{CONF}''$ can be transformed into the CNF by applying the rules $x \Rightarrow y \equiv \neg x \vee y$ and $\neg(x \wedge y) \equiv \neg x \vee \neg y$.

Encoding constraint. First we describe an important STG transformation allowing to capture the current value of each signal in the STG's marking. For each signal $z \in Z$, a pair of complementary places, p^0_z and p^1_z, tracing the value of z is added as follows. For each z^+–labelled transition t, $p^0_z \in {}^\bullet t$ and $p^1_z \in t^\bullet$, and for each z^-–labelled transition t', $p^1_z \in {}^\bullet t'$ and $p^0_z \in t'^\bullet$. Exactly one of these two places is marked at the initial state, accordingly to the initial value of signal z. One can show that at any reachable state of an STG augmented with such places, p^0_z (respectively, p^1_z) is marked iff the value of z is 0 (respectively, 1). Thus, if a transition labelled by z^+ (respectively, z^-) is enabled then the value of z is 0 (respectively, 1), which in turn guarantees the consistency of the augmented STG. Such a transformation can be done completely automatically (one can easily determine the initial values of all the signals from the unfolding prefix). For a consistent STG, it does not restrict the behaviour and yields an STG with an isomorphic state graph; for a non-consistent STG, the transformation restricts the behaviour and may lead to (new) deadlocks. In what follows, we assume that the tracing places are present in the STG, and denote $P^0_Z \stackrel{\text{df}}{=} \{p^0_z \mid z \in Z\}$, $P^1_Z \stackrel{\text{df}}{=} \{p^1_z \mid z \in Z\}$, and $P_Z \stackrel{\text{df}}{=} P^0_Z \cup P^1_Z$.

The role of encoding constraints, $\mathcal{CODE}'$ and $\mathcal{CODE}''$, is to ensure that the signal codes of the final markings of configurations C' and C'' are equal. To build a formula establishing the value code_z of each signal $z \in Z$ at the final state of C', we observe that $\mathsf{code}_z = 1$ iff $p^1_z \in \mathit{Mark}(C')$, i.e., iff $b \in \mathit{Cut}(C')$ for some

p_z^1–labelled condition b (note that the places in P_Z cannot contain more than one token). The latter can be captured by the constraint:

$$\bigwedge_{z \in Z} (\mathsf{code}_z \iff \bigvee_{b \in B_z} \mathsf{cut}'_b) ,$$

where $B_z \stackrel{\text{df}}{=} \{B \setminus E^\bullet_{cut} \mid b \text{ is an instance of } p_z^1\}$. We then define $\mathcal{CODE}'$ as the conjunction of the last formula and

$$\bigwedge_{z \in Z} \bigwedge_{b \in B_z} (\mathsf{cut}'_b \iff \bigwedge_{e \in {}^\bullet b} \mathsf{conf}'_e \wedge \bigwedge_{e \in b^\bullet \setminus E_{cut}} \neg \mathsf{conf}'_e) ,$$

which ensures that $b \in Cut(C')$ iff the event 'producing' b has fired, but no event 'consuming' b has fired. (Note that since $|{}^\bullet b| \leq 1$, $\bigwedge_{e \in {}^\bullet b} \mathsf{conf}_e$ in this formula is either the constant 1 or a single variable.) One can see that if C' is a configuration and $\mathcal{CODE}'$ is satisfied then the value of signal z at the final state of C' is given by code_z. $\mathcal{CODE}''$ is defined similarly.

The use of the same variables code in both $\mathcal{CODE}'$ and $\mathcal{CODE}''$ ensures that the encodings of the final states of C' and C'' are the same, if both constraints are satisfied.

It is straightforward to build the CNF of $\mathcal{CODE}'$:

$$\bigwedge_{z \in Z} \Bigg((\neg \mathsf{code}_z \vee \bigvee_{b \in B_z} \mathsf{cut}'_b) \wedge \bigwedge_{b \in B_z} (\mathsf{code}_z \vee \neg \mathsf{cut}'_b) \wedge$$

$$\bigwedge_{b \in B_z} \Bigg(\bigwedge_{e \in {}^\bullet b} (\neg \mathsf{cut}'_b \vee \mathsf{conf}'_e) \wedge \bigwedge_{e \in b^\bullet \setminus E_{cut}} (\neg \mathsf{cut}'_b \vee \neg \mathsf{conf}'_e) \wedge (\mathsf{cut}'_b \vee \bigvee_{e \in {}^\bullet b} \neg \mathsf{conf}'_e \vee \bigvee_{e \in b^\bullet \setminus E_{cut}} \mathsf{conf}'_e) \Bigg) \Bigg) ,$$

and the CNF of $\mathcal{CODE}''$ can be built similarly.

Next-state constraint. The role of this constraint is to ensure that $Nxt_z(C') \neq Nxt_z(C'')$. Since all the other constraints are symmetric w.r.t. C' and C'', one can rewrite it as $Nxt_z(C') = 0 \wedge Nxt_z(C'') = 1$. Moreover, it follows from the definition of Nxt_z that $Nxt_z(C) \equiv \neg Code_z(C) \iff Out_z(C)$, and so the next-state constraint can be rewritten as the conjunction of $Code_z(C') \iff Out_z(C')$ and $\neg Code_z(C'') \iff Out_z(C'')$.

We observe that an output signal z is enabled by $Mark(C')$ iff there is a z^+- or z^-–labelled event $e \notin C'$ 'enabled' by C', i.e., such that $C' \cup \{e\}$ is a configuration (note that e can be a cut-off event). We then define the formula $\mathcal{NEXTZERO}'$, ensuring that $Nxt_z(C') = 0$, as the conjunction of

$$\mathsf{code}'_z \iff \bigvee_{e \in E_z} \mathsf{en}'_e \quad \text{and} \quad \bigwedge_{e \in E_z} (\mathsf{en}'_e \iff \bigwedge_{f \in {}^\bullet({}^\bullet e)} \mathsf{conf}'_f \wedge \bigwedge_{f \in ({}^\bullet e)^\bullet \setminus E_{cut}} \neg \mathsf{conf}'_f) ,$$

where $E_z \stackrel{\text{df}}{=} \{e \in E \mid \text{e is an instance of } z^\pm\}$. The former conjunct ensures that $Code_z(C') \iff Out_z(C)$ (it takes into account that z is enabled by the final state of C' iff at least one its instance is enabled by C') and the latter one states for each instance e of z that e is enabled by C' iff all the events 'producing' tokens in ${}^\bullet e$ are in C' but no events 'consuming' tokens from ${}^\bullet e$ (including e itself) are in C'.

The formula $\mathcal{NEXTONE}''$, ensuring that $Nxt_z(C'') = 1$, is defined as the conjunction of

$$\neg\mathsf{code}''_z \iff \bigvee_{e\in E_z} \mathsf{en}''_e$$

and a constraint 'computing' en''_e, which is similar to that for $\mathcal{NEXTZERO}'$. Now the next-state constraint can be expressed as $\mathcal{NEXTZERO}' \wedge \mathcal{NEXTONE}''$.

The CNF of $\mathcal{NEXTZERO}'$ is

$$(\neg\mathsf{code}'_z \vee \bigvee_{e\in E_z} \mathsf{en}'_e) \wedge \bigwedge_{e\in E_z} (\mathsf{code}'_z \vee \neg\mathsf{en}'_e) \wedge$$

$$\bigwedge_{e\in E_z}\Big(\bigwedge_{f\in{}^\bullet({}^\bullet e)} (\neg\mathsf{en}'_e \vee \mathsf{conf}'_f) \wedge \bigwedge_{f\in({}^\bullet e)^\bullet\setminus E_{cut}} (\neg\mathsf{en}'_e \vee \neg\mathsf{conf}'_f) \wedge (\mathsf{en}'_e \vee \bigvee_{f\in{}^\bullet({}^\bullet e)} \neg\mathsf{conf}'_f \vee \bigvee_{f\in({}^\bullet e)^\bullet\setminus E_{cut}} \mathsf{conf}'_f)\Big),$$

and the CNF of $\mathcal{NEXTONE}''$ can be built similarly.

Translation to SAT. Finally, the problem of detection of CSC^z conflicts can be formulated as the SAT problem for the formula

$$\mathcal{CSC}^z \stackrel{\text{df}}{=} \mathcal{CONF}' \wedge \mathcal{CONF}'' \wedge \mathcal{CODE}' \wedge \mathcal{CODE}'' \wedge \mathcal{NEXTZERO}' \wedge \mathcal{NEXTONE}'' ,$$

and the CSC problem is reduced to checking the CSC^z condition for each output signal z. In principle, the CSC problem can also be reduced to a *single* SAT instance [39], but according to our experiments the method presented here tends to be more efficient.

Computing All Cores. The method for resolution of CSC conflicts described in Section 5.2 requires to compute all conflict cores. This can be done by computing all the solutions of CSC^z for all output signals z using the incremental SAT approach. However, as the same complementary set can correspond to multiple conflict pairs, this approach is unnecessarily expensive. A better approach would be to eliminate all the solutions corresponding to a newly computed complementary set $\mathcal{CS}$ each time it is computed, by appending new clauses to the formula. This can be done as follows. For each event $e \in E \setminus E_{cut}$ we create a variable cs_e, and the following constraint is added to the formula:

$$\Big(\bigwedge_{e\in E\setminus E_{cut}} (\mathsf{cs}_e \iff (\mathsf{conf}'_e \oplus \mathsf{conf}'_e)) \Big) \wedge \bigvee_{e\in E\setminus E_{cut}} \begin{cases} \neg\mathsf{cs}_e \text{ if } e \in \mathcal{CS} \\ \mathsf{cs}_e \text{ otherwise} . \end{cases}$$

Note that the first part of this constraint is the same for all the computed complementary sets, and thus can be generated just once. The CNF of this constraint is

$$(\neg\mathsf{conf}'_e \vee \mathsf{conf}''_e \vee \mathsf{cs}_e) \wedge (\mathsf{conf}'_e \vee \neg\mathsf{conf}''_e \vee \mathsf{cs}_e) \wedge$$
$$(\mathsf{conf}'_e \vee \mathsf{conf}''_e \vee \neg\mathsf{cs}_e) \wedge (\neg\mathsf{conf}'_e \vee \neg\mathsf{conf}''_e \vee \neg\mathsf{cs}_e) \wedge \bigvee_{e \in E \setminus E_{cut}} \begin{cases} \neg\mathsf{cs} \text{ if } e \in \mathcal{CS} \\ \mathsf{cs} \text{ otherwise} . \end{cases}$$

The Case of Prefixes without Structural Conflicts. In many cases the performance of the proposed method can be improved by exploiting specific properties of the Petri net underlying an STG Γ. For instance, if Γ is free from dynamic choices (in particular, this is the case for marked graphs) then the union of any two configurations of its unfolding is also a configuration. This observation can be used to reduce the search space. Indeed, according to Proposition 2 below, it is then enough to look only for those cases when the configurations C' and C'' being tested are ordered in the set-theoretical sense.

Proposition 2 ([39]). *Let $\langle C', C''\rangle$ be a conflict pair of configurations in the unfolding of a consistent STG Γ satisfying $C' \not\subseteq C''$, $C'' \not\subseteq C'$ and $C' \cup C''$ is a configuration. Then $C \stackrel{\text{df}}{=} C' \cap C''$ is such that either $\langle C, C'\rangle$ or $\langle C, C''\rangle$ is a conflict pair.*

Note that freeness from structural conflicts can easily be detected: it is enough to check that $|b^\bullet| \leq 1$, for all conditions b of the prefix.

Since we do not know in advance whether $C' \subseteq C''$ or $C'' \subseteq C'$ (and the order does matter because the suggested implementation of the next-state constraint breaks the symmetry), a new Boolean variable, $\mathsf{v}_\subseteq$, is introduced. If its value is 1 then the former possibility is checked, otherwise the latter possibility is tried out. This is captured by the constraint

$$\bigwedge_{e \in E \setminus E_{cut}} ((\mathsf{v}_\subseteq \rightarrow (\mathsf{conf}'_e \rightarrow \mathsf{conf}''_e)) \wedge (\neg\mathsf{v}_\subseteq \rightarrow (\mathsf{conf}''_e \rightarrow \mathsf{conf}'_e))) ,$$

which should be added to the formula. Note that it can easily be transformed into the CNF by applying the rule $x \rightarrow y \equiv \neg x \vee y$.

Experimental Results. We implemented our method using the zChaff SAT solver [48]. All the experiments were conducted on a PC with a *Pentium*^{TM} IV/2.8GHz processor and 512M RAM.

A few classes of benchmarks have been attempted (the STGs with names containing the occurrence of 'Csc' satisfy the CSC property, the others exhibit CSC conflicts). The first group of examples comes from the real design practice. They are as follows:

- LazyRing and Ring – Asynchronous Token Ring Adapters described in [10, 44]. LazyRingCsc and RingCsc have been obtained by resolving CSC conflicts in these test cases.
- Dup4ph, Dup4phCsc, Dup4phMtr, Dup4phMtrCsc, DupMtrMod, DupMtrModUtg, and DupMtrModCsc – control circuits for the Power-Efficient Duplex Communication System described in [28].

Table 4. Experimental results: checking CSC.

Problem	Net $\|S\|$	Net $\|T\|$	Net In/Out	States	Prefix $\|B\|$	Prefix $\|E\|$	Prefix $\|E_{cut}\|$	Time, [s] Pfy	Time, [s] Clp	Time, [s] Sat
Real-Life STGs										
LazyRing	35	32	5/6	160	87	66	5	1	<1	<1
LazyRingCsc	42	37	5/7	187	88	71	5	1	<1	<1
Ring	147	127	11/17	16508	763	498	59	694	<1	<1
RingCsc	185	172	11/18	16320	650	484	55	837	15	<1
Dup4ph	133	123	12/15	169	144	123	11	13	<1	<1
Dup4phCsc	135	123	12/15	171	146	123	11	13	<1	<1
Dup4phMtr	109	96	10/12	121	117	96	8	8	<1	<1
Dup4phMtrCsc	114	105	10/16	149	122	105	8	9	<1	<1
DupMtrMod	129	100	10/11	345	199	132	10	89	<1	<1
DupMtrModUtg	116	165	10/11	323	344	218	65	286	<1	<1
DupMtrModCsc	152	115	10/17	321	228	149	13	116	<1	<1
CfSymCscA	85	60	8/14	6672	1341	720	56	153	16	2
CfSymCscB	55	32	8/8	690	160	71	6	6	<1	<1
CfSymCscC	59	36	8/10	2416	286	137	10	11	<1	<1
CfSymCscD	45	28	4/10	414	120	54	6	3	<1	<1
CfAsymCscA	128	112	8/26	147684	1808	1234	62	1551	439	11
CfAsymCscB	128	112	8/24	147684	1816	1238	62	2602	471	10
Marked Graphs										
PpWk(2,3)	23	14	0/7	$5{\cdot}2^5 = 160$	41	23	1	<1	<1	<1
PpWk(2,6)	47	26	0/13	$5{\cdot}2^5 = 10240$	119	62	1	5	<1	<1
PpWk(2,9)	71	38	0/19	$5{\cdot}2^5 > 6{\cdot}10^5$	233	119	1	43	<1	<1
PpWk(2,12)	95	50	0/25	$5{\cdot}2^5 > 4{\cdot}10^7$	383	194	1	494	1	<1
PpWkCsc(2,3)	24	14	0/7	$2^7 = 128$	38	20	1	<1	<1	<1
PpWkCsc(2,6)	48	26	0/13	$2^{13} = 8192$	110	56	1	4	<1	<1
PpWkCsc(2,9)	72	38	0/19	$2^{19} > 5{\cdot}10^5$	218	110	1	43	3	<1
PpWkCsc(2,12)	96	50	0/25	$2^{25} > 3{\cdot}10^7$	362	182	1	2076	264	<1
PpWk(3,3)	34	20	0/10	$13{\cdot}2^7 = 1664$	63	35	1	1	<1	<1
PpWk(3,6)	70	38	0/19	$13{\cdot}2^{16} > 8{\cdot}10^5$	183	95	1	103	<1	<1
PpWk(3,9)	106	56	0/28	$13{\cdot}2^{25} > 4{\cdot}10^8$	357	182	1	2121	12	<1
PpWk(3,12)	142	74	0/37	$13{\cdot}2^{34} > 2{\cdot}10^{11}$	585	296	1	*mem*	1031	<1
PpWkCsc(3,3)	36	20	0/10	$2^{10} = 1024$	57	29	1	1	<1	<1
PpWkCsc(3,6)	72	38	0/19	$2^{19} > 5{\cdot}10^5$	165	83	1	44	3	<1
PpWkCsc(3,9)	108	56	0/28	$2^{28} > 2{\cdot}10^8$	327	164	1	7936	2285	<1
PpWkCsc(3,12)	144	74	0/37	$2^{37} > 10^{11}$	543	272	1	*mem*	*time*	<1
STGs with Arbitration										
PpArb(2,3)	48	32	2/13	$291{\cdot}2^4 = 4656$	110	66	2	7	<1	<1
PpArb(2,6)	72	44	2/19	$291{\cdot}2^{10} > 2{\cdot}10^5$	218	120	2	57	<1	<1
PpArb(2,9)	96	56	2/25	$291{\cdot}2^{16} > 10^7$	362	192	2	1726	<1	<1
PpArb(2,12)	120	68	2/31	$291{\cdot}2^{22} > 10^9$	542	282	2	11493	<1	<1
PpArbCsc(2,3)	48	32	2/13	$207{\cdot}2^4 = 3312$	110	66	2	3	<1	<1
PpArbCsc(2,6)	72	44	2/19	$207{\cdot}2^{10} > 2{\cdot}10^5$	218	120	2	41	2	<1
PpArbCsc(2,9)	96	56	2/25	$207{\cdot}2^{16} > 10^7$	362	192	2	316	153	<1
PpArbCsc(2,12)	120	68	2/31	$207{\cdot}2^{22} > 8{\cdot}10^8$	542	282	2	*mem*	12745	<1
PpArb(3,3)	71	48	3/19	$1647{\cdot}2^6 > 10^5$	188	114	3	97	<1	<1
PpArb(3,6)	107	66	3/28	$1647{\cdot}2^{15} > 5{\cdot}10^7$	368	204	3	1726	<1	<1
PpArb(3,9)	143	84	3/37	$1647{\cdot}2^{24} > 2{\cdot}10^{10}$	602	321	3	*mem*	<1	<1
PpArb(3,12)	179	102	3/46	$1647{\cdot}2^{33} > 10^{13}$	890	465	3	*mem*	<1	<1
PpArbCsc(3,3)	71	48	3/19	$297{\cdot}2^8 = 76032$	118	114	3	43	1	<1
PpArbCsc(3,6)	107	66	3/28	$297{\cdot}2^{17} > 3{\cdot}10^7$	368	204	3	1186	379	<1
PpArbCsc(3,9)	143	84	3/37	$297{\cdot}2^{26} > 10^{10}$	602	321	3	27512	*time*	<1
PpArbCsc(3,12)	179	102	3/46	$297{\cdot}2^{35} > 10^{13}$	890	465	3	*mem*	*time*	<1

– CfSymCscA, CfSymCscB, CfSymCscC, CfSymCscD, CfAsymCscA, and CfAsymCscB – control circuits for the Counterflow Pipeline Processor described in [72].

Some of these STGs, although built by hand, are quite large in size. The results for this group are summarised in the first part of Table 4. Two other groups, PpWk(m,n) and PpArb(m,n), contain scalable examples of STGs modelling m pipelines weakly synchronised without arbitration (in PpWk(m,n)) and with arbitration (in PpArb(m,n)). (See [40] for a more detailed description.) The former offers the possibility of studying the effect of the optimisation described in Section 5.6 (all STGs in the PpWk(m,n) series are marked graphs, and so their prefixes contain no structural conflicts). These benchmarks come in pairs: for each test case satisfying the CSC property there is a very similar one exhibiting CSC conflicts. This allowed us to test the algorithm on almost identical specifications with and without encoding conflicts. The results for these two groups are summarised in the last two parts of Table 4.

The meaning of the columns is as follows (from left to right): the name of the problem; the number of places, transitions, and input and output signals in the original STG; the number of conditions, events and cut-off events in the complete prefix; the number of reachable states in the STG; the time spent by a special version of the Petrify tool, which did not attempt to resolve the encoding conflicts it had identified; the time spent by the integer programming algorithm proposed in [38]; and the time spent by the proposed method. We use 'mem' if there was a memory overflow and 'time' to indicate that the test had not stopped after 15 hours. We have not included in the table the time needed to build complete prefixes, since it did not exceed 0.1sec for all the attempted STGs.

Although performed testing was limited in scope, one can draw some conclusions about the performance of the proposed algorithm. In all cases the proposed method solved the problem relatively easily, even when it was intractable for the other approaches. In some cases, it was faster by several orders of magnitude. The time spent on all of these benchmarks was quite satisfactory – it took just 11 seconds to solve the hardest one. Overall, the proposed approach was the best, especially for hard problem instances.

5.7 Logic Synthesis Based on Unfolding Prefixes

In Section 5.6, the CSC conflict detection problem was solved by reducing it to SAT. More precisely, given a finite and complete prefix of an STG's unfolding, one can build for each output signal z a formula CSC^z which is satisfiable iff there is a CSC^z conflict. Here we modify that construction in the way described below. We assume a given consistent STG satisfying the CSC property, and consider in turn each output signal z.

Let C' and C'' be two configurations of the unfolding of a consistent STG, z be an output signal, and X is some set of signals. C' and C'' are in *Complete State Coding conflict for z w.r.t. X (CSC^z_X conflict)* if $Code_x(C') = Code_x(C'')$ for all

$x \in X$ and $Nxt_z(C') \neq Nxt_z(C'')$. The notion of CSC^z_X is a generalisation of the notion of CSC^z conflict (indeed, the latter can be obtained from the former by choosing X to be the set of all signals in the STG). X is a *support* of an output signal z if no two configurations of the unfolding are in CSC^z_X conflict. In such a case the next-state value of z at each reachable state of the STG is determined without ambiguity by the encoding of this state restricted to X, i.e., z can be implemented as a gate with the support X. A support X of z is *minimal* if no set $X' \subset X$ is a support of z. In general, a signal can have several distinct minimal supports.

The starting point of the proposed approach is to consider the set $\mathcal{NSUPP}^z$ of all sets of signals which are *non-supports* of z. Within the Boolean formula $\mathcal{CSC}^z_{\mathsf{nsupp}}$, which we are going to construct, non-supports are represented by variables $\mathsf{nsupp} \stackrel{\text{df}}{=} \{\mathsf{nsupp}_x \mid x \in Z\}$, and, for a given assignment A, the set of signals $X = \{x \mid A(\mathsf{nsupp}_x) = 1\}$ is identified with the projection $A|_{\mathsf{nsupp}}$. The key property of $\mathcal{CSC}^z_{\mathsf{nsupp}}$ is that $\mathcal{NSUPP}^z = Proj^{\mathcal{CSC}^z_{\mathsf{nsupp}}}_{\mathsf{nsupp}}$, and so it is possible to use the incremental SAT approach to compute $\mathcal{NSUPP}^z$. However, for our purposes it will be enough to compute the maximal non-supports $\mathcal{NSUPP}^z_{\max} \stackrel{\text{df}}{=} \max_{\subseteq} \mathcal{NSUPP}^z$ which can then be used for computing the set

$$\mathcal{SUPP}^z_{\min} \stackrel{\text{df}}{=} \min_{\subseteq}\{X \subseteq Z \mid X \not\subseteq X', \text{ for all } X' \in \mathcal{NSUPP}^z_{\max}\}$$

of all the minimal supports of z (another incremental SAT run will be needed for this).

$\mathcal{SUPP}^z_{\min}$ captures the set of all possible supports of z, in the sense that any support is an extension of some minimal support, and vice versa, any extension of any minimal support is a support. However, the simplest equation is usually obtained for some minimal support, and this approach was adopted in our experiments. Yet, this is not a limitation of our method as one can also explore some or all of the non-minimal supports, which can be advantageous, e.g., for small circuits and/or when the synthesis time is not of paramount importance (this would sometimes allow to find a simpler equation). On the other hand, not all minimal supports have to be explored: if some minimal support has many more signals compared with another one, the corresponding equation will almost certainly be more complicated, and so too large supports can safely be discarded. Thus, as usual, there is a trade-off between the execution time and the degree of design space exploration, and our method allows one to choose an acceptable compromise. Typically, several 'most promising' supports are selected, the equations expressing Nxt_z as a function of signals in these supports are obtained (as described below), and the simplest among them is implemented as a logic gate.

Suppose now that X is one of the chosen supports of z. In order to derive an equation expressing Nxt_z as a function of the signals in X, we build a Boolean formula $\mathcal{EQN}^z_X$ which has a variable code_x for each signal $x \in X$ and is satisfiable iff these variables can be assigned values in such a way that there is a configuration C in the prefix such that $Code_x(C) = \mathsf{code}_x$, for all $x \in X$. Now, using the incremental SAT approach one can compute the projection of the set of reachable encodings onto X (differentiating the stored solutions according to

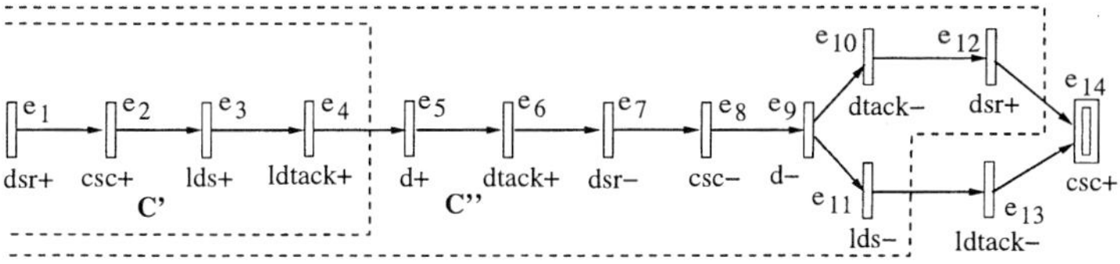

inputs: $dsr, ldtack$; **outputs:** $lds, d, dtack$; **internal:** csc

Fig. 23. An STG unfolding illustrating a $CSC^{csc}_{\{dsr,ldtack\}}$ conflict between configurations C' and C''. Note that e_{14} is not enabled by C'' (since $e_{13} \notin C''$), and thus $Nxt_{csc}(C') = 1 \neq Nxt_{csc}(C'') = 0$. The order of signals in the binary encodings is: *dsr, ldtack, dtack, lds, d, csc.*

the value of the next-state function for z), and feed the result to a Boolean minimiser.

To summarise, the proposed method is executed separately for each output signal z and has three main stages: (I) computing the set $\mathcal{NSUPP}^z_{\max}$ of maximal non-supports of z; (II) computing the set $\mathcal{SUPP}^z_{\min}$ of minimal supports of z; and (III) deriving an equation for a chosen support X of z. In the sequel, we describe each of these three stages in more detail.

It should be noted that the size of the truth table for Boolean minimisation and the number of times a SAT solver is executed in our method can be exponential in the number of signals in the support. Thus, it is crucial for the performance of the proposed algorithm that the support of each signal is relatively small. However, in practice it is anyway difficult to implement as an atomic logic gate a Boolean expression depending on more than, say, eight variables. (Atomic behaviour of logic gates is essential for the speed-independence of the circuit, and a violation of this requirement can lead to hazards [14, 18].) This means that if an output signal has only 'large' supports then the specification must be changed (e.g., by adding new internal signals) to introduce 'smaller' supports. Such transformations are related to the *technology mapping* step in the design cycle for asynchronous circuits (see, e.g., [18]); we do not consider them here.

Computing Maximal Non-supports. Suppose that we want to compute the set of all maximal non-supports of an output signal z. At the level of a branching process, a CSC^z_X conflict can be represented as an unordered *conflict pair* of configurations $\langle C', C''\rangle$ whose final states are in CSC^z_X conflict, as shown in Fig. 23.

As already mentioned, our aim is to build a Boolean formula $\mathcal{CSC}^z_{\mathsf{nsupp}}$ such that $Proj^{\mathcal{CSC}^z_{\mathsf{nsupp}}}_{\mathsf{nsupp}} = \mathcal{NSUPP}^z$, i.e., after assigning arbitrary values to the variables nsupp, the resulting formula is satisfiable iff there is a CSC^z_X conflict, where $X \stackrel{\text{df}}{=} \{x \mid \mathsf{nsupp}_x = 1\}$.

The target formula $\mathcal{CSC}^z_{\mathsf{nsupp}}$ is very similar to the formula $\mathcal{CSC}^z$ built in Section 5.6, with the following changes. For each signal $x \in Z$, instead of a variable code_x we create two Boolean variables, code'_x and code''_x, tracing the values of $Code_x(C')$ and $Code_x(C'')$ respectively; $\mathcal{CODE}'$ and $\mathcal{CODE}''$ are amended accordingly. Moreover, we create for each signal $x \in Z$ a variable nsupp_x indicating whether x belongs to a non-support.

Now we need a ensure that $\mathsf{code}'_x = \mathsf{code}''_x$ whenever $\mathsf{nsupp}_x = 1$. This can be expressed by the following constraint:

$$\bigwedge_{x \in Z} \left(\mathsf{nsupp}_x \Rightarrow (\mathsf{code}'_x \iff \mathsf{code}''_x) \right) ,$$

with the CNF

$$\bigwedge_{x \in Z} \left((\neg \mathsf{code}'_x \vee \mathsf{code}''_x \vee \neg \mathsf{nsupp}_x) \wedge (\mathsf{code}'_x \vee \neg \mathsf{code}''_x \vee \neg \mathsf{nsupp}_x) \right) .$$

This completes the construction of $\mathcal{CSC}^z_{\mathsf{nsupp}}$. For example, its satisfying assignment (except the variables cut' and cut'') for the $CSC^{csc}_{\{dsr,ldtack\}}$ conflict depicted in Fig. 23 is as follows: $\mathsf{conf}' = 1111000000000$, $\mathsf{conf}'' = 1111111111110$, $\mathsf{code}' = 110101$, $\mathsf{code}'' = 110000$, $\mathsf{nsupp} = 110000$, $\mathsf{en}'_{e_2} = \mathsf{en}'_{e_8} = \mathsf{en}'_{e_{14}} = 0$, $\mathsf{en}''_{e_2} = \mathsf{en}''_{e_8} = \mathsf{en}''_{e_{14}} = 0$.

Now the problem of computing the set $\mathcal{NSUPP}^z_{\max}$ of maximal non-supports of z can now be formulated as a problem of finding the maximal elements of the projection $Proj^{\mathcal{CSC}^z_{\mathsf{nsupp}}}_{\mathsf{nsupp}}$. It can be solved using the incremental SAT approach, as described in Section 5.5.

Computing Minimal Supports. Let $\mathcal{NSUPP}^z_{\max}$ be the set of maximal non-supports computed in the first stage of the method. Now we need to compute the set $\mathcal{SUPP}^z_{\min}$ of the minimal supports of z. This can be achieved by computing the set of minimal assignments for the Boolean formula

$$\bigwedge_{\mathsf{nsupp}^* \in \mathcal{NSUPP}^z_{\max}} \left(\bigvee_{x \in Z : \mathsf{nsupp}^*_x = 0} \mathsf{supp}_x \right) ,$$

which is satisfied by an assignment A iff for all maximal non-supports nsupp^* in $\mathcal{NSUPP}^z_{\max}$, $A \not\leq \mathsf{nsupp}^*$. This again can be done using the incremental SAT approach, as described in Section 5.5. Note that this Boolean formula is much smaller than that for the first stage of the method (it contains at most $|Z|$ variables), and thus the corresponding incremental SAT problem is much simpler.

Deriving an Equation. Suppose that X is a (not necessarily minimal) support of z. We need to express Nxt_z as a Boolean function of signals in X. This can be done by generating a truth table for z as a Boolean function of signals in X, and then applying Boolean minimisation.

The set of encodings appearing in the first column of the truth table coincides with the projections of the formula

$$\mathcal{EQN}^z_X \stackrel{\text{df}}{=} \mathcal{CONF}' \wedge \mathcal{CODE}'_X$$

onto the set of variables $\{\mathsf{code}_x \mid x \in X\}$, where $\mathcal{CODE}'_X$ is $\mathcal{CODE}'$ restricted to the set of signals X (i.e., all the conjunctions of the form $\bigwedge_{x \in Z} \cdots$ are replaced by $\bigwedge_{x \in X} \cdots$). It also can be computed using the incremental SAT approach, as described in Section 5.5. Note that at each step of this computation, the SAT solver returns information not only about the next element of the projection, but also the values of all the other variables in the formula. That is, along with the restriction of some reachable encoding onto the set X we have an information about a configuration C via which it can be reached. Thus, the value of Nxt_z on this element of the projection can be computed simply as $Nxt_z(C)$. This essentially completes the description of our method.

Optimisations. In [40] we describe optimisations which can significantly reduce the computation effort required by our method. In particular, we suggest a heuristic helping to compute a part of a signal's support without running the SAT solver, based on the fact that any support for an output z must include all the *triggers* of z, i.e., those signals whose firing can enable z. (The information about triggers can be derived from the finite and complete prefix.) Moreover, one can speed up the computation in the case of prefixes without structural conflicts, as described in Section 5.6.

Experimental Results. We implemented our method using the zChaff SAT solver [48] and the Espresso Boolean minimiser [5], and the benchmarks from Section 5.6 satisfying the CSC property were attempted. All the experiments were conducted on a PC with a *Pentium*^{TM} IV/2.8GHz processor and 512M RAM.

The experimental results are summarised in Table 5, where the meaning of the columns is as follows: the total number of equations obtained by our method (this is equal to the total number of minimal supports for all the output signals and gives a rough idea of the explored design space); the time spent by the Petrify tool; and the time spent by the proposed method. We use 'mem' if there was a memory overflow and 'time' to indicate that the test had not stopped after 15 hours. (Table 4 provides additional data about the benchmarks.)

Although the performed testing was limited in scope, one can draw some conclusions about the performance of the proposed algorithm. In all cases the proposed method solved the problem relatively easily, even when it was intractable for Petrify. In some cases, it was faster by several orders of magnitude. The time spent on all these benchmarks was quite satisfactory – it took less than 50 seconds to solve the hardest one (CfAsymCscA); note however, that in that case a total of 450 equations were obtained, i.e., more than 9 equations per second.

Table 5. Experimental results.

Real-Life STGs			
Problem	Eqns (SAT)	Time, [s] PFY	SAT
LazyRingCsc	14	1	<1
RingCsc	63	850	3
Dup4phCsc	48	20	<1
Dup4phMtrCsc	46	13	<1
DupMtrModCsc	165	125	1
CfSymCscA	60	163	16
CfSymCscB	34	10	<1
CfSymCscC	18	13	<1
CfSymCscD	16	3	<1
CfAsymCscA	450	1448	48
CfAsymCscB	93	2323	17

Marked Graphs			
Problem	Eqns (SAT)	Time, [s] PFY	SAT
PpWkCsc(2,3)	7	<1	<1
PpWkCsc(2,6)	13	4	<1
PpWkCsc(2,9)	19	44	<1
PpWkCsc(2,12)	25	2082	<1
PpWkCsc(3,3)	10	1	<1
PpWkCsc(3,6)	19	43	<1
PpWkCsc(3,9)	28	7380	<1
PpWkCsc(3,12)	37	*time*	1

STGs with Arbitration			
Problem	Eqns (SAT)	Time, [s] PFY	SAT
PpArbCsc(2,3)	18	4	<1
PpArbCsc(2,6)	24	42	<1
PpArbCsc(2,9)	30	315	<1
PpArbCsc(2,12)	36	3840	1
PpArbCsc(3,3)	29	45	<1
PpArbCsc(3,6)	38	1001	<1
PpArbCsc(3,9)	47	24941	1
PpArbCsc(3,12)	56	*mem*	2

It is important to note that these improvements in memory and running time come *without any reduction in quality of the solutions*. In fact, our method is *complete*, i.e., it can produce all the valid complex-gate implementations of each signal. However, in our implementation we restricted the algorithm to only minimal supports. Nevertheless, the explored design space was quite satisfactory: as the 'Eqns' column in Table 5 shows, in many cases our method proposed quite a few alternative implementations for signals. In fact, among the list of solutions produced by our tool there was always a solution produced by Petrify (with, perhaps, only minor differences due to the non-uniqueness of the result of Boolean minimisation). Overall, the proposed approach turned out to be clearly superior, especially for hard problem instances.

5.8 Conclusion and Future Work

We have proposed a complex-gate design flow for asynchronous circuits based on STG unfolding prefixes comprising: (i) a SAT-based algorithm for detection of encoding conflicts; (ii) a framework for visualisation and resolution of encoding conflicts; and (iii) an algorithm for derivation of Boolean equations for the gates implementing the circuit based on the incremental SAT approach.

Note that in all the test cases (Table 4) the size of the complete prefix was relatively small. This can be explained by the fact that STGs usually contain a lot of concurrency but relatively few choices, and thus the prefixes are in many cases not much bigger then the STGs themselves. For the scalable benchmarks, one can observe that the complete prefixes exhibited polynomial (in fact, quadratic) growth, whereas the number of reachable states grew exponentially. As a result, the unfolding-based method had a clear advantage over that based on state graphs, both in terms of memory usage and running time. The experimental results demonstrated that the devised algorithms could handle quite large specifications in relatively short time, obtaining high-quality solutions. Moreover, the proposed approach is applicable to all bounded Petri nets, without any structural restrictions such as Marked Graph of Free-Choice constraint.

An important observation one can make is that the combination 'unfolder & solver' turns out to be quite powerful. It has already been used in a number of papers (see, e.g., [30, 38]). Most of 'interesting' problems for safe Petri nets are $\mathcal{PSPACE}$-complete [23], and unfolding such a net allows to reduce this complexity class down to $\mathcal{NP}$ (or even $\mathcal{P}$ for some problems, e.g., checking consistency). Though in the worst case the size of a finite and complete unfolding prefix can be exponential in the size of the original Petri net, in practice it is often relatively small. In particular, according to our experiments, this is almost always the case for STGs. A problem formulated for a prefix can usually be translated into some canonical problem, e.g., an integer programming one [38], a problem of finding a stable model of a logic program [30], or SAT as here. Then an appropriate solver can be used for efficiently solving it.

The presented framework for interactive refinement aimed at resolution of encoding conflicts is based on the visualisation of conflict cores, which are sets of transitions 'causing' state encoding conflicts. Cores are represented at the level of the unfolding prefix, which is a convenient model for understanding the behaviour of the system due to its simple branching structure and acyclicity.

The advantage of using cores is that only those parts of STGs which cause encoding conflicts, rather than the complete list of CSC conflicts, are considered. Since the number of cores is usually much smaller than the number of encoding conflicts, this approach saves the designer from analysing large amounts of information. Resolution of encoding conflicts requires the elimination of cores by introducing additional signals into the STG. The refinement contains several interactive steps aimed at helping the designer to obtain a customised solution. The case studies demonstrate the positive features of the interactive refinement process.

Heuristics for signal insertion based on the height map and exploiting the intersections of cores use the most essential information about encoding conflicts, and thus should be quite efficient. In fact, the conflict resolution procedure can be automated either partially or completely. However, in order to obtain an optimal solution, a semi-automated resolution process should be employed. For example, the tool might suggest the areas for insertion of new signal transitions, which are to be used as guidelines. Yet, the designer is free to intervene at any stage and choose an alternative location, in order to take into account the design constraints.

We view these results as encouraging. In future work we intend to include also the technology mapping step into the described design flow, as well as incorporate other methods for resolving encoding conflicts (concurrency reduction [17], timing assumption [18], etc.) into the proposed framework for visualisation and resolution of encoding conflicts.

6 Other Related Work and Future Directions

There has been a large amount of research in hardware design using Petri nets in the last few years. This chapter has covered only the main advances made recently in logic synthesis from STGs, and some of them, such as the topic of STG decomposition, only briefly.

The reader is however encouraged to look broader and for that we briefly list here a number of relevant and interesting developments.

- **STG Decomposition.** The idea of reducing complexity in logic synthesis from STG by STG decomposition is not new. For example, in [14] the contraction method was introduced in which the logic equations for output signal were derived from the projections of the state graph on the set of relevant signals forming the support of the derived function. This idea has been recently developed further in [71], in order remove some restrictions on the class of the STG (live and safe free choice). It also approaches the decomposition problem in a powerful equivalence framework, which is a bisimulation with angelic nondeterminism. Another attempt in this direction, perhaps in a more practical context of the HDL-based design flow was recently reported in [77].
- **Implementability Checking in Polynomial Time.** Another important source of complexity reduction is a search for polynomial algorithms for various stages in asynchronous logic synthesis for restricted STG classes, in particular for free-choice nets. Such an algorithm has been developed in [24].
- **Optimisation in Direct Mapping from STG.** The advantages of the direct mapping of Petri nets to circuits can be exploited in the STG level, although at extra cost in circuit area. The direct mapping does not however affect performance negatively. In fact in many cases, direct mapping offers solutions where the latency between input and output signal transitions is minimal. New techniques of translating STGs into circuits using David cells, structured into 'tracker' and 'bouncer', also include optimisation of the logic size [61].
- **Synthesis from STGs in Restricted Bases.** While logic synthesis of speed-independent circuits in complex gates provides a satisfactory solution for modern CMOS design technologies, in the future it may not be reliable enough to guarantee correct operation. The effects of delays in wires

and parametric instabilities may require a much more conservative approach to the implementation of control circuits. In this respect, advances in the synthesis of circuits that are monotonic [62], i.e., having no "zero-delay" inverters on the inputs, and free from isochronic forks [63] are important.
- **Synthesis with Relative Timing Assumptions.** Unlike the above, sometimes designing circuits under conservative assumptions can lead to significant wastage of area, speed and power. More optimistic considerations can be made about delays in the system, for example based on the knowledge of actual delays in the data path or in the environment, or due to information about relative speeds of system components. Use of relative timing has been investigated under the notion of lazy transition systems [15].
- **Synthesis from Delay-Insensitive Process Specifications.** A potential way to automatic compilation of HDLs based on communicating processes to asynchronous circuits may be via an important semantical link between delay-insensitive (DI) process algebras and Petri nets. Such a link has been established and developed to the level of tool support in [33]. A particularly interesting contribution has been the definition of the DI process decomposition which helps avoiding CSC conflicts in the STG that is constructed automatically from a process-algebraic specification [34].

Acknowledgements

We would like to thank Alex Bystrov, Michael Kishinevsky, Alex Kondratyev, Maciej Koutny, Luciano Lavagno and Agnes Madalinski for contributing to this research at various stages. This research was partially supported by EU Framework 5 ACiD-WG and EPSRC grants GR/M99293, GR/M94366 (MOVIE) and GR/R16754 (BESST).

References

1. A. Allan, et. al., 2001 Technology Roadmap for Semiconductors, *Computer*, January 2002, pp. 42-53.
2. K. van Berkel. *Handshake Circuits: an Asynchronous Architecture for VLSI Programming*, volume 5 of International Series on Parallel Computation. Cambridge University Press, 1993.
3. E. Best and B. Grahlmann. PEP – more than a Petri Net Tool. *Proc. of Tools and Algorithms for the Construction and Analysis of Systems (TACAS'96)*, Springer-Verlag, Lecture Notes in Computer Science 1055, Springer-Verlag (1996) 397-401.
4. I. Blunno and L. Lavagno. Automated synthesis of micro-pipelines from behavioral VERILOG HDL, *Proc. of IEEE Symp. on Adv. Res. in Async. Cir. and Syst. (ASYNC 2000)*, IEEE CS Press, pp. 84–92.
5. R. Brayton, G. Hachtel, C. McMullen and A. Sangiovanni-Vincentelli: *Logic Minimisation Algorithms for VLSI Synthesis.* Kluwer Academic Publishers (1984).
6. A. Bystrov and A. Yakovlev. Asynchronous Circuit Synthesis by Direct Mapping: Interfacing to Environment, *Proc. ASYNC'02*, Manchester, April 2002.

7. J. Carmona and J. Cortadella. Input/Output Compatibility of Reactive Systems. In *Fourth International Conference on Formal Methods in Computer-Aided Design (FMCAD)*, Portland, Oregon, USA, November 2002. Springer-Verlag.
8. J. Carmona and J. Cortadella. ILP Models for the Synthesis of Asynchronous Control Circuits. In *Proc. International Conf. Computer-Aided Design (ICCAD)*, San Jose, California, USA, November 2003.
9. J. Carmona, J. Cortadella, and E. Pastor. A structural encoding technique for the synthesis of asynchronous circuits. *Fundamenta Informaticae*, pages 135–154, April 2001.
10. C. Carrion and A. Yakovlev: Design and Evaluation of Two Asynchronous Token Ring Adapters. Tech. Rep. CS-TR-562, School of Comp. Sci., Univ. of Newcastle (1996).
11. Daniel M. Chapiro, *Globally-Asynchronous Locally-Synchronous Systems*. PhD thesis, Stanford University, October 1984.
12. T. Chelcea, A. Bardsley, D. Edwards and S.M. Nowick. A burst-mode oriented back-end for the BALSA synthesis system, *Proc. of Design, Automation and Test in Europe (DATE'02)*, IEEE CS Press, pp. 330-337.
13. T.-A. Chu, C. K. C. Leung, and T. S. Wanuga. A design methodology for concurrent VLSI systems. In *Proc. International Conf. Computer Design (ICCD)*, pages 407-410. IEEE Computer Society Press, 1985.
14. T.-A. Chu: *Synthesis of Self-Timed VLSI Circuits from Graph-Theoretic Specifications*. PhD Thesis, MIT/LCS/TR-393 (1987).
15. J. Cortadella, M.Kishinevsky, S.M. Burns, K.S. Stevens, A. Kondratyev, L. Lavagno, A. Taubin, A. Yakovlev. Lazy Transition Systems and Asynhronous Circuit Synthesis with Relative Timing Assumptions. *IEEE Trans. of CAD*, Vol. 21, No. 2, Feb. 2002, pages 109-130.
16. J. Cortadella, M. Kishinevsky, A. Kondratyev, L. Lavagno and A. Yakovlev: A Region-Based Theory for State Assignment in Speed-Independent Circuits. *IEEE Transactions on Computer-Aided Design of Integrated Circuits and Systems* 16(8) (1997) 793–812.
17. J. Cortadella, M. Kishinevsky, A. Kondratyev, L. Lavagno and A. Yakovlev: Automatic Handshake Expansion and Reshuffling Using Concurrency Reduction. Proc. of *HWPN'98*, (1998) 86–110.
18. J. Cortadella, M. Kishinevsky, A. Kondratyev, L. Lavagno, and A. Yakovlev: *Logic Synthesis of Asynchronous Controllers and Interfaces*. Springer Verlag (2002).
19. W. J. Dally and J. W. Poulton: *Digital Systems Engineering*. Cambridge University Press (1998).
20. R. David. Modular design of asynchronous circuits defined by graphs. *IEEE Transactions on Computers*, 26(8):727–737, August 1977.
21. J. Desel and J. Esparza. Reachability in cyclic extended free-choice systems. *TCS 114, Elsevier Science Publishers B.V.*, 1993.
22. D. Edwards and A. Bardsley. BALSA: An asynchronous hardware synthesis language. *The Computer Journal*, 45(1):12-18, 2002.
23. J. Esparza: Decidability and Complexity of Petri Net Problems – an Introduction. In: *Lectures on Petri Nets I: Basic Models*, W. Reisig and G. Rozenberg (Eds.). LNCS 1491 (1998) 374–428.
24. J. Esparza. A Polynomial-Time Algorithm for Checking Consistency of Free-Choice Signal Transition Graphs, Proc. of the 3rd Int. Conf. Applications of Concurrency to System Design (ACSD'03), IEEE CS Press, June 2003, pp. 61-70.
25. J. Esparza, S. Römer and W. Vogler: An Improvement of McMillan's Unfolding Algorithm. *FMSD* 20(3) (2002) 285–310.

26. D. Ferguson and M. Hagedorn, The Application of NULL Convention Logic to Microcontroller/Microconverter Product, Second ACiD-WG Workshop, Munich, 2002. URL: http://www.scism.sbu.ac.uk/ccsv/ACiD-WG/Workshop2FP5/Programme/.
27. S. Furber, Industrial take-up of asynchronous design, Keynote talk at the Second ACiD-WG Workshop, Munich, 2002. URL: http://www.scism.sbu.ac.uk/ccsv/ACiD-WG/Workshop2FP5/Programme/.
28. S. B. Furber, A. Efthymiou, and M. Singh: A Power-Efficient Duplex Communication System. Proc. of *AINT'00*, TU Delft, The Netherlands (2000) 145–150.
29. F. García Vallés and J.M. Colom. Structural analysis of signal transition graphs. In D. Holdt In B. Farwer and M.O. Stehr, editors, *Proceedings of the Workshop Petri Nets in System Engineering (PNSE?97). Modelling, Verification and Validation*, pages 123–134, Hamburg (Germany). September 25–26, September 1997. Published as report n 205 of the Computer Science Department of the University of Hamburg.
30. K. Heljanko: Using Logic Programs with Stable Model Semantics to Solve Deadlock and Reachability Problems for 1-Safe Petri Nets. *Fundamentae Informaticae* 37(3) (1999) 247–268.
31. K. Heljanko, V. Khomenko, and M. Koutny: Parallelization of the Petri Net Unfolding Algorithm. Proc. of *TACAS'2002*, LNCS 2280 (2002) 371–385.
32. L.A. Hollaar. Direct implementation of asynchronous control units. *IEEE Transactions on Computers*, C-31(12):1133–1141, December 1982.
33. H. K. Kapoor, M. B. Josephs and D. P. Furey: Verification and Implementation of Delay-Insensitive Processes in Restrictive Environments. Proc. of *ICACSD'04*, IEEE Comp. Soc. Press (2004) to appear.
34. H. K. Kapoor and M. B. Josephs: Automatically decomposing specifications with concurrent outputs to resolve state coding conflicts in asynchronous logic synthesis. Proc. of DAC'04, 2004 (to appear).
35. V. Khomenko and M. Koutny: LP Deadlock Checking Using Partial Order Dependencies. Proc. of *CONCUR'2000*, LNCS 1877 (2000) 410–425.
36. V. Khomenko and M. Koutny: Towards An Efficient Algorithm for Unfolding Petri Nets. Proc. of *CONCUR'2001*, LNCS 2154 (2001) 366–380.
37. V. Khomenko, M. Koutny, and V. Vogler: Canonical Prefixes of Petri Net Unfoldings. Proc. of *CAV'2002*, LNCS 2404 (2002) 582–595. Full version: Acta Informatica 40(2) (2003) 95-118.
38. V. Khomenko, M. Koutny and A. Yakovlev: Detecting State Coding Conflicts in STGs Using Integer Programming. Proc. of *DATE'02*, IEEE Comp. Soc. Press (2002) 338–345.
39. V. Khomenko, M. Koutny, and A. Yakovlev: Detecting State Coding Conflicts in STG Unfoldings Using SAT. Proc. of *ICACSD'03*, IEEE Comp. Soc. Press (2003) 51–60. Full version: to appear in Special Issue on Best Papers from ICACSD'2003, Fundamenta Informaticae.
40. V. Khomenko, M. Koutny, and A. Yakovlev: Logic Synthesis Avoiding State Space Explosion. Proc. of *ICACSD'04*, IEEE Comp. Soc. Press (2004) to appear. Full version: Tech. Rep. CS-TR-813, School of Comp. Science, Univ. of Newcastle. URL: http://homepages.cs.ncl.ac.uk/victor.khomenko/home.formal/papers/papers.html.
41. D. J. Kinniment, B. Gao, A. Yakovlev and F. Xia: Towards asynchronous A-D conversion. Proc. of *ASYNC'00*, IEEE Comp. Soc. Press (2000) 206–215.
42. Michael Kishinevsky, Alex Kondratyev, Alexander Taubin, and Victor Varshavsky. *Concurrent Hardware: The Theory and Practice of Self-Timed Design*. Series in Parallel Computing. John Wiley & Sons, 1994.
43. A. Kondratyev and K. Lwin. Design of asynchronous circuits using synchronous CAD tools. *IEEE Design and Test of Computers*, 19(4):107-117, 2002.

44. K. S. Low and A. Yakovlev: Token Ring Arbiters: an Exercise in Asynchronous Logic Design with Petri Nets. Tech. Rep. CS-TR-537, School of Comp. Sci., Univ. of Newcastle (1995).
45. A. Madalinski, A. Bystrov, V. Khomenko, and A. Yakovlev: Visualisation and Resolution of Coding Conflicts in Asynchronous Circuit Design. Proc. of *DATE'03*, IEEE Comp. Soc. Press (2003) 926–931. Full version: Special Issue on Best Papers from DATE'2003, IEE Proceedings: Computers & Digital Techniques 150(5) (2003) 285–293.
46. K. L. McMillan: Using Unfoldings to Avoid State Explosion Problem in the Verification of Asynchronous Circuits. Proc. of *CAV'92*, LNCS 663 (1992) 164–174.
47. G. De Micheli. *Synthesis and Optimisation of Digital Circuits*, McGraw-Hill, 1994.
48. S. Moskewicz, C. Madigan, Y. Zhao, L. Zhang and S. Malik: CHAFF: Engineering an Efficient SAT Solver. Proc. of *DAC'01*, ASME Technical Publishing (2001) 530–535.
49. T. Murata. Petri Nets: Properties, analysis and applications. *Proceedings of the IEEE*, pages 541–580, April 1989.
50. E. Pastor, J. Cortadella, A. Kondratyev, and O. Roig. Structural methods for the synthesis of speed-independent circuits. *IEEE Transactions on Computer-Aided Design*, 17(11):1108–1129, November 1998.
51. S.S. Patil and J.B. Dennis. The description and realization of digital systems. In *Proceedings of the IEEE COMPCON*, pages 223–226, 1972.
52. M.A. Peña and J. Cortadella, Combining process algebras and Petri nets for the specification and synthesis of asynchronous circuits, *Proc. of IEEE Symp. on Adv. Res. in Async. Cir. and Syst. (ASYNC'96)*, IEEE CS Press, pp. 222-232.
53. C. A. Petri. *Kommunikation mit Automaten.* PhD thesis, Bonn, Institut für Instrumentelle Mathematik, 1962. (technical report Schriften des IIM Nr. 3).
54. P. Riocreux: *Private communication.* UK Asynchronous Forum (2002).
55. L. Y. Rosenblum and A. V. Yakovlev. Signal graphs: from self-timed to timed ones. In *Proceedings of International Workshop on Timed Petri Nets*, pages 199–207, Torino, Italy, July 1985. IEEE Computer Society Press.
56. A. Schrijver. *Theory of Linear and Integer Programming.* John Wiley & Sons, 1998.
57. A. Semenov: *Verification and Synthesis of Asynchronous Control Circuits Using Petri Net Unfolding.* PhD Thesis, University of Newcastle upon Tyne (1997).
58. D. Shang, F. Xia and A. Yakovlev. Asynchronous Circuit Synthesis via Direct Translation, *Proc. Int. Symp. on Cir. and Syst. (ISCAS'02)*, Scottsdale, Arizona, May 2002.
59. Manuel Silva, Enrique Teruel, and José Manuel Colom. Linear algebraic and linear programming techniques for the analysis of place/transition net systems. *Lecture Notes in Computer Science: Lectures on Petri Nets I: Basic Models*, 1491:309–373, 1998.
60. J. Sparsø and S. Furber, Edt., *Principles of Asynchronous Circuit Design: A Systems Perspective.* Kluwer Academic Publishers, 2001
61. D. Sokolov, A. Bystrov and A. Yakovlev. STG optimisation in the direct mapping of asynchronous circuits, Proc. Design and Test in Europe (DATE), March 2003, 932-937.
62. N. Starodoubtsev, S. Bystrov, M. Goncharov, I. Klotchkov and A. Smirnov. Towards Synthesis of Monotonic Circuits from STGs, In Proc. of 2ndd Int. Conf. Applications of Concurrency to System Design (ACSD'01), IEEE CS Press, June 2001, pp. 179-180.
63. N. Starodoubtsev, S. Bystrov, and A. Yakovlev. Monotonic circuits with complete acknowledgement, Proc. of ASYNC'03, Vancouver, IEEE CS Press, pp. 98-108.

64. Ivan E. Sutherland. Micropipelines. *Communications of the ACM,* 32(6):720-738, June 1989.
65. A. Valmari. A stubborn attack on state explosion. *Formal Methods in System Design*, 1(4):297–322, 1992.
66. P. Vanbekbergen. *Synthesis of Asynchronous Control Circuits from Graph-Theoretic Specifications.* PhD thesis, Catholic University of Leuven, 1993.
67. P. Vanbekbergen, F. Catthoor, G. Goossens and H. De Man: Optimised Synthesis of Asynchronous Control Circuits form Graph-Theoretic Specifications. Proc. of *ICCAD'90*, IEEE Comp. Soc. Press (1990) 184–187.
68. V. I. Varshavsky and V. B. Marakhovsky. Asynchronous control device design by net model behavior simulation. In J. Billington and W. Reisig, editors, *Application and Theory of Petri Nets 1996*, volume 1091 of *Lecture Notes in Computer Science*, pages 497–515. Springer-Verlag, June 1996.
69. V. I. Varshavsky, editor. *Self-Timed Control of Concurrent Processes: The Design of Aperiodic Logical Circuits in Computers and Discrete Systems.* Kluwer Academic Publishers, Dordrecht, The Netherlands, 1990.
70. Thomas Villiger, Hubert Kslin, Frank K. Grkaynak, Stephan Oetiker, and Wolfgang Fichtner. Self-timed ring for globally-asynchronous locally-synchronous systems. Proc. *International Symposium on Advanced Research in Asynchronous Circuits and Systems*, pp. 141-150. IEEE Computer Society Press, May 2003.
71. W.Vogler and R. Wollowski. Decomposition in asynchronous circuit design. In J. Cortadella, A. Yakovlev, and G. Rozenberg, editors, *Concurrency and Hardware Design*, volume 2549 of *Lecture Notes in Computer Science*, pages 152-190. Springer-Verlag, 2002.
72. A. Yakovlev: Designing Control Logic for Counterflow Pipeline Processor Using Petri nets. *FMSD* 12(1) (1998) 39–71.
73. A. Yakovlev, S. Furber and R. Krenz, Design, Analysis and Implementation of a Self-timed Duplex Communication System, CS-TR-761, Dept. Computing Science, Univ. of Newcastle upon Tyne, March 2002. URL: `http://www.cs.ncl.ac.uk/people/alex.yakovlev/home.informal/some_papers/duplex-TR.ps`.
74. A. Yakovlev and A. Koelmans. Petri nets and Digital Hardware Design *Lectures on Petri Nets II: Applications. Advances in Petri Nets, Lecture Notes in Computer Science*, vol. 1492, Springer-Verlag, 1998, pp. 154-236.
75. A. Yakovlev and A. Petrov: Petri Nets and Asynchronous Bus Controller Design. Proc. of *ICATPN'90*, (1990) 244–262.
76. A. Yakovlev, V. Varshavsky, V. Marakhovsky and A. Semenov. Designing an asynchronous pipeline token ring interface, *Proc. of 2nd Working Conference on Asynchronous Design Methdologies, London, May 1995*, IEEE Comp. Society Press, N.Y., 1995, pp. 32-41.
77. T. Yoneda and C. Myers. Synthesis of Speed Independent Circuits based on Decomposition, Proceedings of ASYNC 2004, Heraklion, Greece, IEEE CS Press, April 2004.
78. L. Zhang and S. Malik: The Quest for Efficient Boolean Satisfiability Solvers. Proc. of *CAV'02*, LNCS 2404 (2002) 17–36.

Teaching Coloured Petri Nets: Examples of Courses and Lessons Learned

Søren Christensen and Jens Bæk Jørgensen

Department of Computer Science, University of Aarhus
Aabogade 34, DK-8200 Aarhus N, Denmark
{schristensen,jbj}@daimi.au.dk

Abstract. In this paper, we describe and discuss three different courses in which Coloured Petri Nets (CPN) is used: (1) an introductory course on distributed systems and network protocols; (2) an advanced course on CPN; (3) a course on industrial application of CPN. Courses (1) and (2) are taught at the Department of Computer Science, University of Aarhus and course (3) is given for professional software engineers. For each course, we briefly present contents, format, and role of CPN. Then we describe a number of lessons learned from teaching the three courses. We have two aims in mind: In the first place, we want to share our specific experiences with other teachers. Secondly, we want to contribute to a more general discussion and exchange of ideas on Petri nets and education.

1 Introduction

Coloured Petri Nets (CPN) [10] has had a place in the curriculum at the Department of Computer Science, University of Aarhus (henceforth abbreviated with the Danish acronym *DAIMI* [32]) for the last twenty years. The main reason is that formal modelling languages like CPN are suitable for many educational activities within computer science. Another contributing factor is the presence of professor Kurt Jensen [28], whose PhD work around 1980 defined the first version [9] of the CPN language and laid the foundation for the research of the CPN Group [26] at DAIMI.

In the 1980's, CPN was used at DAIMI as a general system description language in the introductory first-year course taken by 150-200 students each year. One of the main purposes of using CPN was to teach students that making abstract system descriptions (or models) is an important activity in computer science. A number of lectures on CPN were given, the students read some introductory material, and they were required to solve exercises on CPN. Examples of exercises were to model the flow of customers through the local canteen and to model a traffic light. These were non-trivial exercises, especially because at that time, tool support for CPN (and other kinds of Petri nets) was scarce. Models were drawn on paper and simulations were carried out by playing the token game with coins or drawing pins on sheets of paper. Another main purpose of using CPN was to introduce the students to formal semantics of programming

J. Desel, W. Reisig, and G. Rozenberg (Eds.): ACPN 2003, LNCS 3098, pp. 402–412, 2004.

languages via CPN, e.g., the semantics of various language constructs of Pascal [12].

Around 1990, DAIMI began to offer advanced courses on the CPN language itself. There were two kinds of courses: (1) application-oriented courses, where students constructed and analysed fairly large CPN models (made possible by the emergence of the Design/CPN tool [27] in 1989); (2) theoretical courses, where students immersed in the mathematical foundation of CPN and typically pursued formal verification methods like state spaces or place invariants.

In the 1990's, CPN made up about half of the curriculum in a third-year course on distributed systems; the mid-1990's incarnation of that course is described in detail in the paper [6]. Two textbooks were used: Jensen's CPN book [10] and Tanenbaum's book on distributed operating systems [24]. The main emphasis of the CPN part of the course was on CPN modelling and use of the Design/CPN tool as vehicles for design and analysis of distributed systems, but students were also thoroughly introduced to the mathematical foundation and to formal verification methods of CPN.

We, the authors of this paper, are members of the CPN Group at DAIMI. We have seen CPN in education from different perspectives, starting with our first encounter in the early and mid 1980's, when we were students at DAIMI. From around 1990, we have experienced CPN from the other side of the table: as teachers in various computer science courses. This paper is based on our experiences with using CPN in education. We have two aims: In the first place, we want to share our specific experiences with other teachers, and hopefully provide some kind of inspiration (and perhaps save computer science students at other universities from teachers making the same errors as we did). Secondly, we want to contribute to a more general discussion and exchange of ideas on Petri nets and education.

In Sect. 2, we describe three courses we have taught and which have used CPN. In Sect. 3, we report some lessons learned. We draw some conclusions in Sect. 4.

2 Examples of Courses

Currently, CPN is used at DAIMI in two different courses:

- The *distributed systems course*: a third-year introductory course on distributed systems and network protocols [33].
- The *advanced course on CPN*: a graduate course in which CPN is studied as a language in its own right.

We describe these two courses more thoroughly in this section. In addition, we describe:

- The *industrial application of CPN course*: a course on application of CPN held for a group of engineers from a software company.

2.1 Distributed Systems Course

Since the late 1990's, CPN has been used as an ingredient of a third-year introductory course on distributed systems and network protocols. The course runs over 15 weeks and is attended by close to 100 students. The course gives a credit of 10 ECTS points and consists of two parts of equal size. The first part introduces basic concepts and design techniques for distributed systems. The second part introduces the basic ideas behind computer networks and network protocols, including a detailed coverage of the Internet protocols. The textbooks currently used are Coulouris et al's on distributed systems [7] and Stallings' on networks and network protocols [23].

The format of the course is a combination of lectures for all students and tutorials where the students are divided into classes of approximately 20 students, and where an older student is available as teaching assistant. In average, there are three hours of lectures and three hours of tutorials each week. Students are expected to use a total of up to 15 hours on the course each week.

Three two-hour lectures on CPN are given early in the course. The first informally introduces the basic concepts of CPN. The second gives some practical hints on the construction of CPN models and it introduces basics of the Standard ML language [17]; some elementary Standard ML programming skills are necessary in order to properly apply the CPN tool, which is introduced in the third lecture in an extensive tool demonstration. The CPN literature we have used over the years is excerpts from Jensen's book [10] (chapter 1 and parts of chapter 3) and the practitioner's guide to CPN [15]. The tools which have been applied are Design/CPN and the newer CPN Tools [21, 30].

CPN is used to give the students a better understanding of distributed systems and network protocols than can be provided by the textbooks alone. Thus, the role of CPN is to be a supplement to the main curriculum. We feel that the exercises are a weak part of the two textbooks we use. Therefore, we need additional exercises and CPN helps us to achieve this. CPN exercises are put forward throughout the course. An example of a CPN exercise aiming at making the students better understand fundamental mechanisms of distributed file systems as described in the textbook (chapter 8 of Coulouris et al's book [7]) is: (1) Make a CPN model of caching in Sun's Network File System (NFS); (2) based on your model, discuss the problem of cache consistency and advantages and drawbacks of the NFS solution.

In addition to smaller exercises, we have asked the students to solve a larger mandatory project on CPN over a time period of three weeks. The project varies from year to year. In 2003, the students were asked to design a protocol ensuring reliable communication over an unreliable communication channel with appropriate use of timers, sequence numbers, retransmissions, etc.

CPN is a suitable language for this course because it allows the students to explicitly describe their interpretation of the highly prose-based and always slightly ambiguous and sometimes even vague presentation of algorithms, protocols etc. from the textbooks. In particular, CPN facilitates the students' comprehension of traditionally hard-to-understand issues related to concurrency, resource sharing, synchronisation, and conflicts. CPN models give a solid foundation for discussions between students and between students and teaching assistants.

2.2 Advanced Course on CPN

CPN is the subject of an advanced course, which runs over 12-15 weeks and is typically attended by 10-15 students. The course gives a credit of 10 ECTS points and comprises three parts. The second part is rather special; it consists in participation in an international workshop on CPN being held in Aarhus (for the 2002 incarnation of that workshop, see [31]). Thus, from the students' perspective, the workshop is an integrated part of the course. The first part consists of introductory lectures and student presentations of the scientific papers on CPN, which are accepted for the workshop. Together, the first and the second parts occupy about a month of calendar time, in which the students work intensively with the course. The third part, which takes approximately two months, consists in carrying out a project. The literature used in the course is excerpts from volumes 1 and 2 of Jensen's books [10, 11], the practitioner's guide [15], and workshop proceedings (which in 2002 were [8]).

The students choose between two categories of projects: One category is application-oriented project, i.e., creation and analysis of CPN models of domains of interest for the students. An example of such a project from 2002 is a group of students who worked together with the large Danish company Danfoss to model and investigate the behaviour of control software for an industrial embedded system. The other category comprises theoretical projects. Examples of such projects are study of methods for verification by means of state space analysis. The students read relevant literature and sometimes do small practical exercises using various tools (depending on availability and quality of such tools). All students interested in this subject study basic state space analysis. Subsequently, some choose to pursue more advanced methods, e.g., state space analysis using equivalence classes or symmetries [11], state space analysis using stubborn sets [25], or state space analysis by the sweepline method [5].

2.3 Industrial Application of CPN Course

In addition to teaching CPN to computer science students, we have given several CPN courses for software engineers from the industry over the years. These courses are tailored for particular companies and vary in contents and format. Examples of CPN projects where a course has been an integrated part are analysis of audio/video transmission protocols at Bang and Olufsen as described in [4], design of alarm systems at Dalcotech [20], and analysis of car control systems at Peugeot Citroen [18].

Typically, between two and six engineers participate in the course, which runs over a total of six full days, divided into two parts each comprising three days in one week and three days in another week. The course is very application-oriented and no introduction to the formal, mathematical foundation of CPN is given. The attendees use most of their time doing practical hands-on exercises in small groups.

In the first part, the basic CPN concepts and a CPN tool are introduced. The first day covers the most fundamental CPN concepts (corresponding to chapter 1 of Jensen's book [10]). The second day covers hierarchical CPN models (corresponding to parts of chapter 3 of Jensen's book [10]) and on the third day, CPN models with time are introduced. The introductions are not traditional lecture-like presentations, but integrated parts of extensive tool demonstrations. It is always shown how to create, edit, and simulate CPN models (it is often shown how to carry out simple state space analysis as well). The basic functionality of the tool is explained and step-by-step instructions on how to carry out modelling tasks are given. Throughout the first part, the engineers do practical hands-on exercises with the tool. As example, on the first day, they make various modifications of a small model of a simple communications protocol, e.g., modify a stop-and-wait protocol to become a more general sliding-window protocol.

On the last day of the first part, much time is allocated for discussion and determination of the more specific contents of the second part. It is crucial that the engineers make this choice themselves. They identify problems in their domain which they would like to address using CPN. CPN instructors often start to outline model drafts. In the time between the first and the second part, the engineers and the CPN instructors continue to think about how CPN can be used for the particular problem that the engineers want to solve. This thinking is crucial preparation for the second part in which the engineers spend most of the time creating and analysing larger domain-specific models. The CPN instructors are available to help the engineers, who, thus, are in a good position to work efficiently. Quite often, one engineer and one CPN instructor sit together in front of a computer and build models together.

3 Lessons Learned

In this section, we describe and discuss a number of lessons we have learned from teaching the three courses described above. The sources include course evaluation forms that students fill out after having attended a course and feedback from teaching assistants.

3.1 On Literature

As mentioned in the previous section, in the distributed systems course and the industrial application course, we have tried two possibilities for introductory literature on CPN: excerpts of Jensen's book [10] and the practitioner's guide [15].

In the distributed systems course, we have experienced that Jensen's book work better than the practitioner's guide. Students seem to prefer the thorough, step-wise introduction of the basic CPN concepts given in Jensen's book via easily understandable place-transition nets (PT nets) and small examples of CPN models. In the industrial application course, we have better experiences with the practitioner's guide, which is written to be directly appealing to industrial software engineers. It is non-formal, emphasises the application aspects of CPN, and advocates CPN as a way to address common software development problems. As an example, many industrial software engineers have scalability as a main concern; some may be reluctant to use of formal methods at all because they have seen approaches that do not scale well (e.g., when they studied computer science some years ago). One of the main purposes of the practitioner's guide is to demonstrate that CPN scales well to the size of problems that the software industry is dealing with. Therefore, the practitioner's guide introduces the basic CPN concept via a quite large example model and without PT nets, which do not scale well to modelling of industrial systems.

In the advanced course, the students who attend are particularly interested in CPN and want a thorough and broad coverage of the language. Therefore, we use both Jensen's books [10, 11] for a well-founded introduction of the basic concepts and analysis methods, and the practitioner's guide to set the stage for large-scale modelling. In addition, we use workshop proceedings from the current year. Typically, the range of subjects covered by the papers in these proceedings is quite broad. Reading the papers gives the students an introduction to research in CPN and to the process of writing and publishing scientific papers. Workshop papers often describe early results and work in progress. They may later mature into conference and journal papers, after more research, writing, and rewriting. Therefore, it is often possible for students to find errors and shortcomings and to propose constructive improvements. It is useful for the students to see the authors present their papers at the workshop and to compare this with the presentation that they (the students) gave themselves earlier in the course. And the students are in an excellent position to ask questions and to engage in discussions.

3.2 On Tools

The choice of which tool to use in a particular course may have a high impact on the quality of the course as experienced by the students.

In the distributed systems course, in which relatively many unexperienced users use the tool, it must be easy to learn, stable, and well documented. In the advanced course, the demands to the tools are lower for a number of reasons. In the first place, the students are older and more mature than the students, who attend the distributed systems course. Secondly, the students have a particular interest in CPN, and thirdly, they have a higher willingness to accept the inherent limitations of research prototypes of tools. In the industrial application course, the requirements to the tool are very high because industrial software engineers will inevitably compare it with top-quality commercial tools that they are used to from their everyday development work.

For the last couple of years, the choice between the new CPN Tools [21, 30] and the older Design/CPN tool has been difficult in all three courses. The trade-off between stability and being easy to learn has not been easy. Design/CPN has a quite steep learning curve; there are a number of obstacles causing troubles for unexperienced users like young students or software engineers previously unfamiliar with CPN. As an example, Design/CPN does not have a fully incremental syntax check which often makes it difficult to debug models. On the other hand, for the last many years, Design/CPN has been a stable and well-tested tool with many useful and nice features (and a bit old-fashioned user interface). CPN Tools alleviates many of the problems that are present with Design/CPN. In particular, it seems to be faster to become a proficient CPN Tools user than a Design/CPN user. However, sometimes we have been too eager to use new versions of CPN Tools. They have not always been tested well enough and have occasionally caused frustration for students (who do not want to spend their valuable time as alpha or beta testers of a tool, which is not sufficiently mature for a large group of unexperienced users). In 2003, CPN Tools had reached a maturity that ensured a successful use by approximately 100 students in the distributed systems course.

In all the CPN related courses we have taught, we have experienced that it is important that a long extensive tool demo is given. The students who do not attend the demo often have had severe problems getting started with the tool.

3.3 On Teacher Skills

The required level of CPN skills for the teacher or teachers varies between the three courses discussed in this paper. It must of course be solid, but is lowest for the distributed systems course, in which only basic CPN is taught and only relatively small exercises are put forward. The skill level required to run the industrial application course is higher: It is necessary that the teachers are experienced in building large CPN models and have the ability to understand the domain of interests for the engineers. The advanced course demands the highest skill level: It can probably only be taught properly by teachers who are themselves CPN researchers.

The teaching assistants for the distributed systems course are appointed by the Faculty of Science, University of Aarhus. We have taught instances of the course where the teaching assistants were not sufficiently proficient with either CPN itself or with the applied CPN tool. That problem immediately propagated on to the students, who were not being appropriately helped. To solve the problem, we now staff the weekly tutorials in the weeks in which CPN are introduced with older students who we know are well experienced with CPN and CPN tools (e.g., recruited among the CPN Group's PhD students and student programmers). In this way, help is readily available, and students do not have to spend excessive amounts of time trying to figure out themselves the peculiarities of CPN and CPN Tools (including the Standard ML programming language, which is new to the majority of the students).

In the advanced course and the industrial application course, we have always hand-picked teaching assistants to ensure that they were sufficiently experienced CPN users.

3.4 On Student Motivation

In the advanced course and the industrial application course, the attendees themselves have usually chosen that they want to learn about CPN. Therefore, we always teach highly motivated people in these two courses. But in the much broader distributed systems course, a number of students have seen CPN as a small and irritating "appendix" that they did not have to take too seriously. This has caused these students to more or less ignore CPN in the weeks where it was introduced. As a consequence, it turned out to be very difficult and sometimes even impossible for them to solve the CPN exercises put forward in conjunction with the main curriculum. To address this problem, we now require that all students carry out a larger mandatory CPN project early in the course.

In the advanced course, we have experienced that it is very motivating for students to participate in an international workshop giving them an opportunity to meet other students and researchers from foreign universities and companies. We have also experienced that it is motivating for some students to collaborate with an industrial partner like Danfoss.

In the industrial application course, it is important that focus is on the domain-specific problems that the engineers are facing. Therefore, as we saw, an example model of something that the engineers is familiar with from their everyday work is always a central ingredient in the course; we have not met many industrial software engineers, who find the dining philosophers or similar toy examples very appealing. Moreover, it is essential to present CPN as a useful supplement to the software development techniques that the engineers are already using; software engineers do typically not take a CPN course because they (or their managers) want to make dramatic changes to their company's software development practices. They want to improve what they are already good at. Today, this means that CPN often must be presented as a supplement to UML [19, 22], the de-facto modelling language of the software industry. CPN models may supplement, e.g., UML use cases [13], class diagrams, sequence diagrams, and collaboration diagrams. CPN may be seen as vehicle to make strong descriptions of behaviour and as an alternative to UML state machines and activity diagrams [14].

3.5 On Integration with Main Curriculum

The discussion in this section only applies to courses in which CPN is one component amongst others; not for courses exclusively on CPN. In the scope of this paper, this means the distributed systems course, where a number of students have criticized us for not integrating CPN well enough with the main curriculum. One of their arguments is that CPN is not sufficiently used at the lectures.

We partly agree with the students. As mentioned earlier, CPN mainly play a role in the tutorials and there are a number of reasons for not using CPN more extensively at the lectures. In the first place, the course is about distributed systems and network protocols; CPN is merely a vehicle for gaining a better understanding, description, and discussion of concepts and problems of the main curriculum. More extensive use of CPN could cause the course to become a "CPN Modelling of Distributed Systems and Network Protocols" course. Secondly, the lectures would deviate more from the textbooks than we feel they should. Thirdly, there is already plenty of material for the students to digest, so expanding the use of CPN would force us to reduce something else, which we do not believe is a good decision for this particular course.

However, finding the right balance between CPN and the rest of the curriculum is a non-trivial task that we are currently working on and have not yet solved to our full satisfaction.

4 Conclusions

In this paper, we have described three CPN related courses which we have taught and a number of lessons we have learned. Naturally, many of the lessons are of a general kind, applicable not only to the particular courses described here, but to many other kinds of courses as well: All teachers put an effort into finding good literature, making sure to have appropriate skills, and worrying about student motivations; many computer science teachers also have to deal with tools.

At this point in our writing, we would have liked to make a qualified comparative discussion of the experiences of colleagues who have taught Petri nets related courses at other universities. Berthelot and Petrucci report on experiences with education in relation to modelling, simulating, and verifying a train system using CPN and Design/CPN [2]. However, we have not been able to find many papers discussing Petri nets and education. Therefore, we encourage others to publish their experiences; we would like improve our teaching and to gain inspiration from an exchange of ideas with other teachers of Petri nets related courses.

If we take a broader perspective, going from teaching Petri nets to teaching computer science in general, there is a host of sources for more information, e.g., proceedings from the Innovation and Technology in Computer Science Education (ITiCSE) conferences, see, e.g., [3], and the ACM Curricula Recommendations [29]. Putting this paper into a broader perspective using such sources is future work.

We believe that over the years, DAIMI students and industrial software engineers have developed many useful skills from courses in which CPN has been used. This may well continue in the future. However, due to the general growth in computer science knowledge, the competition for a place in computer science curricula and for attention from the software industry is getting harder. At DAIMI, in the 1990's, CPN constituted about half of the curriculum of the distributed systems course. The amount of CPN has been decreasing; in 2003, CPN made up about 10-15 percent of the curriculum. In general, "exotic" subjects like CPN are at risk of having to leave the curriculum in order to accommodate something else, in particular in the undergraduate courses. Two examples on relatively recent additions to the undergraduate curriculum at DAIMI are a course on web technology and a course on security – that is tough competition. When use of Petri nets is considered in broader courses, there seems to always be good alternatives like process algebras (e.g., CCS [16]) and timed automata [1] in theoretical courses on concurrency and verification, and UML in software engineering oriented courses. In summary, we believe that it is worthwhile to think about good arguments to justify Petri nets in computer science curricula.

Acknowledgements

We thank Kurt Jensen, Lars M. Kristensen, and Thomas Mailund for allowing us to draw from their teaching experiences as a supplement to our own; we also thank them for discussions and helpful comments on this paper.

References

1. R. Alur and D. Dill. Automata for Modelling Real-Time Systems. *Theoretical Computer Science*, 126(2):183–236, 1994.
2. G. Berthelot and L. Petrucci. Specification and Validation of a Concurrent System: an Educational Project. *Software Tools for Technology Transfer*, 3(4):372–381, 2001.
3. R. Boyle and G. Evangelidis (eds.). Proceedings of the 8th Annual Conference on Innovation and Technology in Computer Science Education, 2003. Thessaloniki, Greece.
4. S. Christensen and J.B. Jørgensen. Analysing Bang & Olufsen's BeoLink Audio/Video System Using Coloured Petri Nets. In P. Azema and G. Balbo, editors, *Proceedings of the 18th International Conference on Application and Theory of Petri Nets*, volume 1248 of *LNCS*, pages 387–406, Toulouse, France, 1997. Springer-Verlag.
5. S. Christensen, L.M. Kristensen, and T. Mailund. A Sweep-Line Method for State Space Exploration. In T. Margaria and W. Yi, editors, *Proceedings of the 7th International Conference on Tools and Algorithms for the Construction and Analysis of Systems (TACAS 2001)*, volume 2031 of *LNCS*, pages 450–464, Genova, Italy, 201. Springer-Verlag.
6. S. Christensen and K.H. Mortensen. Teaching Coloured Petri Nets – A Gentle Introduction to Formal Methods in a Distributed Systems Course. In P. Azema and G. Balbo, editors, *Proceedings of the 18th International Conference on Application and Theory of Petri Nets*, LNCS, pages 290–309, Toulouse, France, 1997. Springer-Verlag.
7. G. Coulouris, J. Dollimore, and T. Kindberg. *Distributed Systems – Concepts and Design*. Addison-Wesley, 2001.
8. K. Jensen (ed.). Proceedings of the 3rd CPN Workshop, CPN'02. Technical report DAIMI PB-560, Department of Computer Science, University of Aarhus, 2002.
9. K. Jensen. Coloured Petri Nets and the Invariant Method. *Theoretical Computer Science*, 14:317–336, 1981.
10. K. Jensen. *Coloured Petri Nets – Basic Concepts, Analysis Methods and Practical Use. Volume 1, Basic Concepts.* Monographs in Theoretical Computer Science. An EATCS Series. Springer-Verlag, 1992.
11. K. Jensen. *Coloured Petri Nets – Basic Concepts, Analysis Methods and Practical Use. Volume 2, Analysis Methods.* Monographs in Theoretical Computer Science. An EATCS Series. Springer-Verlag, 1994.
12. K. Jensen and E.M. Schmidt. Pascal Semantics by a Combination of Denotational Semantics and High-level Petri Nets. In G. Rozenberg, editor, *Advances in Petri Nets*, volume 222 of *LNCS*, pages 297–329. Springer-Verlag, 1985.
13. J.B. Jørgensen and C. Bossen. Requirements Engineering for a Pervasive Health Care System. In *Proceedings of the IEEE International Requirements Engineering Conference (RE'03)*, pages 55–64, Monterey Bay, California, 2003. IEEE.

14. J.B. Jørgensen and S. Christensen. Executable Design Models for a Pervasive Healthcare Middleware System. In J.M. Jézéquel, H. Hussmann, and S. Cook, editors, *Proceedings of the 5th International Conference on the Unified Modeling Language (UML'02)*, volume 2460 of *LNCS*, pages 140–149, Dresden, Germany, 2002. Springer-Verlag.
15. L.M. Kristensen, S. Christensen, and K. Jensen. The Practitioner's Guide to Coloured Petri Nets. *International Journal on Software Tools for Technology Transfer*, 2(2):98–132, 1998.
16. R. Milner. *Communication and Concurrency.* Prentice Hall, 1989.
17. R. Milner, R. Harper, and M. Tofte. *The Definition of Standard ML.* MIT Press, 1990.
18. G. Monvelet, S. Christensen, H. Demmou, M. Paludetto, and J. Porras. Analysing a Mechatronic System with Coloured Petri Nets. *Software Tools for Technology Transfer*, 2(2):160–167, 1998.
19. OMG Unified Modeling Language Specification, Version 1.4. Object Management Group (OMG); UML Revision Taskforce, 2001.
20. J.L. Rasmussen and M. Singh. Designing a Security System by Means of Colored Petri Nets. In J. Billington and W. Reisig, editors, *Proceedings of the 17th International Conference on Application and Theory of Petri Nets*, volume 1091 of *LNCS*, pages 400–419, Osaka, Japan, 1996. Springer-Verlag.
21. A.V. Ratzer, L. Wells, H.M. Lassen, M. Laursen, J.F. Qvortrup, M.S. Stissing, M. Westergaard, S. Christensen, and K. Jensen. CPN Tools for Editing, Simulating, and Analysing Coloured Petri Nets. In W. van der Aalst and E. Best, editors, *Proceedings of the 24th International Conference on Application and Theory of Petri Nets*, volume 2679 of *LNCS*, pages 450–462, Eindhoven, The Netherlands, 2003. Springer-Verlag.
22. J. Rumbaugh, I. Jacobson, and G. Booch. *The Unified Modeling Language Reference Manual.* Addison-Wesley, 1999.
23. W. Stallings. *Data & Computer Communications, Sixth Edition.* Prentice Hall, 2000.
24. A.S. Tanenbaum. *Modern Operating Systems.* Prentice Hall, 1992.
25. A. Valmari. A Stubborn Attack on State Explosion. *Formal Methods in System Design*, 1:297–322, 1992.
26. Home page of the CPN Group at the University of Aarhus. www.daimi.au.dk/CPNets.
27. Home page of Design/CPN. www.daimi.au.dk/designCPN.
28. Home page of Kurt Jensen. www.daimi.au.dk/~kjensen.
29. Home page of ACM Curricula Recommendations. www.acm.org/education/curricula.html.
30. Home page of CPN Tools. www.daimi.au.dk/CPNtools.
31. Home page of 3rd CPN Workshop, CPN'02. www.daimi.au.dk/CPnets/Workshop02.
32. Home page of Department of Computer Science, University of Aarhus. www.daimi.au.dk.
33. Home page of Distributed Systems Course. www.daimi.au.dk/dDist (in Danish).

Unbounded Petri Net Synthesis

Philippe Darondeau

IRISA, campus de Beaulieu, F35042 Rennes Cedex

Abstract. We address the problem of deciding uniformly for graphs or languages of a given class whether they are generated by unlabelled Place-Transition nets whose sets of reachable markings may be infinite.

1 Introduction

Initialized Petri nets may be seen alternatively as graph generators or as language generators. In the first case, the generated graph is the reachable state graph of the net, considered up to isomorphisms of graphs (*i.e.* any set in bijection with the set of reachable markings may be used equivalently to represent the vertices of this graph). In the second case, the generated language is the set of firing sequences of the net (we will not introduce in this paper any labelling of transitions nor any special subset of accepting states or markings). The Petri net synthesis problem consists in deciding uniformly for a fixed class of graphs or languages whether a given member of this class has a Petri net generator and in producing such a generator if it exists. For classes of graphs or languages where the decision is not possible, a connected problem is to produce from a given object a Petri net generator which approximates it at best.

The Petri net synthesis problem may be addressed for several classes of nets, including notably the Elementary Nets and the Place-Transition Nets. Synthesis was dealt with originally by Ehrenfeucht and Rozenberg in the context of finite graphs and *Elementary Nets* [22] [23]. As the number of (simple) elementary nets with a fixed set of transitions is finite, the decision problem has an obvious solution in this context. The goal of the cited authors was to put forward a graph theoretic and axiomatic solution. The seminal idea which they introduced for this purpose is the concept of *regions* of a graph. The regions of a graph are particular subsets of vertices. The regions of a graph edge-labelled on T correspond bijectively with the simulations of this graph by elementary nets of the *atomic* form $(\{p\}, T, F, M_0)$. A finite and reachable rooted graph (loopfree, deterministic and simple) is simulated by an elementary net if all walks in the graph are matched by similar firing sequences of the net, such that two walks ending at the same vertex are always simulated by firing sequences ending at the same marking. Each simulation induces thus a (unique) map from vertices to markings. The regions of the graph are the inverse images of the marking $p = 1$ under arbitrary simulations of the graph by atomic nets $(\{p\}, T, F, M_0)$. Ehrenfeucht and Rozenberg gave a purely graph theoretic characterization of these regions. Their logical structure was studied further in [11] and [12].

J. Desel, W. Reisig, and G. Rozenberg (Eds.): ACPN 2003, LNCS 3098, pp. 413–438, 2004.

Given any graph edge-labelled on T, one may *synthesize* an elementary net from this graph by gluing together on their common transitions all the simulating atomic nets $(\{p_l\}, T, F_l, M_{l,0})$. A graph is generated by some elementary net *if and only if* it is generated by the elementary net constructed in this way. It follows that the family of graphs with elementary net generators may be characterized by two axioms: *i)* for any two distinct vertices v and v', some region contains either v or v' (but not both); *ii)* for any $t \in T$ and for any vertex v, if no edge labelled with t leaves the vertex v, then the vertex v is outside some region that contains all sources of edges labelled with t and none of their targets.

Synthesis algorithms based on the above characterization were proposed in [20], [11], and [17]. In the context of Elementary Nets, synthesis is an NP-complete problem [3]. Efficient heuristic algorithms have been implemented in the tool PETRIFY, with application to Asynchronous Circuit Design [16]. On the side of theory, a categorical version of the correspondence between Elementary Graphs (finite and reachable rooted graphs, loopfree, deterministic and simple, satisfying axioms *(i)* and *(ii)*) and Elementary Nets was given in [38]. The latter work sheds additional light on synthesis: it indicates that morphisms of nets may also be synthesized from morphisms of graphs (*i.e.* net synthesis is functorial). For more on the synthesis of Elementary Nets, we refer the reader to [22] [23], to the papers mentionned above, and to the survey [6]. A closely related topic is the synthesis of labelled one-safe nets from Asynchronous Transition Systems, which was explored in [44] and [8].

The concept of regions, which was introduced in the context of Elementary Nets, was quickly adapted to *Place-Transition Nets.* In this different context, the regions of a graph edge-labelled on T are in bijective correspondence with the simulations of this graph by P/T-nets of the *atomic* form $(\{p\}, T, F, M_0)$. A rooted, reachable and deterministic graph is simulated by a P/T-net if all walks in the graph are matched by similar firing sequences of the net, such that two walks ending at the same vertex are simulated by firing sequences ending at the same marking. Each simulation of a graph by an atomic P/T-net $(\{p\}, T, F, M_0)$ induces thus a (unique) map from vertices to non-negative integers. Regions are no longer subsets of vertices. They are multisets of vertices: a region assigns to each vertex v the weight defined by this induced map. Multiset regions may still be given a graph-theoretic characterization, but their logical structure is unclear. This is compensated for by nice algebraic properties: the (multiset) sum of two regions is a region, and the (multiset) difference of two regions, when it is defined, is also a region. We shall intensively exploit this linear algebraic structure in the body of the paper.

Regions as multisets were introduced independently by several groups of researchers. Slightly different definitions of regions were given, depending on the amount of concurrency embedded in the classes of labelled graphs considered. Concurrency by steps was considered in [33] and [37] (the first two papers in which multiset regions were defined), and regions served there to characterize respectively the subclass of Local Trace Languages with P/T-net generators and the subclass of Step Transition Systems with P/T-net generators. Another form of concurrency (pairwise independence, which is weaker than step independence) was considered in [21], where regions served to characterize the subclass of Automata with Concurrency Relations that may be generated from P/T-nets. Multiset regions of ordinary graphs, *i.e.* graphs without concurrency, were defined in [9], where they served to characterize the subclass of finite graphs with P/T-net generators. All characterizations are expressed by two axioms akin to Ehrenfeucht and Rozenberg's axioms *(i)* and *(ii)*. The adaptation of the axiom *(i)* is immediate: a multiset region separates two vertices v and v' if they have different weights in this region. The adaptation of the axiom *(ii)* is not so immediate and it depends on the exact definition of regions that is used. Roughly, the modified axiom requires from any vertex v that, if no edge leaving v bears the label t, then some region assigns to the vertex v a weight strictly lower than the weights of all sources of edges labelled with t.

While the accent was set on categorical correspondences between graphs and nets in [37] and [21], it was set in [9] on the algebraic and combinatorial properties of regions. It was shown in the latter reference that the *minimal* regions of a finite graph provide all the information needed to determine whether this graph has a P/T-net generator. The decision problem was however left unsolved. It was actually shown in [2] that the synthesis problem is decidable for pure and bounded P/T-nets and for finite graphs or for regular languages. The main principle for the decision is to compute a *finite* set of regions that generate all other regions (the regions of a finite graph form a module, and the bounded regions of a regular language do the same). The synthesis algorithm is polynomial in the size of the graph, and it is exponential in the size of the regular expression. This algorithm was extended in [5] to bounded P/T-nets (which may be impure) and to finite Step Transition Systems. The synthesis algorithm for bounded P/T-nets has been implemented in the tool SYNET, with tentative applications to the distributed realization of protocols [4]. Another application of the bounded P/T-net synthesis is the computation of Petri net supervisory controllers [27]. All state-avoidance problems in one-safe Petri nets may in fact be solved in this way. For more on the synthesis of bounded P/T-nets and applications, we refer the reader to the above mentioned papers or to the survey [6].

The remaining sections of the paper are devoted to the algorithmic synthesis of unbounded P/T-nets, a topic which was not covered in [6] because it was studied afterwards. Section 2 deals with the synthesis of unbounded P/T-nets from languages. Section 3 deals with the synthesis of unbounded P/T-nets from infinite graphs. The last section summarizes the results and indicates some directions for future work. The paper is self-contained. No familiarity with the synthesis of Elementary Nets nor with the synthesis of bounded Place-Transition Nets is assumed. The presentation of net synthesis given here is simpler than the general presentation given in [6], but it ignores most of the results reported there. The presentation below is based on the work of the author and his colleagues from IRISA (see references in the bibliography).

2 Net Synthesis from Languages

Let us recall the definition of Place-Transition nets (or P/T-nets for short).

Definition 1 (P/T-nets). *A P/T-net is a triple $N = (P, T, F)$ where P and T are finite disjoint sets of* places *and* transitions*, respectively, and F is a function, $F : (P \times T) \cup (T \times P) \to \mathbb{N}$. A* marking *of N is a map $M : P \to \mathbb{N}$. The* state graph *of N is a labelled graph, with markings as vertices, where there is an edge labelled with transition t from M to M' (in notation: $M[t\rangle M'$) if and only if, for every place $p \in P$, $M(p) \geq F(p,t)$ and $M'(p) = M(p) - F(p,t) + F(t,p)$. The* reachable state graph *of an* initialized *P/T-net $\mathcal{N} = (P, T, F, M_0)$, with* initial marking *M_0, is the restriction of its state graph induced by the subset of vertices that may be reached inductively from M_0. The net $\mathcal{N}$ is* unbounded *if its reachable state graph is infinite. The* language *of an* initialized *P/T-net is the set of sequences $w \in T^*$ that label walks from the root M_0 of this graph. Thus, the language of $\mathcal{N}$ is the set $\{w \in T^* \mid M_0[w\rangle\}$ where $M_0[w\rangle$ means that the sequence w may be fired inductively from M_0.*

Example 1. Let $P = \{1,2\}$ with $M_0(1) = 0$ and $M_0(2) = 1$. Let $T = \{a, b\}$ with $F(a,1) = 1 = F(1,b)$ and $F(2,b) = 1$. Let F evaluate to 0 for all the remaining arguments. The reachable state graph of the specified net is the infinite graph shown below. The language of this net is the regular language $a^* + aa^*ba^*$. □

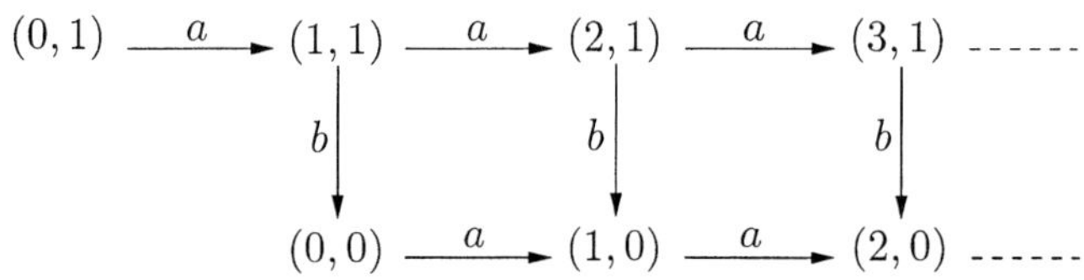

The P/T-net synthesis problem for a class of languages is the problem whether one can decide *uniformly* from any language $\mathcal{L}$ in this class whether it coincides with the language of some (initialized) P/T-net, and construct such a net when it exists. Uniformity means that the same constructive procedure should apply to all languages in the considered class. For instance, the P/T-net synthesis problem has a (positive) solution for $\mathcal{L} = a^* + aa^*ba^*$ in the singleton class $\{\mathcal{L}\}$, but this does not mean that the P/T-net synthesis problem has a solution for $\mathcal{L} = a^* + aa^*ba^*$ in the class of regular languages over two letters a and b.

We propose in this section a uniform procedure that computes, for any class of semi-linear languages closed under right quotients with letters, the least over-approximation of a language in the class by the language of a P/T-net. We propose moreover a uniform procedure that solves the P/T-net synthesis problem for classes of semi-linear languages closed under right quotients with letters and under the *max* operation (w.r.t. the order prefix). We show that the synthesis problem is decidable for the regular or deterministic context-free languages, whereas it is undecidable for the context-free languages and for the languages of High-level Message Sequence Charts (or HMSCs for short). We argue finally about the practical relevance of approximating languages by P/T-net languages.

Before we describe the common principles under the two procedures, let us add two remarks about example 1. First, infinitely many different P/T-nets have the language $a^* + aa^*ba^*$. Therefore, one cannot require from an effective synthesis procedure to produce all of them. Second, any P/T-net with the language $a^* + aa^*ba^*$ has an infinite number of reachable markings. To see this, assume the opposite. Then the transition a should act as the identity on the markings of some initialized P/T-net $(P, \{a, b\}, F, M_0)$ generating this language. As $M_0[ab\rangle$, necessarily $M_0[b\rangle$, a contradiction. This remark shows that the synthesis of *unbounded* P/T-nets is a relevant problem for all reasonable classes of languages.

Henceforth in this section $T = \{t_1, \ldots, t_n\}$ is a fixed alphabet of transitions, and P/T-nets have always the set of transitions T. The languages $\mathcal{L}$ under consideration are always subsets of T^*. As we are mainly interested in languages of P/T-nets and these languages are non-empty and prefix-closed, it will always be assumed that $\mathcal{L}$ is non-empty and prefix-closed, *i.e.* $(\forall w \in \mathcal{L})\ w = u \cdot v \Longrightarrow u \in \mathcal{L}$ where $\cdot$ denotes the concatenation product in T^*. In particular, the empty word ε is always in $\mathcal{L}$. In the sequel, initialized P/T-nets are called P/T-nets for short. The language of the P/T-net $\mathcal{N}$ is denoted $L(\mathcal{N})$.

2.1 The Regions of a Language

The two essential facts on which is based the synthesis of P/T-nets from languages are stated in the (almost obvious) propositions 1 and 2 below.

Definition 2 (Atomic subnets). *A P/T-net $\mathcal{N} = (P, T, F, M_0)$ is a* subnet *of $\mathcal{N}' = (P', T, F', M_0')$ if $P \subseteq P'$ and F and M_0 are the induced restrictions of F' and M_0' (respectively on $(P \times T) \cup (T \times P)$ and on P). The net (P, T, F, M_0) is* atomic *if $|P| = 1$. An* atomic subnet *of $\mathcal{N}'$ is a subnet of $\mathcal{N}'$ which is atomic.*

Proposition 1. *The language of a P/T-net is the intersection of the languages of its atomic subnets.*

Definition 3 (P/T-regions). *Given a word $w \in T^*$, an atomic P/T-net $\mathcal{N} = (\{p\}, T, F, M_0)$ is a* P/T-region *of w if $w \in L(\mathcal{N})$. Given a language $\mathcal{L}$, an atomic P/T-net $\mathcal{N}$ is a P/T-region of $\mathcal{L}$ if it is a P/T-region of every word $w \in \mathcal{L}$.*

Proposition 2. *$\mathcal{L}$ is the language of a P/T-net if and only if the set of P/T-regions of $\mathcal{L}$ contains a finite subset $\{\mathcal{N}_1, \ldots, \mathcal{N}_m\}$ such that, for every $t \in T$ and for every $w \in \mathcal{L}$, if $w \cdot t \notin \mathcal{L}$, then some $\mathcal{N}_l$* is not *a P/T-region of $w \cdot t$. When this condition is satisfied, $\mathcal{L} = L(\mathcal{N})$ where $\mathcal{N}$ is the P/T-net with the set of atomic subnets $\{\mathcal{N}_1, \ldots, \mathcal{N}_m\}$.*

Example 2. The P/T-net described in the example 1 has two atomic subnets $\mathcal{N}_1$ and $\mathcal{N}_2$, where 1 is the unique place of $\mathcal{N}_1$ and 2 is the unique place of $\mathcal{N}_2$. Both nets are regions of the language $a^* + aa^*ba^*$. The word $\varepsilon \cdot b$ is not in this language, but it does not belong either to $L(\mathcal{N}_1)$, thus $\mathcal{N}_1$ is not a region of this word. Similarly, $\mathcal{N}_2$ is not a region of any word $a^k ba^l \cdot b$ for $k > 0$ and $l \geq 0$. Finally observe that the language $a^* + aa^*ba^*$ has an infinite set of regions. For instance, all the atomic P/T-nets $(\{p\}, \{a, b\}, F, M_0)$ such that $F(p, a) = 0$ and $F(p, b) = 0$ are regions of this language. □

In view of proposition 2, the feasability of a procedure for the decision of the P/T-net synthesis problem (with respect to a fixed class of languages) depends on the feasability of two subproblems. First, one should compute an effective representation of the set of regions of a language, notwithstanding the fact that this set is always *infinite*. Second, one should decide whether some *finite* subset of regions suffices to *reject* all minimal words (with respect to the order prefix) in the complement of the given language, even though these *unwanted* words may form an *infinite* set (*e.g.*, the set $b + aa^*bb$ in example 1).

The above problems cannot be solved without specific assumptions on the considered classes of languages. Fortunately, the first problem has an easy solution for semi-linear languages.

2.2 A Procedure for Computing Generating Regions

Let us recall two definitions.

Definition 4 (Commutative image). *The* commutative image *of a word $w \in T^*$ is the n-vector $[w]$ whose respective entries $[w]_i$ count for each $i \in [1, n]$ the occurrences of the letter t_i in w. The commutative image of a language $\mathcal{L} \subseteq T^*$ is the set $[\mathcal{L}] = \{[w] \mid w \in \mathcal{L}\}$.*

Definition 5 (Semi-linear subset). *Let $\mathcal{M} = (\mathcal{M}, \cdot, 1)$ be a monoid. A subset of $\mathcal{M}$ is* linear *if it may be expressed as $m \cdot \mathcal{F}^*$ where $m \in \mathcal{M}$, $\mathcal{F}$ is a finite subset of $\mathcal{M}$, and $\mathcal{F}^*$ is the least submonoid of $\mathcal{M}$ containing $\mathcal{F}$. A finite union of linear subsets of $\mathcal{M}$ is called a* semi-linear *subset.*

A language $\mathcal{L}$ is said to be *semi-linear* if its commutative image $[\mathcal{L}]$ is a semi-linear subset of $\mathbb{N}^n$, the commutative monoid where the product $\cdot$ is the addition of n-vectors and where the neutral element 1 is the all-zeroes n-vector.

Example 3. For $\mathcal{L} = a^* + aa^*ba^*$, where we let $a = t_1$ and $b = t_2$ for convenience, $[\mathcal{L}] = <1,0>^* + <1,0> \cdot <1,0>^* \cdot <0,1> \cdot <1,0>^*$. By commutativity of the product (*i.e.* addition) in $\mathbb{N}^n$, $[\mathcal{L}] = <1,0>^* + <1,1> \cdot <1,0>^*$, hence this set is semi-linear (in regular expressions, + denotes set union). □

The considerations in the above example may be generalized to all regular expressions. It should therefore be clear that for any language $\mathcal{L}$, $[\mathcal{L}]$ is semi-linear if and only if $[\mathcal{L}] = [\mathcal{R}]$ for some regular language $\mathcal{R}$. A celebrated theorem by Parikh shows that this condition holds for the context-free languages (see section 6.9 in [31] for the construction of $\mathcal{R}$ from a context-free grammar generating $\mathcal{L}$).

In order to achieve our goals, we shall actually require a little more than the semi-linearity of $[\mathcal{L}]$. Namely, we require that all right derivatives $\mathcal{L}/t_j$ are semi-linear, where $t_j \in T$ and $\mathcal{L}/t_j = \{v \in T^* \mid v \cdot t_j \in \mathcal{L}\}$. Under this stronger requirement (which is met by context-free languages), one can effectively compute a finite representation of the infinite set of P/T-regions of $\mathcal{L}$. Moreover, this representation yields for free a P/T-net $\mathcal{N}$ whose language $L(\mathcal{N})$ is the *least* net language larger than $\mathcal{L}$. The construction is explained in the rest of the section.

Recall that a P/T-region of $\mathcal{L}$ is an atomic P/T-net $\mathcal{N} = (\{p\}, T, F, M_0)$ such that $\mathcal{L} \subseteq L(\mathcal{N})$. An atomic P/T-net $\mathcal{N}$ as above may be represented equivalently as a $(2n+1)$-vector $< M_0(p), F(p,t_1), \ldots, F(p,t_n), F(t_1,p), \ldots, F(t_n,p) >$. We claim that a $(2n+1)$-vector $\boldsymbol{x} =< x_0, x_1, \ldots, x_n, x_{n+1}, \ldots, x_{2n} >$ defines a region of $\mathcal{L}$ if and only if all its entries x_k are non-negative integers, and for each (non empty) word $v \cdot t_j$ in $\mathcal{L}$

$$x_0 + \sum_{i=1}^{n} [v]_i \times (x_{(n+i)} - x_i) \geq x_j \tag{1}$$

Actually, if the vector $\boldsymbol{x}$ is seen as an atomic P/T-net, the above inequality may be read as $M[t_j\rangle$ where M is the marking of the net reached after firing the sequence of transitions v, *assuming that* v *may be fired.* Since $\mathcal{L}$ is prefix-closed, this will necessarily be the case if similar inequalities hold for all the non-empty prefixes $u \cdot t_k$ of v. Let us now use the assumption that all derivatives $\mathcal{L}/t_j$ are semi-linear. Thus, for each $t_j \in T$, the set $[\mathcal{L}/t_j]$ is a finite union of linear sets $\boldsymbol{e} \cdot \mathcal{F}^*$, where $\boldsymbol{e} \in \mathbb{N}^n$ and $\mathcal{F}$ is a finite subset of $\mathbb{N}^n$. For each $t_j \in T$ and for each linear set $\boldsymbol{e} \cdot \mathcal{F}^*$ in $[\mathcal{L}/t_j]$, the collection of instances of 1 generated from words $v \in \mathcal{L}/t_j$ such that $[v] \in \boldsymbol{e} \cdot \mathcal{F}^*$ may be replaced equivalently with the *finite* linear system:

$$\sum_{i=1}^{n} \boldsymbol{e}[i] \times (x_{(n+i)} - x_i) \geq x_j - x_0 \tag{2}$$

$$\sum_{i=1}^{n} \boldsymbol{f}[i] \times (x_{(n+i)} - x_i) \geq 0 \tag{3}$$

where f ranges over the finite set $\mathcal{F}$. Let us justify this claim. For any vector $\boldsymbol{x}$ which is a solution of the finite linear system, the inequality 1 is obviously satisfied for all $v \in \mathcal{L}/t_j$ such that $[v] \in \boldsymbol{e} \cdot \mathcal{F}^*$. Conversely, the conjunction of all such inequalities entails 2 and 3. To see that it entails 3, suppose for a contradiction that 3 does not hold for some $\boldsymbol{f} \in F$. Then, for h large enough, $\sum_{i=1}^{n} ((\boldsymbol{e}+h\boldsymbol{f})[i]) \times (x_{(n+i)} - x_i) < x_j - x_0$. As $[\boldsymbol{e}+h\boldsymbol{f}] = [v]$ for some $v \in \mathcal{L}/t_j$ and the inequality 1 cannot hold for the considered v, a contradiction has been reached. Therefore, the set of P/T-regions of $\mathcal{L}$, seen as vectors $\boldsymbol{x} \in \mathbb{N}^{2n+1}$, is the set of solutions of a finite system of linear inequalities (T is finite, and each set $[\mathcal{L}/t_j]$ is a finite union of linear subsets). Moreover, all inequalities in this system are *homogeneous*, *i.e.* they may be written equivalently in the form $\sum_{k=0}^{2n} \alpha_k \times x_k \geq 0$ (where the α_k are constants in $\mathbb{Z}$).

Example 4. For $\mathcal{L} = a^* + aa^*ba^*$, one obtains $\mathcal{L}/a = \mathcal{L}$ and $\mathcal{L}/b = aa^*$. Therefore if we let $a = t_1$ and $b = t_2$, $[\mathcal{L}/t_1] =< 1,0 >^* + < 1,1 > \cdot < 1,0 >^*$ and $[\mathcal{L}/t_2] =< 1,0 > \cdot < 1,0 >^*$. The P/T-regions $\mathcal{L}$ are the solutions of the system

$$0 \geq x_1 - x_0$$
$$(x_3 - x_1) + (x_4 - x_2) \geq x_1 - x_0$$
$$(x_3 - x_1) \geq x_2 - x_0$$
$$(x_3 - x_1) \geq 0$$

The atomic net $\mathcal{N}_1 = (\{1\}, T, F, M_0)$ given by $x_0 = M_0(1) = 0$, $x_1 = F(1,a) = 0$, $x_2 = F(1,b) = 1$, $x_3 = F(a,1) = 1$, and $x_4 = F(b,1) = 0$ is a P/T-region of $\mathcal{L}$, and similarly is the atomic net $\mathcal{N}_2 = (\{2\}, T, F, M_0)$ given by $x_0 = M_0(2) = 1$, $x_1 = F(2,a) = 0$, $x_2 = F(2,b) - 1$, $x_3 - F(a,2) = 0$, and $x_4 = F(b,2) = 0$. □

Let $\mathcal{S}$ be the finite system of linear inequalities in the variables $x_0, \ldots, x_{2n}$ which defines the regions of $\mathcal{L}$, augmented with inequalities $x_k \geq 0$ for all $k \in [0, 2n]$. If one lets the variables x_k range over the set $\mathbb{Q}$ of rational numbers, the solutions of $\mathcal{S}$ in $\mathbb{Q}^{2n+1}$ form a cone with a finite set of generators $\boldsymbol{x}_1 \ldots \boldsymbol{x}_m$ (see [41]). This means that a rational vector $\boldsymbol{x}$ is a solution of $\mathcal{S}$ if and only if $\boldsymbol{x} = \sum_{l=1}^{m} q_l \boldsymbol{x}_l$ for some non-negative rational coefficients q_l. Moreover, one can effectively compute a minimal set of generators $\boldsymbol{x}_1 \ldots \boldsymbol{x}_m$, *e.g.* using Chernikova's algorithm [15]. A finite representation of the set of P/T-regions of $\mathcal{L}$ is then obtained.

A vector $\boldsymbol{x}$ *is a P/T-region of* $\mathcal{L}$ *if and only if* $\boldsymbol{x} \in \mathbb{N}^{2n+1}$ *and* $\boldsymbol{x} = \sum_{l=1}^{m} q_l \, \boldsymbol{x}_l$ *for some non-negative rational coefficients* q_l.

Henceforth in the section, we assume that $\boldsymbol{x}_1 \ldots \boldsymbol{x}_m$ are vectors of integers (this may be assumed w.l.o.g. since $\boldsymbol{x}_l$ may be replaced equivalently with $q_l \, \boldsymbol{x}_l$ for any non-negative q_l), and we call them the *generating regions* of $\mathcal{L}$.

Example 5. For $\mathcal{L} = a^* + aa^*ba^*$, where $a = t_1$ and $b = t_2$, the generating regions are the columns of the following table (we let $\boldsymbol{x} =< m_0, ^\bullet a, ^\bullet b, a^\bullet, b^\bullet >$ for convenience):

m_0	1	0	0	1	1	0	1
$^\bullet a$	0	0	0	1	1	0	0
$^\bullet b$	0	0	0	1	0	1	1
$a^\bullet$	0	1	0	1	1	1	0
$b^\bullet$	0	0	1	1	0	0	0

The last two regions correspond to the atomic nets $\mathcal{N}_1$ and $\mathcal{N}_2$ already seen. □

Let $\mathcal{N}$ be the P/T-net formed by gluing together on transitions $t_j \in T$ the atomic P/T-nets $\mathcal{N}_1 \ldots \mathcal{N}_m$ defined by the generating regions $\boldsymbol{x}_1 \ldots \boldsymbol{x}_m$ of $\mathcal{L}$.

$L(\mathcal{N})$ *is the* least *net language larger than* $\mathcal{L}$.

Let us establish this claim. As each atomic net $\mathcal{N}_l$ ($l \in [1, m]$) is a region of $\mathcal{L}$, it should be clear from definition 3 and proposition 1 that $L(\mathcal{N})$ is larger than $\mathcal{L}$.

Now let $\mathcal{N}'$ be any P/T-net such that $\mathcal{L} \subseteq L(\mathcal{N}')$. Suppose for a contradiction that some word w belongs to $L(\mathcal{N}) \setminus L(\mathcal{N}')$. We may assume w.l.o.g. that w is minimal w.r.t. the order prefix among the words in that case. Then necessarily, $w = v \cdot t_j$ for some $t_j \in T$ and $v \in L(\mathcal{N}) \cap L(\mathcal{N}')$. As $v \in L(\mathcal{N}')$ and $v \cdot t_j \notin L(\mathcal{N}')$, the inequality 1 does certainly not hold for some $2n+1$-vector $\boldsymbol{x}'$ representing an atomic subnet of $\mathcal{N}'$. Since $\mathcal{L} \subseteq L(\mathcal{N}')$, this atomic subnet is a region of $\mathcal{L}$, thus $\boldsymbol{x}' = \sum_{l=1}^{m} q_l \, \boldsymbol{x}_l$ for some *non-negative* rational coefficients q_l. Therefore, the inequality 1 does not hold for some vector $\boldsymbol{x} = \boldsymbol{x}_l \in \{\boldsymbol{x}_1, \ldots, \boldsymbol{x}_m\}$. It follows that $v \cdot t_j \notin L(\mathcal{N}_l)$, and hence $v \cdot t_j \notin L(\mathcal{N})$, a contradiction.

The above construction may be applied to any class of languages with semi-linear right derivatives. This is the case of every semi-linear full TRIO (by definition, a full TRIO is closed under homomorphisms, inverse homomorphisms, and intersection with regular languages). Examples are the regular languages, the context-free languages, the simple matrix languages of fixed degree [35], the languages of flip-pushdown automata with a fixed number of reversals [32], and the full slip AFLs described in [30]. This is also the case of two other classes of languages generated by parallel systems, namely the languages of parallel communicating grammar systems with terminal transmission and with fully synchronized mode [25], and the languages of HMSCs [14].

2.3 A Procedure for the Decision of the Net Synthesis Problem

Deciding whether a given language $\mathcal{L}$ has a P/T-net generator amounts to deciding whether $\mathcal{L} = L(\mathcal{N})$ where $\mathcal{N}$ is the net constructed from the generating regions of $\mathcal{L}$. We propose now a decision procedure that works under additional requirements of semi-linearity on the considered class of languages. Namely, we require that the complements in $\mathcal{L}$ of the right derivatives are also semi-linear. This requirement is significant: the assumption that all sets $[\mathcal{L}/t_j]$ are semi-linear does *not* entail that all sets $[\mathcal{L} \setminus (\mathcal{L}/t_j)]$ are semi-linear (although $[\mathcal{L}]$ must be semi-linear in this case).

For convenience of notation, let $\mathcal{L} \ominus t_j = \mathcal{L} \setminus (\mathcal{L}/t_j)$, thus $\mathcal{L} \ominus t_j$ is the set of the words $v \in \mathcal{L}$ such that $v \cdot t_j \notin \mathcal{L}$. Clearly, $\mathcal{L} = L(\mathcal{N})$ if and only if $v \cdot t_j$ is *not* in $L(\mathcal{N})$ whenever $t_j \in T$ and $v \in (\mathcal{L} \ominus t_j)$. Seeing that $\mathcal{N}$ was built up from the atomic nets defined by the generating regions of $\mathcal{L}$, $v \cdot t_j$ is not in $L(\mathcal{N})$ if and only if

$$x_0 + \sum_{i=1}^{n} [v]_i \times (x_{(n+i)} - x_i) < x_j \tag{4}$$

for some generating region $\boldsymbol{x} \in \{\boldsymbol{x}_1, \ldots, \boldsymbol{x}_m\}$. Let $\boldsymbol{y} = [v]$ thus $\boldsymbol{y} \in [\mathcal{L} \ominus t_j]$, and let $\boldsymbol{y} =< y_1, \ldots, y_n >$, then relation 4 may be rewritten to the linear inequality

$$x_0 + \sum_{i=1}^{n} y_i \times (x_{(n+i)} - x_i) < x_j \tag{5}$$

When the x_k are fixed constants in $\mathbb{N}$ (for $k \in [0, 2n]$) and the y_i are variables in $\mathbb{N}$ (for $i \in [1, n]$), the formula 5 amounts to a Presburger formula (it may be expressed equivalently as a comparison between two sums), hence it defines an effective semi-linear subset of $\mathbb{N}^n$ [29]. For each $l \in [1,m]$ and for each $j \in [1,n]$, let $Y_{l,j}$ be the semi-linear subset of $\mathbb{N}^n$ which is defined with formula 5 for the constants $x_k = \boldsymbol{x}_l[k]$ (*i.e.* for $\boldsymbol{x} = \boldsymbol{x}_l$). Now, $\mathcal{L} = L(\mathcal{N})$ if and only if, for all t_j,

$$[\mathcal{L} \ominus t_j] \subseteq \cup_{l=1}^{m} Y_{l,j} \tag{6}$$

As we assumed that all sets $[\mathcal{L} \ominus t_j]$ are semi-linear, and the semi-linear subsets of $\mathbb{N}^n$ form an effective boolean algebra [28], one can compute $[\mathcal{L} \ominus t_j] \setminus \cup_{l=1}^{m} Y_{l,j}$ and decide whether this set is empty. Therefore, one can decide whether $\mathcal{L} = L(\mathcal{N})$. We have thus obtained a decision procedure for the P/T-net synthesis problem.

Assuming that all sets $[\mathcal{L}/t_j]$ and $[\mathcal{L} \setminus (\mathcal{L}/t_j)]$ are semi-linear, one can decide whether the language $\mathcal{L}$ has a P/T-net generator.

When the decision is successful, it may occur that $\mathcal{L} = L(\mathcal{N}')$ for some proper subnet $\mathcal{N}'$ of the net $\mathcal{N}$ constructed from all the generating regions of $\mathcal{L}$. The procedure may be adapted in order to produce directly some minimal subnet $\mathcal{N}'$ of $\mathcal{N}$ such that $\mathcal{L} = L(\mathcal{N}')$. The subsets of $\{\boldsymbol{x}_1, \ldots, \boldsymbol{x}_m\}$ should be explored in increasing order until discovering some subset $\{\boldsymbol{x}_{l_1}, \ldots, \boldsymbol{x}_{l_p}\}$ large enough to make relation 6 valid for all $j \in [1,n]$ when l ranges over $\{l_1, \ldots, l_p\}$. The solution net $\mathcal{N}'$ is then constructed from the atomic nets defined by $\boldsymbol{x}_{l_1} \ldots \boldsymbol{x}_{l_p}$.

Example 6. For $\mathcal{L} = a^* + aa^*ba^*$, one obtains $(\mathcal{L} \ominus a) = \emptyset$ and $(\mathcal{L} \ominus b) = \varepsilon + aa^*ba^*$. Let $a = t_1$ and $b = t_2$, then $[\mathcal{L} \ominus t_2] = < 0,0 > + < 1,1 >< 1,0 >^*$. For the two regions $\boldsymbol{x}_1 = < 0,0,1,1,0 >$ and $\boldsymbol{x}_2 = < 1,0,1,0,0 >$ (see example 5), the sets $Y_{1,2}$ and $Y_{2,2}$ are defined by the respective formulas $y_1 - y_2 < 1$ and $1 - y_2 < 1$. Clearly, $< 0,0 > \in Y_{1,2}$ and for any non-negative integer h, $< 1+h, 1 > \in Y_{2,2}$. Therefore, $\mathcal{L}$ is the language of the net formed of the two atomic P/T-nets $\mathcal{N}_1$ and $\mathcal{N}_2$ from example 4. Seeing that $< 0,0 > \notin Y_{2,2}$ and $< 2,1 > \notin Y_{1,2}$, this net is a minimal net generator for $\mathcal{L}$. □

It remains to show classes of languages in which our working assumptions hold, *i.e.* where $[\mathcal{L}/t_j]$ and $[\mathcal{L} \ominus t_j]$ are semi-linear for every prefix-closed language $\mathcal{L}$ and for every $t_j \in T$. For any language $\mathcal{L}$ of T^*, define

$$max(\mathcal{L}) = \{u \in \mathcal{L} \,|\, (\forall v \in T^*)\, u \cdot v \in \mathcal{L} \Longrightarrow v = \varepsilon\}$$

Then, for any prefix-closed language $\mathcal{L}$ of T^*, $v \in \mathcal{L}$ and $v \cdot t_j \notin \mathcal{L}$ if and only if $v \cdot t_j \in max(\mathcal{L} \cdot t_j)$. Therefore, $\mathcal{L} \ominus t_j = (max(\mathcal{L} \cdot t_j))/t_j$. It follows that the sets $[\mathcal{L} \ominus t_j]$ are semi-linear in every class of semi-linear languages with the following properties:

i) the class is closed under right products and right quotients with letters,
ii) the class is closed under max.
Property *(i)* holds in any full TRIO, but property *(ii)* does not! As the following example shows, it does not hold *e.g.* for the context-free languages.

Example 7. ([18]) Define context-free languages on the alphabet $\{a,b,c,d,e\}$ as follows. First, let $A=\{a^n b\, c^m \mid n \neq m\}$, $B = b\, c^*$, $C=\{c^n b\, c^m \mid n\neq m\}$. Next, let $D = a^* B^* B B\, b\, d$, $E = AB^* B\, b\, d\, e + a^* B^* b\, C B^* b\, d\, e$, and $L = D+E$. Then $max(L) = E + F$, where $F = \{a^n (b\, c^n)^m b\, d \mid n \geq 0 \wedge m \geq 2\}$. Assume that $[max(L)]$ is semi-linear. Since $[E]$ is semi-linear and $[E]$ and $[F]$ are disjoint, $[F] = [max(L)] \setminus [E]$ and this set is semi-linear, hence it may be defined by a Presburger formula. Now $[F]$ is the set of the integer vectors of the form $< n, m+1, n \times m, 1, 0 >$. As multiplication cannot be defined in Presburger arithmetic, a contradiction has been reached, hence $[max(L)]$ is not semi-linear.

We know actually only two classes of semi-linear languages with properties *(i)* and *(ii)*: the regular languages (a full TRIO) and the deterministic context-free languages (which do not form a full TRIO). The deterministic context-free languages are indeed closed under right products and quotients with regular languages (see [31]), and they are closed under the *max* operation (see [34]).

The P/T-net synthesis problem is decidable for regular or deterministic *context-free languages.*

2.4 Two Undecidable Cases

We show that it is undecidable *i)* whether an arbitrary context-free language has a P/T-net generator, and *ii)* whether the language of an arbitrary HMSC has a P/T-net generator. The proofs for the two facts are similar.

We consider first context-free languages. Given any context-free language $\mathcal{L}$ of T^*, let $\mathcal{N}$ be the P/T-net defined by the generating regions of $\mathcal{L}$ (see section 2.2). Then $\mathcal{L} \neq T^*$ *if and only if* $\mathcal{L} \neq L(\mathcal{N})$ *or* $L(\mathcal{N}) \neq T^*$. The complement of a (*deterministic*) P/T-net language may be generated by a *labelled* P/T-net with a finite subset of *final* partial markings [39]. The reachability of partial markings is decidable [36]. Therefore one can decide whether $L(\mathcal{N}) = T^*$. If one could decide whether $\mathcal{L} = L(\mathcal{N})$, one could decide whether $\mathcal{L} = T^*$. Now it is undecidable for an arbitrary context-free language $\mathcal{L}$ of T^* whether $\mathcal{L} = T^*$ (see *e.g.* [34]). Therefore, the P/T-net synthesis problem is undecidable for context-free languages.

We consider now HMSC languages. In this case, the alphabet T has a specific structure. On the one hand, it is equipped with a map $\ell : T \rightarrow [1, K]$ that assigns a specific *location* to each transition. On the other hand, the transitions in T are divided into message *emissions* (emit towards location k), message *receptions* (receive from location k), and *internal* transitions. An HMSC $\mathcal{H}_1$ with K locations and with internal transitions only may be simulated by an HMSC $\mathcal{H}_2$ with $2K$ locations and with no internal transitions, such that $L(\mathcal{H}_1)$ is a P/T-net language *if and only if* $L(\mathcal{H}_2)$ is a P/T-net language: each internal transition at location k may be simulated by an emission from k to $k+K$ plus a matching reception at $k + K$. We shall assume below, for simplicity, that all transitions in T are internal.

Let $T = T_1 \cup T_2$ where $\ell(t) = k$ for $t \in T_k$, and let $E_k = T_k \setminus \{\$_k\}$ where $\$_k$ is a distinguished symbol in T_k. Whenever a relation $\mathcal{R} \subseteq (E_1^* \times E_2^*)$ is accepted by a finite automaton over the product monoid $E_1^* \times E_2^*$, the relation $\mathcal{R} \cdot (\$_1, \$_2)$ is accepted by a finite and trim automaton over $T_1^* \times T_2^*$. This trim automaton may be seen as an HMSC over T. For any *rational* relation $\mathcal{R} \subseteq (E_1^* \times E_2^*)$, there exists therefore an HMSC $\mathcal{H}$ over T with the language

$$L(\mathcal{H}) = pref\{w \mid \exists (u, v) \in \mathcal{R} : w \in (u \cdot \$_1) ⧢ (v \cdot \$_2)\}$$

where $pref$ denotes prefix closure and ⧢ is the shuffle operation. Let ⧢ be defined on languages of T^* by an additive extension of the latter. There obviously exists a P/T-net $\mathcal{N}'$ such that $L(\mathcal{N}') = pref\,((T_1^* \cdot \$_1) ⧢ (T_2^* \cdot \$_2))$. It follows from the definition of $L(\mathcal{H})$ that $L(\mathcal{H}) = L(\mathcal{N}')$ *if and only if* $\mathcal{R} = (E_1^* \times E_2^*)$.

Now let $\mathcal{N}$ be the P/T-net constructed from the generating regions of $L(\mathcal{H})$. Then $\mathcal{R} \neq (E_1^* \times E_2^*)$ *if and only if* $L(\mathcal{H}) \subsetneq L(\mathcal{N})$ *or* $L(\mathcal{N}) \subsetneq L(\mathcal{N}')$. From the results in [39] and [36] one can decide whether $L(\mathcal{N}) = L(\mathcal{N}')$. If one could decide whether $L(\mathcal{H}) = L(\mathcal{N})$, one could decide whether $\mathcal{R} = (E_1^* \times E_2^*)$. Now, provided that each subalphabet E_k contains at least two letters, it is undecidable for an arbitrary rational relation $\mathcal{R}$ whether $\mathcal{R} = (E_1^* \times E_2^*)$, see [26] or [10]. Therefore, the P/T-net synthesis problem is undecidable for HMSC languages.

The P/T-net synthesis problem is undecidable for context-free languages and for HMSC languages.

2.5 Comments and Complements

Many classes of semi-linear languages extend the context-free languages, or are based on rational relations. The undecidability results presented in section 2.4 apply in both cases. In order to extend the decidability results established in section 2.3, it appears suitable to focus on sub-classes of languages generated with *deterministic* automata, as we have done for the context-free languages. This way is still open for HMSC languages, since deterministic generators have not yet been thoroughly investigated in this context.

As concerns applications, one may argue that least over-approximations of languages by P/T-nets are often more suitable than exact realizations. Two cases in support of this thesis are discussed below.

Let us come again to HMSCs. As these are intended to serve at an early stage of design of distributed systems, collections of scenarios defined by HMSCs are usually seen as *incomplete specifications* of a system. P/T-net synthesis may be used to build a *prototype* of the specified system, *i.e.* a distributed scale model that may be run and model-checked before designing software. Now model-checking is undecidable for *general* HMSCs (see [1] or [14]). Therefore, one should accept that a prototype system may have a language larger than the language of the specifying HMSC. Approximating HMSC languages by P/T-net languages as indicated in section 2.2 is justified, because the model-checking of P/T-nets w.r.t. linear-time μ-calculus is decidable [24]. Moreover, relations $L(\mathcal{N}) \subseteq \mathcal{R}$ and $\mathcal{R} \subseteq L(\mathcal{N})$ are decidable for arbitrary P/T-nets $\mathcal{N}$ and regular languages $\mathcal{R}$ (because $\mathcal{R}$ and $\complement\mathcal{R}$ have labelled P/T-net generators with final markings). A matter not yet discussed is distribution. Recall that P/T-regions may be seen as vectors $\boldsymbol{x} =< x_0, \ldots, x_{2n} >$ in $\mathbb{N}^{2n+1}$, where $T = \{t_1, \ldots, t_n\}$. By simply imposing on vectors $\boldsymbol{x}$, for some $\mathcal{I} \in \mathcal{P}([1,n])$, the additional constraint $(\exists I \in \mathcal{I})(\forall i \in I)\,(x_i = 0)$, *distributable* P/T-nets may be produced by the synthesis procedure. In a distributable P/T-net (see [4]), the transitions have locations, and an input place is never shared by transitions with different locations. Because competitions for tokens are local, distributable P/T-nets may be cut to local subnets communicating by asynchronous message passing. Distributed prototypes of HMSCs may be obtained in this way.

Another field of application is *supervisory control*. Let us briefly recall the framework defined by Ramadge and Wonham [40]. A *plant* is a finite automaton over an alphabet A with two orthogonal partitions: $A = A_c \cup A_{uc}$ where the transitions in A_c and A_{uc} are respectively *controllable* and *uncontrollable*, and $A = A_o \cup A_{uo}$ where the transitions in A_o and A_{uo} are respectively *observable* and *unobservable*. Let $\mathcal{R}_p$ be the language of the plant, and let $\mathcal{R}_l \subseteq \mathcal{R}_p$ be a regular subset of *legal* firing sequences. For the sake of simplicity, assume that $\mathcal{R}_p$ and $\mathcal{R}_l$ are prefix-closed and $A_c \subseteq A_o$ (unobservable transitions are uncontrollable). A *controller* is then a (finite or infinite) automaton that defines a prefix-closed language $\mathcal{K}$ of A_o^*. The problem is to search for some $\mathcal{K}$ in a given class of languages such that $\{u \in \mathcal{R}_p \,|\, \pi_o(u) \in \mathcal{K}\} \subseteq \mathcal{R}_l$ where π_o projects A^* on A_o^*. An *admissible* controller $\mathcal{K}$ should moreover satisfy

$$\begin{aligned} &\forall t \in (A_o \cap A_{uc}) \quad \forall u \in \mathcal{R}_p \quad \forall v \in \mathcal{K} \\ &v = \pi_o(u) \;\wedge\; (u \cdot t) \in \mathcal{R}_p \quad \Longrightarrow \quad (v \cdot t) \in \mathcal{K} \end{aligned}$$

Deciding whether *maximally permissive* admissible controllers exist reduces to deciding whether for some $\mathcal{K}$,

$$\smile \;\cap\; \pi_o(\mathcal{R}_l) \;\subseteq\; \mathcal{K} \;\subseteq\; \smile$$

where $\smile$ is the largest subset of A_o^* containing no observation sequence $v = \pi_o(u)$ such that $uw \in \mathcal{R}_p$ and $uw \notin \mathcal{R}_l$ for some uncontrollable sequence $w \in A_{uc}^*$. Thus, if $\cdot/\cdot$ denotes quotient of languages, $\smile$ is the complement in A_o^* of the set

$$\frown \;=\; \pi_o\left(\,(\mathcal{R}_p \cap \complement\,\mathcal{R}_l)\,/A_{uc}^*\,\right)$$

As $\frown$ is a regular set, $\smile$ is regular, hence $\mathcal{L} = \smile \;\cap\; \pi_o(\mathcal{R}_l)$ is regular. The problem amounts to deciding whether there exists some $\mathcal{K}$ in the specified class of languages such that $\mathcal{L} \subseteq \mathcal{K} \subseteq \smile$ where $\mathcal{L}$ and $\smile$ are two regular languages. This problem may be posed w.r.t. the class of P/T-net languages. The solution is to compute $\mathcal{N}$ from $\mathcal{L}$ as shown in section 2.2, such that $\mathcal{K} = L(\mathcal{N})$ is the least P/T-net language larger than $\mathcal{L}$, and then to check whether $L(\mathcal{N}) \subseteq \smile$ (this is decidable). Maximally permissive P/T-net controllers are then obtained.

3 Net Synthesis from Infinite Graphs

The P/T-net synthesis problem for a class of graphs is the problem whether one can decide uniformly from any graph in this class whether it is isomorphic to the reachable state graph of some initialized P/T-net, and construct such a net when it exists. In this section, $T = \{t_1, \ldots, t_n\}$ is a fixed alphabet, all P/T-nets have the set of transitions T, and all graphs have directed edges with labels in T.

Let $\mathcal{G} = (V, E, v_0)$ denote a graph with respective sets of *vertices* and *edges* V and E, where $v_0 \in V$ is the *root* and $E \subseteq (V \times T \times V)$. An edge $(v, t, v') \in E$ has a *source* v, a *label* t, and a *target* v'. We consider deterministic and reachable graphs exclusively, *i.e.* we assume that every vertex v can be reached by some walk from v_0 to v, and that distinct edges with a common source have distinct labels. A *morphism* of graphs $\sigma : \mathcal{G} \to \mathcal{G}'$, where $\mathcal{G}' = (V', E', v'_0)$, is a map $\sigma : V \to V'$ such that $\sigma(v_0) = \sigma(v'_0)$ and $(\sigma(v), t, \sigma(v')) \in E'$ for every edge $(v, t, v') \in E$. Note that there is *at most one* morphism from $\mathcal{G}$ to $\mathcal{G}'$. Let $\mathcal{G} \leq \mathcal{G}'$ when this morphism exists. It is easily seen that $\leq$ is an order relation and that two graphs $\mathcal{G}$ and $\mathcal{G}'$ are isomorphic ($\mathcal{G} \cong \mathcal{G}'$) *if and only if* $\mathcal{G} \leq \mathcal{G}'$ and $\mathcal{G}' \leq \mathcal{G}$. Let $G(\mathcal{N})$ denote the reachable state graph of the P/T-net $\mathcal{N}$. The problem is to decide from a given graph $\mathcal{G}$ whether $\mathcal{G} \cong G(\mathcal{N})$ for some P/T-net $\mathcal{N}$.

This problem is a strengthening of the problem dealt with in section 2. Indeed, $\mathcal{G} \cong G(\mathcal{N}) \implies L(\mathcal{G}) = L(\mathcal{N})$ where $L(\mathcal{G})$ is the set of sequences in T^* labelling walks from v_0 to arbitrary vertices v in $\mathcal{G}$. The converse implication does not hold. We show in this section that the P/T-net synthesis problem for graphs may be solved by a modification of the techniques already presented. The leading idea is to replace the relation of language inclusion $\subseteq$ used in section 2 with the order relation $\leq$ on graphs. The development given hereafter mimics the development given in this earlier section.

3.1 The Regions of a Graph

To begin with, let us observe that any finite family of graphs $\mathcal{G}_l = (V_l, E_l, v_{l,0})$, $l \in [1, m]$, has a *greatest lower bound* $\bigwedge_l \mathcal{G}_l$. This greatest lower bound is a graph (V, E, v_0) where $V \subseteq (V_1 \times \ldots \times V_m)$ and $v_0 = (v_{1,0}, \ldots, v_{m,0})$. Moreover V and E are the least sets such that $v_0 \in V$ and the following *closure* axiom is satisfied:

if $v = (v_1, \ldots, v_m)$ is in V and for some $t \in T$, $(v_l, t, v'_l) \in E_l$ for all $l \in [1, m]$, then $v' = (v'_1, \ldots, v'_m)$ is in V and (v, t, v') is in E.

Proposition 3. *The reachable state graph of a P/T-net is isomorphic to the greatest lower bound of the reachable state graphs of its atomic subnets.*

The proposition follows immediately from the firing rule of nets. In view of this fundamental property of reachable state graphs, the modified definition of P/T-regions which is proposed hereafter supplies a basis for the synthesis of P/T-nets from graphs.

Definition 6 (P/T-regions of a graph). *A* P/T-region *of $\mathcal{G}$ is any atomic P/T-net $\mathcal{N} = (\{p\}, T, F, M_0)$ such that $\mathcal{G} \leq G(\mathcal{N})$.*

Example 8. Consider the following graph $\mathcal{G}$:

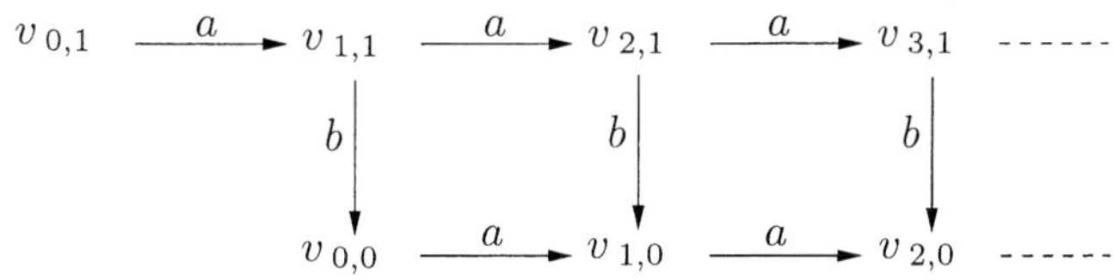

The atomic nets $\mathcal{N}_1$ and $\mathcal{N}_2$ defined in example 4 are two P/T-regions of $\mathcal{G}$. The respective graphs $G(\mathcal{N}_1)$ and $G(\mathcal{N}_2)$ are shown below, with $G(\mathcal{N}_1)$ on the left hand side.

The inequalities $\mathcal{G} \leq G(\mathcal{N}_1)$ and $\mathcal{G} \leq G(\mathcal{N}_2)$ are established by the respective morphisms $\sigma_1(v_{i,j}) = i$ and $\sigma_2(v_{i,j}) = j$. □

The next proposition follows from proposition 3 and the (obvious) fact that $G(\mathcal{N}) \leq G(\mathcal{N}_l)$ for every atomic subnet $\mathcal{N}_l$ of $\mathcal{N}$.

Proposition 4. *$\mathcal{G} \cong G(\mathcal{N})$ for some P/T-net $\mathcal{N}$ if and only if $\mathcal{G} \cong \bigwedge_l G(\mathcal{N}_l)$ for some* finite *collection $\{\mathcal{N}_1, \ldots, \mathcal{N}_m\}$ of P/T-regions of $\mathcal{G}$.*

Definition 6 and proposition 4 are too abstract and they should be refined. We aim in the sequel at equivalent statements with better algorithmic contents. Because $\mathcal{G} \leq G(\mathcal{N}) \Rightarrow L(\mathcal{G}) \subseteq L(\mathcal{N})$, every region of a graph $\mathcal{G}$ is a region of the language $L(\mathcal{G})$. In example 8, all regions of $L(\mathcal{G})$ are regions of $\mathcal{G}$, but this is not true in general. For instance, if $\mathcal{G}$ has edges (v_0, a, v_1) and (v_0, b, v_1), a P/T-net $(\{p\}, T, F, M_0)$ such that $F(a,p) - F(p,a) \neq F(b,p) - F(p,b)$ may be a region of $L(\mathcal{G})$ but it cannot be a region of $\mathcal{G}$. This distinction is clarified below.

Definition 7. *Given $\mathcal{G} = (V, E, v_0)$ and $w \in L(\mathcal{G})$, let ∂w denote the vertex at the end of the walk with label w from the root v_0. Two words w and w' of $L(\mathcal{G})$ are said to* converge *in $\mathcal{G}$ if $\partial w = \partial w'$, and they are said to* diverge *otherwise.*

Proposition 5. *Given graphs $\mathcal{G}_1$ and $\mathcal{G}_2$, $\mathcal{G}_1 \leq \mathcal{G}_2$ if and only if $L(\mathcal{G}_1) \subseteq L(\mathcal{G}_2)$ and every pair of words that converges in $\mathcal{G}_1$ converges in $\mathcal{G}_2$.*

Proof. The two conditions are clearly necessary to the existence of a morphism of graphs from $\mathcal{G}_1$ to $\mathcal{G}_2$. Conversely, when both conditions are satisfied, the map σ defined with $\sigma(\partial_1 w) = \partial_2 w$, where $w \in L(\mathcal{G}_1)$ and ∂_1 and ∂_2 are interpreted w.r.t. $\mathcal{G}_1$ and $\mathcal{G}_2$, respectively, is a morphism of graphs. □

Corollary 1. *Let $\mathcal{N} = (\{p\}, T, F, M_0)$ be a region of $L(\mathcal{G})$, then $\mathcal{N}$ is a region of $\mathcal{G}$ if and only if every pair of words that converges in $\mathcal{G}$ converges in $G(\mathcal{N})$.*

By definition, $\mathcal{G} \leq G(\mathcal{N}_l)$ for every P/T-region $\mathcal{N}_l$ of $\mathcal{G}$, hence $\mathcal{G} \leq \bigwedge_l G(\mathcal{N}_l)$ for every finite collection $\{\mathcal{N}_1, \ldots, \mathcal{N}_m\}$ of P/T-regions of $\mathcal{G}$. By proposition 5, the converse inequality $\bigwedge_l G(\mathcal{N}_l) \leq \mathcal{G}$ holds *if and only if* $\bigcap_l L(\mathcal{N}_l) \subseteq L(\mathcal{G})$ and every pair of words that diverges in $\mathcal{G}$ diverges in $G(\mathcal{N}_l)$ for some l. Proposition 4 may therefore be restated equivalently as follows.

Proposition 6. *$\mathcal{G} \cong G(\mathcal{N})$ for some P/T-net $\mathcal{N}$ if and only if there exists a finite collection $\{\mathcal{N}_1, \ldots, \mathcal{N}_m\}$ of P/T-regions of $L(\mathcal{G})$ such that:*

i) *every pair of words that converges in $\mathcal{G}$ converges in $G(\mathcal{N}_l)$ for all $l \in [1, m]$,*
ii) *for every $t \in T$ and for every $w \in L(\mathcal{G})$, if $w \cdot t \notin L(\mathcal{G})$, then $w \cdot t \notin L(\mathcal{N}_l)$ for some $l \in [1, m]$,*
iii) *every pair of words that diverges in $\mathcal{G}$ diverges in $G(\mathcal{N}_l)$ for some $l \in [1, m]$.*

When these conditions are satisfied, $\mathcal{G} \cong G(\mathcal{N})$ where $\mathcal{N}$ is the P/T-net with the set of atomic subnets $\{\mathcal{N}_1, \ldots, \mathcal{N}_m\}$.

A comparison between proposition 6 and proposition 2 indicates that two new problems should be solved if one wants to decide on the P/T-net synthesis problem for classes of graphs. First, the computation of the generating regions defined in section 2.2 should be accomodated to the constraints induced by the requirement *(i)* in the above proposition. Second, the procedure defined in section 2.3 should be augmented so as to decide whether *both* requirements *(ii)* and *(iii)* in the proposition can be satisfied by the generating regions. The two problems are examined in sequence in the sections below.

3.2 A Procedure for Computing Generating Regions

Let us introduce two definitions.

Definition 8. *Given a graph $\mathcal{G} = (V, E, v_0)$, a prefix-closed language $\mathcal{L} \subseteq L(\mathcal{G})$* spans $\mathcal{G}$ *if $(\forall v \in V)\ (\exists w \in \mathcal{L})\ \ v = \partial w\ \ (in\ \mathcal{G})$.*

Definition 9. *For any vector $\psi \in \mathbb{N}^{2n}$, let ψ_L and ψ_R denote the respective vectors in $\mathbb{N}^n$ such that ψ decomposes to (ψ_L, ψ_R) through the isomorphism $\mathbb{N}^{2n} \cong (\mathbb{N}^n \times \mathbb{N}^n)$. For any pair of words w_L and w_R in T^*, let $[w_L, w_R]$ denote the (unique) vector $\psi \in \mathbb{N}^{2n}$ such that $\psi_L = [w_L]$ and $\psi_R = [w_R]$.*

In order to compute effectively from a graph $\mathcal{G} = (V, E, v_0)$ a finite set of P/T-regions generating all regions of this graph, we require that $\mathcal{G}$ should be spanned by some (prefix-closed) language $\mathcal{L}$ such that:

for every edge label $t_j \in T$,

$$\Psi_j = \{\, [w, w'] \mid w, w' \in \mathcal{L} \wedge (\partial w, t_j, \partial w') \in E\}$$

is a semi-linear subset of $\mathbb{N}^{2n}$.

Example 9. Let $\mathcal{G}$ be the infinite graph depicted below. It is easily seen that this graph is spanned by the prefix-closed language $\mathcal{L} = (ab)^*a^* + (ab)^*b + (ab)^*bb$.

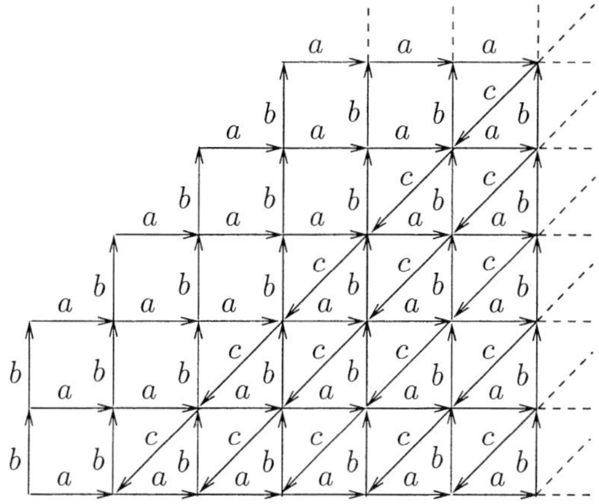

Let $a = t_1$, $b = t_2$, $c = t_3$ and define the following vectors in $\mathbb{N}^3 \times \mathbb{N}^3 \cong \mathbb{N}^6$ (where ; is used in place of , for better readability):

$$\begin{aligned} \mathbf{0} &= < 0,0,0\,;0,0,0 > \\ \boldsymbol{\delta}_{ab} &= < 1,1,0\,;1,1,0 > \\ \boldsymbol{\delta}_{a} &= < 1,0,0\,;1,0,0 > \\ \boldsymbol{\delta}_{b} &= < 0,1,0\,;0,1,0 > \\ \boldsymbol{\delta}_{bb} &= < 0,2,0\,;0,2,0 > \end{aligned}$$

Let $\Psi = \mathbf{0} \cdot (\boldsymbol{\delta}_{ab} + \boldsymbol{\delta}_a)^* + \boldsymbol{\delta}_b \cdot (\boldsymbol{\delta}_{ab})^* + \boldsymbol{\delta}_{bb} \cdot (\boldsymbol{\delta}_{ab})^*$, thus Ψ is a semi-linear set. For $t \in \{a,b,c\}$, the respective sets $\Psi_t = \{\, [w,w'] \,|\, w,w' \in \mathcal{L} \wedge (\partial w, t, \partial w') \in E\}$ may be given the semi-linear expressions:

$$\Psi_a = < 0,0,0\,;1,0,0 > \cdot \Psi \tag{7}$$

$$\Psi_b = < 0,0,0\,;0,1,0 > \cdot (\boldsymbol{\delta}_{ab} + \boldsymbol{\delta}_a)^* + < 0,0,0\,;0,1,0 > \cdot \boldsymbol{\delta}_b \cdot (\boldsymbol{\delta}_{ab})^* \tag{8}$$

$$\Psi_c = < 2,1,0\,;1,0,0 > \cdot (\boldsymbol{\delta}_{ab} + \boldsymbol{\delta}_a)^* \tag{9}$$

The requested condition is fulfilled. □

We define now a procedure that computes the generating P/T-regions of a graph from a language $\mathcal{L}$ spanning this graph and from the associated semi-linear sets Ψ_j ($j \in [1,n]$). Recall that an atomic P/T-net $\mathcal{N} = (\{p\}, T, F, M_0)$ may be represented as a $(2n+1)$-vector $\boldsymbol{x} =< x_0, x_1, \ldots, x_n, x_{n+1}, \ldots, x_{2n} >$ where $x_0 = M_0(p)$ and for all $j \in [1,n]$, $x_j = F(p,t_j)$ and $x_{n+j} = F(t_j,p)$. We claim that a $(2n+1)$-vector $\boldsymbol{x} =< x_0, x_1, \ldots, x_n, x_{n+1}, \ldots, x_{2n} >$ defines a region of $\mathcal{G}$ if and only if all its entries x_k are non-negative integers, and the following inequalities and equations hold for all $t_j \in T$ and $\psi \in \Psi_j$:

$$\sum_{i=1}^{n} \psi_L[i] \times (x_{(n+i)} - x_i) \geq x_j - x_0 \tag{10}$$

$$\sum_{i=1}^{n} \psi_R[i] \times (x_{(n+i)} - x_i) = x_{(n+j)} - x_j + \sum_{i=1}^{n} \psi_L[i] \times (x_{(n+i)} - x_i) \tag{11}$$

In order to see that both conditions are necessary, let $\mathcal{N}$ be the atomic P/T-net defined by the vector $\boldsymbol{x}$. If $\mathcal{N}$ is a region of $\mathcal{G}$, then $L(\mathcal{G}) \subseteq L(\mathcal{N})$ and therefore $\mathcal{L} \subseteq L(\mathcal{N})$. The inequality 10 states that whenever $u \in \mathcal{L}$ and $u \cdot t_j \in L(\mathcal{G})$, if u can be fired in $\mathcal{N}$ then $u \cdot t_j$ can be fired in $\mathcal{N}$. This must be true since $L(\mathcal{G})$ cannot be included in $L(\mathcal{N})$ otherwise. Under the same assumptions, the equation 11 states that whenever $u \cdot t_j$ and v converge in $\mathcal{G}$ for some v in $\mathcal{L}$, $u \cdot t_j$ and v converge in $G(\mathcal{N})$. The corollary 1 states that this also must be true.

In order to establish the claim, it remains to show that whenever the conditions 10 and 11 hold for a vector $\boldsymbol{x}$, the atomic P/T-net $\mathcal{N}$ defined by $\boldsymbol{x}$ is a region of the graph $\mathcal{G}$. By corollary 1, it suffices to prove that $L(\mathcal{G}) \subseteq L(\mathcal{N})$ and that all pairs of words of $L(\mathcal{G})$ that converge in $\mathcal{G}$ converge in $G(\mathcal{N})$.

Proposition 7. *$\mathcal{L} \subseteq L(\mathcal{N})$ and moreover, the pairs of words of $\mathcal{L}$ that converge in $\mathcal{G}$ converge also in $G(\mathcal{N})$.*

Proof. As $\mathcal{L}$ is prefix-closed, $\mathcal{L} \subseteq L(\mathcal{N})$ follows from 10 by induction on words. Consider $v, w \in \mathcal{L}$ such that $\partial v = \partial w$ in $\mathcal{G}$. If $v = w$, they do converge in $G(\mathcal{N})$. If $v \neq w$, at least one of them is non-empty. Assume w.l.o.g. that $v = u \cdot t_j$ with $t_j \in T$. As $\mathcal{L}$ is prefix-closed, $u \in \mathcal{L}$. Therefore, $\psi = [u, v]$ and $\psi' = [u, w]$ are vectors in Ψ_j. It follows from the equation 11 that $\sum_{i=1}^{n} [v]_i \times (x_{(n+i)} - x_i) = \sum_{i=1}^{n} [w]_i \times (x_{(n+i)} - x_i)$. Therefore, v and w converge in $G(\mathcal{N})$. □

Proposition 8. *For all $v' \in L(\mathcal{G})$ and for all $w \in \mathcal{L}$ such that $\partial v' = \partial w$ in $\mathcal{G}$:*

i) *$v' \in L(\mathcal{N})$, and*
ii) *v' and w converge in $G(\mathcal{N})$.*

Proof. Since $L(\mathcal{G})$ is prefix-closed, one may use an induction on words. As $\varepsilon \in \mathcal{L}$, the basis of the induction is clear from proposition 7. For the induction step, let $v' = u' \cdot t_j$ where $t_j \in T$. Choose $u, v \in \mathcal{L}$ such that $\partial u = \partial u'$ and $\partial v = \partial v'$ (since $\mathcal{L}$ spans $\mathcal{G}$, such words must exist). As $v' \in L(\mathcal{G})$, $(\partial u', t_j, \partial v')$ is an edge of $\mathcal{G}$, and this edge is equal to $(\partial u, t_j, \partial v)$. By proposition 7, $u, v \in L(\mathcal{N})$ since $u, v \in \mathcal{L}$. Hence $u \cdot t_j \in L(\mathcal{N})$, in view of the inequality 10, and $u \cdot t_j$ and v converge in $G(\mathcal{N})$, in view of the equation 11. From the induction hypothesis, $u' \in L(\mathcal{N})$, and u and u' converge in $G(\mathcal{N})$. Therefore, $u' \cdot t_j \in L(\mathcal{N})$, and $u' \cdot t_j$ and $u \cdot t_j$ converge in $G(\mathcal{N})$. This entails that $v' \in L(\mathcal{N})$, and v' and v converge in $G(\mathcal{N})$. As $\partial v = \partial v' = \partial w$ and $v, w \in \mathcal{L}$, v and w converge in $G(\mathcal{N})$, by proposition 7. Therefore, v' and w converge in $G(\mathcal{N})$. □

Corollary 2. *$\mathcal{N}$ is a P/T-region of $\mathcal{G}$.*

Proof. Seeing that $\mathcal{L}$ spans $\mathcal{G}$, proposition 8 entails that whenever two words v', v'' of $L(\mathcal{G})$ converge in $\mathcal{G}$, they converge in $G(\mathcal{N})$. □

Using the assumption that all sets Ψ_j are semi-linear, the (possibly) infinite collection of linear homogeneous constraints that derive as instances of (10) or (11) for some $\psi \in \Psi_j$ ($j \in [1, n]$) may be reduced to a finite linear system.

As regards the inequality (10), which is similar to the inequality (1), the reduction follows the same lines as in section 2.2 (the former set $[\mathcal{L}/t_j]$ is replaced with the semi-linear set $\Sigma_j = \{\psi_L \mid \psi \in \Psi_j\}$).

As regards the equation (11), let $\Delta_j = \{\psi_R - \psi_L \mid \psi \in \Psi_j\}$ for each $j \in [1, n]$. Since Ψ_j is a semi-linear subset of $\mathbb{N}^{2n}$, and in view of the definition 5, Δ_j is a finite union of linear sets $\boldsymbol{e} \cdot \mathcal{F}^*$, where $\boldsymbol{e} \in \mathbb{Z}^n$ and $\mathcal{F}$ is a finite subset of $\mathbb{Z}^n$. For each linear subset $\boldsymbol{e} \cdot \mathcal{F}^*$ of Δ_j, the set of constraints that derive as instances of the equation 11 for some $\psi \in \boldsymbol{e} \cdot \mathcal{F}^*$ may be replaced equivalently with the finite linear system:

$$\sum_{i=1}^{n} \boldsymbol{e}[i] \times (x_{(n+i)} - x_i) = x_{(n+j)} - x_j \tag{12}$$

$$\sum_{i=1}^{n} \boldsymbol{f}[i] \times (x_{(n+i)} - x_i) = 0 \tag{13}$$

where f ranges over the finite set $\mathcal{F}$. Therefore, the collection of instances of the equation 11 for all $j \in [1, n]$ and for all $\psi \in \Psi_j$ reduces to a finite system.

Let $\mathcal{S}$ be the finite linear system in the variables $x_0, \ldots, x_{2n}$ formed of the reduced systems defined above, plus inequalities $x_k \geq 0$ for all $k \in [0, 2n]$. One can compute as was explained in section 2.2 a finite and minimal set of solutions $\boldsymbol{x}_1 \ldots \boldsymbol{x}_m$ of $\mathcal{S}$ in $\mathbb{N}^{2n+1}$, called the *generating regions* of the graph $\mathcal{G}$, such that the regions of this graph may be characterized as follows:

A vector $\boldsymbol{x}$ is a P/T-region of $\mathcal{G}$ if and only if $\boldsymbol{x} \in \mathbb{N}^{2n+1}$ and $\boldsymbol{x} = \sum_{l=1}^{m} q_l \, \boldsymbol{x}_l$ for some non-negative rational coefficients q_l.

Example 10. Let us compute the generating regions of the graph $\mathcal{G}$ from example 9. In order to enhance the readability, let $\boldsymbol{x} = < m_0, {}^\bullet a, {}^\bullet b, {}^\bullet c, a^\bullet, b^\bullet, c^\bullet >$ where $t_1 = a$, $t_2 = b$, and $t_3 = c$. The respective sets Δ_1, Δ_2, and Δ_3 are the singleton sets defined with $\Delta_1 = \Delta_a = \{< 1, 0, 0 >\}$, $\Delta_2 = \Delta_b = \{< 0, 1, 0 >\}$, and $\Delta_3 = \Delta_c = \{< -1, -1, 0 >\}$. The finite system derived from equation 11 is:

$$\begin{array}{l} a^\bullet - {}^\bullet a = a^\bullet - {}^\bullet a \\ b^\bullet - {}^\bullet b = b^\bullet - {}^\bullet b \\ {}^\bullet a - a^\bullet + {}^\bullet b - b^\bullet = c^\bullet - {}^\bullet c \end{array}$$

Two trivial equations may be dropped. The linear constraints generated from the instances of the inequality 10, where ψ_L ranges over the respective semi-linear sets $\Sigma_j = \{\psi_L \mid \psi \in \Psi_j\}$ $(j \in [1, 3])$, are as follows.
$\Sigma_1 = \Sigma_a = \langle 0,0,0 \rangle \cdot (\langle 1,1,0 \rangle + \langle 1,0,0 \rangle)^* + \langle 0,1,0 \rangle \cdot \langle 1,1,0 \rangle^* + \langle 0,2,0 \rangle \cdot \langle 1,1,0 \rangle^*$

produces the constraints

$$\begin{array}{l} 0 \geq {}^\bullet a - m_0 \\ b^\bullet - {}^\bullet b \geq {}^\bullet a - m_0 \\ 2\,(b^\bullet - {}^\bullet b) \geq {}^\bullet a - m_0 \\ a^\bullet - {}^\bullet a \geq 0 \\ a^\bullet - {}^\bullet a + b^\bullet - {}^\bullet b \geq 0 \end{array}$$

$\Sigma_1 = \Sigma_b = \langle 0,0,0 \rangle \cdot (\langle 1,1,0 \rangle + \langle 1,0,0 \rangle)^* + \langle 0,1,0 \rangle \, \langle 1,1,0 \rangle^*$ adds two constraints

$$0 \geq {}^\bullet b - m_0$$
$$b^\bullet - {}^\bullet b \geq {}^\bullet b - m_0$$

$\Sigma_3 = \Sigma_c = \langle 2,1,0 \rangle \cdot (\langle 1,1,0 \rangle + \langle 1,0,0 \rangle)^*$ brings finally one more constraint

$$2\,(a^\bullet - {}^\bullet a) + b^\bullet - {}^\bullet b \geq {}^\bullet c - m_0$$

The generating regions computed by Chernikova's algorithm arc thc following:

m_0	2	1	1	1	0	2	1	0	0	2	1	1	2	1	1
${}^\bullet a$	0	1	0	1	0	0	0	0	0	0	1	1	0	0	0
${}^\bullet b$	1	0	1	1	0	2	0	0	0	1	0	1	2	1	0
${}^\bullet c$	3	1	1	1	1	3	1	2	1	0	0	0	0	0	0
$a^\bullet$	1	1	0	1	1	1	0	1	0	1	1	1	1	0	0
$b^\bullet$	0	0	1	1	0	1	0	0	1	0	0	1	1	1	0
$c^\bullet$	3	1	1	1	0	3	1	1	0	0	0	0	0	0	0

Many generating regions are useless, as we will show later on. □

We claim that the P/T-net $\mathcal{N}$ built up from the atomic subnets $\mathcal{N}_1 \ldots \mathcal{N}_m$ defined by the generating regions $\boldsymbol{x}_1 \ldots \boldsymbol{x}_m$ is the best net-approximation of the graph $\mathcal{G}$ in the following sense:

$$\mathcal{G} \leq G(\mathcal{N}) \quad \textit{and} \quad \forall \mathcal{N}' \quad \mathcal{G} \leq G(\mathcal{N}') \Longrightarrow G(\mathcal{N}) \leq G(\mathcal{N}')$$

The relation $\mathcal{G} \leq G(\mathcal{N})$ is easily established, as $\mathcal{G} \leq G(\mathcal{N}_l)$ for all $l \in [1,m]$ (by definition of the regions of a graph) and $G(\mathcal{N}) = \bigwedge_l G(\mathcal{N}_l)$ (by proposition 3). The two propositions below aim at establishing the second part of the claim.

Proposition 9. $\forall \mathcal{N}' \quad \mathcal{G} \leq G(\mathcal{N}') \Longrightarrow L(\mathcal{N}) \subseteq L(\mathcal{N}')$

Proof. Assuming the converse, let $w \in L(\mathcal{N}) \cap L(\mathcal{N}')$ and $t_j \in T$ such that $w \cdot t_j \in L(\mathcal{N})$ and $w \cdot t_j \notin L(\mathcal{N}')$. Necessarily, $\sum_i [w]_i \times (\boldsymbol{x}'[n+i] - \boldsymbol{x}'[i]) < \boldsymbol{x}'[j] - \boldsymbol{x}'[0]$ for some $(2n+1)$-vector $\boldsymbol{x}'$ representing an atomic subnet of $\mathcal{N}'$. As $\mathcal{G} \leq G(\mathcal{N}')$ and $G(\mathcal{N}') \leq G(\mathcal{N}'')$ for every subnet $\mathcal{N}''$ of $\mathcal{N}'$, this atomic subnet of $\mathcal{N}'$ is a region of $\mathcal{G}$. Therefore, $\boldsymbol{x}' = \sum_{l=1}^{m} q_l \, \boldsymbol{x}_l$ for some non-negative rational coefficients q_l. Owing to the sign of the coefficients, it must be true for some $l \in [1,m]$ that $\sum_i [w]_i \times (\boldsymbol{x}_l[n+i] - \boldsymbol{x}_l[i]) < \boldsymbol{x}_l[j] - \boldsymbol{x}_l[0]$. But this inequality entails $w \cdot t_j \notin L(\mathcal{N}_l)$ and hence $w \cdot t_j \notin L(\mathcal{N})$, a contradiction. □

Proposition 10. *If $\mathcal{G} \leq G(\mathcal{N}')$, then two words of $L(\mathcal{N})$ converge in $G(\mathcal{N}')$ whenever they converge in $G(\mathcal{N})$.*

Proof. Assuming the converse, let w, w' converge in $G(\mathcal{N})$ and diverge in $G(\mathcal{N}')$. Necessarily, $\sum_i [w]_i \times (\boldsymbol{x}'[n+i] - \boldsymbol{x}'[i]) \neq \sum_i [w']_i \times (\boldsymbol{x}'[n+i] - \boldsymbol{x}'[i])$ for some

$(2n+1)$-vector $\boldsymbol{x}'$ representing an atomic subnet of $\mathcal{N}'$. Like in the proof of the former proposition, $\boldsymbol{x}' = \sum_{l=1}^{m} q_l\, \boldsymbol{x}_l$, and it must be true for some $l \in [1,m]$ that $\sum_i [w]_i \times (\boldsymbol{x}_l[n+i] - \boldsymbol{x}_l[i]) \neq \sum_i [w']_i \times (\boldsymbol{x}_l[n+i] - \boldsymbol{x}_l[i])$. As a consequence, w and w' diverge in $G(\mathcal{N}_l)$ and hence in $G(\mathcal{N})$, a contradiction. □

In view of these propositions, the second part of the claim follows from the proposition 5. Optimal net-approximations may therefore be computed in any class of graphs (V, E, v_0) spanned by languages $\mathcal{L}$ such that, for every $t_j \in T$, the *transition relation* $\mathcal{T}_j = \{\, (w, w') \,|\, w \in \mathcal{L} \wedge w' \in \mathcal{L} \wedge (\partial w, t_j, \partial w') \in E \,\}$ is semi-linear (*i.e.* $\{\, [w, w'] \,|\, (w, w') \in \mathcal{T}_j \,\}$ is semi-linear). A trivial example is the class of finite graphs. Two other examples are the classes of *labelled* domains induced by recognizable sets of Mazurkiewicz traces, or by Finite Automata with Concurrency Relations [13]. In both cases, the language of a labelled domain $\mathcal{G}$ is a regular language $\mathcal{L}$, and $\{\, [w, w'] \,|\, (w, w') \in \mathcal{T}_j \,\} = \{\, [w, w \cdot t_j] \,|\, (w \cdot t_j) \in \mathcal{L} \}$, hence the transition relations are semi-linear. Optimal net-approximations may also be computed in any class of graphs where the transition relations $\mathcal{T}_j$ may be defined with finite 2-tape automata (this is the case for deterministic pushdown graphs [42][43]), or more generally with (non-deterministic) 2-tape pushdown automata (this particular use of 2-tape pda's is a suggestion of ours).

3.3 A Procedure for the Decision of the Net Synthesis Problem

We show in this section that under additional conditions on $\mathcal{G}$ and the spanning language $\mathcal{L}$, one can decide whether $\mathcal{G}$ has a P/T-net generator, *i.e.* whether $\mathcal{G} \cong G(\mathcal{N})$ where $\mathcal{N}$ is the net constructed in section 3.2.

Because $\mathcal{G} \leq G(\mathcal{N})$ and $G(\mathcal{N})$ is the least net-approximation of $\mathcal{G}$, the graph $\mathcal{G}$ has a P/T-net generator *if and only if* $G(\mathcal{N}) \leq \mathcal{G}$. By proposition 5, the following two conditions are necessary and sufficient:

i) for every $w \in L(\mathcal{G})$ and $t_j \in T$, if $w \cdot t_j \notin L(\mathcal{G})$, then $w \cdot t_j \notin L(\mathcal{N})$,
ii) every pair of words of $L(\mathcal{N})$ that diverges in $\mathcal{G}$ diverges in $G(\mathcal{N})$.

The condition *(i)* reads as $L(\mathcal{N}) \subseteq L(\mathcal{G})$, hence when it holds, $L(\mathcal{N}) = L(\mathcal{G})$ because $\mathcal{G} \leq G(\mathcal{N}) \Longrightarrow L(\mathcal{G}) \subseteq L(\mathcal{N})$. We thus retrieve the respective conditions *(ii)* and *(iii)* stated in proposition 6 (the atomic subnets of $\mathcal{N}$ are the generating regions $\mathcal{N}_l$ of $\mathcal{G}$, hence they are regions of $L(\mathcal{G})$ and they satisfy condition *(i)* in prop. 6). Recalling that $\mathcal{L} \subseteq L(\mathcal{G})$, the above conditions may be simplified to:

i') for every $w \in \mathcal{L}$ and $t_j \in T$, if $w \cdot t_j \notin L(\mathcal{G})$, then $w \cdot t_j \notin L(\mathcal{N})$,
ii') every pair of words of $\mathcal{L}$ that diverges in $\mathcal{G}$ diverges in $G(\mathcal{N})$.

(i') entails (i) : Let $w \in L(\mathcal{G})$ and $w \cdot t_j \notin L(\mathcal{G})$. As $\mathcal{L}$ spans $\mathcal{G}$, $\partial w = \partial u$ for some u in $\mathcal{L}$. As $w \cdot t_j \notin L(\mathcal{G})$, $u \cdot t_j \notin L(\mathcal{G})$, hence $u \cdot t_j \notin L(\mathcal{N})$. As $\mathcal{G} \leq G(\mathcal{N})$ and w and u converge in $\mathcal{G}$, they lead to the same marking of $\mathcal{N}$, hence $w \cdot t_j \notin L(\mathcal{N})$.
(ii') entails (ii) : Let w and w' be words of $L(\mathcal{G})$ such that $\partial w \neq \partial w'$. As $\mathcal{L}$ spans $\mathcal{G}$, $\partial w = \partial u$ and $\partial w' = \partial u'$ for some u and u' in $\mathcal{L}$. As $\partial u \neq \partial u'$, u and u' lead

to distinct markings of $\mathcal{N}$. As $\mathcal{G} \leq G(\mathcal{N})$, w and u lead to the same marking of $\mathcal{N}$, and similarly do w' and u', hence w and w' diverge in $G(\mathcal{N})$.

In order to decide whether $\mathcal{G} \cong G(\mathcal{N})$, we add the following requirements on $\mathcal{G}$ and its spanning language $\mathcal{L}$:

1. $Dis = \{\, [w, w'] \mid w \in \mathcal{L} \wedge w' \in \mathcal{L} \wedge \partial w \neq \partial w' \,\}$ should be semi-linear,
2. $Inh_j = \{\, [w] \mid w \in \mathcal{L} \wedge w \cdot t_j \notin L(\mathcal{G}) \,\}$ should be semi-linear for all $j \in [1, m]$.

Assuming these requirements are fulfilled, we propose a decision procedure. Recall that $\mathcal{N}$ has m atomic subnets $\mathcal{N}_1 \ldots \mathcal{N}_m$, *viz.* the generating regions of $\mathcal{G}$, represented with $(2n+1)$-vectors $\boldsymbol{x}_1 \ldots \boldsymbol{x}_m$. Thus, for any $w \in \mathcal{L}$ and for any $l \in [1, m]$, the marking reached after firing w in the atomic subnet $\mathcal{N}_l = (\{p\}, T, F, p = m_0)$ is defined with $p = m_0 + \sum_i [w]_i \times (\boldsymbol{x}_l[n+i] - \boldsymbol{x}_l[i])$.

The condition *(ii')* is satisfied *if and only if* $Dis \subseteq \cup_{l=1}^{m} Dis_l$ where:

$$Dis_l = \{\, \psi \in \mathbb{N}^{2n} \mid \sum_{i=1}^{n} (\psi_R[i] - \psi_L[i]) \times (\boldsymbol{x}_l[n+i] - \boldsymbol{x}_l[i]) \neq 0 \,\}$$

Now, for fixed l, all coefficients $(\boldsymbol{x}_l[n+i] - \boldsymbol{x}_l[i])$ are constants in $\mathbb{Z}$, hence the above formula is a Presburger formula and Dis_l is a semi-linear subset of $\mathbb{N}^{2n}$. Such subsets form an effective boolean algebra. Therefore, when Dis is semi-linear, one can decide whether the condition *(ii')* is satisfied.

The condition *(i')* is satisfied *if and only if*, for all $j \in [1, m]$, $Inh_j \subseteq \cup_{l=1}^{m} Inh_j^l$ where:

$$Inh_j^l = \{\, \psi \in \mathbb{N}^{n} \mid \sum_{i=1}^{n} \psi[i] \times (\boldsymbol{x}_l[n+i] - \boldsymbol{x}_l[i]) < \boldsymbol{x}_l[j] - \boldsymbol{x}_l[0]$$

As Inh_j^l is defined by a Presburger formula, Inh_j^l is a semi-linear subset of $\mathbb{N}^{2n}$. Therefore, when all inhibitor sets Inh_j are semi-linear, one can decide whether the condition *(i')* is satisfied.

The P/T-net synthesis problem is decidable in classes of graphs spanned by languages $\mathcal{L}$ such that the set Dis and all sets Ψ_j and Inh_j are semi-linear

---

Like in section 2, it may occur that $\mathcal{G} \cong G(\mathcal{N}')$ for some proper subnet $\mathcal{N}'$ of $\mathcal{N}$. Minimal nets $\mathcal{N}'$ may be derived from minimal subsets of generating regions such that $Dis \subseteq \cup_l Dis_l$ and $Inh_j \subseteq \cup_l Inh_j^l$ for l ranging over indices of regions in these subsets.

Example 11. For the graph $\mathcal{G}$ in the example 9, $\mathcal{L} = (ab)^*a^* + (ab)^*b + (ab)^*bb$, and $Dis = \{\psi \mid \psi_L \in [\mathcal{L}] \wedge \psi_R \in [\mathcal{L}] \wedge \psi_L \neq \psi_R\}$ where $[\mathcal{L}]$ is the commutative image of $\mathcal{L}$. As $\mathcal{L}$ is a regular language, $[\mathcal{L}]$ is a semi-linear set, hence Dis is a semi-linear set (because it is defined by a Presburger formula). Consider now the two regions of $\mathcal{G}$ represented respectively by the columns 1 and 9 in the table at the end of the example 10. The respective sets Dis_1 and Dis_9 are given by the semi-linear expressions:

$Dis_1 = \{< n_a, n_b, n_c \,;\, n'_a, n'_b, n'_c > \mid n_a - n_b \neq n'_a - n'_b\}$

$Dis_9 = \{< n_a, n_b, n_c \,;\, n'_a, n'_b, n'_c > \mid n_b - n_c \neq n'_b - n'_c\}$

Clearly, for any $\psi =< n_a, n_b, n_c \,;\, n'_a, n'_b, n'_c >\in Dis$, $n_c = 0$ and $n'_c = 0$ because $\mathcal{L} \subseteq \{a, b\}^*$, and $n_a \neq n'_a$ or $n_b \neq n'_b$. If $n_b \neq n'_b$ then $\psi \in Dis_9$. If $n_a \neq n'_a$ and $n_b = n'_b$ then $\psi \in Dis_1$. Thus, $Dis \subseteq Dis_1 \cup Dis_9$.

The respective inhibitor sets $Inh_1 = Inh_a$, $Inh_2 = Inh_b$, and $Inh_3 = Inh_c$ are given by the semi-linear expressions:

$Inh_1 = \emptyset$
$Inh_2 = \langle 0, 2, 0\rangle \cdot \langle 1, 1, 0\rangle^*$
$Inh_3 = \langle 0, 0, 0\rangle \cdot \langle 1, 0, 0\rangle^* + (\langle 0, 0, 0\rangle + \langle 0, 1, 0\rangle + \langle 0, 2, 0\rangle) \cdot \langle 1, 1, 0\rangle^*$

As $Inh_2^1 = \{\langle n_a, n_b, n_c\rangle \mid n_a - n_b < -1\}$, it follows that $Inh_2 \subseteq Inh_2^1$.

Now $Inh_3^1 = \{\langle n_a, n_b, n_c\rangle \mid n_a - n_b < 1\}$, and $Inh_3^9 = \{\langle n_a, n_b, n_c\rangle \mid n_b - n_c < 1\}$.

Clearly, $\langle 0, 0, 0\rangle \cdot \langle 1, 0, 0\rangle^* \subseteq Inh_3^9$, and $Inh_3 \subseteq Inh_3^1 \cup Inh_3^9$.

The graph $\mathcal{G}$ is therefore isomorphic to the reachable state graph of the net built from the atomic nets defined by the respective vectors $\boldsymbol{x}_1 = \langle 2, 0, 1, 3, 1, 0, 3\rangle$ and $\boldsymbol{x}_9 = \langle 0, 0, 0, 1, 0, 1, 0\rangle$. This net is shown in the figure below. □

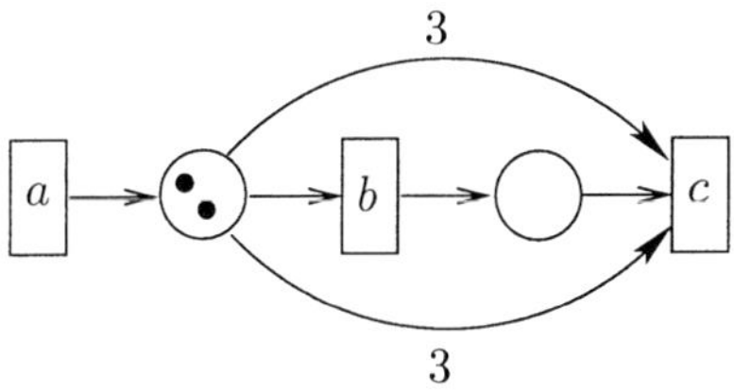

A well known class of graphs where the requirements 1 and 2 are fulfilled is the class of the deterministic pushdown graphs. This assertion is not trivial and it follows from the results establihed by Sénizergues in his unpublished work [43]. Therefore, the general decision procedure presented in this section may be considered as an extension of the specific procedure proposed in [19] for the deterministic pushdown graphs. This extension owes much to Sénizergues's view of graphs with an automatic structure. Building on his ideas, a wide class of graphs where the P/T-net synthesis problem is decidable was proposed in [7].

4 Conclusion

In this paper, we focussed on the problem whether a language or an infinite graph may be realized exactly by an unbounded P/T-net, a problem which was ignored in [6]. We have shown that this problem is decidable under strong requirements of semi-linearity, met by deterministic pushdown languages and graphs, and by graphs in wider families. We have shown that the exact net-realization problem is undecidable for pushdown languages and for HMSC languages. These negative results, and the strong constraints imposed for deciding on the synthesis problem when this is possible, indicate that approximate net-realizations of languages or graphs is often the best one can expect. We have shown that (least) over-approximations by nets may be computed under mild assumptions of semi-linearity on languages or graphs. It was argued that such approximations are particularly adequate in the context of supervisory control problems.

It might be objected that the procedures we have proposed are too limited, since the P/T-nets produced by these procedures have always semi-linear sets of reachable markings. We are conscious of this limitation, but we do not see how it could be removed consistently with our approach.

If one agrees that the exact realization of languages or graphs by nets is not the central problem, there are two ways for further research. One is to search for approximate realizations of languages or graphs by nets, as was proposed in this paper. A second way is to change the data of the P/T-net synthesis problem, by taking *sets* of graphs or languages as inputs, in place of individuals. Then, the problem is to search for a net $\mathcal{N}$ such that $G(\mathcal{N})$ or $L(\mathcal{N})$ belongs to the given set. This problem has been solved in [7] for sets of graphs defined by path-automatic specifications, a combination of modal transition systems and automatic graphs. We are currently working on a similar problem in the context of languages.

References

1. Alur, R., Yannakakis, M.: Model Checking of Message Sequence Charts. Proc. Concur, LNCS **1664** (1999) 114-129.
2. Badouel, E., Bernardinello, L., Darondeau, Ph.: Polynomial Algorithms for the Synthesis of Bounded Nets. Proc. CAAP, LNCS **915** (1995) 647-679.
3. Badouel, E., Bernardinello, L., Darondeau, Ph.: The Synthesis Problem for Elementary Net Systems is NP-complete. *Theoretical Computer Science* **186** (1997) 107-134.
4. Badouel, E., Caillaud, B., Darondeau, Ph.: Distributing finite automata through Petri net synthesis. *Formal Aspects of Computing* **13** (2002) 447-470.
5. Badouel, E., Darondeau, Ph.: On the Synthesis of General Petri Nets. Inria Research Report 3025 (1996).
6. Badouel, E., Darondeau, Ph.: Theory of regions. In W. Reisig and G. Rozenberg, editors, *Lectures on Petri Nets I: Basic Models*, Advances in Petri Nets, LNCS **1491** (1998) 529-586.
7. Badouel, E., Darondeau, Ph.: The Synthesis of Petri Nets from Path-Automatic Specifications. draft, submitted (2003).
8. Bednarczyk, M., Borzyszkowski, A.: On Concurrent Realization of Reactive Systems and their Morphisms. In H. Ehrig, G. Juhas, J. Padberg and G. Rozenberg, editors, *Unifying Petri Nets*, Advances in Petri Nets, LNCS **2128** (2001) 346-379.
9. Bernardinello, L., De Michelis, G., Petruni, K., Vigna, S.: On the Synchronic Sructure of Transition Systems. In J. Desel, editor, *Structures in Concurrency Theory*, Workshops on Computing, Springer Verlag (1996) 11-31.
10. Berstel, J.: *Transductions and Context-Free Languages*. Teubner Verlag (1979).
11. Bernardinello, L.: Synthesis of Net Systems. Proc. ATPN, LNCS **691** (1993) 11-31.

12. Bernardinello, L., Ferrigato, C., Pomello, L.: An Algebraic Model of Observable Properties in Distributed Systems. *Theoretical Computer Science* **290** (2003) 637-678.
13. Bracho, F., Droste, M.: Labelled Domains and Automata with Concurrency Relations. *Theoretical Computer Science* **135**,2 (1994) 289-318.
14. Caillaud, B., Darondeau, Ph., Helouet, L., Lesventes, G.: HMSCs as Partial Specifications ... with PNs as Completions. In F. Cassez, C. Jard, B. Rozoy, M.D. Ryan, editors, *Modeling and Verification of Parallel Processes*: 4th summer school / MOVEP'2000, LNCS **2067** (2001) 125-152.
15. Chernikova, N.: Algorithm for finding a general formula for the non-negative solutions of a system of linear inequalities. *USSR Computational Mathematics and Mathematical Physics* **5**,2 (1965) 228-233.
16. Cortadella, J., Kishinevsky, M., Kondratyev, A., Lavagno, L., Yakovlev, A.: Hardware and Petri Nets: Application to Asynchronous Circuit Design. Proc. ATPN, LNCS **1825** (2000) 1-15.
17. Cortadella, J., Kishinevsky, M., Lavagno, L., Yakovlev, A.: Deriving Petri Nets from Finite Transition Systems. *IEEE Transactions on Computers* **47**,8 (1998) 859-882.
18. Darondeau, Ph.: Deriving Petri Nets from Formal Languages. Proc. CONCUR, LNCS **1466** (1998) 533-548.
19. Darondeau, Ph.: On the Petri net realization of context-free graphs. *Theoretical Computer Science* **258** (2001) 573-598.
20. Desel, J., Reisig, W.: The Synthesis Problem of Petri Nets. *Acta Informatica* **33** (1996) 297-315.
21. Droste, M., Shortt, R.M.: Petri Nets and Automata with Concurrency Relations - an Adjunction. In M. Droste and Y. Gurevich, editors, *Semantics of Programming Languages and Model Theory* (1993) 69-87.
22. Ehrenfeucht, A., Rozenberg, G.: Partial (Set) 2-Structures; *Part I:* Basic Notions and the Representation Problem. *Acta Informatica* **27** (1990) 315-342.
23. Ehrenfeucht, A., Rozenberg, G.: Partial (Set) 2-Structures; *Part II:* State Spaces of Concurrent Systems. *Acta Informatica* **27** (1990) 343-368.
24. Esparza, J.: Decidability of model-checking for infinite-state concurrent systems. *Acta Informatica* **34**,2 (1997) 85-107.
25. Fernau, H.: Parallel Communicating Grammar Systems with Terminal Transmission. *Acta Informatica* **37** (2001) 511-540.
26. Fischer, P.C., Rosenberg, A.L.: Multitape One-Way Nonwriting Automata. *Journal of Computer and System Sciences* **2** (1968) 88-101.
27. Ghaffari, A., Rezg, N., Xie, X.: Live and Maximally Permissive Controller Synthesis using Theory of Regions. In B. Caillaud, Ph. Darondeau, L. Lavagno and X. Xie, editors, *Synthesis and Control of Discrete Event Systems*, Kluwer Academic Publishers (2002)
28. Ginsburg, S., and Spanier, E.H.: Bounded Algol-like Languages. *Transactions of the American Mathematical Society* (1964) 333-368.
29. Ginsburg, S., and Spanier, E.H.: Semigroups, Presburger formulas, and Languages. *Pacific Journal of Mathematics* **16** (1966) 285-296.
30. Ginsburg, S., and Spanier, E.H.: AFL with the Semilinear Property. *Journal of Computer and System Sciences* **5** (1971) 365-396.
31. Harrisson, M.A.: *Introduction to Formal Language Theory.* Addison-Wesley (1978).
32. Holzer, M., Kutrib, M.: Flip-Pushdown Automata: k+1 Pushdown Reversals are Better than k. IFIG Research Report 0206, Universitaet Giessen (2002).

33. Hoogers, P.W., Kleijn, H.C.M., Thiagarajan, P.S.: A Trace Semantics for Petri Nets. Proc. ICALP, LNCS **623** (1992) 595-604.
34. Hopcroft, J., Ullman, J.: *Introduction to Automata Theory, Languages, and Computation.* Addison Wesley (1979).
35. Ibarra, O.: Simple Matrix Languages. *Information and Control* **17** (1970) 359-394.
36. Mayr, E.: An Algorithm for the General Petri Net Reachability Problem. *SIAM Journal on Computing* **13** (1984) 441-460.
37. Mukund, M.: Petri Nets and Step Transition Systems. *International Journal of Foundations of Computer Science* **3**,4 (1992) 443-478.
38. Nielsen, M., Rozenberg, G., Thiagarajan, P.S.: Elementary Transition Systems. *Theoretical Computer Science* **96**,1 (1992) 3-33.
39. Pelz, E.: Closure Properties of Deterministic Petri Nets. Proc. Stacs, LNCS **247** (1987) 373-382
40. Ramadge, P.J, Wonham, W.M.: Supervisory Control of a Class of Discrete Event Processes. *SIAM Journal of Control and Optimization* **25**,1 (1987) 206-230.
41. Schrijver, A.: *Theory of Linear and Integer Programming.* John Wiley (1986).
42. Sénizergues, G.: Definability in weak monadic second order logic of some infinite graphs. Dagstuhl seminar on Automata theory: Infinite computations, Wadern **28** (1992) 16.
43. Sénizergues, G.: *unpublished work communicated to the author*
44. Vogler, W.: Concurrent Realization of Asynchronous Transition Systems. Proc. ATPN, LNCS **1639** (1999) 284-303.

Petri Nets and Software Engineering

Giovanni Denaro and Mauro Pezzè

Università degli Studi di Milano-Bicocca,
Dipartimento di Informatica Sistemistica e Comunicazione,
I-20126 Milano, Italy
{denaro,pezze}@disco.unimib.it

Abstract. Software engineering and Petri net theory are disciplines of different nature. Research on software engineering focuses on a problem domain, i.e., the development of complex software systems, and tries to find a coherent set of solutions to cope with the different aspects of the problem, while research on Petri nets investigates applications and properties of a specific model (Petri nets).

When Petri nets can solve some problems of software development, the two disciplines meet with mutual benefits: software engineers may find useful solutions, while Petri net experts may find new stimuli and challenges in their domain.

Petri nets and software engineering have similar age: Karl Adam Petri wrote his thesis in 1962, while the term "software engineering" was coined in 1968 at a NATO conference held in Germany. The two disciplines met several times in the past forty years with alternate fortune. Presently, software engineering and Petri nets do not find many meeting points, as witnessed by the scarce references to Petri nets in software engineering journals and conferences and vice versa, but software engineering is facing many new challenges and the Petri net body of knowledge is extending with new results.

This paper attempts to illustrate the many dimensions of software engineering, to point at some aspects of Petri nets that have been or can be exploited to solve software engineering problems, and to identify new software engineering challenges that may be solved with Petri net results. This paper does not have the ambition of completely surveying either discipline, but hopes to help scientists and practitioners in identifying interesting areas where software engineers and Petri net experts can fruitfully collaborate[1].

1 Introduction

Software engineering presents several problems that can be attacked with many different techniques and methodologies. Software engineers do not focus on a particular technique or model to solve all problems, but select the solutions that best fit the requirements for each different problem and context. Solutions that

[1] This work has been partially funded by the European Union through the EU IST project SegraVis.

J. Desel, W. Reisig, and G. Rozenberg (Eds.): ACPN 2003, LNCS 3098, pp. 439–466, 2004.

are excellent in a specific context and at a given time, may be sub-optimal in other domains, may not suite well other problems, or may become obsolete in other moments. Software specification and design are typical examples: structure analysis based solutions that were very popular in the eighties, became less and less popular in the nineties, and are now substituted by object oriented based solution; client-server solutions that may solve well many classes of problems, may be ignored in contexts that benefit from other equally good solutions. A quick scan of software engineering handbooks, conferences, and journals would clearly give a variegate picture from the methods and techniques viewpoint.

Disciplines like software engineering that focus on problems and search for the best solutions regardless of the underlying methods or techniques can be identified as *problem-oriented* disciplines. The main characteristic of these disciplines is the presence of many complex problems and the co-existence of alternative solutions, none of which optimal per se. Problem-oriented disciplines are eclectic, since problems may be solved in many different ways with radically different techniques, and fickle, since techniques can be adopted and abandoned as the field evolves [1].

Conversely, the research on Petri nets focuses on a "solution": Petri nets. The research on Petri nets is not driven by a problem domain that asks for successful solutions, but is rather driven by a theory that is studied for solving problems of different nature. Research on Petri nets investigates the various possibilities presented by the theory, and proposes the theory to solve problems in different domains. Advances in Petri nets can be used for attacking problems in computer science, chemistry, biology, hardware design, software specification, distributed computing, multimedia and so forth. Disciplines like Petri net research that focus on theory and offer it for different application domains can be identified as *solution-oriented* disciplines. Solution-oriented disciplines are homogeneous and have a well-defined theory and a stable set of tools.

The meeting of problem and solution-oriented disciplines may bring enormous benefits to both fields: problem-oriented disciplines may find efficient solutions to key problems, while solution-oriented disciplines may find new stimuli in the field. Unfortunately meeting of different disciplines is difficult: few scientists understand different fields well enough to be able to see the potentialities for cross fertilization, and blind attempts to investigate new fields to search for novel solutions are often frustrated by skepticism and lack of successes. However, when problem- and solution-oriented disciplines meet, the whole scientific community can benefit from scientific and technological progresses. This is happening for example in the meeting of biology and research on algorithms that is opening enormous opportunities in bioinformatics.

Software engineering and Petri nets met several time in the past and the meeting seeded interesting ideas in both fields. Useful applications of Petri nets have been proposed in requirement engineering (e.g., [2]), reverse engineering (e.g., [3, 4]), design of user interfaces (e.g., [5]), modeling and analysis of safety critical systems (e.g., [6]), distributed systems (e.g., [7, 8]), real time systems (e.g., [9–12]), multimedia systems (e.g., [13–17]), software process management (e.g., [18, 19]), and software performance evaluation (e.g., [20]). However, the cross fertilization has never stabilized an the two fields are passing a period of scarce communication.

Software engineering is characterized by many dimensions that assume different relevance from different perspectives and are difficult to summarize and frame. Figure 1 suggests three main dimensions: *product development*, *process support* and *application domain*. Software engineers must find an adequate process support to fit the different characteristics of the product for the specific application domain. Each dimension includes many elements with mutually dependent choices hard to concert in a successful project.

The first dimension considered in the figure is related to the development of products, i.e., concerns with the development of software, and is characterized by the specific aspects of the software system, the development phases and the activities performed during development, and the involved stakeholders.

The *aspects* of software systems are the relations among components of the system from different perspectives. They include *structure* and *architecture*, i.e., relations among components, *functions*, i.e., relations among values, *behavior*, i.e., relations among processes over time, and *non-functional properties*, i.e., relations between the system and its environment. The distinct aspects can be instantiated in many ways, but instantiations are not independent: the system structure may strongly impact on the behavior of the system, which may impact on non-functional properties or functions, and so on. For example, a pipeline architecture may limit concurrency that may result in low performances. Thus, factoring aspects independently can be hard and not always possible.

Software development includes different *phases* that span from *requirements analysis* and *specification* to *design*, *implementation* and *test*. Each phase copes with specific problems at distinct abstraction levels, and uses suitable tools and techniques. Phases are not independent. Many activities performed in different phases overlap and influence each other. The distinction of phases over time, as postulated by the waterfall model, is merely conventional and does not reflect the complex intertwining among phases, which characterize real life processes. The real situation is better represented for example by the process model shown in Figure 2, which captures the effort allocation among phases and process iterations. The vertical slices of the figure show how effort is concurrently allocated to different phases. Each vertical slice corresponds to a process iteration that involves all phases. Each process iteration produces a complete version of the software that improves the former version: inception and elaboration iterations produce early prototypes, construction iterations produce beta versions and release candidates, transition iterations produces software evolutions.

Each development phase requires many *activities*, *analysis*, *abstraction*, *modeling*, *construction*, *refinement*, *documenting*, *testing*, *comprehension*, *refactoring*, *reverse engineering*, etc... System activities must adapt to the different phases and may require different tools and techniques, depending on the phase in which they are performed and the aspects of the developed system. For example, modern modeling methodologies offer many different models suited to distinct de-

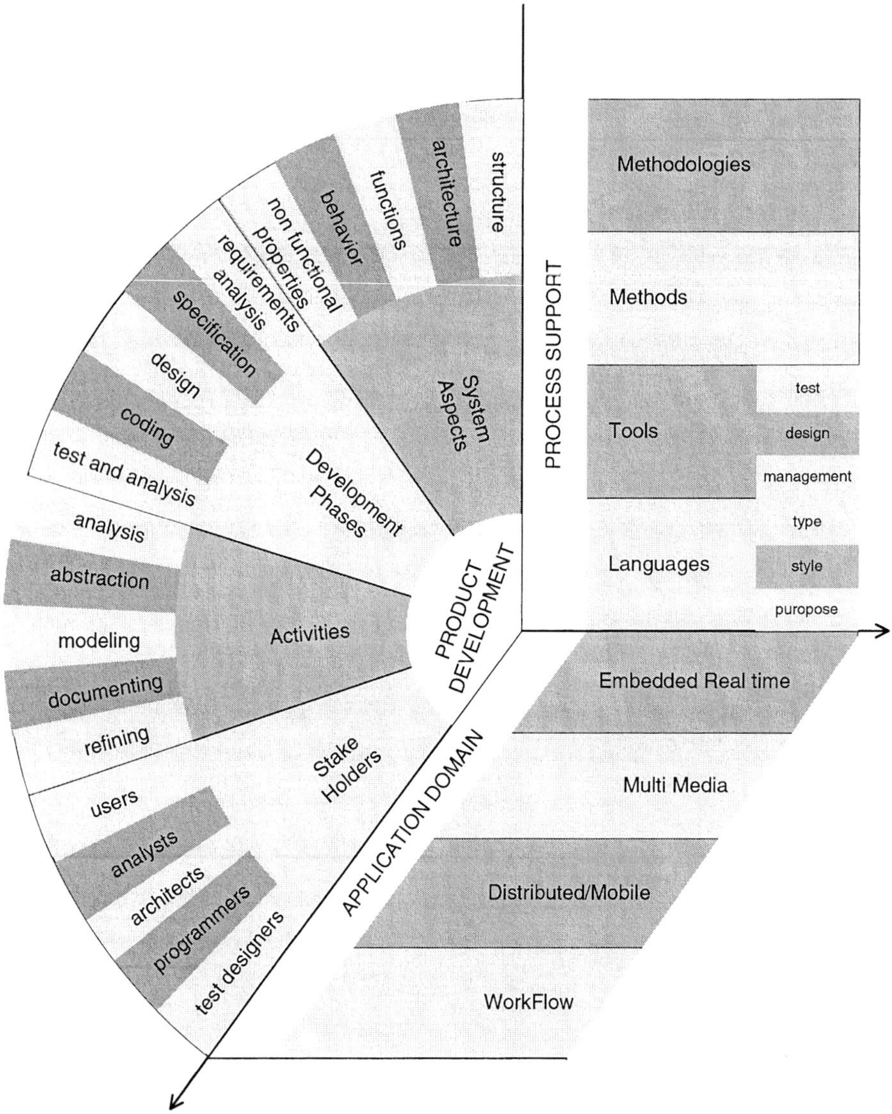

Fig. 1. Software engineering dimensions

velopment stages and to different aspects (Figure 4 at page 447 illustrates this issue in the case of the Unified Modeling Language.)

Software development involves many *stakeholders* who are involved with different roles, speak different languages, have different expectations, and focus on different problems: *users*, *analysts*, *software architects*, *developers*, *test designers*, *managers*, *marketing analysts*, etc... Software engineering must cope with the different needs and must provide a suitable means to coordinate stakehold-

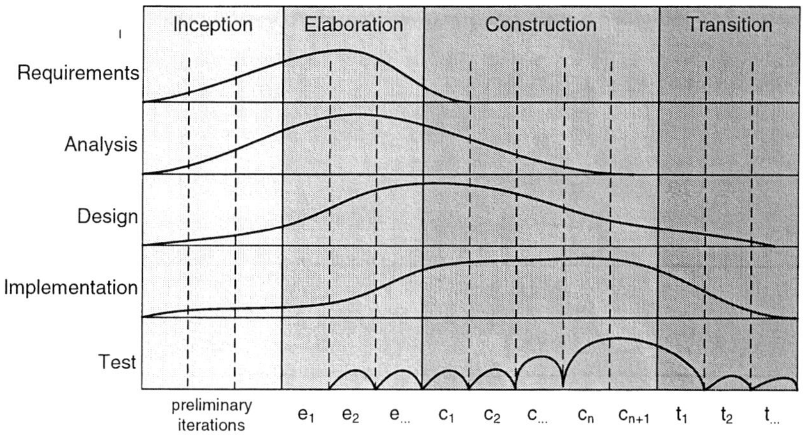

Fig. 2. The unified process model

ers. Lack of communication and comprehension among stakeholders can impact on the overall costs and even on the success of the whole project. For example, difficulties of analysts to understand the user language can lead to ill-designed requirements, while difficulties of users to read design models can lead to poor validation in the early stages, resulting in a final system that does not meet the user requirements. Expertise and needs of stakeholders impact on activities, on phases, and on the way system aspects influence the overall process. For example, familiarity of users with specific notations may influence the organization of system analysis and validation, the presence of an independent quality team may impact on the planning of activities and phases, the lack of familiarity with the application domain and the programming languages required by the user may require specific training, and so forth.

The second dimension considered in the figure is the software process: a suitable blend of methodologies, languages and tools that support the activities of the stakeholders through the development phases of the different aspects of the system.

Software development may involve many *languages* at different stages of development, e.g., specification, design, development, as well as within the same development phase, e.g., class diagrams, statecharts, interaction diagrams during specifications, or different programming languages for different subsystems. Languages involved in the development process are of different *type* (*operational*, *descriptive*, *executable*, etc...) and *style* (*textual*, *visual*, *diagrammatic*, *hybrid*, etc...). Languages affect other factors of the development process. For example, the presence of code generators or tools for analysis for a given language changes the effort required in specific phases, while strong user requirements, due for instance to requests of certification agencies or the legacy of the application, may impact on the organization of activities and roles.

Human activities are supported and complemented by many *tools*, which are responsible for shaping the process: *planning and monitoring*, *configuration management*, *design and specification*, *analysis*, *test case generators* tools are essential in mature development processes. Availability and functionalities offered by tools may determine choices of methodologies, activities, people and phases.

The third dimension considered in the figure concerns the application domains that emphasize different *software characteristics* that further impact on the development process. The development of *interactive*, *reactive*, *embedded*, *real-time*, *distributed*, *mobile*, *batch* systems may require different techniques, tools, languages, phases, and people.

Thus large variety of dimensions and choices that characterize problem oriented disciplines adds a critical dimension to the problems to be solved. Finding solutions to single problems is not sufficient: single techniques must be suitably blended within a general context and changes in one solution may affect many other solutions to different problems. For example, techniques for test and analysis may require different approaches to specification, design and coding, they may change the overall organization of the different development phases, they may require new skills and training, they may be based on new tools that may in turn impact on organization, methodologies, phases, and so forth.

Petri nets, as any solution-oriented discipline, cannot cope with all software engineering problems, but they can help for an unexpected variety of problems that involve all dimensions of software engineering. They can and have been successfully used during software development for modeling and analyzing behavior as well as non functional aspects both in the specification and design phases; They can support analysis, abstraction, modeling and documenting activities; They can provide a means for communication among users and analysts. They have been proposed as specification language for analysis as well as a means for modeling and enacting software processes. They have been used in many application domains that include real-time, workflow, multimedia, and distributed systems. Figure 3 summarizes the software engineering dimensions that can benefit from Petri nets.

Surveying either of the two disciplines would be impossible in the length of a single paper and it is out of the scope of this paper. Main goal of this paper is to illustrate how Petri nets can and have been used through the three outlined dimensions of software engineering, i.e., as models for software development, for describing and enacting software processes, and for solving problems in the specific domain of embedded real time systems.

2 The Role of Models in Software Engineering

The unexpectedly wide spectrum of applicability of Petri nets in software engineering derives from the central role of models in this discipline. Engineering software means describing and reasoning about the problem domain, the software solution, and the process evolution: analysts must capture the problem domain

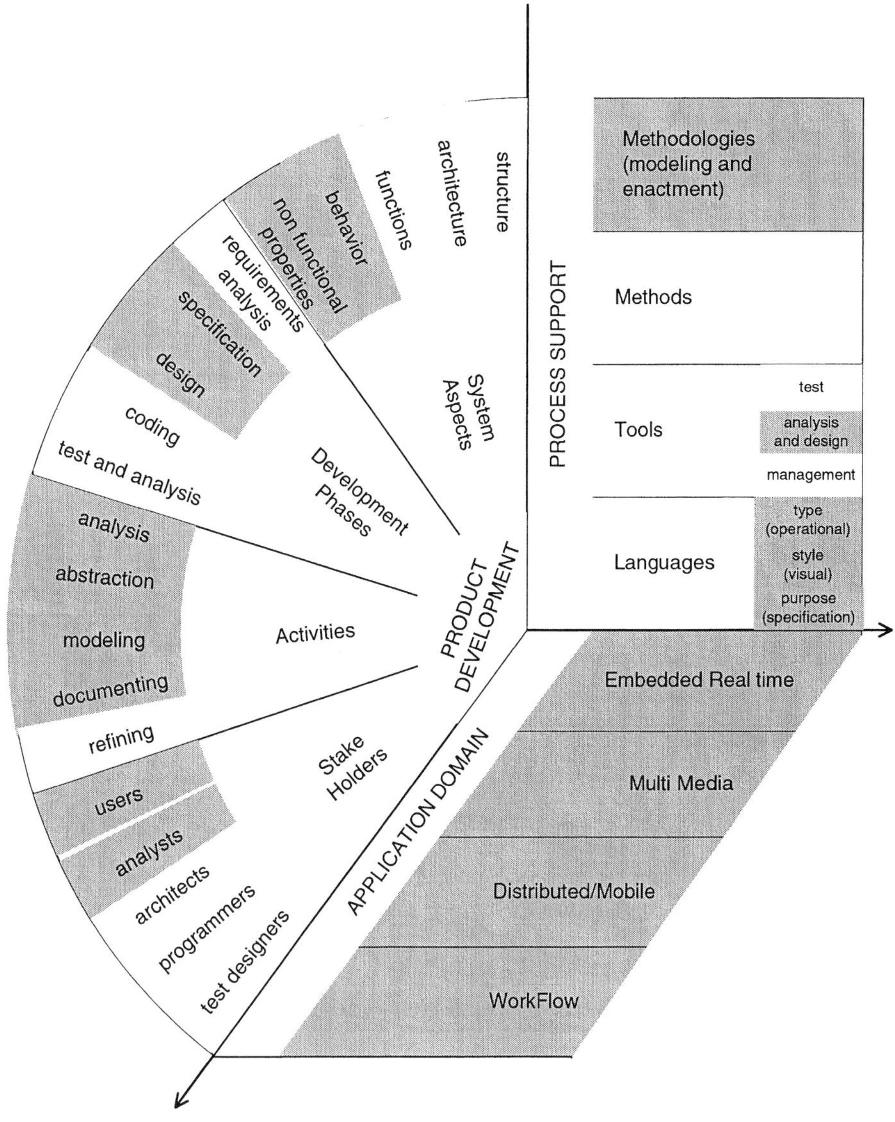

Fig. 3. Software engineering dimensions that can be supported by Petri nets are highlighted with grey background

to understand what has to be solved, designers must describe the system in terms of its architecture, programmers and test designers must understand the data and the control flow through the program, architects must capture the structure of systems to engineer and evolve applications, managers must build a cost model to plan and monitor the process.

Models are essential for communicating and reasoning about systems and must adapt to the people and the properties of interest. Models must capture the relevant system aspects in the different design phases; They must abstract from details that hide the overall picture; They must provide a common language for the different actors; They must support the analysis of the properties of interest. No model suites all phases, aspects, activities, stakeholders, and characteristics of software. The development of a single product usually requires the construction and analysis of many different models. The requirements of a good model depend from the goal of the model: models used for communication among people must be easily understandable for all involved specialists; models used for reasoning about properties must support efficient analysis of the target properties. For example, a detailed data flow model of a program can hardly be used for discussing software requirements or design strategies, but may be excellent for identifying anomalies in the code; conversely, use cases or interaction scenarios provide little help for analyzing program properties, but are often used to discuss the system requirements among software specialists, and between software specialists and domain experts.

During software development we need to discuss and reason about all aspects of the systems: structure, function, behavior, non-functional properties. These aspects cover a wide spectrum of elements, relations and views of the system. Capturing such a variety of elements with a single language requires enormous flexibility and generality that is hardly available in a single language. Universal languages, e.g., natural languages, provide such wide-spectrum coverage, but introduce ambiguities that reduce the possibility of analyzing properties. Modern methodologies, e.g. UML [21], are grounded on sets of complementary languages that cover different aspects for supporting communication and analysis of many aspects at different levels. Sets of languages help describing different aspects at different abstraction levels. In the case of UML, use case and sequence diagrams can be used in the early analysis phases to discuss early requirements with domain experts, class diagrams, collaboration diagrams and Statecharts can support modeling of behaviors during the detailed design of the system, component and deployment diagrams can model design and implementation details, as in Figure 4.

Models are used for communicating design decisions among different stakeholders. To this end, languages must be comprehensible to the involved people and must suite goals such as documentation, analysis, testing, early validation and problem understanding. For example, requirements analysis languages must provide a means for communication among analysts, users, architects, test designers, as well as software specialists, managers and marketing staff. Different attitudes and purposes inspired a large variety of languages that span from textual to visual and diagrammatic, from informal to formal, from detailed to abstract, etc...

Models are used for systems with different characteristics and requirements, e.g., interactive, reactive, embedded, real-time, control, workflow, distributed, mobile, multimedia, web-based systems. Different languages provide means for

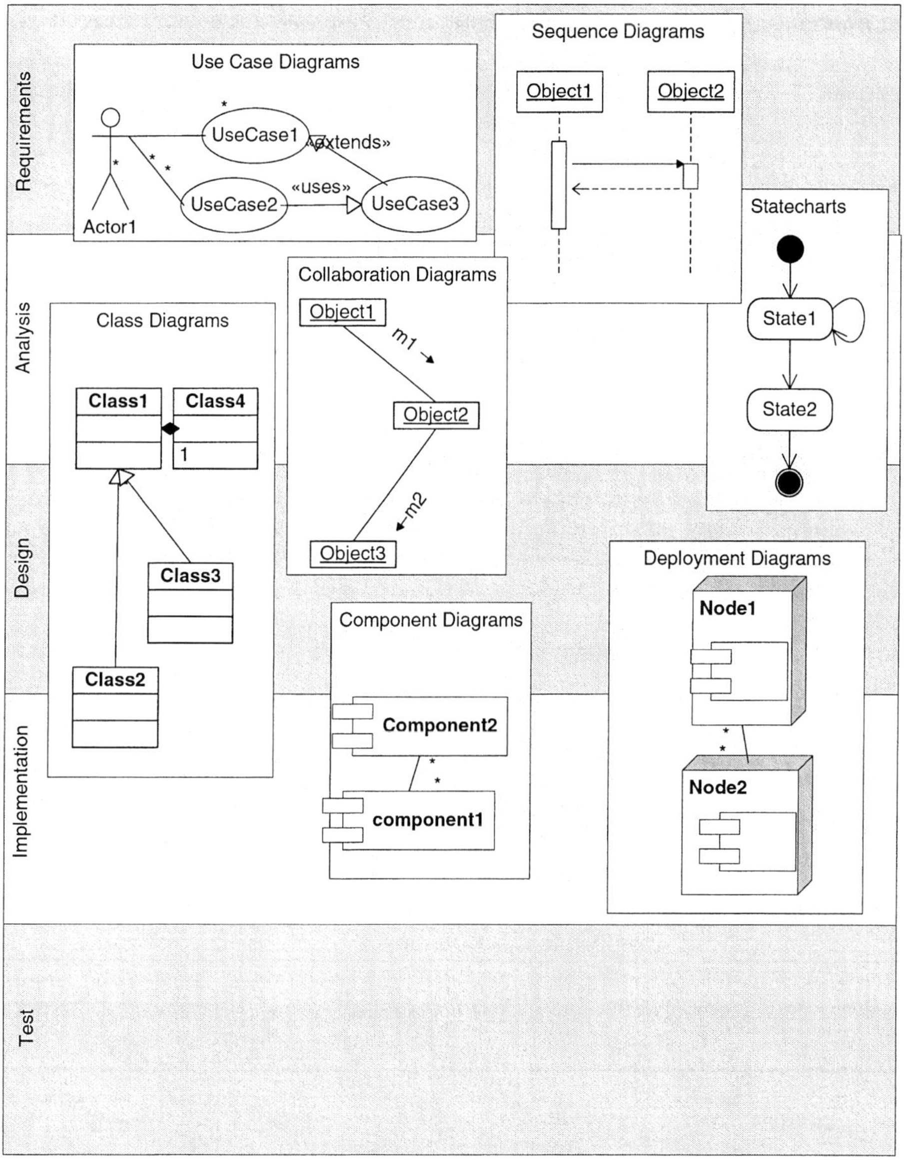

Fig. 4. Support of UML diagram to the software process

dealing with distinct characteristics: for example, Statecharts have been designed for reactive systems, Petri nets and process algebras for concurrent systems, UML-RT for real time systems.

Software systems are extremely complex and can change both during and after development. Constructing a complete and consistent model of the system is almost impossible and never cost effective. Modeling languages must support flexibility, adaptability and instrumentability of specification and design. Software developers need a suitable blend of precision, for supporting analysis, and incompleteness that stems from lack of knowledge of the application domain and evolving design.

3 Petri Nets for Specification and Verification

The variety of uses of models in software engineering requires modeling languages with properties that may be very different if not contradicting. It is difficult to imagine a single language that satisfies all requirements and needs, rather, software engineers tend to uses different modeling languages through the many phases of software development, as well as within the same phase. Modern methodologies, e.g., UML, are based on sets of modeling languages with different characteristics that are integrated in a unifying framework as in Figure 4.

The main challenge in software engineering is rarely to invent yet another modeling language, but more often it is to identify modeling languages suitable to the specific needs, and to integrate them in a coherent framework.

As any other modeling language, Petri nets cannot satisfy all needs of software engineering, but present features that can be appealing in many context within the software development process. The ability of easily modeling concurrency and synchronization aspects, the intuitive graphic notation, the formal semantics that supports powerful analysis capabilities and the availability of several supporting tools make Petri nets an appealing candidate in many situations. However, despite these advantages and the success stories in several application domains, Petri nets are not widely used in software engineering, and successful methodologies often suggest alternative modeling languages, e.g., SDL or Statecharts.

Goal of this paper is not to discuss the mutual diffusion of alternative modeling language, nor to identify the remote causes of relative successes and failures that may change in a few years, as happened many times in the still young history of software engineering. Rather, in this section, we will try to understand limits of Petri nets in coping with software engineering problems aiming at providing directions for further investigation.

Petri nets are available in many variants and extensions that span from place/transition nets to high-level nets and timed Petri nets. Here we focus on untimed models, leaving timed extensions to the next section, where we discuss the usage of Petri nets in the domain of embedded real-time systems.

Place/transition nets provides an essential model of concurrency that can be very useful in addressing specific problems. The absence of constructs for dealing with data and negative conditions results in powerful analysis mechanisms, but limits the applicability to many interesting software engineering problems, and affects scalability.

Figure 5 illustrates the limits of place/transition nets from the modeling viewpoint. The place/transition net of Figure 5 (a) captures the essence of the problem (a process A that can either synchronize with process B or be interrupted by an asynchronous interrupt or by the expiration of a timeout), but cannot model intuitively the duration of the timeout, or the conditions that may govern the timeout or the interrupt handler: if all places *interrupt*, *process A*, and *process B* are marked, the transition that fires is chosen non deterministically among *interrupt handler*, *timeout*, or *synchronization*. We cannot easily specify that transition *timeout* fires only if the token in place *process A* has a given age, or that transition *interrupt handler* fires only if the interrupt has a given priority or is of a given type. Figure 5 (b) illustrates the problems of scaling the description to the presence of different handlers for different interrupts. We can model several instances of interrupts of the same nature increasing the number of tokens, but we cannot distinguish the single tokens, and thus, to keep track of the identity of tokens we need to use different subnets.

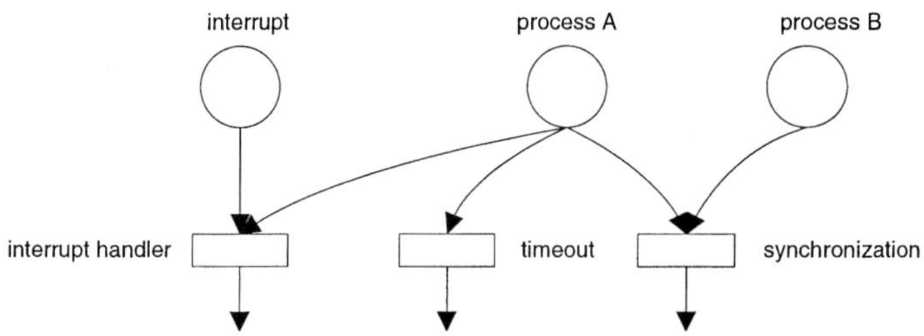

(a) Process A synchronizes with process B unless timed out or interrupted.

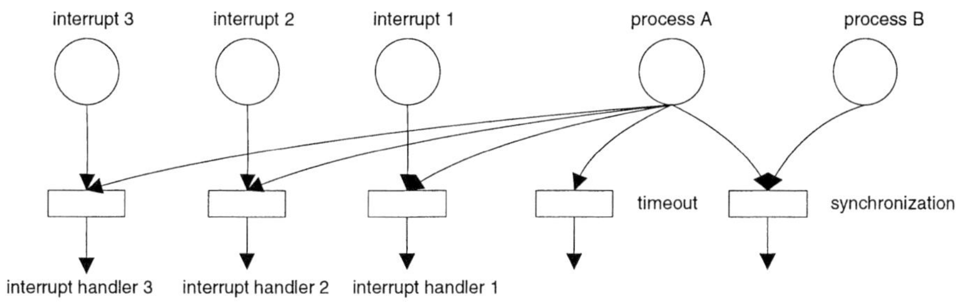

(b) Presence of different interrupt handlers for different signals.

Fig. 5. Limits of place/transition nets as modeling language. (Transitions and places are labeled only for easy referencing)

Place/transition nets have been extended in several ways to overcome their modeling limits: inhibitor arcs, priority, time and predicates help solving different problems. For example, we can use predicates to distinguish different interrupt handling routines, like in Figure 6 that shows a colored Petri net model for three types of interrupt handlers that access resources "L" and "M" in different ways.

The top rectangle groups declarations of colors and variables. In the example, we have two colors: "IH", corresponding to tokens of type "Interrupt Handlers", and “R”, corresponding to tokens of type “Resources”. We have three types of handlers: “i”, “j” and “k”, and two types of resources “l” and “m”. Variable “x” of type “IH” is used in the expressions that annotate arcs to indicate an interrupt handler. Places are annotated with the type of tokens they can contain, and with a marking and an initialization expression. The marking is expressed as a number in a circle and an expression nearby. Place ”L” is initially market with three tokens of type “l” and Place “M” is marked with 2 tokens of type “m”. The figure is a subnet of a larger model, place A is marked by the firing of “ancestor” transitions. The initialization expression is expressed as an underlined expression beside places and helps initializing the net. In the figure, the initial marking corresponds to the effect of applying the initialization expression. Arcs are annotated with expressions that indicate the number and type of tokens flowing on the arcs. The firing of transition “T1” removes a token of type “IH” from place “A” and either a token of type “l” from place “L” if the considered interrupt is of type “i” or “j”, or two tokens of type “l” if the interrupt is of type “k”, and produces an “IH” token in place “B” with the same identity of the token removed from the input place.

The special keyword “empty” indicates that no tokens of that type flows on the arc. For example, handlers of type “j” and “k” release resources of type “l” after the firing of transition “T2” (one and two resources, respectively), while handlers of type “i” returns the resource of type “l” only after the firing of transition “T3”. Similarly, handlers of type “i” and “k” perform actions “T1”, “T2” and “T3”, while handlers of type “j” stop after action “T2” and continue with “T4”.

We can see that colored Petri nets allow identifying different handlers, and to model handlers of different types without affecting the complexity of the net structure, capturing identity and actions with colors and annotations.

The information captured by structure and annotations can be balanced in different ways. For example, we could compact all states of the interrupt handlers in a single state and use colors and predicates to describe the evolution of the computation.

The various extensions of Petri nets and in particular colored Petri nets (or, more generally, high-level Petri nets) are very useful for modeling many software engineering problems, and find some important applications. Unfortunately, adding modeling capabilities and “compacting” the structure solves only some of the limits of the modeling language. Software engineers need flexible, adaptable and scalable modeling notations: they need to change and adapt the notation to the different abstraction levels, to the different stakeholders involved in the development, to the different application domains, and to the evolution of requirements during and after development. They need to cope with large problems and they look for models that help mastering complexity and size. Petri nets, as many other formal methods, rarely provide the flexibility, adaptability, modularity and scalability required in software development. Research in Petri nets moved in two main directions: adding either modeling power or user-friendly interfaces to Petri nets.

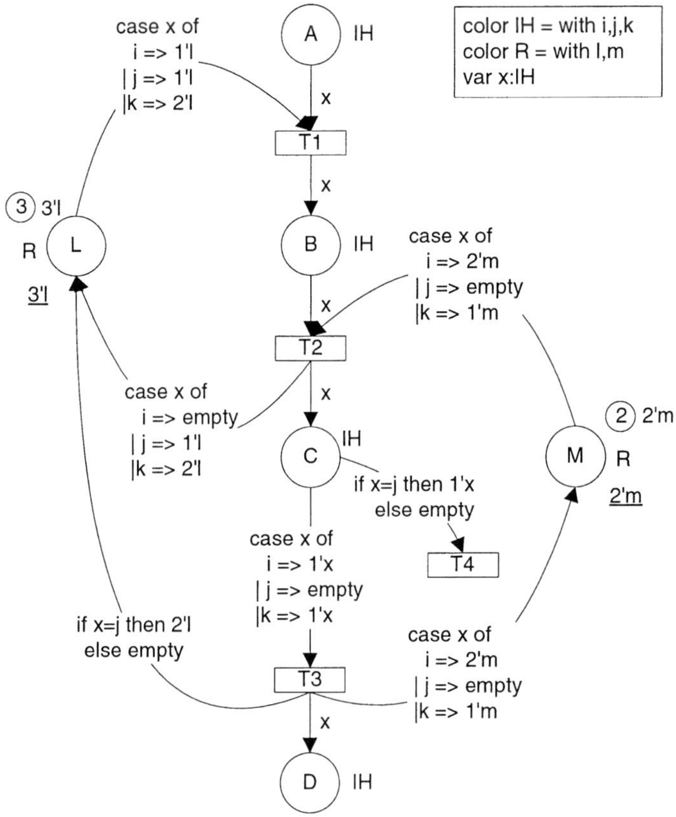

Fig. 6. A colored Petri net model for a set of interrupt handling routines that access resources "L" and "M"

Petri nets have been augmented in many additional ways with hierarchy and object oriented constructs to add flexibility, adaptability and scalability, i.e., the *modeling power* sought by software engineers. Although the different extensions provide useful capabilities and find interesting applications, none of them has a prominent role in software engineering yet. All these useful attempts move in a single direction, ignoring the complexity of the software engineering domain. Software engineering must consider modeling power, precision, analyzability, but also costs, understandability, tool support, and familiarity with the notation. Similarly to programming languages, modeling languages succeed when they present an appealing balance among the different needs. Stochastic Petri nets represent a notable case: they do not present specific features that make them more modular, flexible, adaptable or scalable than other Petri net extensions, but they address a specific problem, namely performance evaluation. When performance is prominent, and then the tradeoff among costs, training, understandability, flexibility, adaptability and scalability is unbalanced towards analysis, software engineers do not hesitate to use stochastic Petri nets in the mosaic of notations adopted in the software development. The (limited, but notable) success of stochastic Petri nets in software engineering may indicate a direction to pursue to increase the applicability of Petri nets, or, from a pessimistic viewpoint, a limit of their applicability: finding specific problems where Petri nets can provide an advantageous solution and adapting Petri nets to the identified problems.

Recognizing that the main advantage of Petri nets, as well as other formal methods, relies not in their intuitive modeling power as communication means, but in their powerful analysis capabilities suggests a different research direction that has been pursued by many scientists: finding "user-friendly" interfaces. The approach is somehow similar to high level programming languages that provide a useful abstraction of the underlying machine and hide the complexity of machine languages, but allow programmers to execute their code, i.e., to take full advantage of the underlying machine language. Similarly, several scientists have been worked on "dual-language" approaches where successful user-friendly specification and design notations are paired with formal models that support powerful analysis capabilities. The approach has been investigated with many specification notations and formal models, including Petri nets. The straightforward operational semantics of Petri nets that supports different types of analysis including "partial" analysis that can be obtained from example by executing partial specifications, provide a strong advantage over other formal models, whose semantics cannot be paired with many specification notations as easily as Petri nets. Moreover, the huge body of knowledge on Petri nets and the immediate modeling of concurrency aspects makes them more appealing than other formal models with operational semantics.

Many scientists defined "compilers" from different specification notations (recently UML, but in the past structured analysis, SDL, etc...) to Petri nets. "Traditional style compilers" that freeze a notation and provide a specific semantics through a fix mapping to Petri nets forget the tradeoff among the variety of aspects to be considered in software engineering: while the primary need of executing the final code overcomes many other requirements, and thus makes acceptable the use of programming languages with fixed and precise semantics, flexibility, understandability and adaptability often overcome the need of analyzability in specifications, thus making it difficult to accept "frozen specification notations". Many projects tried for example to find "the" semantics of structured analysis and provided tools for automatically capturing the identified semantics with a mapping to Petri nets. The resulting frameworks allow for formally analyzing structured analysis, but limit the freedom of the analysts or the architects, who cannot adapt the notation to the specific needs of the end-users, of the application domains or of the changes in requirements. Few attempts survived a few pilot projects.

Several scientists pursue an interesting alternative that consists of providing flexible semantics, i.e., mappings from specification notations to formal models that can be adapted to new needs and requirements. Mappings are given as sets of rules that can be adjusted to meet different needs that result in different interpretation of the same syntactic element or in modifications of the notation.

Flexible approaches seem a better tradeoff for software engineering, but their success is still bound to the ability of identifying a clear advantage in terms of analysis capabilities added with Petri nets.

Software engineering is dealing with new problems that derive from the rapid spread of new applications: pervasive computing, mobile applications, heterogeneous environments, software components that are reused in new unforeseen frameworks, context awareness and new constraints derived from resource bounds like screen size (palm devices), variable bandwidth (mobile computing) present new challenges that may not be easily addressable with traditional techniques. The software engineering community is actively seeking new solutions and ideas to address these new problems. Petri nets as many other modeling languages may provide useful support to some new challenges, thus starting a new time for collaboration among the two communities.

4 Petri Nets for Embedded Real-Time Systems

Petri nets can be used to address the needs of specific application domains. Here we survey embedded real-time systems, which seem particularly well suited for time and stochastic extensions of Petri nets. We will try to summarize the state of art and the future trend in this important domain with respect to possible uses of Petri nets.

In many application domains, software is *embedded* in larger systems. The software is the heart of the systems: it sends control signals and receives feedback. The behavior of these systems is time dependent: the correctness of the software cannot be expressed merely in terms of functional relations between inputs and outputs, but depends on the instants at which the results are produced. A functionally correct result produced too late may be wrong. For example, a drive-by-wire system that computes the correct maneuver for avoiding an obstacle too late, e.g., after crashing into the obstacle, is obviously wrong regardless of the produced value. Results produced too early may be wrong as well. For example the signal for controlling the delivery of power to an electrical engine cannot be produced too early, otherwise the engine may reach a wrong speed at a wrong time.

Missing deadlines can have different consequences for different systems. In some cases, it can be tolerated if it does not happen too frequently, while in other cases, results must be always available within the deadlines. Although the distinction is not sharp, we often classify real-time systems as *hard* and *soft*. Hard real time systems do not tolerate missing deadlines. An approximate results produced within the deadline may be preferable to an exact result produced too late. This is the case of many control systems that must send control signals when they are needed by the controlled systems: we prefer a vehicle to break a bit too suddenly, because the control software computes an approximate control signal, but avoids the collisions, to a vehicle that crashes because the ideal signal is computed after the collision.

Soft real time systems can tolerate some late results. For example, voice packets must be received with specific frequencies for reproducing the correct voice signal, but a small percentage of late packets can be ignored without appreciable degradation of the reproduced voice signal.

Hard and soft real time constraints often coexists in the same systems: the GPS signal used by the drive-by-wire system as well as by the position display on the driver console owns hard real time constraints in one case, since a late signal to the drive-by-wire system may cause the vehicle to crash, and soft real time constraints in the other, since a late signal to the driver display may not be even perceivable to the end user.

The distinction between hard and soft real time system is important to identify suitable analysis techniques: performance analysis may be enough for soft real time systems, but is rarely sufficient for hard real time systems.

Embedded real time systems can be composed of several concurrent subsystems. They include at least the controller and the controlled system, but more often, both the software controller and the controlled system include several concurrent subsystems. Different components are often of different nature, the behavior of the components of the controlled system is usually time-continuous, while the behavior of the controlling software is usually time-discrete. The controlled system and the controlling software interact through special purpose devices (sensors and actuators) that may be responsible of failures. Moreover, the timing of the system depends on elements that are usually not considered in "traditional" systems: hardware, operating system, and middleware. Abstracting from such elements may not be possible for not trivial real-time systems.

Embedded real time systems present many new challenges. Modeling and analyzing systems in the early development phases requires models and analysis technique that can capture the subtle intertwining between functional and timing aspects, and that can model both continuous and discrete timing. The correspondence between requirement specifications and code must take into account limited availability of resources and constraints that can derive from the hardware and software platform. Analysis techniques must cope with new properties that include timing and safety properties.

Petri nets were originally proposed for modeling concurrent systems abstracting away from timing aspects. Extensions of Petri nets for dealing with time have been studied since the early seventies. We can identify two different approaches: *timed Petri nets* that augment Petri nets with deterministic time, and *stochastic Petri nets* that augment Petri nets with time probabilities.

Timed Petri nets have been proposed as early as 1974 by Ramchandani [22], and by Merlin and Farber in 1976 [9]. Since the early proposals, Petri nets have been extended with time in several ways, by adding time to either places or transitions or both, by interpreting time as firing delay or firing duration, by adding a single time value or a time set (interval) to transitions or places. Different models satisfy different needs, but they are all substantially equivalent from the semantic viewpoint.

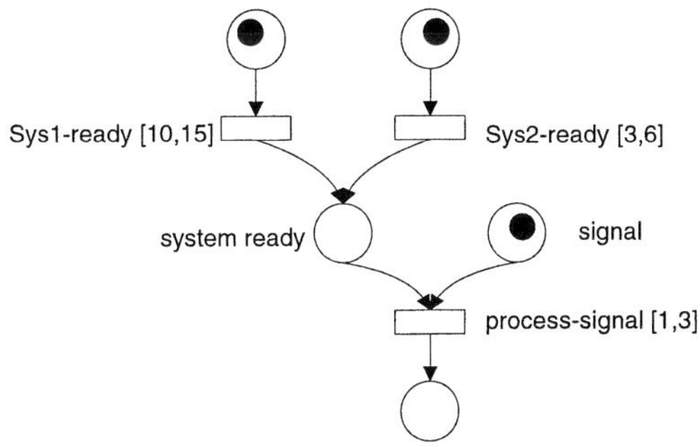

Fig. 7. A simple timed Petri net. Transitions are augmented with pairs of numeric values that represent the minimum and maximum firing time relative to the enabling time, i.e., the instant at which all input places are marked. Labels are added only for referencing. Marking is represented by black dots in places

Extending Petri nets with time can greatly affect the semantics. While *weak time semantics* does not affect the locality of enabling, *strong time semantics* violates the locality principle. Informally, *weak time semantics* considers the time constraints as instants at which the modeled events will happen, if they happen, while *strong time semantics* considers the time constraints as instants at which the events must happen.

Let us consider for example the timed Petri net of Figure 7 that represents a simple systems where an incoming signal can be handled by two different signal handlers (*Sys1* and *Sys2*). *Sys1* required from 10 to 15 time units to become ready, *Sys2* requires from 3 to 6 time units. The signal is processed in 1 to 3 time units.

If we consider each transition "locally", i.e., ignoring relations among firings that may derive from time constraints, both transitions *Sys1-ready* and *Sys2-ready* are enabled. If transition *Sys1-ready* fires at time e.g. 10, transition process signal is enabled in the interval $\langle 11, 13 \rangle$, i.e., between 1 and 3 time units after the enabling at time 10. The firing of transition *process-signal* at time e.g. 12 consumes the tokens. Thus, the token in place *signal* is not available any more. If we now consider again transition *Sys1-ready* enabled between 3 and 6, it can fire e.g. at time 5. The considered sequence of firings can be ordered to obtain a monotonically non-decreasing sequence with respect to time: *Sys1-ready* at time 5, *Sys2-ready* at time 10, *process-signal* at time 12, obtaining a legal firing sequence according to weak time semantics. This is true in general: each firing sequence obtained by considering transitions locally is equivalent to a legal time monotonically non-decreasing firing sequence according to weak time semantics. This means that analysis performed on the underlying Petri net produces results that are valid also for the timed net.

In the considered example, the obtained sequence represents the case in which *Sys1* becomes available for handling the signal at time 5, but does not handle the signal for some reasons that are not explicitly captured by the model. The signal is handled later by *Sys2* that becomes available at time 10.

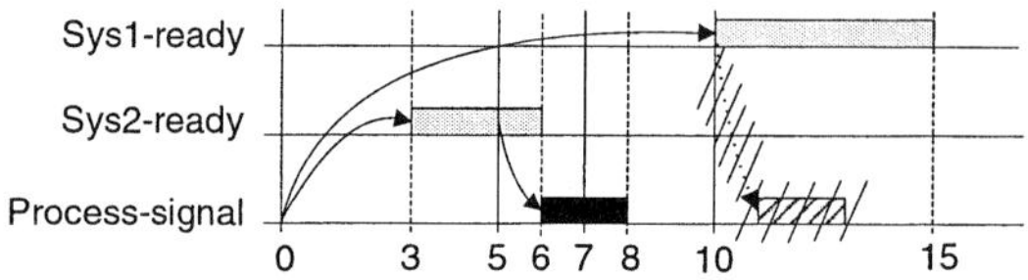

Fig. 8. Weak vs. strong time semantics

Unfortunately, strong time semantics does not have this nice property. The same firing sequence cannot be ordered according to strong time semantics, as illustrated in Figure 8. Initially both transitions *Sys1* and *Sys2* are "locally" enabled, but transition *Sys1* must fire before its deadline (time 6). The firing of transition *Sys2* e.g. at time 5 enables transitions *process-signal* within the time interval $\langle 6, 8\rangle$. Transition *process-signal* must fire before time 6. Transition *Sys1* can fire only after the firing of transition *process-signal*, e.g., at time 10. Since transition *process-signal* is forced to fire before time 8, removing the token from place *signal*, the token produced by the firing of transition *Sys1* cannot enable transition *process-signal*, differently from the case of weak time semantics.

The model used in Figure 7 that associates firing fixed time intervals to transitions cannot capture all aspects of real time systems. In particular, we cannot model complex intertwining between timing and functional aspects. Let us assume for example that the processing of the signal depends on the load of the system that handles it, and that the choice of the handling system depends on the characteristics of the incoming signal. The fixed time interval associated to transition *process-signal* that indicates the minimum and maximum firing time as constants can approximate the modeled system by indicating an upper and a lower bound, and cannot express complex conditions for selecting the signal handler depending on the nature of the signal.

Complex intertwining between timing and functional aspects can be modeled by merging timed and high-level Petri nets (HLTPN). HLTPNs associate data (and timestamps) to the tokens, and predicates, actions and time functions to transitions, as shown in Figure 9.

Soft real time systems can be modeled and analyzed with stochastic Petri nets that were first proposed in the late seventies by Sifakis [23] and later extended by many scientists. Stochastic Petri nets augment transitions with a distribution of probability that the transition will fire. Figure 10 shows a simple example of generalized stochastic Petri nets: a processing task that may or may not require instrumentation, and may or may not require elaboration. The two choices are represented with the two pairs of conflicting transitions *need instrumentation*, *instrumentation ok*, and *need processing*, *system ok*. Black transitions indicate immediate transitions, i.e. transitions that fire immediately, representing instantaneous decisions, while white transitions indicate timed transitions, representing the termination of actions with given durations. Transitions are associated with a priority π and a weight W. Immediate transitions fire first, while timed transitions fire only when no immediate transition is enabled. Pri-

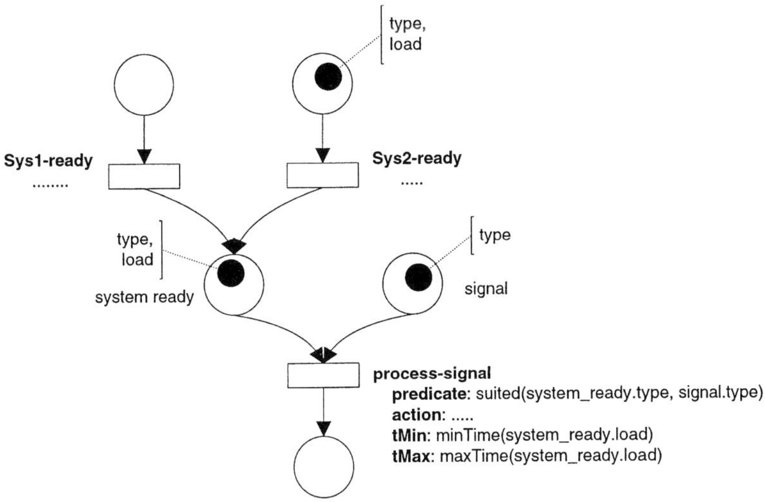

Fig. 9. A simple HLTPN. Data associated with tokens are represented with types. Predicates, actions and time intervals associated with transitions are partially given only for transition *process-signal*. The predicate requires the evaluation of a boolean function *suited* that computes the suitability of the system to process the signal. The time interval can be computed by evaluating functions *minTime* and *maxTime* that compute the minimum and maximum firing time according to the load of the system

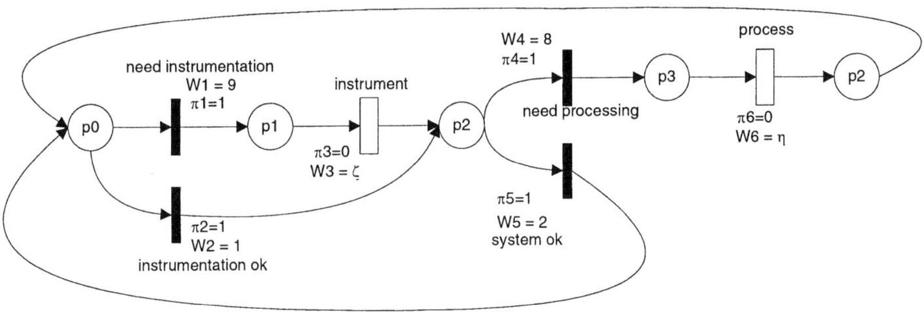

Fig. 10. A simple generalized stochastic Petri net

ority defines a (partial) order of firings among transitions of the same kind. The weight indicates the frequency of firings for immediate transitions with equal priority, and a distribution of probability that describes the firing time of timed transitions.

In the example, transitions *need instrumentation* and *instrumentation ok* have the same priority and fire with a ration of 9:1, while transitions *need processing* and *system ok*, also with equal priority, fires with a ration 8:2.

Stochastic Petri nets support powerful performance analysis that determines their success in important application domains. Timed Petri nets support time reachability analysis.

Real time systems are quickly evolving: the spread of SoC (System on a Chip), the vanishing distinction of hardware and software components, the increasing use of COTS (Components-Of-The-Shelf) in complex real time systems, the introduction of Internet connectivity introduce new challenges that call for new methodologies and techniques and open new potentialities to Petri nets as well as other formal methods.

5 Software Processes

Complex software is developed by a set of specialists that use many techniques and tools, and collaborate over a long period of time to design, develop and maintain a suitable product. Often, nobody knows all the details of a software product, but different actors share partial views of the system. For example, managers and analysts may have an abstract view of some aspects of the overall system, but may not know all implementation details, while architects, designers, programmers and test engineers may have a detailed view of part of the software, but not be familiar with other parts.

People, techniques and tools must be suitably coordinated and organized over time to assure the success of a project, i.e., the developed of the right product within time, resource and environmental constraints. The overall organization of the activities required to develop, test and maintain a software product is called a software process.

A software process consists of a set of interacting software engineering activities aimed at producing (and maintaining) a software product. A key property of a software process is *visibility*, i.e., the ability of examining progresses and results. Process visibility gives the possibility to monitor and steer the process towards its goals. Visibility is often achieved by identifying different phases and associating activities and phases with the production of intermediate artifacts, such as, requirements specifications, design specifications, code and quality reports, which are often associated with process milestones.

Large projects span over many months. Requirements are seldom clear at the beginning of the project. Usually a first core of requirements is detailed and expanded through the process following the increasing understanding of the problem and the solution, and adapting to the evolution of the domain. Systems are rarely developed as single monolithic products. More often, systems are developed incrementally though several iterations that produce many prototypes and releases that increasingly approximate the final product. Process phases can seldom be organized as separate development steps, as postulated by the waterfall process model. They often overlap with complex interaction patterns that must be suitable organized and monitored. Software projects usually involve separated teams that work concurrently on different phases of the process.

The large variety of situations results in many different requirements that cannot be fully captured by a standard process. Each project has its own properties, and requires a specific process. The definition and implementation of a software process is a complex activity that can and shall be suitably programmed and executed [24]. Software processes must be suitably described to guide and coordinate the key activities, and to push forward repeatability and controllability of the processes. Rigorous software process descriptions enable the development of tools to enact process descriptions thus automating coordination of activities, tools and people. The definition of precise process models is also referred to as *software process programming.*

The goal of software process programming is the creation of *process-centered software engineering environments* (PSEEs), i.e., information systems that support the enactment of software processes. The core of a PSEE is a *Process Modeling Language* (PML), i.e., the language used to describe the target processes. Ambriola et al. outline the main requirements for a PML [25], which can be summarized as follows:

Modeling concurrency: Concurrency is intrinsic in software processes. PMLs must be able to clearly capture the concurrency of activities and their synchronization.

Modeling products: The artifacts produced during software development are complex and strictly interrelated. For example, a test report that points out a failure of a software system must be related to the tested software version and to the revealing test case, which in turn is related to a number of implementation artifacts (e.g., test drivers and stubs). PMLs must take into account the structure of the artifacts involved in the process and their mutual relationships.

Managing tool integration: Software process activities are supported by several tools (e.g., editors, compilers, debuggers, configuration management systems, and so forth). Mechanisms to control at a fine-grained level the tools involved with the process are essential for managing the interactions between a PSEE and its users in an effective way. PMLs must provide both active and reactive mechanisms to allow the PSEE to send messages to external tools and to react to messages from external tools, respectively.

Supporting process enactment: Software process descriptions must be interpreted to provide automatic support to processes when they are executed (process enactment). PMLs must produce operational descriptions with well-defined non-ambiguous semantics.

Supporting analysis: PMLs must support verification of important properties, e.g., absence of deadlocks, against process descriptions.

Supporting evolution: Software process improvement is an important issue in software engineering and has been the target of important industrial and research activities in the last years. For example, the SEI Capability Maturity Model (CMM [26]) defines a framework for assessing the level of maturity of a software process. This framework consists of five maturity levels. At the fist level, software processes are chaotic and uncontrolled, activities

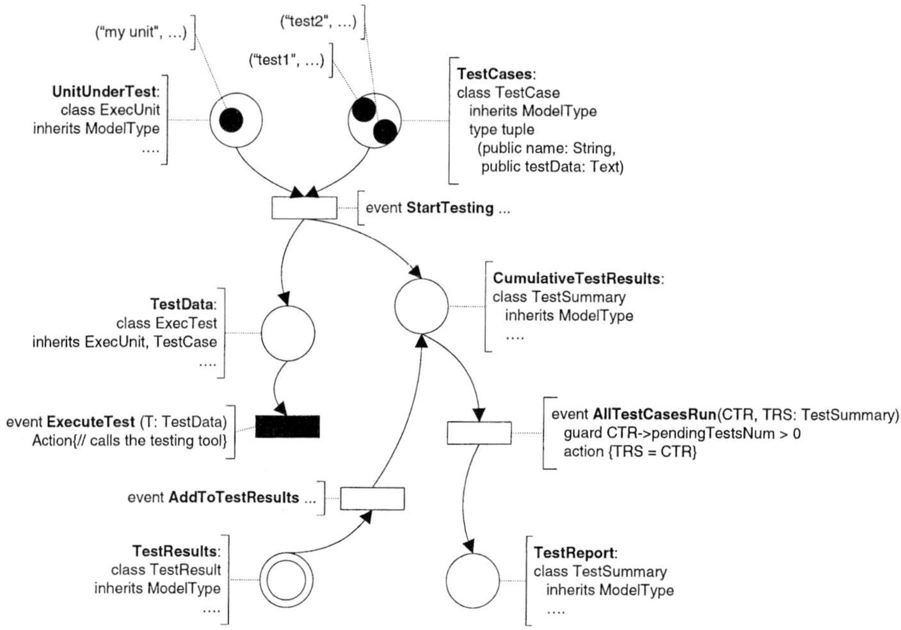

Fig. 11. Description of a generic unit test session in SLANG

are carried without any explicit guideline and there is no description of the process. The introduction of methods and technologies such as configuration management, quantitative measurement, quality control and process description, is expected to gradually increase the maturity of a software process and correspondingly its capability of dealing successfully with complex software projects. At level 5, software processes are continuously improved based on the experience and data accumulated over time. Mature software processes must support evolution of the the process descriptions. To this end PMLs must possess reflexive features allowing to modify process descriptions either off-line or on-the-fly.

Petri nets present several nice properties that make them particularly appealing as PML: They have a precise semantics; They provide an intuitive way of modeling concurrency and non-determinism; Their marking supports easily modeling of the process state, thus facilitating the representation of milestones and conditional choices; Their operational semantics allows to easily represent and analyze process enactment; Their intuitive visualization provides an excellent communication means; There exists a large body of theory that support analysis of many properties under different assumptions; There are many supporting tools.

Basic Petri nets have been extended and specialized in several ways to cover all requirements of a PML, by adding reflexivity, features for modeling process artifacts, and mechanisms for managing tool integration. Extensions of Petri nets to cope with process modeling aspects are illustrated in the example of Figure 11 that shows a model of a generic unit test session in SLANG (the PML of the SPADE PSEE [19]). The availability of both the unit under test and a set of corresponding test cases triggers the testing activity that starts setting up the testing environment and starting the tracking of cumulative test results. The test results are incrementally cumulated while executing the test cases. When all test cases have been executed, a test report is generated.

SLANG extends high-level Petri nets, using tokens to represent the process data. SLANG tokens are structured objects whose types are defined in the traditional object-oriented style, i.e., as a data structure that can be accessed through a set of exported operations. The net places are associated with a type (a.k.a. class) and they can contain only objects of the associated type. All SLANG types are organized in a type hierarchy as specified by the inheritance relation. The object oriented-paradigm makes it possible to describe the structure of software artifacts. For example, in the figure, the test cases are described as tuples with two fields: a string that identifies the test case by name and a text that describes the test data.

The transitions that in SLANG are called *events* are associated with guard predicates that control their execution, and actions that that describes the effect of the firings on the tokens. A transition can fire if enabled by the associated guard predicate evaluated on the tokens in the input places. Its firing removes the tokens from the input places and produces tokens in the output places according to the associated action. For example, in the figure, the transition *AllTestCases-Run* is not enabled (guarded) until there are still test cases to execute. Its firing produces the test report from the incrementally generated test summary.

SLANG uses special *black transitions* and *user places* to integrate CASE tools. Black transitions send asynchronous messages to external tools as part of their action. User places (double circles) change their content as a result of an event that happens in the user environment. For example, in the figure, the event *ExecuteTest* is a black transition whose action calls an external testing tool for executing the test cases. After the execution of each test case, the external tool will produce a token in the user place *TestResults*, thus allowing the process to progress.

SLANG provides reflexive features for dynamically modifying a process description. Figure 12 shows a SLANG type hierarchy that includes both the predefined SLANG types and the user defined types for the previous examples. All types that participate to a process description derive from the predefined type *Token* that defines the set of properties common to all tokens. All user defined types inherit (directly or indirectly) from the predefined type *ModelType* which is a direct descendent of *Token*. The set of predefined types includes three additional types: *Activity*, *Meta type* and *Active copy* that allow to access the activity definitions, the user type definitions and the instantiated copies of a process during enactment, respectively. The specification of how to modify the process can be part of the process description itself.

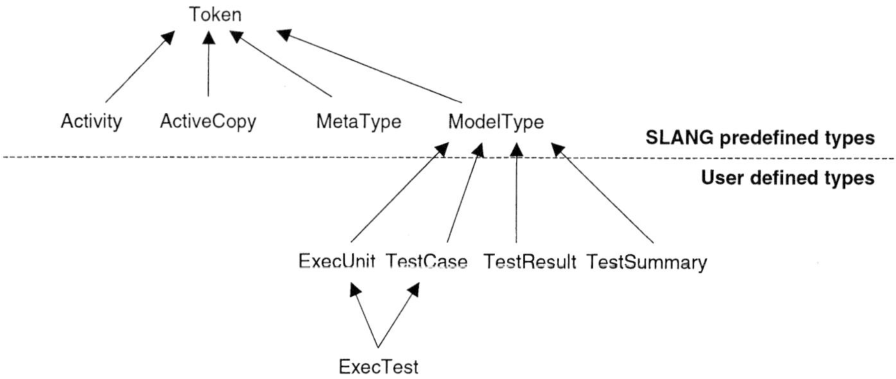

Fig. 12. A SLANG type hierarchy

Thirty years of research solved many problems in software process modeling, but some are still open: tolerability to inconsistencies and incompleteness, non-intrusiveness of PSEE, inconsistency management are some of the problems where Petri nets can find new applications.

6 Further Readings

The literature on software engineering and Petri nets is immense and finding a good compass is hard. Here we try to indicate some doors to access the enormous body of knowledge for identifying areas of common interest for software engineers and Petri net experts following the schema of this paper.

A good way for understanding problems and dimensions of software engineering is the volume "Future of Software Engineering" published in 2000 [27]. The introduction illustrates the many dimensions of the discipline, while the many papers present the current trends of the most important areas.

Modeling languages have been widely studied and it is difficult to identify a good survey. Interested readers can find a general overview in software engineering handbooks, e.g., in [28], [29] or [30]. A good survey of formal methods is given by Wing [31], while an interesting discussion on the role of formal methods in software engineering is proposed by Saiedian [32]. The different models that comprise the UML approach are illustrated in many book, e.g., [21].

A comprehensive overview of Petri nets is given in [33]. Colored Petri nets together with a sample of industrial applications are presented in Jensen's books [34–36]. The volume edited by Agha and De Cindio discusses the use of Petri nets in the object oriented framework [37].

Approaches for mapping various notations to Petri nets have been proposed by many authors: [38–42]. Rule based mappings have been proposed by Paige [43] and Baresi et al. [44].

Stochastic Petri nets are presented in the books by Bause and Kritzinger and by Ajmone Marsan et al. [45, 46]. Timed Petri nets are discussed in the classic paper by Merlin and Farber [9], while high-level timed Petri nets (ER nets, in the paper) and weak and strong time semantics have been introduced by Ghezzi et al. [11]. Time Petri nets are overviewed also in in the book by Nissanke [12]. Time reachability analysis is discussed by Berthomieu and Diaz [10] for timed Petri nets, and by Ghezzi et al. for high-level timed Petri nets [47].

The term "software process" was first proposed by Osterwiel in his seminal paper [24]. Various PSEE are discusses in many papers, e.g., [48, 49, 19, 18]) The possibility of using Petri nets as a PML for describing workflows of business processes and many results related to the use of Petri nets for this purpose have been described in Chapter 1[2] of this book.

7 Conclusions

The meeting of problem- and solution-oriented disciplines can lead to important progresses in both areas. Petri nets provide an excellent means for modeling concurrent aspects and have been extended in many ways to cope with many problems. Petri nets have been successfully applied many times to several software engineering problems. However, the two disciplines do not go through a period of particularly strong cross fertilization. This paper tried to overview some aspects of software engineering, pointing to aspects where Petri nets have been or can be proposed as solutions to critical problems. We hope to have provided few ideas to foster new fruitful collaborations between the two disciplines.

Acknowledgment

This paper springs from a course held at the Advanced Course on Petri Nets (ACPN) 2003. A preliminary version of this material has been presented as a tutorial at the 24th International Conference on Application and Theory of Petri Nets (ICATPN 2003) held in Eindhoven in June 2003 jointly with Gregor Engels, who greatly contributed to the preparation of the tutorial and the organization of the content. We would like to thank Gregor for the invaluable preliminary work that helped shaping the course and the derived paper.

References

1. Young, M.: Neat models of messy problems: Notes on the interplay between solution- and problem-centered disciplines, and more particularly on the interaction between Petri net research and software engineering research. International Journal of Computer Systems Science and Engineering **16** (2001)
2. Oswald, H., Esser, R., Mattmann, R.: An environment for specifying and executing hierarchical petri nets. In: Proceedings of the 12th International Conference on Software Engineering. (1990) 164–172

[2] This is a reference to the Chapter of this book authored by van der Aalst and related to workflow modeling.

3. Duri, S., Buy, U., Devarapalli, R., Shatz, S.: Application and experimental evaluation of state space reduction methods for deadlock analysis in ada. ACM Transactions on Software Engineering and Methodology **3** (1994) 340–380
4. Jahnke, J., Schafer, W., Zundorf, A.: Generic fuzzy reasoning nets as a basis for reverse engineering relational database applications. In: Proceedings of the European Conference on Software Engineering 1997, Springer Verlag (1997)
5. Bastide, R., Palanque, P.: A Petri net based environment for the design of event-driven interfaces. Lecture Notes in Computer Science **935** (1995)
6. Leveson, N., Stolzy, J.: Safety analysis using petri nets. IEEE Transactions on Software Engineering **SE-13** (1987) 386–397 19 refs.
7. Azema, P., Juandle, G., Sanchis, E., Montbernard, M.: Specification and verification of distributed systems using PROLOG interpreted Petri nets. In: Proceedings of the 7th International Conference on Software Engineering, IEEE Computer Society Press (1984) 510–519
8. Suzuki, T., Shatz, S.M., Murata, T.: A protocol modeling and verification approach based on a specification language and petri nets. IEEE Transactions on Software Engineering **16** (1990) 523–536
9. Merlin, P., Faber, D.J.: Recoverability of communication protocols. IEEE Transactions on Communication **24** (1976) 1036–1043
10. Berthomieu, B., Diaz, M.: Modeling and verification of time dependent systems using time Petri nets. IEEE Transactions on Software Engineering **17** (1991) 259–273
11. Ghezzi, C., Mandrioli, D., Morasca, S., Pezzè, M.: A unified High-Level Petri Net formalism for time-cricical systems. IEEE Transactions on Software Engineering **17** (1991) 160–172
12. Nissanke, N.: Realtime Systems. International series in computer science. Prentice Hall (1997)
13. T.D.C., L., A., G.: Synchronisation and storage models for multimedia objects. IEEE Journal on Selected Areas in Communications **8** (1990) 413–427
14. Wahl, T., Rothermel, K.: Representing time in multimedia systems. In: Proceedings of the International Conference on Multimedia Computing and Systems, Boston, USA. (1994)
15. Willrich, R., Saqui-Sannes, P.D., Senac, P., Diaz, M.: Multimedia authoring with hierarchical timed stream Petri nets and Java. Multimedia Tools and Applications **16** (2002) 7–27
16. Engels, G., Sauer, S.: Object-Oriented Modeling of Multimedia Applications. In: Handbook of Software Engineering and Knowledge Engineering (S. K. Chang ed.). Volume 2. World Scientific (2002) 21–52
17. Vazirgiannis, M.: Interactive multimedia documents: modeling, authoring, and implementation experiences. Volume 1564 of Lecture Notes in Computer Science. Springer-Verlag (1999)
18. Emmerich, W., Gruhn, V.: FUNSOFT Nets: a Petri-Net based Software Process Modeling Language. In Ghezzi, C., Roman, G., eds.: Proceedings of the 6th ACM/IEEE Int. Workshop on Software Specification and Design (IWSSD), Como, Italy, IEEE Computer Society Press (1991) 175–184
19. Bandinelli, S., Fuggetta, A., Ghezzi, C., Lavazza, L.: SPADE: An environment for software process analysis, design, and enactment. In Nuseibeh, B., Finkelstein, A., Kramer, J., eds.: Software Process Modelling and Technology. John Wiley and Sons (1994) 33–70

20. Florin, G., Natkin, S.: Generalization of queueing network product form solutions to stochastic Petri nets. IEEE Transactions on Software Engineering **17** (1991) 99–107
21. Booch, G., Rumbaugh, J., Jacobson, I.: The Unified Modeling Language User Guide. 1 edn. Addison-Wesley, Reading, Massachusetts, USA (1999)
22. Ramchandani, C.: Analysis of asynchronous concurrent systems by timed Petri nets. Technical Report Project MAC Tech. Rep. 120, Massachusetts Institute of Technology (1974)
23. Sifakis, J.: Use of Petri nets for performance evaluation. Acta Cybernetica **4** (1978) 185–202
24. Osterweil, L.: Software processes are software too. In: Proceedings of the 9th International Conference on Software Engineering, IEEE Computer Society Press (1987) 2–13
25. Ambriola, V., Conradi, R., Fuggetta, A.: Assessing Process-Centered Software Engineering environments. ACM Transactions on Software Engineering and Methodology **6** (1997) 283–328
26. Paulk, M., Curtis, B., Chrissis, M., Weber, C.: Capability maturity model, version 1.1. IEEE Software **10** (1993)
27. Finkelstein, A., ed.: The future of software engineering (part of the Proceedings of the 22th International Conference on Software Engineering). ACM Press, NY (2000)
28. Ghezzi, C., Jazayeri, M., Mandrioli, D.: Fundamentals of Software Engineering. 2 edn. Prentice Hall, Englewood Cliffs (1999)
29. Sommerville, I.: Software Engineering. 6th edn. Addison-Wesley (2001)
30. van Vliet, H.: Software Engineering: Principles and Practice. John Wiley & Sons, Chichester (1993)
31. Wing, J.: A specifier's introduction to formal methods. Computer **23** (1990) 8, 10–22, 24
32. Saiedian, H.: An invitation to formal methods. IEEE Computer **29** (1996) 16–30
33. Murata, T.: Petri nets: properties, analysis, and applications. In: Proceedings of the IEEE. Volume 77. (1989) 541–580
34. Jensen, K.: Coloured Petri Nets: Basic Concepts, Analysis Methods and Practical Uses, vol. 1: Basic Concepts. EATCS Monographs on Theoretical Computer Science. Springer-Verlag (1992)
35. Jensen, K.: Coloured Petri Nets: Basic Concepts, Analysis Methods and Practical Uses, vol. 2: Analysis Methods. EATCS Monographs on Theoretical Computer Science. Springer-Verlag (1995)
36. Jensen, K., ed.: Coloured Petri Nets: Basic Concepts, Analysis Methods and Practical Uses, vol. 3: Practical Uses. Number 1217 in Lecture Notes in Computer Science. Springer-Verlag (1997)
37. Agha, G., Cindio, F.D., Rozenberg, G.: Concurrent object-oriented programming and Petri Nets: advances in Petri Nets. Volume 2001. Springer-Verlag Inc., New York, NY, USA (2001)
38. France, R.: Semantically extended data flow diagrams: A formal specification tool. IEEE Transactions on Software Engineering **18** (1992) 329–346
39. Fencott, P., Galloway, A., Lockyer, M., O'Brien, S.: Formalising the semantics of Ward/Mellor SA/RT essential models using a process algebra. Lecture Notes in Computer Science **873** (1994)

40. Petersohn, C., Huizing, C., Peleska, J., de Roever, W.: Formal semantics for Ward and Mellor's transformation schemas and its application to fault tolerant systems. International Journal of Computer Systems Science and Engineering **13** (1998) 131–136
41. Shi, L., Nixon, P.: An improved translation of SA/RT specification model to high-level timed Petri nets. In Gaudel, J.W.M.C., ed.: FME '96: Industrial Benefit and Advances in Formal Methods. Volume 1051 of LNCS., Springer Verlag (1996) 518–537
42. Richter, G., Maffeo, B.: Toward a rigorous interpretation of ESML-extended systems modeling language. IEEE Transactions on Software Engineering **19** (1993) 165–180
43. Paige, R.F.: A meta-method for formal method integration. Lecture Notes in Computer Science **1313** (1997)
44. Baresi, L., Orso, A., Pezzè, M.: Introducing formal specification methods in industrial practice. In: Proceedings of the 19th International Conference on Software Engineering (ICSE '97), NY, ACM (1997) 56–67
45. Bause, F., Kritzinger, P.: Stochastic Petri Nets - An Introduction to the Theory. Advanced Studies in Computer Science. Vieweg Verlag (1996)
46. Marsan, M.A., Balbo, G., Conte, G., Donatelli, S., Franceschinis, G.: Modelling with Generalized Stochastic Petri Nets. Wiley Series in Parallel Computing. John Wiley and Sons (1995)
47. Ghezzi, C., Morasca, S., Pezzè, M.: Validating timing requirements of time basic net specifications. Journal of Systems and Software **27** (1994) 97–117
48. Montangero, C., Ambriola, V.: OIKOS: Constructing process-centred SDEs. In A. Finkelstein, J. Kramer, B.N., ed.: Software Process Modelling and Technology. John Wiley and Sons (1994) 131–151
49. Conradi, R., Hagaseth, M., Larsen, J., Nguyen, M., Munch, B., Westby, P., Zhu, W., Jaccheri, M., Liu, C.: EPOS: Object-oriented cooperative process modeling. In Nuseibeh, B., Finkelstein, A., Kramer, J., eds.: Software Process Modelling and Technology. John Wiley and Sons (1994) 33–70

Model Validation in Controller Design

Jörg Desel, Vesna Milijic, and Christian Neumair

Lehrstuhl für Angewandte Informatik
Katholische Universität Eichstätt–Ingolstadt
Ostenstr. 28, 85072 Eichstätt, Germany
{joerg.desel,vesna.milijic,christian.neumair}@ku-eichstaett.de

Abstract. This work considers model construction and validation in controller design. The problem we are interested in is to derive a formal model of a controlled automation system from a semi-formal description of the uncontrolled plant and various requirements concerning the plant and the processes of the controlled system. These requirements are originally formulated on many different abstraction levels, partly employing formal notations, partly using just natural language and partly consisting of mixtures of both. Moreover, they are often incomplete, contain errors, contradict each other and assume some domain knowledge which is typically not explicitly stated. So a crucial part of the model construction process is the formalization of the plant and of the requirements as well as validation of the derived models. We suggest a simulation-based method which employs formal and graphical representations of process models and specifications and which involves an iterative process of formalization and validation of requirements. The approach is based on particular Petri nets, called signal nets, as formal process models and partially ordered runs as their semantics. This contribution also reports on a case study from the automotive industry.

1 Introduction

This contribution is on model based development of software systems that are supposed to run in a technical environment. More precisely, we deal with the development of such systems which is based on formal process models. We use a tailored variant of Petri nets together with a process net semantics.

Model based system development can only lead to a valuable system if the underlying models faithfully represent the requirements. The requirements include information about the existing or the planned environment of the system as well as the desired system behavior within this environment. These statements hold true for a wide range of systems. In this work we concentrate on computer systems which are supposed to function in a given technical environment. These include automation systems composed of a plant and a control restricting the plant's behavior. In particular, we consider embedded systems in cars. In this setting, the aim is to develop a control algorithm such that the controlled system matches the requirements. Unfortunately, the requirement specification is often formulated on many different abstraction levels, partly employing formal

J. Desel, W. Reisig, and G. Rozenberg (Eds.): ACPN 2003, LNCS 3098, pp. 467–495, 2004.

notations, partly using just natural language and partly consisting of mixtures of both. Moreover, it is usually incomplete, contains errors, is contradictory and assumes some domain knowledge which is not explicitly stated.

Since the general aim is to develop the controller software, one possible approach would be to start with generating a formal specification of this software. This software has to run within the environment. Therefore, a formal specification of this environment, namely the plant, is necessary as well. This specification is not easy to obtain because the user is interested in the overall behavior. Thus he will only provide information concerning the controlled system, i.e. the composition of plant and control. Moreover, the precise behavior of the plant might be unknown as well. Faulty assumptions on the plant specification will lead to faulty or incomplete control specifications, which eventually leads to controller software that matches the specification but does not satisfy the user's needs.

Therefore, we proceed differently; we aim at a model of the entire system, including both the plant and the control. This model can be viewed as a specification of the total system. A given control software matches the specification if its behavior together with the plant precisely corresponds to the behavior of the model. The model of the entire system is generated from the different specification items that are given in different form mentioned above. The crucial steps in model construction are the appropriate formalization of the requirements (and their validation) and the correct generation of the model from the formal specifications.

Model construction is used in controller design for the examination of specifications w.r.t. feasibility and for creation of reference models for the final system that are used for verification and tests. These models are also very useful as a basis for model-based test case generation. So we view model construction, formalization and validation as one important early phase in the process of system development.

This work will present an approach for model construction for controlled systems that employs different formalization / validation steps and a synthesis procedure to obtain the model from the specifications in a systematic way. It also presents a case study developed with the car manufacturing company Audi (see also [8]) and reports on experiences with applying this method.

The basis of our formal modelling language are signal nets [11, 14, 15], an extension of Petri nets. In order to adapt our modelling language to industrial relevance, features for modularity, interaction between modules and differentiation between controllable, observable and internal events had to be integrated. Extensions also concern a timing concept for representing real time aspects and real-valued sensor data employing concepts of High-Level nets. The approach is based on simulation and verification. By simulation we mean construction and inspection of partially ordered causal runs, represented again by signal nets.

The paper is organized as follows: In the forthcoming section we describe what we mean by validation of models, in contrast to system validation. We also distinguish validation from verification and formalization from specification. Section three is devoted to the steps of our approach in a general setting. In section four, our formal modelling language is presented. The causal simulation of our extension of signal nets, its advantages and some words about algorithmic aspects is the topic of section five. Section six illustrates applying this approach to the industrial case study, also providing net models and partially ordered runs. Finally, experiences from the case study are outlined in section seven.

The first sections of this paper are strongly based on [2] and [5], where more details can be found. Different aspects of the approach were also adapted to and presented in various different communities (see [1], [3], and see [4] for using part of the concept for education purpose).

2 Model Validation

This section is devoted to a general discussion of the term "model validation" in system design. Validation is usually related to systems. We adapt its meaning to models. The usual definition of validation of a system in relation to verification and evaluation reads as follows:

> **Validation.** Validation is the process determining that the system fulfills the purpose for which it was intended. So it should provide an answer to the question *"Did we build the right system?"* In the negative case, validation should point out which aspects are not captured or any other mismatch between the system and the actual requirements.

> **Verification.** Verification is the automated or manual creation of a proof showing that the system matches the specification. A corresponding question is *"Did we build the system right?"* In the negative case, verification should point out which part of the specification is not satisfied and possibly give hints why this is the case, for example by providing counter examples. Nowadays, *model checking* is the most prominent technique used for automated verification. *Proof techniques* can be viewed as manual verification methods.

> **Evaluation.** Evaluation concerns the questions *"Is the system useful?"*, *"Will the system be accepted by the intended users?"* It considers those aspects of the system within its intended environment that are not formulated or cannot be formulated in terms of formal requirements specifications. The question *"How is the performance of the system?"* might also belong to this category, if the system's performance is not a matter of specification.

This contribution is about validation of *models*, namely process models. So replacing the term "system" in the above definitions by "process model" should provide the definitions we need. Models are used as specifications of systems. Unfortunately, replacing "system" by "specification" in the definitions does not make much sense. So we need a more detailed investigation of the role of models and of validation in model-based system development.

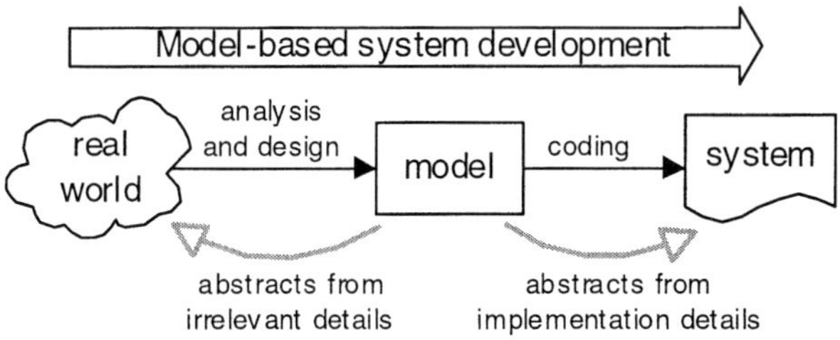

Fig. 1. Model based system development

In Figure 1, the model is an abstract representation of both, the relevant part of the "real world" and the actual system implementation. It abstracts from irrelevant details of the considered part of the "real world", and it abstracts from implementation details of the system. Verification mainly concerns the relation between the model and the system implementation, validation concerns the relation between the model and the "real world", whereas evaluation directly relates the system and the "real world".

The above view ignores that the system to be implemented will have to function within an environment, which also belongs to the "real world". So the left hand side and the right hand side of the picture cannot be completely separated; they are linked via the "real world". Figure 2 shows a more faithful representation of the situation.

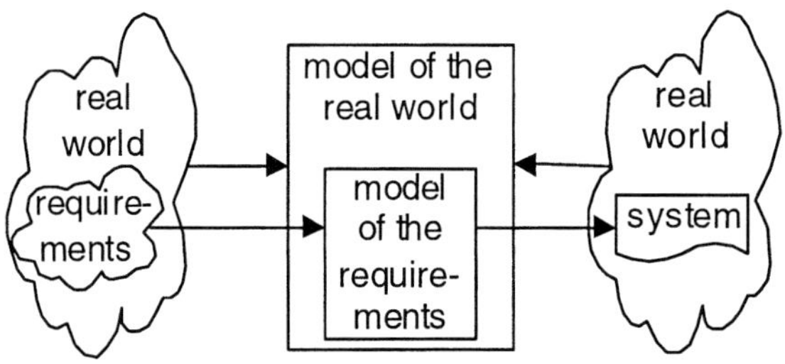

Fig. 2. Capturing the embedding in the real world

Notice that the word "system" is used with different meanings: the "real world" (environment plant), the software system to be implemented (control) and the composition of both (the controlled plant). In the sequel we mainly use the term for the environment together with (part of) the control.

A more detailed view of the model distinguishes *requirements specification* and *design specifications* on the level of the model.

The model of the real world is obtained by analysis of the domain and *formalization* of its relevant aspects. The requirements specification models the requirements and is derived by *formalization* of the requirements that exist within the "real world". The design specification can be viewed as a model of the system im-

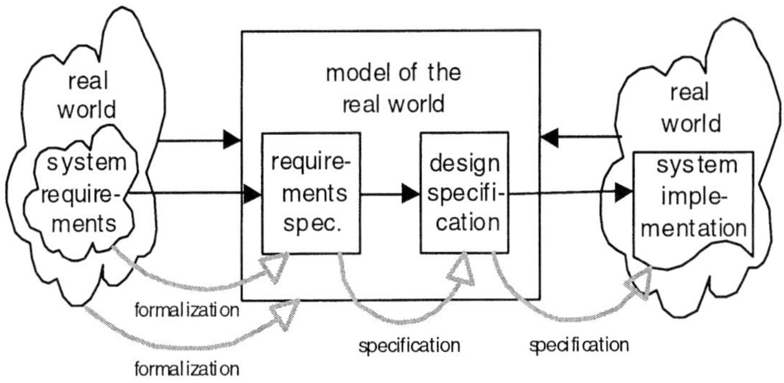

Fig. 3. Capturing requirements and design specifications

plementation, without considering implementation details though. This model has to satisfy all properties formulated in the requirements specification. The transformation from the requirements specification to the design specification is a nontrivial task. Finally, there should be a more or less direct transformation from the design specification to the system implementation. This implementation of the system is also said to be *specified* by the design specification.

Now let us consider the reverse direction. It is a matter of *verification* to check whether the design specification actually matches the requirement specification. It can also be *verified* whether the system implementation reflects the design specification. The correctness of the formalization transformations can only be checked by *validation*. So "formalization" and "validation" is a related pair of terms in the same sense as "specification" and "verification". Finally, requirements that are not captured in the model can only be checked by *evaluation* of the system implementation within the "real world".

In Figure 4, the arrow annotated by "evaluation" points to the "real world" including the system requirements whereas the lower arrow annotated by "validation" addresses only the "real world" without system requirements.

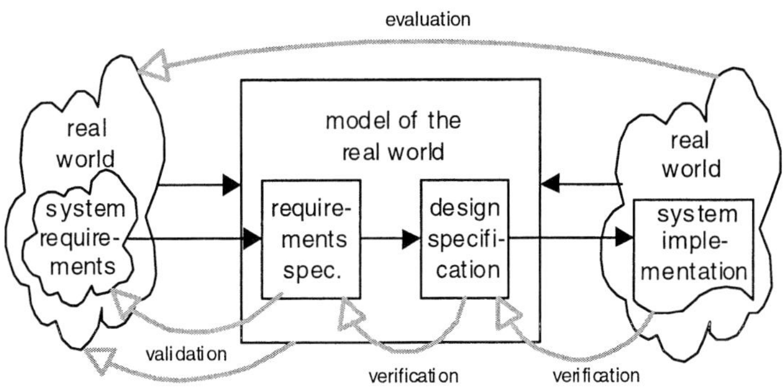

Fig. 4. The position of validation, verification and evaluation

In our context of controller design, the plant is part of the real world (the environment, respectively) and the control plays the role of the system implementation. Formalizing the description of the plant will yield a formal process model whereas the formalization of the requirements have to be interpreted on this process model, or, respectively, on its behavior. Both formalization steps have corresponding validation steps that are supported in our approach.

3 The Approach

How can we derive a valid formal model from a semi-formal description of a controlled system and of its desired behavior? There is no general answer to this question, since modelling is a creative process. Creating a model always means to formalize concepts that have not been formulated that precise before. Therefore, misunderstandings, errors, missing assumptions etc. can not be avoided in general. The best we can expect is to provide means for detecting these errors as soon as possible.

We concentrate on process models that have a dynamic behavior and can thus be executed. So for each process model there is the notion of a run, i.e., one of its executions. Our basic assumption is that the domain expert (the user of our approach) knows well what the correct runs of the desired system should look like but might have problems in formalizing an appropriate specification of this set of runs. We will use *causal* runs, given by partially ordered sets of events and local system states. A definition of causal runs and their graphical representation is deferred to the next sections.

As mentioned in the previous section, formalization tasks appear at different steps: First, a given or planned system that serves as the environment or plant has to be modelled. Second, the requirements of the controlled system has to be specified. Both aspects deserve additional validation procedures. Given a valid model of the plant and a valid specification of the controlled system, the following step is to design the control algorithm and to verify its correctness with respect to the specification. This step is not within the scope of our approach (see [15]). However, it will turn out that some verification means can also be used for validation purposes.

We first consider the problem of modelling a given system (the environment). The behavior of the system should precisely correspond to the behavior of the model. Assuming that we have a version of this model, our approach generates the runs of the model, visualizes this behavior in an appropriate way and presents the result to the expert. This model is often derived directly from the system's structure and architecture. If the behavior of the system rather than its structure is known, then a first version of the system model is constructed from the runs by *folding* appropriate representations of runs (this procedure is given in [3] for workflow models).

The simulation of the system model either shows that the model can be accepted or that it does not yet match the system. In the latter case, the model is changed according to identified modelling errors and the procedure is repeated.

Only when the simulated runs of the model coincide with the required runs, the model can be used to obtain information about the system. The procedure for model validation can be complemented by verification means: If some behavioral properties of the system are known then the model should satisfy according properties as well. Since this verification step is sometimes hard to conduct, there is an intermediate solution for properties that all runs should satisfy: Simulation is paired with verification of the simulated runs. This requires an analysis method for runs, which is also the kernel of the formalization of other requirements, to be discussed next.

We now consider the formalization and validation of requirements. That is, we assume to have a valid model of the environment (the plant) and add requirements that have to be satisfied by the controlled system, i.e., that have to be guaranteed by the desired control. In our approach, we only consider required properties that can be formulated as properties of runs (generally, all properties of a Linear Time Temporal Logic). These requirements are formalized, validated and implemented step by step. In the first step, we begin with some of the requirements and analyze simulated runs of the existing model with respect to these requirements. The result is a distinction of those runs that satisfy the requirements and those that do not. This way, the user gets information about his requirement specification in terms of runs ("did you really want to rule out precisely those runs that failed the test?"). Figure 5 illustrates this step. After an iterative reformulation of the first requirements the simulation based approach should eventually yield a valid specification of this requirement. Thereafter the system is modified in such a way that it satisfies this requirement. For some requirement specifications, there is an automated procedure for this task. In general, however, there is some freedom in how to implement the requirement. The implementation of the requirement is either verified by appropriate verification techniques or checked again by simulation.

After the first step, a second requirement can be formalized, validated and implemented, based on the modified model, in the same way (see Figure 6), and so on. Notice, however, that the implementation of the new requirement should not violate a previously implemented requirement. As long as all requirements only restrict the set of possible runs, this problem does not occur. But, if liveness

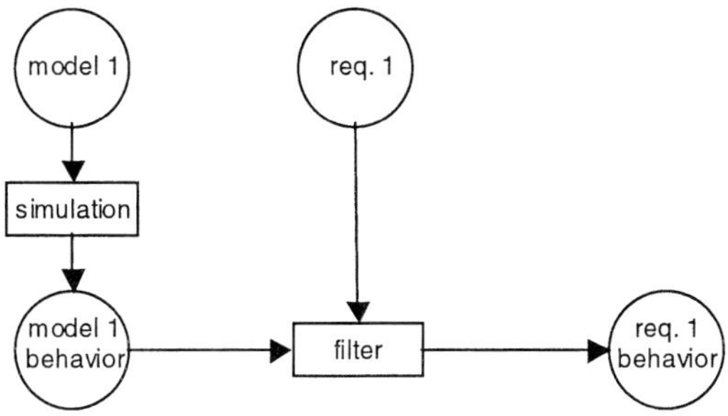

Fig. 5. A first step in requirements validation

properties (requiring that something eventually happens) and safety properties (requiring that something bad does not happen) are added in arbitrary order, then previous steps might have to be repeated.

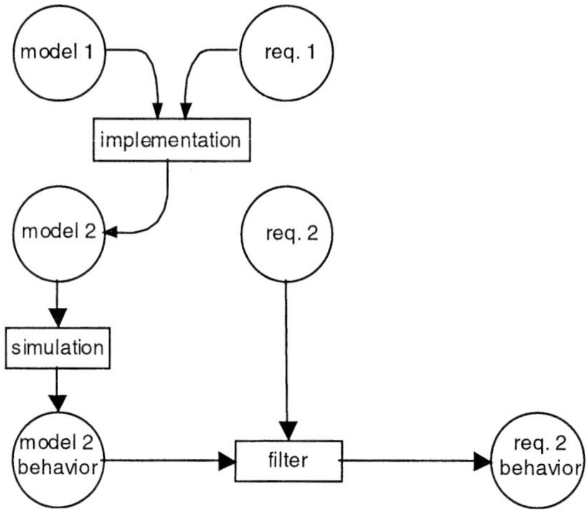

Fig. 6. A second step in requirements validation

4 The Modelling Approach

4.1 The Modelling Language

To model a system, we use signal nets [6, 7, 11, 15, 16] that are an extension of Petri nets. A signal net is, like a Petri net, a graph with two types of nodes. Our modelling language extends signal nets to principles like modularity, interaction between modules and differentiation between controllable, observable and internal actions. Also a timing concept for representing real time aspects and concepts of High-Level nets for representing real-valued sensor data are included.

Places. Each place may contain tokens from a place-specific defined set of token types, called domain. The same token can appear more than once in a place. To depict output-places, where tokens represent data to be read from outside, grey background color is used. We distinguish two kinds of places, namely low-level and high-level places.

- *Low-level places* are represented by a simple circle. The domain of low-level places contains a single token type: a black token. The number of black tokens represents the actual state of the place. If these places represent conditions then in any reachable state they contain at most one black token: one token means that the condition is fulfilled, no token means it is not. This will be the case in the examples given in section six.

- *High-level places* are represented by a double-framed circle. The domain of these places is an arbitrary nonempty set of token types. The state is symbolized by the number of tokens which belong to each token type. In the examples, high-level places will only contain one token and the domain will be always the set of real numbers. Hence, to simplify matters, the state of a high-level place will be symbolized by a real number.

Transitions, drawn as rectangles, represent actions. Actions that do not underly the control algorithm, called uncontrollable, like user driven actions or errors will be symbolized by a darker grey background. Transitions with a light grey background characterize actions that generate new values, for example for sensor data. These transitions and the white colored transitions are controllable.

Arcs. Nodes can be connected by different kinds of arcs.

- *Flow arcs* are black arcs that either connect a place with a transition or a transition with a place. They may have a time label. In the case they connect high-level places and transitions or vice versa, they are labelled by a variable.
- *Read arcs* are double-sided black arcs between transitions and places. If they connect high-level places and transitions, they are labelled by a variable and may also have a time label.
- *Write arcs* are double-sided grey arcs between transitions and high-level places that are labelled by two variables and possibly a time label.
- *Synchronization arcs* connect two transitions and are graphically represented by jagged arcs.

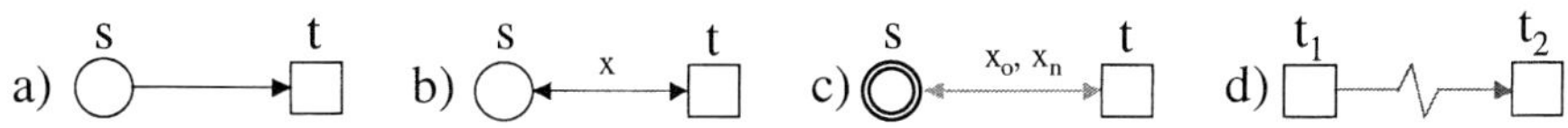

Fig. 7. a) Flow arc b) Read arc c) Write arc d) Synchronization arc

A state of a signal net is determined by its current distribution of tokens in the places, also called *marking*. We denote the initial distribution of tokens by *initial marking*. The *dynamic behavior* of a signal net is given by the firing of transitions. The surrounding arcs determine whether a transition may fire and how its firing changes the marking.

Transitions that are not connected to high-level places are called *low-level transitions*. If every low-level input place, connected to the low-level transition by a flow arc or a test arc, contains a token, this transition is enabled and may fire.

A transition which is connected by labelled arc with at least one high-level place is called *high-level transition*. Each variable of the labelled arc can be substituted by a value of the domain of the connected high-level place (so in our case by a real number). A high-level transition may have a *firing condition*,

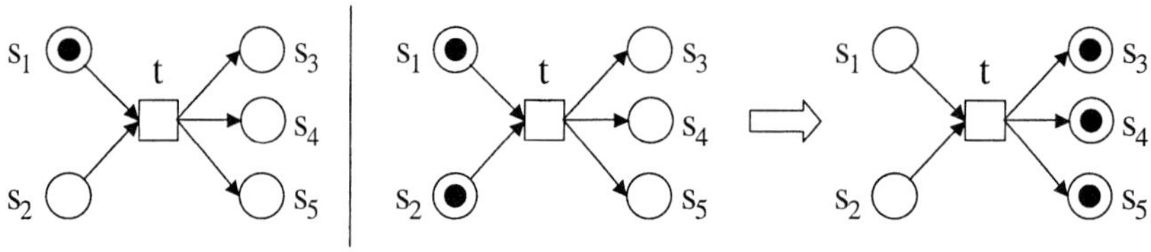

Fig. 8. On the left hand side you can see a net where transition t cannot fire because of the empty place s_2. The right hand side shows a net where transition t is able to fire and the resulting state of the net after firing of t

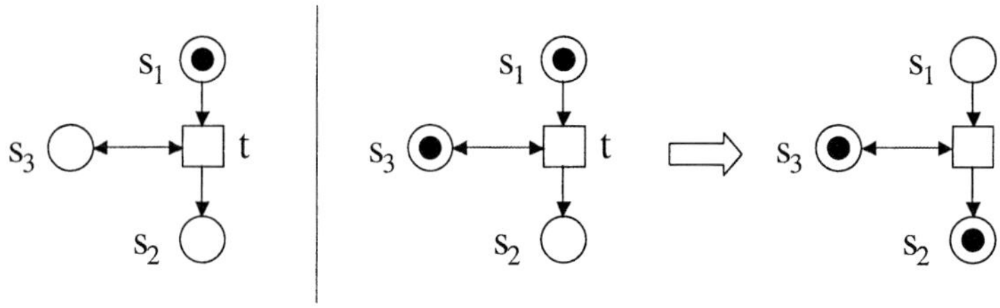

Fig. 9. On the left hand side, transition t cannot fire because of the empty place s_3. On the right hand side, firing is possible

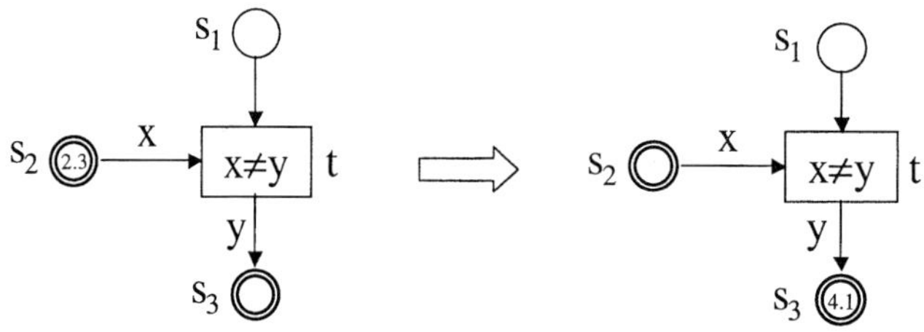

Fig. 10. This net contains two high-level places (s_2 and s_3). Place s_2 contains the token 2.3. Firing transition t substitutes x and y. While x has to be substituted by 2.3 (as the token of place s_2), variable y can be substituted by an arbitrary value but not the same as x. On the right side you can find the resulting net after firing with $x = 2.3$ and $y = 4.1$

i.e., a Boolean term that includes the variables from the labelled arcs connected with the transition. To fire a high-level transition, values for the variables at the surrounding arcs have to be substituted in the following way: The variable of an arc, resp. the first variable in the case of a write arc, leading from a high-level place to the transition is substituted by a token of this place. The substituted values must fulfill the firing condition of the transition. Moreover, each low-level input place, has to contain a token. The firing of a transition (high- or low-level) deletes a token from each low-level input place connected with a flow arc and produces a token in each low-level output place connected with a flow arc. Firing a high-level transition deletes the current value in each input high-level place and produces in every output high-level place the token determined by the substitution of the arc variable. In addition, firing a high-level transition deletes the current value in every high-level place connected with the transition by a write arc, and the value that substitutes the second variable of the write arc is produced. Read arcs do not change the token of the place they are connected with. If an enabled transition is connected to another enabled

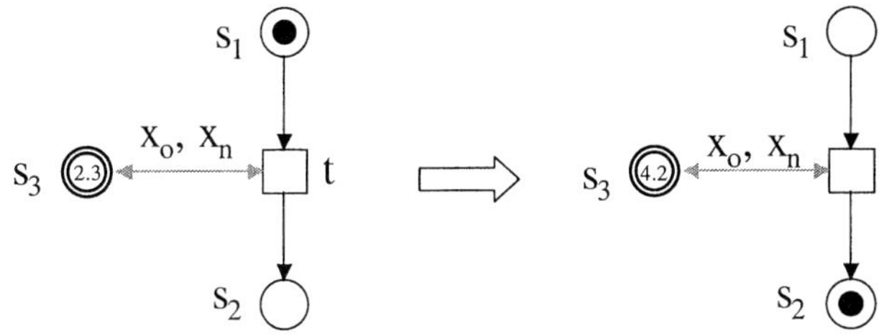

Fig. 11. This example shows a net with a write arc before and after firing of transition t with the substitution $x_a = 2.3$ and $x_n = 4.2$

transition by a synchronization arc, both transitions will fire at the same time (the first transition is synchronizing the second transition). A transition with an incoming synchronization arc will never fire without the synchronization signal, whereas a synchronizing transition can also fire alone. Our extension of signal nets also provides a concept of time. An enabled transition fires immediately after the time on the label of every arc with time label ingoing to the transition has expired after enabledness. A time label of an arc leading from a transition to a place causes that firing the transition produces a token in the place after the time of the time label has passed. The time label of a write arc between a place and a transition means that by firing the transition the old value is replaced by the new value after the time on the label has passed.

Controllable transitions that have only ingoing flow and write arcs without time label underly the progress assumption. This means that for each set of transitions which are pairwise in conflict, one will eventually fire. Two transitions, which are both enabled, are in conflict if after firing one transition the other one is no more enabled.

4.2 Causal Semantics

We now concentrate on process models, i.e., on specifications of runs of a system. Each process model has a dynamic behavior, given by its set of runs. In a run, actions of the system can occur. We will distinguish actions from action occurrences and call the latter events. In general, an action can occur more than once in a single run. Therefore, several events of a run might refer to the same action. Runs and events of our extension of signal nets can be defined in several ways. We will discuss sequential runs, given by occurrence sequences and causal runs, given by process nets.

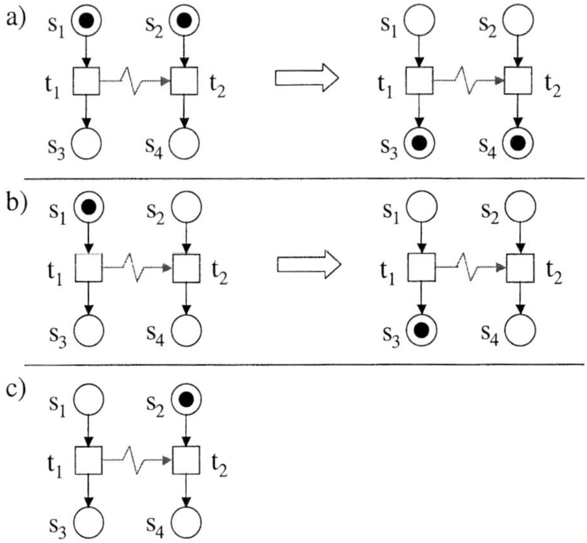

Fig. 12. a) In this net, transitions t_1 and t_2 are enabled to fire. Firing transition t_1 synchronizes transition t_2, which then also fires. The net state before and after firing is presented. b) In the second case, only transition t_1 can fire and does not synchronize transition t_2 as the place s_2 contains no token. c) Transition t_1 cannot fire because of place s_1 contains no token. So transition t_2 will not fire because it is not synchronized by transition t_1

There are basically two different techniques to describe the behavior of our signal net model: A single run can either be represented by a sequence of action names, representing subsequent events, or by a causally ordered set of events.

The first technique is formally described by occurrence sequences. It constitutes the sequential semantics. The main advantage of sequential semantics is formal simplicity. Sequential semantics generalizes well-known concepts of sequential systems. Every occurrence sequence can be viewed as a sequence of global system states and transformations leading from a state to a successor state.If transitions fire synchronously due to synchronization arcs, we combine the names of these transitions and regard them as one event. In sequential semantics, a run is represented by a sequence of events such that causal dependencies are respected; if an event causally depends on another event, then these events will not appear in the reverse order in an occurrence sequence.

The second technique employs process nets representing causal runs. It constitutes the causal semantics of our extension of signal nets. Also process nets are extended signal nets. One of the main advantages of causal semantics is its explicit representation of causal dependency, represented by paths of directed arcs in process nets. Consequently, concurrent events are events that are not connected by a path in a process net.

A causal run consists of a set of events (representing the firing of one or a set of synchronized transitions each), symbolizing action occurrences of the system.

An action can only occur in certain system states, i.e. its pre-conditions have to be satisfied. The occurrence of the action leads to a new system state where some post-conditions of the action start to hold. An event is therefore causally dependent on certain pre-conditions and might lead to new conditions that are causal prerequisites for other events. Combining events with their explicitly modelled pre- and post-conditions yields a causal run, formally represented by a process net.

Our extensions of signal nets make it necessary to take some further notes on causal runs concerning causal dependencies of events.

- A read arc either tests if a condition is fulfilled (low-level place) or reads some value (high-level place) but it does not change the token in the connected place. Hence, it is possible that more than one transition connected with the same place by read arcs can simultaneously access the place. Thus, read arcs do not influence the causal dependencies between events.
- In our modelling language, time aspects are modelled by time labels on some arcs, which either cause that transitions fire immediately after a certain time has passed or that tokens are produced after a certain time period. These time aspects have no effect on the formal system's behavior, i.e., they do not influence dependencies of events in a process net.
- Finally, remember that we consider the firing of synchronous transitions as one event.

In a process net, each token is produced by at most one transition occurrence, and it is consumed (remember that read arcs just test, but do not consume) by at most one transition occurrence. Hence, conditions of process nets are not branched w.r.t. flow and write arcs.

The immediate causal dependency of events is represented by the flow and write arcs of a process net. No two elements can be mutually causally dependent, in other words, the flow and write relation has no cycles. So the causal relation is a partial order that we call causal order. Two different events are causally ordered if and only if they are connected by a chain of directed flow or write arcs. Otherwise, they are not ordered but occur concurrently.

A condition of a process net represents the appearance of a token on a place of the original net and is therefore drawn as a copy of the place labelled by the name of the place. In case of high-level places, the copy also includes the current value of the high-level place.

As an event represents the occurrence of at least one transition, it is depicted as a copy of a transition of the original net. If the event represents the occurrence of a single transition, it is labelled by the name of the transition. If an event represents the occurrence of a set of synchronous transitions, it is labelled by all elements of the set of names of these transitions. Consequently, there are no synchronization arcs in a process net.

Since events represent transition occurrences, the pre- and post-sets of these transitions are respected. The initial state of the process net is the characteristic mapping of the set of conditions that are minimal with respect to the causal order, i.e., these conditions carry one token each, and all other conditions are initially unmarked. We assume that every event has at least one pre-condition and at least one post-condition. By this assumption, all minimal elements are conditions. Finally, the initial state of the process net corresponds to the initial marking of the system net, i.e., each initial token of the system net is represented by a (marked) minimal condition of the process net. Each process net represents a single causal run of a system net.

Using acyclic graphs to define partially ordered runs is common for many computation models. The specific property of process nets is that each process net is formally a signal net with our extensions and that there is a close connection between a process net representing a run and the extended signal net modelling the system; the events of a process net are annotated by respective names of actions of the system. More precisely, mappings from the net elements of the process net to the net elements of the original net representing the system formalize the relations between events of a process net and transitions of a system net and between conditions of a process net and places of a system net.

Sequential and causal runs have strong relations. Sequences of event occurrences of a process net closely correspond to transition sequences of the system net. Therefore, roughly speaking, the set of occurrence sequences of an extended signal net coincides with the set of occurrence sequences of its process nets when only the labels of events of these latter sequences are considered.

5 Simulation by Construction of Runs

By simulation we understand the generation of runs of the process model. For a valid model, each run should represent a corresponding run of the system, and for each system run there should exist a corresponding run of the model. Validation by simulation means generating and inspecting runs of the model with respect to the desired runs of the modelled system. Since neither the system nor its runs are given formally, only domain experts can do this comparison. So this task requires a good and easy understanding of the generated runs of the model.

Usually, the user is supported by a graphical representations of runs: The extended signal net is represented graphically and sequential runs are depicted by subsequent occurrences of transitions of the net. We suggest to construct and visualize causal runs given by partially ordered *process nets* instead. We argue that we gain two major advantages, namely expressiveness and efficiency.

Every sequence of events, i.e. transition occurrences, defines a total order on these events. A transition can either occur after another transition because there is a causal dependency between these occurrences or the order is just an arbitrarily chosen order between concurrent transition occurrences. Hence, an occurrence sequence gives little information on the causal structure of the system run. Interesting aspects of system behavior such as the flow of control, possible parallel behavior etc. are directly represented in process nets, but they are hidden in sequences of events. Causal runs provide full information about these causal dependencies.

The number of event occurrence sequences of a single run grows dramatically when a system exhibits more concurrency. Each of these occurrence sequences represents the very same causal system run. Hence, the simulation of more than one of these sequences yields no additional information on the causal behavior of the system. The gain of efficiency is most evident when all runs of a system can be simulated, i.e. when there is only a finite number of finite runs. In the case of arbitrary large runs, a set of process nets allows to represent a larger significant part of the behavior than a comparable large set of occurrence sequences.

Simulation of a system model means construction of a set of (different) runs. In general, each causal run corresponds to a nonempty set of occurrence sequences. Taking the sequence of labels of events in occurrence sequences of process nets yields all occurrence sequences of the system net.

In previous publications, we have described the simulation algorithms [2, 5], i.e. an algorithm constructing runs. Crucial aspects are a compact representation of similar runs, completeness with respect to all possible alternatives and in particular termination conditions for potentially infinite runs.

As described in the third section, we have to provide means to analyze the constructed runs with respect to specified requirements. These specifications are formulated on the level of the system net in a graphical way (see [2]), adopting the well-known fact transitions [10] and introducing analogous graphical representations for other properties.

As the specifications are interpreted on runs, we developed algorithms for analysis of process nets. It turned out that the particular structure of these nets lead to significant advantages with respect to efficiency, compared to occurrence sequences, at least for some important classes of requirement specifications.

6 The Case Study

In the context of a new production run of cars, this case study with the car manufacturing company Audi was concerned with the control system of the fuel gauge of a car, which is surprisingly complex. The value of the fuel gauge is sometimes determined by various sensors and sometimes calculated by means of consumption and an earlier calculated fill level. Numerous parameters like ignition state, movement of the car, car position and engine on/off are relevant. Because of the special shape of the tank only the values of some sensors – depending on the current fill level – account for the calculation of the fuel gauge. If some sensor fails, a plausibility test avoids that the values of this sensor are used for the calculation. Though the consumption based calculation is rather exact, the summation of minor measurement errors below a certain threshold may lead to a significant difference between real and calculated fuel gauge. Already these problems indicate that complex algorithms are necessary for the control of this technical system which includes continuous and discrete elements.

Starting point of the case study was an informal, mainly textual document where functionality of the fuel gauge control system was described. The following tasks arose in the project:

1. Adaption of appropriate techniques for modelling and model synthesis,
2. development of models of control for the fuel measurement,

3. simulation and validation of the models,
4. feasibility study for similar tasks but with larger scale,
5. specification of the requirements of software tools to be developed.

6.1 Procedure

When starting the project, we expected to receive a basic model of the uncontrolled system and a set of informal and formal requirements [12, 13]. However, actually the input of Audi corporation was a completely different one: informal descriptions of scenarios were given, that had to be realized by the algorithm to be modelled, including some implicit requirements. The basic model of the plant was partly explained by corresponding automotive components (e.g. the functionality of the four tank sensors) and partly assumed as well-known (e.g. possible states of ignition or of car movement). Some of these implicit assumptions could easily become completed, others made it necessary to query at Audi corporation. Interpreting the description of the scenarios, ambiguities were detected. Even by Audi corporation, some ambiguities could not be cleared up instantly. In general, the (not surprising) assumption was affirmed, that modelling strongly depends on feedback of experts and cannot solely be done by the given documents. So, in the procedure, the first steps of modelling and especially of a repeated validation turned out to be of high importance [2, 5].

In the following, we present our intended procedure. In our case study, the later phases of model based generation of test cases are presented only exemplarily.

Procedure

Steps one to six refer only to single modules.

1. Extraction and validation of the model of the plant.
2. Extraction, formulation and validation of the requirements.
3. Modelling of runs from the scenarios comprising the model of the plant.
4. Generation of the complete system by folding the runs. This complete system includes the given runs but may also have some other (desired or undesired) ones.
5. Elimination of undesired runs by implementation of the requirements found in step two. It has to be guarantied that the given runs of step three are still possible.
6. Validation of the single modules and verification of the requirements.
7. Integration of the modules by composition at certain restrictive interfaces.
8. Verification of the complete system.
9. Application as reference model.
10. Analysis for efficient generation of relevant test data.
11. Application as test reference by simulation, in parallel to the real control.

6.2 Model of Calculation of Fuel Gauge while *Ignition off*

In the following three subsections we present the case study. Due to lack of space not every detail can be explained. Each subsection contains a short description of the given specification of the corresponding model. Then the model, including both plant and control is presented. A description of its behavior is visualized by different process nets.

In each model, the plant is represented by dark grey colored transitions and the places connected with these transitions. For simplicity, those parts of the plant, which are not necessary for the control algorithm in the current model are omitted. If a complete model is desired, the models can be connected by identifying common parts of the plant.

The calculation of the fuel gauge while *ignition off* consists of two measurement phases. The first measurement phase starts by turning the ignition off and returns the mean value *MW old.* Turning the ignition on starts the second measurement phase and returns the mean value *MW new.* If the difference between the two mean values exceeds 4 liters, refuelling is recognized. Then the *displayed fuel* is recalculated by adding this difference to the *old value.* The following requirements must be fulfilled for calculating the fuel gauge:

1. If the period of time while the ignition is off is too short to finish measurement phase one, there will be no new calculation of the fuel gauge.
2. The result of measurement phase 1 outlasts if ignition is turned on for a short time.

In the left part of the model the behavior of the ignition is presented by its physical $(1, 2)$ and internal $(1', 2')$ transitions. When the ignition is off for 6 seconds, measurement phase one starts. 4 seconds later the first phase ends by firing transition 5 what produces a value in the place *MW old.* Turning the ignition on initiates the second measurement phase (transition 6), if measurement phase one has already been finished (indicated by a token in the place *synchronization*). If the ignition stays turned on for 0.4 seconds, *MW new* is calculated. Ignition on and a difference between *MW old* and *MW new* greater than 4 liters are preconditions for a change of the fuel gauge value: the transition with firing condition $|x - y| \geq 4L$ fires and the difference is added to the former value in place *old value.* If the difference is less than 4 liters, no refuelling is recognized: the transition with firing condition $|x - y| < 4L$ fires and the marking of the place *displayed fuel* remains unchanged.

At this point the calculation is declared finished (token in place *calculation completed*), to enable – when the ignition is turned off next time – a new calculation of the fuel gauge value.

The following three pictures show relevant process nets of the model: first, a complete calculation and change of the fuel gauge value is done (Fig. 14). In the second process, calculation is aborted because of turning on ignition before ending of first measurement phase (Fig. 15). The last process shows an interrupted calculation: after the first measurement phase, when the second is started by turning ignition on, the driver could turn off again ignition before the second

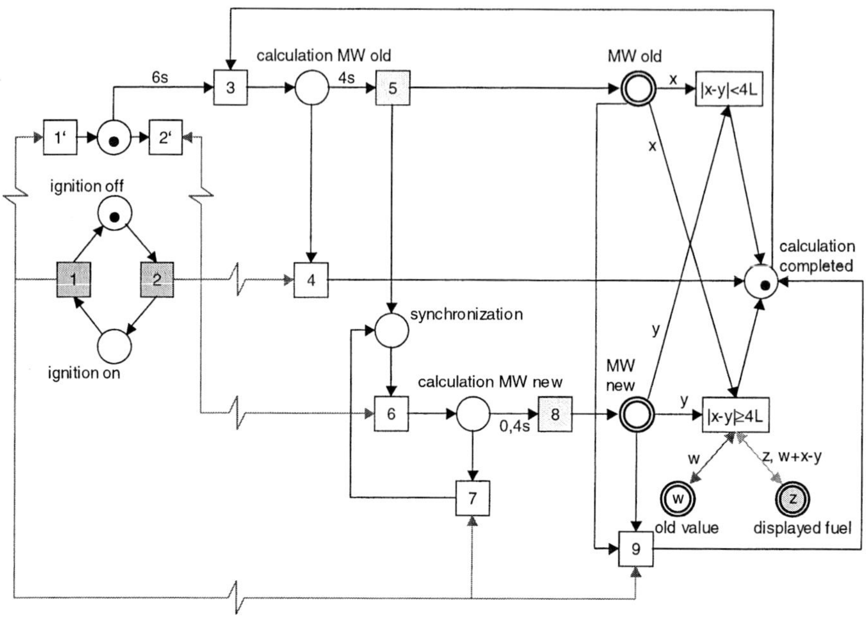

Fig. 13. Signal net model of the fuel gauge while the ignition is off

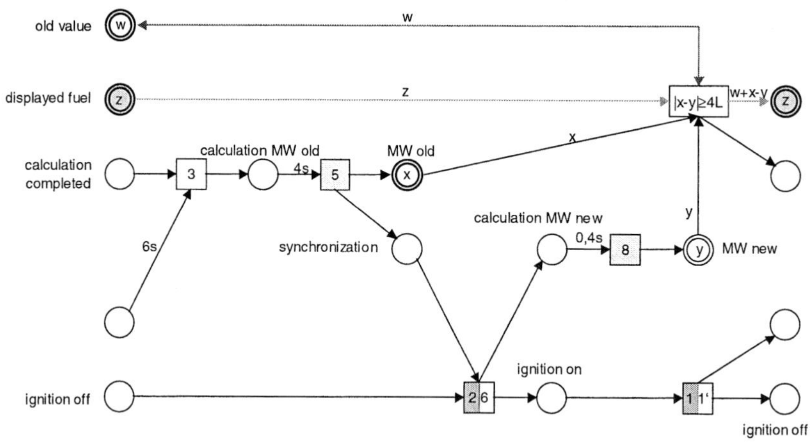

Fig. 14. Process net of a complete calculation of the fuel gauge value

measurement phase can happen. In this case, the result of measurement phase one will be remembered until ignition is turned on and the second measurement phase is executed (Fig. 16).

After some iteration steps, we came up with the following formalization of the requirement, which holds for all simulated runs.

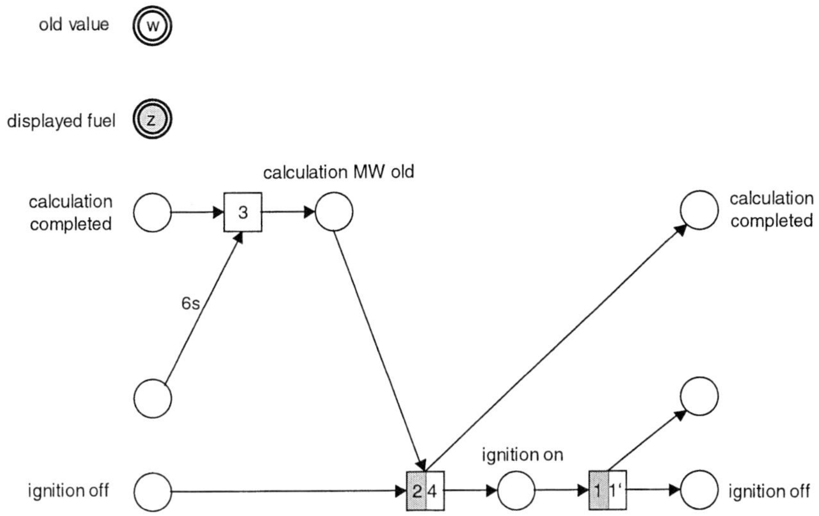

Fig. 15. Process net of an abort of the calculation during the first measurement phase

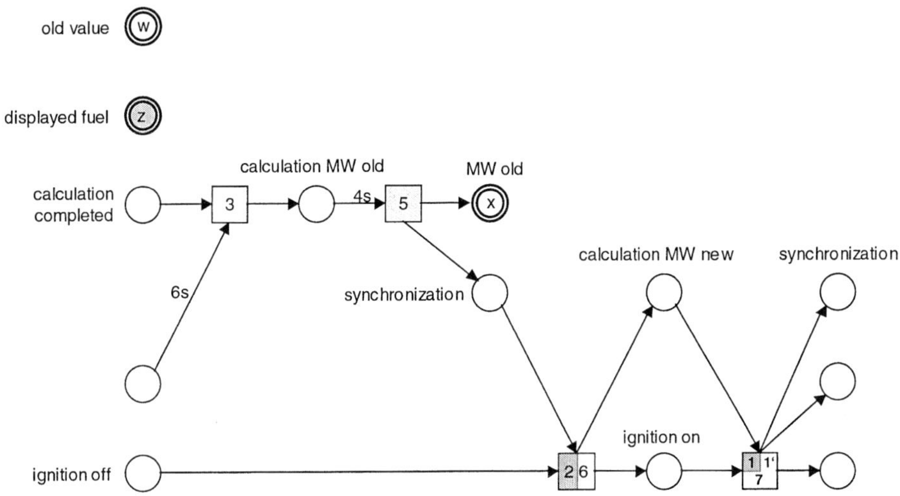

Fig. 16. Process net of an interrupted calculation during the second measurement phase

If exactly the places MW old, ignition off, displayed fuel, old value and synchronization are marked and then transitions 2 *and* 1 *fire sequentially within 0.4 seconds, then the marking in place MW old stays unchanged.*

The verification of this formalized requirement will be exemplarily shown.

A sufficient precondition is that always exactly one of the three places *calculation MW old*, *MW old* and *calculation completed* is marked. This property can be formally proved by so called place invariants. As in our requirement place

MW old is marked, the other two places are not marked. So transitions 3 and 4 cannot fire while *MW old* remains marked.

The only enabled, not synchronized transition is transition 2 which fires by assumption. As transition 6 is enabled it will be synchronized by transition 2. Therefore place *calculation MW new* is marked. By assumption, transition 1 fires within 0.4 seconds after transition 2. Before transition 8 can fire, the place *calculation MW new* is marked for 0.4 seconds. So transition 7 is enabled and synchronized when transition 1 fires. Thus the token in *calculation MW new* is consumed and a token in place *synchronization* is produced, recovering the initial marking.

6.3 Model of Calculation of Fuel Gauge while *Ignition on* and *Vehicle Stops*

The calculation of the fuel gauge while *ignition on* and *vehicle stops* also requires two measurement phases. Measurement phase one starts if the vehicle stops for 8 seconds while the ignition is on. This phase lasts 4 seconds and returns the mean value *MW old.* Subsequently, measurement phase 2 returns repeatedly a new mean value *MW new.* If refuelling is recognized once (difference between *MW old* and *MW new* greater than 4 liters), the fuel level is repeatedly recalculated by the *MW new* values by adding the difference of the mean values *MW new* and *MW old* to the *old value.* Among others, the following (informal) requirements must be fulfilled:

1. Recognizing refuelling once, the time period for detecting the mean values of *MW new* is shortened from 2 seconds to 0.4 seconds.
2. The calculation of the mean value *MW new* lasts until the vehicle moves again or the ignition is turned off.

The left part of the model shows the physical and internal states and the possible changes of states of the ignition and of the vehicle (*vehicle stops*, *vehicle moves*). To start measurement phase 1 (transition 5), the ignition has to be turned off for 8 seconds and the vehicle must not move. After 4 seconds the calculation of *MW old* is completed (transition 6). Afterwards the calculation of the mean value of *MW new* starts (transition 7) and is completed after 0.4 seconds (transition 8), if the ignition stays turned on during this time period. Directly after the calculation of *MW new* (time label 0 seconds), it is checked if the difference between *MW old* and *MW new* is greater than or equal to 4 liters. If this is the case, refuelling is recognized (transition with firing condition $|x-y| \geq 4L$ fires). If this difference is less than 4 liters, no refuelling is recognized (transition with firing condition $|x - y| < 4L$ fires), and then every two seconds a new mean value *MW new* is calculated (transition 9) and the above procedure is repeated again. Once refuelling has been recognized a new mean value *MW new* is yet calculated every 0.4 seconds (transition 10) and then immediately (time label 0 seconds) the value of the fuel gauge is actualized (transition 11). This happens by adding the difference of the mean values to the *old value* (transition 11) until

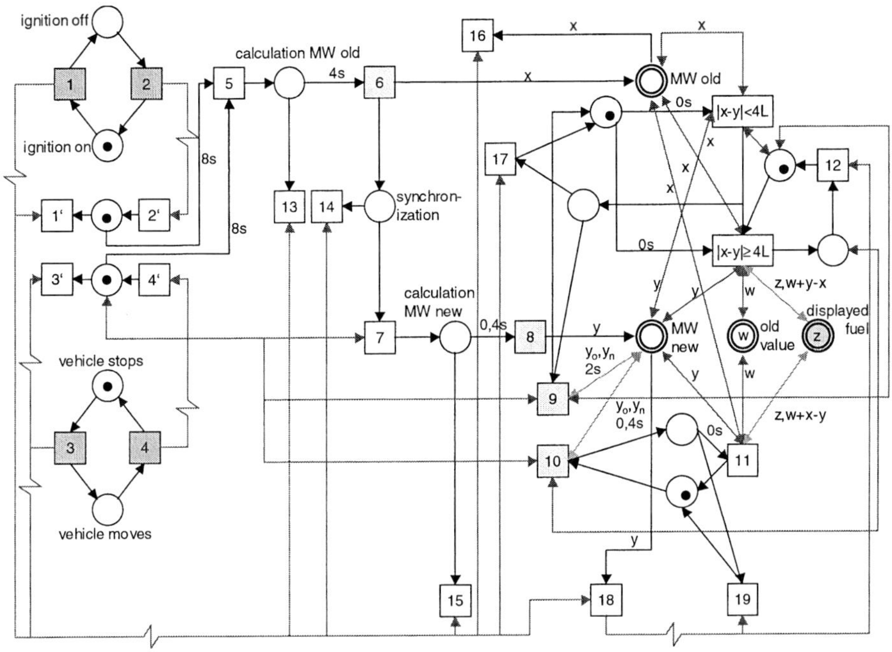

Fig. 17. Signal net model of fuel gauge while *ignition on* and *vehicle stops*

either the ignition is turned off (transition 1) or the vehicle moves (transition 3). In these cases the recalculation of the value of the fuel gauge is stopped and the initial marking in the right part of the model is restored again. This allows a new calculation of the fuel level if the car stops the next time for 8 seconds while the ignition is on.

As an example, the following figures show scenarios given by partial ordered runs of the above model.

We provide a valid formalization of requirement 1 and outline its verification:

After firing of the transition with firing condition $|x - y| \geq 4L$, transition 9 must not fire but transition 10. It fires alternately to transition 11 until either transition 1 or transition 3 fires.

When the calculation of the fuel gauge starts, the place between transition 12 and the transition with the firing condition $|x - y| \geq 4L$ is initially marked. This marking is a precondition for firing transition 9 (calculation of the *MW new* values every two seconds). Analogously, a precondition for firing transition 10 is that the place between the transition with the firing condition $|x - y| \geq 4L$ and transition 12 is marked. As always exactly one of both places contains one token (this property can be formally proved by so called place invariants), it follows: if transition 9 can fire, then transition 10 is not enabled and otherwise. So after firing the transition with the firing condition $|x-y| \geq 4L$, only transition 10 and 11 can alternately fire until transition 12 which is synchronized by transition 1 and 3, fires.

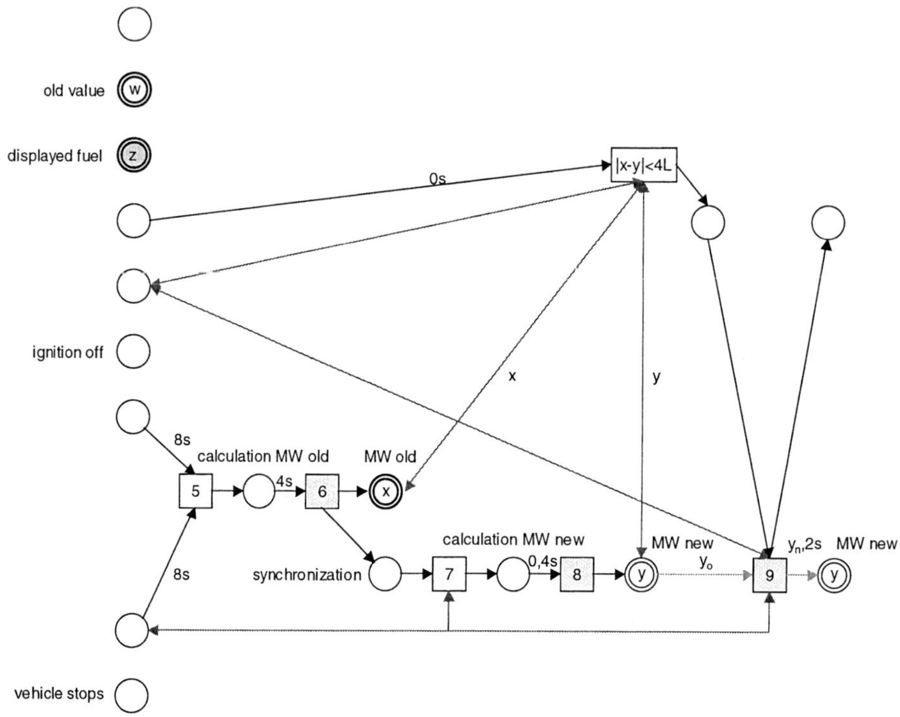

Fig. 18. Process net describing that no refuelling is recognized

6.4 Model of Error Treatment for One Sensor

A sensor can be in one of three different states: *sensor ok*, *sensor broken* or *sensor shorted*. For the sensor value (called ADC-value) the parameters Su (lower threshold), So (upper threshold), Au (initial stop of the fuel gauge) and Ao (back stop of the fuel gauge) are defined. If the ADC-value is beyond the thresholds, a sensor error has to be recognized (sensor broken if the ADC-value is too large, sensor shorted if the ADC-value is too small). If the ADC-value is between lower threshold and initial stop resp. back stop and upper threshold, the stop values are taken into account for the calculation of the fuel gauge level.

The model should fulfill the following requirements:

1. Error treatment for sensors only occurs while the ignition is on.
2. When the ignition is turned on, initially no sensor error is assumed.
3. If the ADC-value is too large ($> So$) for 20 seconds, a break should be recognized.
4. If the ADC-value is too small ($< Su$) for 20 seconds, a short circuit should be recognized.
5. If the ADC-value is again within the thresholds for 4 seconds, from error state should be changed to normal sensor state.

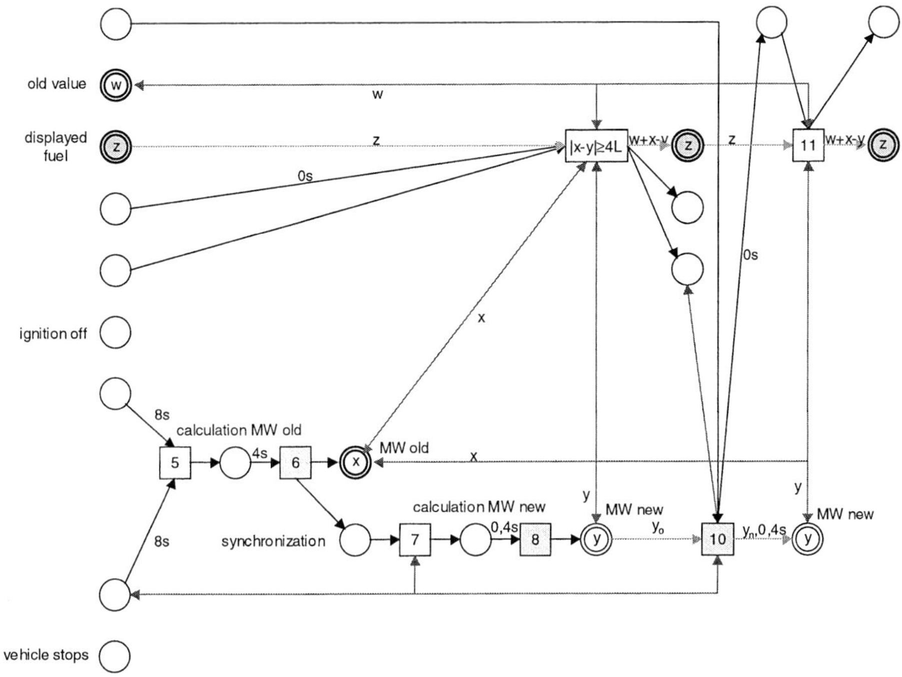

Fig. 19. Process net of the calculation of a new fuel gauge value, because refuelling is recognized

The signal-net model consists of four parts: in the left part, the possible physical states (*sensor ok*, *sensor broken*, *sensor shorted*) and the possible changes of states of the sensor are modelled. Only if the sensor is ok, the actual sensor value x (given by transition 1) is used as ADC-value. Otherwise if *sensor broken* is given, a maximal value $max > So$ is used and in case of *sensor shorted* a minimal value $0 < Su$. In the right lower part the ignition is modelled. The middle and right part show the algorithm for the error treatment of the sensor. In the middle part the treatment of the ADC-value is done which either can be normal (place *ADC-value: ok*), to small (place *ADC-value: too small*), or to large (place *ADC-value: too large*) depending on the measured sensor value in the place *sensor value*. For this purpose the transitions are labelled by the corresponding firing conditions. In the right part, depending on the classification of the ADC-value, the internal detection of a sensor error and the kind of the sensor error is described. This detection only occurs if ignition is on (place *ignition on*). As an example the following figures show scenarios given by partial ordered runs of the above model.

Finally, requirement 3 is formalized and verified:

If for at least 20 seconds the places sensor broken *and* ignition on *are simultaneously marked, the place* break *will be marked after these 20 seconds.*

Obviously requirement 1 has to be integrated as a basic condition for an error treatment of a sensor. The following considerations are necessary for the

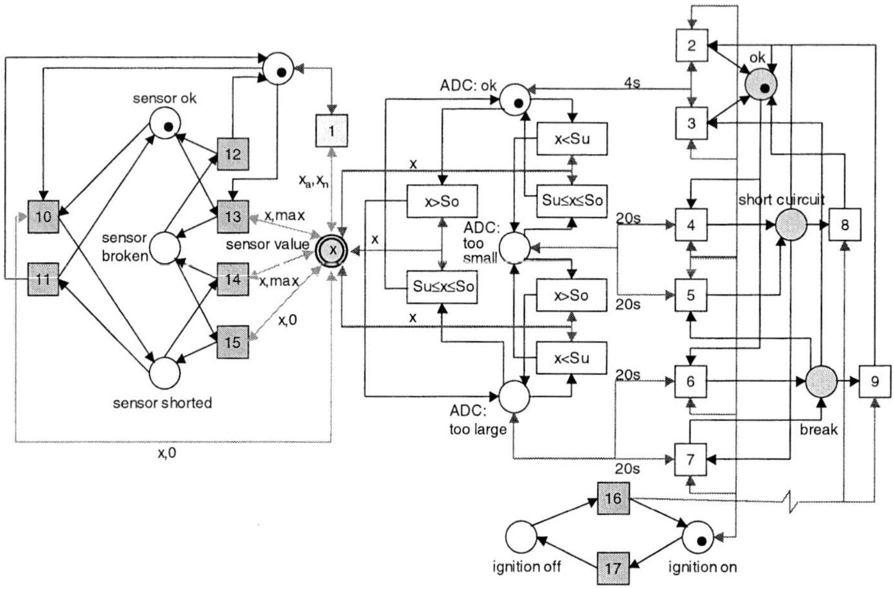

Fig. 20. Signal net model of error treatment for one sensor

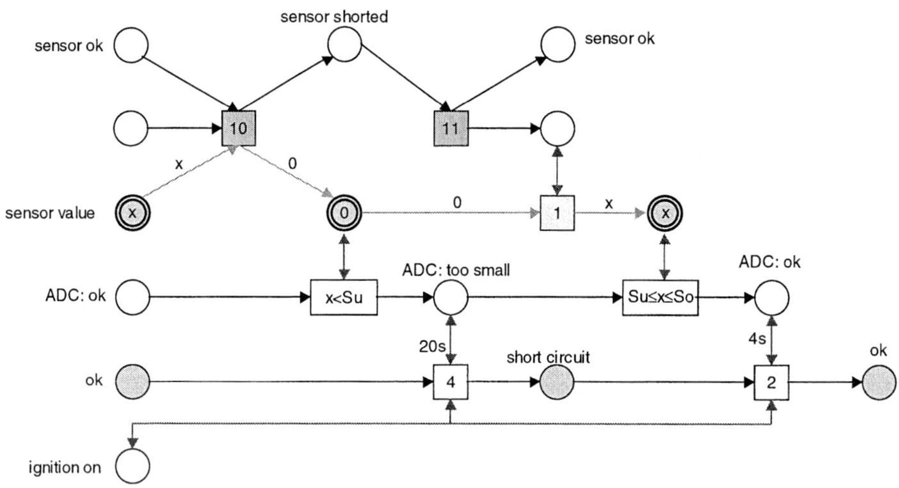

Fig. 21. Process net describing that a short circuit is recognized first, then the sensor is ok again

verification of requirement 3: in both, the middle and the right part of the model always exactly one place is marked with one token (so the corresponding sets of places are place invariants). Furthermore, in both parts of the model always exactly one transition is enabled what implies that there is no conflict between

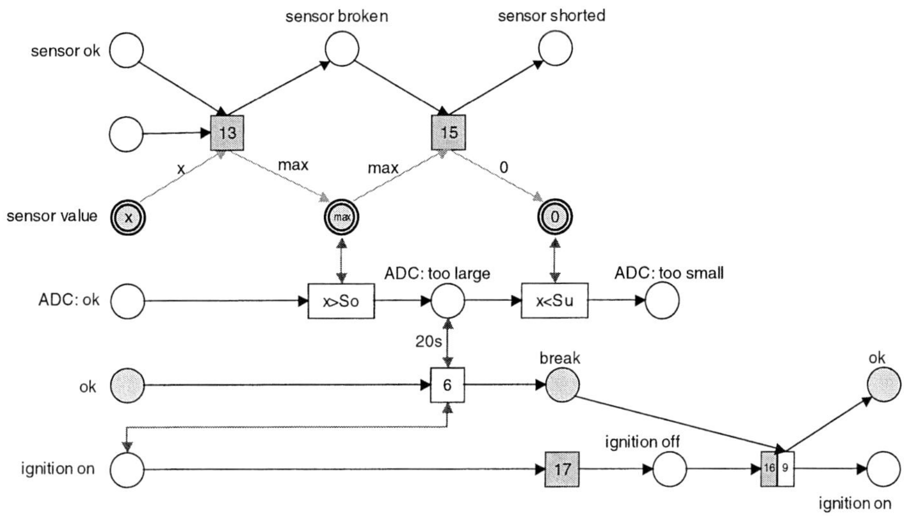

Fig. 22. Process net describing that break is recognized first, a following short circuit is not recognized as the ignition is turned off/on

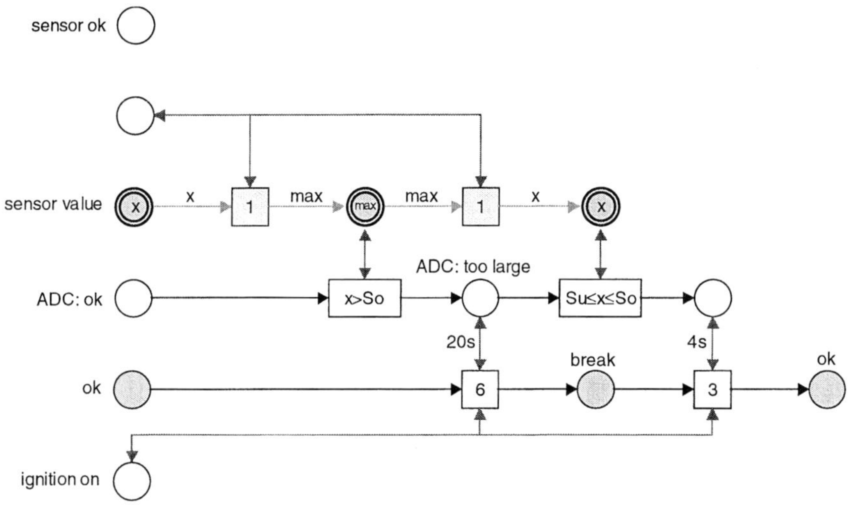

Fig. 23. Process net describing that break is recognized first, then the sensor is ok again

these transitions. To mark the place *sensor broken* either transition 13 or 14 has to fire. In particular, the place *sensor value* is then marked with the value *max*. As the place *sensor broken* stays marked for 20 seconds (as postulated in the requirement), during this period of time no transition in the left part fires. So for at least 20 seconds, the place *sensor value* is marked by the value *max*. Because of our previous considerations, exactly one transition of the middle part of the model with the firing condition $x > So$ can fire. This transition fires because of the progress assumption as it is in no conflict with any other transition of the other parts of the model (there are no flow arcs between the different parts of the model). Afterwards the place *ADC-value too large* is marked and the token stays there for at least 20 seconds (as for this period of time there is no transition enabled which could change this marking). Now the place *break* of the right part of the model is marked and so either no transition of this part is enabled or exactly one of transition 6 or 7 is enabled (depending on the former ADC-value). Because of the progress assumption, one of these transitions fires after 20 seconds and marks the place *break*.

6.5 Generation of Test Cases

We illustrate, exemplarily for the model of error treatment of a sensor, the generation of test cases resp. test vectors:

The values of the input components of a test vector are supplied by the grey resp. light grey colored transitions and correspond to the marking of certain places: these places represent sensor data resp. interfaces to components which deliver sensor data or control parameters. In this model (*sensor ok*, *ignition on*, *max*) is an input vector. The first component represents the state of the sensor delivered by transitions 10 to 15, the second component the state of the ignition (delivered by transitions 16 and 17) and the third component the sensor value (delivered by transition 1). The value of the output components corresponds to the marking of the grey colored places. Suitable values for the test vectors resp. sequences of test vectors can be determined by the model based range of values. If necessary, additional information has to be requested if the range of values does not matter for the algorithm (e.g. the maximal fill level that can be displayed by a sensor; it is determined by the dimension of the tank and the scale of the sensor and is important for plausibility test but not for other algorithms). It is not surprising that for the determination of local test vectors from a model there cannot be extracted more than it was explicitly put in. The advantages of model based determination of test vectors rest in more global aspects. The consideration of dependency resp. independency of input and output values that can be won by analyzing the model reduces the number of necessary combinations of test values extensively. For example, it can be derived by the model that *ignition on* and a incorrect *sensor value* are necessary to detect a sensor error. Therefore not all input combinations of (*sensor ok*, *sensor broken*, *sensor shorted*) and (*ignition on*, *ignition off*) have to be tested as for *ignition off* the first parameter plays no role.

This approach is based on the assumption that the actual realized control algorithm has the same dependencies as the modelled algorithm. Both should be equivalent concerning the specification, but this equivalence can be only supported by tests. So, in the best case, the model based generation of test vectors can give hints for relevant test data. But it cannot be excluded that such errors in the control, which could be detected by other test data, remain hidden. Such test data can only be delivered by further information about the realized algorithm, which cannot be derived from the specification or the model which is only based on the specification.

7 Experiences and Conclusions

After a general discussion of model validation, we have presented an approach for the systematic construction of formal models of embedded systems. In particular, we considered controlled automation systems that consist of an uncontrolled plant and a control software. The main purpose of the model is to specify the behavior of the controlled system, which implies requirements for the controller software. The approach considers the generation of initial system models (the plant) and of formal specifications of the requirements for the controlled system, which are implemented in the model step by step. The approach is based on the assumption that the user knows the desired runs of the system but tends to make errors when formalizing specifications for these runs. Thus, the core of the approach is a simulation based technique to generate runs from specifications and to visualize these runs for inspection by the user. We have argued that causal concurrent runs have important advantages in relation to sequential runs because they better capture relevant aspects of the behavior, allow a more efficient representation of behavior and allow for more efficient analysis methods with respect to system requirements.

We further presented our extension of signal nets as the modelling language to be used and discussed as an industrial case study the fuel gauge control of a car. We illustrated the extended signal net models and some of the causal runs and gave an idea of how to verify given requirements.

The main lesson we have learnt from this case study is the following. The assumption that users start with a vague description of the plant and several requirements and look for the controlling algorithm is only partly justified in this application area. Instead, very precise knowledge about the plant is available. This information has to be transformed in our modelling language which sometimes causes problems and needs feedback, because of hidden assumptions. The semi-formal requirement specification hardly includes an enumeration of safety and liveness properties the controlled system has to satisfy. These requirements are implicitly given and often go without saying (for example, the tank should not become empty without prior warning of the driver). Instead, the modelling work was based on desired scenarios of the style "what happens if...". In our terminology, a set of runs in a semi-formal style was provided, formalized in our approach, and validated by the experts. These runs have interfaces to the model of the plant. Each model of the control algorithm that supports these runs will also support additional, different runs and shows that situations can arise for which no scenarios were provided. Our simulation approach identifies these situations and offers runs to the user that are possible due to the respective model, this way enforcing the users to complete the necessary requirements. Thereafter, the formalized requirements are validated.

To draw a conclusion, the feedback from the users, engineers from Audi corporation, indicates that they considered our approach very useful. We were able to solve most of the problems posed by the users and, perhaps more importantly, we proved that the documents provided by Audi corporation contained much more ambiguities and errors than expected by the users.

The concepts presented in this paper are partly implemented in the VIPtool [9] that was developed by our group, see http://www.informatik.ku-eichstaett.de/projekte/vip. Main features of the tool are a graphical net editor, a simulation engine that generates causal runs, a visualization module that presents runs in a nice and readable way and moreover depicts the relation between process net elements and system net elements.

References

1. Desel, J.: Validation of System Models Using Partially Ordered Runs. In Szczerbicka, H. (Ed.): *Modelling and Simulation: A Tool for the Next Millenium*, Proc. of the 13th European Simulation Multiconference ESM'99, Warschau, Society for Computer Simulation, 295–302, 1999.
2. Desel, J.: Validation of Process Models by Construction of Process Nets. In van der Aalst, W., Desel, J., Oberweis A. (Eds.): *Business Process Management*, LNCS 1806. Springer-Verlag, 110–128, 2000.
3. Desel, J.: Simulation of Petri Net Processes. In Kozk, ., Huba, M. (Eds.): *Proc. of the IFAC Conference on Control System Design*, Bratislava, 14–25, 2000.
4. Desel, J.: Teaching System Modeling, Simulation and Validation. In Joines, J.A., Barton, R.R., Kang, K., Fishwick, P.A. (Eds.): *Proc. of the 2000 Winter Simulation Conference*, WSC'00, Orlando, 1669–1675, 2000.
5. Desel, J.: Model Validation - A Theoretical Issue? In Esparza, J., Lakos, Ch. (Eds.): *Proceedings of the 23rd International Conference on Applications and Theory of Petri Nets 2002*, ICATPN 2002, LNCS 2360. Springer-Verlag, 23–43, 2002.
6. Desel J., Juhás G., Lorenz R.: Process Semantics and Process Equivalence of NCEM. In *Proc 7. Workshop Algorithmen und Werkzeuge für Petrinetze AWPN 2000*, Fachberichte Informatik, Universität Koblenz - Landau, 7–12, 2000.
7. Desel J., Juhás G., Lorenz R.: Input/Output Equivalence of Petri Modules. In *Proc. of the 6th Biennial World Conference on Integrated Design and Process Technology* IDPT 2002, Pasadena, California, 2002.
8. Desel J., Juhás G., Lorenz R., Milijic V., Neumair Ch., Schieber, R.: Modellierung von Steuerungssystemen mit Signal-Petrinetzen – eine Fallstudie aus der Automobilindustrie. In Schnieder, E. (Ed.): *8. Fachtagung Entwurf komplexer Automatisierungssysteme 2003*, Proceedings of EKA 2003, Braunschweig, 273–297, 2003.
9. Desel J., Juhás G., Lorenz R., Neumair Ch.: Modelling and Validation with VipTool In Proc. of the International Conference on Business Process Management, LNCS 3080, Springer-Verlag, 2003.
10. Genrich, H., Thieler-Mevissen, G.: The Calculus of Facts. *Mathematical Foundations of Computer Science*, Springer-Verlag, 588–595, 1976.
11. Hanisch H.-M., Lüder A.: A Signal Extension for Petri nets and its Use in Controller Design. *Fundamenta Informaticae*, 41(4), 415–431, 2000.
12. Hanisch H.-M., Lüder A., Rausch M.: Controller Synthesis for Net Condition/Event Systems with a Solution for Incomplete State Obersvation. *European Journal of Control*, (3), 280–291, 1997.

13. Hanisch H.-M., Thieme J., Lüder A.: Towards a Synthesis Method for Distributed Safety controllers Based on Net Condition/Event Systems. *Journal of Intelligent Manufacturing*, (5), 357–368, 1997.
14. Juhás G., Lorenz R.: Modelling with Petri Modules. In Caillaud, B., Darondeau, Ph., Lavagno, L., Xie, X. (Eds.) *Synthesis and Control of Discrete Event Systems*, 125–138, Kluwer, 2002.
15. Juhás G., Lorenz R., Neumair Ch.: Modelling and Control with Modules of Signal Nets. In Desel J., Reisig W., and Rozenberg G. (Eds.): *Lectures on Concurrency and Petri Nets*, LNCS, Springer, 2004 (In this volume).
16. Sreenivas R. S., Krogh B. H.: Petri Net Based Models for Condition/Event Systems. In *Proceedings of 1991 American Control Conference*, vol. 3, 2899–2904, Boston, MA, 1991.

Graph Grammars and Petri Net Transformations

Hartmut Ehrig and Julia Padberg

Technical University Berlin, Germany
Institute for Software Technology and Theoretical Computer Science
{ehrig,padberg}@cs.tu-berlin.de

Abstract. The aim of this paper is a tutorial introduction to graph grammars and graph transformations on one hand and to Petri net transformations on the other hand. In addition to an introduction to both areas the paper shows how they have influenced each other. The concurrency concepts and semantics of graph transformations have been generalized from those of Petri net using the fact that the token game of Petri nets can be considered as a graph transformation step on discrete graphs. On the other hand each Petri net can be considered as a graph, such that graph transformations can be used to change the net structure of Petri nets. This leads to a rule based approach for the development of Petri nets, where the nets in different development stages are related by Petri net transformations.

1 Introduction

The main idea of graph grammars is the rule-based modification of graphs where each application of a graph rule leads to a graph transformation step. Graph grammars can be used on one hand to generate graph languages in analogy to the idea to generate string languages by Chomsky grammars in formal language theory. On the other hand graphs can be used to model the states of all kinds of systems which allows to use graph transformations to model state changes of these systems. This allows to apply graph grammars and graph transformation systems to a wide range of fields in computer science and other areas of science and engineering. A detailed presentation of different graph grammar approaches and application areas of graph transformations is given in the 3 volumes of the *Handbook of Graph Grammars and Computing by Graph Transformation* [32, 10, 15].

The intention of the first part of this paper is to give a tutorial introduction to the basic concepts and results of one specific graph transformation approach, called double-pushout approach, which is based on pushout constructions in the category of graphs and graph morphisms. Although this approach is based on a categorical concept, we do not require that the reader is familiar with category theory: In fact, we introduce the concept of a pushout in the category of graphs from an intuitive point of view, where a pushout of graphs corresponds to the gluing of two graphs via a shared subgraph.

In Section 2 of this paper we give a general overview of graph grammars and graph transformations including the main approaches considered in literature.

J. Desel, W. Reisig, and G. Rozenberg (Eds.): ACPN 2003, LNCS 3098, pp. 496–536, 2004.

The basic concepts of the double-pushout approach are introduced in Section 3 using as example the Pacman game considered as a graph grammar. Concepts and results concerning parallel and sequential independence as well as parallelism of graph transformations are introduced in Section 4. The main results are the local Church-Rosser Theorem and the Parallelism Theorem. The relationship between graph grammars and Petri nets is discussed in Section 5 of this paper. First we show how the basic concepts of both areas correspond to each other. Then we give an overview of the concurrent semantics of graph transformations, which has been developed in analogy to the corresponding theory of Petri nets.

In the second part of this paper we give an introduction to concepts and results of Petri net transformations. This area of Petri nets has been introduced about 10 years ago in order to allow in addition to the token game of Petri nets, where the net structure of fix, also the change of the nets structure [31, 28] This allows the stepwise development of Petri nets using a rule-based approach in the sense of graph transformations, where the net structure of a Petri net is considered as a graph. An intuitive introduction to Petri net transformations is given in Section 6 using the stepwise development of Petri nets for a baggage handling system as an example.

In Section 7 we show how the basic concepts of graph transformation - introduced in Section 3 for the double-pushout approach - can be extended to Petri net transformations in the case of place/transition nets. In addition we discuss a general result concerning the compatibility of horizontal structuring and transformation of Petri nets, which has been used in the example of Section 6. Moreover we give an overview of results as well for other Petri net classes, which kind of Petri net transformations are preserving interesting properties like safety and liveness.

The conclusion in Section 8 summarizes the main ideas of this paper and further aspects concerning graph grammars and Petri net transformations.

2 General Overview of Graph Grammars and Graph Transformation

The research area of graph grammars or graph transformations is a discipline of computer science which dates back to the early seventies. Methods, techniques, and results from the area of graph transformations have already been studied and applied in many fields of computer science such as formal language theory, pattern recognition and generation, compiler construction, software engineering, concurrent and distributed systems modeling, database design and theory, logical and functional programming, AI, visual modeling, etc.

The wide applicability is due to the fact that graphs are a very natural way of explaining complex situations on an intuitive level. Hence, they are used in computer science almost everywhere, e.g. as data and control flow diagrams, entity relationship and UML diagrams, Petri nets, visualization of software and hardware architectures, evolution diagrams of nondeterministic processes, SADT diagrams, and many more. Like the *token game* for Petri nets, a graph transformation brings dynamic to all these descriptions, since it can describe the evolution of graphical structures. Therefore, graph transformations become attractive as a *modeling and programming paradigm* for complex-structured software and graphical interfaces. In particular, graph rewriting is promising as a comprehensive framework in which the transformation of all these very different structures can be modeled and studied in a uniform way.

Before we go into more detail let us discuss the basic question

2.1 What Is Graph Transformation?

In fact, graph transformation has at least three different roots

- from Chomsky grammars on strings to graph grammars
- from term rewriting to graph rewriting
- from textual description to visual modeling.

Altogether we use the notion graph transformation to comprise the concepts of graph grammars and graph rewriting. In any case, the main idea of graph transformation is the rule-based modification of graphs as shown in Figure 1.

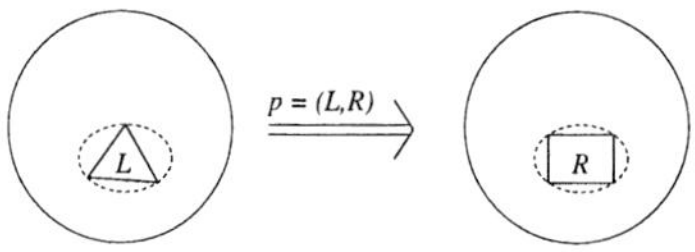

Fig. 1. Rule-based Modification of Graphs

The core of a rule or production $p = (L, R)$ is a pair of graphs (L, R), called left hand side L and right hand side R. Applying the rule $p = (L, R)$ means to find a match of L in the source graph and to replace L by R leading to the target graph of the graph transformation. The main technical problem is how to connect R with the context in the target graph. In fact, there are different solutions how to handle this problem leading to different graph transformation approaches, which are summarized below.

2.2 Overview of Different Approaches

The main graph grammar and graph transformation approaches developed in literature so far are presented in the *Handbook of Graph Grammars and Computing by Graph Transformation vol 1: Foundations* [32].

1. The *node label replacement approach*, mainly developed by Rozenberg, Engelfriet and Janssens, allows replacing a single node as left hand side L by an arbitrary graph R. The connection of R with the context is determined by embedding rules depending on node labels.

2. The *hyperedge replacement approach*, mainly developed by Habel, Kreowski and Drewes, has as left hand side L a labeled hyperedge, which is replaced by an arbitrary hypergraph R with designated attachment nodes corresponding to the nodes of L. The gluing of R with the context at corresponding attachment nodes leads to the target graph.
3. The *algebraic approaches* are based on pushout and pullback constructions in the category of graphs, where pushouts are used to model the gluing of graphs. The double pushout approach, mainly developed by Ehrig, Schneider and the Berlin- and Pisa-groups, is introduced in Sections 3-5 in more detail.
4. The *logical approach*, mainly developed by Courcelle and Bouderon, allows expressing graph transformation and graph properties in modanic second order logic.
5. The *theory of 2-structures* was initiated by Rozenberg and Ehrenfeucht as a framework for decomposition and transformation of graphs.
6. The *programmed graph replacement approach* by Schuerr used programs in order to control the nondeterministic choice of rule applications.

2.3 Aims and Paradigms for Graph Transformation

Computing was originally done on the level of the *von Neumann Machine* which is based on machine instructions and registers This kind of low level computing was considerably improved by assembler and high level imperative languages. From the conceptual - but not yet from the efficiency point of view - these languages were further improved by functional and logical programming languages. This newer kind of computing is mainly based on term rewriting, which - in the terminology of graphs and graph transformations - can be considered as a concept of tree transformations. Trees, however, do not allow sharing of common substructures, which is one of the main reasons for efficiency problems concerning functional and logical programs. This leads to consider graphs rather than trees as the fundamental structure of computing.

The main idea is to advocate graph transformations for the whole range of computing. Our concept of *Computing by Graph Transformations* is not limited to programming but includes also specification and implementation by graph transformations as well as graph algorithms and computational models and computer architectures for graph transformations.

This concept of Computing by Graph Transformations has been developed as basic paradigm in the ESPRIT Basic Research Actions COMPUGRAPH and APPLIGRAPH as well as in the TMR Network GETGRATS in the years 1990-2002. It can be summarized in the following way:

Computing by graph transformation is a fundamental concept for

- programming
- specification
- concurrency
- distribution
- visual modeling.

The aspect to support visual modeling by graph transformation is one of the main intentions of the ESPRIT TMR Network SEGRAVIS (2002-2006). In fact, there is a wide range of applications to support visual modeling techniques, especially in the context of UML, by graph transformation techniques. A state of the art report for applications, languages and tools for graph transformation on one hand and for concurrency, parallelism and distribution on the other hand is given in volumes 2 and 3 of the *Handbook of Graph Grammars and Computing by Graph Transformation* [10] and [15]

3 Introduction to the DPO-Approach

As mentioned already in the general overview there are several algebraic graph transformation approaches based on pushout and pullback constructions in the category of graphs. The most prominent one is the double-pushout approach, short DPO-approach, initiated by Ehrig, Pfender and Schneider in [17]. The main idea is to model graph transformation by two gluing constructions for graphs and each gluing construction by a pushout. Roughly spoken, a production is given by $p = (L, K, R)$, where L and R are the left and right hand side graphs and K is a common interface of L and R. Given a production $p = (L, K, R)$ and a context graph D, which includes also the interface K, the source graph G of a graph transformation $G \Rightarrow H$ via p is given by the gluing of L and D via K, written $G = L +_K D$, and the target graph H by the gluing of R and D via K, written $H = R +_K D$. More precisely we will use graph morphisms $K \to L, K \to R$ and $K \to D$ to express how K is included in L, R, and D respectively. This allows to define the gluing constructions $G = L +_K D$ and $H = R +_K D$ as pushout constructions **(1)** and **(2)** leading to a double pushout in Figure 2.

$$\begin{array}{ccccc} L & \longleftarrow & K & \longrightarrow & R \\ \downarrow & \textbf{(1)} & \downarrow & \textbf{(2)} & \downarrow \\ G & \longleftarrow & D & \longrightarrow & H \end{array}$$

Fig. 2. DPO-Graph Transformation

Before we present more technical details of the DPO-approach, let us point out that it is based on graphs and total graph morphisms. In fact there is a slightly more general approach using graphs and partial graph morphism, where a graph transformation can be expressed by a single pushout. This approach has been initiated by Raoult and fully worked out by Lwe leading to the single pushout approach, short SPO-approach. A detailed presentation and comparison of both approaches is given in volume 1 of the handbook [32]. The DPO-approach has been generalized from graphs to any other kind of high-level structures. This leads to the theory of high-level replacement systems initiated in [12], which can be applied to Petri nets leading to net transformation systems considered in Section 6 and Section 7 of this paper.

3.1 Graphs and Graph Morphisms

A *directed, labeled graph* G, short graph, over fixed sets of colors Ω_E and Ω_V for edges and vertices is given by

$$G \quad = \quad \Omega_E \xleftarrow{le} E \overset{s}{\underset{t}{\rightrightarrows}} V \xrightarrow{lv} \Omega_V$$

Fig. 3. Directed Labeled Graph G

where E and V are the sets of edges and vertices of G, s and t are the source and target functions, and l_e and l_v are the edge and vertex label functions respectively.

An example for such a graph G is the Pacman graph PG in Figure 4, where the color $*$ for the edges is omitted in PG.

- $\Omega_V = \{$  $\}$
- $\Omega_E = \{*\}$
- Identities of nodes and edges are not shown explicitly

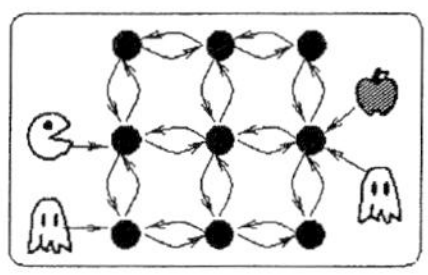

Fig. 4. Pacman Graph PG and Color Sets

A *graph morphism* $f : G \to G'$ consists of a pair of functions $f = (f_E : E \to E', f_V : V \to V')$, which is compatible with source, target, and label functions of G and G', i.e. $f_V \cdot s = s' \cdot f_E, f_V \cdot t = t' \cdot f_E, l'_e \cdot f_E = l_e$ and $l'_v \cdot f_v = l_v$.

The diagram schema for graph morphisms and an example for a graph morphism is given in Figure 5.

The category **Graph** has graphs as objects and graph morphisms as morphisms.

Let us point out that there are also several other notions of graphs and graph morphisms which are suitable for the DPO-approach of graph transformation; especially typed graphs and attributed graphs, where the color sets are replaced by a type graph and a type algebra respectively.

3.2 Graph Productions and Graph Grammars

A *graph production* $p = L \overset{l}{\leftarrow} K \overset{r}{\rightarrow} R$ consists of graphs L, K and R and (injective) graph morphisms $l : K \to L$ and $r : K \to R$ mapping the interface graph K to the left hand side L and the right hand side R respectively.

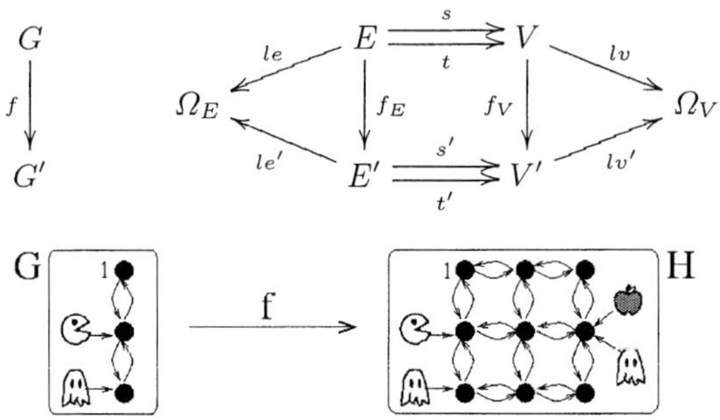

Fig. 5. Graph Morphism $f : G \to G'$

A *graph grammar* $GG = \{S, P, \Omega\}$ consists of a start graph, a set P of graph productions as defined above, and a pair of color sets $\Omega = (\Omega_E, \Omega_V)$, where S and the graphs in P are labeled over Ω. An example is the Pacman graph grammar

$PGG = \{PG, \{moveP, moveG, kill, eat\}\}$,

where the start graph PG is given in Figure 4 and the graph productions *moveP, moveG, kill, eat* in Figure 6. The production *moveG* is similar to *moveP*, where pacman is replaced by the ghost. These productions allow pacman resp. the ghost to move along an arc of the grid of the pacman graph PG. The productions *eat* resp. *kill* allow pacman to eat an apple resp. the ghost to kill pacman, provided that pacman and the apple resp. ghost are on the same node of the grid.

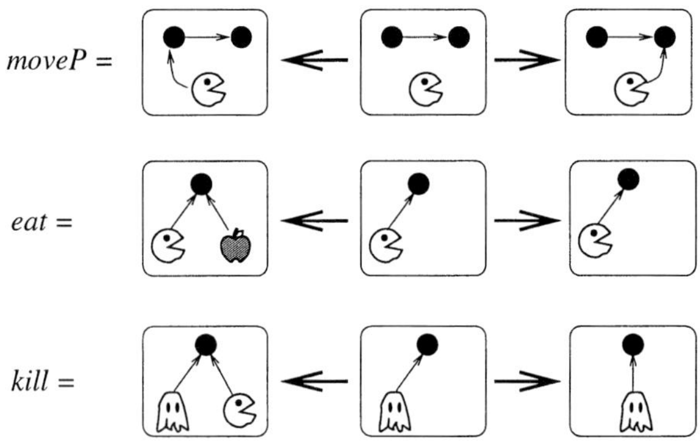

Fig. 6. Graph Productions of the Pacman Graph Grammar

Similar to Chomsky grammars it is also possible to distinguish between terminal and nonterminal color sets. In our case we have only terminal color sets. A graph grammar without distinguished start graph is also called *graph transformation system*.

3.3 Graph Transformation, Derivation and Graph Language

Given a graph production $p = L \stackrel{l}{\leftarrow} K \stackrel{r}{\rightarrow} R$, a graph G and a graph morphism $m : L \rightarrow G$, called *match* of L in G, then there is a *graph transformation*, also called *direct derivation*, if a double-pushout diagram as shown in Figure 7 can be constructed, where (1) and (2) are pushouts in the category **Graph**.

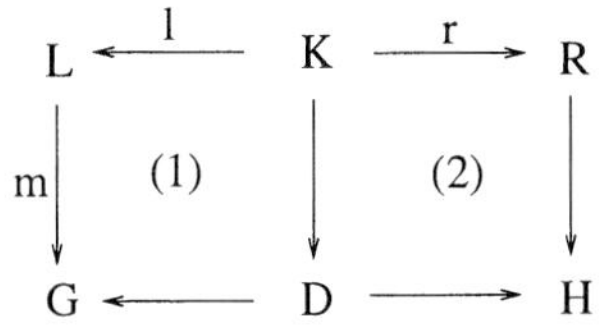

Fig. 7. Graph Transformation with Pushouts (1) and (2)

A *graph transformation* as given in Figure 7 is denoted by $G \stackrel{p,m}{\Longrightarrow} H$, or $G \Longrightarrow H$ via (p, m), where G is the source graph and H the target graph. In the next section we will show that pushouts can be interpreted as gluing constructions. Given a production and a match $m : L \rightarrow G$ means that we require to be able to construct a context graph D such that G is the gluing of L and D along K in pushout (1) and H is the gluing of R and D along K in pushout (2) of Figure 7, written

$G = L +_K D$ and $H = R +_K D$.

The morphism $R \rightarrow H$ in Figure 7 is called *comatch* of the graph transformation. A *graph transformation sequence*, also called *derivation*, is given by a finite sequence of graph transformations

$$G_0 \stackrel{p_1,m_1}{\Longrightarrow}, G_1 \stackrel{p_2,m_2}{\Longrightarrow} \ldots\ldots \stackrel{p_n,m_n}{\Longrightarrow} G_n.$$

The *graph language* generated by a graph grammar $GG = \{S, P, \Omega\}$ is the set of all graphs derivable from the start graph S with productions in P.

An example of a graph transformation using the production *moveP* is given in Figure 8, where pacman is moving from node 1 in graph G to node 2 in graph $H = G_1$. Moreover, Figure 8 shows a graph transformation sequence, where after this first step the productions *moveP* again, and also *eat* and *kill* are applied.

In Figure 8 it is intuitively clear that G is the gluing of L and D along K and H the gluing of R and D along K. Vice versa, given the production *moveP* and the match $m : L \rightarrow G$ the context graph D can be constructed by removing from G all items of L, which are not in the interface K. In our case it is only the arc from pacman to node 1, which has to be removed. This is the first step in the explicit construction of a graph transformation. It leads to a pushout (1) in Figure 7 if a suitable *gluing condition* is satisfied which will be explained below. The second step is the gluing of R and D along K leading to pushout (2) in Figure7.

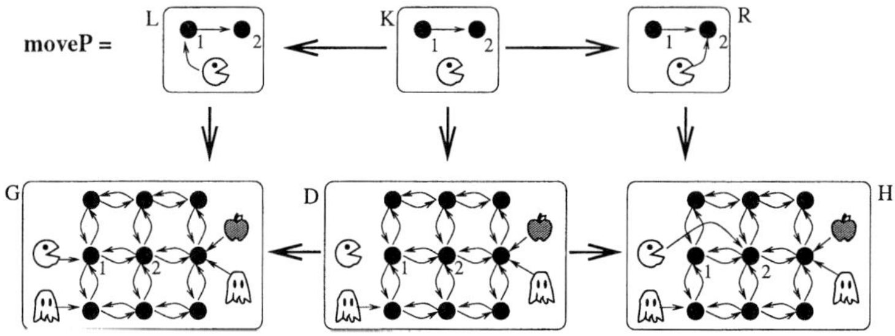

$$G = G_0 \overset{moveP}{\Longrightarrow} G_1 \overset{moveP}{\Longrightarrow} G_2 \overset{eat}{\Longrightarrow} G_3 \overset{kill}{\Longrightarrow}$$

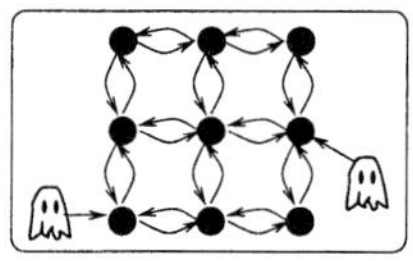

Fig. 8. A Sample Graph Transformation and Derivation

In general the construction of a graph transformation $G \overset{p,m}{\Longrightarrow} H$ from a production $p = L \overset{l}{\leftarrow} K \overset{r}{\rightarrow} R$ and a match $m : L \rightarrow G$ is given in two steps, where the first step requires that the gluing condition (see 3.5 below) is satisfied:

STEP 1 (DELETE): Delete $m(L - l(K))$ from G leading to a context graph D (if the gluing condition is satisfied), s.t. G is the gluing of L and D along K, i.e. $G = L +_K D$ in (1) of Figure 7.

STEP 2 (ADD): Add $R - r(K)$ to D leading to a graph H, s.t. H is the gluing of R and D along K, i.e. $H = R +_K D$ in (2) of Figure 7.

3.4 Gluing Construction and Pushout

The idea of the gluing construction of graphs makes sense also for other kinds of structures, where the idea is to construct the union of structures along a common substructure. For structures given by geometrical figures this kind of union or gluing is shown in Figure 9.

In the framework of category theory the idea of the gluing construction can be formalized by the notion of a *pushout*: Given objects (e.g. sets, graphs or structures) A,B, and C and morphisms (e.g. functors, graph or structure morphisms) $f : A \rightarrow B$ and $g : A \rightarrow C$ an object D together with morphisms $h : B \rightarrow D$ and $k : C \rightarrow D$ is called *pushout* of f and g if we have $h \circ f = k \circ g$ (i.e. diagram (1) in Figure 10 commutes) and the following universal property is satisfied:

For all objects D' and morphisms $h' : B \rightarrow D', k' : C \rightarrow D'$ with $h' \circ f = k' \circ g$ (i.e. the outer diagram in Figure 10 commutes) we have a unique morphism $d : D \rightarrow D'$ s.t. $d \circ h = h'$ and $d \circ k = k'$ (i.e. diagrams (2) and (3) commute).

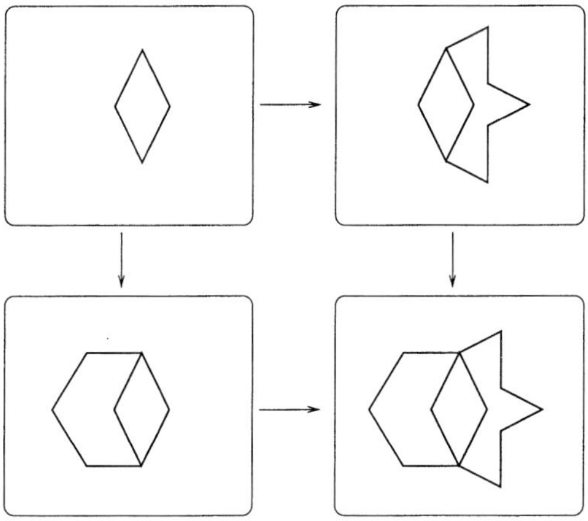

Fig. 9. Gluing Construction for Geometrical Figures

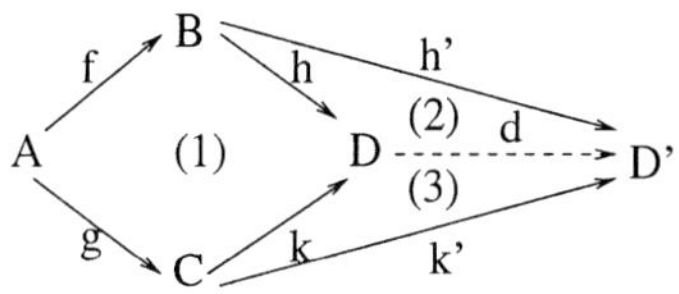

Fig. 10. Universal Pushout Property

In the category **Sets** of sets and functions the pushout object D is given by the quotient set

$D = B + C/\equiv$, short $D = B +_A C$,

where $B + C$ is the disjoint union of B and C and $\equiv$ the equivalence relation generated by $f(a) \equiv g(a)$ for all $a \in A$. In fact D can be interpreted as the gluing of B and C along A: Starting with the disjoint union $B + C$ we glue together the elements $f(a) \in B$ and $g(a) \in C$ for each $a \in A$.

In the category **Graph** the pushout graph D can be constructed componentwise for the set of edges and the set of vertices using the pushout construction in **Sets** discussed above. This shows that also the pushouts in **Graph** can be interpreted as a gluing construction (see Figure 8). In general, the pushout graph $D = (D_E, D_U, s_D, t_D, le_D, lv_D)$ is given as follows:

- $D_E = B_E +_{A_E} C_E$
- $D_V = B_V +_{A_V} C_V$
- $s_D(e) = \begin{cases} [s_B(e')] & ; \text{ if } e = h_E(e') \\ [s_C(e'')] & ; \text{ if } e = k_E(e'') \end{cases}$
- $t_D(e) = \begin{cases} [t_B(e')] & ; \text{ if } e = h_E(e') \\ [t_C(e'')] & ; \text{ if } e = k_E(e'') \end{cases}$

– $le_D(e) = \begin{cases} [le_B(e')] & ; \text{ if } e = h_E(e') \\ [le_C(e'')] & ; \text{ if } e = k_E(e'') \end{cases}$

– $lv_D(v) = \begin{cases} [lv_B(v')] & ; \text{ if } v = h_V(v') \\ [lv_C(v'')] & ; \text{ if } v = k_V(v'') \end{cases}$

In fact the pushout construction is well-defined and unique up to isomorphism. This means that the graph D can also be replaced by any other graph $\overline{D}$, which is isomorphic to D, i.e. there is a bijective graph morphism $f : D \to \overline{D}$.

Uniqueness of pushouts up to isomorphism is a general property of pushouts in arbitrary categories. Moreover, it is a general property that pushouts can be composed horizontally and vertically leading again to pushouts.

3.5 Gluing Condition and Pushout Complement

In order to construct a graph transformation from a given graph production $p = (L \xleftarrow{l} K \xrightarrow{r} R)$ and a match $m : L \to G$ as shown in Figure 7 we have to construct first a graph D and graph morphisms $K \to D$ and $D \to G$ s.t. diagram (1) in Figure 7 becomes a pushout in the category **Graph**. In this case D is called *pushout complement* of $l : K \to L$ and $m : L \to G$. See also the left diagram in Figure 11. In general, however, the pushout complement may not exist, or may not be unique up to isomorphism. In Figure 11 we show two examples in the category **Sets**, where in the middle there is no pushout complement D for given functions $l : K \to L$ and $m : L \to G$. On the right hand side we have two different non-isomorphic pushout complements D and D'.

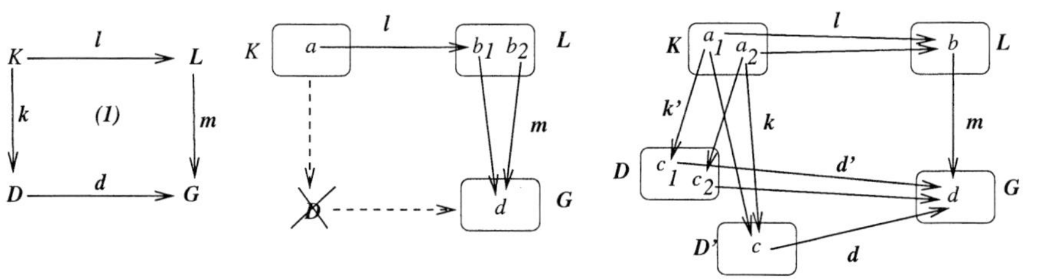

Fig. 11. Construction, Non-Existence and Non-Uniqueness

In the category **Sets** and **Graphs** we have uniqueness of the pushout complement up to isomorphism $l : K \to L$ is injective. For the existence of the pushout complement we need a *Gluing Condition*. Given an injective graph morphism $l : K \to L$ and a match $m : L \to G$ we can construct a pushout complement D leading to the pushout (1) in Figure 11 if and only if the following *Gluing condition* is satisfied that requires that the *boundary* of the match $m : L \to G$ is included in the gluing part $l(K)$ of L. More formally, we have:

Gluing Condition:

$$BOUNDARY \subseteq GLUING$$

where *BOUNDARY* and *GLUING* are subgraphs of L defined by

- $GLUING = l(K)$
- $DANGLING = \{x \in L_V \mid \exists e \in G_E - m_E(L_E) : (m_V(x) = s_G(e) \text{ or } m_V(x) = t_G(e))\}$
- $IDENTIFICATION = \{x \in K \mid \exists y \in K : (x \neq y \text{ and } m(x) = m(y))\}$, where $x \in K$ means $x \in K_V$ with $m = m_V$ or $x \in K_E$ with $m = m_E$, and
- $BOUNDARY = DANGLING \cup IDENTIFICATION$

This means that the boundary of the match m given by the graph *BOUNDARY* consists of a dangling and an identification part. In the identification part we have all those nodes and edges which are identified by the match m. The dangling part consists of those nodes $x \in L$ so that $m_V(x)$ is adjacent to an edge $e \in G_E$, which is not part of the match $m(L)$. These edges are called *dangling edges* because they lack either the source or the target node in the set theoretical complement $G - m(L) = (G_E - m_E(L_E), G_V - m_V(L_V))$. For brevity we call the nodes in *DANGLING* dangling nodes.

Now we can construct the pushout complement graph D in Figure 11 in the diagram to the left by $D = (D_E, D_V, s_D, t_D, l_{eD}, l_{vD})$ with

- $D_E = (G_E \setminus m_E(L_E)) \cup m_E(l_E(K_E))$
- $D_V = (G_V \setminus m_V(L_V)) \cup m_V(l_V(K_V))$
- $T_C = (T_N \setminus m_T(T_L)) \cup m_T(l_T(T_K))$
- s_D, t_D, l_{eD},and l_{vD} are defined by the restriction of s_G, t_G, l_{eG},and l_{vG} respectively.

Finally the graph morphisms in the diagram to the left in Figure 11 $d : D \to G$ and $h : K \to D$ are given by the inclusion $D \subseteq G$ and by $k(x) = m \circ l(x)$ for nodes and edges $x \in K$.

In our pacman graph grammar *PGG* considered above we can have only injective matches $m : L \to G$, because the pacman graph *PG* in Figure 4 and Figure 8 has no loops. This implies that the identification part of the gluing condition is always satisfied. But also the dangling part is satisfied for all productions and all matches, because all nodes of the left-hand side L of each production are gluing nodes. Hence especially all dangling nodes of L are gluing nodes. In the graph transformation shown in Figure 8 both nodes 1 and 2 are dangling and also gluing nodes.

4 Concepts of Parallelism

In this section we present main concepts and results for parallelism of graph transformations. We start with the concepts of parallel and sequential independence leading to a local Church-Rosser Theorem which corresponds to the concept of concurrency by interleaving. However, using the concept of parallel productions and derivations, the DPO-approach also allows to model true concurrency. The Parallelism Theorem shows equivalence of true concurrency and interleaving in our framework. Finally the Parallelism Theorem allows to formulate shift equivalence leading to canonical parallel derivations.

4.1 Parallel and Sequential Independence

Two graph transformations $G \Rightarrow H_1$ via (p_1, m_1), $G \Rightarrow H_2$ via (p_2, m_2) are called *parallel independent* if the matches $m_1 : L_1 \to G$ and $m_2 : L_2 \to G$ only overlap in gluing items which are preserved by both graph transformations, i.e.

$$m_1(L_1) \quad \cap \quad m_2(L_2) \quad \subseteq \quad m_1(l_1)K_1)) \quad \cap \quad m_2(l_2)K_2))$$

for $p_i = (L_i \stackrel{l_i}{\leftarrow} K_i \stackrel{r_i}{\longrightarrow} R_i)$ and $i = 1, 2$.

In Figure 12 we show two productions $p_1 = move\ P$ and $p_2 = move\ G$ with matches $m_1 : L_1 \to PG$ and $m_2 : L_2 \to PG$ which satisfy the conditions for parallel independence. In fact, the matches overlap exactly in node 1 which is gluing node for both productions and hence preserved by the corresponding derivations. The first derivation $PG \Rightarrow H_1$ via (*move* P, m_1) is explicitly shown in fig 8 with $PG = G$ and $H_1 = H$. Moreover, the match $m_2 : L_2 \to PG$ can be extended to a match $m_2 : L_2 \to H_1$ leading to a derivation $H_1 \Rightarrow X$ via (*move* G, m_2'). In fact, the two derivations $PG \Rightarrow H_1$ via (*move* P, m_1) and $H_1 \Rightarrow X$ via $(moveG, m_2')$ are sequential independent in the sense defined below.

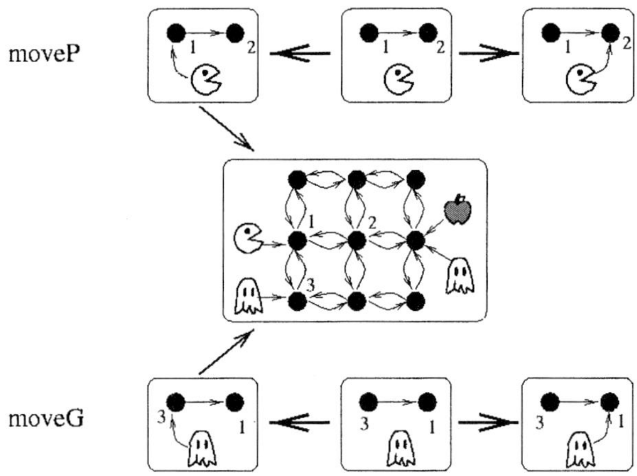

Fig. 12. Parallel Independence

Two graph transformations $G \Rightarrow H_1$ via (p_1, m_1) (with comatch $m_1' : R_1 \to H_1$) and $H_1 \Rightarrow X$ via (p_2, m_2) are called *sequential independent* if the comatch $m_1' : R_1 \to H_1$) and the match $m_2' : L_2 \to H_2$) only overlap in gluing items, i.e.

$$m_1'(R_1) \quad \cap \quad m_2(L_2) \quad \subseteq \quad m_1'(r_1(K_1)) \quad \cap \quad m_2(l_2(K_2))$$

Parallel and sequential independence of graph transformations are suitable conditions to allow interleaving of graph transformations as shown in the following theorem:

4.2 Local Church-Rosser Theorem

The following conditions for graph transformations are equivalent and each of them is leading to the diamond of parallel and sequential graph transformations in Figure 13, called *local Church-Rosser property.*

1. $G \Rightarrow H_1$ via (p_1, m_1) and $G \Rightarrow H_2$ via (p_2, m_2) are parallel independent
2. $G \Rightarrow H_1$ via (p_1, m_1) and $H_1 \Rightarrow X$ via (p_2, m_2') are sequential independent
3. $G \Rightarrow H_2$ via (p_2, m_2'') and $H_2 \Rightarrow X$ via (p_1, m_1') are sequential independent.

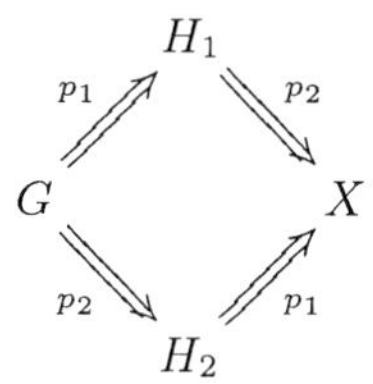

Fig. 13. Local Church-Rosser Property

An explicit proof of the local Church-Rosser Theorem is given in [14]. It is based on suitable composition and decomposition properties of pushouts.

In the following we will see that parallel independence of graph transformations also allows to construct a parallel derivation $G \Rightarrow X$ via a parallel production $p_1 + p_2$.

4.3 Parallel Productions and Parallel Derivations

Given productions $p_i = (L_i \stackrel{l_i}{\leftarrow} K_i \stackrel{r_i}{\rightarrow} R_i)$ for $i = 1, 2$ the *parallel production* $p_1 + p_2$ is given by

$$p_1 + p_2 = (L_1 + L_2 \stackrel{l_1+l_2}{\leftarrow} K_1 + K_2 \stackrel{r_1+r_2}{\rightarrow} R_1 + R_2)$$

where $L_1 + L_2, l_1 + l_2$ etc. is the disjoint union of graphs and graph morphisms respectively. This corresponds to the coproduct of objects and morphisms in the category **Graphs**.

An example for the parallel production *move P + move G* is shown in Figure 14.

Parallel independence of *move P* and *move G* in Figure 12 implies according to the following Parallelism Theorem a parallel derivation $G \Rightarrow X$ via $(p_1 + p_2, m)$, where the match $m : L_1 + L_2 \to G$ is a non-injective graph morphism induced by $m_1 : L_1 \to G$ and $m_2 : L_2 \to G$. The nodes 4 and 1 in Figure 14 are identified with node 1 in Figure 12. In the derived graph X pacman is on node 2 and the ghost on node 1.

In general, a derivation with a parallel production is called *parallel derivation.*

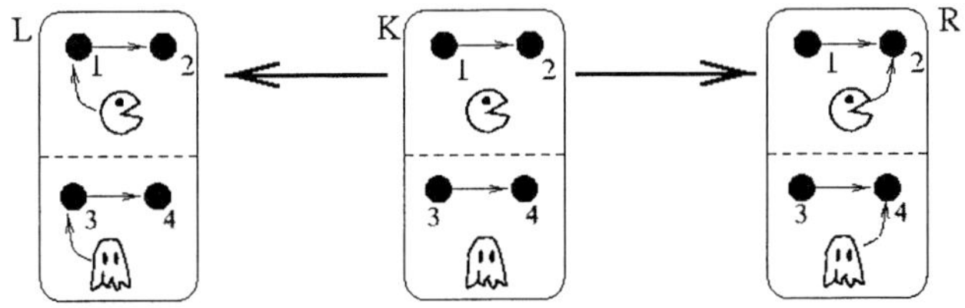

Fig. 14. Parallel Production *move P + move G*

4.4 Parallelism Theorem

The following conditions for graph transformations are equivalent;

1. $G \Rightarrow H_1$ via (p_1, m_1) and $G \Rightarrow H_2$ via (p_2, m_2) are parallel independent
2. $G \Rightarrow X$ via $(p_1 + p_2, m)$ is a parallel derivation, where $(p_1 + p_2)$ is the parallel production of p_1 and p_2 and $m_1 : L_1 + L_2 \rightarrow G$ is the match induced by $m : L_1 \rightarrow G$ and $m_2 : L_2 \rightarrow G$.

Together with the Local Church-Rosser Theorem we obtain the parallelism diamond shown in Figure 15.

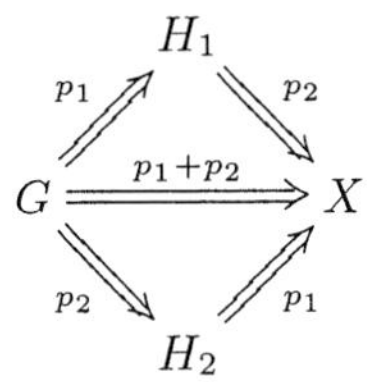

Fig. 15. Parallelism Diamond

If p_1 and p_2 are sequentially independent in a derivation sequence $G_1 \overset{p_1}{\Longrightarrow} G_2 \overset{p_2+p_3}{\Longrightarrow} G_3$ then this sequence is *shift equivalent* to a derivation sequence $G_1 \overset{p_1+p_2}{\Longrightarrow} G_2' \overset{p_3}{\Longrightarrow} G_3$ and we obtain the *shift relation* shown in Figure 16.

$$G_1 \overset{p_1}{\Longrightarrow} G_2 \overset{p_2+p_3}{\Longrightarrow} G_3 \quad \sqsubseteq_{shift} \quad G_1 \overset{p_1+p_2}{\Longrightarrow} G_2' \overset{p_3}{\Longrightarrow} G_3$$

Fig. 16. Shift Relation

Shift equivalence on parallel derivations is the closure of the shift relation under parallel and sequential composition. The shift relation is well-founded. The minimal derivations with respect to shift relation are called *canonical derivations*. Canonical derivations are unique representations of shift equivalent parallel derivation classes (see [3] for more details).

5 Graph Grammars, Petri Nets and Concurrent Semantics

In this section we discuss the relationship between graph grammars and Petri nets. Both of them are well-known as specification formalisms for concurrent and distributed systems. First we show how the token game of place-transition nets can be modeled by double pushouts of discrete labeled graphs. This allows to relate basic notions of place-transition nets like marking, enabling, firing, steps and step sequences, to corresponding notions of graph grammars and to transfer semantical concepts from Petri nets to graph grammars. Since a marking of a net on one hand and a graph of a graph grammar on the other hand correspond to the state of a system to be modeled, graph grammars can be seen to generalize place-transition nets by allowing more structured states. In the second part of this section we give a short overview of the concurrent semantics of graph transformations presented in [3] of the Handbook of Graph grammars volume 3, which is strongly influenced by corresponding semantical constructions for Petri nets in [36]. Finally let us point out that we discuss the modification of the net structure of Petri nets using graph transformations in the next chapter.

5.1 Correspondence of Notions between Petri Nets and Graph Grammars

The firing of a transition in a place-transition net can be modeled by a double pushout in the category of discrete graphs labeled over the places of the transitions. Let us consider the transition firing as token game in Figure 17.

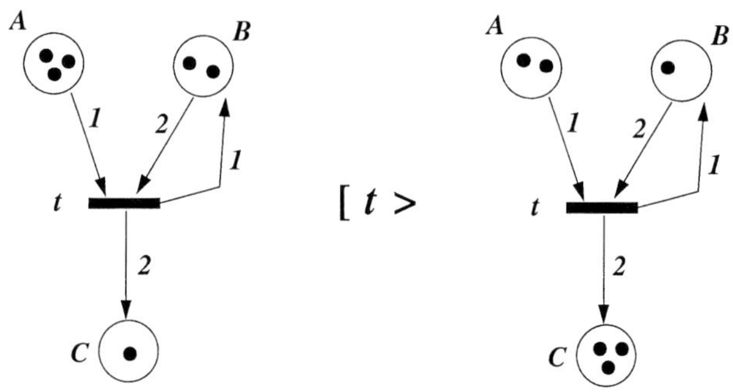

Fig. 17. Transition Firing as Token Game

The transition t in Figure 17 requires in the pre-domain one token on place A and two tokens on place B and produces in the post-domain one token on B and two tokens on place C. This corresponds to the production in the upper row of Figure 18, where the left hand side consists of three nodes labeled A, B and B and the right hand sie of three nodes labeled B, C and C. The empty interface of the production means that no node is preserved by the production, which corresponds to the token game in place-transition nets. In fact, the transition t in Figure 17 consumes two tokens and produces one token on place B. Preservation of tokens in the framework of Petri nets can be modeled by contextual nets, and transition with context places can be modeled by productions with nonempty interface.

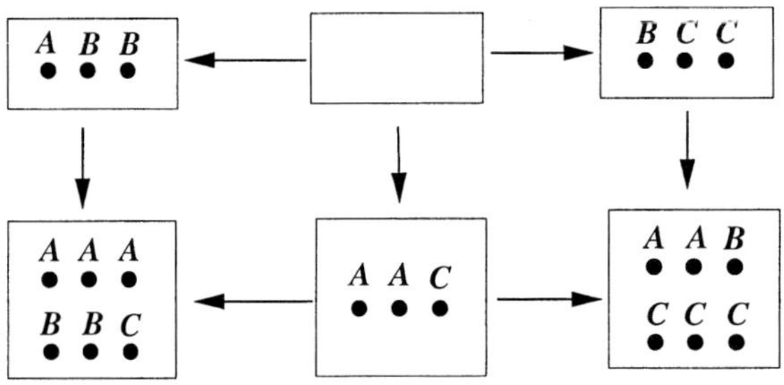

Fig. 18. Transition Firing as Double Pushout

The marking of the left-hand side net in Figure 17 corresponds to the discrete graph to the left in the lower row of Figure 18, while the marking after firing of the transition in Figure 17 corresponds to the discrete graph to the right.

The discrete graph in the middle of Figure 18 is the result of the deleting step of the double pushout and that on the right in the lower row is the result of the adding step. This shows that the firing step in Figure 17 corresponds exactly to a direct derivation in the double-pushout approach. This correspondence of notions between place/transition nets and graph grammars is shown in Table 1 in more detail. In fact, enabling of a transition at a marking corresponds to applicability of a production to a graph, concurrency of transitions corresponds to parallel independent productions applied with non-overlapping matches, conflicts correspond to parallel dependent direct derivations with overlapping matches, a parallel transition step of concurrent transitions corresponds to a parallel direct derivation, and finally a step sequence to a parallel derivation.

5.2 Concurrent Semantics of Graph Transformation

For sequential systems it is often sufficient to consider an input/output semantics and thus the appropriate semantic domain is usually a suitable class of functions from the input to the output domains. When concurrent or distributed features are involved, instead, typically more information about the actual computation of the system has to be recorded in the semantic domain. For instance, one may want to know which steps of computation are independent (concurrent), which are causally related and which are the (non-deterministic) choice points. This information is necessary, for example, if one wants to have a compositional semantics, allowing to reduce the complexity of the analysis of concurrent systems

Table 1. Correspondence of Notions

Petri Nets	**Graph Grammars**
tokens	nodes
places	node labels
marking	discrete, labeled graph
transition enabled at a marking	production applicable to a graph
firing	direct derivation
firing sequence	derivation
concurrent transitions	parallel independent productions
conflict	parallel dependence
step	parallel direct derivation
step sequence	parallel derivation

built form smaller parts, or if one wants to allocate a computation on a distributed architecture. Roughly speaking, *non-determinism* can be represented either by collecting all the possible different computations in a set, or by merging the different computations in a unique *branching* structure where the choice points are explicitly represented. On the other hand, *concurrent* aspects can be represented by using a *truly concurrent* approach, where the casual dependencies among events are described directly in the semantics using a partially ordered structure. Alternatively, an *interleaving* approach can be adopted, where concurrency is reduced to non-determinism, in the sense that the concurrent execution of events is represented as the non-deterministic choice among the possible interleavings of such events.

Let us first have a look to the area of Petri nets, where a well-established theory has been developed already.

Petri nets have been equipped with rich, formal computation-based semantics, including both interleaving and truly concurrent models. In many cases such semantics have been defined by using well-established categorical techniques, often involving adjunctions between suitable categories of nets and corresponding categories of models. Let us point out especially the semantics of safe place-transition nets presented as a chain of adjunctions by Winskel [36].

To propose graph transformation systems as a suitable formalism for the specification of concurrent/distributed systems that generalizes Petri nets, we are naturally led to the attempt of equipping them with a satisfactory semantic framework, where the truly concurrent behavior of grammars can be suitably described and analyzed. The basic result for interleaving and concurrent semantics of graph transformation are the local Church-Rosser Theorem and the Parallelism Theorem presented in the previous section. In the following we present the main ideas of trace, process and event structure semantics for graph transformations. For a more detailed overview we refer to the handbook article [3].

The *trace semantics* for graph transformations is based on parallel derivation sequences introduced in the previous section. Derivation traces are defined as equivalence classes of parallel derivations with respect to the shift equivalence, which is the closure of the shift relation (see Figure 16) under parallel and sequential composition. Abstraction equivalence is a suitable refinement of the isomorphism relation on parallel derivations, which allows to obtain a well-defined concatenation of derivation traces. This leads to a category $\mathbf{Tr}(\mathbf{G})$ of derivation traces of a graph grammar G, which can be considered as the trace semantics of G.

The *process semantics* for graph transformations is based on the notion of a graph process, which is a suitable generalization of a Petri net process. In fact, the idea of occurrence nets and concatenable net processes has been generalized to occurrence graph grammars and concatenable graph processes. The mapping of an occurrence graph grammar O to the original graph grammar G determines for each derivation of O a corresponding derivation of G, such that all derivations of O correspond to the full class of shift-equivalent derivations. This means that the graph process, defined by the occurrence graph grammar O together with the mapping from O to G, can be considered as an abstract representation of the shift-equivalence class. Hence the graph process plays a role similar to the canonical derivation introduced in the previous section. The process semantics for graph transformations is defined by the category $\mathbf{CP}(\mathbf{G})$ of abstract graphs as objects and concatenable processes of G as morphisms.

The *event structure semantics* for graph transformations allows to construct an event structure for a graph grammar G which - in contrast to trace and process semantics - allows to reflect the intrinsic non-determinism of a grammar. Event structures and domains are well-known semantical models not only for Petri nets, but also for other specification techniques for concurrent and distributed systems. The domain $\mathbf{Dom}(\mathbf{G})$ of a graph grammar is a partially ordered set, where the elements of $\mathbf{Dom}(\mathbf{G})$ are derivation traces starting at the start graph G_S of G, and we have $d_1 \leq d_2$ for derivation traces $d_1 : G_S \Rightarrow G_1$ and $d_2 : G_S \Rightarrow G_2$, if there is a derivation trace $d : G_1 \Rightarrow G_2$ with $d \circ d_1 = d_2$. Roughly spoken an event e in the event structure $\mathbf{ES}(\mathbf{G})$ of the graph grammar G corresponds to the application of a basic production $p(e)$ in a derivation trace $d(e) : G_S \Rightarrow G$. Moreover, we have a partial order $\leq$ and a conflict relation $\sharp$ in $\mathbf{ES}(\mathbf{G})$, where roughly spoken $e_1 \leq e_2$ means $d(e_1) \leq d(e_2)$, and $e_1 \sharp e_2$ means that there is no derivation trace $d : G_S \Rightarrow G$ including both $p(e_1)$ and $p(e_2)$. In the first case e_1 and e_2 are casually related and in the second case they are in conflict.

In the handbook article [3], where all these semantics are presented in detail, it is also shown how these different graph transformation semantics are related with each other (see 1.-3. below).

1. The trace semantics $\mathbf{Tr}(\mathbf{G})$ and the process semantics $\mathbf{CP}(\mathbf{G})$ are equivalent in the sense that both categories are isomorphic.
2. For consuming graph grammars G the event structure semantics $\mathbf{ES}(\mathbf{G})$ and the domain semantics $\mathbf{Dom}(\mathbf{G})$ are conceptually equivalent in the sense that one can be recovered from the other. A grammar G is called *consuming* if each production of the grammar deletes at least one node or edge. This correspondence is a consequence of a well-known general result concerning the equivalence of prime event structures and domains, where the configurations

of a prime event structure are the elements of the domain. A configuration of a prime event structure is a subset of events, which is left-closed and conflict free. In our case the configurations of **ES**(**G**) correspond to the derivation traces in **Dom**(**G**).

3. As a consequence of results 1 and 2 above we obtain the following intuitive characterization of events and configurations from **ES**(**G**) in terms of processes: Configurations correspond to processes, which have as source graph the start graph of the grammar. Events are one-to-one with a subclass of such processes having a production which is the maximum w.r.t. the casual ordering.
4. In the case of Petri nets Winskel has shown in [36] that the category of safe place-transition nets is related by a chain of adjoint functors to the categories of domains and prime event structures. Motivated by this chain of adjunctions Baldan has shown in his dissertation [2] that there is a chain of functors between the category of graph grammars and prime event structures, which is based on the trace and event structure semantics **Tr**(**G**) and **ES**(**G**) discussed above. In fact, all but one steps in this chain of functors have been shown already to be adjunctions as in the case of Petri nets.

6 Introduction to Petri Net Transformations

In the second part of this contribution we investigate Petri net transformations.

Note that there is a shift of paradigm. In graph transformation systems graph productions are used to model the behavior. Obviously, this is not required for Petri nets as the token game already models the behavior. In the area of Petri nets the transformations are used to describe changes of the Petri net structure. So, we can describe the stepwise development of nets, and have a formal foundation for the evolution of Petri nets. The main advantages of Petri net transformations are:

- the rule-based approach
- compatibility with structuring and marking graph semantics
- extension to refinement

There already have been a few approaches to describe transformations of Petri nets formally (e.g. in [5, 6, 33, 7, 35]). The intention has been mainly on reduction of nets to support verification, and not on the development process itself.

First we discuss briefly the formal foundation of Petri net transformations as an instantiation of so called high-level replacement systems. This is a generalization of the DPO-approach from graphs to arbitrary specification techniques, that can be instantiated especially to graphs and different classes of Petri nets (see Figure 19). Subsequently we give an extensive example, stepwise developing the baggage handling system of an airport. Finally we discuss the relevance of net transformations as means for the rule-based modification and refinement of nets.

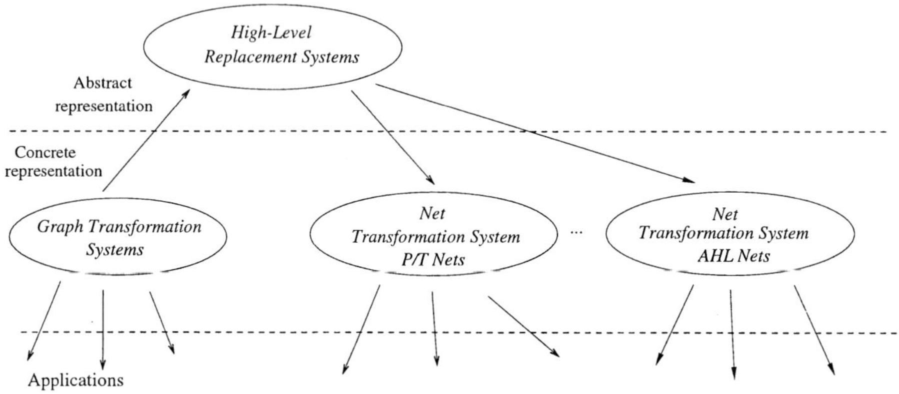

Fig. 19. Generalization and Instantiation

6.1 Formal Foundation Based on High-Level Replacement Systems

In this section we sketch the abstract frame work, that comprises the transformations of graphs in the previous and of Petri nets in the next sections. High-level replacement systems can be considered as a general description of replacement systems, where a left-hand side of the rule is replaced by a right-hand side in the presence of an interface. Historically, rules and transformations of Petri nets have been introduced as an instantiation of high-level replacement systems [12, 13, 28].

These kinds of replacement systems have been introduced in [13] as a categorical generalization of graph transformations in the DPO-approach. High-level replacement systems are formulated for an arbitrary category **Cat** with a distinguished class $\mathcal{M}$ of morphisms, called $\mathcal{M}$-morphisms. Figure 20 illustrates the main idea for some arbitrary specification or structure. The rule given in the upper line describes that a black triangle is replaced by a long dotted rectangular, if there is a light grey square below the triangle. The transformation is given by the bottom line, where the replacement specified by the rule is carried out.

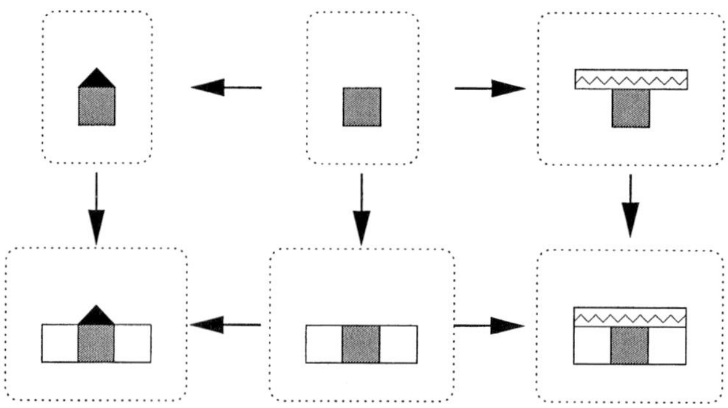

Fig. 20. Abstract Example

High-level replacement systems are a categorical generalization of the algebraic approach to graph transformation systems with double pushouts. They allow formulating the same notions as for graph transformation systems, but not only for graphs but for objects of arbitrary categories. That means, instead of replacing one graph by another one, now one object is replaced by another one. Due to the categorical formulation of high-level replacement systems the focus is not on the structure of the objects but on the properties of the category.

To achieve the results known in graph transformation systems, the instantiated category of a high-level replacement system has to satisfy certain HLR-conditions. In [24] an elegant reformulation of some HLR-conditions [26] is given in terms of adhesive categories.

In [27] we have extended the theory of high-level replacement systems where rules and transformations are required to preserve some desired properties of the specification. To do so, rules and transformations are equipped with an additional morphism that has to preserve or reflect specific properties. At this abstract level we merely can assume suitable classes of morphisms and then guarantee that these morphisms lead to property preserving rules and transformations. In Section 7.6 (and much more detailed in [29]) we give a glance how this approach works for Petri nets.

6.2 Example: Baggage Handling System

In this example we illustrate the rule-based modification of place/transition nets. Rules describe the replacement of a left-hand side net by a right-hand side net. The application of the rule yields a transformation where in the source net the subnet corresponding to the left-hand side is replaced by the subnet corresponding to the right-hand side. At this level as well as in this example there are no statements about the properties of the modified nets. Nevertheless based on the transformation we illustrate here, we already have developed a theory, called rule-based refinement, where the transformations are extended to introduce, preserve, or reflect net specific properties. In [29] a comprehensive survey can be found, in Section 7.6 we discuss this theory briefly.

This example concerns the sorting, screening and moving of baggage at an airport. The physical basis of the baggage handling system consists of check-in counter, conveyor belts, sorter, screening devices, a baggage claim carousel, storages, and loading stations. The conveyor belts are transportation belts, that are starting and ending at some fixed point (as check-in, sorter, loading station, baggage claim carousel, etc). The baggage handling system comprises three check-in counters, the primary sorter, the early baggage, the lost baggage as well as the unclaimed baggage storage, the secondary sorter, two loading stations, the baggage claim sorter with two baggage claim carousels and all the conveyor belts in between. Mainly, there are the following cases to handle:

1. *Check-in:* The baggage has to be moved from the check-in counter to the right loading station of a departing carrier. It has to pass a security check (screening the baggage). At the check-in the baggage items are placed manually into the transport system.

2. *Baggage Claim:* At the loading station a carrier is unloaded and the baggage items are placed manually into the transport system. The baggage has to be moved from the loading station to the right baggage claim carousel.
3. *Transfer:* The baggage has to be moved from the loading station of the arriving carrier to the loading station of the connecting flight carrier. The baggage is moved to the secondary sorter and subsequently either to the right loading station for the connecting flight or to the early baggage storage.
4. *Storing Baggage:* For baggage checked in early and for long waits between connecting flights there must be a storage, called early baggage storage. Moreover, misled or lost baggage has to be identified and is then handled manually. Baggage that is not claimed at the baggage claim carousel has to be stored as well.

To model the above given baggage handling system we can use low-level or high-level Petri nets. High-level net allow modeling the data explicitly, but for the purpose of this paper it is sufficient to use low-level nets. In fact, the basic principles for net transformations are the same for low-level and high-level nets. Subsequently we model the baggage handling system with place/transition nets, which requires some abstraction of the data – for example the baggage tags or the flight numbers are not modeled. Especially, we have modeled baggage as tokens, hence it cannot be distinguished. The choice what happens to baggage is accordingly no longer depended from the data, i.e. the baggage tag, but is done at random.

The place/transition net in Figure 21 is an abstraction of the above specified baggage handling system. The baggage handling system is an open system; baggage enters and leaves the system. We have modeled this using transitions without pre-domain for entering baggage and using transitions without post-domain for leaving baggage. Therefore we have the empty initial marking.

In the net in Figure 21 neither the conveyor belts nor the screening nor the lost or unclaimed baggage storage are modeled explicitly. Subsequently we present a step-by-step development of our first abstraction in Figure 21 that adds the lacking features. We want to add the representation of the conveyor belts by places, as well as the explicit modeling of the screening. Extending the net in this way yields a larger net. So we decompose the net into subnets in order to continue using the smaller subnets. Subsequently we introduce the subnets for the lost baggage storage and for the unclaimed baggage storage.

Introducing Conveyor Belts and Screening: The conveyor belt is not yet explicitly modeled. The transitions *t4* to *t14* represent conveyor belts. There are three different possibilities: A simple conveyor belt connecting two devices of the baggage handling system (e.g. between the sorters and the early baggage storage or the loading station) a conveyor belt connecting several devices (between check-in and the primary sorter), and a complex conveyor belt including a screening device with an optional manual check of unsafe baggage (between primary and secondary sorter). The baggage is considered by the screening device either as safe or as unsafe. If it is safe, then it is left on the conveyor belt. If it is considered

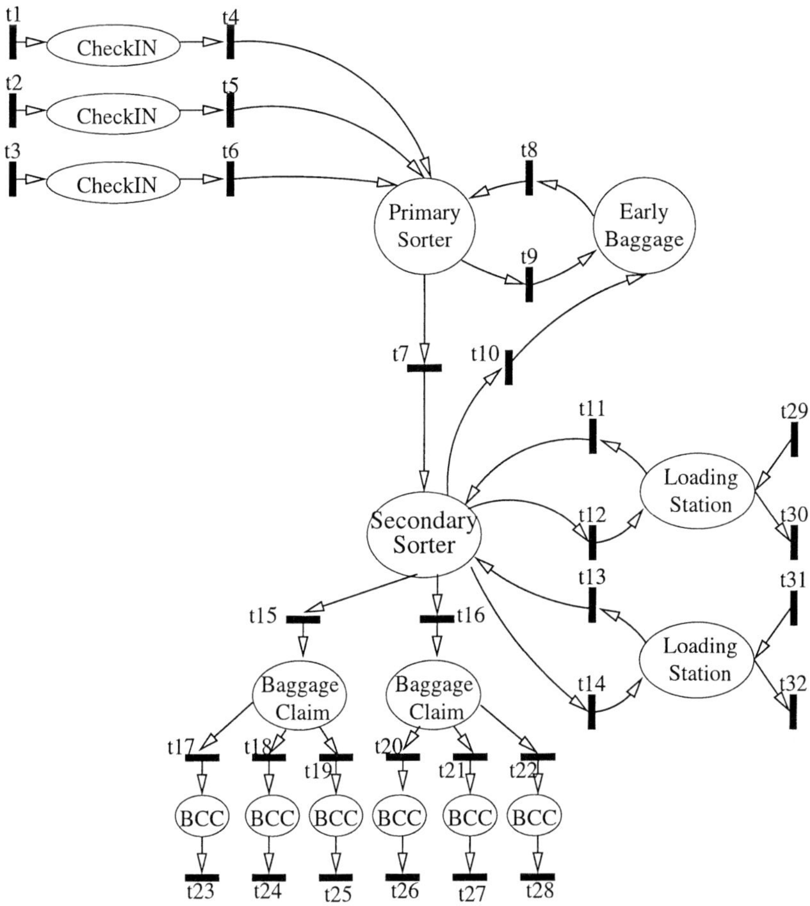

Fig. 21. Baggage Handling System: Net **B0**

to be unsafe, it is taken off the conveyor belt, is checked manually, and either it is taken out of the baggage handling system or it is put back into the subsequent device (sorter, storing, or loading).

For these three cases there are three rules available for the replacement of the corresponding transitions by subnets containing an explicit place **ConveyorBelt**. In the first case it is modeled by the rule $r1 = (L1 \leftarrow K1 \rightarrow R1)$ in Figure 22. This rule states that a transition ***X*** is deleted (including the adjacent arcs) and is replaced by transitions ***T1*** and ***T2*** and the place **ConveyorBelt**.

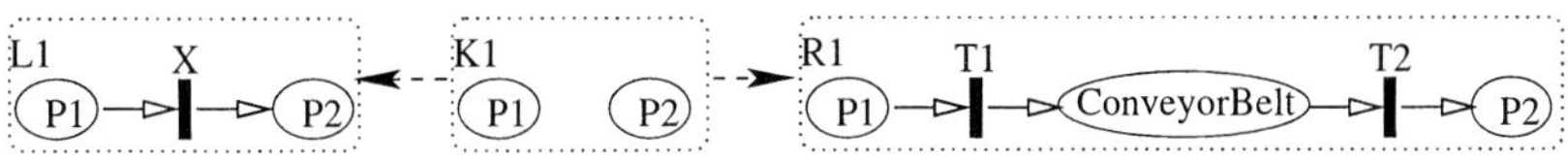

Fig. 22. Introducing Conveyor Belts (Rule $r1$)

In the second case we recursively replace transitions $\boldsymbol{X}$ by the already existing **ConveyorBelt** in Figure 23.

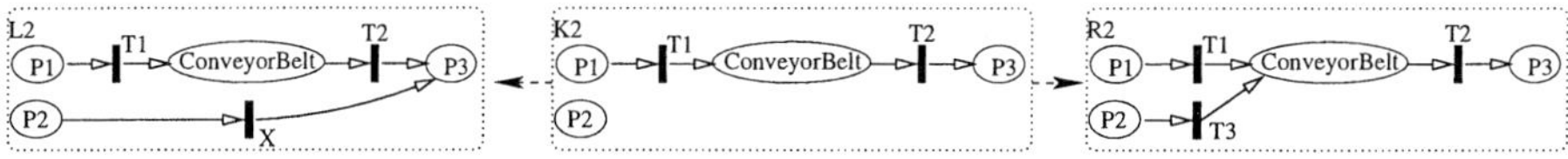

Fig. 23. Recursive Introduction of Conveyor Belts (Rule $r2$)

Introducing the screening is modeled in Figure 24 adding the new places **ConveyorBelt** and **ManualCheck** for the handling of unsafe baggage. The original transition $\boldsymbol{X}$ is deleted, the two new places and the transitions in between are added. The transition $\boldsymbol{T5}$ denotes the removal of the unsafe baggage.

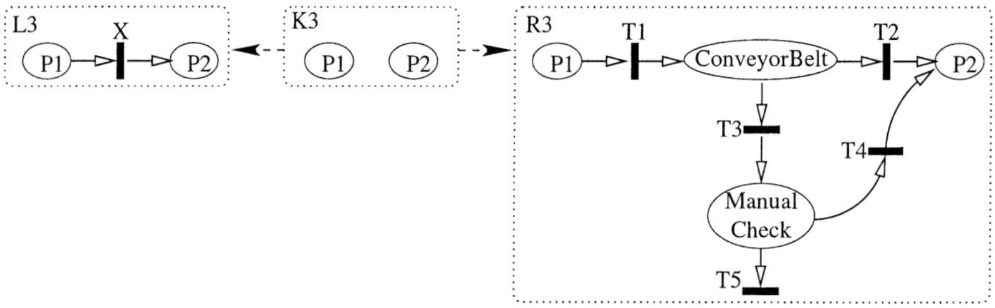

Fig. 24. Introducing Screening Devices (Rule $r3$)

These rules can be applied several times with different matches. First we investigate the application of rule $r3$ with match m in Figure 25 with $m(\boldsymbol{X}) = \boldsymbol{t7}$. Applying rule $r3$ with match m we have again the two steps as for the application of a graph production (see 3.3):

STEP 1 (DELETE): Delete $m(L3 - l(K3))$ from **B0** leading to a context net **D** (if the gluing condition is satisfied), s.t. B is the gluing of $L3$ and D along $K3$, i.e. $B0 = L3 +_{K3} D$ in (1) of Figure 25.

STEP 2 (ADD): Add $R3 - r(K3)$ to D leading to the net **B1**, s.t. **B1** is the gluing of $R3$ and **D** along $K3$, i.e. $B1 = R3 +_{K3} D$ in (2) of Figure 25.

Then we obtain the transformation in Figure 25 consisting of two pushouts, where the context net **D** is the net **B0** without the transition $\boldsymbol{t7}$ and the resulting net **B1** has additional places **ConveyorBelt** and **ManualCheck** with the corresponding transitions and arcs. In Figure 25 we have indicated the changes by a light grey ellipse.

Next we apply rule $r1$ using the matches $m1_i : L1 \rightarrow B1$ mapping the transition $\boldsymbol{X}$ to one of those transitions in **B1** that represent a conveyor belt and mapping the places **P1** and **P2** the adjacent places. So, we have $m1_i(\boldsymbol{X}) = \boldsymbol{t_i}$ for $i \in \{8, ..., 14\}$ leading to the nets $B2, ..., B8$ that are not given explicitly. At last we replace the transitions $\boldsymbol{t4}$, $\boldsymbol{t5}$, and $\boldsymbol{t6}$ by conveyor belts. We use match $m1_4 : L1 \rightarrow B8$ with $m1_4(\boldsymbol{T}) = \boldsymbol{t_4}$ and transform $B8 \overset{(r1, m1_4)}{\Longrightarrow} B9$. Subsequently we can apply rule $r2$ using the matches $m2_5(\boldsymbol{T}) = \boldsymbol{t_5}$ and $m2_6(\boldsymbol{T}) = \boldsymbol{t_6}$.

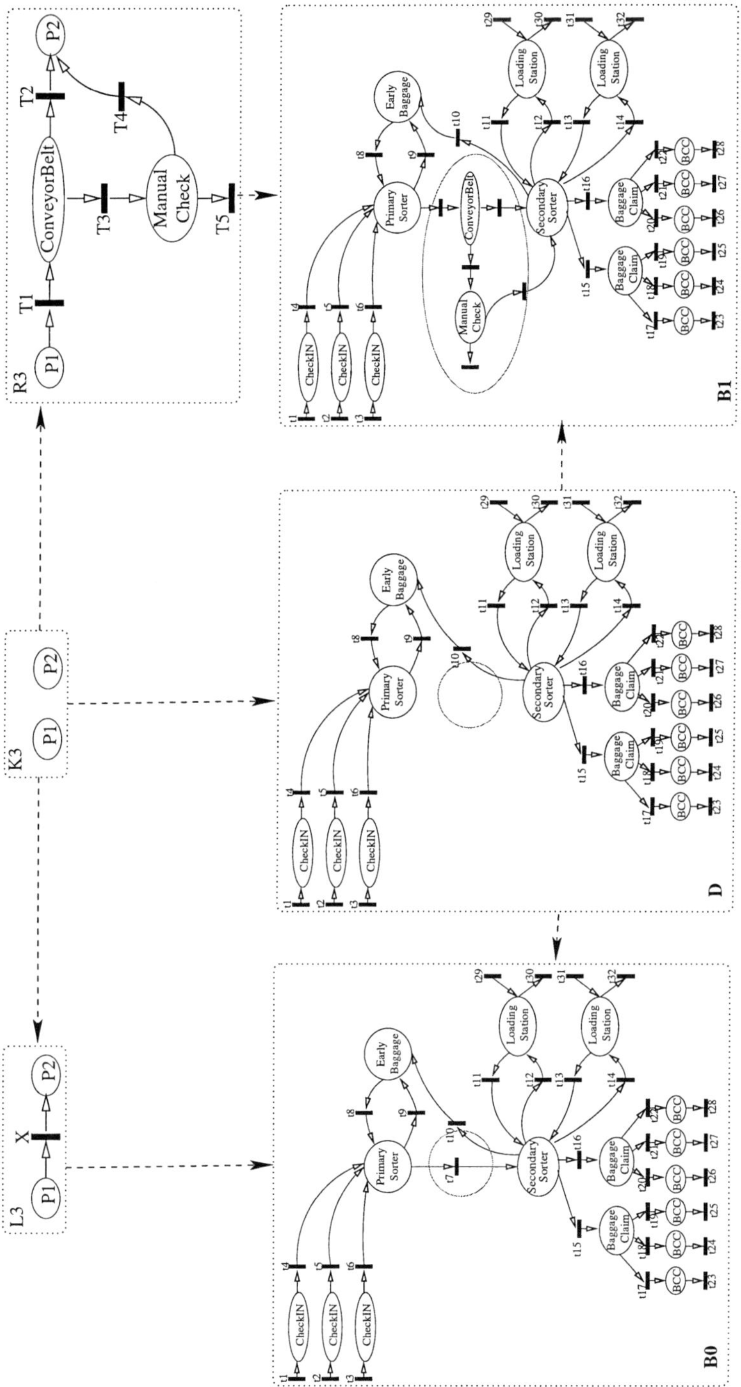

Fig. 25. Transformation $\mathbf{B0} \overset{(r1,m)}{\Longrightarrow} \mathbf{B1}$

This results in the following transformation sequence:

$$\mathbf{B0} \overset{(r3,m)}{\Longrightarrow} \mathbf{B1} \overset{(r1,m1_8)}{\Longrightarrow} B2 \overset{(r1,m1_9)}{\Longrightarrow} B3 \overset{(r1,m1_{10})}{\Longrightarrow} B4 \overset{(r1,m1_{11})}{\Longrightarrow} B5 \overset{(r1,m112)}{\Longrightarrow} B6$$
$$\overset{(r1,m1_{13})}{\Longrightarrow} B7 \overset{(r1,m1_{14})}{\Longrightarrow} B8 \overset{(r1,m1_4)}{\Longrightarrow} B9 \overset{(r2,m2_5)}{\Longrightarrow} B10 \overset{(r2,m2_6)}{\Longrightarrow} \mathbf{B11}$$

Note that the nets typed bold face are illustrated in some Figure, e.g. the net **B11** is depicted in Figure 26.

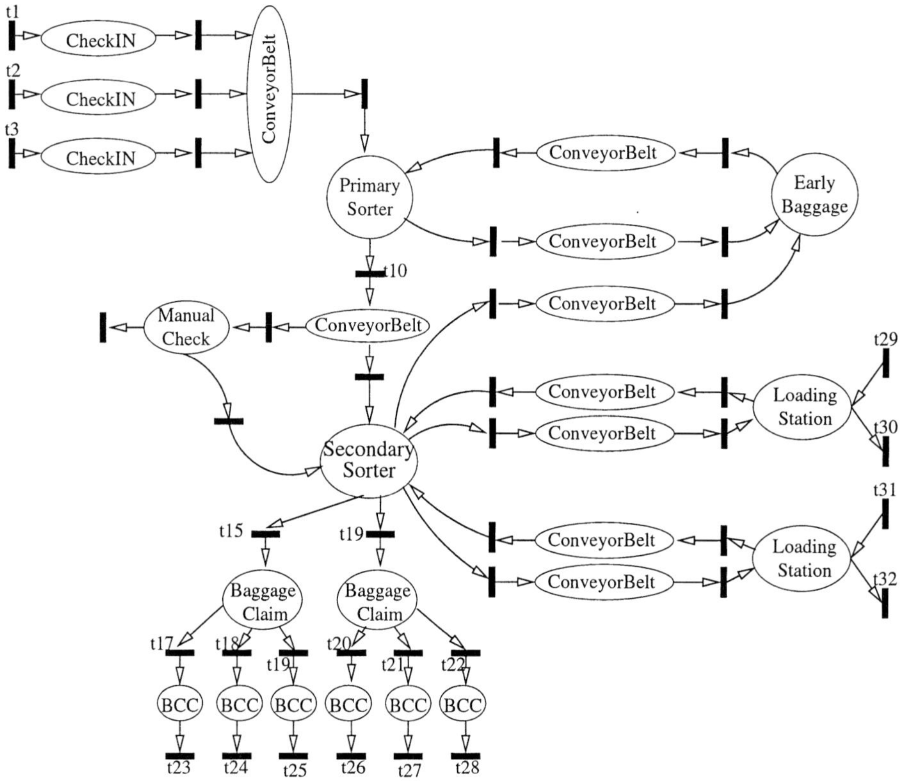

Fig. 26. Net **B11**: After Introducing all Conveyor Belts

Decomposition of the Net: During stepwise development a net usually reaches at some point a size, where it becomes to large and has to be decomposed.

We assume the net **B11** has become too large, so that some structuring is required. In Figure 27 the place/transition net **B11** is decomposed into two subnets **S1** and **S2** and one interface net **I**, consisting of place **SecondarySorter**. The subnets can be glued together using the union construction (see 7.4) and then yield the original net **B11**: We have the embedding of **I** into **S1** and **S2**. The union describes the gluing of the subnets along the interface, hence we have

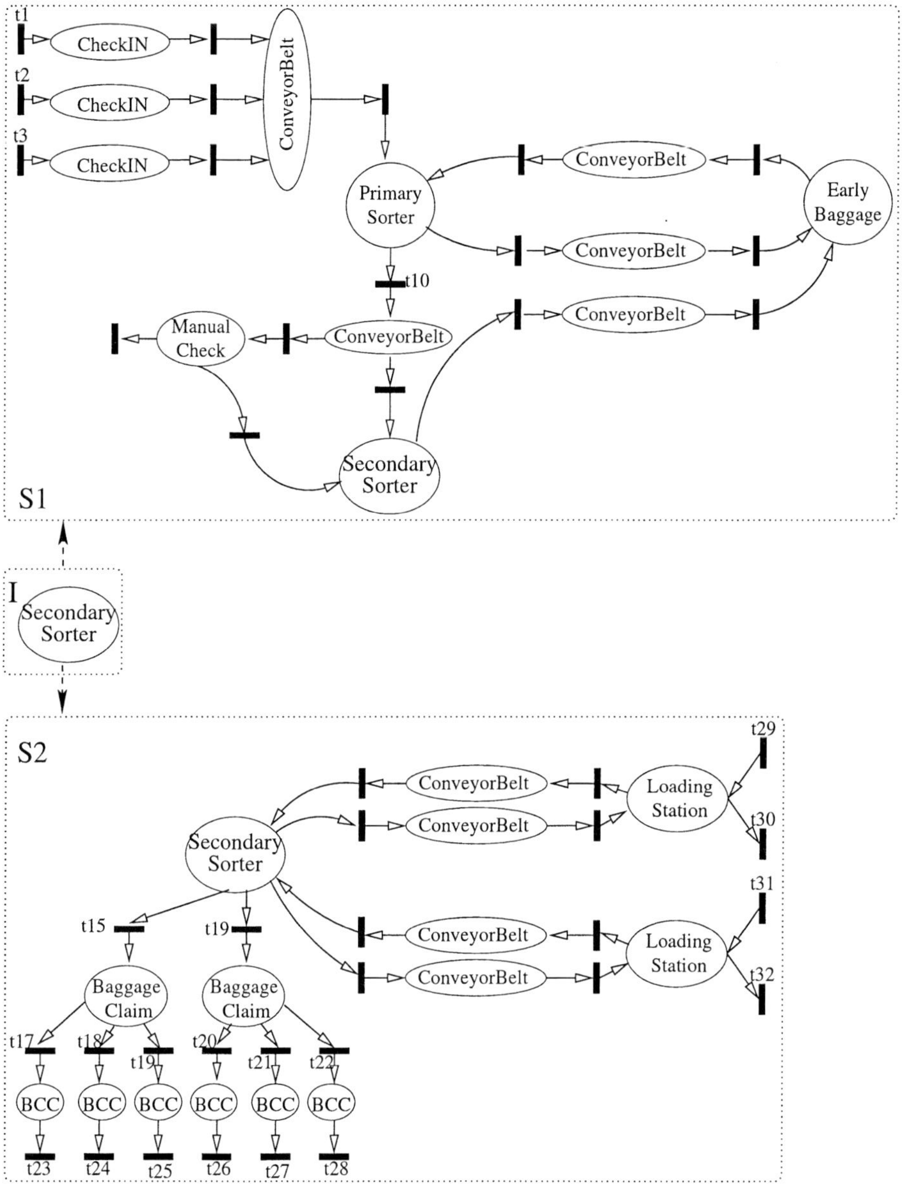

Fig. 27. Decomposition Using Union

the the union $\mathbf{S1} +_{\mathbf{I}} \mathbf{S2} = \mathbf{B11}$ [1]. Now we can modify the subnets independently of each other provided that specific independence conditions are satisfied.

---

[1] In this case the interface net consists of one place only, so that the union corresponds to the usual place fusion of nets. But the general union construction allows having arbitrary subnets as interfaces.

Introducing Lost Baggage Storage: If the baggage is misled or the connecting or the departing carrier are missed, then the baggage is stored in the lost baggage storage. There it is handled manually, that is it is re-tagged and put back into the primary sorter. This is expressed at an abstract level in rule $r4$ in Figure 28.

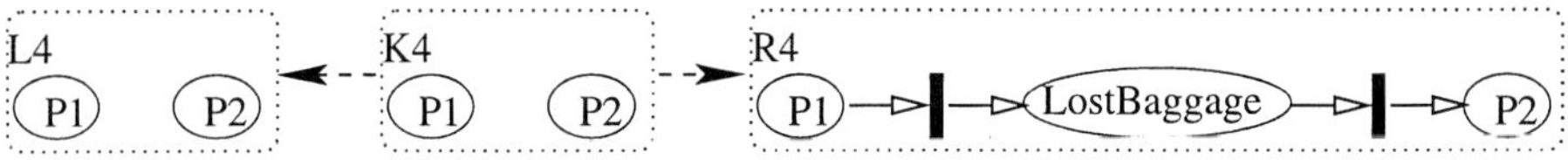

Fig. 28. Introducing Lost Baggage Storage (Rule $r4$)

The application of rule $r4$ to subnet **S1** using the match $m4 : L4 \rightarrow S1$ with $m4(\textsf{P1}) = \textsf{SecondarySorter}$ and $m4(\textsf{P2}) = \textsf{PrimarySorter}$ yields the net $S3$. Applying rule $r1$ twice, subsequently adds the corresponding conveyor belts and we have the transformation sequence $S1 \overset{(r4,m4)}{\Longrightarrow} S3 \overset{r1}{\Longrightarrow} S4 \overset{r1}{\Longrightarrow}$ **S5**. **S5** is depicted in Figure 29.

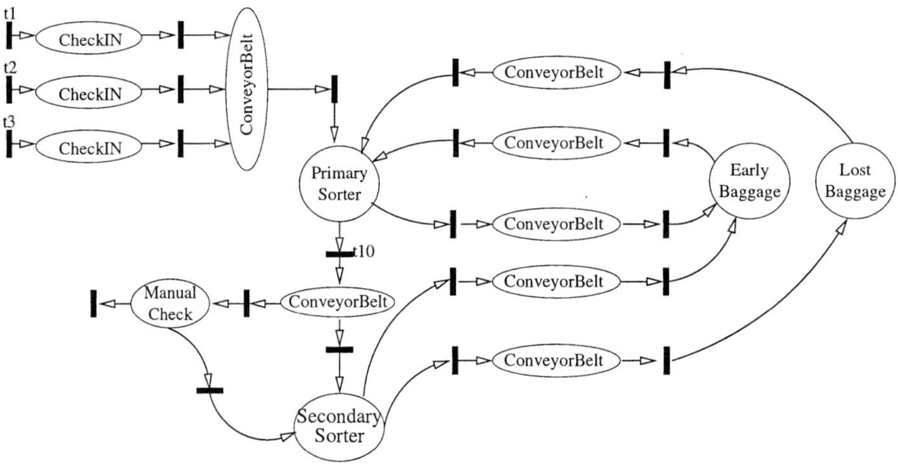

Fig. 29. The Resulting Subnet **S5**

Introducing Unclaimed Baggage Storage. If the baggage is not claimed at the baggage claim it is collected and stored in the unclaimed baggage storage. We use two rules In Figure 30 to introduce the place **UnclaimedBaggage** and the adjacent transitions recursively.

Applying first rule $r5$ and then five times rule $r6$ we obtain the following transformation sequence $\mathbf{S2} \overset{r5}{\Longrightarrow} S6 \overset{5\star(r6)}{\Longrightarrow}$ **S7**, where the resulting subnet **S7** is given in Figure 31.

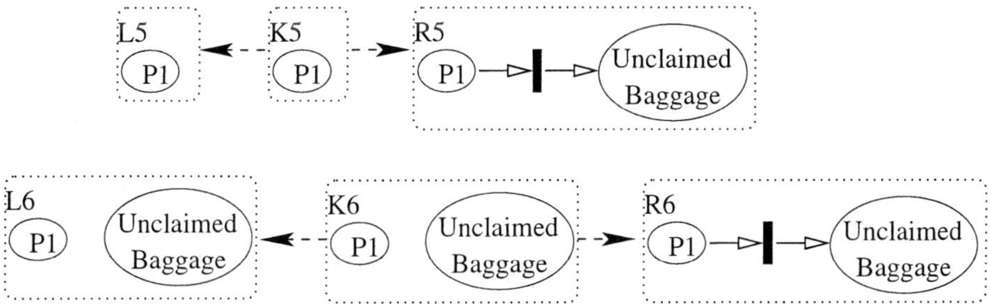

Fig. 30. Introducing Unclaimed Baggage Storage (Rules $r5$ and $r6$)

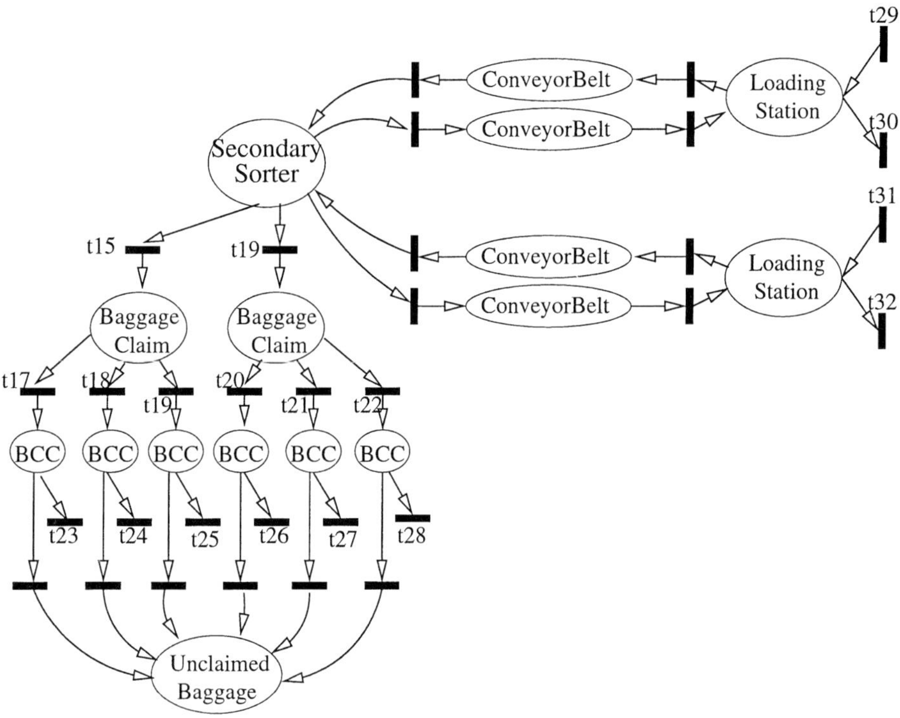

Fig. 31. The Resulting Subnet **S7**

The Union Theorem and the Parallelism Theorem together now guarantee that the resulting net **B14** in Figure 32 of the union $\mathbf{S5} +_{\mathbf{I}} \mathbf{S7} = \mathbf{B14}$ is the same as the result of the following transformation sequence $\mathbf{B11} \stackrel{r4}{\Longrightarrow} B12 \stackrel{r5}{\Longrightarrow} B13 \stackrel{5\star(r6)}{\Longrightarrow} \mathbf{B14}$ according to the case without the decomposition. This is quite obvious if the interface net consists of one place only. In case of more complex interface nets this result can be only achieved if some independence conditions are satisfied. This condition states in principle that nothing from the interface net may be deleted.

6.3 Relevance of Petri Net Transformations

The above example illustrates only some of the possibilities and advantages of net transformations. The usual argument in favor of formal techniques, to have precise notions and valid results clearly holds for this approach as well.

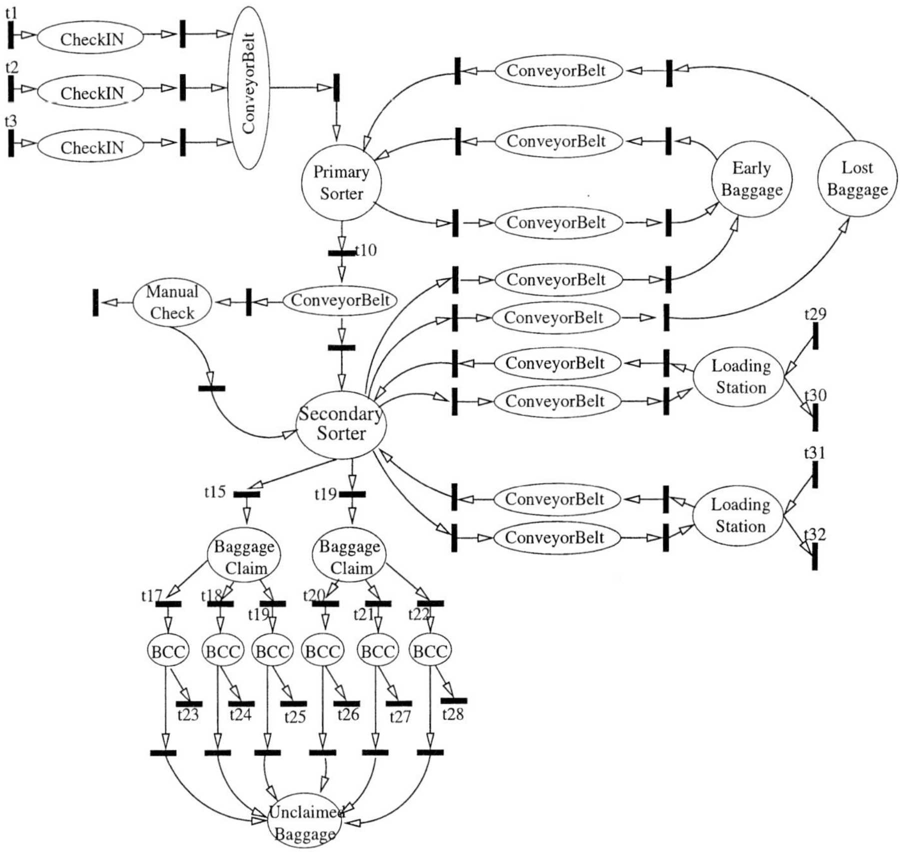

Fig. 32. The Resulting Net **B14**

Moreover, we have already investigated net transformations in high-level Petri net classes (see Section 7.6) that are even more suitable for system modeling than the place/transition nets in our example. The impact for system development is founded in what results from net transformations:

- *Stepwise Development of Models*
  The model of a complex software system may reach a size that is difficult to handle and may compromise the advantages of the (formal) model severely. The one main counter measure is breaking down the model into sub-models, the other is to develop the model top-down. In top-down development the first model is a very abstract view of the system and step by step more modeling details and functionality are added. In general however, this results

in a chain of models, that are strongly related by their intuitive meaning, but not on a formal basis.

Petri net transformations fill this gap by supporting the step-by-step development of a model formally. Rules describe the required changes of a model and their application yields the transformations of the model. Especially the repeated use of a rule ensures a uniform change of a subnet that appears as multiple copies in the model (e.g. replacing one transition by the explicit place **ConveyorBelt** and its adjacent transitions).

Moreover, the representation of change in a visual way using rules and transformations is very intuitive and does not require a deeper knowledge of the theory.

- *Distributed Development of Models*

  Decomposing a model, that is too large, is an important technique for the development of complex models. To combine the advantages of a horizontal structuring with the advantages of step-by-step development techniques for ensuring the consistency of the composed model are required. Then a distributed step-by-step development is available, that allows the independent development of sub-models.

  The theory of net transformations comprises horizontal structuring techniques and ensures compatibility between these and the transformations. In our example we have employed the union construction for the decomposition, and have subsequently developed the subnets independently of each other. The theory allows much more complex decompositions, where the independence of the sub-models is not as obvious as in the given example. So, the formal foundation for the distributed development of complex models is given.

- *Incremental Verification*

  Pure modification of Petri nets is often not sufficient, since the net has some desired properties that have to be ensured during further development. Verification of each intermediate model requires a lot of effort and hence is cost intensive. But refinement can be considered as the modification of nets preserving desired properties. Hence the verification of properties is only required for the net, where they can be first expressed. In this way properties are introduced into the development process and are preserved from then on.

  Rule-based refinement modifies Petri nets using rules and transformations so that specific system properties are preserved. For a brief discussion see Section 7.6.

- *Foundation for Tool Support*

  A further advantage is the formal foundation of rule-based refinement and/or rule-based modification for the implementation of tool support. Due to the theory of Petri net transformations we have a precise description, how rules and transformation work on Petri nets. Tool support is for the practical use the main precondition. The user should get tool support for defining and applying rules. The tool should assist the choice as well as the execution of rules and transformations.

– *Variations of the Development Process*
  Another area, where transformations are very useful, concerns variations in the development process. Often a development is not entirely unique, but variations of the same development process lead to variations in the desired models and resulting systems. These variations can be expressed by different rules yielding different transformations, that are used during the step-by-step development. In our example we can obtain various different baggage handling systems, depending on the rules we use. We can have a system where each conveyor belt is equipped with screening device, if we always use rule $r3$ instead of rule $r1$.

7 Concepts of Petri Net Transformations

In this section we give the precise definitions of the notions that we have already used in our example. For notions and results beyond that we give a brief survey in Section 7.6 and refer to literature.

7.1 Place/Transition Nets and Net Morphisms

Let us first present a notation of place/transition net that is suitable for transformations is the algebraic approach.

These nets are given in the algebraic style as introduced in [25]. A place/transition net $N = (P, T, pre, post)$ is given by the set of places P, the set of transitions T, and two mappings $pre, post : T \to P^{\oplus}$, the pre-domain and the post-domain.

$$T \underset{post}{\overset{pre}{\rightrightarrows}} P^{\oplus} .$$

$P^{\oplus}$ is the free commutative monoid over P that can also be considered as the set of finite multisets over P. The pre- (and post-) domain function maps each transition into the free commutative monoid over the set of places, representing the places and the arc weight of the arcs in the pre-domain (respectively in the post-domain). An element $w \in P^{\oplus}$ can be presented as a linear sum $w = \sum_{p \in P} \lambda_p \cdot p$ or as a function $w : P \to \mathbb{N}$. We can extend the usual operations and relations as $\oplus$, $\ominus$, $\leq$, and so on.

Based on the algebraic notion of Petri nets [25] we use simple homomorphisms that are generated over the set of places. These morphisms map places to places and transitions to transitions. The pre-domain of a transition has to be preserved, that is even if places may be identified the number of tokens that are taken, remains the same. This is expressed by the condition $pre_2 \circ f_T = f_P^{\oplus} \circ pre_1$.

A morphism $f : N_1 \to N_2$ between two place/transition nets $N_1 = (P_1, T_2, pre_1, post_1)$ and $N_2 = (P_2, T_2, pre_2, post_2)$ is given by $f = (f_P, f_T)$ with $f_P : P_1 \to P_2$ and $f_T : T_1 \to T_2$ so that $pre_2 \circ f_T = f_P^{\oplus} \circ pre_1$ and $post_2 \circ f_T = f_P^{\oplus} \circ post_1$. The diagram schema for net morphisms is given in Figure 33.

Several examples of net morphisms can be found in Figure 25 where the dashed arrows denote injective net morphisms.

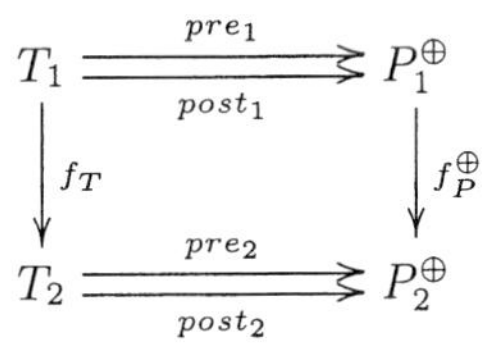

Fig. 33. Net Morphism

7.2 Rules and Transformations

The category **PT** consists of place/transition nets as objects and place/transition net morphisms as morphisms. In order formalize rules and transformations for nets in the DPO-approach we first state the construction of pushouts in the category **PT** of place/transition nets. For any span of two morphisms $N_1 \leftarrow N_0 \rightarrow N_2$ the pushout can be constructed. The construction is based on the pushouts for the sets of transitions and sets of places in the category **Set** of sets and is similar to the pushout construction for graphs (see 3.4).

Given the morphisms $f : N_0 \rightarrow N_1$ and $g : N_0 \rightarrow N_2$ then the pushout N_3 with the morphisms $f' : N2 \rightarrow N_3$ and $g' : N_1 \rightarrow N_3$ is constructed (see Figure 34) as follows:

- $T_3 = T_1 +_{T_0} T_2$ with f'_T and g'_T as pushout of f_T and g_T in **Set**.
- $P_3 = P_1 +_{P_0} P_2$ with f'_P and g'_P as pushout of f_P and g_P in **Set** as well.

- $pre_3(t) = \begin{cases} [pre_1(t_1)] & ; \text{ if } g'_T(t_1) = t \\ [pre_2(t_2)] & ; \text{ if } f'_T(t_2) = t \end{cases}$
- $post_3(t) = \begin{cases} [post_1(t_1)] & ; \text{ if } g'_T(t_1) = t \\ [post_2(t_2)] & ; \text{ if } f'_T(t_2) = t \end{cases}$

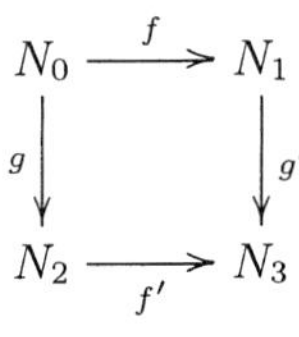

Fig. 34. Pushout of Nets

We introduce rules, that correspond to graph productions in the DPO-approach. Rules describe the replacement of the left-hand side net by the right-hand side net in the presence of an interface net.

- A rule $r = (L \xleftarrow{k_1} K \xrightarrow{k_2} R)$ consists of place/transition nets L, K and R, called left-hand side, interface and right-hand side net respectively, and two injective net morphisms $K \xrightarrow{k_1} L$ and $K \xrightarrow{k_2} R$.

– Given a rule $r = (L \xleftarrow{k_1} K \xrightarrow{k_2} R)$ a direct transformation $N_1 \xRightarrow{r} N_2$, from N_1 to N_2 is given by two pushout diagrams (1) and (2) in Figure 35.
The morphisms $m : L \to N_1$ and $n : R \to N_2$ are called match and comatch, respectively. The net C is called pushout complement or the context net.

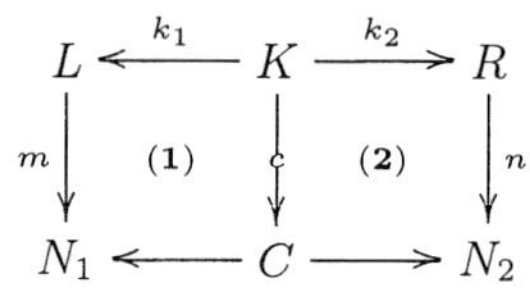

Fig. 35. Net Transformation

The illustration of a transformation can be found for our example in Figure 25, where the rule $r1$ is applied to the net **B0** with match m. The first pushout denotes the gluing of the nets $L3$ and D along the net $K3$ resulting in net **B0**. The second pushout denotes the gluing of net $R3$ and net D along $K3$ resulting in net $B1$.

7.3 Gluing Condition and the Construction of the Context Net

Given a rule r and a match m as depicted in Figure 35, then we construct in a first step the pushout complement provided the gluing condition holds. This leads to the pushout (1) in Figure 35. In a second step we construct the pushout of c and k_2 leading to N_2 and the pushout (2) in Figure 35.

The gluing condition correspond exactly to the gluing condition in the graph case (see 3.5). Using the same interpretation as in the graph case, but the notation from Figure 35 we have the following:

Gluing Condition for Nets:

$$BOUNDARY \subseteq GLUING$$

where $BOUNDARY$ and $GLUING$ are subnets of L defined by

- $GLUING = k_1(K)$
- $DANGLING = \{p \in P_L \mid \exists t \in T_1 - m_T(T_L) : (m_P(p) \in pre_1(t) \text{ or } m_P(p) \in post_1(t))\}$
  where the notation $p \in pre_1(t)$ means $pre_1(t) = \sum_{p \in P_1} \lambda_p \cdot p$ with $\lambda_p > 0$, similar for $post_1$,
- $IDENTIFICATION = \{x \in K \mid \exists y \in K : (x \neq y \text{ and } m(x) = m(y))\}$, where $x \in K$ means $x \in P_K$ with $m = m_P$ or $x \in T_K$ with $m = m_T$, and
- $BOUNDARY = DANGLING \cup IDENTIFICATION$

Now the context net C is the pushout complement C in Figure 35 that is constructed by:

- $P_C = (P_1 \setminus m_P(P_L)) \cup m_P(k_{1P}(P_K))$
- $T_C = (T_1 \setminus m_T(T_L)) \cup m_T(k_{1T}(T_K))$
- $pre_C = pre_{1|T_C}$ and $post_C = post_{1|T_C}$

Note that the pushout complement C leads to the pushout (1) in Figure 35 and that it is unique up to isomorphism.

In our example of the development of the baggage handling system the gluing condition is satisfied in all cases, since the matches are all injective and places are not deleted by our rules.

7.4 Union Construction

The union of two Petri nets sharing a common subnet, that may be empty, is defined by the pushout construction for nets.

The union of place/transition nets N_1, N_2 sharing an interface net I with the net morphisms $f : I \to N_1$ and $g : I \to N_2$ is given by the pushout (1) in Figure 36. Subsequently we use the short notation $N = N_1 +_I N_2$ or $N_1, N_2 > \stackrel{I}{\Longrightarrow} N$.

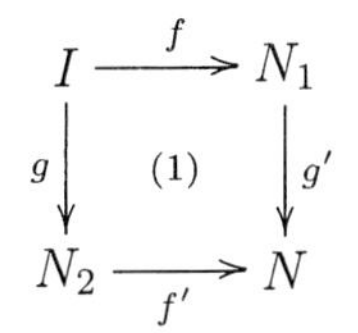

Fig. 36. Union of Nets

In our example we use the union construction to describe the decomposition in Figure 27. The interface net I is mapped by morphisms to the subnets $S1$ and $S2$.

7.5 Union Theorem

The Union Theorem states the compatibility of union and net transformations: Given a union $N_1 +_I N_2 = N$ and net transformations $N_1 \stackrel{r_1}{\Longrightarrow} M_1$ and $N_2 \stackrel{r_2}{\Longrightarrow} M_2$ then we have a parallel rule $r_1 + r_2$ (analogously to a parallel production, see 4.3) and a parallel net transformation $N \stackrel{r_1+r_2}{\Longrightarrow} M$. $M = M_1 +_I M_2$ is then the union of M_1 and M_2 with the shared interface I, provided that the given net transformations preserve the interface I.

The Union Theorem is illustrated in Figure 37:

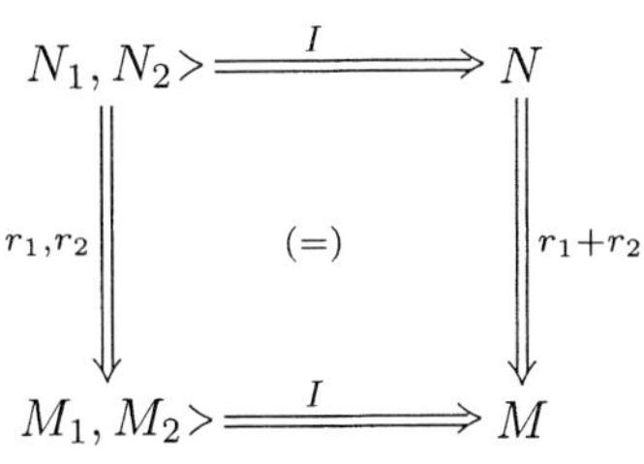

Fig. 37. Diagram for the Union Theorem

Note that the compatibility requires an independence condition stating that nothing from the interface net I may be deleted by one of the transformations of the subnets. This is obviously the case in our example, since the interface consists of one place only and the rules do not delete any places.

7.6 Further Results

We briefly introduce the main net classes we have studied up to now, and subsequently we present some main results.

- Place/transition nets in the algebraic style have already been introduced in the previous Section.
- Coloured Petri nets [20–22] are widely known and very popular. Their practical relevance is very high, due to the very successful tool Design/CPN [19].
- Algebraic high-level nets are available in quite a few different notions e.g. [34, 30, 28]. We use a notion that reflects the paradigm of abstract data types into signature and algebra. An algebraic high-level net (as in [28]) is given by $N = (SPEC, P, T, pre, post, cond, A)$, where $SPEC = (S, OP, E)$ is an algebraic specification in the sense of [16], P is the set of places, T is the set of transitions, $pre, post : T \to (T_{OP}(X) \times P)^{\oplus}$ are the pre- and post-domain mappings, $cond : T \to \mathcal{P}_{fin}(EQNS(SIG))$ are the transition guards, and A is a $SPEC$ algebra.

Horizontal Structuring. Union and fusion are two categorical structuring constructions for place/transition nets, that merge two subnets or two different nets into one.

The Union is introduced in the previous section. Now let us consider fusion: Given a net F that occurs in two copies in the net N_1, represented by two morphisms $F \overset{f}{\underset{f'}{\rightrightarrows}} N_1$ the fusion construction leads to a net N_2, where both occurrences of F in N_1 are merged. If F consists of places $p_1, .., p_n$ then each of the places occurs twice in net N_1, namely as $f(p_1), ..., f(p_n)$ and $f'(p_1), ..., f'(p_n)$. N_2 is obtained from net N_1 fusing both occurrences $f(p_i)$ and $f'(p_i)$ of each place p_i for $1 \leq i \leq n$.

The Union Theorem is presented in the previous section. The Fusion Theorem [27] is expressed similarly: Given a rule r and a fusion $F \rightrightarrows N_1$ then we obtain the same result whether we derive first $N_1 \overset{r}{\Longrightarrow} N_1'$ and then construct the fusion $F \rightrightarrows N_1'$ resulting in N_2' or whether we construct the fusion $F \rightrightarrows N_1$ first, resulting in N_2 and then perform the transformation step $N_2 \overset{r}{\Longrightarrow} N_2'$. Similar to the Union Theorem a certain independence condition is required. Both theorems state that Petri nets transformations are compatible with the corresponding structuring technique under suitable independence conditions. Roughly spoken these conditions guarantee that the interface net I and respectively the fusion net F are preserved by all net transformations.

Parallelism. In Section 4 the concepts of parallelism have been discussed for graphs. The main theorems hold for Petri net transformations as well.

The Church-Rosser Theorem states a local confluence in the sense of formal languages. The required condition of parallel independence means that the matches of both rules overlap only in parts that are not deleted. Sequential independence means that those parts created by the first transformation step are not deleted in the second. The Parallelism Theorem states that sequential or parallel independent transformations can be carried out either in arbitrary sequential order or in parallel. In the context of step-by-step development these theorems are important as they provide conditions for the independent development of different parts or views of the system. More details for horizontal structuring or parallelism are given in see [28] or [27].

Refinement. The extension of High-level replacement systems to rules and transformations preserving properties has the following impact on Petri nets: Rule-based refinement comprises the transformation of Petri nets using rules while preserving certain net properties. For Petri nets the desired properties of the net model can be expressed, e.g in terms of Petri nets (as liveness, boundedness etc.), in terms of logic (e.g. temporal logic, logic of actions etc.) in terms of relation to other models (e.g. bisimulation, correctness etc.) and so on. We have investigated the possibilities to preserve liveness of Petri nets and safety properties in the sense of temporal logic.

Summarizing, we have for place/transition nets, algebraic-high level nets and Coloured Petri nets the following results for rule-based refinement presented in table 2. For more details see [29].

8 Conclusion

In the first part of this paper (Sections 2 - 5) we have given a tutorial introduction to the basic notions of graph grammars and graph transformations including the relationship to corresponding notions of Petri nets. In the second part (Section 6 and Section 7) we have shown how to use Petri nets transformations for the stepwise development of systems and have included a detailed example of a baggage handling system. The main idea of Petri transformations is to extend the classical theory of Petri nets by a rule-based technique that allows studying the changes of the Petri net structure.

In our general overview of graph grammars and transformations in Section 2 we have already pointed out that there is a large variety of different approaches and application areas. The practical use of graph transformations is supported by several tools. The algebraic approach to graph transformations (presented in Sections 3 - 5) is especially supported by the graph transformation environment AGG (see the homepage of [1]). AGG includes an editor for graphs and graph grammars, a graph transformation engine, and a tool for the analysis of graph transformations. the AGG system as well as some other tools are available on a CD which is part of volume 2 of the *Handbook of Graph Grammars and Computing by Graph Transformation* [10]. This volume provides also

Table 2. Achieved results

Notion/Results	PT-nets	AHL-nets	CPNs
Rules, Transformations	√	√	√
Safety property preserving transformations with transition-gluing morphisms	√	√	√
place-preserving morphisms	√	√	√
Safety property introducing transformations	√	√	√
Liveness preserving transformations	√	?	?
Liveness introducing transformations	√	?	?
Church Rosser I + II Theorem	√	√	√
Parallelism Theorem	√	√	√
Union	√	√	√
Fusion	√	√	√
Union Theorems I+II	√	√	√
Fusion Theorem	√	√	√

an excellent introduction to several application areas for graph transformations. Concurrency aspects of graph grammars, which are briefly discussed in Section 5, are presented in much more detail in volume 3 of the handbook [3]. This volume includes also an introduction to high-level replacement systems with application to algebraic specification and Petri nets including the theoretical foundations of Petri net transformations [11].

On top the graph transformation system AGG there is the GenGED environment (see the homepage of [18]) that supports the generic description of visual modeling languages for the generation of graphical editors and the simulation of the behavior of visual models. Especially, Petri net transformations can be expressed using GenGED, e.g. for the animation of Petri nets [9, 4]. In this framework, the animation view of a system modeled as a Petri net consists of a domain-specific layout and an animation according to the firing behavior of the Petri net. This animation view can be coupled to other Petri net tools [8] using the Petri Net Kernel [23] a tool infrastructure for editing, simulating and analyzing Petri nets of different net classes and for integration of other Petri net tools.

References

1. AGG Homepage. http://tfs.cs.tu-berlin.de/agg.
2. P. Baldan. *Modelling Concurrent Computations: From Contextual Petri Nets to Graph Grammars.* PhD thesis, University of Pisa, 2000.

3. P. Baldan, A. Corradini, U. Montanari, F. Rossi, H. Ehrig, and M. Löwe. Concurrent Semantics of Algebraic Graph Transformations. In G. Rozenberg, editor, *The Handbook of Graph Grammars and Computing by Graph Transformations, Volume 3: Concurrency, Parallelism and Distribution*. World Scientific, 1999.
4. R. Bardohl and C. Ermel. Scenario Animation for Visual Behavior Models: A Generic Approach Applied to Petri Nets. In G. Juhas and J. Desel, editors, *Proc. 10th Workshop on Algorithms and Tools for Petri Nets (AWPN'03)*, 2003.
5. G. Berthelot. Checking properties of nets using transformations. *Advances in Petri Nets 1985*, Lecture Notes in Computer Science 222: pages 19–40. Springer 1986.
6. G. Berthelot. Transformations and decompositions of nets. In Brauer, W., Reisig, W., and Rozenberg, G., editors, *Petri Nets: Central Models and Their Properties, Advances in Petri Nets*, Lecture Notes in Computer Science 254, pages 359–376. Springer, 1987.
7. R. David and H. Alla, editors. *Petri Nets and Grafcet*. Prentice Hall (UK), 1992.
8. C. Ermel, R. Bardohl, and H. Ehrig. Specification and implementation of animation views for Petri nets. In DFG Research Group *Petri Net Technology, Proc. of 2nd International Colloquium on Petri Net Technology for Comunication Based Systems*, 2001.
9. C. Ermel, R. Bardohl, and H. Ehrig. Generation of animation views for Petri nets in GenGED. In H. Ehrig, W. Reisig, G. Rozenberg, and H. Weber, editors, *Advances in Petri Nets: Petri Net Technologies for Modeling Communication Based Systems*, Lecture Notes in Computer Science 2472. Springer, 2003.
10. H. Ehrig, G. Engels, H.-J. Kreowski, and G. Rozenberg, editors. *Handbook of Graph Grammars and Computing by Graph Transformation, Volume 2: Applications, Languages and Tools*. World Scientific, 1999.
11. H. Ehrig, M. Gajewsky, and F. Parisi-Presicce. *High-level replacement systems with applications to algebraic apecifications and Petri nets*, chapter 6, pages 341–400. Number 3: Concurrency, Parallelism, and Distribution in Handbook of Graph Grammars and Computing by Graph Transformations. World Scientific, 1999.
12. H. Ehrig, A. Habel, H.-J. Kreowski, and F. Parisi-Presicce. Parallelism and concurrency in high-level replacement systems. *Math. Struct. in Comp. Science*, 1:361–404, 1991.
13. H. Ehrig, A. Habel, H.-J. Kreowski, and F. Parisi-Presicce. Parallelism and concurrency in high-level replacement systems. *Math. Struct. in Comp. Science*, 1:361–404, 1991.
14. H. Ehrig. Introduction to the algebraic theory of graph grammars (A survey). In *Graph Grammars and their Application to Computer Science and Biology*, pages 1–69. Lecture Notes in Computer Science 73. Springer, 1979.
15. H. Ehrig, H.-J. Kreowski, U. Montanari, and G. Rozenberg, editors. *Handbook of Graph Grammars and Computing by Graph Transformation, Volume 3: Concurrency, Parallelism, and Distribution*. World Scientific, 1999.
16. H. Ehrig and B. Mahr. *Fundamentals of Algebraic Specification 1: Equations and Initial Semantics*, volume 6 of *EATCS Monographs on Theoretical Computer Science*. Springer Verlag, Berlin, 1985.
17. H. Ehrig, M. Pfender, and H.J. Schneider. Graph grammars: an algebraic approach. In *14th Annual IEEE Symposium on Switching and Automata Theory*, pages 167–180. IEEE, 1973.
18. GenGED Homepage. http://tfs.cs.tu-berlin.de/genged.
19. K. Jensen, S. Christensen, P. Huber, and M. Holla. *Design/CPN. A Reference Manual*. Meta Software Cooperation, 125 Cambridge Park Drive, Cambridge Ma 02140, USA, 1991.

20. K. Jensen. *Coloured Petri nets. Basic Concepts, Analysis Methods and Practical Use*, volume 1: Basic Concepts. Springer Verlag, EATCS Monographs in Theoretical Computer Science edition, 1992.
21. K. Jensen. *Coloured Petri Nets - Basic Concepts, Analysis Methods and Practical Use*, volume 2: Analysis Methods. Springer Verlag, EATCS Monographs in Theoretical Computer Science edition, 1994.
22. K. Jensen. *Coloured Petri Nets - Basic Concepts, Analysis Methods and Practical Use*, volume 3: Practical Use. Springer Verlag, EATCS Monographs in Theoretical Computer Science edition, 1997.
23. E. Kindler and M. Weber. The Petri net kernel – an infrastructure for building Petri net tools. *Software Tools for Technology Transfer*, 3(4):486–497, 2001.
24. S. Lack and P. Sobociski. Adhesive categories. In *Proc. FOSSACS 04*, 2004. to appear.
25. J. Meseguer and U. Montanari. Petri Nets are Monoids. *Information and Computation*, 88(2):105–155, 1990.
26. J. Padberg. Survey of high-level replacement systems. Technical Report 93-8, Technical University of Berlin, 1993.
27. J. Padberg. Categorical approach to horizontal structuring and refinement of high-level replacement systems. *Applied Categorical Structures*, 7(4):371–403, December 1999.
28. J. Padberg, H. Ehrig, and L. Ribeiro. Algebraic high-level net transformation systems. *Mathematical Structures in Computer Science*, 5:217–256, 1995.
29. J. Padberg and M. Urbášek. Rule-based refinement of Petri nets: A survey. In H. Ehrig, W. Reisig, G. Rozenberg, and H. Weber, editors, *Advances in Petri Nets: Petri Net Technologies for Modeling Communication Based Systems*, Lecture Notes in Computer Science 2472. Springer, 2003.
30. W. Reisig. Petri Nets and Algebraic Specifications. *Theoretical Computer Science*, 80:1–34, 1991.
31. L. Ribeiro, H. Ehrig, and J. Padberg. Formal development of concurrent systems using algebraic high-level nets and transformations. In *Proc. VII Simpósio Brasileiro de Engenharia de Software*, pages 1–16, Tech-report no. 93-13, TU Berlin, 1993.
32. G. Rozenberg, editor. *Handbook of Graph Grammars and Computing by Graph Transformations, Volume 1: Foundations.* World Scientific, 1997.
33. Vanio M. Savi and Xiaolan Xie. Liveness and boundedness analysis for petri nets with event graph modules. In Jensen, K., editor, *13th International Conference on Application and Theory of Petri Nets 1992, Sheffield, UK*, Lecture Notes in Computer Science 616, pages 328–347. Springer, 1992.
34. J. Vautherin. Parallel system specification with coloured Petri nets. In G. Rozenberg, editor, *Advances in Petri Nets 87*, pages 293–308. Lecture Notes in Computer Science 266. Springer Verlag, 1987.
35. W.M.P. van der Aalst. Verification of workflow nets. In P. Azéma and G. Balbo, editors, *Application and Theory of Petri Nets*, Lecture Notes in Computer Science 1248, pages 407–426. Springer, 1997.
36. G. Winskel. Petri nets, algebras, morphisms, and compositionality. *Information and Computation*, 72:197–238, 1987.

Message Sequence Charts

Blaise Genest[1], Anca Muscholl[1], and Doron Peled[2]

[1] LIAFA, Université Paris VII & CNRS
2, pl. Jussieu, case 7014, 75251 Paris cedex 05, France
[2] Department of Computer Science, University of Warwick
Coventry, CV4 7AL, United Kingdom

Abstract. *Message sequence charts* (MSC) are a graphical notation standardized by the ITU and used for the description of communication scenarios between asynchronous processes. This survey compares MSCs and communicating finite-state automata, presenting two fundamental validation problems on MSCs, model-checking and implementability.

1 Introduction

Modeling and validation, whether formal or ad-hoc, are important steps in system design. Over the last couple of decades, various methods and tools were developed for decreasing the amount of design and development errors. A common component of such methods and tools is the use of *formalisms* for *specifying* the behavior and requirements of the system. Experience has shown that some formalisms, such as finite-state machines, are particularly appealing, due to their convenient mathematical properties. In particular, the expressive power of finite-state machines is identical to *regular languages*, an important and well-studied class of languages. Although their expressiveness is restricted, finite-state machines are used for the increasingly successful automatic verification of software and hardware, also called *model-checking* [8, 10]. One of the biggest challenges in developing new validation technology based on finite-state machines is to make this model popular among system engineers.

The *Message Sequence Charts* (MSC) model has become popular in software development throughout its visual representation, depicting the involved processes as vertical lines, and each message as an arrow between the source and the target processes, according to their occurrence order. An international standard [1], and its inclusion in the UML standard, has increased the popularity. The standard has also extended the notation to *Message Sequence Graphs* (MSGs), which consist of finite transition systems, where each state embeds a single MSC. Encouraged by the success of the formalism among software developers, techniques and tools for analyzing MSCs and MSGs have been developed.

In this survey we describe the formal analysis of MSCs and MSGs. The class of systems that can be described using this formalism does not directly correspond to a well-studied class such as regular languages. It turns out that MSGs are incomparable with the class of finite-state communication protocols. One thus needs to separately study the expressiveness of MSG languages, and

J. Desel, W. Reisig, and G. Rozenberg (Eds.): ACPN 2003, LNCS 3098, pp. 537–558, 2004.

adapt the validation algorithms. Several new algorithms are suggested in order to check MSG properties, mostly related to an automatic translation from MSG specification into skeletons of concurrent programs. Our survey concentrates on the following subjects:

Expressiveness: Comparing the expressive power of MSGs to the expressive power of other formalisms, in particular communicating finite-state machines.

Verification: The ability to apply automatic verification algorithms on MSGs, and the various formalisms used to define properties of MSGs.

Implementability: The ability to obtain an automatic translation from MSG specification into skeletons of code.

Generalizations and Restrictions: Various extensions and restrictions of the standard notation are suggested in order to capture further systems, and on the other hand, to obtain decidability of important decision procedures.

Very recently, several MSC-based specification formalisms have been proposed, such as *Live Sequence Charts* [17], *Triggered MSCs* [30], *Netcharts* [25] and *Template MSCs* [12]. The motivation behind these models is to increase the expressiveness of the notation, and to make their usage by designers even more convenient.

2 Message Sequence Graphs and Communicating Finite-State Machines

We present in this section two specification formalisms for communication protocols, *Message Sequence Charts* and *Communicating Finite-State Machines.*

Message Sequence Charts (MSC for short) is a scenario language standardized by the ITU [1]. They are simple diagrams depicting the activity and communications in a distributed system. The entities participating in the interactions are called instances (or processes). They are represented by vertical lines, on which the behavior of each single process is described by a sequence of events. Message exchanges are depicted by arrows from the sender to the receiver. In addition to messages, atomic events, timers, local/global conditions can also be represented.

Definition 1 *A Message Sequence Chart (MSC for short) is a tuple* $M = \langle \mathcal{P}, E, \mathcal{C}, \ell, m, < \rangle$ *where:*

- $\mathcal{P}$ *is a finite set of processes,*
- E *is a finite set of events,*
- $\mathcal{C}$ *is a finite set of names for messages and local actions,*
- $\ell : E \to \mathcal{T} = \{p!q(a), p?q(a), p(a) \mid p \neq q \in \mathcal{P}, a \in \mathcal{C}\}$ *labels an event with its* type*: in process* p*, either a send* $p!q(a)$ *of message* a *to process* q*, or a receive* $p?q(a)$ *of message* a *from process* q*, or a local event* $p(a)$*. The labeling* ℓ *partitions the set of events by type (send, receive, or local),* $E = S \biguplus R \biguplus L$*, and by process,* $E = \biguplus_{p \in \mathcal{P}} E_p$*.*

- $m : S \to R$ *is a bijection matching each send to the corresponding receive. If* $m(s) = r$, *then* $\ell(s) = p!q(a)$ *and* $\ell(r) = p?q(a)$ *for some processes* $p, q \in \mathcal{P}$ *and some message name* $a \in \mathcal{C}$.
- $< \subseteq E \times E$ *is an acyclic relation between events consisting of:*
  1. *a total order on* E_p, *for every process* $p \in \mathcal{P}$, *and*
  2. $s < r$, *whenever* $m(s) = r$.

The event labeling ℓ implicitly defines the process $pr(e)$ for each event e as $pr(e) = p$ if $e \in E_p$ (equivalently, $\ell(e) \in \{p!q(a), p?q(a), p(a)\}$ for some $q \in \mathcal{P}, a \in \mathcal{C}$). Since point-to-point communication is usually FIFO (first-in-first-out) we make in the following the same assumption for MSCs. That is, we assume that whenever $m(s_1) = r_1$, $m(s_2) = r_2$ holds with $pr(s_1) = pr(s_2)$, $pr(r_1) = pr(r_2)$ and $s_1 < s_2$, then we also have $r_1 < r_2$.

The example in figure 1 is an MSC M with messages sent between two processes p_1, p_2. It corresponds to a scenario of the alternating bit protocol, in which the sender p_1 is forced to resend the message to the receiver p_2, since p_2's acknowledgments arrive too late.

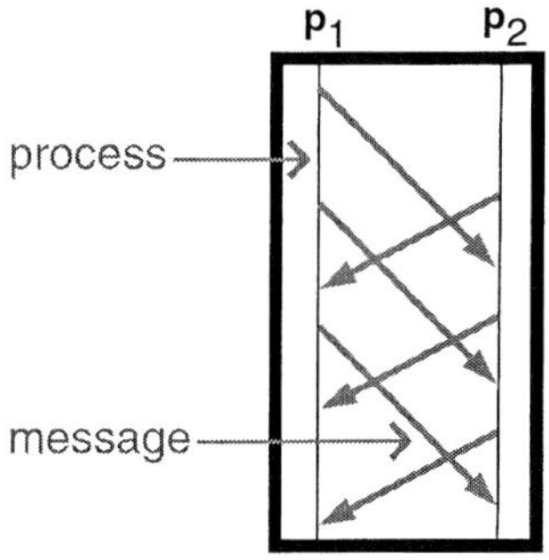

Fig. 1. MSC execution of the alternating bit protocol.

The relation $<$ is called the *visual* order on the MSC, since it corresponds to its graphical representation. It is comprised of the process ordering and the message ordering, pairwise between send and matching receive. Since $<$ is required to be acyclic, its reflexive-transitive closure $<^*$ is a partial order on the set E of events, which we will denote for simplicity also by $\leq$. Any extension of $\leq$ to a total order on E is called a linearization of M. We denote by $\text{Lin}(M)$ the set of all *labeled linearizations* of an MSC M, $\text{Lin}(M) = \{\ell(e_1) \cdots \ell(e_n) \mid e_1 \cdots e_n \text{ is a linearization of } M\}$.

Since the specification of a communication protocol consists of many scenarios, either in positive or in negative form, a high-level description is needed for combining them together and defining infinite sets of (finite or infinite) scenarios. The Z.120 standard description introduces high-level MSCs using non-deterministic branching, concatenation and iteration of finite MSCs. The semantics is provisional, that is, the high-level MSC usually describes *possible* behaviors of the system. Formally, a *Message Sequence Graph* (MSG for short) $G = \langle V, R, v^0, V_f, \lambda\rangle$ consists of a finite transition system (V, R, v^0, V_f) with set of nodes V and set of transitions $R \subseteq V \times V$, initial node $v^0 \in V$ and terminal nodes $V_f \subseteq V$. In pictures, the initial node is marked by an incoming arrow, and final nodes by outgoing arrows. Each node v is labeled by the finite MSC $\lambda(v)$. For instance, the MSG in figure 2 describes the possible runs of a protocol for connecting a user U with a server S through a firewall F. After a connection request (initial node A) either the server accepts the user and the firewall grants the access (final node B), or else the server's accept arrives too late (after the firewall denied the access, node C). This negative behavior can repeat (loop between A and C) and leads eventually to an error (final node D).

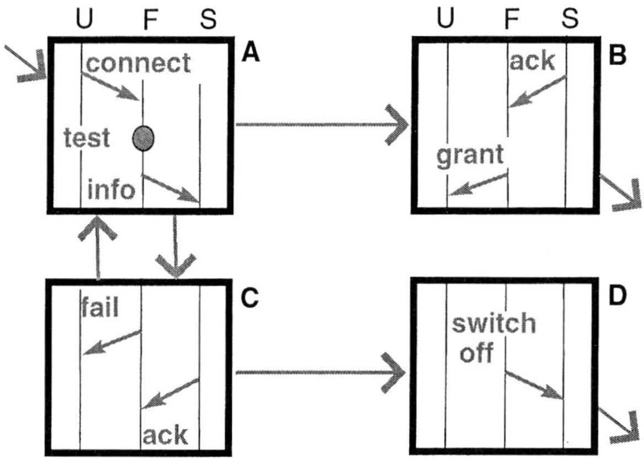

Fig. 2. Communication protocol represented by an MSG.

An *execution* of an MSG G is the labeling $\lambda(v_0)\lambda(v_1)\cdots\lambda(v_k)$ of some accepting path $v^0 = v_0, v_1, \ldots, v_k \in V_f$ of G, i.e., $(v_i, v_{i+1}) \in R$ for every $0 \leq i < k$. For example, $ACAB$ in figure 2 is the execution of G in which the connection fails once, but the second request succeeds. The set of executions of G is denoted by $\mathcal{L}(G)$, the set of linearizations of executions of G is denoted by $\mathrm{Lin}(G)$. The *size* of a MSG G (denoted $|G|$) is the sum of the sizes of its nodes.

Of course, the semantics of MSGs depends on the definition of the MSC product. We consider the usual *weak product* of MSCs, that concatenates MSCs along the process lines. Let $M_1 = \langle \mathcal{P}, E_1, \mathcal{C}_1, \ell_1, m_1, <_1\rangle$ and $M_2 = \langle \mathcal{P}, E_2, \mathcal{C}_2, \ell_2, m_2, <_2\rangle$ be MSCs over the same set of processes $\mathcal{P}$. The product $M_1 M_2$ is the MSC $\langle \mathcal{P}, E_1 \biguplus E_2, \mathcal{C}_1 \cup \mathcal{C}_2, \ell_1 \cup \ell_2, m_1 \cup m_2, <\rangle$ over the disjoint union of events, with the visual order given by:

$$< = <_1 \cup <_2 \cup \{(e, f) \in E_1 \times E_2 \mid pr(e) = pr(f)\}\,.$$

Note that there is no synchronization between different processes when moving from one node to the next one (*weak product*). Hence, it is possible that one process is still involved in some actions of M_1, while another process has advanced to an event of M_2.

A related standardized specification notation for telecommunication applications is SDL (Specification and Description Language, ITU Z.100). SDL is dedicated to the design of real-time, distributed systems and involves complex features as hierarchy, procedure calls and abstract data types. The basic theoretical model behind SDL are nested communicating finite-state machines. We recall the definition of (flat) *communicating finite-state machines* (CFM for short).

A CFM $\mathcal{A} = (\mathcal{A}_p)_{p\in\mathcal{P}}$ consists of finite-state machines $\mathcal{A}_p$ associated with processes $p \in \mathcal{P}$, which communicate over unbounded, error-free, FIFO channels. The content of a channel is a word over a finite alphabet $\mathcal{C}$. With each pair $(p,q) \in \mathcal{P}^2$ of distinct processes we associate a channel $C_{p,q}$. Each finite-state machine $\mathcal{A}_p$ is described by a tuple $\mathcal{A}_p = (S_p, A_p, \rightarrow_p, F_p)$ consisting of a set of local states S_p, a set of actions A_p, a set of final states F_p and a transition relation $\rightarrow_p \subseteq S_p \times A_p \times S_p$. The computation begins in an initial state $s^0 \in \prod_{p\in\mathcal{P}} S_p$. The actions of $\mathcal{A}_p$ are either local actions or sending/receiving a message. We use the same notations as for MSCs. Sending message $a \in \mathcal{C}$ from process p to process q is denoted by $p!q(a)$ and it means that a is appended to the channel $C_{p,q}$. Receiving message a by p from q is denoted by $p?q(a)$ and it means that a must be the first message in $C_{q,p}$, which will be then removed from $C_{q,p}$. A local action a on process p is denoted by $l_p(a)$. We denote a run of the CFM as *successful*, if each process p finishes the execution in some final state and all channels are empty. The set of successful runs of $\mathcal{A}$ is denoted $\mathcal{L}(\mathcal{A})$. The *size* of $\mathcal{A}$ is $\sum_p |\mathcal{A}_p|$ and is denoted $|\mathcal{A}|$.

Note that each successful run of a CFM defines an MSC. Conversely, with each MSC $M = \langle \mathcal{P}, E, \mathcal{C}, \ell, m, < \rangle$ we can associate an equivalent CFM, by defining the behavior of process p as the (ordered) sequence of events E_p. However, the two formalisms MSG and CFM are incomparable in general, as discussed in the next section.

3 Comparing MSG and CFM

Comparing the expressivity of MSG and CFM is interesting for at least two reasons. First, both formalisms are heavily used in protocol design, sometimes for specifying different parts of a system at different stages of the design process. Second, MSCs are usually intended as early requirements, for a rough description of the desired/undesired behavior. Thus, the question whether the described behavior can be turned into a protocol (*implementability/realizability problem*) is an important validation step in the design process.

A qualitative comparison between MSG and CFM concerns two important parameters, *control* and *channels*. Control in a CFM is inherently local, since it corresponds to local transition functions. The control structure of an MSG is global, since the branching from a node concerns all processes occurring in the future execution. The global control mechanism of an MSG is actually imposed by the visual character of the diagram graph, in which MSCs are composed sequentially. One problem arising from the global control is that an MSG G

might be non-implementable, i.e., no CFM $\mathcal{A}$ exists with $\mathcal{L}(\mathcal{A}) = \mathcal{L}(G)$. For a simple example, consider the MSG G consisting of a single node v with a self-loop, labeled by a message from p_1 to p_2 and another message from p_3 to p_4. Since the MSCs in $\mathcal{L}(G)$ must contain equally many messages from p_1 to p_2 and from p_3 to p_4, there can be no equivalent CFM.

We turn now to the second parameter, namely channels. Although none of the models impose any (universal) bound on the channel capacity, validation tasks such as model-checking tend to be "more" decidable for MSGs than for CFMs that are Turing complete, see [9]. The reason is that MSGs have *existentially-bounded* channels, i.e., for each MSG G there exists an integer b such that every MSC in $\mathcal{L}(G)$ can be executed with channels of size at most b. Formally, a set X of MSCs is called *existentially-bounded* if there exists some b such that every MSC $M \in X$ has some linearization $w \in \mathrm{Lin}(M)$ satisfying the following property: for every pair of distinct processes p, q and every prefix v of w, it holds that $0 \le \sum_{a \in C} |v|_{p!q(a)} - \sum_{a \in C} |v|_{q?p(a)} \le b$. For an MSG G the bound b is linear in the maximal size of the MSCs labeling the nodes of G. For an example of property that is undecidable for CFM (but not for MSG) one can consider the question whether a CFM generates at least one MSC [9]. A less trivial example is *pattern-matching*: given an MSC M and an MSG G, we ask whether there is some execution $N \in \mathcal{L}(G)$ and a factorization $N = N_1 M N_2$, where N_1, N_2 are both MSCs. The pattern-matching algorithm described in [12, 13] uses heavily the fact that MSGs are existentially-bounded (with an priori known bound).

More generally, some CFMs cannot be transformed into MSGs since MSGs are *finitely generated*. That is, for any MSG G there exists a finite set X of finite MSCs such that any execution $M \in \mathcal{L}(G)$ can be written as a (finite or infinite) product $M = M_1 M_2 \cdots M_k$ of factors from X, $M_i \in X$ for all i. A typical example of CFM that is not finitely generated corresponds to the alternating bit protocol. The executions of this protocol include the family of MSCs that generalize the pattern of the MSC shown in figure 1 with n crossing messages for every n. None of these MSCs M can be decomposed as $M = M_1 M_2$ with M_1, M_2 non-empty MSCs, since including a send in M_1 forces to add another send on the other process (the one preceding the corresponding receive). More generally, an MSC M is called *atomic* (or *atom*), if for any decomposition $M = M_1 M_2$ where both M_1, M_2 are MSCs, at most one is non-empty. For another example of atomic MSC, consider the MSC M_3 in figure 4. The set of atoms generating the MSC executions of an MSG G is denoted $\mathrm{At}(G)$. It is a finite set and it represents a canonic set of generators of $\mathcal{L}(G)$. Moreover, it can computed by a simple linear-time algorithm, see [19].

On the potentially infinite alphabet At of atomic MSCs, we can define an independence (commutation) relation $I \subseteq \mathrm{At} \times \mathrm{At}$ by letting $A\, I\, A'$ iff $pr(A) \cap pr(A') = \emptyset$. Notice that $A\, I\, A'$ implies that A, A' commute, $AA' = A'A$, and that the decomposition of any MSC into atoms is unique up to commuting adjacent atoms A, A' with $A\, I\, A'$.

Returning to the alternating bit example, it is easily seen that the set of linearizations $\mathrm{Lin}(M)$ of the represented MSC M is regular. Note that in this particular example *every* linearization of M has channel bound at most 3. We call a set X of MSCs *universally-bounded* if for every MSC $M \in X$, *every* linearization $w \in \mathrm{Lin}(M)$, every prefix v of w and every pair of distinct processes p, q we have $0 \leq \sum_{a \in C} |v|_{p!q(a)} - \sum_{a \in C} |v|_{q?p(a)} \leq b$. Notice also that such a universal channel bound for an MSG G does not suffice for $\mathrm{Lin}(G)$ being a regular set. Hence, even if the specification is given as finite-state automaton $\mathcal{A}$, we cannot automatically transform $\mathcal{A}$ into an equivalent MSG G. This led to an extension of the MSG formalism, namely to *Compositional Message Sequence Graphs* (CMSG, for short) [16]. A compositional MSC (CMSC, for short) is defined as an MSC, except that the message function m is partially defined. A send that does not belong to the domain of the message function m, or a receive not belonging to the range of m, are called unmatched events. The product of two CMSC $M_1 M_2$ is defined as for MSC, but in addition the k-th unmatched send of M_1 is matched with the k-th unmatched receive of M_2 (if they exist) in such a way that the FIFO property is satisfied by matched events. Hence, the product of CMSCs is only partially defined. Moreover, it is not associative, hence we define a product $M_1 M_2 \cdots M_k$ as parenthesized from left to right.

It is not very difficult to see that any CFM can be transformed into an equivalent CMSG of exponential size. The rough idea is that nodes correspond to pairs (state,event), where state is a global state of the CFM and event is an event enabled in state. There is a transition from (state1,event1) to (state2,event2) if state2 is obtained from state1 by an event1-transition (that modifies state1 according to the local transition relation). It is easy to check that any CMSC execution of the CFM with no unmatched receive is an execution of the CMSG, and vice-versa. For instance, the CFM generating the alternating bit protocol from example 1 can be transformed into the CMSG in figure 3. (For CMSCs we draw unmatched events by the solid end of a half-dotted message arrow, that suggests the type of the matching event.)

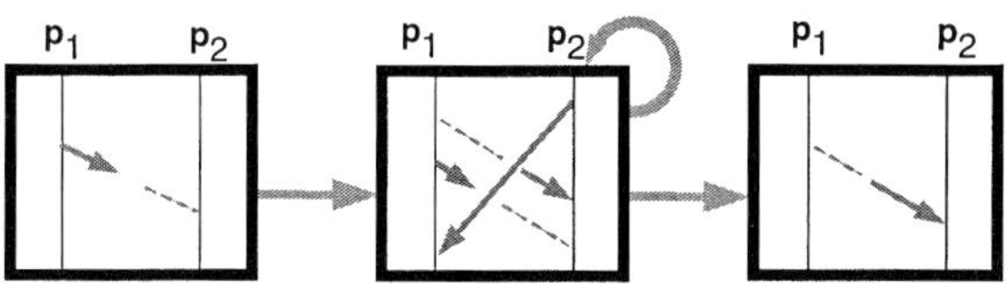

Fig. 3. CMSG depicting the alternating bit protocol.

Theorem 1. *Any CFM can be transformed into an equivalent CMSG of exponential size.*

4 Validating MSC Specifications: Model-Checking and Implementation

MSG specifications are used very early in the design process. Revealing design errors before implementing is of primary importance. This has motivated the design of algorithms that check specific properties of MSGs such as *race conditions* [4, 28] and detecting *non-local choice* [7, 19, 18]. *Model-checking* MSG specifications has been considered w.r.t. properties expressed as MSG [27], automata [6] and partial-order logics [29, 24]. Another test that may reveal the incompleteness of an MSG specification is the one for implementability. Here, we want to know whether the specification can be transformed into a state-based, distributed model as CFM. As discussed in section 4.2, the definition of implementability is not canonical, and the results strongly depend on the variant we consider.

4.1 Model-Checking

In the common model-checking approach (see for recent textbooks [8, 10]) we usually describe bad execution sequences using the same formalism as for specifying the system (e.g., finite automata over infinite words). Then we need to check the emptiness of the intersection between the bad sequences and the system, and counter-examples can be obtained if the intersection is non-empty. In the MSC setting we cannot use complementation as with finite automata. First, the complement of an MSG is not finitely generated, thus it can never be represented by an MSG. Secondly, even if we take the complement w.r.t. the MSCs generated by the same set of atoms, the complement cannot be represented by an MSG in general. This is similar to the fact that the complement of a rational trace language is not rational, in general [11]. Therefore, we consider two variants of model-checking, *positive* and *negative* model-checking. In both cases we specify the property P we want to check, as well as the system S itself, by MSGs. For *negative* model-checking we view P as a set of bad MSC executions and we ask whether $\mathcal{L}(P) \cap \mathcal{L}(S) = \emptyset$. For *positive* model-checking we view P as a set of good MSC executions and we ask whether $\mathcal{L}(S) \subseteq \mathcal{L}(P)$.

In the general setting of MSG specifications, both model-checking variants are undecidable [6, 27]. This holds even if the property P is given by a finite-state automaton or an LTL formula [6]:

Theorem 1 *Given a finite-state automaton P and an MSG graph G, it is undecidable whether $\mathcal{L}(P) \cap \mathcal{L}(G) = \emptyset$.*

The proof for theorem 1 is a straightforward reduction from Post's correspondence problem (PCP). Recall that an instance of PCP consists of pairs of words $(x_i, y_i)_{1 \leq i \leq k}$ over the alphabet $\{0, 1\}$. Then we ask for a non-empty sequence of indices $i_1, \ldots, i_n$ such that $x_{i_1} \cdots x_{i_n} = y_{i_1} \cdots y_{i_n}$.

The MSG G consists of $(k+2)$ nodes $v^0, v_1, \ldots, v_k, v^f$. Node v^0 (v^f, resp.) is initial (final, resp.), and labeled by the empty MSC. Node v_i is labeled by a sequence of messages from p_1 to p_2 labeled 0 or 1 such that the sequence of labels equals x_i, and a message from p_3 to p_4 labeled by i. There is a transition from v^0 to each of v_i, from each v_i to v^f, and one from v^f to v_0. The automaton P accepts precisely the set $(X_1 + \cdots + X_k)^+$, where each X_i is a finite word defined as follows: Let $y_i = a_1 \cdots a_m$, then $X_i = p_1!p_2(a_1)p_2?p_1(a_1) \cdots p_1!p_2(a_m)p_2?p_1(a_m)\, p_3!p_4(i)p_4?p_3(i)$.

Clearly, since there is no synchronization between the process pairs $\{p_1, p_2\}$ and $\{p_3, p_4\}$, both the MSG and the automaton describe MSCs with two parallel threads, one over the PCP words (x for G, y for P) and the other over the corresponding indices. The non-empty intersection between G and P reveals then a PCP solution.

Remark. Note that the undecidability proof above does not rely on the unboundedness of channels, since G is existentially-bounded. Actually the construction can be slightly modified such that G becomes universally-bounded, by adding an acknowledgment after each message. The true reason for undecidability is concurrency, since G and P use different linearizations of the same partial orders of MSC. □

Several decidable variants of model-checking have been considered in subsequent papers. Some of them are obtained by restricting the properties we want to check, others are obtained by restricting the system specification. However, several variants are based on a similar idea. Suppose for instance that the property P is given by a *linearization-closed* finite-state automaton $\mathcal{A}$. That is, for every word $w \in \mathcal{L}(\mathcal{A})$, the automaton $\mathcal{A}$ also accepts every linearization $v \in \mathrm{Lin}(M)$ of the MSC M defined by w. In this case it suffices to consider *representative linearizations* of the system MSG G: We choose for every node v of G some linearization of the MSC labeling v, say l_v. Then we define a finite-state automaton $\mathcal{A}(G)$ from G by replacing the label of v by l_v. Thus, states of $\mathcal{A}(G)$ are labeled by words. It is easy to see now that $\mathcal{L}(G) \cap \mathcal{L}(\mathcal{A}) \neq \emptyset$ if and only if $\mathcal{L}(\mathcal{A}(G)) \cap \mathcal{L}(\mathcal{A}) \neq \emptyset$, and $\mathcal{L}(G) \subseteq \mathcal{L}(\mathcal{A})$ if and only if $\mathcal{L}(\mathcal{A}(G)) \subseteq \mathcal{L}(\mathcal{A})$.

Among the model-checking variants that led to algorithmic solutions we refer to the following ones:

- *Model-checking with gaps* [28]: The property P is given by an MSG, but its semantics differs from the semantics of the system G. An execution M of P is matched *with gaps* by an execution M' of G if there is an embedding ϕ of the events of M in the set of events of M' such that the visual order is preserved: whenever $e < f$ in M, we have $\phi(e) <' \phi(f)$ in M'. This problem has been shown to be NP-complete (even if P is an acyclic MSG).
  The main reason for decidability of model-checking with gaps is that gaps lead to very restricted languages, for which we can compute a sort of linearization-closure.
- *Using partial-order specifications* [29, 24]: Here, the property P is given by a partial-order logic, which makes it linearization-closed. In [29] a logic derived from a fragment of TLC [5] is proposed for MSGs. Basically, this logic corresponds to CTL interpreted over partial-order graphs of MSCs, where the edge relation is the immediate successor relation (on each process, resp. for send/receive pairs). It is shown in [29] how to construct an exponential-size automaton from the specification, hence model-checking is PSPACE w.r.t. the specification (and only linear in the size of the system). In [24] the specification formalism is MSO, interpreted over partial-order graphs of MSCs. Here, the complexity is non-elementary, as it is already in the word case.

A further approach leading to a decidable model-checking problem is to syntactically restrict the MSGs, see sections 5 and 6 for details.

4.2 Implementability

As previously mentioned, the MSG formalism is useful as a specification notation, but it does not provide directly a protocol model. Such a model is usually state-based and distributed, whereas MSGs provide an implicit global control over the behavior of the processes. This allows for specifications that are not implementable because of global choices (see for instance figure 4 which is discussed below). Being able to generate an implementation for an MSC specification also allows to perform tests on the level of requirements, hence it is not longer required to generate code before testing.

The protocol model generally used is CFM over the same set of processes as the MSG specification. But we still have some choice for the semantics of the implementation. For instance, we could allow for an implementation with more (or less) behavior than the MSG. The most natural notion is that the implementation is equivalent to the MSG: An MSG G is *implementable*, if some CFM $\mathcal{A}$ exists such that $\mathcal{L}(\mathcal{A}) = \mathcal{L}(G)$. Furthermore, we allow the implementation to contain additional data in messages. That is, the message contents of the CFM come from a finite set $\mathcal{C}' = \mathcal{C} \times D$, where $\mathcal{C}$ is the set of message contents of the MSG and D is some finite set. Then, the equality $\mathcal{L}(\mathcal{A}) = \mathcal{L}(G)$ is required up to the additional data D. A further, even more relaxed notion of implementability, would also allow for additional messages. Notice that this would make every MSG implementable, since the additional messages can be used for synchronizing all processes after each node. We do not allow additional messages, since in many applications they are neither desired nor possible (e.g., applications where acknowledgments cannot be provided).

The first notion of implementation, which we denote as *standard implementation*, has been proposed in [2, 3]. The standard implementation of the MSG $G = \langle V, R, v^0, V_f, \lambda \rangle$ over the process set $\mathcal{P}$ does not add any data and it is fully determined by the MSG, being defined process by process: The automaton $\mathcal{A}_p$ for process p generates the projection of $\mathcal{L}(G)$ on the events of process p.

We call an MSG *standard-implementable* if it is implementable w.r.t. the standard version of implementability. Notice that this notion is actually too weak, since it captures just a small subset of implementable specifications. The simplest counter-example (see figure 4) is a set of two MSCs over the processes p_1, p_2 where the first MSC M_1 has a message from p_1 to p_2, followed by one from p_2 to p_1. In the second MSC M_2 we have first a message from p_2 to p_1, then one from p_1 to p_2. These two MSCs are not standard-implementable since we can combine the projection of M_1 on p_1 with the projection of M_2 on p_2 and we obtain the MSC M_3. This set is not implementable even with additional data. Changing slightly this example we obtain one which is not standard-implementable, but is implementable with additional data. For this, we just add at the beginning of both M_1, M_2 a first unlabeled message from p_1 to p_2, see figure 5. Then the non-implementability argument given previously still works. However, with

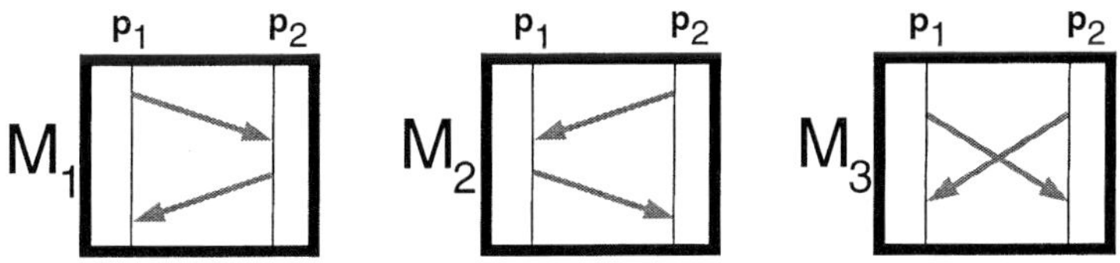

Fig. 4. The set $\{M_1, M_2\}$ is not implementable (it does not contain the implied MSC M_3).

additional data we use the initial message for letting p_1 decide on the outcome M_1 or M_2, and inform p_2.

Two striking weaknesses of the standard notion is that not even simple MSGs are standard-implementable, as seen from the example above. Furthermore, for the restricted class of regular MSGs defined in section 5, the question of standard-implementability is undecidable. However, regular MSGs are implementable with additional data, see section 5 for more details.

Nevertheless, the results of [2, 3] show that standard implementability becomes decidable at least for regular MSGs if one looks for *deadlock-free* implementations, only (called *safe realizability* in [2, 3]), albeit with high algorithmic complexity. A CFM is called a *deadlock-free implementation* of an MSG G if $\mathcal{L}(\mathcal{A}) = \mathcal{L}(G)$ and every configuration of $\mathcal{A}$ that has no successor, is such that all processes have reached a final state and all channels are empty. Deadlock-freeness is of course required in practice, since real-life protocols should not be aborted in some unclean state. We will recall the various results on the implementability problem in sections 5 and 6.

5 Regular MSC Specifications

Regular MSGs have been proposed in the context of model-checking, as a subclass for which both variants of model-checking are decidable [6, 27]. It is a syntactic restriction that ensures that the set of all linearizations, i.e., the set $\mathrm{Lin}(G)$, is regular. Regular MSGs provided to be a theoretically robust class, in terms of logical and automata-theoretic characterizations. In particular, regular MSGs can be implemented with additional data by CFM with universally-bounded channels. However,the CFM implementation is not deadlock-free, in general.

A set X of finite MSCs is called *regular* if $\mathrm{Lin}(X)$ is a regular string language over the alphabet $\mathcal{T}$ of event types [20]. Moreover, there is a syntactic condition ensuring that an MSG G generates a regular set $\mathcal{L}(G)$ of MSCs. This condition roughly means that communication in a loop must be acknowledged to all active processes. Formally, we need to define the *communication graph* of an MSC M: it is a directed graph over the set of communicating processes in M with an edge from process p to process q whenever M contains a message from p to q. An MSG G is called a *regular MSG* (locally-synchronized in [27], bounded in [6]) if any MSC labeling a loop of G has a strongly connected communication graph. This condition is co-NP complete [27].

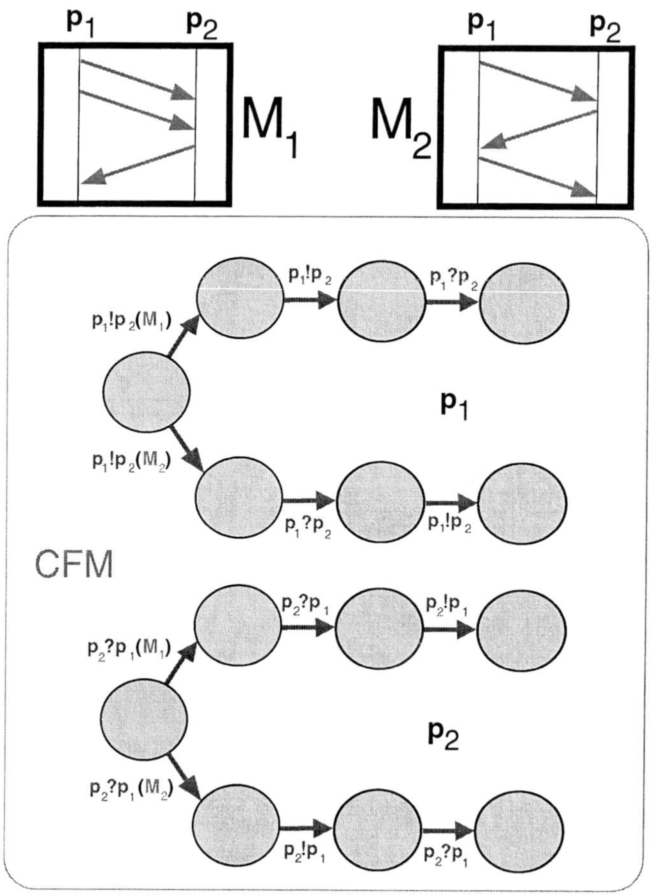

Fig. 5. CFM implementing the set $\{M_1, M_2\}$ with additional data.

For an example, consider the MSG in figure 2. It is a regular MSG, since every loop involves only A, C, and the communication graph of AC is strongly connected (the firewall is connected with both user and server by bidirectional arcs).

Putting together the results from [6, 27, 20] we have the following relationship between regular sets of MSCs and regular MSGs:

Theorem 2 *1. For every regular MSG G the set $\mathcal{L}(G)$ of generated MSCs is regular [6, 27].*

2. For every regular and finitely generated set X of MSCs there exists a regular MSG G with $X = \mathcal{L}(G)$ [20].

The main interest in regular MSGs was to obtain a subclass of MSC specifications with a decidable model-checking problem:

Theorem 3 *[6, 27] The negative model-checking problem $\mathcal{L}(G) \cap \mathcal{L}(H) \neq \emptyset$ where G is a regular MSG, is PSPACE-complete. The positive model-checking problem $\mathcal{L}(G) \subseteq \mathcal{L}(H)$ where H is a regular MSG, is EXPSPACE-complete.*

The theorem above shows that model-checking MSGs is rather expensive, which is actually not very surprising when we deal with concurrent models. The reason is MSGs are more compact than finite-state automata. The upper bounds in the theorem above are based on the fact that if G is a regular MSG then we can compute a finite automaton of *exponential size* generating $\mathrm{Lin}(G)$.

Regular MSC languages also have nice characterizations in the logical and communicating automata framework. The logic used in [21, 24] is MSO with atomic propositions $\ell(e) = t \in \mathcal{T}$, $e \leq f$ and $e \in E$ that have the usual interpretation, as type labeling, partial order of the MSC and membership in a second order variable E.

Theorem 4 *[21, 22] Let X be a universally-bounded set of MSCs. The following assertions are equivalent:*

1. *X can be implemented by a (deterministic) CFM with additional data.*
2. *There exists an MSO formula ϕ such that X is the set of bounded MSCs satisfying ϕ.*

In the first part of the theorem above the implementation is not deadlock-free, since the constructed CFM uses global final states for accepting X. On the other hand, as we mentioned in section 5, the standard implementation is not really helpful when applied to regular MSGs (the upper bound is due to [3], and the lower bound to [23]):

Theorem 2. *[3, 23] It is undecidable to know whether a regular MSG is standard-implementable. It is EXPSPACE-complete to know whether a regular MSG is standard-implementable without deadlocks.*

Remark. The undecidability result in theorem 2 heavily depends on the fact that channels are FIFO. Without FIFO, standard implementability for regular MSGs becomes decidable [26]. □

6 Globally-Cooperative MSGs

As seen in section 5, model-checking for regular MSGs is decidable and of tractable complexity (PSPACE for the basic variant). However, the situation is far from being ideal. Notice first that some trivial protocols cannot be represented by regular MSGs. For instance, the protocol where process p_1 can send any number of messages to process p_2. The reason is that regular MSGs have universally bounded channels, which restricts severely their expressive power. Second, for real life communication protocols one can usually find a (sufficiently large) bound b so that any run of the protocol can be executed with channels each bounded by b. Similarly to MSGs, we call such protocols *existentially b-bounded.* Whereas an algorithm exists to check whether a CFM or a finite-state machine is existentially b-bounded for a given b, its complexity depends severely (exponentially) on b. Hence, in practice we cannot hope to be able to fix a sufficiently large bound b that takes care of all executions. The last problem is that we cannot obtain in general an automaton generating all linearizations of executions for models that are strictly more expressive than regular MSGs. Instead, we can try to use representative linearizations rather than all linearizations, requiring that the set of representative linearizations is b-bounded, with b as small as possible.

Definition 1 *An MSG G is called* globally-cooperative *(*gc-MSG *for short) if every loop of G has a weakly connected communication graph.*

Thus, an MSG G is a gc-MSG if any MSC M labeling a loop cannot be written as $M = M_1||M_2$ with M_1, M_2 non-empty MSCs with no common process. It is co-NP complete to know whether an MSG is a gc-MSG. For an example of a gc-MSG, see figure 6, or suppose that we add a self-loop on node A in figure 2. The MSG thus obtained is not regular anymore, but it is a gc-MSG. Clearly, every regular MSG is also a gc-MSG. Moreover, it can be noted that regular MSGs correspond exactly to gc-MSGs with universally-bounded channels.

The representative linearizations that we use for model-checking are the linearizations that execute atoms one by one. More precisely, for any $M \in \mathcal{L}(G)$ we consider only linearizations in $\mathrm{Lin}(M)$ of the form $w = w_1 \cdots w_n$, where $M = A_1 \cdots A_n$ is some decomposition of M into atoms A_i and $w_i \in \mathrm{Lin}(A_i)$ for all i. Let us denote by $\mathrm{Lin}^a(G) \subseteq \mathrm{Lin}(G)$ the set of such linearizations of MSCs of $\mathcal{L}(G)$. For an example, let G be the graph consisting of a single node with a self-loop, labeled by a message from p_1 to p_2. Let $s = p_1!p_2$ and $r = p_2?p_1$, then $\mathrm{Lin}^a(G) = (sr)^*$. Of course, $\mathrm{Lin}(G)$ is not regular, it corresponds to the Dyck language over one pair of brackets.

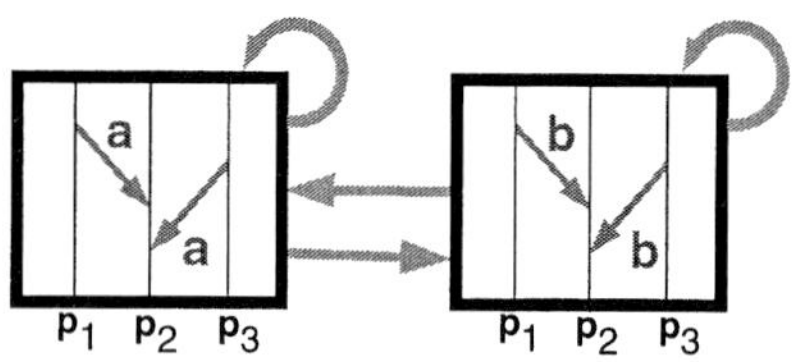

Fig. 6. Globally-cooperative MSG.

We can use representative linearizations for model-checking as follows. Let G, H be two MSGs. Then it is easy to see that $\mathcal{L}(G) \cap \mathcal{L}(H) = \emptyset$ if and only if $\mathrm{Lin}^a(G) \cap \mathrm{Lin}^a(H) = \emptyset$ (respectively, $\mathcal{L}(G) \subseteq \mathcal{L}(H)$ if and only if $\mathrm{Lin}^a(G) \subseteq \mathrm{Lin}^a(H)$. Recall that for any MSG G, atoms of $\mathcal{L}(G)$ are finite and finitely many. Hence the set of representative linearizations $\mathrm{Lin}^a(G)$ is b-bounded, where b is such that G is existentially b-bounded. For getting a regular set of representative linearizations $\mathrm{Lin}^a(G)$ we impose a theoretically well-known restriction, that of *loop-connectedness.*

We can change slightly the graph of a gc-MSG G by replacing each node v labeled by some non-empty MSC M by a path of new nodes $v_1, \ldots, v_k$ where v_i is labeled by A_i and $M = A_1 \cdots A_k$ is some decomposition of M into atoms A_i. The new graph G' can be seen as an automaton with states labeled over the alphabet of atoms $\mathrm{At}(G)$. The property of G being a gc-MSG translates to G' being loop-connected, which is a well-known property from the theory of Mazurkiewicz traces. It means that every loop of G' is labeled by a sub-alphabet of $\mathrm{At} = \mathrm{At}(G)$ that is connected w.r.t. the symmetric dependence $D = (\mathrm{At} \times \mathrm{At}) \setminus I$, that is $A\, D\, A'$ if A and A' share at least one process. With this restriction it is well-known that the closure under commutation I of the regular set generated by G' is regular, and an automaton generating the closure can be effectively computed [11, 27]. From this automaton we obtain $\mathrm{Lin}^a(G)$ and an automaton generating it simply by replacing every atom $A \in \mathrm{At}(G)$ by some linearization of A. Since the size of the automaton generating $\mathrm{Lin}^a(G)$ is exponential in the size of G we obtain:

Theorem 5 *[15] Given a gc-MSG G and an arbitrary MSG H, it is PSPACE-complete to decide whether $\mathcal{L}(G) \cap \mathcal{L}(H) = \emptyset$. The positive model-checking problem $\mathcal{L}(G) \subseteq \mathcal{L}(H)$ where H is a gc-MSG, is EXPSPACE-complete.*

Notice that the complexity of model-checking gc-MSGs is not higher than for regular MSGs. Moreover, the situation for gc-MSGS is better, since we do not have to compute all linearizations, but a smaller subset that has the additional property of being b-bounded for a small b, yielding an algorithm that is faster in practice than the one given for regular MSGs. A regular MSG G is universally B-bounded with a B that can be exponential in the size of G.

We turn now to the implementation problem. The situation here enforces the idea that that universal channel bounds are not needed. For the safe variant of the standard implementability problem, i.e., where the implementation is not allowed any additional data but must be deadlock free, the complexity is the same for regular MSGs and for gc-MSGs:

Theorem 6 *[3, 23] Given a gc-MSG G, it is EXPSPACE-complete to decide whether there exists a deadlock-free CFM $\mathcal{A}$ with $\mathcal{L}(G) = \mathcal{L}(\mathcal{A})$.*

Again, this result is not really practical, given the high complexity. Moreover, in practice it might be the case that the standard implementation does not work for some gc-MSG G, but that G is still implementable with a little more data. For an example see figure 8 in section 7.

In the case where one allows data to be added to messages but deadlocks are not allowed, there are gc-MSGs that cannot be implemented. For instance, consider the gc-MSG G in figure 6 with two nodes with self-loops, and two edges between them. Both nodes are labeled by MSCs with two messages, one from p_1 to p_2 and one from p_3 to p_2. The first node has its messages carrying the data a, while the second node carries the data b. Both nodes are initial and final. In any CFM implementation processes p_1 and p_3 should decide to send either both a or both b, but this is impossible with no additional synchronization (messages). Hence, this protocol cannot be implemented without deadlocks. It remains open whether every gc-MSG can be implemented with additional data and allowing deadlocks. The conjecture in [15] is that this is always possible.

7 Choice and Implementability

Deadlock-free implementability being a key feature required for communicating protocols, tractable algorithms that help implementing an MSG with additional data are needed. One reasonable way of doing this is first to exhibit a non-trivial subclass of MSGs that is always implementable with additional data and no deadlocks. Then we want to test whether an MSG can be represented inside our subclass, preserving the MSC language.

As mentioned before, the reason for non-implementability of an MSG is the global control, whereas the choice in a CFM must be done locally. The idea is then to define MSGs that have only local choices, that is any node is controlled by a single process [7, 19].

Definition 2 *An MSG* $G = \langle V, R, v^0, V_f, \lambda\rangle$ *is called* local-choice *(lc-MSG for short) if each MSC labeling any node* v *of* G *is a* triangle, *that is it has a single minimal event* $\min(v)$ *in the partial order* $\leq$. *Moreover,* $\min(w)$ *belongs to the process set of node* v*, whenever* $(v, w) \in R$.

Figure 7 shows an lc-MSG G. Note that G is equivalent to the MSG in figure 2, which is not an lc-MSG. Checking that an MSG is local-choice can be done in polynomial time.

It is not very hard to translate a lc-MSG into a deadlock-free CFM, using linear additional data. The idea is to use a *leader process* and to let the current leader choose the current_node to be executed and the next_leader. The node is chosen among the nodes that follow the node being executed, and that begin with a minimal event belonging to the leader. The next_leader should be chosen among the minimal processes of nodes that follow the chosen node.

In the procedure polling_state below process p waits for a message informing it about the next node to execute and the next leader:

```
void polling_state()
{ while (true) {
    if p receives a message (a,v,q) then
          { current_node=v; next_leader=q; return;} } }
```

Initially, the current node is initialized by letting next_leader $= pr(\min((v^0))$. Before executing its event from current_node, process p goes to a polling state, unless it is the leader process. Here is the algorithm for process p:

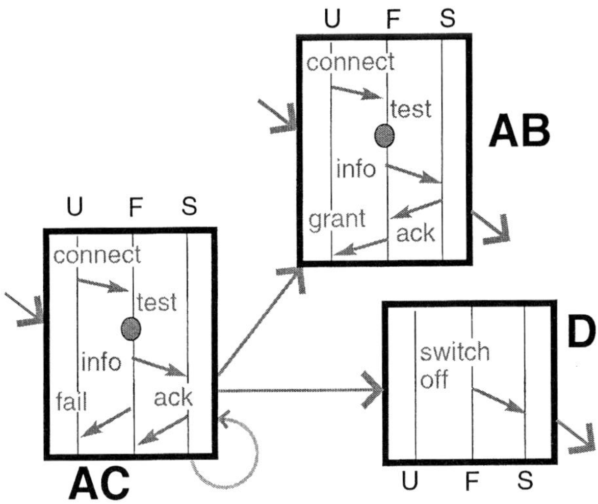

Fig. 7. Local-choice MSG.

```
initialization();
 while (true)
 { if (p ≠ next_leader) polling_state();
   else {current_node=guess(current_node);
         next_leader=guessp(current_node);}
   execute_path(current_node); }
```

The algorithm execute_path(current_node) above makes that process p executes its events from current_node, if any. In this case each message sent by p contains the additional data (current_node, next_leader).

Theorem 7 *[15] Every lc-MSG G is implementable by a deadlock-free CFM with additional data which is of size linear in* $|G|$.

Note that in a triangle (see definition 2), every process but the minimal process begins by a receive. A process that is chosen to be the leader is always informed, since it occurs in the node where it is chosen.

It is important to see that if additional data is forbidden, then there are lc-MSG that are not implementable, even when allowing deadlocks. Consider for example an lc-MSG with three nodes 1, 2, 3, see figure 8. The initial node consists of a message from process p_1 to p_2. Then either node 2 is executed, with process p_1 sending to process p_3, or node 3 is executed with process p_2 sending to process p_4. Since processes p_1 and p_2 do not know which one will be next (no additional data, same past), both can begin, thus both nodes 2 and 3 can start. The execution must stop, and there is no distributed way to know whether the protocol went fine or not. Hence without additional data this protocol is not implementable at all. [19] proposes a sufficient condition for the standard-implementability of lc-MSGs.

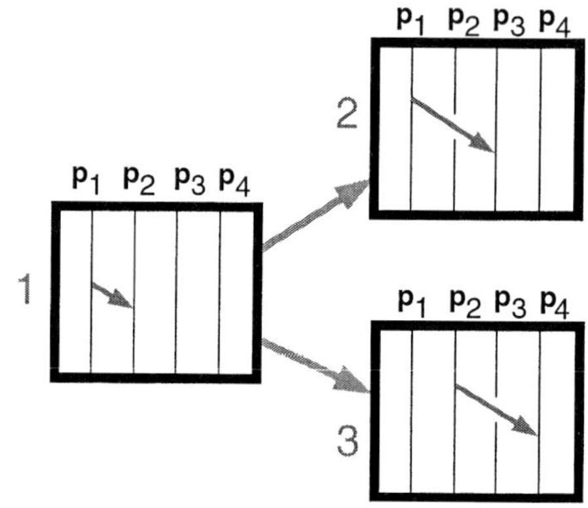

Fig. 8. Lc-MSG not implementable without additional data.

While local-choice is defined syntactically, we show that it corresponds to a semantic property. Moreover, one can test whether an MSG is transformable into a lc-MSG. Triangles are of huge importance here. We first define a generic lc-MSG H_n over triangles of size bounded by n. The MSG H_n has for each triangle T of size at most n, one node v_T labeled by T. There is an edge $v_T \to v_{T'}$ if $pr(\min(T')) \in pr(T')$.

Proposition 1 *[14] An MSG G is equivalent to some lc-MSG iff there exists some n such that $\mathcal{L}(G) \subseteq \mathcal{L}(H_n)$. If this is the case, then we can obtain a lc-MSG equivalent to G, of size exponential in $|G|$ and n.*

If the test in proposition 1 on G answers yes, then an equivalent lc-MSG can be constructed by synchronizing G and H_n.

While it is PSPACE to test whether $\mathcal{L}(G) \subseteq \mathcal{L}(H_n)$ by theorem 9, the value of n is not bounded so far. For testing, we need a bound. We use for this the following three structural properties of lc-MSG G.

1. Every MSC M in $\mathcal{L}(G)$ is a triangle.
2. There is a bound b s.t. for every MSC in $\mathcal{L}(G)$ containing a factor $(U||V)$ with U, V MSCs (that is, U, V share no process), either $|U| < b$ or $|V| < b$. This implies that for an MSG to be equivalent to some lc-MSG, it is necessary to be a gc-MSG.
3. There is a bound b s.t. for every MSC in $\mathcal{L}(G)$ of the form URV with R an MSC of size at least b, there exists triangle T that is a suffix of RV, such that $\min(T)$ belongs to R.

Obviously, an MSG G that is equivalent to some lc-MSG satisfies these three properties. The important point is that the converse holds, too. It allows us to state:

Theorem 8 *[14] Testing whether an MSG G is equivalent to some lc-MSG is in PSPACE. Moreover, if the answer is positive, then an equivalent lc-MSG of doubly exponential size can be constructed.*

Proof. We can check whether G is a gc-MSG in co-NP [15]. Checking the first property above is in polynomial time. Checking the second property for gc-MSG is in co-NP. If true, the test provides a bound b that is polynomial in $|G|$.

Checking the third property for gc-MSGs is in PSPACE. If true, the test provides a bound b that is exponential in $|G|$.

We can then compute an equivalent lc-MSG building the product $H_b \times \mathrm{Lin}^a(G)$. As b is exponential in $|G|$ and H is exponential in b, the result is at most doubly exponential in $|G|$. □

One important question is whether local-choice is expressive enough, else the test to know whether an MSG is equivalent to some lc-MSG would almost certainly lead to a negative answer. Comparing lc-MSGs to regular MSGs, lc-MSGs tend to be more useful in practice. In particular, the restriction of universally-bounded channels of regular MSGs is not required for lc-MSGs. Moreover, lc-MSGs can be implemented without deadlock, while this is not the case for regular MSGs. A drawback of lc-MSGs is the fact that they exclude long parallel MSCs, while this is possible with regular MSGs (albeit not in the same loop of the graph). Actually, it would not be difficult to cut a protocol into parallel ones, and implement each one using lc-MSGs.

Since lc-MSGs form a subclass of gc-MSGs, one can hope that they are easier to model-check than gc-MSGs. In order to improve the model-checking algorithm, triangles can be used as generators instead of atoms. For a given lc-MSG each node v labeled by a triangle T can be sliced into two nodes labeled by triangles R, S, as long as $T = RS$ satisfies $pr(\min(w)) \in S$ for every $v \to w$. Notice that by the definition of a triangle, we have that $pr(\min(S)) \in R$. Let $T_1 \cdots T_n, T'_1 \cdots T'_{n'}$ be sequences of triangles labeling two paths ρ, ρ' in lc-MSGs G, H sliced in this way. Then there exist k, X s.t. $T_i = T'_i$ for all $i < k$, and $T'_k = XT_{k+1} \cdots T_n$, $T_k = XT'_{k+1} \cdots T'_{n'}$. Hence, $T_{k+1} \cdots T_n$ is smaller than the largest node of G. The same applies for $T'_{k+1} \cdots T'_{n'}$. This idea allows to do model-checking very similarly to word automata.

Theorem 9 *[15] Given two lc-MSGs G, H, the negative model-checking question $\mathcal{L}(G) \cap \mathcal{L}(H) = \emptyset$ can be answered in quadratic time. The positive model-checking question $\mathcal{L}(G) \subseteq \mathcal{L}(H)$ with H an lc-MSG and G an arbitrary MSG, is PSPACE-complete.*

8 Conclusions

The MSC/MSG standard is a popular notation for concurrent system specification, in particular for communication protocols. Stemming from its successful use by software engineers, new techniques and tools have been developed for MSC/MSG analysis. The finite states model was designed by researchers. Although this model has many mathematical properties, it is not always easy to transfer its related technology to the software developers. The MSC notation, on the other hand, has gained first popularity with the software developers. Consequently, this notation does not fit directly the main classes of formal languages.

This calls for studying the expressiveness of the notation and developing new validation and implementation methods.

It is evident from the collection of results surveyed here that one of the main challenges in studying MSCs/MSGs is how to achieve the appropriate expressiveness, while maintaining decidability with respect to automatic verification. This calls for developing various extensions and restrictions on the allowed class of MSCs/MSGs.

The MSC/MSG standard provides an alternative for the communicating automata model. In particular, the main compositional operator for the former is sequential composition, while the main way to connect communicating automata is using parallel composition. Although sequential composition is often considered simpler than the parallel one, it is evident that this is not the case here. The reason is that the sequential composition is asynchronous, relating partial orders. In particular, the parallel composition of two MSCs (i.e., that share no process) is expressed when we compose them sequentially (as is the case in classical Mazurkiewicz trace theory [11]). This is also manifested by the high complexity results on MSG decision procedures. Note however that subclasses as lc-MSGs have the same complexity as finite-state machines.

The theory of MSCs is related to models of true concurrency, including partial orders and Mazurkiewicz traces. While these theories flourished in the recent decades, their practical use was limited, due to the high complexity they generally possess, when compared to the finite-state machine model. The MSC model provides an important use of these true concurrency models. The intuitive nature of these models is manifested by the use of the MSC as a popular visual notation for concurrency.

References

1. ITU-TS recommendation Z.120, 1996.
2. R. Alur, K. Etessami, and M. Yannakakis. Inference of Message Sequence Charts. In *Proceedings of the 22nd International Conference on Software Engineering, Limerick (Ireland)*, pages 304–313. ACM, 2000.
3. R. Alur, K. Etessami, and M. Yannakakis. Realizability and verification of MSC graphs. In *Proceedings of the 28th International colloquium on Automata, Languages and Programming (ICALP'01), Crete (Greece) 2001*, number 2076 in Lecture Notes in Computer Science, pages 797–808. Springer, 2001.
4. R. Alur, G. H. Holzmann, and D. A. Peled. An analyzer for message sequence charts. *Software Concepts and Tools*, 17(2):70–77, 1996.
5. R. Alur, D. Peled, and W. Penczek. Model-checking of causality properties. In *Proceedings of the 10th Annual IEEE Symposium on Logic in Computer Science (LICS '95)*, pages 90–100. IEEE, 1995.
6. R. Alur and M. Yannakakis. Model checking of message sequence charts. In *Proceedings of the 10th International Conference on Concurrency Theory CONCUR'99, Eindhoven (The Netherlands)*, number 1664 in Lecture Notes in Computer Science, pages 114–129. Springer, 1999.

7. H. Ben-Abdallah and S. Leue. Syntactic detection of process divergence and non-local choice in message sequence charts. In *Proceedings of the Tools and Algorithms for the Construction and Analysis of Systems, Third International Workshop, TACAS'97*, number 1217 in Lecture Notes in Computer Science, pages 259–274, Enschede, The Netherlands, 1997. Springer.
8. B. Berard, M. Bidoit, A. Finkel, F. Laroussinie, A. Petit, L. Petrucci, and P. Schnoebelen. *Systems and Software Verification. Model-Checking Techniques and Tools.* Springer, 2001.
9. D. Brand and P. Zafiropulo. On communicating finite-state machines. *Journal of the ACM*, 30(2):323–342, 1983.
10. E. Clarke, O. Grumberg, and D. Peled. *Model Checking.* MIT Press, 2000.
11. V. Diekert and G. Rozenberg, editors. *The Book of Traces.* World Scientific, Singapore, 1995.
12. B. Genest, M. Minea, A. Muscholl, and D. Peled. Specifying and verifying partial order properties using template MSCs. In *Proceedings of the 7th International Conference on Foundations of Software Science and Computation Structures (FoSSaCS'04)*, Lecture Notes in Computer Science. Springer, 2004.
13. B. Genest and A. Muscholl. Pattern matching and membership for Hierarchical Message Sequence Charts. In *Proceedings of the LATIN 2002: Theoretical Informatics, 5th Latin American Symposium, Cancun, Mexico*, number 2286 in Lecture Notes in Computer Science, pages 326–340. Springer, 2002.
14. B. Genest and A. Muscholl. The structure of local choice in High-Level Message Sequence Charts (HMSC) Internal report LIAFA, 2003 Available at http://www.crans.org/~genest/report_lc.ps.
15. B. Genest, A. Muscholl, H. Seidl, and M. Zeitoun. Infinite-state High-level MSCs: Model-checking and realizability. In *Proceedings of the 29thInternational colloquium on Automata, Languages and Programming (ICALP'02), Malaga (Spain), 2002*, number 2380 in Lecture Notes in Computer Science, pages 657–668. Springer, 2002.
16. E. Gunter, A. Muscholl, and D. Peled. Compositional Message Sequence Charts. In *Proceedings of the Tools and Algorithms for the Construction and Analysis of Systems, 7th International Workshop (TACAS'01)*, number 2031 in Lecture Notes in Computer Science, pages 496–511. Springer, 2001. Journal version to appear in the International Journal on Software Tools for Technology Transfer.
17. D. Harel and R. Marelly. Come, Let's Play: Scenario-Based Programming Using LSCs and the Play-Engine. Springer 2003.
18. L. Hélouët and C. Jard. Conditions for synthesis of communicating automata from HMSCs. In *5th International Workshop on Formal Methods for Industrial Critical Systems*, Berlin, 2000.
19. L. Hélouët and P. Le Maigat. Decomposition of Message Sequence Charts. In *Proceedings of the 2nd Workshop on SDL and MSC (SAM2000), Col de Porte, Grenoble*, pages 46–60, 2000.
20. J. G. Henriksen, M. Mukund, K. Narayan Kumar, and P. Thiagarajan. On Message Sequence Graphs and finitely generated regular MSC languages. In *Proceedings of the 27th International colloquium on Automata, Languages and Programming (ICALP'00), Geneva (Switzerland), 2000*, number 1853 in Lecture Notes in Computer Science, pages 675–686. Springer, 2000.

21. J. G. Henriksen, M. Mukund, K. Narayan Kumar, and P. Thiagarajan. Regular collections of Message Sequence Charts. In *Proceedings of the 25th Symposium on Mathematical Foundations of Computer Science (MFCS'00), Bratislava (Slovakia), 2000*, number 1893 in Lecture Notes in Computer Science, pages 405–414. Springer, 2000.
22. D. Kuske. A Further Step towards a Theory of Regular MSC Languages. In *Proceedings of the 19th Annual Symposium on Theoretical Aspects of Computer Science (STACS'02), Juan-les-Pins (France), 2002*, number 2285 in Lecture Notes in Computer Science, pages 489–500. Springer, 2002.
23. M. Lohrey. Safe realizability of High-level Message Sequence Charts. In *Proceedings of the Concurrency Theory, 13th International Conference (CONCUR'02)*, number 2421 in Lecture Notes in Computer Science, pages 177–192. Springer, 2002.
24. P. Madhusudan. Reasoning about sequential and branching behaviours of Message Sequence Graphs. In *Proceedings of the 28th International colloquium on Automata, Languages and Programming (ICALP'01), Crete (Greece) 2001*, number 2076 in Lecture Notes in Computer Science, pages 809–820. Springer, 2001.
25. M. Mukund, K. Narayan Kumar, and P.S. Thiagarajan. Netcharts: bridging the gap between HMSCs and executable specifications. In In *Proceedings of the Concurrency Theory, 14th International Conference (CONCUR'03)*, number 2761 in Lecture Notes in Computer Science, pages 296–310. Springer, 2003.
26. R. Morin. Recognizable Sets of Message Sequence Charts In *Proceedings of the 19th Annual Symposium on Theoretical Aspects of Computer Science (STACS'02), Juan-les-Pins (France), 2002*, number 2285 in Lecture Notes in Computer Science, pages 523–540. Springer, 2002.
27. A. Muscholl and D. Peled. Message sequence graphs and decision problems on Mazurkiewicz traces. In *Proceedings of the 24th Symposium on Mathematical Foundations of Computer Science (MFCS'99), Szklarska Poreba (Poland) 1999*, number 1672 in Lecture Notes in Computer Science, pages 81–91. Springer, 1999.
28. A. Muscholl, D. Peled, and Z. Su. Deciding properties of Message Sequence Charts. In *Proceedings of the 1st International Conference on Foundations of Software Science and Computation Structures (FoSSaCS'98), Lisbon, Portugal, 1998*, number 1378 in Lecture Notes in Computer Science, pages 226–242. Springer, 1998.
29. D. Peled. Specification and verification of Message Sequence Charts. In *Proceedings of Formal Techniques for Distributed System Development, FORTE/PSTV 2000, Pisa, Italy*, pages 139–154, 2000.
30. B. Sengupta and R. Cleaveland. Triggered Message Sequence Charts. In *Proceedings of SIGSOFT 2002/FSE-10*, pages 167–176, ACM Press, 2002.

Model-Based Development of Executable Business Processes for Web Services

Reiko Heckel[1,2] and Hendrik Voigt[2]

[1] Faculty of Computer Science, University of Dortmund, Germany
[2] Faculty of Computer Science, Electrical Engineering and Mathematics
University of Paderborn, Germany
{reiko,hvoigt}@upb.de

Abstract. In order to implement business processes, the composition of simpler services provided by different independent participants requires a high degree of standardization and flexibility. For this purpose, platform-independent XML-based languages like the *Business Process Execution Language for Web Services (BPEL4WS)* are suitable. XML documents are in fact human readable, but in general they are hard to produce and to understand by business experts which are, however, most qualified for defining business processes. We present a *model-based development method* based on an intuitive and adequate modelling notation, an automatic transformation of process models to their XML-based encoding, and techniques to analyze processes. In this context the *Unified Modeling Language (UML)* as standard notation for modelling software, *graph transformation* as meta language for defining model transformations, and a semantic interpretation of process models in terms of *Communicating Sequential Processes (CSP)* are used.

1 Introduction

A *Web service* is a software component that can be dynamically discovered, linked, and invoked by its clients via XML-based protocols. This software-oriented definition of the term can be contrasted with a business-oriented view, considering a Web service as a *business process*, implemented by the composition (and coordination) of simpler services provided by other businesses.

The composition of services provided by different independent parties, at both development time or runtime, requires a high degree of standardization and flexibility. Therefore, rather than hard-coding business processes in platform-specific programming languages which depend on certain compilers and runtime environments, platform-independent XML-based languages like the *Business Process Execution Language for Web Services (BPEL4WS)* [1] are advocated. Such processes in XML representation can, at least in theory, be adapted at runtime, exchanged between different services, and executed on different standardized interpreters.

However, even if XML documents are text files and therefore, in principle, human readable, the XML representation of a processes is hard to produce and to

J. Desel, W. Reisig, and G. Rozenberg (Eds.): ACPN 2003, LNCS 3098, pp. 559–584, 2004.

understand even by an experienced programmer. It resembles, in a linear form, the abstract syntax tree of a program without providing the usual front-end notation. What is more, in their role as business processes, Web service processes should be defined by business experts which are not typically programmers.

Therefore a *model-based development method* is required based on

1. an intuitive and adequate *modelling notation*, to allow precise specifications of processes at the conceptual level
2. an *automatic transformation of process models* to their XML-based encoding, to avoid the costly and error-prone task of deriving the implementation manually
3. *techniques to analyze processes* at the model level for syntactic and semantic properties, to avoid "debugging" the XML code

These problems and requirements are prototypical for a wide variety of languages and platforms, in the Web services domain and elsewhere. Therefore, instead of defining and implementing languages, transformations, and analysis tools for every single problem, reusable solutions are required.

In this paper, we will present an approach based on the combination of three such solutions: the *Unified Modeling Language (UML)* [8] as standard notation for modelling software, *graph transformation* [12] as meta language for defining model transformations, and a semantic interpretation of process models in terms of *Communicating Sequential Processes (CSP)* [6] which offers a language to express semantic consistency properties and tool support for analysis.

In the following section we will discuss the complementary roles of these techniques in general and outline their application to the model-based development of Web service processes.

2 Defining a Model-Based Development Method

An outline of our approach can best be given in terms of the triangle in Fig. 1, whose vertices are the languages by which processes may be represented, and whose edges represent uni- or bi-directional transformations between these representations.

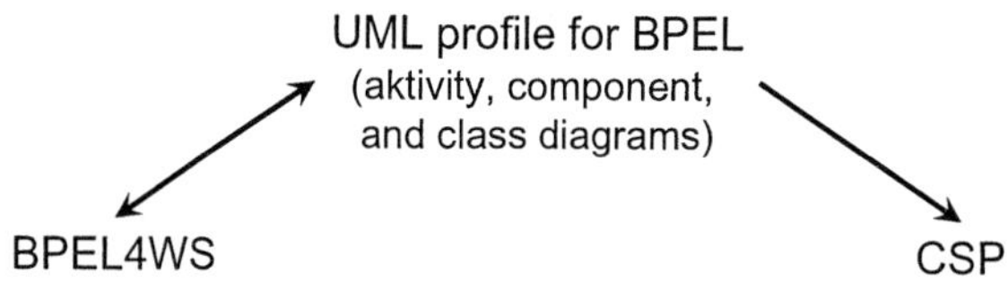

Fig. 1. Outline of the approach: languages and transformations

The UML, as a general-purpose modelling language, provides a rich set of concepts to model all kinds a software system. However, to address the more specific aspects of a particular application domain or implementation platform, the language needs to be specialized and extended. For this purpose, the standard [8] foresees the extension mechanism of *profiles*, a compromise between desirable flexibility of the language and necessary compatibility with existing tools. We shall use a profile to tailor, in particular, UML activity diagrams to the specification of BPEL4WS processes. In conjunction with these, class diagrams and component diagrams shall be used to describe, respectively, data types and software architecture relevant to the process.

Besides the concrete visual representation of a UML model, an abstract representation is required to capture its *semantically relevant* structure. This *abstract syntax* of UML models is defined by means of a *meta model*, i.e., a class diagram with well-formedness constraints expressed in the Object Constraint Language (OCL) [7]. The meta model specifies the collection of all legal *abstract syntax graphs*—its instances—each of which represents a legal model. This graph-based internal representation of UML models, which is typical of visual languages in general, shall be needed when defining the transformation of models.

These transformations do, in fact, represent the core features of a model-based development approach. In our case, they occur in two places: the transformation into BPEL4WS, the implementation language, and into CSP, the language for behavioral analysis. In many situations, two-way transformations are required, e.g., to support a *round-trip engineering* approach, where not only models are transformed into implementations (forward engineering), but also vice versa (reverse engineering), thus allowing incremental changes at both levels.

Our tool for describing (potentially bi-directional) transformations between models and other (typically textual) languages is the approach of *pair grammars* [10], i.e., a coupling of context-free grammars which allows to generate a sentence in the target language after parsing a given sentence in the source language. Since at least one of the languages will be a graphical one, context-free *graph* grammars [5] shall be employed which generalize context free grammars on strings by describing languages whose sentences are graphs.

For a mapping specification to be manageable and reusable, a modular approach is important which is structured in terms of the fundamental concepts of the domain. In this case, whenever a concept is added or modified, the corresponding transformation rules can be exchanged without affecting the rest of the mapping specification. For the domain of executable business processes, or workflow models, a corresponding concept analysis has produced an established list of *workflow patterns* [16], a subset of which is supported by UML activity diagrams. In fact, it turns out that these workflow patterns, interpreted over either activity diagrams or BPEL4WS processes, provide us with the pairs of context-free rules making up the pair grammar that specifies the translation.

The model-based analysis of processes represents the final ingredient of our approach. Depending on the representation on which the analysis is performed, we distinguish between *syntactic* and *semantic* analysis. The former is often restricted to the evaluation of well-formedness constraints on (the abstract syntax graph of) the model which reveal inconsistencies in structural dependencies and typing. However, syntactic analysis also includes the manual review of models based on semi-formal error patterns, a method that is quite successful in revealing behavioral problems and that may be the only method available in the presence of semi-formal models.

Formal analysis of behavioral properties, however, can hardly be done at the syntactic level, but requires a mapping of models into a semantic domain providing (1) a representation of the behavior to be analyzed, (2) means to express the desired properties, and (3) techniques and tools to check if these properties hold [4]. We have chosen the semantic domain of CSP [6] for this purpose, whose refinement relations are the basis for expressing properties over processes while tool support is provided by the FDR2 model checker [11].

The paper is structured according to the triangle in Fig. 1. The following section is devoted to the modelling of processes in the UML. Then, Sections 4 and 5 deal, respectively, with the mapping between UML and BPEL4WS, and the analysis of processes, including the mapping to CSP. Section 6 concludes the paper and summarizes the results.

3 Modelling BPEL4WS Processes in the UML

In this section, we describe how UML diagrams can be used to model Web service processes. We put special emphasis on the behavioral aspect given by BPEL4WS process interactions. As already discussed in [15], visual and more high-level modelling languages like UML have important advantages in comparison to low-level XML-based specification languages. Among others, they allow a better abstraction from implementation details and are therefore better understandable.

In particular UML use case-, component-, class-, and activity diagrams are suitable for modelling Web service processes in the context of business processes. In the following, we will demonstrate this by a sample model of an online shop.

Use case diagrams describe the business segment of our example. As shown in Fig. 2, the use case itself symbolizes the business process, in this case the service provided by the shop. The participants in the use case represent the roles of the partners that interact with the process. In the example, a buyer, a delivery service, and an invoice service interact with the online shop service.

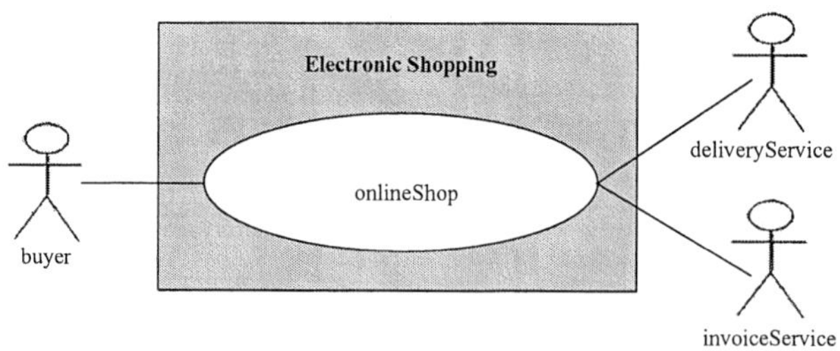

Fig. 2. Use Case Diagram

Component diagrams are used to refine the dependencies described in the use case diagram. In doing so, the component symbol is used for depicting a Web service. Both, the participants of the business process and the business process itself are modelled as Web services (see Fig. 3). In order to establish possible points of interactions, *port types* are added to the diagram as interfaces (the circular symbols, PT is used as abbreviation for port type). If a Web service *provides* a port type, the interface is connected to the component symbol by a solid line. If a Web service *requires* a port type, this is modelled by a dashed arrow (a UML *dependency*) to the corresponding interface.

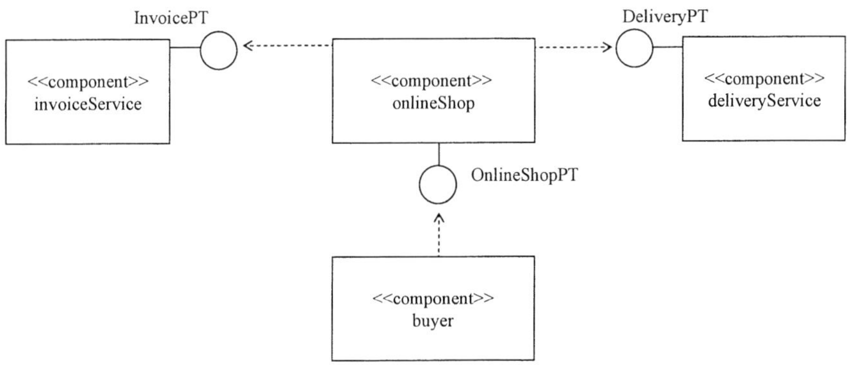

Fig. 3. Component Diagram

For example in Fig. 3, the online shop service provides exactely one port type OnlineShopPT and requires the port types DeliveryPT and InvoicePT for processing the corresponding data.

Class diagrams are used to provide further details of the different port types by defining their operations and involved parameters. In Fig. 4, the three port types from the component diagram are refined. In this simplified example, each port type provides only one operation which is mainly used to submit and receive the processing data to and from the partners. Likewise, the necessary messages and parameter types could be modelled by the class diagram.

Protocols and business processes for Web services are modelled with activity diagrams. Besides control-flow elements (decision, fork, join, etc.), the activity diagrams contain the necessary basic activities for interacting with the partner services. The activities are stereotyped like *receive*, *reply*, or *invoke* depending on their function as defined in the BPEL4WS specification [1]. A triplet consisting of partner, port type and operation follows the stereotype. This triplet assigns one of the available port type operations to each activity.

Fig. 5 shows the process of the online shop service. The first part is used to accept an order provided by the buyer through OnlineShopPT. After the data has been received, the online shop concurrently invokes an invoice and a delivery service with the required data. In order to simplify the example, we have

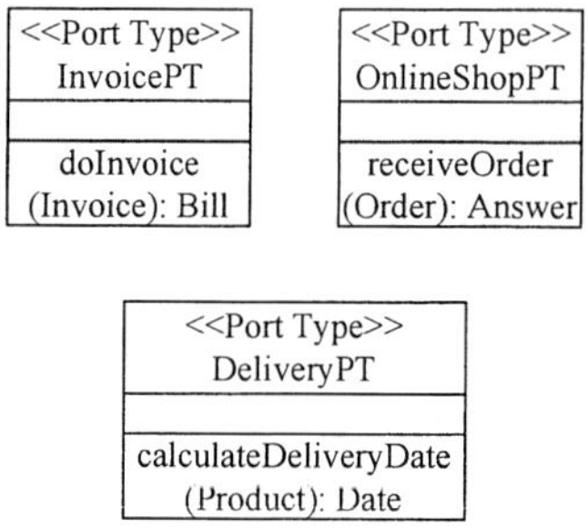

Fig. 4. Class Diagram

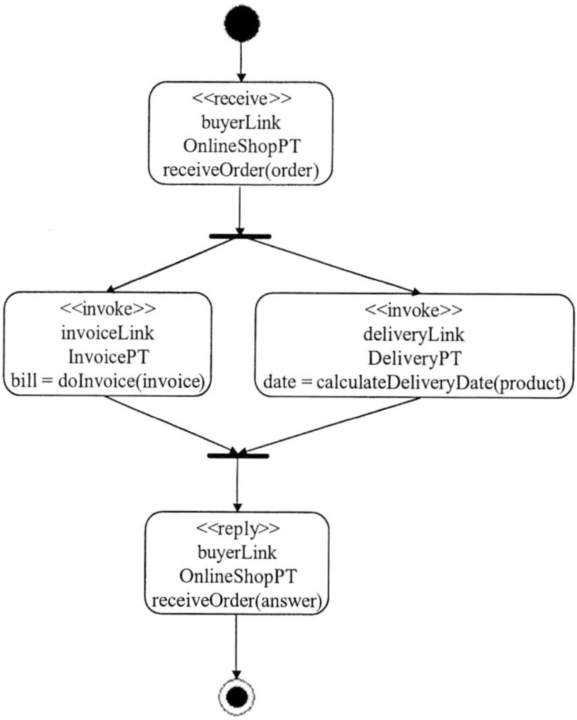

Fig. 5. Activity Diagram: Online Shop

omitted the details of assigning variables to establish the connection between the received order and further activities. In addition, the decision process for generating the answer for the buyer is hidden. The last activity in the row, which can be identified by the stereotype *receive* in Fig. 5, sends the corresponding answer to the buyer.

The protocols of the partners are modelled with activity diagrams, too. This step is essential for further analysis, because inconsistent behavior between the participants has to be discovered to ensure a correct execution of the interactions.

Therefore, the following three activity diagrams in Fig. 6 represent the processes on the Buyer, Invoice Service and Delivery Service part, respectively. In this example, a quick comparison of their activities with the activities of the Online Shop shows that the processes are behaviorally compatible.

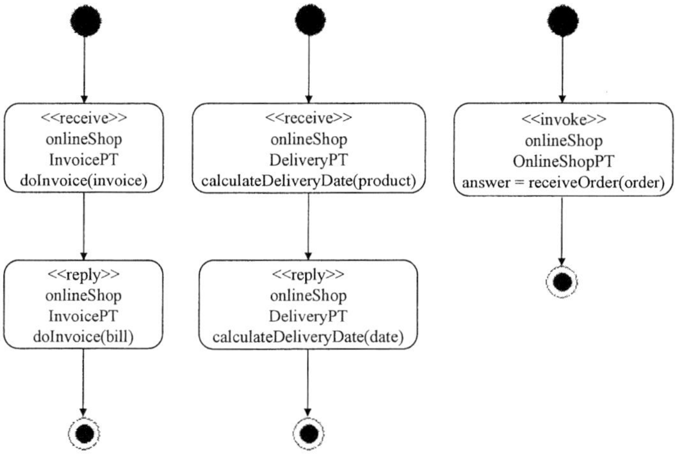

Fig. 6. Activity Diagram: Buyer, Invoice Service and Delivery Service

For more complex interactions, the consistency of all involved partner processes cannot be checked as easily as in this example. For this reason, Section 5 discusses more advanced methods to maintain certain consistency properties. Before, however, the transformation of process models into BPEL4WS processes by means of pair grammars is described.

4 Mappings between UML and BPEL4WS

A formal definition of the translation between UML models and an executable process language is important, not only to automate the translation, but also to define and ensure consistency at the model level (cf. Section 3).

Translations between string and graph representations of programs and data may be formally defined by means of *pair grammars* [10]. A pair grammar consists of two context-free grammars describing, respectively, the source and the target language, together with a correspondence between their rules and non-terminals. In this way, it defines a correspondence between source and target sentences which represents a mapping between the two languages. By virtue of their symmetric nature, pair grammars allow the definition of bi-directional mappings between UML activity diagrams and BPEL4WS processes.

To support the translation of *graphical* languages, pair grammars are based on context-free *graph* grammars, i.e., formal grammars similar to ordinary context-free grammars, except that the language defined is a set of graphs rather than a set of strings. Context-free grammars, in fact, form a subclass of context-free graph grammars. In this paper, an informal introduction to the most important concepts of pair grammars is given. For an exhaustive and formal treatment we refer to [10].

Our presentation is based on *edge replacement graph grammars*, one of the simplest forms of context-free graph grammars which form, in their generalized form of *hyperedge replacement (HR) graph grammars* [5], one of the two major approaches in the literature. Here, context-freeness means that the left-hand side of a rule is given by a single edge (or hyperedge, i.e., an edge attached to an arbitrary number of vertices) representing a nonterminal which is replaced by the graph that forms the right-hand side of the rule. The obvious alternative consists in replacing a nonterminal node, leading to the family of *node-replacement* approaches [2].

4.1 Pair Grammars

First, we introduce the basic notions of graphs and context-free graph grammars. A *graph* consists of vertices and edges such that each edge has a source and a target vertex in the graph, respectively. In accordance with [5], in our graphs *labelled edges* carry the relevant information, while nodes just represent the points where the edges are attached.

Definition 1 (edge-labelled graphs). *Let C be a fixed set of edge labels. A (directed edge-labelled)* graph $G = \langle G_V, G_E, src^G, tar^G, lab\rangle$ *over C has a set of vertices G_V, a set of edges G_E, two functions $src^G : G_E \to G_V$ and $tar^G : G_E \to G_V$ associating to each edge its source and target vertex, and a labelling function $lab : E_V \to C$ associating with every edge its label.*

Definition 2 (edge replacement (ER) graph grammars). *Let $N \subseteq C$ be a set of nonterminal labels. A production rule over N is a pair $p = A \stackrel{x,y}{\longrightarrow} R$ where $A \in N$ and R is a graph over C with distinguished vertices $x, y \in R_V$.*

An edge replacement (ER) graph grammar $\mathcal{G} = \langle C, N, P, S\rangle$ *consists of the sets of labels and nonterminals introduced above, a set P of productions over N, and a start symbol $S \in N$.*

Edge replacement graph grammars subsume context free grammars by representing strings of terminal and nonterminal symbols as chains of correspondingly labelled edges. Every application of a rule derived in this way from a context-free grammar rule takes out one nonterminal edge and replace it with a path, gluing the source vertex of the path to the (former) source vertex of the edge, and analogously for the target.

In the general case of ER grammar rules, an edge in graph is replaced by a graph with two attachment points x, y (that we think of as "source" and "target" vertices) which are glued to the source and target vertex of the replaced edge, respectively.

Full hyperedge replacement grammars generalize this by allowing an arbitrary number of attachment points for the graphs to be inserted and, consequently, for the edge to be replaced. We have limited ourselves to ER graph grammars here for simplicity, and because they are sufficient to illustrate the concept of grammar-based translation of graphical languages.

The definition of pair grammars, originally formulated for a simple kind of node-replacement graph grammars, has been tailored to our purposes.

Definition 3 (pair grammar). *A pair grammar is a quadruple* $Q = \langle C, N, PP, S\rangle$ *where* C *and* N *are sets of labels and nonterminals as before,* $S \in N$ *is a start symbol, and* PP *is a finite set of triples* (p_1, h, p_2)*, where*

1. $p_1 = A_1 \stackrel{x_1,y_1}{\longrightarrow} R_1$ *and* $p_2 = A_2 \stackrel{x_2,y_2}{\longrightarrow} R_2$ *are ER rules over* C *and* N *as above such that* $A_1 = A_2$*,*
2. h *is a nonterminal edge pairing of* R_1 *and* R_2*, i.e., a bijection between their nonterminally labelled edges such that* $e_1 h e_2$ *implies* $label(e_1) = label(e_2)$*.*

The language defined by a pair grammar Q consists of ordered pairs of graphs from the left and right language, respectively, of Q. The pair grammar defines how these graph pairs may be generated in parallel form the same start symbol. At each intermediate stage in the generation we have a pair of graphs, each containing some nonterminal nodes, and a correspondence between these nonterminal nodes. At each rewriting, a corresponding pair of nonterminal nodes, one in each graph, is rewritten according to a rule of the pair grammar, and a new correspondence is set up between nonterminal nodes in the resulting graphs using the nonterminal pairing of the grammar rule.

4.2 UML–BPEL4WS Mapping

Let us illustrate the notions introduced so far by means of a mapping between UML activity diagrams and BPEL4WS processes, specified by the pair grammar whose rules are shown in Fig. 7 through 10.

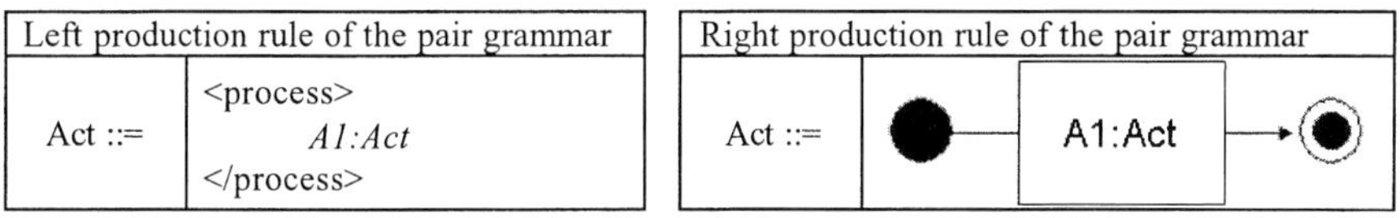

Fig. 7. Pattern 0: Start and End

Production rules of this pair grammar combine ordinary context-free grammar rules for BPEL4WS processes and truly graphical rules for (a subset of) UML activity diagrams. In order to regard these diagrams as graphs that can be generated by edge rewriting, activities shapes as well as fork / join bars are interpreted as terminal edges. Nodes (presented as little circles) are introduced whenever two activities are connected by a transition. As the only nonterminal label, *Act* stands for an arbitrary diagram with one entry and one exit transition. Thus our sets of labels are defined by $C = \{activity, bar, Act\}$ and $N = \{Act\}$, such that *Act* is the only possible start symbol.

Vertices have no labels, but we distinguish attachment points as filled circles with name x or y. Nonterminal edges (i.e., with label *Act*) are denoted as boxes connected to their source and pointing to their target vertex.

The bijection h between nonterminal edges of the right-hand sides of left and right rules is given by identical names for corresponding edges. According to Def. 3, corresponding edges as well as left hand sides must be of the same label. For example, in the rule of Fig. 8, symbols $A1 : Act$ through $A1 : Act$ in the upper BPEL4WS production correspond to the edges with the same name in the lower UML production rule.

Left production rule of the pair grammar	
Act ::=	<sequence> *A1:Act* ... *An:Act* </sequence> with n ∈ N.

Right production rule of the pair grammar

Act ::=

Fig. 8. Pattern 1: Sequence

It is interesting to note how the definition of the individual rules is inspired by the workflow patterns [16]. For example, in the case of *sequence* or *parallel split and synchronization*, the interpretation of each pattern in both BPWL4WS and activity diagrams yields the right hand sides of the two corresponding rules.

The mapping specified by the pair grammar shall be applied to our online shop example. Recall Fig. 5 specifying a business process of the shop. In order to execute this process it shall be translated into the BPEL4WS. The operational idea is to start parsing the sentence of the source language (UML) and to generate the sentence in the target language (BPEL4WS) along the resulting derivation tree.

Parsing the source language. As result of parsing the activity diagram, its syntactic structure is represented by the derivation tree in Fig. 11. In the first step of the construction of this diagram, the basic activities (rectangles with rounded corners) are replaced with nonterminal edges. It follows the detection of structured activities based on the patterns of sequence (the outermost box $a6 : Act$) and parallelism (box $a5 : Act$).

Thus, the parsing yields a hierarchical decomposition of the diagram which can be seen as a tree with the innermost boxes (terminal edges) as leafs and the outermost box (the start symbol) as root. This structure resembles derivation trees of ordinary context-free grammars. In particular, it abstracts from the ordering of independent derivation steps, i.e., boxes that are not nested in one another, like $a1, a2, a3, a4$ in Figure 11, can be processed in any order or in parallel.

When the graph representing the activity diagram is reduced to the start symbol, the parsing process is finished successfully.

Generating the target language. The next step is to generate the corresponding BPEL4WS sentence, invoking the rules of the right grammar following the structure of the derivation tree. This second phase begins with the start symbol, i.e., the tree is computed bottom up, but evaluated top down.

After the first step, one derives the box with the *¡process¿* tag (see the left production rule in Figure 7). Since the correspondence between the rules of the two languages is fixed by the pairing, there is no other option but to start with this pattern, which represented the last step in the generation of the tree.

Next, the structure of the process is refined by introducing sequence and flow instructions (compare Fig. 8 and 9). First, the sequence is inserted because this is the next outermost structure. Subsequently, the parallel part is generated. Alternatively, one could have substituted the first activity in the flow of the sequence, because this step is independent of the generation of the parallel activities.

Left production rule of the pair grammar	
Act ::=	<flow> *A1:Act* ... *An:Act* </flow> with n ∈ N.

Right production rule of the pair grammar	
Act ::=	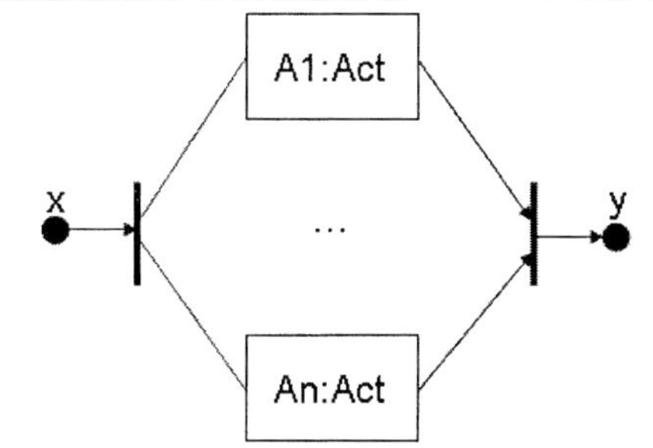

Fig. 9. Pattern 2: Parallel Split and Synchronization

We emphasize the role of the correspondence between the nonterminals on the right sides. Symbols $A1 : Act$ to $An : Act$ are associated with the edges $A1 : Act$ to $An : Act$, so that the "content" of edge Ai can determine the replacement of the corresponding nonterminal.

At this stage, the considered BPEL-process looks like this:

```
<process>
  <sequence>
    A1:Act
    <flow>
      A2:Act
      A3:Act
    </flow>
    A4:Act
  </sequence>
</process>
```

In the last step, all nonterminals are substituted by terminals. As shown in Figure 10, all variables of the right production rule must be replaced with the corresponding terminals for partner, port type, and so on. After generating all basic activities, the transformation is completed. The resulting BPEL4WS process is listed below.

```
<process>
  <sequence>
    <receive name="receiveOrder"
      partnerLink="ns:buyerLink"
      portType="ns:onlineShopPT"
      operation="ns:receiveOrder"
      variable="order"
      createInstance="yes"/>
    <flow>
      <invoke name="invokeBank"
        partnerLink="ns:invoiceLink"
        portType="ns:InvoicePT"
        operation="ns:doInvoice"
        inputVariable="invoice"
        outputVariable="bill"/>
      <invoke name="invokeDeliverer"
        partnerLink="ns:deliveryLink"
        portType="ns:DeliveryPT"
        operation="ns:calculateDeliveryDate"
        inputVariable="product"
        outputVariable="date"/>
    </flow>
    <receive name="replyOrder"
      partnerLink="ns:buyerLink"
      portType="ns:onlineShopPT"
      operation="ns:receiveOrder" variable="answer"/>
  </sequence>
</process>
```

Left production rule of the pair grammar	
Act ::=	<invoke partnerLink="PLname" portType="PTname" operation="OPname" inputVariable="input" outputVariable="output"> </invoke>

Right production rule of the pair grammar

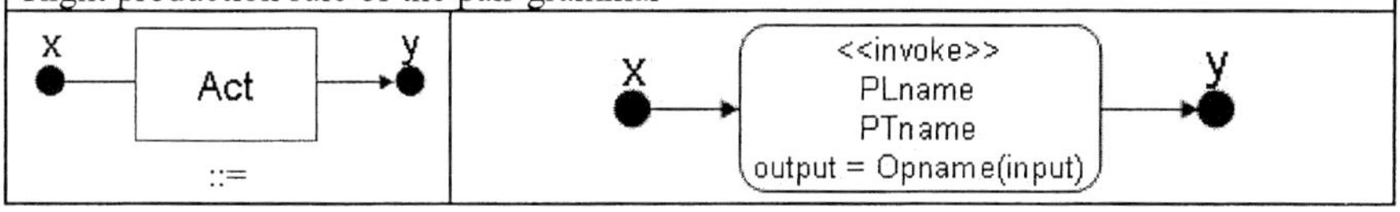

Fig. 10. Pattern: Invoke

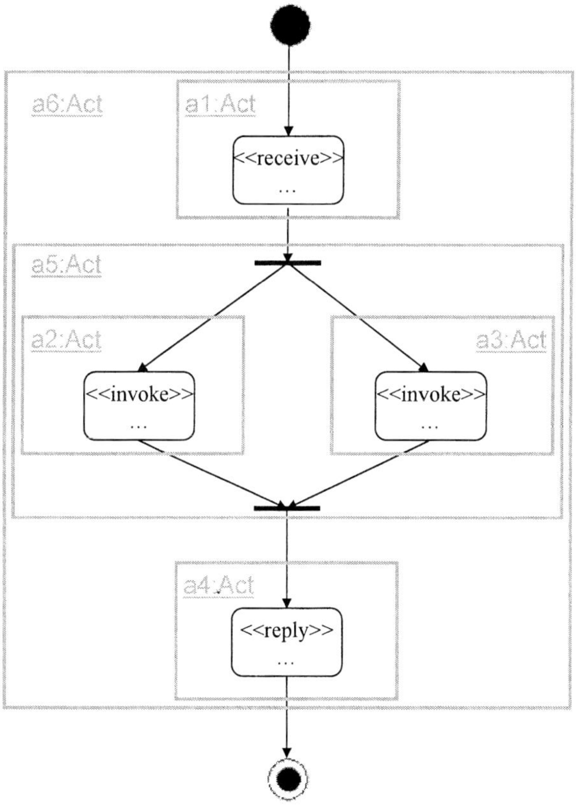

Fig. 11. Decomposition of the activity diagram

4.3 Properties of the Mapping

A basic requirement for an automated translation, even before semantic considerations are made, is that there should be a unique sentence of the target language generated for every given sentence of the source language. Uniqueness may be relaxed by some notion of semantic equivalence, but in many cases such notion is not readily available, or very hard to verify.

A purely syntactic criterion is the notion of *(un)-ambiguity* of the pair grammar which is based on corresponding notions for the underlying grammars.

Definition 4 (ambiguity). *An ER graph grammar $\mathcal{G}$ is ambiguous iff there exists a graph G in the generated language which has two distinct derivation trees.*

That means, a grammar is *un*ambiguous if every sentence can be parsed in essentially one way, up to the ordering of independent steps which are abstracted from in the parse tree.

It is difficult to prove unambiguity in general, but there exists a simpler sufficient condition based on the idea of critical pairs in rewriting: Consider the grammar as a reduction system, applying its rules from right to left in order to reduce the given graph to the start symbol. Now, unambiguity is ensured if there is never a true conflict between the application of two reduction rules. A conflict is evident in an overlapping of the left-hand sides of two reduction rules (i.e., the right-hand sides of two production rules) if (cf. [9])

- they intersect in anything else than attachment vertices, and
- both rules are applicable to reduce the graph formed by the overlapping

The second condition is violated if one of the reduction rules attempts to delete a vertex that is connected to an edge originating from the other rule. In this case, the resulting structure is no longer a graph since the edge misses its source or target vertex. Hence, in a critical pair, an overlap in a non-attachment node entails that all edges connected to this node in both rules are also in the intersection. In turn, an overlap of an edge obviously entails an overlap of its source and target node.

That means, the intersection includes all nodes and edges of both rules indirectly reachable from a non-attachment node or edge in the intersection. Since the right-hand sides of our rules are connected, this implies that the overlap is complete whenever a non-attachment node or an edge is involved.

For a pair grammar we consider unambiguity for both directions of the translation. Generally, it requires that the source grammar is unambiguous, so the parse tree is uniquely determined, and that for every source production there is exactly one target production, so the tree uniquely induces a derivation in the target. Depending on who is source and who is target, this results in the notions of left and right unambiguity.

Definition 5 (unambiguity of pair grammars). *A pair grammar $\mathcal{Q}$ is left (right) unambiguous if the left (right) grammar of $\mathcal{Q}$ is unambiguous, and Q contains no two distinct rules with identical left (right) rules.*

Q is unambiguous if it is both left and right unambiguous.

The pair grammar presented in Section 4.2 is unambiguous. However, there exists another possibility for describing sequences in BPEL4WS, besides the one shown in Fig. 8. The alternative shown in Fig. 12 uses the *flow* construct, specifying the desired temporal dependencies by means of links between activities.

Left production rule of the pair grammar	

Act ::=	<flow>
	    <links>
	        <link name="L_1"/>
	    </links>
	    A1:Act
	        <source linkName="L_1"/>
	    ...
	    An:Act
	        <target linkName="L_{n-1}"/>
	</flow>
	with $n \in N$.

Right production rule of the pair grammar

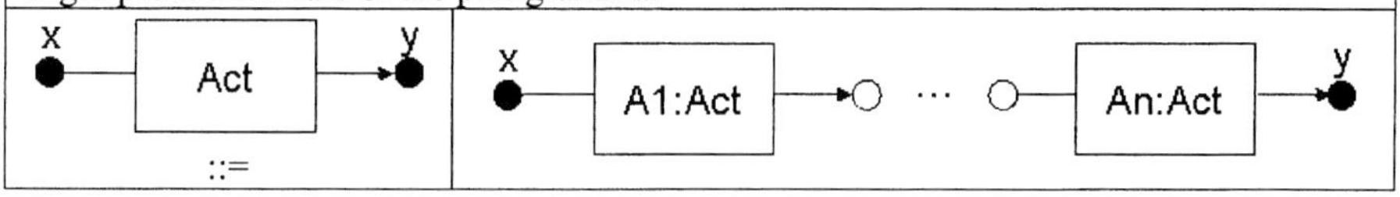

Fig. 12. Pattern 1: Sequence (via BPEL4WS flow construct)

The resulting pair grammar is no longer right unambiguous because there are two rule pairs sharing the same right rule. Indeed, since we have two choices to implement a sequence, the result of the translation is no longer unique. Hence, for right-to-left mappings, one of the two rules should be disregarded.

On the other hand, the alternative sequence rule is useful for the left-to-right from BPEL4WS to UML because, when reverse engineering a process we cannot assume a certain style of implementation, but have to handle the full spectrum of language constructs. Fortunately, the extended pair grammar is still left unambiguous.

This example shows that, while in potentially bi-directional, it could be necessary to tailor the mapping description to one or the other direction in order to achieve unambiguity.

5 Model-Based Analysis

Model-based development tends to create a variety of artifacts that describe the system to be built from different viewpoints and at different levels of abstraction. This allows developers to concentrate on the concern of present interest, reducing complexity by hiding other not so relevant concerns, but it also creates consistency problems between the different descriptions.

Besides this general, domain and language-independent cause for consistency problems, there are more specific reasons resulting from the choice of a specific development method, target language, or application domain. Typical for Web services is, for example, the consistency between descriptions of *required* and *provided* services, which need to be matched before services can be composed. This includes, e.g., the compatibility of their signatures to ensure type safety, and of their interaction protocols to avoid deadlocks.

Other consistency problems are inherited from the target language of development, in our case BPEL4WS, which entails particular restrictions for processes formulated in that language. Type checking rules for BPEL4WS require, for example, that all operations used in activities of the process must be declared in the appropriate port types. This induces a dependency between the class diagram containing the interfaces from which the port types are derived and the activity diagram where the process is modelled. Thus, the precise notion of consistency that needs to be applied at the model level depends on both the restrictions at the implementation level and the mapping of models to implementations as described in Section 4.

In this section, we are dealing with model-based analysis of consistency problems. Both subject and result of the analysis are given at the model level because, obviously, it would limit the applicability of a method if developers were forced to work on different representations of processes, e.g., to eliminate a fault directly in an XML-based business process language, or to analyze a process in a process algebra. This, again, emphasizes the need for automated mappings between different representations.

For dealing with consistency in UML-based development processes, we apply a general methodology for specifying and analyzing consistency [4]. In short, the methodology consists of the following steps.

1. Consistency problems must be identified in a given UML-based development process, documented and categorized into
   - problems of *syntactic* (e.g., structural) nature that can be formulated and solved at the level of models;
   - problems of *semantic* (e.g., behavioral) nature that require a separate semantic representation.
2. For each semantic consistency problem, a suitable semantic domain must be chosen and a semantic mapping of models into this domain must be designed.
3. For both syntactic and semantic problems, consistency conditions must be stated as constraints, either over the abstract syntax or the semantic representation of models.

In the following two subsections we consider, in turn, the syntactic and the semantic case.

5.1 Syntactic Analysis

In our sample process, for illustration we identify syntactic consistency problems between *component diagrams* and *class diagrams*, as well as between *activity diagrams* and *class diagrams*.

For component and class diagrams a consistency problem occurs whenever an interface is used in a component diagram that is not declared in the class diagram. This consistency problem can be considered syntactic because a syntactic condition can be formulated using OCL, requiring that each interface used is declared in the class diagram.

With regard to activity diagrams and class diagrams, a similar kind of consistency problem is illustrated in Fig. 13: An activity contains references to the partner component, interface and operation which must be in a certain relation (e.g., the interface declares the operation and is implemented or used by the component playing the partner role). Considering the class diagram in Fig. 4, it is obvious, that the OnlineShopPT does not support a receiveInstruction operation.

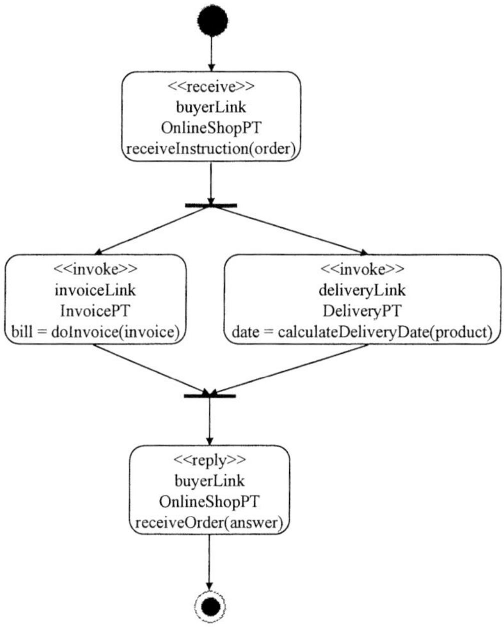

Fig. 13. Activity Diagram: Syntactic consistency problem

When modelling Web service processes, we also have to take into account language-specific consistency properties, as for instance: In a BPEL4WS process a process instance must not simultaneously enable two *receive* actions for the same partner, port type, and operation. If two receive actions for the same partner, port type, and operation are, in fact, simultaneously enabled, e.g., in two concurrent threads of the process, then a standard fault must be thrown by the process interpreter that complies with the BPEL4WS specification. In such a case, the processing of the current scope is terminated. Possible results are the invocation of compensation handlers (if defined) or the abortion of the entire process. In any case, the chance for a successful completion of the process decreases.

In order to avoid such conflicts, we can provide sufficient static conditions at the model level, e.g., by means of a semi-formal error pattern as illustrated in Fig. 14. *ActivityA* and *ActivityB* are place holders for sub-processes of arbitrary structure. Such a pattern, which must not occur in a process, can serve as a guideline for the developer or as a documentation for a formal analysis of this property, if available.

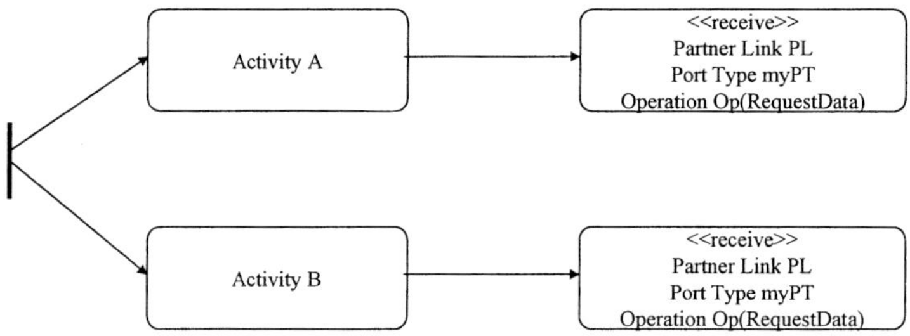

Fig. 14. Conflict potentials in parallel sections

Another example of an error pattern is shown in Fig. 15. It reflects, at the level of models, the fact that BPEL4WS requires control flows that are based on links to be acyclic.

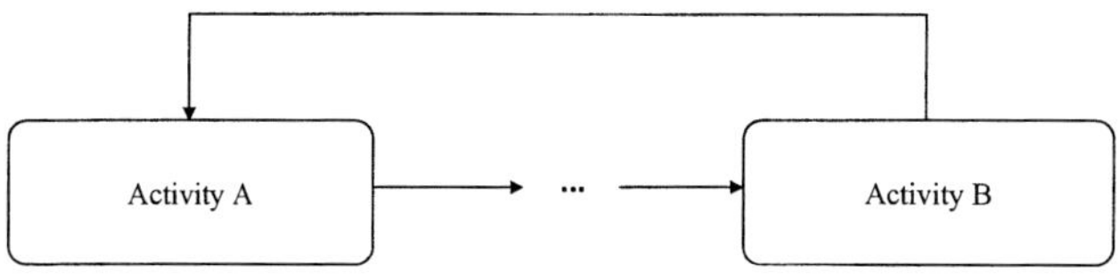

Fig. 15. Cycles

5.2 Semantical Analysis

Next, we deal with consistency conditions that are formulated and analyzed in a separate semantic domain. In our example, different interaction protocols of the participants are combined and conclusions about their compatibility are given. Thus, we focus on the behavioral aspect. In particular, if the business processes have a complex control flow, the concrete or abstract syntax of models is not suitable for such problems. Therefore we choose the process algebra CSP [6] as semantic domain for analysis.

Since we cannot assume that developers are familiar with CSP, and in order to avoid mistakes in the translation, an automated mapping of models into CSP is required, as well as a mapping of analysis results back into UML models.

Translation from UML to CSP. In principle, the translation from UML activity diagrams into the semantic domain CSP is based on the same concepts as the translation shown in Section 4. However, in this case a pair grammar is not fully satisfactory, because one has to generate complex *Pre* and *Post* processes for managing events accurately in CSP. These are interconnected by an *Environment.* In order to control these processes, the concept of a *Global Scheduler* is adopted. Furthermore, the *Control* process guarantees the correct termination of the whole system. Both basic and structured activities (i.e., nodes like split and join that describe the control flow) are coded as separate CSP processes denoted as *Activity processes.* Thus Activity processes emulate the proper control flow as well as the sending and receiving of events. In order to execute the Activity processes independently, they are combined by interleaving.

Processes may be composed by operators which require synchronization on some events. Each component must be willing to participate in a given event before the whole can make the transition. In this regard, suitable communication and synchronization *alphabets* consisting of events must be defined. The composition of processes is itself a process, allowing a hierarchical description of a system. The process structure can be represented as a tree. The root node represents the process as a whole (cf. *System_Control*). According to the number of sub-components of the node branches are added. Fig. 16 visualizes the process structure, whereby rectangles indicate the parallel composition symbol including the alphabet in question and ovals display CSP processes. The several atomic Pre, Post and Activity processes are hidden, because their occurrence strongly depends on the underlying example.

In the following, we focus on the generation of the Activity processes. As already mentioned, the concept of pair grammars is suitable for demonstrating the basic idea. Now the grammar for UML activity diagrams is paired with the one for CSP processes. This is shown in Fig. 17.

On each left-hand side of the considered rules the type of the expected nonterminal is used. In this way, these nonterminals are paired. For the grammar for UML activity diagrams we refer to Section 4, because it has already been discussed. The right-hand sides of the rules for the CSP grammar include terminals like *if* and *else*, and nonterminals like $ActivityA_1$, whereby *Activity* indicates the type and A_1 indicates the variable. We have chosen this alternative representation for a better readability of the CSP process. Below we explain the *Activity process* in short without deepening the language CSP too much.

$ActivityA_1$ is the name of the considered *Activiy process.* The first event in the flow activates this process (compare $act\_ActivityA_1$). Afterwards t reads out the channel $transition_x$, whereas the value 1 respectively 0 indicates a positive respectively negative precondition. If t has in fact the value 1, then the corresponding transition process is instructed to reset itself. After this the transition for the following Activity process is set and $ActivityA_1$ is disabled and ends with a recursive invocation on itself, so it is available in the next step of the Global Scheduler.

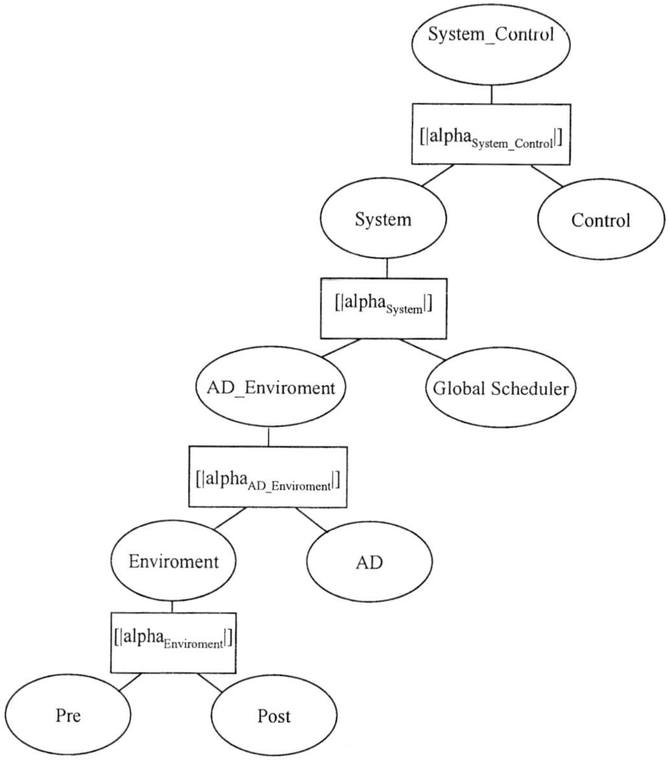

Fig. 16. Combing processes in CSP

For UML activity diagrams and CSP, we do not require a bi-directional translation, because we do not assume that business processes are formulated as CSP processes. For analyzing CSP processes we use the model checker FDR2 [11]. As results of a check one obtains a trace, which the process is, or is not, willing to execute. These traces can be transformed into a sequence diagram. Hence, a developer is able to work solely at the model level. Moreover, the complex structure of the established CSP process is the reason for customizing the concept of pair grammars in this regard. For completely describing this translation, we would have to upgrade from pair grammars to so-called triple graph grammars [13]. Beside lifting the restriction to context-free grammars, triple graph grammars allow to store auxiliary data, accumulated during the translation, inside a third intermediate graph, which also keeps track of the relation between source and target. This feature is important to determine the alphabet of a process by collecting data on operations and partners occurring in the process. However, in this paper we stick to the pair grammar representation which is still sufficient to convey the basic ideas.

After defining a translation from UML activity diagrams into CSP, we can finally turn to the actual semantical analysis. Basically, we distinguish between

Left production rule of the pair grammar

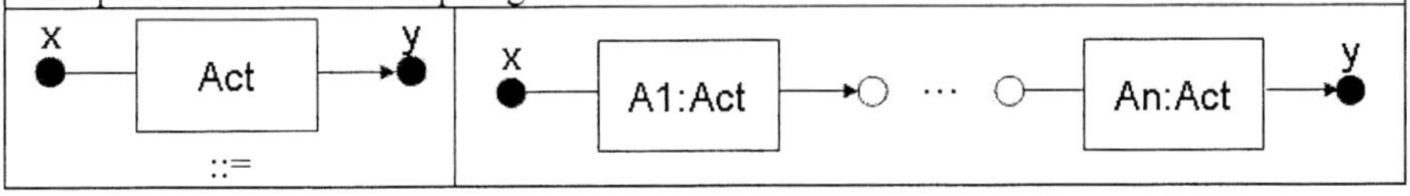

Right production rule of the pair grammar

Act ::=

```
ActivityA1 =
      act_ActivityA1 ->
      transistion_x?t ->
      if (t == 1) then
            ini_Transition_x ->
            post_Transition_ActivityA1_ActivityA2 ->
            ackn_ActivityA1 -> ActivityA1
      else
            ackn_ActivityA1 -> ActivityA1

...

ActivityAn =
      act_ActivityAn ->
      transistion_ActivityAn-1_ActivityAn?t ->
      if (t == 1) then
            ini_Transition_ActivityAn-1_ActivityAn ->
            post_Transition_y ->
            ackn_ActivityAn -> ActivityAn
      else
            ackn_ActivityAn -> ActivityAn
```

Fig. 17. Pattern 1: Sequence

classical requirements for concurrent processes and requirements for business processes.

Classical concurrency properties. At first we consider classical requirements like *deadlock* or *livelock*. Concerning the property of *deadlock freedom*, we need to provide a consistency concept for activity diagrams. At first we give a definition for deadlock.

Definition 6 (deadlock). *A set of processes is* deadlocked *if each process in the set is waiting for an event that only another process in the set can cause.*

As an example, consider the modified activity diagram of the Buyer shown in Fig. 18 and the one of the Online Shop Service, illustrated in Fig. 5. Both processes expect a signal from each other which gives rise to a deadlock. This circumstance is independent of changes in their interfaces.

In general, whenever a deadlock occurs, processes can not be completed successfully. Hence, we take into account suitable measures to avoid such conflicts. The tool FDR2 supports the detection of deadlocks. However, due to the complex structure of the CSP process implementing a business process, CSPs definition of a deadlock, which requires that a process does not communicate at all, is not applicable here. Instead, we have to check for a livelock in order to capture the notion of a business process deadlock in CSP, see [17] for details.

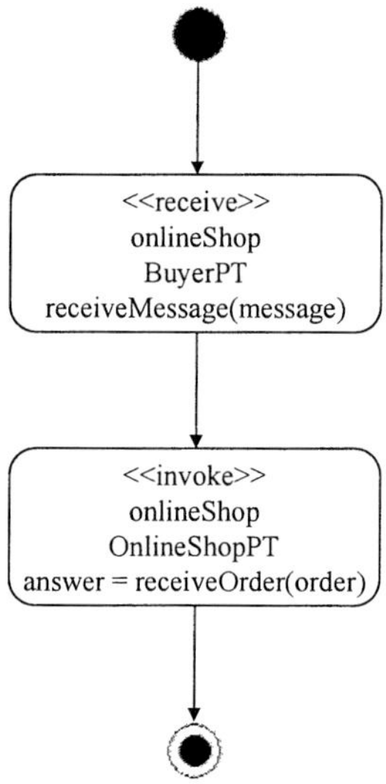

Fig. 18. A deadlock situation

Requirements for business processes. Now we turn to requirements which must be formulated depending on the context of a business process. The verification of so called security properties is based on the comparison of CSP processes to a set of traces. This set defines sequences of events and by doing so secure states are specified. In this sense, a CSP process is in fact secure if its provided traces are in the set of traces of the security property. In addition we want to check, if a concrete activity diagram covers several scenarios. These scenarios can be formalized as UML diagrams. In this context UML behavior diagrams are of special importance (compare UML sequence and statechart diagrams). In order to compare all these different diagram types, a sufficient transformation into the semantic domain CSP must be established. In [14], Stehr picks the translation of sequence and statechart diagrams out as a central theme. In this survey we have already demonstrated, how activity diagrams can be translated into CSP.

Such a scenario can be derived from different use cases.

- By modelling sequence diagrams one has the opportunity to define permitted respectively prohibited examples. This means that a developer identifies concrete scenarios, which are checked regarding a concrete business process.
- Concrete executions of existing business process instances can be monitored. By doing so sequence diagrams can be formulated by assigning exchanged messages to objects.

In both cases, one has to establish the relationship between the different message and event names, respectively, used in the diagrams. In general, sequence diagrams are much less formalized than activity diagrams. This task can only be handled by the developer himself. Agreeing on the same name space is a precondition for meaningful analysis.

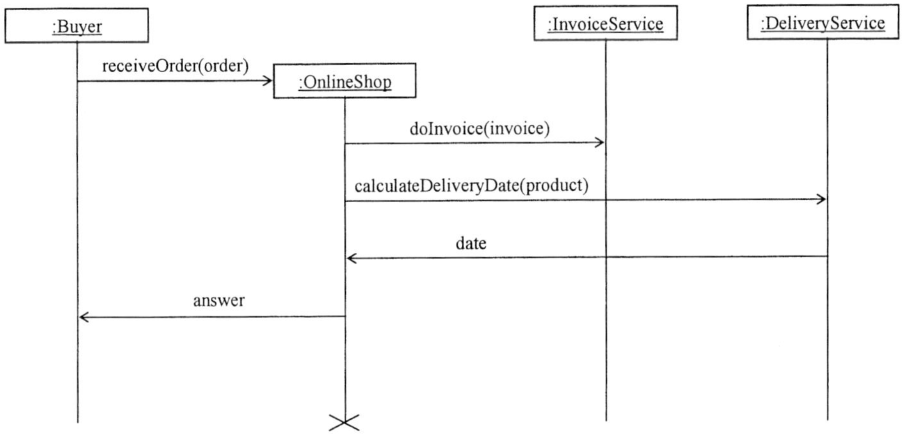

Fig. 19. Property formulated as a UML sequence diagram

Messages	Events
receiveOrder(order)	post_Event_buyerLink_OnlineShopPT_receiveOrder_Request
doInvoice(invoice)	post_Event_invoiceLink_InvoicePT_doInvoice_Request
	post_Event_invoiceLink_InvoicePT_doInvoice_Response
calculateDeliveryDate (product)	post_Event_deliveryLink_DeliveryPT_calculateDeliveryDate_Request
date	post_Event_deliveryLink_DeliveryPT_calculateDeliveryDate _Response
answer	post_Event_buyerLink_OnlineShopPT_receiveOrder_Response

Fig. 20. Assigning messages of the sequence diagram to the events of the CSP process

An example of such an assignment is shown in Fig. 20. We emphasize that the events of the CSP process are not introduced in this context, because the underlying example consists of over 1400 lines of code. For the complete example we again refer to [17].

The result of a check shows whether the trace of the sequence diagram is in the set of traces of the activity diagram. In this example, a required event does not have any correspondence in the sequence diagram. And in fact, this event must be executed in the underlying business process. Hence, the CSP processes (and thus the business processes visualized as activity diagrams) do not conform to the defined property.

Further on, complete subsystems can be compared to each other. In doing so, we want to check, if a subsystem might be replaced by another. This analysis is only based on CSP processes, which are representations of UML activity diagrams. For this purpose the FDR2 trace refinement checker is suitable.

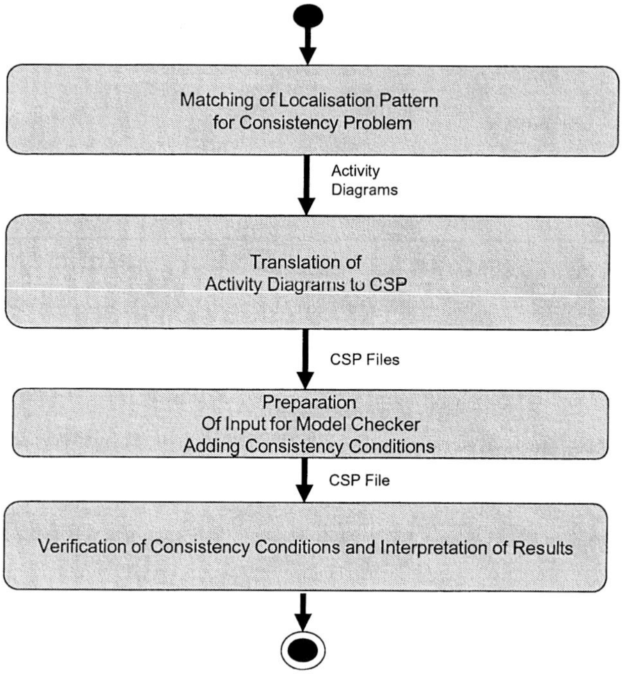

Fig. 21. A sample consistency check

Defining consistency checks. On the basis of a consistency concept, consistency checks can be defined in order to validate that a model is consistent. A consistency check must therefore validate the consistency conditions of a consistency concept. Within our approach, such a check may involve the translation of a model into a semantic domain, the verification of consistency conditions by a model checker, and an interpretation of the results.

Informally, the specification of such a consistency check can be visualized by an activity diagram extended by mechanisms for modelling object flow. In the following, we will sketch the definition of a consistency check for the consistency problem type of activity diagrams, ensuring their deadlock freedom.

In Fig. 21, the consistency check for activity diagrams is shown (with object flow visualized by arrows). Within the first activity, a UML localization pattern is used for locating and identifying, within a larger UML model, those activity diagrams relevant for the consistency check. These are then given to the translation activities. Within the translation activities, the translation to CSP is performed. Resulting CSP files are then assembled to a single file which can be handed over directly to the model checker.

This concept is implemented in the ConWork tool developed at the university of Paderborn, which allows to define flexible consistency checks based on rule-based translations of UML diagrams into CSP [3].

6 Conclusions

With the wide integration of Web services into software development, modelling of Web service processes is gaining increasing importance. In order to be beneficial, a modelling approach should take into account the characteristics of Web service processes. Currently, the Unified Modeling Language is the accepted industrial standard for modelling object-oriented systems. In this paper, we have discussed how the UML can be applied for modelling Web service processes. Furthermore we have introduced several UML models for suitable abstractions of both structure and behavior of Web service processes. Then we have focused on a bi-directional translation between UML activity diagrams and BPEL4WS. Thus we provide a framework for forward and reverse engineering based on the considered translation concept. As consistency is not established by the language definition of UML, it must be ensured by the software engineer applying UML for modelling Web service processes. In order to prove consistency conditions in regard to the UML models, we categorized possible inconsistency types into syntactical and semantical problems depending on the language that is suitable for solving these. In this context the analysis of different interaction protocols participating in a given business process is of particular interest. For this task we chose CSP as semantic domain for further analysis. On this account we provide a translation from UML into CSP and propose an approach, how results of the model checker FDR2 can be visualized as UML models. The visual modelling language UML facilitates an adequate abstraction of implementation details and supports a better understanding of consistency analysis.

Future work includes the definition of a generic development process for Web service processes and the elaboration of consistency management within this development process. For that purpose, we must automate the translation between UML activity diagrams and BPEL4WS as well as UML activity diagrams into a semantic domain such as CSP. Currently, we are developing tool support for this task based on the Consistency Workbench [3]. This tool allows the software engineer to define translations of UML models into a semantic domain and define consistency checks as workflows, like visualized in Fig. 21.

References

1. T. Andrews, F. Curbera, H. Dholakia, Y. Goland, J. Klein, F. Leymann, K. Liu, D. Roller, D. Smith, S. Thatte, I. Trickovic, and S. Weerawarana. Business Process Execution Language for Web Services, Version 1.1, May 2003. `http://www-106.ibm.com/developerworks/webservices/library/ws-bpel/`.
2. J. Engelfriet and G. Rozenberg. Node replacement graph grammars. In Rozenberg [12], pages 1 – 94.
3. G. Engels, R. Heckel, and J. M. Küster. The consistency workbench: A tool for consistency management in uml-based development. In *UML 2003 - The Unified Modeling Language. Modeling Languages and Applications. 6th International Conference, San Francisco, USA*, LNCS. Springer, 2003.

4. G. Engels, J.M. Küster, L. Groenewegen, and R. Heckel. A methodology for specifying and analyzing consistency of object-oriented behavioral models. In V. Gruhn, editor, *Proc. European Software Engineering Conference (ESEC/FSE 01), Vienna, Austria*, volume 1301 of *LNCS*, pages 327–343. Springer Verlag, 2001.
5. A. Habel. *Hyperedge Replacement: Grammars and Languages*, volume 643 of *Lecture Notes in Computer Science*. Springer-Verlag, Berlin, 1992.
6. C. Hoare. Communicating sequential processes. *Communicat. Associat. Comput. Mach.*, 21(8):666–677, 1978.
7. Object Management Group. Object constraint language (OCL) 2.0, 2003. `http://www.omg.org/uml`.
8. Object Management Group. Unified modelling language(UML) 2.0, 2003. `http://www.omg.org/uml`.
9. D. Plump. Hypergraph Rewriting:Critical Pairs and Undecidability of Confluence. In M. Plasmeijer and M.C. van Eekelen, editors, *Term Graph Rewriting*, pages 201–214. Wiley, 1993.
10. T. W. Pratt. Pair grammars, graph languages and string-to-graph translations. *Journal of Computer and System Sciences*, 5:560–595, 1971.
11. A.W. Roscoe. *The Theory and Practice of Concurrency*. Prentice-Hall, 1997.
12. G. Rozenberg, editor. *Handbook of Graph Grammars and Computing by Graph Transformation, Volume 1: Foundations*. World Scientific, 1997.
13. A. Schür. Specification of graph translators with triple graph grammars. In Tinhofer, editor, *Proceedings WG'94 International Workshop on Graph-Theoretic Concepts in Computer Science*, pages 151–163. LNCS 903, Springer–Verlag, 1994.
14. J. Stehr. *Semantical Consistency Check of UML Behavior Diagrams for Modelling Embedded Systems [in German]*. Diploma thesis, University of Paderborn, 2003.
15. S.Thöne, R.Depke, and G.Engels. Process-Oriented, Flexible Composition of Web Services with UML. In *Proc. of ER-Workshop on Conceptual Modeling Approaches for e-Business (eCOMO 2002); Tampere, Finland*. LNCS, 2002.
16. W.M.P. van der Aalst, A.H.M. ter Hofstede, B. Kiepuszewski, and A.P. Barros. *Distributed and Parallel Databases*. 2003.
17. H. Voigt. *Model-based Analysis of Executable Business Processes for Web Services [in German]*. Diploma thesis, University of Paderborn, 2003.

Modelling and Control with Modules of Signal Nets

Gabriel Juhás, Robert Lorenz, and Christian Neumair*

Lehrstuhl für Angewandte Informatik
Katholische Universität Eichstätt, 85071 Eichstätt, Germany
{gabriel.juhas,robert.lorenz,christian.neumair}@ku-eichstaett.de

Abstract. We present a modular formalism and methodology for modelling and control of discrete event systems, such as flexible manufacturing systems. The formalism is based on Petri net modules which communicate via signals. Two kinds of signals are employed, namely active signals, which force occurrence of (enabled) events (typically switches), and passive signals which enable/prohibit occurring of events (typically sensors). Modelling with such modules appears to be very natural from engineering perspective, enables hierarchical structuring, and support locality principle.
Further, we discuss the role of both kinds of signals in control tasks and we focus on the control aspects in general. We present a methodology for synthesis of controlled behavior for systems modelled by modules of signal sets. Given an uncontrolled system (a plant) modelled by a module of a signal net, and a control specification given as a regular language representing the desired signal output behavior of this system, we show how to synthesize the maximal permissive and non-blocking behavior of the plant respecting the control specification. Finally, we show how to synthesize the controller (as a module of a signal net) forcing the plant to realize the controlled behavior.

1 Introduction

Petri Nets are already widely used for modelling and control of Discrete event systems [10, 22], because of their modelling power, graphical expression, strong theoretical background, very developed analytical methods, tools, and many other features. However, there are still some features which are not directly supported by Petri Nets (at least in their basic version), but are, on the other hand, quite natural for engineers working with real applications. For example, to cover control tasks, Petri nets were extended by adding external conditions, which are necessary for enabling occurrence of transitions [10]. In the following paragraphs we are trying to identify some of features which are important for applications and are not directly supported by Petri nets. Based on this discussion we are presenting an extension of Petri nets, which can still benefit from all strong advantages that Petri nets bring, but also enables to deal with the discussed unsupported features in an effective way.

Petri nets are in principle distributed, however they do not support modularity. Modularity is quite natural and important in engineering. In complex application, models are usually built in several steps and are described on several levels of abstraction. Almost

* Supported by DFG: Project "SPECIMEN".

J. Desel, W. Reisig, and G. Rozenberg (Eds.): ACPN 2003, LNCS 3098, pp. 585–625, 2004.

each system is a part of a bigger system, such as a robot is a part of a manufacturing cell, as well as almost each system itself is composed from subsystems. This fact gives an importance to principle of compositionality. Thinking on one level of abstraction one does not need to reason about all details of subsystems which were taken into consideration in a sublevel. It is usually sufficient to consider just those parts of subsystems, which are in contact with environment, i.e. "input/output" parts and to consider the "inside" of the subsystems being a "black box". Such approach supports local changes in the whole system, it enables a replacement of one module by another with the same "input/output" functionality. A typical example of a modular approach in control applications are block diagrams. It would be very nice to have such a modular approach based on Petri nets. There are already developed many compositional frameworks for Petri nets, mostly based on gluing common places and/or transitions. However, because the subject of engineering are mostly complex systems, it is desirable that the composition of modules preserve the structure of modules.

In classical control theory it is given a system which can interfere with environment via inputs and outputs. The aim of its control is to ensure desired behaviour by giving the system right inputs in order to get the right outputs. The central idea in control theory is, that system and control build a so called *closed loop* (or *feedback loop*), which means, roughly speaking, that the control gives inputs to the system based on the system outputs which are observed by the control. In this paper, we are interested in control of discrete event systems, where the dynamic behaviour of a system is described by occurrence of discrete events changing the states of the system. The crucial question to be answered when choosing a formalism for modelling control systems is how to formalize "giving inputs and observing outputs". In order to answer the question, let us discuss a very simple example. Consider a switch which can turn on a light. Turning on the switch forces the bulb to light, however, only if the bulb is not damaged. And of course the bulb can not start to light, if no switch is turned on. In other words, in the previous situation the switch is the *actuator* of the bulb. In this example the switch plays the role of the control, while the light (bulb with cables etc.) plays the role of the system to be controlled. Thus, inputs to the system can be actuators representing conditional assymmetric synchronization - events of the control are trying to force events in the system. This is a typical situation in control of discrete event systems: a product line will not start without pressing a control button, or a mobile phone will not call a number without pressing appropriate buttons, but a printer is not printing without paper even if the "print" button was pressed. The other typical interaction between control and discrete event systems are sensors readings: An event in the system can occur only if a sensor in the control is in a certain state and vice versa. Thus events can be enabled/prohibited via states of sensors.

Thus, an event of a system can have two kinds of inputs: Actuators, which try to force the event, or sensors, which can prohibit the event. Events associated to inputs are called *controllable*. Of course there can be uncontrollable events in the system. Regarding for example a printer, a "paper jam" event can occur without any influence from the control. The following two kinds of outputs can be observed: Either the occurrence of an event (via actuators) or the fact that a state is reached (via sensors). Event resp. states associated to outputs are called *observable*. Of course there can be unobservable events resp. states. As mentioned, it would be natural to model control of a discrete events systems by influencing its behaviour by actuators and sensors in order to observe desired outputs as decribed above.

However, the solution in the discrete event control community, which is now quite accepted, is to use only the sensors. More exactly, in supervisory control [2, 16] the events of the system to be controlled are divided as above into controllable and uncontrollable. But the controllable events can only be enabled/prohibited by a supervisor. Thus, in supervisory control actuators can only be modelled indirectly using the "sensor principle" by prohibiting all controllable events, except the event which is actuated ([1], pp. 185 - 202). For example, modelling a switch and a light, one needs to prohibit all controllable events except the event "a bulb starting to light" to model the situation when the switch turns on. In fact, in case of supervisory control, the control means to restrict the behaviour of the system to fulfil the control specification. As mentioned in [1], pp. 185 - 202, "sometimes it is desirable to have a controller which not only disables controllable events but also chooses one among the enabled ones. This event can be interpreted as a command given to the plant." The solution to such cases is given by a construction of "an inplementation", which is a special supervisor, enabling at most one controllable event at a time. There arises the natural question, why not directly model actuators?

We would like to have an extension of Petri nets, which support input/output structuring using actuator and sensors, modularity and compositionality in an intuitive graphical way.

So, as a modelling formalism, we use modules communicating by means of the above described signals. This formalism is based on the work [18], where automata were used to describe the internal behavior of a system, and the paper [19], where Petri nets were used for this purpose. We call these models, i.e. Petri nets equipped with two kinds of signals, *signal nets*. Nowadays, this concept is successfully used in modelling and control of discrete event systems by a growing community. There are several dialects of these nets and several different names, such as net condition/event systems [8, 7, 9] or signal nets [20]. In this paper we are using the name signal nets. One reason is that the name condition/event nets is used in the Petri net context for a well known basic net class. A signal net is a Petri net enriched by *event signals*, which force the occurrence of (enabled) events (typically switches), and *condition signals* which enable/prohibit the occurring of events (typically sensors). Adding input and output signals to a signal net, one gets a *module of a signal net*. Modules of signal nets can be composed by connecting their respective input and output signals.

There are several related works employing modules of signal nets in control of discrete event systems. In [8, 7, 9] effective solutions for particular classes of specifications, such as forbidden states, or simple desired and undesired sequences of events, are described. Recently, an approach for control specifications given by cycles of observable events was presented in [15]. However, in [15] the actuators are used only to observe events of the controlled system, but surprisingly, for control actions only condition signals (for prohibiting events) are taken. In our paper, we adapt the framework of supervisory control providing a methodology for control of discrete event systems using **both** concepts, namely actuators and sensors. Such a methodology with the slogan "forcing and prohibiting instead of only prohibiting" would be more appropriate for the class of discrete event systems, where actuators and commands are used in practice. In addition, we consider a general class of control specification in form of a language over steps of event outputs (steps of observable events). We consider steps (i.e. sets) of outputs, rather then simple outputs, because some outputs can be simultaneously synchronized by an event of the system. We allow also steps containing an input with some outputs. Such a situation describes that an input signal is trying to synchronize a controllable event of the system, which is also observable. So the controller can immediately (i.e. in the same step) observe whether the input signal has forced the event to occur or not. However, since the control is assumed to send inputs based on observed outputs (as stated in the beginning), we do not allow the symmetric situation: observable events can not synchronize inputs in the same step.

As it was already mentioned, in case of supervisory control, the behavior of the DES can not be forced by the supervisor: control means to restrict the behavior of the system to fulfill the control specification. Formally (see e.g. [2]), there is given a regular prefix closed language over the set of system events. This language represents the uncontrolled behavior of the system. Control specification is given in form of a regular subset of this language and is representing the desired behavior. Moreover, some states in an automaton representing the uncontrolled behavior are marked. The sequences (words) of events leading to these states describe completed tasks. Remember also, that the events of the system are divided into controllable events, which can be enabled/prohibited by the control, and uncontrollable events. The basic aim of supervisory control is to find a supervisor, which will prohibit the controllable events in such a way, that the behavior of the system is restricted to its maximal regular sublanguage, which still respects the control specification, and is moreover non-blocking (every sequence of this language can be completed to a marked state, i.e. no dead- or livelocks occur in unmarked states). Such a supervisor is called minimally restrictive nonblocking supervisor.

In our framework we identify which input signals have to be sent to the module of the plant in order to observe only such sequences of (steps of) output signals, which are prefixes of the control specification, and every sequence of (steps of) output signals can be completed to a sequence of output signals belonging to the control specification. The presented solution is maximal in the sense, that we match all sequences of (steps of) outputs which can be achieved by sending appropriate inputs without being in danger to observe a sequence of (steps of) outputs which is not a prefix of a sequence in the control specification, or a sequence of (steps of) outputs which can not be completed to a sequence in the control specification (i.e. which is blocking). The maximality is achieved under the paradigm, that no output signal of the plant can synchronize an input signal of the plant (as already stated above). In other words, we construct a language over steps of input and output signals of the module of the plant, which represents the maximally permissive nonblocking behavior and fulfills the control specification. Finally, we show that for such a behavior there exists a control module (of a signal net), which will in composition with the plant module realize this behavior. As the main result we will construct such a control module.

The paper is organized as follows: In Section 2 we present *modules of signal nets* with definition of step semantics, composition rules and input/output behavior. In Section 3 we outline our control framework implementing the "forcing and prohibiting"-paradigm by means of modules of signal nets. It is compared in detail to classical supervisory control. The Section splits into two parts. In Subsection 4.1 we synthesize the maximally permissive nonblocking behavior of a module of a signal net (representing the plant) respecting a given regular specification language. Finally, in Subsection 4.2 we present the construction of the controller as a module of a signal net.

2 Modules of Signal Nets

As mentioned in the introduction we present an extension of Petri nets which allows to model actuators, sensors and modularity, and still has all the benefits that Petri nets bring. We assume the underlying Petri nets to be elementary Petri nets (1-safe Petri nets) equipped with the so called *first consume, then produce* semantics (since we want to allow loops, e.g. [13]). The first step in the extension is to add two kinds of signals, namely active signals, which force the occurrence of (enabled) events (typically switches or actuators), and passive signals which enable/prohibit the occurrence of events (typically sensors). These signals are expressed using two kind of arcs. A Petri net extended with such signals is simply called a *signal net*.

Active signals are represented using arcs connecting transitions and can be interpreted in the following way: An active signal arc, also called *event arc*, leading from a transition t_1 to a transition t_2 specifies that if transition t_1 occurs and transition t_2 is enabled to occur then the occurrence of t_2 is forced (synchronized) by the occurrence of t_1, i.e. transitions t_1 and t_2 occur in one (synchronized) step. If t_2 is not enabled, t_1 occurs without t_2, while an occurrence of t_2 without t_1 is not allowed. Taking an example, an event turning on a switch would be modelled via the transition t_1, while the event lighting the bulb would be modelled via transition t_2.

In general (synchronized) steps of transitions are build inductively in the above way. Every step starts at a unique transition, which is not synchronized itself. Notice that this implies that event arcs build no cycles. Consider a transition t which is synchronized by several transitions $t_1, \ldots, t_n$, $n \geqslant 2$. Then two situations can be distinguished. For simplicity consider the case $n = 2$.

If the transitions t_1 and t_2 do not build a synchronized step themselves, either t_1 or t_2 can synchronize transition t in the above sense, but never transitions t_1 and t_2 can occur in one synchronized step. As an example you can think of several switches to turn on a light on (see Figure 1, part (a)).

If the transitions t_1 and t_2 build a synchronized step themselves, then there are two dialects in literature to interpret such a situation: In the first one ([8, 7, 9]) both transitions t_1, t_2 have to agree to synchronize t. Thus the only possible step of transitions involving t has to include transitions t_1 and t_2, too. We call this dialect AND-semantics (see Figure 1, part (b)).

In the second one ([4]) the occurrence of at least one of the transitions t_1 and t_2 synchronizes transition t, if t is enabled. It is also possible, that t_1, t_2 and t occur in one synchronized step. We call this dialect OR-semantics (see Figure 1, part (c)).

In general the relation given by event arcs builds a forest of arbitrary depth. In this paper we introduce the most general interpretation, where both semantics are pos-

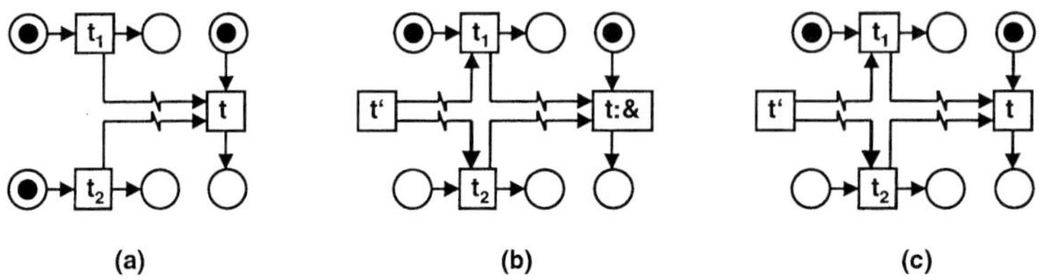

Fig. 1. In Figure (a) the enabled steps are $\{t_1, t\}$ and $\{t_2, t\}$. Figure (b) shows a signal net in AND-semantics: here the only enabled step is $\{t', t_1\}$, i.e. t is not synchronized. In Figure (c) the same net is shown in OR-semantics: here we have the enabled step $\{t', t_1, t\}$, i.e. t is synchronized.

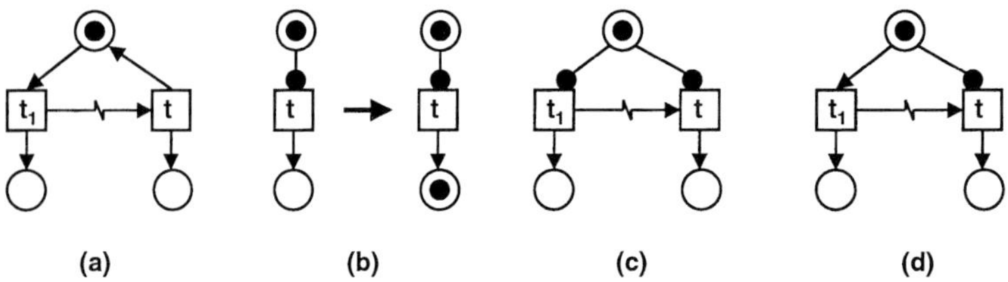

Fig. 2. Figure (a) shows an enabled step $\{t_1, t\}$. The left part of Figure (b) shows an enabled transition t, which tests a place to be marked. The occurrence of t leads to the marking shown in the right part of Figure (b). Figures (c) and (d) again present situations of an enabled step $\{t_1, t\}$.

sible and are interpreted locally backward. That means we distinguish between OR- and AND-synchronized transitions. An OR-synchronized transition demands to be synchronized by at least one of its synchronizing transitions, whereas an AND-synchronized transition demands to be synchronized by all of its synchronizing transitions. Since we allow loops w.r.t. single transitions, we also allow loops w.r.t. steps of transitions (see Figure 2, part (a)).

Passive signals are expressed by so called *condition arcs* (also called read arcs or test arcs in the literature) connecting places and transitions. A condition arc leading from a place to a transition models the situation that the transition can only occur if the place is in a certain state but this state remains unchanged by the transition's occurrence (read operation) (see Figure 2, part (b)). Of course several transitions belonging to a synchronized step can test a place to be in a certain state via passive signals simultaneously, since the state of this place is not changed by their occurrence (see Figure 2, part (c)).

We also allow that a transition belongs to a synchronized step of transitions testing a place to be in a certain state via a passive signal, whereas the state of this place is changed by the occurrence of another transition in this step. That means we use the so called *a priori* semantics ([12]) for the occurrence of steps of transitions, where testing of states precedes changing of states by occurrence of steps of transitions (see Figure 2, part (d)).

Definition 1 (Signal nets). *A signal net is a six-tuple* $N = (P, T, F, CN, EN, m_0)$ *where*

P *denotes the finite set of* places,
$T = T_{AND} \dot{\cup} T_{OR}$ *the distinct union of the finite sets of* AND-synchronized transitions T_{AND} *and* OR-synchronized transitions T_{OR} *(*$P \cap T = \emptyset$*),*
$F \subseteq (P \times T) \cup (T \times P)$ *the* flow relation,
$CN \subseteq (P \times T)$ *the set of* condition arcs *(*$CN \cap (F \cup F^{-1}) = \emptyset$*),*
$EN \subseteq (T \times T)$ *the acyclic set of* event arcs *(*$EN^{+} \cap id_T = \emptyset$*), and*
$m_0 \subseteq P$ *the* initial marking.

Places, transitions and the flow relation are drawn as usual using circles, boxes and arrows. To distinguish between AND- and OR-synchronized transitions, AND-synchronized transitions are additionally labelled by the symbol "&". Event arcs and condition arcs are visualized using arcs of a special form given in Figure 1 and Figure 2.

For a place or a transition x we denote
$^{\bullet}x = \{y \mid (y, x) \in F\}$ the *preset* of x,
$x^{\bullet} = \{y \mid (x, y) \in F\}$ the *postset* of x.

For a transition t we denote
$^{+}t = \{p \mid (p, t) \in CN\}$ the *positive context* of t,
$^{\leadsto}t = \{t' \mid (t', t) \in EN\}$ the *synchronization set* of t,
$t^{\leadsto} = \{t' \mid (t, t') \in EN\}$ the *synchronized set* of t.

Given a set $\xi \subseteq T$ of transitions, we extend the above notions to: $^{\bullet}\xi = \bigcup_{t \in \xi} {^{\bullet}t}$ and $\xi^{\bullet} = \bigcup_{t \in \xi} t^{\bullet}$, $^{\leadsto}\xi = \bigcup_{t \in \xi} {^{\leadsto}t}$, $\xi^{\leadsto} = \bigcup_{t \in \xi} t^{\leadsto}$.

Definition 2 (Enabling of transitions). *A transition* $t \in T$ *is* enabled *at a marking* $m \subseteq P$*, if* $^{\bullet}t \cup {^{+}t} \subseteq m$ *and* $(t^{\bullet} \setminus {^{\bullet}t}) \cap m = \emptyset$*.*

The following definition introduces a notion of steps of transitions which is different to the usual one used in Petri nets. A step denotes a set of transitions connected by event arcs, which occur synchronously. A transition, which is not synchronized by another transition, is a step. Such transitions are called *spontanuous*. In general, steps are sets of transitions such that for every non-spontaneous OR-synchronized transition in this step at least one of it's synchronizing transitions belongs also to this step, and for every AND-synchronized transition in this step all of it's synchronizing transitions belong also to this step.

Definition 3 (Steps). *Given a signal net* N*, steps are sets of transitions* ξ *defined inductively by*

- *If* $t \in T$ *with* $^{\leadsto}t = \emptyset$ *(*t *is spontaneous), then* $\xi = \{t\}$ *is a step.*
- *If* ξ *is a step, and* $t \in T \setminus \xi$ *is a transition, then* $\xi \cup \{t\}$ *is a step, if either* $t \in T_{OR}$ *and* $^{\leadsto}t \cap \xi \neq \emptyset$*, or* $t \in T_{AND}$ *and* $^{\leadsto}t \subseteq \xi$*.*

Now we introduce how a step is enabled to occur. A step ξ is said to be potentially enabled at a marking m if every transition $t \in \xi$ is enabled at m and no transitions $t_1, t_2 \in \xi$ are in conflict, except for possible loops $p \in {^{\bullet}\xi} \cap \xi^{\bullet}$ w.r.t. ξ, where $p \in m$ is required. From all steps potentially enabled at a marking only those are enabled which are maximal with this property.

Definition 4 ((Potentially) enabling of steps). *A step* ξ *is* potentially enabled *in a marking* m *if*

- *All $t \in \xi$ are enabled: $^\bullet t \cup {}^+t \subseteq m$ and $(t^\bullet \setminus {}^\bullet\xi) \cap m = \emptyset$ and*
- *No pair of transitions $t, t' \in \xi$ is in conflict: $^\bullet t \cap {}^\bullet t' = t^\bullet \cap (t')^\bullet = \emptyset$.*

The step ξ is enabled, *if ξ is potentially enabled, and there is not a potentially enabled step $\eta \supsetneq \xi$ (ξ is maximal).*

Definition 5 (Occurrence of steps and follower markings). *The occurrence of an enabled step ξ yields the follower marking $m' = (m \setminus {}^\bullet\xi) \cup \xi^\bullet$. In this case we write $m[\xi\rangle m'$.*

Definition 6 (Reachable markings, occurrence sequences). *A marking m is called reachable from the initial marking m_0 if there is a sequence of markings $m_1, \ldots, m_k = m$ and a sequence of steps $\xi_1, \ldots, \xi_k$, such that $m_0[\xi_1\rangle m_1, \ldots, m_{k-1}[\xi_k\rangle m_k$. Such a sequence of steps is called an occurrence sequence.*

Adding some inputs and outputs to signal nets, i.e. adding condition and event arcs coming from or going to an environment, we get modules of signal nets with input and output structure.

Definition 7 (Modules of signal nets). *A module of a signal net is a triple $M = (N, \Psi, c_0)$, where $N = (P, T, F, CN, EN, m_0)$ is a signal net, and $\Psi = (\Psi^{sig}, \Psi^{arc})$ is the input/output structure, where*
$\Psi^{sig} = C^{in} \cup E^{in} \cup C^{out} \cup E^{out}$ is a finite set of input/output signals, and
$\Psi^{arc} = CI^{arc} \cup EI^{arc} \cup CO^{arc} \cup EO^{arc}$ is a finite set of arcs connecting input/output signals with the elements of the net N. Namely,
C^{in} resp. E^{in} denotes the set of condition resp. event inputs,
C^{out} resp. E^{out} the set of condition resp event outputs,
$CI^{arc} \subseteq C^{in} \times T$ resp. $EI^{arc} \subseteq E^{in} \times T$ the set of condition resp event input arcs,
$CO^{arc} \subseteq P \times C^{out}$ resp. $EO^{arc} \subseteq T \times E^{out}$ the set condition resp event output arcs,
$c_0 \subseteq C^{in}$ the initial state of the condition inputs.

We extend the notions of preset, postset, positive context, synchronization set and synchronized set to the elements of Ψ^{sig} in the obvious way. An example of a module of a signal net, with $C^{in} = \{ci\}$, $E^{in} = \{j, k\}$, $C^{out} = \{co\}$ and $E^{out} = \{u, v\}$ is shown in the Figure 3.

Two modules can be composed by identifying some inputs of the one module M_1 with appropriate outputs of the other module M_2 and vice versa with a composition mapping Ω. The connections of the nets to the involved identified inputs and outputs are replaced by direct signal arcs respecting the identification (see Figure 7), such that

- the initial markings are compatible with the initial states of the condition inputs, and
- no cycles of event arcs are generated.

The composition of M_1 and M_2 w.r.t. Ω is denoted by $M_1 *_\Omega M_2$.

Definition 8 (Composition of modules of signal nets). *Let $M_1 = (N_1, \Psi_1, c_{01})$, $M_2 = (N_2, \Psi_2, c_{02})$ be modules of signal nets with input/output structures $\Psi_i = (\Psi_i^{sig}, \Psi_i^{arc})$ and initial markings m_{0i} ($i = 1, 2$).*

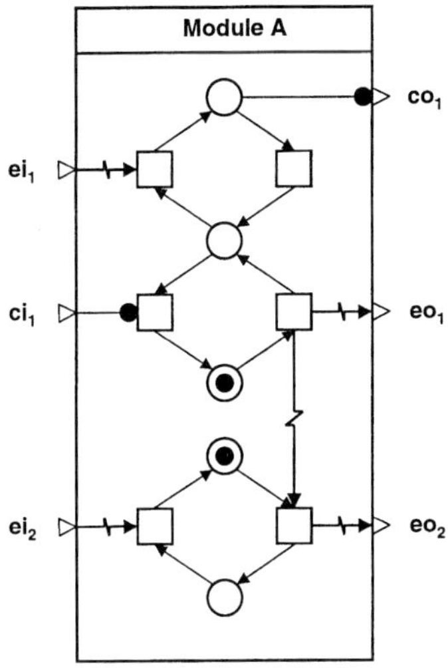

Fig. 3. A module of a signal net with condition inputs $C^{in} = \{ci\}$, event inputs $E^{in} = \{j, k\}$, condition outputs $C^{out} = \{co\}$ and event outputs $E^{out} = \{u, v\}$.

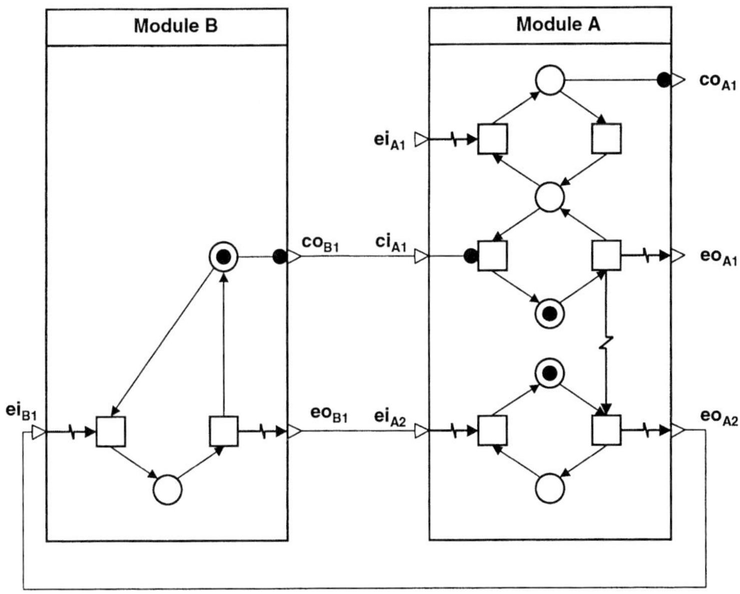

Fig. 4. The composition of two modules.

Let $Q \subseteq \Psi_1^{sig}$ and $\Omega : Q \rightarrow \Psi_2^{sig}$ be an injective mapping, such that the initial markings are compatible with the initial states of the condition inputs:
$(p, co) \in CO_1^{arc} \wedge \Omega(co) \in c_{02} \Rightarrow p \in m_{01}$ *and*
$(p, co) \in CO_2^{arc} \wedge \Omega^{-1}(co) \in c_{01} \Rightarrow p \in m_{02}$.

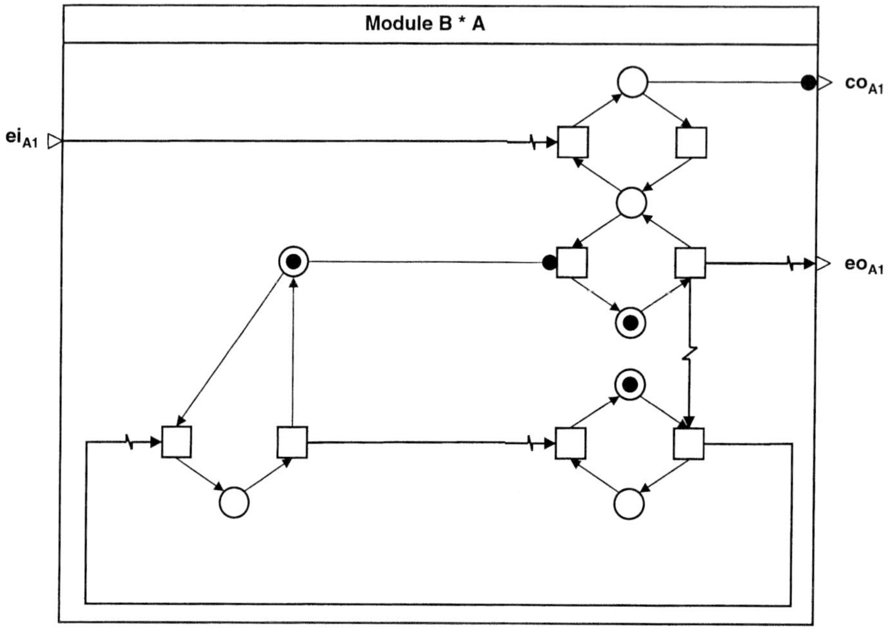

Fig. 5. The result of the composition of the modules from Figure 4.

Ω has to satisfy:
$\Omega(E_1^{in} \cap Q) \subseteq E_2^{out}$, $\Omega(E_1^{out} \cap Q) \subseteq E_2^{in}$, $\Omega(C_1^{in} \cap Q) \subseteq C_2^{out}$, *and* $\Omega(C_1^{out} \cap Q) \subseteq C_2^{in}$.
Finally, no cycles of event arcs should be generated.

Then the composition $M = M_1 *_\Omega M_2$ of M_1 and M_2 w.r.t. Ω *is the module* $M = (N, \Psi, c_0)$ *with* $N = (P_1 \cup P_2, T_1 \cup T_2, F_1 \cup F_2, CN, EN, m_{01} \cup m_{02})$ *and* $\Psi = (\Psi^{sig}, \Psi^{arc})$*, where involved inputs, outputs and corresponding signal arcs are deleted, i.e.*

$$\begin{aligned}
\Psi^{sig} &= (\Psi_1^{sig} \setminus Q) \cup (\Psi_2^{sig} \setminus \Omega(Q), \\
\Psi^{arc} &= (\Psi_1^{arc} \setminus (({}^\bullet Q \times Q) \cup (Q \times Q^\bullet))) \cup \\
&\quad (\Psi_2^{sig} \setminus (({}^\bullet\Omega(Q) \times \Omega(Q)) \cup (\Omega(Q) \times \Omega(Q)^\bullet))), \\
c_0 &= (c_{01} \setminus Q) \cup (c_{02} \setminus \Omega(Q)),
\end{aligned}$$

and new signal arcs are added according to Ω in the following way:

$$\begin{aligned}
CN = {} & CN_1 \cup CN_2 \cup \\
& \{(p,t) \mid \exists co \in C_1^{out} : (p, co) \in CO_1^{arc} \wedge (\Omega(co), t) \in CI_2^{arc}\} \cup \\
& \{(p,t) \mid \exists ci \in C_1^{in} : (ci, t) \in CI_1^{arc} \wedge (p, \Omega(ci)) \in CO_2^{arc}\}, \\
EN = {} & EN_1 \cup EN_2 \cup \\
& \{(t,t') \mid \exists eo \in E_1^{out} : (t, eo) \in EO_1^{arc} \wedge (\Omega(eo), t') \in EI_2^{arc}\} \cup \\
& \{(t,t') \mid \exists ei \in E_1^{in} : (ei, t') \in EI_1^{arc} \wedge (t, \Omega(ei)) \in CO_2^{arc}\}
\end{aligned}$$

Remark 1. For each new arc (t, t') in a composed module $M_1 *_\Omega M_2$ w.r.t. an event output $eo \in E_1^{out}$ with $(t, eo) \in EO_1^{arc}$ and $(\Omega(eo), t') \in EI_2^{arc}$ we say that t' *replaces* eo and t *replaces* $\Omega(eo)$. A similar notion is used also for new arcs w.r.t. event inputs and condition inputs and outputs.

In order to define the behavior of a module, observe: transitions connected by an event input to the environment are not able to occur spontaneously, but need to be synchronized by the event input in order to occur. Similar a transition connected by an condition input to the environment is only able to occur, if the condition input is activated. Therefore we are interested in the behavior of the module w.r.t. a given environment. In the most general case this environment is assumed to be maximal permissive in the sense, that there is no causal restriction in sending event inputs and activating condition inputs. We will model such an environment also as a module $\mathcal{E}$ of a signal net and then compose the environment module appropriately with the original module M. $\mathcal{E}$ realizes a maximally permissive environment in the following sense:

- at any moment $\mathcal{E}$ can send event inputs to M: so each event signal of M is modelled in $\mathcal{E}$ by a corresponding always enabled transition;
- at any moment $\mathcal{E}$ can enable and disable condition inputs of M: so each condition input of M is modelled in $\mathcal{E}$ by a corresponding place, which can be marked and unmarked by associated transitions;
- $\mathcal{E}$ can observe outputs of M: every output of M is modelled in $\mathcal{E}$ by a corresponding transition, which synchronized in the case of an event output, and enabled in the case of an condition output;
- in $\mathcal{E}$ no synchronization between its transitions is allowed: in particular, inputs should not be sent in steps from $\mathcal{E}$ to M, and outputs M should only be observed by $\mathcal{E}$ and not synchronize inputs of M via $\mathcal{E}$.

Definition 9 (Maximally permissive environment). *Let $M = (N, \Psi, c_0)$ be a module with $\Psi = (\Psi^{sig}, \Psi^{arc})$. Define the maximally permissive environment module $\mathcal{E} = (N_\mathcal{E}, \Psi_\mathcal{E}, c_{0\mathcal{E}})$, $\Psi_\mathcal{E} = (\Psi_\mathcal{E}^{sig}, \Psi_\mathcal{E}^{arc})$, w.r.t. M by $EN_\mathcal{E} = CN_\mathcal{E} = \emptyset$ and*

$$\begin{aligned}
P_\mathcal{E} &= \{p_{ci.on} \mid ci \in C^{in}\},\\
T_\mathcal{E} &= \{t_c \mid c \in C^{out}\} \cup \\
&\qquad \{t_{ci.on} \mid ci \in C^{in}\} \cup \{t_{ci.off} \mid ci \in C^{in}\} \cup \\
&\qquad \{t_e \mid e \in E^{out} \cup E^{in}\},\\
F_\mathcal{E} &= \{(t_{ci.on}, p_{ci.on}) \mid ci \in C^{in}\} \cup \{(p_{ci.on}, t_{ci.off}) \mid ci \in C^{in}\},\\
m_{0\mathcal{E}} &= \{p_{ci.on} \mid ci \in C^{in} \cap c_0\},\\
C_\mathcal{E}^{in} &= \{ci_c \mid c \in C^{out}\},\\
C_\mathcal{E}^{out} &= \{co_c \mid c \in C^{in}\},\\
E_\mathcal{E}^{in} &= \{ei_e \mid e \in E^{out}\},\\
E_\mathcal{E}^{out} &= \{eo_e \mid e \in E^{in}\},\\
CI_\mathcal{E}^{arc} &= \{(ci_c, t_c) \mid c \in C^{out}\},\\
CO_\mathcal{E}^{arc} &= \{(p_c, co_c) \mid c \in C^{in}\},
\end{aligned}$$

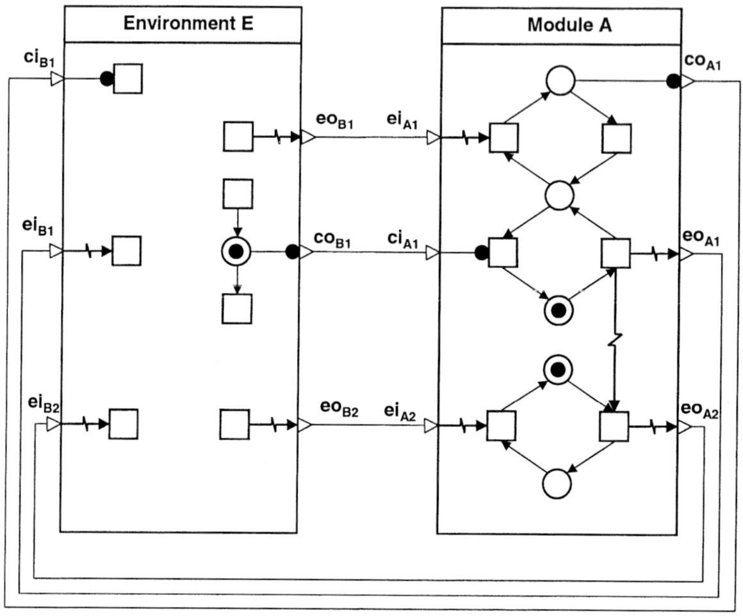

Fig. 6. The composition of the module in Figure 3 with its maximally permissive environment module.

$$EI_{\mathcal{E}}^{arc} = \{(ei_e, t_e) \mid e \in E^{out}\},$$
$$EO_{\mathcal{E}}^{arc} = \{(t_e, eo_e) \mid e \in E^{in}\}.$$

The composition of M with its maximally permissive environment $\mathcal{E}$ is called the standalone of M (observe that this composition has empty input/output structure) (as an example see Figure 7).

Definition 10 (Standalones). *Let M be a module of a signal net and $\mathcal{E}$ be the maximally permissive environment module of M. The* standalone *of M is the composition module $M_S = (N_S, \Psi_S) = M *_\Omega \mathcal{E}$ w.r.t. the following composition mapping $\Omega : \Psi^{sig} \to \Psi_{\mathcal{E}}^{sig}$:*

$$\Omega(e) = ei_e \, for \, e \in E^{out},$$
$$\Omega(e) = eo_e \, for \, e \in E^{in},$$
$$\Omega(c) = ci_c \, for \, c \in C^{out},$$
$$\Omega(c) = co_c \, for \, c \in C^{in}.$$

Definition 11 (Behavior of modules of signal nets). *Let M be a module of a signal net and let $M_S = (N_S, \Psi_S)$ be the standalone of M. The set L_M of all occurrence sequences of N_S is called the behavior of the module M.*

L_M represents the set of all possible sequences of steps of input signals, output signals and inner transitions of M under the assumptions: Output signals of M can

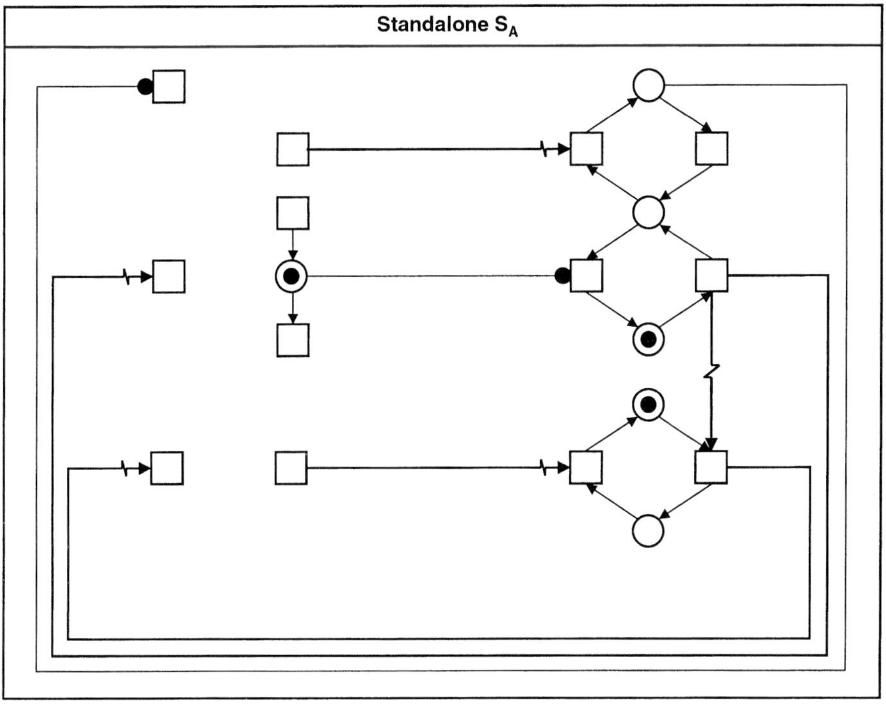

Fig. 7. The standalone of the module of a signal net in Figure 3.

not synchronize input signals of M via the maximally permissive environment module. Several input signals of M can not be sent in steps from the maximally permissive environment module.

Thus, modules of signal nets are a Petri net extension supporting input/output structuring, modularity and compositionality in an intuitive graphical way. They are used in many applications in the area of design, modelling and control of discrete event systems, such as flexible manufacturing systems and control of traffic systems for more than ten years, see e.g. [8, 7, 9, 19]. This fact gives a motivation for a more detailed theoretical investigation of this extension of Petri nets. In this section we have provided a proper formal foundation for this modelling framework, including definitions of input/output structure and composition of modules. In [14], we have concentrated on a definition of an equivalence w.r.t. input/output behaviour, which is preserved by the composition of modules. It is a crucial concept for hierarchical modelling, which enables to replace a module with a more abstract/concrete module with the same "input/output" functionality.

In the following sections we discuss the role of both kinds of signals in control tasks and we focus on the control aspects in general.

3 Controller Synthesis

As mentioned in the introduction, in classical control the aim is to influence the behavior of a system by a control via sensors and actuators in order to get a specified desired behavior. In principle there are two possibilities to express a desired behavior (see [2] for an actual survey, and [1,21] for recent developments):

- the event based approach used in the seminal work of Ramadge and Wonham on supervisory control of discrete event systems (DES) [16]. In this framework the desired behavior is given in the form of legal sequences of events.
- the state based approach ([10]), where the desired behavior is derived from a set of legal resp. forbidden states.

Considering discrete event systems (DES) in both approaches the main problem is that the considered modelling formalisms (languages, automata, Petri nets) do not provide a mechanism for asymmetric synchronization intended by actuators.

For example in classical supervisory control this problem is solved by modelling actuators via prohibiting all other possible events ([5]). As a consequence, the behavior of the DES cannot be forced by the control, now called supervisor, but only be restricted. Formally there is given a regular prefix closed language over a fixed set of events representing the uncontrolled behavior of the DES and a regular subset of this language representing the restricted desired behavior. In the most general case one distinguishes between controllable events (which can be prohibited by the supervisor) and uncontrollable events, and between observable events (which can be observed by the supervisor) and unobservable events. The question is, which controllable events should be prohibited by the supervisor after observing a certain sequence of observable events in order to disable all undesired behavior in a *minimal restrictive way*.

We present an alternative to the existing approaches to control of DES with *direct modelling of actuators*. Our formalism is suitable for both kind of specifying the desired behavior. Because in literature the event based approach is more developed than the state based approach in the sense that it allows more general specifications ([23]), we concentrate in this paper on a event based specification of the desired behavior.

In particular, we specify the desired behavior by sequences of event output signals. Therefore we consider modules, modelling the plant, without condition output signals (which correspond to states). Notice that a condition output signal c in a behavior specification could be replaced by two event output signals $c.on$ and $c.off$ synchronized by transitions marking and emptying the place in ^+c, respectively.

Our framework could be easily adapted to behavior specifications which include input signals: In this case one could additionally consider specifications of the form *"After sending input i, we want to observe a sequence of outputs w"* or *"input i always synchronizes output o"*. The restriction to sequences of output signals is only for sake of simplicity.

Throughout this section we consider a module $\mathcal{P}$ of a signal net as a model of an uncontrolled plant and its maximally permissive environment $\mathcal{E}$. As in the previous section T denotes the set of transitions of $\mathcal{P}$. We additionally fix the set I of transitions of $\mathcal{E}$ corresponding to event input signals together with transitions switching condition input signals on/off, and the set O of transitions of $\mathcal{E}$ corresponding to event output signals:

$$I = \{t_e \mid e \in E^{in}\} \cup \{t_{ci.on} \mid ci \in C^{in}\} \cup \{t_{ci.off} \mid ci \in C^{in}\},$$
$$O = \{t_e \mid e \in E^{out}\}.$$

We consider the inside of $\mathcal{P}$ as a black box: We only can send input signals to $\mathcal{P}$ and meanwhile observe sequences of output signals. In particular, the behavior of the DES (represented by $\mathcal{P}$) is forced, not only restricted from outside. Of course this approach leads to formal and technical differences to the classical supervisory control approach:

Mainly, all events of $\mathcal{P}$ are assumed to be uncontrollable and unobservable. Controllable are only the input signals, modelled by the set of transitions I of $\mathcal{E}$, and observable are, beside the input signals, exactly the output signals, modelled by the set of transitions O of $\mathcal{E}$.

Remember that we specify a desired behavior of $\mathcal{P}$ by a set of desired sequences only of output signals (in difference to supervisory control, where the specification is over all events, observable and unobservable, controllable and uncontrollable ones). Observe that, since the event arc relation produces a step semantics, we observe sequences of steps of output signals.

The aim of control synthesis is to find a control module $\mathcal{C}$ which appropriately composed (by a composition mapping Ω) with $\mathcal{P}$ fulfills: Each occurrence sequence of the underlying signal net of $\mathcal{C} *_{\Omega} \mathcal{P}$ respects the desired behavior in the sense that the projection of this occurrence sequence onto the set of transitions of $\mathcal{C}$ which replaces output signals of $\mathcal{P}$ (see remark 1) belongs to the desired behavior.

We synthesize such a control module $\mathcal{C}$ in two steps. First we define conditions of *controllability* of a subbehavior of the behavior $L_{\mathcal{P}}$ of the plant module $\mathcal{P}$ (analogously to [16]) and show how to compute the maximal controllable subbehavior of $L_{\mathcal{P}}$ respecting the desired behavior (if it exists), see subsection 4.1. Second we show that for every controllable subbehavior of $L_{\mathcal{P}}$ there is a control module $\mathcal{C}$, which in composition with the plant module $\mathcal{P}$ realizes this controllable subbehavior. As the main result of the second step we will construct such a control module by adding new net structure to the maximally permissive environment module $\mathcal{E}$ in $\mathcal{E} *_{\Omega} \mathcal{P}$ (see Figure 7), see subsection 4.2.

Since sets of occurrence sequences of signal nets are regular languages[1] over an alphabet of steps, we assume the desired behavior to be a regular language. In the following section we provide a short introduction to the theory of regular languages, which will be used in the subsections 4.1 and 4.2.

4 Regular Languages

We need the following language theoretic notations ([11] and [2]). For a finite set A we denote

- $2^A = \{B \mid B \subseteq A\}$ the set of all subsets of A,
- ϵ the empty word,
- $A^* = \{a_1 \dots a_n \mid n \in \mathbb{N}_0,\, a_1, \dots, a_n \in A\} \cup \{\epsilon\}$ the set of all finite words over the alphabet A.

In paragraph 3 we are solely concerned with regular languages $L \subseteq A^*$ with $A = 2^X$ for finite sets X. In the following we will briefly introduce some representations of regular languages and some operations on regular languages.

[1] Observe that we use elementary nets, which have a finite reachability graph.

Finite Automata. A regular language L can be represented as the language $L(G)$ of a (deterministic) finite automaton $G = (S, A, \delta, F, s_0)$, where

- S is the set of states.
- $\delta : S \times A \to S$ is the transition function: $\delta(s, a) = s'$ means that the automaton reaches state s' when reading a in state s.
- s_0 is the initial state.
- $F \subseteq S$ is the set of accepting states.

The transition function is extended in the obvious way to $\delta : S \times A^* \to S$: $\delta(s, w) = s'$ means that the automaton reaches state s' when reading w in state s. A word $w \in A^*$ belongs to $L(G)$ if and only if $\delta(s_0, w) \in F$. The states of G can be denoted as equivalence classes over A^*:

$$[w]_G = \{v \in A^* \mid \delta(s_0, v) = \delta(s_0, w)\}.$$

A finite automaton can be viewed as a labelled transition system (S, Σ, A), where S is the set of states, $\Sigma = \{s \xrightarrow{a} s' \mid \delta(s, a) = s'\}$ the set of transitions and A the set of (event) labels.

Regular Expressions. A regular language L can be represented as the language $L(\alpha)$ of a regular expression α over A. Regular expressions are build inductively from the elementary regular expressions $\alpha = x$ with $x \in A \cup \{\epsilon, \emptyset\}$, where $L(\alpha) = L(x) = \{x\}$ for $x \neq \emptyset$ and $L(\emptyset) = \emptyset$, by:

- *union*:
  α, β regular expressions $\Rightarrow \alpha + \beta$ regular expression with $L(\alpha + \beta) = L(\alpha) \cup L(\beta)$.
- *concatenation*:
  α, β regular expressions $\Rightarrow \alpha\beta$ regular expression with $L(\alpha\beta) = L(\alpha)L(\beta) = \{uv \mid u \in L(\alpha),\ v \in L(\beta)\}$.
- *iteration*:
  α regular expressions $\Rightarrow \alpha^*$ regular expression with $L(\alpha^*) = (L(\alpha))^* = \{u_1 \ldots u_n \mid u_i \in L(\alpha),\ n \in \mathbb{N}\}$.

Operations on Regular Languages. There is the so called *prefix relation* $\leqslant \subset A^* \times A^*$:

$$u \leqslant v \Leftrightarrow \exists x \in A^* : \ ux = v.$$

In this case u is called a *prefix* of v. For $x \neq \epsilon$ we call u a *proper prefix* of v. Beside the set operations $\cap$, $\cup$ and $\setminus$ there are the following operations on languages:

- cocatenation L_1L_2 (already mentioned).
- iteration L^* (already mentioned).
- prefix closure $\overline{L} = \{v \in A^* \mid v \text{ is prefix of a word in } L\}$.
- postfix closure $post(L) = \{v \in A^* \mid \text{ a word in } L \text{ is prefix of } v\} = LA^*$.
- minimal words $min(L) = \{v \in L \mid \text{ no proper prefix of } v \text{ is in } L\}$.
- quotient $L_1/L_2 = \{v \in A^* \mid \exists w \in L_2 : \ vw \in L_1\}$.

- projection $P_B(L) = \{P_B(w) = P_B(x_1) \dots P_B(x_n) \mid w = x_1 \dots x_n \in L\}$, where $P_B(\epsilon) = \epsilon$ and

$$P_B(x) = \begin{cases} x \text{ for } x \in B, \\ \epsilon \text{ otherwise.} \end{cases}$$

- pumping $P_B^{-1}(L) = \{w \in (A \cup B)^* \mid P_B(w) \in L\}$.

The set of regular languages is closed under all these operations.

The Hiding Operator. For the alphabets of the form $A = 2^X$ as we consider, we need a more sophisticated projection operator, called *hiding* operator, to hide characters from subsets $Y \subseteq X$. We define the *hiding operator* λ_Y *w.r.t.* Y by:

- For a character $a \in A$:
  $\lambda_Y(a) = a \setminus Y$ if $a \setminus Y \neq \emptyset$, and $\lambda_Y(a) = \epsilon$ otherwise.
- For a word $w \in A^*$:
  $\lambda_Y(w) = \lambda_Y(a_1) \dots \lambda_Y(a_n)$ if $w = a_1 \dots a_n$, and $\lambda_Y(w) = \epsilon$ if $w = \epsilon$.
- For a language $L \subseteq A^*$:
  $\lambda_Y(L) = \{\lambda_Y(w) \mid w \in L\}$.

The hiding operator defines equivalence classes over A^* in the following way: For a $w \in A^*$ denote

$$[w]_Y = \{v \in A^* \mid \lambda_Y(w) = \lambda_Y(v)\}.$$

The set of regular languages is closed under the hiding and corresponding pumping operations. This can be seen by constructing from a given regular expression α two regular expressions $\lambda_Y(\alpha)$ and $ext_Y(\alpha)$ such that $L(\lambda_Y(\alpha)) = \lambda_Y(L(\alpha))$ and $L(ext_Y(\alpha)) = \lambda_Y^{-1}(L(\alpha))$:

Definition 12. *Let* X, Y *be two finite sets and* α *be a regular expression over the alphabet* $A = 2^X$.

(a) We construct a regular expression $\lambda_Y(\alpha)$ *over the alphabet* 2^Y *by replacing every character* $a \in 2^X$ *in* α *by* $\lambda_Y(a)$.

(b) We construct a regular expression $ext_Y(\alpha)$ *over the alphabet* $2^{X \cup Y}$ *by replacing every character* $a \in 2^X$ *in* α *by*

$$ext_Y(a) = \left(\sum_{b \in 2^Y} b\right)^* \left(\sum_{b \in 2^Y} a \cup b\right) \left(\sum_{b \in 2^Y} b\right)^*,$$

and by replacing the character ϵ *by*

$$ext_Y(\epsilon) = \left(\sum_{b \in 2^Y} b\right)^*.$$

Observe that by construction

$$
\begin{aligned}
ext_Y(\alpha_1 + \alpha_2) &= ext_Y(\alpha_1) + ext_Y(\alpha_2), \\
\lambda_Y(\alpha_1 + \alpha_2) &= \lambda_Y(\alpha_1) + \lambda_Y(\alpha_2), \\
ext_Y(\alpha_1\alpha_2) &= ext_Y(\alpha_1)ext_Y(\alpha_2), \\
\lambda_Y(\alpha_1\alpha_2) &= \lambda_Y(\alpha_1)\lambda_Y(\alpha_2), \\
ext_Y(\alpha^*) &= (ext_Y(\alpha))^*, \\
\lambda_Y(\alpha^*) &= (\lambda_Y(\alpha))^*.
\end{aligned}
$$

Lemma 1. *Let X, Y be finite sets and α be a regular expression over the alphabet $A = 2^X$.*

(a) Each $w \in \left(2^Y\right)^*$ satisfies

$$w \in \lambda_Y(L(\alpha)) \Leftrightarrow w \in L(\lambda_Y(\alpha)).$$

(b) Each $w \in \left(2^{X \cup Y}\right)^*$ satisfies

$$\lambda_Y(w) \in L(\alpha) \Leftrightarrow w \in L(ext_Y(\alpha)).$$

Proof. We will only show '(b):⇒'. The proofs of the other cases are similar argumentations using again structural induction over the construction rules of regular expressions: Let $\alpha = x \in 2^X \cup \{\epsilon\}$ (these are the constants of a regular expression over 2^X), and let $w \in (2^{X \cup Y})^*$ satisfying $\lambda_Y(w) = x$. Then w is of the form $w = x_1 \ldots x_n$ ($x_i \in 2^{X \cup Y}$), such that there exists an index k satisfying $\lambda_Y(x_k) = x$ and $\lambda_Y(x_i) = \epsilon$ for $i \neq k$. It follows immediately from the construction above, that $w \in L(ext_Y(\alpha))$. Let α_1 and α_2 be regular expressions over the alphabet 2^X satisfying the induction hypothesis, and let $\beta_1 = ext_Y(\alpha_1)$ and $\beta_2 = ext_Y(\alpha_2)$ be the corresponding extensions according to the above construction. We get for $w \in (2^{X \cup Y})^*$:

(i) Assume $\lambda_Y(w) \in L(\alpha)$ for $\alpha = \alpha_1 + \alpha_2$:
In this case $\lambda_Y(w) \in L(\alpha_1)$ or $\lambda_Y(w) \in L(\alpha_2)$. By induction hypothesis $w \in L(\beta_1)$ or $w \in L(\beta_2)$. This implies $w \in L(\beta_1 + \beta_2) = L(ext_Y(\alpha))$.

(ii) Assume $\lambda_Y(w) \in L(\alpha)$ for $\alpha = \alpha_1\alpha_2$:
There are $w_1, w_2 \in (2^{X \cup Y})^*$ satisfying $w = w_1w_2$ and $\lambda_Y(w_i) \in L(\alpha_i)$ ($i = 1, 2$). By induction hypothesis
$w = w_1w_2 \in L(\beta_1)L(\beta_2) = L(\beta_1\beta_2) = L(ext_Y(\alpha))$.

(iii) Assume $\lambda_Y(w) \in L(\alpha)$ for $\alpha = (\alpha^1)*$:
There are $w_1 \ldots w_n \in (2^{X \cup Y})^*$ satisfying $w = w_1 \ldots w_n$ and $\lambda_Y(w_i) \in L(\alpha_1)$ ($i = 1, \ldots, n$). By induction hypothesis $w = w_1 \ldots w_n \in (L(\beta_1))^* = L(\beta_1^*) = L(ext_Y(\alpha))$.

□

4.1 The Behavior of the Controlled Plant

We will formulate our approach language theoretically similarly as it is done in classical supervisory control. We will see, that despite the mentioned differences, some algorithms of classical supervisory control can at least be adapted to our framework. While omitting therefore most details of these algorithms, out paper remains selfcontained, i.e. can be understood without previous knowledge of supervisory control.

We search for a sublanguage K (of occurrence sequences) of the language $L_{\mathcal{P}}$ representing the behavior of the controlled plant, which can be realized by a composition of the plant module $\mathcal{P}$ and a control module $\mathcal{C}$. This implies the following requirements on K:

- If an occurrence sequence in K can be extended by a step of output transitions or unobservable transitions to an occurrence sequence in $L_{\mathcal{P}}$, then also this extended occurrence sequence should be in K. This follows the paradigm: "*What cannot be prevented, should be legal*".
- According to the unobservability of some events, some occurrence sequences in $L_{\mathcal{P}}$ cannot be distinguished by the control. As a consequence, following the paradigm "*what cannot be distinguished, cannot call for different control actions*", if an input is sent to the plant after a sequence w of steps has occurred, then the same input has to be send after occurrence of any other sequence, which is undistinguishable to w

Observe that the first condition correspond to the classical one in supervisory control. The second one is due to our step semantics, where an input can synchronize different unobservable and output transitions depending on the state of $\mathcal{P}$, in combination with the notion of *observability* in supervisory control. Such a sublanguage K is called controllable w.r.t. $L_{\mathcal{P}}$, I and O (figure 8):

Definition 13 (Controllable Language). *A prefix closed, regular sublanguage K of $L_{\mathcal{P}}$ is said to be* controllable w.r.t. $L_{\mathcal{P}}$, I and O, *if*

$$\forall w \in K, \forall o \in 2^{OUT} : \; wo \in L_{\mathcal{P}} \Rightarrow wo \in K, \tag{1}$$

$$\begin{aligned} \forall vj \in K,\, j \cap I \neq \emptyset, \\ \forall j',\, j \cap I = j' \cap I, \\ \forall v' \in K,\, \lambda_T(v) = \lambda_T(v') : \;\; v'j' \in L_{\mathcal{P}} \Rightarrow v'j' \in K. \end{aligned} \tag{2}$$

If the sets $L_{\mathcal{P}}$, I and O are clear, we simply call K controllable.

Observe that for each controllable sublanguage K of $L_{\mathcal{P}}$ two undistinguishable words in $L_{\mathcal{P}}$ are either both in K, or both not in K. This property is also called *normality* in supervisory control. In case each controllable event is also observable (as in our framework) every controllable sublanguage of $L_{\mathcal{P}}$ can be proved to be *normal.*

Lemma 2. *Let $K \subseteq L_{\mathcal{P}}$ be controllable. Then for each $w \in L_{\mathcal{P}}$ either $[w]_T \cap L_{\mathcal{P}} \subseteq K$ or $[w]_T \cap L_{\mathcal{P}} \cap K = \emptyset$.*

Proof. We prove the lemma by contradiction: let $v, w \in L_{\mathcal{P}}$ with $\lambda_T(w) = \lambda_T(v)$ and

$$w \in K \quad \text{and} \quad v \notin K.$$

First observe $v \neq \epsilon$, since $\epsilon \in K$. Assume without loss of generality

$$v = v'x \quad \text{with} \quad v' \in K.$$

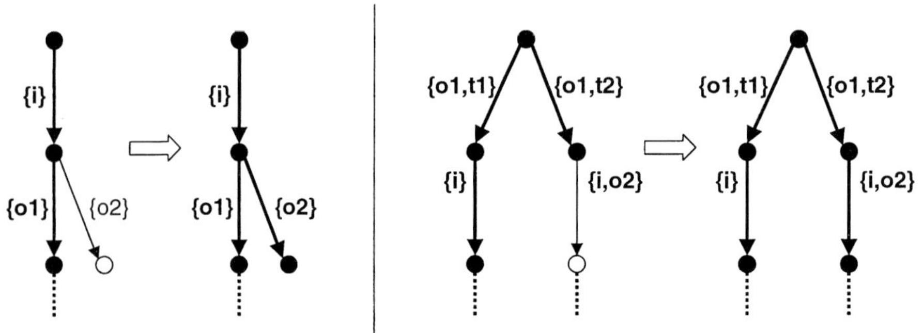

(a) Controllability (1) **(b)** Controllability (2)

Fig. 8. A step (sequence) consisting of transitions from T (I resp. O) are denoted by (indexed) t's (i's resp. o's). Part (a) (controllability condition (1)): After sending the input i, the output o_2 cannot be avoided. Part (b) (controllability condition (2)): The steps $\{o_1, t_1\}$ and $\{o_1, t_2\}$ cannot be distinguished from the control, because only the output o_1 is visible. Therefore sending the input i should be allowed either in both cases or in none case. Hereby it does not play a role, what effect has this input to the plant (i.e. whether it synchronizes another output or not).

By condition (1) $x \cap I \neq \emptyset$. From $\lambda_T(w) = \lambda_T(v)$ we deduce

$$w = w'y \quad \text{with} \quad \lambda_T(w') = \lambda_T(v') \wedge x \cap I = y \cap I.$$

This contradicts condition (2). □

Observe that the property of normality can also be written in the form

$$\lambda_T^{-1}(\lambda(K)) \cap L_{\mathcal{P}} = K.$$

In supervisory control there exists moreover the notion of observability of sublanguages $K \subseteq L_{\mathcal{P}}$, which essentially states that if the supervisor cannot distinguish between sequences of events (according to some unobservable events), these sequences need the same control action. Observe, that this concept is directly integrated into the above definitions.

As mentioned we are searching for a controllable K, which additionally respects L_c and is maximal with this property:

Definition 14 (Maximally Permissive Controllable Language). *Let L_c be a regular language over the alphabet 2^O and let $K \subseteq L_{\mathcal{P}}$ be controllable w.r.t $L_{\mathcal{P}}$, I and O satisfying*

$$\lambda_{T \cup I}(K) \subseteq \overline{L_c}.$$

We say that K is maximally permissive controllable *w.r.t. L_c, $L_{\mathcal{P}}$, I and O, if there exists no language K' satisfying $K \subsetneq K' \subseteq L$, which is controllable w.r.t. $L_{\mathcal{P}}$, I and O and fulfills $\lambda_{T \cup I}(K') \subseteq \overline{L_c}$.*

If the sets L_c, $L_{\mathcal{P}}$, I and O are clear, we simply call K maximally permissive controllable.

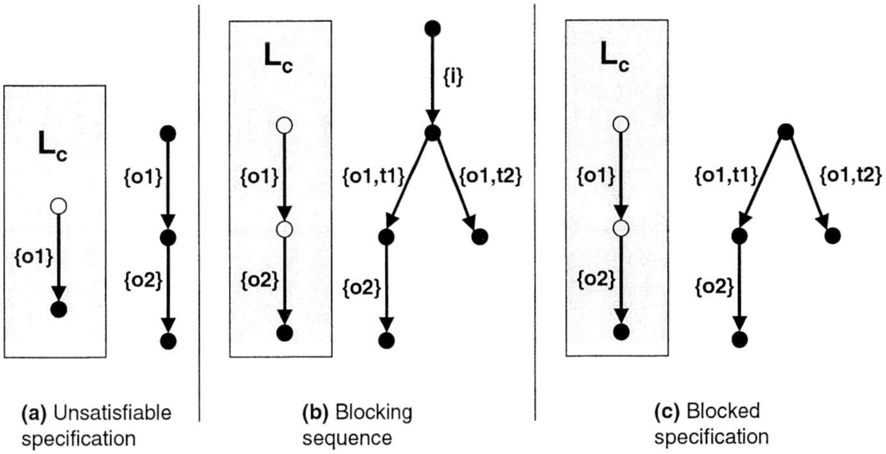

Fig. 9. A step (sequence) consisting of transitions from T (I resp. O) are denoted by (indexed) t's (i's resp. o's). The desired behavior is represented by finite automata, where accepting states are black. Part (a) (condition (3)): The occurrence sequence $\{o_1\}\{o_2\}$ is not a prefix of a word in L_c. So, L_c is unsatisfiable. Part (b) (condition (4)): The occurrence sequence $\{i\}\{o_1, t_2\}$ cannot be completed to a word respecting L_c. This blocking situation can only be avoided by not sending the input i before. Part (c) (condition (5)): The occurrence sequence $\{o_1, t_2\}$ cannot be completed to a word respecting L_c and cannot be avoided, i.e. L_c is only blocking satisfiable.

It is possible to get the result $K = \{\epsilon\}$ as maximally permissive controllable language, what means that the maximal behavior respecting the specification is empty, but there happens nothing wrong without inputs from outside. If even without any input the specification can be violated, we call L_c unsatisfiable (figure 9 (a)).

Definition 15. *L_c is said to be* unsatisfiable (w.r.t. $L_\mathcal{P}$, I and O), *if*

$$\exists w \in (2^{O \cup T})^* : w \in L_\mathcal{P} \wedge \lambda_T(w) \notin \overline{L_c}. \tag{3}$$

If this is not the case, we call L_c satisfiable (w.r.t. $L_\mathcal{P}$, I and O).

Consider a maximally permissive controllable language K: By definition every occurrence sequence in K is a prefix of an occurrence sequence respecting L_c. But it can happen there are such occurrence sequences that cannot be extended within K to an occurrence sequence respecting L_c, i.e. the desired behavior is blocked. We require additionally K to be nonblocking (figure 9 (b)):

Definition 16 (Nonblocking Language). *Let $K \subseteq L$ be maximally permissive controllable w.r.t L_c, $L_\mathcal{P}$, I and O and let $M \subseteq K$ be controllable w.r.t. $L_\mathcal{P}$, I and O satisfying*

$$\forall r \in M : \exists x \in (2^{O \cup I \cup T})^* \textit{ with } rx \in M,\, \lambda_{I \cup T}(rx) \in L_c. \tag{4}$$

We say that M is nonblocking controllable *w.r.t. L_c, $L_\mathcal{P}$, I and O. If it is maximal with this property, M is called* maximally permissive nonblocking controllable language *w.r.t. L_c, $L_\mathcal{P}$, I and O.*

If the sets L_c, $L_\mathcal{P}$, I and O are clear, we simply call K maximally permissive nonblocking controllable language.

In classical supervisory control some states in an automaton representing the behavior of the uncontrolled plant are marked. The sequences (words) of events leading to these states describe completed tasks. Nonblocking control requires that every sequence of the language of the controlled plant can complete to a marked state (no dead- or livelock occurs in unmarked states).

In comparison we define nonblocking w.r.t. the given (not prefix closed) specification L_c, thus specifying desired tasks also by L_c and not by an extra given set of states in a fixed automaton representing the uncontrolled behavior. As we will see later, one can construct an automaton recognizing $L_\mathcal{P}$, where the set words respecting L_c correspond to certain marked states.

Assume L_c is satisfiable and let K be the maximally permissive controllable. If $K \neq \{\epsilon\}$, it is possible to get $M = \{\epsilon\}$ as maximally permissive *nonblocking* controllable language, but only if $\epsilon \in L_c$. In this case the maximal nonblocking behavior is empty, but without inputs from outside no blocking state can be reached. If even without any input a blocking state can be reached, we call L_c blocked (figure 9 (c)).

Definition 17. *Let $M \subseteq L_\mathcal{P}$ be prefix closed. We refer to the the condition*

$$\exists w \in (2^{OUT})^* : w \in M \wedge (\forall x \in (2^{OUIUT})^* : wx \in M \Rightarrow \lambda_{T \cup I}(wx) \notin L_c), \quad (5)$$

as blocking condition *w.r.t. M.*

Remark 2. If the condition (5) is fulfilled w.r.t. a language M, L_c is said to be *blocked* w.r.t. M.

Lemma 3. *If L_c is blocked w.r.t. $L_\mathcal{P}$, then there is no nonblocking controllable sublanguage of $L_\mathcal{P}$.*

In the next two paragraphs we synthesize the maximally permissive nonblocking controllable language M, if it exists. First we examine the case, when L_c is prefix closed. In this case the maximally permissive controllable sublanguage of $L_\mathcal{P}$ is already nonblocking. In particular safefy properties can be formalized via a prefix closed specification L_c.

Safety Properties

Safety properties specify undesired behavior, that should not happen (for example forbidden states of the system). If some undesired behavior is realized by an occurrence sequence, the whole possible future of this occurrence sequence is undesired, too. In other words: If an occurrence sequence realizes no undesired behavior, all prefixes also do so. That means, safety properties can be formalized by prefix closed specification languages. On the other hand every prefix closed specification language can be regarded as the formalization of safety properties.

The searched (control) language M as defined in the last paragraph is computed in several steps. First we define the (potentially safe) language L_{psafe} as the set of all occurrence sequences of $L_\mathcal{P}$ respecting L_c.

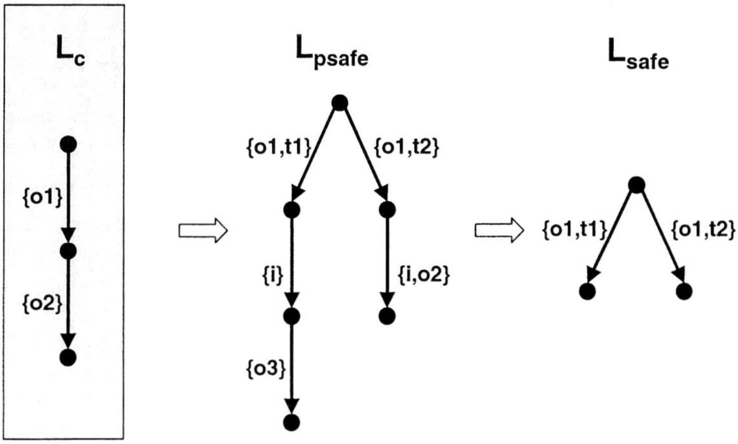

Fig. 10. The languages are represented by finite automata, where accepting states are black (observe that all languages are prefix closed). Sending the input i after observing the output o_1 can cause the not avoidable output o_3. This gives an occurrence sequence not respecting L_c. Therefore allways after observing o_1 the input i should not be sent.

Definition 18. *We define*

$$L_{psafe} = \{w \in L_{\mathcal{P}} \mid \lambda_{I \cup T}(w) \in L_c\} = \lambda_{I \cup T}^{-1}(L_c) \cap L_{\mathcal{P}},$$
$$L_{unsafe} = \{w \in L_{\mathcal{P}} \mid \lambda_{I \cup T}(w) \notin L_c\} = L_{\mathcal{P}} \setminus L_{psafe}.$$

Observe that

$$L_c \text{ unsatisfiable } \Leftrightarrow \exists w \in L_{unsafe} \cap (2^{OUT})^*.$$

L_{psafe} is only a first approximation to M, since it is in general not controllable. In particular it may contain occurrence sequences which are not closed under extensions by outputs (condition (1) in definition 13). Such occurrence sequences must be cut at the last possible input (the last possibility of control), if there is one. Due to condition (2) (of controllability) such an input must be avoided at all undistinguishable places. The words ending with these inputs are collected in the language L_{danger}. Deleting the futures of occurence sequences in L_{danger} from L_{psafe} gives the language L_{safe}, which we will prove below to be the searched language M (figure 10).

Definition 19. *We define*

$$\begin{aligned} L_{danger} = \{vj \in L_{psafe} \mid\ & j \cap I \neq \emptyset, \\ & \exists v' : \lambda_T(v') = \lambda_T(v), \\ & \exists j' : j' \cap I = j \cap I, \\ & \exists y \in (2^{T \cup O})^* : v'j'y \in L_{unsafe})\}. \end{aligned}$$

$$L_{safe} = L_{psafe} \setminus post(L_{danger}).$$

If L_c is satisfiable, every word from L_{unsafe} has a nonempty prefix in L_{danger}. It is obvious from the definitions of L_{psafe}, L_{unsafe}, L_{danger} and L_{safe} that every set $[w]_T \cap L_{\mathcal{P}}$ is either subset of or disjoint to these languages (see also lemma 2).

The main result of this subsection is the following theorem:

Theorem 1. *L_{safe} is maximally permissive nonblocking controllable w.r.t. L_c, $L_{\mathcal{P}}$, I and O, if L_c is satisfiable w.r.t. $L_{\mathcal{P}}$, I and O.*

Before proving this theorem we give an algorithm to compute L_{safe}: It is essentially shown, that L_{safe} can be constructed by appropriate operations on regular languages. We want to remark here that for computing the maximally permissive controllable language also the more sophisticated framework presented in [2] could be adapted (since our different notion of controllability is still compatible with the union operation $\cup$). In [2] can also be found some hints to the complexity of the computation.

By definition L_{psafe} and L_{unsafe} are regular (see section 4):

Lemma 4. *L_{psafe} and L_{unsafe} are regular.*

Next we show that L_{danger} is regular. We will give L_{danger} as a simple formula over the regular languages $(2^{OUT})^*$ and L_{unsafe}. First observe that the regular language

$$L^{real}_{danger} = (L_{unsafe}/(2^{OUT})^*) \cap L_{\mathcal{P}},$$

where the symbol "/" denotes the quotient operation on languages, is the set of those words $vj \in L_{danger}$, which themselves can be extended by an $y \in (2^{OUT})^*$ to a word in L_{unsafe}. The remaiming words in L_{danger} are of the form $v'j'$ with $\lambda_T(v') = \lambda_T(v)$ and $j' \cap I = j \cap I$ for a word $vj \in L^{real}_{danger}$. We get these words by means of a special defined hiding operator $\overline{\lambda}$ defined by

$$v \in (2^{OUIUT})^*, \ x \in 2^{OUIUT} : \ \overline{\lambda}(vx) = \lambda_T(v)\lambda_{T\cup O}(x).$$

Obviously the operators $\overline{\lambda}$ and $(\overline{\lambda})^{-1}$ preserve the regularity of languages, since this is the case for the hiding operator λ as argued in lemma 1 (section 4). We get

Lemma 5. *L_{danger} and L_{safe} are regular.*

Proof. L_{danger} is regular, since it can be constructed by regularity preserving operations in the following way:

$$L_{danger} = (\overline{\lambda})^{-1}(\overline{\lambda}(L^{real}_{danger})) \cap L_{psafe}.$$

Then also $L_{safe} = L_{psafe} \setminus post(L_{danger})$ is regular as a formula over regular languages. □

The main theorem 1 now is shown in two steps by the following lemmata.

Lemma 6. *Let L_c be satisfiable. Then L_{safe} is controllable.*

Proof. We show both conditions of controllability by contradiction:

(i) Condition (1):
Assume there is an word $w \in L_{safe}$ and a step $o \in 2^{OUT}$ satisfying $wo \in L_{\mathcal{P}}$, but $wo \notin L_{safe}$. There are two cases:
  - $wo \in L_{psafe}$:
    Then $wo \in post(L_{danger})$. This imlies obviously $w \in post(L_{danger})$, what contradicts $w \in L_{safe}$.
  - $wo \notin L_{psafe}$:
    Then by definition $wo \in L_{unsafe}$. Since L_c is satisfiable w.r.t. $L_{\mathcal{P}}$, I and O, wo has a prefix in L_{danger}. This again contradicts $w \in L_{safe}$.

(ii) Condition (2):
Assume there are words $v', vj \in L_{safe}$ and a step j' with $j \cap I = j' \cap I \neq \emptyset$ and $\lambda_T(v') = \lambda_T(v)$ satisfying $v'j' \in L_{\mathcal{P}}$, but $v'j' \notin L_{safe}$. For such vj and $v'j'$ we have according to the definition of L_{danger}:

$$vj \in post(L_{danger}) \Leftrightarrow v'j' \in post(L_{danger}).$$

From this it follows $vj \notin L_{safe}$ analogously to the first case. A contradiction.

□

Lemma 7. *Let L_c be satisfiable. There is no language $K \subseteq L_{\mathcal{P}}$ satisfying $L_{safe} \subsetneq K$, which is controllable and fulfills $\lambda_{I \cup T}(K) \subseteq \overline{L_c}$.*

Proof. We choose a $w \in K \setminus L_{safe}$ and construct from w a word $w' \in K$ satisfying $\lambda_{I \cup T}(w') \notin \overline{L_c}$. As $w \in L_{\mathcal{P}}$, there are two cases:

- $w \notin L_{psafe}$:
  Then $w \in L_{unsafe}$ and thus $\lambda_{I \cup T}(w) \notin L_c$.
- $w \in L_{psafe}$:
  Then $w \in post(L_{danger})$, i.e. w has a prefix $vj \in L_{danger}$. That means, there is are words $v' \in L_{\mathcal{P}}, y \in (2^{OUT})^*$ and a step j' with $j \cap I = j' \cap I$ and $\lambda_T(v) = \lambda_T(v')$ such that $v'j'y \in L_{unsafe}$, i.e. $\lambda_{I \cup T}(v'j'y) \notin L_c$. Since K is controllable, $v'j'$ also belongs to K (condition (2)) and consequently $v'j'y \in K$ (condition (1)).

□

It follows immedeately from the above proof, that L_{safe} is the unique maximally permissive language, analogously to related results in supervisory control.

Nonblocking Control

More general properties as for example the full execution of certain tasks cannot be formalized by a regular language L_c which is prefix closed. Of course a maximally permissive controllable language K w.r.t. a not prefix closed L_c should contain occurrence sequences of the standalone of $L_{\mathcal{P}}$ which represent prefixes of words in L_c, but only such ones, which can be extended to a word in L_c within K, i.e. which are nonblocking.

We now search for a sublanguage L_{nbsafe} of L_{safe}, which is controllable and respecting L_c, nonblocking and maximal with these two properties according to definition 16. As mentioned, in our framework every controllable event is also observable. Therefore, we are able to adapt a result in supervisory control ([2], subsection 3.7.5), which states (under the assumption that every controllable event is also observable): If there is at least one controllable language respecting L_c which is nonblocking, then there is a unique maximal one.

In order to compute L_{nbsafe}, we collect all blocking occurrence sequences of L_{safe} in the set $L_{blocking}$ (observe that every future of a blocking occurrence sequence is blocking, too). We have to cut all occurrence sequences in this set at the last possible input, if there is one. Due to condition (2) (of controllability) such an input must be avoided at all undistinguishable places. The prefixes ending with these inputs are collected in the language $L_{badchoice}$. Deleting the futures of occurrence sequences in $L_{badchoice}$ possibly produces new blocking words. Therefore we have to iterate this procedure. We define (figure 11)

Definition 20. *Let $M \subseteq L_{safe}$. Denote*

$$\begin{aligned} M_{blocking} &= \{w \in M \mid \nexists x \in (2^{O \cup I \cup T})^* : wx \in M \wedge \lambda_{I \cup T}(wx) \in L_c\}, \\ &= M \setminus ((M \cap \lambda_{I \cup T}^{-1}(L_c))/(2^{O \cup I \cup T})^*), \end{aligned}$$

$$\begin{aligned} M_{badchoice} = \{vj \in M \mid\ & j \cap I \neq \emptyset, \\ & \exists v' : \lambda_T(v') = \lambda_T(v), \\ & \exists j' : j' \cap I = j \cap I, \\ & \exists y \in (2^{T \cup O})^* : v'j'y \in min(M_{blocking})\}. \end{aligned}$$

The language $M_{blocking}$ is regular by definition, if M is regular. In analogy to L_{danger}, then $M_{badchoice}$ is regular, too. Observe that

$$L_c \text{ blocked w.r.t. } M \Leftrightarrow \exists w \in M_{blocking} \cap (2^{O \cup T})^*.$$

We are now prepared to state the algorithm to compute L_{nbsafe}:

Input: Language $M^0 = L_{safe}$, Integer $k = 0$.

Step 1:
Compute $M^k_{blocking}$.

Step 2:
If $M^k_{blocking} \cap (2^{O \cup T})^* \neq \emptyset$: **return** "*$L_{nbsafe}$ does not exist*".
If $M^k_{blocking} = \emptyset$: **return** M^k.

Step 3:
Compute $M^k_{badchoice}$.
$M^{k+1} = M^k \setminus post(M^k_{badchoice})$.
Set $k = k + 1$.
Goto Step 1.

Starting with $M^0 = L_{safe}$ the algorithm iteratively deletes blocking words by cutting them at the last possible inputs (and by additionally cutting all undistinguishable words). This is done until either no new blocking words are produced (in which case L_{nbsafe} is found) or an L_c is blocked w.r.t. the actually computed language (in which

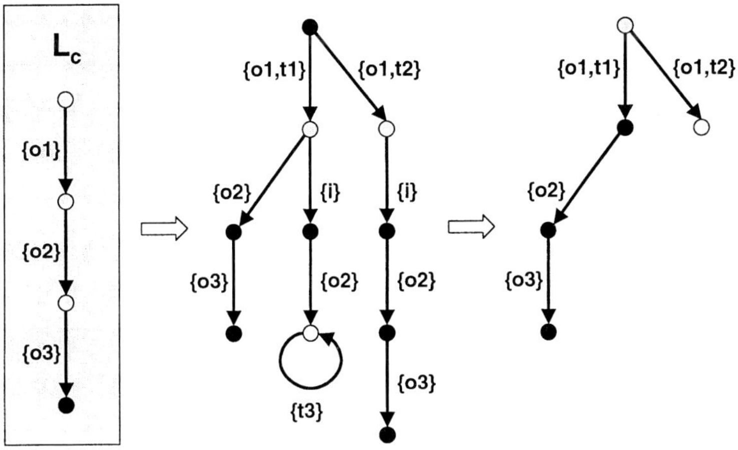

Fig. 11. The black states in the automaton representing L_c are the accepting states. The white states in the other automata (representing languages computed in the nonblocking algorithm) are blocking w.r.t. the given L_c. In the grey states there is the last possibility not to send an input in order to avoid the blocking situation. The input must be avoided at all undistinguishable states. This can cause new blocking situations, which can be even not avoidable.

case no controllable nonblocking language exists). All computed languages M^k are controllable and normal, but possibly not nonblocking. Observe that if $M^k_{blocking} \cap (2^{OUT})^* = \emptyset$, then each word in $M^k_{blocking}$ has a nonempty prefix in $M^k_{badchoice}$ and therefore does not belong to M^{k+1}.

Before stating the main result, namely that this algorithm returns L_{nbsafe} if and only if a maximally permissive nonblocking controllable sublanguage exists, we have to verify, that the algorithm allways terminates. For completeness we will give a sketch of the proof below. A detailed proof can be found for example in [3]: the algorithm presented there only slightly differs from ours. The following procedure is repeated: *It first iteratively deletes blocking words by cutting them at the last possible inputs, without additionally cutting all undistinguishable words. This is also done until no new blocking word is found. The resulting language is controllable and nonblocking, but not normal. Then all cuts done so far are also realized for all undistinguishable words, which yields a controllable and normal language, which is possibly not nonblocking.* The whole procedure in repeated until the resulting language is nonblocking. Both algorithms have the same output.

The main idea for showing the termination is to find a deterministic finite automaton $G = (S, (2^{I \cup OUT})^*, \delta, F, s_0)$ recognizing L_{safe}, such that *deleting words* from $post(M^k_{badchoice})$ (in the algorithm) corresponds to *deleting edges* in G. A nessecary *and* sufficient condition for this is that the states of G distinguish words in $M^k_{badchoice}$ from words not in $M^k_{badchoice}$, i.e. (see also Figure 13)

$$\delta(s_0, w) = \delta(s_0, v) \Rightarrow (w \in M^k_{badchoice} \Leftrightarrow v \in M^k_{badchoice}). \tag{6}$$

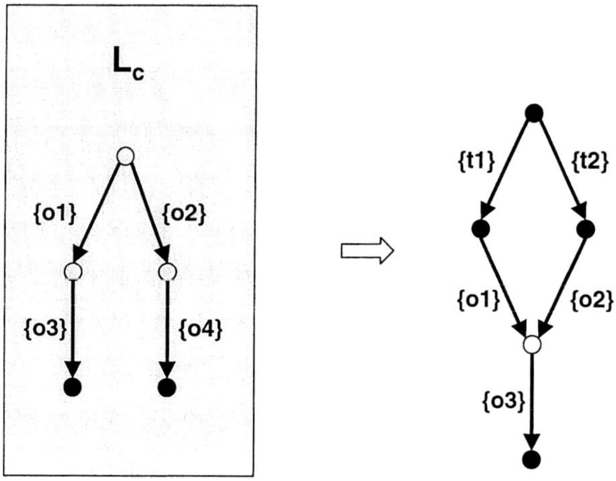

Fig. 12. In general two words $v = \{t_2\}\{o_2\} \notin M^k_{blocking}$ and $w = \{t_1\}\{o_1\} \in M^k_{blocking}$ can have the same follower state in an automaton A recognizing L_{safe}. For the automaton G implementing the nonblocking algorithm however we require that one can distinguish between "*blocking*-states" and "*not blocking*-states". Such states can be splitted appropriately by synchronizing A with an automaton recognizing $\lambda^{-1}_{T\cup I}(L_c)$.

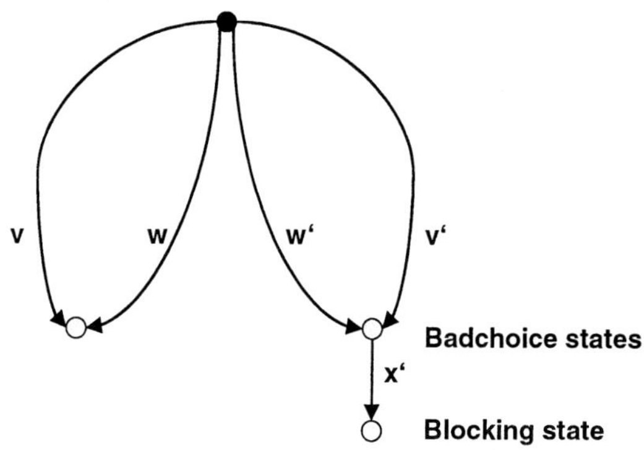

Fig. 13. In general two words $v \notin M^k_{badchoice}$ and $w \in M^k_{badchoice}$ can have the same follower state in an automaton A recognizing L_{safe}. For the automaton G implementing the nonblocking algorithm however we require that one can distinguish between "*badchoice*-states" and "*not badchoice*-states": If there is a w' undistinguishable from w leading to a blocking state, there must also be a word v' undistinguishable from v leading to a blocking state. G is even choosen such that v' and w' have the same follower state.

Taking $v, w \in L_{safe}$ with $\delta(s_0, w) = \delta(s_0, v)$ and $w \in M^k_{badchoice}$, we know that there is a $w' \in M^k_{badchoice}$ with $\lambda_T(w) = \lambda_T(w')$ and there is a $x' \in (2^{OUT})^*$ with $w'x' \in M^k_{blocking}$. To prove (6) it suffices to find $v' \in M^k_{badchoice}$ with $\lambda_T(v) = \lambda_T(v')$ such that $\delta(s_0, v') = \delta(s_0, v)$ and $v'x' \in M^k_{blocking}$ (Figure 13). In other words, the possible futures of words undistinguishable from v and the possible futures of words undistinguishable from w should be the same.

In general a finite automaton A recognizing L_{safe} does not fulfill this property, i.e. does not distinguish by its states words, for which the possible futures of their undistinguishable words are not the same. Such states have to be split appropriately by synchronizing A with another automaton B. B can be constructed from A in such a way that a state $[w]_B$ of B is defined as the set of exactly those states of A which are follower states of words undistinguishable from w:

$$s \in [w]_B \Rightarrow \exists w' \in L_{safe} : \lambda_T(w) = \lambda_T(w') \wedge s = [w']_A.$$

Formally B can be constructed from A in three steps. For this let $(S_A, \Sigma_A, (2^{I \cup O \cup T}))$ be the labelled transition system associated to A:

(1) Replace each edge $s \xrightarrow{x} s' \in \Sigma_A$ by $s \overset{\lambda_T(x)}{\rightarrow} s'$ (this yields in particular $s \xrightarrow{\epsilon} s'$ for $x \subseteq T$). The result is a so called nondeterministic ϵ-automaton. By definition of such automata ([11]) the possible follower states of a word $z \in (2^{I \cup O})^*$ in this automaton are exactly the follower states of words $w \in L_{safe}$ in A with $\lambda_T(w) = z$.
(2) Compute the deterministic finite automaton simulating the nondeterministic ϵ-automaton from (1) by the well-known subset construction ([11]): Then exactly the sets of possible follower states of a word $z \in (2^{I \cup O})^*$ in the above ϵ-automaton define the states of this deterministic automaton.
(3) Pump the automaton from (2) by steps of unobservable transitions from 2^T in the following way: For all states s and all $x \subseteq T$ (loop) transitions $s \xrightarrow{x} s$ are added. For all transition $s \xrightarrow{y} s'$ with $y \subseteq I \cup O$ and forall $x \subseteq T$ transitions $s \overset{x \cup y}{\rightarrow} s'$ are added.

Finally we have to require the automaton A to distinguish words in $M^k_{blocking}$ from words not in $M^k_{blocking}$ by its states (which is not the case in general, see Figure 12):

$$\forall w, \forall v \in [w]_A : w \in M^k_{blocking} \Leftrightarrow v \in M^k_{blocking}, \tag{7}$$

This can be achieved by building A as the synchronized product of the minimal automata recognizing L_{safe} and $\lambda^{-1}_{I \cup T}(L_c)$. Then A fulfills

$$\forall v \in [w]_A : \lambda_{I \cup T}(w) \in L_c \Leftrightarrow \lambda_{I \cup T}(v) \in L_c. \tag{8}$$

It can be seen as follows, that then property (7) is also satisfied: Take $v \in [w]_A$, $v \neq w$. Assume $w \notin M^k_{blocking}$. There is a $x \in (2^{I \cup O \cup T})^*$, such that $wx \in M^k$ and $\lambda_{I \cup T}(wx) \in L_c$. Since $vx \in [wx]_A$ it follows from property (8), that $\lambda_{I \cup T}(vx) \in L_c$, i.e. $v \notin M^k_{blocking}$.
Let us state the main theorem of this subsection:

Theorem 2. *There exists a maximally permissive nonblocking controllable sublanguage of L_{safe}, if and only if the previous algorithm returns a language L_{nbsafe}. In this case L_{nbsafe} is this searched sublanguage.*

Proof. Let $L_{safe} = M^0, \ldots, M^{N_0}$ be the sequence of languages the algorithm has computed until it has stopped.

We first show the "only if"-part:
Assume the previous algorithm outputs "L_{nbsafe} *does not exist*". We have to show, that there is no maximally permissive nonblocking controllable sublanguage of L_{safe}. We will show this by contradiction: Assume there is a controllable language $M \subseteq L_{safe}$ with $M_{blocking} = \emptyset$. Observe that

$$M \cap M^0_{blocking} \subseteq M_{blocking},$$

i.e. the assumption in particular implies

$$M \cap M^0_{blocking} = \emptyset. \tag{9}$$

By **Step 1** M^{N_0} fulfills

$$\exists v_0 \in M^{N_0}_{blocking} \cap (2^{OUT})^*.$$

From the controllability of M (condition (1)) we deduce $v_0 \in M$. From (9) it follows $v_0 \notin M^0_{blocking}$. That means v_0 can be extended by some word $y \neq \epsilon$ (remark that $\lambda_{T\cup I}(v_0) \notin L_c$!) to a word $v_0 y \in M^0$ with $\lambda_{I\cup T}(v_0 y) \in L_c$. By the assumption one of these extensions $v_0 y_0$ must be in M:

$$v_0 y_0 \in M \quad \text{and} \quad \lambda_{I\cup T}(v_0 y_0) \in L_c.$$

Since $v_0 \in M^{N_0}_{blocking}$, we have moreover $v_0 y_0 \notin M^{N_0}$. By construction (Step 2) there must be an index $N_1 < N_0$, such that

$$v_0 y_0 \in post(M^{N_1}_{badchoice}).$$

Let $v_0 x i \in M^{N_1}_{badchoice}$ for a prefix xi of y_0 with $i \cap I \neq \emptyset$. By definition of $M^{N_1}_{badchoice}$ there is a $v_1 = v_0' x' i' y \in M^{N_1}_{blocking}$ with

(a) $\lambda_T(v_0' x') = \lambda_T(v_0 x)$,
(b) $i' \cap I = i \cap I$, and
(c) $y \in (2^{OUT})^*$.

Remember now that all prefixes of $v_0 y_0$, in particular $v_0 x$ and $v_0 x i$, belong to M. Since M is assumed to be controllable, M contains all words in $L_{\mathcal{P}} \cap [v_0 x]_T$ (lemma 2). In particular $v_0' x' \in M$ (property (a)). From the condition (2) (of controllability) and (b) we get further $v_0' x' i' \in M$, and therefore $v_1 = v_0' x' i' y \in M$ (condition (1) (of controllability) and (c)).

By repeating this construction we get an strictly decreasing sequence of natural numbers $N_0 > N_1 > \ldots$ and associated words $v_0, v_1, \ldots \in M$, such that $v_i \in M^{N_i}_{blocking}$, $i = 0, 1, \ldots$. Finally $N_k = 0$ for some k, which implies $v_k \in M_{blocking}$, what contradicts our assumption.

Next we consider the "if"-part:
By construction $M^{N_0} = L_{nbsafe}$ is controllable and nonblocking. It remains to show that it is maximally permissive with these two properties. We show this statement by contradiction. Assume another language M to be controllable and nonblocking controllable satisfying $L_{nbsafe} \subsetneq M \subseteq L_{safe}$. In particular, by assumption $M_{blocking} = \emptyset$.

There is a $x \in M \setminus M^{N_0}$. As M^{N_0} and M are prefix closed we can assume (without loss of generality) that x is of the form

$$x = wj \in M,\ w \in M^{N_0},\ j \cap I \neq \emptyset.$$

Since $wj \notin M^{N_0}$, for some step $N_1 < N_0$ it holds $wj \in M^{N_1}_{badchoice}$. By definition of $M^{N_1}_{badchoice}$ there is a $v_0 = w'j'y \in M^{N_1}_{blocking}$ with

(a) $\lambda_T(w') = \lambda_T(w)$,
(b) $j' \cap I = j \cap I$, and
(c) $y \in (2^{OUT})^*$.

As above, since M is assumed to be controllable, we follow $v_0 \in M$. By assumption, as in the 'only if'-part, $v_0 \notin M^0_{blocking}$. Therefore v_0 must have an extension within M to a word respecting L_c. Let v_0y_0 be this extension of v_0. Proceed now as in the "only if"-part. □

4.2 Synthesis of Control Modules

In this subsection we show how to synthesize a control module $\mathcal{C}$ from a given behavior $L_{cb} \subseteq L_{\mathcal{P}}$ of $\mathcal{P}$ and to compose this module with $\mathcal{P}$, such that the resulting composed module has exactly this behavior up to transitions of $\mathcal{C}$ which are not in $I \cup O$, whenever possible. Of course, as a first necessary condition, we have to require L_{cb} to be a prefix closed regular language, since the set of occurrence sequences of a module has this property. Formally a control module $\mathcal{C}$ w.r.t. such a language L_{cb} is defined as follows:

Definition 21. *Let $\mathcal{C}$ be a module of a signal net with the set of transitions $T_{\mathcal{C}}$ and denote $U = T_{\mathcal{C}} \setminus (I \cup O)$. Then $\mathcal{C}$ is the* control module *of $\mathcal{P}$ w.r.t. L_{cb}, if there is a composition mapping Ω, such that the set of all occurrence sequences $L_{\mathcal{CP}}$ of the module $\mathcal{C} *_{\Omega} \mathcal{P}$ satisfies $\lambda_U(L_{\mathcal{CP}}) = L_{cb}$.*

We claim that for the existence of such a control module it is sufficient to require L_{cb} to be *controllable* (see definition 13). This gives the main theorem of this subsection:

Theorem 3. *If $L_{cb} \subseteq L_{\mathcal{P}}$ is a regular, controllable, prefix closed language, then there is a control module $\mathcal{C}$ of $\mathcal{P}$ w.r.t. L_{cb}.*

In practice this statement can be applied to $L_{cb} = L_{safe}$ or $L_{cb} = L_{nbsafe}$. We prove the theorem by constructing $\mathcal{C}$. The main idea is to synthesize $\mathcal{C}$ by adding new net structure to $\mathcal{E}$ (see Figure 7). In particular $\mathcal{C}$ is composed with $\mathcal{P}$ via the connections (given by Ω) between $\mathcal{P}$ and $\mathcal{E}$.

For the construction we use a deterministic finite automaton $A = (S, 2^{I \cup O}, \delta, F, s_0)$ recognizing $\lambda_T(L_{cb})$. We denote $\Sigma = \{s \xrightarrow{x} s' \mid \delta(s,x) = s'\}$ the set of edges of A and $l : \Sigma \to 2^{I \cup O}$, $l(s \xrightarrow{x} s') = x$, the labelling of Σ.

Remember that L_{cb} is prefix closed. Without loss of generality we assume that

(a) All states are accepting states: $F = S$. Just omit all edges leading to non-accepting states.

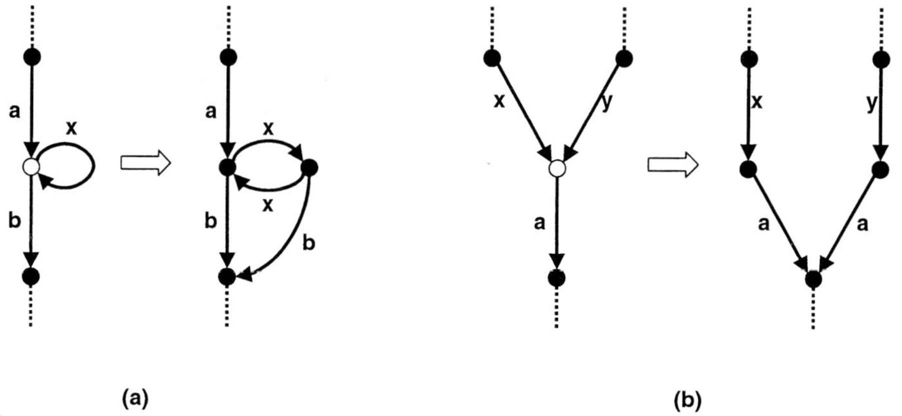

Fig. 14. Part (a): splitting states to avoid loops. Part (b): splitting states to distinguish words according to their last character.

(b) There are no loops in A: $s \xrightarrow{x} s \notin \Sigma$. If this is not the case for a state of A you can think of splitting this state into two copies and thus transforming the loop into a cycle of length 2 (see Figure 14, (a)).
(c) The states of A distinguish words according to their last character: $s' \xrightarrow{x} s, s'' \xrightarrow{y} s \in \Sigma \Rightarrow x = y$. As long as this is not the case for a state s of A, i.e. $x \neq y$, you can think of splitting s into two copies, one for words ending with x and one for words ending with y (see Figure 14, (b)). Hence we get $l(s') = l(s \xrightarrow{x} s') = x$ for $s \xrightarrow{x} s' \in \Sigma$.

Formally (b), (c) can be achieved by synchronizing A with appropriate other finite automata.

We will construct a signal net $N = (P, U \cup I \cup O, F, EN, CN, m_0)$, where P is the set of places, $U \cup I \cup O$ is the set of transitions, F is the flow arc relation, EN is the event arc relation, CN is the context arc relation, and m_0 is the initial marking. N together with input/output structure of $\mathcal{E}$, will give the searched module $\mathcal{C}$. For simplicity we will use two kinds of context arcs: Usual positive context arcs, called *condition arcs* in the context of signal nets, which test places for presence of tokens, and negative context arcs, also called *inhibitor arcs* in literature ([12]), which test places for absence of tokens. It is well known that in elementary nets negative context arcs can be equivalently replaced by a structure using positive context arcs and so-called *co-places* ([12], see Figure 15). So CN splits into the set of condition arcs CN^+ and inhibitor arcs CN^-. We modify some notions for the enabling of steps w.r.t. CN^-: For a transition t we denote $^+t = \{p \mid (p,t) \in CN^+\}$ and $^-t = \{p \mid (p,t) \in CN^-\}$. Given a set $\xi \subseteq T$ of transitions, we extend the above notions to: $^+\xi = \bigcup_{t \in \xi} {}^+t$ and $^-\xi = \bigcup_{t \in \xi} {}^-t$.

Definition 22 (Potentially enabled for negative context). *A step* ξ *is* potentially enabled *in a marking* m *if*

– *All* $t \in \xi$ *are enabled:* $^\bullet t \cup {}^+t \subseteq m$, $^-t \cap m = \emptyset$ *and* $(t^\bullet \setminus {}^\bullet\xi) \cap m = \emptyset$ *and*
– *All pairs* $t, t' \in \xi$ *are not in conflict:* $^\bullet t \cap {}^\bullet t' = t^\bullet \cap (t')^\bullet = \emptyset$.

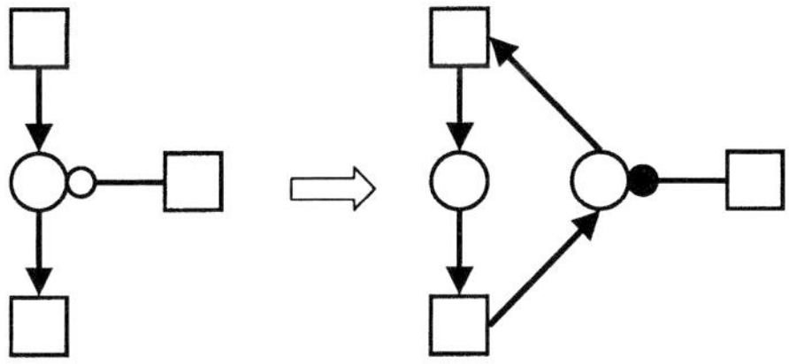

Fig. 15. Introducing coplaces to translate inhibitor arcs into condition arcs.

We set

- $P = \{p_s \mid s \in S\}$.
- $U_{OR} = \{t_s^{empty} \mid s \in S\} \cup \{t_s^{fill} \mid s \in S\} \cup \{t_{s,i} \mid \exists s \xrightarrow{x} s' \in \Sigma : i \in x \cap I\}$.
- $U_{AND} = \{t_{s \xrightarrow{x} s'} \mid s \xrightarrow{x} s' \in \Sigma\}$.

Our aim is to identify each state $s \in S$ of A with a unique set of places $P_s \subseteq P$, and each edge $s \xrightarrow{x} s' \in \Sigma$ of A with a unique set of transitions $\xi_{s \xrightarrow{x} s'} \subseteq U \cup I \cup O$, such that

- $\xi_{s \xrightarrow{x} s'}$ is enabled if and only if exactly the places in P_s are marked,
- $\xi_{s \xrightarrow{x} s'}$ together with an appropriate set of transitions of T build a step in $\mathcal{C} *_\Omega \mathcal{P}$, and
- ${}^\bullet\xi_{s \xrightarrow{x} s'} = P_s$ and $\xi^\bullet_{s \xrightarrow{x} s'} = P_{s'}$.

The idea is the following: Assume that $\mathcal{C}$ is in the state P_s. Then for each $s \xrightarrow{x} s' \in \Sigma$ it should be possible to send the input $i \in x \cap I$ (if there is one) to the plant synchronizing a step in the plant (in the case of event inputs) or switching a condition input on/off. In the first case this step in the plant should synchronize the step of outputs $x \cap O$ sent from the plant. If there is no input $i \in x \cap I$, there should be a step in the plant synchronizing the step of outputs $x \cap O$ sent from the plant. Let x be such a set, and let $i \in x \cap I$. Since there are in general also states in which i is not allowed to be sent to the plant, we model the transition $t_{s,i}$ in such a way that

- $t_{s,i}$ is enabled exactly under the marking P_s via condition and inhibitor arcs (Figures 18 and 20), and
- $t_{s,i}$ synchronizes the transition i (Figure 18).

A transition $t_{s \xrightarrow{x} s'}$ is intended to simulate the step of signals x in the control module $\mathcal{C}$, if $\mathcal{C}$ is in state P_s. Therefore $t_{s \xrightarrow{x} s'}$

- is enabled exactly under the marking P_s via condition and inhibitor arcs (Figures 16 and 20),
- synchronizes the transition t_s^{empty} which is intended to empty exactly the places in P_s (Figure 16),
- synchronizes all transitions $t_{s''}^{fill}$, which are intended to mark the places $p_{s''}$ in the follower marking $P_{s'}$ (Figure 16), and
- is synchronized by all outputs in $x \cap O$ and by $t_{s,i}$ for $i \in x \cap I$ (Figures 17 and 18).

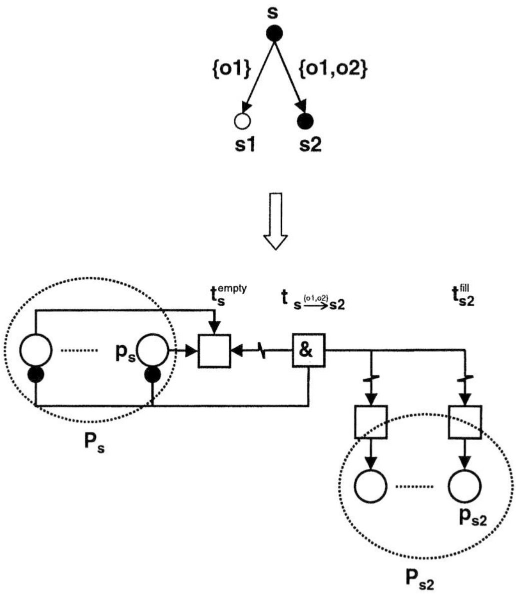

Fig. 16. For clearness we model only the transition $s \overset{\{o_1,o_2\}}{\longrightarrow} s_2$ of the above behavior (the black states). The transition $t_{s\overset{\{o_1,o_2\}}{\longrightarrow}s_2}$ simulates the edge $s \overset{\{o_1,o_2\}}{\longrightarrow} s_2$. $t_{s\overset{\{o_1,o_2\}}{\longrightarrow}s_2}$ synchronizes t_s^{empty} to empty the marking P_s, and synchronizes exactly the transitions t_p^{fill} with $p \in P_{s_2}$ (observe in particular $p_s \in P_s$ and $p_{s_2} \in P_{s_2}$). Moreover $t_{s\overset{\{o_1,o_2\}}{\longrightarrow}s_2}$ tests the places in P_s to be marked. The event arc connections to input and output transitions are omitted (see Figure 17).

The second part of the last condition is necessary, since in general in the same state the same step of outputs can occur spontaneous or can be initiated by an input. Observe that, if C is in state P_s, a step of outputs $x \cap O$ synchronizes beside the transition $t_{s\overset{x}{\longrightarrow}s'}$ also each transition $t_{s\overset{y}{\longrightarrow}s''}$ with $y \subset x$. See figure 19.

We are now able to define the sets $\xi_{s\overset{x}{\longrightarrow}s'}$ and $P_{s'}$:

- From the event arc relation we deduce (Figure 19) $\xi_{s\overset{x}{\longrightarrow}s'} = \{t_{s,i} \mid i \in x \cap I\} \cup x \cup \{t_{s\overset{y}{\longrightarrow}s''} \mid y \subseteq x\} \cup \{t_s^{empty}\} \cup \{t_{s''}^{fill} \mid p_{s''} \in P_{s'}\}$.
- Because $t_{s\overset{x}{\longrightarrow}s'}$ synchronizes $t_{s'}^{fill}$) the place $p_{s'}$ belongs to $P_{s'}$. For every $s \in S$ with $s \overset{x}{\longrightarrow} s' \in \Sigma$ the step of inputs and outputs x also synchronizes beside the transition $t_{s\overset{x}{\longrightarrow}s'}$ each transition $t_{s\overset{y}{\longrightarrow}s''}$ with $y \subseteq x$. Therefore we get in such cases $P_{s''} \subseteq P_{s'}$. This procedure has to be applied recursively. Therefore we define $P_{s'}$ to be the smallest set satisfying (Figure 20)
  (i) $p_{s'} \in P_{s'}$.
  (ii) $\forall s \overset{x}{\longrightarrow} s', s \overset{y}{\longrightarrow} s'' \in \Sigma$ $(s' \neq s'')$ with $y \subset x$: $P_{s''} \subseteq P_{s'}$.

Altogether we get formally:

- $F = \{(\mathbf{p}, \mathbf{t}_s^{\mathbf{empty}}) \mid s \in S, p \in P_s\} \cup \{(\mathbf{t}_s^{\mathbf{fill}}, \mathbf{p_s}) \mid s \in S\}$,
- $CN^+ = \{(\mathbf{p}, \mathbf{t}_{\mathbf{s}\overset{\mathbf{x}}{\longrightarrow}\mathbf{s}'}) \mid s \in S, p \in P_s\} \cup \{(\mathbf{p}, \mathbf{t_{s,i}}) \mid s \in S, p \in P_s\}$,

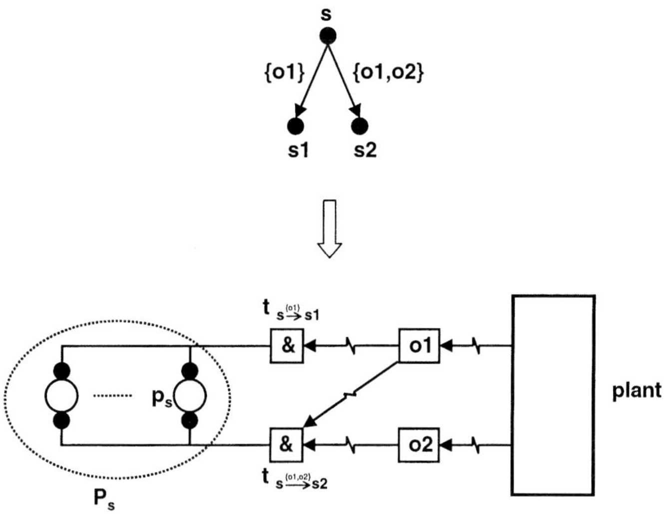

Fig. 17. The transition $t_{s \xrightarrow{\{o_1,o_2\}} s_2}$ simulates the edge $s \xrightarrow{\{o_1,o_2\}} s_2$. It is therefore synchronized by the output transitions o_1 and o_2. The occurrence of o_1 together with o_2 synchronizes also $t_{s \xrightarrow{\{o_1\}} s_1}$, since $t_{s \xrightarrow{\{o_1\}} s_1}$ is also enabled under P_s. The occurrence of o_1 without o_2 only synchronizes $t_{s \xrightarrow{\{o_1\}} s_1}$. For clearness the connections to empty- and fill-transitions and places in the follower marking are omitted (see Figure 16), but observe that this implies $p_{s_1} \in P_{s_1} \subset P_{s_2}$. (see also Figure 20).

- $CN^- = \{(\mathbf{p}, \mathbf{t}_{s \xrightarrow{x} s'}) \mid \exists \bar{s} \in S :\ P_s \subset P_{\bar{s}} \wedge p \in P_{\bar{s}} \setminus P_s\} \cup \{(\mathbf{p}, \mathbf{t}_{s,i}) \mid \exists \bar{s} \in S :\ P_s \subset P_{\bar{s}} \wedge p \in P_{\bar{s}} \setminus P_s\}$, and
- $EN = \{(\mathbf{t}_{s,i}, \mathbf{i}) \mid s \in S,\ i \in I\} \cup \{(\mathbf{t}_{s,i}, \mathbf{t}_{s \xrightarrow{x} s'}) \mid s \in S,\ i \in x \cap I\} \cup \{(\mathbf{o}, \mathbf{t}_{s \xrightarrow{x} s'}) \mid s \in S,\ o \in x \cap O\} \cup \{(\mathbf{t}_{s \xrightarrow{x} s'}, \mathbf{t}^{\mathbf{fill}}_{s''}) \mid s \xrightarrow{x} s' \in \Sigma,\ p_{s''} \in P_{s'}\} \cup \{(\mathbf{t}_{s \xrightarrow{x} s'}, \mathbf{t}^{\mathbf{empty}}_{s}) \mid s \in S\}$,

Remark 3. Observe that $p_{s'} \in P_s$ implies either $s' = s$ or $l(s') \subset l(s)$. This implies that for all $s, s' \in S$: $p_s \notin P_{s'}$ and/or $p_{s'} \notin P_s$.

Lemma 8. *(a) The mapping $\phi : S \to 2^P$, $\phi(s) = P_s$ is injective.*
(b) $(\xi_{s \xrightarrow{x} s'} \setminus x)^\bullet = P_{s'}$ and ${}^\bullet(\xi_{s \xrightarrow{x} s'} \setminus x) = P_s$.
(c) $\xi_{s \xrightarrow{x} s'} \setminus x$ is potentially enabled in P_s.
(d) $\xi_{s \xrightarrow{x} s'}$ is maximal w.r.t. U and P_s: For each transition $t \in U \setminus \xi_{s \xrightarrow{x} s'}$ with ${}^{\leadsto}t \cap \xi_{s \xrightarrow{x} s'} \neq \emptyset$ the set $\xi_{s \xrightarrow{x} s'} \cup \{t\}$ is not potentially enabled in P_s.

Proof.
ad (a): see remark 3.

ad (b): Follows from

$$(\xi_{s \xrightarrow{x} s'} \setminus x)^\bullet = \bigcup_{p_{s''} \in P_{s'}} (t^{fill}_{s''})^\bullet, \quad {}^\bullet(\xi_{s \xrightarrow{x} s'} \setminus x) = {}^\bullet(t^{empty}_{s}).$$

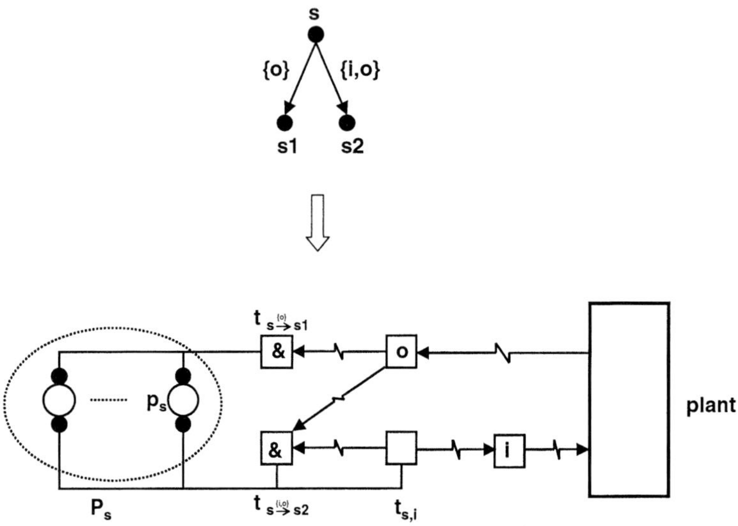

Fig. 18. Here the label of the edge $s \overset{\{i,o\}}{\rightarrow} s_2$ contains an input transition i. This input transition is synchronized by the spontaneous transition $t_{s,i}$, which is exactly enabled under P_s. If i synchronizes the output transition o via transitions in the plant, the transition $t_{s \overset{\{i,o\}}{\rightarrow} s_2}$ is synchronized together with the transition $t_{s \overset{\{o\}}{\rightarrow} s_1}$ (analogously to Figure 17). It is also possible that the plant sends the output o without the input i. In this case $t_{s \overset{\{o\}}{\rightarrow} s_1}$ is synchronized alone. Observe that no cycles of event arcs are produced.

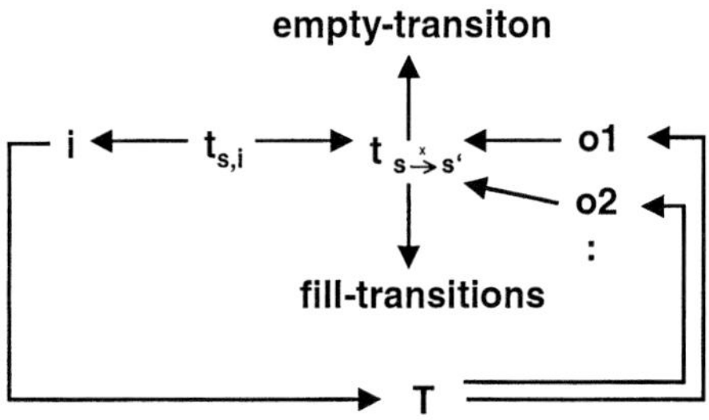

Fig. 19. The event arc relation w.r.t. an edge $s \overset{\{x\}}{\rightarrow} s' \in E$. The event arc from i to T is optional: it exists only, if i replaces an event input signal, and not in the case i is an on- or off-transition of an condition imput signal.

ad (c): By definition of F, CN^- and CN^+, the only sets of places of the form ${}^\bullet t$, $t^\bullet$ or ${}^+t$ for $t \in \xi_{s \overset{x}{\rightarrow} s'} \setminus x$ which are not empty are:

$$^+t_{s \overset{x}{\rightarrow} s'} = {}^+t_{s,i} = P_s, \ {}^\bullet t_s^{empty} = P_s, \ (t_{s''}^{fill})^\bullet = \{p_{s''}\}.$$

This gives ${}^\bullet t \cup {}^+t \subseteq P_s$. Moreover, ${}^-t \cap P_s = \emptyset$ by definition. Finally, from (b) we get $(t^\bullet \setminus {}^\bullet\xi_{s \overset{x}{\rightarrow} s'}) \cap P_s = (t^\bullet \setminus P_s) \cap P_s = \emptyset$. By this the first part of the *potentially enabled* definition 22 is fulfilled.

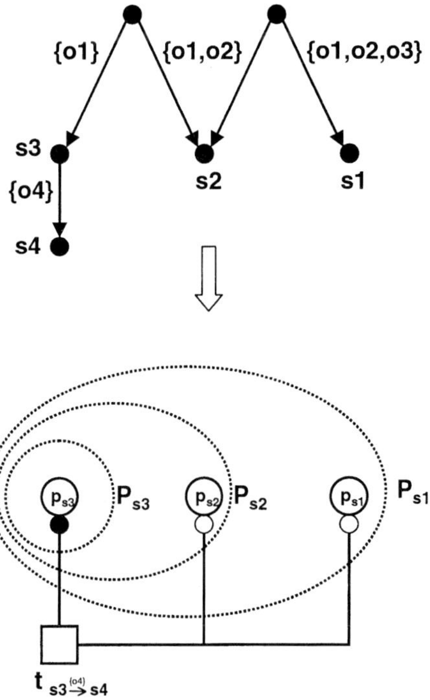

Fig. 20. By definition we get $P_{s_3} \subset P_{s_2} \subset P_{s_1}$. The transition $t_{s_3 \overset{\{o_4\}}{\rightarrow} s_4}$ has to test the places in P_{s_3} for the presence of tokens and the places in $P_{s_1} \setminus P_{s_3}$ for the absence of tokens.

It remains to verify that there are no conflicts w.r.t. pre- resp. postsets in $\xi_{s \xrightarrow{x} s'} \setminus x$: The only transition in $\xi_{s \xrightarrow{x} s'} \setminus x$ with nonempty preset is t_s^{empty}. So there are no conflicts w.r.t. presets. The only transitions with nonempty postsets in $\xi_{s \xrightarrow{x} s'} \setminus x$ are of the form $t_{s''}^{fill}$ for $p_{s''} \in P_{s'}$. All postsets of such transitions consist of a unique place, and so are pairwise distinct.

ad (d): Each transition $t \in U \setminus \xi_{s \xrightarrow{x} s'}$ with ${}^{\rightsquigarrow}t \cap \xi_{s \xrightarrow{x} s'} \neq \emptyset$ is of the form (see figure 19) $t = t_{\overline{s} \xrightarrow{x} \overline{s}'}$. There are two cases:

- $\overline{s} \neq s$:
  From $P_s \neq P_{\overline{s}}$ we deduce that t is not enabled. Therefore $\xi_{s \xrightarrow{x} s'} \cup \{t\}$ is not potentially enabled.
- $\overline{s} = s$ and $l(\overline{s}') \not\subseteq l(s')$:
  That means ${}^{\rightsquigarrow}t \not\subseteq \xi_{s \xrightarrow{x} s'}$. Since $t \in U_{AND}$, $\xi_{s \xrightarrow{x} s'} \cup \{t\}$ is not a step. □

We are now going to prove $\lambda_U(L_{\mathcal{CP}}) = L_{cb}$. We need some additional notions: For an occurrence sequence $w = x_1 \dots x_n$ of $\mathcal{E} *_{\Omega} \mathcal{P}$ we denote

- $w_i = x_1 \dots x_i$,
- m_i the marking of $\mathcal{E} *_{\Omega} \mathcal{P}$ after the occurrence of w_i,
- $s_i = \delta(s_0, \lambda_T(w_i))$ the state in A after executing $\lambda_T(w_i)$.

Observe $s_\epsilon = s_0$ and $x_i \cap (I \cup O) = \emptyset \Leftrightarrow s_{i-1} = s_i$. For $x_i \cap (I \cup O) \neq \emptyset$ we get $l(s_i) = \lambda_T(x_i)$. Define

$$\eta_i = \begin{cases} x_i \cup \xi_{s_{i-1} \xrightarrow{x} s_i} & \text{if } x = x_i \cap (I \cup O) \neq \emptyset, \\ x_i & \text{else.} \end{cases}$$

For a set of transitions σ and a marking m of $\mathcal{C} *_\Omega \mathcal{P}$ we denote

- $\sigma^\mathcal{C} = \lambda_T(\sigma)$ and $m^\mathcal{C} = m \cap P$ the $\mathcal{C}$-parts, and
- $\sigma^{\mathcal{EP}} = \lambda_U(\sigma)$ and $m^{\mathcal{EP}} = m \setminus P$ the $\mathcal{E} *_\Omega \mathcal{P}$-parts.

Observe that $\eta_i^{\mathcal{EP}} = x_i$, $\eta_i^\mathcal{C} = \xi_{s_{i-1} \xrightarrow{x} s_i}$ resp. $= \emptyset$, $(m_i \cup P_{s_i})^\mathcal{C} = P_{s_i}$ and $(m_i \cup P_{s_i})^{\mathcal{EP}} = m_i$. Between the net structure in module $\mathcal{E} *_\Omega \mathcal{P}$ and the additional net structure in $\mathcal{C} *_\Omega \mathcal{P}$ there are only event arc connections. There are no events arc connections between transitions in T and transitions in U. Therefore:

- σ is potentially enabled in $\mathcal{C} *_\Omega \mathcal{P}$ under the marking m *if and only if* both $\sigma^\mathcal{C}$ is potentially enabled in $\mathcal{C}$ under the marking $m^\mathcal{C}$ and $\sigma^{\mathcal{EP}}$ is potentially enabled in $\mathcal{E} *_\Omega \mathcal{P}$ under the marking $m^{\mathcal{EP}}$
- σ is maximal w.r.t. $T \cup U$ and m in the sense of lemma 8 (d) *if and only if* both $\sigma^\mathcal{C}$ is maximal w.r.t. U and $m^\mathcal{C}$ and $\sigma^{\mathcal{EP}}$ is maximal w.r.t. T and $m^{\mathcal{EP}}$ in the sense of lemma 8 (d).

Putting this together, observe that σ is an enabled step in $\mathcal{C} *_\Omega \mathcal{P}$ under the marking m *if and only if* σ is a step and $\sigma^\mathcal{C}$ and $\sigma^{\mathcal{EP}}$ are potentially enabled and maximal as above.

Lemma 9. $\lambda_U(L_{\mathcal{CP}}) \supseteq L_{cb}$.

Proof. We show by induction on the length of $w = x_1 \ldots x_n \in L_{cb}$:

(A) The occurrence sequence $\eta = \eta_1 \ldots \eta_n$ is enabled in $\mathcal{C} *_\Omega \mathcal{P}$ under the marking $m_0 \cup P_{s_0}$.
(B) The occurrence of η gives the follower marking $m_n \cup P_{s_n}$.
(C) $x_1 \ldots x_n x_{n+1} \in L_{cb}$ implies that the step η_{n+1} is enabled in $\mathcal{C} *_\Omega \mathcal{P}$ under the marking $m_n \cup P_{s_n}$.

First let $w = \epsilon$ $(n = 0)$: (A) and (B) are clear. Ad (C): Observe that in each case the $\mathcal{C}$- and $\mathcal{EP}$-parts of η_1 are potentially enabled and maximal in the above sense. According to the above considerations, it remains to show that η_1 is a step. We distinguish three cases:

(i) $x_1 \subseteq T$:
$\eta_1 = x_1$ is clearly a step.
(ii) $x_1 \subseteq T \cup O$ and $x_1 \cap O \neq \emptyset$:
x_1 builds a step in $\mathcal{E} *_\Omega \mathcal{P}$ which includes the set of transitions $x = l(s_1)$. From the transitions in x the transition $t_{s_0 \xrightarrow{x} s_1}$ is synchronized. The transition $t_{s_0 \xrightarrow{x} s_1}$ synchronizes the empty and fill transitions of $\xi_{s_0 \xrightarrow{x} s_1}$ (see figure 19).
(iii) $x_1 \cap I \neq \emptyset$:
analogously to (ii).

Let $w = x_1 \dots x_{n-1}x_n$. By assumption the statements $(A) - (C)$ are valid for w_{n-1}. This implies (A) for w. Statement (B) is clear (since the markings are unions of the $\mathcal{C}$- and $\mathcal{EP}$-parts). (C) can be seen analogously to the above argumentation.

Lemma 10. $\lambda_U(L_{\mathcal{CP}}) \subseteq L_{cb}$.

Proof. To see the statement, the controllability of L_{cb} will play the crucial role: Observe that $\lambda_U(L_{CP}) \subseteq L_{\mathcal{P}}$. Assume there is $\sigma = \sigma_1 \dots \sigma_n \sigma_{n+1} \in \lambda_U(L_{CP}) \setminus L_{cb}$. Without loss of generality $\sigma_1 \dots \sigma_n \in L_{cb}$. Denote $s = \delta(s_0, \lambda_T(\sigma_1 \dots \sigma_n))$ and $s' = \delta(s_0, \lambda_T(\sigma))$.

(i) $\sigma_{n+1} \subseteq T \cup O$:
According to the first condition (1) of controllability it follows $\sigma \in L_{cb}$. A contradiction.

(ii) $\sigma_{n+1} \cap I \neq \emptyset$:
Let $\sigma_{n+1} \cap I = \{i\}$. The sequence $\lambda_T(\sigma_1) \dots \lambda_T(\sigma_n)\lambda_T(\sigma_{n+1}) = \lambda_T(\sigma_1) \dots \lambda_T(\sigma_n)l(s')$ is a path in A from s_0 to s'. That means there is a word $x_1 \dots x_n x_{n+1} \in L_{cb}$ with $\lambda_T(x_i) = \lambda_T(\sigma_i)$ $(i = 1, \dots, n+1)$. In particular we have $\lambda_{T \cup O}(x_{n+1}) = \lambda_O(l(s'))$, since every step contains at most one input. According to the second condition (2) of controllability it follows $\sigma \in L_{cb}$. A contradiction.

□

5 Conclusion

In this paper we have presented a methodology for synthesis of the controlled behavior of discrete event systems employing actuators which try to force events and sensors which can prohibit event occurrences. As a modelling formalism, we have used modules of signal nets. The signal nets offer a direct way to model typical actuators behavior. Another advantage of such modules consists in supporting input/output structuring, modularity and compositionality in an intuitive graphical way.

In the paper we were not focusing on complexity issues. It is known that the complexity of the supervisory control problem is in general PSPACE-hard, and sometimes even undecidable ([21], pp. 15 - 36). To get efficient algorithms one has to restrict the setting in some way, for example by considering only very special kinds of specifications.

As the main result of the paper, we have shown how to synthesize the control module from the behavior of the controlled plant under the paradigm, that outputs of the plant cannot force inputs of the plant via the control module. This paradigm of course (structurally) restricts the class of modules which can be used as control modules. It would be interesting to discuss a generalization of this concept, where the composition of a control module with a plant module is not restricted. That means, in any pair of composed modules both modules can be considered as the control module symmetrically, or even more generally both modules can be considered to control each other. For sake of simplicity, we have restricted the control specification over set of outputs. We are presently working on extension of our methodology for the control specifications including input signals. The methodology for the specifications over observable states (i.e. condition output signals) is also an interesting subject of the further research.

The presented approach considers only Petri nets on a very elementary level. For complex industrial-size systems, these nets tend to be either very large or too abstract. In particular, data and time aspects can not be modelled in a natural way. Therefore, we are working on extension of modules of signal nets by special high-level Petri net features.

References

1. B. Caillaud, P. Darondeau, L. Lavagno and X. Xie (Eds.). Synthesis and Control of Discrete Event Systems Kluwer Academic Press, 2002.
2. C. G. Cassandras and S. Lafortune. Introduction to Discrete Event Systems. Kluwer, 1999.
3. H. Cho and S.I. Marcus. On supremal languages of classes of sublanguages that arise in supervisor synthesis problems with partial observation. em Mathematics of Control, Signals, and Systems, Vol. 2, No. 2, pp. 47-69, 1989.
4. J. Desel, G. Juhás and R. Lorenz. Input/Output Equivalence of Petri Modules. In *Proc. of IDPT 2002*, Pasadena, USA, 2002.
5. P. Dietrich, R. Malik, W.M. Wonham and B.A. Brandin. Omlementation Consideration in Supervisory Control. In [1].
6. H.-M. Hanisch, A. Lüder: Modular Modeling of Closed-Loop Systems, Colloquium on Petri Net Technologies for Modelling Communication Based Systems, Berlin 1999, pp. 103-126.
7. H-M. Hanisch and A. Lüder. A Signal Extension for Petri nets and its Use in Controller Design. *Fundamenta Informaticae*, 41(4) 2000, 415–431.
8. H.-M. Hanisch, A. Lüder, M. Rausch: Controller Synthesis for Net Condition/Event Systems with Incomplete State Observation, European Journal of Control, Nr. 3, 1997, S. 292-303.
9. H.-M. Hanisch, J. Thieme and A. Lüder. Towards a Synthesis Method for Distributed Safety controllers Based on Net Condition/Event Systems. *Journal of Intelligent Manufacturing*, 5, 1997, 8, 357-368.
10. L.E. Holloway, B.H. Krogh and A. Giua. A Survey of Petri Net Methods for Controlled Discrete Event Systems. *Discrete Event Dynamic Systems: Theory and Applications*, 7, 1997), 151–190.
11. J.E. Hopcroft, R. Motwani and J.D.Ullman. Introduction to Automata Theory, Languages, and Computation. Addison Wesley, 2001.
12. R. Janicki and M. Koutny. Semantics of Inhibitor Nets. *Information and Computations*, 123, pp. 1–16, 1995.
13. G. Juhás. On semantics of Petri nets over partial algebra. In J. Pavelka, G. Tel and M. Bartosek (Eds.) *Proc. of 26th Seminar on Current Trends in Theory and Practice of Informatics SOFSEM'99*, Springer, LNCS 1725, pp. 408-415, 1999.
14. G. Juhás and R. Lorenz. Modelling with Petri Modules. In [1].
15. L.E. Pinzon, M.A. Jafari, H.-M. Hanisch and P. Zhao Modelling admissible behavior using event signals submitted
16. P.J. Ramadge, W.M. Wonham: The Control of Discrete Event Systems. Proceedings of the IEEE, 77 (1989) 1, S. 81-98.
17. G. Rozenberg, and J. Engelfriet. Elementary Net Systems. In W. Reisig and G. Rozenberg (Eds.) *Lectures on Petri Nets I: Basic Models*, Springer, LNCS 1491, pp. 12-121, 1998.
18. R.S. Sreenivas und B.H. Krogh. On Condition/Event Systems with Discrete State Realizations. *Discrete Event Dynamic Systems: Theory and Applications*, 2, 1991, 1, 209–236.
19. R. S. Sreenivas and B. H. Krogh. Petri Net Based Models for Condition/Event Systems. In *Proceedings of 1991 American Control Conference*, vol. 3, 2899–2904, Boston, MA, 1991.

20. P. H. Starke. Das Komponieren von Signal-Netz Systemen. In Proc 7. Workshop Algorithmen und Werkzeuge für Petrinetze AWPN 2000, Universität Koblenz - Landau, pages 1–6, 2000.
21. P. Darondeau and S. Kumagai (Eds.). Proceedings of the *Workshop on Discrete Event System Control*. Satellite Workshop of ATPN 2003.
22. M.C. Zhou und F. DiCesare. *Petri Net Synthesis for Discrete Event Control of Manufacturing Systems*. Kluwer Adacemic Publishers, Boston, MA, 1993.
23. Zhonghua Zhang and W.M. Wonham. STCT: An Efficient Algorithm for Supervisory Control Design. In [1].

Application of Coloured Petri Nets in System Development

Lars Michael Kristensen*, Jens Bæk Jørgensen, and Kurt Jensen

Department of Computer Science, University of Aarhus
IT-parken, Aabogade 34, DK-8200 Aarhus N, Denmark
{lmkristensen,jbj,kjensen}@daimi.au.dk

Abstract. Coloured Petri Nets (CP-nets or CPNs) and their supporting computer tools have been used in a wide range of application areas such as communication protocols, software designs, and embedded systems. The practical application of CP-nets has also covered many phases of system development ranging from requirements to design, validation, and implementation. This paper presents four case studies where CP-nets and their supporting computer tools have been used in system development projects with industrial partners. The case studies have been selected such that they illustrate different application areas of CP-nets in various phases of system development.

1 Introduction

System development and engineering [73] is a complex task involving a multitude of activities such as analysis, requirement engineering, design, implementation, and testing. Several approaches to system development have been suggested and described in the literature such as the classical waterfall approach [44] and the newer, iterative Rational Unified Process (RUP) [61]. One universal technique that can be used across many of the activities in system development is *modelling*. The act of constructing a model of the system to be developed is typically done in early phases of system development, and is also known from other disciplines, e.g., when engineers construct bridges and architects design buildings. The main benefit of modelling is that it provides insight about the properties of the system prior to implementation. This allows many issues about the system to be resolved in the requirements and design phase rather than in the implementation phase. Many modelling languages have been suggested and are being used for system development. The most prominent example is the Unified Modeling Language (UML) [69,78] which is the de-facto standard modelling language of the software industry and which supports modelling of the structure and behaviour of systems.

CP-nets [47,48,50,58] is a graphical modelling language suited for modelling concurrency, synchronisation, and communication in systems. Prototypical application domains of CP-nets and Petri nets are communication protocols, data

* Supported by the Danish Natural Science Research Council.

J. Desel, W. Reisig, and G. Rozenberg (Eds.): ACPN 2003, LNCS 3098, pp. 626–685, 2004.

networks, embedded systems, and other types of reactive systems. CP-nets and Petri nets are, however, also applicable more generally for modelling systems where concurrency and communication are key characteristics. Examples of this are business process/workflow modelling and manufacturing systems.

The CPN modelling language combines Petri nets and programming languages. Petri nets [24, 77] provide the foundation of the graphical notation and the semantical foundation for modelling concurrency, synchronisation, and communication in systems. The functional programming language Standard ML [86] provides the primitives for compactly modelling the sequential aspects of systems (such as data manipulation) and for creating compact and parameterisable models. CP-nets have a module concept allowing CPN models to be organised into several modules (called pages). The module concept is hierarchical, allowing a module to have a number of submodules and allowing a set of modules to be composed to form new modules. This enables the modeller to work both top-down and bottom-up when constructing CPN models. CPN models can be timed, meaning that the time taken by different events in the system can be modelled. This means that CP-nets can be used to investigate both logical and functional properties such as absence of deadlocks, and performance properties such as execution times and queue lengths.

The CPN modelling language is supported by two computer tools: CPN Tools and Design/CPN. The Design/CPN tool [25] was developed in 1989 and is now being replaced by the next generation of tool support: CPN Tools [22]. The CPN computer tools support construction of CPN models including syntax check, type checking, and simulation (execution) of CPN models. Editing and simulation of the CPN models are done directly on the graphical representation of CP-nets. It is also possible to animate the system behaviour using a number of graphical libraries [13,75]. These libraries can be used on top of the CPN models to display graphics specific to the application domain. The basic idea in this behavioural animation is to have the CPN model display the evolution of the system using other graphical means such as, e.g., message sequence charts [9, 13].

The CPN computer tools support state space (reachability) analysis [48] of CPN models. The basic idea in state spaces is to calculate all reachable states and state changes of the system and represent these as a directed graph. The state space of a CPN model can be used to verify a number of properties of the system under consideration. A number of state space reduction methods [15, 16, 49] are also available in the computer tools for alleviating the state explosion problem [88], i.e., the fact that the number of reachable states can be large for complex systems. The computer tools also allow the performance of the system to be analysed based on simulation.

This paper presents four projects where CP-nets and their supporting computer tools have been used in system development. The four projects make it evident that CP-nets can be used in many phases of system development. CP-nets is however not a modelling language designed to replace other modelling languages (such as UML). In our view it should be used as a supplement to existing modelling languages and methodologies. CP-nets are suited for modelling and analysing behaviour in concurrent and distributed systems – an aspect where many other modelling languages, and in particular UML, are weak. While UML sequence- and collaboration diagrams are widely used to describe examples of system behaviour, the UML diagrams available for modelling behaviour in a general way, i.e., UML state machines and activity diagrams, are more rarely used. They have a number of limitations, and, in many cases, there are substantial technical reasons to prefer CP-nets over, e.g., UML state machines. The latter lack a well-defined execution semantics, do not support modelling of multiple instances of classes, and do not scale well to large systems [30,55]. CP-nets may be seen as a convenient supplement to the well-established UML diagram types such as sequence diagrams and class diagrams. On the other hand, CP-nets are not suited for giving purely static descriptions of system architecture and structure.

Another characteristic of the CPN modelling language is that it is general instead of domain specific, i.e., it is not aimed directly at modelling a specific class of systems, but aimed towards a very broad class of systems that can be characterised as concurrent and distributed. This is also evident in that the CPN language has few, but powerful modelling primitives that make it possible to model systems and concepts at different levels of abstraction. This is both a weakness and a strength of the CPN modelling language. The capability of CP-nets to model systems at different levels of abstraction is one of the keys to making formal analysis (e.g., state space analysis) of such models tractable, as large and very detailed models will usually be intractable for state space analysis. Finding the different abstraction levels that are useful at different points in systems development and more generally finding the right abstraction level is one of the arts of modelling. Finally, the CPN modelling language is able to describe large and complex systems. The use of a full programming language (Standard ML) gives CP-nets a scalability at the modelling level that cannot be found in low-level Petri nets.

Below we give a brief introduction to the four projects presented in this paper. The presented CPN models have all been constructed in joint projects between the CPN group [23] at the University of Aarhus and industrial partners.

Modelling Scenarios in Ad Hoc Networking. This joint project [57] with Ericsson Telebit A/S [33] was concerned with network architectures for integrating stationary core networks and mobile ad-hoc networks. The presented CPN model was developed in an early phase of the project to specify the network architecture itself and the mobility and communication scenarios to be supported by the communication protocols to be developed in later phases. CPN modelling was hence used to formalise the problem domain and for specifying requirements for the later implementation. This application of CP-nets is presented in Sect. 2.

Modelling Requirements in Pervasive Healthcare. This joint project [53] with Systematic Software Engineering A/S [84] and Aarhus County Hospital was concerned with specifying the business processes at Aarhus County Hospital and their support by a new IT system. The CPN model was used to engineer requirements for the system. Input from nurses was crucial in this process. The project demonstrated how application-specific graphics driven by underlying CPN models can be used to visualise system behaviour and to discuss requirements with people who are not familiar with the CPN modelling language. This application of CP-nets in presented in Sect. 3.

State Space Analysis of an Audio/Video Protocol. This joint project [14] with Bang and Olufsen A/S [5] was concerned with the design of the communication protocols to be used in the next generation of the B & O BEOLINK system. The presented CPN model was used to specify the new lock management protocol, and state space analysis was used to validate and analyse the protocol. The project took place in 1995-1996 when only very basic state space analysis was available in the CPN computer tools. Since then, a number of new state space methods have been developed and implemented in the CPN computer tools. A revised CPN model of the lock management protocol is presented in Sect. 4, together with the application of the state space methods currently available in the CPN computer tools.

Implementation of a Planning Tool. This joint project [94] with the Australian Defence Science and Technology Organisation (DSTO) [4] was concerned with the development of the Course of Action Scheduling Tool (COAST). CPN modelling has been used to conceptualise and formalise the planning domain to be supported by the COAST tool. Furthermore, the constructed CPN model has been extracted in executable form from the CPN computer tools and embedded into the server of the COAST tool together with a number of state space analysis algorithms. This project demonstrated how a constructed CPN model can be used for the implementation of a computer tool by effectively bridging the gap between the design specified as a CPN model and the implementation of the system. This application of CP-nets is presented in Sect. 5.

The four projects presented in this paper can be read in any order, but we have ordered their presentation according to the typical phases in system development starting with analysis and requirements, moving on to design and validation, and finally implementation. For readers with only limited or no prior knowledge of Petri nets we recommend reading Sect. 2 first as it also gives some introduction to the basic constructs in the CPN modelling language. We sum up the conclusions in Sect. 6 and give references to further reading on CP-nets.

2 Modelling Scenarios in Ad-Hoc Networking

The overall topic of this joint research project with Ericsson Telebit A/S [57] presented in this section is the use of the Internet Protocol v6 (IPv6) [42] for ad-hoc networking [71]. An ad-hoc data network is a collection of (typically) mobile nodes, such as laptops, personal digital assistants (PDAs), and mobile phones, capable of establishing a communication infrastructure for their common use. Ad-hoc networking differs from conventional data networks in that the network of nodes operates in a fully self-configuring and distributed manner, i.e., there is no central network management, control, or components at the network layer. Furthermore, there is no preexisting infrastructure, such as base stations and routers, available. One of the challenges in ad-hoc networking is to design the routing protocols in such a way that they are able to quickly adapt to the frequent changes in network topology due to node mobility and nodes leaving/joining the network. Ad-hoc networking has a number of application areas, such as sensor networks, rescue operations in remote areas, mobile conferencing, home networking, and wireless personal area networks. Routing protocols for ad-hoc networking are under development by the IETF Mobile Ad-hoc Networks working group [35]. The main focus of the project is the integration of routing protocols for conventional wired data networks e.g, OSPF, RIP, and BGP [83]) with routing protocols for ad-hoc networks (e.g, DSR, AODV, and OLSR [71]).

Figure 1 shows the IPv6 based network architecture considered in the project. The network architecture consists of an IPv6 core network connecting a number of mobile ad-hoc networks (MANETs) on the edge of the core network. The network architecture is aimed at supporting communication between nodes residing in different ad-hoc networks and communication between nodes in the ad-hoc networks and stationary nodes in the core network. Communication between nodes in the same ad-hoc network is facilitated by the ad-hoc network itself. Another important aspect of the network architecture is mobility. *Macromobility* is concerned with the movement of nodes from one ad-hoc network to another ad-hoc network, and the movement of an entire ad-hoc network from one point of attachment to the core network to another point of attachment. *Micromobility* is concerned with the movement of the nodes within an ad-hoc network which changes the topology of the ad-hoc network.

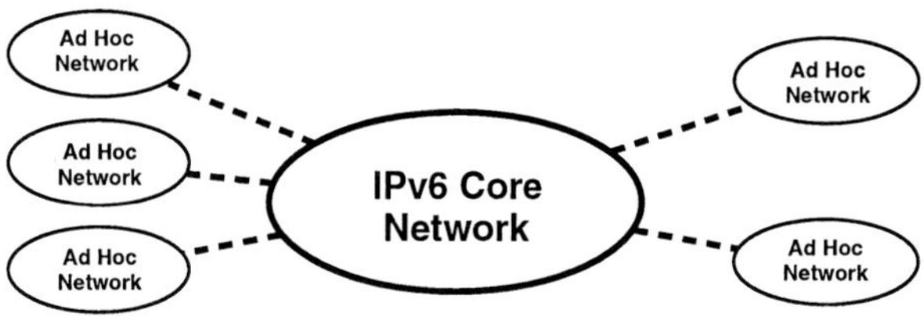

Fig. 1. IPv6 based networking architecture.

CPN modelling was used in the first phase of the project to develop the network architecture shown in Fig. 1 and to capture in a rigorous way the communication and mobility scenarios that must be supported. Capturing these requirements was done by constructing a CPN model that described mobility and communication in the above networking architecture. In the following, we give a detailed description of this CPN model.

2.1 CPN Modelling of Mobility and Communication

Figure 2 shows the *hierarchy page* of the CPN model. The hierarchy page provides an overview of the *pages* (modules) constituting the CPN model and their relationship. Each node in Figure 2 represents a page in the CPN model, and is labelled with a page name and a page number. As an example, the page node at the top left of Figure 2 is named Scenarios and has page number 1. Page Scenarios is the most abstract page in the CPN model. An arc between two nodes indicates that the destination page is a subpage (submodule) of the source page. The arc label(s) specifies the name of the *substitution transition*(s) representing the corresponding subpage at the source page. Substitution transitions are explained in more detail later.

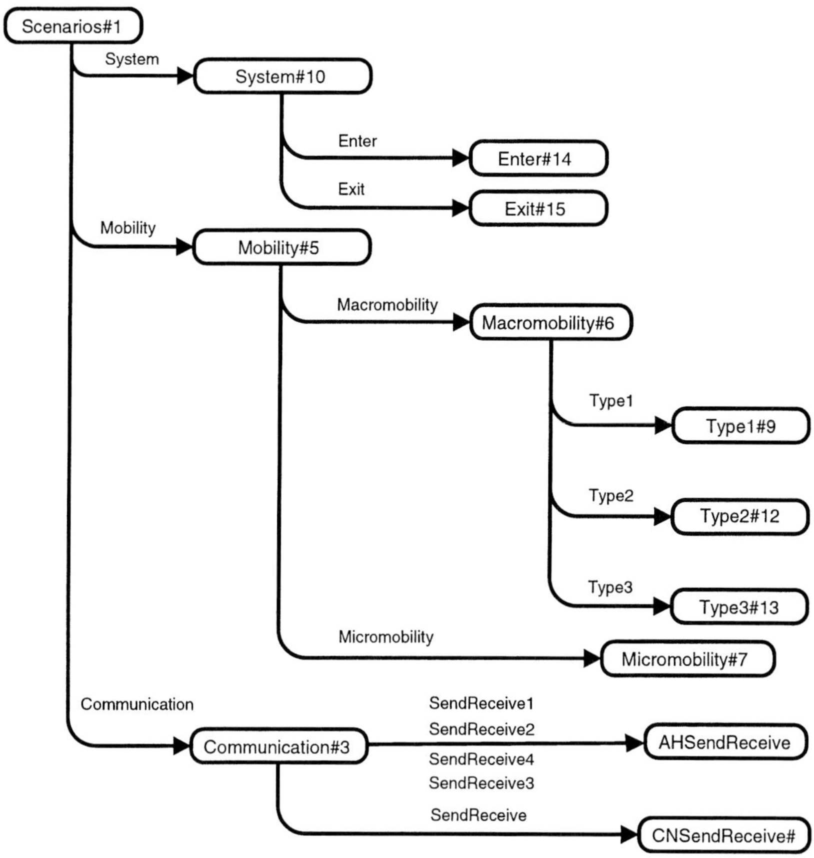

Fig. 2. Hierarchy page - overview of CPN model.

The CPN model consists of three main parts. Page System and its two subpages model the system scenarios which are concerned with ad-hoc nodes entering and leaving the system. Page Mobility and its five subpages model the mobility scenarios, i.e., the movement of the nodes in the ad-hoc networks. Page Communication and its two subpages, AHSendReceive and CNSendReceive, model the communication between nodes in the system.

Figure 3 depicts page Scenarios which is the most abstract part of the CPN model. It corresponds to the Scenarios page node in Fig. 2. The rectangles in Figure 3 are *substitution transitions* as indicated by the associated HS-tag (in the lower left corner of each rectangle). Each substitution transition has an associated subpage modelling the compound behaviour represented by the substitution transition in more detail. The name of the subpage is given in the dashed box next to the HS-tag. The communication scenarios are modelled by the substitution transition Communication which has page Communication (see Fig. 2) as its associated subpage. The mobility scenarios are modelled by the substitution transition Mobility which has page Mobility as its associated subpage. The system scenarios are modelled by the substitution transition System which has page System as its associated subpage.

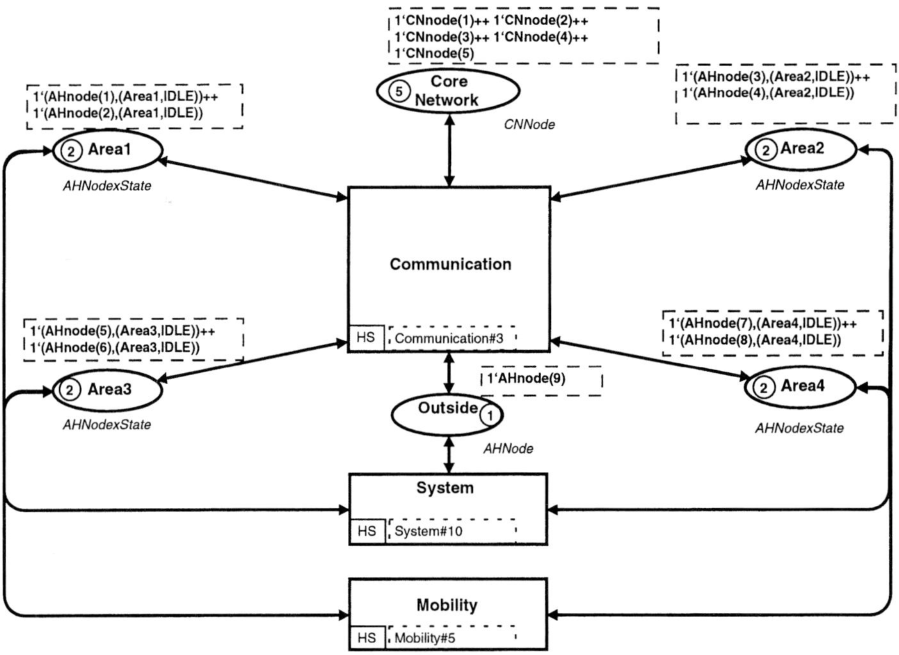

Fig. 3. The Scenarios page - top level page in the CPN model.

The ellipses in Fig. 3 are called *places* and are used to model the state of the system. The state of a CPN model is called a *marking* and is a distribution of *tokens* on the places of the CPN model. Each of the places Area1, Area2, Area3, and Area4 correspond to areas where ad-hoc networks can exist. In our scenarios, ad-hoc networks can exist in four areas. Nodes that are part of the ad-hoc network in a given area are modelled as tokens residing on the corresponding place. The place CoreNetwork is used for modelling the nodes in the core network. The place Outside is used for modelling the ad-hoc nodes currently outside of the system.

The kind of tokens that may reside on a place is determined by the *colour set* of the place. A colour set in a CPN model is similar to a type in a programming language, and the values in a colour set are referred to as *colours*. The colour set of a place is typically written below the place and is declared using the Standard ML programming languages. As an example, place Area1 has the colour set AHNodexState. The declarations of the colour sets used in Fig. 3 are listed in Fig. 4 and will be explained below.

```
val AHn = 9;
color AHInt = int with 1..AHn;
color AHNode = union AHnode : AHInt;

color Macrostate = with IDLE | MACROMOVE;
color Area = with Area1 | Area2 | Area3 | Area4;
color State = product Area * Macrostate;

color AHNodexState = product AHNode * State;

val CNn = 5;
color CNInt = int with 1..CNn;
color CNNode = union CNnode : CNInt;
```

Fig. 4. Colour sets used in Fig. 3.

The symbolic constant AHn is used to specify the total number of ad-hoc nodes in the system. Colour sets are declared using the keyword color. The colour set AHInt denotes the set of integers in the range from 1 to AHn. The colour set AHNode is used to model the ad-hoc nodes. An ad-hoc node is specified as a value (colour) with the form AHnode(i) where $1 \leq i \leq$ AHn. The colour set Macrostate is used to model the internal state of an ad-hoc node with respect to movement from one area to another area. The state may either be IDLE indicating that the node is currently not on the move from one area to another area, or MACROMOVE indicating that the node is currently on the move from one area to another area. The state of an ad-hoc node is modelled by the colour set State which is the cartesian product of the colour sets Area and Macrostate. Hence, the state of an ad-hoc node specifies the area that the ad-hoc node is currently in, and whether the ad-hoc node is currently moving from one area to another. The area places in Fig. 3 all have the colour set AHNodexState. Hence, tokens residing on these places represents ad-hoc nodes. Place Outside on page Scenarios has the colour set AHNode. The reason for this is that the state of the ad-hoc node is not important when the node is outside the system. The colour set CNNode is used to model the nodes in the core network. A core network node is specified as a value (colour) with the form CNnode(ci) where $1 \leq$ bi $\leq$ CNn. Place CoreNetwork in Fig. 3 has the colour set CNNode.

The small circles and associated dashed boxes in Fig. 3 show the *current marking* of the CPN model. The small circle positioned inside a place indicates the number of tokens on the given place in the current marking. In the marking shown, there are two ad-hoc nodes in each of the four areas, and ad-hoc node 9 is currently outside the system. There are fives nodes in the core network. The dashed boxes positioned next to the places specify the colours of the individual tokens residing on that place. The marking of a place is a *multi-set* of tokens over the colour set of the place, i.e., there can be multiple appearances of the same token. The text inside the dashed boxes specifies the multi-set of tokens residing on the place using ++ to denote union (pronounced and) and ' (pronounced of) to specify coefficients, i.e., the number of occurrences of tokens with that value. As an example, on place Area1 in the marking shown in Fig. 3 there is one token *of* colour (AHnode(1),(Area1,IDLE)) *and* one token *of* colour (AHN-ode(1),Area1,IDLE). The CPN model contains an initialisation step responsible for the initial distribution of tokens on the CPN model. It is, however, possible for the modeller to also manually specify the *initial marking* of the CPN model.

The transitions and places in Fig. 3 are connected by double-deaded *arcs*. Some of these arcs have been partly positioned on top of each other to improve readability of the figure. A place connected to a substitution transition is called a *socket place*, and a socket place is associated to a *port place* on the subpage associated with the substitution transition. This is called a *port-socket* assignment. This association has the effect that the port and the socket places will always have identical markings. Note that a place may be a socket place for several substitution transitions. The dynamics of a CPN model consists of *occurrences* of transitions (ordinary, not substitution transitions) which add and remove tokens to/from the places of the CPN model, thereby changing the current marking. An arc leading to a place from a substitution transition means that transitions on the subpage associated with the substitution transitions will add tokens on this place. Similarly, an arc leading from a place to a substitution transition means that transitions on the subpage will remove tokens from this place. A double-deaded arc is a shorthand for an arc in each direction. The basic idea in the CPN model is to capture mobility scenarios of the network architecture by moving tokens corresponding to ad-hoc nodes from one area place to another area place. Similarly, communication scenarios will be modelled by moving tokens in the CPN model corresponding to packets.

2.2 Modelling Mobility

Figure 5 depicts page Mobility which is the most abstract page in the part of the CPN model specifying mobility. Two types of mobility are considered: macromobility and micromobility. Recall that macromobility is concerned with the mobility of ad-hoc nodes between ad-hoc networks. In the CPN model we consider only the macromobility case of one ad-hoc node moving from one ad-hoc network to another ad-hoc network. The case of an entire ad-hoc network moving can be viewed as the individual movement of all of the nodes in the ad-hoc network. Micromobility is concerned with the movement of ad-hoc nodes within an ad-hoc network. The two types of mobility are modelled by the subpages of the substitution transitions Macromobility and Micromobility, respectively. The four places Area1-4 are port places of this page - indicated by the P-tags positioned next to them. The I/O-tag specifies that they are input and output port places. This means that tokens may be added and removed to/from these places. Each of the area places are associated with the identically named socket place in Fig. 3. The places Area1-4 are also socket places since they are connected to the Macromobility and Micromobility substitution transitions.

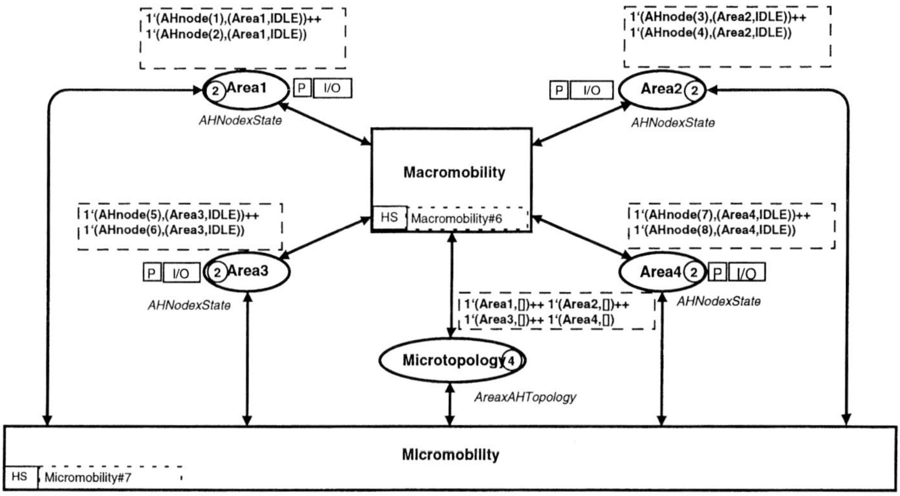

Fig. 5. The Mobility page.

The macromobility scenarios are specified by considering the movement of ad-hoc nodes between the places Area1-4. The place Microtopology is used to represent the current topology of the ad-hoc networks. The definition of the colour set AreaxAHTopology is given in Fig. 6.

```
color AHNodexAHNode = product AHNode * AHNode;
color AHTopology = list AHNodexAHNode;

color AreaxAHTopology = product Area * AHTopology;
```

Fig. 6. Declaration of colour set AreaxAHTopology.

The topology of an ad-hoc network is a pair consisting of the ad-hoc network and a list of pairs specifying the current set of links between the nodes in the ad-hoc network. For example, a pair (AHnode(6),AHnode(5)) captures that ad-hoc node 5 can be reached from ad-hoc node 6, but not necessarily the other way around as links may be unidirectional. In the current marking of place

Microtopology shown in Fig. 5, no ad-hoc nodes are able to reach each other in any area, and hence the topology in each area is specified as the empty list []. The substitution transition Macromobility is connected to the place Microtopology by a double arc. When a node moves from one area to another area, all existing links to nodes in the area being moved from disappear.

Micromobility. Figure 7 depicts page Micromobility specifying the micromobility. The micromobility scenarios are abstractly modelled by viewing the ad-hoc network as a directed graph where edges represent connectivity. Hence, we have abstracted from the physical location of the nodes in the ad-hoc networks. The nodes in the ad-hoc networks are represented as tokens on the area places. The current topology of the ad-hoc network is represented by the tokens on place Microtopology. All five places on this page are connected via port-socket relationships to the identically named places on page Mobility (see Fig. 5). The two rectangles AddLink and DeleteLink are ordinary transitions. Transition AddLink models that a new link between two ad-hoc nodes arises, and transition DeleteLink models that an existing link between two ad-hoc nodes disappears.

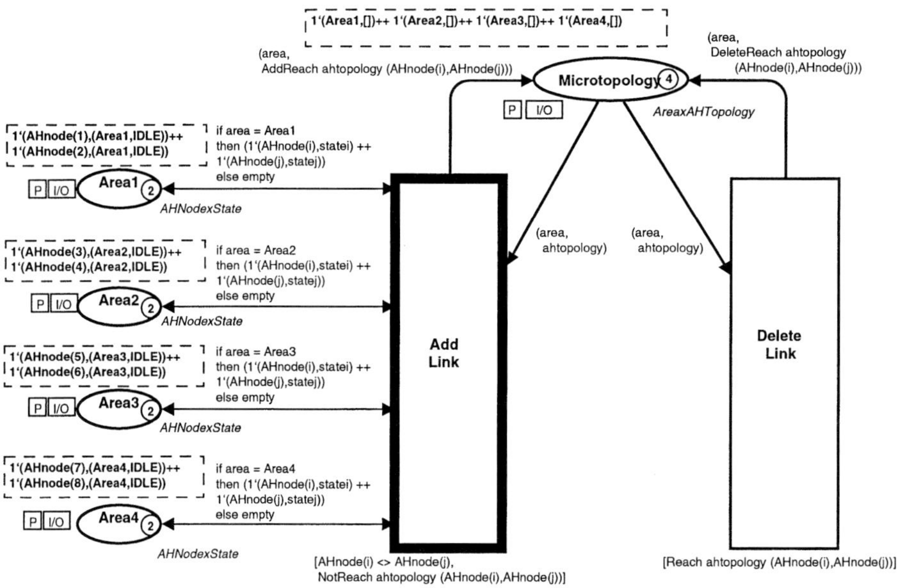

Fig. 7. The Micromobility page.

The actions of a CPN model consist of occurrences of *enabled* transitions removing tokens from places connected to incoming arcs and adding tokens to places connected to outgoing arcs of the transition. The transition AddLink is enabled in the marking shown in Fig. 7. This is indicated by the thick border of the transition. Transition AddLink has five input places and five output places. A transition is required to be *enabled* before it may occur. A transition is enabled if sufficient tokens with adequate colours exist in each of its input places. When a transition occurs, it removes tokens from input places and adds tokens to output places. The exact multi-set of tokens required for a transition to be enabled and removed from input places when it occurs, and the exact multi-set of tokens added to output places of the transition are determined by assigning value to the *variables* of the transition, and by evaluating the *arc expressions*, i.e., the inscriptions positioned next to the arcs. Arc expressions are written in the Standard ML language.

To evaluate the arc expressions on the surrounding arcs of a transition, a *binding* of the transition must be created. A binding is an assignment of data values to the variables of the transition. Figure 8 shows the declaration of the variables appearing in the surrounding arcs of the NewReach transition in Fig. 7. The definition of the colour sets have previously been given in Fig. 4.

```
var area             : Area
var i,j              : AHInt;
var statei, statej : State;
var ahtopology      : AHTopology;
```

Fig. 8. Variables used on page Micromobility shown in Fig. 7.

A binding of a transition is enabled in the current marking if when evaluating each of the arc expressions on input arcs, the resulting multi-set of tokens is a subset of the multi-set of tokens currently present in the corresponding input place. An enabled binding of the transition AddLink is the following which lists the value assigned to each variable of the transition:

$$< \text{area=Area1,i=1, statei=IDLE, j=2, statej=IDLE, ahtopology=[]} >$$

This binding corresponds to the event that ad-hoc node 1 is now able to reach ad-hoc node 2. Evaluating the input arc expression from place Area1 in this binding yields the multi-set: 1'(AHnode(1),IDLE) ++ 1'(AHnode(2),IDLE). The result of evaluating the input arc expression from place Microtopology yields the multi-set 1'(Area1,[]). The remaining input arc expressions all yield the empty multi-set since the variable area is bound to the value Area1. This binding is enabled since each of the multi-sets of tokens are present on the corresponding input places, and because the *guard* (shown in square bracket below the transition) of the NewReach transition is satisfied. A guard is a boolean expression that must evaluate to true in the binding in order for the transition to be enabled. The guard expresses the condition that the two ad-hoc nodes determined by the binding of the variables i and j must be distinct, and there must not already exist a link between ad-hoc node i and j. The latter requirement is checked by the function NotReach which is a function implemented in Standard ML. The implementation of the NotReach function is shown in Fig. 9. The implementation of the NotReach function uses the built-in Standard ML function [79] List.all to check whether the edge between ad-hoc node i and j already exists in the list ahtopology corresponding to the current topology. We explain the other functions listed in Fig. 9 shortly.

```
fun NotReach ahtopology (ahnode1,ahnode2) =
    (List.all (fn edge => edge <> (ahnode1,ahnode2)) ahtopology)

fun AddReach ahtopology (ahnode1,ahnode2) = (ahnode1,ahnode2)::ahtopology

fun Reach ahtopology (ahnode1,ahnode2) =
    (List.exists (fn edge => edge = (ahnode1,ahnode2)) ahtopology)

fun DeleteReach ahtopology (ahnode1,ahnode2) =
    List.filter (fn edge => edge <> (ahnode1,ahnode2)) ahtopology
```

Fig. 9. Function used in arc expression on page Micromobility in Fig. 7.

If the above enabled binding of transition AddLink *occurs*, it will remove the multi-set of tokens from input places of the transition obtained by evaluating the input arc expressions, and add the multi-set of tokens to each output place obtained by evaluating the corresponding output arc expression. Since the AddLink transition is connected to the area places with double arcs, the same multi-set of tokens will be removed and added for each of these places. Hence, the marking of these places will remain unchanged. The marking of place Microtopology will change as the token (Area1,[]) will be removed and a new token will be added as described by the arc expression from AddLink to Microtopology. This arc expression uses the function AddReach to add the edge AHnode(i),AHnode(j) to the microtopology in area 1. The AddReach function uses the list constructor :: to insert the edge (AHnode(i),AHnode(j)) at the head of the list ahtopology representing the current topology. Figure 10 shows the marking of page Micromobility after the occurrence of the above binding of the AddLink transition. The marking of the place Microtopology has changed so that Ahnode(1) is now able to reach AHnode(2) in area 1. The transition AddLink is also enabled in other bindings. In fact, it is enabled in bindings corresponding to all the possible edges that can arise between nodes given the current location of nodes in the four areas.

In the marking shown in Fig. 10 both transitions are enabled. Transition DeleteLink is enabled with the binding:

$$< \text{area=Area1,i=1, j=2, ahtopology=[(AHnode(1),AHnode(2))]} >$$

The guard of the DeleteLink transition uses the function Reach to ensure that the transition is only enabled in bindings corresponding to links that exists in the area. The implementation of Reach is given in Fig. 9, and it uses the list library function List.exists to ensure that the link to be removed is an existing list in the current topology of the ad-hoc network. The output arc expression to

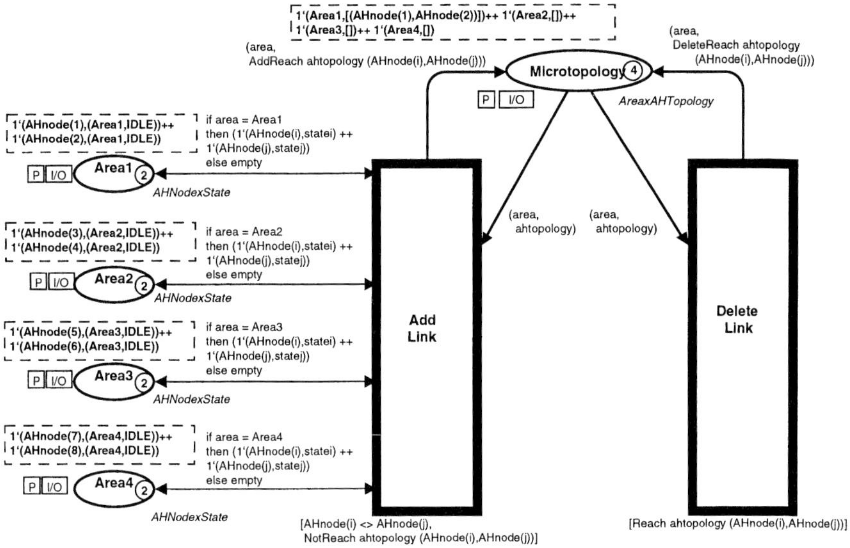

Fig. 10. The Micromobility page - after occurrence of AddLink.

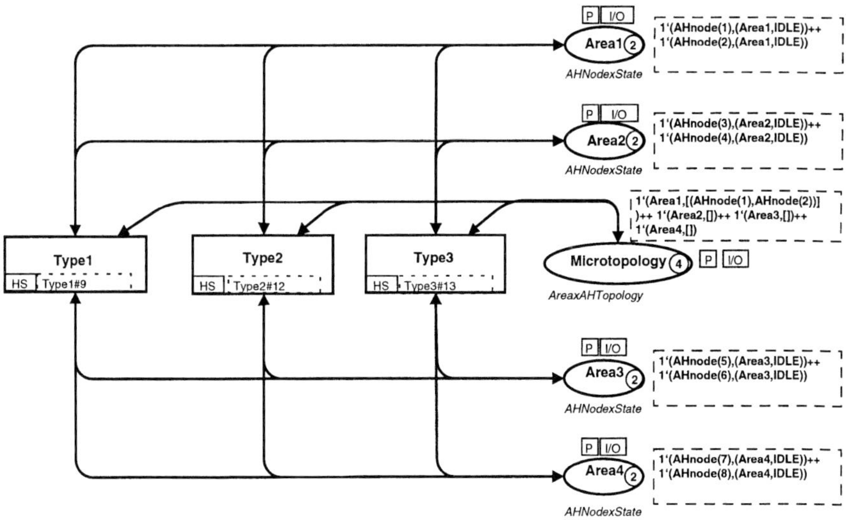

Fig. 11. The Macromobility page.

Microtopology uses the DeleteReach function to delete the edge in the list describing the topology in the area where the link disappears. If transition DeleteReach occurs in the above binding, it will result in the marking shown in Fig. 7. This means that it will remove the link which was added when AddLink occurred.

Macromobility. Figure 11 depicts page Macromobility specifying the macromobility scenarios. Three types of macromobility are considered and modelled by the subpages of the accordingly named substitution transitions. All arcs in Fig. 11 are double-headed arcs, but they have been positioned on top of each other to reduce the number of crossing arcs. Each of the three types of macro-mobility is described below.

Type 1: This type specifies the movement of an ad-hoc node from one ad-hoc network to another ad-hoc network. The subpage Type1 modelling this type is shown in Fig. 12. The transition InstantMove represents the instantaneous move from one ad-hoc network to another ad-hoc network, i.e., at the same moment as the node leaves the ad-hoc network in one area it joins the ad-hoc network in another area. The declarations used are listed in Fig. 13. The value bound to the variable i (on the arcs between the area places and InstantMove) of type Int corresponds to the ad-hoc node that moves. When the InstantMove transition occurs, the variable to will be bound to the area which is being moved to and the variable from will be bound to the area being moved from. The microtopology of the area being moved from is also updated by changing the corresponding token

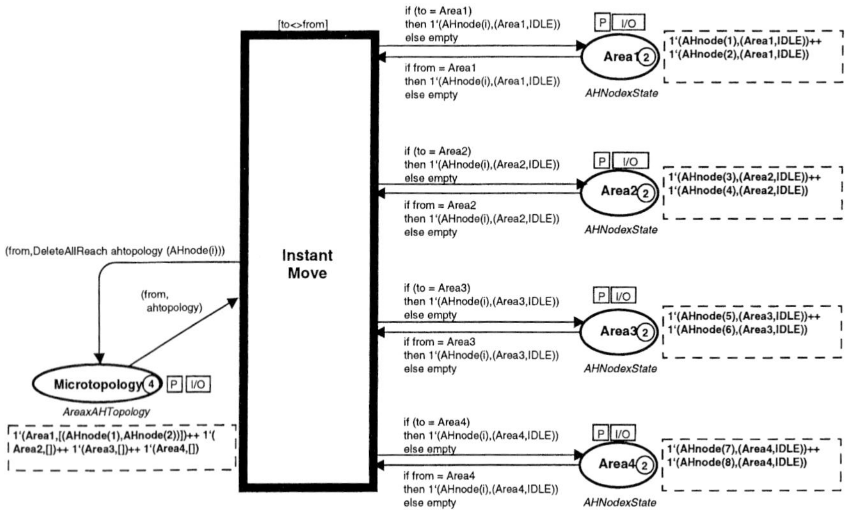

Fig. 12. Macromobility – Type 1.

```
var area,to,from : Area;
var ahtopology   : AHTopology;

fun DeleteAllReach ahtopology ahnode =
    List.filter
    (fn (snode,dnode) => (snode <> ahnode) andalso (dnode <> ahnode))
    ahtopology
```

Fig. 13. Declarations for macromobility – Type 1.

on place Microtopology. The function DeleteAllReach uses the built-in function List.filter to delete all edges in the microtopology related to the ad-hoc node that moves. An ad-hoc node has to be in its IDLE state to move from one area to another area. This ensures that the ad-hoc node is not currently moving according to one of the other types of macromobility types described below. The guard of the transition ensures that it is only enabled when the variables to and from are bound to different areas, i.e., the binding corresponds to movement of nodes between distinct areas.

The following binding is an example of an enabled binding of the transition InstantMove in the marking shown in Fig. 14. It corresponds to the movement of ad-hoc node 1 from area 1 to area 2:

$$< \text{i=1,from=Area1,to=Area2,ahtopology=[AHnode(1),AHnode(2)]} >$$

The transition is enabled in bindings corresponding to all the possible movement of nodes between areas. An occurrence of the above binding results in the marking shown in Fig. 14 where the token corresponding to ad-hoc node 1 is now positioned on the place corresponding to area 2 and the link between ad-hoc nodes 1 and 2 in area 1 no longer exists. The movement of ad-hoc nodes is also evident on page Scenarios shown in Fig. 15.

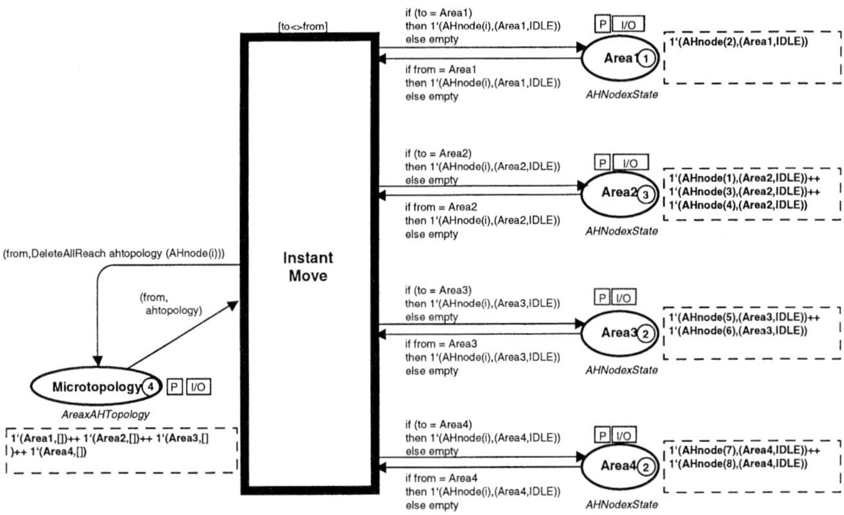

Fig. 14. Macromobility Type 1 - after occurrence of InstantMove.

Type 2: This type specifies the movement of an ad-hoc node from one ad-hoc network to another ad-hoc network. The difference between type 2 and type 1 is that there is a period of time in which the nodes moving are not part of any of the ad-hoc networks in Area1-4. The page for type 2 mobility is similar to the one for type 1 and is therefore omitted.

Type 3: This type specifies the movement of an ad-hoc node from one ad-hoc network to another with the addition that there is a period of time in which

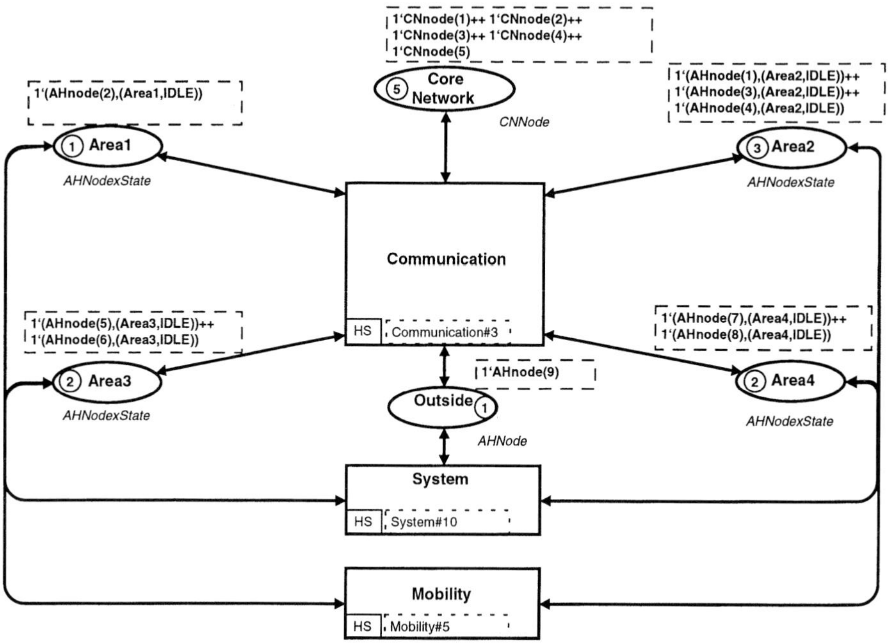

Fig. 15. The Scenarios page - after occurrence of InstantMove.

the node moving is part of both the ad-hoc network being moved from and the ad-hoc network being moved to. The page for type 3 mobility is similar to the one for type 1 and is therefore omitted.

2.3 Modelling Communication

Figure 16 depicts page Communication which is the most abstract page modelling the communication. The page models that each of the nodes (ad-hoc and core network nodes) may send and receive packets. Packets in transit between ad-hoc nodes are represented as tokens on place Routing. At the abstraction level of the CPN model, there is no distinction made between communication internally in an ad-hoc network and between nodes in different ad-hoc networks. The CPN model simply specifies the requirement that the packet must be delivered to the appropriate node - no matter in which ad-hoc network the node currently resides. Place Routing hence abstractly represents the routing functionality that will have to be implemented to get the packets from the source to the destination. How this is done is a design and implementation issue. The transition Drop Packet models that packets for ad-hoc nodes currently outside the system will be dropped.

The declarations used for modelling communication between nodes are listed in Fig. 17. The colour set Packet is used modelling packets. A packet is abstractly represented as having a source and destination. As an example, a packet sent from AHnode(1) to AHnode(2) will be represented as a token with value (colour)

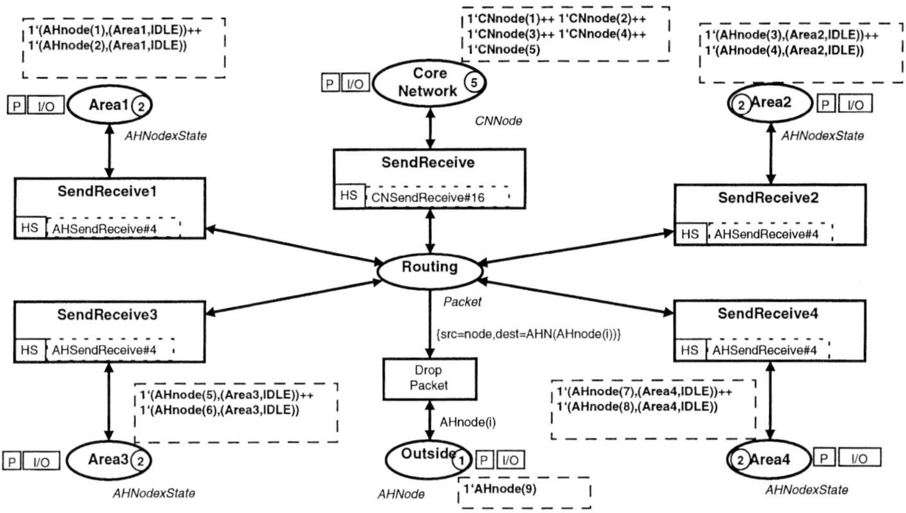

Fig. 16. The Communication page.

```
color Node   = union CNN : CNNode + AHN : AHNode;
color Packet = record src : Node * dest : Node;
var node : Node;
```

Fig. 17. Declarations for modelling communication.

{src = AHN(AHnode(1)), dest = AHN(AHnode(2))}. Hence for specification of requirements, we abstract from the actual content of packets.

Page AHSendReceive modelling the sending and receiving of packets by ad-hoc nodes is shown in Fig. 18. It is the subpage of each of the four SendReceive1-4 substitution transitions in Fig. 16. This means that there will be four *instances* of this page when the CPN model is executed, one for each of the substitution transitions. The marking and enabling of transitions on these instances will be independent of each other. The instance depicted in Fig. 18 corresponds to the instance associated with the substitution transition SendReceive1 in Fig. 16. The transition SendPacket models the transmission of a packet from ad-hoc node i to a node assigned to the variable node. The transition ReceivePacket models the reception of a packet by ad-hoc node i. An occurrence of this transition will remove the token corresponding to the packet being received from place Routing.

An occurrence of the SendPacket transition in Fig. 18 in a binding with: i=1, state=IDLE,dest=AHN(AHnode(3)) results in the marking shown in Fig. 19. The corresponding marking of page Communication is shown in Fig. 20. The ReceivePacket transition on the instance of the AHSendReceive page corresponding to the substitution transition SendReceive2 will now be enabled in binding corresponding to ad-hoc node 3 receiving the packet. The reception of the packet will result in the corresponding token being removed from place Routing.

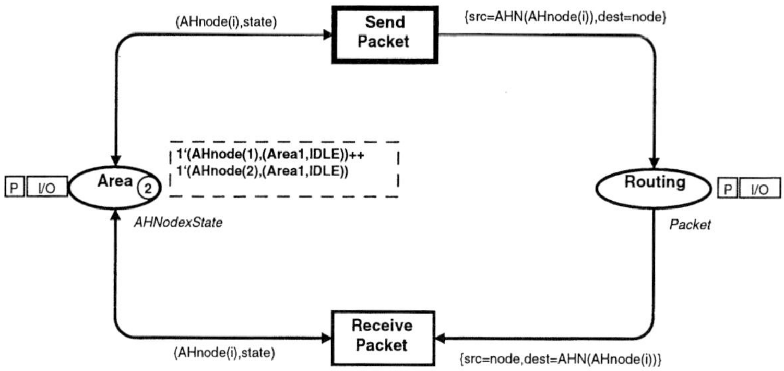

Fig. 18. The AHSendReceive page - instance for SendReceive1.

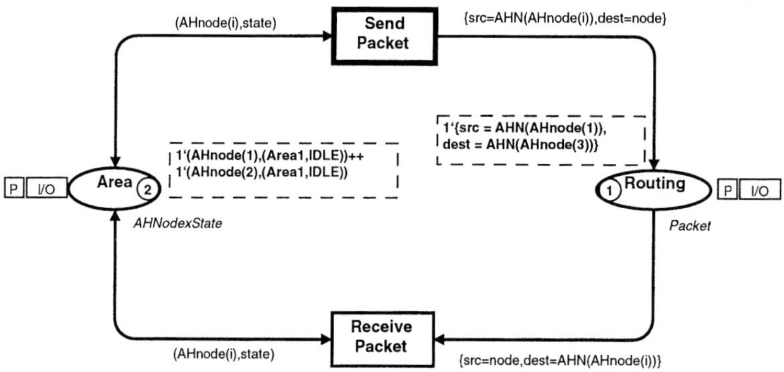

Fig. 19. The AHSendReceive page - after occurrence of SendPacket.

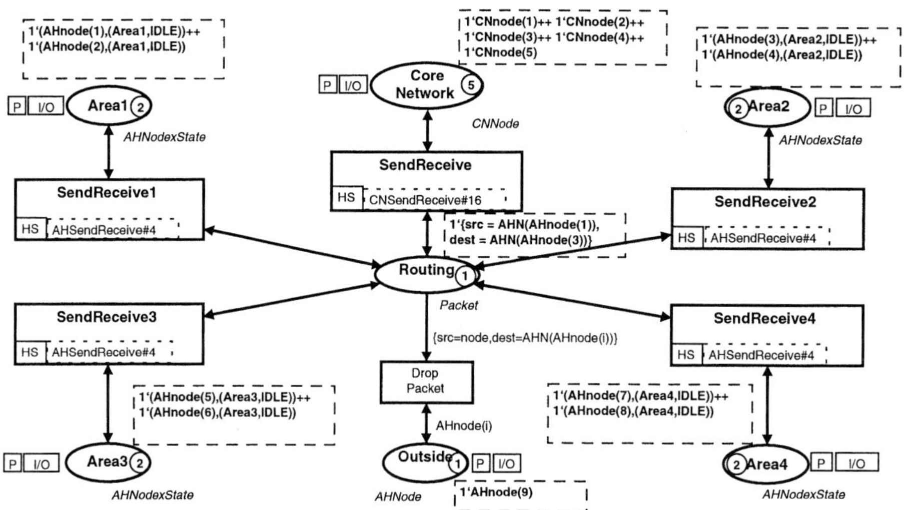

Fig. 20. Marking of the Communication page - packet in transit.

The modelling of send and receive for nodes in the core network is similar to the ad-hoc nodes. Hence, we do not give a detailed explanation of page CNSendReceive.

2.4 Conclusions on Modelling Ad-Hoc Networking Scenarios

The CPN model developed describes the abstract network architecture and associated communication and mobility scenarios considered in the project. A key point of the CPN model is that it captures the communication and the mobility aspects in a single model. The CPN model also allows derivation of combined scenarios involving simultaneously communication and mobility. As such, the CPN model can be seen as a formal documentation of the network architecture and its communication and mobility requirements. The CPN model is also suitable for generation of communication and mobility test-cases against which the later protocol designs can be checked. A number of such interesting scenarios were derived using simulation of the CPN model. The plan is to use these scenarios as test-cases for the protocols to be developed in later phases of the project. Finally, and probably most importantly, the development of the CPN model has served as an important tool for stimulating discussion of the network architecture and requirements.

The graphical layout of the CPN model currently mimics the network architecture. This was chosen since it is easier to visualise the behaviour of the system directly at the level of the CPN model. This has been useful when presenting the CPN model to people without CPN knowledge. A more compact CPN model with tokens representing ad-hoc networks instead of tokens representing ad-hoc nodes could be developed. This would make it possible to model an arbitrary number of areas where ad-hoc networks can exists and it could be considered a more direct way of modelling mobility of an entire ad-hoc network. The CPN model would, however, lose some of its graphical appeal and hence possibly other means of graphics showing mobility and communication would have to be added to the CPN model. This approach seems more suitable for later CPN modelling of the actual protocol designs.

The CPN model does not have an explicit representation of the connection between the core network and the ad-hoc networks since the purpose of the CPN model was to abstractly specify the communication and mobility requirements and scenarios related to nodes in the ad-hoc networks. The operation of gateways integrating the IPv6 core network routing protocols and the ad-hoc routing protocols is at a lower level of abstraction than the current CPN model. The purpose of the presented CPN model was to describe the scenarios and hence capture requirements in an implementation-independent manner.

3 Modelling Requirements in Pervasive Health Care

The *pervasive health care system (PHCS)* [12] is envisioned in a joint project between Aarhus County Hospital, the software company Systematic Software

Engineering A/S [84], and the Centre for Pervasive Computing [10] at the University of Aarhus. In this section, we describe how CP-nets are applied in *requirements engineering* for PHCS. The section is based on previous descriptions given in [52–54], and has benefitted from efforts of many participants in the pervasive health care research project [72].

The aim of PHCS is to improve the electronic patient record (EPR) [1], which is currently being deployed at the hospitals in Aarhus, Denmark. EPR is a comprehensive health care IT system with a budget of approximately 15 million US dollars; it will eventually have 8-10,000 users.

EPR solves obvious problems occurring with paper-based patient records such as being not always up-to-date, only present in one location at a time, misplaced, and sometimes even lost. However, the version of EPR currently being deployed is a desktop PC based system which is not very practical for hospital work, since the users like nurses and doctors are often on the move and away from their offices (and, thus, desktop PCs). Moreover, users are frequently interrupted. Therefore, the desktop PC based EPR potentially induces at least two central problems for its users [6]. The first problem is *immobility*: in contrast to a paper-based record, an electronic patient record accessed only from desktop PCs cannot be easily transported. The second problem is *time-consuming login and navigation*: EPR requires user identification and login to ensure information confidentiality and integrity, and to start using the system for clinical work, a logged-in user must navigate, e.g., to find a specific document for a given patient.

The motivation for PHCS is to address these problems. In the ideal situation, the users should have access to the IT system wherever they need it, and it should be easy to resume a work process which has previously been interrupted.

3.1 The Pervasive Health Care System

Use of personal digital assistants (PDAs), with which nurses and doctors could access EPR using a wireless network, is a possible solution to the immobility problem. That approach has been considered, but is not ideal, e.g., because of well-known characteristics of PDAs like small screens and limited memory, and because it does not fully address the time-consuming login and navigation problem. PHCS is a more ambitious solution which to a larger extent takes advantage of the possibilities of *pervasive computing* [81,90]. Three basic design principles are exploited.

The first principle is *context-awareness* [80]. This means that PHCS is able to register and react upon certain changes of context. More specifically, nurses, patients, beds, medicine trays, and other items are equipped with radio frequency identity (RFID) tags [74], enabling the presence of such items to be detected automatically by involved context-aware computers, e.g., located by the medicine cabinet and by the patient beds.

The second design principle is that PHCS is *propositional*, in the sense that it makes qualified propositions, or guesses. Context changes may result in automatic generation of buttons that appear at the task-bar of computers. Users may explicitly accept a proposition by clicking a button – and implicitly ignore or reject it by not clicking. The presence of a nurse holding a medicine tray for patient P in front of the medicine cabinet is a context that triggers automatic generation of a button Medicine plan:P, because in many cases, the intention of the nurse is now to navigate to the medicine plan for P. If the nurse clicks the button, she is logged in and taken to P's medicine plan. It is, of course, impossible always to guess the intention of a user from a given context, and without the propositional principle, automatic shortcutting could become a nuisance since guesses would sometimes be wrong.

The third design principle is that PHCS is *non-intrusive*, i.e., not interfering with or interrupting hospital work processes in an undesired way. Thus, when a nurse approaches a computer, it should react to her presence in such a way that a second nurse, who may currently be working on the computer, is not disturbed or interrupted. The last two design principles cooperate to ensure satisfaction of a basic mandatory user requirement: important hospital work processes have to be executed as conscious and active acts by responsible human personnel, not automatically by a computer.

Figure 21 outlines PHCS (with an interface that is simplified and translated into English for the purpose of this paper). The current context of the system is that nurse Jane Brown is engaged in pouring medicine for patient Bob Jones for the giving to take place at 12 a.m. The medicine plan on the display shows which medicine has been prescribed (indicated by 'Pr'), poured ('Po'), and given ('G') at the current time. In this way, it can be seen that Advil and Tylenol have been poured for the 12 a.m. giving, but Comtrex not yet. Moreover, the medicine tray for another patient, Tom Smith, stands close to the computer, as can be seen from the task-bar buttons.

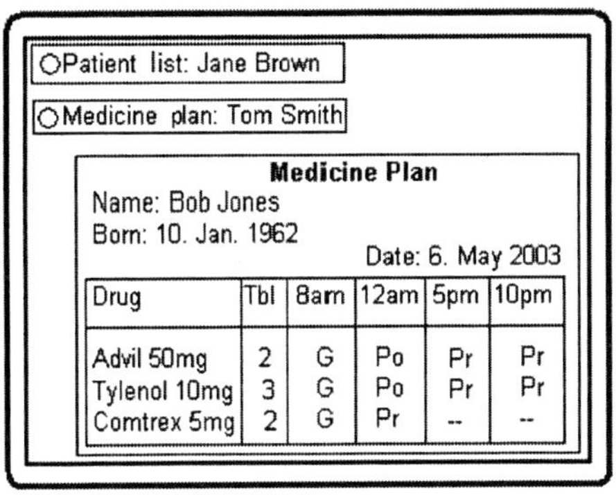

Fig. 21. PHCS – outline.

3.2 Medicine Administration

To aid requirements engineering for PHCS, CPN models of envisioned new work processes and of their proposed computer support were created. The scope of this section is the work process *medicine administration*, which is described below.

Assume that nurse N wants to pour medicine into a medicine tray and give it to patient P. First, N goes to the room containing the medicine cabinet (the medicine room). Here is a context-aware computer on which the buttons Login:N and Patient list:N appear on the task-bar when N approaches. If the second button is clicked, N is logged in and a list of those patients of which she is in charge is displayed on the computer. A medicine tray is associated with each patient. When N takes P's tray nearby the computer, the button `Medicine plan:P` will appear on the task-bar, and a click will make P's medicine plan appear on the display. N pours the prescribed medicine into the tray and acknowledges this in PHCS. When N leaves the medicine room, she is automatically logged out. N now takes P's medicine tray and goes to the ward where P lies in a bed, which is supplied with a context-aware computer. When N approaches, the buttons Login:N, Patient list:N, and Medicine plan:P will appear on the task-bar. If the last button is clicked, the medicine plan for P is displayed. Finally, N gives the medicine tray to P and acknowledges this in PHCS. When N leaves the bed area, she is automatically logged out again.

The given description captures just one specific combination of sub work processes. There are numerous other scenarios to take into account, e.g., medicine may be poured for one or more patients, for only one round of medicine giving, all four regular rounds of a 24 hours period, or for ad hoc giving; a nurse may have to fetch trays left at the wards prior to pouring; a nurse may approach the medicine cabinet without intending to pour medicine, but only to log into EPR (via PHCS) or to check an already filled medicine tray; two or more nurses may do medicine administration at the same time. To support a smooth medicine administration work process, the requirements for PHCS must deal with all these scenarios and many more. A CPN model, with its fine-grained and coherent nature, is able to support that.

3.3 Medicine Administration CPN Model

The medicine administration CPN model consists of 11 pages with a total of 54 places and 29 transitions. An overview of the model in terms of the hierarchy page is given in Fig. 22. The graph shows how the work process medicine administration is decomposed in sub-work processes.

We give an impression of the model by describing the page shown in Fig. 23. The page models the pouring and checking of trays and is represented by the node PourChkTrays in Fig. 22. The medicine cabinet computer is in focus. It is modelled by a token on the Medicine cabinet computer place. This place has colour set COMPUTER, whose elements are 4-tuples (compid,display,taskbar,users) consisting of a computer identification, its display (main screen), its task-bar buttons, and its current users. In the initial marking, the computer has a blank display, no task-bar buttons, and no users.

The colour set NURSE is used to model nurses. A nurse is represented as a pair (nurse,trays), where nurse identifies the nurse and trays is a container data structure holding the medicine trays that this nurse currently has in possession.

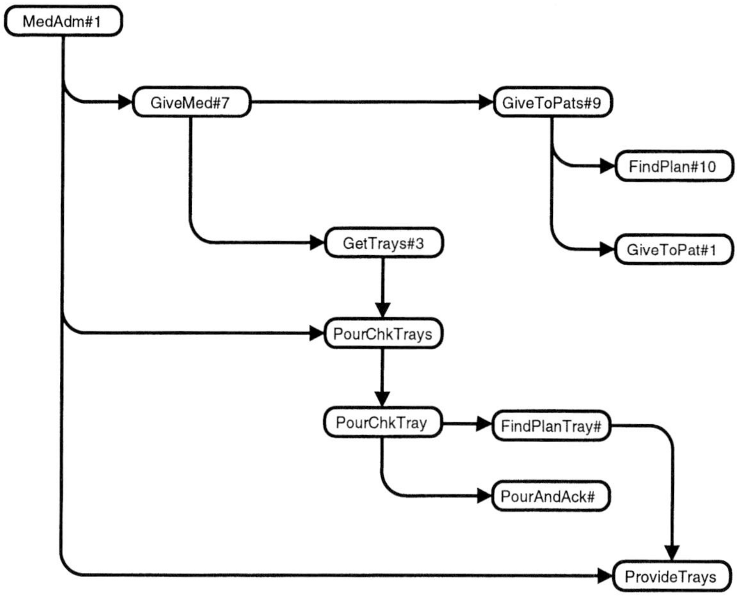

Fig. 22. Medicine administration CPN model: hierarchy page.

Initially, the nurses Jane Brown and Mary Green are ready (represented as tokens in the Ready place) and have no trays.

Occurrence of the Approach medicine cabinet transition models that a nurse changes from being ready to being busy nearby the medicine cabinet. At the same time, two buttons are added to the task-bar of the medicine cabinet computer, namely one login button for the nurse and one patient list button for the nurse. In the CPN model, these task-bar buttons are added by the function addMedicineCabinetButtons appearing on the arc from the transition Approach medicine cabinet to the place Medicine cabinet computer.

The possible actions for a nurse who is by the medicine cabinet are modelled by the three transitions Pour/check tray, Enter EPR via login button, and Leave medicine cabinet. Often, a nurse at the medicine cabinet wants to pour and/or check some trays. How this pouring and checking is carried out is modelled on the subpage PourChkTray, which is the subpage of the substitution transition Pour/check tray.

The Enter EPR via login button transition models that a nurse clicks on the login button and makes a general-purpose login to EPR. It is outside the scope of the model to describe what the nurse subsequently does – the domain of the model is specifically medicine administration, not general EPR use. The transition has a guard which checks if a nurse is allowed to log into EPR. When a nurse logs in, the login button for that nurse is removed from the task-bar of the computer, modelled by the removeLoginButton function. Moreover, the nurse is added to the set of current users by the function addUser.

The Leave medicine cabinet transition models the effect of a nurse leaving: it is checked whether the nurse is currently logged in, modelled by the function

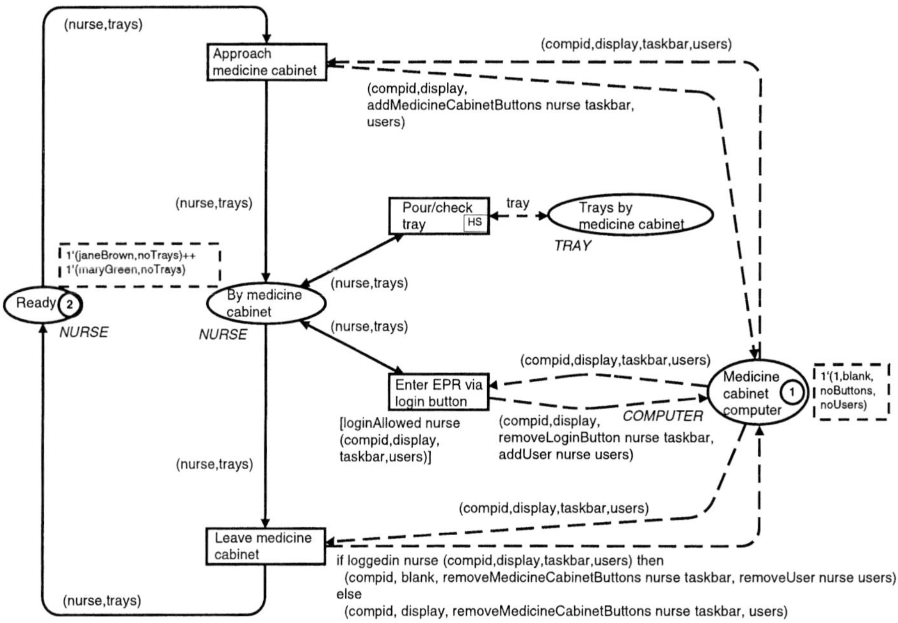

Fig. 23. Medicine administration CPN model: PourChkTrays page.

loggedin appearing in the if-then-else expression on the arc going from Leave medicine cabinet to the Medicine cabinet computer place. If the nurse is logged in, the medicine cabinet computer automatically returns to a blank screen, removes the nurse's task-bar buttons (removeMedicineCabinetButtons), and logs her off (removeUser). If she is not logged in, the buttons generated because of her presence are removed, but the state of the computer is otherwise left unaltered. In any case, the token corresponding to the nurse is put back on the Ready place.

3.4 Medicine Administration Animation

An animation built on top of the CPN model is shown in Fig. 24. The animation is an interface to the CPN model, i.e., the animation is consistent with the CPN model and reflects the markings, transition occurrences, and marking changes that appear when the CPN model is executed. The animation hides the technicalities of CP-nets, e.g., concepts like places, transitions, tokens, enabling, occurrence, etc. In this way, the animation supports communication between users and system developers, by reducing the semantic distance [26] between the CPN model and the conception by the users of future work processes and their proposed computer support. The limitations of formal specifications as a means of communication in general, and, thus, the need for an animation, are widely recognised, see, e.g. [95].

The link between the CPN model and the animation is that the transitions of the CPN model are calling drawing functions related to the animation when

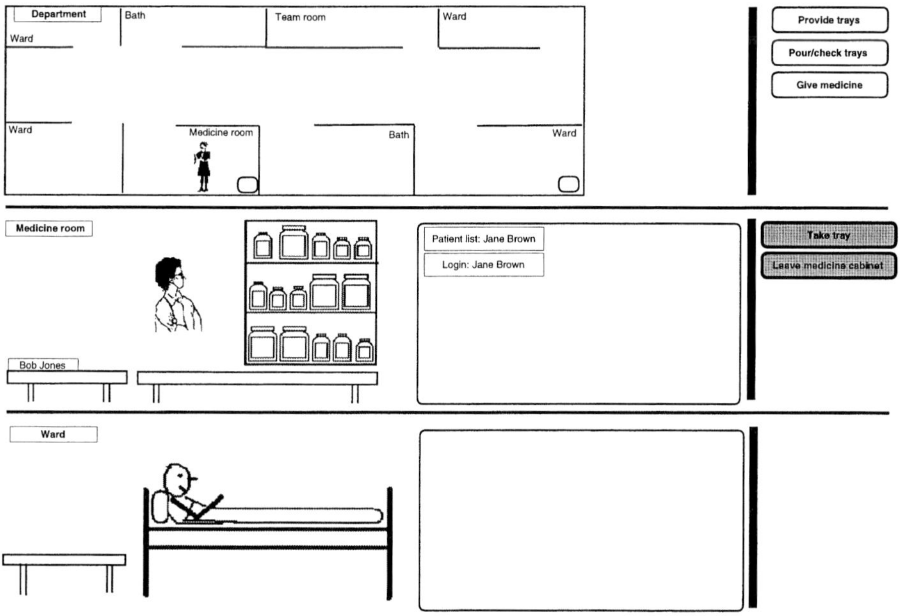

Fig. 24. Medicine administration animation.

they occur. Occurrence of a transition in this way triggers that graphical objects like nurse icons are created, moved, deleted, etc. in the animation.

The animation runs in three windows. The Department window (at the top of Fig. 24) shows the layout of a hospital department with wards, the medicine room, the so-called team room (the nurses' office), and two bathrooms. The Medicine room window (in the middle of Fig. 24) shows the medicine cabinet, pill boxes, tables, medicine trays, and the computer screen (enlarged). The Ward window (at the bottom of Fig. 24) shows a patient, a bed, a table, and the computer screen. Thus, the Department window gives an overview, and the other windows zoom in on areas of interest.

The animation is interactive in the sense that the animation user is prompted to make choices. In Fig. 24, the animation shows a situation where nurse Jane Brown is in the medicine room, shown in the Department window and the Medicine room window, sufficiently close to produce two task-bar buttons at the computer. The animation user must make choices in order to drive the animation further. Specifically, by selecting one of the buttons to the right in the Medicine room window, the animation user can choose to take a tray or leave the medicine room. Also, the animation user can select one of the task-bar buttons at the computer. These four choices correspond to enabled transitions in the CPN model. As examples, if the animation user pushes the Leave medicine cabinet button, it forces the transition with the same name in the CPN model (cf. Fig. 23) to occur. The result of the occurrence is experienced by the animation user who sees Jane Brown walking away from the medicine cabinet and the removal of the task-bar buttons on the computer screen, which were generated because of

Jane Brown's presence. If the animation user pushes the Take tray button and then selects Bob Jones' medicine tray, it is moved close to the computer, and a medicine plan button for Bob Jones appears on the task-bar. If this button is pushed, the computer will display a screen similar to the one shown in Fig. 21.

3.5 CPN in Requirements Engineering for PHCS

The PHCS project started in early 2001. The first activities were domain analysis in the form of ethnographic field work, and a series of vision workshops with participation of nurses, doctors, computer scientists, and an anthropologist. An outcome of this analysis was natural-language descriptions of work processes and their proposed computer support. The first version of the CPN model presented in this section was based on these prose descriptions. The CPN model and the animation were extended and modified in a number of iterations, each version based on feedback on the previous versions. The animation has served as a basis for discussions in evaluation workshops with participation of nurses from hospitals in Aarhus and personnel from the involved software company.

Through construction and use of the CPN model and the animation, in particular at the evaluation workshops, we have gained some experiences with CP-nets in requirements engineering. In the terminology of [89], we have seen that for PHCS, the CPN model and the animation have been an effective means for *specification*, *specification analysis*, *elicitation*, and *negotiation and agreement*. Each of these concepts will be discussed in more detail below.

Specification and Specification Analysis. Our specification has a sound foundation because of the formality and unambiguity of the CPN model. From the CPN model of medicine administration, requirements are precisely described by the transitions modelling manipulation of the involved computers. Each transition connected to the places modelling computers, e.g., the place Medicine cabinet computer shown in Fig. 23, must be taken into account. The following are examples of requirements induced by the transitions on the page of Fig. 23:

1. (R1) When a nurse approaches the medicine cabinet, the medicine cabinet computer must add a login button and a patient list button for that nurse to the task-bar (transition Approach medicine cabinet).
2. (R2) When a nurse leaves the medicine cabinet, if she is logged in, the medicine cabinet computer must return to a blank display, remove the nurse's login button and patient-list button from the task-bar, and log her out (transition Leave medicine cabinet).
3. (R3) When a nurse selects her login button, she must be added as a user of EPR, and the login button must be removed from the task-bar of the computer (transition Enter EPR via login button).

Specification analysis is well supported through simulation that allows experiments and trial-and-error investigations of various scenarios for the new envisioned work process. Specification analysis may also be supported through formal verification. However, the CPN model of medicine administration is too large and complex to make, e.g., verification by exploration of the full state space possible in practice. In general, we believe that the full state space of a CPN model made to support requirements engineering typically will be very large. The reason is that often, in the view of the users who should be actively involved in the requirements engineering process, a representation of a work process and its proposed computer support must include many details. This conflicts with modelling the work process in a more coarse-grained, abstract way, with a corresponding smaller state space. Therefore, verification of CPN models supporting requirements engineering is an application area where strong methods for state space reduction, condensation, and exploration are highly needed.

Elicitation. Elicitation includes the discovery of new requirements and the gain of a better understanding of known requirements. Elicitation is, like specification analysis, well supported through simulation. Simulation spurs elicitation by triggering many questions. Simulation of a CPN model typically catalyses the participants' cognition and generates new ideas. Interaction with an executable model that is a coherent description of multiple scenarios most likely brings up questions, and issues appear that the participants had not thought about earlier. Examples of questions (Qs) that have appeared during simulation of the CPN model for medicine administration and corresponding answers (As) are:

1. (Q1) What happens if two nurses are both close to the medicine cabinet computer? (A1) The computer generates login buttons and patient list buttons for both of them.
2. (Q2) What happens when a nurse carrying a number of medicine trays approaches a bed? (A2) In addition to a login button and a patient list button for that nurse, only one medicine plan button is generated – a button for the patient associated with that bed.
3. (Q3) Is it possible for one nurse to acknowledge pouring of medicine for a patient while another nurse at the same time acknowledges giving of medicine for that same patient? (A3) No, that would require a more fine-grained concurrency control exercised over the patient records.

Questions like Q1, Q2, and Q3 may imply changes to be made to the CPN model, because sometimes emergence of a question indicates that the current version of the CPN model does not reflect the work process properly. As a concrete example, in an early version of the medicine administration CPN model, the leaving of any nurse from the medicine cabinet resulted in the computer display being blanked off. To be compliant with the non-intrusive design principle for PHCS, the leaving of a nurse who is not logged in, should of course not disturb another nurse who might be working at the computer, and the CPN model had to be changed accordingly.

Negotiation and Agreement. Leaving practical issues such as being widely accepted by involved stakeholders aside, negotiation and agreement may be eased via CPN models. In large projects, negotiation about requirements inevitably takes place during the project. In many cases, this has strong economical consequences, because a requirements specification for a software system may be an essential part of a legal contract between a customer, e.g., a hospital, and a software company. Therefore, it is important to be able to determine which requirements were included in the initial agreement. Questions like Q1, Q2, and Q3 above may easily be subject to dispute. However, if the involved parties have the agreement that medicine administration should be supported, and have the overall stipulation that the formal and unambiguous CPN model is the authoritative description, many disagreements can quickly be settled.

3.6 Conclusions on Modelling Requirements to the PHCS

In this section, we have demonstrated that CPN models are able to support various common requirements engineering activities. However, of course, CP-nets are not a panacea. Use of CP-nets does not address, e.g., how to carry out the necessary initial domain analysis, interviews with users, etc. Moreover, the purpose of the presented CPN model is solely to describe the requirements of an IT system, relative to the work processes to be supported. A number of other requirements issues are not addressed properly by the CPN model, e.g., performance and availability issues.

The CPN model and the animation of the medicine administration work process can be seen as an alternative to or supplement to UML use cases [20,45]. Use cases model work processes to be supported by a new IT system, and a set of use cases is interpreted as functional requirements for that system.

A main motivation for our choice of requirements engineering approach for PHCS was to build on top of prose descriptions of work processes and proposed computer support, consolidated as UML use cases, with which the stakeholders of PHCS were already familiar via EPR. A key observation, done many times before, is that UML use cases have a number of weaknesses and shortcomings, e.g., [82] points out a number of problems under headlines like *use case modelling misses long-range logical dependency* and *use case dependency is non-logical and inconsistent*. Various remedies have been proposed, see, e.g., [2,3].

Having an executable representation of a work process, instead of a static representation in terms of a UML use case, supports specification analysis and elicitation as we discussed. This is possible via the CPN model itself, but can only be done properly by people who are able to read and understand the formal model. In practice, this often means only the system developers. The animation enables users like nurses and doctors to be actively engaged in specification analysis and elicitation, which is crucial. User participation increases the probability that a system is ultimately built that fits with the future users' work processes.

4 State Space Analysis of an Audio/Video Protocol

Bang & Olufsen [5] is a Danish manufacturer of audio/video systems. The project described in this section was originally conducted in 1995-1996 [14] and was concerned with the design of the next generation of the BeoLink system. The BeoLink system makes it possible to connect audio and video devices in a home via a dedicated network. The CPN modelling and analysis focused on the design of the *lock management protocol* in the BeoLink system. This protocol is used to grant devices exclusive access to services in the system, such as being able to use the loud speakers when playing music. The lock management protocol is based on the notion of a *key*, and a device is required to possess the key to access services in the system. When the system is switched on, exactly one key must be generated by the devices currently in the system. Furthermore, this key must be generated within 2 seconds for the system to be properly working. Special devices in the system called *audio* and *video masters* are responsible for generating the key when the system is switched on.

A CPN model modelling BeoLink systems with 1-4 devices was constructed in the original project and analysed using the state space method of CP-nets. The CPN model constructed in the project was timed, meaning that the time taken by the various events in the lock management protocol was reflected in the CPN model. This was needed since the correctness of the lock management protocol depends on timing. When the project was conducted, the CPN computer tools had only support for ordinary state spaces, i.e., state spaces in their most basic form. Since the ordinary state space of the timed CPN model was infinite, this meant that only the initialisation phase of the lock management protocol could be validated. The initialisation phase is concerned with generating the key when the system is switched on. Since then, a number of more powerful state space methods have been developed and implemented in the CPN computer tools.

In this section we give a brief presentation of a revised and more compact CPN model of the BeoLink system able to capture any number of devices. This is followed by a demonstration of how the more elaborate set of state space analysis methods currently available can be used to verify the full lock management protocol.

4.1 The Revised BeoLink CPN Model

Figure 25 shows the hierarchy page of the BeoLink CPN model. The subpage network models the network connecting the devices in the system. Page device and its subpages model the lock management protocol entities in each device. The subpages on the right, from reqkey down to fltimeo2 correspond to the different functional blocks of the lock management protocol. The subpage keyuser models the behaviour of devices as seen from the lock management protocol.

Figure 26 shows the BeoLink page. The substitution transition Network represents the network connecting the devices in the system. The substitution transition Device models the devices in the system. The CPN model provides a *folded* representation of the behaviour of the devices. This is achieved by encoding the identity of the devices as part of the colour of tokens. This makes it possible to capture any number of devices without having to make changes to the net structure of the CPN model, and without having an instance of the subpages of the substitution transition Device for each of the devices in the system. This

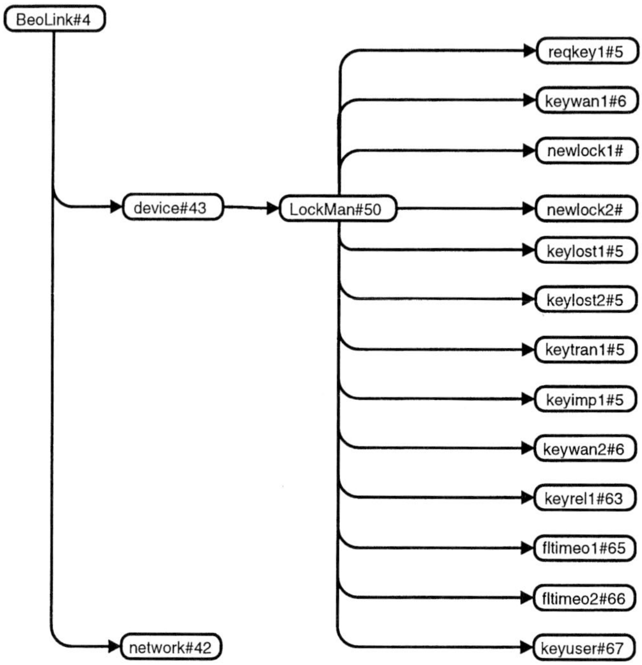

Fig. 25. Hierarchy page for the CPN BeoLink model.

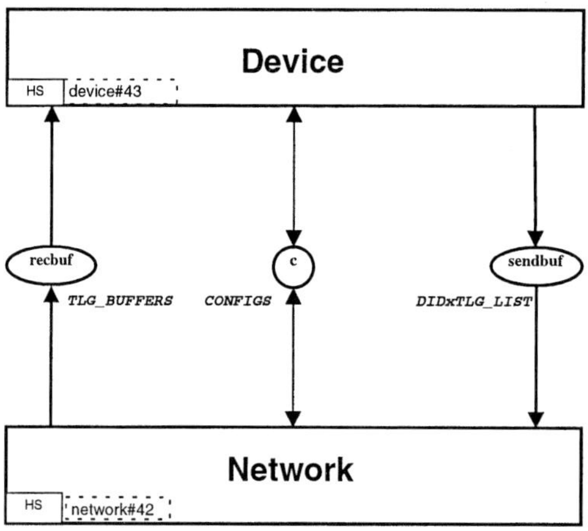

Fig. 26. The BeoLink page.

way of compactly representating any number of devices, and which makes the CPN model parametric will become evident when we present the keyuser page.

The socket places recbuf and sendbuf in Fig. 26 connecting the two substitution transitions, model send and receive message buffers between the devices and the network. Messages in the lock management protocol are called *telegrams* and are abbreviated TLG. Each device has a buffer for outgoing and incoming

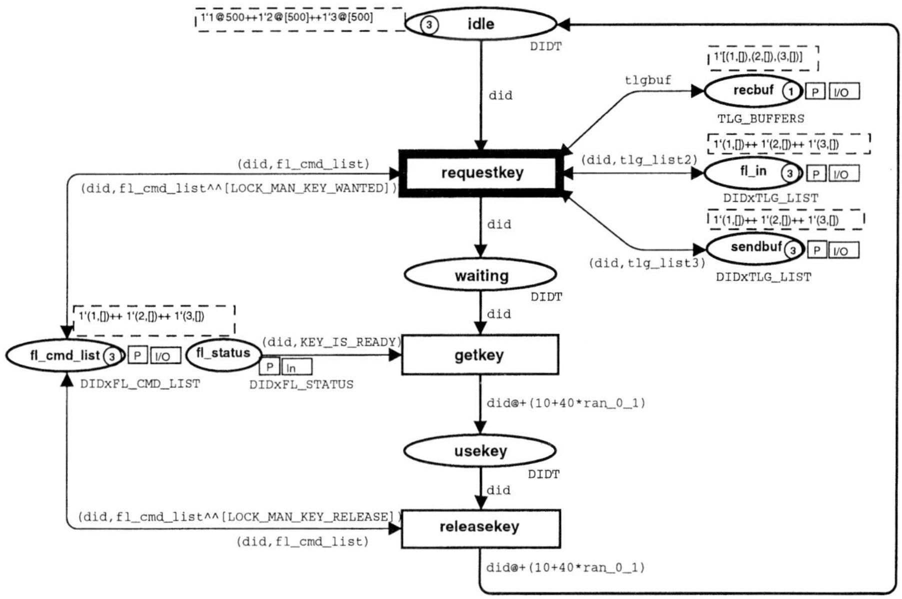

Fig. 27. The keyuser page - initial marking.

telegrams. The place c is used for configuration of the CPN model and will not be explained in any detail.

The behaviour of devices, as seen from the lock management protocol, is modelled by page keyuser shown in Fig. 27. Each device has a cyclic control flow where the device is initially idle (modelled by place idle), then it asks for the key (modelled by the transition requestkey), and it enters a state where it is waiting for the key (modelled by place waiting). Granting of the key to a device is modelled by the transition getkey which causes the device to enter a state where it is using the key (modelled by the place usekey). When the device is finished using the key, it will release the key and return to the idle state where it may then ask for the key again. The places fl_status, fl_cmd_list, and fl_in are used to model the internal state of a device. The places sendbuf and recbuf are linked to the accordingly named places on page BeoLink via port/socket relationship. The markings of these five places are also changed by the different functional blocks of the lock management protocol.

Figure 27 shows a marking of the CPN model with three devices all in their idle state, as represented by the three tokens on place idle. A device is simply identified by a number. In the marking shown in Fig. 27 any of the three devices may ask for the key corresponding to the requestkey transition being enabled in three different bindings depending on the device identifier assigned to the variable did of colour set DIDT. The domain of the DIDT colour set is the set of device identifiers. If the transition occurs in a binding with did $= 1$, the token with colour 1 will be removed from place idle and added to place waiting. Figure 28 shows a marking of page keyuser where device 1 uses the key, whereas devices 2 and 3 have requested but have not been granted the key.

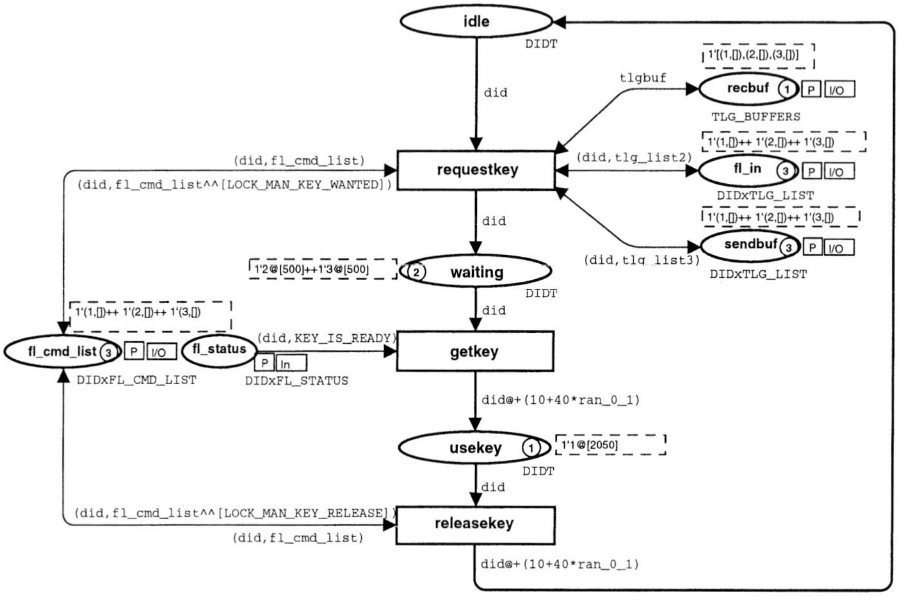

Fig. 28. The keyuser page - device 1 using the key.

The CPN model of the BeoLink system is timed. This means that the CPN model captures the time taken by the different events in the protocol. The time concept of CP-nets is discrete and is based on the introduction of a *global clock* used to represent *current model time*. Furthermore, in addition to a data value, tokens in a timed CPN model may carry time stamps. The time stamp of a token describes the earliest model time at which the token can be consumed, i.e., removed by the occurrence of a transition. The time stamps of tokens are written as part of the current marking. As an example, the three tokens on place idle in Fig. 27 all have time stamp 500. This can be seen from the number in square brackets written after the @ sign in the box showing the details of the tokens residing on that place. To model that an event corresponding to the occurrence of a transition takes r time units, the tokens added to output places of the transition are given a time stamp that is r time units larger than the model time at which the transition occurs. The time units to add to the current model time when tokens are produced by the occurrence of a transition are specified using the @+ operator. As an example, the transition getkey uses the @+ operator in the arc expression on the output arc leading to the place usekey. The time units to add to the current model time is specified by the expression 10+40*ran_0_1 where ran_0_1 is a variable that can be bound to either 0 or 1. This models that the event of obtaining the key take either 10 or 50 time units.

The execution of a timed CPN model is time driven. The CPN model remains at a given model time as long as there are enabled transitions at that model time. When no more transitions are enabled at the current model time, the global clock is incremented to the earliest next model time at which transitions are enabled. The model time in the marking shown in Fig. 27 is 500. Hence, transition requestkey is enabled since the time stamps of the tokens on place idle are less than or equal to current model time. The model time in the marking shown in Fig. 28 is 2036. This is the reason why the releasekey transition is not enabled, since the time stamp of the token residing on place usekey is 2050. In the marking shown, transitions are enabled in the other pages of the CPN model.

4.2 Full State Spaces

The basic idea of state spaces is to calculate all reachable states and state changes of the system and represent these as a directed graph. The state space of a CPN model has a node for each of its reachable markings, i.e., markings that can be reached by occurrences of transitions starting from the initial marking. The outgoing arcs of a node n in the state space correspond to the set of enabled *binding elements* in that marking. A binding element is a pair consisting of a transition and an assignment of values to the variables of the transition. The destination node of an arc originating in node n is the node representing the marking resulting from the occurrence of the binding element corresponding to the arc in the marking represented by node n.

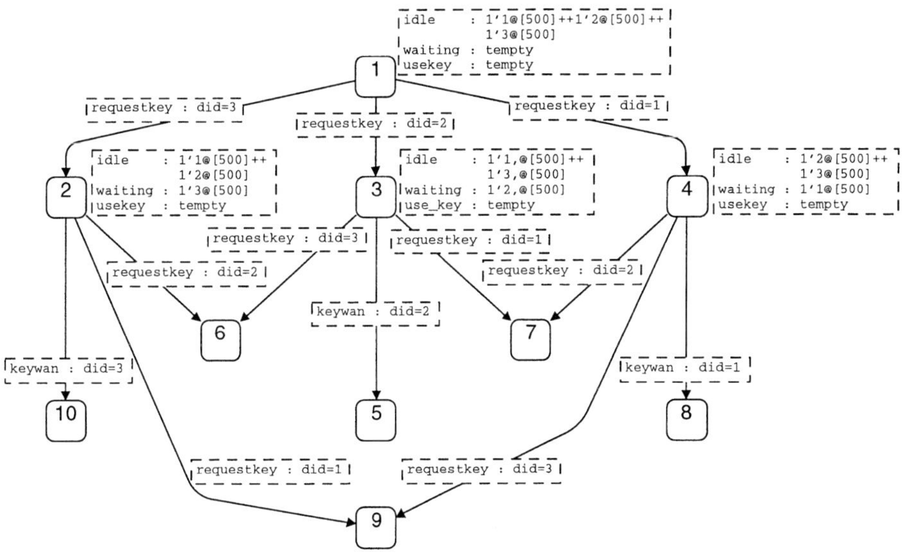

Fig. 29. Initial fragment of state space.

Figure 29 shows an initial fragment of the state space for the BEOLINK system. Node 1 corresponds to the marking previously shown in Fig. 27. Figure 29 shows the nodes in the state space that are reachable by at most two occurrences of transitions starting from node 1. In the marking corresponding to node 1, there are three enabled binding elements corresponding to the three outgoing arcs from node 1. This three outgoing arcs correspond to all three devices being able to request the key in the marking corresponding to node 1. The dashed boxes shown next to nodes 1-4 list the tokens present on selected places on the keyuser page in the marking represented by the node. The constant tempty denotes the empty set of tokens in a timed CPN model. These boxes have been omitted for some nodes to make the figure readable. The dashed boxes positioned on top of the arcs describes the enabled binding element to which the arc corresponds. The transition keywan is on another page in the CPN model.

Figure 29 was created using the support in the CPN computer tools for drawing selected parts of a state space. The CPN computer tools make it possible to generate the state space manually as well as automatically. The state space can be generated either depth-first or breadth-first. From a constructed state space it is possible to automatically verify a number of properties of the system such as absense of deadlocks and other safety properties. The CPN computer tools contain a number of query functions that allow the analyst to investigate and verify the system using state spaces. The three main correctness criteria of the lock management protocol are listed below.

1. (C1) Key generation. When the system is booted, a key is eventually generated. The key is to be generated within 2.0 seconds.
2. (C2) Mutual exclusion. At any time during the operation of the system at most one key exists.
3. (C3) Key access. Any given device always has the possibility of obtaining the key, i.e., no device is ever excluded from getting access to the key.

In the original analysis conducted in [14] only the first property was verified. The remaining properties could not be verified due to the state space of the timed CPN model being infinite. The key generation property was investigated by considering a partial state space, i.e., a finite fragment of the full state space. The partial state space was obtained by not generating successors for states where the key had been generated or where the model time had passed two seconds. It was then checked that in all markings for which successor states had not been generated, a key was present in the system. Table 1 lists some statistics showing the number of states in the partial state space and the CPU time it took to generate the partial state space. Configurations written with the form $VM : n$ are configurations with a video master and a total of n devices. Similarly, configurations with one audio master and a total of n devices are written with the form $AM : n$. CPU time is written on the form $h : mm : ss$ where h is

Table 1. Statistics for partial state space of initialisation phase.

Config	Nodes	Time
AM : 3	1,839	0:00:07
AM : 4	22,675	0:02:42
AM : 5	282,399	1:47:44
VM : 3	1,130	0:00:04
VM : 4	13,421	0:01:26
VM : 5	164,170	0:58:28

hours, *mm* is minutes, and *ss* is seconds. The results using partial state spaces and the revised CPN model were obtained on a HP Unix Workstation with 1 Gb of memory.

4.3 Timed Condensed State Spaces

A main problem with state spaces in their most basic form is that they are infinite for timed CPN models of cyclic/reactive systems. The problem is that the absolute notion of time as represented by the global clock and the time stamps of tokens are carried over into the state space. The BEOLINK system is an example of a cyclic system since the devices are executing a loop where they request the key, are granted the key, and finally release the key. As a concrete example, consider the marking of the keyuser page shown in Fig. 30. This marking is similar to the marking previously shown in Fig. 28, except that all devices have had the key once and device 1 now possesses the key again. The markings in Fig. 28 and Fig. 30 are, however, represented by two nodes in the state space because the time stamps of the tokens and the value of the global clock differ. Intuitively, the markings are, however, similar.

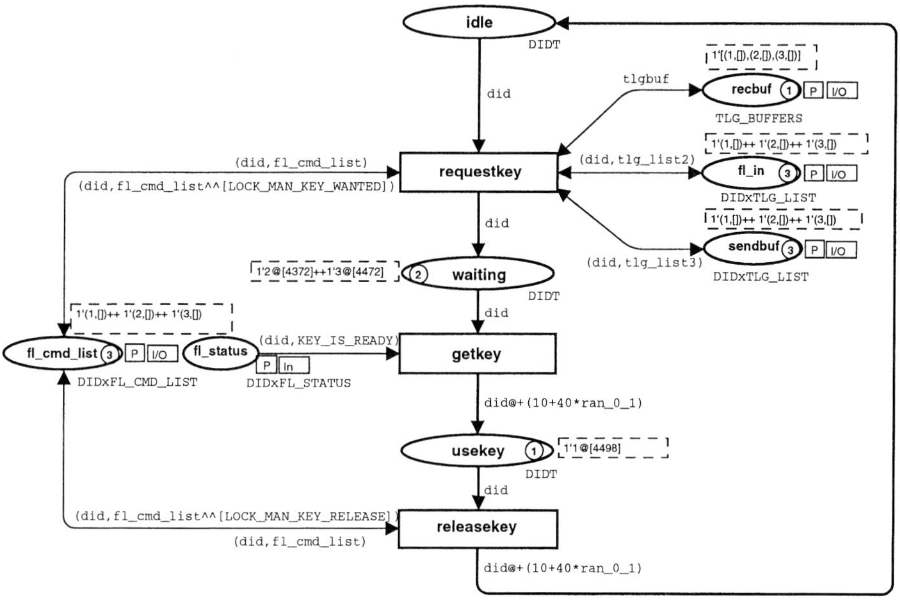

Fig. 30. The keyuser page - all devices have used the key once.

Timed condensed state spaces [16] have been developed to overcome this problem, and use equivalence on the states to factor out the absolute notion of time. In this way, the infinite state space can be condensed into a finite state space. The *condensed state space* can be computed using a variant of the standard algorithm for state space construction, but without first constructing the full state space. The basic idea is to normalise each state encountered during state space generation by:

1. (N1) Setting all time stamps on tokens that are less than current model time to zero (as their time stamp cannot influence enabling).
2. (N2) Subtracting the current model time from all time stamps of tokens that are greater than current model time.
3. (N2) Setting the current model time to 0.

It have been proven [16] that all properties of the system expressible in the real-time temporal logic RCCTL* [31] are preserved in the condensed state space. This set of properties includes all standard dynamic properties of CP-nets. Table 2 shows statistics for the condensed state space for the full BeoLink system. The results were obtained on a HP Unix Workstation with 1Gb of memory. It was not possible to generate the time condensed state space for more than 3 devices with the available amount of memory. Using the condensed state space it is now also possible to verify properties C2 and C3 from Sect. 4.2. Property C2 can be expressed as the property that in no reachable marking is there more than one token on place usekey (see Fig. 27), and property C3 can be expressed as the property that from any reachable marking and for any device it is always possible to reach a marking where the token corresponding to this device is on place usekey. These two properties can be expressed using the query functions in the CPN state space tool and answers was computed in a few seconds.

Table 2. Statistics for the time condensed state spaces.

Config	Nodes	Arcs	Time
AM : 2	346	399	0:00:03
AM : 3	27,246	37,625	0:04:10
VM : 2	274	310	0:00:02
VM : 3	10,713	14,917	0:01:34

4.4 The Symmetry Method

Many concurrent systems possess a certain degree of symmetry. For example, many concurrent systems are composed of similar components whose identities are interchangeable from a verification point of view. This symmetry is also reflected in the state spaces of such systems. The basic idea in the *symmetry method* [17, 19, 32, 43, 48, 49] is to represent symmetric states and symmetric binding elements using *equivalence classes*. State spaces can be reduced by factoring out this symmetry, and the symmetry-reduced state space is typically orders of magnitude smaller than the full state space. Furthermore, the same set of dynamic properties can be verified and analysed based on the symmetry-reduced state space without unfolding to the full state space.

The devices in the BeoLink system that are not audio or video masters are symmetric, in the sense that they behave in the same way. They are only distinguishable by their device identity. This symmetry is also reflected in the state space (see Fig. 29). Consider, for instance, the two states 2 and 3 that correspond to states in which exactly one non-master device (device 1 is the audio master in the considered configuration) has requested the key. These two states are symmetric in the sense that node 2 can be obtained from node 3 by swapping the identity of device 2 and 3. Similarly, the two states represented by node 5 and node 10 can be obtained from each other by interchanging the identity of devices 2 and 3. These two states correspond to states in which one device has requested the key and the lock management protocol has registered the request. Furthermore, it can be observed that two symmetric states such as state 2 and state 3 have symmetric sets of enabled binding elements, and symmetric sets of successor states. This property can be extended to finite and infinite occurrence sequences of transitions.

Figure 31 shows the initial fragment of the symmetry-reduced state space for the BeoLink system obtained by considering two states equivalent if one can be obtained from the other by a permutation of the identity of the non-master devices. The nodes and arcs now represent equivalence classes of markings and binding elements, respectively. The equivalence class of states represented by a node is listed in brackets in the inscription of the node, e.g., node 2 represents the states 2 and 3 from Fig. 29. A similar notation is used for binding elements. The basic idea in symmetry-reduced state spaces is to represent these equivalence classes by picking a representative for each equivalence class. The symmetries used to reduce the state space are required to be symmetries actually present in the CPN model. The CPN model is, therefore, required to satisfy a set of static properties relative to the set of symmetries to be used for the reduction [48].

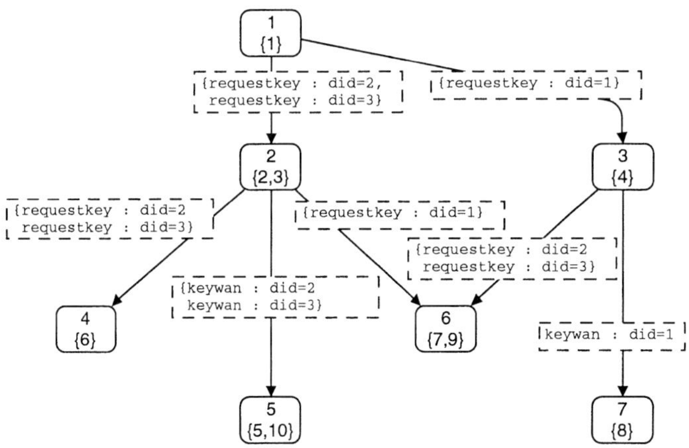

Fig. 31. Initial fragment of symmetry-reduced state space.

Table 3 shows the results when using the symmetry method on the initialisation phase of the BeoLink system. The size of the full state space for the

Table 3. Statistics for symmetry reduced state space - initialisation phase.

	Full State Spaces		**Symmetry Reduced**			
Config	Nodes	Time	Nodes	Time	Factor	$(n-1)!$
AM : 3	1,839	0:00:07	53.0 %	100.0 %	1.9	2
AM : 4	22,675	0:02:42	19.3 %	40.1 %	5.2	6
AM : 5	282,399	1:47:44	5.6 %	10.4 %	17.8	24
AM : 6	3,417,719	-	1.4 %	2:13:29	71.4	120
VM : 3	1,130	0:00:04	53.2 %	100.0 %	1.9	2
VM : 4	13,421	0:01:26	19.4 %	40.6 %	5.1	6
VM : 5	164,170	0:58:28	5.6 %	10.1 %	17.6	24
VM : 6	1,967,159	-	1.4 %	1:10:35	71.4	120
VM : 7	22,892,208	-	0.3 %	≈15 hours	333.3	840

AM:6, VM:6, and VM:7 configurations has been calculated from the symmetry-reduced state space by computing the size of each equivalence class. The results were obtained on a HP Unix Workstation with 1Gb memory. Calculation of the symmetry-reduced state space is based on calculating canonical representatives for each equivalence class [63]. This means that whenever a state is generated, this state is transformed into a canonical representative for its equivalence class. It is then checked whether this canonical representative is already included in the state space. The numbers in the Nodes column for the symmetry-reduced state space are relative to the number of nodes in the full state space, i.e., the number of nodes in the symmetry-reduced state space divided by the number of nodes in the full state space. The numbers in the Time column are also relative to the generation of the full state space for those configurations where the full state space could be generated. The Factor column gives the number of nodes in the full state space divided by the number of nodes in the symmetry reduced state space. The column $(n-1)!$ lists the factorial of $n-1$ where n is the number of devices in the configuration. When there are n devices in the configuration, there are $n-1!$ possible permutations of the non-master devices. Hence, $(n-1)!$ is the theoretical upper limit on the reduction factor that can be obtained for a configuration with n devices. It can be seen that the computation time becomes large for 7 devices. This is due to the calculation of canonical representative being costly. It has been proven [17] that computing canonical representative for equivalence classes is at least as hard as the *graph isomorphism problem* for which no polynomial time algorithm is known.

Table 4 lists the statistics for the symmetry-reduced state space of the full BeoLink system. Here we have used the symmetry method and the time condensed state space simultaneously. The number of nodes for the $AM : 4$ and $VM : 4$ configurations in the time condensed state space has been computed from the symmetry reduced state space.

4.5 The Sweep-Line Method

The amount of available main memory is often the limiting factor in the use of state spaces. During construction of the state space, the set of markings en-

Table 4. Statistics for symmetry reduced state space - full system.

Config	Time Equivalence		Symmetry Reduced			
	Nodes	Time	Nodes	Time	Factor	$(n-1)!$
AM : 2	346	0:00:03	100.0 %	100.0 %	1	1
AM : 3	27,246	0:04:10	50.1 %	52.0 %	1.9	2
AM : 4	12,422,637	-	16.7 %	≈25 hours	5.9	6
VM : 2	274	0:00:02	100. %	100.0 %	1	1
VM : 3	10,713	0:01:34	50.6 %	50.0 %	1.9	2
VM : 4	3,557,441	-	16.7 %	7:10:21	5.9	6

countered are kept in memory to recognise already visited marking and thereby ensure that the state space generation terminates. The basic idea of the sweep-line method [15, 59] is to exploit a certain kind of *progress* exhibited by many systems. Exploiting progress makes it possible to explore all the reachable markings of a CPN model, while only storing small fragments of the state space in memory at a time. This means that the peak memory usage is reduced. The sweep-line method was used in [38] for verification of transactions in the Wireless Application Protocol (WAP) with a reduction in peak memory usage to 20%. The sweep-line method is aimed at on-the-fly verification of safety properties, e.g., determining whether a reachable marking exists satisfying a given state predicate. Hence, it can be used to verify properties C1 and C2 of the BeoLink system, but not property C3.

The Basic Sweep-Line Method. For the BeoLink system, one source of progress is the time in the system (the model time) as represented by the value of the *global clock* in the CPN model. The global clock in a timed CP-net [48] has the property that for two markings M and M', where M' is a successor marking of M, the value of the global clock in M is less than or equal to the value of the global clock in M'. This progress can be formalised as a *progress measure* mapping a marking into the corresponding value of the global clock. The progress measure based on the global clock is a *monotonic progress measure.*

Figure 32 shows how the markings/nodes in the state space fragment from Fig. 29 can be ordered according to this notion of progress. The intuition of the ordering is the following: markings in one layer all have the same value of the progress measure (the global clock), and markings in higher numbered layers are markings where the system has progressed further than in markings in lower numbered layers. Layer 0 contains markings in which the global clock has value 0. Layer 1 contains markings where the global clock is 500 time units.

A marking in a given layer has successor markings either in the same layer or in a layer that represents further progress, but never in a layer that represents less progress. Markings in Layer 0 can thus never be reached by markings in Layer 1. If we calculate the markings one layer at a time, moving from one layer to the next when all markings in the first layer have been calculated and not before, we can think of it as a sweep-line moving through the state space. At any

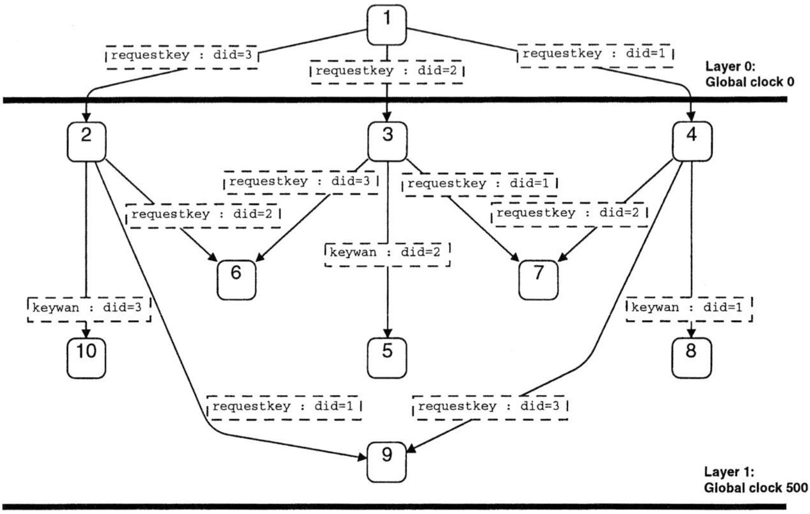

Fig. 32. Initial fragment of full state space – arranged by progress.

one point during state space generation, the sweep-line corresponds to a single layer—all the states in the layer are "on" the sweep-line—and all new markings calculated are either on the sweep-line or in front of the sweep-line. Table 5 lists the statistics for the application of the sweep-line method on the initialisation phase of the BeoLink system and using the global clock as the progress measure. The figures in the Sweep-Line Method column are given relative to the numbers in the Full State Spaces columns. The results were obtained on a Pentium II PC with 160 Mb of memory.

Table 5. Application of the sweep-line method – initialisation phase.

	Full State Spaces		**Sweep-Line Method**	
Config	Nodes	Time	Peak Nodes	Time
AM: 3	1,839	0:00:11	100.0 %	100.0 %
AM: 4	22,675	0:05:32	22.8 %	84.0 %
AM: 5	282,399	5:03:53	12.4 %	39.4 %
VM: 3	1,130	0:00:06	100.0 %	100.0 %
VM: 4	13,421	0:02:40	38.5 %	106.0 %
VM: 5	164,170	2:30:27	21.3 %	45.4 %

The Generalised Sweep-Line Method. While the basic idea behind the sweep-line described above is intuitive and simple, it has the obvious drawback that it only works on systems exhibiting this kind of monotonic progress. While a lot of systems have a certain degree of progress, it is usually not strictly monotonic. There will be occasional occurrences of binding elements from high-progress markings to low-progress markings. The generalised sweep-line method [59] solves this problem by introducing multiple sweeps of the state space. Each sweep follows only binding elements that result in markings with unchanged or increasing progress measure and collects information about *regress-arcs* that result in markings with decreasing progress measure. The markings at the end of regress-arcs are then marked as persistent meaning that they cannot be deleted again, and they are used as starting point for a subsequent sweep. The generalised sweep-line method visits all the reachable markings, but may visit some markings multiple times.

To apply the sweep-line method for the full BeoLink system we first need to obtain a finite state space using the time condensed state spaces as described in Sect. 4.3. This, however, has the drawback that the value of the global clock becomes 0 in all markings. Hence, the progress measure defined above based on the global clock will map all markings into 0, resulting in no peak memory reduction when we apply the sweep-line method. It is however possible to define a non-monotonic progress measure for the BeoLink system based on the control flow of the devices. Recall that the devices have a cyclic control flow where they are first idle, then they request the key, and finally they obtain the key. When they have used the key they return to the idle state. This is a kind of local progress starting from the idle state progressing towards the state where they have the key. This ordering on the states of the devices can be used to define a non-monotonic progress measure. Details of such a progress measure can be found in [59]. With this progress measure, the marking shown in Fig. 28 will have a higher progress value than the marking shown in Fig. 27. When a device releases the key and moves to the idle state, then this will result in a regress-arc in the state space. Table 6 lists statistics for the application of the generalised sweep-line method to the full BeoLink system using the progress measure informally defined above. The experiments were conducted on a Pentium II PC with 160 Mb of memory. It can be seen that some states are explored multiple times which causes a time penalty, but the sweep-line method still achieves a reduction in peak memory usage to about 10 %. The relatively large time penalty is due to an inefficient implementation of deletion of states in the Design/CPN Sweep-Line Library [37]. A more efficient algorithm for deletion of states has been developed in [60].

Table 6. Application of the sweep-line method – full system.

	Time Equivalence		**Sweep-Line Method**		
Config	Nodes	Time	Nodes Explored	Peak Nodes	Time
AM:2	346	00:02	102.6 %	18.8 %	200.0 %
AM:3	27,246	06:54	104.1 %	9.7 %	327.8 %
VM:2	274	00:02	103.3 %	15.0 %	200.0 %
VM:3	10,713	02:19	106.3 %	9.7 %	207.2 %

Above we have seen that it is possible to combine time condensed state spaces with both the symmetry method and the sweep-line method. It is also possible to use the sweep-line method and the symmetry method simultaneously. This combination was investigated in [7] where it was demonstrated by means of experimental results that using the two methods simultaneously leads to better reduction than when either method is used in isolation.

4.6 Conclusions on Audio/Video Protocol and State Space Analysis

The revised state space analysis of the BeoLink system illustrates the use of the state space reduction methods that have been developed and implemented in the CPN computer tools in recent years. In addition to time condensed state spaces, the symmetry method, and the sweep-line method, several other methods have been developed to combat the state explosion problem. Examples of these include partial order reduction methods [70,87,93], the unfolding method [34,65], and methods based on Binary Decision Diagrams (BDDs) [66]. Until now, the above methods have only been used in practice on low-level Petri nets or by unfolding the high-level Petri net into the equivalent low-level Petri net. For CPN models constructed in industrial projects which often have variables from infinite domains, approaches based on unfolding to low-level are not feasible. Some work has been done on developing a version of the stubborn set method for CP-nets without having to rely on unfolding to low-level Petri net [56]. Methods that appear more promising for being included in the CPN computer tools include the *bit-state hashing method* [40,41] and the *state space caching method* [39,46] which both are based on ideas similar to the sweep-line method, i.e., deleting information about states during the state space exploration. In general, the CPN computer tools must support a suite of state space reduction methods since these reduction methods exploit different characteristics of the modelled system to achieve the reduction. As a consequence, only some reduction methods will work on a given CPN model. The protocol verification technique used, e.g., in [38] based on language comparison to verify a protocol specification against a service specification is another candicate for inclusion into the CPN computer tools.

The support for state space methods in the CPN computer tools differs from other tools, such as SPIN [85], in its support for drawing selected parts of a state space and the support for a query language not based on temporal logic and model checking [18] but on functions to traverse the state space and extract information from the nodes and arcs. While it seldom makes sense to draw the full state space of a system, practical experience has shown that being able to visualise small fragments of the state space is an efficient way of investigating local behavior of the system in detail. A main reason for supporting a query language that allows the user to write traversals of the state space is that it provides better support for analysis. With the available query functions it is possible to compute quantitative values such as e.g., minimum and maximum number of tokens on a place rather than just yes/no answers as supported by temporal logic. Furthermore, query functions for typical dynamic properties of CPN models is also available. Instantiating these query functions is much more convenient for practitioners than writing the equivalent formulas in temporal logic. Support for CTL model checking [11] is, however, available as a library to the state space tool.

Another feature of the state space tool that has shown to be valuable in the practical use of state space methods is the support for generation of a predefined *state space report.* The state space report contains information regarding a set of standard dynamic properties of CP-nets and can be generated fully automatically and then inspected by the user. Generation of the state space report is usually the first activity in state space analysis, and many errors and problems in a design are often detectable from the state space report.

5 Implementation of a Planning Tool

This project [94] is concerned with the development of the Course of Action Scheduling Tool (COAST) by the Australian Defence Science and Technology Organisation (DSTO) [4]. A Course of Action (COA) (also referred to as a plan) consists of a set of tasks. The key capability of COAST is the computation of task schedules called *line of operations* (LOPs) and is aimed at supporting the planner in COA Development and COA Analysis which are two of the main activities in a military planning process. The basic idea in COAST is to use a CPN model for modelling the execution of tasks according to the pre- and postconditions of tasks, imposed synchronisations, and available resources. The LOPs are then obtained by generating a state space for the CPN model and extracting paths in the state space leading from the initial marking to certain markings representing *end-states.*

The COAST planning tool has been developed in close cooperation with DSTO researchers, the Computer System Engineering Centre at University of South Australia [21], and planners at the Australian Defence Force. The latter group is the envisioned user of the tool. The role of CP-nets in the development of COAST has been threefold. Firstly, CPN modelling has been used in the development and specification of the underlying framework. Secondly, the constructed CPN model has been used directly in the implementation of COAST by embedding it into the COAST server which constitutes the computation back-end of COAST. Hence, CP-nets provide the semantical foundation by formalising and implementing the abstract conceptual framework underlying the tool. Finally, the analysis capabilities of COAST are based on state space methods.

5.1 An Example Plan

In this section we give a brief overview of the conceptual framework of the COAST and present a small example plan used as a running example throughout this section. The framework underlying COAST is based on four key concepts:

Tasks are the basic units in a plan and have associated preconditions and effects describing the conditions required for a task to start and the effect of executing the task. A task also includes a specification of the resources required to execute the task, and may also have a specified duration. Tasks also have other attributes, but these are omitted in this presentation.

Conditions are used to describe the explicit logical dependencies between tasks via preconditions and effects. As an example, a task T1 may have an effect used as a precondition of a task T2. Hence, T2 logically depends on T1 in the sense that it cannot be started until T1 has been executed.

Resources are used by tasks during their execution. Resources typically represent planes, ships, and personal required to executed a task. Resources may be available only at certain times due to e.g., service intervals. Resources may be lost in the course of executing a task.

Synchronisation can be used to capture that a set of tasks must begin or end simultaneously, that there has to be a specific amount of time between the start and end of certain tasks, and that a task can only start after a certain point in time. A set of tasks that are required to begin at the same time is said to be *begin-synchronised.* A set of tasks required to end at the same time is said to be *end-synchronised.* End-synchronisations can cause the duration of tasks to be extended.

Table 7 lists an example of a plan with 6 tasks. The table specifies for each task its preconditions, its effects, the required resources, and the duration of the tasks. In addition to the information provided in the table, the set {T5,T6} of tasks are begin-synchronised and the set {T4,T5,T6} of tasks are end-synchronised. The assigned resources are: 4'R1++3'R2++3'R3++1'R5++1'R6 (written as a multi-set). Figure 33 provides a graphical illustration of the dependencies between tasks using dashed lines to indicate begin-synchronisations and end-synchronisations.

Table 7. A example plan with 6 tasks.

Task	Preconditions	Effects	Resources	Duration
T1	-	E1	4'R1	2
T2	E1	E2	2'R2 ++ 2'R3	4
T3	E1	E3	2'R2 ++ 2'R3	7
T4	E1	E4	1'R2 ++ 1'R3	-
T5	E2	E5	1'R5	7
T6	E3	E6	1'R6	7

For this example, we are interested in the possible ways (if any) that the set of tasks can be sequenced such that a state satisfying conditions E4, E5, and E6 can be reached given the assigned resources and synchronisation constraints. Figure 34 illustrates one such possible *line of operation* (LOP) by giving the start and end time of tasks such that the desired end-state is reached.

5.2 Engineering COAST

Figure 35 shows the client-server software architecture of COAST. The COAST client, which includes a domain-specific graphical user interface, is implemented

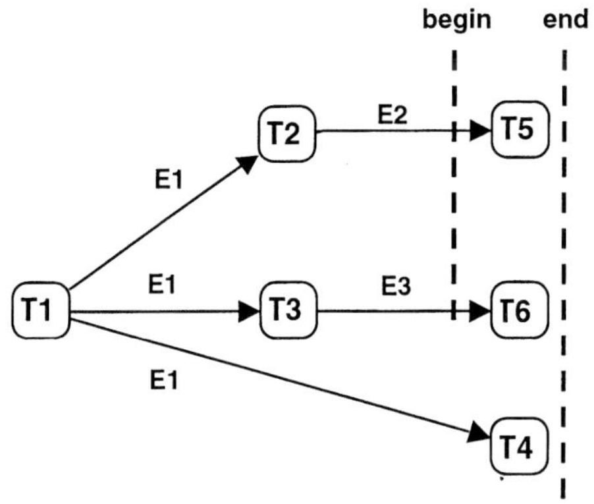

Fig. 33. Illustration of dependencies between tasks in the example plan.

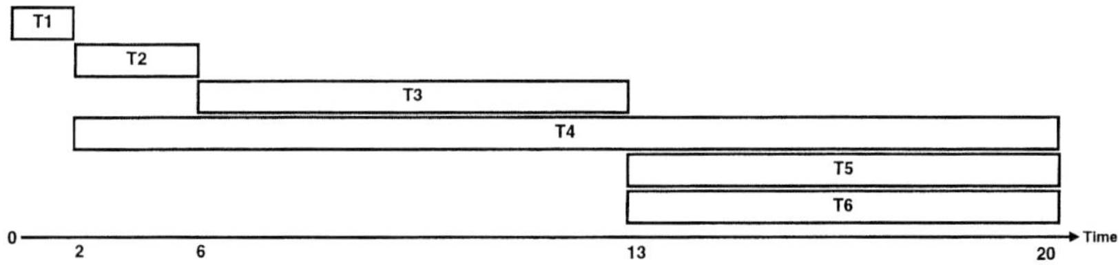

Fig. 34. One possible line of operation for the example plan.

in Java, whereas the COAST server is implemented in Standard ML (SML) via the embedded CPN model which forms the core of the COAST server. Communication between the client and the server is based on Comms/CPN and Comms/JAVA [36], a library supporting TCP/IP communication between CPN models and external applications. A SML session layer has been implemented on top of Comms/CPN. This layer allows the client to invoke functions available in the server and receive the corresponding results. The SML Session layer is implemented by allowing the client to submit SML code to the server for evaluation. The received SML code is then executed by the server, and results are sent back to the client. The SML code sent to the server corresponds to the invocation of the SML functions made available by the server by the COA Analysis module. The COAST client consists of two main parts: an Editor for creating and editing plans, and an Analyser for the analysis of plans. The COAST server consists of three main parts. The Initialisation module allows the CPN model to be initialised according to the plan to be analysed. The Simulation Code module for executing the CPN model which consists of the simulation code generated by the CPN computer tools for executing CPN models. The State Space Code and COA Analysis modules support the generation of state spaces and LOPs. The State Space Code module consists of the code for generation of state spaces in the CPN computer tools.

Figure 36 is a snapshot from the COAST client illustrating how the user views the plans in the editor. There are four main windows. A window displaying the

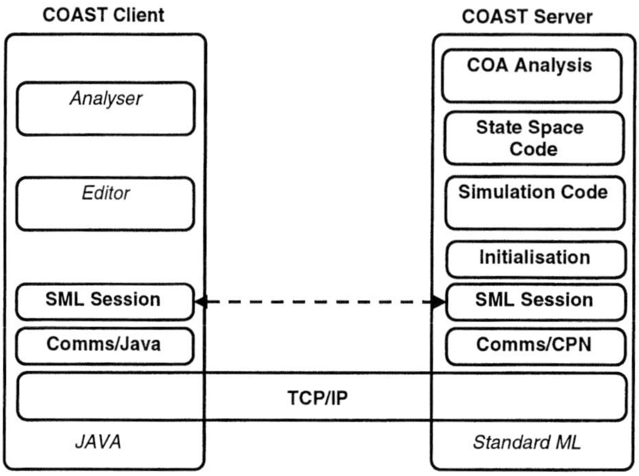

Fig. 35. Architectural overview of the COAST tool.

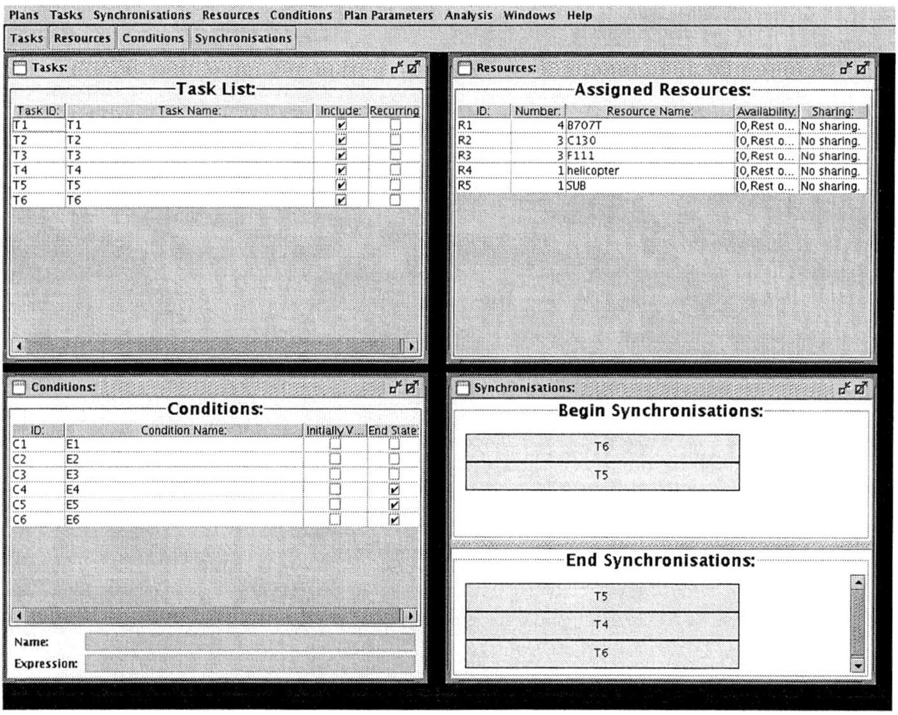

Fig. 36. Snapshot from the COAST editor.

set of tasks, a window showing the assigned resources, and a window showing the conditions, and a window showing the synchronisations.

Figure 37 shows an example of how LOPs are reported to the user in the Analyser part of the COAST client. The window gives a specification of the LOP

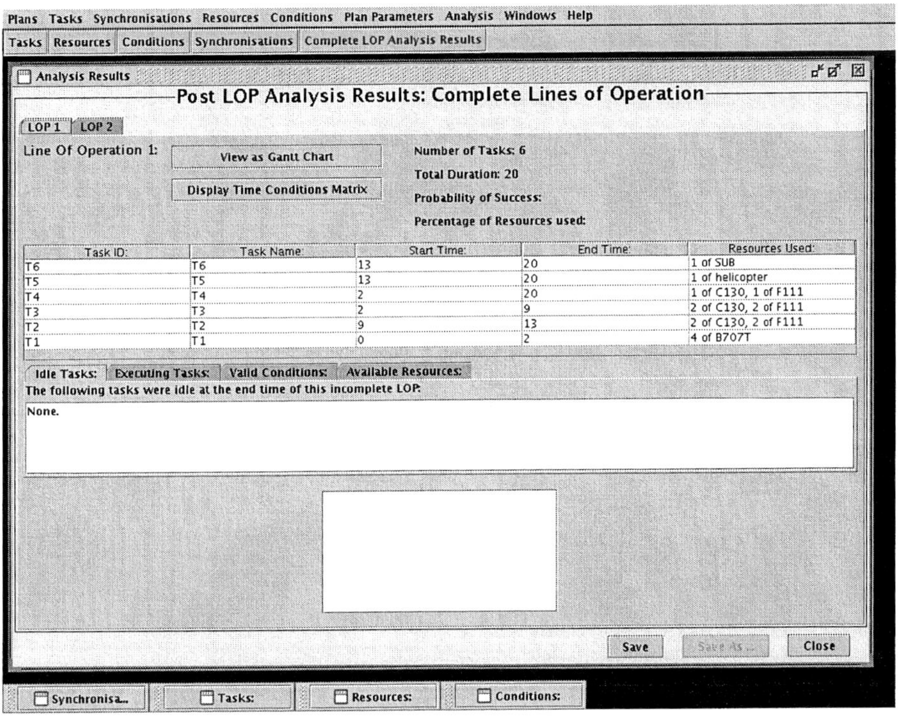

Fig. 37. Snapshot from the COAST analyser.

for the example plan corresponding to the one previously shown in Fig. 34. That the COAST server uses a CPN model as a basis for the scheduling analysis is fully transparent to the analyst using the COAST client.

5.3 The CPN Model

The CPN model has been parameterised with respect to the set of tasks, resources, conditions, and task synchronisations. This ensures that a given set of tasks, resources, and task synchronisation can be analysed by setting the initial marking of the CPN model accordingly, i.e., no changes to the structure of the CPN model are required to analyse a different set of tasks. Figure 38 shows the hierarchy page for the CPN model. The page CoastServer is the top level page in the CPN model which consists of three main parts. Page Execute (left) and its subpages model the execution of tasks, i.e, start, termination, abortion, and failure of tasks according to the set of tasks, resources, conditions, and synchronisation in the plan. Page Environment and its subpages model the environment in which tasks execute, and is responsible for managing the availability of resources over time, change of conditions over time, and task failures. Page Initialisation and its subpages are used for the initialisation of the model according to the concrete set of tasks, synchronisation, and resources in a plan. The CPN model

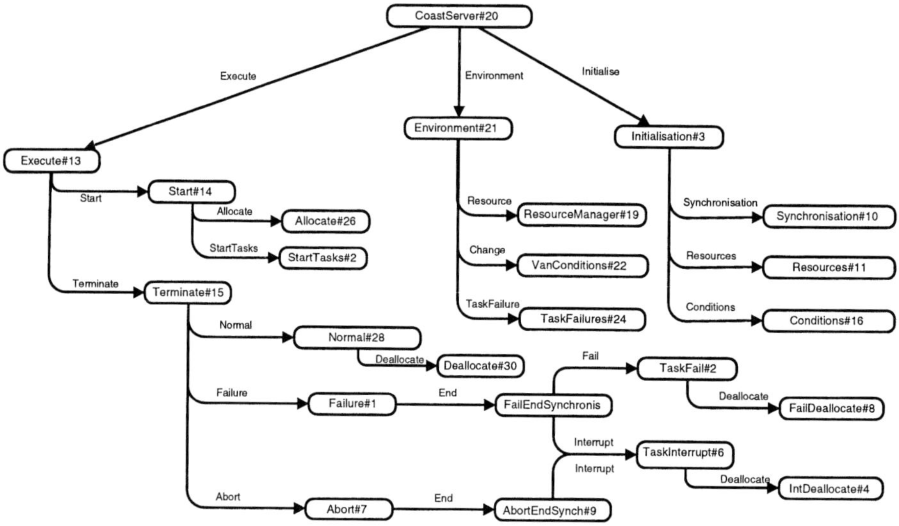

Fig. 38. Hierarchy page of the COAST CPN model.

is timed since capturing the time taken by executing a task is an important part of the computation of LOPs.

The COAST server was obtained from the CPN model by first generating the standard simulation and state space code. SML files implementing Comms/CPN and the SML session layer were then loaded together with the functions implementing the server and the LOP generation algorithms. The resulting executable file constitutes the COAST server and includes the functions required to execute the CPN model and to conduct state space analysis. The COAST server is totally detached from the GUI of the CPN computer tools. When the COAST server is started, it will wait for an incoming TCP connection, and once the COAST client has established a connection, it can start invoking functions on the COAST server and thereby conduct the task scheduling analysis.

Figure 39 shows the top level page of the CPN model with the three main parts of the CPN model represented as the substitution transitions Initialise, Execute, and Environment. The marking shown is the marking of the CPN model after initialisation of the CPN model with the example plan from Table 7. Place Tasks contains six tokens corresponding to the six tasks in the example plan. Place Conditions contains one token which is a list containing the conditions in the plan and their truth value. It can be seen that all conditions are initially false. Place Resources contains two tokens. There is one token consisting of a list describing the current set of idle (available) resources, and one token consisting of a list describing the resources that have been lost until now. Since the colour of the tokens on the places Resources and Tasks are of a complex colour set, we have not shown the detailed colours of the tokens but only the number of tokens. As an example, the colour set Task modelling tasks is record type with more than 15 fields.

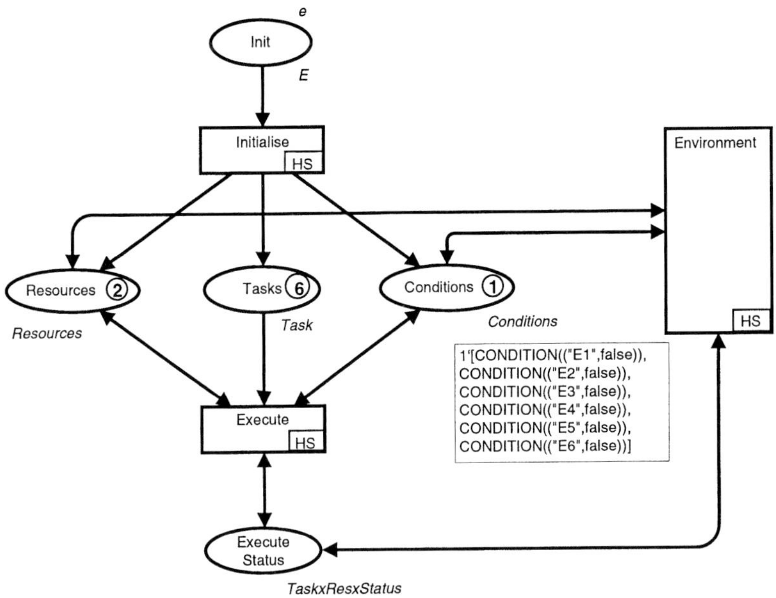

Fig. 39. The CoastServer page after initialisation.

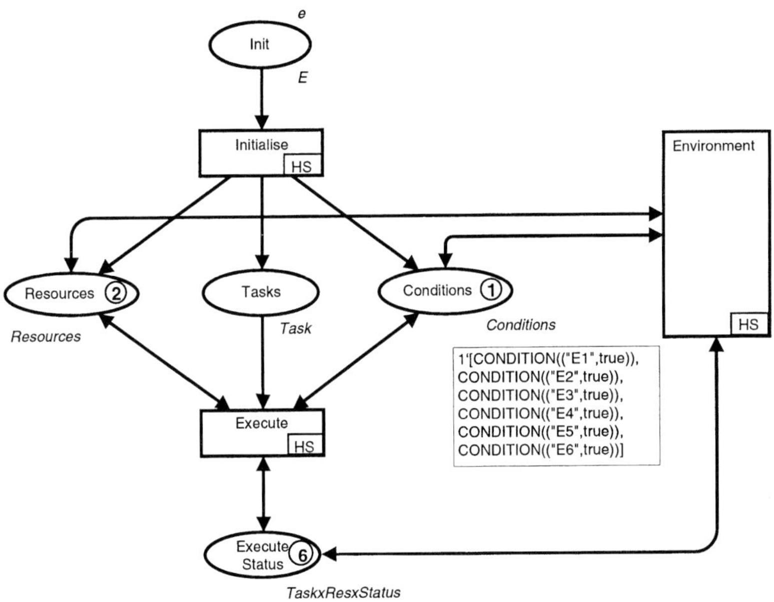

Fig. 40. The CoastServer page – all tasks executed.

Figure 40 shows the top level page of the CPN model in a marking where all six tasks in the example plan have been executed. All six tokens have been removed from place Tasks since the tasks have now been executed. The marking shown corresponds to a desired end-state since the conditions E4, E5, and E6 are now all satisfied as can be seen from the marking of place Conditions.

5.4 Line of Operation Generation

The main analysis capability of COAST is the generation of LOPs. A LOP is a specification of start and end times for the tasks in the plan. The LOP generation implemented in the COAST server consists of two phases. In the first phase, the state space is generated relative to the plan to be analysed. Successors are not generated for states that qualify as desired end-states according to the conditions specified by the user. In the second phase, LOPs are computed by traversing the constructed state space. The LOPs are determined from the paths in the state space, and they are divided into two classes. *Complete LOPs* are LOPs that lead from the initial marking to a marking representing a desired end-state. The *incomplete LOPs* are LOPs that lead to markings representing undesired end-states, i.e., markings without enabled transitions that do not satisfy the conditions specified by the user. When incomplete LOPs are reported, the user will typically investigate the causes of these using queries about tasks, conditions, and resources in different states. In that sense, COAST also supports the planner in identifying errors and inconsistencies in the plan under analysis.

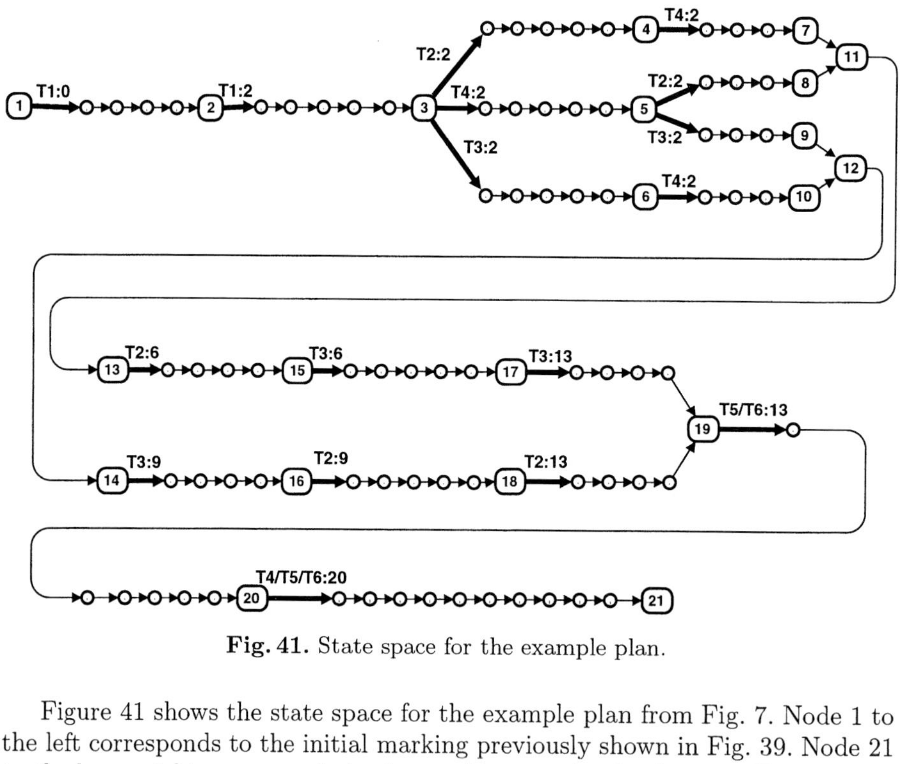

Fig. 41. State space for the example plan.

Figure 41 shows the state space for the example plan from Fig. 7. Node 1 to the left corresponds to the initial marking previously shown in Fig. 39. Node 21 to the lower right corresponds to the marking previously shown in Fig. 39. The thick arcs in the state space correspond to start and termination of tasks. The other arcs correspond to internal events in the CPN model related to the start and termination of tasks. The thick arcs have labels of the form $Ti : t$ where i specifies the task number and t specifies the time at which the event takes place. As an example, task T1 starts at time 0 as specified by the label on the outgoing arc from node 1, and terminates at time 2 as specified by the label on the outgoing arc from node 2.

The computation of LOPs is based on a breadth-first traversal of the state space starting from the initial marking. The basic idea is to compute the LOPs leading to each marking encountered during the traversal of the state space, where the LOPs in a given marking are computed from the LOPs associated with its predecessor markings. The LOPs associated with a given marking are then deleted once the LOPs have been computed for all its successor markings. The algorithm exploits the fact that the state space of the CPN model is acyclic for any plan, and that the paths leading to a given marking in the state space all have the same length measured in occuring binding elements.

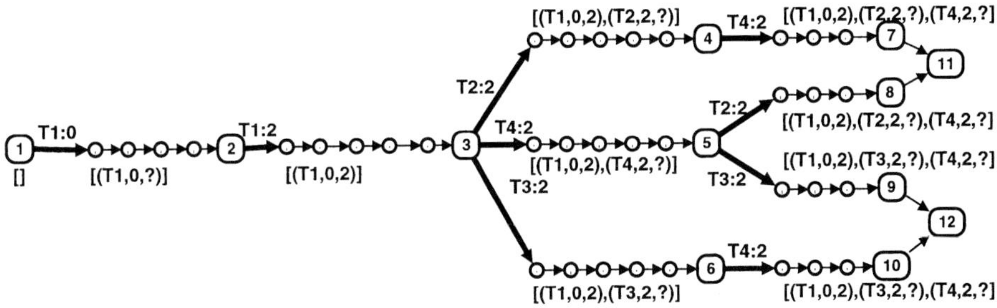

Fig. 42. LOP generation after initial marking has been processed.

We now illustrate how the algorithm operates. Figure 42 shows the LOPs information associated with each marking in the first part of the state space. The LOP associated with the initial marking is the empty LOP represented as the empty list []. LOPs for the successor marking of the initial marking are now computed. Since the arc leading to the successor marking corresponds to the start of a task, the LOP is augmented with information about the time at which T1 was started. This results in the LOP: [(T1,0,?)] being associated with the successor marking of the initial marking. The LOP remains the same until the arc corresponding to the termination of T1 at time 2 is reached. In this case, the termination time of T1 can be recorded in the LOP. This results in associating the LOP [T1,0,2] with the successor marking of node 2. The LOP now associated with the succesor of node 2 is propagated forward. The LOP generation now proceeds and when node 3 is reached, the LOPs are propagated along three branches corresponding to the three successor markings of node 3. The LOP generation will now continue until the nodes 7, 8, 9, and 10 are reached. Here the LOPs associated with nodes 7 and 8 will be merged and associated with node 11 since the start and termination time of each of the tasks in the LOPs are identical. Similarly, the LOPs associated with node 9 and 10 will be merged and associated with node 12. The breadth-first traversal will now continue until eventually the situation shown in Fig. 43 is reached where the two complete LOPs leading to the desired end-state have been computed.

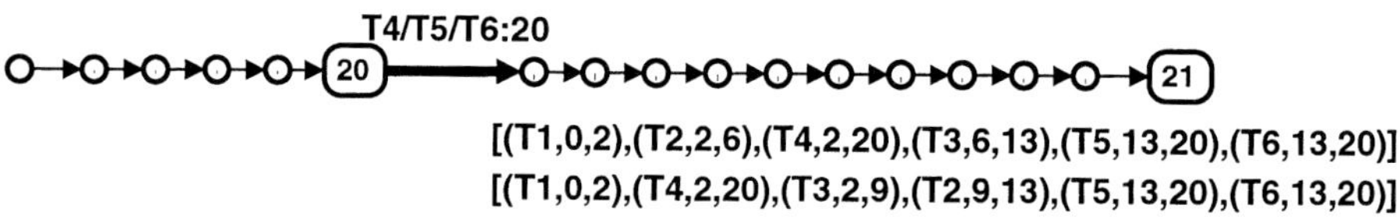

Fig. 43. Termination of the LOP generation.

Typical planning problems to which COAST is applied consist of 15 to 25 tasks resulting in state spaces with 10,000 to 20,000 nodes and 25,000 to 35,000 arcs. Such state spaces can be generated in less than 2 minutes on a standard PC. The state spaces are relatively small because the conditions, available resources, and imposed synchronisations in practice strongly limit the possible orders in which the tasks can be executed.

5.5 Conclusions on the Development of COAST

The development of the COAST tool is an example of how the usual gap between design as specified by a CPN model and the final implementation of a system can be overcome. The CPN model that was constructed to develop the conceptual and semantical foundation of COAST is being used directly in the final implementation of the COAST server. The project also demonstrates the value of having a full programming language environment in the form of the Standard ML compiler integrated in the CPN computer tools. The use of Standard ML as part of the CPN computer tools was crucial in several ways in the development of COAST. It allowed a highly compact and parameterisable CPN model to be constructed, and it allowed the CPN model to become the implementation of the COAST server. The parameterisation is important to ensure that the COAST server is able to analyse any set of tasks, resources, and synchronisations without having to make changes to the CPN model. Having a full programming language available also made it possible to extend the COAST server with the specialised algorithms required to extract the task schedules from a generated state space.

6 Conclusions and Future Directions

In this paper we have presented four projects where the CPN modelling language and computer tools have been put into practical use in system development projects. The project on modelling communication and mobility scenarios for ad-hoc networking illustrates how quite abstract CPN models can be used in an early phase of system development to determine the boundaries of the project and specify requirements. The pervasive health care project illustrated how CP-nets can be used to construct an executable use case in the form of an animated

CPN model. An informal use case described in prose was augmented with notions of execution, formality, and animation. We have illustrated the use of state space methods for the analysis of the BeoLink system and for obtaining lines of operation in the COAST tool. The revised case study on the BeoLink system demonstrates that significant progress has been made in recent years on the support for state space analysis. The work presented on the COAST tool also shows how a CPN model can be integrated into an application making the use of CP-nets transparent to the user and overcoming the usual gap between design and implementation. Another example of automatic code generation from CPN models can be found in [67]. The paper [62] describes an approach to making a tailored graphical user interface on top of a CPN model using web technology.

In general, many CPN projects have been carried out and documented in papers and reports. As examples, the proceedings of the CPN workshops 1998-2002 [51], and the two special issues of the Software Tools for Technology Transfer journal [27, 28] contain many papers on practical use of CP-nets. The most comprehensive overview of application and industrial use of CP-nets can be found on the web pages [22,23,25] that are maintained by our research group. Together, all these projects provide solid evidence that CP-nets have good potential to be used in the software industry. On the other hand, as evidenced by the recent survey [68], in general formal methods (like CP-nets) are only rarely used in the software industry. An interesting direction for future research is to try to increase the use of Petri nets in the software industry. That is for obvious reasons attractive for us as Petri net researchers. However, it may also be attractive for many parts of the software industry. It is widely recognised that today's mainstream software development methods and tools are not always adequate for solving the range of difficult problems that software developers are facing.

The choice of formal modelling language to be used in a system development project is non-trivial, and many aspects must be taken into account, e.g., available tool support and background of the involved system developers. Choosing CP-nets has a number of virtues. CP-nets has a sound, mathematically well-founded execution semantics, is well-proven, and has proper tool support. This includes support for creating animations of CPN models, which has been used in a number of projects, see, e.g., [76] for an alarm system, [64] for mobile phones, and [8] for ISDN services.

Even though we see a number of advantages of using Petri nets, other researchers and practitioners may have other opinions and preferences. If we want to advocate wide-spread and long-term use of Petri nets in system development in a company, we have to convince not only the software developers, but also higher-level decision makers like business and project managers. In conversations with the latter, we must stress the key business question: How does my company save time and money by using Petri nets? Sometimes, we should perhaps talk about reducing time to market, increasing return of investment, and limitation of risks, instead of about, e.g., nice theoretical properties like formal semantics. We should also promote Petri nets as a supplement to existing software development practices, not as something fundamentally new. In particular, with the success of UML, the software industry has in large scale adopted modelling as a valuable discipline in everyday software development. Many software developers appreciate UML (in particular the static parts of UML such as class diagrams) as a productive asset to help them in their work. Those who have also tried to model behavioural aspects in UML might have encountered problems with UML state machines and activity diagrams. Therefore, for many developers, the motivation to use a supplementary modelling language together with UML may be quite high. In this way, the success of UML can be seen as a good chance to establish Petri nets more broadly in the software industry.

The reader interested in getting started with CPN modelling is referred to the paper [58] and the book [47]. The paper introduces the CPN modelling language using a simple communication protocol, whereas the book contains several smaller examples and also the formal definition of CP-nets. Readers interested in getting started using the state space method is referred to the book [48], the introductory paper [58], and the examples found on the web pages [22, 25]. The reader interested in the recent work on state space methods is referred to [16] for the time condensed state spaces, [15, 59] for the sweep-line method, and [29] for the symmetry method. CPN models can also be analysed using simulation, and the papers [91,92] describe how quantitative measures such as throughput and delay of the system can be obtained using simulation-based performance analysis and the CPN computer tools.

The web pages for the CPN computer tools [22, 25] contain several tutorials and small examples of CPN models useful for getting started using CPN modelling and the CPN computer tools. A license for the CPN computer tools can be obtained free of charge, and a licence form is available electronically from our web-pages [22]. Mailing lists have also been established for users of the CPN computer tools.

References

1. Aarhus Amt Electronic Patient Record. `www.epj.aaa.dk`.
2. D. Amyot, R.J.A. Buhr, T. Gray, and L. Logrippo. Use Case Maps for the Capture and Validation of Distributed Systems Requirements. In *Proc. of 4th IEEE International Symposium on Requirements Engineering*, pages 44–53. IEEE Computer Society, 1999.
3. A.I. Antón, R.A. Carter, A. Dagnino, J.H. Dempster, and D.F. Siege. Deriving Goals from a Use-Case Based Requirements Specification. *Requirements Engineering Journal*, 6:63–73, 2001. Springer-Verlag.
4. Australian Defence Science and Technology Organisation. `www.dsto.defence.gov.au`.
5. Bang & Olufsen. `www.bang-olufsen.com`.
6. J.E. Bardram and C. Bossen. Moving to get aHead: Local Mobility and Collaborative Work. In *Proc. of 8th European Conference on Computer-supported Cooperative Work*, pages 355–374. Kluwer Academic Publishers, 2003.
7. J. Billington, G. Gallasch, L.M. Kristensen, and T. Mailund. Exploiting Equivalence Reduction and the Sweep-Line Method for Detecting Terminal States. *IEEE Transactions on Systems, Man, and Cybernetics. Part A: Systems and Humans*, 2004. To appear.

8. C. Capellmann, S. Christensen, and U. Herzog. Visualising the Behaviour of Intelligent Networks. In *Services and Visualisation, Towards User-Friendly Design*, volume 1385 of *Lecture Notes in Computer Science*, pages 174–189. Springer-Verlag, 1998.
9. ITU (CCITT). Recommendation Z.120: MSC. Technical report, International Telecommunication Union, 1992.
10. Centre for pervasive computing. `www.pervasive.dk`.
11. A. Cheng, S. Christensen, and K.H. Mortensen. Model Checking Coloured Petri Nets Exploiting Strongly Connected Components. In *Proc. of the International Workshop on Discrete Event Systems, WODES96*. Institution of Electrical Engineers, Computing and Control Division, Edinburgh, UK, 1996.
12. H.B. Christensen and J.E. Bardram. Supporting Human Activities – Exploring Activity-Centered Computing. In *Proc. of 4th Ubicomp Conference*, volume 2498 of *Lecture Notes in Computer Science*. Springer-Verlag, 2002.
13. S. Christensen. *Message Sequence Charts. User's Manual*, January 1997.
14. S. Christensen and J.B. Jørgensen. Analysis of Bang and Olufsen's BeoLink Audio/Video System Using Coloured Petri Nets. In *Proc. of 18th International Conference on Application and Theory of Petri Nets*, volume 1248 of *Lecture Notes in Computer Science*, pages 387–406. Springer-Verlag, 1997.
15. S. Christensen, L.M. Kristensen, and T. Mailund. A Sweep-Line Method for State Space Exploration. In *Proc. of 7th International Conference on Tools and Algorithms for the Construction and Analysis of Systems*, volume 2031 of *Lecture Notes in Computer Science*, pages 450–464. Springer-Verlag, 2001.
16. S. Christensen, L.M. Kristensen, and T. Mailund. Condensed State Spaces for Timed Petri Nets. In *Proc. of 22nd International Conference on Application and Theory of Petri Nets*, volume 2075 of *Lecture Notes in Computer Science*, pages 101–120. Springer-Verlag, 2001.
17. E. Clarke, E.A. Emerson, S. Jha, and A.P. Sistla. Symmetry Reductions in Model Checking. In *Proc. of 10th International Conference on Computer-Aided Verification*, volume 1427 of *Lecture Notes in Computer Science*, pages 147–159. Springer-Verlag, 1998.
18. E. Clarke, O. Grumberg, and D. Peled. *Model Checking*. The MIT Press, 1999.
19. E.M. Clarke, R. Enders, T. Filkorn, and S. Jha. Exploiting Symmetries in Temporal Logic Model Checking. *Formal Methods in System Design*, 9(1/2):77–104, 1996.
20. A. Cockburn. *Writing Effective Use Cases*. Addison-Wesley, 2000.
21. Computer Systems Engineering Centre at University of South Australia. `www.unisa.edu.au/eie/csec/`.
22. CPN Tools. `www.daimi.au.dk/CPNtools`.
23. The CPN Group at University of Aarhus. `www.daimi.au.dk/CPnets`.
24. J. Desel and W. Reisig. Place/Transition Petri Nets. In *Lecture on Petri nets I: Basic Models*, volume 1491 of *Lecture Notes in Computer Science*, pages 122–173. Springer-Verlag, 1998.
25. Design/CPN. `www.daimi.au.dk/designCPN`.
26. N. Dulac, T. Viguier, N. Leveson, and M.-A. Storey. On the Use of Visualization in Formal Requirements Specification. In *Proc. of 7th IEEE International Symposium on Requirement Engineering*, pages 71–80. IEEE Computer Society, 2002.
27. K. Jensen (ed.). International Journal on Software Tools for Technology Transfer, Vol. 2, No. 2. Special section on Coloured Petri nets, 1998.
28. K. Jensen (ed.). International Journal on Software Tools for Technology Transfer, Vol. 3, No. 4. Special section on Coloured Petri nets, 2001.

29. L. Elgaard. *The Symmetry Method for Coloured Petri Nets*. PhD thesis, Department of Computer Science, University of Aarhus, July 2002.
30. M. Elkoutbi and R.K. Keller. User Interface Prototyping Based on UML Scenarios and High-Level Petri Nets. In *Proc. of 21st International Conference on Application and Theory of Petri Nets*, volume 1825 of *Lecture Notes in Computer Science*. Springer-Verlag, 2000.
31. E. A. Emerson, A.K. Mok, A.P Sistla, and J. Srinivasan. Quantitative Temporal Reasoning. In *Proc. of 2nd International Workshop on Computer-Aided Verification*, volume 531 of *Lecture Notes in Computer Science*, pages 136–145. Springer-Verlag, 1990.
32. E.A. Emerson and A.P. Sistla. Symmetry and Model Checking. *Formal Methods in System Design*, 9(1/2):105–131, 1996.
33. Ericsson Telebit A/S. `www.ericssontelebit.dk`.
34. J. Esparza. Model Checking using Net Unfoldings. *Science of Computer Programming*, 23:151–195, 1994.
35. Internet Engineering Task Force. Mobile ad-hoc networks. `www.ietf.org/html.charters/manet-charter.html`.
36. G. Gallasch and L. M. Kristensen. Comms/CPN: A Communication Infrastructure for External Communication with Design/CPN. In *Proc. of the 3rd Workshop and Tutorial on Practical Use of Coloured Petri Nets and the CPN Tools*, pages 79–93. Department of Computer Science, University of Aarhus, 2001. DAIMI PB-554.
37. G.E. Gallasch, L.M. Kristensen, and T. Mailund. Sweep-Line State Space Exploration for Coloured Petri Nets. In *Proc. of 4th Workshop and Tutorial on Practical Use of Coloured Petri Nets and the CPN Tools*, pages 101–120. Department of Computer Science, University of Aarhus, 2002. DAIMI PB-560.
38. S. Gordon, L.M. Kristensen, and J. Billington. Verification of a Revised WAP Wireless Transaction Protocol. In *Proc. of 23rd International Conference on Application and Theory of Petri Nets*, volume 2360 of *Lecture Notes in Computer Science*, pages 182–202. Springer-Verlag, 2002.
39. G.J. Holzmann. Tracing Protocols. *Bell System Technical Journal*, 64:2413–2434, 1985.
40. G.J. Holzmann. *Design and Validation of Computer Protocols*. Prentice-Hall International Editions, 1991.
41. G.J. Holzmann. An Analysis of Bitstate Hashing. *Formal Methods in System Design*, 13(3):287–305, 1998.
42. C. Huitema. *IPv6: The New Internet Protocol*. Prentice-Hall, 1998.
43. C.N. Ip and D.L. Dill. Better Verification Through Symmetry. *Formal Methods in System Design*, 9(1/2):41–75, 1996.
44. M. Jackson. *System Development*. Prentice-Hall, 1983.
45. I. Jacobson, M. Christerson, P. Jonsson, and G. Övergaard. *Object-Oriented Software Engineering: A Use Case Driven Approach*. Addison-Wesley, 1992.
46. C. Jard and T. Jeron. Bounded-memory Algorithms for Verification On-the-fly. In *Proc. of 3rd International Workshop on Computer-Aided Verification*, volume 575 of *Lecture Notes in Computer Science*, pages 192–202. Springer-Verlag, 1991.
47. K. Jensen. *Coloured Petri Nets - Basic Concepts, Analysis Methods and Practical Use. - Volume 1: Basic Concepts*. Springer-Verlag, 1992.
48. K. Jensen. *Coloured Petri Nets - Basic Concepts, Analysis Methods and Practical Use. - Volume 2: Analysis Methods*. Springer-Verlag, 1995.
49. K. Jensen. Condensed State Spaces for Symmetrical Coloured Petri Nets. *Formal Methods in System Design*, 9(1/2):7–40, 1996.

50. K. Jensen. *Coloured Petri Nets - Basic Concepts, Analysis Methods and Practical Use. - Volume 3: Practical use.* Springer-Verlag, 1997.
51. K. Jensen, editor. *Proceedings Workshop and Tutorial on Practical Use of Coloured Petri Nets and CPN Tools.* Available via `www.daimi.au.dk/CPnets`, 1998-2002.
52. J.B. Jørgensen. Coloured Petri Nets in Development of a Pervasive Health Care System. In *In Proc. of 24th International Conference on Application and Theory of Petri Nets*, volume 2679 of *Lecture Notes in Computer Science*, pages 256–275. Springer-Verlag, 2003.
53. J.B. Jørgensen and C. Bossen. Requirements Engineering for a Pervasive Health Care System. In *Proc. of 11th IEEE International Requirements Engineering Conference*, pages 55–64. IEEE Computer Sociery, 2003.
54. J.B. Jørgensen and C. Bossen. Executable Use Cases: Requirements for a Pervasive Health Care System. *IEEE Software*, March/April 2004. To appear.
55. J.B. Jørgensen and S. Christensen. Executable Design Models for a Pervasive Healthcare Middleware System. In *In Proc. of the 5th UML Conference*, volume 2460 of *Lecture Notes in Computer Science*, pages 140–149. Springer-Verlag, 2002.
56. L. M. Kristensen and A. Valmari. Finding Stubborn Sets of Coloured Petri Nets Without Unfolding. In *Proceedings of 19th International Conference on Application and Theory of Petri Nets*, volume 1420 of *Lecture Notes in Computer Science*, pages 104–123. Springer-Verlag, 1998.
57. L.M. Kristensen. Ad-hoc Networking and IPv6: Modelling and Validation. `www.pervasive.dk/projects/IPv6/IPv6_summary`.
58. L.M. Kristensen, S. Christensen, and K. Jensen. The Practitioner's Guide to Coloured Petri Nets. *International Journal on Software Tools for Technology Transfer*, 2(2):98–132, 1998.
59. L.M. Kristensen and T. Mailund. A Generalised Sweep-Line Method for Safety Properties. In *Proc. of Formal Methods Europe*, volume 2391 of *Lecture Notes in Computer Science*, pages 549–567. Springer-Verlag, 2002.
60. L.M. Kristensen and T. Mailund. A Compositional Sweep-Line State Space Exploration Method. In *Proc. of Formal Techniques for Networked and Distributed Systems*, volume 2529 of *Lecture Notes in Computer Science*, pages 327–343. Springer-Verlag, 2002.
61. P. Kruchten. *The Rational Unified Process: An Introduction.* Addison-Wesley, 1999.
62. B. Lindstrøm. Web-Based Interfaces for Simulation of Coloured Petri Net Models. *International Journal on Software Tools for Technology Transfer*, 3(4):405–416, 2001.
63. L. Lorentsen and L. M. Kristensen. Exploiting Stabilizers and Parallelism in State Space Generation with the Symmetry Method. In *Proceedings of International Conference on Application of Concurrency in System Design*, pages 211–220. IEEE Computer Society, 2001.
64. L. Lorentsen, A-P Tuovinen, and J. Xu. Modelling Features and Feature Interactions of Nokia Mobile Phones Using Coloured Petri Nets. In *Proc. of the 23rd International Conference on Application and Theory of Petri Nets*, Lecture Notes in Computer Science. Springer-Verlag, 2002.
65. K. L. McMillan. A Technique of State Space Search Based on Unfolding. *Formal Methods in System Design*, 6(1):45–65, 1995.
66. K.L. McMillan. *Symbolic Model Checking.* Kluwer Academic Publishers, 1993.
67. K.H. Mortensen. Automatic Code Generation Method Based on Coloured Petri Net Models Applied on an Access Control System. In *Proceedings of 21st International Conference on Application and Theory of Petri Nets*, volume 1825 of *Lecture Notes in Computer Science*, pages 367–386. Springer-Verlag, 2000.

68. C.J. Neill and P.A. Laplante. Requirements Engineering: The State of the Practice. *IEEE Software*, 20(6):61–69, 2003.
69. OMG Unified Modeling Language Specification, Version 1.4. Object Management Group (OMG); UML Revision Taskforce, 2001.
70. D. Peled. All from One, One for All: On Model Checking Using Representatives. In *Proc. of 5th International Conference on Computer-Aided Verification*, volume 697 of *Lecture Notes in Computer Science*, pages 409–423. Springer-Verlag, 1993.
71. C.E. Perkins. *Ad Hoc Networking*. Addison-Wesley, 2001.
72. Pervasive Healthcare. `www.healthcare.pervasive.dk`.
73. S. Lawrence Pfleeger. *Software Engineering: Theory and Practice*. Prentice-Hall, 2nd edition, 2001.
74. Radio Frequency Identification. `www.rfid.org`.
75. J. L. Rasmussen and M. Singh. *Mimic/CPN. A Graphical Simulation Utility for Design/CPN. User's Manual*. `www.daimi.au.dk/designCPN`.
76. J.L. Rasmussen and M. Singh. Designing a Security System by Means of Coloured Petri Nets. In *Proc.of 17th International Conference on Application and Theory of Petri Nets*, volume 1091 of *Lecture Notes in Computer Science*, pages 400–419. Springer-Verlag, 1996.
77. W. Reisig. *Petri Nets*, volume 4 of *EACTS Monographs in Theoretical Computer Science*. Springer-Verlag, 1985.
78. J. Rumbaugh, I. Jacobson, and G. Booch. *The Unified Modeling Language Reference Manual*. Addison-Wesley, 1999.
79. The Standard ML Basis Library. `http://www.smlnj.org/doc/basis/pages/sml-std-basis.html`.
80. M. Satyanarayanan. Challenges in Implementing a Context-Aware System. In *Pervasive Computing*, volume 1(3). IEEE, 2002.
81. M. Satyanarayanan, editor. *Pervasive Computing*, volume 1(1). IEEE, 2002.
82. A.J.H. Simons and I. Graham. 30 Things That Go Wrong in Object Modelling with UML 1.3. In H. Kilov, B. Rumpe, and I. Simmonds, editors, *Behavioral Specifications of Businesses and Systems*. Kluwer Academic Publishers, 1999.
83. W. Stallings. *Data & Computer Communications*. Prentice Hall, 6th edition, 2000.
84. Systematic Software Engineering A/S. `www.systematic.dk`.
85. The SPIN Tool. `netlib.bell-labs.com/netlib/spin/whatispin.html`.
86. J.D. Ullman. *Elements of ML Programming*. Prentice-Hall, 1998.
87. A. Valmari. A Stubborn Attack on State Explosion. In *Proc. of 2nd International Workshop on Computer-Aided Verification*, volume 531 of *Lecture Notes in Computer Science*, pages 156–165. Springer-Verlag, 1990.
88. A. Valmari. The State Explosion Problem. In *Lectures on Petri Nets I: Basic Models*, volume 1491 of *Lecture Notes in Computer Science*, pages 429–528. Springer-Verlag, 1998.
89. A. van Lamsweerde. Requirements Engineering in the Year 00: A Research Perspective. In *Proc. of the 22nd International Conference on Software Engineering*. ACM Press, 2000.
90. M. Weiser. The Computer for the 21st Century. In *Scientific American*, volume 265 (3). Scientific American, Inc., 1991.
91. L. Wells. Performance Analysis Using Coloured Petri Nets. In *Proc. of the Tenth IEEE International Symposium on Modeling, Analysis and Simulation of Computer and Telecommunication Systems*, pages 217–221. IEEE Computer Society, 2002.

92. L. Wells, S. Christensen, L.M. Kristensen, and K. Mortensen. Simulation Based Performance Analysis of Web Servers. In *Proc. of the 9th International Workshop on Petri Nets and Performance Models*, pages 59–68. IEEE Computer Society, 2001.
93. P. Wolper and P. Godefroid. Partial Order Methods for Temporal Verification. In *Proc. of 4th International Conference on Concurrency Theory*, volume 715 of *Lecture Notes in Computer Science*, pages 233–246. Springer-Verlag, 1993.
94. L. Zhang, L.M. Kristensen, C. Janczura, G. Gallasch, and J. Billington. A Coloured Petri Net based Tool for Course of Action Development and Analysis. In *Proc. of Workshop on Formal Methods Applied to Defence Systems*, volume 12 of *Conferences in Research and Practice in Information Technology*, pages 125–134. Australian Computer Society, 2001.
95. M.K. Zimmerman, K. Lundqvist, and N. Leveson. Investigating the Readability of State-Based Formal Requirements Specification Languages. In *Proc. of 24th International Conference on Software Engineering*, pages 33–43. ACM Press, 2002.

Bigraphs for Petri Nets

Robin Milner

University of Cambridge, The Computer Laboratory,
J J Thomson Avenue, Cambridge CB3 0FD, UK

Abstract. A simple example is given of the use of bigraphical reactive systems (BRSs). It provides a behavioural semantics for condition-event Petri nets whose interfaces are named condition nodes, using a simple form of BRS equipped with a labelled transition system and its associated bisimilarity equivalence. Both of the latter are derived from the standard net firing rules by a uniform technique in bigraphs, which also ensures that the bisimilarity is a congruence. Furthermore, this bisimilarity is shown to coincide with one induced by a natural notion of *experiment* on condition-event nets, defined independently of bigraphs.
The paper is intended as a bridge between Petri net theory and bigraphs, as well as a pedagogical exercise in the latter.

1 Introduction

This paper conducts a simple exercise in bigraphical reactive systems (BRSs) [4], consisting of a behavioural study of condition-event Petri nets [12]. The exercise has two very different purposes. The first is pedagogical: condition-event nets can be modelled as a *link-graph* reactive system (LRS), which is a simple form of BRS, so they illustrate the use of bigraphs while avoiding some of their complexity. The other purpose is to promote future research: since bigraphs model systems that can reconfigure both their placing and their linking, the exercise illustrates a framework in which Petri nets may be generalised to deal with mobile informatic systems.

The exercise involves the interpretation of condition-event nets in terms of *bisimilarity* [8]. As in process calculi, it may sometimes be useful to employ an abstract model of the behaviour of a Petri net in which two nets are regarded as equivalent if they cannot be distinguished by certain forms of *experiment*. If an experiment e can be carried out on a system in state g, changing its state to g', we write

$$g \xrightarrow{e} g'$$

and call it a *labelled transition* between the two states. If we fix a vocabulary of labels e and define the possible transitions for each one, we have a *(labelled) transition system* (TS) $\mathcal{L}$. Then a symmetric binary relation $\mathcal{R}$ between two system states is said to be a *bisimulation* (for $\mathcal{L}$) if

whenever $f\mathcal{R}g$ and $f \xrightarrow{e} f'$, there exists
a state g' such that $g \xrightarrow{e} g'$ and $f'\mathcal{R}g'$.

In other words: given two related processes, whatever one of them can do, the other can also do without losing the relationship. The bisimulation property is preserved by union

J. Desel, W. Reisig, and G. Rozenberg (Eds.): ACPN 2003, LNCS 3098, pp. 686–701, 2004.

of relations, so there is a largest bisimulation which is the union of all bisimulations, and it is easily found to be an equivalence relation. We call it *bisimilarity* (for $\mathcal{L}$).

All this holds for any interpretation of 'experiment'. We call bisimulations (and bisimilarity) *weak* or *strong*, and denote the equivalence by $\approx$ or $\sim$, according to whether or not a single experiment e can be accompanied by any finite amount of internal activity. In the strong case we consider each individual internal action as an experiment, even though it is indistinguishable from any other such action. Both weak and strong bisimilarity abstract away from the causal behaviour of systems, but the weak form is more generous in turning a blind eye to internal activity.

We now consider what might be an experiment on a condition-event net. There are various ways to make parts of a net externally accessible, in order to observe – or induce – behaviour of the net from the outside. Authors (some of whom are cited in the next section) have considered making accessible certain *events* or *actions*, or alternatively certain *conditions* or *states*. This exercise is of the latter kind; we allow an experimenter to change certain conditions from holding to not holding or vice versa, by removing or adding a token. This choice was made because it makes the exercise simple, but the alternatives may well yield to a similar approach.

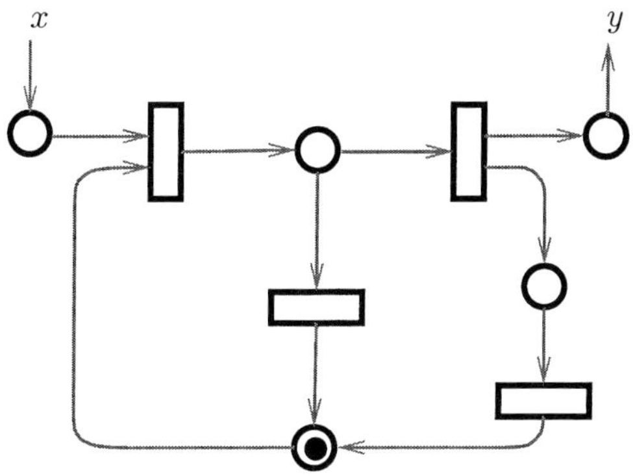

Consider the above net, for example. At different times the experimenter will be able to add or remove a token at x or at y. In general, given a state g, i.e. a *marking* of the net, the transition $g \xrightarrow{+x} g'$ or $g \xrightarrow{-x} g'$ represents the addition or subtraction of a token at x. Since we are dealing with condition-event nets, in any given state exactly one of these experiments is possible for each accessible condition. A third kind of transition, $g \xrightarrow{\tau} g'$, represents an internal event involving no external participation.

These three kinds of transition are the basis of a TS; we shall it call it $\mathcal{L}_\mathsf{p}$, and its induced bisimilarity $\sim_\mathsf{p}$. In the rest of this paper we shall compare this TS and its bisimilarity with another one, which arises from setting up condition-event nets as an LRS and then deriving a TS $\mathcal{L}_\mathsf{g}$ by a construction [5,4] that is uniform over all LRSs (and BRSs). We shall find that the labels of $\mathcal{L}_\mathsf{g}$ differ from those of $\mathcal{L}_\mathsf{p}$, but that the two bisimilarities $\sim_\mathsf{p}$ and $\sim$ coincide. This gives us confidence that the dynamics of nets may be faithfully presented in bigraph theory.

2 Related Work on Petri Nets

In the introduction we declared two goals: first, to give a simple tutorial in bigraphs; second, to treat Petri nets in the bigraphical framework, thus perhaps easing the extension of the net model to admit mobility. We shall tackle both goals by means of a simple case study. In this section we briefly describe how the study relates to existing work in Petri net behaviour, with reference to some recent papers on that topic.

Pomello, Rozenberg and Simone [10] give a comprehensive survey of behavioural equivalences for Petri nets. They cover those based on observation both of actions and of states, and range from fine equivalences respecting causality to coarser ones, for example the failures equivalence from CSP, the coarsest which respects deadlock. The study of congruence is reported as being rather incomplete at that date (1992).

Nielsen, Priese and Sassone [9] characterise some behavioural congruences on nets. Given semantic function $\mathcal{B}$ that assigns an abstract behaviour to each net, they consider the congruence $\approx$ it induces upon nets; this is defined by

$$N_0 \approx N_1 \overset{\text{def}}{\Leftrightarrow} \mathcal{B}(C[N_0]) = \mathcal{B}(C[N_1]) \text{ for every context } C\ .$$

This definition presupposes a precise notion of *context.* An important contribution of their paper is to define such a notion, by means of a set of *combinators* upon nets. They are then able to characterise the congruences, for each of four semantic functions $\mathcal{B}$, by showing that for each pair N_0, N_1 there is a single easily identified context that is sufficient to determine whether or not $N_0 \approx N_1$.

Priese and Wimmel [11] continue this programme; they enrich the net combinators, and consider a wider range of semantic functions.

The Petri Box calculus of Best, Devillers and Hall [1], like the previous two, emphasises combinators and algebra. By identifying certain net-patterns as operators, it presents a modular semantics of nets in terms of equivalence classes of Boxes (a special class of nets). A main result of the paper is agreement between this denotational semantics and a structured operational semantics of Box expressions.

This brief summary does not do justice to the four papers, which represent well the progress towards a modular treatment of Petri nets. But it helps us to identify differences with bigraph theory, which suggest contributions that can be made by the latter. The first difference is that, since bigraphs and their contexts are the arrows of a category, whenever a class of agents (e.g. nets) is encoded in bigraphs the contexts and combinators are already determined; they need not be defined specifically for each class. The second difference is that the semantic function on bigraphical agents is defined not by specific means, but as the quotient by a generic equivalence relation that pertains to all bigraphical systems. Finally, many such equivalences – including bisimulation (which we use in this paper) but also others – are guaranteed by bigraphical theory to be congruences.

In this brief discussion we have tried to explain the way in which bigraphs aim at a theory shared by different models of concurrency. Much work is needed to determine how far they can achieve this aim. Success can be measured in two ways: by the range of different models that can be satisfactorily treated in bigraphs, and by the depth of the theory thus shared among them. The present paper begins to evaluate these measures with particular reference to Petri nets.

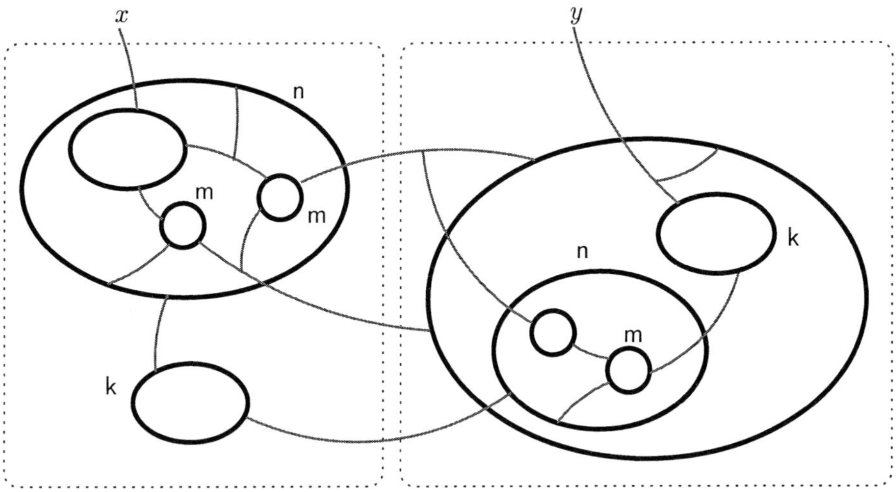

Fig. 1. A typical bigraph

3 Bigraphs and Link Graphs

Figure 1 shows a typical bigraph. The ovals and circles are *nodes*. Associated with each node is a *control* which indicates what kind of node it is. Here we show three controls, k, m and n; the controls of other nodes are not shown. Each control has an *arity*, a finite ordinal indicating the number of *ports* on that kind of node; here k, m and n have arity 2, 3 and 2 respectively.

A bigraph is so called because its nodes are structured in two ways. The first structure is *placing*; the nodes are nested inside one another, giving an ordered set of trees, i.e. a forest. In our example there are two trees; each has a *root* – not itself a node – represented by a dotted rectangle. The second structure is *linking*; the ports of the bigraphs are partitioned into *links*, shown by curved lines. A link may be *open* or *closed*; each open link has a distinct *name* (here x or y). Names allow bigraphs to be joined via their open links.

The two structures are totally independent; note here how the links cross node boundaries and even link different trees in the forest.

In other applications of bigraphs the nesting of nodes plays an important role in the way bigraphs reconfigure themselves; both placing and linking may vary dynamically. But in our present application the placing vanishes, so we shall work only with *link graphs*, i.e. the linkage structure. In following sections we shall explain only those parts of link-graph theory that we need.

Link graphs. It is common in graph theory to distinguish between *concrete* and *abstract* graphs. In the former the nodes and edges have identity, and we distinguish two graphs that differ only by a bijection between their nodes and edges; in the latter we equate them. For link graphs we are interested in both kinds; for applications we usually want the abstract ones, but the concrete ones provide us a convenient means to develop the theory. Here we shall work mainly with the concrete link graphs; at the end we point out how the results, once derived, transfer to the abstract ones.

We treat concrete link graphs, then, as the morphisms of a *supported precategory*. This is like a category except that each morphism has a finite set, its *support*, and the composition of two morphisms is defined only if their supports are disjoint. The support of a composite morphism is the union of the supports of its components. Identity morphisms have empty support. Two morphisms F and F' are *support equivalent*, written $F \simeq F'$, if they differ only by a bijection between their supports.

Working with supported precategories is hardly different from working with categories; in this paper the reader can rest assured that any concept familiar from the latter means practically the same for the former. More discussion of this point can be found in the concluding section.

In the supported precategory of (concrete) link graphs, the objects are finite sets $X, Y, \ldots$ of *names*. A link graph $H : X \to Y$ has *inner face* X and *outer face* Y. An example appears in Figure 2; think of H as a *context* in which to embed a link graph G with outer face X. The *points* of a link graph are its ports and its inner names, so H has eleven points: three ports for each a-node, two for the b-node and three inner names $X = \{x_1, x_2, x_3\}$. (Note that the points do not include the *outer* names.) The *links* constitute a partition of the points, and to each *open* link (which we mentioned already) is assigned a distinct outer name; for H, these are $Y = \{y_1, y_2\}$. Note that H has two open and three closed links.

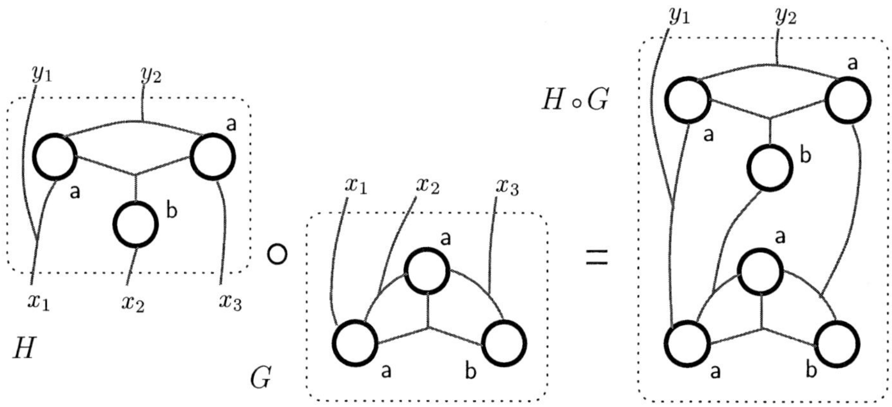

Fig. 2. The composite $H \circ G$ of link graphs $G : \emptyset \to X$ and $H : X \to Y$

The support of a link graph consists of its nodes and its closed links (the latter corresponding to the edges of a classical graph). Their identity is not shown in the diagram, but when we show a composition of two link graphs we assume disjoint supports. Figure 2 shows the composition of $G : \emptyset \to X$ and $H : X \to Y$; each open link in G is joined to the link in H that contains the corresponding inner name, and then that name is erased. The outer and inner names of a link graph need not be disjoint. The identities are those link graphs $\mathsf{id}_X : X \to X$ with empty support in which each outer name $x \in X$ is assigned to the link whose only point is the inner name x.

A link graph with empty inner face, such as G in the diagram, is called *ground*; ground link graphs, and more generally ground bigraphs, are used to represent *agents*, such as a condition-event net with no missing pieces. We typically use lower case letters $f, g, h, \ldots$ for ground link graphs.

Algebra. Let us review briefly how complex link graphs may be built from simpler ones. As well as composition, we use *tensor product*: if F_1 and F_2 have disjoint supports, disjoint inner faces X_1 and X_2, and also disjoint outer faces Y_1 and Y_2, then their tensor product

$$F_1 \otimes F_2 \colon X_1 \cup X_2 \to Y_1 \cup Y_2$$

is formed by placing them side-by-side. The unit for $\otimes$ is just $\mathsf{id}_\emptyset$. Using composition and tensor product we can build all link graphs from the *atomic* ones (those with a single node) with the help of *wirings* (those with no nodes). If k is a control with arity n and $\boldsymbol{x}$ a sequence of n distinct names then a k-*atom* with ports named $x_1, \ldots, x_n$ is denoted by $\mathsf{k}_{\boldsymbol{x}}$. All wirings can be built from two elementary kinds: a *linker* $y/\boldsymbol{x}$ and a *closure* $/x$. These three elementary link graphs are as follows, when $\boldsymbol{x} = x_1x_2x_3$:

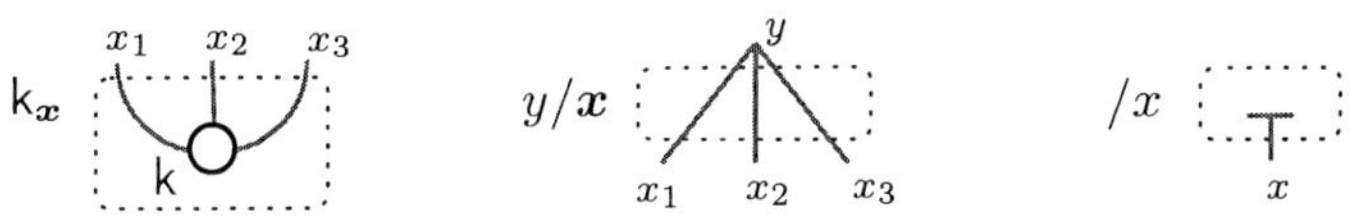

For example, suppose the outer face of F is $\{xyz\}$. We may want to replace x and y by v, leaving z unaffected; or we may want to do the same but close off the link z. In each case we can form $\omega \circ F$, where the wiring ω is respectively

$$\omega = v/xy \otimes \mathsf{id}_z \quad \text{or} \quad \omega = v/xy \otimes /z\,.$$

More generally, the algebra of link graphs consists of expressions built from the elements using $\circ$, $\otimes$ and identities, and satisfying some simple equations. In this paper we shall use a little algebra, but rely more upon diagrams.

One abbreviation will come in handy. If F has outer face $\{xyz\}$ and G has inner face $\{xy\}$, then we may write $G \circ F$ instead of $(G \otimes \mathsf{id}_z) \circ F$. In other words, we sometimes omit identities in composition when no confusion arises.

Sorting. For many purposes, it is useful to enrich link graphs by imposing a *sorting*, i.e. a discipline of *sorts* (or *types*). We set up a sorting in three stages:

1. Specify a set $S = \{\alpha, \beta, \ldots\}$ of sorts.
2. Declare for each control with arity n an ordered list of n sorts. This determines a sort for every port in a link graph.
3. Enrich interfaces $X, Y, \ldots$ by assigning a sort to each name.

We may then define a *well-sorted* link graph to be one that satisfies certain constraints. Here we are interested especially in *many-one* sorting, in which there are just two sorts α and β. Each link may have any number of α-points, but β-points are constrained follows:

- A closed link has exactly one β-point;
- An open link with a β-name has exactly one β-point;
- An open link with an α-name has no β-points.

As an example, for link graphs with controls a and b we may declare that every port of an a-node has sort α, and every port of an b-node has sort β. It can be checked for Figure 2 that G, H and $H \circ G$ are well-sorted if X has the sorting $\{x_1 \colon \alpha, x_2 \colon \alpha, x_3 \colon \beta\}$ and Y has the sorting $\{y_1 \colon \alpha, y_2 \colon \alpha\}$. The reader may like to look ahead and see the rôle of sorting in representing condition-event nets; it ensures that each port on an event node will be connected to at most one pre- or post-condition node.

Dynamics. To equip link graphs with behaviour, we first specify a subclass of the interfaces called *agent interfaces*. If X is such an interface we call any $f \colon \emptyset \to X$ an *agent*; these are the link graphs whose behaviour we want to define. For this purpose we specify a set of *reaction rules*, each being a pair (r, r') of ground link graphs with the same outer face. In each rule we call r the *redex* and r' the *reactum*. Then we specify the *reaction relation* $\longrightarrow\!\!\!\triangleright$ over agents to be the smallest such that

$$D \circ r \longrightarrow\!\!\!\triangleright D \circ r'$$

for every reaction rule (r, r') and every context D for which the compositions are defined and are agents. We also require both the rule-set and the reaction relation to be closed under support-equivalence. In the next section we shall set up the firing rules of condition-event nets as reaction rules.

What we have defined so far is called a *reactive system*. In process calculi it has become usual to refine this to a *(labelled) transition system* (TS), with transitions of the form $f \xrightarrow{\ell}\!\!\!\triangleright f'$, where the *labels* ℓ are specific to each calculus. Intuitively, ℓ represents the contribution that f may make to a reaction; typically this contribution is incomplete, so the transition makes precise the idea that both and agent and its environment may contribute to a reaction. In terms of these TSs, one may define bisimilarity and other equivalences and preorders over agents; a test of a good TS is that these behavioural relations are *congruential*, i.e. preserved by insertion into any context.

In bigraph theory we adopted a proposal by Leifer and Milner [5] to derive TSs uniformly over all bigraphical reactive systems, in a way that guarantees congruential behavioural relations. For link graphs, it works as follows. We consider a label L to be a (link graph) context into which an agent may be inserted in order to enable a reaction to occur; that is, we define the transition

$$f \xrightarrow{L}\!\!\!\triangleright f'$$

to mean that the equation $L \circ f = D \circ r$ holds for some context D and reaction rule (r, r'), and moreover that $f' \simeq D \circ r'$. Think of this as inserting f into a (small) context L so that an instance of the redex r occurs in the composite $L \circ f$, and then replace this occurrence by r'.

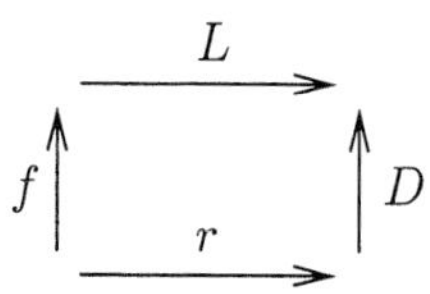

But if we were to allow *all* such contexts as labels L, there would be an unwieldy multitude of labels. Indeed, a moment's reflection reveals that if L is a transition label, then so would be *any larger context* $C \circ L$! To avoid this, the theory limits labels L to be the (in some sense) *smallest* for which the equation $L \circ f = D \circ r$ holds for some D, given f and r. It turns out that strong bisimilarity is congruential for any TS so defined, and we believe that this extends to other behavioural relations. By *smallest*, we mean that the above diagram should not only commute but should also be an *idem pushout* (IPO), a weaker version of the more familiar *pushout*.

We need not explain IPOs here, because our precategory of condition-event nets actually has pushouts where we need them. We shall not show how to construct pushouts for link graphs; we shall just exhibit the resulting TS and then work with it. The construction can be found in the Technical Report by Jensen and Milner [4]. We should note that pushouts – even IPOs – exist only for *concrete* bigraphs, not for abstract ones. Intuitively, support provides a means of defining exactly which nodes and edges are shared between two link graphs.

4 Condition-Event Nets as Link Graphs

We are now ready to set up condition-event nets as link graphs[1]. There are many ways to do it; we choose one that appears to give a smooth treatment. We shall use the example from the introduction as an illustration:

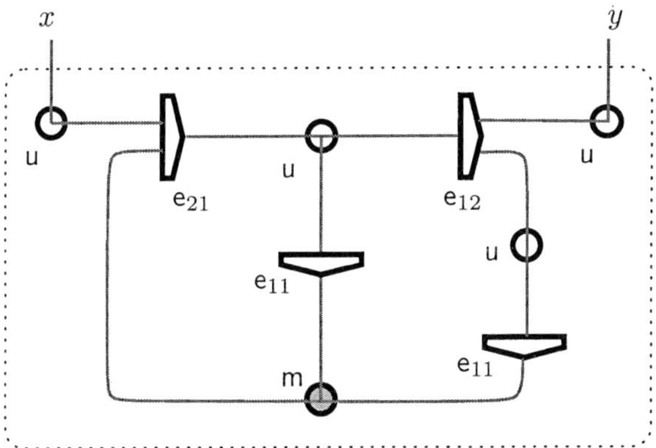

We choose three kinds of control: m ('marked') and u ('unmarked') for holding and non-holding conditions, and e_{hk} for events with h pre- and k post-conditions. The shape and colour of each node will save us from writing controls in diagrams. Conditions have arity 1; we site the single port of a condition node in its centre. An e_{hk}-node has $h + k$ ports; h *pre-ports* for pre-conditions, k *post-ports* for post-conditions. You may like to check that the above net has two open and three closed links.

[1] Terminology can become confused when discussing two different formalisms. In particular, Petri nets and bigraphs differ in their use of the terms 'transition' and 'place'. Fortunately, in this paper we are concerned only with *condition-event* nets, not *place-transition* nets, so we are able to avoid confusion.

We adopt the many-one sorting described above. Specifically, there are two sorts, γ for condition ports and η for event ports. An interface assigns one of these sorts to each of its names. When a net satisfies the many-one sorting constraints from the previous section (with η and γ for α and β) we call it *well-sorted*. Thus, in a well-sorted net, each condition has a single link to all its pre- and post-events, and each event port is linked to at most one condition. Let us denote the precategory of well-sorted condition-event nets by ´CE; the accent means that we are dealing with *concrete* link graphs.

In general an interface may contain both γ-names and η-names. But in the example you will notice that both x and y are γ-names, because each names a link containing a condition. In fact we shall confine our attention to the subprecategory ´CE_γ whose interfaces contain only γ-names, and whose nets are well-sorted. We call these *γ-nets*, for short. The ground γ-nets are our agents; note in particular that an agent contains all the pre- and post-conditions of its events.

The reader should note that our encoding of condition-event nets into γ-nets is not surjective, even up to support equivalence. The reason is that, in an encoded condition-event net, each pre-condition of a single event is linked to exactly one of its pre-ports (and similarly for post-conditions). This constraint is illustrated thus:

There are γ-nets that violate this constraint; it is not imposed by many-one sorting. This situation arises because in ´CE_γ we have equipped events with several ports, for technical reasons. But these spurious γ-nets need not disturb us, for it can be shown that a spurious one never arises from a genuine one as the result of a transition.

Let us now add dynamics to ´CE_γ, making it a reactive system. To do this, we introduce the usual Petri-net firing rules as reaction rules (r, r'), one for each e_{hk}. Figure 3 shows the rule for $h = 1, k = 2$. Note that r and r' are indeed agents. Note also that all the links of r are open; this means for every occurrence of r in an agent f there is a context D such that $f = D \circ r$. You may be concerned that we have given *particular* names to the interface of our reaction rules; this is no constraint, because by using wirings we can rename – or close – these names at will.

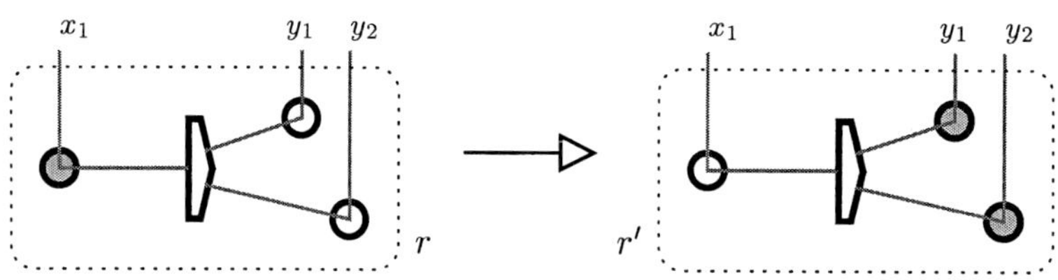

Fig. 3. A link-graph reaction rule for condition-event nets

We are now ready to examine the behaviour of γ-agents. Recall from the introduction that we already have a TS for them, namely $\mathcal{L}_\mathsf{p}$, defined without any help from link graph theory; the labels ℓ in its transitions $f \xrightarrow{\ell} f'$ take one of the forms $+x$,$-x$ or τ.

We shall assume each transition relation $\xrightarrow{\ell}$ to be closed under support equivalence. Denote by $\sim_{\mathsf{p}}$ the strong bisimilarity induced by $\mathcal{L}_{\mathsf{p}}$.

To compare this with the strong bisimilarity $\sim$ induced by link graph theory, let us now define the latter equivalence accurately. First recall that in the TS $\mathcal{L}_{\mathsf{g}}$, each L-transition $f \xrightarrow{L} f'$ is such that, for some D and reaction rule (r, r'), the pair (L, D) is a pushout for (f, r), and $f' \doteq D \circ r'$. This ensures that $\mathcal{L}_{\mathsf{g}}$ also is closed under support equivalence. Then, recalling the introduction, the equivalence $\sim$ is the largest symmetric relation such that

$$\begin{array}{l}\text{whenever } f \sim g \text{ and } f \xrightarrow{L} f', \text{ with } L \circ g \text{ defined,}\\ \text{there exists } g' \text{ such that } g \xrightarrow{L} g' \text{ and } f' \sim g'.\end{array}$$

(The condition that $L \circ g$ be defined is needed because we are working in a precategory.) Unlike $\sim_{\mathsf{p}}$, the bisimilarity $\sim$ is guaranteed by link graph theory to be a congruence, i.e. preserved by insertion into any context.

Our first task is to characterise the labels of $\mathcal{L}_{\mathsf{g}}$. We omit the detailed analysis. It turns out that (up to isomorphism in $'\mathrm{CE}_\gamma$) each label is either an identity, or an open γ-net with exactly one e-node, linked to zero or more m-nodes as preconditions and u-nodes as post-conditions. An identity label just signifies that the agent makes a transition with no assistance from its environment. In fact $f \xrightarrow{\mathsf{id}} f'$ iff $f \longrightarrow f'$; this justifies our use of the same arrow for both reactions and transitions.

Figure 4 shows a non-identity label. It is not quite a redex; it requires its client agent

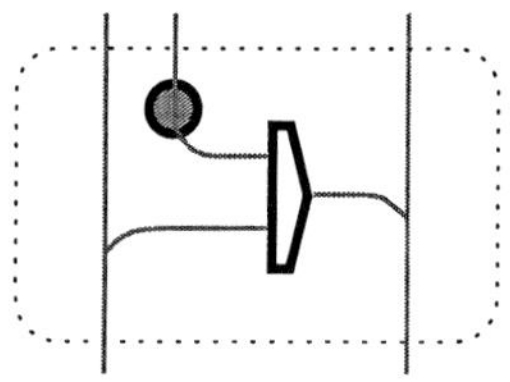

Fig. 4. A typical label in $\mathcal{L}_{\mathsf{g}}$

to provide one marked precondition and one unmarked postcondition. Figure 5 shows the anatomy of a transition $f \xrightarrow{L} f'$ with this label. Note that f' takes the form $\overline{L} \circ \overline{f}$; we call $\overline{L}$ and $\overline{f}$ the *residuals* of L and f respectively. We see that a single transition may change the marking of several named conditions of f, however far apart they may lie in f. Any other agent g with the same interface as f will have a similar transition, provided only that it has the same initial marking of its named conditions.

The two TSs $\mathcal{L}_{\mathsf{p}}$ and $\mathcal{L}_{\mathsf{g}}$ are significantly different, so it is not immediately clear that they will induce the same bisimilarity. We prove that they do so in the next section.

5 Coincidence of Bisimilarities

In $'\mathrm{CE}_\gamma$ we have two TSs on condition-event nets: $\mathcal{L}_{\mathsf{p}}$ defined directly with labels ℓ of the form $+x$, $-x$ or τ, and $\mathcal{L}_{\mathsf{g}}$ derived in link graph theory, with labels L consisting of

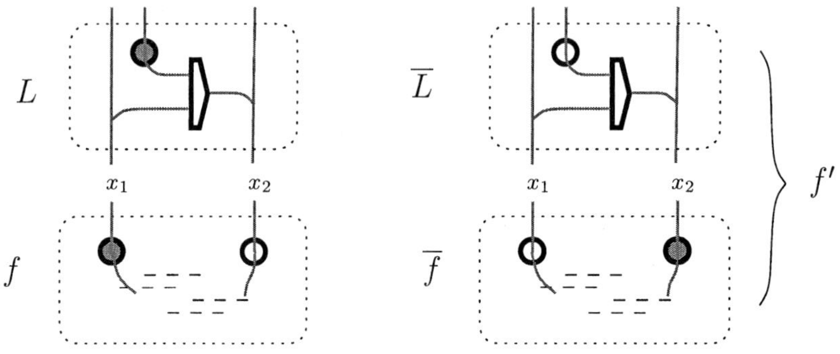

Fig. 5. Anatomy of a transition $f \xrightarrow{L} f'$ in $\mathcal{L}_{\mathrm{g}}$

link graph contexts having at most a single event node. The bisimilarities for the two TSs are $\sim_{\mathsf{p}}$ and $\sim$ respectively.

We shall first show that $\sim \subseteq \sim_{\mathsf{p}}$. This asserts that if we can distinguish two γ-nets f and g by using 'experiments' ℓ like $+x$ and $-x$, then we can also do so using 'experiments' L that are elementary link graph contexts. So, among the labels L generated by our theory (see Figure 4), we need to find those that can do the job of the experiments $+x$ and $-x$.

It turns out that such labels need only involve events with one pre- and one post-condition; we call them *input* and *output probes* respectively. They are denoted by in_{xz} and out_{xz}, and are shown in the first column of Figure 6. The second column shows the *spent* probes, the residuals of the probes. The third column shows the spent probes with their conditions closed; they are defined by $\overline{\mathsf{in}}_x \stackrel{\text{def}}{=} /z \circ \overline{\mathsf{in}}_{xz}$ and $\overline{\mathsf{out}}_x \stackrel{\text{def}}{=} /z \circ \overline{\mathsf{out}}_{xz}$. They may be called *twigs* because, up to the equivalence $\sim$, they can be broken off. The

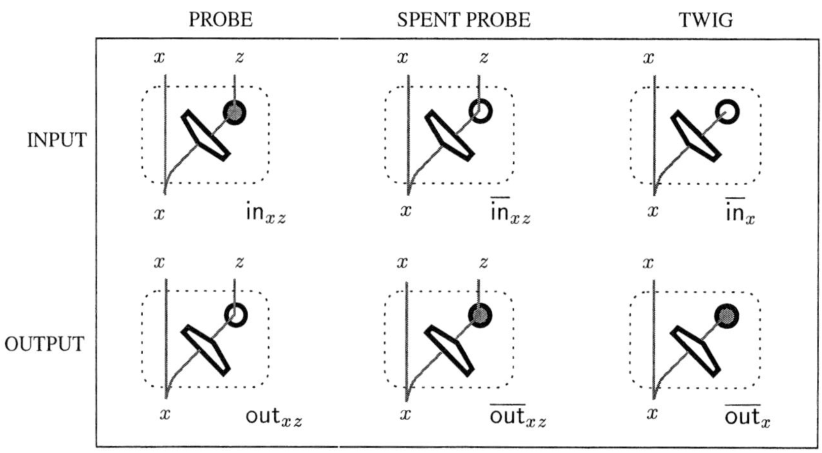

Fig. 6. Probes: labels in $\mathcal{L}_{\mathrm{g}}$ for observing conditions in a γ-net

intuition is simply that a twig occurring anywhere in a net can never fire. In fact we have a lemma, proved easily in link graph theory:

Lemma 1. *For any γ-agent f having x in its outer face,* $\overline{\mathsf{in}}_x \circ f \sim \overline{\mathsf{out}}_x \circ f \sim f$.

Now to prove that $\sim \subseteq \sim_{\mathsf{p}}$ it is enough to show that $\sim$ is an $\mathcal{L}_{\mathsf{p}}$-bisimulation. For this, suppose that $f \sim g$, and let $f \xrightarrow{\ell} f'$ in $\mathcal{L}_{\mathsf{p}}$. We must find g' such that $g \xrightarrow{\ell} g'$ and $f' \sim g'$. If $\ell = \tau$ this is easy, because then our assumption implies that $f \longrightarrow f'$, and hence $f \xrightarrow{\mathsf{id}} f'$ in $\mathcal{L}_{\mathsf{g}}$; but then by bisimilarity in $\mathcal{L}_{\mathsf{g}}$ we have $g \xrightarrow{\mathsf{id}} g' \sim f'$, and by reversing the reasoning for f we get that $g \xrightarrow{\tau} g'$ and we are done.

Now let $\ell = +x$ (the case for $-x$ is dual), so that $f \xrightarrow{+x} f'$. This means that f has an unmarked condition named x, so that in $\mathcal{L}_{\mathsf{g}}$ we have

$$f \xrightarrow{\mathsf{in}_{xz}} f'' \stackrel{\text{def}}{=} \overline{\mathsf{in}}_{xz} \circ f' .$$

Hence by bisimilarity in $\mathcal{L}_{\mathsf{g}}$ we have

$$g \xrightarrow{\mathsf{in}_{xz}} g'' = \overline{\mathsf{in}}_{xz} \circ g'$$

where $f'' \sim g''$ and g' is the residual of g'' under the transition. This residual g' differs from g only in having a marked condition named x that was unmarked in g, and hence we also have $g \xrightarrow{+x} g'$ in $\mathcal{L}_{\mathsf{p}}$. It remains only to show that $f' \sim g'$. We deduce this using the congruence of $\sim$ and Lemma 1:

$$\begin{aligned} f' \sim \overline{\mathsf{in}}_x \circ f' &= /z \circ \overline{\mathsf{in}}_{xz} \circ f' = /z \circ f'' \\ \sim /z \circ g'' &= /z \circ \overline{\mathsf{in}}_{xz} \circ g' = \overline{\mathsf{in}}_x \circ g' \\ \sim g' , \end{aligned}$$

and so we have proved

Lemma 2. $\sim \subseteq \sim_{\mathsf{p}}$.

To complete our theorem we must prove the converse, $\sim_{\mathsf{p}} \subseteq \sim$. It would be enough to prove that $\sim_{\mathsf{p}}$ is an $\mathcal{L}_{\mathsf{g}}$-bisimulation; but this is false. Instead we have to consider the closure of $\sim_{\mathsf{p}}$ under all contexts, namely

$$\mathcal{S} \stackrel{\text{def}}{=} \{ (C \circ f, C \circ g) \mid f \sim_{\mathsf{p}} g \} .$$

In fact it will be enough to prove that $\mathcal{S}^{\hat{=}}$, the closure of $\mathcal{S}$ under support equivalence, is a bisimulation. We get the required result by considering the case $C = \mathsf{id}$.

So let us assume that $f \sim_{\mathsf{p}} g$, and that $C \circ f \xrightarrow{M} f''$ in $\mathcal{L}_{\mathsf{g}}$. Then there is a reaction rule r and context D such that (M, D) forms a pushout for $(C \circ f, r)$, as shown in the left-hand diagram of Figure 7, and $f'' \hat{=} D \circ r'$. We now take the pushout (L, F) for (f, r), and properties of pushouts yield the right-hand diagram, in which the upper square is also a pushout. So there is a transition $f \xrightarrow{L} f'$, where $f' \hat{=} F \circ r'$; note also that $f'' \hat{=} C' \circ f'$.

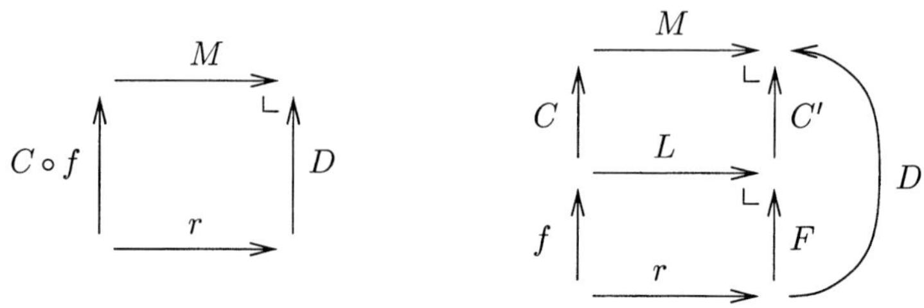

Fig. 7. Pushouts underlying transitions of $C \circ f$ and f

Now consider the anatomy of this transition, exemplified in Figure 5. We know that the residual $\overline{f}$ differs from f only in the changed marking of zero or more named conditions. It follows therefore that in $\mathcal{L}_{\mathsf{p}}$ there is a sequence of transitions

$$f \xrightarrow{\ell_1} f_1 \ldots \xrightarrow{\ell_n} f_n = \overline{f} \qquad (n \geq 0)$$

where $\ell_i \in \{+x_i, -x_i\}$; each transition marks or unmarks a single named condition. Moreover $f' = \overline{L} \circ \overline{f}$. Since $f \sim_{\mathsf{p}} g$ there exists a similar sequence

$$g \xrightarrow{\ell_1} g_1 \ldots \xrightarrow{\ell_n} g_n = \overline{g}$$

with $\overline{f} \sim_{\mathsf{p}} \overline{g}$. This implies that g has the same initial marking as f for the named conditions involved in the transitions. But we know that $L \circ g$ is defined (since we assumed $M \circ C \circ g = C' \circ L \circ g$ to be defined), so in $\mathcal{L}_{\mathsf{g}}$ there is a transition $g \xrightarrow{L} g' \stackrel{\text{def}}{=} \overline{L} \circ \overline{g}$. Its underlying pushout is shown in the left-hand diagram of Figure 8. Also it has an underlying reaction rule (s, s'), with $g' \simeq G \circ s'$.

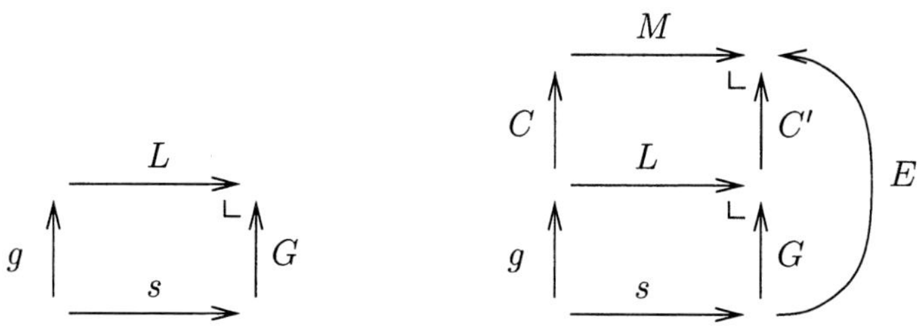

Fig. 8. Pushouts underlying transitions of g and $C \circ g$

Now we form the right-hand diagram of Figure 8 by replacing this pushout for the lower square in right-hand diagram of Figure 7. Since both small squares are pushouts, so is the large square; therefore it underlies an $\mathcal{L}_{\mathsf{g}}$-transition

$$C \circ g \xrightarrow{M} g'' \stackrel{\text{def}}{=} E \circ s' .$$

To complete our proof we need only show that the pair (f'', g'') lies in $\mathcal{S}^{\hat{=}}$. We already know that $f'' \hat{=} C' \circ f' = C' \circ \overline{L} \circ \overline{f}$. We can now compute

$$g'' = E \circ s' = C' \circ G \circ s' \hat{=} C' \circ g' = C' \circ \overline{L} \circ \overline{g}\ ,$$

and hence $(f'', g'') \in \mathcal{S}^{\hat{=}}$ since $\overline{f} \sim_{\mathsf{p}} \overline{g}$. It follows that $\sim_{\mathsf{p}} \subseteq \sim$.

So we have proved:

Theorem 1. (coincidence of bisimilarities) $\sim = \sim_{\mathsf{p}}$ *in* ´CE_γ.

The reader will remember that we have worked in *concrete* link graphs ´CE_γ in order to ensure the existence of IPOs (in fact pushouts); these were needed to define a transition system in a way that ensures congruence of bisimilarity. Having done this, we can now tranfer both the transitions and the bisimilarity to the corresponding category CE_γ of *abstract* link graphs, which has no IPOs. Note that CE_γ is indeed a category, not just a precategory, because support no longer places a constraint upon composition.

If G is a concrete link graph, let $[G]$ denote the corresponding abstract one – essentially the support-equivalence class of G. Then we define the transition system $[\mathcal{L}_{\mathsf{g}}]$ in CE_γ to be the smallest set such that

$$\text{if } g \xrightarrow{L} g' \text{ in } \mathcal{L}_{\mathsf{g}} \text{ then } [g] \xrightarrow{[L]} [g'] \text{ in } [\mathcal{L}_{\mathsf{g}}]\ .$$

Similarly we define $[\mathcal{L}_{\mathsf{p}}]$ in CE_γ; this is even simpler because its labels ℓ are not subject to support equivalence.

These two abstract transition systems induce corresponding bisimilarities in CE_γ; we shall again denote them by $\sim$ and $\sim_{\mathsf{p}}$. They are simply related to those in ´CE_γ. We conclude by stating this relationship, omitting the proof; it also has the consequence that the assertion of Theorem 1 for concrete link graphs is matched for abstract link graphs.

Corollary 1. (coincidence of bisimilarities in abstract condition-event nets)

1. $f \sim g$ *in* ´CE_γ *iff* $[f] \sim [g]$ *in* CE_γ.
2. $f \sim_{\mathsf{p}} g$ *in* ´CE_γ *iff* $[f] \sim_{\mathsf{p}} [g]$ *in* CE_γ.
3. *In* CE_γ *bisimilarity* $\sim$ *is a congruence and coincides with* $\sim_{\mathsf{p}}$.

6 Discussion

This exercise has shown that a particular class of Petri nets, condition-event nets, can be modelled and analysed in link graphs. It has not shown that this modelling is canonical, nor that it extends to other net disciplines. I hope that the relatively simplicity of the present analysis may provoke interest in these questions.

This would not only determine how far the present theory of bigraphs goes for Petri nets; it may also suggest improvements and variations of bigraph theory. Indeed it was by trying to analyse other concurrency models – especially the π-calculus of Milner, Parrow and Walker [7] and the mobile ambients of Cardelli and Gordon [2] – that bigraphs evolved from their predecessor action calculi, Milner [6]. A large concern in defining bigraphs has been to admit theoretical analysis (such as we have illustrated) which could not be provided so well for action calculi. Furthermore, Jensen and Milner [3] have recently shown that the behavioural theory of a version of the π-calculus can be exactly mirrored in bigraphs.

Even within the present exercise, interesting points emerge. In the present encoding of Petri nets, we stratify the event controls e_{hk} by their arities; this limits the number of pre- and post-conditions that can be connected to a given event in any context. In contrast, Nielsen et al [9] use a *recursion* combinator that connects a given condition to a given event. Thus the algebraic combinators provided uniformly by link graphs may not coincide with those designed for a particular application, and the comparison of the two requires further examination.

Another interesting outcome is the mismatch between the transition system $\mathcal{L}_\mathsf{g}$ generated by link-graph theory and the simple specific transition system $\mathcal{L}_\mathsf{p}$, despite the coincidence of the bisimilarities they induce. It is clear that the labels in $\mathcal{L}_\mathsf{g}$ are redundant, in the sense that the same phenomenon may be detected by more than one experiment. This is not surprising, because the labels are generated from each reaction rule *separately*; no attempt has yet been made to discover to what extent the labels from a *family* of reaction rules duplicate each other's discriminating power. The exercise suggests that in further development the theory of BRSs we should try to identify general properties of rule-sets that lead to such redundancies; this could yield more economical transition systems.

Our formulation of bigraphs uses precategories for two reasons. First, they provide concrete bigraphs with RPOs, which categories do not. Second, more generally, they provide a very direct way to distinguish different *occurrences* within the same bigraph. Although manipulation with precategories is not troublesome, they are not standard in category theory. Sassone and Sobocinski [13] have provided a valuable link with a more standard categorical concept, 2-categories; using these they have been able to recover exactly the RPO theory and congruence theory. Whether 2-categories will ease the further development of bigraphical theory, as presented in Jensen and Milner [4], is a topic for further research. But there is already advantage in an alternative formulation.

Finally, although no proposal is made here about how to enrich Petri nets with mobility, the present exercise offers a microcosm in which to test such proposals.

In summary, Petri nets and bigraphs may be able to enrich one another.

Acknowledgement

I am grateful to Mogens Nielsen, Vladimiro Sassone and Thiagarajan for helpful comments on the ideas in this paper.

References

1. Best, E., Devillers, R. and Hall, J.G., The box algebra: a model of nets and process expressions. 20th International Conference on Application and Theory of Petri Nets, LNCS 1639 (1999) 344–363.
2. Cardelli, L. and Gordon, A.D., Mobile ambients. Foundations of System Specification and Computational Structures, LNCS 1378 (2000) 140–155.

3. Jensen, O.-H. and Milner, R., Bigraphs and transitions. In Proc. 30th SIGPLAN-SIGACT Symposium on Principles of Programming Languages (2003).
4. Jensen, O.-H. and Milner, R., Bigraphs and mobile processes. Technical Report UCAM-CL-TR-570, University of Cambridge Computer Laboratory (2003). Also available at `http://www.cl.cam.ac.uk/users/rm135`, together with an index and slides.
5. Leifer, J.J. and Milner,R., Deriving bisimulation congruences for reactive systems. Proc. CONCUR 2000, 11th International Conference on Concurrency Theory (2000) 243–258.
6. Milner, R., Calculi for interaction. Acta Informatica 33 (1996) 707–737,
7. Milner, R., Parrow, J. and Walker D., A calculus of mobile processes, Parts I and II. Journal of Information and Computation 100 (1992) 1–77.
8. Park, D., Concurrency and automata on infinite sequences. In LNCS 104, Springer Verlag (1980).
9. Nielsen, M., Priese, L. and Sassone, V., Characterizing behavioural congruences for Petri nets. Proc. CONCUR'95, LNCS 962 (1995) 175–189.
10. Pomello, L., Rozenberg, G. and Simone, C., A survey of equivalence notions for net-based systems. Advances in Petri Nets '92, LNCS 609 (1992) 410–472.
11. Priese, L. and Wimmel, H., A uniform approach to true-concurrency and interleaving semantics for Petri nets. Theoretical Computer Science 206 (1998) 219–206.
12. Reisig, W., Petri Nets: an Introduction. Springer Verlag, Berlin (1985).
13. Sassone, V. and Sobocinski, P., Deriving bisimulation congruences: 2-categories vs. precategories. In Proc. FOSSACS '03, LNCS 2620 (2003) 409–424.

Notes on Timed Concurrent Constraint Programming

Mogens Nielsen[1,2,*] and Frank D. Valencia[1,2,**]

[1] **BRICS** University of Aarhus, Denmark
[2] Dept. of Information Technology, Uppsala University
{mn,fvalenci}@brics.dk

Abstract. A *constraint* is a piece of (partial) information on the values of the variables of a system. *Concurrent constraint programming* (ccp) is a model of concurrency in which agents (also called processes) interact by telling and asking information (constraints) to and from a shared store (a constraint). *Timed* (or *temporal*) ccp (tccp) extends ccp by agents evolving over time. A distinguishing feature of tccp, is that it combines in one framework an *operational and algebraic* view from process algebra with a *declarative* view based upon temporal logic. Tccp has been widely used to specify, analyze and program reactive systems.

This note provides a comprehensive introduction to the background for and central notions from the theory of tccp. Furthermore, it surveys recent results on a particular tccp calculus, `ntcc` , and it provides a classification of the expressive power of various tccp languages.

1 Introduction

Saraswat's *concurrent constraint programming* (ccp) [45] is a well-established formalism for concurrency based upon the shared-variables communication model where interaction arises via constraint-imposition over shared-variables. In ccp, agents can interact by *adding* (or *telling*) partial information to a medium, a so-called *store*. Partial information is represented by *constraints* (i.e., first-order formulae such as $x > 42$) on the shared variables of the system. The other way in which agents can interact is by *asking* partial information to the store. This provides the synchronization mechanism of the model; asking agents are suspended until there is enough information in the store to answer their query.

As other models of concurrency, ccp has been extended to capture aspects such as mobility [8, 12, 37], stochastic behavior [13], and most prominently time [5, 14, 40, 42]. *Timed* ccp extends ccp by allowing agents to be constrained by time requirements.

Modal extensions of logic study time in logic reasoning, and in the same way mature models of concurrency have been extended with explicit notions of time. For instance, neither Milner's CCS [25], Hoare's CSP [19], nor Petri Nets [33], in their original form, were concerned explicitly with temporal behavior, but they all have been extended to

* The contribution of M. Nielsen to this work was suported by Basic Research in Computer Science, Centre of the Danish National Research Foundation.

** The contributation of F. Valencia to this work was supported by the **PROFUNDIS** Project.

J. Desel, W. Reisig, and G. Rozenberg (Eds.): ACPN 2003, LNCS 3098, pp. 702–741, 2004.

incorporate an explicit notion of time, e.g. Timed CCS [53], Timed CSP [35], and Timed Petri Nets [54].

A distinctive feature of timed ccp is that it combines in one framework an *operational and algebraic* view based upon process calculi with a *declarative* view based upon temporal logic. So, processes can be treated as computing agents, algebraic terms and temporal formulae, and the combination in one framework of the alternative views of processes, allows timed ccp to benefit from the large body of techniques of well established theories.

Furthermore, timed ccp allows processes to be (1) expressed using a vocabulary and concepts appropriate to the *specific domain* (of some application under consideration), and (2) read and understood as temporal logic *specifications*. This feature is suitable for timed concurrent systems, since they often involve *specific domains* (e.g., controllers, databases, reservation systems) and have time-constraints *specifying* their behavior. Several timed extensions of ccp have been developed as settings for the modeling, programming and specification of timed systems [5, 14, 40, 43].

Organization. This note provides an overview of timed ccp with its basic background and various approaches explored in the literature. Furthermore, the note offers an introduction to a particular timed ccp process calculus called `ntcc` . In Sections 2 and 3 we give a basic background on ccp and timed ccp. Section 4 is devoted to present the developments of the timed ccp calculus `ntcc` [30]. In Section 5 we describe in detail several timed ccp languages and provide a classification of their expressive power. Finally, in Section 6 we discuss briefly some related and future work on timed ccp.

2 Background: Concurrent Constraint Programming

In his seminal PhD thesis [39], Saraswat proposed concurrent constraint programming as a model of concurrency based on the shared-variables communication model and a few primitive ideas taking root in logic. As informally described in the next section, the ccp model elegantly combines logic concepts and concurrency mechanisms.

Concurrent constraint programming traces its origins back to Montanari's pioneering work [28] leading to constraint programming and Shapiro's concurrent logic programming [46]. The ccp model has received a significant theoretical and implementational attention: Saraswat, Rinard and Panangaden [45] as well as De Boer, Di Pierro and Palamidessi [6] gave fixed-point denotational semantics to ccp, whilst Montanari and Rossi [36] gave a (true-concurrent) Petri-Net semantics (using the formalism of contextual nets); De Boer, Gabrielli et al [7] developed an inference system for proving properties of ccp processes; Smolka's Oz [48] as well as Haridi and Janson's AKL [17] programming languages are built upon ccp ideas.

The ccp Model. A concurrent system is specified in the ccp model in terms of *constraints* over the variables of the system. A constraint is a first-order formula representing *partial information* about the values of variables. As an example, for a system with variables x and y taking natural numbers as values, the constraint $x + y > 16$ specifies possible values for x and y (those satisfying the inequation). The ccp model is parameterized by a *constraint system*, which specifies the constraints of relevance for the kind of system under consideration, and an *entailment relation* $\models$ between constraints (e.g, $x + y > 16 \models x + y > 0$).

During a ccp computation, the state of the system is specified by an entity called the *store* in which information about the variables of the system resides. The store is represented as a constraint, and thus it may provide only partial information about the variables. This differs fundamentally from the traditional view of a store based on the Von Neumann memory model, in which each variable is assigned a uniquely determined value (e.g., $x = 16$ and $y = 7$), rather than a set of possible values.

The notion of store in ccp suggests a model of concurrency with a central memory. This is, however, only an abstraction which simplifies the presentation of the model. The store may be distributed in several sites according to the sharing of variables (see [39] for further discussions about this matter). Conceptually, the store in ccp is the *medium* through which agents interact with each other.

A ccp process can update the state of the system only by adding (or *telling*) information to the store. This is represented as the (logical) conjunction of the store representing the previous state and the constraint being added. Hence, updating does not change the values of the variables as such, but constrains further some of the previously possible values.

Furthermore, ccp processes can synchronize by querying (or *asking*) information from the store. Asking is blocked until there is enough information in the store to *entail* (i.e., answer positively) the query, i.e. the ask operation determines whether the constraint representing the store entails the query.

A ccp computation terminates whenever it reaches a point, called a *resting* or a *quiescent* point, in which no more information can be added to the store. The output of the computation is defined to be the final store, also called the *quiescent store*.

Example 1. Consider the simple ccp scenario illustrated in Figure 1. We have four agents (or processes) wishing to interact through an initially empty store. Let us name them, starting from the upper leftmost agent in a clockwise fashion, A_1, A_2, A_3 and A_4, respectively.

In this scenario, A_1 may move first and tell the others through the store the (partial) information that the temperature value is greater than 42 degrees. This causes the addition of the item "temperature>42" to the previously empty store.

Now A_2 may ask whether the temperature is exactly 50 degrees, and if so it wishes to execute a process P. From the current information in the store, however, the exact value of the temperature can not be entailed. Hence, the agent A_2 is blocked, and so is the agent A_3 since from the store it cannot be determined either whether the temperature is between 0 and 100 degrees.

However, A_4 may tell the information that the temperature is less than 70 degrees. The store becomes "temperature $> 42 \wedge$ temperature < 70", and now process A_3 can execute Q, since its query is entailed by the information in the store . The 2 agent A_2 is doomed to be blocked forever unless Q adds enough information to the store to entail its query. □

In the spirit of process calculi, the language of processes in the ccp model is given by a small number of primitive operators or combinators. A typical ccp process language contains the following operators:

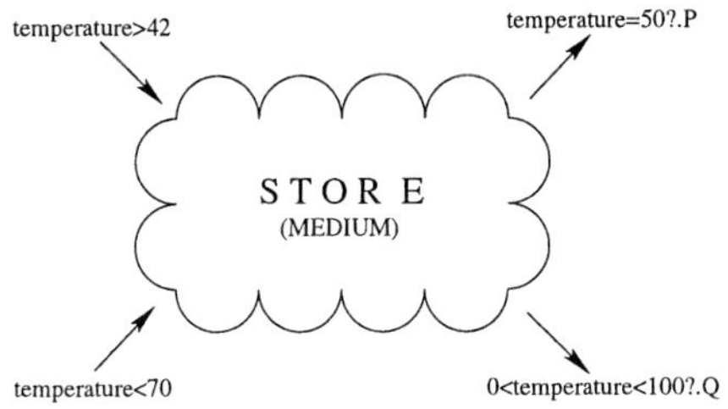

Fig. 1. A simple ccp scenario

- *A tell operator*, telling constraints (e.g., agent A_1 above).
- *An ask operator*, prefixing another process, its continuation (e.g. the agent A_2 above).
- *Parallel composition*, combining processes concurrently. For example the scenario in Figure 1 can be specified as the parallel composition of A_1, A_2, A_3 and A_4.
- *Hiding* (also called *restriction* or *locality*), introducing local variables, thus restricting the interface through which a process can interact with others.
- *Summation*, expressing a nondeterministic combination of agents to allow alternate courses of action.
- *Recursion*, defining infinite behavior.

It is worth pointing out that without summation, the ccp model is deterministic, in the sense that the final store is always the same, independently of the execution order (scheduling) of the parallel components [45].

3 Timed Concurrent Constraint Programming

The first timed ccp model was introduced by Saraswat et al [40] as an extension of ccp aimed at programming and modeling timed, reactive systems. This tcc model elegantly combines ccp with ideas from the paradigms of Synchronous Languages [2, 15].

The tcc model takes the view of reactive computation as proceeding *deterministically* in discrete time units (or time *intervals*). In other words, time is conceptually divided into discrete intervals. In each time interval, a deterministic ccp process receives a stimulus (i.e. a constraint) from the environment, it executes with this stimulus as the *initial store*, and when it reaches its resting point, it responds to the environment with the final store. Furthermore, the resting point determines a residual process, which is then executed in the next time interval.

This view of reactive computation is particularly appropriate for programming reactive systems such as robotic devices, micro-controllers, databases and reservation systems. These systems typically operate in a cyclic fashion; in each cycle they receive and input from the environment, compute on this input, and then return the corresponding output to the environment.

The tcc model extends the standard ccp with fundamental operations for programming reactive systems, e.g. *delay* and *time-out* operations. The delay operation forces the execution of a process to be postponed to the next time interval. The time-out (or weak *pre-emption*) operation waits during the current time interval for a given piece of information to be present and if it is not, triggers a process in the *next time interval.*

In spite of its simplicity, the tcc extension to ccp is far-reaching. Many interesting temporal constructs can be expressed, see [40] for details, As an example, tcc allows processes to be "clocked" by other processes. This provides meaningful pre-emption constructs and the ability of defining *multiple forms of time* instead of only having a unique global clock.

The tcc model has attracted a lot of attention recently. Several extensions have been introduced and studied in the literature. One example can be found in [43], adding a notion of strong pre-emption: the time-out operations can trigger activity in the current time interval. Other extensions of tcc have been proposed in [14], in which processes can evolve continuously as well as discretely.

The tccp framework, introduced in [5] by Gabrielli et al, is a fundamental representative model of nondeterministic timed ccp. In [5] the authors advocate the need of nondeterminism in the context of timed ccp. In fact, they use tccp to model interesting applications involving nondeterministic timed systems (see [5]).

It would be hard to introduce all the tcc extensions in detail, and hence we focus in the following on the `ntcc` calculus, which is a generalization of the tcc model introduced in [30] by Palamidessi and the present authors. The calculus is built upon few basic ideas but it captures several aspects of timed systems. As tcc, `ntcc` can model unit delays, time-outs, pre-emption and synchrony. Additionally, it can model *unbounded but finite delays, bounded eventuality, asynchrony* and *nondeterminism*. The applicability of the calculus has been illustrated with several examples of discrete-time systems involving , mutable data structures, robotic devices, multi-agent systems and music applications [38].

The major difference between tccp model from [5] and `ntcc` is that the former extends the original ccp while the latter extends the tcc model. More precisely, in tccp the information about the store is carried through the time units, thus the semantic setting is completely different. The notion of time is also different; in tccp each time unit is identified with the time needed to ask and tell information to the store. As for the constructs, unlike `ntcc`, tccp provides for arbitrary recursion and does not have an operator for specifying unbounded but finite delays.

4 The `ntcc` Process Calculus

This section gives a formal introduction to the `ntcc` model. We introduce the syntax and the semantics of the `ntcc` process language, and illustrate the expressiveness by modeling robotic devices. Furthermore, we shall present some of the reasoning techniques provided by `ntcc` focusing on

1. *Behavioural equivalences*, which are characterized operationally, relating process behavior much like the behavioral equivalences for traditional process calculi (e.g., bisimilarity and trace-equivalence).
2. A *denotational semantics* which interprets a given process as the set of sequences of input/output behaviours it can potentially exhibit while interacting with arbitrary environments.

3. A *process logic* expressing specifications of behaviors of processes, and an associated *inference system* providing proofs of processes fulfilling specifications.

Informal Description of `ntcc` Processes

We shall begin with an informal description of the process calculus with examples. These examples are also meant to give a flavour of the range of application of `ntcc` .

As for the tcc model, the `ntcc` model is parameterized by a *constraint system.* A constraint system provides a *signature* from which syntactically denotable objects called *constraints* can be constructed, and an *entailment relation* $\models$ specifying interdependencies between these constraints.

We can set up the notion of constraint system by using first-order logic. Let us suppose that Σ is a signature (i.e., a set of constants, functions and predicate symbols) and that Δ is a consistent first-order theory over Σ (i.e., a set of sentences over Σ having at least one model). Constraints can be thought of as first-order formulae over Σ. We can then decree that $c \models d$ if the implication $c \Rightarrow d$ is valid in Δ. This gives us a simple and general formalization of the notion of constraint system as a pair (Σ, Δ).

In the examples below we shall assume that, in the underlying constraint system, Σ is the set $\{=, <, 0, 1 \ldots\}$ and Δ is the set of sentences over Σ valid for the natural numbers.

We now proceed to describe with examples the basic ideas underlying the behavior of `ntcc` processes. For this purpose we shall model simple behavior of controllers such as Programmable Logic Controllers (PLC's) and RCX bricks.

PLC's are often used in timed systems of industrial applications [9], whilst RCX bricks are mainly used to construct autonomous robotic devices [21]. These controllers have external input and output ports. One can attach, for example, sensors of light, touch or temperature to the input ports, and actuators like motors, lights or alarms to the output ports. Typically PLC's and RCX bricks operate in a cyclic fashion. Each cycle consists of receiving an input from the environment, computing on this input, and returning the corresponding output to the environment.

Our processes will operate similarly. Time is conceptually divided into *discrete intervals (or time units).* In a particular time interval, a process P_i receives a *stimulus* c_i from the environment. The stimulus is some piece of information, i.e., a constraint. The process P_i executes with this stimulus as the initial store, and when it reaches its resting point (i.e., a point in which no further computation is possible), it *responds* to the environment with a resulting store d_i. Also the resting point determines a residual process P_{i+1}, which is then executed in the next time interval.

The following sequence illustrates the stimulus-response interactions between an environment that inputs $c_1, c_2, \ldots$ and a process that outputs $d_1, d_2, \ldots$ on such inputs as described above.

$$P_1 \overset{(c_1,d_1)}{\Longrightarrow} P_2 \overset{(c_2,d_2)}{\Longrightarrow} \ldots P_i \overset{(c_i,d_i)}{\Longrightarrow} P_{i+1} \overset{(c_{i+1},d_{i+1})}{\Longrightarrow} \ldots \tag{1}$$

Telling and Asking Information. The `ntcc` processes communicate with each other by posting and reading partial information about the variables of system they model. The basic actions for communication provide the *telling* and *asking* of information. A tell action adds a piece of information to the common store. An ask action queries the store to decide whether a given piece of information is present. The store is a constraint itself. In this way, addition of information corresponds to logical conjunction, and determining the presence of information corresponds to logical entailment.

The tell and ask processes have the syntactic forms respectively

$$\mathbf{tell}(c) \text{ and } \mathbf{when}\ c\ \mathbf{do}\ P. \tag{2}$$

The only action of a tell process $\mathbf{tell}(c)$ is to add, within a time unit, c to the current store d. The store then becomes $d \wedge c$. The addition of c is carried out even if the store becomes inconsistent, i.e., $(d \wedge c) = \mathtt{false}$, in which case we can think of such an addition as generating a *failure*.

Example 2. Suppose that $d = (\mathtt{motor}_1\mathtt{\_speed} > \mathtt{motor}_2\mathtt{\_speed})$. Intuitively, d tells us that the speed of motor one is greater than that of motor two. It does not tell us the specific speed values. The execution in store d of process

$$\mathbf{tell}(\mathtt{motor}_2\mathtt{\_speed} > 10)$$

causes the store to become $(\mathtt{motor}_1\mathtt{\_speed} > \mathtt{motor}_2\mathtt{\_speed} > 10)$ in the current time interval, thus increasing the information we know about the system.

Notice that in the underlying constraint system $d \models \mathtt{motor}_1\mathtt{\_speed} > 0$, therefore the process

$$\mathbf{tell}(\mathtt{motor}_1\mathtt{\_speed} = 0)$$

in store d causes a failure. □

The process $\mathbf{when}\ c\ \mathbf{do}\ P$ performs the action of asking c. If during the current time interval c can eventually be inferred from the store d (i.e., $d \models c$) then P is executed within the same time interval. Otherwise, $\mathbf{when}\ c\ \mathbf{do}\ P$ is precluded from execution (i.e., it becomes permanently inactive).

Example 3. Suppose that $d = (\mathtt{motor}_1\mathtt{\_speed} > \mathtt{motor}_2\mathtt{\_speed})$ is the store. The process

$$P = \mathbf{when}\ \mathtt{motor}_1\mathtt{\_speed} > 0\ \mathbf{do}\ Q$$

will execute Q in the current time interval since $d \models \mathtt{motor}_1\mathtt{\_speed} > 0$, by contrast the process

$$P' = \mathbf{when}\ \mathtt{motor}_1\mathtt{\_speed} > 10\ \mathbf{do}\ Q$$

will not execute Q, unless more information is added to the store during the current time interval entailing $\mathtt{motor}_1\mathtt{\_speed} > 10$.

□

Nondeterminism. As argued above, partial information allows us to model behavior for alternative values of variables. In concurrent systems it is often convenient to model behavior for *alternative courses* of action, i.e., nondeterministic behavior.

We generalize the processes of the form **when** c **do** P described above to guarded-choice summation processes of the form

$$\sum_{i \in I} \textbf{when } c_i \textbf{ do } P_i \tag{3}$$

where I is a finite set of indices. The expression $\sum_{i \in I} \textbf{when } c_i \textbf{ do } P_i$ represents a process that, in the current time interval, *nondeterministically* chooses a process P_j ($j \in I$) whose corresponding constraint c_j is entailed by the store. The chosen alternative, if any, precludes the others. If no choice is possible during the current time unit, all the alternatives are precluded from execution. In the following example we shall use "+" for binary summations.

Example 4. Often RCX programs operate in a set of simple stimulus-response rules of the form **IF** E **THEN** C. The expression E is a condition typically depending on the sensor variables, and C is a command, typically an assignment. In [11] these programs respond to the environment by choosing a rule whose condition is met and executing its command.

If we wish to abstract from the particular implementation of the mechanism that chooses the rule, we can model the execution of these programs by using the summation process. For example, the program operating in the set

$$\left\{ \begin{array}{l} (\textbf{IF } \texttt{sensor}_1 > 0 \ \ \textbf{THEN } \texttt{motor}_1\texttt{\_speed} := 2), \\ (\textbf{IF } \texttt{sensor}_2 > 99 \ \textbf{THEN } \texttt{motor}_1\texttt{\_speed} := 0) \end{array} \right\}$$

corresponds to the summation process

$$P = \begin{array}{l} \textbf{when } \texttt{sensor}_1 > 0 \ \ \textbf{do tell}(\texttt{motor}_1\texttt{\_speed} = 2) \\ + \\ \textbf{when } \texttt{sensor}_2 > 99 \ \textbf{do tell}(\texttt{motor}_1\texttt{\_speed} = 0). \end{array}$$

In the store $d = (\texttt{sensor}_1 > 10)$, the process P causes the store to become $d \wedge (\texttt{motor}_1\texttt{\_speed} = 2)$ since $\textbf{tell}(\texttt{motor}_1\texttt{\_speed} = 2)$ is chosen for execution and the other alternative is precluded. In the store $\texttt{true}$, P cannot add any information. In the store $e = (\texttt{sensor}_1 = 10 \wedge \texttt{sensor}_2 = 100)$, P causes the store to become either $e \wedge (\texttt{motor}_1\texttt{\_speed} = 2)$ or $e \wedge (\texttt{motor}_1\texttt{\_speed} = 0)$. □

Parallel Composition. Given P and Q we denote their parallel composition by the process

$$P \parallel Q \tag{4}$$

In one time unit processes P and Q operate concurrently, "communicating" via the common store by telling and asking information.

Example 5. Let P be defined as in Example 4 and

$$Q = \begin{array}{l} \textbf{when } \texttt{motor}_1\texttt{\_speed} = 0 \textbf{ do tell}(\texttt{motor}_2\texttt{\_speed} = 0) \\ + \\ \textbf{when } \texttt{motor}_2\texttt{\_speed} = 0 \textbf{ do tell}(\texttt{motor}_1\texttt{\_speed} = 0). \end{array}$$

Intuitively Q turns off one motor if the other is detected to be off. The parallel composition $P \parallel Q$ in the store $d = (\texttt{sensor}_2 > 100)$ will, in one time unit, cause the store to become $d \wedge (\texttt{motor}_1\texttt{\_speed} = \texttt{motor}_2\texttt{\_speed} = 0)$. □

Local Behavior. Most process calculi have a construct to restrict the interface through which processes can interact with each other, thus providing for the modeling of *local (or hidden)* behavior. We introduce processes of the form

$$(\mathbf{local}\, x)\, P \tag{5}$$

The process $(\mathbf{local}\, x)\, P$ declares a variable x, private to P. This process behaves like P, except that all the information about x produced by P is hidden from external processes and the information about x produced by other external processes is hidden from P.

Example 6. In modeling RCX or PLC's one uses "global" variables to represent ports (e.g., sensor and motors). However, one often also uses variables, which represent some local (or private) computational data.

Suppose that R is a given process modeling some controller task. Furthermore, suppose that R uses a variable z, which is set at random to a value $v \in \{0, 1\}$ in the process P, i.e.

$$P = \Big(\sum_{v \in \{0,1\}} \mathbf{when}\ \mathtt{true}\ \mathbf{do}\ \mathbf{tell}(z = v)\Big) \parallel R$$

representing the behavior of R under P's random assignment of z.

We may want to declare z in P to be local since it does not represent an input or output port. Moreover, notice that if we need to run two copies of P, i.e., process $P \parallel P$, a failure may arise as each copy can assign a different value to z. Therefore, the behavior of R under the random assignment to z can be best represented by $P' = (\mathbf{local}\, z)\, P$. In fact, if we run two copies of P', no failure can arise from the random assignment to the z's as they are private to each P'. □

The processes hitherto described generate activity within the current time interval only. We now turn to constructs that can generate activity in future time intervals.

Unit Delays and Time-Outs. As in the Synchronous Languages [2] we have constructs whose actions can delay the execution of processes. These constructs are needed to model time dependency between actions, e.g., actions depending on preceding actions.

The unit-delay operators have the form

$$\mathbf{next}\, P \ \text{ and } \ \mathbf{unless}\ c\ \mathbf{next}\, P \tag{6}$$

The process $\mathbf{next}\, P$ represents the activation of P in the next time interval. The process $\mathbf{unless}\, c\, \mathbf{next}\, P$ is similar, but P will be activated only if c cannot be inferred from the resulting (or final) store d in the current time interval, i.e., $d \not\models c$. The "unless" processes add time-outs to the calculus, i.e., they wait during the current time interval for a piece of information c to be present and if it is not, they trigger activity in the next time interval.

Notice that $\mathbf{unless}\, c\, \mathbf{next}\, P$ is not equivalent to $\mathbf{when}\, \neg c\, \mathbf{do}\, \mathbf{next}\, P$ since $d \not\models c$ does not necessarily imply $d \models \neg c$. Notice also that $Q = \mathbf{unless}\ \mathtt{false}\ \mathbf{next}\, P$ is not the same as $R = \mathbf{next}\, P$, since R (unlike Q) always activates P in the next time interval, even if the store entails $\mathtt{false}$.

Example 7. Let us consider the following process:

$$P = \textbf{when}\ \texttt{false}\ \textbf{do next tell}(\texttt{motor}_1\texttt{\_speed} = \texttt{motor}_2\texttt{\_speed} = 0).$$

P turns the motors off by decreeing that $\texttt{motor}_1\texttt{\_speed} = \texttt{motor}_2\texttt{\_speed} = 0$ in the next time interval if a failure takes place in the current time interval. Similarly, the process

$$\textbf{unless}\ \texttt{false}\ \textbf{next}\,(\textbf{tell}(\texttt{motor}_1\texttt{\_speed} > 0) \parallel \textbf{tell}(\texttt{motor}_2\texttt{\_speed} > 0))$$

makes the motors move at some speed in the next time unit, unless a failure takes place in the current time interval. □

Asynchrony. We now introduce a construct that, unlike the previous ones, can describe arbitrary (finite) delays. The importance of this construct is that it allows us to model asynchronous behavior across the time intervals.

We use the operator "$\star$" which corresponds to the unbounded but finite delay operator for synchronous CCS [26]. The process

$$\star\, P \tag{7}$$

represents an arbitrary long but finite delay for the activation of P. Thus, $\star\,\textbf{tell}(c)$ can be viewed as a message c that is eventually delivered but there is no upper bound on the delivery time.

Example 8. Let $S = \star\,\textbf{tell}(\texttt{malfunction}(\texttt{motor}_1\texttt{\_status}))$. The process S can be used to specify that $\texttt{motor}_1$, at some unpredictable point in time, is doomed to malfunction □

Infinite Behavior. Finally, we need a construct to define infinite behavior. We shall use the operator "!" as a delayed version of the replication operator for the π−calculus [27]. Given a process P, the process

$$!\,P \tag{8}$$

represents $P \parallel (\textbf{next}\,P) \parallel (\textbf{next next}\,P) \parallel \cdots \parallel !\,P$, i.e., unboundedly many copies of P, but one at a time. The process $!\,P$ executes P in one time unit and persists in the next time unit.

Example 9. The process R below repeatedly checks the state of $\texttt{motor}_1$. If a malfunction is reported, R tells that $\texttt{motor}_1$ must be turned off.

$$R = !\,\textbf{when}\ \texttt{malfunction}(\texttt{motor}_1\texttt{\_status})\ \textbf{do tell}(\texttt{motor}_1\texttt{\_speed} = 0)$$

Thus, $R \parallel S$ with $S = \star\,\textbf{tell}(\texttt{malfunction}(\texttt{motor}_1\texttt{\_status}))$ (Example 8) eventually tells that $\texttt{motor}_1$ is turned off. □

Some Derived Forms

We have informally introduced the basic process constructs of ntcc and illustrated how they can be used to model or specify system behavior. In this section we shall illustrate how they can be used to obtain some convenient derived constructs.

In the following we shall omit "**when** true **do**" if no confusion arises. The "blind-choice" process $\sum_{i \in I}$ **when** true **do** P_i, for example, can be written as $\sum_{i \in I} P_i$. We shall use $\prod_{i \in I} P_i$, where I is finite, to denote the parallel composition of all the P_i's. We use $\mathbf{next}^n(P)$ as an abbreviation for $\mathbf{next}(\mathbf{next}(\ldots(\mathbf{next}\, P)\ldots))$, where **next** is repeated n times.

Inactivity. The process doing nothing whatsoever, **skip** can be defined as an abbreviation of the empty summation $\sum_{i \in \emptyset} P_i$. This process corresponds to the inactive processes $\mathbf{0}$ of CCS and $STOP$ of CSP. We should expect the behavior of $P \parallel \mathbf{skip}$ to be the same as that of P under any reasonable notion of behavioral equivalence.

Abortion. Another useful construct is the process **abort** which is somehow to the opposite extreme of **skip**. Whilst having **skip** in a system causes no change whatsoever, having **abort** can make the whole system fail. Hence **abort** corresponds to the $CHAOS$ operator in CSP. In Section 4 we mentioned that a tell process causes a failure, at the current time interval, if it leaves the store inconsistent. Therefore, we can define **abort** as $!\,\mathbf{tell}(\mathtt{false})$, i.e., the process that once activated causes a constant failure. Therefore, any reasonable notion of behavioral equivalence should not distinguish between $P \parallel \mathbf{abort}$ and **abort**.

Asynchronous Parallel Composition. Notice that in $P \parallel Q$ both P and Q are forced to move in the current time unit, thus our parallel composition can be regarded as being a synchronous operator. There are situations where an asynchronous version of "$\parallel$" is desirable. For example, modeling the interaction of several controllers operating concurrently where some of them could be faster or slower than the others at responding to their environment.

By using the star operator we can define a *(fair) asynchronous* parallel composition $P \mid Q$ as

$$(P \parallel \star Q) \;+\; (\star P \parallel Q)$$

A move of $P \mid Q$ is either one of P or one of Q (or both). Moreover, both P and Q are eventually executed (i.e. a fair execution of $P \mid Q$). This process corresponds to the asynchronous parallel operator described in [26].

We should expect operator "$\mid$" to enjoy properties of parallel composition. Namely, we should expect $P \mid Q$ to be the same as $Q \mid P$ and $P \mid (Q \mid R)$ to be the same as $(P \mid Q) \mid R$. Unlike in $P \parallel \mathbf{skip}$, however, in $P \mid \mathbf{skip}$ the execution of P may be arbitrary postponed, therefore we may want to distinguish between $P \mid \mathbf{skip}$ and P. Similarly, unlike in $P \parallel \mathbf{abort}$, in $P \mid \mathbf{abort}$ the execution of **abort** may be arbitrarily postponed.

Bounded Eventuality and Invariance. We may want to specify that a certain behavior is exhibited within a certain number of time units, i.e., *bounded eventuality*, or during a certain number of time units, i.e., *bounded invariance*. An example of bounded eventuality is "the light must be switched off within the next ten time units" and an example of bounded invariance is "the motor should not be turned on during the next sixty time units".

The kind of behavior described above can be specified by using the bounded versions of $!\, P$ and $\star\, P$, which can be derived using summation and parallel composition in the obvious way. We define $!_I P$ and $\star_I P$, where I is a closed interval of the natural numbers, as an abbreviation for

$$\prod_{i \in I} \mathbf{next}^i P \text{ and } \sum_{i \in I} \mathbf{next}^i P$$

respectively. Intuitively, $\star_{[m,n]} P$ means that P is eventually active between the next m and $m+n$ time units, while $!_{[m,n]} P$ means that P is always active between the next m and $m+n$ time units.

4.1 The Operational Semantics of ntcc

Following the informal description of ntcc above, we now proceed with a formal definition. We shall begin by formalizing the notion of constraint system and the syntax of ntcc . We shall then give meaning to the ntcc processes by means of an operational semantics. The semantics, which resembles the reduction semantics of the π-calculus [27], provides *internal* and *external* transitions describing process evolutions. The internal transitions describe evolutions within a time unit, and they are considered to be unobservable. The external transitions describe evolution across the time units, and they are considered to be observable.

Constraint Systems. For our purposes it will suffice to consider the notion of constraint system based on first-order logic, following e.g. [47].

Definition 1 (Constraint System). *A* constraint system (cs) *is a pair* (Σ, Δ) *where* Σ *is a signature of function and predicate symbols, and* Δ *is a decidable theory over* Σ *(i.e., a decidable set of sentences over* Σ *with a least one model).*

Given a constraint system (Σ, Δ), let $(\Sigma, \mathcal{V}, \mathcal{S})$ be its underlying first-order language, where $\mathcal{V}$ is a countable set of variables $x, y, \ldots$, and $\mathcal{S}$ is the set of logic symbols $\neg, \wedge, \vee, \Rightarrow, \exists, \forall$, $\mathtt{true}$ and $\mathtt{false}$. *Constraints* $c, d, \ldots$ are formulae over this first-order language. We say that c *entails* d in Δ, written $c \models d$, iff $c \Rightarrow d$ is true in all models of Δ. The relation $\models$, which is decidable by the definition of Δ, induces an equivalence $\approx$ given by $c \approx d$ iff $c \models d$ and $d \models c$.

Convention 1 *Henceforth,* $\mathcal{C}$ *denotes the set of constraints modulo* $\approx$ *under consideration in the underlying constraint system.*

Let us now give some examples of constraint systems. The classical example is the Herbrand constraint system [39].

Definition 2 (Herbrand Constraint System). *The Herbrand constraint system is such that:*

- Σ *is a set with infinitely many function symbols of each arity and equality* $=$.
- Δ *is given by Clark's Equality Theory with the schemas*
  $f(x_1,\ldots,x_n){=}f(y_1,\ldots,y_n) \Rightarrow x_1 = y_1 \wedge \ldots \wedge x_n = y_n$
  $f(x_1,\ldots,x_n){=}g(y_1,\ldots,y_n) \Rightarrow \texttt{false}$, *if* f, g *are distinct symbols*
  $x = f(\ldots x \ldots) \Rightarrow \texttt{false}$.

The importance of the Herbrand constraint system is that it underlies conventional logic programming and many first-order theorem provers. Its value lies in the Herbrand Theorem, which reduces the problem of checking unsatisfiability of a first-order formula to the unsatisfiability of a quantifier-free formula interpreted over finite trees.

Another widely used constraint system is the finite-domain constraint system **FD** defined in [18]. In **FD** variables are assumed to range over finite domains and, in addition to equality, we may have predicates that restrict the range of a variable to some finite set. The following is a simplified finite-domain constraint system.

Definition 3 (A Finite-Domain Constraint System). *Let* $n > 0$. *Define* $\mathbf{FD}[n]$ *as the constraint system such that:*

- Σ *is given by the constants symbols* $0, 1, \ldots., n-1$ *and the equality* $=$.
- Δ *is given by the axioms of equational theory* $x = x$, $x = y \Rightarrow y = x$, $x = y \wedge y = z \Rightarrow x = z$, *and* $v = w \Rightarrow \texttt{false}$ *for each two different constants* v, w *in* Σ.

Intuitively $\mathbf{FD}[n]$ provides a theory of variables ranging over a finite domain of values $\{0, \ldots, n-1\}$ with syntactic equality over these values.

The following is a somewhat more complex finite-domain constraint system.

Definition 4 (Modular Arithmetic Constraint System). *Let* $n > 0$. *Define* $\mathbf{A}[n]$ *as the constraint system such that:*

- Σ *is given by* $\{0, 1, \ldots., n-1, \mathrm{succ}, \mathrm{pred}, +, \times, =, >\}$.
- Δ *is the set of sentences valid in arithmetic modulo* n.

The intended meaning of $\mathbf{A}[n]$ is the natural numbers interpreted as in arithmetic modulo n. Due to the familiar operations it provides, we shall often assume that $\mathbf{A}[n]$ is the underlying constraint system in our examples and applications.

Other examples of constraint systems include: Rational intervals, Enumerated type, the Kahn constraint system and the Gentzen constraint system (see [45] and [39] for details).

Process Syntax and Semantics

Following the informal description above, the process constructions in the ntcc calculus are given by the following syntax:

Definition (Processes, *Proc*). *Processes* P, Q, $\ldots \in$ *Proc are built from constraints* $c \in \mathcal{C}$ *and variables* $x \in \mathcal{V}$ *in the underlying constraint system by:*

$$\begin{array}{rcl} P, Q, \ldots & ::= & \mathbf{tell}(c) \mid \sum_{i \in I} \mathbf{when}\ c_i\ \mathbf{do}\ P_i \mid P \parallel Q \mid (\mathbf{local}\, x)\, P \\ & \mid & \mathbf{next}\, P \mid \mathbf{unless}\ c\ \mathbf{next}\, P \mid \star\, P \mid\ !\, P \end{array}$$

The informal semantic meaning provided above of the constructs is formalized in terms of the following structural operational semantics (SOS) of ntcc . This semantics defines *transitions* between process-store *configurations* of the form $\langle P, c\rangle$, with stores represented as constraints and processes quotiented by the congruence $\equiv$ below.

Let us define precisely what we mean by the term "congruence" of processes, a key concept in the theory of process algebra. First, we need to introduce the standard notion of *process context*. Informally speaking, a process context is a process expression with a single hole, represented by $[\cdot]$, such that placing a process in the hole yields a well-formed process. More precisely,

Definition 5 (Process Context). Process contexts C *are given by the syntax*

$$\begin{array}{rll} C ::= & [\cdot] & \mid \mathbf{when}\ c\ \mathbf{do}\ C + M \mid C \parallel P \mid P \parallel C \mid (\mathbf{local}\ x)\ C \\ \mid & \mathbf{next}\ C & \mid \mathbf{unless}\ c\ \mathbf{next}\ C \quad \mid \star\ C \quad \mid !\,C \end{array}$$

where M stands for summations. The process $C[Q]$ results from the textual substitution of the hole $[\cdot]$ in C with Q.

An equivalence relation is a congruence if it respects all contexts:

Definition 6 (Process Congruence). *An equivalence relation $\cong$ on processes is said to be a process* congruence *iff for all contexts C, $P \cong Q$ implies $C[P] \cong C[Q]$.*

We can now introduce the structural congruence $\equiv$. Intuitively, the relation $\equiv$ describes irrelevant syntactic aspects of processes. It states that $(Proc/\equiv, \parallel, \mathbf{skip})$ is a commutative monoid.

Definition 7 (Structural Congruence). *Let $\equiv$ be the smallest congruence over processes satisfying the following axioms:*

1. $P \parallel \mathbf{skip} \equiv P$
2. $P \parallel Q \equiv Q \parallel P$
3. $P \parallel (Q \parallel R) \equiv (P \parallel Q) \parallel R.$

We extend $\equiv$ to configurations by decreeing that $\langle P, c\rangle \equiv \langle Q, c\rangle$ iff $P \equiv Q$.

Convention 2 *Following standard notation, we extend the syntax with a construct* $\mathbf{local}\,(x, d)\ \mathbf{in}\ P$, *to represent the evolution of a process of the form* $\mathbf{local}\ x\ \mathbf{in}\ Q$, *where d is the local information (or store) produced during this evolution. Initially d is "empty", so we regard* $\mathbf{local}\ x\ \mathbf{in}\ P$ *as* $\mathbf{local}\ (x, \mathtt{true})\ \mathbf{in}\ P$.

The transitions of the SOS are given by the relations $\longrightarrow$ and $\Longrightarrow$ defined in Table 1. The *internal* transition $\langle P, d\rangle \longrightarrow \langle P', d'\rangle$ should be read as "P with store d reduces, in one internal step, to P' with store d' ". The *observable transition* $P \xrightarrow{(c,d)} R$ should be read as "P on input c, reduces in one *time unit* to R and outputs d".

Intuitively, the observable reduction is obtained from a sequence of internal reductions starting in P with initial store c and terminating in a process Q with final store d. The process R, which is the one to be executed in the next *time interval* (or time unit), is obtained by removing from Q what was meant to be executed only during the

Table 1. Rules for internal reduction $\longrightarrow$ (upper part) and observable reduction $\Longrightarrow$ (lower part). $\gamma \not\longrightarrow$ in OBS holds iff for no γ', $\gamma \longrightarrow \gamma'$. $\equiv$ and F are given in Definitions 7 and 8.

$$\text{TELL}\ \frac{}{\langle \mathbf{tell}(c), d\rangle \longrightarrow \langle \mathbf{skip}, d \wedge c\rangle} \qquad \text{SUM}\ \frac{d \models c_j \ \ j \in I}{\langle \sum_{i\in I} \mathbf{when}\ c_i\ \mathbf{do}\ P_i, d\rangle \longrightarrow \langle P_j, d\rangle}$$

$$\text{PAR}\ \frac{\langle P, c\rangle \longrightarrow \langle P', d\rangle}{\langle P \parallel Q, c\rangle \longrightarrow \langle P' \parallel Q, d\rangle} \qquad \text{LOC}\ \frac{\langle P, c \wedge \exists_x d\rangle \longrightarrow \langle P', c'\rangle}{\langle (\mathbf{local}\ x, c)\ P, d\rangle \longrightarrow \langle (\mathbf{local}\ x, c')\ P', d \wedge \exists_x c'\rangle}$$

$$\text{UNL}\ \frac{}{\langle \mathbf{unless}\ c\ \mathbf{next}\ P, d\rangle \longrightarrow \langle \mathbf{skip}, d\rangle}\ \text{if } d \models c$$

$$\text{REP}\ \frac{}{\langle !\, P, d\rangle \longrightarrow \langle P \parallel \mathbf{next}\, !\, P, d\rangle} \qquad \text{STAR}\ \frac{}{\langle \star\, P, d\rangle \longrightarrow \langle \mathbf{next}^{\,n} P, d\rangle}\ \text{if } n \geq 0$$

$$\text{STR}\ \frac{\gamma_1 \longrightarrow \gamma_2}{\gamma_1' \longrightarrow \gamma_2'}\ \text{if } \gamma_1 \equiv \gamma_1' \text{ and } \gamma_2 \equiv \gamma_2'$$

$$\text{OBS}\ \frac{\langle P, c\rangle \longrightarrow^* \langle Q, d\rangle \not\longrightarrow}{P \overset{(c,d)}{\Longrightarrow} R}\ \text{if } R \equiv F(Q)$$

current time interval. Notice that the store d is not transferred to the next time interval, i.e. information in d can only be transfered to the next time unit by P itself.

Most of the rules in Table 1 should be straightforward from the informal description of the intended semantics given above. For detailed comments we refer to [30], and here we only comment on two of the rules: the rule for local variables LOC and OBS (covering the seemingly missing rules for "next" and "unless" processes).

Consider the process

$$Q = (\mathbf{local}\, x, c)\, P$$

in Rule LOC. The global store is d and the local store is c. We distinguish between the *external* (corresponding to Q) and the *internal* point of view (corresponding to P). From the internal point of view, the information about x, possibly appearing in the "global" store d, cannot be observed. Thus, before reducing P we should first hide the information about x that Q may have in d. We can do this by existentially quantifying x in d. Similarly, from the external point of view, the observable information about x that the reduction of internal agent P may produce (i.e., c') cannot be observed. Thus we hide it by existentially quantifying x in c' before adding it to the global store corresponding to the evolution of Q. Additionally, we should make c' the new private store of the evolution of the internal process for its future reductions.

Rule OBS says that an observable transition from P labeled with (c, d) is obtained from a terminating sequence of internal transitions from $\langle P, c\rangle$ to a $\langle Q, d\rangle$. The process R to be executed in the next time interval is equivalent to $F(Q)$ (the "future" of Q). $F(Q)$ is obtained by removing from Q summations that did not trigger activity and any local information which has been stored in Q, and by "unfolding" the sub-terms within "next" and "unless" expressions.

Definition 8 (Future Function). *Let $F : Proc \rightharpoonup Proc$ be defined by*

$$F(Q) = \begin{cases} \textbf{skip} & \textit{if } Q = \sum_{i \in I} \textbf{when } c_i \textbf{ do } Q_i \\ F(Q_1) \parallel F(Q_2) & \textit{if } Q = Q_1 \parallel Q_2 \\ (\textbf{local}\, x)\, F(R) & \textit{if } Q = (\textbf{local}\, x, c)\, R \\ R & \textit{if } Q = \textbf{next}\, R \textit{ or } Q = \textbf{unless}\, c\, \textbf{next}\, R \end{cases}$$

*Remark 1. F need no to be total since whenever we need to apply F to a Q (*OBS *in Table 1), every* $\textbf{tell}(c)$, $\star R$ *and* $!R$ *in Q will occur within a "next" or "unless" expression.*

Example 10. Recall Example 9. Processes R and S were defined as:

$$R = !\,\textbf{when}\; c\; \textbf{do tell}(e)$$
$$S = \star\, \textbf{tell}(c)$$

where $c = \texttt{malfunction}(\texttt{motor}_1\texttt{\_status})$ and $e = (\texttt{motor}_1\texttt{\_speed} = 0)$.

Let $P = R \parallel S$, $S' = \textbf{tell}(c)$ and $R' = \textbf{when}\; c\; \textbf{do tell}(e)$. One can verify that for an *arbitrary* $m > 0$, the following is a valid sequence of observable transitions starting with P:

$$R \parallel S \overset{(\texttt{c,c}\wedge\texttt{e})}{\Longrightarrow} R \parallel \textbf{next}^{\,m} S' \overset{(\texttt{true,true})}{\Longrightarrow} R \parallel \textbf{next}^{\,m-1} S' \overset{(\texttt{true,true})}{\Longrightarrow} \ldots$$
$$\ldots \overset{(\texttt{true,true})}{\Longrightarrow} R \parallel S' \overset{(\texttt{true,c}\wedge\texttt{e})}{\Longrightarrow} R \overset{(\texttt{true,true})}{\Longrightarrow} \ldots.$$

Intuitively, in the first time interval the environment tells c (i.e., c is given as input to P) thus R', which is created by $!R$, tells e. The output is then $c \wedge e$. Furthermore, S creates an S' which is to be triggered in an arbitrary number of time units $m + 1$. In the following time units the environment does not provide any input whatsoever. In the $m + 1$-th time unit S' tells c and then R' tells e. □

4.2 Observable Behavior

In this section we recall some notions introduced in [31] of *process observations*. We assume that what happens within a time unit cannot be directly observed, and thus we abstract from internal transitions, and focus on observations in terms of external transitions.

Notation 1 *Throughout this paper $\mathcal{C}^\omega$ denotes the set of infinite sequences of constraints in the underlying set of constraints $\mathcal{C}$. We use $\alpha, \alpha', \ldots$ to range over $\mathcal{C}^\omega$.*

Let $\alpha = c_1.c_2.\ldots$ and $\alpha' = c'_1.c'_2.\ldots$. We use the notation $P \overset{(\alpha,\alpha')}{\Longrightarrow}{}^{\omega}$ to denote the existence of an infinite sequence of observable transitions (or *run*): $P = P_1 \overset{(c_1,c'_1)}{\Longrightarrow} P_2 \overset{(c_2,c'_2)}{\Longrightarrow} \ldots$.

IO and Output Behavior. Consider a run of P as above. At the time unit i, the environment *inputs* c_i to P_i, which then responds with an output c'_i. As observers, we can see that on α, P responds with α'. We refer to the set of all (α, α') such that $P \overset{(\alpha,\alpha')}{\Longrightarrow}{}^{\omega}$

as the *input-output (io) behavior* of P. Alternatively, if $\alpha = \texttt{true}^\omega$, we interpret the run as an interaction among the parallel components in P *without the influence of any (external) environment*; as observers what we see is that P produces α on its own. We refer to the set of all α' such that $P \xRightarrow{(\texttt{true}^\omega,\alpha')}{}^\omega$ as the *output* behavior of P.

Quiescent Sequences and SP. As a third alternative, we may observe the quiescent input sequences of a process. These are sequences of input on which P can run without adding any information; we observe whether $\alpha = \alpha'$ whenever $P \xRightarrow{(\alpha,\alpha')}{}^\omega$.

In [30] it is shown that the set of quiescent sequences of a given P can be characterized as *the set of infinite sequences that P can possibly output under arbitrary environments*; the strongest postcondition (sp) of P.

Summing up, we have the following notions of observable behavior.

Definition 9 (Observable Behavior). *The behavioral observations that can be made of a process are:*

1. *The* input-output (or stimulus-response) behavior *of P, written, $io(P)$, defined as*
$$io(P) = \{(\alpha,\alpha') \mid P \xRightarrow{(\alpha,\alpha')}{}^\omega\}.$$
2. *The* (default) output behavior *of P, written $o(P)$, defined as*
$$o(P) = \{\alpha' \mid P \xRightarrow{(\texttt{true}^\omega,\alpha')}{}^\omega\}.$$
3. *The* strongest postcondition *behavior of P, written $sp(P)$, defined as*
$$sp(P) = \{\alpha \mid P \xRightarrow{(\alpha',\alpha)}{}^\omega \textit{ for some } \alpha'\}.$$

Given these notions of observable behaviors, we have the following naturally induced equivalences and congruences (recall the notion of congruence given in Definition 6.)

Definition 10 (Behavioral Equivalences). *Let $l \in \{io, o, sp\}$. Define $P \sim_l Q$ iff $l(P) = l(Q)$. Furthermore, let $\approx_l$ the congruence induced by $\sim_l$, i.e., $P \approx_l Q$ iff $C[P] \sim_l C[Q]$ for every process context C.*

We shall refer to equivalences defined above as *observational equivalences*. Notice, that they identify processes whose internal behavior may differ widely. Such an abstraction from internal behavior is essential in the theory of several process calculi; most notably in weak bisimilarity for CCS [25].

Example 11. Let a, b, c, d and e mutually exclusive constraints. Consider the processes P and Q below:

$$\underbrace{\textbf{when } a \textbf{ do next} \begin{array}{c}\textbf{when } b \textbf{ do next tell}(d) \\ + \\ \textbf{when } c \textbf{ do next tell}(e)\end{array}}_{P} \; , \; \underbrace{\begin{array}{c}\textbf{when } a \textbf{ do next when } b \textbf{ do next tell}(d) \\ + \\ \textbf{when } a \textbf{ do next when } c \textbf{ do next tell}(e)\end{array}}_{Q}$$

The reader may verify that $P \sim_o Q$ since $o(P) = o(Q) = \{\texttt{true}^\omega\}$. However, $P \not\sim_{io} Q$ nor $P \not\sim_{sp} Q$ since if $\alpha = a.c.\texttt{true}^\omega$ then $(\alpha,\alpha) \in io(Q)$ and $\alpha \in sp(Q)$ but $(\alpha,\alpha) \notin io(P)$ and $\alpha \notin sp(P)$. □

Congruence and Decidability Issues. In [30] it is proven that none of the three observational equivalences introduced in Definition 10 are congruences. However, $\sim_{sp}$ is a congruence if we confine our attention to the so-called *locally-independent* fragment of the calculus, i.e. the fragment without non-unary summations and "unless" operations, whose guards depend on local variables.

Definition 11 (Locally-Independent Processes). *P is* locally-independent *iff for every* **unless** c **next** Q *and* $\sum_{i\in I}$ **when** c_i **do** Q_i *($|I| \geq 2$) in P, neither c nor the c_i's contain variables in $bv(P)$ (i.e., the bound variables of P).*

The locally-independent fragment is indeed very expressive. Every summation process whose guards are either all equivalent or mutually exclusive can be encoded in this fragment [51]. Moreover, the applicability of this fragment is witnessed by the fact all the `ntcc` applications we are aware of [30,31,51] can be model as locally-independent processes. Also, the (parameterless-recursion) tcc model can be expressed in this fragment as, from the expressiveness point of view, the local operator is redundant in tcc with parameterless-recursion [29]. Furthermore, it allows us to express infinite-state processes (i.e., there are processes that can evolve into infinitely many other processes). Hence, it is rather surprising that $\sim_{sp}$ is decidable for the local-independent fragment as recently proved in [52]. In 5 below we shall present a number of other seemingly surprising decidability results for other fragments of `ntcc` .

⋆ ⋆ ⋆

4.3 Denotational Semantics

In the previous section we introduced the notion of strongest-postcondition of `ntcc` processes in operational terms. In the following we show the abstract denotational model of this notion, first presented in [32].

The denotational semantics is defined as a function $[\![\cdot]\!]$ associating with each process a set of infinite constraint sequences, $[\![\cdot]\!] : Proc \rightarrow \mathcal{P}(\mathcal{C}^\omega)$. The definition of this function is given in Table 2. Intuitively, $[\![P]\!]$ is meant to capture the set of all sequences P can possibly output. For instance, the sequences associated with **tell**(c) are those for which the first element is stronger than c (see DTELL, Table 2). Process **next** P has not influence on the first element of a sequence, thus $d.\alpha$ is a possible output if α is a possible output of P(see DNEXT, Table 2). The other cases can be explained analogously.

From [7] we know that there cannot be a $f : Proc \rightarrow \mathcal{P}(\mathcal{C}^\omega)$, compositionally defined, such that $f(P) = sp(P)$ for all P. Nevertheless, as stated in the theorem below, Palamidessi et al [32] showed that the sp denotational semantics matches its operational counter-part for the locally-independent fragment 11.

Theorem 1 (Full Abstraction, [32]). *For every* `ntcc` *process P, $sp(P) \subseteq [\![P]\!]$ and if P is locally-independent then $[\![P]\!] \subseteq sp(P)$.*

The full-abstraction result above has an important theoretical value; i.e., for a significant fragment of the calculus we can abstract away from operational details by working

Table 2. Denotational semantics of ntcc . Symbols α and α' range over the set of infinite sequences of constraints $\mathcal{C}^\omega$; β ranges over the set of finite sequences of constraints $\mathcal{C}^*$. Notation $\exists_x \alpha$ denotes the sequence resulting by applying $\exists_x$ to each constraint in α.

DTELL:	$[\![\mathbf{tell}(c)]\!] = \{d.\alpha \mid d \models c\}$
DSUM:	$[\![\sum_{i\in I} \mathbf{when}\ c_i\ \mathbf{do}\ P_i\ ]\!] = \bigcup_{i\in I} \{d.\alpha \mid d \models c_i \text{ and } d.\alpha \in [\![P_i]\!]\} \cup \bigcap_{i\in I}\{d.\alpha \mid d \not\models c_i\}$
DPAR:	$[\![P \parallel Q]\!] = [\![P]\!] \cap [\![Q]\!]$
DLOC:	$[\![(\mathbf{local}\, x)\, P]\!] = \{\alpha \mid \text{there exists } \alpha' \in [\![P]\!] \text{ s.t. } \exists_x \alpha' = \exists_x \alpha\}$
DNEXT:	$[\![\mathbf{next}\, P]\!] = \{d.\alpha \mid \alpha \in [\![P]\!]\}$
DUNL:	$[\![\mathbf{unless}\, c\, \mathbf{next}\, P]\!] = \{d.\alpha \mid d \models c\} \cup \{d.\alpha \mid d \not\models c \text{ and } \alpha \in [\![P]\!]\}$
DREP:	$[\![!\, P]\!] = \{\alpha \mid \text{ for all } \beta, \alpha' \text{ s.t. } \alpha = \beta.\alpha', \text{ we have } \alpha' \in [\![P]\!]\}$
DSTAR:	$[\![\star\, P]\!] = \{\beta.\alpha \mid \alpha \in [\![P]\!]\}$

with $[\![P]\!]$ rather than $sp(P)$. Furthermore, an interesting corollary of the full-abstraction result is that $\sim_{sp}$ is a congruence, if we confine ourselves to locally-independent processes.

4.4 LTL Specification and Verification

Processes in ntcc denote observable behavior of timed systems. As with other such formalisms, it is often convenient to express specifications of such behaviors in logical formalisms. In this section we present the ntcc logic first introduced in [32]. We start by defining a linear-time temporal logic (LTL) expressing temporal properties over infinite sequences of constraints. We then define what it means for a process to satisfy a specification given as a formula in this logic. Finally, we present an inference system aimed at proving processes satisfying specifications.

A Temporal Logic. The ntcc LTL expresses properties of infinite sequences of constraints, and we shall refer to it as **CLTL**.

Definition 12 (CLTL Syntax). *The formulae $F, G, ... \in \mathcal{F}$ are built from constraints $c \in \mathcal{C}$ and variables $x \in \mathcal{V}$ in the underlying constraint system by:*

$$F, G, \ldots := c \mid \dot{\mathtt{true}} \mid \dot{\mathtt{false}} \mid F \dot{\wedge} G \mid F \dot{\vee} G \mid \dot{\neg} F \mid \dot{\exists}_x F \mid \circ F \mid \Box F \mid \Diamond F$$

Here c is a constraint (i.e., a first-order formula in the underlying constraint system) representing a *state formula* c. The symbols $\dot{\mathtt{true}}, \dot{\mathtt{false}}, \dot{\wedge}, \dot{\vee}, \dot{\neg}, \dot{\exists}$ represent linear-temporal logic true, false, conjunction, disjunction, negation and existential quantification. As clarified later, the dotted notation is introduced since in **CLTL** these operators may have different interpretations from the symbols $\mathtt{true}, \mathtt{false}, \wedge, \vee, \neg, \exists$ in the underlying constraint system. The symbols $\circ$, $\Box$, and $\Diamond$ denote the temporal operators *next*, *always* and *sometime*.

The standard interpretation structures of linear temporal logic are infinite sequences of states [22]. In the case of `ntcc` , it is natural to replace states by constraints, and hence our interpretations are elements of $\mathcal{C}^\omega$.

The **CLTL** semantics is given in Definition 14. Following [22] we introduce the notion of *x-variant.*

Notation 2 *Given a sequence $\alpha = c_1.c_2.\ldots$, we use $\exists_x\alpha$ to denote the sequence $\exists_x c_1 \exists_x c_2 \ldots$. We shall use $\alpha(i)$ to denote the $i-th$ element of α.*

Definition 13 **(x-variant).** *A constraint d is an x-*variant *of c iff $\exists_x c = \exists_x d$. Similarly α' is an x-*variant *of α iff $\exists_x\alpha = \exists_x\alpha'$.*

Intuitively, d and α' are x-variants of c and α, respectively, if they are logically the same except for information about x. For example, $x = 0 \wedge y = 0$ is an x-variant of $x = 1 \wedge y = 0$.

Definition 14 **(CLTL Semantics).** *We say that $\alpha \in \mathcal{C}^\omega$ satisfies (or that it is a model of) the* **CLTL** *formula F, written $\alpha \models_{\mathrm{CLTL}} F$, iff $\langle\alpha, 1\rangle \models_{\mathrm{CLTL}} F$, where:*

$$\begin{array}{lll}
\langle\alpha, i\rangle \models_{\mathrm{CLTL}} \dot{\mathtt{true}} & & \langle\alpha, i\rangle \not\models_{\mathrm{CLTL}} \dot{\mathtt{false}} \\
\langle\alpha, i\rangle \models_{\mathrm{CLTL}} c & \textit{iff} & \alpha(i) \models c \\
\langle\alpha, i\rangle \models_{\mathrm{CLTL}} \dot{\neg} F & \textit{iff} & \langle\alpha, i\rangle \not\models_{\mathrm{CLTL}} F \\
\langle\alpha, i\rangle \models_{\mathrm{CLTL}} F \dot{\wedge} G & \textit{iff} & \langle\alpha, i\rangle \models_{\mathrm{CLTL}} F \textit{ and } \langle\alpha, i\rangle \models_{\mathrm{CLTL}} G \\
\langle\alpha, i\rangle \models_{\mathrm{CLTL}} F \dot{\vee} G & \textit{iff} & \langle\alpha, i\rangle \models_{\mathrm{CLTL}} F \textit{ or } \langle\alpha, i\rangle \models_{\mathrm{CLTL}} G \\
\langle\alpha, i\rangle \models_{\mathrm{CLTL}} \circ F & \textit{iff} & \langle\alpha, i+1\rangle \models_{\mathrm{CLTL}} F \\
\langle\alpha, i\rangle \models_{\mathrm{CLTL}} \Box F & \textit{iff} & \textit{for all } j \geq i \ \langle\alpha, j\rangle \models_{\mathrm{CLTL}} F \\
\langle\alpha, i\rangle \models_{\mathrm{CLTL}} \Diamond F & \textit{iff} & \textit{there is a } j \geq i \textit{ such that } \langle\alpha, j\rangle \models_{\mathrm{CLTL}} F \\
\langle\alpha, i\rangle \models_{\mathrm{CLTL}} \dot{\exists}_x F & \textit{iff} & \textit{there is an } x\textit{-variant } \alpha' \textit{ of } \alpha \textit{ such that } \langle\alpha', i\rangle \models_{\mathrm{CLTL}} F.
\end{array}$$

Define $[\![F]\!]=\{\alpha \mid \alpha \models_{\mathrm{CLTL}} F\}$. We say that F is **CLTL** valid *iff $[\![F]\!] = \mathcal{C}^\omega$, and that F is* **CLTL** satisfiable *iff $[\![F]\!] \neq \emptyset$.*

State Formulae as Constraints. Let us comment briefly on the role of constraints as state formulae in our logic. A temporal formula F expresses a property of sequences of constraints. As a state formula, c expresses a property, which is satisfied by those $e.\alpha'$ such that $e \models c$. Hence, the state formula `false` (and consequently $\Box$ `false`) is satisfied by $\mathtt{false}^\omega$. On the other hand, the temporal formula $\dot{\mathtt{false}}$ has no model whatsoever.

Similarly, the models of the temporal formula $c \dot{\vee} d$ are those $e.\alpha'$ such that either $e \models c$ or $e \models d$ holds. Therefore, the formula $c \dot{\vee} d$ and the atomic proposition $c \vee d$ may have different models since, in general, one can verify that $e \models c \vee d$ may hold while neither $e \models c$ nor $e \models d$ hold – e.g. take $e = (x = 1 \vee x = 2)$, $c = (x = 1)$ and $d = (x = 2)$.

In contrast, the formula $c \dot{\wedge} d$ and the atomic proposition $c \wedge d$ have the same models since $e \models (c \wedge d)$ holds if and only if both $e \models c$ and $e \models d$ hold.

The above discussion tells us that the operators of the constraint system should not be confused with those of the temporal logic. In particular, the operators $\vee$ and

Table 3. A proof system for linear-temporal properties of `ntcc` processes.

$$\text{LTELL: } \mathbf{tell}(c) \vdash c \qquad\qquad \text{LPAR: } \frac{P \vdash F \quad Q \vdash G}{P \parallel Q \vdash F \dot{\wedge} G}$$

$$\text{LSUM: } \frac{\forall i \in I \;\; P_i \vdash F_i}{\sum_{i\in I} \mathbf{when}\ c_i\ \mathbf{do}\ P_i \vdash \dot{\bigvee}_{i\in I} (c_i \dot{\wedge} F_i) \dot{\vee} \dot{\bigwedge}_{i\in I} \dot{\neg} c_i} \qquad \text{LLOC: } \frac{P \vdash F}{(\mathbf{local}\, x)\, P \vdash \dot{\exists}_x F}$$

$$\text{LNEXT: } \frac{P \vdash F}{\mathbf{next}\, P \vdash \bigcirc F} \qquad\qquad \text{LUNL: } \frac{P \vdash F}{\mathbf{unless}\ c\ \mathbf{next}\, P \vdash c \dot{\vee} \bigcirc F}$$

$$\text{LREP: } \frac{P \vdash F}{!P \vdash \Box F} \qquad\qquad \text{LSTAR: } \frac{P \vdash F}{\star P \vdash \Diamond F}$$

$$\text{LCONS: } \frac{P \vdash F}{P \vdash G} \quad \text{if } F \dot{\Rightarrow} G$$

$\dot{\vee}$. This distinction does not make our logic intuitionistic. In fact, classically (but not intuitionistically) valid statements such as $\dot{\neg} A \dot{\vee} A$ and $\dot{\neg}\dot{\neg} A \dot{\Rightarrow} A$ are also valid in our logic.

Process Verification

We are now ready to define what it means for a process P to satisfy a specification F.

Definition 15 (Verification). P satisfies F, *written* $P \models_{\mathrm{CLTL}} F$, *iff* $sp(P) \subseteq [\![F]\!]$.

Thus, the intended meaning of $P \models_{\mathrm{CLTL}} F$ is that every sequence P can possibly output on inputs from arbitrary environments satisfies the temporal formula F. For example, $\star\,\mathbf{tell}(c) \models \Diamond c$, since in every infinite sequence output by $\star\,\mathbf{tell}(c)$ on arbitrary inputs, there must be an element entailing c.

Following the discussion above, notice that $P = \mathbf{tell}(c) + \mathbf{tell}(d) \models (c \dot{\vee} d)$ as every constraint e output by P entails either c or d. In contrast, $Q = \mathbf{tell}(c \vee d) \not\models (c \dot{\vee} d)$ in general since Q can output a constraint e which entails $c \vee d$, but neither c nor d.

4.5 Proof System for Verification

[32] introduces a *proof (or inference) system* for assertions of the form $P \vdash F$, where $P \vdash F$ is intended to be the "counterpart" of $P \models F$ in the sense that $P \vdash F$ should approximate $P \models_{\mathrm{CLTL}} F$ as closely as possible (ideally, they should be equivalent). The system is presented in Table 3.

Definition 16 **($P \vdash F$).** *We say that* $P \vdash F$ *iff the assertion* $P \vdash F$ *has a proof in the system in Table 3.*

Inference Rules. Let us briefly comment on (the soundness of) some of the inference rules of the proof system. The inference rule for the tell operator is given by

$$\text{LTELL: } \mathbf{tell}(c) \vdash c$$

Rule LTELL gives a proof reflecting the fact that every output of $\mathbf{tell}(c)$ on arbitrary input, indeed satisfies the atomic proposition c, i.e., $\mathbf{tell}(c) \models_{\text{CLTL}} c$.

Consider now the rule for the choice operator:

$$\text{LSUM: } \frac{\forall i \in I \quad P_i \vdash F_i}{\sum_{i \in I} \mathbf{when}\ c_i\ \mathbf{do}\ P_i \vdash \dot{\bigvee}_{i \in I} (c_i \mathbin{\dot\wedge} F_i) \mathbin{\dot\vee} \dot{\bigwedge}_{i \in I} \dot\neg c_i}$$

Rule LSUM can be explained as follows. Suppose that for $P = \sum_{i \in I} \mathbf{when}\ c_i\ \mathbf{do}\ P_i$ we are given a proof that each P_i satisfies F_i, i.e. (inductively) $P_i \models_{\text{CLTL}} F_i$. Then we may conclude that every output of P on arbitrary input will satisfy either: (a) some of the guards c_i and their corresponding F_i (i.e., $\dot{\bigvee}_{i \in I}(c_i \mathbin{\dot\wedge} F_i)$), or (b) none of the guards (i.e., $\dot{\bigwedge}_{i \in I} \dot\neg c_i$).

The inference rule for parallel composition is defined as

$$\text{LPAR: } \frac{P \vdash F \qquad Q \vdash G}{P \parallel Q \vdash F \mathbin{\dot\wedge} G}$$

The soundness of this rule can be justified as follows. Assume that each output of P, under the influence of arbitrary environments, satisfies F. Assume the same about Q and G. In $P \parallel Q$, the process Q can be thought as one of those arbitrary environment under which P satisfies F. Then $P \parallel Q$ must satisfy F. Similarly, P can be one of those arbitrary environment under which Q satisfies G. Hence, $P \parallel Q$ must satisfy G as well. We therefore have grounds to conclude that $P \parallel Q$ satisfies $F \mathbin{\dot\wedge} G$.

The inference rule for the local operator is

$$\text{LLOC: } \frac{P \vdash F}{(\mathbf{local}\, x)\, P \vdash \dot\exists_x F}$$

The intuition is that since the outputs of $(\mathbf{local}\, x)\, P$ are outputs of P with x hidden then if P satisfies F, $(\mathbf{local}\, x)\, P$ should satisfy F with x hidden, i.e., $\dot\exists_x F$.

The following are the inference rules for the temporal `ntcc` constructs:

$$\text{LNEXT: } \frac{P \vdash F}{\mathbf{next}\, P \vdash \circ F} \qquad \text{LUNL: } \frac{P \vdash F}{\mathbf{unless}\ c\ \mathbf{next}\, P \vdash c \mathbin{\dot\vee} \circ F}$$

$$\text{LREP: } \frac{P \vdash F}{!\, P \vdash \Box F} \qquad \text{LSTAR: } \frac{P \vdash F}{\star\, P \vdash \Diamond F}$$

Assume that $P \vdash F$, i.e. (inductively) $P \models_{\text{CLTL}} F$. Rule LNEXT reflects that we may then conclude that also the process $\mathbf{next}\, P$ satisfies $\circ F$. Rule LUNL is similar, except that P can also be precluded from execution, if some environment provides c. Thus

unless c **next** P satisfies either c or $\circ F$. Rule LREP says that if F is satisfied by P, then executing P in each time interval will imply F to be satisfied in each time interval, i.e. $!\,P$ satisfies $\Box F$. Rule LSTAR reflects that if P is executed in some time interval, then in that time interval F is satisfied, and hence $\star\, P$ satisfies $\Diamond F$.

Finally, we have the rule:

$$\text{LCONS:}\ \frac{P \vdash F}{P \vdash G} \quad \text{if } F \dot{\Rightarrow} G$$

Notice that this rule refers to some unspecified way of inferring validity of **CLTL** formulae. We shall return to this point shortly. Rule LCONS simply says that if P satisfies a specification F then it also satisfies any weaker specification G. We shall also refer to LCONS as *the consequence rule.*

Notice that the inference rules reveal a pleasant correspondence between `ntcc` operators and the logic operators. For example, parallel composition and locality corresponds to conjunction and existential quantification. The choice operator corresponds to some special kind of conjunction. The next, replication and star operators correspond to the next, always, and eventuality temporal operator.

The Proof System at Work. Let us now give a simple example illustrating a proof in inference system.

Example 12. Recall Example 9. We have a process R which was repeatedly checking the state of $\texttt{motor}_1$. If a malfunction is reported, R would tell that $\texttt{motor}_1$ must be turned off. We also have a process S stating that motor $\texttt{motor}_1$ is doomed to malfunction. Let $R = !\,\textbf{when}\ c\ \textbf{do}\ \textbf{tell}(e)$ and $S = \star\,\textbf{tell}(c)$ with the constraints $c = \texttt{malfunction}(\texttt{motor}_1\texttt{\_status})$ and $e = (\texttt{motor}_1\texttt{\_speed} = 0)$. We want to provide a proof of the assertion: $R \parallel S \vdash \Diamond\, e$. Intuitively, this means that the parallel execution of R and S satisfies the specification stating that $\texttt{motor}_1$ is eventually turned off. The following is a derivation of the above assertion.

$$\dfrac{\dfrac{\dfrac{\dfrac{}{\textbf{when}\ c\ \textbf{do}\ \textbf{tell}(e) \vdash (c \,\dot{\wedge}\, e)\,\dot{\vee}\,\dot{\neg}\, c}\ \text{LSUM}}{\textbf{when}\ c\ \textbf{do}\ \textbf{tell}(e) \vdash c \dot{\Rightarrow} e}\ \text{LCONS}}{R \vdash \Box\,(c \dot{\Rightarrow} e)}\ \text{LREP} \qquad \dfrac{\dfrac{}{\textbf{tell}(c) \vdash c}\ \text{LTELL}}{S \vdash \Diamond\, c}\ \text{LSTAR}}{\dfrac{R \parallel S \vdash \Box\,(c \dot{\Rightarrow} e)\,\dot{\wedge}\,\Diamond c}{R \parallel S \vdash \Diamond\, e}\ \text{LCONS}}\ \text{LPAR}$$

More complex examples of the use of the proof system for proving the satisfaction of processes specification can be found in [30]—in particular for proving properties of mutable data structures. □

Let us now return to the issue of the relationship between $\vdash$ and $\models_{\text{CLTL}}$.

Theorem 2 (Relative Completeness, [30]). *If P is locally-independent then $P \vdash F$ iff $P \models_{\text{CLTL}} F$.*

Notice that this is indeed a "relative completeness" result, in the sense that, as mentioned earlier, one of our proof rules refer to the validity of temporal implication. This means that our proof system is complete, if we are equipped with an oracle that is guaranteed to provide a proof or a confirmation of each valid temporal implication. Because of this, one may wonder about the decidability of the validity problem for our temporal logic. We look at this issue next.

Decidability Results. In [52] it is shown that the verification problem (i.e., given P and F whether $P \models_{\mathrm{CLTL}} F$) is decidable for the locally-independent fragment of ntcc and negation-free **CLTL** formulae. Please recall that the locally-independent fragment of ntcc admits infinite-state processes. Also note that **CLTL** is first-order. Most first-order LTL's in computer science are not recursively axiomatizable, let alone decidable [1].

Furthermore, [52] proves the decidability of the validity problem for implication of negation-free **CLTL** formulae. This is done by appealing to the close connection between ntcc processes and LTL formulae to reduce the validity of implication to the verification problem. More precisely, it is shown that given two negation-free formulae F and G, one can construct a process P_F such that $sp(P_F) = [\![F]\!]$ and then it follows that $P_F \models_{\mathrm{CLTL}} G$ iff $F \Rightarrow G$. As a corollary of this result, we obtain the decidability of *satisfiability* for the negation-free first-order fragment of **CLTL**

A theoretical application of the theory of ntcc is presented in [52], stating a new positive decidability result for a first-order fragment of Pnueli's first-order **LTL** [22]. The result is obtained from a reduction to **CLTL** satisfiability, and thus it also contributes to the understanding of the relationship between (timed) ccp and (temporal) classic logic.

5 A Hierarchy of Timed CCP Languages

In the literature several timed ccp languages have been introduced, differing in their way of expressing infinite behavior. In this section we shall introduce a few fundamental representatives of mechanisms introducing infinite behavior, expressed as variants of the ntcc calculus. We shall also characterize their relative expressiveness following [29].

Since timed CCP languages are deterministic we shall confine our attention to the *deterministic* processes of ntcc as described in [30]. These are all the star-free processes with all summations having at most one guard. On top of this fragment we consider the following variants:

- rep: deterministic ntcc ; infinite behavior given by replication.
- $\mathtt{rec_p}$: obtained from deterministic ntcc replacing replication by *parametric recursion*. In $\mathtt{rec_p}$ each procedures body *has no free variables* other than its formal parameters.
- $\mathtt{rec_i}$: same as $\mathtt{rec_p}$, but where the actual parameters in recursive calls are *identical* to the formal parameters; i.e., we do not vary the parameters in the recursive calls.

- $\texttt{rec}_\texttt{d}$: obtained by using *parameterless recursion*, but *including* free variables in procedure bodies with *dynamic scope*.
- $\texttt{rec}_\texttt{s}$: same as $\texttt{rec}_\texttt{d}$ but with *static scope*.

In the following, the expressive power of these process languages is compared with respect to the notion of input-output behavior, as introduced in Section 4.2. More precisely, one language is considered at least as expressive as another, if any input-output behavior expressed by a process in the latter can be expressed also by a process in the former. The comparison results can be summarized as follows:

- $\texttt{rec}_\texttt{p}$ and $\texttt{rec}_\texttt{d}$ are equally expressive, and strictly more expressive than the other languages,
- $\texttt{rep}$, $\texttt{rec}_\texttt{s}$ and $\texttt{rec}_\texttt{i}$ are equally expressive.

In fact, [29] shows a strong separation result between the languages $\texttt{rec}_\texttt{p}/\texttt{rec}_\texttt{d}$ and $\texttt{rep}/\texttt{rec}_\texttt{s}/\texttt{rec}_\texttt{i}$: the input-output equivalence is undecidable for the languages in the first class, but decidable for the languages in the second class.

The undecidability results holds even if we fix an underling constraint system with a finite domain having at least one element. The undecidability result is obtained by a reduction from Post's correspondence problem [34] and an input-output preserving encoding between $\texttt{rec}_\texttt{p}/\texttt{rec}_\texttt{d}$.

The decidability results hold for arbitrary constraint systems, and follow from Büchi automata [3] representation of ntcc processes and input-output preserving encodings between the languages in $\texttt{rep}/\texttt{rec}_\texttt{s}/\texttt{rec}_\texttt{i}$.

The expressiveness gaps illustrated above may look surprising to readers familiar with the π-calculus [27], since it is well known that the π-calculus correspondents of $\texttt{rep}, \texttt{rec}_\texttt{i}$ and $\texttt{rec}_\texttt{p}$ all have the same expressive power. The reason for these differences can be attributed to the fact that the π-calculus has some powerful mechanisms (such as mobility), which compensate for the weakness of replication and the lower forms of recursion.

We start by formally defining our five classes of process languages.

5.1 Replication

We shall use rep to denote the deterministic fragment of ntcc . The processes in the deterministic fragment are those star-free processes in which the cardinality of every summation index set is at most one. Thus, the resulting syntax of process in rep is given by:

$$\begin{array}{rll} P, Q, \ldots ::= \mathbf{skip} & \mid \mathbf{tell}(c) \mid \mathbf{when}\ c\ \mathbf{do}\ P \mid P \parallel Q \mid (\mathbf{local}\ x)\ P \\ \mid\ \mathbf{next}\ P & \mid\ \mathbf{unless}\ c\ \mathbf{next}\ P & \mid\ !\ P \end{array} \tag{9}$$

Infinite behavior in rep is provided by using replication. This way of expressing infinite behavior is also considered in [43]. To be precise, [43] uses the **hence** operator. However, **hence** P is equivalent to **next** $!\,P$ and, similarly $!\,P$ is equivalent to $P \parallel$ **hence** P.

5.2 Recursion

Infinite behavior in tcc languages may also be introduced by adding recursion, as e.g. in [40,41,49]. Consider the process syntax obtained from replacing replication $!P$ with *process* (or *procedure*) calls $A(y_1, \ldots, y_n)$, i.e.:

$$\begin{array}{rll} P, Q, \ldots ::= \textbf{skip} & \mid \textbf{tell}(c) \mid \textbf{when}\ c\ \textbf{do}\ P \mid P \parallel Q \mid (\textbf{local}\ x)\ P \\ \mid \textbf{next}\ P \mid & \textbf{unless}\ c\ \textbf{next}\ P \qquad \mid A(y_1, \ldots, y_n) \end{array} \tag{10}$$

The process $A(y_1, \ldots, y_n)$ is an *identifier* with arity n. We assume that every identifier has a (recursive) *process (or procedure) definition* of the form $A(x_1, \ldots, x_n) \stackrel{\texttt{def}}{=} P$ where the x_i's are pairwise distinct, and the intuition is that $A(y_1, \ldots, y_n)$ behaves as P with y_i replacing x_i for each i.

We declare $\mathcal{D}$ to be the set of recursive definitions under consideration. We shall often use the notation $\boldsymbol{x}$ as an abbreviation of $x_1, x_2, \ldots, x_n$ if n is unimportant or obvious. We shall sometimes say that $A(\boldsymbol{y})$ is an *invocation* with *actual parameters* $\boldsymbol{y}$, and given $A(\boldsymbol{x}) \stackrel{\texttt{def}}{=} P$ we shall refer to P as its *body* and to $\boldsymbol{x}$ as its *formal parameters*.

Finite Dependency and Guarded Recursion Following [40], we shall require, for all the forms of recursion defined next, the following: (1) any process to depend only on finitely many definitions and (2) recursion to be "next" guarded. For example, given $A(\boldsymbol{x}) \stackrel{\texttt{def}}{=} P$, every invocation $A(\boldsymbol{y})$ in P must occur within the scope of a "next" or "unless" operator operator. This avoids non-terminating sequences of internal reductions (i.e., non-terminating computation within a time interval). Below we give a precise formulation of (1) and (2).

Given $A_1(\boldsymbol{x}_1) \stackrel{\texttt{def}}{=} P_1$ and $A_2(\boldsymbol{x}_2) \stackrel{\texttt{def}}{=} P_2$, we say that A_1 (directly) *depends* on A_2, written $A_1 \leadsto A_2$, if there is an invocation $A_2(\boldsymbol{y})$ in P_1. Requirement (1) can be then formalized by requiring the strict ordering induced by $\leadsto^*$ (the reflexive and transitive closure of $\leadsto$)[1] to be well founded.

To formalize (2), suppose that $A_1 \leadsto A_2 \leadsto \ldots \leadsto A_n \leadsto A_{n+1} = A_1$, where $A_i(\boldsymbol{x}_1) \stackrel{\texttt{def}}{=} P_i$. We shall require that for at least one i, $1 \leq i \leq n$, the occurrences of A_{i+1} in P_i are within the scope of a "next" or an "unless" operator.

Parametric Recursion

We consider a further restriction for the case of recursion involving parameters. *All the free variables in definitions' bodies must be formal parameters*; more precisely, for each $A(x_1, \ldots, x_n) \stackrel{\texttt{def}}{=} P$, we decree that $fv(P) \subseteq \{x_1, \ldots, x_n\}$.

We shall use $\texttt{rec}_\texttt{p}$ to denote the tcc language with recursion with the above syntactic restriction. The operational rules for $\texttt{rec}_\texttt{p}$ are obtained from Table 1 by replacing the rule for replication REP with the following rule for recursion:

$$\text{REC}\ \frac{}{\langle A(\boldsymbol{y}), d\rangle \longrightarrow \langle P[\boldsymbol{y}/\boldsymbol{x}], d\rangle}\ A(\boldsymbol{x}) \stackrel{\texttt{def}}{=} P \tag{11}$$

[1] The relation $\leadsto^*$ is a pre-ordering. By induced strict ordering we mean the strict component of $\leadsto^*$ modulo the equivalence relation obtained by taking the symmetric closure of $\leadsto^*$.

As usual $P[y_1, \ldots, y_n/x_1, \ldots, x_n]$, with all the x_i's being pairwise distinct, is the process that results from syntactically replacing every free occurrence of x_i by y_i using α-conversion wherever needed to avoid capture.

Identical Parameters Recursion. An interesting tcc language considered in [40] arises from $\mathtt{rec_p}$ by restricting the parameters not to change through recursive invocations. In the π-calculus this restriction does not cause any loss of expressive power since such form of recursion can encode general recursion (see [27]).

An example satisfying the above restriction is $R_P(\boldsymbol{x}) \stackrel{\mathrm{def}}{=} P \parallel \mathbf{next}\, R_P(\boldsymbol{x})$. Here the actual parameters of the invocation in the body of the definition are the same as the formal parameters of R_P. An example not satisfying the restriction is $R'_P(\boldsymbol{x}) \stackrel{\mathrm{def}}{=} P \parallel \mathbf{next}\,(\mathbf{local}\, \boldsymbol{x})\, R'_P(\boldsymbol{x})$. Here the actual parameters are bound and therefore different from those of the formal parameters.

One can formalize the identical parameters restriction on a set of mutually recursive definitions as follows. Suppose that $A_1 \rightsquigarrow A_2$ and $A_2 \rightsquigarrow^* A_1$ with $A_1(\boldsymbol{x}_1) \stackrel{\mathrm{def}}{=} P_1$ and $A_2(\boldsymbol{x}_2) \stackrel{\mathrm{def}}{=} P_2$ in the underlying set of definitions $\mathcal{D}$. Then for each invocation $A_2(\boldsymbol{y})$ in P_1 we should require $\boldsymbol{y} = \boldsymbol{x}_2$ and $\boldsymbol{y} \notin bv(P_1)$. In other words the actual parameters of the invocation A_2 in P_1 (i.e., $\boldsymbol{y}$) should be syntactically the same as its formal parameters (i.e., $\boldsymbol{x}_2$). Furthermore, they should not be bound in P_1 to avoid cases such as $R'_P(\boldsymbol{x})$ above.

The processes of tcc with identical parameters are those of $\mathtt{rec_p}$ that satisfy this requirement. We shall refer to this language as $\mathtt{rec_i}$.

Parameterless Recursion

Tcc with parameterless recursion have been studied in [40]. All identifiers have arity zero, and hence, for convenience, we omit the "( )" in $A(\,)$.

Given a parameterless definition $A \stackrel{\mathrm{def}}{=} P$, requiring all variables in $\mathit{fv}(P)$ to be formal parameters, as in $\mathtt{rec_p}$, would mean that the body P has no free variables, and the resulting class of process languages would be expressively weak. Hence, we now suggest to allow free variables in procedure bodies.

Now. assuming that the operational rules for parameterless recursion are the same as for $\mathtt{rec_p}$, what are the resulting scope rules for free variables in procedure bodies? Traditionally, one distinguishes between *dynamic* and *static* scoping, as illustrated in the following example.

Example 13. Consider a constant identifier A with the following definition

$$\begin{aligned} A \stackrel{\mathrm{def}}{=}\ & \mathbf{tell}(x = 1) \\ & \parallel \mathbf{next}\,(\mathbf{local}\, x)\,(A \parallel \mathbf{when}\ x = 1\ \mathbf{do}\ \mathbf{tell}(z = 1)) \end{aligned}$$

In the case of dynamic scoping, an outside invocation A causes the execution $\mathbf{tell}(z = 1)$ in the second time interval. The reason is that $(\mathbf{local}\, x)$ binds the x resulting from the unfolding of the A inside the definition's body. In fact, the telling of $x = 1$, in the second time unit, will not be visible in the store. In the case of static scoping, $(\mathbf{local}\, x)$

does not bind the x of the unfolding of A because such an x is intuitively a "global" variable, and hence $\mathbf{tell}(z = 1)$ will not be executed. In fact, the telling of $x = 1$, will also be visible in the store in the second time interval. □

Parameterless Recursion with Dynamic Scoping. The rule LOC in Table 1 combined with REC causes the parameterless recursion to have dynamic scoping[2]. As illustrated in the example below, the idea is that since $(\mathbf{local}\, x)\, P$ reduces to a process of the form $(\mathbf{local}\, x)\, Q$, the free occurrences of x in the unfolding of invocations in P get bounded.

Example 14. Consider A as defined in Example 13. Let us abbreviate the definition of A as $A \stackrel{\text{def}}{=} \mathbf{tell}(x = 1) \parallel P$. Also let $Q = \mathbf{skip} \parallel P$. We have the following reduction of $(\mathbf{local}\, x)\, A$ in store $\mathtt{true}$.

$$\dfrac{\dfrac{\dfrac{\dfrac{}{\langle \mathbf{tell}(x = 1), \mathtt{true}\rangle \longrightarrow \langle \mathbf{skip}, x = 1\rangle}\ \text{TELL}}{\langle \mathbf{tell}(x = 1) \parallel P, \mathtt{true}\rangle \longrightarrow \langle Q, x = 1\rangle}\ \text{PAR}}{\langle A, \mathtt{true}\rangle \longrightarrow \langle Q, x = 1\rangle}\ \text{REC}}{\langle (\mathbf{local}\, x, \mathtt{true})\, A, \mathtt{true}\rangle \longrightarrow \langle (\mathbf{local}\, x, x = 1)\, Q, \mathtt{true}\rangle}\ \text{LOC}$$

Thus, $(\mathbf{local}\, x)\, A$ in store $\mathtt{true}$ reduces to $(\mathbf{local}\, x, x = 1)\, (\mathbf{skip} \parallel P)$ in store $\mathtt{true}$. Notice that the free x in A's body become local to $(\mathbf{local}\, x, x = 1)\, (\mathbf{skip} \parallel P)$, i.e, it now occurs in the local store but not in the global one. □

We shall refer to the language allowing only parameterless recursion with free-variables in the procedure bodies as $\mathtt{rec}_{\mathtt{d}}$; parameterless recursion with dynamic scoping.

Remark 2. It should be noticed that, unlike in $\mathtt{rec}_{\mathtt{p}}$, we cannot freely α-convert processes in $\mathtt{rec}_{\mathtt{d}}$ without changing behavior. For example, we could α-convert the process $(\mathbf{local}\, x)\, A$ in the above example into $(\mathbf{local}\, z)\, A$ (since $A[z/x]$ is syntactically equal to A) but the behavior of $(\mathbf{local}\, z)\, A$ would not be the same as that of $(\mathbf{local}\, x)\, A$.

Parameterless Recursion with Static Scoping. From the previous section it follows that static scoping as in [40] requires an alternative to the rule for local behavior LOC .

The rule LOC′ defines locality for the parameterless recursion with static scoping language henceforth referred to as $\mathtt{rec}_{\mathtt{s}}$.

$$\text{LOC}' \quad \frac{\langle P[y/x], d\rangle \longrightarrow \langle P', d'\rangle}{\langle (\mathbf{local}\, x)\, P, d\rangle \longrightarrow \langle P', d'\rangle} \ \text{if } y \text{ is fresh} \tag{12}$$

As in [24], we use the notion of *fresh variable* meaning that it does not occur elsewhere in a process definition or in the store. It will be convenient to presuppose that the set of variables $\mathcal{V}$ is partitioned into two infinite sets $\mathcal{F}$ and $\mathcal{V} - \mathcal{F}$. We shall assume that the fresh variables are taken from $\mathcal{F}$ and that no input from the environment or

[2] Rules LOC and REC are basically the same in ccp, hence the observations made in this section regarding dynamic scoping apply to ccp as well.

processes, other than the ones generated when applying LOC$'$, can contain variables in $\mathcal{F}$.

The fresh variables introduced by LOC$'$ are not to be visible from the outside. We hide these fresh variables, as suggested in [43], by using existential quantification in the output constraint of observable transitions. More precisely, we replace in Table 1 the rule for the observable transitions OBS with the rule

$$\text{OBS}' \quad \frac{\langle P, c\rangle \longrightarrow^* \langle Q, d\rangle \not\longrightarrow}{P \overset{(c,\exists_{\mathcal{F}} d)}{\Longrightarrow} F(Q)} \tag{13}$$

where $\exists_{\mathcal{F}} d$ represents the constraint resulting from the existential quantification in d of free occurrences of variables in $\mathcal{F}$.

In order to see why LOC$'$ causes static scoping in $\texttt{rec}_\texttt{s}$, suppose that P in Rule LOC$'$ in Equation 12 contains an invocation A where $A \stackrel{\text{def}}{=} R$. When replacing x with y in P, A remains the same since $A[y/x]$ is A. Furthermore, since y is chosen from $\mathcal{F}$, there will be no capture of free variables in R when unfolding A. This causes the scoping to be static. Let us illustrate this by revisiting the previous example.

Example 15. Let A, P and Q as in the previous example. We have the following reduction of $(\mathbf{local}\, x)\, A$ in store $\texttt{true}$.

$$\cfrac{\cfrac{\cfrac{\cfrac{}{\langle \mathbf{tell}(x=1), \texttt{true}\rangle \longrightarrow \langle \mathbf{skip}, x=1\rangle}\ \text{TELL}}{\langle \mathbf{tell}(x=1) \parallel P, \texttt{true}\rangle \longrightarrow \langle Q, x=1\rangle}\ \text{PAR}}{\langle A, \texttt{true}\rangle \longrightarrow \langle Q, x=1\rangle}\ \text{REC}}{\langle (\mathbf{local}\, x)\, A, \texttt{true}\rangle \longrightarrow \langle Q, x=1\rangle}\ \text{LOC}'$$

Thus, $(\mathbf{local}\, x)\, A$ in store $\texttt{true}$ reduces to $\mathbf{skip} \parallel P$ in store $(x = 1)$ making the free x in A's body visible in the "global" store . □

Remark 3. Notice that, as in $\texttt{rec}_\texttt{d}$, in $\texttt{rec}_\texttt{s}$ we do not need α-conversion since in the reductions of $\texttt{rec}_\texttt{s}$ we only use syntactic replacements of variables by fresh variables.

5.3 Summary of TCC Languages

We have described five classes of tcc languages with infinite behabior, based on the literature. We adopt the following convention.

Convention 3 *We shall use $\mathcal{L}$ to designate the set of tcc languages*

$$\{\texttt{rep}, \texttt{rec}_\texttt{p}, \texttt{rec}_\texttt{i}, \texttt{rec}_\texttt{d}, \texttt{rec}_\texttt{s}\}.$$

Furthermore, we shall index sets and relations involving tcc processes with the appropriate tcc language name to make it clear what is the language under consideration. We shall omit the index when it is unimportant or clear from the context.

For example, $\longrightarrow_{\texttt{rec}_\texttt{p}}$ and $\overset{(.,.)}{\Longrightarrow}_{\texttt{rec}_\texttt{p}}$ refer to the (internal and observable) reduction of $\texttt{rec}_\texttt{p}$. Similarly, $Proc_{\texttt{rec}_\texttt{p}}$ denotes the set of processes in $\texttt{rec}_\texttt{p}$, $\sim_{io}^{\texttt{rec}_\texttt{p}}$ denotes the input-output equivalence (Definition 10) for processes in $Proc_{\texttt{rec}_\texttt{p}}$, and $\approx_{io}^{\texttt{rec}_\texttt{p}}$ denotes congruence induced by $\sim_{io}^{\texttt{rec}_\texttt{p}}$.

5.4 The TCC Equivalences

In this section we relate the equivalences and their congruences for the various tcc languages. Each behavioral equivalence (and congruence) for a tcc language ℓ is obtained by taking the ntcc transitions given in Definition 9 (and thus in Definition 10) to be those of ℓ (i.e., replace $\xRightarrow{(\cdot,\cdot)}$ with $\xRightarrow{(\cdot,\cdot,)}_{\ell}$).

The theorem below states the relationship among the equivalences.

Theorem 3 (Equivalence Results, [29]). *For each $\ell \in \mathcal{L}$,*

1. *If $\ell = \mathtt{rec_s}$ then $\approx^{\ell}_{io} = \approx^{\ell}_{o} \subset \sim^{\ell}_{io} \subset \sim^{\ell}_{o}$.*
2. *If $\ell \neq \mathtt{rec_s}$ then $\approx^{\ell}_{io} = \approx^{\ell}_{o} = \sim^{\ell}_{io} \subset \sim^{\ell}_{o}$.*

The theorem says the input-output and output congruences coincide for all languages. It also states that the input-output behavior is a congruence for every tcc language but $\mathtt{rec_s}$. This reveals a distinction between $\mathtt{rec_s}$ and the other tcc languages and, in fact, between $\mathtt{rec_s}$ and the standard model of concurrent constraint programming [45].

In the following sections we shall classify the tcc languages based on the decidability of their input-output equivalence.

5.5 Undecidability Results

In [29] it is shown that $\sim^{\mathtt{rec_p}}_{io}$ is undecidable for processes with an underlying finite-domain constraint system. Recall that a finite-domain constraint system $\mathbf{FD}[n]$ (see Definition 3) provides a theory of variables ranging over a finite domain of values $D = \{0, 1, \ldots, n-1\}$ with syntactic equality over these values. We shall also prove a stronger version of this result establishing that $\sim^{\mathtt{rec_p}}_{io}$ is undecidable even for the finite-domain constraint system with one single constant $\mathbf{FD}[1]$, i.e., $|D| = 1$. In sections 5.7 we shall give an input-output preserving constructive encoding from $\mathtt{rec_p}$ into the parameterless recursion language $\mathtt{rec_d}$, thus proving also the undecidability of $\sim^{\mathtt{rec_d}}_{io}$.

Theorem 4 (Undec. of $\sim^{\mathtt{rec_p}}_{io}$, [29]). *The problem of deciding given $P, Q \in Proc_{\mathtt{rec_p}}$ in a finite-domain constraint system, whether or not $P \sim^{\mathtt{rec_p}}_{io} Q$, is undecidable.*

We find it convinient to outline the proof of the above theorem given in [29] since it decribes very well the computational power of $\mathtt{rec_p}$. The proof is a reduction from Post's correspondence problem (PCP) [34].

Definition 17 (PCP). *A Post's Correspondence Problem (PCP)* instance *is a tuple (V, W), where $V = \{v_0, \ldots, v_n\}$ and $W = \{w_0, \ldots, w_n\}$ are two lists of non-empty words over the alphabet $\{0, 1\}$. A* solution *to this instance is a sequence of indexes $i_0, \ldots, i_m$ in $I = \{0, \ldots, n\}$ with $i_0 = 0$ s.t.*

$$v_{i_0}.v_{i_2} \ldots v_{i_m} = w_{i_0}.w_{i_2} \ldots w_{i_m}.$$

PCP is the following problem: given a PCP instance (V, W), does it have a solution?

The Post's Correspondence Problem is known to be undecidable [34]. We reduce PCP to the problem of deciding input-output equivalence between $\mathtt{rec_p}$ processes, thus proving Theorem 4.

The Post's Correspondence Problem Reduction. Let (V, W) be a PCP instance where $V = \{v_0, \ldots, v_n\}$ and $W = \{w_0, \ldots, w_n\}$ are sets of non-empty words. Let $\mathbf{FD}[m]$ (Definition 3) be the underlying constraint system where $m = max(|V|, 2)$ (i.e., we need at least two constants in the encoding below).

For each $i \in I = \{0, \ldots, |V| - 1\}$, we shall a define process $A_i(a, b, index, x)$ which intuitively behaves as follows:

1. It waits until is told that $a = 1$ to start writing v_i, one symbol per time unit. Each such a symbol, say s, will be written in x by telling $x = s$. Similarly, it waits until $b = 1$ to start writing w_i, one symbol per time unit. Each such a symbol will also be written in x.
2. It spawns a process $A_j(a', b', index, x)$ when the environment inputs an index $index = j$ in I.
3. It sets $a = 0$ and $a' = 1$ when it finishes writing v_i, i.e., $|v_i|$ time units later after it started writing v_i (this way it announces that its job of writing v_i is done, and allows A_j to start writing v_j). Similarly, it sets $b = 0$ and $b' = 1$ when it finishes writing w_i.
4. It aborts unless the environment provides an *index* in I. It also aborts if an inconsistency arises: Either two symbols (one from a V word and another from a W word) are written in x in the same time unit and they do not match (thus generating `false`), or the environment itself inputs `false`.

Thus, intuitively the A_i's keep writing V and W words, as the environment dictates, as long as the symbols match and the environment keeps providing indexes in I at each time unit.

Auxiliary Constructs We use the following constructs:

$$\begin{aligned} W_{c,P}(\boldsymbol{x}) &\stackrel{\text{def}}{=} \textbf{when } c \textbf{ do } P \parallel \textbf{unless } c \textbf{ next } W_{c,P}(\boldsymbol{x}) \\ R_Q(\boldsymbol{y}) &\stackrel{\text{def}}{=} Q \parallel \textbf{next } R_Q(\boldsymbol{y}) \end{aligned}$$

where $fv(P) \cup fv(c) = \{\boldsymbol{x}\}$ and $fv(Q) = \{\boldsymbol{y}\}$. We use the more readable notation **wait** c **do** P and **repeat** Q for $W_{c,P}(\boldsymbol{x})$ and $R_Q(\boldsymbol{y})$, respectively. We also define **whenever** c **do** P as an abbreviation of **repeat when** c **do** P.

We now define $A_i(a, b, index, x)$ for each $i \in I$ according to Items 1-4. The local variable $ichosen$ is used as flag to check whether the environment inputs an index.

$$\begin{aligned} & A_i(a, b, index, x) \stackrel{\text{def}}{=} (\textbf{local}\, a'\, b'\, ichosen)\,(\\ & \quad \textbf{wait } a = 1 \textbf{ do } V_i \\ & \parallel \textbf{wait } b = 1 \textbf{ do } W_i \\ & \parallel \textstyle\prod_{j \in I} \textbf{when } index = j \textbf{ do } (\textbf{tell}(ichosen = 1) \\ & \qquad\qquad\qquad \parallel \textbf{next } A_j(a', b', index, x)) \\ & \parallel Abort\ ) \end{aligned}$$

The process V_i writes, one by one, the v_i symbols in x (notation $v_i(n)$ denotes the $n-$th element of v_i). Furthermore it sets $a = 0$ and $a' = 1$ when it finishes writing v_i.

The process W_i is defined analogously.

$$V_i = \prod_{0 \leq k < |v_i|} \mathbf{next}^{\,k}\mathbf{tell}(x = v_i(k)) \parallel \mathbf{next}^{\,|v_i|}(\mathbf{tell}(a = 0) \parallel \mathbf{tell}(a' = 1))$$

$$W_i = \prod_{0 \leq k < |w_i|} \mathbf{next}^{\,k}\mathbf{tell}(x = w_i(k)) \parallel \mathbf{next}^{\,|w_i|}(\mathbf{tell}(b = 0) \parallel \mathbf{tell}(b' = 1))$$

The process *Abort* aborts, according to Item 4 above, by telling `false` thereafter (thus creating a constant inconsistency).

$$\begin{array}{l} Abort = \\ \parallel \mathbf{unless}\ ichosen = 1\ \mathbf{next\,repeat\,tell}(\mathtt{false}) \\ \parallel \mathbf{when}\ \mathtt{false}\ \mathbf{do\,repeat\,tell}(\mathtt{false}) \end{array}$$

Let us now define a process $B_i(a, b, index, x, ok)$ for each $i \in I$ that behaves exactly like $A_i(a, b, index, x)$, but in addition it outputs $ok = 1$ whenever it stops writing v_i and w_i exactly in the same time interval[3]. This happens when both a and b are set to zero in the same unit and it will imply that a solution of the form $v_{i_0}.\ldots.v_i = w_{i_0}.\ldots.w_i$ for the PCP (V, W) has been found.

$$\begin{array}{l} B_i(a, b, index, x, ok) \stackrel{\mathrm{def}}{=} (\mathbf{local}\ a'\ b'\ ichosen)\ (\\ \quad \mathbf{wait}\ a = 1\ \mathbf{do}\ V_i \\ \parallel \mathbf{wait}\ b = 1\ \mathbf{do}\ W_i \\ \parallel \prod_{j \in I} \mathbf{when}\ index = j\ \mathbf{do}\ (\mathbf{tell}(ichosen = 1) \\ \qquad\qquad\qquad\qquad \parallel \mathbf{next}\ B_j(a', b', index, x, ok)) \\ \parallel Abort \\ \parallel \mathbf{whenever}\ a = 0 \land b = 0\ \mathbf{do\ tell}(ok = 1)) \end{array}$$

Since we require the first index in a solution for PCP (V, W) to be 0, we define two processes $A(index, x)$ and $B(index, x, ok)$ which trigger A_0 and B_0 as follows.

$$\begin{array}{l} A(index, x) \stackrel{\mathrm{def}}{=} (\mathbf{local}\ a\ b)\ (\\ \quad \mathbf{tell}(a = 1) \parallel \mathbf{tell}(b = 1) \parallel A_0(a, b, index, x)) \end{array}$$

$$\begin{array}{l} B(index, x, ok) \stackrel{\mathrm{def}}{=} (\mathbf{local}\ a\ b)\ (\\ \quad \mathbf{tell}(a = 1) \parallel \mathbf{tell}(b = 1) \parallel B_0(a, b, index, x, ok)) \end{array}$$

One can verify that the only difference between the processes $A(index, x)$ and $B(index, x, ok)$ is that the latter eventually tells $ok = 1$ iff there is a solution to the PCP (V, W).

Since the PCP problem is undecidable, from the lemma above it follows that given $P, Q \in Proc_{\mathtt{rec_p}}$ in a finite-domain constraint system, the question of whether $P \sim_{io}^{\mathtt{rec_p}} Q$ or not is undecidable. This proves Theorem 4. □

[3] The reader may wonder why the A_i's do not have the formal parameter ok as well. This causes no problem here, but you can think of A as having a dummy ok formal parameter if you wish

Undecidability over Fixed Finite-Domains

Actually [29] gives a stronger version of the above theorem; input-output equivalence in undecidable in $\texttt{rec}_\texttt{p}$ even if we fix the underlying constraint system to be $\mathbf{FD}[1]$, which is the finite-domain constraint system whose only constant is 0.

Theorem 5 ([29]). *Fix* $\mathbf{FD}[1]$ *to be the underlying constraint system. The question of whether* $P \sim_{io}^{\texttt{rec}_\texttt{p}} Q$ *or not is undecidable.*

From Theorems 5 and 3, we also have that the input-output and default output congruences are undecidable for $\texttt{rec}_\texttt{p}$ over a fixed finite-domain constraint system.

Theorem 6. *The input-output and output congruences* $\approx_{io}^{\texttt{rec}_\texttt{p}}$ *and* $\approx_{o}^{\texttt{rec}_\texttt{p}}$ *are undecidable for processes in the finite-domain constraint system* $\mathbf{FD}[1]$.

Notice that $\mathbf{FD}[1]$ is a very simple constraint system (i.e., only equality and one single constant). So, the undecidability results for other constraint systems providing theories with equality and an at least one constant symbol follow from Theorem 5. This includes almost all constraint system of interest (e.g. the Herbrand constraint system [39], the Kahn constraint system [45], Enumerated Types [39] and modular arithmetic [32]).

5.6 Decidability Results

In sharp contrast to the undecidability result for $\texttt{rec}_\texttt{p}$, the equivalence of $\texttt{rep}$ processes is decidable even for *arbitrary constraint systems* [29].

Theorem 7. *The following equivalences for processes in* $\texttt{rep}$ *over arbitrary constraint system are decidable:*

1. *The input-output equivalence* $\sim_{io}^{\texttt{rep}}$, *default output equivalence* $\sim_{o}^{\texttt{rep}}$ *and strongest-postcondition equivalence* $\sim_{sp}^{\texttt{rep}}$.
2. *The output congruences* $\approx_{io}^{\texttt{rep}}$ *and* $\approx_{o}^{\texttt{rep}}$.

In section 5.7 we shall show via constructive encodings that $\texttt{rep}$, $\texttt{rec}_\texttt{i}$, $\texttt{rec}_\texttt{s}$ have the same expressive power. We then conclude that the corresponding equivalences for $\texttt{rec}_\texttt{i}$ and $\texttt{rec}_\texttt{s}$ are also decidable. These decidability results in $\texttt{rep}$ with arbitrary constraint system are to be contrasted to the undecidability results in $\texttt{rec}_\texttt{p}$ with the simple finite-domain constraint system $\mathbf{FD}[1]$.

5.7 Classification of the Timed CCP Languages

In this section we discuss the relation between the various tcc languages, and we classify them on the basis of their expressive power.

Figure 2 shows the sub-language inclusions and the encodings preserving the input-output behaviour between the various tcc versions. To complete the picture, we have included the class $\texttt{rec}_0$ denoting the language with neither parameters nor free variables in the bodies of definitions. Classes I, II, III represent a partition based on the

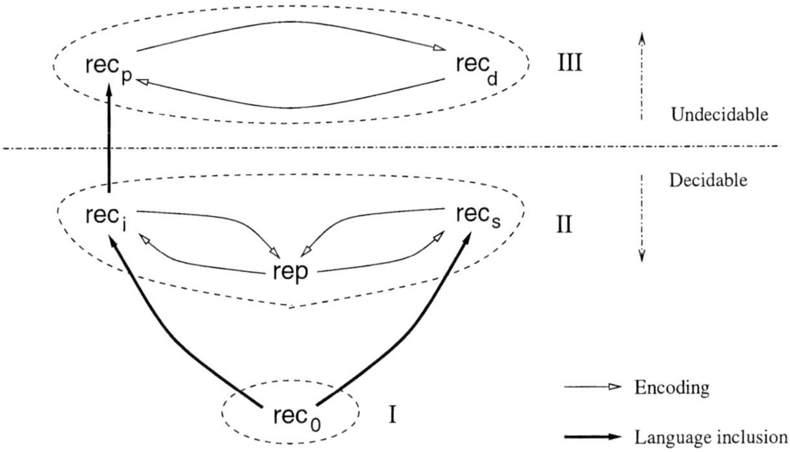

Fig. 2. Classification of the various tcc languages: The tcc hierarchy.

expressive power: two languages are in the same class if and only if they have the same expressive power. We will first discuss the separation results, and then the equivalences.

Given the input-output preserving encodings in [29], which we will recall in the next section, the separation between Classes II and III is already suggested by the results in Sections 5.6 and 5.5. From the proof of Theorem 4 it follows that $\texttt{rec}_\texttt{p}$ is capable of expressing the "behavior" of Post's correspondence problems, and hence clearly capable of expressing output behavior not accepted by Büchi automata. It turns out that the output (and input-ouput) behavior of every process in rep can be represented as a language accepted by a Büchi automata [29].

The separation between Classes I and II, on the other hand, follows from the fact that without parameters or free variables the recursive calls cannot communicate with the external environment, hence in $\texttt{rec}_0$ a process can produce information on variables for a finite number of time intervals only.

The Encodings

Let us recall briefly the input-output preserving encodings among the various tcc languages in [29]. Henceforth, $[\![\cdot]\!] : \ell \to \ell'$ will represent the encoding function from class ℓ to class ℓ' We shall say that $[\![\cdot]\!]$ is *homomorphic* wrt to the parallel operator if $[\![P \parallel Q]\!] = [\![P]\!] \parallel [\![Q]\!]$, and similarly for the other operators.

Notation 3 *We shall use the following notation:*

- *We use call*(x) *as abbreviation of* $x = 1$ *and declare, for each identifier* A*, a fresh variable* z_A *uniquely associated to it.*
- *We denote by* $I(P)$ *the set of identifiers on which* P *depends, i.e. the transitive closure of* $\rightsquigarrow$ *of the identifiers occurring in* P *(see Section 5.2).*
- *We often use* $\mathcal{D}_\ell$ *to denote the set of recursive definitions under consideration for processes in* ℓ*. As usual we omit* ℓ *when it is clear from the context.*

Encoding $\mathtt{rec_s} \rightarrow \mathtt{rep}$. Here the idea is to simulate a procedure definition by a replicated process that activates (the encoding of) its body P each time it is called. The activation can be done by using a construct of the form **when** c **do** P. The call, of course, will be simulated by **tell**(c).

The key case is the local operator, since we do not want to capture the free variables in the bodies of procedures. Thus, we need to α-convert by renaming the local variables with fresh variables.

First we need two auxiliary encodings $[\![\cdot]\!]_{\mathcal{D}}$ and $[\![\cdot]\!]_0$: given by :

$$[\![A \stackrel{\mathtt{def}}{=} P]\!]_{\mathcal{D}} = !\mathbf{when}\ call(z_A)\ \mathbf{do}\ [\![P]\!]_0$$

$$[\![A]\!]_0 = \mathbf{tell}(call(z_A))$$

$$[\![(\mathbf{local}\, x)\, P]\!]_0 = (\mathbf{local}\, y)\,([\![P[y/x]]\!]_0) \quad \text{where } y \text{ is fresh}$$

with $[\![\cdot]\!]_0$ being homomorphic on all the other operators of $\mathtt{rec_s}$.

We are now ready to give our encoding of $\mathtt{rec_s}$ into $\mathtt{rep}$.

Definition 18. *The encoding* $[\![\cdot]\!] : \mathtt{rec_s} \rightarrow \mathtt{rep}$ *is given by:*

$$[\![A]\!] = (\mathbf{local}\, \boldsymbol{z})\,([\![P]\!]_0 \parallel \prod_{i=1}^{n} [\![A_i(\boldsymbol{x}_i) \stackrel{\mathtt{def}}{=} P_i]\!]_{\mathcal{D}})$$

with $I(P) = \{A_1, \ldots, A_n\}$ *and* $\boldsymbol{z} = z_{A_1} \ldots z_{A_n}$.

Encoding $\mathtt{rec_i} \rightarrow \mathtt{rep}$. This encoding is similar to the encoding in the previous section, except that now we need to encode the passing of parameters as well. Let us give some intuition first.

A call $A(\boldsymbol{y})$, where $A(\boldsymbol{x}) \stackrel{\mathtt{def}}{=} P$, can occur in a process or in the definition of identifier B (possibly A itself). Consider the case in which there is no mutual dependency between A and B or A is a call in a process. Then, the actual parameters of A may be different from the formal ones (i.e., $\boldsymbol{y} \neq \boldsymbol{x}$). If so, we need to model the call by providing a copy of the replicated process that encodes the definition of A and by making the appropriate parameter replacements.

Now, consider the case in which there is a mutual dependency between A and B (i.e. if also A depends on B). From the restriction imposed on (the mutual) recursion of $\mathtt{rec_i}$ (see Section 5.2), we know that the actual parameters must coincide with the formal ones (i.e., $\boldsymbol{y} = \boldsymbol{x}$) and therefore we do not need to make any parameter replacement. Neither do we need to provide a copy of the replicated processes as it will be available at the top level.

As for the previous encoding, we first define the auxiliary encodings $[\![\cdot]\!]_{\mathcal{D}}$ and $[\![\cdot]\!]_0$:

$$[\![A(\boldsymbol{x}) \stackrel{\mathtt{def}}{=} P]\!]_{\mathcal{D}} = !\mathbf{when}\ call(z_A)\ \mathbf{do}\ [\![P]\!]_0$$

$$[\![A(\boldsymbol{y})]\!]_0 = \mathbf{tell}(call(z_A)) \quad \text{if } \boldsymbol{y} = \boldsymbol{x} \text{ and } A(\boldsymbol{x}) \stackrel{\mathtt{def}}{=} P \in \mathcal{D}$$

$$\begin{aligned}[\![A(\boldsymbol{y})]\!]_0 = {} & (\mathbf{local}\, z_A)\,(\\ & \mathbf{tell}(call(z_A)) \parallel [\![A(\boldsymbol{x}) \stackrel{\text{def}}{=} (P[\boldsymbol{y}/\boldsymbol{x}])]\!]_{\mathcal{D}}) \\ & \text{if } \boldsymbol{y} \neq \boldsymbol{x} \text{ and } A(\boldsymbol{x}) \stackrel{\text{def}}{=} P \in \mathcal{D}\end{aligned}$$

with $[\![\cdot]\!]_0$ homomorphic on all the other operators of $\mathtt{rec_i}$.

It worth noticing that if we did not have the restriction on the recursion in $\mathtt{rec_i}$ mentioned above, the encoding $[\![.]\!]_{\mathcal{D}}$ would not be well-defined. E.g., consider the definition $A(x) \stackrel{\text{def}}{=} \mathbf{next}\,(\mathbf{local}\, y)\; A(y)$ which violates the restriction, and try to compute $[\![A(x) \stackrel{\text{def}}{=} (\mathbf{local}\, y)\; A(y)]\!]_{\mathcal{D}}$.

We are now ready to give our encoding of $\mathtt{rec_i}$ into $\mathtt{rep}$.

Definition 19. *The encoding* $[\![\cdot]\!] : \mathtt{rec_i} \rightarrow \mathtt{rep}$ *is given by:*

$$[\![A(\boldsymbol{y})]\!] = (\mathbf{local}\, \boldsymbol{z})\, ([\![P]\!]_0 \parallel \prod_{i=1}^{n} [\![A_i(\boldsymbol{x}_i) \stackrel{\text{def}}{=} P_i]\!]_{\mathcal{D}})$$

with $I(P) = \{A_1, \ldots, A_n\}$ *and* $\boldsymbol{z} = z_{A_1} \ldots z_{A_n}$.

Encoding $\mathtt{rep} \rightarrow \mathtt{rec_i}$**.** This encoding is rather simple. The idea is to replace $!\,P$ by a call to a new process identifier R_P, defined as a process that expands P and then calls itself recursively in the next time interval. The free variables of $!\,P$, $\boldsymbol{x}$, are passed as (identical) parameters.

Definition 20. *The encoding* $[\![\cdot]\!] : \mathtt{rep} \rightarrow \mathtt{rec_i}$ *is given by:*

$$\begin{aligned}[\![!\,P]\!] = {} & R_P(\boldsymbol{x}) \\ & \text{where } R_P(\boldsymbol{x}) \stackrel{\text{def}}{=} [\![P]\!] \parallel \mathbf{next}\, R_P \in \mathcal{D}_{\mathtt{rec_i}}, \boldsymbol{x} = fv(P).\end{aligned}$$

with $[\![\cdot]\!]$ *homomorphic on all the other operators of* $\mathtt{rep}$.

Encoding $\mathtt{rec_d} \rightarrow \mathtt{rec_p}$**.** Intuitively, if the free variables are treated dynamically, then they could equivalently be passed as parameters.

Definition 21. *The encoding* $[\![\cdot]\!] : \mathtt{rec_d} \rightarrow \mathtt{rec_p}$ *is given by*

$$\begin{aligned}[\![A]\!] = {} & A(\boldsymbol{x}) \\ & \text{where } A \stackrel{\text{def}}{=} P \in \mathcal{D}_{\mathtt{rec_d}} \\ & \text{and } A(\boldsymbol{x}) \stackrel{\text{def}}{=} [\![P]\!] \in \mathcal{D}_{\mathtt{rec_p}}, \boldsymbol{x} = fv(P)\end{aligned}$$

with $[\![\cdot]\!]$ *homomorphic on all the other operators of* $\mathtt{rec_d}$

Encoding $\mathtt{rec_p} \rightarrow \mathtt{rec_d}$**.** The idea is to establish the link between the formal parameters $\boldsymbol{x}$ and the actual parameters $\boldsymbol{y}$ by telling the constraint $\boldsymbol{x} = \boldsymbol{y}$. However, this operation has to be encapsulated within a $(\mathbf{local}\, \boldsymbol{x})$ in order to avoid confusion with other potential occurrences of $\boldsymbol{x}$ in the same context of the call.

Definition 22. *The encoding* $[\![\cdot]\!] : \mathtt{rec_p} \to \mathtt{rec_d}$ *is given by*

$$[\![A(\boldsymbol{y})]\!] = (\mathbf{local}\,\boldsymbol{x})\,(A \parallel E_{\boldsymbol{y}/\boldsymbol{x}})$$
$$\textit{where } A(\boldsymbol{x}) \stackrel{\mathrm{def}}{=} P \in \mathcal{D}_{\mathtt{rec_p}},\ A \stackrel{\mathrm{def}}{=} [\![P]\!] \in \mathcal{D}_{\mathtt{rec_d}},$$
$$\textit{and } \ E_{\boldsymbol{y}/\boldsymbol{x}} \stackrel{\mathrm{def}}{=} \mathbf{tell}(\boldsymbol{y} = \boldsymbol{x}) \parallel \mathbf{next}\,E_{\boldsymbol{y}/\boldsymbol{x}} \in \mathcal{D}_{\mathtt{rec_d}}$$

with $[\![\cdot]\!]$ *homomorphic on all the other operators of* $\mathtt{rec_d}$.

Encoding $\mathtt{rep} \to \mathtt{rec_s}$**.** Here we take advantage of the automata representation of the input-output behavior of rep processes given in [29]. Basically, the idea is to use the recursive definitions as equations describing these input-output automata.

Let P be an arbitrary process in rep. Let us recall the automaton $M_P = A_P^{io}$ in [29] representing the input-output behavior of P on the inputs of relevance for P. The start state of M_P is P. Let T_P be the set of transitions of M_P. Each transition from Q to R with label (c,d), written $\langle Q,(c,d),R\rangle \in T_P$, represents an observable transition $Q \stackrel{(c,d)}{\Longrightarrow} R$.

So, for each state Q of M_P we define an identifier A_Q as follows:

$$A_Q \stackrel{\mathrm{def}}{=} \prod_{\langle Q,(c,d),R\rangle\in T_P} \mathbf{when}\ c\ \mathbf{do}\ (\mathbf{tell}(d) \parallel O(\sqcup c, R))$$
$$\text{with } \sqcup c = \bigvee_{e\in\{c' \mid c'\neq c,\ c'\models c,\ \langle Q,(c',d'),R'\rangle\in T_P\}} e$$

where $O(\sqcup c, R)$ takes the form $\mathbf{unless}\ \sqcup c\ \mathbf{next}\ A_R$ if $c \neq \mathtt{false}$, otherwise it takes the form $\mathbf{next}\ A_R$.

Intuitively, A_Q expresses that if we are in state Q and c is the strongest constraint entailed by the input, then the next state will be R and the output will be d, with $\langle Q,(c,d),R\rangle \in T_P$.

Definition 23. *The encoding* $[\![\cdot]\!] : \mathtt{rep} \to \mathtt{rec_s}$ *is defined as* $[\![P]\!] = A_P$.

6 Related Work and Concluding Remarks

Saraswat el al proposed a proof system for tcc [40], based on an intuitionistic logic enriched with a next operator. The system is complete for hiding-free and finite processes. Also Gabrielli et al [4] introduced a proof system for the tccp model (see Section 3). The underlying second-order linear temporal logic in [4] can be used for describing input-output behavior. In contrast, the ntcc logic can only be used for the strongest-postcondition, but also it is semantically simpler and defined as the standard first-order linear-temporal logic of [22].

The decidability results for the ntcc equivalences here presented are based on reductions from ntcc processes into finite-state automata [29,31,52]. The work in [43] also shows how to compile tcc into finite-state machines thus providing an execution model of tcc.

In [49] Tini explores the expressiveness of tcc languages, focusing on the capability of tcc to encode synchronous languages. In particular, Tini shows that Argos [23] and a version of Lustre restricted to finite domains [16] can be encoded in tcc.

In the context of tcc, Tini [50] introduced a notion of bisimilarity with a complete and elegant axiomatization for the hiding-free fragment of tcc. The notion of bisimilarity has also been introduced for `ntcc` by Valencia in his PhD thesis [51].

On the practical side, Saraswat el al introduced Timed Gentzen [41], a particular tcc-based programming language for reactive-systems implemented in PROLOG. More recently, Saraswat el al released jcc [44], an integration of timed (default) ccp into the JAVA programming language. Rueda et al [38] demonstrated that essential ideas of computer generated music composition can be elegantly represented in `ntcc`. Hurtado and Muñoz [20] in joint work with Fernández and Quintero [10] gave a design and efficient implementation of an `ntcc`-based reactive programming language for LEGO RCX robots [21]—the robotic devices chosen in Section 4 as motivating examples.

Future Work. Timed ccp is still under development and certainly much remain to be explored. In order to contribute to the development of timed ccp as a well-established model of concurrency, a good research strategy could be to address those issues that are central to other mature models of concurrency. In particular, the analysis and formalization of the `ntcc` behavioral equivalences, which at present time are still very immature (e.g., axiomatizations of process equivalences and automatic tools for behavioral analysis).

Furthermore, the decision algorithms for `ntcc` verification and satisfiability, are very inefficient, and of theoretical interest only. For practical purposes, it is important to conduct studies on the design and implementation of efficient algorithms for verification.

Acknowledgments

We owe much to Catuscia Palamidessi for her contributions to the development of the `ntcc` calculus.

References

1. M. Abadi. The power of temporal proofs. *Theoretical Computer Science*, 65:35–84, 1989.
2. G. Berry and G. Gonthier. The ESTEREL synchronous programming language: design, semantics, implementation. *Science of Computer Programming*, 19(2):87–152, 1992.
3. J. R. Buchi. On a decision method in restricted second order arithmetic. In *Proc. Int. Cong. on Logic, Methodology, and Philosophy of Science*, pages 1–11. Stanford University Press, 1962.
4. F. de Boer, M. Gabbrielli, and M. Chiara. A temporal logic for reasoning about timed concurrent constraint programs. In *TIME 01*. IEEE Press, 2001.
5. F. de Boer, M. Gabbrielli, and M. C. Meo. A timed concurrent constraint language. *Information and Computation*, 161:45–83, 2000.
6. F. de Boer, A. Di Pierro, and C. Palamidessi. Nondeterminism and infinite computations in constraint programming. *Theoretical Computer Science*, 151(1):37–78, 1995.
7. F. S. de Boer, M. Gabbrielli, E. Marchiori, and C. Palamidessi. Proving concurrent constraint programs correct. *ACM Transactions on Programming Languages and Systems*, 19(5), 1997.

8. J.F. Diaz, C. Rueda, and F. Valencia. A calculus for concurrent processes with constraints. *CLEI Electronic Journal*, 1(2), 1998.
9. H. Dierks. A process algebra for real-time programs. In *FASE*, volume 1783 of *LNCS*, pages 66–81. Springer Verlag, 2000.
10. D. Fernádez and L. Quintero. *VIN: An ntcc visual language for LEGO Robots*. BSc Thesis, Universidad Javeriana-Cali, Colombia, 2003. http://www.brics.dk/~fvalenci/ntcc-tools.
11. J. Fredslund. The assumption architecture. Progress Report, Department of Computer Science, University of Aarhus, 1999.
12. D. Gilbert and C. Palamidessi. Concurrent constraint programming with process mobility. In *Proc. of the CL 2000*, LNAI, pages 463–477. Springer-Verlag, 2000.
13. V. Gupta, R. Jagadeesan, and P. Panangaden. Stochastic processes as concurrent constraint programs. In *Symposium on Principles of Programming Languages*, pages 189–202, 1999.
14. V. Gupta, R. Jagadeesan, and V. A. Saraswat. Computing with continuous change. *Science of Computer Programming*, 30(1–2):3–49, 1998.
15. N. Halbwachs. Synchronous programming of systems. *LNCS*, 1427:1–16, 1998.
16. N. Halbwachs, P. Caspi, P. Raymond, and D. Pilaud. The synchronous data-flow programming language LUSTRE. *Proc. of the IEEE*, 79(9):1305–1320, 1991.
17. S. Haridi and S. Janson. Kernel andorra prolog and its computational model. In *Proc. of the International Conference on Logic Programming*, pages 301–309. MIT Press, 1990.
18. P. Van Hentenryck, V. Saraswat, and Y. Deville. Constraint processing in `cc(fd)`. In *Proceedings of the 2000 ACM symposium on Applied computing 2000*, volume 910 of *LNCS*. Springer-Verlag, 1995.
19. C. A. R. Hoare. *Communications Sequential Processes*. Prentice-Hall, Englewood Cliffs (NJ), USA, 1985.
20. R. Hurtado and M. Muñoz. *LMAN: An ntcc Abstract Machine for LEGO Robots*. BSc Thesis, Universidad Javeriana-Cali, Colombia, 2003. http://www.brics.dk/~fvalenci/ntcc-tools.
21. H. H. Lund and L. Pagliarini. Robot soccer with LEGO mindstorms. *LNCS*, 1604:141–151, 1999.
22. Z. Manna and A. Pnueli. *The Temporal Logic of Reactive and Concurrent Systems, Specification*. Springer, 1991.
23. F. Maraninchi. Operational and compositional semantics of synchronous automaton compositions. In *CONCUR '92*, volume 630 of *LNCS*, pages 550–564. Springer-Verlag, 1992.
24. N.P. Mendler, P. Panangaden, P. J. Scott, and R. A. G. Seely. A logical view of concurrent constraint programming. *Nordic Journal of Computing*, 2(2):181–220, 1995.
25. R. Milner. *Communication and Concurrency*. International Series in Computer Science. Prentice Hall, 1989. SU Fisher Research 511/24.
26. R. Milner. A finite delay operator in synchronous ccs. Technical Report CSR-116-82, University of Edinburgh, 1992.
27. R. Milner. *Communicating and Mobile Systems: the π-calculus*. Cambridge University Press, 1999.
28. U. Montanari. Networks of constraints: Fundamental properties and applications to picture processing. *Information Science*, 7, 1974.
29. M. Nielsen, C. Palamidessi, and F. Valencia. On the expressive power of concurrent constraint programming languages. In *Proc. of PPDP'02*, pages 156–167. ACM Press, 2002.
30. M. Nielsen, C. Palamidessi, and F. Valencia. Temporal concurrent constraint programming: Denotation, logic and applications. *Nordic Journal of Computing*, 9(2):145–188, 2002.
31. M. Nielsen and F. Valencia. *Temporal Concurrent Constraint Programming: Applications and Behavior*, chapter 4, pages 298–324. Springer-Verlag, LNCS 2300, 2002.
32. C. Palamidessi and F. Valencia. A temporal concurrent constraint programming calculus. In *Proc. of CP'01*. Springer-Verlag, LNCS 2239, 2001.

33. C.A. Petri. Fundamentals of a theory of asynchronous information flow. In *Proc. IFIP Congress '62*, 1962.
34. E. L. Post. A variant of a recursively unsolvable problem. *Bulletin of the American Mathematical Society*, 52:264–268, 1946.
35. G.M. Reed and A.W. Roscoe. A timed model for communication sequential processes. *Theoretical Computer Science*, 8:249–261, 1988.
36. F. Rossi and U. Montanari. Concurrent semantics for concurrent constraint programming. In *Constraint Programming: Proc. 1993 NATO ASI*, pages 181–220, 1994.
37. J.H. Réty. Distributed concurrent constraint programming. *Fundamenta Informaticae*, 34(3), 1998.
38. C. Rueda and F. Valencia. Proving musical properties using a temporal concurrent constraint calculus. In *Proc. of the 28th International Computer Music Conference (ICMC2002)*, 2002.
39. V. Saraswat. *Concurrent Constraint Programming*. The MIT Press, Cambridge, MA, 1993.
40. V. Saraswat, R. Jagadeesan, and V. Gupta. Foundations of timed concurrent constraint programming. In *Proc. of LICS'94*, pages 71–80, 1994.
41. V. Saraswat, R. Jagadeesan, and V. Gupta. Programming in timed concurrent constraint languages. In *Constraint Programming*, NATO Advanced Science Institute Series, pages 361–410. Springer-Verlag, 1994.
42. V. Saraswat, R. Jagadeesan, and V. Gupta. Default timed concurrent constraint programming. In *Proc. of POPL'95*, pages 272–285, 1995.
43. V. Saraswat, R. Jagadeesan, and V. Gupta. Timed default concurrent constraint programming. *Journal of Symbolic Computation*, 22(5–6):475–520, 1996.
44. V. Saraswat, R. Jagadeesan, and V. Gupta. jcc: Integrating timed default concurrent constraint programming into java. http://www.cse.psu.edu/~saraswat/jcc.html, 2003.
45. V. Saraswat, M. Rinard, and P. Panangaden. The semantic foundations of concurrent constraint programming. In *POPL '91*, pages 333–352, 1991.
46. E. Shapiro. The Family of Concurrent Logic Programming Languages. *Computing Surveys*, 21(3):413–510, 1990.
47. G. Smolka. A Foundation for Concurrent Constraint Programming. In *Constraints in Computational Logics*, volume 845 of *LNCS*, 1994. Invited Talk.
48. G. Smolka. The Oz programming model. In Jan van Leeuwen, editor, *Computer Science Today*, volume 1000 of *LNCS*, pages 324–343. Springer-Verlag, 1995.
49. S. Tini. On the expressiveness of timed concurrent constraint programming. *Electronics Notes in Theoretical Computer Science*, 1999.
50. S. Tini. An axiomatic semantics for the synchronous language gentzen. In *FOSSACS'01*, volume 2030 of *LNCS*. Springer-Verlag, 2001.
51. F. Valencia. *Temporal Concurrent Constraint Programming*. PhD thesis, BRICS, University of Aarhus, 2003.
52. F. Valencia. Timed concurrent constraint programming: Decidability results and their application to LTL. In *Proc. of ICLP'03*. Springer-Verlag, LNCS, 2003.
53. W. Yi. *A Calculus for Real Time Systems*. PhD thesis, Chalmers Institute of Technology, Sweden, 1991.
54. W. M. Zuberek. Timed petri nets and preliminary performance evaluation. In *Proc. of the 7th Annual Symposium on Computer Architecture*, pages 88–96. ACM and IEEE, 1980.

Petri Nets and Manufacturing Systems: An Examples-Driven Tour*

Laura Recalde, Manuel Silva, Joaquín Ezpeleta, and Enrique Teruel

Dep. Informática e Ingeniería de Sistemas, I3A
Universidad de Zaragoza, María de Luna 1, E-50018 Zaragoza, Spain
{lrecalde,silva,ezpeleta,eteruel}@unizar.es

Abstract. There exists ample literature on Petri nets and its potential in the modelling, analysis, synthesis and implementation of systems in the manufacturing applications domain (see for example [54, 15, 18]; besides, in [66] an important bibliography is presented). This paper provides an examples-driven perspective. Nevertheless, not only complete examples from the application domain are considered. Manufacturing systems are frequently large systems, and conceptual complexity often appears because of some particular "local" constructions.

The examples considered in this selected tour try to introduce in a progressive way some applied concepts and techniques. The starting point is an assembly cell, for which models concerning several phases of the design life-cycle are presented. Afterwards, some pull control and kanban management strategies are modelled. Then, two coloured models of production lines are presented. After that, a manufacturing system with two cells is modelled, and the difficulty of the practical analysis is shown. For very populated manufacturing systems or systems with high cadence, relaxation of discrete event models leads to hybrid and continuous approximations, an example of which will be shortly introduced.

1 Motivation and Objectives

Petri Nets (PNs) constitute a well known *paradigm* for the design and operation of many systems allowing a discrete event view [53]. The purpose of this work is to present, in a tutorial style, some examples in which manufacturing systems are modelled and analysed. Several books about PNs and the design and operation of manufacturing systems have been published at the end of the last century [17, 15, 65, 56, 44, 66]. In the sequel, the reader is assumed to be introduced to the main concepts in Petri Nets [50, 42].

Basically a case study driven perspective is provided in this work. Nevertheless, not only full examples from the application domain are considered. Manufacturing systems are frequently large systems, and conceptual complexity appears because of some particular constructions that appear in part of the system.

* Partially supported by projects FEDER and CICYT TIC2001-1819, and DPI2003-06376.

J. Desel, W. Reisig, and G. Rozenberg (Eds.): ACPN 2003, LNCS 3098, pp. 742–788, 2004.

The examples considered in this selected tour try to progressively present some applied concepts and techniques.

The starting point (Sect. 2) is a manufacturing cell in which some conveyors move parts that, processed into two different machines (*M1* and *M2*), are assembled and evacuated. The internal movements of the parts in the cell are executed by an industrial robot. Moreover, due to a relatively high rate of failures of a machine (*M1*), a buffer allows a partial decoupling with respect to the assembly machine (hence, also with respect to *M2*). This store (or buffer) acts like condensers in RC circuits: filtering high frequency perturbations (i.e., attenuating the effect of frequent short failures that usually lead to many small unavailability periods). From an abstract perspective, this introductory example shows some interesting interleaving among *cooperation* (here, the assembly of two different kinds of parts) and *competition* (for the shared resource: the robot) relationships. In general terms, the intricate interleaving of these two kinds of relationships leads to the kernel of the conceptual complexity to master the behaviour of discrete event systems (DES). The presentation of this introductory example is focused on the advantages of using different models of the same PN modelling paradigm in order to deal with the different phases of the design and operation that appear during the life cycle of the process.

In general terms, the control of manufacturing systems often uses some preestablished strategies. Among them the *push* strategy (from the input to the output: from the raw parts to the finished products), *pull* (from the output backwards to the input: from the demand to the input of raw parts) and *kanban*, that may represent many different kinds of tradeoffs between the above mentioned basic strategies, are specially relevant. The purpose of Sect. 3 is to show that this kind of control mechanisms (or management strategies) can be appropriately modelled by means of PNs (see, for example, [11]). Analysis and optimisation of the obtained models can be done, but this topic is not considered in detail in this section, since the main purpose is to show the practical modelling power of the PN formalisms. This paper is mainly devoted to aspects related to modelling, analysis and control design, and not on other topics, like simulation or implementation issues, that although interesting and useful are not developed here. However, simulation will be used in this particular section to illustrate the comparison of different control techniques.

In many manufacturing systems a significant part of the apparent complexity may derive from the existence of several subsystems having identical (similar) behaviours, or from many parts having similar processing plans. Under these conditions (i.e., having significant symmetries among components), the use of high level PNs may be of interest. For this purpose two different examples are presented. The first one (Sect. 4) concerns a French manufacturing line for car assembly. The basic model is constructed in a very systematic way, by merging a coloured PN model of the stations where manufacturing operations are performed and a coloured PN model for the transportation system. The problem with this basic model is that deadlocks may appear. A quite simple solution is presented, being directly implementable in PN terms, just by adding a place (i.e., a constraint) appropriately marked. A step further is done through the presentation of a closed line corresponding to an ovens production factory sited in Zaragoza (Sect. 5).

In order to approach the limits of the actual knowledge in the theory and application of PNs to manufacturing examples, two additional cases are introduced in Sect. 6. In the first one (Sect. 6.1), a model of a Flexible Manufacturing System (FMS) (held in the Department of Computer Science and Systems Engineering of the University of Zaragoza) is established [25]. Even if modelling can be done in this case in a "straightforward" way, analysis "requires", in the actual state of the art, some manipulations allowing the computation of sequentialised views for the different *process plans*. In other words, it is not a direct application of theory that brings some solutions, but an indirect-pragmatically oriented engineering approach. Going in the same direction, in Sect. 6.2 modelling with object nets is done: this leads to a powerful modelling approach [62]. Unfortunately, it usually happens that the higher the abstraction level the formalism allows, the more complicated its analysis becomes. However, it is always possible to apply simulation techniques, which can give insight of some system behaviours.

Discrete event "views" may be very convenient in many cases for manufacturing systems. Nevertheless, in some other cases, either because of computational complexity problems (due to state explosion) or because the system presents a "regular" high cadence behaviour or is highly populated, fluidification or continuisation may be of interest [3, 51, 52]. A hybrid (partially continuised) model of this category is presented in Sect. 7. For systems in which some parts are "naturally perceived as continuous", a different PN interpretation leads to hybrid modelling (PrTr-DAE). In the present state of knowledge, this last approach uses simulation as the main analysis technique (besides the application of standard analysis techniques for the study of the underlying discrete model). Hybrid models analysis techniques should much improve in the future. Finally, some concluding remarks close this work.

2 Life Cycle and an Introductory Example: An Assembly Manufacturing Cell

This introductory example deals with a system in which the process plan is quite easy: Parts "*A*" and "*B*" should be produced (at machines *M1* and *M2*, respectively) and later assembled (a *rendez-vous*) in machine *M3* to obtain a final product that leaves the manufacturing cell. In this trivial cooperative system, two additional elements are introduced. First, relatively important failures and repairs are taken into account for *M1*. With the idea in mind of partially decoupling these accidents with respect to the operation of downstream machines (here *M3*), a *buffer* (inventory place, deposit) is introduced. If *M1* fails, the downstream machine, *M3*, may continue working for a while consuming the parts already in the buffer. If the upstream machine *M1* is repaired before the buffer is emptied, the failure will not affect the downstream line (here *M3*, only). Since *M3* is an assembly machine, its stopping condition will propagate to the upstream line (here *M2*). The buffer is a passive element. At this point, the full system only exhibits cooperative activities. A typical competition relationship is introduced by means of the movement of parts inside the system. In this case a robot feeds *M1* and *M2* (from the conveyor belt), feeds the buffer (from *M1*), and moves parts *A* (from the buffer) and *B* (from *M2*) to *M3*. Therefore, all these activities are in mutual exclusion (*mutex*). Thus this introductory example (Fig. 1, that will be explained more in detail in Sect. 2.1) has *cooperation* and *competition* relationships. If the competition for the use of the robot is ignored, the cooperative parts can be described by a *free-choice* net system [57]. The addition of the robot-idle place transforms the net into a *simple* or *asymmetric choice.*

2.1 Basic Autonomous Model: Dealing with Basic Relationships at the Net Level

The net in Fig. 1 models both the *plant* and the *work plan*, from a coordination viewpoint. In the initial state, all the machines and the robot are idle, and the buffer is empty. The only enabled transitions are those that represent the start of the loading operation of either *M1* or *M2*, but only one of them can occur (i.e., there is a *conflict* situation). The autonomous model leaves undetermined which one will occur, it only states that these are the possibilities. Assume *M1* is to be loaded, what is represented by the occurrence of transition *t1*. Then the marking changes: one token is removed from each input place of the transition (*R idle* and *M1 idle*) and one token is put into the output place (*M1 loading*). Notice that tokens were required from two input places, meaning that the loading operation requires that both the machine and the robot are ready: it is a *synchronisation* of both. Now the only enabled transition is the one representing the end of the loading operation, but the autonomous model leaves undetermined when will this happen, it only states that it can only happen whenever loading is in course (which allows to represent *sequencing*). At the firing, the token is removed from *M1 loading* and tokens are put in *M1 working* and *R idle*. In this new marking, both output transitions of *M1 working* are enabled in conflict (it may either complete the work or fail), and also the start of the loading of *M2* is enabled. This latter transition and a transition from *M1* can occur simultaneously, or in any order (their enabling is independent), what allows to faithfully model *concurrency*. Notice the correspondence of subnets and subsystems (*M1*, *M2*, *M3*, *B1*, and *R*), and the natural representation of their mutual interactions. (It goes without saying that operation places could be refined to show the detailed sequence of operations in each machine, etc.)

We have depicted as bars those transitions that represent *control events*, while transitions depicted as boxes represent the *end of an operation*, or the *occurrence of a failure*. At the present stage of autonomous systems, these drawing conventions, and also the various labels, are literature: the dynamics of the model is not affected by these details, which are intended to make clearer the "physical" meaning of the model.

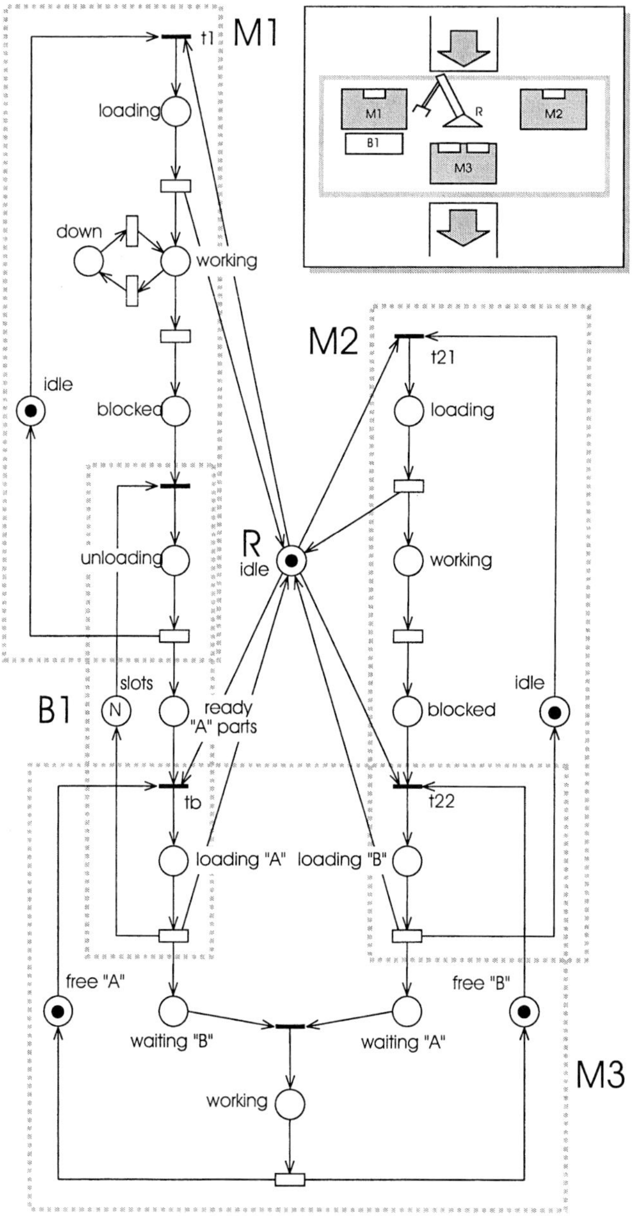

Fig. 1. An autonomous place/transition system that formally describes the logic behaviour of a manufacturing cell.

This autonomous model can be used for documentation/understanding purposes, and also to formally analyse the non-deterministic possible behaviours. Classical PN analysis techniques allow to efficiently decide that this system model is bounded (i.e., finite state space), live (i.e., no action can become unattainable), and reversible (i.e., from any state the system can evolve to its initial state).

Classical (and basic) reduction rules [49] allow to transform the model into a marked graph:

1. Every path *start loading* $\longrightarrow$ *loading* $\longrightarrow$ *end loading* is a *macrotransition.* Therefore it can be reduced to a single *load* transition, preserving the (projected) language, hence liveness, boundedness, reversibility, etc.
2. After the previous step, place *R idle* self-loops around the four *load* transitions, and can be removed preserving the language (i.e., it was an *implicit* place).
3. The places *working* and *down* in *M1* and their connecting transitions form a *macroplace.*

The resulting marked graph is strongly connected. Therefore, it is structurally bounded (i.e., it is bounded for any initial marking, not just for the one that is shown here), and it does not contain unmarked circuits, so it is live and reversible.

2.2 The Performance Evaluation Model: Stochastic T-Timed Interpretation and Analysis

If the purpose of the model is to evaluate the performance of the manufacturing cell, or to investigate different scheduling policies, then *timing information* (e.g., duration of operations, mean time between failures, etc.) can be incorporated to the model, for instance specifying the delay in the firing of transitions. Diverse timing specifications are possible (e.g., stochastic, deterministic, time intervals, etc.), each one best suited for a particular purpose or degree of detail required. In Fig. 2 the firing delays are specified by their mean times.

In a preliminary design stage, where the issue is machine selection and dimensioning of the system, a stochastic timing specification, such as that of *generalised stochastic PNs* [1], is best suited. In the example we assume that the distribution of time delays corresponding to operations and movements is *phase-type*, namely *Erlang-3*, while failures and repairs follow *exponential* distributions. All other transitions are *immediate*, they fire as soon as they are enabled (so they are *prioritary* w.r.t. timed transitions). Conflicts between timed transitions are solved by *race* policy, while conflicts between immediate ones are solved in a probabilistic fashion).

It was seen in Sect. 2.1 that this system is reversible. Therefore, the reachability graph is strongly connected, and this allows to deduce ergodicity of the stochastic process and irreducibility of the underlying Markov chain.

Markovian performance analysis can be used to assist in the dimensioning of *B1*, or to analyse its impact. With given failure and repair rates for *M1*, throughput is plotted versus buffer size in Fig. 3.

Economic considerations (in terms of throughput, required investment, and work in progress) would allow to optimise the buffer size. The plots in Fig. 4

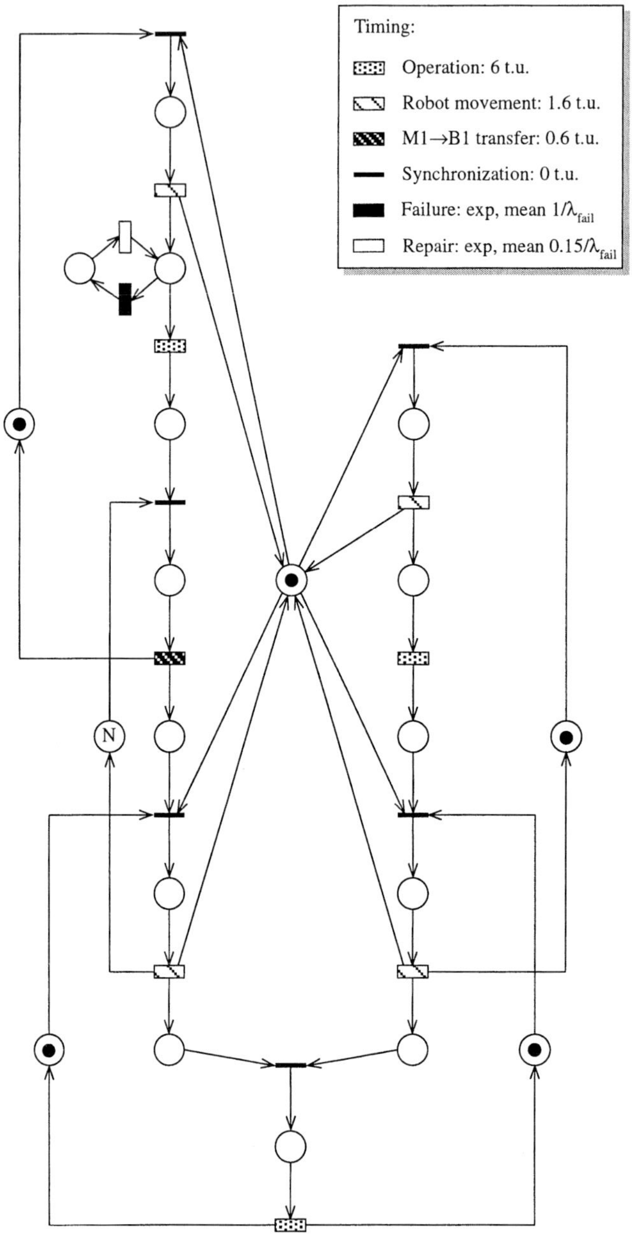

Fig. 2. A timed place/transition system that allows performance evaluation and optimisation of a manufacturing cell.

show how the effect of the buffer varies depending on the nature of the failures. Keeping the failure/repair ratio constant (i.e., the % of unavailability of the machine due to a failure is constant), different situations can be observed:

- Very unfrequent failures with very long repair times (left side of the plot). The throughput is reduced, and is insensible to the buffer size, because the repair time exceeds largely the time to empty the buffer.
- On the other extreme, in the case of very frequent slight failures, a relatively small buffer is able to filter out the high frequency perturbations represented by the failures, and the throughput is equal to the throughput in the case of no failures.
- When the order of magnitude of repair times are similar to the time required to empty the buffer, its size is most critical in order to increase the throughput.

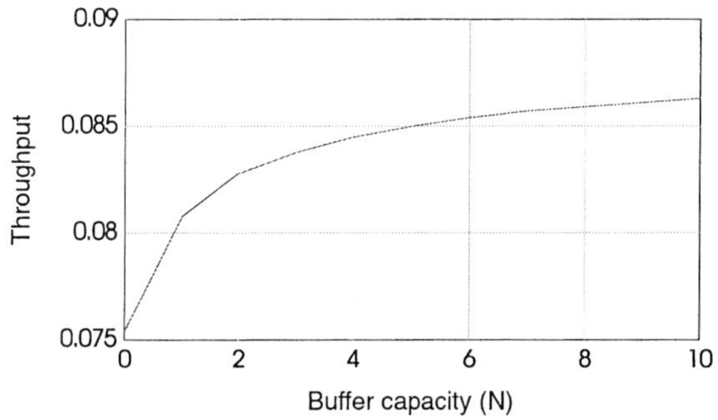

Fig. 3. Performance evaluation of the cell in Fig. 1 with respect to buffer capacity.

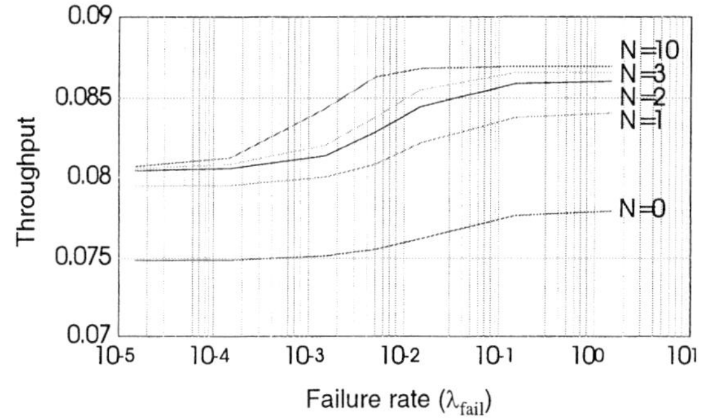

Fig. 4. Performance evaluation of the cell in Fig. 1 with respect to failure rate.

Notice that for the case $N = 0$ the model in Fig. 1 should be changed, removing *B1*. That is, the "*unloading*" operation should be merged with the "*loadingA*" and place *slots* removed since it becomes implicit. Then, *M1* becomes essentially identical to *M2*, except for the presence of failures. It results in a more tight coupling of the machines that leads to a significantly lower throughput.

2.3 On the Optimal Scheduling: Performance Control

Assume that, after the optimisation of the design that involved performance evaluation, the capacity of the buffer is fixed to two. Although the plant parameters are fixed, the actual performance of the system may vary depending on how it is controlled. The scheduler is in charge of controlling the evolution by enabling/disabling the transitions that initiate robot load operations (i.e., these are the *controllable* transitions here).

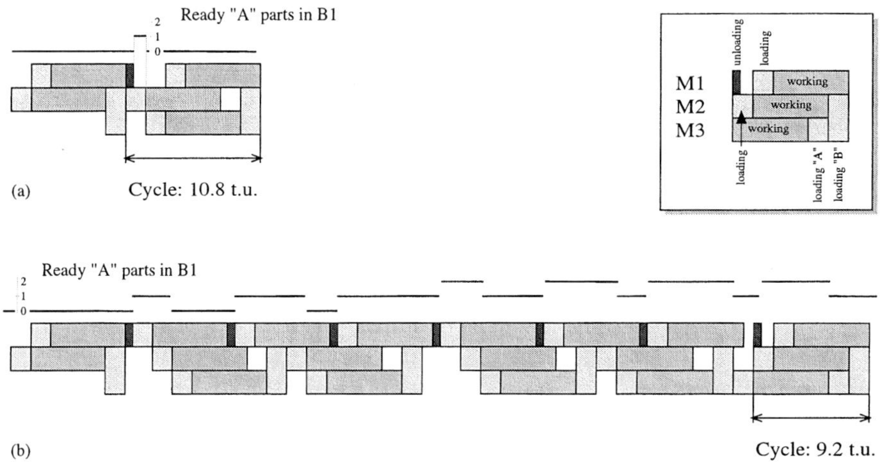

Fig. 5. Effect of different scheduling policies in the manufacturing cell of Fig. 1.

Fig. 5 shows the Gantt charts of two possible scheduling policies assuming deterministic timing and disregarding failures. In Fig. 5(a) operations are scheduled as soon as possible, solving eventual conflicts in the allocation of the robot by fixed priorities (*M2* is prioritary over *M1*). A periodic regime is quickly reached, in which:

- The cycle time is 10.8 (i.e., throughput *without failures* is 0.0926).
- The buffer contains at most one part, so parts are not accumulated to be used in the event of a failure.

The Gantt chart in Fig. 5(b) shows an evolution in which the scheduler prevents interrupting *M1* until it gets blocked, and prevents interrupting *M2* and *M3* from then on. This policy fills up the buffer to be prepared for eventual failures and achieves a cycle time of 9.2 (i.e., throughput 0.1087) in normal operation, thus the buffer allows to increase productivity in more than 11%. Let us check that this policy can be proved to be optimal.

As already mentioned, let us consider the system without failures (i.e., removing the failure-repair loop). One way of reasoning to obtain an optimal schedule for this system is as follows: the skeleton of the system is clearly a strongly connected marked graph provided with a monitor place (idle state for the robot).

Thus the unique T-semiflow is $\mathbf{x} = \mathbf{1}$ (i.e., a vector of 1s). This means that all the transitions, in particular the four immediate in which the robot starts to work, should be fired in the same proportion in any "long enough" sequence. Even more, the steady state should be defined by repeating sequences in which *t1*, *tb*, *t21* and *t22* (i.e., all the transitions before the "*loading*" places) appear once. Since those transitions are the only ones that may be in conflict, the scheduling problem reduces to choosing the relative order in which they should be fired. Given the repetitive behaviour of the steady state, in principle any transition can be taken as the first, thus there exist at most $3! = 6$ possibilities to explore. Assume *t22* is fired first. In this case nothing opposes to take *t21* as the second one to fire, because there is a marked place (*M2idle*) connecting the end of the first loading operation with the start of the second one (in other words, by choosing *t*21 as the second one no constraint is added). Therefore, the question now is to choose between *t1* and *tb*. Before going to that question, let us observe that firing an appropriate transient sequence the buffer can be filled, at least partially. In doing that, the firing of *t1* and *tb* are "decoupled" by a finite sequence, i.e., both can be fired in any order, while keeping the goal of computing an optimal schedule. If, after *t21*, transition *t1* is fired, the cycle of use of the shared resource (the robot) is finished by firing *tb* (and later *t22* for a new cycle).

A general upper bound of the throughput (lower for the cycle time) of the original system can be computed by means of a linear programming problem [9]. For this particular case, the lower bound for the cycle time is 9.2 time units. Looking at Fig. 5(b) it is clear that this lower bound can be reached with the previous ordering. However, an alternative procedure can be used to prove it.

Introducing places $\{p2, p3, p4\}$ to put an order in the use of the robot: *t21*-*p2*-*t1*, *t1*-*p3*-*tb*, *tb*-*p4*-*t22* (observe that *p1*, for *t22*-*p1*-*t21*, is equal to *M2idle*, and so it is already present and marked), the place representing the idle state of the robot becomes concurrently implicit [55], thus it can be removed for any time interpretation, and a marked graph is found (see Fig. 6). Under deterministic timing the exact cycle time for any marked graph can be computed by means of the same linear programming problem mentioned above [8]. The obtained value for this case is once again 9.2, thus under deterministic timing and no failures, the set of added constraints, places $\{p2, p3, p4\}$, constitute an optimal scheduler. The reason is that adding that constraints (places *p2*, *p3* and *p4*) the lower bound for the cycle time is now known to be reachable.

2.4 The Controller: The Marking Diagram Interpretation and Fault-Tolerant Implementation

Controlling an existing manufacturing system (MS) means constraining its evolution in order to guarantee the desired logic behaviour or/and to optimise its performances at operation. If the plant to be controlled is modelled as a PN, the control decides the firing or not of enabled transitions. Usually, not every transition can be disabled (e.g., a failure, the completion of an operation, etc.), so transitions can be classified as *controllable* or *uncontrollable*. Controllable points

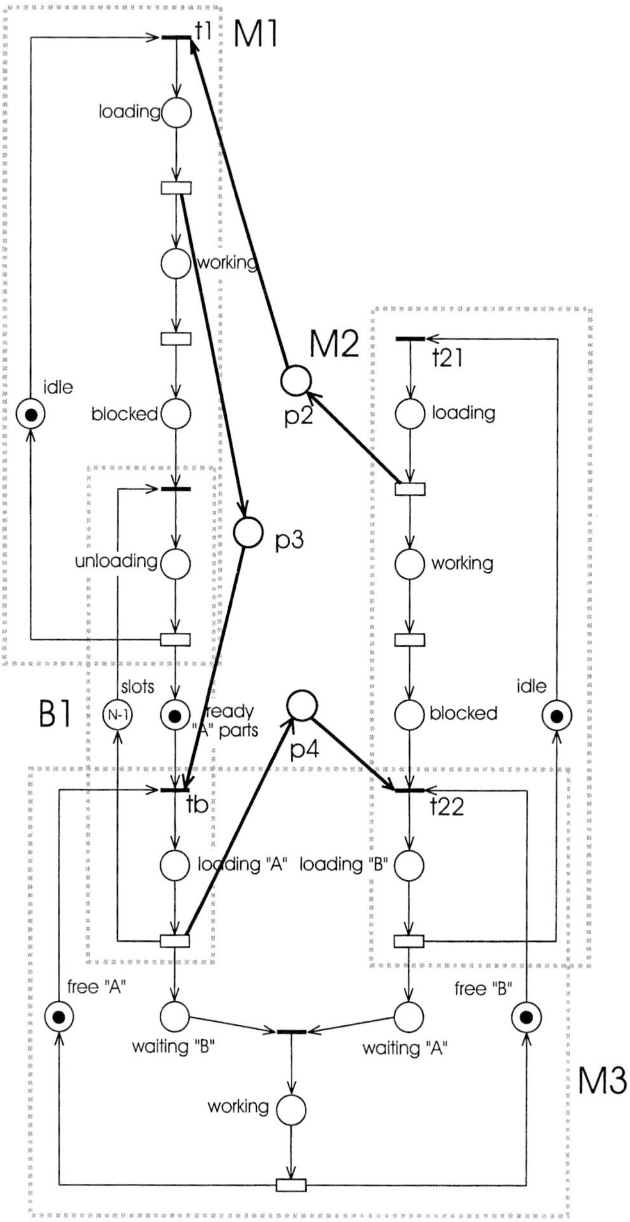

Fig. 6. Implementation of a scheduler that leads to the minimum cycle time.

are those at which the decision maker (e.g., a scheduler) influences the behaviour of the system.

Typically, concerning the logic behaviour, it is important to avoid undesirable or forbidden states, such as deadlocks, or to guarantee certain mutual exclusions, while performance control aims to maximise throughput or a more general cost function (e.g., involving also work in progress, machine utilisations, etc.), by determining the firing epoch for transitions (scheduling). PNs with an appropriate timed interpretation are very well suited to the modelling of scheduling problems in parallel and distributed systems. PNs allow to model within a single formalism the *functional*, *temporal*, and *resource* constraints. These determine the enabled transitions, and then the scheduling problem is reducing the indeterminism by deciding *when* to fire *which* transitions among the enabled ones. In scheduling theory [12] it is conventionally assumed that tasks are to be executed *only once*. Periodic or cyclic schedules [34] are seldom treated by the theory despite they abound in practice. PN scheduling techniques allow to face these problems. The same as for the analysis, enumerative, net-driven, and net-based approaches can be found in the literature. The computational complexity of scheduling problems leads in practice to sub-optimal solutions obtained using heuristics, artificial intelligence techniques, etc.

Usually, the control receives inputs from the plant, besides of emitting signals to it, so it operates in closed loop (the plant and the control are composed in parallel, in discrete event systems terminology). The same as PN can be used to model and analyse an MS, its control can often be represented within the PN formalism, perhaps incorporating an appropriate interpretation.

Coming back to the manufacturing example, if the model is meant as a specification for a logic controller, the firing of transitions must be related to the corresponding external events or inputs, and the outputs that must be emitted have to be specified. The inputs, which condition the evolution of the controller, may come from plant sensors (e.g., when *R* finishes loading *M2* it emits a signal `loaded_M2`) or from other levels in the control hierarchy (e.g., when the scheduler decides — in view of the state of the system and the production requirements — that *M1* should be loaded, it sends `sched_M1`). The outputs may command the actuators (e.g., `START_M3` initiates the assembly sequence in *M3*) or send information to other levels in the control hierarchy (e.g., `REPAIR!` raises an alarm to call the attention of maintenance staff, or an interrupt that activates automatic recovery; `B1_CONT(m)` updates the number of ready "A" parts in the production database, etc.). The PN model in Fig. 7 captures this information. Following appropriate conventions in the specification (e.g., those imposed in the definition of Grafcet [15]), a model similar to this one could be used directly as a logic controller program.

Once a suitable PN model for a controller has been obtained it has to be *implemented*. Basically an implementation is a physical device which emulates the behaviour expressed by the model. One advantage of using PNs as a specification formalism is their independence w.r.t. the precise technology (pneumatic, electronic, etc.) and techniques (hardwired, microprogrammed, etc.) of the final implementation. Presently, in MS control, programmed implementations are the most usual, running on a wide range of computer systems (e.g., industrial PC's, programmable logic controllers, etc.).

The (programmed) implementation is affected by the selected PN *formalism* (low or high level, different interpretations of the firing rule), the *algorithmic ap-*

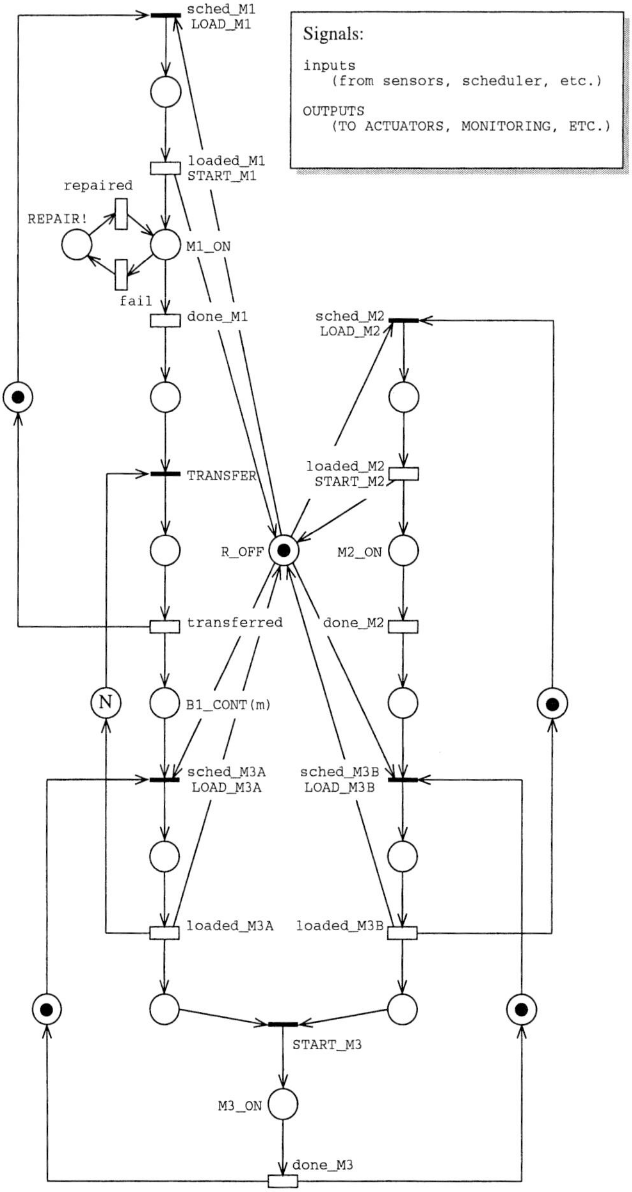

Fig. 7. A marking diagram that specifies the behaviour of the logic controller of a manufacturing cell.

proach (interpreted, where the PN model is a data structure, or compiled, where a program is obtained from the given PN; centralised or parallel/distributed schemas), and the *computer architecture* (high or low level programming language; single or multi processor).

For the case of local controllers specified by low level PNs with input and output signals (like that shown in Fig. 7), a usual choice are interpreted implementations ("token players") [61, 48]. The basic schema is a cyclic program that reads the inputs, computes the evolution of the marking, and generates the outputs once and again. A major issue is the efficient computation of enabled transitions. An example of an efficient technique for this purpose are *representing places* (see, for instance, [13]). The idea is to appropriately select one input place per transition (its *representing place*). It is always possible (perhaps after some net transformations) to classify places as either representing or *synchronisation places*, where each of the former is the representing place of all its output transitions. The marked representing places are kept in a list (we assume safeness for simplicity), that is updated at each transition firing. In each cycle, only the output transitions of marked representing places are tested for enabledness, eventually checking the marking of some synchronisation places. A possible selection of representing places for the net in Fig. 7 are all but *R idle*, *slots*, *ready* "A" *parts*, *waiting* "A", and *free* "B" (thus, these would be the synchronisation places).

The inherent parallelism captured by a PN model is somehow dismissed in centralised implementations. Diverse parallel and distributed implementations have been proposed (see, for instance, [13]). The structure theory of PNs allows to identify certain components in a given net that are useful for distributing or parallelising the implementation. Particularly, live and safe state machine components lead to cyclic sequential processes that can be directly implemented, for instance, as Ada tasks. In such case, other places can be represented as global variables, semaphores, etc. Coming back to the example, we easily identify *M1* and *M2* as sequential tasks, *M3* can be decomposed into two synchronised sequential tasks, *slots* and *ready* "A" *parts* are semaphores, and *R idle* is a mutual exclusion semaphore.

In the implementation of higher control levels, some convergence has appeared between the fields of PNs and artificial intelligence (see, for instance, [40], [60]). In this sense, transitions play the role of *rules* while the *working memory* can be split into several nodes corresponding to the respective input places. With respect to classical PNs implementations, the search for enabled transitions is carried out by the *matching phase* in the rule system, which can take advantage from the partition into local working memories. For the selection phase transitions can be grouped into *conflict sets* by inspecting the net structure, and each one can be provided with a particular resolution strategy.

An important issue when designing a control system is that of *safety*. Formal modelling and analysis tools are needed to engineer safe computer-controlled systems. For this task it is necessary to consider both the control system and its environment, for which PNs are a suitable formalism [37]. When faults can happen the controller should be able to detect them and even react appropriately degrading system's performance as little as possible.

Let us briefly concentrate here on the detection and recovery of faults in the controller itself. Several techniques have been proposed to produce safe and/or *fault-tolerant* PNs based controllers. We illustrate next one of these techniques which are supported by PNs theory: the *spy/observer* schema.

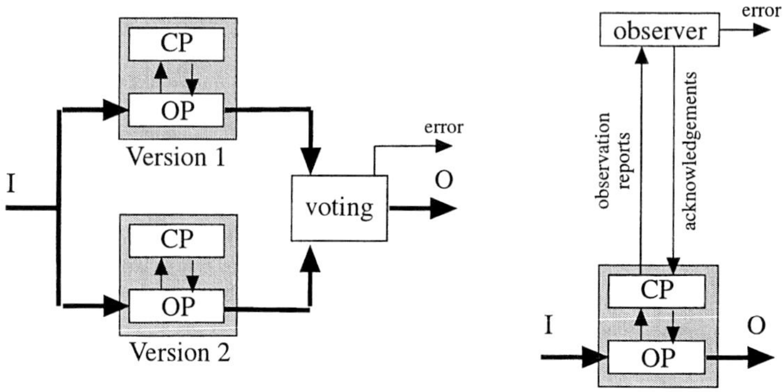

Fig. 8. Duplication versus observation.

In general, *N-version* programming techniques, that is, the controller is replicated and a voting mechanism is introduced [4], can be used. A less expensive schema is based on the idea of an *observer* [5] or *spy* [63], which accepts "normal" behaviours seen through some *observable*, or *check*, points. In Fig. 8 duplication and observation schemas are compared. The observable points are transitions whose firing is reported to the spy/observer (transitions are classified as observable or non-observable, dually to the classification into controllable and uncontrollable). The spy/observer can be modelled as a PN equivalent to the original one w.r.t. observable transitions (non observable transitions are considered silent and can be reduced). In the final implementation, the code corresponding to the spy is merged with the code of the proper controller. An observer is also employed in [19] for formal validation.

Coming back to the example, considering as observable all the synchronisation transitions in the net (i.e., those corresponding to the initiation of robot operations, initiation of a transfer from *M1* to *M2*, and initiation of an assembly in *M3*) the corresponding spy is shown in Fig. 9. (Notice that this spy is obtained applying the same reduction rules that were applied for the analysis.)

3 Modelling Some Classical Management Strategies in Manufacturing: Pull Control and Kanban

The primary goal of many manufacturing systems can be expressed in terms of the maximisation of the production rate, the minimisation of the *work-in-process* (WIP) inventory, and minimisation of the delivering delay (difference between the date of a demand and the date of serving it). The above criteria usually leads to some contradictory situations. For example, minimising WIP usually lead to higher delivering delays, what may even represent losing some selling opportunities (impatient clients).

Among the many imaginable strategies for the management of production systems, *push control* is based on the idea of "advancing" tasks relative to production as much as possible. Thus the behaviour of the production plant is

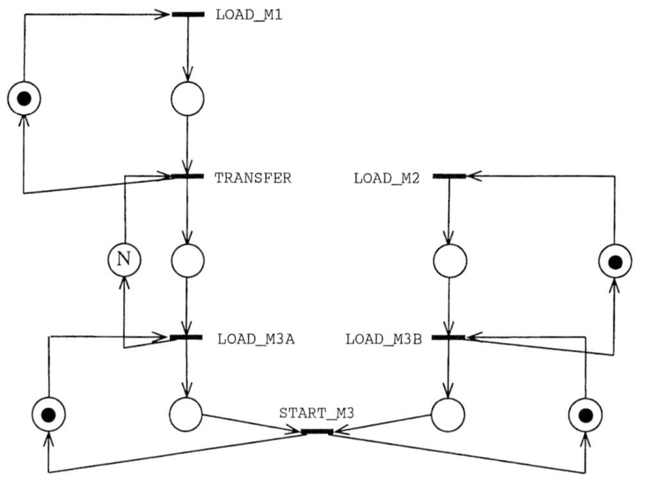

Fig. 9. A spy for the net in Fig. 1.

"externally" constrained by the raw materials available, and by the capacity of buffers for storing finished products. Under this strategy, raw materials "push the production", and delivering delays are minimised at the expense of, eventually, important WIP costs. In many cases push-type behaviours use demand forecasts to generate the production plans. On the contrary, under the basic *pull control* strategy, the customers demands trigger the production, i.e., "pull the production". Thus the WIP cost is reduced to a minimum, at the expense of more important delays for delivering, i.e., at the expense of decreasing the quality of customer service.

In the manufacturing arena, it is well known that *just in time* (JIT) approaches lead to low WIP costs. In order to conciliate the above mentioned contradictory performances, many hybrid push/pull control algorithms have been proposed in the literature. *Kanban systems* allow to deal with different kinds of those strategies, trying to smooth and balance material flows by using several appropriately controlled intermediate inventories. In essence kanbans are cards that circulate between a machine (or sequence of machines) and a downstream buffer. When a withdrawal operation liberates a position of an intermediate buffer, a card is recirculated in order to allow the production of a new part to compensate "the previous loss" in the inventory site. The number of kanbans around a machine(s)-buffer subsystem determines the buffer size. In a kanban controlled system, production of parts is triggered in response to "intermediate demands". As already mentioned in the cell manufacturing example of Sect. 2, the parts in any intermediate buffer try to "protect" the operation of downstream machines from possible interruptions of upstream machines. If the repairing time of the machine under failure is "not too big", the buffer will not empty and the failure will not affect the downstream machine. Therefore intermediate buffers "can be perceived" as condensers in electrical circuits or resorts in mechanical systems, allowing relatively uncoupled behaviours on production lines subsystems. A certain number of questions arise in order to optimise the production: Where to put the intermediate buffers?, How large?, Which strategies should be used for control?, etc.

The point here is that at a general level, Petri nets –with some timed interpretation, for example, Generalised Stochastic Petri Nets [1]– can be used to model different designs and control strategies. By using appropriate performance evaluation models, the optimisation of the strategy used to control the material flow (i.e., making the more appropriate decisions), even the tuning of its parameters, can be formally studied.

Single-output assembly manufacturing systems have usually, from the output point of view, a tree-like topology. In the manufacturing domain, it is usual to represent machines as circles and buffers as triangles (Fig. 10). The (output) root of the tree represents the finished goods buffer. In order to simplify the presentation, let us assume a single level assembly stage and two previous manufacturing stages (Fig. 11).

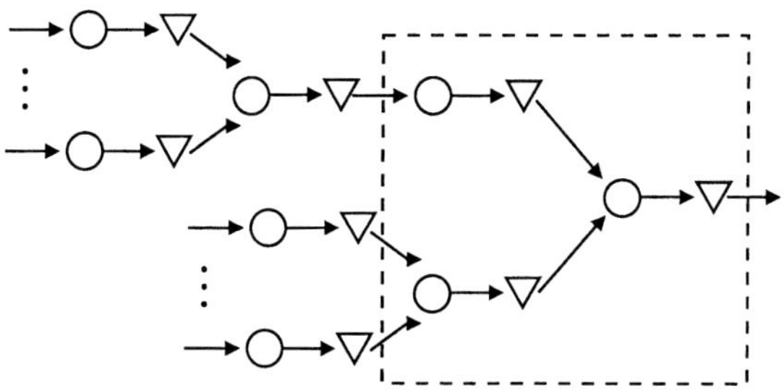

Fig. 10. Topology of an assembly manufacturing system: machines are shown as circles and buffers as triangles.

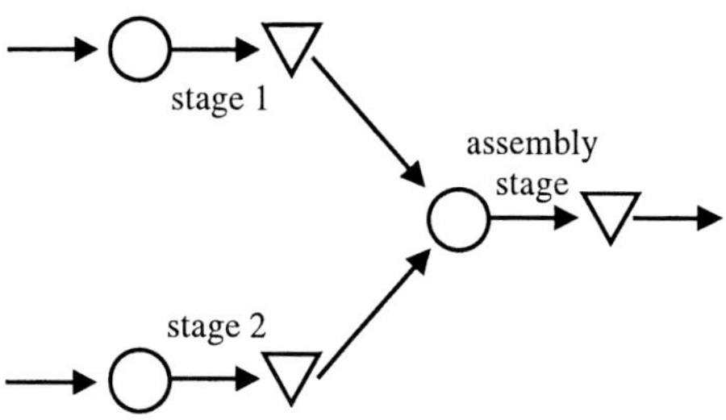

Fig. 11. Two manufacturing stages (with their buffers) followed by an assembly stage (with the finished products buffer).

The basic schema of a production stage can be easily described in PNs terms by means of the connected marked graph in Fig. 12(a). According to that, production stages are composed of a raw parts container (*raw*) synchronised with

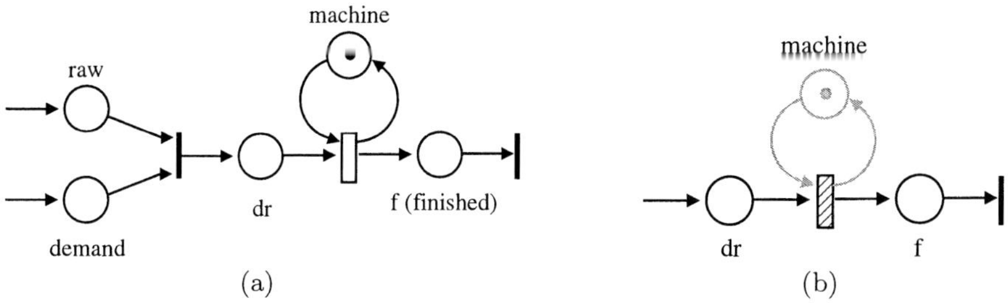

Fig. 12. Basic schema of a production stage.

a demand for production (*demand*), followed by the waiting queue and machine working place (*dr*), and the place representing the single machine (*machine*); and finally its output buffer of finished parts (*f*). The transition in the self-loop of the machine is timed (processing time of a part). Thus the utilisation rate of the machine is given by the probability of non null marking in place *dr* (at least one part needs to be processed).

It is common in certain cases to assume that there are always enough raw parts. This means that place *raw* can be removed because it is not a constraint any more (it is implicit: i.e., it is never the unique that forbids the firing of its output transition). In doing so, because the transition between places *demand* and *dr* is immediate, both places can be merged into a single one (we keep the name *dr*). In Fig. 12(b), the simplified model is presented. It will be a basic building block for the models of this section. In order to simplify the drawing of nets, in the sequel place *machine* will be removed, while it is assumed that the firing semantics of the corresponding transitions is *single server* [8]. Transitions with single servers semantics will be graphically denoted here as dashed timed transitions. Observe that at this level it is assumed that the machines do not fail.

A basic pull control system (*base stock control system*, BSCS [11]) is presented in Fig. 13. It consists of two production stages (with $k1$ and $k2$ parts finished in stage 1 and stage 2, respectively), feeding an assembly stage (initially with $k3$ finished parts). When a customer's demand appears, places *dr1* and *dr2* receive a (new) token, in order to produce another part for each stage. Customers demand allows to serve finished parts, represented by tokens in place *f3*, initially marked with $k3$ tokens. A main problem in this basic schema is that the limitation of the WIP is not assured in any of the three stages (two for production and one for assembly, in the present case). It is not difficult to see that *under saturation of customers demands* (i.e., under the hypothesis that there exists an infinite number of customers demands), the production cycle time (the inverse of the throughput) is bounded by the slower of the three machines:

$$\theta = max\{\theta 1, \theta 2, \theta 3\}$$

Simultaneous kanban control system (SKCS) and *independent kanban control system* (IKCS) are modelled in Fig. 14 and Fig. 15. As happened before, in

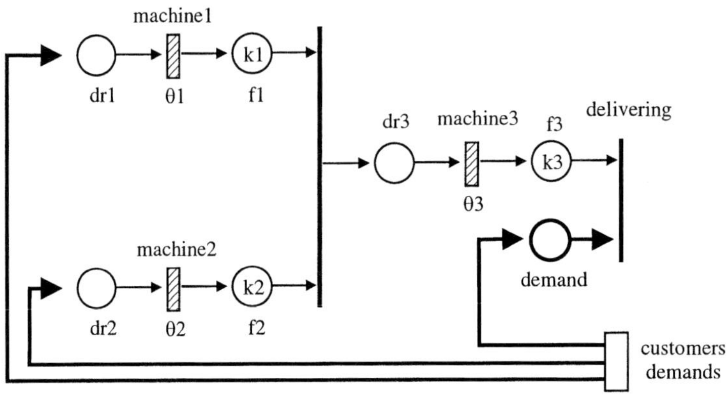

Fig. 13. Production of parts A and B (stages 1 and 2) and final assembly (stage 3), with a basic stock (pull) control system (BSCS) and assuming single server semantics.

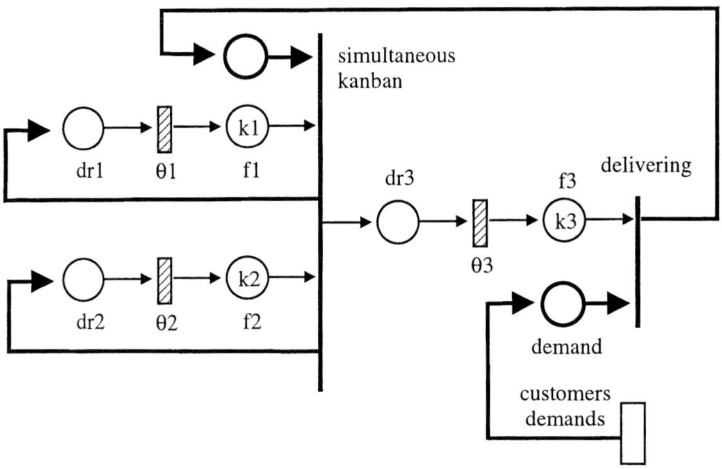

Fig. 14. Simultaneous kanban control system (SKCS).

both cases two production stages are followed by an assembly stage. Even under saturation of customers demands, the capacity of the stages are $k1, k2$ and $k3$, respectively, while the production cycle time under deterministic timing is once again θ, i.e., defined by the slower machine (because all ki are greater than zero). Under stochastic timing, θ is a lower bound for the cycle time (i.e., $1/\theta$ is an upper bound for the throughput).

The difference among SKCS and IKCS is that the first one feeds simultaneously the assembly stage and the new production order for the (two) previous stages. In the second case, separate kanbans feed stages 1 and 2, while feeding the assembly stage is automatic, when appropriate parts exists (in *b1* and *b2*).

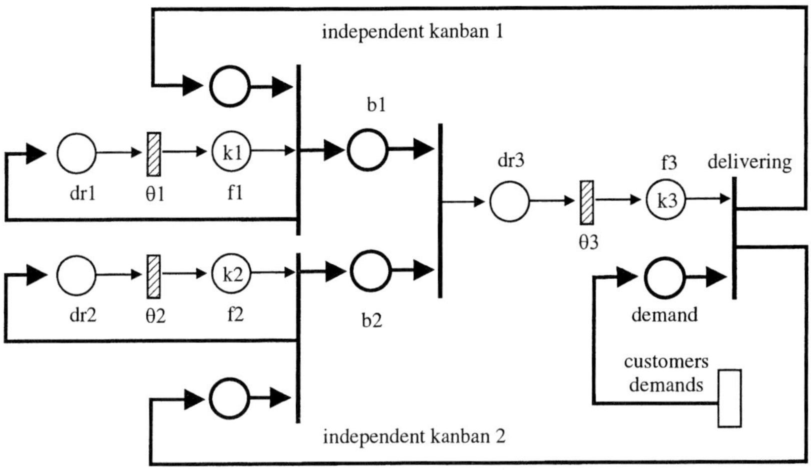

Fig. 15. Independent kanban control system (IKCS): Kanbans are independently generated for machine 1 and machine 2.

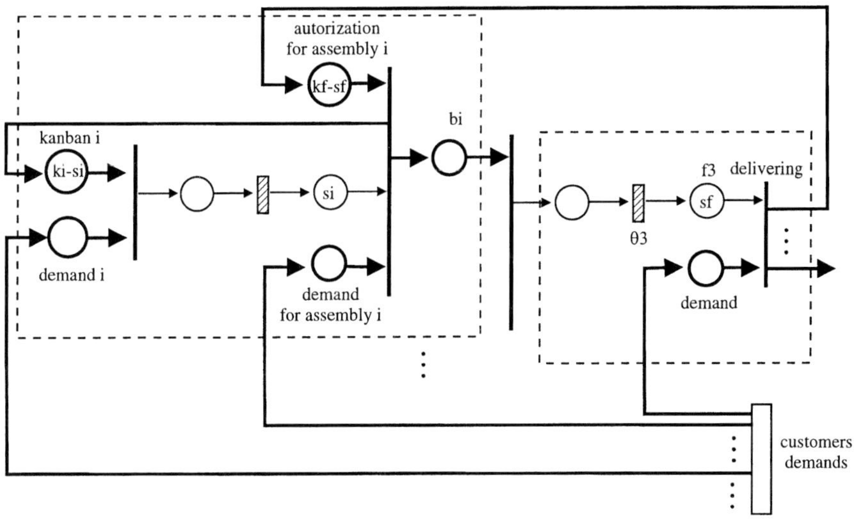

Fig. 16. Independent extended kanban control system (IEKCS).

Obviously, in transient behaviours, the independent case can be better than the simultaneous one.

A more elaborated kanban system is presented in Fig. 16. It is the so called *independent extended kanban control system* (IEKCS) [11]. Under saturation of customers demands it behaves exactly like the above schemes (SKCS and IKCS). Nevertheless, in this case different kanbans send simultaneously requests for the production of primary parts (in stage 1 and stage 2), for an assembly to be done, and for the delivery of a finished part. This may lead to some interesting behaviours, potentially reducing the WIP, while keeping a good reactivity to demands.

These control policies have been simulated assuming in all cases that $\theta 1 = 0.5$, $\theta 2 = 1$, $\theta 3 = 0.4$, $k1 = 1$, $k2 = 1$, $k3 = 2$ and, for IEKCS, $s1 = s2 = 0$ and $s3 = 1$. A burst of 5 simultaneous demands is simulated at 15 t.u. The results for the different control systems in Figs. 13-16 are represented in Fig. 17, where (a) shows the marking of place *demand* (unsatisfied demand), (b) shows the marking of place *f3* (complete products in stock), and (c) shows the throughput of the assembly station. Because the "delivering" transition is immediate, the unsatisfied demand at 15 t.u. is equal to 5 minus the products in stock: 2 for BSCS, 3 for SKCS and IKCS, and 4 for IEKCS. In this case, BSCS, SKCS and IKCS need more or less the same time to "satisfy the demand" (the marking of the place *demand* returns to zero), while IEKCS is the last one. However, the stock of complete products in absence of demand is much larger under BSCS (3), than under IEKCS (1). With respect to the throughput, SCKS, IKCS and IEKCS work on demand, so the throughput is zero before the demand. Under

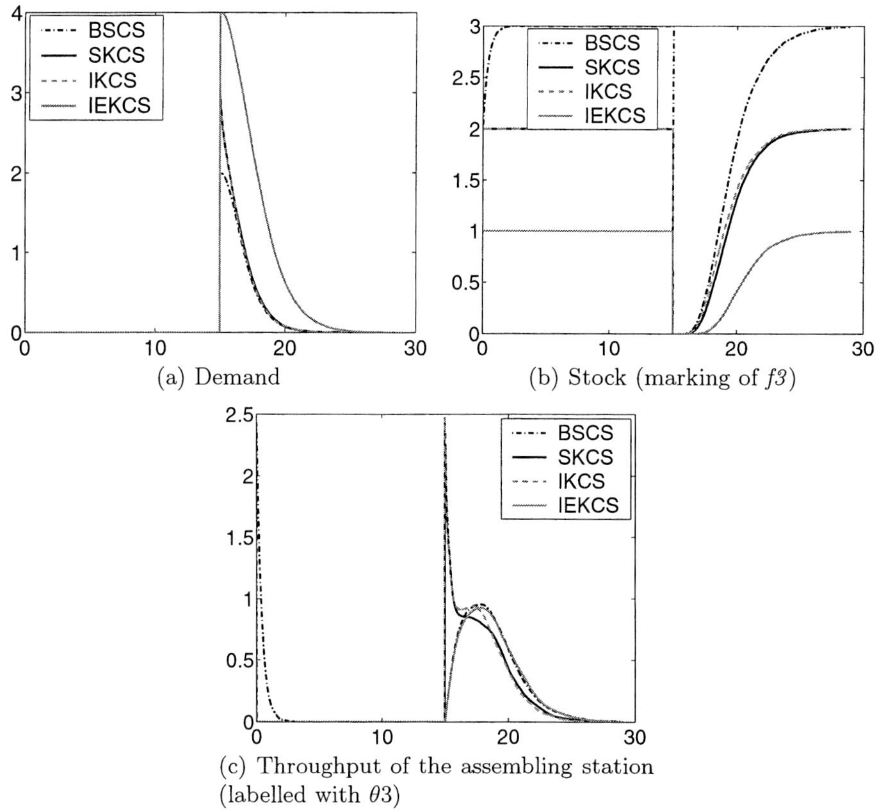

(a) Demand

(b) Stock (marking of *f3*)

(c) Throughput of the assembling station (labelled with $\theta 3$)

Fig. 17. Simulation of the different control policies in Figs. 13-16.

BSCS, a first outburst of the production appears, since the intermediate stocks *f1* and *f2* are used to produce the final assembly. In other words, the system tries to complete as much products as it can, instead of keeping stocks of intermediate elements. That is the reason why although the stock under BSCS is 3 and under IEKCS is only 1, it does not take three more times to satisfy the demand in the latter case.

Many other schemes of this type can be imagined. The important point at this level is that modelling with PNs is frequently quite straightforward (if control strategies do not depend too much on particular data), and analysis can provide useful information about the behaviour of the intended control strategy.

4 A Coloured Model for a Car Manufacturing Line

A relatively frequent characteristic of production systems is the existence of *symmetries* due to the presence of subsystems that behave "in a similar way". Coloured PNs allow to exploit these symmetries and generate a more compact model. Coloured Petri nets can also be extended, as in [30, 31], or abstraction on the formalism (i.e., the underlying PN model) can be done in application oriented interfaces, as in [64]. Here just basic coloured Petri nets will be used to model some examples.

4.1 A Car Manufacturing System

The following example shows a coloured PN model of a realistic MS (part of a flexible workshop of a car factory), taken from a case study [39].

The FMS shown in Fig. 18 consists of:

- Several workstations (*S1* to *Sn*). All the workstations behave in a similar way: car bodies to be processed are loaded in table L (input buffer of capacity one), then transferred to table P (actual processing), and then transferred to table U for unloading (output buffer of capacity one). For simplicity, we disregard the nature of the precise operations performed in the station, and therefore, we represent a model of a generic workstation. A station behaves as a pipeline with three stages: L, P, and U, represented by the corresponding places, which can be active simultaneously. The complementary places FL, FP, and FU represent, when marked, that the respective stage is free. The colour domain of all these places is $\{1, \ldots, n\}$ for the stations. A token of colour i in place P represents that workstation *Si* is processing. Transferring a processed part from table P to table U in workstation *Si* requires one i-token in P and FU, and puts one i-token in U and FP.
- An unidirectional transport system, consisting of several roller tables (*T1* to *Tn*). Car bodies enter the system in table *T1* and leave it from *Tn*, after being processed in one station (the one decided by the scheduler). The model for this transport system consists of two places, T and FT, for the occupied and free tables, and transitions to represent the input or output

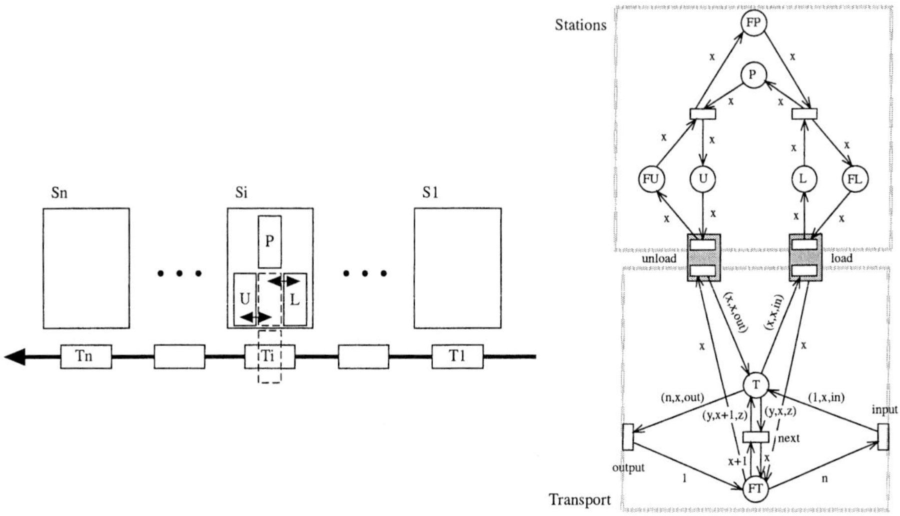

Fig. 18. A flexible workshop that processes car bodies in several stations, and its coloured PN model.

of a car body, a movement to the next table, and the load or unload of a station. The colour domain of FT is $\{1, \ldots, n\}$ for the tables, and the colour domain of T is $(\{1, \ldots, n\}, \{1, \ldots, n\}, \{in, out\})$, where the first field identifies the table, the second one the destination station of the car body, and the third one the status of the car body (*in* when not yet processed and *out* when ready to leave the cell). Notice that, at the firing of transition *input*, a destination station is assigned to the incoming car body. In net terms, this means solving a conflict among the different firing modes of the input transition. The destination is determined by the scheduler, possibly taking into account the state of the system and the production requirements. That is, the scheduler (placed at a higher level) controls the behaviour of the coordination model represented by the coloured PN.

The complete net model is obtained merging the *load* and *unload* transitions of the submodels for the workstations and the transport system. The loading of *Si* from *Ti* is represented by the firing of transition *load* in mode i: it consumes a token (i, i, in) from T and an i-token from FL and puts i-tokens in L and FT. Similarly for the unloading, where the "status" colour of the token deposited in T is *out* indicating that the car body in the corresponding table has been processed.

4.2 On the Control of the Production Line

Besides avoiding deadlocks, let us consider a control policy to improve the performance.

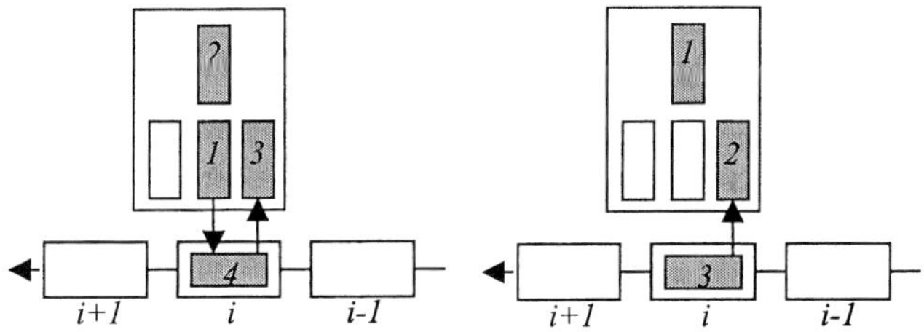

Fig. 19. (a) Complete deadlock (b) Temporary deadlock.

Analysis of this system proves the existence of deadlocks: when all the tables in a given station are occupied and a car body is waiting in the corresponding table of the transport system to enter this station, a deadlock is reached, see Fig. 19(a). The deadlock can be avoided by making sure that no more than three car bodies scheduled for the same station are present in the system at any time. This can be enforced by limiting the number of firings of *input* in a given mode w.r.t. the number of firings of *output* in that mode. This is implemented by place O (for orders) in Fig. 20(a), whose colour domain is $\{1, \ldots, n\}$ for the destination stations, marked with three tokens of each colour.

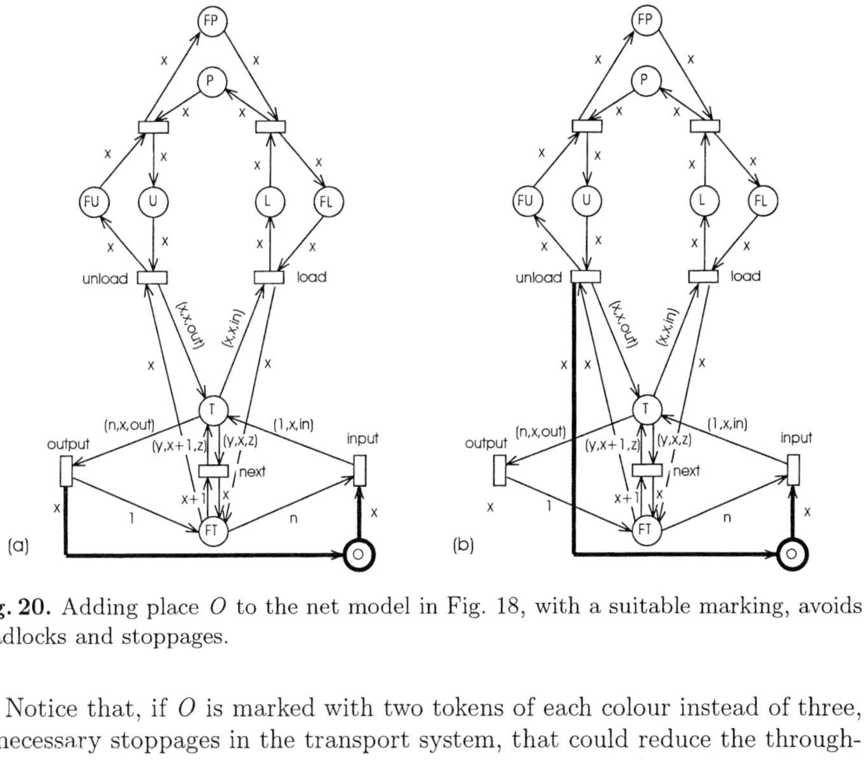

Fig. 20. Adding place O to the net model in Fig. 18, with a suitable marking, avoids deadlocks and stoppages.

Notice that, if O is marked with two tokens of each colour instead of three, unnecessary stoppages in the transport system, that could reduce the throughput, are avoided. These stoppages appear when a car body waits in front of

Table 1. Throughput comparison for the system in Fig. 20(a), if place O is marked with two or three tokens of each colour.

Mean processing time	Observe the output			Observe the unload		
	Throughput		Increase	Throughput		Increase
	Three orders	Two orders		Three orders	Two orders	
1	0.2971	0.2984	0.45 %	0.2969	0.3002	1.11 %
5	0.2434	0.2763	13.54 %	0.2378	0.2809	18.14%
10	0.1669	0.2173	30.24 %	0.1617	0.2210	36.66%
15	0.1227	0.1671	36.17 %	0.1189	0.1690	42.12%
20	0.0964	0.1331	38.07 %	0.0935	0.1341	43.45%
50	0.0418	0.0578	38.51 %	0.0406	0.0579	42.70%

its destination station because this station is processing and the load table is occupied, see Fig. 19(b). We cannot proceed to load the third car body until processing is completed, the processed car body is transferred to the table U, and the car body in table L is transferred to table P. In the meanwhile, other car bodies may be prevented from advancing to their destination beyond that station.

The first columns in Table 1 (observe the output) compare the steady state throughput of these two control policies for different processing times in a three cells workshop. All the cells are assumed to be equal, and the car bodies are sent to all of them with the same probability. The transitions are assumed to follow exponential distributions, of mean one for all the transport operations (both inside and outside the cells). It can be seen that, if the processing is fast with respect to the transport, the two policies are more or less equivalent. However, if the processing takes “much time”, the throughput is better under the most restrictive policy. Intuitively, since the processing needs more time than the transportation, it is better to be sure that the parts can advance till the processing station.

Finally, in the above control it was assumed that the scheduler controls transition *input* and observes just transition *output*. If also the occurrences of transition *unload* were observed, it might be possible to improve the performance of the control policy by allowing a limited number of *unprocessed* orders in the system (see Fig. 20(b)).

Table 1 compares the results of both control policies for the previous example. It shows that if the number of orders allowed in the system for each machine is 2, the throughput increases slightly when the *unload* transition is observed. However, if three orders are allowed, the throughput decreases. Intuitively, with at most three orders for each machine the system was already saturated, and allowing a greater number of car bodies only makes it worse.

5 On a Production Line for Ovens

This section describes a new manufacturing system where the set of production orders compete for a set of physical resources. The system is quite similar to the one in the previous section. Here, the attention is focused on how to obtain the coloured Petri net model, by first modelling the plant layout taking into account the possible ways parts can flow through the system and then imposing to each flowing part the execution of its associated process plan, which needs of model refinement. Finally, it will be shown how to prevent deadlocks and how the deadlock related control approach can be improved taking a more abstract point of view.

5.1 System Description

Fig. 21(a) depicts the structure of a flexible manufacturing cell for the production of microwave ovens (a more detailed description can be found in [24]). The cell has an entry station, `EntryStation`, an exit station, `ExitStation` and n workstations, $w_0, w_1, \ldots, w_{n-1}$. These workstations are loaded and unloaded by a circular conveyor belt with a continuous movement in a unique direction. The manufacturing of each oven is made according to its process plan. There are several scales and models of ovens with their respective process plans. The components of an oven arrive at `EntryStation` after having been previously pre-assembled; once an oven reaches that point, it is fixed to a pallet that will be inserted into the transport system when possible. One of such loaded pallets must visit a set of workstations, according to the process plan of the part it contains, and then leave the system through the `ExitStation`. The pallet goes then to the pallet store, to be reused. The system has a total of K pallets.

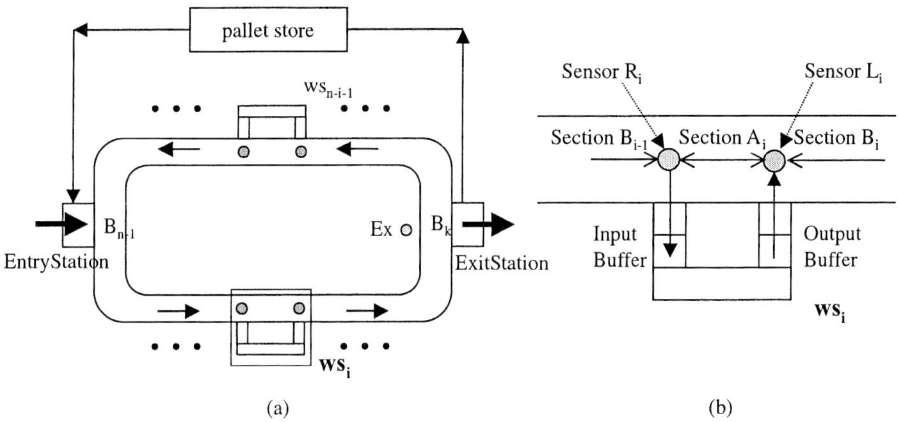

Fig. 21. a) General plant representation of a cell for the manufacturing of microwave ovens. b) Detailed view of the structure of a workstation and its related sections.

As detailed in Fig. 21(b), each workstation w_i has an input buffer I_i and an output buffer O_i. Both consist of two roller tables, each with capacity for one pallet. The pallets in each buffer follow a FIFO policy. A workstation can operate with one pallet at a time. In order to control the system, the conveyor belt has a set of sensors distributed as shown in Fig. 21(b): $R_0, L_0, \ldots, R_{n-1}, L_{n-1}$ and Ex. Associated to these detection points there are mechanisms that, under the control of the workshop coordination system, allow to carry out the following transfer operations, schematised by means of arrows in Fig. 21: introduction of a pallet from `EntryStation`, exit of a pallet from the `Ex` point towards `ExitStation`, loading of a pallet in workstation w_i by transferring it from position R_i to the input buffer of w_i, I_i, unloading of a pallet from the output buffer of w_i, O_i, to point L_i of the conveyor belt. Each A_i or B_i section will have its own capacity, which corresponds to the number of pallets the section can hold.

5.2 A Coloured Petri Net Model of the Coordination System

A first approach to the modelling of material flow is shown in Fig. 22. Let us explain the main elements in the model.

The transport system: The set of states a pallet can be in the transport system is modelled by means of places B, R, A, L. Place B models the set of B sections. Place A models the set of A sections, while places R and L model sensor points between sections B_{i-1} and section A_i and between sections A_i and B_i, respectively. The colour domain of all these places is $WS = \{w_0, \ldots, w_{n-1}\}$, the set of workstations. The initial marking of each one of these places is the multi-set 0, which means that, at the initial state, no pallet is inside the system. Transitions t_{in} and t_{out} model the actions by which a pallet with a new oven enters the system and a pallet with a terminated oven leaves the system, respectively. Ordinary (non-coloured) place AP models the set of free pallets, whose initial marking is K, the number of available pallets. In the system, it is assumed that `EntryStation` loads pallets into section B_{n-1} and that `ExitStation` unloads pallets from section B_k.

Places BC and AC, whose colour domain is also WS, model the capacities of B_i and A_i sections, respectively. The initial marking of BC is the multi-set $\sum_{i=0}^{n-1} b_i \cdot w_i$, being b_i the capacity of section B_i. Analogously, the initial marking of AC is the multi-set $\sum_{i=0}^{n-1} a_i \cdot w_i$, being a_i the capacity of section A_i. Places CR and CL represent that only one pallet can be in sensor points R_i and L_i, respectively. The initial marking of both places is $\sum_{i=0}^{n-1} 1 \cdot w_i$.

Transition t_{br} models a pallet reaching an R_i sensor (the function labelling the arc (t_{br}, R), $w@1$, represents the addition of 1, modulo the number of sections, n). Transition t_{ra} models a pallet entering an A_i section, transition t_{al} models a pallet reaching an L_i sensor. Finally, transition t_{lb} models that a pallet reaches a B_i section.

Transition t_{ls} (t_{us}) models a pallet being loaded into (unloaded from) a workstation.

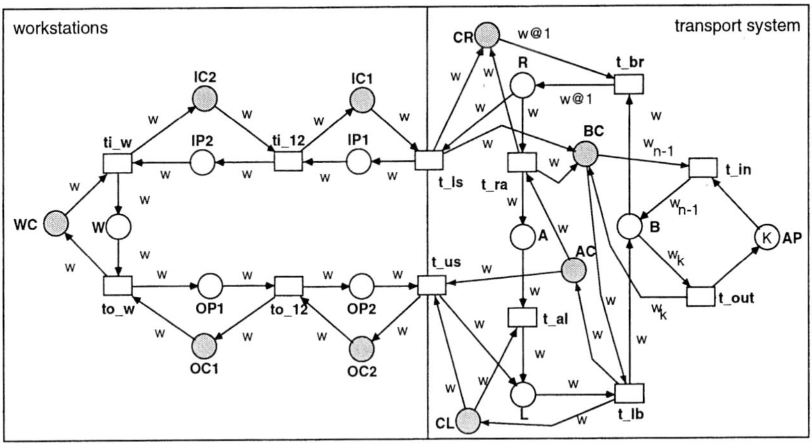

Fig. 22. A coloured Petri net model of flow of pallets in the system in Fig. 21.

The set of workstations: A pallet loaded into a workstation, by means of the firing of transition t_{ls}, must, successively, visit the two input buffer positions (places $IP1$ and $IP2$), to be processed in the workstation (place W), and visit the two output buffer positions (places $OP1$ and $OP2$). The initial marking of any of these places is the multi-set 0: there is no pallet in any workstation.

Places $IC1, IC2, WC, OC1$ and $OC2$ impose the capacity constraints of being able to have at most one pallet in each one of the components of a workstation. The initial marking of any of these places is $AW = \sum_{i=0}^{n-1} 1 \cdot w_i$.

It is important to notice that, even if all the transitions in the model represent system actions that change the system state, from the control point of view two kinds of transitions are considered:

- Transitions whose firing is observable but not controllable. This is the case of $\{t_{br}, t_{ra}, t_{al}, t_{lb}, ti_{12}, to_{12}\}$. Since the conveyor has a continuous movement the firing of one of such transitions will be realised when a pallet reaches or leaves the corresponding sensor. The events can be noticed and thus the system state can be updated in the model.
- Transitions whose firing is decided and executed by the control system (controllable transitions). These are the transitions that can be controlled in order to ensure that every incoming part will be processed according to its associated process plan, and also to impose some control policy in order to ensure some desired properties, as deadlock freeness or to impose some scheduling policies. This set of transitions is composed of $\{t_{ls}, t_{us}, t_{in}, t_{out}, ti_w, to_w\}$.

5.3 Inclusion of the Process Plans

Each oven that enters the system must execute its associated process plan, which consist of a sequence of operations to be executed in the system workstations.

This sequence is described by means of a sequence of pairs (o, w), where o defines the operation to be executed, and w the workstation where such operation must be done. The sequence of operations for an oven has been pre-established by the system controller before loading the oven into the system. In the specification level considered here, which concentrates on the material flow control, it is possible to make abstraction of the operations to be executed, describing the process plan as the ordered sequence of workstations to be visited by the oven. Therefore, a process plan will have the following form: $p = (w_p^1; w_p^2; \ldots; w_p^{n_p})$, where each w_p^i, $i \in \{1 \ldots n_p\}$, belongs to WS.

There exists a set of predefined process plans $PP \subset WS^+$. Each part that enters the system has an associated process plan belonging to PP. The first element in the ordered sequence of workstations in the process plan corresponds to the first workstation to be visited. In order to identify the state in the processing of a part in the system, tuples of the form $(p, i) \in PP \times \mathbb{N}$ will be used: p identifies the process plan, while i identifies the position in the process plan sequence of the next workstation to be visited. For instance, when an oven whose associated process plan is $p = (w_p^1; w_p^2; \ldots; w_p^{n_p})$ enters the system, it will be identified by means of the token $(p, 1)$, meaning that w_p^1 is the next workstation to be visited. When the oven is processed in w_p^1, the tuple identifying the oven will be $(p, 2)$; when terminated, it will be identified by means of $(p, n_p + 1)$.

According to this codification of the processing state of an oven in the system, the model in Fig. 22 must be transformed. Since the system layout is still the same, only colour domains and functions in the arcs have to be changed. If in the initial model a token in place A, for instance, was of the form w, just indicating the concrete A-section where the pallet was, now a token in such place will be of the form (p, i, w) indicating that there is a pallet in w A-section, containing an oven whose associated process plan is p and that has to next visit workstation w_p^i. Accordingly, the colour domain of places modelling physical locations that can contain pallets with ovens is $PP \times \mathbb{N} \times WS$.

Notice that, in order to forbid a pallet to enter a workstation that is not its next destination, predicate $[w_i^p = w]$ has been associated to transition t_{ls}. Also, predicate $[i = n_p + 1]$ has been associated to transition t_{out} so that only pallets containing ovens whose process plan has been completely executed can be unloaded from the system. Notice also that the firing of transition to_w transforms a token of the form (p, i, w) into $(p, i + 1, w)$, which corresponds to changing the next destination workstation for the considered oven.

The resulting model is shown in Fig. 23, *where places modelling resource capacity constraints have not been represented*, for the sake of clarity. In any case, they are exactly the same as in Fig. 22.

5.4 Preventing Deadlocks. A First Solution

If the control model in Fig. 23 is directly implemented, the system can reach deadlock situations. Let us consider, for instance, a reachable state in which a workstation w_i is full (input and output buffers are full and the workstation is also processing an oven) and also the transport system is full of pallets that must

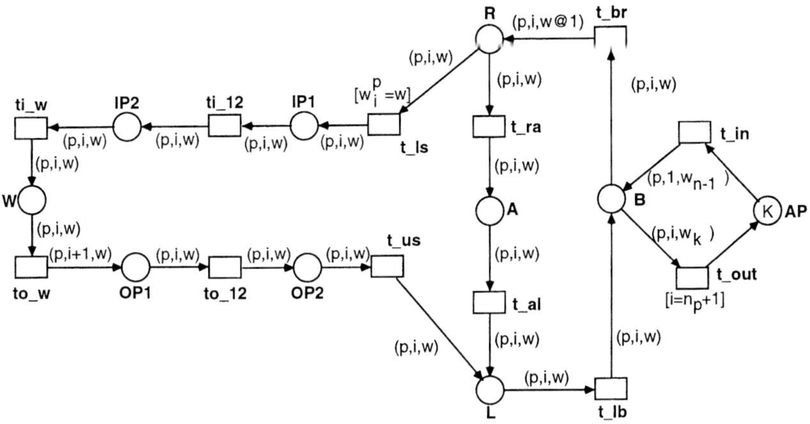

Fig. 23. A coloured Petri net model of the system in Fig. 21 once the process plans are considered (capacity constraints have not been represented, for the sake of clarity).

enter workstation w_i. In this situation, no new pallet can enter the system, no pallet in the conveyor can be loaded into workstation w_i and no pallet can leave it since the conveyor is full. All the deadlock situations are related to states in which full stations require to unload pallets to the transport system, which is full of pallets that must enter a full workstation.

An easy way of preventing such situations consists in ensuring that no more than five pallets inside the system need to visit a given workstation. This is the deadlock control implemented in the following. The implementation is based on the following function, called *workstation requirements*, and defined as follows. Let $p = (w_p^1; w_p^2; \ldots; w_p^{n_p})$ be a process plan, and let $i \in \{1, \ldots, n_p + 1\}$ be an index associated to p. For the tuple (p, i) the following multi-set of workstations is defined: $wr(p,i) = \sum_{j=0}^{n-1} \lambda_{p_i}^j \cdot w_j$, where $\lambda_{p_i}^j$ is 1 if $w_j \in \{w_p^i, w_p^{i+1}, \ldots, w_p^{n_p}\}$ (in the case of $i = n_p + 1$ the addition is made over an empty set of workstations, and it is assumed to be the empty multi-set). Notice that, in fact, $wr(p,i)$ is the characteristic function of the workstations to be visited by the oven from the index i until the associated production plan is terminated. Notice also that if $i_1 < i_2$, then $wr(p, i_1) \geq wr(p, i_2)$.

In order to implement such control policy in the Petri net model place DPS (Deadlock Prevention Solution) is added, whose colour domain is WS and whose initial marking is the multi-set $\sum_{i=0}^{n-1} 5 \cdot w_i$ (Fig. 24 shows the Petri net elements to be added to the model in Fig. 23). For a pallet that enters the system (firing transition t_{in}) with an oven whose associated process plan is p, the set of possible workstations the pallet must visit is "reserved". This is implemented by means of the function $wr(p, 1)$ labelling the arc (DPS, t_{in}). Moreover, each time a pallet leaves a workstation, if this oven does not need to visit that workstation again in the future, the reservation must be released. This is implemented by means of the arc (t_{us}, DPS). As noticed previously, the label $wr(p, i-1) - wr(p, i)$ is properly defined since $i - 1 < i$. Notice also that the control is related to transitions t_{in} and t_{us}, which are both controllable.

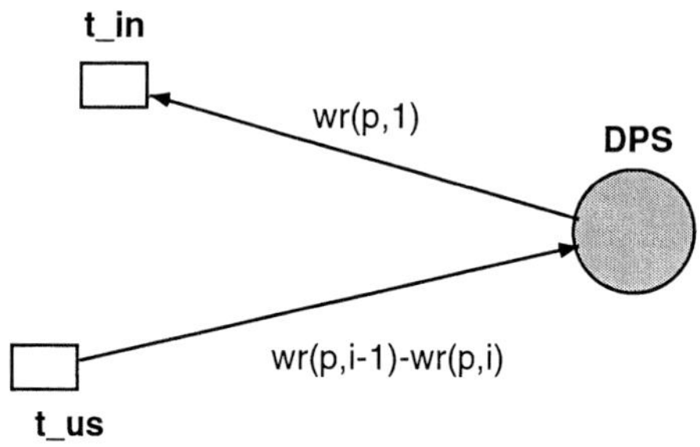

Fig. 24. The implementation of a deadlock prevention solution for the considered system.

5.5 Preventing Deadlocks. A More Accurate Solution

The solution for deadlock prevention just proposed is of the same type as in Sect. 4. However, taking a detailed look at an abstract view of the underlying non-coloured model a more accurate solution can be adapted. Let us, for instance, consider a process plan $p = (w_1; w_2)$. Taking into account that with an adequate control every pallet in the transport system can reach any workstation and also that every free position in the transport system can be used for the downloading of any workstation, the ordinary Petri net in Fig. 25 is an abstract view of the processing of a part whose process plan is p.

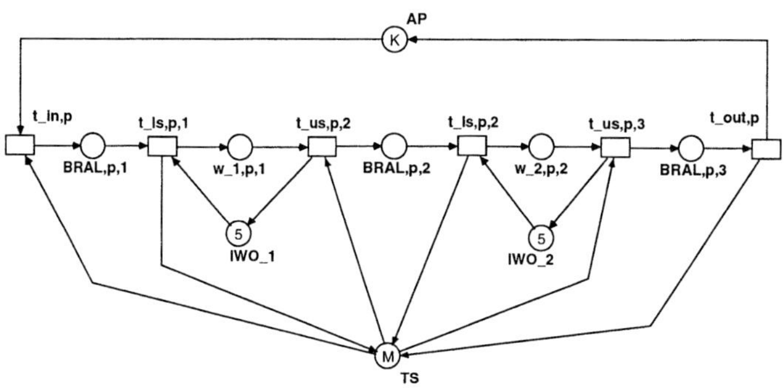

Fig. 25. An abstract point of the processing of a part whose associated process plan is $p = (w_1, w2)$.

The meanings of the different elements in the model are the following. Place TS in an abstraction of the whole transport system; its initial marking is $M = \sum_{i=0}^{n-1} a_i + b_i$, the total number of available locations for parts in the conveyor. Place IWO_1, whose initial marking is 5, models the total capacity of workstation w_1, considering in it the input buffer, the output buffer and the workstation itself. Places "$BRAL, p, *$" model the different states of a part of type p in the transport system. Transitions "$t_{ls}, p, *$" model the different firings of transition t_{ls} when the processing of a part of type p advances. Analogously for transitions "$t_{us}, p, *$". Transition t_{in}, p (t_{out}, p) models the loading (unloading) of part of type p into (from) the system. Considering the models of all the involved process plans, a final model will be obtained by means of the fusion of the places in the models of the process plans corresponding to the capacities of the resources they share.

The resulting Petri net belongs to a class of resource allocation systems (RAS) which have been intensively studied in the literature, and for which a wide set of different approaches for deadlock prevention and avoidance have been developed. [23, 58] use an structure-based approach to synthesise the deadlock-freeness related control. In both cases, the Petri net structure (siphons) is used to characterise deadlock problems and also to obtain generalised mutual exclusion solutions that forbid deadlock related stated. These mutual exclusion constraints are implemented by means of the addition to the former uncontrolled model of new places and arcs. Any of the solutions can be used to control the system here considered. The implementation can be done as in [22], in an analogous way as in the previous subsection, by means of the addition of a control place (as is the case of place DPS previously used) and some related labelled arcs.

The use of any of these last approaches will yield, in general, more permissive solutions than using the approach in section 5.4 (the less states of the uncontrolled system a control policy allows, the less permissive it is). However, they have the drawback that since the control is based on a deep use of the abstract unfolded model and the competition relations among the involved process plan models, the addition of new process plans will require the re-computation of the necessary control, making the approach less adaptable to changes in the production than using the approach in section 5.4.

6 Additional Examples: On Modelling and Analysis

Among the advantages of formal modelling are primarily the rational, non-ambiguous, "complete" description of behaviour and the capability of analysis. In the actual state of the art, analysis is not always straightforward, even "efficient" techniques may not be known.

In some cases, analysing the "natural" model an engineer produces is not an easy task. This is due to the fact that the resulting model can be complex. Analysis techniques (mainly those techniques that do not use the reachability graph or simulation, such as structure-based techniques or transformation techniques, for instance) have some limitations for general Petri net models, becoming more difficult when using high level Petri nets. In this section two new practical cases are described. The first one uses ordinary Petri net models, but there are not techniques able to control the natural model (deadlock-freeness related control is once again the objective). This problem is then solved by the transformation of the initial model into one with an equivalent behaviour, and for which control techniques exist. The second case uses a different modelling approach, based on the *Nets-within-Nets* paradigm as used in [62]. This paradigm falls into the object-oriented modelling approach.

6.1 Modelling and Deadlock Avoidance for a Two Cells Manufacturing System

The objective is to model and control, avoiding deadlock states, the manufacturing system in the Department of Computer Science and Systems Engineering of the University of Zaragoza. To do that, ordinary Petri nets have been selected as the modelling tool. It could have been modelled also using coloured PNs, as the previous examples. However, since the technique that is being used for the control needs a non-coloured model, it has been decided to use ordinary nets instead of building a coloured model and unfolding it afterwards.

The system and the modelling approach. Figure 26 depicts the plant of the manufacturing cell, consisting of six machines (M1 to M6) that process the components, one buffer with place to store up to 16 intermediate products, and two robots (R1 and R2). The process is organised in two rings, with the buffer connecting them. A final product (Fig. 27) is composed of a base on which three cylinders are set. The base may be black or white, and there are three types of cylinders: cylinders that are composed of a case, a piston, a spring, and a cover (called "complete" cylinders), cylinders with just a case and a cover (called "hollow" cylinders), and cylinders in one piece (called "solid" cylinders). The cases and the solid cylinders may be red, black or metallic. Bases, pistons, springs, covers, cases, and solid cylinders are considered as the raw materials. An unbounded amount of raw material is assumed to feed the system. A set of 330 different products can be composed using these materials.

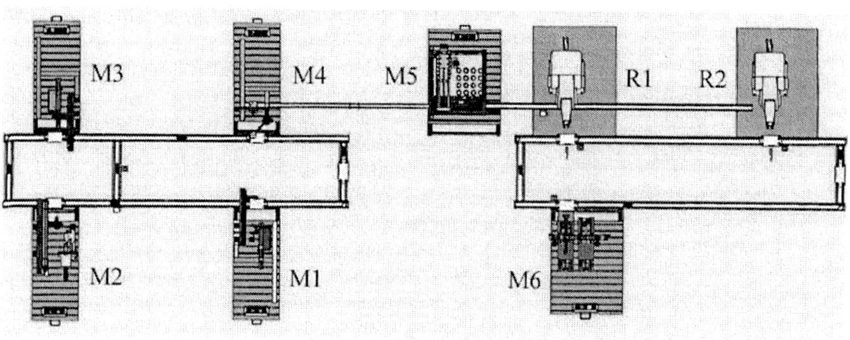

Fig. 26. A plan of the physical system.

The processing goes as follows: machine M1 takes a case from a feeder, and verifies that it corresponds to the order, that is, if the colour is correct and whether it is a case or a solid cylinder. If it is not correct, then it is discarded, otherwise, it is put on a pallet, and the kind of processing that the part needs is written on the pallet. If it is a solid cylinder, a switch is activated to carry it directly to M4. Otherwise it goes to M2. Machine M2 puts the piston and the spring, if the cylinder needs them, and then the part goes to M3, which adds the cover. In M4 the parts are verified, the pallets are released and the parts are put on a conveyor that moves them to the entrance of the buffer. Machine M5 can temporarily store the cylinders in the buffer. When needed to assemble the final product, M5 puts them in a conveyor that takes them to robot R1. Machine M6 puts a base of the right colour on a pallet, and it is carried to robot R1. The robot takes the three cylinders one by one and puts them on the base. The product is then complete, and goes to robot R2, which takes it out of the system.

Fig. 27. The kind of products that the system in Fig. 26 produces.

The adopted modelling approach is as follows. Each possible *production order* (corresponding to a type of product) has been modelled by means of a Petri net. Then, a set of places, modelling the capacity constraints of the physical resources involved in the production process (robots, intermediate store, pallets, etc.), have been modelled.

Figure 28 shows the Petri net model of one of the products in the system here considered: a product made of three complete cylinders is shown. Place `IDLE` represents the state in which the production order has not been started, the rest of "tagged" places model the system resources (resource places), while the "non-tagged" places model the different states of the component elements inside the system (state places). In the example the resources are of two kinds. On the one hand there are machines, robots, and space in the intermediate buffer (i.e, physical constraints). On the other, there are constraints that are not strictly necessary but are advisable for the correct evolution of the system, for example not to allow more than one pallet on each conveyor segment, that make the conveyor segment to be considered as a resource with capacity one. The final model will be obtained by means of the composition, by fusion of the common places modelling system resources, of the models corresponding to the whole set of products.

Deadlock avoidance control. In order to have a completely automated system, the objective now is to synthesise the control necessary to ensure that no deadlocks can appear. As in Sect. 5.4, the system falls into the class of Resource

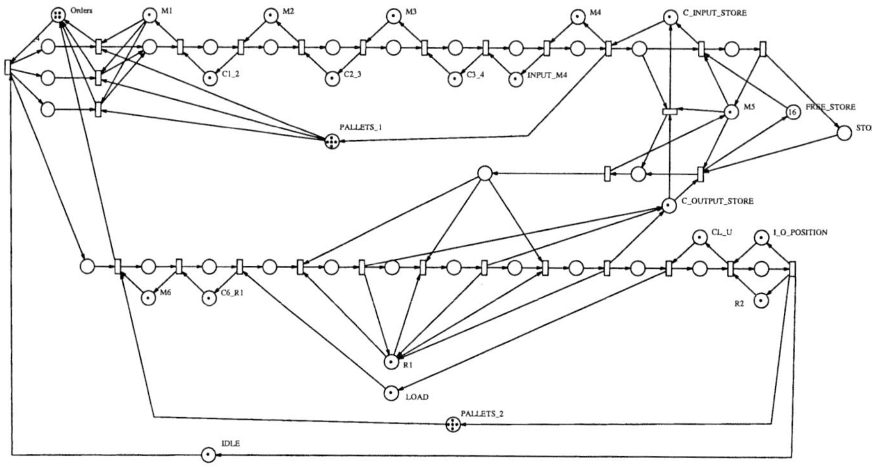

Fig. 28. A non-sequential RAS modelling the assembly of a product made of three complete cylinders and a base.

Allocation Systems: it is composed of a set processes which in their execution must compete for the set of system resources. The complexity of dealing with deadlocks strongly depends on the system structure. Different classes of RAS systems have been defined in the literature. The features that distinguish these classes refer to the process structure (wether the process is sequential or concurrent and whether routing flexibility is allowed or not, mainly) and the way in which resources are allowed to be used and allocated/released (one-by-one or as multi-sets). These characteristics define the class of Petri nets the model belongs to. In the case of a process with a sequential nature (sequential RAS), a state machine can be used to model it (places modelling constraints capacities imposed by the physical or logical resources have then to be added); in the case of non sequential processes, more sophisticated Petri net models are needed, including fork/joint transitions (non-sequential RAS). In systems where resources are allowed to be allocated/released as multi-sets, weights will appear in the arcs related to places modelling resources, which means that the model will belong to the class of generalised Petri nets. These elements will directly influence the analysis and synthesis capabilities of the Petri net model.

An "easy" way of applying deadlock related control is based on the computation of the reachability graph of the system model, to detect the deadlock states and then to forbid them somehow. However, computing the reachability graph of the whole system was not possible, because of its enormous size (for instance, the reachability graph of just one production order as the one in Fig. 28 has 2442 states, while the reachability graph with two production orders being concurrently executed had 241951 states; computing the reachability graph in the case of three production orders was not possible). Therefore, some deadlock prevention/avoidance strategy based on the model structure instead of the reachability graph is needed.

In the case of sequential RAS many different solutions can be found in the literature, adopting different points of view. See, for instance, [23, 35, 43, 26] as a very short list of solutions. However, in our concrete case, there exist transitions with more than one input state place (see Fig. 28), which make our system to belong to the non-sequential RAS class. Adopting a Petri net perspective [47, 28] propose deadlock avoidance solutions for sub-classes of assembly systems. However, the present system falls out of these classes.

In the sequel, a different engineering strategy is adopted: to transform the problem into one with known and applicable solutions. If a deadlock avoidance strategy is adopted, any resource-related state change in the system must be controlled in such a way that only if the reached state is proved to be *safe* (safe means that it can be ensured that all the active processes can be terminated) the change is allowed, otherwise it is forbidden. This means that the application of a deadlock avoidance method imposes a kind of "sequentialisation" in the system behaviour. Therefore, and concentrating on the execution of a production order, substituting its model by the state machine corresponding to the reachability graph of the production model itself is just a change in the model, but not in the behaviour. Notice that doing so a sequential RAS model for the system is obtained. Resource places of the initial model are added to this state machine (they are implicit places and can be added without changing the behaviour) and the final system model is obtained by means of the composition by fusion of the places modelling system resources of the sequential models of the set of products. The considered model belongs to the class of systems for which a deadlock avoidance method is proposed in [26], which can be, then, applied to control the considered system.

The control is based on an adaptation of the Banker's algorithm [20, 33]. In order to consider a given state as safe, the Banker's algorithm looks for an ordering in the set of active processes such that the first process can terminate using the resources granted to it plus the free ones, the second process can terminate using the resources it holds plus the ones free upon the hypothetical termination of the first process, and so on. The basic step is to know if a given process is able to terminate using a given set of available resources. The solution in [26] is a two steps algorithm. First, mark those state places of the state machine modelling the considered process and that require no more resources than the free ones plus the ones in use by the process itself. Second, look for a path of marked state places joining the place corresponding to the state the process is in and the final state.

One important issue when applying deadlock avoidance approaches is the time used to decide whether a given state is safe, since the procedure must be called every time a state change engages new resources. Implementing the control method the following results have been obtained. In the case of the non-sequential RAS in Fig. 28, the corresponding sequential model (the reachability graph of the net in that figure) has 2442 state places, 7814 transitions, using each state up to 22 types of resources. Checking if an active process was able to terminate using the free resources has been implemented. Its takes about

0.003 CPU seconds using a Pentium(4) processor at 1.7 GHz under Microsoft Windows 2000 operating system (this computation uses a Depth First Search algorithm, which is linear in the size of the unfolded system). If the whole system is considered, and given that no more than 26 components can stay at the same time in the system (considering the 10 pallets plus the 16 storage places in Fig. 27) and that a direct implementation of the algorithm in [26] grows in a quadratic way with respect to the number of active production orders, the time to know if a system state is safe takes about 2 CPU–seconds in the worst case.

In order to obtain more efficient solutions some approaches are currently being studied trying to solve the problem for non-sequential RAS using directly the initial model structure. A solution for a class non-sequential RAS, where processes must have a tree-like structure can be found in [27].

6.2 Beyond the State of the Art for the Analysis: Modelling with Object Nets

The aim of this section is to show a different approach for the modelling of production systems. It is based on the clear and intuitive characteristic that in a production system, among other elements, there are two main components. On the one hand, the *system architecture*, which corresponds to the distribution of the physical elements in the plant. Usually, this structure is rather static, and not easily changeable. On the other hand, the set of *process plans* corresponding to the different types of products to be produced in the system. These plans can be seen as logical constraints to be imposed to the free flow of parts in the system. In many cases the set of process plans can change (new process plans are required to face demands of new products, while others disappear, corresponding to products with very low demand). Therefore, doing a separated consideration of that elements when designing the system control software makes easier to adapt it to changes in the set of products the system is able to deal with.

A way of doing that was proposed in [22], where the final model was a coloured Petri net in which the system architecture provided the net skeleton (the set of places, transition and arcs) while the set of part flow restrictions imposed by the process plans were modelled by means of the colour domains of places and transitions and the functions labelling the arcs. This has also been the approach followed in the previous sections. In this section a different approach is going to be adopted. It is based on the *Nets-within-Nets* paradigm, as used, for instance in [62], which support a modelling of systems by Petri nets following the paradigm of Object Oriented Modelling. Applications of the paradigm to the case of manufacturing systems can be seen in [29, 41, 38].

Roughly speaking, one of such models is composed of a *System Net* and one or more *Object Nets* which can be seen as token objects of the system net. Both, the system net and the object nets are Petri nets. A token in the system net can be either a reference to an object net or a black token. Each object net state represents the state of the element it models. Changes in such state can be produced by its own internal dynamics (*autonomous* occurrences), but can also be due to some interactions with the system net. On the other hand, some transitions in the system net can influence the internal state of object nets, but others just move object nets between different locations of the system nets (*transport* occurrences).

Therefore, in the definition of an elementary object system, besides the system net, the set of object nets and the initial marking, a set of *interactions* must be considered. The interactions define how the system net and the object nets must synchronise their activities. These concepts directly apply for the modelling of manufacturing systems. The model of the physical system will correspond to the system net, while each part will be modelled by means of an object net.

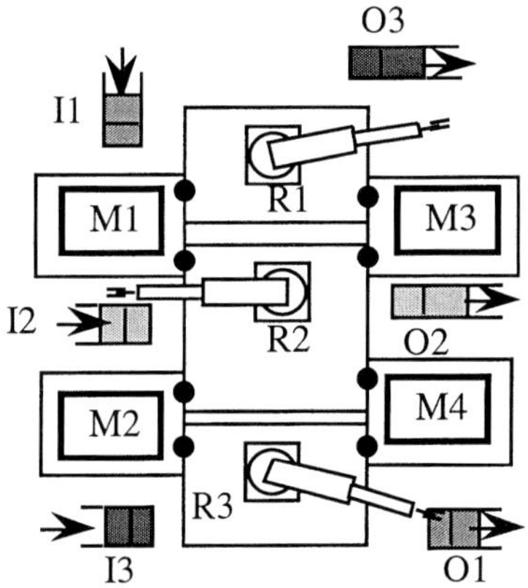

Fig. 29. A manufacturing cell composed of four machines and three robots. Black dots represent the possibility of part flow between two resources.

The objective of this section is not the introduction of the Nets-within-Nets paradigm, but just to show that it is very well adapted to model production systems. To do that, let us apply it to the same example used in [23, 62]. Figure 29 depicts a manufacturing cell composed of four machines, $M1, M2, M3$ and $M4$ (each one can process two products at a time) and three robots $R1$, $R2$ and $R3$ (each one can hold a product at a time). There are three loading points (named $I1, I2, I3$) and three unloading points (named $O1, O2, O3$). The action area for robot $R1$ is $I1, O3, M1, M3$, for robot $R2$ is $I2, O2, M1, M2, M3, M4$ and for robot $R3$ is $M2, M4, I3, O1$.

Every raw product arriving to the cell belongs to one of the three following types: $W1$, $W2$ and $W3$. The type of product characterises the process to be made in the cell as follows: 1) a raw product of type $W1$ is taken from $I1$ and, once it has been manufactured, is moved to $O1$. The sequences of operations for this type are either $(M1, op1); (M2, op2)$ (execute $op1$ in $M1$ and then $op2$ in $M2$) or $(M3, op1); (M4, op2)$ (execute $op1$ in $M3$ and then $op2$ in $M4$). 2) a raw product of type $W2$ is taken from $I2$, manufactured in $M2$ (operation $op5$) and then routed towards $O2$. 3) a raw product of type $W3$ is taken from $I3$, manufactured in $M4$ (operation $op4$) and then in $M3$ (operation $op3$) and,

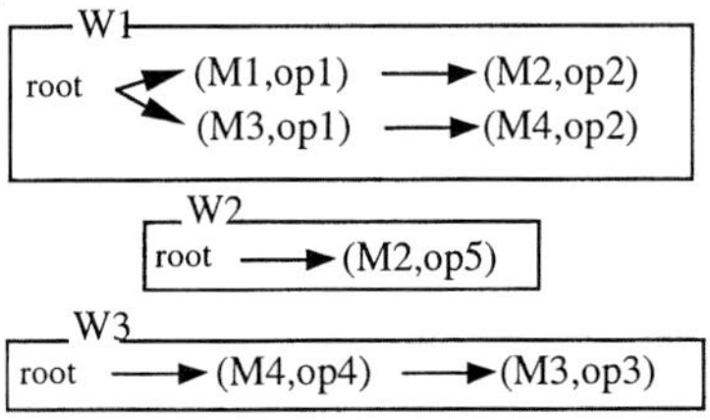

Fig. 30. Three directed acyclic graphs specifying three different types of parts to be processed in the cell depicted in Fig. 29.

finally, routed towards $O3$. Figure 30 represents, by means of directed acyclic graphs, the possible operation sequences for such set of types of parts.

Analogously as in the example in 5.3, the (uncontrolled) Petri net in Fig. 31 represents the possible flow of parts in the considered system. In order to be able to ensure that each part in the system will be produced according to its corresponding process plan, some control has to be added to this skeleton model, which will correspond to the system net in the Nets-within-Nets model (the meaning of places named $W1r, W2r, W3r$ and $W1t, W2t, W3t$ will be explained later).

Figure 32 shows three object nets corresponding to the three types of parts to be produced in the considered system (since in this example all the transitions in the object nets must interact with the system, transition names in Fig. 32 are not represented, just the interactions, for the sake of clarity). Let us explain one of these models. The Petri net labelled $W2$ in Fig. 32 corresponds to a part type $W2$ (in fact, each $W2$-type part will be modelled by one instance of such net). The token in place $p2_1$ models the raw material for one of such products before being loaded into the system. This state is changed when that raw material enters the system. According to the system net in Fig. 31, this is done by the firing of transition $I2$. Therefore, firing such (system) transition must also make the token in $p2_1$ to move to place $p2_2$, which is imposed by the interaction $\langle i11 \rangle$. Place $p2_2$ models a part of type $W2$ inside the system and that must be processed in $M2$. The transition joining $p2_2$ and $p2_3$ is used to model the fact that such part enters $M2$, which in the system net corresponds to transition $R2M2$. Interaction $\langle i13 \rangle$ takes that into account. Interaction $\langle i15 \rangle$ is used to move the part from $M2$ to the robot $R2$. Finally, interaction $\langle i12 \rangle$ is needed to model the unloading of such part from the system.

In the system net in Fig. 31 tokens in place $W1r$ are instances of object net $W2$ in Fig. 32, and correspond to raw parts of type $W2$ (there are $K2$ of such net instances). Once terminated, these object nets will be in place $W1t$, which "collects" terminated products of type $W2$.

Any further refinement in the model is easy to be done. Let us suppose also the different operations each machine is able to do need to be considered. For instance, machine $M3$ is able to carry out operations $op1$ and $op3$. Figure 33(a) shows how place pi_M3 in the net in Fig. 31 could be refined in order to consider

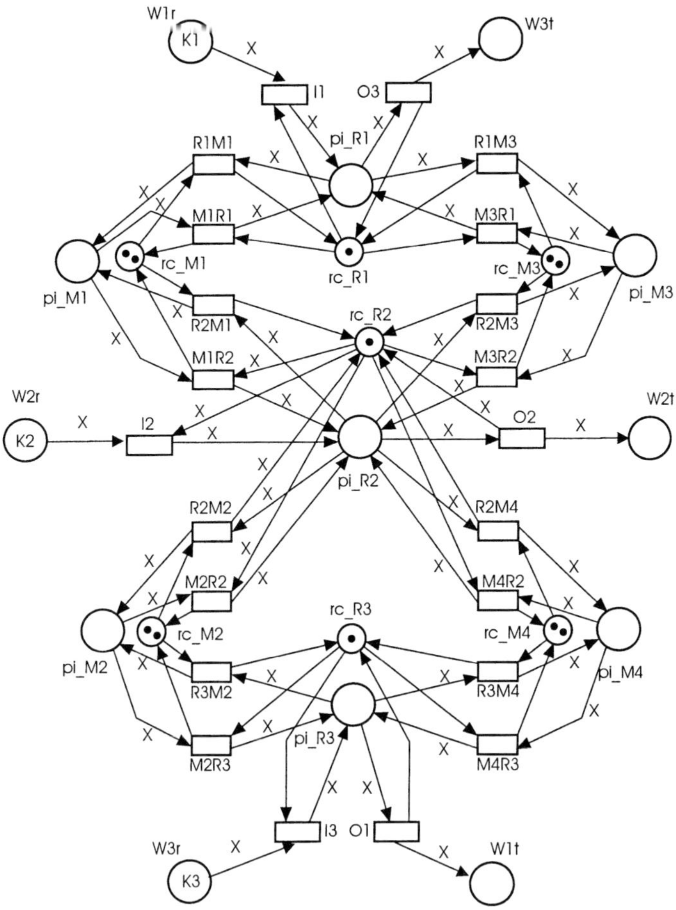

Fig. 31. Petri net model of the part flow in the cell depicted in Fig. 29.

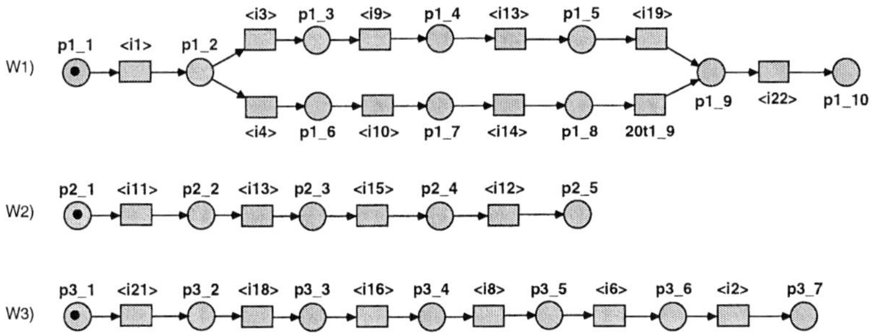

Fig. 32. Three object nets modelling the three types of parts to be processed in the system in Fig. 29. Transition names are not presented, only the interactions with the system net.

the operations it is able to do (capacity of $M3$ is not represented for the sake of clarity). On the other hand, Fig. 33(b) shows how the place $p3_5$ of the object net corresponding to the processing of parts of type $W3$ in Fig. 32 could be refined so that the process plan it models takes into account that the operation $op3$ has to be done in $M3$ for such parts (notice that transitions $M3_1$ and $M3_2$ correspond to transport occurrences).

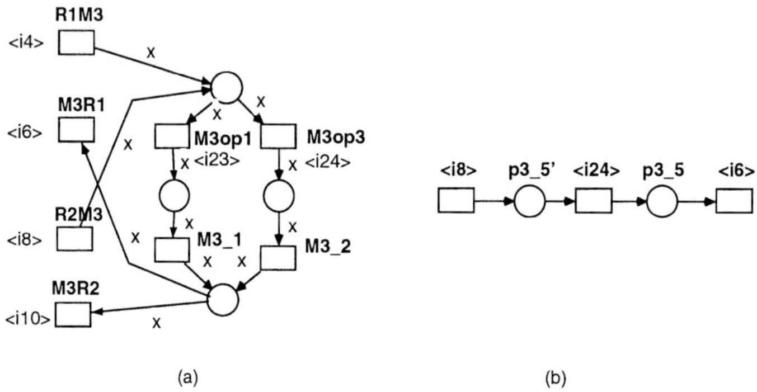

Fig. 33. A refined model for machine $M3$ and how it affects the object net modelling $W3$ parts.

High level Petri net-based formalisms provide very useful tools for the modelling, analysis and control of complex concurrent systems. However, the higher the abstraction level the formalism allows, the more complicated its analysis becomes. This is the case of coloured Petri nets, for instance (structure-based techniques are not as general as in the case of ordinary Petri nets) and also the case for Nets-within-Nets models. It is always possible to apply simulation techniques, which can give insight of some system behaviours allowing the system designer to easily test different system configurations in order to have arguments to choose one or another. In the case of Nets-within-Nets, the tool Renew [36] is a good environment for modelling and simulation.

7 From Discrete Event Models towards Hybrid Models

In the last years a new kind of models based on Petri nets has appeared. They differ from the previous ones in that they are not discrete event models, but hybrid models. That is, the state is not only represented by discrete variables, but it is partly relaxed into continuous variables (in the extreme case, even all the variables may be continuous in piecewise continuous systems).

These hybrid models have been defined in many different ways. For example, (discrete) Petri nets may be combined with differential algebraic equations associating them either to places (Pr/Tr Petri nets) [10] or to markings (DAE Petri nets) [59]. Another possibility is to partially relax the integrality condition in the firing of the transitions, i.e., continuise or fluidify the firing, as in Hybrid Petri nets [3, 52]. This means that the marking of the places around these transitions

Table 2. The four cases for possible continuisation of a transition [52].

Clients	Servers	Semantics of the transition
few (D)	few (D)	Discrete transition
few (D)	many (C)	Discrete transition (servers become *implicit places*)
many (C)	few (D)	Continuous finite server semantics (bounds to firing speed)
many (C)	many (C)	Continuous infinite servers semantics (speed is enabling-driven)

is no longer guaranteed to be integer (with the possible exception of self-loop arcs). When a total fluidification is done the result is a Continuous Petri net [14, 51]. This kind of hybrid models can be used both to represent systems whose "more reasonable view" is hybrid, or as an approximation of discrete systems under high traffic conditions. The idea of continuisation of discrete models is not new and has been employed in many different fields, for example, population dynamics [46], manufacturing systems [16, 32], communication systems [21], etc. In the following we will concentrate on Hybrid Petri nets. This is not the place to present the state of the art of the analysis of continuous and hybrid PNs (see for example [45, 51]), but just to point out that (partial) fluidification of the untimed model does not preserve in general liveness properties of the discrete model.

In timed models, in order to associate a time semantics to the fluidification of a transition, it should be taken into account that a transition is like an station in Queuing Networks, thus "the meeting point" of clients and servers. Assuming that there may be many or few of each one of them, fluidification can be considered for clients, for servers or for both. Table 2 represents the four theoretically possible cases. If there were few clients, the transition should be considered discrete.

Basically, the idea is to use a first order (or deterministic) approximation of the discrete case [45], assuming that the delays associated to the firing of transitions can be approximated by their mean values. A similar approach is used, for example, in [6]. This means that in continuous transitions the firing is approximated by a continuous flow, whose exact value depends on the semantics being used. The two basic semantics defined for continuous transitions (see Table 2) are *infinite servers* (or *variable speed*) and *finite servers* (or *constant speed*) [3, 45]. Under finite servers semantics, the flow of t_i has just an upper bound, $\boldsymbol{\lambda}[t_i]$ (the number of servers times the speed of a server). Then $\mathbf{f}(\tau)[t_i] \leq \boldsymbol{\lambda}[t_i]$ (knowing that at least one transition will be in saturation, that is, its utilisation will be equal to 1). Under infinite servers semantics, the flow through a timed transition t is the product of the speed, $\boldsymbol{\lambda}[t]$, and the enabling of the transition, i.e., $\mathbf{f}[t] = \boldsymbol{\lambda}[t] \cdot \text{enab}(t, \mathbf{m}) = \boldsymbol{\lambda}[t] \cdot \min_{p \in {}^\bullet t}\{\mathbf{m}[p]/\mathbf{Pre}[p,t]\}$.

It should be pointed out that finite server semantics, equationally modelled by bounding the firing speed of continuised transitions, corresponds at conceptual level to a *hybrid* behaviour: fluidification is applied only to clients, while servers are kept as discrete, i.e., counted as a finite number (the firing speed is bounded by the product of the speed of a server and the number of servers in the station).

On the other hand, infinite servers semantics really relax clients and servers, being the firing speed driven by the enabling degree of the transition. In this case, even if the fluidification is total, the model is hybrid in the sense that it is a piecewise linear system, in which switching among the embedded linear systems is not externally driven as in [7], but internally through the minimum operators.

The following example is taken from [2, 3]. It models a station in a Motorola production system. This station can produce two kinds of parts, *c1* and *c2*, whose processing corresponds to the left and right part of the figure, respectively. The parts arrive in batches of 30000 and 20000 parts at times 0 and 1000. After the arrival of a bach, parts are downloaded into a buffer at a speed of 1 part per time unit. The processing does not start immediately, but waits until at least 500 parts of type *c1* or 600 parts of type *c2*) have been downloaded. At that point some set up is done on the machine, which takes 300 time units for parts *c1* and 360 for *c2*, before the processing starts. When all the parts in the batch have been processed, the machine is liberated. Pieces are removed in batches of the input size.

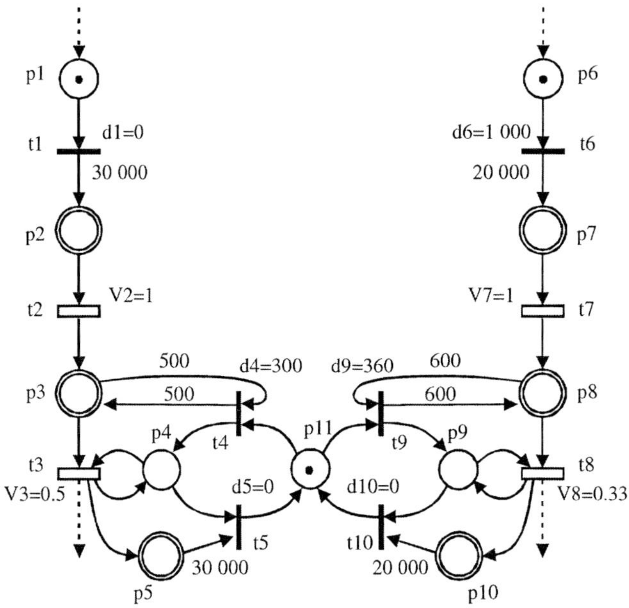

Fig. 34. Hybrid Petri net modelling the behaviour of a production system.

A model of this system can be seen in Fig. 34. Although it is a discrete system, the model is not discrete, but hybrid. The transitions represented as bars in the figure are discrete (the usual transitions in Petri nets), while those represented as boxes are continuous. Analogously, the circles drawn with a simple line are discrete, while those with the double line are continuous.

In this example, since the size of the batches is quite large, the firing of transitions t_2, t_3, t_7 and t_8 can be approximated by a continuous flow. This kind of approximation (when applicable) may simplify the study of the system. For example, in [2] it is reported that for this system the simulation time reduces from 454 sec. to 0.15, that is, it is divided by 3000!

Basic understanding of hybrid systems, and analysis and synthesis techniques need much improvement before they can be effectively used [51, 52]. Moreover, it should be pointed out that there exist some "natural" limits to the properties that can be studied. For example, mutual exclusion (in the marking of places or in the firing of transitions), and the difference between home space and reversibility cannot be studied in general [51]. Additionally, basic properties like deadlock-freeness of the autonomous continuised model is neither necessary, nor sufficient for the discrete case [51]. However, the use of hybrid models as partial relaxations of discrete models is a quite new and promising approach.

References

1. M. Ajmone Marsan, G. Balbo, G. Conte, S. Donatelli, and G. Franceschinis. *Modelling with Generalized Stochastic Petri Nets.* Wiley, 1995.
2. H. Alla, J.B. Cavaille, M. Le Bail, and G. Bel. Les systémes de production par lot: une approche discret-continu utilisant les réseaux de Petri Hybrides. In *Proc. of ADPM'92*, Paris, France, January 1992.
3. H. Alla and R. David. Continuous and hybrid Petri nets. *Journal of Circuits, Systems, and Computers*, 8(1):159–188, 1998.
4. A. Avizenis and J. P. Kelly. Fault tolerance by design diversity: Concepts and experiments. *Computer*, 17(8):67–80, 1984.
5. J. M. Ayache, P. Azema, and M. Diaz. Observer, a concept for on line detection for control errors in concurrent systems. In *Proc.* 9th *IEEE Int. Sypm. Fault-Tolerant Computing*, pages 79–86, Madison, WI, USA, June 1992.
6. F. Balduzzi, A. Giua, and G. Menga. First-order hybrid Petri nets: A model for optimization and control. *IEEE Trans. on Robotics and Automation*, 16(4):382–399, 2000.
7. A. Bemporad, A. Giua, and C. Seatzu. An iterative algorithm for the optimal control of continuous-time switched linear systems. In M. Silva, A. Giua, and J.M. Colom, editors, *WODES 2002: 6th Workshop on Discrete Event Systems*, pages 335–340, Zaragoza, Spain, 2002. IEEE Computer Society.
8. J. Campos, G. Chiola, J. M. Colom, and M. Silva. Properties and performance bounds for timed marked graphs. *IEEE Trans. on Circuits and Systems-I: Fundamental Theory and Applications*, 39(5):386–401, 1992.
9. J. Campos, G. Chiola, and M. Silva. Ergodicity and throughput bounds of Petri net with unique consistent firing count vector. *IEEE Trans. on Software Engineering*, 17(2):117–125, 1991.
10. R. Champagnat, R. Valette, J.C. Hochon, and H. Pingaud. Modeling, simulation and analysis of batch production systems. *Discrete Event Dynamic Systems: Theory and Application*, 11(1/2):119–136, 2001.
11. C. Chaouiya and Y. Dallery. Petri net models of pull control systems for assembly manufacturing systems. In *Procs. of the* 2nd *Int. Workshop on Manufacturing and Petri Nets, ICATPN*, pages 85–103, Toulouse, France, 1997.

12. P. Chretienne, E. G. Coffman, J. K. Lengstra, and Z. Liu, editors. Wiley, 1995.
13. J. M. Colom, M. Silva, and J. L. Villarroel. On software implementation of Petri nets and colored Petri nets using high-level concurrent languages. In *Proc. 7th European Workshop on Application and Theory of Petri Nets*, pages 207–241, Oxford, England, July 1986.
14. R. David and H. Alla. Continuous Petri nets. In *Proc. of the 8th European Workshop on Application and Theory of Petri Nets*, pages 275–294, Zaragoza, Spain, 1987.
15. R. David and H. Alla. *Petri Nets and Grafcet.* Prentice-Hall, 1992.
16. R. David, X. Xie, and Y. Dallery. Properties of continuous models of transfer lines with unreliable machines and finite buffers. *IMA Journal of Mathematics Applied in Business and Industry*, 6:281–308, 1990.
17. A. Desrochers and R. Y. Al-Jaar. *Applications of Petri Nets in Manufacturing Systems.* IEEE Press, 1994.
18. Alan A. Desrochers and Robert Y. Al-Jaar. *Applications Of Petri Nets In Manufacturing Systems. Modeling, Control, And Performance Analysis.* IEEE Press, 1995.
19. M. Diaz, G. Juanole, and J. P. Courtiat. Observer — a concept for formal on-line validation of distributes systems. *IEEE Trans. on Software Engineering*, 20(12):900–913, 1994.
20. E. W. Dijkstra. Cooperating sequential processes. In F. Genuys, editor, *Programming Languages.* Academic Press, 1968.
21. E.I. Elwadi and D. Mitra. Statistical multiplexing with loss priorities in rate-based congestion control of high-speed networks. *IEEE Transactions on Communications*, 42(11):2989–3002, 1994.
22. J. Ezpeleta and J.M. Colom. Automatic synthesis of colored Petri nets for the control of FMS. *IEEE Transactions on Robotics and Automation*, 13(3):327–337, June 1997.
23. J. Ezpeleta, J.M. Colom, and J. Martínez. A Petri net based deadlock prevention policy for flexible manufacturing systems. *IEEE Trans. on Robotics and Automation*, 11(2):173–184, April 1995.
24. J. Ezpeleta and J. Martínez. Formal specification and validation in production plants. In *Proceedings of the 3th. International Conference on Computer Integrated Manufacturing*, pages 64–73, Rensselaer Polytechnic Institute, Troy (New York), May 1992. IMACS.
25. J. Ezpeleta and L. Recalde. A deadlock avoidance approach for non-sequential resource allocation systems. *IEEE Trans. on Systems, Man, and Cybernetics*, 2004. Accepted.
26. J. Ezpeleta, F. Tricas, F. García- Vallés, and J.M. Colom. A Banker's solution for deadlock avoidance in FMS with routing flexibility and multi–resource states. *IEEE Transactions on Robotics and Automation*, 18(4):621–625, August 2002.
27. J. Ezpeleta and R. Valk. A polynomial solution for deadlock avoidance in assembly systems modelled with petri nets. In *Proceedings of the Multiconference on Computational Engineering in Systems Applications (CESA2003)*, pages 1–8, Lille (France), July, 9–11 2003.
28. M.P. Fanti, B. Maione, and B. Turchiano. Design of supervisors to avoid deadlock in flexible assembly systems. *The International Journal of Flexible Manufacturing Systems*, 14:157–175, 2002.
29. B. Farwer, D. Moldt, and F. Garcí-Vallés. An approach to modelling fms with dynamic object petri nets. In *Proceedings of the IEEE International Conference on Systems, Man and Cybernetics*, Hammamet (Tunisia), October 2002.

30. J. C. Gentina, J. P. Bourey, and M. Kapusta. Coloured adaptive structured Petri nets. *Computer-Integrated Manufacturing*, 1(1):39–47, 1988.
31. J. C. Gentina, J. P. Bourey, and M. Kapusta. Coloured adaptive structured Petri nets (II). *Computer-Integrated Manufacturing*, 1(2):103–109, 1988.
32. S. B. Gershwin. *Manufacturing Systems Engineering.* Prentice-Hall, 1994.
33. A. N. Habermann. Prevention of systems deadlocks. *Communications of the ACM*, 12(7):373–385, July 1969.
34. C. Hanen and A. Munier. Cyclic scheduling problems: An overview. In Chretienne et al. [12].
35. YiSheng Huang, MuDer Jeng, and Xiaolan Xie. A deadlock prevention policy for flexible manufacturing systems using siphons. In *Proc. of the 2001 IEEE International Conference on Robotics and Automation*, pages 541–546, Seoul (Korea), May 2001.
36. O. Kummer and F. Wienberg. Renew. the reference net workshop. *Petri Net Newsletter*, (56):12–16, 1999.
37. N. G. Leveson and J. L. Stolzy. Safety analysis using Petri nets. *IEEE Trans. on Software Engineering*, 13(3):386–397, 1987.
38. E. López-Mellado and J.G. Morales-Montelongo. Agent-based distributed controllers for discrete manufacturing systems. In *Proceedings of the Multiconference on Computational Engineering in Systems Applications (CESA2003)*, pages 1–7, Lille (France), July, 9–11 2003.
39. J. Martínez, P. Muro, and M. Silva. Modeling, validation and software implementation of production systems using high level Petri nets. In M. Silva and T. Murata, editors, *Invited Sessions: Petri Nets and Flexible Manufacturing. IEEE Int. Conf. on Robotics and Automation*, pages 1180–1185, Raleigh, NC, USA, April 1987.
40. J. Martínez, P. Muro, M. Silva, S. F. Smith, and J. L. Villarroel. Merging artificial intelligence techniques and Petri nets for real time scheduling and control of production systems. In R. Huber et al., editors, *Artificial Intelligence in Scientific Computation*, pages 307–313. Scientific Publishing Co., 1989.
41. D. Moldt and J. Ezpeleta. A proposal for flexible testing of deadlock control strategies in resource allocation systems. In *Proceedings of the International Conference on Computational Intelligence for Modelling Control and Automation (CIMCA'03)*, pages 586–595, Vienna, Austria, February 2003.
42. T. Murata. Petri nets: Properties, analysis and applications. *Proceedings of the IEEE*, 77(4):541–580, 1989.
43. J. Park and S. Reveliotis. Deadlock avoidance in sequential resource allocation systems with multiple resource acquisitions and flexible routings. *IEEE Transactions on Automatic Control*, 46(10):1572–1583, October 2001.
44. J. M. Proth and X. Xie. *Petri Nets. A Tool for Design and Management of Manufacturing Systems.* Wiley, 1996.
45. L. Recalde and M. Silva. Petri Nets fluidification revisited: Semantics and steady state. *APII-JESA*, 35(4):435–449, 2001.
46. E. Renshaw. A survey of stepping-stone models in population dynamics. *Adv. Appl. Prob.*, 18:581–627, 1986.
47. E. Roszkowska and R. Wojcik. Problems of process flow feasibility in FAS. In K. Leiviska, editor, *IFAC CIM in Process and manufacturing Industries*, pages 115–120, Espoo, Finland, 1992. Oxford: Pergamon Press.
48. M. Silva. *Las Redes de Petri: en la Automática y la Informática.* AC, 1985.
49. M. Silva. Interleaving functional and performance structural analysis of net models. In M. Ajmone Marsan, editor, *Application and Theory of Petri Nets 1993*, volume 691 of *Lecture Notes in Computer Science*, pages 17–23. Springer, 1993.

50. M. Silva. Introducing Petri nets. In *Practice of Petri Nets in Manufacturing*, pages 1–62. Chapman & Hall, 1993.
51. M. Silva and L. Recalde. Petri nets and integrality relaxations: A view of continuous Petri nets. *IEEE Trans. on Systems, Man, and Cybernetics*, 32(4):314–327, 2002.
52. M. Silva and L. Recalde. On fluidification of Petri net models: from discrete to hybrid and continuous models. In *IFAC Conference on Analysis and Design of Hybrid Systems, ADHS03*, pages 9–20, Saint-Malo, France, June 2003.
53. M. Silva and E. Teruel. A systems theory perspective of discrete event dynamic systems: The Petri net paradigm. In P. Borne, J. C. Gentina, E. Craye, and S. El Khattabi, editors, *Symposium on Discrete Events and Manufacturing Systems. CESA '96 IMACS Multiconference*, pages 1–12, Lille, France, July 1996.
54. M. Silva and E. Teruel. Petri nets for the design and operation of manufacturing systems. *European Journal of Control*, 3(3):182–199, 1997.
55. M. Silva, E. Teruel, and J. M. Colom. Linear algebraic and linear programming techniques for the analysis of net systems. In G. Rozenberg and W. Reisig, editors, *Lectures in Petri Nets. I: Basic Models*, volume 1491 of *Lecture Notes in Computer Science*, pages 309–373. Springer, 1998.
56. M. Silva, E. Teruel, R. Valette, and H. Pingaud. Petri nets and production systems. In G. Rozenberg and W. Reisig, editors, *Lectures in Petri Nets. II: Applications*, volume 1492 of *Lecture Notes in Computer Science*, pages 85–124. Springer, 1998.
57. E. Teruel, J. M. Colom, and M. Silva. Choice-free Petri nets: A model for deterministic concurrent systems with bulk services and arrivals. *IEEE Trans. on Systems, Man, and Cybernetics*, 27(1):73–83, 1997.
58. F. Tricas, F. García-Vallés, J.M. Colom, and J. Ezpeleta. An iterative method for deadlock prevention in FMS. In R. Boel and G. Stremersch, editors, *Discrete Event Systems: Analysis and Control. Proc. of the Workshop On Discrete Event Systems 2000*, pages 139–148, Ghent, Belgium, Aug 2000. Kluwer Academic Publishers.
59. C. Valentin-Roubinet. Modeling of hybrid systems: DAE supervised by Petri nets. the example of a gas storage. In *Proc. of ADPM'98*, pages 142–149, Reims, France, March 1998.
60. R. Valette and M. Courvoisier. Petri nets and artificial intelligence. In R. Zurawski and T. Dillon, editors, *Modern Tools for Manufacturing Systems*, pages 385–405. Elsevier, 1993.
61. R. Valette, M. Courvoisier, J. M. Bigou, and J. Albukerque. A Petri nets based programmable logic controller. In *IFIP 1st Int. Conf. on Computer Applications in Production and Engineering*, Amsterdam, Holland, April 1983.
62. R. Valk. Petri nets as token objects - an introduction to elementary object nets. *Lecture Notes in Computer Science: 19th Int. Conf. on Application and Theory of Petri Nets, ICATPN'98, Lisbon, Portugal, June 1998*, 1420:1–25, June 1998.
63. S. Velilla and M. Silva. The spy: A mechanism for safe implementation of highly concurrent systems. In *Real Time Programming 1988, 15th IFAC/IFIP Workshop*, pages 95–102, Valencia, Spain, May 1988. Pergamon.
64. J. L. Villarroel, J. Martínez, and M. Silva. GRAMAN: A graphic system for manufacturing system design. In S. Tzafestas, A. Eisinberg, and L. Carotenuto, editors, *IMACS Symp. on System Modelling and Simulation*, pages 311–316. Elsevier, 1988.
65. N. Viswanadham and Y. Narahari. *Performance Modeling of Automated Manufacturing Systems*. Prentice-Hall, 1992.
66. Mengchu Zhou and Kurapati Venkatesh. *Modeling, Simulation, and Control of Flexible Manufacturing Systems : A Petri Net Approach*, volume 6 of *Series in Intelligent Control and Intelligent Automation*. World Scientific, 1999.

Communicating Transaction Processes: An MSC-Based Model of Computation for Reactive Embedded Systems

Abhik Roychoudhury and Pazhamaneri Subramaniam Thiagarajan

School of Computing, National University of Singapore, Singapore 117543
{abhik,thiagu}@comp.nus.edu.sg

Abstract. Message Sequence Charts (MSC) have been traditionally used to depict execution scenarios in the early stages of design cycle. MSCs portray inter-object interactions. Synthesizing intra-object executable specifications from an MSC-based description is a non-trivial task. Here we present a model of computation called *Communicating Transaction Processes* (CTP) based on MSCs from which an executable specification can be extracted in a straightforward manner. Our model describes a network of communicating processes in which the processes interact via common action labels. Each action is a non-atomic interaction described as a guarded choice of MSCs. Thus our model achieves a separation of concerns: the high-level network of processes depicting intra-process computations and control flow, while the common non-atomic communication actions capture inter-process interaction via MSCs. We show how to extract an ordinary Petri net from a CTP model thereby leading to a standard operational semantics. We also discuss the connection of our formalism to Live Sequence Charts, an extension of MSCs which also has an executable semantics.

1 Introduction

Message Sequence Charts (MSCs) are an attractive visual formalism which are used in the early design stages of reactive systems. They portray scenarios that arise from component interactions and hence can be used to capture requirements and test cases. MSCs and a related mechanism called HMSCs (High-level Message Sequence Charts) have been standardized [26] for specifying telecommunication software. A version of MSC called Sequence Diagram is a behavioral diagram type used in the Unified Modeling Language (UML) [10].

In all these settings, MSCs are used to capture system requirements. To move towards an implementation, one must obtain an executable specification which is related in some fashion to the MSC-based requirements. The key difficulty here, as identified in [14], is that the inter-object interactions described in form of MSCs must be related to -or synthesized as- executable specifications given in terms of intra-object behaviors, say, one state-chart for each object. This is a difficult problem and it has been studied in various limited contexts [1, 14, 17, 20].

J. Desel, W. Reisig, and G. Rozenberg (Eds.): ACPN 2003, LNCS 3098, pp. 789–818, 2004.

In this paper, we propose using MSCs to construct executable specifications in a more direct fashion. The main idea is to use traditional methods to capture the control flow of the system components while using MSCs to describe the *non-atomic* component interactions. Among the various possibilities for describing the control flow in a multi-component system, we choose here the well-known model of synchronized product of transition systems; a network of labeled transition systems that synchronize on common actions. With suitable modifications one could easily use other related models as well.

We impose two restrictions on the control flow; a minor technical one that we will come to later but also a major one which requires that branchings in the control flow is effected by the components in a *local* fashion. In Petri net terms, this is the so called free choice property [9]. In particular, this restriction ensures that choices regarding which interactions to take part in are made by the components in a local fashion.

Starting with a network of labeled transition systems that synchronize on common actions, we refine each common abstract action γ involving a set of agents into a *transaction scheme* T_γ. Each such scheme is a guarded choice of MSCs. The life lines of the MSCs in T_γ will be from the set of agents participating in the common action γ. Each guarded MSC in T_γ, called a *transaction* will represent one possible interaction and will involve a complex flow of data and control signals. *When* a transaction scheme is to be executed is determined by the control flow in the high level product transition system. As to *which* transaction in T_γ will be chosen to be executed is determined by the guards which are propositional formulas built out of atomic propositions. The truth values of these atomic propositions, and hence those of the guards, will capture abstracted properties of the values of the variables associated with the agents. A central feature of the model is that both the control flow and the evaluation of the guards (which then leads to the execution of a specific transaction within a transaction scheme) are done in a distributed and asynchronous manner. In broad terms, this is our Communicating Transaction Processes (CTP) model.

Our model is in line with the emerging consensus that system-level design methods for embedded systems should be based on models of computation in which there is a clean separation of computational and communication features [11, 3, 12]. The CTP formalism basically uses finite state machines with data paths to model computational -and the attendant control flow- aspects while deploying guarded choices of MSCs to capture complex interactions between the different computational threads.

Our strategy of striking a balance between control flow and component interactions yields a model which is flexible, powerful and at the same time amenable to formal analysis and synthesis. Indeed, the problem of extracting an executable specification from the CTP model becomes very manageable and amenable to automation as we show in section 3. Our main point of reference for this work is the formalism of Live Sequence Charts [8] and more specifically the Play-in/Play-out approach [16] in which the component interactions are elaborated in a powerful way using the LSC language while the control flow information is completely suppressed. On the other hand, in models such as Petri nets and distributed transitions systems, the focus is on a detailed presentation of control flow while the only mechanisms for capturing component interactions are the atomic notions of synchronizing transitions and shared buffers.

An alternative way to use MSCs to capture system behavior is via HMSCs. However, an HMSC is just a presentation of a *collection* of MSCs. The problem of extracting an executable specification from an HMSC is a non-trivial one. There are a variety of choices available for the executable specification mechanism such as state charts [20], Petri nets [5] and networks of automata communicating through FIFOs [17, 1]. Many versions of this synthesis problem -i.e. deriving an intra-object executable specification from an HMSC- are not even decidable [17, 1, 5]. In contrast, as we shall show, we can extract an executable specification in the form of a finite Petri net from a CTP model effectively and in a manner that can be automated quite easily.

In the next section we introduce the CTP model while illustrating its main features with simple examples. In Section 3, we provide the operational semantics of the CTP model in terms of ordinary Petri nets. The key step in this process is converting the transaction schemes into an executable mechanism called event structures. In Section 4, we present a more detailed example based on the AMBA bus protocol in order to highlight the communicational aspects of the CTP model supported by the use of transaction schemes based on MSCs. In Section 5, we discuss behavioral properties and the means for determining these properties. In particular we present the notion of well-formed transaction schemes and illustrate its importance. In the subsequent section, we provide a more detailed comparison with the closely related formalism of LSCs. Section 7 reports our current efforts for building an experimental framework to enable the use of the CTP model to support the specification, verification and implementation of reactive embedded systems. The concluding section provides additional pointers to future research.

2 The CTP Model

Being based on MSCs, the CTP model captures non-atomic inter-process communications. However, in order to be amenable to efficient distributed implementation, this is combined with notations for describing intra-process control flow. As a starting example, consider the specification shown in Figure 1[1]. Each process repeatedly interacts with the other process and then performs some internal computational action. Note that the inter-process interaction and the internal actions have been separated into distinct units. A number of processes P will be involved in the execution of a chart. A process p which takes part in such an execution might next participate in a chart involving a different set of processes, say Q.

[1] We adopt the usual MSC convention that horizontal and downward sloping arrows denote message send-receives between two processes. Further, a □ symbol on a single vertical line denotes an internal action (such as actions a and b in Figure 1). We denote a control state of a process as a circle.

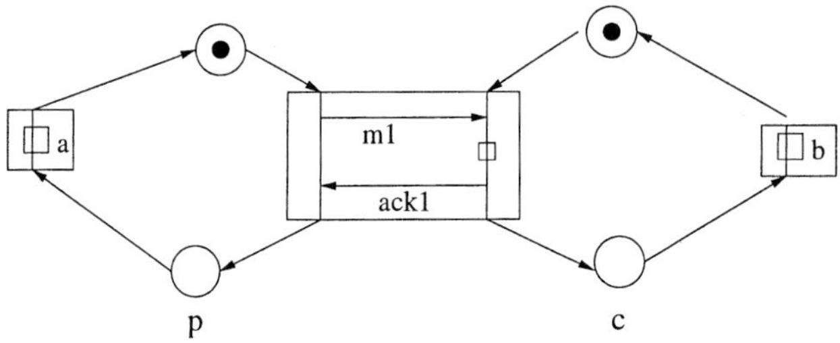

Fig. 1. Inter-process communication and intra-process control flow

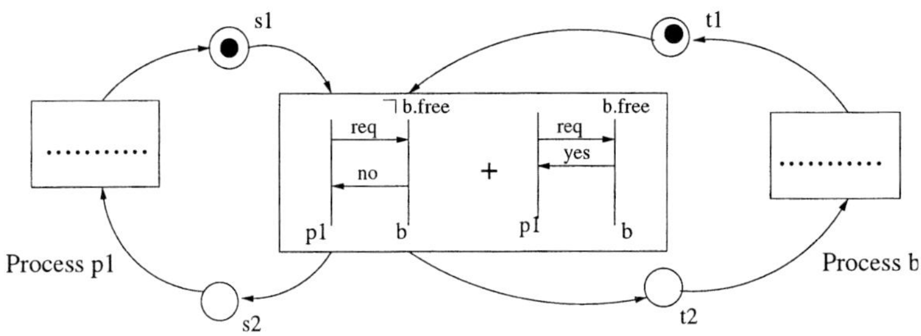

Fig. 2. Choice of Inter-process communication

The organization of interactions into the various units is up to the convenience of the designer. For example, in Figure 1 we could have made the actions a and b also to be part of the chart involving processes p and c. Note also that the example shown in Figure 1 is essentially a Petri net where the local control states in each process are the places of the net denoted by circles. Each top-level transition of this net is, in general, a collection of Message Sequence Charts at the refined level. A particular execution of the high-level transition, is an abstraction of the activity in which one of the charts associated with the high-level transition is chosen and executed. In the example of Figure 1 each net transition has a single chart associated with it. (An internal action is a degenerate chart involving just one process executing just one action). The choice as to which chart is executed -in case more than one chart is associated with a transition- is based on the value of the local variables of the processes. This is illustrated in Figure 2 where the choice is determined by the value of the variable $free$ belonging to process b). If $b.free$ holds once control reaches $s1$ and $t1$ respectively, we must execute the right-hand chart of Figure 2.

In general, the choice of which chart is executed at a particular net transition is a distributed one. Let the charts contained in a particular net transition be as shown in Figure 3. If $p1.data$ holds then chart 1 is ruled out. However, still we do not know whether chart 2 or chart 3 will be executed. This will depend on the value of variable $free$ in process b. As shown in Figure 3, each MSC associated with a net transition has a guard (which we will also refer to as a pre-condition). This guard is a distributed one in that it will in general involve

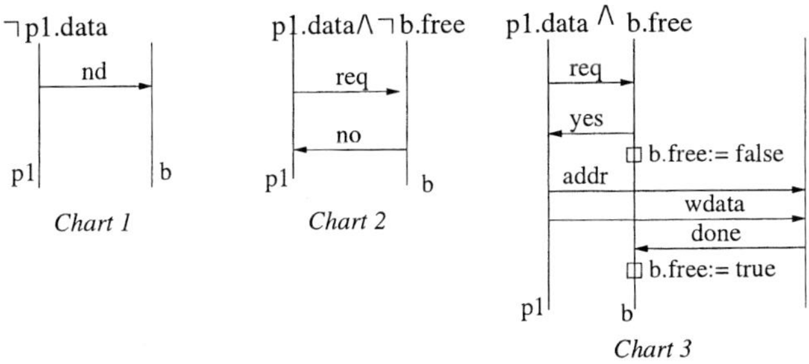

Fig. 3. Distributed nature of choice in a net transition

propositions belonging to different processes participating in the MSC. At the end of the execution of a chart, the truth values of the various propositions will be set to new values in general.

2.1 The Definition of the CTP Model

A product transition system is a network of sequential transition systems that synchronize on common actions. The CTP model is obtained by taking a restricted class of product transition systems and refining the common actions into collections of guarded MSCs called *transaction schemes*.

Fix a finite set of process names $\mathcal{P}$ with p, q ranging over $\mathcal{P}$. Fix also a finite set of labels Γ and a family $\{\Gamma_p\}_{p \in \mathcal{P}}$ with each Γ_p a subset of Γ and $\bigcup \Gamma_p = \Gamma$. This induces the function *loc* which assigns to each label in Γ the set of agents that participate in the execution of that action. This function is given by: $loc(\gamma) = \{p \mid \gamma \in \Gamma_p\}$. If $loc(\gamma) = \{p\}$ then γ will be called p-local action. The members of Γ will be treated as abstract action labels in the first step where we define the control flow model. In the second step they will be interpreted as transaction schemes and further elaborated. Γ_p is the set of (abstract) actions that the process p will participate in.

Anticipating the need to build guards in the second step, we also fix AP_p a finite set of atomic propositions, one for each p and set $AP = \bigcup_{p \in \mathcal{P}} AP_p$. If $P \subseteq \mathcal{P}$ then we let $AP_P = \bigcup_{p \in P} AP_p$. By convention, we shall write $AP_{\mathcal{P}}$ as AP. Each subset of AP_P will be called P-valuation. If $P = \{p\}$ is a singleton we will write p-valuation.

For each p let $TS_p = \langle S_p, \Gamma_p, \longrightarrow_p, init_p, V_{p,in} \rangle$ be a finite-state transition system over Γ_p with an initial p-valuation. In other words, S_p is a finite set of states, $init_p \in S_p$ is the initial state, $\longrightarrow_p \subseteq S_p \times \Gamma_p \times S_p$ denotes the transition relation and $V_{p,in} \subseteq AP_p$ is the initial valuation of atomic propositions in AP_p. In this paper we will be only interested in control flows in which the choices as to which transaction scheme that p will take part in is decided locally by p (free choice). Further, to avoid notational clutter, we will require that each member of Γ_p is the label of at most one transition in TS_p. These two restrictions on TS_p can be formalized as follows.

(1) if $s \xrightarrow{\gamma}_p s_1$, $s \xrightarrow{\gamma'}_p s_2$ and $s_1 \neq s_2$, then γ and γ' are p-local actions. Thus, $loc(\gamma) = loc(\gamma') = \{p\}$.
(2) If $s_1 \xrightarrow{\gamma}_p s_2$ and $s_3 \xrightarrow{\gamma}_p s_4$ then $s_1 = s_3$ and $s_2 = s_4$.

Definition 1 Product Transition System *A product transition system over* $(\{\Gamma_p, AP_p\})_{p\in\mathcal{P}}$ *is denoted as* $\{TS_p\}_{p\in\mathcal{P}}$ *where each*

$$TS_p = \langle S_p, \Gamma_p, \longrightarrow_p, init_p, V_{p,in}\rangle$$

is as specified above. As usual, the behavior of this product transition system is defined to be the global transition system $\langle S, \Longrightarrow, init, V_{in}\rangle$ *where:*

- $S = \prod_{p\in\mathcal{P}} S_p$
- $init = \prod_{p\in\mathcal{P}} init_p$
- $s \overset{\gamma}{\Longrightarrow} s'$ *iff* $s(p) \xrightarrow{\gamma}_p s'(p)$ *if* $p \in loc(\gamma)$ *and* $s(p) = s'(p)$ *otherwise. The notation* $s(p)$ *denotes local state of process* p *in global control state* s.
- $V_{in} = \bigcup_{p\in\mathcal{P}} V_{p,in}$

Next, we need to define transaction schemes. We begin with the standard notion of MSCs which we shall view, in the present context, as certain kinds of labeled partial orders. Their visual representation will be as shown in the various examples already. We shall use Σ_p to denote the set of actions executed by the process p. It consists of actions of the form $\langle p!q, m\rangle$, $\langle p?q, m\rangle$ and $\langle p, a\rangle$ where M is an alphabet of messages and *Act* is an alphabet of internal actions. The communication action $\langle p!q, m\rangle$ stands for p sending the message m to q and $\langle p?q, m\rangle$ stands for p receiving the message m from q. On the other hand, $\langle p, a\rangle$ is an internal action of p with a being the member of *Act* being executed. We set $\Sigma = \bigcup_{p\in\mathcal{P}} \Sigma_p$. We also denote the set of channels $Chan$ given by $Chan = \{(p,q) \mid p \neq q\}$.

Turning now to the definition of MSCs, we define Σ-labeled poset to be a structure $Ch = (E, \leq, \lambda)$ where $(E, \leq)$ is a poset and $\lambda : E \rightarrow \Sigma$ is a labeling function. For $X \subseteq E$ we define $\downarrow(X) = \{e' \mid e' \leq e \text{ for some } e \in X\}$. When $X = \{e\}$ is a singleton we shall write $\downarrow(e)$ instead of $\downarrow(\{e\})$. We say that X is *downclosed* in case $X = \downarrow(X)$. For $p \in \mathcal{P}$, we set $E_p = \{e \mid \lambda(e) \in \Sigma_p\}$. These are the events that p takes part in. Further, $E_{p!q} = \{e \mid e \in E_p \text{ and } \lambda(e) = \langle p!q, m\rangle \text{ for some } m \in M\}$. Similarly, $E_{p?q} = \{e \mid e \in E_p \text{ and } \lambda(e) = \langle p?q, m\rangle \text{for some } m \in M\}$. We define for any channel $c = (p,q)$, the communication relation R_c as: $(e, e') \in R_c$ iff $\mid \downarrow(e) \cap E_{p!q} \mid = \mid \downarrow(e') \cap E_{q?p} \mid$ and $\lambda(e) = \langle p!q, m\rangle$ and $\lambda(e') = \langle q?p, m\rangle$ for some message m.

An MSC (over $(\mathcal{P}, M, Act)$) is a Σ-labeled poset $Ch = (E, \leq, \lambda)$ which satisfies:

(1) $\leq_p$ is a linear order for each p where $\leq_p$ is $\leq$ restricted to $E_p \times E_p$.
(2) Suppose $\lambda(e) = \langle p?q, m\rangle$. Then $\mid \downarrow(e) \cap E_{p?q} \mid = \mid \downarrow(e) \cap E_{q!p} \mid$.
(3) For every p, q with $p \neq q$, $\mid E_{p?q} \mid = \mid E_{q!p} \mid$.
(4) $\leq = (\leq_{\mathcal{P}} \cup R_{Chan})^\star$ where $\leq_{\mathcal{P}} = \bigcup_{p\in\mathcal{P}} \leq_p$ and $R_{Chan} = \bigcup_{c\in Chan} R_c$.

This definition assumes a FIFO discipline for each channel. Other variations can also be dealt with easily. In what follows, we let $agents(Ch)$ denote the set of agents participating in the MSC $Ch = (E, \leq, \lambda)$ and define it as $agents(Ch) = \{p \mid E_p \neq \emptyset\}$.

Definition 2 Transaction Scheme *A Transaction Scheme γ is a finite collection of guarded Message Sequence Charts $\{[I^i : Ch^i]\}_{i=1}^k$. Each Ch^i is an MSC over $(\mathcal{P}, M, Act)$. Each I^i is of the form $\bigwedge_{p \in agents(Ch^i)} I_p^i$ where I_p^i is a propositional logic formula built from the propositions in AP_p.*

For each chart Ch^i in a transaction scheme, we have only mentioned a pre-condition. We have not specified the valuations of atomic propositions upon exiting from a chart. However, send and receive actions have a well-defined meaning. We can also assume that the internal actions are expressed in a standard imperative language. The operational semantics of this imperative language then lends a meaning to the internal actions. Consequently each event in a chart will have a well-defined effect on the truth-values of the local atomic propositions and as a sum total of these effects, we can associate with each chart an output valuation O^i. If more than one output valuation is possible, we can consider them as different transactions. Hence in what follows, we will assume that a transaction scheme is of the form $\{[I^i : Ch^i : O^i]\}_{i=1}^k$ over $(\mathcal{P}, M, Act)$.

Finally, we can now define a Communicating Transaction Processes (CTP) system model as follows.

Definition 3 CTP System Model *A CTP model is a product transition system $\{TS_p\}_{p \in \mathcal{P}}$ over $(\Gamma, \mathcal{P})$ where Γ is a finite set of transaction schemes over $(\mathcal{P}, M, Act)$. Further, for each $\gamma \in \Gamma$, $agents(\gamma) = loc(\gamma)$.*

Here $loc(\gamma)$ is as before where γ is viewed as an abstract action label in high level product transition system; $agents(\gamma)$ is the set of agents participating in some transaction associated with the transaction scheme γ. Let $\gamma = \{[I^i : Ch^i : O^i]\}_{i=1}^k$. Then $agents(\gamma) = \bigcup_{i=1,2,\ldots n} agents(Ch^i)$. Thus the restriction in the above says that the processes taking part in a high level transition in the control flow model are the same as the processes taking part in the transaction scheme associated with this high level transition. This restriction still allows the designer to reorganize the distribution of transactions across the various transaction schemes. Indeed, in the extreme case can one collapse the whole model into a single messy transaction scheme with just one control state for each process! A subclass of CTPs can be obtained by requiring $loc(\gamma) = agents(Ch^i)$ for each i above. In such CTPs one cannot arbitrarily rearrange the transactions.

2.2 A Simple Example

Consider two processors communicating with a shared memory via a bus. The bus controller serves as an arbiter for bus access and serializes the bus access requests by the two processors. The memory controller provides data to the processors for read requests and commits data for write requests. Two of the

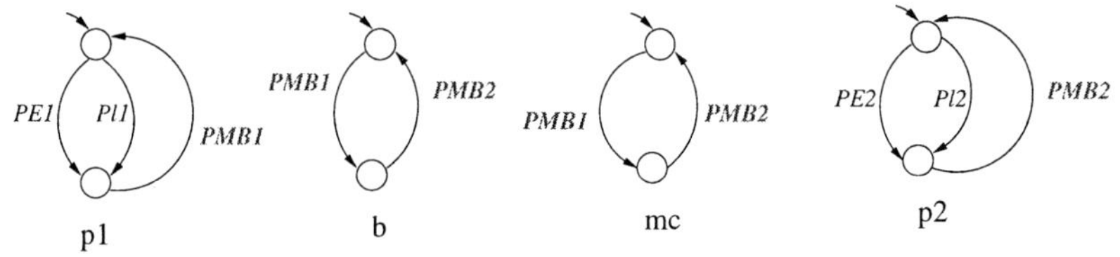

Fig. 4. CTP system model of Multiprocessor example (common actions are shown in bold)

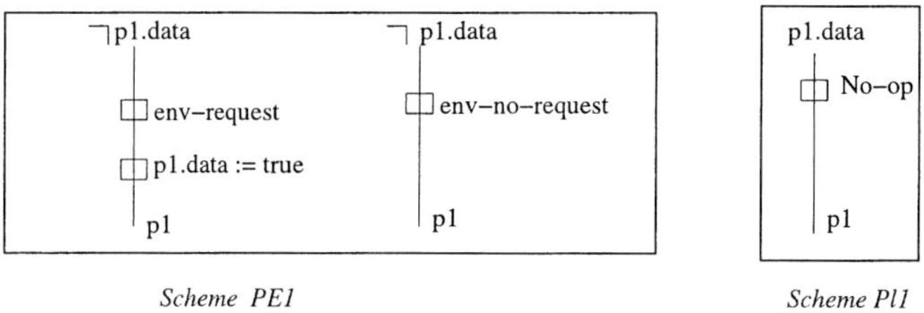

Fig. 5. Local Choices and Environment Interaction in Transaction Schemes of Fig. 4

simple schemes of this system are shown in Figure 5 whereas the high-level control flow is as shown in Figure 4. The two processors are denoted by processes $p1$ and $p2$; b is the bus controller and mc is the memory controller.

The schemes *Pl1* and *PE1* are local schemes in which only $p1$ participates (refer Figure 5). They represent local choices. Scheme *Pl1* is executed when processor $p1$ has data to transfer ($p1.data$ is true). Thus, this scheme consists of a single degenerate MSC. The MSC consists of a single internal action which is a no-op. If processor $p1$ has no data to transfer, then scheme *PE1* is executed. This scheme consists of two MSCs. The choice of which chart is executed is made by the environment. If the environment (*i.e.* the application running on the processor) has data to transfer then $p1.data$ is set; otherwise it remains reset. In this simple example, whether *Pl1* or *PE1* is executed, the process $p1$ next participates in the same transaction scheme, namely, *PMB1*. In general however, this branching in the control flow could lead to different transaction schemes being chosen.

Since the processors have similar behavior, the scheme *PMB1* is identical to *PMB2* except that process $p1$ is replaced by $p2$. (Similar remarks hold for *PE1* and *PE2* as well as *Pl1* and *Pl2*) The scheme *PMB1* is the one shown earlier in Figure 3. This scheme involves a decision by the bus controller b about granting bus access to $p1$. In Figure 3, $p1.data$ holds when $p1$ has data to transfer; $b.free$ holds when the bus is free for transfer. After the transfer the bus is set free. In this simple example, we have assumed that the bus is released after every access, and only write transfers are shown. We have also used our formalism to model more complex interactions such as burst transfers and split transfers, as discussed in Section 4.

3 The Petri Net Semantics

Our goal here is to provide an operational semantics for the CTP model in terms of Petri nets. A key step in our semantics is to combine the different guarded transactions within a transaction scheme into a single entity. This entity will consist of a parallel composition of computation trees; one computation tree for each process that participates in the transaction scheme. Finite labeled *event structures* [21] can be conveniently used for representing such a parallel composition. We define:

Definition 4 Event Structure *An* event structure *is a triple* $ES = (E, \leq, \#)$ *where* E *is a set of events,* $\leq \subseteq E \times E$ *is a partial ordering* causality *relation and* $\# \subseteq E \times E$ *is a* conflict *relation which is required to satisfy the following conditions: (a)* $\#$ *is irreflexive and symmetric, and (b) conflict is inherited via causality, that is* $(e_1 \# e_2 \wedge e_2 \leq e_3) \Rightarrow e_1 \# e_3$.

The idea is that in any execution if an event e occurs and $e' \leq e$ then e' must have occurred earlier in the same execution. On the other hand two events that are in conflict are mutually exclusive. They can never both occur in the same execution. Consequently, if e and e' are mutually exclusive and e'' causally depends on e' then e and e'' are mutually exclusive as well. This is captured by the fact that conflict is inherited via causality.

As a related notion, we define a Σ-labeled event structure to be a structure $ES = (E, \leq, \#, \Lambda)$ where $(E, \leq, \#)$ is an event structure and $\Lambda : E \to \Sigma$ is a labeling function. In diagrams, as illustrated in Figure 6, it will be convenient to represent the causality and conflict relation in a minimal fashion. To this end, we define the immediate causality relation $\lessdot$ and the immediate conflict relation $\#_\mu$ via:

- $e \lessdot e'$ iff $e < e'$ and for every e'', if $e \leq e'' \leq e'$ then $e = e''$ or $e'' = e'$.
- $e \#_\mu e'$ iff $(\downarrow(e) \times \downarrow(e')) \cap \# = \{(e, e')\}$.

Thus two events are in immediate conflict if they are in conflict and their being in conflict can not be attributed to an earlier conflict that is inherited via the causality relation.

The event structure corresponding to the transaction scheme in Figure 2 is shown in Figure 6. The minimal causal relationship is captured by unidirectional arrows. The minimal conflict relation $\#_\mu$ is captured by curved bidirectional arrows. The way the minimal and maximal events of this event structure are connected to the input and output control states of the transaction scheme are also shown.

A (labeled) event structure is accompanied by a natural dynamics. A state is a set of events that have occurred so far along an execution. States are usually referred as configurations. Formally, a configuration c of the event structure $ES = (E, \leq, \#)$ is subset of E which is downclosed and conflict-free. In other words $\downarrow(c) = c$ and $(c \times c) \cap \# = \emptyset$. The empty set is a configuration; it is the initial configuration.

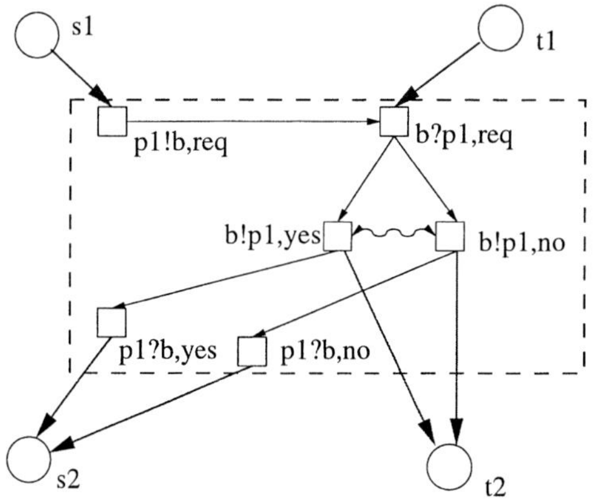

Fig. 6. Event Structure for Transaction scheme in Figure 2 (shown in dashed box)

Let $\mathcal{C}_{ES}$ be the set of (finite) configurations of ES. The event e is *enabled* at the configuration c if e is not in c and $c \cup \{e\}$ is also a configuration. This leads to the transition relation $\longrightarrow_{ES} \subseteq \mathcal{C}_{ES} \times E \times \mathcal{C}_{ES}$ where $c \xrightarrow{e}_{ES} c'$ iff e is enabled at c and $c' = c \cup \{e\}$. Thus we can associate the transition system $TS_{ES} = (\mathcal{C}_{ES}, \longrightarrow_{ES}, \emptyset)$ with the event structure ES. These ideas extend in the expected manner to labeled event structures.

3.1 Constructing Event Structures

In order to define our operational semantics, we first recall that $AP = \bigcup_{p \in \mathcal{P}} AP_p$ is the set of atomic propositions. Let γ be a *transaction scheme* (refer Definition 2) of the form $\gamma = \{I^i : Ch^i : O^i\}_{i=1}^n$ where each I^i is a propositional formula built out of AP, each $Ch^i = (E^i, \leq^i, \lambda^i)$ is a chart over $(\mathcal{P}, M, Act)$ and each O^i is a subset of AP. We let $\gamma^i = [I^i : Ch^i : O^i]$ for each i and call I^i, the *input guard*, Ch^i the *body* and O^i the *output valuation* of the transaction T^i. We will assume without loss of generality that the sets $\{E^i\}_{i=1,\ldots,n}$ are pairwise disjoint.

We construct the labeled event structure $ES_\gamma = (E, \leq \#, \lambda)$ to be associated with a transaction scheme γ as follows. The set of events E is obtained from the event sets E^i $(i = 1, \ldots, n)$ but after identifying events that have isomorphic pasts. Consequently we start with a set X whose elements will be of the form (e, i, P, V_P) where $e \in E^i$, $P = \{p \mid \exists e' \in E^i_p \, \text{and} \, e' \leq^i e\}$ and V_P is a P-valuation such that $V_P \models \bigwedge_{p \in P} I^i_p$. Note that E^i_p is the set of events in E^i in which p participates, that is, for any event $e \in E^i_p$, the label $\lambda^i(e)$ is of the form $\langle p!q, m\rangle$ or $\langle p?q, m\rangle$ or $\langle p, a\rangle$.

Actually, the second and third components in (e, i, P, V_P) are redundant but we will carry them for convenience. Next let $x = (e, i, P, V_P)$ and $y = (d, j, Q, V_Q)$ be in X. Then $x \equiv y$ iff $\downarrow(e)$ in Ch^i is isomorphic to $\downarrow(d)$ in Ch^j in the obvious sense. We shall denote the $\equiv$-equivalence class containing x as $[x]$.

- *Set of Events:* We define E, the set of events of ES_γ to be the $\equiv$-equivalence classes of X. Thus, $E = \{[x] \mid x \in X\}$. Thus for the scheme shown in figure 2, the two events of $p1$ that send a request message req are equivalent as also the two corresponding receive events.
- *Causality Relation:* Let $[x], [y]$ be in E. Then $[x] \leq [y]$ iff there exists (e, i, P, V_P) in $[x]$ and (d, j, Q, V_Q) in $[y]$ such that $i = j$, $e \leq^i d$ and $V_Q \cap AP_P = V_P \cap AP_P$.
- *Conflict Relation:* First we define the relation $\widehat{\#}$ to be the least subset of $E \times E$ which satisfies the following. Suppose $[x], [y] \in E$ are such that $[x] \not\leq [y]$ and $[y] \not\leq [x]$. Furthermore, there exist (e, i, P, V_P) in $[x]$ and (d, j, Q, V_Q) in $[y]$ such that $e \in E_p^i$ and $d \in E_p^j$ for some p but $i \neq j$. Then $[x]\widehat{\#}[y]$. We now define the conflict relation $\#$ as the least subset of $E \times E$ which (a) contains $\widehat{\#}$, (b) is a symmetric relation, and (c) inherits through causality, that is, $[x]\#[y]$ and $[y] \leq [z]$ implies $[x]\#[z]$.
- *Labeling Function:* Finally, the labeling function Λ is given by:

$$\Lambda([(e, i, P, V_P)]) = \lambda^i(e)$$

Lemma 1 *$ES_\gamma = (E, \leq, \#, \Lambda)$ is a labeled event structure.*

Proof: Due to the isomorphism condition imposed in the definition of $\equiv$, it is easy to observe that $\leq$ is a partial ordering relation. From the definition of the relation $\#$ it is symmetric and is inherited via $\leq$. We need to show that it is irreflexive. Assume for contradiction that there exists $[x]$ such that $[x]\#[x]$. In this case it is not difficult to see there exist $[y]$ and $[z]$ such that $[y]\widehat{\#}[z]$ and $[y] \leq [x]$ and $[z] \leq [x]$. This implies there exist (e, i, P, V_P) in $[y]$ and $(e1, i, P1, V_{P1})$ in $[x]$ such that $e \leq^i e1$. Further, there exist (d, j, Q, V_Q) in $[z]$ and $(d1, j, Q1, V_{Q1})$ in $[x]$ such that $d \leq^j d1$. Then by the definition of the $\equiv$ relation, it follows that there exists $(d', i, Q', V_{Q'})$ in $[z]$ such that $d' \leq^i e1$. But then from the definition of $\widehat{\#}$ it follows that there exists p such that $e \in E_p^i$ and $d' \in E_p^i$. This leads to $e \leq^i d'$ or $d' \leq^i e$ which in turn leads to $[y] \leq [z]$ or $[z] \leq [y]$ contradicting $[y]\widehat{\#}[z]$. The fact that the labeling function is well-defined is obvious. □

3.2 The Petri Net Semantics

We construct the Petri net semantics for the CTP model in three steps. First we convert each labeled event structure yielded by a transaction scheme into an acyclic net (without an initial marking). We then merge these nets with the high level control flow net. As a last step we refine local control states and the transitions to expose information about the valuations of the atomic propositions.

From Event Structure to Acyclic Net. First, let γ be a transaction scheme and $ES_\gamma = (E, \leq, \#, \lambda)$ be its event structure representation. For $e \in E$ we set

$proc(e) = p$ if there exists (x, i, P, V_P) in e such that $x \in E_p^i$. It is easy to see that $proc$ is a well-defined function, and it is also easy to check that if $e \widehat{\#} e'$, $proc(e) = proc(e')$.

Now, we define the net associated with our event structure. Before doing so, note that the *minimal* causality relation of event structure ES_γ is denoted as $\lessdot$; the minimal conflict relation is denoted as $\#_\mu$. From the construction of ES_γ it follows easily that that if $e \#_\mu e'$ then $proc(e) = proc(e')$. Furthermore, $\#_\mu$ is transitive (and symmetric). Hence in what follows, while writing $\#$ instead of $\#_\mu$ for convenience, we will let $[e]_\#$ denote the set of events given by $[e]_\# = \{e\} \cup \{e' \mid e \#_\mu e'\}$.

We define the net representation $ES_\gamma = (E, \leq, \#, \lambda)$ as the acyclic net $N_\gamma = (B_\gamma, E_\gamma, F_\gamma)$ where:

(1) The set of transitions $E_\gamma = E$.

(2) The set of places B_γ and the flow relation F_γ are the least sets which satisfy:
  - (i) Suppose $e \lessdot e'$ and $proc(e) \neq proc(e')$. Then $(e, e') \in B_\gamma$, $(e, (e, e')) \in F_\gamma$, and $((e, e'), e') \in F_\gamma$.
  - (ii) Let $e \lessdot e'$ and $proc(e) = proc(e')$. Then $(e, [e']_\#) \in B_\gamma$ and $(e, (e, [e']_\#)) \in F_\gamma$ and $((e, [e']_\#), e'') \in F_\gamma$ for every e'' in $[e']_\#$.

The net representation of the event structure of Figure 6 is shown in Figure 7.

Merging the Control Flow. Let $\mathrm{TP} = \{TS_p\}_{p \in \mathcal{P}}$ be a CTP over $(\Gamma, \mathcal{P})$ where Γ is a finite set of transaction schemes over $(\mathcal{P}, M, Act)$. Let $TS_p = (S_p, \Gamma_p, \longrightarrow_p, init_p, V_{p,in})$ be the transition system associated with transaction process p (note that $V_{p,in}$ is the initial p-valuation). For each transaction scheme γ in Γ let ES_γ be its event structure representation and $N_\gamma = (B_\gamma, E_\gamma, F_\gamma)$, the net associated with ES_γ.

For convenience we will denote the set of pre and post control states of the transaction scheme γ as ${}^\bullet\gamma$ and $\gamma^\bullet$ respectively and define these sets as:
${}^\bullet\gamma = \{s \mid \gamma \in \Gamma_p \text{ and } s \xrightarrow{\gamma}_p s' \text{ for some } s, s' \in S_p\}$.
$\gamma^\bullet = \{s' \mid \gamma \in \Gamma_p \text{ and } s \xrightarrow{\gamma}_p s' \text{ for some } s, s' \in S_p\}$.

We can now carry out the second step in providing the operational semantics. The *control flow Petri net* of TP is the Petri net

$$\mathrm{CFN}_{\mathrm{TP}} = (S_{\mathrm{TP}}, T_{\mathrm{TP}}, F_{\mathrm{TP}}, M_{in,\mathrm{TP}})$$

where:

- $S_{\mathrm{TP}} = \bigcup\{S_p \mid p \in \mathcal{P}\} \cup \bigcup\{B_\gamma \mid \gamma \in \Gamma\}$.
- $T_{\mathrm{TP}} = \bigcup\{E_\gamma \mid \gamma \in \Gamma\}$
- $F_{\mathrm{TP}} = \bigcup_{\gamma \in \Gamma}(F_\gamma) \cup$
  $\{(s, e) \mid e \in min(E_\gamma), s \in {}^\bullet\gamma \cap S_p, proc(e) = p\} \cup$
  $\{(e, s') \mid e \in max(E_\gamma), s' \in \gamma^\bullet \cap S_p, proc(e) = p\}$)
- $M_{in,\mathrm{TP}}(z) = 1$ if there exists p s.t. $z = init_p$ (the initial state of some p). Otherwise $M_{in,\mathrm{TP}}(z) = 0$.

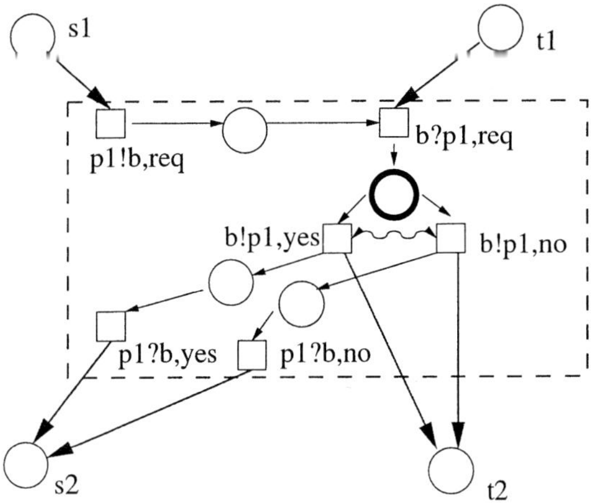

Fig. 7. Acyclic Net for Transaction scheme in Figure 2 (shown in dashed box)

By $min(E_\gamma)$ ($max(E_\gamma)$) we mean the set of minimal (maximal) elements under the causality relation of the event structure ES_γ.

The control flow net of a CTP description captures the behaviors in the individual processes with one major caveat. For events which are in minimal conflict, it does not expose the valuations of the atomic propositions which resolve the conflict. As an example, consider the event structure of Figure 6 and its net representation shown in Figure 7. Now, consider the place marked in bold in Figure 7; this place has two outgoing flow arcs leading to two events in minimal conflict. Here, the control flow net contains infeasible behaviors not allowed by the transaction scheme of Figure 2. This is because the control flow net of Figure 7 does not capture the condition which needs to be evaluated to decide which of the two conflicting events is executed (in this case, the condition is $b.free$). The simplest solution is to annotate the flow arcs with this condition (*i.e.*, the two arcs should be annotated with $b.free$ and $\neg b.free$)[2]. However adding such annotations does not give an executable model of the allowed behaviors for a CTP. To construct such an executable model, we need to systematically expose the data dependencies, that is, the valuation of atomic propositions in the places and transitions of the control flow net. This is now done by constructing a Petri net corresponding to any CTP specification.

Constructing the Petri Net. Let CFN $= (S_{\text{TP}}, T_{\text{TP}}, F_{\text{TP}}, M_{in,\text{TP}})$ be the control flow net of TP, a CTP. Then PN_{TP} is the Petri net representation of TP and it is the Petri net $\text{PN}_{\text{TP}} = (S, T, F, M_{in})$ where S, T and F are the least set of elements satisfying the following conditions:

[2] One could capture this easily using Colored Petri nets [19] but this would entail an additional intermediate description.

- Suppose s is in S_p. Then (s, V_p) is in S where V_p is a p-valuation. Next let γ be in Γ and $N_\gamma = (B_\gamma, E_\gamma, F_\gamma)$ be the net associated with ES_γ and $(x, y) \in B_\gamma$. Now suppose $(e, i, P, V_P) \in x$. Then $((x, y), V_P)$ is in S.
- Let γ be in Γ and $N_\gamma = (B_\gamma, E_\gamma, F_\gamma)$ be the net associated with ES_γ and $x \in E_\gamma$. Suppose $(e, i, P, V_P) \in x$. Then (x, V_P) is in T.
- Suppose $(s, x) \in F_{\mathrm{TP}}$ with $s \in S_p$ for some p and $(e, i, \{p\}, V_p) \in x$. Then $((s, V_p), (x, V_p))$ is in F. Also, suppose $(x, s') \in F_{\mathrm{TP}}$ with $s' \in S_p$ for some p and $(e, j, Q, V_Q) \in x$ and O^j is the output valuation of the transaction $[I^j : Ch^j : O^j]$. Then $((x, V_Q), (s', V_p))$ is in F where $V_p = O^j \cap AP_p$. Finally, let $((x, y), V_P) \in S$. Then $((x, V_P), ((x, y), V_P))$ is in F. Furthermore, $(((x, y), V_P), (y, V_Q)) \in F$ provided $proc(x) \neq proc(y)$ and (y, V_Q) is in T and $V_Q \cap AP_P = V_P$. In case $proc(x) = proc(y)$ then $(((x, y), V_P), (z, V_Q)) \in F$ provided $z \in y$ and $V_Q \cap AP_P = V_P$. In general, y denotes a set of events in minimal conflict and belonging to the same process.
- $M_{in}(z) = 1$ if $z = (init_p, V_{p,in})$ for some p. Otherwise $M_{in}(z) = 0$.

For example, the Petri net fragment for the transaction scheme of Figure 2, together with the refined representation of its surrounding control places will be as shown in Figure 8. Here for convenience we have assumed that the output guard for both the transactions is $b.free$.

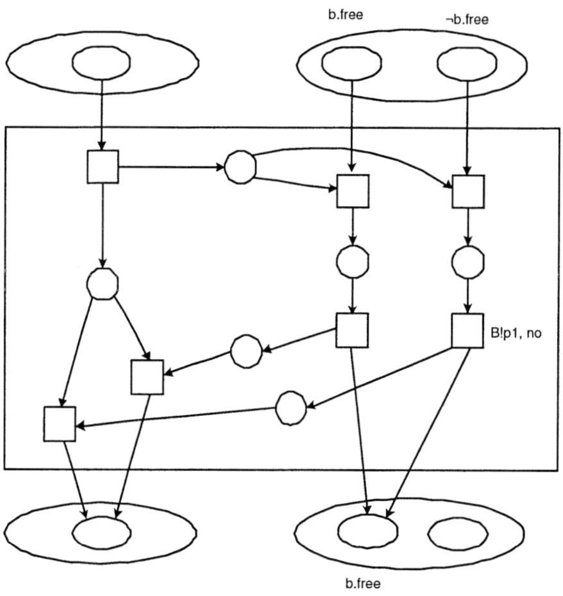

Fig. 8. Petri net fragment for Transaction scheme in Figure 2

This concludes the construction of the Petri net to be associated with a CTP. The execution semantics of a CTP is then just the usual execution semantics of its associated Petri net.

4 Specifying the AMBA Bus Protocol

In this section, we present a non-trivial example to show the use of the CTP as a modeling language. In particular, we model the data communication between two components via a bus. We call the originator of the data communication the *master* and the receiver of the communication the *slave*. Our model consists of five processes executing in parallel: the master component (called P_m), interface of the master component (called I_m), the bus controller (called BC), interface of the slave component (called I_s) and the slave component (called P_s). The master and slave components (P_m and P_s) are often processors or co-processors. The high level transition systems of the individual processes are shown in Figure 9. In the diagram, the constituent guarded transactions of the various transaction schemes have not been shown.

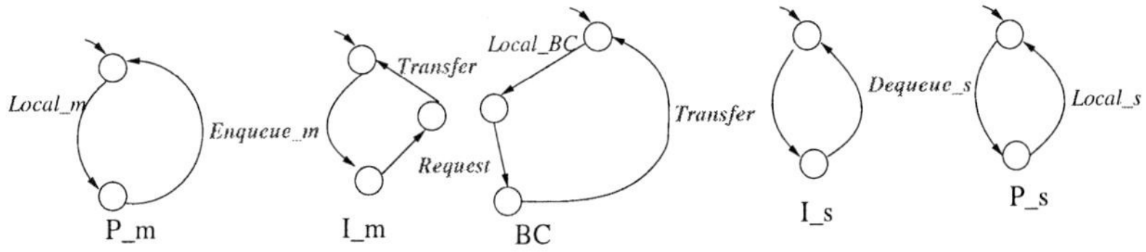

Fig. 9. CTP model of interfaces between two embedded co-processors. Common actions are shown in bold. Wherever possible, labels of repeated occurrences of a common action have been shared to reduce visual clutter.

To develop our example, we fix: (1) a specific bus protocol, (2) storage capabilities of the interfaces, I_m and I_s (3) interaction between the components and interfaces. We choose the popular AMBA bus protocol used in ARM system-on-chip designs [2]. We assume that each interface contains a bounded queue to hold data in transit. The interaction between a component and its interface then involves enqueueing and dequeuing these queues. In particular, our choice of the component-interface protocol is drawn from the interface modules developed in the European COSY project [7]. These interfaces were originally designed for data transfer between co-processors connected to a common bus running the PI-Bus protocol. Here instead we shall be using the AMBA bus protocol.

The transaction schemes $Local_m$, $Local_{BC}$ and $Local_s$ have only one participating process: namely P_m, BC and P_s respectively. They represent internal computations of these processes and we do not describe them here. The other three transaction schemes denote the following interactions. $Enqueue_m$ involves enqueueing of data by the master process P_m into the queue of the master interface I_m. Similarly, $Dequeue_s$ denotes the dequeuing of data from the queue of the slave interface I_s by slave processor P_s. The scheme *Request* denotes request for bus access by the master to the bus controller, and subsequent granting of bus access (if any). Finally, the scheme *Transfer* denotes the transfer of data from master interface I_m into slave interface I_s over the bus.

We now describe the transaction scheme $Transfer$ where

$$agents(Transfer) = \{I_m, I_s, BC\}$$

In particular, this will show our formal specification of the AMBA bus protocol. Conditions on the local variables of each of these processes are used to decide which chart of $Transfer$ is executed in a particular execution. We will freely use values of these local variables in our charts. The events in the charts pass these values between variables of different processes, thereby modeling data transfer.

Local Variables. We present the local variables of the processes I_m, I_s and BC in Figure 10. We wish to note that $maxwait$ denotes a predefined fixed positive constant, and $\mathcal{D}$ denotes the data type of the data being transmitted from master component P_m to slave component P_s. Furthermore, *Addr* denotes the range of addresses manipulated by I_m and I_s.

Process	Local Variables
I_m	mq : Queue of $(Addr, \mathcal{D})$ $data\_sent, wait\_data : \mathcal{D}$ $wait\_addr : Addr$ $grant_m$: boolean
I_s	sq : Queue of $(Addr, \mathcal{D})$ $addr\_rcvd : Addr$ $waitcnt : 0 \ldots maxwait$
BC	$gnt_m, split_m$: boolean

Fig. 10. Local Variables in the Interface Example

The master and slave interfaces I_m and I_s each contain a queue mq and sq. The master queue mq receives data from P_m and passes it to the slave interface I_s. The slave queue sq receives data from master interface I_m and passes it to the slave component P_s. The transfer of data between the master and slave interfaces is over a bus, and is thus dictated by the bus protocol. In this case, we consider the AMBA bus protocol which has the following features. This will clarify the need for the various local variables.

Bus Access Protocol. Each transfer is preceded by a grant of bus access by the bus controller to a master. This information is stored by the bus controller BC in the boolean variable gnt_m. Its value is communicated to I_m in the *Request* transaction scheme (not shown here) when I_m requests for bus access. I_m stores this information in $grant_m$. Thus there is clear relationship between $I_m.grant_m$ and $BC.gnt_m$. Similar relationships exist between other local variables of different processes owing to the flow of values via messages.

Pipelined Transfer. Multiple transfers from I_m to I_s are *pipelined.* For example suppose I_m wants to transfer (a_1, d_1),(a_2, d_2), (a_3, d_3) to I_s. This is a request to write d_1 to address a_1, d_2 to address a_2 and d_3 to address a_3. The transfer over the address and data lines proceeds as follows.

Clock cycle:	1	2	3	4
Address :	a_1	a_2	a_3	-
Data :	-	d_1	d_2	d_3

Since in every cycle, the data of the previous cycle's address is transmitted, this needs to be remembered. This information is stored in the local variable *data_sent* of I_m. Similarly, on the slave interface side, the address received in previous cycle is stored in the variable *addr_rcvd* of process I_s.

Transfer with Wait Cycles. The slave interface I_s may not be ready to write data in every cycle *e.g.* the slave queue *sq* may be full. This results in insertion of "*wait cycles*". The number of such wait cycles is stored in the local variable *waitcnt*. In the presence of wait cycles, the transfer can be as follows:

Clock cycle:	1	2	3	4	5	6
Address :	a_1	a_2	a_2	a_2	a_3	-
Data :	-	d_1	d_1	d_1	d_2	d_3

Here, d_1 is transfered after two wait cycles. During these wait cycles, the master interface needs to keep on transmitting a_2 as address and d_1 as data; otherwise the correspondence between address and data is lost. Hence the need for the local variables *wait_addr* and *wait_data* in process I_m.

Split Transfer. If the number of wait cycles equals a threshold *maxwait*, the slave interface I_s informs the bus controller *BC* that it is currently unable to service the master interface I_m. The bus controller *BC* then records that I_m is suspended by setting $split_m$ (which is reset later when I_s is able to serve I_m).

Message Sequence Charts. The transaction scheme *Transfer* is collection of MSCs, one for each of the following mutually exclusive conditions[3].

(1) $\neg grant_m \vee (empty(mq) \wedge waitcnt = 0) \vee (split_m \wedge full(sq))$
(2) $grant_m \wedge \neg split_m \wedge \neg empty(mq) \wedge \neg full(sq) \wedge waitcnt = 0$
(3) $grant_m \wedge \neg split_m \wedge \neg empty(mq) \wedge full(sq) \wedge waitcnt = 0$
(4) $grant_m \wedge \neg split_m \wedge \neg full(sq) \wedge waitcnt > 0$
(5) $grant_m \wedge \neg split_m \wedge full(sq) \wedge waitcnt > 0 \wedge waitcnt < maxwait$
(6) $grant_m \wedge \neg split_m \wedge full(sq) \wedge waitcnt = maxwait$
(7) $grant_m \wedge split_m \wedge \neg full(sq) \wedge waitcnt = maxwait$

In case 1, either the bus is busy ($\neg grant_m$ holds) or the master queue *mq* is empty and $waitcnt = 0$ (*i.e.* new data needs to be dequeued from *mq* which is empty), or the data transfer from I_m has been split, but I_s is still not ready to

[3] These guards are also total, when the relationships between the local variables of various processes are taken into account.

input data (sq is full). In these cases, no data is transmitted, no control signals are exchanged and the chart is a no-op.

In case 2 (shown in Figure 11), the master is granted access to the bus, data is dequeued from the master queue mq, and enqueued into the slave queue sq. This corresponds to "normal" data transfer without wait cycles and split transfer. Each message is of the form Signal_name(Value), such as ADDR(a). Access to mq and sq are through the *Enqueue* and *Dequeue* methods.

The chart for case 3 is shown in Figure 12. This corresponds to the scenario where wait cycles are initiated (note that $waitcnt = 0$) for some transfer, since the queue at I_s is full. Note that the first three actions by I_m in this chart are the same as Figure 11. This illustrates the distributed decision-making performed by agents of a transaction scheme in deciding which chart is to be executed. As long as the slave interface I_s does not execute its internal actions, we cannot decide whether chart for case 2 or case 3 is being executed.

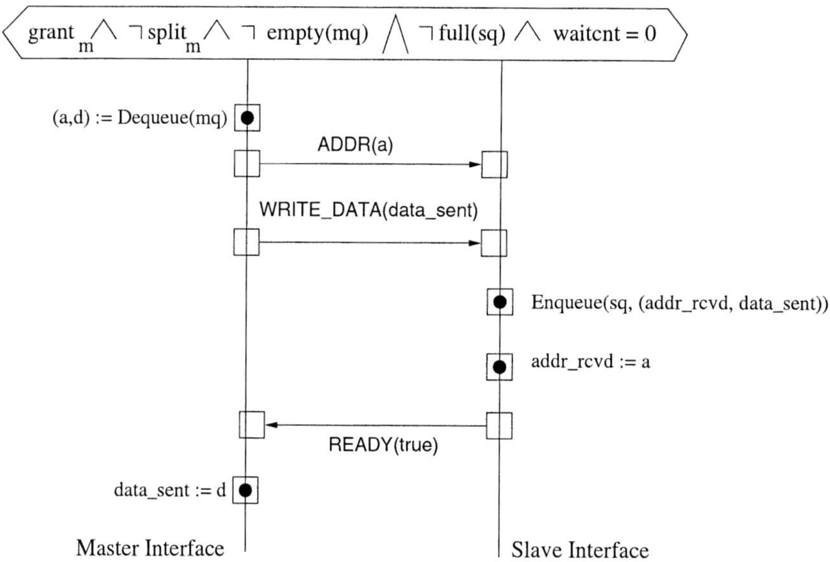

Fig. 11. Normal data transfer between master and slave interface

Case 4 corresponds to $grant_m \wedge \neg split_m \wedge \neg full(sq) \wedge waitcnt > 0$. Thus, the master has been granted bus access (since $grant_m$ holds) and is currently going through a wait cycle (since $waitcnt > 0$). The slave is however ready to input data (since $\neg full(sq)$), that is, the master need not wait any more. Thus, this scenario corresponds to the last wait cycle. The chart is shown in Figure 13.

Case 5 corresponds to $grant_m \wedge \neg split_m \wedge full(sq) \wedge waitcnt > 0 \wedge waitcnt < maxwait$. Here again the master has been granted bus access (since $grant_m$ holds) and is currently going through a wait cycle (since $waitcnt > 0$). The slave

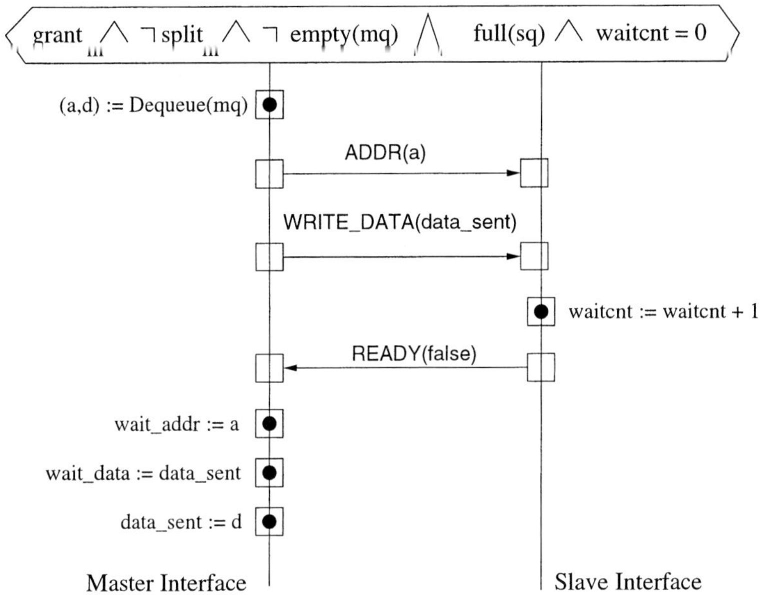

Fig. 12. Initiation of wait cycles

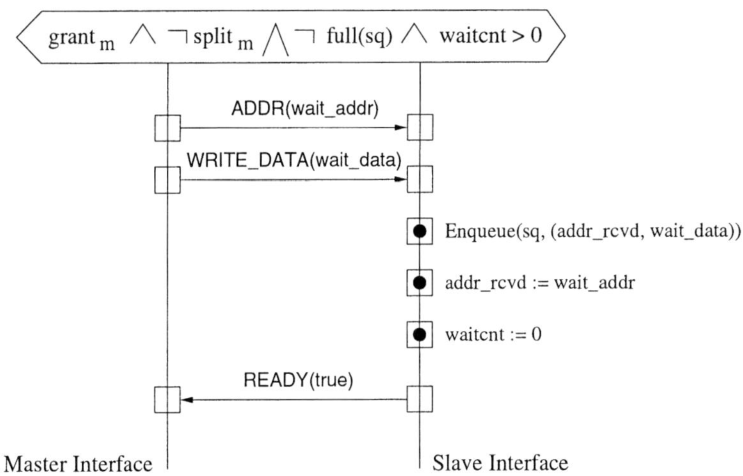

Fig. 13. The last wait cycle

is still not ready to input data (since $full(sq)$). This scenario corresponds to a wait cycle which is not the last. The chart appears in Figure 14.

Case 6 corresponds to $grant_m \wedge \neg split_m \wedge full(sq) \wedge waitcnt = maxwait$. Here the master is going through a wait cycle, but the number of wait cycles has reached the pre-defined threshold *maxwait*. Thus, this requires the slave to

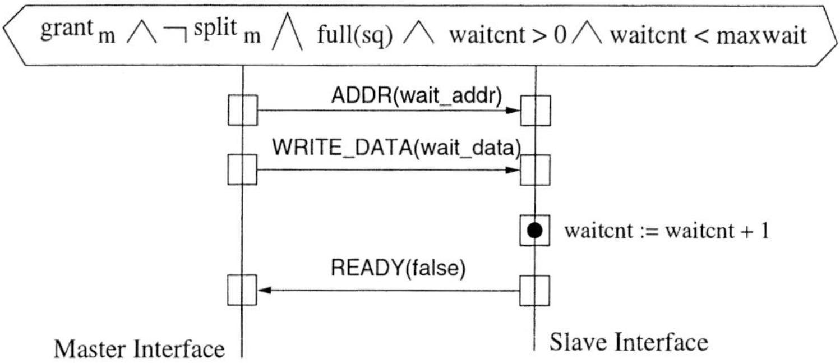

Fig. 14. A wait cycle which is not the last

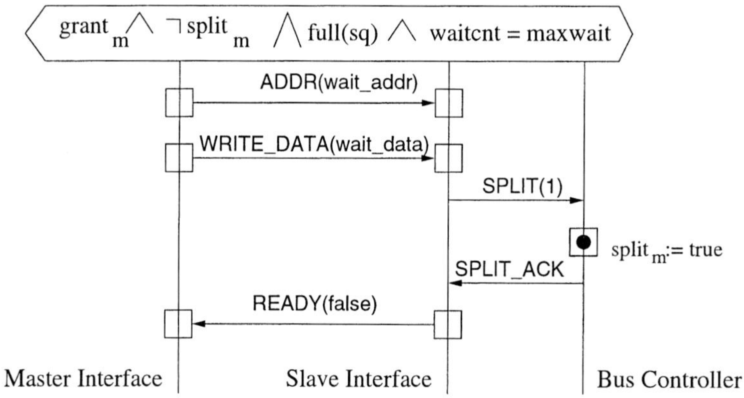

Fig. 15. Initiation of split transfer

initiate *split transfer* by interacting with the bus controller. The chart appears in Figure 15.

Case 7 corresponds to $grant_m \wedge split_m \wedge waitcnt = maxwait \wedge \neg full(sq)$. This means that the transfer from I_m to I_s was previously split, thus $split_m$ holds. However, the slave is currently ready to input data (since $\neg full(sq)$), thereby terminating the split transfer. Thus, this chart will involve exchange of SPLIT and SPLIT_ACK signals along the lines of Figure 15, and the resetting of *waitcnt* to zero.

Remark. As a matter of fact, the AMBA bus protocol is intended for interaction between multiple masters and multiple slaves. Here we have modeled only one master and one slave. However, all the features for multi-component interaction have been, in principle, captured. For example, the suspension of bus access to a master (split transfers) is to allow another master to take over bus access. In our case, even with one master we have modeled this feature via the variable $split_m$.

In future, we plan to explicitly model interaction among multiple masters and slaves, that is, multiple instances of Γ_m and Γ_s. An elegant way of modeling masters and slaves is to treat all masters as one process class, and all slaves as another process class. The individual masters and slaves then correspond to concrete objects of these classes. This requires extending our model to handle objects, classes and subclasses. We are currently pursuing research in this direction.

5 Behavioral Properties of CTPs

In this section, we introduce some important behavioral properties of the CTP model and the techniques currently available for determining these properties.

5.1 Well-Formed Transaction Schemes

For pragmatic reasons, our definition of the CTP model imposes almost no syntactic restrictions. As a result, one can easily specify behaviors which are problematic from both specification and implementation standpoints. For instance, consider the transaction scheme shown in Figure 16 and its associated event structure. If the control flow enables this transaction scheme with the valuation $\{\neg A, B\}$, there will be a deadlock and no event in the associated event structure will execute. On the other hand if the valuation is $\{A, \neg B\}$ then the send events $\langle p!q, m1\rangle$ and $\langle q!p, m2\rangle$ can execute with no order after which there will be a deadlock. Thus local deadlocks can arise due to incomplete specification of the transaction schemes. As a method for detecting and eliminating such undesirable behaviors, we propose the notion of *well-formed* transaction schemes. Intuitively, this notion says that in the locality of a transaction scheme, the maximal executions of the event structure associated with the transaction scheme are precisely the executions of the transactions mentioned in the transaction scheme.

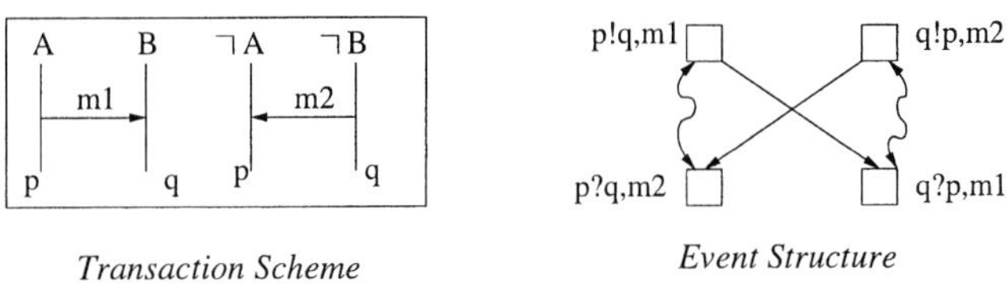

Fig. 16. A Transaction Scheme which is not well-formed

Definition 5 Well-formed Transaction Scheme *Let* $\mathrm{T} = \{T^i = [I^i : Ch^i : O^i]\}_{i=1,2,\ldots,n}$ *be a transaction scheme and* $ES_{\mathrm{T}} = (E, \leq, \#, \Lambda)$ *be its event structure representation. For a configuration (a downclosed conflict-free subset of events) c of* ES_{T}*, we let* $ES_{c,\mathrm{T}}$ *be the sub-event structure induced by c; it is the event structure* $(c, \leq_c, \#_c)$ *where* $\leq_c(\#_c)$ *is* $\leq(\#)$ *restricted to c. Let* $MAXC_{\mathrm{T}}$

be the sub-event structures of ES_T induced by the set of maximal configurations of T. *Then, transaction scheme* T *is said to be* well-formed *iff there exists a bijection $f : \{1, 2, \ldots, n\} \rightarrow MAXC_T$ such that Ch^i is isomorphic to $f(i)$ for each i in $\{1, 2, \ldots, n\}$.*

Each transaction scheme can be effectively analyzed to determine if it is well formed. Clearly the transaction scheme shown in Figure 16 is not well-formed. We are *not* advocating the notion of well-formedness as mandatory but we believe it is a useful criterion using which certain types of incomplete and inconsistent specifications at the level of transaction schemes can be caught. It should also be clear that well-formedness alone will not suffice to guarantee sound implementations. For instance if the behaviors of a transaction scheme exhibit intra-process non-determinism then hardware implementation can be problematic.

5.2 Behavioral Properties

Let TP be a CTP and N_{TP} be its Petri net representation. We shall assume the standard behavioral notions for Petri nets here [9]. We will say that that TP is *transaction-deterministic* if for every reachable marking M of PN_{TP}, if x and y are events belonging to the event structure associated with a transaction scheme in TP and both x and y are enabled at M then $proc(x) \neq proc(y)$. Consequently x and y can occur causally independent of each other at M. Thus transaction-determinism guarantees during the course of executing events taken from a transaction scheme, there will be no conflict. We will also say that TP is *bounded* in case its Petri net is bounded (has only a finite set of reachable markings).

We say that a transaction scheme is *anchored* in case each of its transactions is anchored. A transaction is anchored if its associated MSC, say, $Ch = (E, \leq, \lambda)$ has a least element e_{in} and greatest element e_{fin} and further more,there exists p such that $e_{in}, e_{fin} \in E_p$. Thus the transaction is initiated and terminated by a single agent. An interesting observation here is that if each transaction scheme in a CTP is anchored, then the CTP is bounded.

Via the Petri net semantics, the notions of TP being *live* and *dead-lock free* can also be defined. Clearly, all these properties are decidable since the corresponding problems for Petri nets are decidable. We are currently studying how efficient decision procedures can be developed by exploiting the additional structure provided by the CTP model.

6 Connection to Live Sequence Charts (LSC)

Our CTP formalism serves as a high level executable specification language based on Message Sequence Charts. Recently, Damm and Harel have developed the Live Sequence Charts (LSC) formalism which is also a MSC based modeling language. A powerful execution framework for LSCs based on the so called Play-in/Play-out approach is also being developed by Harel and his collaborators [13, 15]. In this section, we explore the connections between the CTP and LSC formalisms, namely (a) how to interpret LSCs over the CTP model and (b) how to translate CTP models to the LSC language.

The basic feature of LSCs is that it has two types of charts, namely *existential* and *universal* charts. The universal charts are used to specify requirements that *all* the possible system runs must satisfy. A universal chart typically contains a *pre-chart* followed by a main chart to capture the requirement that if along any run, the scenario depicted in the pre-chart occurs then the system *must* also execute the main chart. Existential charts specify sample interactions, typically between the system components and the environment that at least one system run must satisfy. Existential charts can be used to specify system tests and illustrate typical unrestricted runs. The LSC formalism also uses *cold* and *hot* conditions which are in some sense provisional and mandatory guards. If a cold condition holds during an execution then control is intended to pass to the location immediately after the cold condition. If it is false then the chart-context in which this condition occurs is exited. A hot condition, on the other hand, must always be true. If an execution reaches a hot condition which evaluates to false then this signals the violation of requirement and the system is supposed to abort. Thus we can attach the constant *false* condition at the end of chart Ch to capture the requirement that Ch must never occur. In other words, Ch is a forbidden scenario. On the other hand, cold conditions can be used to program if-the-else constructs. Similarly we can also specify events to be hot or cold.

It will be convenient to break down the features of the LSC language into simple units and present them individually. Assuming the notations and terminology developed in the previous section, we define a basic *universal* LSC (over $(\mathcal{P}, M, Act)$) with a *pre-chart* as a structure $[PCh, BCh]$ where:

(1) $PCh = (E_{PCh}, \leq_{PCh}, \lambda_{PCh})$ is a MSC called the *pre-chart*.
(2) $BCh = (E, \leq, \lambda)$ is a MSC called the *body* with $E_{PCh} \cap E = \emptyset$.
(3) $[PCh, BCh]$ denotes $PCh \circ BCh$, the asynchronous concatenation of PCh with BCh. This is in keeping with the asynchronous nature of our CTP model; [8] mentions a variant involving synchronous concatenation. Thus, strictly speaking, we consider an asynchronous version of the LSC formalism [8].
(4) $agents(min(BCh)) \subseteq agents(PCh)$.

In order to explain the last condition given above, recall that $min(BCh)$ is the set of minimal events in BCh. The last condition is intended to ensure that in the asynchronous concatenation of PCh followed by BCh, every event of BCh will have a causal predecessor in PCh. As might be expected, we define the asynchronous concatenation $Ch^1 \circ Ch^2$ of two MSCs $Ch^1 = (E^1, \leq^1, \lambda^1)$ and $Ch^2 = (E^2, \leq^2, \lambda^2)$ with $E^1 \cap E^2 = \emptyset$ as the MSC $Ch = (E, \leq, \lambda)$ where $E = E^1 \cup E^2$ and $\lambda(e) = \lambda^1(e)$ $(\lambda^2(e))$ if e is in E^1 (E^2). Finally $\leq$ is the least partial ordering relation over E which contains $\leq^1$ and $\leq^2$ and satisfies: if $e \in E^1_p$ and $e' \in E^2_p$ for some p then $e \leq e'$.

We define a basic *universal* chart with *pre-condition* as a structure $[P, \varphi, BCh]$ where φ is a propositional formula built out of AP_P called the pre-condition and

BCh is a chart called the body such that $agents(min(BCh)) \subseteq P$. Basic *existential* charts denoted $\langle PCh, BCh \rangle$ with a pre-chart as well as basic existential charts with pre-conditions denoted $\langle P, \varphi, BCh \rangle$ can be defined in a similar fashion. Neither existential nor universal charts with post-charts are interesting. We however define a basic universal (existential) chart with a post-condition as the structure $[BCh, P, \varphi]$ ($\langle BCh, P, \varphi \rangle$) where, as before, φ is a propositional formula built out of AP_P and $agents(max(BCh)) \subseteq P$. Intuitively such charts denote the property that if BCh is executed, φ must (may) hold.

A cold condition is a basic existential chart with a pre-condition whose body chart is empty. A hot condition is a basic universal chart with a post-condition whose body chart is empty. LSCs can now be inductively obtained by starting with the basic charts and allowing the body itself to be an LSC. Viewing the resulting class of LSCs as atomic assertions, one can obtain LSC specifications by forming boolean combinations of these atomic assertions.

6.1 When Does a CTP Model Satisfy a LSC Specification?

We now interpret the basic LSC specifications over CTPs. It is easy to extend this interpretation to more complicated LSC specifications.

Let TP $= \{TS_p\}_{p \in \mathcal{P}}$ be a CTP over $(\Gamma, \mathcal{P})$ where Γ is a finite set of transaction schemes over $(\mathcal{P}, M, Act)$. Let $\mathrm{PN_{TP}}$ be the Petri net representation of TP, constructed from Σ-labeled event structures representing the transaction schemes. Let $\Longrightarrow$ be the labeled transition relation defined over the reachable markings of $\mathrm{PN_{TP}}$ given by: $M \stackrel{\alpha}{\Longrightarrow} M'$ iff there exists a transition t of $\mathrm{PN_{TP}}$ such that t is enabled at M and M' is the resulting marking when t occurs at M. Furthermore, $\Lambda(t) = \alpha$ where Λ is the obvious labeling function that assigns to each transition of $\mathrm{PN_{TP}}$, a label in Σ (the set of labels of events appearing in the transaction schemes). This transition relation $\Longrightarrow$ is extended to Σ-sequences in the obvious way and this extension will be also be denoted as $\Longrightarrow$. Next let $Ch = (E, \leq, \lambda)$ be an MSC. Then λ applied pointwise to a linearization of $(E, \leq)$ yields a member of Σ^*. We let $lin(Ch)$ be the subset of Σ^* obtained this way, and refer to it also, by abuse of terminology, as the linearizations of Ch. We define Σ_{Ch} to be the subset of Σ given by $\Sigma_{Ch} = \{\lambda(e) \mid e \in E\}$. Finally, if $\sigma \in \Sigma^*$ and $\Sigma' \subseteq \Sigma$ then $\lceil_{\Sigma'} (\sigma)$ is the Σ' projection of σ. For an MSC Ch we will often write $\lceil_{Ch}$ instead of $\lceil_{\Sigma_{Ch}}$. We are now prepared to interpret the basic LSC specifications over CTPs.

(1) Let $[PCh, BCh]$ be a basic universal chart with a pre-chart. Then TP satisfies $[PCh, BCh]$ iff *every* reachable marking (s_0, V_0) of $\mathrm{PN_{TP}}$ satisfies the following condition. Suppose $M_0 \stackrel{\sigma_0}{\Longrightarrow} M_1$ such that $\lceil_{PCh} (\sigma_0)$ is in $lin(PCh)$. Then for *every* $M_1 \stackrel{\sigma_1}{\Longrightarrow} M_2$, there exists $M_2 \stackrel{\sigma_2}{\Longrightarrow} M_3$ such that a prefix of $\lceil_{\Sigma'} (\sigma_0\sigma_1\sigma_2)$ corresponds to a member of $lin(PCh \circ BCh)$ where $\Sigma' = \Sigma_{PCh} \cup \Sigma_{BCh}$. Thus, this universal requirement demands that whenever (a linearization of) PCh has been executed then this *must* be followed by an execution of (a linearization of) BCh.

(2) Next suppose $[P, \varphi, BCh]$ is a basic universal chart with a pre-condition. Then TP satisfies $[P, \varphi, BCh]$ iff *every* reachable marking $M_0 = (s_0, V_0)$ of PN_{TP} satisfies the following condition. Suppose $V_0 \models \varphi$. Then for every $M_0 \stackrel{\sigma_0}{\Longrightarrow} M_1$ there exists $M_1 \stackrel{\sigma_1}{\Longrightarrow} M_2$ such that a prefix of $\lceil_{BCh} (\sigma_0\sigma_1)$ is a member of $lin(BCh)$. Hence this universal requirement demands that whenever φ holds then this *must* be followed by an execution of BCh.

(3) Next suppose $\langle PCh, BCh\rangle$ is a basic existential chart with pre-chart. Then TP satisfies $\langle PCh, BCh\rangle$ iff there *exists* a reachable marking M_0 and $M_0 \stackrel{\sigma_0}{\Longrightarrow} M_1$ such that a prefix of $\lceil_{\Sigma'} (\sigma_0)$ contains a member of $lin(PCh \circ BCh)$. As before $\Sigma' = \Sigma_{PCh} \cup \Sigma_{BCh}$. Thus this existential requirement is satisfied if there exists a reachable marking starting from which there is an execution of linearization of PCh followed by an execution of BCh.

(4) Now suppose $\langle P, \varphi, BCh\rangle$ is a basic existential chart with a pre-condition. Then TP satisfies $\langle P, \varphi, BCh\rangle$ iff there exists a reachable marking M_0 and $M_0 \stackrel{\sigma_0}{\Longrightarrow} M_1$ such that $V_0 \models \varphi$ and a prefix of σ_0 corresponds to a member of $lin(BCh)$.

(5) Basic charts with post-conditions are dealt with similarly. For instance, suppose $[BCh, P, \varphi]$ is basic universal chart with a post-condition. Then TP satisfies this requirement iff every reachable marking M_0 satisfies the following condition. Suppose $M_0 \stackrel{\sigma_0}{\Longrightarrow} M_1$ such that $\lceil_{BCh} (\sigma_0)$ has a prefix which is a member of $lin(BCh)$ and σ is the least prefix of σ_0 with this property. Then $V \models \varphi$ where $M_0 \stackrel{\sigma}{\Longrightarrow} M$ and $M = (s, V)$. Thus whenever an execution of BCh takes place then at the resulting marking, the condition φ holds. The semantics of the basic existential chart with a post-condition is defined in a similar way.

One can effectively decide whether or not a bounded CTP TP satisfies an LSC requirement lsc. This is so because from PN_{TP} (the Petri Net corresponding to TP), we can extract a finite Kripke structure. Moreover, it is known that lsc can be effectively transformed into a CTL^* formula [14]. As a result we can apply the known model checking procedure for CTL^* to solve this problem [6]. This however will involve high computational complexity and more efficient decision procedures are needed to solve this problem.

6.2 Translating CTP Models to LSC

We can also translate a CTP model into an LSC specification. Consequently the play engine mechanism developed in the LSC framework [15] becomes readily accessible for simulating CTP models. Furthermore, this translation also makes it clear that the CTP model is a restricted version of the LSC formalism in which only universal charts are used but the intra-object control flow is explicitly specified using traditional mechanisms.

Let $TP = \{TS_p\}_{p \in \mathcal{P}}$ be a CTP over $(\Gamma, \mathcal{P})$ where Γ is a finite set of transaction schemes over $(\mathcal{P}, M, Act)$. Assume as before that each process $p \in \mathcal{P}$ is associated with $TS_p = (S_p, \Gamma_p, \longrightarrow_p, init_p, V_{p,in})$. Recall also that the set of pre and post control states of the transaction scheme T denoted as ${}^\bullet\mathrm{T}$ and

$\mathrm{T}^{\bullet}$. To construct the LSC specification of TP, we will deploy $\bigcup\{S_p \mid p \in \mathrm{P}\}$ also as atomic propositions. Now let T be a transaction scheme of TP with $\mathrm{T} = \{[I^i : Ch^i : O^i] \mid i = 1, 2, \ldots, n\}$. We recall that each I^i is a propositional formula built out of AP, each $Ch^i = (E^i, \leq^i, \lambda^i)$ is a chart over $(\mathcal{P}, M, Act)$ and each O^i is a subset of AP. With T, we associate the LSC specification lsc_{T} given by $lsc_{\mathrm{T}} = lsc_1 \wedge lsc_2 \ldots \wedge lsc_n$ where for each i we have $lsc_i = [BCh_i, post_i]$ with $BCh_i = [pre_i, Ch^i]$. Also, pre_i and $post_i$ are given by $pre_i = \bigwedge_{s \in {}^{\bullet}\mathrm{T}} s \wedge I^i$ and $post_i = \bigwedge_{s \in \mathrm{T}^{\bullet}} s \wedge \bigwedge_{A \in O^i} A$.

Intuitively, T translates to the universal requirement: whenever the pre-control states of T hold and the guard for the i-th transaction holds, then the i-th transaction *must* execute followed by the holding of the post-valuation O^i and the post-control states of T. The actual semantics is given in an asynchronous execution framework.

7 Using the CTP Language

Based on the abstract formal model presented so far, we have designed a simple language, called CTPL, in order to explore the feasibility of using our approach for system-level design of reactive embedded systems. In order to develop CTPL into a full-fledged modeling language, we have had to elaborate several features of the formalism such as (a) syntax/semantics of the internal actions, (b) data types of messages sent and received, (c) local variable declarations in individual processes etc. Here we shall touch upon the major issues. The full syntax of the current version of the language can be obtained from `www.comp.nus.edu.sg/~ctp`.

Language Features. We use a simple imperative language without iterations to describe the internal actions. Thus an internal action is an imperative program with arithmetic and boolean expressions, whose control flow is acyclic. Note that this is not a restriction in the expressive power of the language, since the overall control flow of a process allows (potentially) unbounded iterations. A related issue is that our current modeling language does not allow iterative executions within a transaction scheme. Such an extension would allow one execution of a transaction scheme γ to be specified as a number of iterations of the constituent transactions (where in each iteration, one of the constituent transactions is executed). Exit from the execution of γ happens when the guards of none of the constituent transactions of γ are enabled. In future, we plan to extend CTP (and CTPL) along these lines to support the specification of iterations within a transaction scheme. This will of course naturally define iterations for internal actions as well, since an internal action is simply a degenerate transaction scheme. For the data types of messages as well as local variables of processes, the language implementation currently supports scalar types (such as boolean, integers, user-defined subrange types) as well as vectors (registers) of these scalar types. The guards of transactions are boolean expressions, where the propositions in the boolean expression are allowed to use arithmetic expressions.

Language Implementation. Currently the implementation of CTPL is supported with the following tools and applications.

- a Graphical User Interface for constructing diagrammatic specifications and visualizing them
- a translator for converting visual CTP specifications to a textual format (based on XML)
- a scanner and parser for the textual format generated by translating the visual specifications
- A translator that produces Verilog code from the Intermediate Representation produced by the parser.
- Modeling and deriving of an FPGA-based implementation of an embedded controller for a 16-chamber micro-Polymerase Chain Reaction (μ-PCR) biochip. This real-life embedded controller co-ordinates a complex thermal cycling process requiring highly accurate temperature control and real-time temperature monitoring.

All of these tools are under active development and again, more details can be found at `www.comp.nus.edu.sg/~ctp`.

Integration into Co-design Environments. One of our goals is also to integrate CTPL as a front-end into a hardware-software co-design toolkit. Towards this goal, we are working on translating CTPL into the Metropolis Meta Model (MMM) language [3]. MMM is a common intermediate language for specifying heterogeneous embedded systems, and allows for simulation of system descriptions specified via different models of computation. The Metropolis project provides a SystemC based simulator for generating traces of a system described MMM; this is useful for functionality checking and performance evaluation. Based on experience gained through hand-translations of simple bus protocol examples from CTPL to MMM, we have developed a strategy for designing the translator. In the current version, we are implementing transaction schemes in CTPL via more centralized channels in MMM. The results concerning this effort will be reported elsewhere.

Formal Verification. We are also working on automated verification of CTPL programs via model checking. Towards this end, we have developed a translator from CTP to the input language of the SMV[4] model checker. However, the state machine like input language of SMV has a very different specification style as compared to CTPL and hence the SMV-based verification method does not appear to be the ideal one for CTPL specifications. Consequently, we are also building a translator from CTP to Promela (the input language of the SPIN [18] model checker). Unlike the SMV model checker, Spin allows modeling of explicit control-flow within processes. This is similar in flavor to the process specification style of the CTP modeling language. Asynchronous message passing communication (as used in our MSCs) is also directly supported in Promela via channels.

Synthesis. As for automatic synthesis, we have developed a translator to generate Verilog descriptions from CTPL programs, thereby creating a path to hardware implementation. So far, we have not studied the means for generating software code from CTPL specifications. One possibility would be to convert CTPL specifications to multi-threaded programs (where the processes in a CTPL specification map to threads in the generated program). This may require translating the message passing style of communication espoused in CTPL (via the use of MSCs) into shared variable communication among threads in the target programming language (such as Java). Developing an automated translation scheme for generating multi-threaded Java code from CTPL specifications is a topic of current and future work in our project [25].

8 Discussion

In this paper we have presented CTP, a high level specification language for modeling reactive systems. Our model is based on Message Sequence Charts (which emphasize inter-process communication) and explicit description of intra-process computations and control flow. The main questions to be pursued in this context involves well-formedness checking, formal verification as well synthesizing implementations from such models.

We have constructed a translator that transforms a CTP program into an internal representation of the Petri net representing the behavior of the CTP model. A crucial step in this translation consists of obtaining the event structure representation of each transaction scheme. Using this translator we are currently automating the analysis of transaction schemes for well-formedness. Work is also underway to devise a more efficient and direct procedure for determining the boundedness property. An interesting related problem is the issue of schedulability analysis as formulated in [24], which focuses on ensuring bounded message queues during system execution.

As for formal verification, the Petri net representation of bounded CTP models can be represented as a finite transition system. Indeed, due to the presence of the atomic propositions, this transition system can be viewed as a Kripke structure. Hence dynamic properties of the system being modeled as a CTP can be specified in a temporal logic such as LTL and formal verification of these specifications can be carried out using model checking tool such as Spin [18].

We are also currently exploring the means for extending our model along a number of dimensions, namely: parameterizing each component as an instance of a class together with the parameterization of the transaction schemes; further relaxing the control flow restrictions; and, adding timing constraints. Introduction of classes and objects into our model is particularly interesting since it can allow to symbolically simulate a CTP specification with unboundedly many objects (which form finite number of classes in terms of behaviors). Formalizing these ideas and observations is a topic of future work.

Acknowledgments

Preliminary versions of parts of this paper have appeared in [22] and [23]. We would like to thank the anonymous referees of these two papers for their comments. This research has been supported by an A*STAR (Agency for Science, Technology and Research, Singapore) Research Grant 022 106 0042 funded under the Embedded and Hybrid Systems Program.

We would like to thank Tran Tuan Anh for providing a substantial amount of inputs for all aspects of the research reported here. We would also like to thank the following people for their contributions in the implementation of the CTP language and its usage in system level design: Prakash Chandrasekaran, Kathy Nguyen Dang, Roman Gagarsky, Pankaj Jain, Nikhil Jain and Kamrul Hasan Talukder.

References

1. R. Alur, K. Etessami, and M. Yannakakis. Realizability and verification of MSC graphs. In *International Colloquium on Automata, Languages and Programming (ICALP)*, 2001.
2. ARM Limited. *AMBA On-chip Bus Specification*, 1999.
3. F. Balarin, L. Lavagno, C. Passerone, A. Sangiovanni-Vincentelli, Y. Watanabe, and G. Yang. Concurrent execution semantics and sequential simulation algorithms for the Metropolis Meta-Model. In *International symposium on Hardware/software codesign (CODES)*, 2002.
4. Cadence Berkeley Laboratories, California, USA. *The SMV Model Checker*, 1999. `www-cad.eecs.berkeley.edu/~kenmcmil/smv/`.
5. B. Caillaud, P. Darondeau, L. Helouet, and G. Lesventes. Hmscs as partial specifications ... with pns as completions. In *Modeling and Verification of Parallel Processes 4th Summer School, MOVEP 2000, LNCS 2067*, 2001.
6. E.M. Clarke, O. Grumberg, and D.A. Peled. *Model Checking.* MIT Press, 1999.
7. Codesign, Simulation and Synthesis (COSY) project. *Generic Interface Modules for PI-Bus*, 2001.
8. W. Damm and D. Harel. LSCs: Breathing life into message sequence charts. *Formal Methods in System Design*, 19(1), 2001.
9. J. Desel and J. Esparza. *Free Choice Petri Nets.* Cambridge University Press, 1995.
10. B.P. Douglass. *Doing Hard Time: Developing Real-time Systems using UML, Objects, Frameworks and Patterns.* Addison-Wesley, 1999.
11. D.D. Gajski, J. Zhu, R. Dmer, A. Gerstlauer, and S. Zhao. *SpecC: Specification Language and Methodology.* Kluwer Academic Publishers, 2000.
12. T. Grotker, S. Liao, G. Martin, and S. Swan. *System Design with SystemC.* Kluwer Academic Publishers, 2002.
13. D. Harel and H. Kugler. From play-in scenarios to code: An achievable dream. In *Fundamental Approaches to Software Engineering (FASE), LNCS 1783*, 2000.
14. D. Harel and H. Kugler. Synthesizing state-based object systems from LSC specifications. *International Journal on Foundations of Computer Science*, 13(1), 2002.
15. D. Harel, H. Kugler, R. Marelly, and A. Pnueli. Smart play-out of behavioral requirements. In *International Conference on Formal Methods in Computer Aided Design (FMCAD)*, 2002.

16. D. Harel and R. Marelly. *Come, Let's Play: Scenario-Based Programming Using LSCs and the Play-Engine.* Springer-Verlag, 2003.
17. J.G. Hendriksen, M. Mukund, K.N. Kumar, and P.S. Thiagarajan. Message sequence graphs and finitely generated regular MSC languages. In *International Colloquium on Automata, Languages and Programming (ICALP), LNCS 1853*, 2000.
18. G. J. Holzmann. The model checker SPIN. *IEEE Transactions on Software Engineering*, 23(5):279–295, May 1997.
19. K. Jensen. *Coloured Petri Nets. Basic Concepts, Analysis Methods and Practical Use.* Springer Verlag, 1997.
20. I. Krueger, R. Grosu, P. Scholz, and M. Broy. From MSCs to statecharts. In *International Workshop on Distributed and Parallel Embedded Systesms (DIPES)*, 1998.
21. M. Nielsen, G. Plotkin, and G. Winskel. Petri nets, event structures and domains. *Theoretical Computer Science (TCS)*, 13, 1981.
22. A. Roychoudhury and P.S. Thiagarajan. Communicating transaction processes. In *IEEE International Conference on Applications of Concurrency in System Design (ACSD)*, 2003.
23. A. Roychoudhury and P.S. Thiagarajan. An executable specification language based on message sequence charts. In *Formal Methods at the Crossroads: from Panacea to Foundational Support.* Springer Verlag, LNCS 2757, 2003.
24. M. Sgroi and L. Lavagno. Synthesis of embedded software using free-choice Petri nets. In *ACM Design Automation Conference (DAC)*, 1999.
25. P.S. Thiagarajan et al. Communicating Transaction Processes (CTP) project, 2003. `http://www.comp.nus.edu.sg/~ctp`.
26. Z.120. Message Sequence Charts (MSC'96), 1996.

Object Petri Nets
Using the Nets-within-Nets Paradigm

Rüdiger Valk

Universität Hamburg, Vogt-Kölln-Str.30, D-22527 Hamburg, Germany
valk@informatik.uni-hamburg.de

Abstract. The nets-within-nets paradigm provides an innovative modelling technique by giving tokens themselves the structure of a Petri net. These nets, called *token nets* or *object nets*, also support the object oriented modelling technique as they may represent real world objects with a proper dynamical behaviour. Between object nets and the surrounding net, called *system net*, various interaction mechanisms exist as well as between different object nets. This introduction into the field of object Petri nets starts with small examples and proceeds by giving formal semantics. Some of the examples are modelled within the formalism of the Renew tool. Finally the differences between reference and two kinds of value semantics are discussed.

1 Nets within Nets

Tokens in a Petri net place can be interpreted as objects. In place/transition nets (P/T nets), in most cases these objects represent resources or indicate the state of control. More complex objects are modelled by typed tokens in coloured Petri nets. Object-oriented modelling, however, means that software is designed as the interaction of discrete objects, incorporating both data structure and behaviour [1]. From a Petri net point of view it is quite natural to represent such objects by tokens, that are nets again. We denote this approach as the "nets-within-nets paradigm".

In many applications objects not only belong to a specific environment but are also able to switch to a different one. Examples of such objects are agents, including the classical meaning of persons belonging to a secrete service, as well as software modules in the context of agent-oriented programming. In Fig. 1a) the current environment of agent X is denoted as "location A", which may be a logical state or a physical site. The arrow represents a possible transition to location B. A structurally similar situation is given with a mobile computer switching between "security environments" A and B (Fig. 1b). In Fig. 2a) the object is a task to be executed on a machine A together with a plan for the execution procedure. As before, they can move to machine B. Finally, in Fig. 2b) a workflow is with an employee A, moving to employee B afterwards. Note, that in all of these examples, also the internal state of the object is changed when

J. Desel, W. Reisig, and G. Rozenberg (Eds.): ACPN 2003, LNCS 3098, pp. 819–848, 2004.

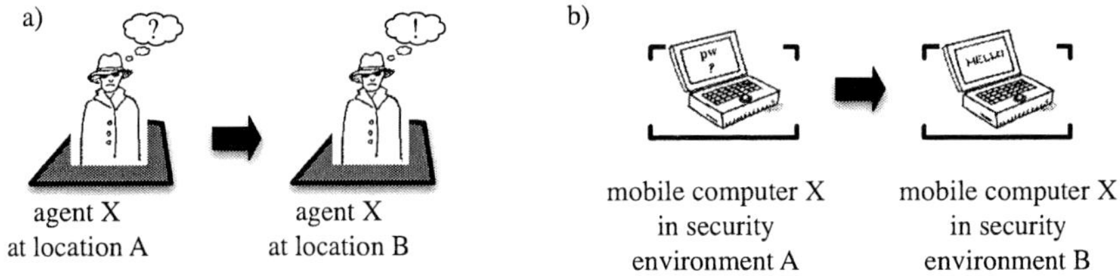

Fig. 1. Moving objects I

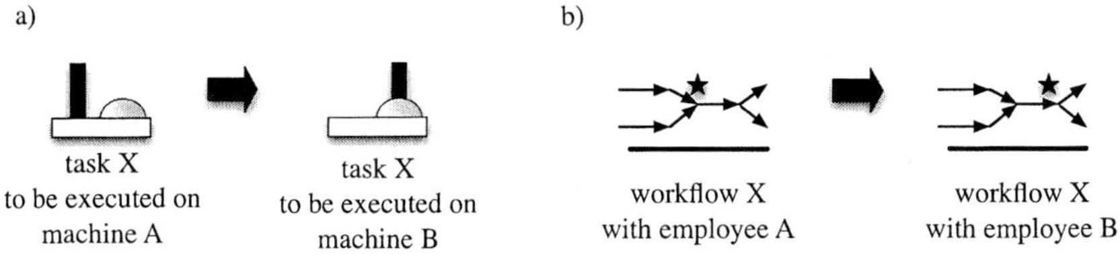

Fig. 2. Moving objects II

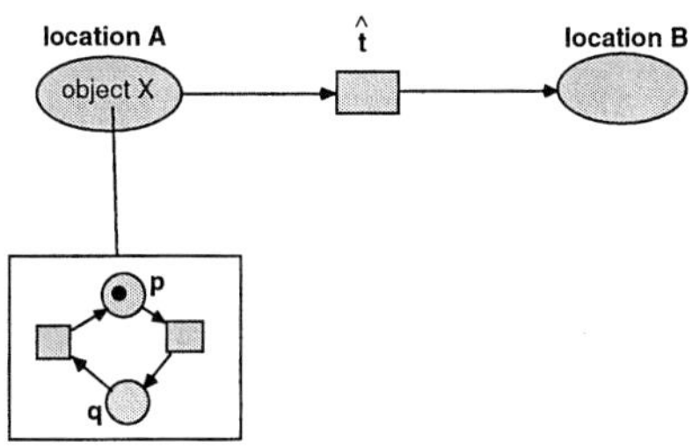

Fig. 3. Object system transition with token net

moving to a different location. Abstracting from these examples, we model the movement or switch of an object X by a transition, which is enabled by the token "object X", as shown in Fig. 3. In addition, as the object has a dynamical behaviour, say alternating states p and q, the token is again a marked net. It is therefore called a "token net". A token net is also called *object net* in distinction to the *system net*, to which it belongs. The whole system is then called an *object net system* or shortly *object system*. In Fig. 4 the movement of the net-token is shown as the firing of transition $\hat{t}$. Obviously, also the token net can fire autonomously without being moved (Fig. 5). Both, transport and autonomous firing can interleave, but are to be considered as concurrent actions. This should be distinguished from a situation, where these transition occurrences are synchronised, i.e. the object moves if and only if some object net transition occurs. Such an action may be triggered by the object net, by the system net, or by both of them. Therefore such a situation is denoted by the neutral term "interaction". Interacting transitions are labelled by a corresponding symbol, such as $\langle i \rangle$ in Fig. 6. Finally in this introduction we discuss two different semantics of

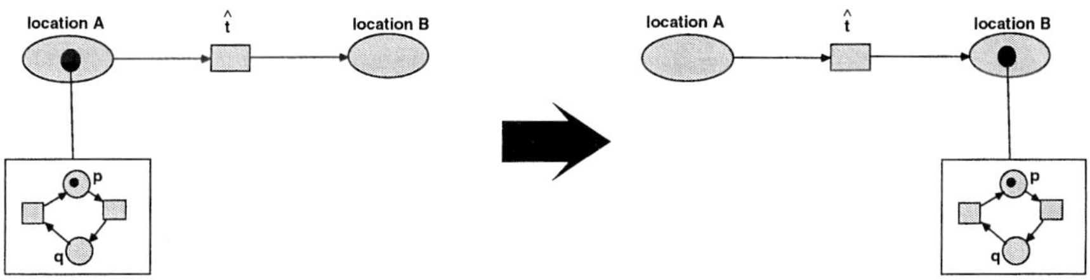

Fig. 4. Transport

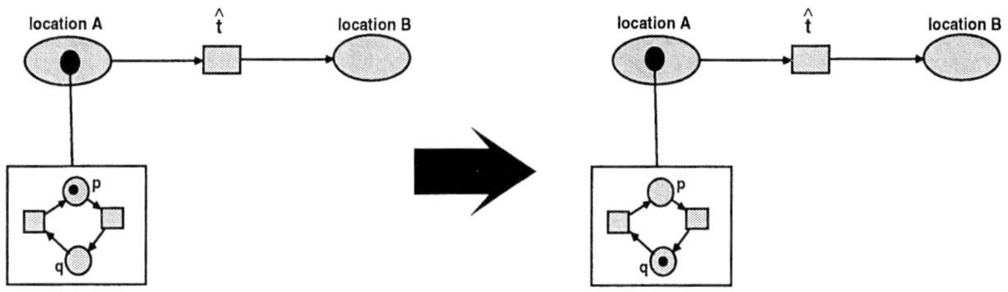

Fig. 5. Autonomous transition

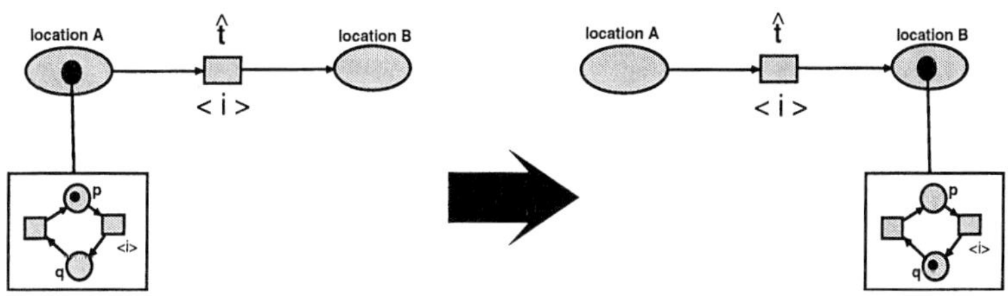

Fig. 6. Interaction

object systems, namely *value semantics* and *reference semantics*. The difference between these semantics becomes obvious when objects (in particular agents) perform concurrent actions at different locations. A single (human) agent can execute independent actions using his two hands. To improve conceivability we prefer to speak of a group of agents, an *agency*. In Fig. 7 such an agency moves from location A to locations B and C. This means that one or more members of the agency are doing so. The abstract net form is given in Fig. 8. The concurrent behaviour of the agency is represented by transitions, labelled $\langle i \rangle$ and $\langle j \rangle$. The question now is, what will be the marking after firing the leftmost transition in the system net? The proposal in Fig. 9 shows references to the object net from both of the output places of the transition. This can be interpreted in such a way, that the members of the agency refer to the same action plan as before, but from different locations. In the graphical representation dashed arrows are used to distinguish references from the lines used before for linking the object nets. As the action plan matches the system net, the concurrent actions labelled $\langle i \rangle$ and $\langle j \rangle$ can be concurrently executed. Later in this paper we discuss how to define semantics for the corresponding join of these action sequences. Distributed systems are characterised by the impossibility of direct access to common data. To

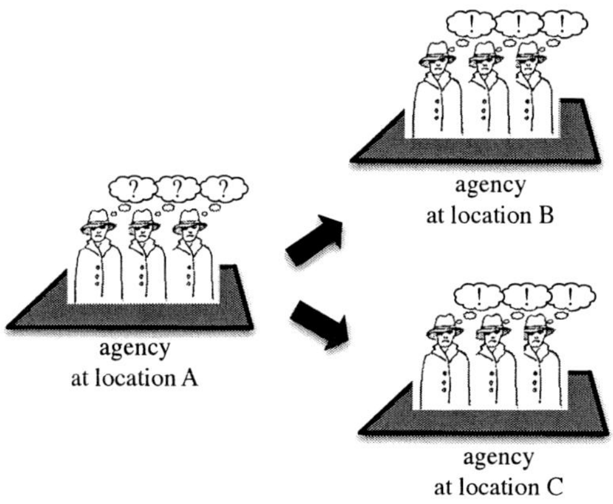

Fig. 7. Creating distributed agencies

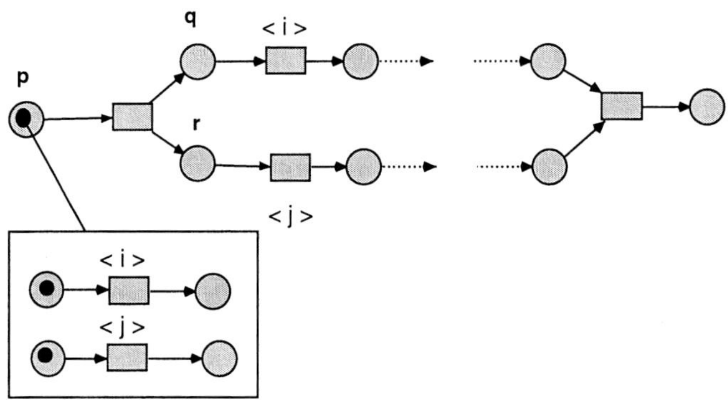

Fig. 8. Creating distributed object nets

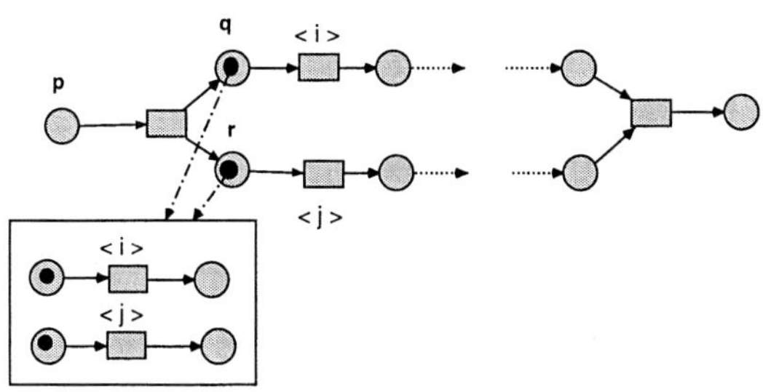

Fig. 9. Reference semantics

meet this paradigm, *value semantics* have been introduced. Instead of references identical copies of the object net are assigned to the output places of the system net transition. This is similar to *call by value* in procedure parameter passing. In the running example, with value semantics, from the marking in Fig. 8 we get

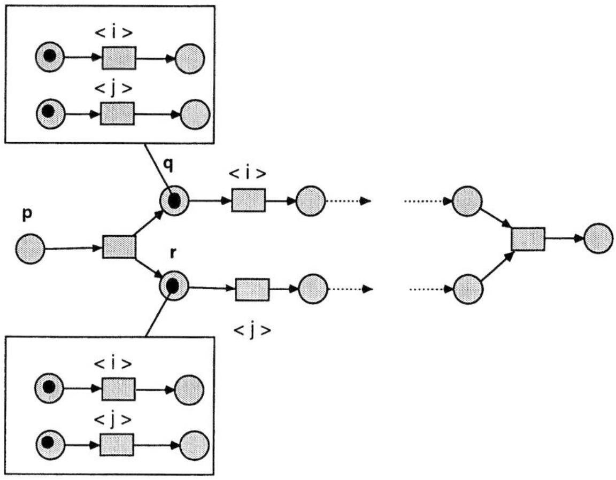

Fig. 10. Value semantics

a successor marking as shown in Fig. 10. Note that normal lines are used again (instead of dashed arrows).

The nets-within-nets concept was first introduced in the nineteen-eighties as *task/flow-nets*, [2–4]. Further results have been published in [5], [6] and [7]. The relation between reference and value semantics is discussed in [8–10]. Reference semantics are carefully studied in the theses of O. Kummer [11] and F. Wienberg [12], in particular in the context of the Renew tool [13]. In the thesis of B. Farwer [14] value semantics is studied within the framework of linear logic (see also [15–19]). Recent work with applications to distributed agents, mobile systems, security problems and socionics can be found in [20–24].

Many references connect Petri net models with object orientation [25–52]. These approaches introduce features of object oriented languages into Petri nets, like classes and inheritance, and partially also refer to the nets-within-nets paradigm.

2 Elementary Object Systems

We now formally introduce *elementary object systems* which form a restricted class of general object systems. We only allow two types (colours) for places, namely objects from a given set of object nets (which do not contain token-nets again) and ordinary black tokens. For general object systems more types are allowed. By this restriction the model remains simple, yet most important features can be introduced. To alleviate the distinction between system and object nets the components of the system net will bear a hat: $\hat{t}, \hat{p}, \widehat{P}, \widehat{T}, \ldots$ etc.

Definition 1. *An* elementary object system *is a tuple* $OS = (SN, \mathcal{ON}_{m^0}, \varrho, \mathbf{R}_0)$ *where* SN *is the* system net, $\mathcal{ON}_{m^0}$ *is a finite set* $\{(ON_1, m_1^0) \ldots, (ON_k, m_k^0)\}$ *of* marked object nets, ϱ *is the* interaction relation *and* $\mathbf{R}_0$ *is the* initial marking, *which are defined as follows. (The sets of places and transitions of all involved nets are assumed to be finite and disjoint.)*

a) *A* system net *is a Petri net* $SN = (\widehat{P}, \widehat{T}, \widehat{W})$ *where*
   1. *the set* $\widehat{P} = P_{ob} \cup P_{bt}$ *of places is divided into disjoint sets of* object places P_{ob} *and* black-token places P_{bt}, $\widehat{T}$ *is the set of* transitions,
   2. *the set of arrows is given as a mapping* $\widehat{W} : (\widehat{P} \times \widehat{T}) \cup (\widehat{T} \times \widehat{P}) \to \mathbb{N}$. *For* $\widehat{W}(x, y) > 0$ *the arrow* (x, y) *is called an* object arrow *if* $\{x, y\} \cap P_{ob} \neq \emptyset$ *and a* black-token arrow *if* $\{x, y\} \cap P_{bt} \neq \emptyset$.

b) *An* object net *is a P/T net* $ON_i = (P_i, T_i, W_i)$ *(cf. the Appendix)*[1].

c) $\varrho \subseteq \widehat{T} \times T$ *is the* interaction relation *where* $T := \bigcup_{i=1}^{k} T_i$.

d) $\mathbf{R}_0$ *specifies the* initial token distribution, *where* $\mathbf{R}_0 : \widehat{P} \to \mathbb{N} \cup Bag(\mathcal{ON})$ *with* $\mathcal{ON} := \{ON_1, \ldots, ON_k\}$. *It has to satisfy the condition* $\mathbf{R}_0(\hat{p}) \in \mathbb{N} \Leftrightarrow \hat{p} \in P_{bt}$.

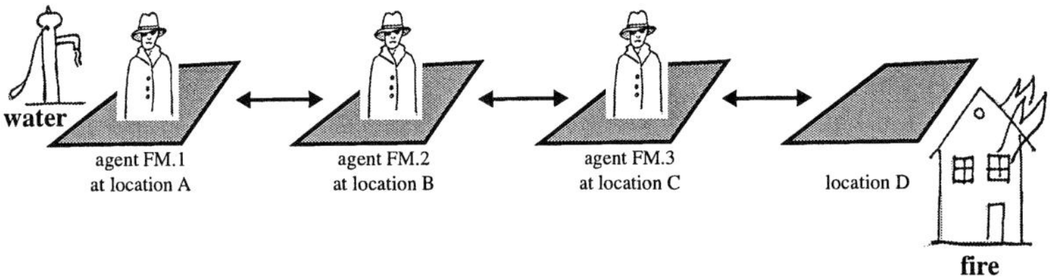

Fig. 11. Parallel fire extinction by agents

In the example of Fig. 12 an object system $OS = (SN, \mathcal{ON}, \varrho, \mathbf{R}_0)$ is shown, where $\mathcal{ON} = \{FM.1, FM.2, FM.3\}$. Black-token arrows of SN can be identified by their labelling from $\mathbb{N}$. Hence `water` is a black-token place, whereas $\{A, B, C, D\}$ are object-places. In the initial marking places A, B and C contain the object net $FM.1$, $FM.2$ and $FM.3$, respectively. They have the same structure and could be generated from a type pattern FM. To keep the formal definition simple, we start with all instances of such patterns already generated. The interaction relation is given by corresponding labels in angle brackets:

$$\varrho = \{(\texttt{refill},\texttt{a.1}), (\texttt{AtoB},\texttt{b.1}), \ldots, (\texttt{refill},\texttt{a.2}), (\texttt{AtoB},\texttt{b.2}), \ldots\}$$

(the labels are <refill> and <approachFire> in the given cases).

The formal behaviour of this example object system will be defined in the next section. It is a modification of the well-known *fire extinction example* of C. A. Petri [53]: tree agents (or firemen) $FM.1$, $FM.2$ and $FM.3$ are in locations A, B and C (Fig. 11). The location D is empty. They can change their position

[1] It is not very important, which class of Petri nets is chosen for object nets. To keep definitions simple, we define them as P/T nets (see Appendix).

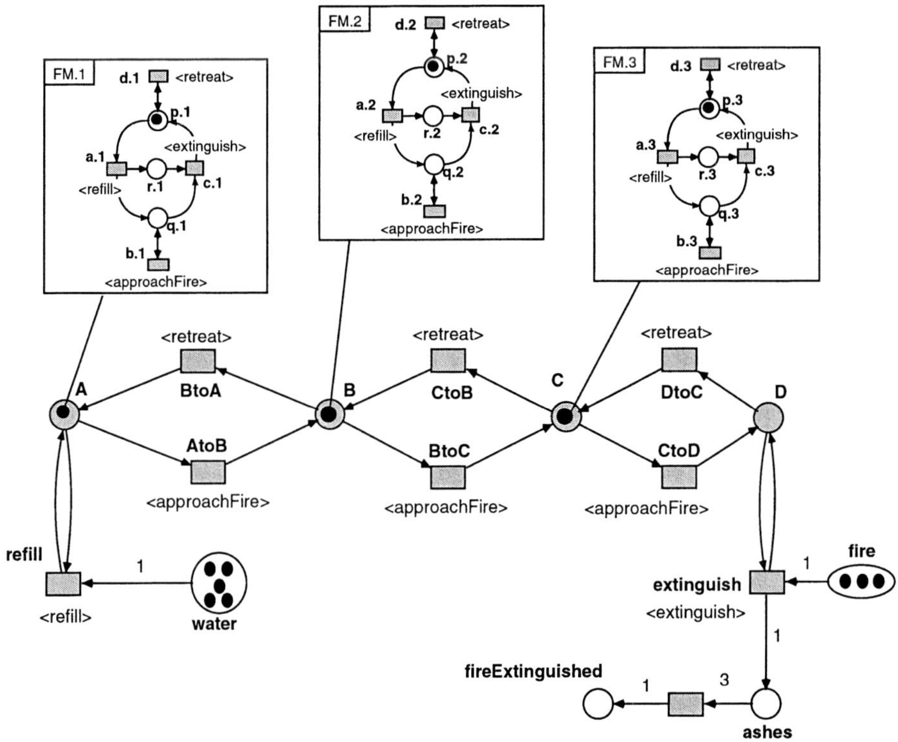

Fig. 12. Elementary object net for parallel fire extinction

to location A, fill their bucket[2] at a water source, go to location D and help to extinguish the fire. They do this quite independently. In a modification, that will be shown later, they will coordinate their actions to form a chain, like in Petri's setting. To show progress up to termination the amount of water is quantified by 5 black tokens, whereas the fire can be extinguished in 3 steps by removing 3 tokens from the place `fire`. The final marking includes the place `fireExtinguished` and the reader is invited to specify the terminal markings, i.e. where the agents terminate.

3 Reference Semantics of Object Systems

We start by introducing the notion of a marking for object systems under reference semantics. Recall that by Definition 1 an object system contains a set $\mathcal{ON}_{m^0} = \{(ON_1, m_1^0) \ldots, (ON_k, m_k^0)\}$ of marked object nets. By omitting the markings we obtain the set of (unmarked) object nets $\mathcal{ON} = \{ON_1, \ldots, ON_k\}$. Hence, in general a marking is given by

[2] The bucket is not modelled here, but in a version given later.

a) a distribution of object nets or black tokens $\mathbf{R} : \widehat{P} \to \mathbb{N} \cup Bag(\mathcal{ON})$ and
b) the vector $\mathbf{M} = (m_1, \ldots, m_k)$ with the current marking of each ON_i $(1 \leq i \leq k)$.

$\mathbf{R}$ specifies for each system net place $\hat{p}$ a number of black tokens (if $\hat{p}$ is a black token place) or a multi-set of unmarked object nets (if $\hat{p}$ is an object place). Since the elements of $\mathcal{ON}$ are unmarked, $\mathbf{R}$ can be thought of as a *reference* to the object nets. If we abbreviate $(m_1, \ldots, m_k)$ by $\mathbf{M}$ and the set of all such vectors by $\mathcal{M}$ we obtain the following Definition 2. By $pr_i(\mathbf{M})$ we denote the i-th component m_i of $\mathbf{M}$ and by $\mathbf{M}_{i \to m}$ the tuple, where the i-th component is substituted by m.

Definition 2. *Given an object system* $OS = (SN, \mathcal{ON}_{m^0}, \varrho, \mathbf{R}_0)$ *we define* $\mathcal{M} := \{\mathbf{M} \mid \mathbf{M} = (m_1, \ldots, m_k) \wedge m_i \in Bag(P_i)\}$. *Then a marking of an elementary object system is a pair* $(\mathbf{R}, \mathbf{M})$ *where* $\mathbf{M} \in \mathcal{M}$ *and* $\mathbf{R} : \widehat{P} \to \mathbb{N} \cup Bag(\mathcal{ON})$ *satisfying* $\mathbf{R}(\hat{p}) \in \mathbb{N} \Leftrightarrow \hat{p} \in P_{bt}$. *Specifying* $\mathbf{M}_0$ *by the initial markings of the marked object nets* $\mathbf{M}_0 := (m_1^0, \ldots, m_k^0)$ *we obtain the initial marking* $(\mathbf{R}_0, \mathbf{M}_0)$ *of* OS.

The occurrence rule for object systems will be introduced in three parts. First we consider the case when an interaction occurs. In this case we assume that a system net transition $\hat{t} \in \widehat{T}$ and an object net transition $t \in T_i$ of some object net ON_i is activated *and* both transitions are related by the interaction relation ϱ. i.e. $(\hat{t}, t) \in \varrho$. This case of the occurrence rule is called an *interaction.*

Definition 3. *(interaction / reference-semantics) Let* $(\mathbf{R}, \mathbf{M})$ *be marking of an object system* $OS = (SN, \mathcal{ON}_{m^0}, \varrho, \mathbf{R}_0)$, $\hat{t} \in \widehat{T}$ *a transition of* SN, $t \in T_i$ *a transition of an object net* $ON_i = (P_i, T_i, W_i) \in \mathcal{ON}$ *such that* $(\hat{t}, t) \in \varrho$. *Then* $(\hat{t}, t)$ *is* activated *in* $(\mathbf{R}, \mathbf{M})$ *if:*

a) $\mathbf{R}(\hat{p}) \geq \widehat{W}(\hat{p}, \hat{t})'ON_i$ *for all* $\hat{p} \in {}^\bullet\hat{t} \cap P_{ob}$ [3],
b) $\mathbf{R}(\hat{p}) \geq W(\hat{p}, \hat{t})$ *for all* $\hat{p} \in {}^\bullet\hat{t} \cap P_{bt}$ [4] *and*
c) t *is activated in* $m = pr_i(\mathbf{M})$ *(see Appendix).*

This is denoted by $(\mathbf{R}, \mathbf{M})[\hat{t}, t\rangle$. *Let be* $m[t\rangle m'$ *(w.r.t.* ON_i*, see Appendix). In this case the successor marking* $(\mathbf{R}', \mathbf{M}')$ *of* OS *is defined by*

a) $\mathbf{R}'(\hat{p}) = \mathbf{R}(\hat{p}) - \widehat{W}(\hat{p}, \hat{t})'ON_i + \widehat{W}(\hat{t}, \hat{p})'ON_i$ *for all* $\hat{p} \in P_{ob}$.
b) $\mathbf{R}'(\hat{p}) = \mathbf{R}(\hat{p}) - W(\hat{p}, \hat{t}) + W(\hat{t}, \hat{p})$ *for all* $\hat{p} \in P_{bt}$.
c) $\mathbf{M}' = \mathbf{M}_{i \to m'}$.

This is denoted by $(\mathbf{R}, \mathbf{M})[\hat{t}, t\rangle(\mathbf{R}', \mathbf{M}')$.

---

[3] $\widehat{W}(p,t)'ON_i$ denotes the multi-set containing one element ON_i with multiplicity $\widehat{W}(p,t)$. Hence $\geq$ denotes the superset relation of multi-sets.
[4] $\geq$ denotes the ordering relation of $\mathbb{N}$.

In our running example of OS from Fig. 12 with $(\mathbf{R}, \mathbf{M}) = (\mathbf{R}_0, \mathbf{M}_0)$, $\hat{t} =$ `refill`, $ON = FM.1$ and $t =$ `a.1` we obtain $(\mathbf{R}, \mathbf{M})[\texttt{refill}, \texttt{a.1}\rangle(\mathbf{R}', \mathbf{M}')$ where $\mathbf{R}' = \mathbf{R}$ and $\mathbf{M}' = \mathbf{M}_{1 \to m_1}$ and $m_1 = \{\texttt{q.1}\}$, i.e. fireman FM.1 fills his bucket. In a further step $(\mathbf{R}', \mathbf{M}')[\texttt{AtoB}, \texttt{b.1}\rangle(\mathbf{R}'', \mathbf{M}'')$ a marking $(\mathbf{R}'', \mathbf{M}'')$ is reached where place B contains two token nets, namely $\mathbf{R}''(B) = 1'FM.1 + 1'FM.2$ and $\mathbf{M}'' = (\{\texttt{q.1}\}, \{\texttt{p.2}\}, \{\texttt{p.3}\})$. In this step fireman $FM.1$ approaches the fire by moving to location B.

If a system net transition is activated without being included in the interaction relation, a chosen object net does not change its current marking. As it changes its location in the system net such an occurrence is called a *transport*. The following definition can be seen as the special case of Definition 3 where the involved object net is not changed, i.e. $\mathbf{M}' = \mathbf{M}$ [5].

Definition 4. *(transport / reference-semantics) Let* $(\mathbf{R}, \mathbf{M})$ *be a marking of an object system* $OS = (SN, \mathcal{ON}_{m^0}, \varrho, \mathbf{R}_0)$, $\hat{t} \in \widehat{T}$ *a transition of* SN, *such that* $\hat{t} \notin \mathrm{dom}(\varrho) := \{\hat{t}_1 \mid \exists t : (\hat{t}_1, t) \in \varrho\}$. *Then* $\hat{t}$ *is* activated *in* $(\mathbf{R}, \mathbf{M})$ *if there is an object net* ON_i *such that*

a) $\mathbf{R}(\hat{p}) \geq \widehat{W}(\hat{p}, \hat{t})'ON_i$ *for all* $\hat{p} \in {}^\bullet\hat{t} \cap P_{ob}$,
b) $\mathbf{R}(\hat{p}) \geq W(\hat{p}, \hat{t})$ *for all* $\hat{p} \in {}^\bullet\hat{t} \cap P_{bt}$

Since we use τ *for the empty action, this is denoted by* $(\mathbf{R}, \mathbf{M})[\hat{t}, \tau\rangle$. *In this case the successor marking* $(\mathbf{R}', \mathbf{M}')$ *is defined by*

a) $\mathbf{R}'(\hat{p}) = \mathbf{R}(\hat{p}) - \widehat{W}(\hat{p}, \hat{t})'ON_i + \widehat{W}(\hat{t}, \hat{p})'ON_i$ *for all* $\hat{p} \in P_{ob}$.
b) $\mathbf{R}'(\hat{p}) = \mathbf{R}(\hat{p}) - W(\hat{p}, \hat{t}) + W(\hat{t}, \hat{p})$ *for all* $\hat{p} \in P_{bt}$.
c) $\mathbf{M}' = \mathbf{M}$

This is denoted by $(\mathbf{R}, \mathbf{M})[\hat{t}, \tau\rangle(\mathbf{R}', \mathbf{M}')$.

In the example of OS from Fig. 12 there is no transport as all system net transitions are labelled for interaction. But we can easily modify the example by deleting all the labels `<retreat>` in both the system net and all the object nets. Then we obtain the same behaviour as before since there is no different possibility to move for the firemen in the corresponding cases (and their marking did not change anyway). As mentioned in Section 1 object nets may change their state without moving:

Definition 5. *(autonomous action / reference-semantics) Let* $(\mathbf{R}, \mathbf{M})$ *be a marking of an object system* $OS = (SN, \mathcal{ON}_{m^0}, \varrho, \mathbf{R}_0)$ *and* $t \in T_i$ *a transition of an object net* $ON_i = (P_i, T_i, W_i) \in \mathbf{R}(\hat{p})$ *for some* $\hat{p} \in \widehat{P}$, *such that* $t \notin range(\varrho) := \{t_1 \mid \exists \hat{t} : (\hat{t}, t_1) \in \varrho\}$ *and* t *is activated in* ON_i. *Then we say that* (τ, t) *is* activated *in* $(\mathbf{R}, \mathbf{M})$ *(denoted* $(\mathbf{R}, \mathbf{M})[\tau, t\rangle$*). The successor marking* $(\mathbf{R}', \mathbf{M}')$ *of* OS *is defined by*

a) $\mathbf{R}' = \mathbf{R}$.
b) $\mathbf{M}' = \mathbf{M}_{i \to m'}$ *if* $m[t\rangle m'$ *for* $pr_i(\mathbf{M}) = m$.

[5] Definitions 3 and 4 could be easily merged. This is not done to emphasise the differences.

This is denoted by $(\mathbf{R},\mathbf{M})[\tau,t\rangle(\mathbf{R}',\mathbf{M}')$.

Definition 6. *For the new alphabet* $\Gamma := (\widehat{T}\cup\{\tau\})\times(T\cup\{\tau\})\backslash(\tau,\tau)$, *where* (τ,τ) *denotes the neutral element of the free monoid* Γ^*, *we define:*

a) $(\mathbf{R},\mathbf{M})[\tau,\tau\rangle(\mathbf{R}',\mathbf{M}')$ *if* $(\mathbf{R},\mathbf{M})=(\mathbf{R}',\mathbf{M}')$ *and*
b) $(\mathbf{R},\mathbf{M})[\widetilde{w}(\hat{t},\alpha)\rangle(\mathbf{R}',\mathbf{M}')$ *if* $\exists(\mathbf{R}'',\mathbf{M}''):(\mathbf{R},\mathbf{M})[\widetilde{w}\rangle(\mathbf{R}'',\mathbf{M}'')$ *and* $(\mathbf{R}'',\mathbf{M}'')[\hat{t},\alpha\rangle(\mathbf{R}',\mathbf{M}')$ *for some* $\widetilde{w}\in\Gamma^*$, $\hat{t}\in\widehat{T}\cup\{\tau\}$ *and* $\alpha\in T\cup\{\tau\}$.

The examples of transition occurrences given before lead to the following occurrence sequence: $(\mathbf{R}_0,\mathbf{M}_0)[(\texttt{refill},\texttt{a.1}),(\texttt{AtoB},\texttt{b.1})\rangle(\mathbf{R}'',\mathbf{M}'')$.

4 Object Interaction, Object Creation and Renew

In this section we extent the definition of elementary object systems to include interaction between objects (with respect to reference semantics). Furthermore we show how these concepts are represented in the Renew tool which includes the creation of object nets.

4.1 Object Interaction

Interaction between object nets is very similar to interaction of system and object nets. But there are good reasons to define them separately. Interacting transitions of different object nets are represented by the interaction relation σ.

Definition 7. *The* object-object-interaction-relation σ *is defined as a set of pairs* (t_i,t_j) *of transitions* t_i *and* t_j *of different object nets* ON_i *and* ON_j. *The relation is supposed to be symmetric (i.e. also contains* (t_j,t_i)*) but irreflexive. Furthermore, to have a simpler formalism it is supposed to be disjoint with the interaction relation* ϱ *in the following sense:* (ϱ,σ) *are* separated *if* $(t_1,t_2)\in\sigma\Rightarrow\varrho t_1 = \varrho t_2=\emptyset$ [6].

As for autonomous occurrences object nets may interact without moving. This is restricted, however, to the case where the object nets are locally "near", which is formalised as *to be in the same place.*

Definition 8. *(object-object-interaction / reference-semantics) Let* $(\mathbf{R},\mathbf{M})$ *be a marking of an object system* $OS=(SN,\mathcal{ON}_{m^0},\varrho,\mathbf{R}_0)$ *and* $\hat{p}\in\widehat{P}$ *a place containing two different object nets* $ON_i=(P_i,T_i,W_i)$ *and* $ON_j=(P_j,T_j,W_j)$, *i.e.* $ON_i+ON_j\le\mathbf{R}(\hat{p})$. *Then* ON_i *and* ON_j can interact *in* $(\mathbf{R},\mathbf{M})$ *if there are transitions* $t_i\in T_i$ *and* $t_j\in T_j$ *such that*

a) t_u *is activated in* $m_u=pr_u(\mathbf{M})$ *for* <u>both</u> $u\in\{i,j\}$.
b) $(t_i,t_j)\in\sigma$.

[6] $\varrho t=\{\hat{t}\mid(\hat{t},t)\in\varrho\}$

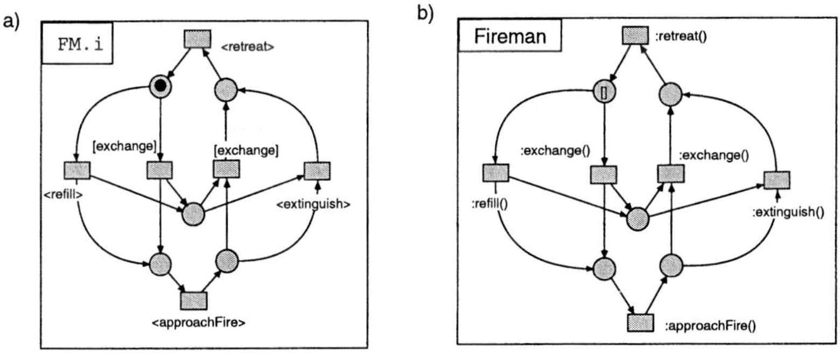

Fig. 13. Fireman with bucket exchange for the net a) of Fig. 12 and b) of Fig. 14

This is denoted $(\mathbf{R},\mathbf{M})[\tau, t_i|t_j\rangle$. *The corresponding successor marking* $(\mathbf{R}',\mathbf{M}')$ *of OS is defined by*

a) $\mathbf{R}' = \mathbf{R}$.
b) $\mathbf{M}' = (\mathbf{M}_{i\to m'})_{j\to m''}$ *if* $m_i[t_i\rangle m'$ *and* $m_j[t_j\rangle m''$.

This is denoted by $(\mathbf{R},\mathbf{M})[\tau, t_i|t_j\rangle(\mathbf{R}',\mathbf{M}')$.

To give an example we substitute the object nets $FM.1$, $FM.2$ and $FM.3$ from Fig. 12 by three copies of the net from Fig. 13a) for $i = 1,2,3$. The modification is the following. Each fireman can proceed only one step by a transition labelled `<retreat>` or `<approachFire>`. Between these steps there has to be a step with `<refill>`, `<extinguish>` or `[exchange]`. Transitions of different object nets labelled by `[exchange]` are in the object nets interaction relation σ which is indicated by brackets `[` and `]`. Such an interaction can occur only in the "rendezvous" places B and C, where they exchange their full and empty buckets. The resulting behaviour is a firemen chain as in Petri's original example: each fireman moves only between two neighbouring places, whereas the buckets move from the water to the fire and back. The place in the middle of $FM.i$ is redundant: if marked the fireman has a full bucket.

4.2 Object Creation and the Renew Tool

The Renew tool allows to design and simulate the example nets in a closely related manner. The system net from Fig. 12 is shown as a Renew model in Fig. 14. There are only little differences. For instance, black tokens are represented by `[]` and integer weights `n` by n such token symbols separated by semi-colons.

The three object nets of the example of Fig. 12 are similar in structure and differ only in the identifiers of places and transitions. The Renew tool supports to define object nets as patterns and to generate instances at simulation runtime. Such a creation is executed by the transition which is placed in the leftmost, upper part of Fig. 14. By the inscription `f1:new Fireman` an instance of the net `Fireman` (see Fig. 13b)) is created with a new identifier `Fireman[n]`, where `n` is

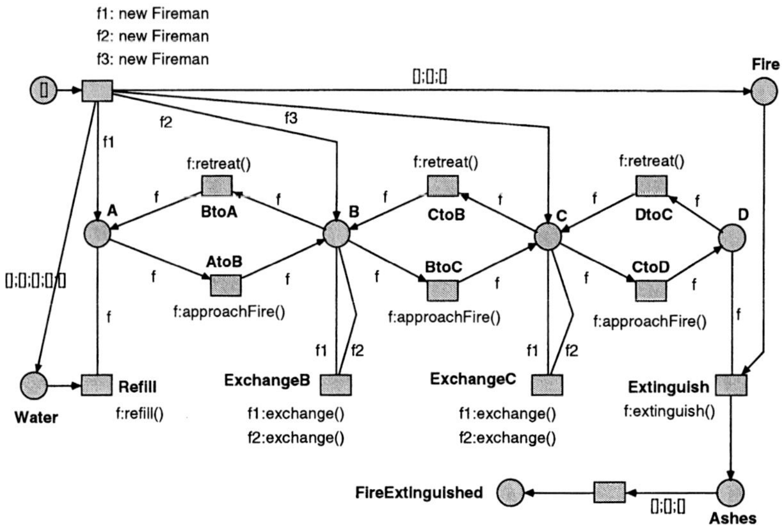

Fig. 14. Fire extinction net modelled with the Renew tool (system net)

an identifier chosen by the tool. Then a reference to this instance is introduced from the place A (by the arc labelled `f1`)[7]. In total, the transition creates references to three different object nets in the places A, B and C, 5 black tokens in `Water` and 3 black tokens in `Fire`.

An inscription like `f:approachFire()` is called a *downlink* and is related to a corresponding *uplink* `:approachFire()` in the net which is referenced by `f` (see Fig. 13). The semantics of this pair is the same as for the interaction relation pair (`AtoB`,`b1`) $\in \varrho$ of Fig. 12. Object-object-interaction is implemented quite different, namely also by down- and uplinks. For such an interaction an extra transition is introduced, like the transition `ExchangeB` in the Renew example net. The downlinks `f1:exchange()` and `f2:exchange()` contain references to nets in the place B (say `Fireman[1]` and `Fireman[2]`) to synchronise two of their transitions labelled by an uplink: `exchange()`. Hence, the behaviour is like the object-object-interaction in the formal definition.

The nets-within-nets paradigm in Renew is not restricted to a 2-level hierarchy: in fact, there is no hierarchy necessary at all. We add a 3-level version of the fire extinction example, where the buckets form an additional level: Fig. 15 and 16. The bucket has two states: `empty` and `full`. The initial state is introduced by the transition with uplink `:new()`, which is executed when the net instance is created (by firing the transition with downlink `b:new Bucket` in the net `Fireman`). It is also interesting to observe how the exchange of a full and an empty bucket is implemented by the transition `ExchangeB` or `ExchangeC`. In the first case the fireman in place B, which is referenced by `f1`, executes the transition with uplink `:exchange(be,bf)` whereas a second one, referenced by

[7] For details see the documentation of the tool at `http://www.renew.de`.

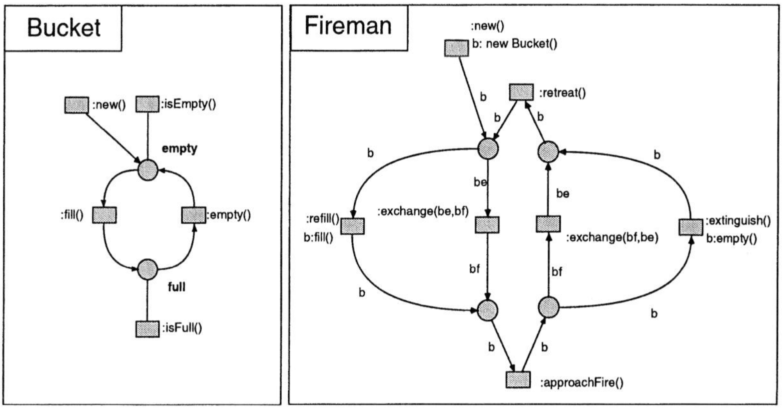

Fig. 15. Three-level fire extinction net modelled with the Renew tool (fireman and bucket net)

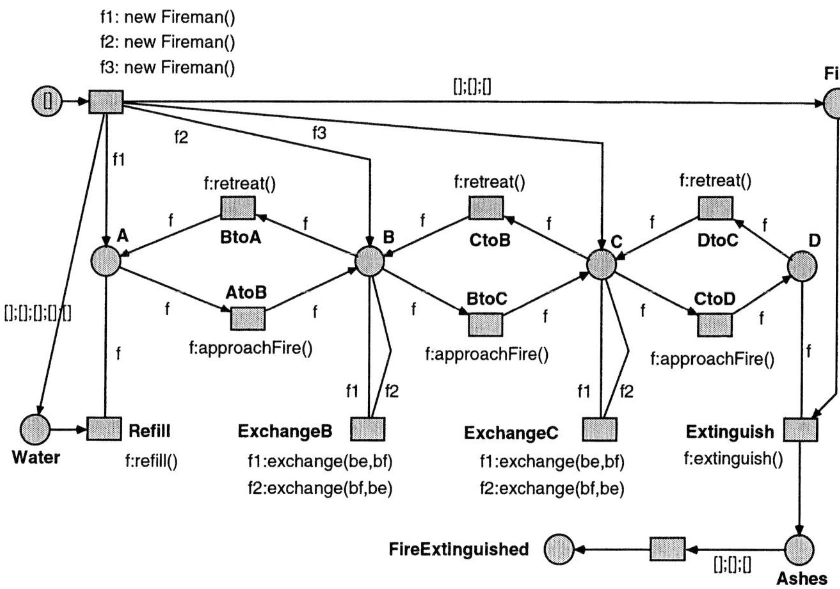

Fig. 16. Three-level fire extinction net modelled with the Renew tool (system net)

`f2`, does the same for `:exchange(bf,be)`. Hence the empty bucket (referenced by `be`) and the full bucket (referenced by `bf`) are exchanged.

Note that the system nets of the 2-level model (Fig. 14) and the 3-level model (Fig. 16) are very similar, showing the abstraction power of the nets-within-nets modelling paradigm.

5 Value Semantics of Object Nets

In this section value semantics, as introduced in the introduction, will be defined formally. We start with a general definition which will be refined in two following subsections to *distributed tokens semantics* and *history process semantics.* To keep definitions simpler, weights of system nets arrows different to 1 are not considered in this section (i.e. $\widehat{W}$ maps to $\{0,1\}$ instaed of $\mathbb{N}$).

5.1 Value Semantics of Object Nets: General Definition

In value semantics each object net instance has its own state (marking), which is formalised by the following set $\mathcal{ON}_m$.

Definition 9. *Given an elementary object system* $OS = (SN, \mathcal{ON}_{m^0}, \varrho, \mathbf{R}_0)$ *as in Definition 1 with* $\mathcal{ON} = \{ON_1, \ldots, ON_k\}$*, we now define* $\mathcal{ON}_m := \{(ON_1, m_1), \ldots, (ON_k, m_k) \mid (m_1, \ldots, m_k) \in \mathbf{M}\}$.

Then a marking of an elementary object system under value semantics *is a mapping*

$$\mathbf{V} : \widehat{P} \to \mathbb{N} \cup Bag(\mathcal{ON}_m)$$

satisfying $\mathbf{V}(\hat{p}) \in \mathbb{N} \Leftrightarrow \hat{p} \in P_{bt}$*. The initial marking* $\mathbf{V}_0$ *has to meet the condition* $\mathbf{V}_0(\hat{p})(ON_i, m_i^0) = \mathbf{R}_0(\hat{p})(ON_i)$*, when* $\hat{p}$ *is an object place and* $\mathbf{V}_0(\hat{p}) = \mathbf{R}_0(\hat{p})$ *otherwise.*

When an interaction occurs with a transition, where several marked object nets are involved at the input side, some kind of unification of their current state (marking) is to be constructed. This corresponds to the collection of partial results (of concurrent computations) to a consistent state, unifying these partial states. The definition of such a function *unify* is left unspecified in the following definition, but made explicit in the subsequent subsections. In a symmetric way, for the output-places a function *distribute* is introduced, which constructs from the state (marking m) a tuple $(m_{\hat{p}_1}, ..., m_{\hat{p}_q})$ of states (markings) for the object nets to be created in the output places $(\hat{p}_1, ..., \hat{p}_q)$ of the transition.

Definition 10. *(interaction / value-semantics) Let* $\mathbf{V}$ *be a marking of an elementary object system* $OS = (SN, \mathcal{ON}_{m^0}, \varrho, \mathbf{R}_0)$, $\hat{t} \in \widehat{T}$ *a transition of* SN, $t \in T$ *a transition of an object net* $ON_i = (P_i, T_i, W_i) \in \mathcal{ON}$ *such that* $(\hat{t}, t) \in \varrho$*. Then* $(\hat{t}, t)$ *is* activated *in* $\mathbf{V}$ *if for each input place* $\hat{p} \in {}^\bullet\hat{t} \cap P_{ob}$ *there is a submultiset* $\mathbf{V}_{\hat{p}} \subseteq \mathbf{V}(\hat{p})$ *such that*

a) $\widetilde{\mathbf{V}}_{\hat{p}} = \widehat{W}(\hat{p}, \hat{t})'ON_i$ *for all* $\hat{p} \in {}^\bullet\hat{t} \cap P_{ob}$ *for all* $p \in {}^\bullet t \cap P_{ob}$,
b) $\mathbf{V}(\hat{p}) \geq W(\hat{p}, \hat{t})$ *for all* $\hat{p} \in {}^\bullet\hat{t} \cap P_{bt}$ *and*
c) $m = \text{unify}(\{m_1 \mid (ON_i, m_1) \in \mathbf{V}_{\hat{p}} \ \wedge \ \hat{p} \in {}^\bullet\hat{t} \cap P_{ob}\})$ *is defined and* t *is activated in* m*, where* unify *is a partial mapping from the set* $2^{2^{P_i}}$ *of all marking sets of* ON_i *to the set* 2^{P_i} *of markings of* ON_i.

This is denoted by $\mathbf{V}[\hat{t}, t\rangle$*. Let be* $m[t\rangle m'$ *(w.r.t.* ON_i*, see Appendix). In this case the successor marking* $\mathbf{V}'$ *of* OS *is defined by* *a)* $\mathbf{V}'(\hat{p}) = \mathbf{V}(\hat{p}) - \mathbf{V}_{\hat{p}} + \widehat{W}(\hat{t},\hat{p})'(ON_i, m_{\hat{p}})$ *for all* $\hat{p} \in P_{ob}$ *where* $m_{\hat{p}}$ *comes from* $(m_{\hat{p}_1}, ..., m_{\hat{p}_q}) \in \text{distribute}(m')$ *and* $\{\hat{p}_1, ..., \hat{p}_q\} = \hat{t}^\bullet$. Distribute *is a mapping from the set* 2^{P_i} *of markings of* ON_i *to the set* $(2^{P_i})^q$ *of all q-tuples of markings of* ON_i*, where* q *is the number of output-places of* $\hat{t}$.

b) $\mathbf{V}'(\hat{p}) = \mathbf{V}(\hat{p}) - W(\hat{p},\hat{t}) + W(\hat{t},\hat{p})$ *for all* $\hat{p} \in P_{bt}$.

This is denoted by $\mathbf{V}[\hat{t}, t\rangle\mathbf{V}'$.

The definitions for transport and autonomous action are similar and omitted here.

5.2 Distributed Tokens Semantics

In distributed tokens semantics the tokens of those object nets, whose copies are distributed to the output-places of a transition are distributed as well, in such a way that they form the original marking when taken all together. Hence instead of Fig. 10 a successor marking like in Fig. 17 is appropriate.

Definition 11. *(interaction / distributed tokens semantics) Distributed tokens semantics is obtained by defining the (total) mappings* unify *and* distribute *of Definition 10 as follows:*

$$\text{unify}\{m_1, ..., m_s\} := \sum_{i=1}^{s} m_i$$

$$\textit{and} \quad \text{distribute}(m') := \{(m_1, ..., m_q) \mid \sum_{i=1}^{q} m_i = m'\}$$

(Recall that markings are multi-sets and the sum is the multi-set addition.)

In Fig. 18 successor markings of Fig. 17 are shown illustrating the application "unify"of Definition 11. The figure contains two markings, namely before and after the occurrence of the rightmost transition. The selection of tokens from the image of the mapping *distribute* is nondeterministic. There are also selections that are "wrong" in the sense that subsequent occurrences are different or impossible. This feature is much like nondeterministic firing of Petri net transitions in general, where conflict solution is left out of consideration.

5.3 History Process Semantics

A different strategy of token distribution is followed by history process semantics. Here all output transitions are supplied with the same information. Then by the subsequent behaviour the appropriate selection is chosen. In order to check whether concurrent executions are consistent instead of markings, processes (occurrence nets) are used as state information. There is a well-developed theory of processes which is not repeated here (see [54] for instance). We mention that there is a partial order on the set of all processes of a net and a well-defined operation "least upper bound", which is used in the following definition.

In Fig. 19 the place $\hat{p}$ contains the initial process of the object net (which is omitted in the place). After the occurrence of the interaction $(\hat{a}, a)$ the output-places $\hat{q}$ and $\hat{r}$ are marked with the corresponding enlarged processes. Finally

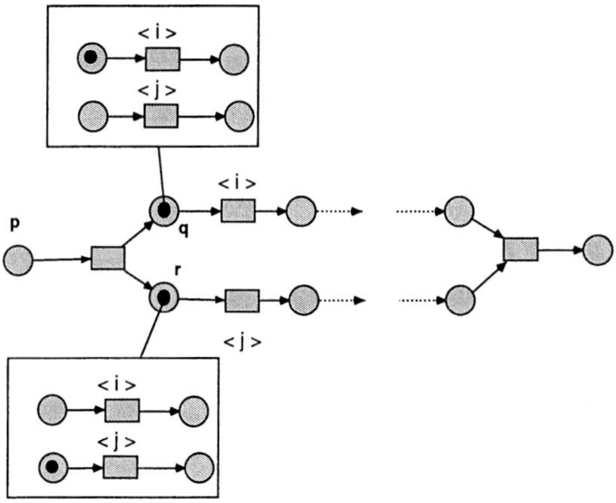

Fig. 17. Successor marking for Fig. 8 with respect to distributed tokens semantics

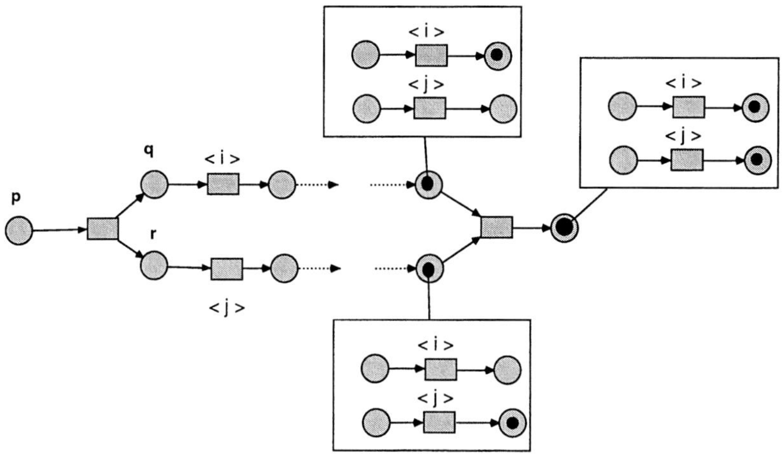

Fig. 18. Successor markings for Fig.17 with respect to distributed tokens semantics

from the processes in $\hat{s}$ and $\hat{s}'$ (see Fig. 20) the least upper bound is constructed and then enlarged by the transition b and added to all output-places of $\hat{e}$.

Definition 12. *(interaction / history process semantics) History process semantics is obtained by defining the mappings* unify *and* distribute *of Definition 3 as follows:* $\text{unify}\{proc_1, ..., proc_s\} := \bigsqcup_{i=1}^{s} proc_i$ *and* $\text{distribute}(proc') := \{(proc', ..., proc')\}$, *where* $\bigsqcup_{i=1}^{s} proc_i$ *is the least upper bound of all processes* $proc_i$ *and* $proc'$ *the least upper bound enlarged by the transition* t [8].

[8] Recall that the transition is not activated, if the least upper bound does not exists.

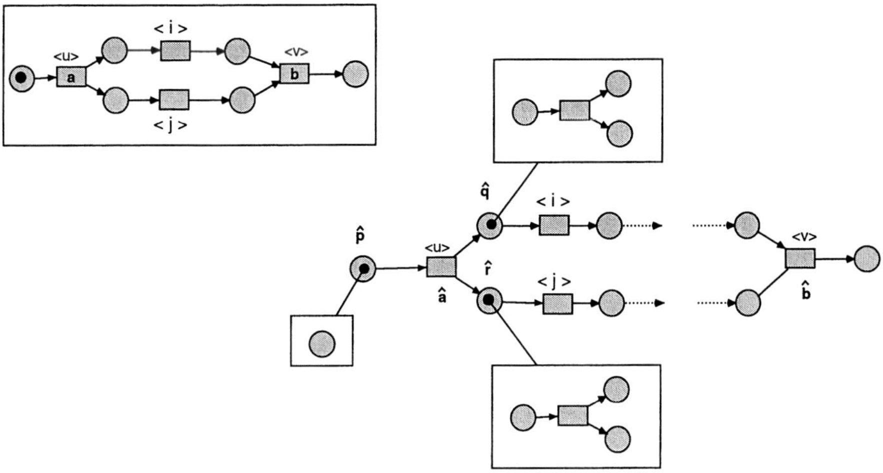

Fig. 19. Initial and first successor markings with respect to history process semantics

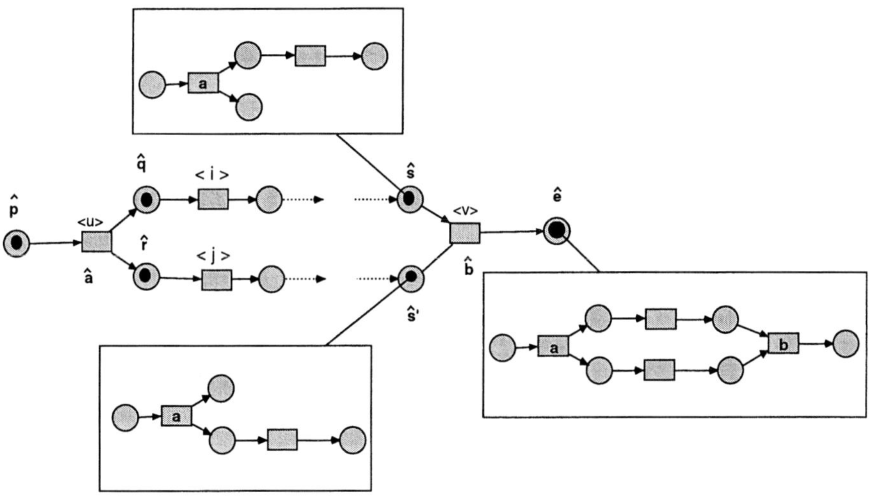

Fig. 20. Successor markings for Fig. 19 with respect to history process semantics

6 Agency under Reference and Value Semantics

In this section we will use a simple example (due to M. Köhler) to explain differences between the introduced semantics. An agent or an agency is designed to first get some money by visiting a bank (3 units of money in our case). Then the agency has to buy flowers uptown (for 1 unit of money) and independently to buy jewels downtown (for 2 units of money). Finally they return from up- and downtown to meet together and deliver their shoppings. We consider 4 different scenarios *system 1* to *system 4* in Fig. 21 to 24. In each of these cases we consider the three introduced semantics *reference semantics* (ref-semantics), *distributed tokens value semantics* (dt-semantics) and *history process value semantics* (hp-semantics). The behaviour is called *correct* if there is an occurrence sequence that marks exactly the terminal places[9] in system and object nets, when started with the given initial marking. As discussed above, in particular with dt-semantics there may be correct *and* incorrect behaviours from the same initial marking.

1. Object system 1 (Fig. 21)
   a) ref-semantics: correct, b) dt-semantics: correct, c) hp-semantics: correct.
2. Object system 2 (Fig. 22)
   a) ref-semantics: correct, b) dt-semantics: not correct (downtown agency member has no money), c) hp-semantics: not correct (downtown agency member has no money).
3. Object system 3 (Fig. 23)
   a) ref-semantics: not correct (2 `<visit_bank>`-actions are impossible), b) dt-semantics: not correct (only one agency member has money.), c) hp-semantics: correct.
4. Object system 4 (Fig. 24)
   a) ref-semantics: not correct (2 `<buy_jewels>`-actions are impossible), b) dt-semantics: not correct (not enough money for both agency members: detected by paying since at least one agency member has not enough money), c) hp-semantics: not correct (not enough money for both agency members: detected by joining since the unify-function is undefined).

Object system 3 (Fig. 23) is of particular interest as this case shows a difference between dt- and hp-semantics. The concept behind hp-semantics is similar to transaction handling in distributed data base systems. Such a transaction is considered *consistent* if it computes consistent results in all sites of the distributed data base system. The `<visit-bank>`-transition is executed uptown *and* downtown, and latter tested on consistency by the last system net transition.

In Fig. 25 two distributed data base systems $DB1$ and $DB2$ are represented by a simple system net. The object net reads the value of data x (either $x = 0$ or $x = 1$) and terminates with transition `end consistency check`, which in fact behaves like a consistency check under hp-semantics.

7 The Garbage Can Example

This section contains a larger example modelled with the Renew tool. Hence, reference semantics is used. It represents a system of agents that behave partially independent and interact in various ways. This example shows how the nets-within-nets paradigm provides a transparent modelling concept for representing objects of the real world by highly independent, but interacting net instances.

The example comes from a project of *socionics* [22], where knowledge from social sciences and multi-agent systems are combined to profit from each other.

[9] A place is called *terminal* if it has no output-transitions.

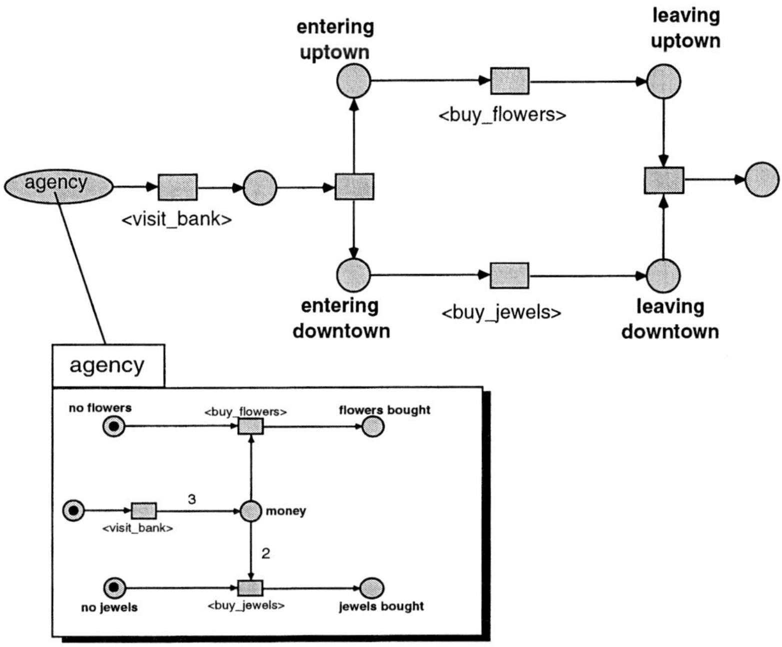

Fig. 21. Object system 1

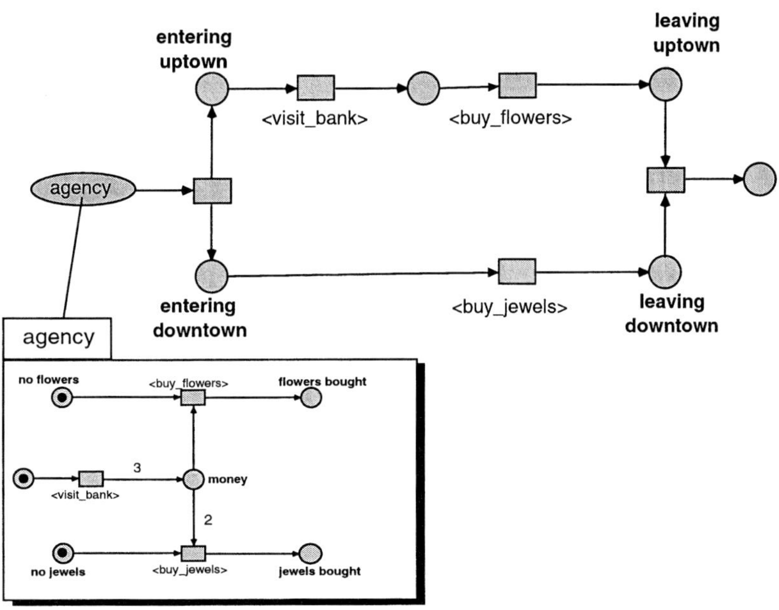

Fig. 22. Object system 2

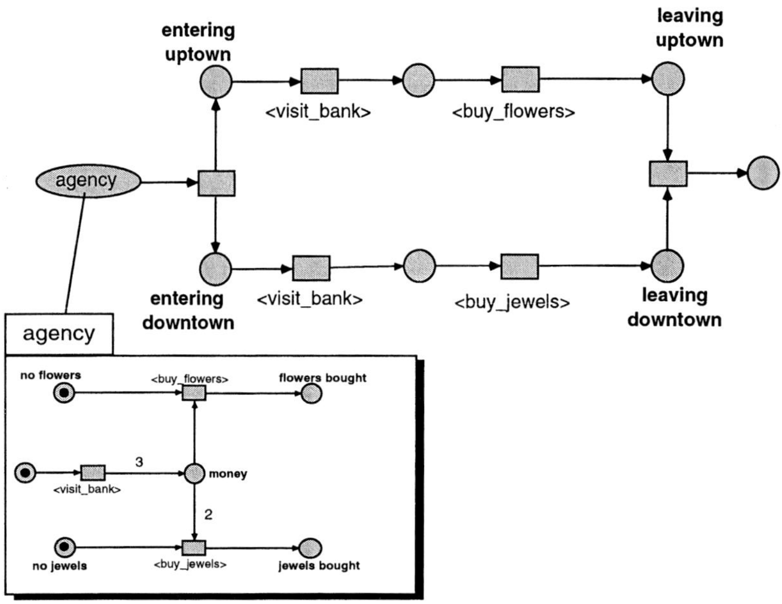

Fig. 23. Object system 3

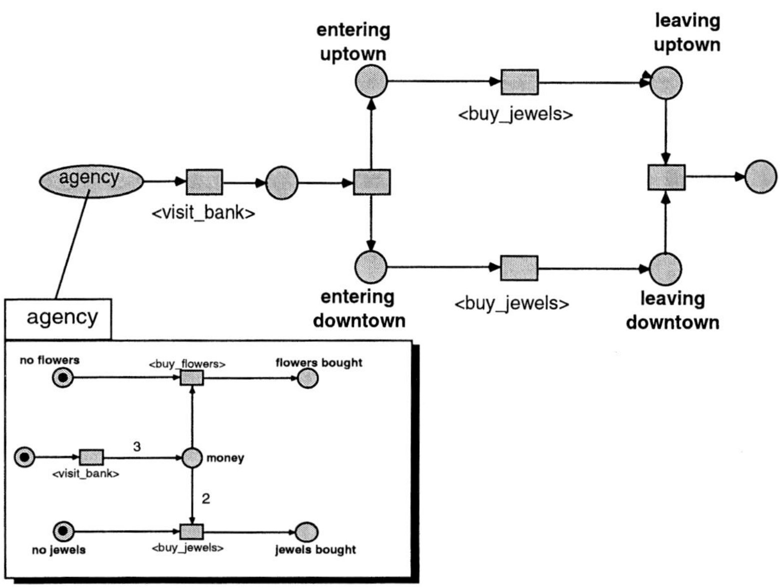

Fig. 24. Object system 4

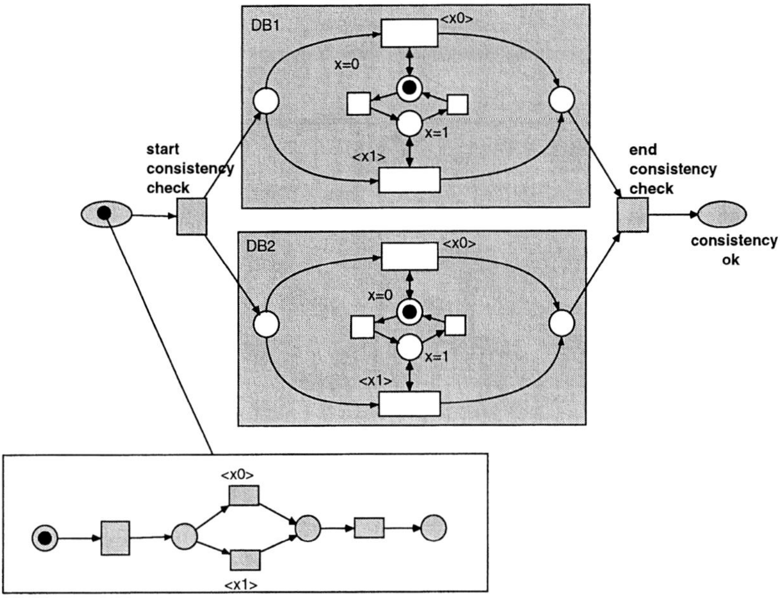

Fig. 25. Consistent distributed reading

It refers to a subfield of organisational theories, where the laws of anarchic behaviour in academic organisations are studied [55]. Following this paper, such organisations - or decision situations - are characterised by three general properties. The first is *problematic preferences.* In the organisation it is difficult to impute a set of preferences to the decision situation that satisfies the standard consistency requirements for a theory of choice. The organisation operates on the basis of a variety of inconsistent and ill-defined preferences. It can be described better as a loose collection of ideas than as a coherent structure; it discovers preferences through action more than it acts on the basis of preferences. The second property is *unclear technology.* Although the organisation manages to survive and even produce, its own processes are not understood by its members. It operates on the basis of simple trial-and-error procedures, the residue of learning from accidents of past experience, and pragmatic inventions of necessity. The third property is *fluid participation.* Participants vary in the amount of time and effort they devote to different domains; involvement varies from one time to another. As a result, the boundaries of the organisation are uncertain and changing; the audiences and decision makers for any particular kind of choice change capriciously.

The authors in [56] distinguish the three notions of *problems*, *solutions*, *participants* and *choice opportunities. Problems* are the concern of people inside and outside the organisation. A *solution* is somebody's product. Despite the dictum that you cannot find the answer until you have phrased the question well, you often do not know what the question is in organisational problem solving until you know the answer. Participants come and go. Substantial variation in participation stems from other demands on the participants' time (rather than from features of the decision under study). *Choice opportunities* are occasions when an organisation is expected to produce behaviour that can be called a decision. Opportunities arise regularly and every organisation has ways of declaring an occasion for choice. Contracts must be signed; people hired, promoted, or fired; money spent; and responsibilities allocated. The dynamic behaviour is the highly concurrent composition of a stream of choices, a stream of problems, rate of flow of solutions and stream of engaged participants. Where they meet and interact in a unpredictable way is called a *garbage can.*

To get an approximate but more concrete impression of the model, the authors of [56] reconsidered the finale of the James Bond movie, "A View to a Kill". Agent 007 poises on the main cable of the Golden Gate Bridge, a woman in distress clinging to his arm, a blimp approaching for rescue: the blimp is a solution, 007 a choice opportunity, and the woman a problem. In the movies's happy ending, the hero is finally picked up, together with the woman, and a solution by resolution takes place; the problem is solved.

Now imagine numerous blimps, women, and heroes, all arriving out of the blue in random sequence. Heroes take their positions on the main cable. Women cling to heroes, blimps hover above the scene. Heroes are able to hold an unlimited number of women, but the blimp's carrying capacity is limited; heroes with too many women cannot be rescued. Blimps retrieve rescuable, i.e., not-too-heavy, heroes. Women in distress are aware of that and switch heroes opportunistically, choosing the hero closest to retrieval. (In our model, however, women choose heroes nondeterministically.)

Since women, as well as blimps, make their choices simultaneously, but independently of each other, a light hero, on the verge of rescue, may suddenly find himself overburdened. Heavy heroes, in turn, may become rescuable all of a sudden as their women desert them. This mechanism, called "fluid participation", creates the possibility of nonsensical solutions or non-solutions. Women may switch heroes too swiftly and end up with an overburdened hero each time; then, problems are not solved. Or heroes are rescued just as all women have left; then, a "decision by flight" is made. Finally, heroes may be salvaged upon arrival at the scene before any woman in distress has a chance to grab their arm; then "decisions by oversight" are said to be made. Nevertheless, decisions by resolution do occur. Fig. 26 shows the system net, called `bridge`, containing the creation of the agents `woman` (by the annotation `woman:new woman`) and, similarly of the agents `hero` and `blimp`. The modelling of different locations for the `heroes` on the bridge is omitted, but could be easily added. The `heroes` can move to the (common) place `heroes on cable` and are able to cling one or more `women`. By the transition `rescue` they are picked up by a `blimp`, which may continue its flight to the hangar by transition `fly`. As example of a runtime shot in Fig. 26 the instances of agents `woman[1]`, `hero[3]`, `hero[4]` and `blimp[1]` are shown. The instance `blimp[1]` is drawn from the pattern `blimp`, as shown in Fig. 28. Within the place `heroes on cable` there are two instances `hero[3]` and `hero[4]` of the class `hero`. Fig. 27 gives the classes of `hero` and `woman`.

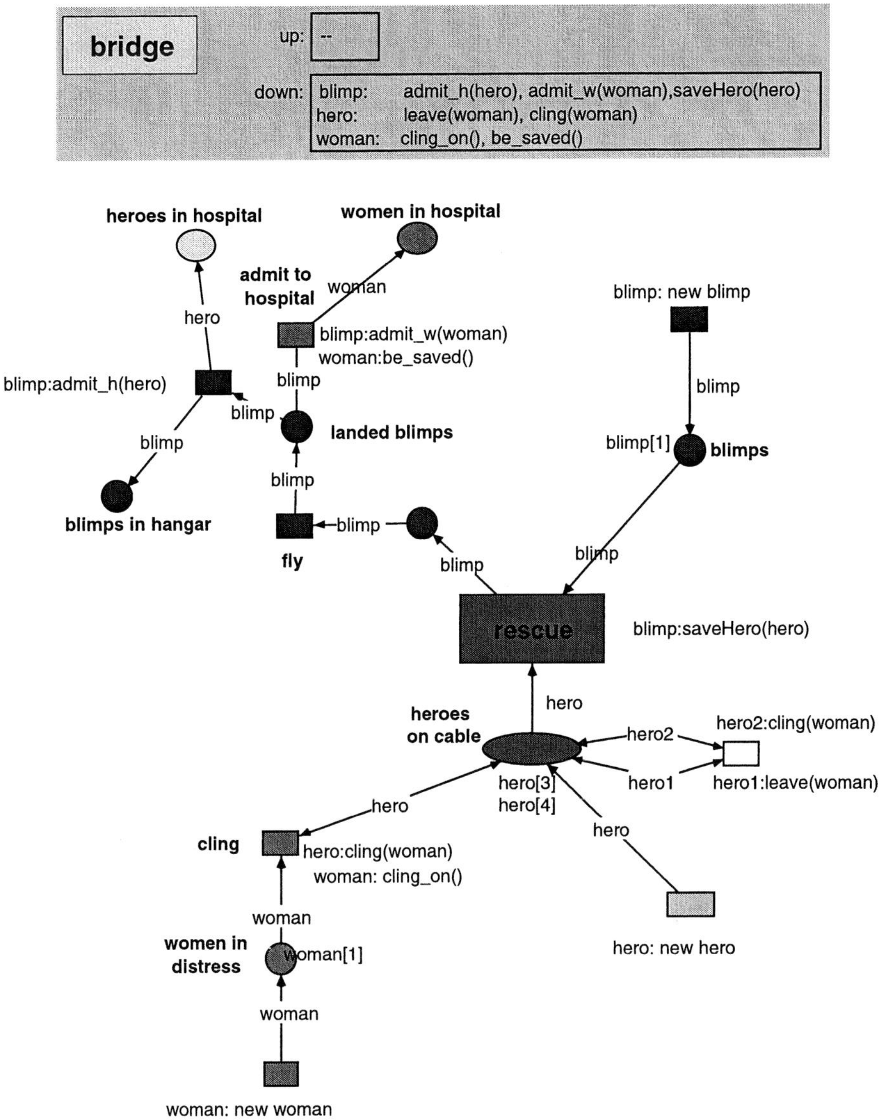

Fig. 26. The garbage can object system: bridge

Each `hero` counts the number of object net instances `woman` picked up by him (see place `counter` in the net `hero`). If there are too many (≥ 3 in our case) transition `rescue` cannot fire for this `blimp` (see leftmost transition of `blimp` in Fig. 28). All down- and uplinks are given in the nets as declarations. This allows for better reading and understanding, but is not supported by the tool. As a modelling style a hierarchy is respected, such that downlinks refer to the next lower level only. By this it is demonstrated how the object-relation *to be contained in* is represented in our modelling style.

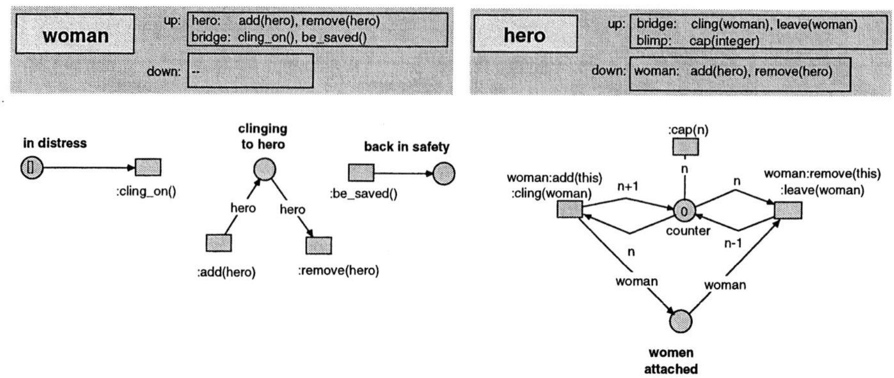

Fig. 27. The garbage can object systems: woman and hero

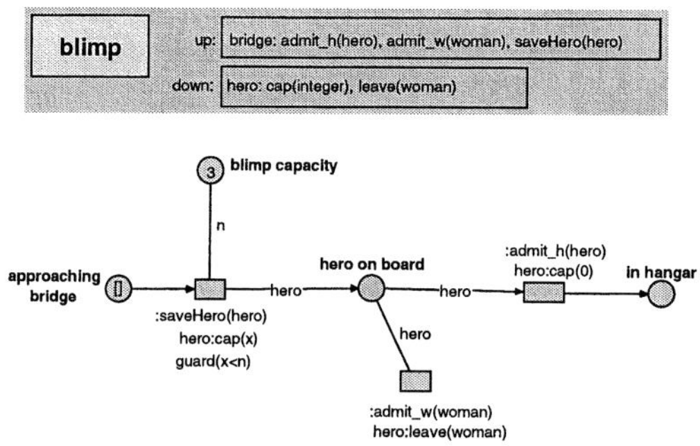

Fig. 28. The garbage can object system: blimp

Note how the change of `woman[4]` from `hero[3]` to `hero[4]` works in this example. `woman[4]` is restricted to communicate only via "her" `hero[3]`, who has to forward the procedure call. `hero[3]` has, in turn, also no direct access to `hero[4]`. Instead, he uses the common level `bridge` by the transition `swap`. This mimics reality, where a communication medium is always necessary (e.g. by sight, by mobile phone, by Internet).

This example of garbage cans can be seen as a prototype to other applications of interacting agents, for instance workflow or flexible manufacturing. In the latter case the bridge stands for the machine configuration, the blimps for a conveyors, heroes and women parts to be processed. Furthermore a "production plan" could be added as a further object net, which takes control over the production order, assembly, disassembly etc.

8 Conclusion and Current Research

We have shown that object Petri nets provide a "natural" modelling method, which is easy to understand and is supported by an appropriate tool. The nets-within-nets concept reduces much of the complexity (e.g. readability, simpler arc inscriptions, modular structure) that would result in modelling the same application by ordinary coloured nets. This is partially a result of the direct representation of object relations like "belongs to" or "is in location". Furthermore these concepts lead to natural representations of typical properties in distributed systems or mobile computation.

Due to space limitations, it was not possible to present formal results and further modelling examples in application domains. We therefore we give some references to related current research. More definitions, results and examples concerning the concept of history process semantics are given in [6] and [9], whereas distributed tokens semantics are studied in [20, 21, 57–60]. The latter group of references contains results on unbounded marking recursion, concurrency notions and decidability properties of object Petri nets. Modelling mobility and security properties is investigated in [61] and [62]. The bucket-chain example is extended to processor failure (fireman failure) and analysed using the MAUDE tool in [63]. There are also results on pattern based workflow design using reference semantics [64] and a proposal for structuring agent interaction protocols [65]. Model checking for object Petri nets via a translation into Prolog is introduced in [66] while some foundations of dynamic Petri net structures can be found in [67]. Fehling's concept of hierarchical Petri nets [68] is extended to a class of object Petri nets in [69]. The monograph [22] (in German) reports numerous results on the use of object Petri nets in socionics (also see [70]). Applications to flexible manufacturing systems can be found in [71] and [72].

Appendix: Basic Concepts

Multi-sets: Let $A \neq \emptyset$ be a set. A *multi-set* s over A is a mapping $s : A \rightarrow \mathbb{N}$, which associates to each element $a \in A$ a non-negative integer coefficient (or multiplicity) $s(a)$. We denote by $Bag(A)$ the set of multi-sets over A. A multi-set will be represented as the symbolic addition of its components: $s = \sum_{a \in A} s(a)'a$. Let s_1 and s_2 be two multi-sets defined over the same set A. The *addition* of multi-sets is defined by $s_1 + s_2 = \sum_{a \in A}(s_1(a) + s_2(a))'a$. On the other hand, $s_1 \geq s_2$ when for each $a \in A$, $s_1(a) \geq s_2(a)$. The *difference* operation extends the corresponding set-operation by $(s_1 - s_2)(a) := max(s_1(a) - s_2(a), 0)$. For short, $\emptyset$ will be used to denote the *empty multi-set.*

Place/Transition nets: A *Place/Transition net (P/T net)* $N = (P, T, W)$ is defined as follows.

1. P and T are finite and disjoint sets of *places* and *transitions*, respectively, and $W : (P \times T) \cup (T \times P) \rightarrow \mathbb{N}$ is the set of weighted arrows.
2. Specifying a marking $m : P \rightarrow \mathbb{N}$ we obtain a *marked P/T net* (N, m_0). Markings are considered as multi-sets over P.
3. For each $t \in T$ let be $PRE(t)$ and $POST(t)$ the multi-sets over P defined by $PRE(t)(p) := W(p, t)$ and $POST(t)(p) := W(t, p)$, respectively. Then $t \in T$ is *activated* in a marking m if $PRE(t) \leq m$ (denoted $m[t\rangle$) and the transition relation is defined by: $m[t\rangle m' :\Leftrightarrow PRE(t) \leq m \quad \wedge \quad m' = m - PRE(t) + POST(t)$. m' is called *successor marking* of m (w.r.t. t).

References

1. Rumbaugh, J., Jacobson, I., Booch, G.: The unified modeling language reference manual: The definitive reference to the UML from the original designers. Addison-Wesley object technology series. Addison-Wesley, Reading, Mass. (1999)
2. Jessen, E., Valk, R.: Rechensysteme: Grundlagen der Modellbildung. Springer-Verlag, Berlin (1987)
3. Valk, R.: Modelling of task flow in systems of functional units. 124, Universität Hamburg, Fachbereich Informatik (1987)
4. Valk, R.: Nets in computer organisation. In Brauer, W., Reisig, W., Rozenberg, G., eds.: Lecture Notes in Computer Science: Petri Nets: Central Models and Their Properties, Advances in Petri Nets 1986, Part I, Proceedings of an Advanced Course, Bad Honnef, September 1986. Volume 254., Springer-Verlag (1987) 377–396
5. Valk, R.: On processes of object Petri nets. Fachbereichsbericht 185, Fachbereich Informatik, Universität Hamburg (1996)
6. Valk, R.: Petri nets as token objects - an introduction to elementary object nets. In Desel, J., Silva, M., eds.: Application and Theory of Petri Nets 1998, Proceedings 15th International Conference, Lisbon, Portugal. Volume 1420 of Lecture Notes in Computer Science., Springer-Verlag (1998) 1–25
7. Valk, R.: Concurrency in communicating object Petri nets. [73] 164–195
8. Valk, R.: Reference and value semantics for object Petri nets. In: Proceedings of Colloquium on Petri Net Technologies for Modelling Communication Based Systems, October 21-22, 1999, Fraunhofer Gesellschaft, ISST (1999) 169–187
9. Valk, R.: Relating different semantics for object Petri nets, formal proofs and examples. Technical Report FBI-HH-B-226, University of Hamburg, Department for Computer Science Report/00 (2000)
10. Valk, R.: Mobile and distributed object versus central referencing. In Grabowski, Jens, Heymer, Stefan, eds.: Proceedings of 10. GI-ITG-Fachgespräch FBT 2000: Formale Beschreibungstechniken für verteilte Systeme, Lübeck, June 2000, Aachen, Shaker Verlag (2000) 7–27
11. Kummer, O.: Referenznetze. Logos Verlag (2002)
12. Wienberg, F.: Informations- und prozesorientierte Modellierung verteilter Systeme auf der Basis von Feature-Structure-Netzen. Dissertation, Universität Hamburg, Fachbereich Informatik (2001)
13. Kummer, O., Wienberg, F., Duvigneau, M.: Renew – The Reference Net Workshop. `http://renew.de/` (2004)
14. Farwer, B.: Linear Logic Based Calculi for Object Petri Nets. Logos Verlag, ISBN 3-89722-539-5, Berlin (2000)
15. Farwer, B.: Modelling protocols by object-based Petri nets. In Czaja, L., ed.: Concurrency Specification and Programming (CSP'01), Proceedings, University of Warsaw (2001) 87–96 published in Fundamenta Informaticae, 2002.
16. Farwer, B.: Comparing concepts of object Petri net formalisms. Fundamenta Informaticae **47** (2001) 247–258
17. Farwer, B., Lomazova, I.: A systematic approach towards object-based Petri net formalisms. In Bjorner, D., Zamulin, A., eds.: Perspectives of System Informatics, Proceedings of the 4th International Andrei Ershov Memorial Conference, PSI 2001, Akademgorodok, Novosibirsk, LNCS 2244. Springer-Verlag (2001) 255–267
18. Farwer, B.: A multi-region linear logic based calculus for dynamic Petri net structures. Fundamenta Informaticae **43** (2000) 61–79

19. Farwer, B.: A linear logic view of object Petri nets. Fundamenta Informaticae **37** (1999) 225–246
20. Köhler, M.: Mobile object net systems: Petri nets as active tokens. Technical Report 320, Universität Hamburg, Fachbereich Informatik, Vogt-Kölln Str. 30, 22527 Hamburg, Germany (2002)
21. Köhler, M., Rölke, H.: Concurrency for mobile object net systems. Fundamenta Informaticae **54** (2003)
22. v. Lüde, R., Moldt, D., Valk, R.: Sozionik: Modellierung soziologischer Theorie. Lit-Verlag (2003)
23. Cabac, L., Moldt, D., Rölke, H.: A proposal for structuring Petri net-based agent interaction protocols. In van der Aalst, W., Best, E., eds.: Proc. of 24nd International Conference on Application and Theory of Petri Nets 2003 (ICATPN 2003), Eindhoven, NL. Volume 2679 of Lecture Notes in Computer Science., Springer-Verlag (2003) 102–120
24. Köhler, M., Moldt, D., Rölke, H.: Modelling mobility and mobile agents using nets within nets. In van der Aalst, W., Best, E., eds.: Proc. of 24nd International Conference on Application and Theory of Petri Nets 2003 (ICATPN 2003), Eindhoven, NL, Berlin Heidelberg New York, to be published in Lecture Notes in Computer Science, Springer-Verlag (2003)
25. Battiston, E., De Cindio, F., Mauri, G.: Objsa nets: A class of high-level nets having objects as domains. In Rozenberg, G., ed.: Advances in Petri Nets 1988. Volume 340 of Lecture Notes in Computer Science., Springer-Verlag (1988) 20–43
26. Battiston, E., Chizzoni, A., Cindio, F.D.: Clown as a testbed for concurrent object-oriented concepts. [73] 131–163
27. Buchs, D., Guelfi, N.: Co-opn: A concurrent object oriented Petri net approach. In: Application and Theory of Petri Nets, 12th International Conference, IBM Deutschland (1991) 432–454
28. Christensen, S., Damgaard Hansen, N.: Coloured Petri nets extended with channels for synchronous communication. Technical Report DAIMI PB–390, Aarhus University (1992)
29. Biberstein, O.: CO-OPN/2: An Object-Oriented Formalism for the Specification of Concurrent Systems. Ph.d. thesis, University of Geneva (1997)
30. Biberstein, O., Buchs, D., Guelfi, N.: Object-oriented nets with algebraic specifications: The co-opn/2 formalism. [73] 73–130
31. Lakos, C.A.: Object Petri nets – definition and relationship to coloured Petri nets. Technical Report 94–3, Computer Science Department, University of Tasmania (1993)
32. Lakos, C.A.: From coloured Petri nets to object Petri nets. In: Proceedings of the 16th International Conference on the Application and Theory of Petri Nets, Turin, Italien. Volume 935 of Lecture Notes in Computer Science., Springer-Verlag (1995) 278–297 `http://www.cs.adelaide.edu.au/users/charles/`.
33. Lakos, C.A.: The role of substitution places in hierarchical coloured Petri nets. Technical Report 93–7, Computer Science Department, University of Tasmania (1993)
34. Lakos, C.A.: Pragmatic inheritance issues for object Petri nets. In: Proceedings of TOOLS Pacific 1995, Melbourne, Australien, Prentice-Hall (1995) 309–321 `http://www.cs.adelaide.edu.au/users/charles/`.
35. Sibertin-Blanc, C.: Cooperative nets. In Valette, R., ed.: Application and Theory of Petri Nets 1994, Proceedings 15th International Conference, Zaragoza, Spain. Volume 815 of Lecture Notes in Computer Science., Springer-Verlag (1994) 471–490

36. Sibertin-Blanc, C.: Cooperative objects: Principles, use and implementation. [73] 216–246
37. Sibertin-Blanc, C.: Syroco: A c++ implementation of cooperative objects. Workshop on Petri Nets and Object-Oriented Models of Concurrency (1995) Überarbeitete Version.
38. Project, T.M.: Project homepage. `http://www.tik.ee.ethz.ch/~moses/` (2002)
39. Češka, M., Janoušek, V., Vojnar, T.: Pntalk – a computerized tool for object oriented Petri nets modelling. In Pichler, F., Moreno-Diaz, R., eds.: 6th International Workshop on Computer Aided Systems Theory (EUROCAST'97), Las Palmas de Gran Canaria. Volume 1333 of Lecture Notes in Computer Science., Springer-Verlag (1997) 591–610
40. Goldberg, A., Robinson, D.: Smalltalk-80: The Language. Addison-Wesley (1989)
41. PNtalk: Project homepage. `http://www.fee.vutbr.cz/UIVT/homes/janousek/pntalk/` (2002)
42. Janoušek, V.: Synchronous interactions of objects in object oriented Petri nets. In: Proceedings of MOSIS'99. (1999) 73–80 `http://www.fee.vutbr.cz/~janousek/`.
43. Philippi, S.: OOPr/T-modelle – ein Pr/T-netz basierter Ansatz zur objektorientierten Modellierung. In Desel, J., Oberweis, A., eds.: 6. Workshop Algorithmen und Werkzeuge für Petrinetze, J.W. Goethe-Universität, Institut für Wirtschaftinformatik, Frankfurt am Main (1999) 36–41
44. Philippi, S.: Seamless object-oriented software development on a formal base. In: Workshop on Software Engineering and Petri-Nets, 21st International Conference on Application and Theory of Petri-Nets, Aarhus. (2000) `http://www.uni-koblenz.de/~philippi/`.
45. Giese, H., Graf, J., Wirtz, G.: Closing the gap between object-oriented modeling of structure and behavior. In France, R., Rumpe, B., eds.: The Second International Conference on The Unified Modeling Language (UML'99). Volume 1723 of Lecture Notes in Computer Science., Springer-Verlag (1999) 534–549
46. OCoN: Project homepage. `http://wwwmath.uni-muenster.de/cs/u/versys/research/ocon/` (2002)
47. Basten, T., van der Aalst, W.M.: Inheritance of behavior. Journal of Logic and Algebraic Programming **47** (2001) 47–145
48. Schöf, S., Sonnenschein, M., Wieting, R.: Efficient simulation of Thor nets. In De Michealis, G., Diaz, M., eds.: Proceeding of the 16th International Conference on Application and Theory of Petri Nets, Turin. Volume 935 of Lecture Notes in Computer Science., Springer-Verlag (1995) 412–431
49. Köster, F., Schöf, S., Sonnenschein, M., Wieting, R.: Modelling of a library with thorns. [73] 375–390
50. Han, Y.: Software Infrastructure for Configurable Workflow Systems; A Model-Driven Approach Based on Higher-Order Object Nets and Corba. Wissenschaft und Technik Verlag, Berlin (1997) Dissertation an der TU Berlin.
51. Lilius, J.: Ob(pn)^2: An object based Petri net programming notation. [73] 247–275
52. Agarwal, R., Bruno, G., Pescarmona, M.: Object-oriented extensions for Petri nets. Petri Net Newsletter **60** (2001) 26–41
53. Petri, C.: Introduction to general net theory. In Brauer, W., ed.: Net Theory and Applications: Proceedings of the Advanced Course on General Net Theory of Processes and Systems, Hamburg, 1979. Volume 84 of Lecture Notes in Computer Science., Springer-Verlag (1979) 1–19
54. Best, E., Fernández, C.: Nonsequential Processes. A Petri Net View. Volume 13. Springer Verlag EATCS Monographs on Theoretical Computer Science (1988)
55. Cohen, M., March, J., Olsen, J.: A garbage can model of organizational choice. Administrative Science Quarterly **17** (1972) 1–25

56. Masuch, M., LaPotin, P.: Beyond Garbage Cans: An AI Model of Organizational Choice. Administrtive Science Quarterly **36** (1989) 38–67
57. Köhler, M., Farwer, B.: Mobile object-net systems and their processes. In: Proceedings of the International Workshop on Concurrency, Specification, and Programming, CS&P 2003. (2003) 134–149
58. Köhler, M.: Mobile object net systems. In: 10. Workshop Algorithmen und Werkzeuge für Petrinetze, Universität Eichstätt (2003) 51–60
59. Köhler, M.: Object Petri nets: Definitions, properties and related models. Technical Report 329, Universität Hamburg, Fachbereich Informatik, Vogt-Kölln Str. 30, 22527 Hamburg, Germany (2003)
60. Köhler, M.: Decidability problems for object Petri nets. In Gesellschaft für Informatik, ed.: Informatiktage 2003. Fachwissenschaftlicher Informatik-Kongreß, Konradin Verlag (2003)
61. Köhler, M., Rölke, H.: Modelling sandboxes for mobile agents using nets within nets. In Busi, N., Martinelli, F., eds.: Workshop on Issues in Security and Petri Nets (WISP'03) at the International Conference on Application and Theory of Petri Nets 2003, University of Eindhoven (2003)
62. Köhler, M., Moldt, D., Rölke, H.: Modelling mobility and mobile agents using nets within nets. In v. d. Aalst, W., Best, E., eds.: Proceedings of the International Conference on Application and Theory of Petri Nets 2003. Volume 2679 of Lecture Notes in Computer Science., Springer-Verlag (2003) 121–140
63. Köhler, M., Rölke, H.: Formal analysis of multi-agent systems: The bucket-chain example. Technical Report to appear, University of Hamburg, Department for Computer Science Report/04 (2004)
64. Moldt, D., Rölke, H.: Pattern based workflow design using reference nets. In van der Aalst, W., ter Hofstede, A., Weske, M., eds.: Proc. of International Conference on BUSINESS PROCESS MANAGEMENT, Eindhoven, NL, Berlin Heidelberg New York, to be published in Lecture Notes in Computer Science, Springer-Verlag (2003)
65. Cabac, L., Moldt, D., Rölke, H.: A proposal for structuring petri net-based agent interaction protocols. In: Lecture Notes in Computer Science: 24th International Conference on Application and Theory of Petri Nets, Eindhoven, Netherlands, June 2003. (2003)
66. Farwer, B., Leuschel, M.: Model checking object Petri nets in Prolog. Technical Report DSSE-TR-2003-4, Declarative Systems and Software Engineering Group, School of Electronics and Computer Science, University of Southampton, SO17 1BJ, UK (2003)
67. Farwer, B., Misra, K.: Dynamic modification of system structures using LLPNs. In: Perspectives Of System Informatics, Proceedings of the 5th International Andrei Ershov Memorial Conference, PSI 2003, Akademgorodok, Novosibirsk, LNCS 2890. Springer-Verlag (2003) 274–293
68. Fehling, R.: A concept of hierarchical Petri nets with building blocks. In Rozenberg, G., ed.: Advances in Petri Nets 1993. LNCS 674, Springer-Verlag (1993) 148–168
69. Farwer, B., Misra, K.: Modelling with hierarchical object Petri nets. Fundamenta Informaticae **55** (2003) 129–147
70. Köhler, M., Moldt, D., Rölke, H., Valk, R.: Structuring of complex socionic systems using reference nets. Technical Report FBI-HH-B-248, University of Hamburg, Department for Computer Science Report/03 (2003)

71. Ezpeleta, J., Moldt, D.: A proposal for flexible testing of deadlock control strategies in resource allocation systems. In Pahlavani, Z., ed.: Proceedings of International Conference on Computational Intelligence for Modelling Control and Automation, in Vienna, Austria, 12-14 February. (2003)
72. Ezpeleta, J., Valk, R.: Modelling assembly systems using object Petri nets and deadlock avoidance. Technical Report to appear, University of Hamburg, Department for Computer Science Report/04 (2004)
73. Agha, G., De Cindio, F., Rozenberg, G., eds.: Advances in Petri Nets: Concurrent Object-Oriented Programming and Petri Nets. Volume 2001 of Lecture Notes in Computer Science. Springer-Verlag (2001)

Author Index

Lecture Notes in Computer Science

For information about Vols. 1–3005

please contact your bookseller or Springer-Verlag

Vol. 3120: J. Shawe-Taylor, Y. Singer (Eds.), Learning Theory. X, 648 pages. 2004.

Vol. 3109: S.C. Sahinalp, S. Muthukrishnan, U. Dogrusoz (Eds.), Combinatorial Pattern Matching. XII, 486 pages. 2004.

Vol. 3105: S. Göbel, U. Spierling, A. Hoffmann, I. Iurgel, O. Schneider, J. Dechau, A. Feix (Eds.), Technologies for Interactive Digital Storytelling and Entertainment. XVI, 304 pages. 2004.

Vol. 3099: J. Cortadella, W. Reisig (Eds.), Applications and Theory of Petri Nets 2004. XI, 505 pages. 2004.

Vol. 3098: J. Desel, W. Reisig, G. Rozenberg (Eds.), Lectures on Concurrency and Petri Nets. VIII, 849 pages. 2004.

Vol. 3096: G. Melnik, H. Holz (Eds.), Advances in Learning Software Organizations. X, 173 pages. 2004.

Vol. 3094: A. Nürnberger, M. Detyniecki (Eds.), Adaptive Multimedia Retrieval. VIII, 229 pages. 2004.

Vol. 3093: S.K. Katsikas, S. Gritzalis, J. Lopez (Eds.), Public Key Infrastructure. XIII, 380 pages. 2004.

Vol. 3092: J. Eckstein, H. Baumeister (Eds.), Extreme Programming and Agile Processes in Software Engineering. XVI, 358 pages. 2004.

Vol. 3091: V. van Oostrom (Ed.), Rewriting Techniques and Applications. X, 313 pages. 2004.

Vol. 3089: M. Jakobsson, M. Yung, J. Zhou (Eds.), Applied Cryptography and Network Security. XIV, 510 pages. 2004.

Vol. 3086: M. Odersky (Ed.), ECOOP 2004 – Object-Oriented Programming. XIII, 611 pages. 2004.

Vol. 3085: S. Berardi, M. Coppo, F. Damiani (Eds.), Types for Proofs and Programs. X, 409 pages. 2004.

Vol. 3084: A. Persson, J. Stirna (Eds.), Advanced Information Systems Engineering. XIV, 596 pages. 2004.

Vol. 3083: W. Emmerich, A.L. Wolf (Eds.), Component Deployment. X, 249 pages. 2004.

Vol. 3079: Z. Mammeri, P. Lorenz (Eds.), High Speed Networks and Multimedia Communications. XVIII, 1103 pages. 2004.

Vol. 3078: S. Cotin, D.N. Metaxas (Eds.), Medical Simulation. XVI, 296 pages. 2004.

Vol. 3077: F. Roli, J. Kittler, T. Windeatt (Eds.), Multiple Classifier Systems. XII, 386 pages. 2004.

Vol. 3076: D. Buell (Ed.), Algorithmic Number Theory. XI, 451 pages. 2004.

Vol. 3074: B. Kuijpers, P. Revesz (Eds.), Constraint Databases and Applications. XII, 181 pages. 2004.

Vol. 3073: H. Chen, R. Moore, D.D. Zeng, J. Leavitt (Eds.), Intelligence and Security Informatics. XV, 536 pages. 2004.

Vol. 3072: D. Zhang, A.K. Jain (Eds.), Biometric Authentication. XVII, 800 pages. 2004.

Vol. 3070: L. Rutkowski, J. Siekmann, R. Tadeusiewicz, L.A. Zadeh (Eds.), Artificial Intelligence and Soft Computing - ICAISC 2004. XXV, 1208 pages. 2004. (Subseries LNAI).

Vol. 3068: E. André, L. Dybkj{\}ae r, W. Minker, P. Heisterkamp (Eds.), Affective Dialogue Systems. XII, 324 pages. 2004. (Subseries LNAI).

Vol. 3067: M. Dastani, J. Dix, A. El Fallah-Seghrouchni (Eds.), Programming Multi-Agent Systems. X, 221 pages. 2004. (Subseries LNAI).

Vol. 3066: S. Tsumoto, R. S lowiński, J. Komorowski, J.W. Grzymala-Busse (Eds.), Rough Sets and Current Trends in Computing. XX, 853 pages. 2004. (Subseries LNAI).

Vol. 3065: A. Lomuscio, D. Nute (Eds.), Deontic Logic in Computer Science. X, 275 pages. 2004. (Subseries LNAI).

Vol. 3064: D. Bienstock, G. Nemhauser (Eds.), Integer Programming and Combinatorial Optimization. XI, 445 pages. 2004.

Vol. 3063: A. Llamosí, A. Strohmeier (Eds.), Reliable Software Technologies - Ada-Europe 2004. XIII, 333 pages. 2004.

Vol. 3062: J.L. Pfaltz, M. Nagl, B. Böhlen (Eds.), Applications of Graph Transformations with Industrial Relevance. XV, 500 pages. 2004.

Vol. 3061: F.F. Ramas, H. Unger, V. Larios (Eds.), Advanced Distributed Systems. VIII, 285 pages. 2004.

Vol. 3060: A.Y. Tawfik, S.D. Goodwin (Eds.), Advances in Artificial Intelligence. XIII, 582 pages. 2004. (Subseries LNAI).

Vol. 3059: C.C. Ribeiro, S.L. Martins (Eds.), Experimental and Efficient Algorithms. X, 586 pages. 2004.

Vol. 3058: N. Sebe, M.S. Lew, T.S. Huang (Eds.), Computer Vision in Human-Computer Interaction. X, 233 pages. 2004.

Vol. 3057: B. Jayaraman (Ed.), Practical Aspects of Declarative Languages. VIII, 255 pages. 2004.

Vol. 3056: H. Dai, R. Srikant, C. Zhang (Eds.), Advances in Knowledge Discovery and Data Mining. XIX, 713 pages. 2004. (Subseries LNAI).

Vol. 3055: H. Christiansen, M.-S. Hacid, T. Andreasen, H.L. Larsen (Eds.), Flexible Query Answering Systems. X, 500 pages. 2004. (Subseries LNAI).

Vol. 3054: I. Crnkovic, J.A. Stafford, H.W. Schmidt, K. Wallnau (Eds.), Component-Based Software Engineering. XI, 311 pages. 2004.

Vol. 3053: C. Bussler, J. Davies, D. Fensel, R. Studer (Eds.), The Semantic Web: Research and Applications. XIII, 490 pages. 2004.

Vol. 3052: W. Zimmermann, B. Thalheim (Eds.), Abstract State Machines 2004. Advances in Theory and Practice. XII, 235 pages. 2004.

Vol. 3051: R. Berghammer, B. Möller, G. Struth (Eds.), Relational and Kleene-Algebraic Methods in Computer Science. X, 279 pages. 2004.

Vol. 3050: J. Domingo-Ferrer, V. Torra (Eds.), Privacy in Statistical Databases. IX, 367 pages. 2004.

Vol. 3049: M. Bruynooghe, K.-K. Lau (Eds.), Program Development in Computational Logic. VIII, 539 pages. 2004.

Vol. 3047: F. Oquendo, B. Warboys, R. Morrison (Eds.), Software Architecture. X, 279 pages. 2004.

Vol. 3046: A. Laganà, M.L. Gavrilova, V. Kumar, Y. Mun, C.K. Tan, O. Gervasi (Eds.), Computational Science and Its Applications – ICCSA 2004. LIII, 1016 pages. 2004.

Vol. 3045: A. Laganà, M.L. Gavrilova, V. Kumar, Y. Mun, C.K. Tan, O. Gervasi (Eds.), Computational Science and Its Applications – ICCSA 2004. LIII, 1040 pages. 2004.

Vol. 3044: A. Laganà, M.L. Gavrilova, V. Kumar, Y. Mun, C.K. Tan, O. Gervasi (Eds.), Computational Science and Its Applications – ICCSA 2004. LIII, 1140 pages. 2004.

Vol. 3043: A. Laganà, M.L. Gavrilova, V. Kumar, Y. Mun, C.K. Tan, O. Gervasi (Eds.), Computational Science and Its Applications – ICCSA 2004. LIII, 1180 pages. 2004.

Vol. 3042: N. Mitrou, K. Kontovasilis, G.N. Rouskas, I. Iliadis, L. Merakos (Eds.), NETWORKING 2004, Networking Technologies, Services, and Protocols; Performance of Computer and Communication Networks; Mobile and Wireless Communications. XXXIII, 1519 pages. 2004.

Vol. 3040: R. Conejo, M. Urretavizcaya, J.-L. Pérez-de-la-Cruz (Eds.), Current Topics in Artificial Intelligence. XIV, 689 pages. 2004. (Subseries LNAI).

Vol. 3039: M. Bubak, G.D.v. Albada, P.M. Sloot, J.J. Dongarra (Eds.), Computational Science - ICCS 2004. LXVI, 1271 pages. 2004.

Vol. 3038: M. Bubak, G.D.v. Albada, P.M. Sloot, J.J. Dongarra (Eds.), Computational Science - ICCS 2004. LXVI, 1311 pages. 2004.

Vol. 3037: M. Bubak, G.D.v. Albada, P.M. Sloot, J.J. Dongarra (Eds.), Computational Science - ICCS 2004. LXVI, 745 pages. 2004.

Vol. 3036: M. Bubak, G.D.v. Albada, P.M. Sloot, J.J. Dongarra (Eds.), Computational Science - ICCS 2004. LXVI, 713 pages. 2004.

Vol. 3035: M.A. Wimmer (Ed.), Knowledge Management in Electronic Government. XII, 326 pages. 2004. (Subseries LNAI).

Vol. 3034: J. Favela, E. Menasalvas, E. Chávez (Eds.), Advances in Web Intelligence. XIII, 227 pages. 2004. (Subseries LNAI).

Vol. 3033: M. Li, X.-H. Sun, Q. Deng, J. Ni (Eds.), Grid and Cooperative Computing. XXXVIII, 1076 pages. 2004.

Vol. 3032: M. Li, X.-H. Sun, Q. Deng, J. Ni (Eds.), Grid and Cooperative Computing. XXXVII, 1112 pages. 2004.

Vol. 3031: A. Butz, A. Krüger, P. Olivier (Eds.), Smart Graphics. X, 165 pages. 2004.

Vol. 3030: P. Giorgini, B. Henderson-Sellers, M. Winikoff (Eds.), Agent-Oriented Information Systems. XIV, 207 pages. 2004. (Subseries LNAI).

Vol. 3029: B. Orchard, C. Yang, M. Ali (Eds.), Innovations in Applied Artificial Intelligence. XXI, 1272 pages. 2004. (Subseries LNAI).

Vol. 3028: D. Neuenschwander, Probabilistic and Statistical Methods in Cryptology. X, 158 pages. 2004.

Vol. 3027: C. Cachin, J. Camenisch (Eds.), Advances in Cryptology - EUROCRYPT 2004. XI, 628 pages. 2004.

Vol. 3026: C. Ramamoorthy, R. Lee, K.W. Lee (Eds.), Software Engineering Research and Applications. XV, 377 pages. 2004.

Vol. 3025: G.A. Vouros, T. Panayiotopoulos (Eds.), Methods and Applications of Artificial Intelligence. XV, 546 pages. 2004. (Subseries LNAI).

Vol. 3024: T. Pajdla, J. Matas (Eds.), Computer Vision - ECCV 2004. XXVIII, 621 pages. 2004.

Vol. 3023: T. Pajdla, J. Matas (Eds.), Computer Vision - ECCV 2004. XXVIII, 611 pages. 2004.

Vol. 3022: T. Pajdla, J. Matas (Eds.), Computer Vision - ECCV 2004. XXVIII, 621 pages. 2004.

Vol. 3021: T. Pajdla, J. Matas (Eds.), Computer Vision - ECCV 2004. XXVIII, 633 pages. 2004.

Vol. 3019: R. Wyrzykowski, J.J. Dongarra, M. Paprzycki, J. Wasniewski (Eds.), Parallel Processing and Applied Mathematics. XIX, 1174 pages. 2004.

Vol. 3018: M. Bruynooghe (Ed.), Logic Based Program Synthesis and Transformation. X, 233 pages. 2004.

Vol. 3017: B. Roy, W. Meier (Eds.), Fast Software Encryption. XI, 485 pages. 2004.

Vol. 3016: C. Lengauer, D. Batory, C. Consel, M. Odersky (Eds.), Domain-Specific Program Generation. XII, 325 pages. 2004.

Vol. 3015: C. Barakat, I. Pratt (Eds.), Passive and Active Network Measurement. XI, 300 pages. 2004.

Vol. 3014: F. van der Linden (Ed.), Software Product-Family Engineering. IX, 486 pages. 2004.

Vol. 3012: K. Kurumatani, S.-H. Chen, A. Ohuchi (Eds.), Multi-Agnets for Mass User Support. X, 217 pages. 2004. (Subseries LNAI).

Vol. 3011: J.-C. Régin, M. Rueher (Eds.), Integration of AI and OR Techniques in Constraint Programming for Combinatorial Optimization Problems. XI, 415 pages. 2004.

Vol. 3010: K.R. Apt, F. Fages, F. Rossi, P. Szeredi, J. Váncza (Eds.), Recent Advances in Constraints. VIII, 285 pages. 2004. (Subseries LNAI).

Vol. 3009: F. Bomarius, H. Iida (Eds.), Product Focused Software Process Improvement. XIV, 584 pages. 2004.

Vol. 3008: S. Heuel, Uncertain Projective Geometry. XVII, 205 pages. 2004.

Vol. 3007: J.X. Yu, X. Lin, H. Lu, Y. Zhang (Eds.), Advanced Web Technologies and Applications. XXII, 936 pages. 2004.

Vol. 3006: M. Matsui, R. Zuccherato (Eds.), Selected Areas in Cryptography. XI, 361 pages. 2004.